Contents

3 Systems of Linear Equations and Problem Solving 147

Frequently Used Symbols and Formulas

A Key to the Icons Appearing in the Exercise Sets

Concept reinforcement exercises, indicated by purple exercise numbers, provide basic practice with the new concepts and vocabulary.

1 Following most examples, students are directed to **TRY EXERCISES**. These selected exercises are identified with a color block around the exercise numbers.

Aha! Exercises labeled Aha! can often be solved quickly with the proper insight.

Calculator exercises are designed to be worked using a scientific or graphing calculator.

Graphing calculator exercises are designed to be worked using a graphing calculator and often provide practice for concepts discussed in the Technology Connections.

Writing exercises are designed to be answered using one or more complete sentences.

Symbols

$=$	Is equal to
$\approx$	Is approximately equal to
$>$	Is greater than
$<$	Is less than
$\geq$	Is greater than or equal to
$\leq$	Is less than or equal to
$\in$	Is an element of
$\subseteq$	Is a subset of
$\lvert x \rvert$	The absolute value of x
$\{x \mid x \ldots\}$	The set of all x such that $x \ldots$
$-x$	The opposite of x
$\sqrt{x}$	The square root of x
$\sqrt[n]{x}$	The nth root of x
LCM	Least Common Multiple
LCD	Least Common Denominator
π	Pi
i	$\sqrt{-1}$
$f(x)$	f of x, or f at x
$f^{-1}(x)$	f inverse of x
$(f \circ g)(x)$	$f(g(x))$
e	Approximately 2.7
Σ	Summation
$n!$	Factorial notation

Formulas

$m = \dfrac{y_2 - y_1}{x_2 - x_1}$	Slope of a line
$y = mx + b$	Slope–intercept form of a linear equation
$y - y_1 = m(x - x_1)$	Point–slope form of a linear equation
$(A + B)(A - B) = A^2 - B^2$	Product of the sum and difference of the same two terms
$\left.\begin{array}{l}(A + B)^2 = A^2 + 2AB + B^2, \\ (A - B)^2 = A^2 - 2AB + B^2\end{array}\right\}$	Square of a binomial
$d = rt$	Formula for distance traveled
$\dfrac{1}{a} \cdot t + \dfrac{1}{b} \cdot t = 1$	Work principle
$s = 16t^2$	Free-fall distance
$y = kx$	Direct variation
$y = \dfrac{k}{x}$	Inverse variation
$x = \dfrac{-b \pm \sqrt{b^2 - 4ac}}{2a}$	Quadratic formula
$P(t) = P_0 e^{kt}, k > 0$	Exponential growth
$P(t) = P_0 e^{-kt}, k > 0$	Exponential decay
$d = \sqrt{(x_2 - x_1)^2 + (y_2 - y_1)^2}$	Distance formula
$\dbinom{n}{r} = \dfrac{n!}{(n - r)! r!}$	$\dbinom{n}{r}$ notation

Resources Designed with You in Mind

This textbook was designed with features and applications to help make learning easier for you, but there's more than just the textbook if you want additional help. At www.mypearsonstore.com, you can check out these and other supplemental materials that can help you pass your math course in style.

Worksheets for Classroom or Lab Practice
(ISBN-10: 0-321-59936-5; ISBN-13: 978-0-321-59936-0)

Need more practice? Each worksheet provides key terms, fill-in-the-blank vocabulary practice, and exercises for each objective. There are two worksheets for every section of the text.

Videos on DVD
(ISBN-10: 0-321-59938-1; ISBN-13: 978-0-321-59938-4)

Miss a lecture? Need some extra help studying the night before an exam? Having a tough time with a certain topic? The Videos on DVD are here to help. Watch an experienced math instructor present important definitions, procedures, and concepts from each section of the book. The instructor will show you how to solve examples and exercises taken straight from the text.

Student's Solutions Manual
(ISBN-10: 0-321-58874-6; ISBN-13: 978-0-321-58874-6)

Looking for more than just the answer in the back of the book? The *Student's Solutions Manual* contains step-by-step solutions for all the odd-numbered exercises in the text as well as step-by-step solutions for Connecting the Concepts, Chapter Review Exercises, and Chapter Test exercises.

Intermediate Algebra

CONCEPTS AND APPLICATIONS

EDITION

8

MARVIN L. BITTINGER
Indiana University Purdue University Indianapolis

DAVID J. ELLENBOGEN
Community College of Vermont

ADDISON-WESLEY

Boston San Francisco New York
London Toronto Sydney Tokyo Singapore Madrid
Mexico City Munich Paris Cape Town Hong Kong Montreal

Editorial Director	Christine Hoag
Editor in Chief	Maureen O'Connor
Executive Project Manager	Kari Heen
Associate Editor	Joanna Doxey
Editorial Assistant	Jonathan Wooding
Production Manager	Ron Hampton
Editorial and Production Services	Martha K. Morong/Quadrata, Inc.
Art Editor and Photo Researcher	The Davis Group, Inc.
Compositor	Pre-Press PMG
Senior Media Producer	Ceci Fleming
Associate Producer	Jennifer Thomas
Software Development	Eileen Moore and Marty Wright
Marketing Manager	Marlana Voerster
Marketing Coordinator	Nathaniel Koven
Prepress Supervisor	Caroline Fell
Manufacturing Manager	Evelyn Beaton
Senior Manufacturing Buyer	Carol Melville
Senior Media Buyer	Ginny Michaud
Text Designer	The Davis Group, Inc.
Cover Designer	Beth Paquin
Cover Photograph	Euxiphipops xanthometapon—Konovalikov Andrey/Shutterstock

Photo Credits
Photo credits appear on page xviii.

Library of Congress Cataloging-in-Publication Data
Bittinger, Marvin L.
  Intermediate algebra : concepts and applications.
  — Eighth ed. / Marvin L. Bittinger, David J. Ellenbogen.
    p. cm.
  Includes indexes.

1. Algebra—Textbooks.   I. Ellenbogen, David.   II. Title.
  QA154.3.B58 2010
  512. 9—dc22   2008024319

1 2 3 4 5 6 7 8 9 10—RRDJC—12 11 10 09 08

© 2010, 2006, 2002, 1998, 1994, 1990, 1986, 1982 Pearson Education, Inc.

Addison-Wesley
is an imprint of

www.pearsonhighered.com

ISBN-13: 978-0-321-55718-6
ISBN-10: 0-321-55718-2

4   Inequalities and Problem Solving   219

5   Polynomials and Polynomial Functions   277

6 Rational Expressions, Equations, and Functions   351

7   Exponents and Radicals   429

10 Conic Sections   649

11 Sequences, Series, and the Binomial Theorem   691

Preface

It is with great pleasure that we introduce you to the eighth edition of *Intermediate Algebra: Concepts and Applications*. Our goal, as always, is to present content that is easy to understand and has the depth required for success in this and future courses. In this edition, faculty will recognize features, applications, and explanations that they have come to rely on and expect. Students and faculty will also find many changes resulting from our own ideas for improvement as well as insights from faculty and students throughout North America. Thus this new edition contains exciting new features and applications, along with updates and refinements to those from previous editions.

Appropriate for a one-term course in intermediate algebra, this text is intended for those students who have a firm background in elementary algebra. It is one of three texts in an algebra series that also includes *Elementary Algebra: Concepts and Applications*, Eighth Edition, by Bittinger/Ellenbogen, and *Elementary and Intermediate Algebra: Concepts and Applications*, Fifth Edition, by Bittinger/Ellenbogen/Johnson.

Approach

Our goal, quite simply, is to help today's students both learn and retain mathematical concepts. To achieve this goal, we feel that we must prepare developmental-mathematics students for the transition from "skills-oriented" elementary and intermediate algebra courses to more "concept-oriented" college-level mathematics courses. This requires that we teach these same students critical thinking skills: to reason mathematically, to communicate mathematically, and to identify and solve mathematical problems. Following are three aspects of our approach that we use to help meet the challenges we all face when teaching developmental mathematics.

Problem Solving

One distinguishing feature of our approach is our treatment of and emphasis on problem solving. We use problem solving and applications to motivate the material wherever possible, and we include real-life applications and problem-solving techniques throughout the text. Problem solving not only encourages students to think about how mathematics can be used, it helps to prepare them for more advanced material in future courses.

In Chapter 1, we introduce our five-step process for solving problems: (1) Familiarize, (2) Translate, (3) Carry out, (4) Check, and (5) State the answer. These steps are then used consistently throughout the text when encountering a problem-solving situation. Repeated use of this problem-solving strategy helps provide students with a starting point for any type of problem they encounter, and frees them to focus on the unique aspects of the particular problem situation. We often use estimation and carefully checked guesses to help with the *Familiarize* and *Check* steps (see pp. 33 and 392).

Applications

Interesting applications of mathematics help motivate both students and instructors. Solving applied problems gives students the opportunity to see their conceptual understanding put to use in a real way. In the eighth edition of *Intermediate Algebra: Concepts and Applications*, we have increased the number of applications, the number of real-data problems, and the number of reference lines that specify the sources of the real-world data. As in the past, art is integrated into the applications and exercises to aid the student in visualizing the mathematics. (See pp. 191, 285, 396, 481, and 524.)

Pedagogy

New!

> TRY EXERCISES

Try Exercises. This icon concludes nearly every example by pointing students to one or more parallel exercises from the corresponding exercise set so that they can immediately reinforce the concepts and skills presented in the examples. For easy identification in the exercise sets, the "Try" exercises have a shaded block on the exercise number. (See pp. 132, 135, and 634.)

New!

Translating for Success and **Visualizing for Success.** These matching exercises help students learn to associate word problems (through translation) and graphs (through visualization) with their appropriate mathematical equations. (See p. 37 (Translating); pp. 123 and 359 (Visualizing).) Each feature contains a corresponding activity in MyMathLab.

Revised!

Connecting the Concepts. Revised and expanded to include new Mixed Review exercises, this midchapter review helps students understand the big picture and prepare for chapter tests and cumulative reviews by relating the concept at hand to previously learned and upcoming concepts. (See pp. 128, 262, and 305.)

Revised!

Study Summary. Found at the end of each chapter and now presented in a two-column format organized by section, this synopsis gives students a fast and effective review of key chapter terms and concepts paired with accompanying examples. (See pp. 138, 343, and 420.)

Revised!

Cumulative Review. This review now appears after every chapter to help students retain and apply their knowledge from previous chapters. (See pp. 275, 351, and 577.)

Algebraic–Graphical Connections. This feature provides students with a way to visualize concepts that might otherwise prove elusive. (See pp. 153, 247, 385, and 500.)

Study Skills. This feature in the margin provides tips for successful study habits that even experienced students will appreciate. Ranging from time management to test preparation, these study skills can be applied in any college course. (See pp. 129, 242, and 390.)

Student Notes. These notes in the margin give students extra explanation of the mathematics appearing on that page. These comments are more casual in format than the typical exposition and range from suggestions for avoiding common mistakes to how to best read new notation. (See pp. 77, 245, and 600.)

Technology Connection. These optional boxes in each chapter help students use a graphing calculator to better visualize a concept that they have just learned. To connect this optional instruction to the exercise sets, certain exercises are marked with a graphing calculator icon 📟 to indicate the optional use of technology. (See pp. 89, 255, and 593.)

Revised!

Concept Reinforcement Exercises. Now with all answers listed in the answer section at the back of the book, these section and review exercises build students' confidence and comprehension through true/false, matching, and fill-in-the-blank exercises at the start of most exercise sets. To help further student understanding, emphasis is given to new vocabulary and notation developed in the section. (See pp. 57, 314, and 443.)

Aha!

Aha! Exercises. These exercises are not more difficult than their neighboring exercises and can be solved quickly, without going through a lengthy computation, if the student has the proper insight. Designed to reward students who "look before they leap," the icon indicates the first time a new insight applies, and then it is up to the student to determine when to use the Aha! method on subsequent exercises. (See pp. 104, 479, and 509.)

Revised!   **Skill Review Exercises.** These exercises, included in Section 1.2 and every section thereafter, review skills and concepts from preceding sections of the text. In most cases, these exercises prepare students for the next section. An introduction to each set directs students to the appropriate sections to review if necessary. On occasion, Skill Review exercises focus on a single topic in greater depth and from multiple perspectives. (See pp. 177, 249, and 527.)

Synthesis Exercises. Synthesis exercises follow the Skill Review exercises at the end of each exercise set. Generally more challenging, these exercises synthesize skills and concepts from earlier sections with the present material, often providing students with deeper insight into the current topic. Aha! exercises are sometimes included as Synthesis exercises. (See pp. 157, 362, and 627.)

 Writing Exercises. These appear just before the Skill Review exercises (two basic writing exercises) and also in the Synthesis exercises (at least two more challenging exercises). Writing exercises aid student comprehension by requiring students to use critical thinking to provide explanations of concepts in one or more complete sentences. Because some instructors may collect answers to writing exercises and because more than one answer can be correct, only answers to writing exercises in the review section are included at the back of the text. (See pp. 107 and 362.)

Collaborative Corner. These optional activities for students to explore together usually appear two to three times per chapter at the end of an exercise set. Studies show that students who study in groups generally outperform those who do not, so these exercises are for students who want to solve mathematical problems together. Additional collaborative activities and suggestions for directing collaborative learning appear in the *Instructor and Adjunct Support Manual*. (See pp. 232, 289, and 562.)

What's New in the Eighth Edition?

We have rewritten many key topics in response to user and reviewer feedback and have made significant improvements in design, art, pedagogy, and an expanded supplements package. Detailed information about the content changes is available in the form of a conversion guide. Please ask your local Pearson sales consultant for more information. Following is a list of the major changes in this edition.

NEW DESIGN

While incorporating a new layout, a fresh palette of colors, and new features, we have a larger page dimension for an open look and a typeface that is easy to read. As always, it is our goal to make the text look mature without being intimidating. In addition, we continue to pay close attention to the pedagogical use of color to make sure that it is used to present concepts in the clearest possible manner.

CONTENT CHANGES

A variety of content changes have been made throughout the text. Some of the more significant changes are listed below.

- Examples and exercises that use real data are updated or replaced with current applications.

- Over 35% of the exercises are new or updated.
- Quick-glance reminders for multistep processes are included next to examples. These appear by one multistep example of each type. (See pp. 256, 367, and 448.)
- In Chapter 2, more emphasis is placed on using the graph of a function to find its domain and range when the function has an infinite domain.
- Inequalities are now graphed on number lines using brackets and parentheses. Interval notation can thus be read directly from the graph of an inequality.
- Domains of radical functions are now discussed in Section 4.1, separately from domains of rational functions in Section 4.2.
- The distance formula is now presented in Section 7.7 as one application of the Pythagorean theorem.
- In Chapter 8, discussion of the solutions of quadratic equations now directly follows the quadratic formula.

ANCILLARIES

The following ancillaries are available to help both instructors and students use this text more effectively.

STUDENT SUPPLEMENTS

New! Chapter Test Prep Video CD

- Watch instructors work through step-by-step solutions to all the chapter test exercises from the textbook. The Chapter Test Prep Video CD is included with each new student text.

New! Worksheets for Classroom or Lab Practice

by Carrie Green
These lab- and classroom-friendly workbooks offer the following resources for every section of the text:

- A list of learning objectives;
- Vocabulary practice problems;
- Extra practice exercises with ample work space.

ISBNs: 0-321-59936-5 and 978-0-321-59936-0

Student's Solutions Manual

by Christine S. Verity

- Contains completely worked-out solutions with step-by-step annotations for all the odd-numbered exercises in the text, with the exception of the writing exercises.
- New! Now contains all solutions to Chapter Review, Chapter Test, and Connecting the Concepts exercises.

ISBNs: 0-321-58874-6 and 978-0-321-58874-6

INSTRUCTOR SUPPLEMENTS

Annotated Instructor's Edition

- Provides answers to all text exercises in color next to the corresponding problems.
- Includes Teaching Tips.
- Icons identify writing and graphing calculator exercises.

ISBNs: 0-321-55947-9 and 978-0-321-55947-0

Instructor's Solutions Manual

by Christine S. Verity

- Contains fully worked-out solutions to the odd-numbered exercises and brief solutions to the even-numbered exercises in the exercise sets.
- Available for download at www.pearsonhighered.com

ISBNs: 0-321-58875-4 and 978-0-321-58875-3

Instructor and Adjunct Support Manual

- Includes resources designed to help both new and adjunct faculty with course preparation and classroom management.
- Offers helpful teaching tips correlated to the sections of the text.

ISBNs: 0-321-56731-5 and 978-0-321-56731-4

Videos on DVD

- A complete set of digitized videos on DVD for use at home or on campus.
- Includes a full lecture for each section of the text, many presented by author team members David J. Ellenbogen and Barbara Johnson.
- Optional subtitles in English are available.

ISBNs: 0-321-59938-1 and 978-0-321-59938-4

InterAct Math® Tutorial Website

www.interactmath.com

- Online practice and tutorial help.
- Retry an exercise with new values each time for unlimited practice and mastery.
- Every exercise is accompanied by an interactive guided solution that gives helpful feedback when an incorrect answer is entered.
- View the steps of a worked-out sample problem similar to those in the text.

Printable Test Bank

by Laurie Hurley

- Contains two multiple-choice tests per chapter, six free-response tests per chapter, and eight final exams.
- Available for download at www.pearsonhighered.com

PowerPoint® Lecture Slides

- Present key concepts and definitions from the text.
- Available for download at www.pearsonhighered.com

TestGen

www.pearsonhighered.com/testgen

- Enables instructors to build, edit, print, and administer tests using a computerized bank of questions developed to cover all text objectives.
- Algorithmically based, TestGen allows instructors to create multiple but equivalent versions of the same question or test with the click of a button.
- Instructors can also modify test bank questions or add new questions.
- Tests can be printed or administered online.

Pearson Math Adjunct Support Center

http://www.pearsontutorservices.com/math-adjunct.html

Staffed by qualified instructors with more than 50 years of combined experience at both the community college and university levels, this center provides assistance for faculty in the following areas:

- Suggested syllabus consultation;
- Tips on using materials packed with the text;
- Book-specific content assistance;
- Teaching suggestions, including advice on classroom strategies.

AVAILABLE FOR STUDENTS AND INSTRUCTORS

MyMathLab® Online Course (access code required)

MyMathLab is a series of text-specific, easily customizable online courses for Pearson Education's textbooks in mathematics and statistics. Powered by CourseCompass™ (our online teaching and learning environment) and MathXL® (our online homework, tutorial, and assessment system), MyMathLab gives you the tools you need to deliver all or a portion of your course online, whether your students are in a lab setting or working from home. MyMathLab provides a rich and flexible set of course materials, featuring free-response exercises that are algorithmically generated for unlimited practice and mastery. Students can also use online tools, such as video lectures, animations, and a multimedia textbook, to independently improve their understanding and performance. Instructors can use MyMathLab's homework and test managers to select and assign online exercises correlated directly to the textbook, and they can also create and assign their own online exercises and import TestGen tests for added flexibility. MyMathLab's online gradebook—designed specifically for mathematics and statistics—automatically tracks students' homework and test results and gives the instructor control over how to calculate final grades. Instructors can also add offline (paper-and-pencil) grades to the gradebook. MyMathLab also includes access to the **Pearson Tutor Center** (www.pearsontutorservices.com). The Tutor Center is staffed by qualified mathematics instructors who provide textbook-specific tutoring for students via toll-free phone, fax, e-mail, and interactive Web sessions. MyMathLab is available to qualified adopters. For more information, visit our website at www.mymathlab.com or contact your sales representative.

MathXL® Online Course (access code required)

MathXL® is a powerful online homework, tutorial, and assessment system that accompanies Pearson Education's textbooks in mathematics or statistics. With MathXL, instructors can create, edit, and assign online homework and tests using algorithmically generated exercises correlated at the objective level to the textbook. They can also create and assign their own online exercises and import TestGen tests for added flexibility. All student work is tracked in MathXL's online gradebook. Students can take chapter tests in MathXL and receive personalized study plans based on their test results. The study plan diagnoses weaknesses and links students directly to tutorial exercises for the objectives they need to study and retest. Students can also access supplemental animations and video clips directly from selected exercises. MathXL is available to qualified adopters. For more information, visit our website at www.mathxl.com, or contact your Pearson sales representative.

MathXL® Tutorials on CD

This interactive tutorial CD-ROM provides algorithmically generated practice exercises that are correlated at the objective level to the exercises in the textbook. Every practice exercise is accompanied by an example and a guided solution designed to involve students in the solution process. Selected exercises may also include a video clip to help students visualize concepts. The software provides helpful feedback for incorrect answers and can generate printed summaries of students' progress.

Acknowledgments

No book can be produced without a team of professionals who take pride in their work and are willing to put in long hours. Laurie Hurley, in particular, deserves extra thanks for her work as developmental editor. Ann Ostberg, Laurie Hurley, Holly Martinez, and Christine Verity also deserve special thanks for their careful accuracy checks, well-thought-out suggestions, and uncanny eye for detail. Thanks to Carrie Green, Laurie Hurley, and Christine Verity for their outstanding work in preparing supplements.

We are also indebted to Chris Burditt and Jann MacInnes for their many fine ideas that appear in our Collaborative Corners and Vince McGarry and Janet Wyatt for their recommendations for Teaching Tips featured in the Annotated Instructor's Edition.

Martha Morong, of Quadrata, Inc., provided editorial and production services of the highest quality imaginable—she is amazing and a joy to work with. Geri Davis, of the Davis Group, Inc., performed superb work as designer, art editor, and photo researcher, and is always a pleasure to work with. Network Graphics generated the graphs, charts, and many of the illustrations. Not only are the people at Network reliable, but they clearly take pride in their work. The many illustrations appear thanks to Bill Melvin—an artist with insight and creativity.

Our team at Pearson deserves special thanks. Acquisitions Editor Randy Welch provided many fine suggestions, remaining involved and accessible throughout the project. Executive Project Manager Kari Heen carefully coordinated tasks and schedules, keeping a widely spread team working together. Associate Editor Joanna Doxey coordinated reviews and assisted in a variety of tasks with patience and creativity. Editorial Assistant Jonathan Wooding responded quickly to all requests, always in a pleasant manner. Production Manager Ron Hampton's attention to detail, willingness to listen, and creative responses helped result in a book that is beautiful to look at. Marketing Manager Marlana Voerster and Marketing Assistant Nathaniel Koven skillfully kept us in touch with the needs of faculty. Our Editor in Chief, Maureen O'Connor, and Editorial Director, Chris Hoag, deserve credit for assembling this fine team.

We also thank the students at Indiana University Purdue University Indianapolis and the Community College of Vermont and the following professors for their thoughtful reviews and insightful comments.

Marie Aratari, *Oakland Community College–Orange Ridge Campus*
Douglas Brozovic, *University of North Texas*
Barbara Burke, *Hawaii Pacific University*
Laura Burris, *Sam Houston State University*
Lisa Carnell, *High Point University*
Sharon Edgmon, *Bakersfield College*
Karen Ernst, *Hawkeye College*
Kathy Garrison, *Clayton College and State University*
Cynthia Harrison, *Baton Rouge Community College*
Tracey L. Johnson, *University of Georgia*
Joanne Kawczenski, *Luzerne County Community College*
Rachel Lamp, *North Iowa Community College*
Kevin J. Leith, *Central New Mexico Community College*
Stephanie Lochbaum, *Austin Community College*
Rob McCarthy, *Community College of Allegheny County—South Campus*
Doug Mace, *Kirtland Community College*
Rhea Meyerholtz, *Indiana State University*
Kausha Miller, *Lexington Community College*
Rebecca Parrish, *Ohio University*
Kay Petrash, *Sam Houston State University*
Debra Pharo, *Northwestern Michigan College*
Terry Reeves, *Red Rocks Community College*
Kathy Rod, *Wharton County Junior College*
Nicole Saporito, *Luzerne Community College*
Elgin Schilhab, *Austin Community College*
M. Terry Simon, *University of Toledo*
Fran Smith, *Oakland Community College*
Donald Soloman, *University of Wisconsin–Milwaukee*

Finally, a special thank-you to all those who so generously agreed to discuss their professional use of mathematics in our chapter openers. These dedicated people all share a desire to make math more meaningful to students. We cannot imagine a finer set of role models.

M.L.B.
D.J.E.

Photo Credits

Algebra and Problem Solving

JEN SCRIBNER
KAYAK OUTFITTER
Machias, Maine

As an office manager, I use math applications daily to balance the checkbook and to figure interest rates and finance charges and the effect those have on our business. As guides, we use triangulation to find where we are in the fog and on land to keep on our correct course and stay safe.

AN APPLICATION

Sea kayaking in Puget Sound is a popular sport in the Seattle area. Stella can maintain a speed of 3.5 mph in a calm sea. The club to which she belongs plans to paddle 6 mi into a 1.9-mph current. How long will it take Stella to make the trip?

Source: Based on information from the University Kayak Club and the University of Washington

This problem appears as Exercise 3 in Section 1.4.

The principal theme of this text is problem solving in algebra. In Chapter 1, we begin with a review of algebraic expressions and equations. The use of algebra as part of an overall strategy for solving problems is presented in Section 1.4. Additional and increasing emphasis on problem solving appears throughout the book.

1.1 Some Basics of Algebra

Algebraic Expressions and Their Use ▪ Translating to Algebraic Expressions ▪
Evaluating Algebraic Expressions ▪ Sets of Numbers

The primary difference between algebra and arithmetic is the use of *variables*. In this section, we will see how variables can be used to represent many different situations. We will also examine the different types of numbers that will be represented by variables throughout this text.

Algebraic Expressions and Their Use

We are all familiar with expressions like

$$95 + 21, \quad 57 \times 34, \quad 9 - 4, \quad \text{and} \quad \frac{35}{71}.$$

In algebra, we use these as well as expressions that include letters, like

$$x + 21, \quad l \cdot w, \quad 9 - s, \quad \text{and} \quad \frac{d}{t}.$$

A letter that can be any one of various numbers is called a **variable**. If a letter always represents a particular number that never changes, it is called a **constant**. If r represents the radius of the earth, in kilometers, then r is a constant. If a represents the age of a baby chick, in minutes, then a is a variable because a changes, or *varies*, as time passes.

An **algebraic expression** consists of variables, numbers, and operation signs. All of the expressions above are examples of algebraic expressions. When an equals sign is placed between two expressions, an **equation** is formed.

Algebraic expressions and equations arise frequently in problem-solving situations. Suppose, for example, that we want to determine by how much the number of single-track downloads has increased from 2004 to 2006.

U. S. Digital Music Market

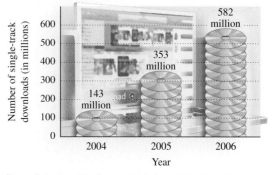

Source: Federation of the Phonographic Industry, Digital Music Report 2007

By using x to represent the increase in downloads, in millions, we can form an equation:

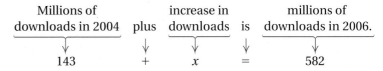

$$143 \quad + \quad x \quad = \quad 582$$

To find a **solution**, we can subtract 143 from both sides of the equation:

$$x = 582 - 143$$
$$x = 439.$$

We see that the number of single-track downloads increased by 439 million from 2004 to 2006.

Translating to Algebraic Expressions

To translate problems to equations, we need to know which words correspond to which symbols:

Key Words

Addition	Subtraction	Multiplication	Division
add	subtract	multiply	divide
sum of	difference of	product of	quotient of
plus	minus	times	divided by
increased by	decreased by	twice	ratio
more than	less than	of	per

When the value of a number is not given, we represent that number with a variable.

Phrase	Algebraic Expression
Five *more than* some number	$n + 5$
Half *of* a number	$\frac{1}{2}t,$ or $\frac{t}{2}$
Five *more than* three *times* some number	$3p + 5$
The *difference* of two numbers	$x - y$
Six *less than* the *product of* two numbers	$rs - 6$
Seventy-six percent *of* some number	$0.76z,$ or $\frac{76}{100}z$

Note that an expression like rs represents a product and can also be written as $r \cdot s$, $r \times s$, or $(r)(s)$. The multipliers r and s are called **factors**.

EXAMPLE   1   Translate to an algebraic expression:

Five less than forty-three percent of the quotient of two numbers.

SOLUTION   We let r and s represent the two numbers.

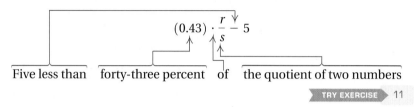

$$(0.43) \cdot \frac{r}{s} - 5$$

Five less than   forty-three percent   of   the quotient of two numbers

TRY EXERCISE   11

Some algebraic expressions contain *exponential notation*. Many different kinds of numbers can be used as *exponents*. Here we establish the meaning of a^n when n is a counting number, $1, 2, 3, \ldots$.

> ### Exponential Notation
>
> The expression a^n, in which n is a counting number, means
>
> $$\underbrace{a \cdot a \cdot a \cdots \cdot a \cdot a}_{n \text{ factors}}.$$
>
> In a^n, a is called the *base* and n is the *exponent*. When no exponent appears, the exponent is assumed to be 1. Thus, $a^1 = a$.

The expression a^n is read "a raised to the nth power" or simply "a to the nth." We read s^2 as "s-squared" and x^3 as "x-cubed." This terminology comes from the fact that the area of a square of side s is $s \cdot s = s^2$ and the volume of a cube of side x is $x \cdot x \cdot x = x^3$.

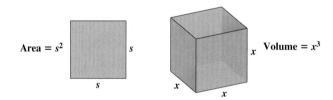

Evaluating Algebraic Expressions

When we replace a variable with a number, we say that we are **substituting** for the variable. The calculation that follows the substitution is called **evaluating the expression**.

Geometric formulas are often evaluated. In the next examples, we use the formula for the area of a square with sides of length s, as well as the formula for the area of a triangle with a base of length b and a height of length h.

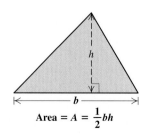

Area $= A = \dfrac{1}{2}bh$

EXAMPLE **2** A square rug has sides of length 5 ft. Find the area of the rug.

SOLUTION We substitute 5 for s and *evaluate* the expression:

$$s^2 = 5^2 \qquad \text{We use color to highlight the substitution.}$$
$$= 5 \cdot 5$$
$$= 25 \text{ square feet (sq ft or ft}^2\text{).}$$ ▶ TRY EXERCISE 25

EXAMPLE **3** The base of a triangular sail is 3.1 m and the height is 4 m. Find the area of the sail.

SOLUTION We substitute 3.1 for b and 4 for h and multiply to evaluate the expression:

$$\tfrac{1}{2} \cdot b \cdot h = \tfrac{1}{2} \cdot 3.1 \cdot 4$$
$$= 6.2 \text{ square meters (sq m or m}^2\text{).}$$ ▶ TRY EXERCISE 29

4 m

3.1 m

Exponential notation tells us that 5^2 means $5 \cdot 5$, or 25, but what does $1 + 2 \cdot 5^2$ mean? If we add 1 and 2 and multiply by 25, we get 75. If we multiply 2 times 5^2, or 25, and add 1, we get 51. A third possibility is to square $2 \cdot 5$ to get 100 and then add 1. The following convention indicates that only the second of these approaches is correct: We square 5, then multiply, and then add.

STUDENT NOTES ————

Note: Step 3 states that when division precedes multiplication, the division is performed first. Thus, $20 \div 5 \cdot 2$ represents $4 \cdot 2$, or 8. Similarly, $9 - 3 + 1$ represents $6 + 1$, or 7.

> **Rules for Order of Operations**
> 1. Simplify within any grouping symbols such as $(\ )$, $[\ \ ]$, $\{\ \}$, working in the innermost symbols first.
> 2. Simplify all exponential expressions.
> 3. Perform all multiplication and division, as either occurs, working from left to right.
> 4. Perform all addition and subtraction, as either occurs, working from left to right.

EXAMPLE **4** Evaluate $5 + 2(a - 1)^2$ for $a = 4$.

SOLUTION

$$\begin{aligned}
5 + 2(a - 1)^2 &= 5 + 2(4 - 1)^2 && \text{Substituting} \\
&= 5 + 2(3)^2 && \text{Working within parentheses first} \\
&= 5 + 2(9) && \text{Simplifying } 3^2 \\
&= 5 + 18 && \text{Multiplying} \\
&= 23 && \text{Adding}
\end{aligned}$$ ▶ TRY EXERCISE 35

Step (3) in the rules for order of operations tells us to divide before we multiply when division appears first, reading left to right. This means that an expression like $6 \div 2x$ means $(6 \div 2)x$.

> *CAUTION!*
> $$6 \div 2x = (6 \div 2)x,$$
> $$6 \div (2x) = \frac{6}{2x}$$
> $6 \div 2x$ *does not mean* $6 \div (2x)$.

EXAMPLE   **5**   Evaluate $9 - x^3 + 6 \div 2y^2$ for $x = 2$ and $y = 5$.

SOLUTION

$$
\begin{aligned}
9 - x^3 + 6 \div 2y^2 &= 9 - 2^3 + 6 \div 2(5)^2 && \text{Substituting} \\
&= 9 - 8 + 6 \div 2 \cdot 25 && \text{Simplifying } 2^3 \text{ and } 5^2 \\
&= 9 - 8 + 3 \cdot 25 && \text{Dividing} \\
&= 9 - 8 + 75 && \text{Multiplying} \\
&= 1 + 75 && \text{Subtracting} \\
&= 76 && \text{Adding}
\end{aligned}
$$

TRY EXERCISE ▶ 37

Sets of Numbers

When evaluating algebraic expressions, and in problem solving in general, we often must examine the *type* of numbers used. For example, if a formula is used to determine an optimal class size, any fraction results must be rounded up or down, since it is impossible to have a fractional part of a student. Three frequently used sets of numbers are listed below.

Natural Numbers, Whole Numbers, and Integers

Natural Numbers (or Counting Numbers)

Those numbers used for counting: $\{1, 2, 3, \ldots\}$

Whole Numbers

The set of natural numbers with 0 included: $\{0, 1, 2, 3, \ldots\}$

Integers

The set of all whole numbers and their opposites:

$$\{\ldots, -4, -3, -2, -1, 0, 1, 2, 3, 4, \ldots\}$$

The dots are called ellipses and indicate that the pattern continues without end.

The integers correspond to the points on the number line as follows:

To describe sets containing the numbers between integers, we use a different kind of set notation.

The set containing the numbers -2, 1, and 3 can be written $\{-2, 1, 3\}$. This way of writing a set is known as **roster notation**. Roster notation was used for the three sets listed above. A second type of set notation, **set-builder notation**, specifies conditions under which a number is in the set. The following example of set-builder notation is read as shown:

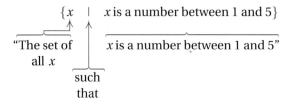

Set-builder notation is generally used when it is difficult to list a set using roster notation.

EXAMPLE 6   Using both roster notation and set-builder notation, represent the set consisting of the first 15 even natural numbers.

SOLUTION

Using roster notation: $\{2, 4, 6, 8, 10, 12, 14, 16, 18, 20, 22, 24, 26, 28, 30\}$

Using set-builder notation: $\{n \mid n \text{ is an even number between 1 and 31}\}$

Note that other descriptions of the set are possible. For example, $\{2x \mid 1 \leq x \leq 15\}$ is a common way of writing this set.

▶ TRY EXERCISE 51

The symbol $\in$ is used to indicate that an element or member belongs to a set. Thus if $A = \{2, 4, 6, 8\}$, we can write $4 \in A$ to indicate that 4 *is an element of A*. We can also write $5 \notin A$ to indicate that 5 *is not an element of A*.

EXAMPLE 7   Classify the statement $8 \in \{x \mid x \text{ is an integer}\}$ as either true or false.

SOLUTION   Since 8 *is* an integer, the statement is true. In other words, since 8 is an integer, it belongs to the set of all integers.

▶ TRY EXERCISE 67

With set-builder notation, we can describe the set of all *rational numbers*.

> ## Rational Numbers
> Numbers that can be expressed as an integer divided by a nonzero integer are called *rational numbers*:
> $$\left\{ \frac{p}{q} \ \middle| \ p \text{ is an integer, } q \text{ is an integer, and } q \neq 0 \right\}.$$

Rational numbers can be written using fraction notation or decimal notation. *Fraction notation* uses symbolism like the following:

$$\frac{5}{8}, \quad \frac{12}{-7}, \quad \frac{-17}{15}, \quad -\frac{9}{7}, \quad \frac{39}{1}, \quad \frac{0}{6}.$$

In *decimal notation*, rational numbers either *terminate* (end) or *repeat* a block of digits.

EXAMPLE 8   When written in decimal form, does each of the following numbers terminate or repeat? **(a)** $\frac{5}{8}$; **(b)** $\frac{6}{11}$.

SOLUTION

a) Since $\frac{5}{8}$ means $5 \div 8$, we perform long division to find that $\frac{5}{8} = 0.625$, a decimal that ends. Thus, $\frac{5}{8}$ can be written as a terminating decimal.

b) Using long division, we find that $6 \div 11 = 0.5454\ldots$, so we can write $\frac{6}{11}$ as a repeating decimal. Repeating decimal notation can be abbreviated by writing a bar over the repeating part—in this case, $0.\overline{54}$.

Many numbers, like π, $\sqrt{2}$, and $-\sqrt{15}$, are not rational numbers. For example, $\sqrt{2}$ is the number for which $\sqrt{2} \cdot \sqrt{2} = 2$. A calculator's representation of $\sqrt{2}$ as 1.414213562 is only an approximation since $(1.414213562)^2$ is not exactly 2. Note that $\sqrt{2}$ does not repeat or terminate when written in decimal form and cannot be written as a fraction.

To see that $\sqrt{2}$ is a "real" point on the number line, it can be shown that when a right triangle has two legs of length 1, the remaining side has length $\sqrt{2}$. Thus we can "measure" $\sqrt{2}$ units and locate $\sqrt{2}$ on a number line.

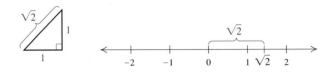

Numbers like π, $\sqrt{2}$, and $-\sqrt{15}$ are said to be **irrational**. Decimal notation for irrational numbers neither terminates nor repeats.

The set of all rational numbers, combined with the set of all irrational numbers, gives us the set of all **real numbers**.

Real Numbers

Numbers that are either rational or irrational are called *real numbers*:

$$\{x \mid x \text{ is rational or } x \text{ is irrational}\}.$$

Every point on the number line represents some real number and every real number is represented by some point on the number line.

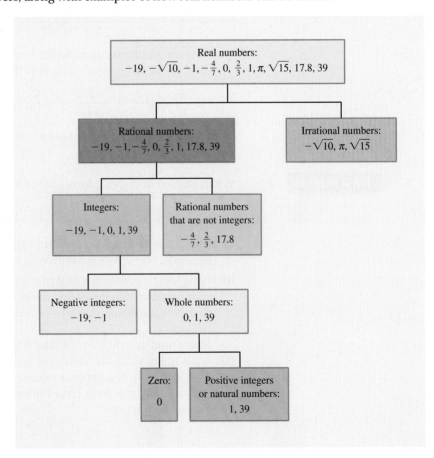

The following figure shows the relationships among various kinds of numbers, along with examples of how real numbers can be sorted.

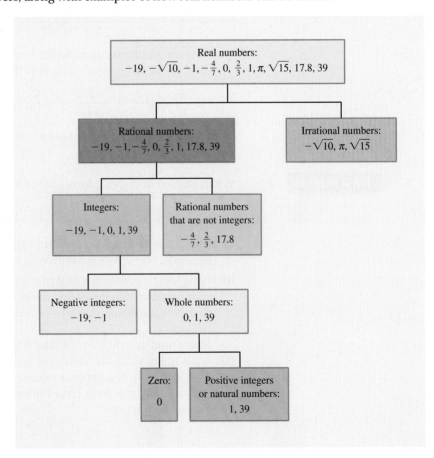

EXAMPLE 9   Which numbers in the following list are **(a)** whole numbers? **(b)** integers? **(c)** rational numbers? **(d)** irrational numbers? **(e)** real numbers?

$$-29, \quad -\tfrac{7}{4}, \quad 0, \quad 2, \quad 3.9, \quad \sqrt{42}, \quad 78$$

SOLUTION

a) 0, 2, and 78 are whole numbers.

b) -29, 0, 2, and 78 are integers.

c) -29, $-\tfrac{7}{4}$, 0, 2, 3.9, and 78 are rational numbers.

d) $\sqrt{42}$ is an irrational number.

e) -29, $-\tfrac{7}{4}$, 0, 2, 3.9, $\sqrt{42}$, and 78 are all real numbers.   **TRY EXERCISE** 63

When every member of one set is a member of a second set, the first set is a **subset** of the second set. Thus if $A = \{2, 4, 6\}$ and $B = \{1, 2, 4, 5, 6\}$, we write $A \subseteq B$ to indicate that *A is a subset of B*. Similarly, if $\mathbb{N}$ represents the set of all natural numbers and $\mathbb{Z}$ the set of all integers, we can write $\mathbb{N} \subseteq \mathbb{Z}$. Additional statements can be made using other sets in the diagram above.

STUDY SKILLS ⎯⎯⎯⎯⎯⎯

Throughout this textbook, you will find a feature called *Study Skills*. These tips are intended to help improve your math study skills. On the first day of class, we recommend that you complete this chart.

Course Information

Instructor: Name _____

  Office hours and location _____

  Phone number _____

  E-mail address _____

Find a student whom you could contact for information or questions:

Name _____

Phone number _____

E-mail address _____

Math lab on campus:

  Location _____

  Hours _____

Tutoring:

  Campus location _____

  Hours _____

Supplements recommended by the instructor:

(See the preface for a complete list of available supplements.)

_____

_____

1.1   EXERCISE SET

↪ *Concept Reinforcement*   *In each of Exercises 1–10, fill in the blank with the appropriate word or words.*

1. A letter representing a specific number that never changes is called a(n) _____.

2. A letter that can be any one of a set of numbers is called a(n) _____.

3. In the expression $7y$, the multipliers 7 and y are called _____.

4. In a^b, the a is called the _____ and the b is called the _____.

5. When all variables in a variable expression are replaced by numbers and a result is calculated, we say that we are _____ the expression.

6. To calculate $4 + 12 \div 3 \cdot 2$, the first operation that we perform is _____.

7. A number that can be written in the form a/b, where a and b are integers (with $b \neq 0$), is said to be a(n) _____ number.

8. A real number that cannot be written as a quotient of two integers is an example of a(n) _____ number.

9. Division can be used to show that $\frac{7}{40}$ can be written as a(n) _____ decimal.

10. Division can be used to show that $\frac{13}{7}$ can be written as a(n) _____ decimal.

To the student and the instructor: *The* **TRY EXERCISES** *for examples are indicated by a shaded block   on the exercise number. Complete step-by-step solutions for these exercises appear online at www.pearsonhighered.com/ bittingerellenbogen.*

Use mathematical symbols to translate each phrase.

11. Five less than some number

12. Ten more than some number

13. Twice a number

14. Eight times a number

15. Twenty-nine percent of some number

16. Thirteen percent of some number

17. Six less than half of a number

18. Three more than twice a number

19. Seven more than ten percent of some number

20. Four less than six percent of some number

21. One less than the product of two numbers

22. One more than the difference of two numbers

23. Ninety miles per every four gallons of gas

24. One hundred words per every sixty seconds

In Exercises 25–28, find the area of a square flower garden with the given length of a side.

25. Side = 6 ft

26. Side = 12 ft

27. Side = 0.5 m

28. Side = 2.5 m

In Exercises 29–32, find the area of a triangular window with the given base and height.

29. Base = 3 ft, height = 5 ft

30. Base = 1.2 m, height = 2 m

31. Base = 3 m, height = 2.4 m

32. Base = 3.5 ft, height = 1.5 ft

To the student and the instructor: Throughout this text, selected exercises are marked with the icon Aha!. *Students who pause to inspect an Aha! exercise should find the answer more readily than those who proceed mechanically. This may involve looking at an earlier exercise or example, or performing calculations in a more efficient manner. These exercises are included to discourage rote memorization. Some Aha! exercises are left unmarked to encourage students to always pause before working a problem.*

Evaluate each expression using the values provided.

33. $4x + y$, for $x = 2$ and $y = 3$

34. $8a - b$, for $a = 5$ and $b = 7$

35. $20 + r^2 - s$, for $r = 5$ and $s = 10$

36. $m^3 + 7 - n$, for $m = 2$ and $n = 8$

37. $2c \div 3b$, for $b = 2$ and $c = 6$

38. $3z \div 2y$, for $y = 1$ and $z = 6$

Aha! **39.** $3n^2p - 3pn^2$, for $n = 5$ and $p = 9$

40. $2a^3b - 2b^2$, for $a = 3$ and $b = 7$

41. $5x \div (2 + x - y)$, for $x = 6$ and $y = 2$

42. $3(m + 2n) \div m$, for $m = 7$ and $n = 0$

43. $[10 - (a - b)]^2$, for $a = 7$ and $b = 2$

44. $[17 - (x + y)]^2$, for $x = 4$ and $y = 1$

45. $[5(r + s)]^2$, for $r = 1$ and $s = 2$

46. $[3(a - b)]^2$, for $a = 7$ and $b = 5$

47. $x^2 - [3(x - y)]^2$, for $x = 6$ and $y = 4$

48. $m^2 - [2(m - n)]^2$, for $m = 7$ and $n = 5$

49. $(m - 2n)^2 - 2(m + n)$, for $m = 8$ and $n = 1$

50. $(r - s)^2 - 3(2r - s)$, for $r = 11$ and $s = 3$

Use roster notation to write each set.

51. The set of letters in the word "algebra"

52. The set of all days of the week

53. The set of all odd natural numbers

54. The set of all even natural numbers

55. The set of all natural numbers that are multiples of 10

56. The set of all natural numbers that are multiples of 5

Use set-builder notation to write each set.

57. The set of all even numbers between 9 and 99

58. The set of all multiples of 5 between 7 and 79

59. $\{0, 1, 2, 3, 4\}$

60. $\{-3, -2, -1, 0, 1, 2\}$

61. $\{11, 13, 15, 17, 19\}$

62. $\{24, 26, 28, 30, 32\}$

*In Exercises 63–66, which numbers in the list provided are **(a)** whole numbers? **(b)** integers? **(c)** rational numbers? **(d)** irrational numbers? **(e)** real numbers?*

63. $-8.7, -3, 0, \dfrac{2}{3}, \sqrt{7}, 6$

64. $-\dfrac{9}{2}, -4, -1.2, 0, \sqrt{5}, 3$

65. $-17, -0.01, 0, \dfrac{5}{4}, 8, \sqrt{77}$

66. $-6.08, -5, 0, 1, \sqrt{17}, \dfrac{99}{2}$

Classify each statement as either true or false. The following sets are used:

$\mathbb{N} =$ the set of natural numbers;
$\mathbb{W} =$ the set of whole numbers;
$\mathbb{Z} =$ the set of integers;
$\mathbb{Q} =$ the set of rational numbers;
$\mathbb{H} =$ the set of irrational numbers;
$\mathbb{R} =$ the set of real numbers.

67. $196 \in \mathbb{N}$

68. $\mathbb{N} \subseteq \mathbb{W}$

69. $\mathbb{W} \subseteq \mathbb{Z}$

70. $\sqrt{8} \in \mathbb{Q}$

71. $\frac{2}{3} \in \mathbb{Z}$

72. $\mathbb{H} \subseteq \mathbb{R}$

73. $\sqrt{10} \in \mathbb{R}$

74. $4.3 \notin \mathbb{Z}$

75. $\mathbb{Z} \not\subseteq \mathbb{N}$

76. $\mathbb{Q} \subseteq \mathbb{R}$

77. $\mathbb{Q} \subseteq \mathbb{Z}$

78. $\dfrac{8}{15} \in \mathbb{H}$

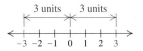

To the student and the instructor: *The icon is used to denote writing exercises. These exercises are meant to be answered with one or more English sentences. Because many writing exercises have a variety of correct answers, these solutions are not listed in the answers at the back of the book.*

79. What is the difference between rational numbers and integers?

80. Oscar insists that $15 - 4 + 1 \div 2 \cdot 3$ is 2. What error is he making?

Synthesis

To the student and the instructor: *Synthesis exercises are designed to challenge students to extend the concepts or skills studied in each section. Many synthesis exercises require the assimilation of skills and concepts from several sections.*

81. Is the following true or false, and why?

$$\{2, 4, 6\} \subseteq \{2, 4, 6\}$$

82. On a quiz, Mia answers $6 \in \mathbb{Z}$ while Giovanni writes $\{6\} \in \mathbb{Z}$. Giovanni's answer does not receive full credit while Mia's does. Why?

Translate to an algebraic expression.

83. The quotient of the sum of two numbers and their difference

84. Three times the sum of the cubes of two numbers

85. Half of the difference of the squares of two numbers

86. The product of the difference of two numbers and their sum

Use roster notation to write each set.

87. The set of all whole numbers that are not natural numbers

88. The set of all integers that are not whole numbers

89. $\{x \mid x = 5n,\ n \text{ is a natural number}\}$

90. $\{x \mid x = 3n,\ n \text{ is a natural number}\}$

91. $\{x \mid x = 2n + 1,\ n \text{ is a whole number}\}$

92. $\{x \mid x = 2n,\ n \text{ is an integer}\}$

93. Draw a right triangle that could be used to measure $\sqrt{13}$ units.

1.2   Operations and Properties of Real Numbers

Absolute Value ■ Inequalities ■ Addition, Subtraction, and Opposites ■ Multiplication, Division, and Reciprocals ■ The Commutative, Associative, and Distributive Laws

In this section, we review addition, subtraction, multiplication, and division of real numbers. We also study important rules for manipulating algebraic expressions. First, however, we must discuss absolute value and inequalities.

Absolute Value

Both 3 and -3 are 3 units from 0 on the number line. Thus their distance from 0 is 3. We use *absolute-value* notation to represent a number's distance from 0. Note that distance is never negative.

> ### Absolute Value
>
> The notation $|a|$, read "the absolute value of a," represents the number of units that a is from zero.

EXAMPLE 1

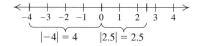

$|-4| = 4$ $|2.5| = 2.5$

Find the absolute value: **(a)** $|-4|$; **(b)** $|2.5|$; **(c)** $|0|$.

SOLUTION

a) $|-4| = 4$ -4 is 4 units from 0.

b) $|2.5| = 2.5$ 2.5 is 2.5 units from 0.

c) $|0| = 0$ 0 is 0 units from itself.

TRY EXERCISE 11

Note that whereas the absolute value of a nonnegative number is the number itself, to find the absolute value of a negative number, we must make it positive.

Inequalities

For any two numbers on the number line, the one to the left is said to be less than, or smaller than, the one to the right. The symbol $<$ means "is less than" and the symbol $>$ means "is greater than." The symbol $\leq$ means "is less than or equal to" and the symbol $\geq$ means "is greater than or equal to." These symbols are used to form **inequalities**.

As shown in the figure below, $-6 < -1$ (since -6 is to the left of -1) and $|-6| > |-1|$ (since 6 is to the right of 1).

$|-1|$ $|-6|$

EXAMPLE 2

Write out the meaning of each inequality and determine whether it is a true statement.

a) $-7 < -2$ **b)** $4 > -1$ **c)** $-3 \geq -2$

d) $5 \leq 6$ **e)** $6 \leq 6$

SOLUTION

Inequality	*Meaning*
a) $-7 < -2$	"-7 is less than -2" is true because -7 is to the left of -2.
b) $4 > -1$	"4 is greater than -1" is true because 4 is to the right of -1.
c) $-3 \geq -2$	"-3 is greater than or equal to -2" is false because -3 is to the left of -2.
d) $5 \leq 6$	"5 is less than or equal to 6" is true if either $5 < 6$ or $5 = 6$. Since $5 < 6$ is true, $5 \leq 6$ is true.
e) $6 \leq 6$	"6 is less than or equal to 6" is true because $6 = 6$ is true.

TRY EXERCISE 23

Addition, Subtraction, and Opposites

We are now ready to review the addition of real numbers.

Addition of Two Real Numbers

1. *Positive numbers*: Add the numbers. The result is positive.
2. *Negative numbers*: Add absolute values. Make the answer negative.
3. *A negative number and a positive number*: If the numbers have the same absolute value, the answer is 0. Otherwise, subtract the smaller absolute value from the larger one.

   a) If the positive number has the greater absolute value, make the answer positive.
   b) If the negative number has the greater absolute value, make the answer negative.

4. *One number is zero*: The sum is the other number.

EXAMPLE   3   Add: **(a)** $-9 + (-5)$; **(b)** $-3.24 + 8.7$; **(c)** $-\frac{3}{4} + \frac{1}{3}$.

SOLUTION

a) $-9 + (-5)$ We add the absolute values, getting 14. The answer is *negative*: $-9 + (-5) = -14$.

b) $-3.24 + 8.7$ The absolute values are 3.24 and 8.7. Subtract 3.24 from 8.7 to get 5.46. The positive number is further from 0, so the answer is *positive*: $-3.24 + 8.7 = 5.46$.

c) $-\frac{3}{4} + \frac{1}{3} = -\frac{9}{12} + \frac{4}{12}$ The absolute values are $\frac{9}{12}$ and $\frac{4}{12}$. Subtract to get $\frac{5}{12}$. The negative number is further from 0, so the answer is *negative*: $-\frac{3}{4} + \frac{1}{3} = -\frac{5}{12}$.   ▶ **TRY EXERCISE** ▶ 37

When numbers like 7 and -7 are added, the result is 0. Such numbers are called **opposites**, or **additive inverses**, of one another. The sum of two additive inverses is the **additive identity**, 0.

The Law of Opposites

For any two numbers a and $-a$,

$$a + (-a) = 0.$$

(The sum of opposites is 0.)

EXAMPLE   4   Find the opposite: **(a)** -17.5; **(b)** $\frac{4}{5}$; **(c)** 0.

SOLUTION

a) The opposite of -17.5 is 17.5 because $-17.5 + 17.5 = 0$.

b) The opposite of $\frac{4}{5}$ is $-\frac{4}{5}$ because $\frac{4}{5} + \left(-\frac{4}{5}\right) = 0$.

c) The opposite of 0 is 0 because $0 + 0 = 0$.   ▶ **TRY EXERCISE** ▶ 53

To name the opposite, we use the symbol "$-$" and read the symbolism $-a$ as "the opposite of a."

CAUTION! $-a$ does not necessarily represent a negative number. In particular, when a is *negative*, $-a$ is *positive*.

EXAMPLE **5**

Find $-x$ for the following: **(a)** $x = -2$; **(b)** $x = \frac{3}{4}$.

SOLUTION

a) If $x = -2$, then $-x = -(-2) = 2$.   The opposite of -2 is 2.

b) If $x = \frac{3}{4}$, then $-x = -\frac{3}{4}$.   The opposite of $\frac{3}{4}$ is $-\frac{3}{4}$.

TRY EXERCISE 59

Using the notation of opposites, we can formally define absolute value.

> **Absolute Value**
>
> $$|x| = \begin{cases} x, & \text{if } x \geq 0, \\ -x, & \text{if } x < 0 \end{cases}$$
>
> (When x is nonnegative, the absolute value of x is x. When x is negative, the absolute value of x is the opposite of x. Thus, $|x|$ is never negative.)

A negative number is said to have a negative "sign" and a positive number a positive "sign." To subtract, we can add an opposite. This can be stated as: "Change the sign of the number being subtracted and then add."

EXAMPLE **6**

Subtract: **(a)** $5 - 9$; **(b)** $-1.2 - (-3.7)$; **(c)** $-\frac{4}{5} - \frac{2}{3}$.

SOLUTION

a) $5 - 9 = 5 + (-9)$ Change the sign and add.

$\qquad\quad = -4$

b) $-1.2 - (-3.7) = -1.2 + 3.7$ Instead of *subtracting negative* 3.7, we *add positive* 3.7.

$\qquad\qquad\qquad\quad = 2.5$

c) $-\frac{4}{5} - \frac{2}{3} = -\frac{4}{5} + \left(-\frac{2}{3}\right)$ Instead of *subtracting* $\frac{2}{3}$, we *add* the opposite, $-\frac{2}{3}$.

$\qquad\qquad = -\frac{12}{15} + \left(-\frac{10}{15}\right)$ Finding a common denominator

$\qquad\qquad = -\frac{22}{15}$

TRY EXERCISE 69

Multiplication, Division, and Reciprocals

Multiplication of real numbers can be regarded as repeated addition or as repeated subtraction that begins at 0. For example,

$$3 \cdot (-4) = 0 + (-4) + (-4) + (-4) = -12 \qquad \text{Adding } -4 \text{ three times}$$

and

$$(-2)(-5) = 0 - (-5) - (-5) = 0 + 5 + 5 = 10. \qquad \text{Subtracting } -5 \text{ twice}$$

When one factor is positive and one is negative, the product is negative. When both factors are positive or both are negative, the product is positive.

To divide, recall that the quotient $a \div b$ (also written a/b) is that number c for which $c \cdot b = a$. For example, $10 \div (-2) = -5$ since $(-5)(-2) = 10$. Thus the rules for division are just like those for multiplication.

TECHNOLOGY CONNECTION

Technology Connections highlight situations in which calculators (primarily graphing calculators) or computers can be used to enrich the learning experience. Most Technology Connections present information in a generic form—consult an outside reference for specific keystrokes.

Graphing calculators use two different keys for subtracting and writing negatives. The key labeled (−) is used to create a negative sign, whereas − is used for subtraction.

1. Use a graphing calculator to check Example 6.
2. Calculate: $-3.9 - (-4.87)$.

> **Multiplication or Division of Two Real Numbers**
>
> 1. To multiply or divide two numbers with *unlike signs*, multiply or divide their absolute values. The answer is *negative*.
> 2. To multiply or divide two numbers having the *same sign*, multiply or divide their absolute values. The answer is *positive*.

EXAMPLE   7   Multiply or divide as indicated: **(a)** $(-4)9$; **(b)** $\left(-\frac{2}{3}\right)\left(-\frac{3}{8}\right)$; **(c)** $20 \div (-4)$; **(d)** $\frac{-45}{-15}$.

SOLUTION

a) $(-4)9 = -36$ Multiply absolute values. The answer is negative.

b) $\left(-\frac{2}{3}\right)\left(-\frac{3}{8}\right) = \frac{6}{24} = \frac{1}{4}$ Multiply absolute values. The answer is positive.

c) $20 \div (-4) = -5$ Divide absolute values. The answer is negative.

d) $\frac{-45}{-15} = 3$ Divide absolute values. The answer is positive.

> **TRY EXERCISE** 81

Note that since

$$\frac{-8}{2} = \frac{8}{-2} = -\frac{8}{2} = -4,$$

we have the following generalization.

> **The Sign of a Fraction**
>
> For any number a and any nonzero number b,
>
> $$\frac{-a}{b} = \frac{a}{-b} = -\frac{a}{b}.$$

Recall that

$$\frac{a}{b} = \frac{a}{1} \cdot \frac{1}{b} = a \cdot \frac{1}{b}.$$

That is, rather than divide by b, we can multiply by $\frac{1}{b}$. Provided that b is not 0, the numbers b and $\frac{1}{b}$ are called **reciprocals**, or **multiplicative inverses**, of each other. The product of two multiplicative inverses is the **multiplicative identity**, 1.

> **The Law of Reciprocals**
>
> For any two numbers a and $\frac{1}{a}$ $(a \neq 0)$,
>
> $$a \cdot \frac{1}{a} = 1.$$
>
> (The product of reciprocals is 1.)

Every number except 0 has exactly one reciprocal. The number 0 has no reciprocal.

EXAMPLE **8**   Find the reciprocal: **(a)** $\frac{7}{8}$; **(b)** $-\frac{3}{4}$; **(c)** -8.

SOLUTION

a) The reciprocal of $\frac{7}{8}$ is $\frac{8}{7}$ because $\frac{7}{8} \cdot \frac{8}{7} = 1$.

b) The reciprocal of $-\frac{3}{4}$ is $-\frac{4}{3}$.

c) The reciprocal of -8 is $\frac{1}{-8}$, or $-\frac{1}{8}$.

> TRY EXERCISE 101

To divide, we can multiply by a reciprocal. We sometimes say that we "invert and multiply."

EXAMPLE **9**   Divide: **(a)** $-\frac{1}{4} \div \frac{3}{5}$; **(b)** $-\frac{6}{7} \div (-10)$.

SOLUTION

a) $-\frac{1}{4} \div \frac{3}{5} = -\frac{1}{4} \cdot \frac{5}{3}$ "Inverting" $\frac{3}{5}$ and changing division to multiplication

$\qquad\qquad = -\frac{5}{12}$

b) $-\frac{6}{7} \div (-10) = -\frac{6}{7} \cdot \left(-\frac{1}{10}\right) = \frac{6}{70}$, or $\frac{3}{35}$ Multiplying by the reciprocal of the divisor

> TRY EXERCISE 109

Thus far, we have never divided by 0 or, equivalently, had a denominator of 0. There is a reason for this. Suppose 5 were divided by 0. The answer would have to be a number that, when multiplied by 0, gave 5. But any number times 0 is 0. Thus we cannot divide 5 or any other nonzero number by 0.

What if we divide 0 by 0? In this case, our solution would need to be some number that, when multiplied by 0, gave 0. But then *any* number would work as a solution to $0 \div 0$. This could lead to contradictions so we agree to exclude division of 0 by 0 also.

Division by Zero

We never divide by 0. If asked to divide a nonzero number by 0, we say that the answer is *undefined*. If asked to divide 0 by 0, we say that the answer is *indeterminate*. Thus,

$\frac{7}{0}$ is *undefined*   and $\frac{0}{0}$ is *indeterminate*.

The rules for order of operations discussed in Section 1.1 apply to *all* real numbers, regardless of their signs.

EXAMPLE **10**   Simplify: **(a)** $(-5)^2$; **(b)** -5^2.

SOLUTION   An exponent is always written immediately after the base. Thus in the expression $(-5)^2$, the base is (-5); in the expression -5^2, the base is 5.

a) $(-5)^2 = (-5)(-5) = 25$ Squaring -5

b) $-5^2 = -(5 \cdot 5) = -25$ Squaring 5 and then taking the opposite

Note that $(-5)^2 \neq -5^2$.

> TRY EXERCISE 115

EXAMPLE 11

Simplify: $7 - 5^2 + 6 \div 2(-5)^2$.

SOLUTION

$$7 - 5^2 + 6 \div 2(-5)^2 = 7 - 25 + 6 \div 2 \cdot 25 \qquad \text{Simplifying } 5^2 \text{ and } (-5)^2$$

$$= 7 - 25 + 3 \cdot 25 \qquad \text{Dividing}$$

$$= 7 - 25 + 75 \qquad \text{Multiplying}$$

$$= -18 + 75 \qquad \text{Subtracting}$$

$$= 57 \qquad \text{Adding}$$

TRY EXERCISE 121

In addition to parentheses, brackets, and braces, groupings may be indicated by a fraction bar, an absolute-value symbol, or a radical sign ($\sqrt{}$).

EXAMPLE 12

Calculate: $\dfrac{12|7 - 9| + 4 \cdot 5}{(-3)^4 + 2^3}$.

SOLUTION   We simplify the numerator and the denominator and divide the results:

$$\frac{12|7 - 9| + 4 \cdot 5}{(-3)^4 + 2^3} = \frac{12|-2| + 20}{81 + 8}$$

$$= \frac{12(2) + 20}{89}$$

$$= \frac{44}{89}. \qquad \text{Multiplying and adding}$$

TRY EXERCISE 127

The Commutative, Associative, and Distributive Laws

When a pair of real numbers is added or multiplied, the order in which the numbers are written does not affect the result.

The Commutative Laws

For any real numbers a and b,

$$a + b = b + a \qquad a \cdot b = b \cdot a.$$
(for Addition)   (for Multiplication)

The commutative laws provide one way of writing *equivalent expressions*.

Equivalent Expressions

Two expressions that have the same value for all possible replacements are called *equivalent expressions*.

Much of this text is devoted to finding equivalent expressions.

EXAMPLE 13   Use a commutative law to write an expression equivalent to $7x + 9$.

SOLUTION   Using the commutative law of addition, we have

$7x + 9 = 9 + 7x$.   Changing the order in which the terms are written

We can also use the commutative law of multiplication to write

$7 \cdot x + 9 = x \cdot 7 + 9$.   Changing the order in which the factors are written

The expressions $7x + 9$, $9 + 7x$, and $x \cdot 7 + 9$ are all equivalent. They name the same number for any replacement of x.   **TRY EXERCISE** 133

The commutative laws allow us to write equivalent expressions by changing order. The *associative laws* also enable us to form equivalent expressions by changing grouping.

The Associative Laws

For any real numbers a, b, and c,

$$a + (b + c) = (a + b) + c; \qquad a \cdot (b \cdot c) = (a \cdot b) \cdot c.$$
(for Addition)   (for Multiplication)

EXAMPLE 14   Write an expression equivalent to $(3x + 7y) + 9z$, using the associative law of addition.

SOLUTION   We have

$(3x + 7y) + 9z = 3x + (7y + 9z)$.   Regrouping the terms

The expressions $(3x + 7y) + 9z$ and $3x + (7y + 9z)$ are equivalent. They name the same number for any replacements of x, y, and z.   **TRY EXERCISE** 137

The *distributive law* that follows provides still another way of forming equivalent expressions. In essence, the distributive law allows us to rewrite the *product* of a and $b + c$ as the *sum* of ab and ac.

The Distributive Law

For any real numbers a, b, and c,

$$a(b + c) = ab + ac.$$

EXAMPLE 15   Obtain an expression equivalent to $5x(y + 4)$ by multiplying.

STUDENT NOTES ──────

The commutative, associative, and distributive laws are used so often in this course that it is worth the effort to memorize them.

SOLUTION   We use the distributive law to get

$$
\begin{aligned}
5x(y + 4) &= 5x \cdot y + 5x \cdot 4 && \text{Using the distributive law} \\
&= 5xy + 5 \cdot 4 \cdot x && \text{Using the commutative law of multiplication} \\
&= 5xy + 20x. && \text{Simplifying}
\end{aligned}
$$

The expressions $5x(y + 4)$ and $5xy + 20x$ are equivalent. They name the same number for any replacements of x and y.   **TRY EXERCISE** 141

When we reverse what we did in Example 15, we say that we are **factoring** an expression. This allows us to rewrite a sum or a difference as a product.

EXAMPLE **16**   Obtain an expression equivalent to $3x - 6$ by factoring.

SOLUTION   We use the distributive law to get

$$3x - 6 = 3(x - 2).$$

> **TRY EXERCISE** 149

In Example 16, since the product of 3 and $x - 2$ is $3x - 6$, we say that 3 and $x - 2$ are **factors** of $3x - 6$. Thus the word "factor" can act as a noun or as a verb.

1.2   EXERCISE SET

👋 *Concept Reinforcement*   *Classify each statement as either true or false.*

1. The sum of two negative numbers is always negative.

2. The product of two negative numbers is always negative.

3. The product of a negative number and a positive number is always negative.

4. The sum of a negative number and a positive number is always negative.

5. The sum of a negative number and a positive number is always positive.

6. If a and b are negative, with $a < b$, then $|a| > |b|$.

7. If a and b are positive, with $a < b$, then $|a| > |b|$.

8. The commutative law of addition states that for all real numbers a and b, $a + b$ and $b + a$ are equivalent.

9. The associative law of multiplication states that for all real numbers a, b, and c, $(ab)c$ is equivalent to $a(bc)$.

10. The distributive law states that the order in which two numbers are multiplied does not change the result.

Find each absolute value.

11. $|-10|$ 12. $|-3|$ 13. $|7|$

14. $|13|$ 15. $|-46.8|$ 16. $|-36.9|$

17. $|0|$ 18. $|3\frac{3}{4}|$ 19. $|1\frac{7}{8}|$

20. $|7.24|$ 21. $|-4.21|$ 22. $|-5.309|$

Write the meaning of each inequality, and determine whether it is a true statement.

23. $-5 \le -4$ 24. $-2 \le -8$

25. $-9 > 1$ 26. $-9 < 1$

27. $0 \ge -5$ 28. $9 \le 9$

29. $-8 < -3$ 30. $7 \ge -8$

31. $-4 \ge -4$ 32. $2 < 2$

33. $-5 < -5$ 34. $-2 > -12$

Add.

35. $4 + 8$ 36. $5 + 7$

37. $(-3) + (-9)$ 38. $(-6) + (-8)$

39. $-5.3 + 2.8$ 40. $9.3 + (-5.7)$

41. $\frac{2}{7} + \left(-\frac{3}{5}\right)$ 42. $\frac{3}{8} + \left(-\frac{2}{5}\right)$

43. $-3.26 + (-5.8)$ 44. $-2.1 + (-7.5)$

45. $-\frac{1}{9} + \frac{2}{3}$ 46. $-\frac{1}{2} + \frac{4}{5}$

47. $-6.25 + 0$ 48. $0 + (-3.69)$

49. $4.19 + (-4.19)$ 50. $-8.35 + 8.35$

51. $-18.3 + 22.1$ 52. $21.7 + (-28.3)$

Find the opposite, or additive inverse.

53. 2.37 54. 6.98 55. -56

56. -11 57. 0 58. $-2\frac{1}{3}$

Find $-x$ for each of the following.

59. $x = 8$ 60. $x = 12$

61. $x = -15$ 62. $x = -18$

63. $x = -4.67$ 64. $x = 3.14$

65. $x = 0$ 66. $x = -7$

Subtract.

67. $10 - 4$

68. $9 - 1$

69. $4 - 10$

70. $1 - 9$

71. $-5 - (-12)$

72. $-3 - (-7)$

73. $-5 - 14$

74. $-9 - 8$

75. $2.7 - 5.8$

76. $3.7 - 4.2$

77. $-\frac{3}{5} - \frac{1}{2}$

78. $-\frac{2}{3} - \frac{1}{5}$

79. $0 - (-5.37)$

80. $0 - 9.09$

Multiply.

81. $(-3)8$

82. $(-5)9$

83. $(-2)(-11)$

84. $(-6)(-7)$

85. $(4.2)(-5)$

86. $(3.5)(-8)$

87. $\frac{3}{7}(-1)$

88. $-1 \cdot \frac{2}{5}$

89. $(-17.45) \cdot 0$

90. 15.2×0

91. $-\frac{2}{3}\left(\frac{3}{4}\right)$

92. $\frac{5}{6}\left(-\frac{3}{10}\right)$

Divide.

93. $\frac{-28}{-7}$

94. $\frac{-18}{-6}$

95. $\frac{-100}{25}$

96. $\frac{40}{-4}$

97. $\frac{73}{-1}$

98. $\frac{-62}{1}$

99. $\frac{0}{-7}$

100. $\frac{0}{-11}$

Find the reciprocal, or multiplicative inverse, if it exists.

101. 8

102. -7

103. $-\frac{5}{7}$

104. $\frac{4}{3}$

105. 0

106. $-\frac{9}{10}$

Divide.

107. $\frac{3}{5} \div \frac{6}{7}$

108. $\frac{2}{3} \div \frac{5}{6}$

109. $-\frac{3}{5} \div \frac{1}{2}$

110. $\left(-\frac{4}{7}\right) \div \frac{1}{3}$

111. $-\frac{2}{9} \div (-8)$

112. $\left(-\frac{2}{11}\right) \div (-6)$

Aha! **113.** $-\frac{12}{7} \div \left(-\frac{12}{7}\right)$

114. $\left(-\frac{2}{7}\right) \div (-1)$

Calculate using the rules for order of operations. If an expression is undefined, state this.

115. -10^2

116. $(-10)^2$

117. $-(-3)^2$

118. $-(-2)^2$

119. $(2 - 5)^2$

120. $2^2 - 5^2$

121. $9 - (8 - 3 \cdot 2^3)$

122. $19 - (4 + 2 \cdot 3^2)$

123. $\frac{5 \cdot 2 - 4^2}{27 - 2^4}$

124. $\frac{7 \cdot 3 - 5^2}{9 + 4 \cdot 2}$

125. $\frac{3^4 - (5 - 3)^4}{8 - 2^3}$

126. $\frac{4^3 - (7 - 4)^2}{3^2 - 7}$

127. $\frac{(2 - 3)^3 - 5|2 - 4|}{7 - 2 \cdot 5^2}$

128. $\frac{8 \div 4 \cdot 6|4^2 - 5^2|}{9 - 4 + 11 - 4^2}$

129. $|2^2 - 7|^3 + 4$

130. $|-2 - 3| \cdot 4^2 - 3$

131. $32 - (-5)^2 + 15 \div (-3) \cdot 2$

132. $43 - (-9 + 2)^2 + 18 \div 6 \cdot (-2)$

Write an equivalent expression using a commutative law. Answers may vary.

133. $6 + xy$

134. $4a + 7b$

135. $-9(ab)$

136. $(7x)y$

Write an equivalent expression using an associative law.

137. $(3x)y$

138. $-7(ab)$

139. $(3y + 4) + 10$

140. $x + (2y + 5)$

Write an equivalent expression using the distributive law.

141. $7(x + 1)$

142. $3(a + 5)$

143. $5(m - n)$

144. $6(s - t)$

145. $-5(2a + 3b)$

146. $-2(3c + 5d)$

147. $9a(b - c + d)$

148. $5x(y - z + w)$

Find an equivalent expression by factoring.

149. $8a + 8b$

150. $5d + 30$

151. $9p - 3$

152. $15x - 3$

153. $7x - 21y + 14z$

154. $6y - 9x - 3w$

Aha! **155.** $255 - 34b$

156. $132a + 33$

157. Describe in your own words a method for determining the sign of the sum of a positive number and a negative number.

158. Explain the difference between the expressions "five is less than x" and "five less than x."

Skill Review

To the student and the instructor: Exercises included for Skill Review cover skills previously studied in the text. Usually these exercises provide preparation for the next section of the text. The section(s) in which these types of exercises first appeared is shown in brackets. Answers to all Skill Review exercises appear at the back of the book.

To prepare for Section 1.3, review evaluating algebraic expressions (Section 1.1).

Evaluate. [1.1]

159. $2(x + 5)$ and $2x + 10$, for $x = 3$

160. $2a - 3$ and $a - 3 + a$, for $a = 7$

Synthesis

161. Explain in your own words why 7/0 is undefined.

162. Write a sentence in which the word "factor" appears once as a verb and once as a noun.

Insert one pair of parentheses to convert each false statement into a true statement.

163. $8 - 5^3 + 9 = 36$

164. $2 \cdot 7 + 3^2 \cdot 5 = 104$

165. $5 \cdot 2^3 \div 3 - 4^4 = 40$

166. $2 - 7 \cdot 2^2 + 9 = -11$

Calculate using the rules for order of operations.

167. $17 - \sqrt{11 - (3 + 4)} \div [-5 - (-6)]^2$

168. $15 - 1 + \sqrt{5^2 - (3 + 1)^2}(-1)$

169. Find the greatest value of a for which $|a| \geq 6.2$ and $a < 0$.

170. Use the commutative, associative, and distributive laws to show that $5(a + bc)$ is equivalent to $c(b \cdot 5) + a \cdot 5$. Use only one law in each step of your work.

171. Are subtraction and division commutative? Why or why not?

172. Are subtraction and division associative? Why or why not?

1.3   Solving Equations

Equivalent Equations • The Addition and Multiplication Principles •
Combining Like Terms • Types of Equations

STUDY SKILLS

Do the Exercises

• When you have completed the odd-numbered exercises in your assignment, you can check your answers at the back of the book. If you miss any, closely examine your work and, if necessary, seek help (see the Study Skills on p. 9).

• Whether or not your instructor assigns the even-numbered exercises, try to do some on your own. Check your answers later with a friend or your instructor.

Solving equations is an essential part of problem solving in algebra. In this section, we review and practice solving basic equations.

Equivalent Equations

In Section 1.1, we saw that the solution of $143 + x = 582$ is 439. That is, when x is replaced with 439, the equation $143 + x = 582$ is a true statement. It is important to know how to find such a solution using the principles of algebra. These principles are used to produce *equivalent equations* from which solutions are easily found.

> **Equivalent Equations**
>
> Two equations are *equivalent* if they have the same solution(s).

EXAMPLE 1   Determine whether $4x = 12$ and $10x = 30$ are equivalent equations.

SOLUTION   The equation $4x = 12$ is true only when x is 3. Similarly, $10x = 30$ is true only when x is 3. Since both equations have the same solution, they are equivalent.

TRY EXERCISE 11

Note that the equation $x = 3$ is also equivalent to the equations in Example 1.

EXAMPLE 2   Determine whether $3x = 4x$ and $3/x = 4/x$ are equivalent equations.

SOLUTION   Note that 0 is a solution of $3x = 4x$. Since neither $3/x$ nor $4/x$ is defined for $x = 0$, the equations $3x = 4x$ and $3/x = 4/x$ are *not* equivalent.

TRY EXERCISE 15

The Addition and Multiplication Principles

Suppose that a and b represent the same number and that some number c is added to a. If c is also added to b, we will get two equal sums, since a and b are the same number. The same is true if we multiply both a and b by c. In this manner, we can produce equivalent equations.

The Addition and Multiplication Principles for Equations

For any real numbers a, b, and c:

a) $a = b$ is equivalent to $a + c = b + c$;
b) $a = b$ is equivalent to $a \cdot c = b \cdot c$, provided $c \neq 0$.

As shown in Examples 3 and 4, either a or b usually represents a variable expression.

EXAMPLE 3   Solve: $y - 4.7 = 13.9$.

SOLUTION

$$y - 4.7 = 13.9$$

$$y - 4.7 + 4.7 = 13.9 + 4.7 \qquad \text{Using the addition principle; adding 4.7 to both sides}$$

$$y + 0 = 13.9 + 4.7 \qquad \text{Using the law of opposites}$$

$$y = 18.6 \qquad \text{The solution of this equation is 18.6.}$$

Check:

$$\begin{array}{c|c} y - 4.7 = 13.9 \\ \hline 18.6 - 4.7 & 13.9 \qquad \text{Substituting 18.6 for } y \\ 13.9 \overset{?}{=} 13.9 & \text{TRUE} \end{array}$$

The solution is 18.6.

TRY EXERCISE 17

In Example 3, why did we add 4.7 to both sides? Because 4.7 is the opposite of -4.7 and we wanted y alone on one side of the equation. Adding 4.7 gave us $y + 0$, or just y, on the left side. This led to the equivalent equation $y = 18.6$, from which the solution, 18.6, is immediately apparent.

EXAMPLE 4   Solve: $\frac{2}{5}x = -\frac{9}{10}$.

SOLUTION   We have

$$\frac{2}{5}x = -\frac{9}{10}$$

$$\frac{5}{2} \cdot \frac{2}{5}x = \frac{5}{2} \cdot \left(-\frac{9}{10}\right) \qquad \text{Using the multiplication principle, we multiply by } \frac{5}{2}, \text{ the reciprocal of } \frac{2}{5}.$$

$$1x = -\frac{45}{20} \qquad \text{Using the law of reciprocals}$$

$$x = -\frac{9}{4}. \qquad \text{Simplifying}$$

The check is left to the student. The solution is $-\frac{9}{4}$.

TRY EXERCISE 19

In Example 4, why did we multiply by $\frac{5}{2}$? Because $\frac{5}{2}$ is the reciprocal of $\frac{2}{5}$ and we wanted x alone on one side of the equation. When we multiplied by $\frac{5}{2}$, we got $1x$, or just x, on the left side. This led to the equivalent equation $x = -\frac{9}{4}$, from which the solution, $-\frac{9}{4}$, is clear.

There is no need for a subtraction principle or a division principle because subtraction can be regarded as adding opposites and division can be regarded as multiplying by reciprocals.

Combining Like Terms

In an expression like $8a^5 + 17 + 4/b + (-6a^3b)$, the parts that are separated by addition signs are called *terms*. A **term** is a number, a variable, a product of numbers and/or variables, or a quotient of numbers and/or variables. Thus, $8a^5$, 17, $4/b$, and $-6a^3b$ are terms in $8a^5 + 17 + 4/b + (-6a^3b)$. When terms have variable factors that are exactly the same, we refer to those terms as **like**, or **similar**, **terms**. Thus, $3x^2y$ and $-7x^2y$ are similar terms, but $3x^2y$ and $4xy^2$ are not. We can often simplify expressions by **combining**, or **collecting**, **like terms**.

EXAMPLE **5**   Write an equivalent expression by combining like terms: $3a + 5a^2 - 7a + a^2$.

SOLUTION

$$3a + 5a^2 - 7a + a^2 = 3a - 7a + 5a^2 + a^2 \qquad \text{Using the commutative law}$$

$$= (3 - 7)a + (5 + 1)a^2 \qquad \text{Using the distributive law. Note that } a^2 = 1a^2.$$

$$= -4a + 6a^2 \qquad \text{TRY EXERCISE } 39$$

Sometimes we must use the distributive law to remove grouping symbols before like terms can be identified and then combined. Remember to remove the innermost grouping symbols first.

EXAMPLE **6**   Simplify: $3x + 2[4 + 5(x - 2y)]$.

SOLUTION

$$3x + 2[4 + 5(x - 2y)] = 3x + 2[4 + 5x - 10y] \qquad \text{Using the distributive law}$$

$$= 3x + 8 + 10x - 20y \qquad \text{Using the distributive law again}$$

$$= 13x + 8 - 20y \qquad \text{Combining like terms}$$

$$\text{TRY EXERCISE } 43$$

The product of a number and -1 is its opposite, or additive inverse. For example,

$$-1 \cdot 8 = -8 \qquad \text{(the opposite of 8).}$$

Thus we have $-8 = -1 \cdot 8$, and in general, $-x = -1 \cdot x$. We can use this fact along with the distributive law when parentheses are preceded by a negative sign or subtraction.

EXAMPLE 7   Simplify $-(a - b)$, using multiplication by -1.

SOLUTION   We have

$$-(a - b) = -1 \cdot (a - b)$$ Replacing $-$ with multiplication by -1

$$= -1 \cdot a - (-1) \cdot b$$ Using the distributive law

$$= -a - (-b)$$ Replacing $-1 \cdot a$ with $-a$ and $(-1) \cdot b$ with $-b$

$$= -a + b, \text{ or } b - a.$$ Try to go directly to this step.

The expressions $-(a - b)$ and $b - a$ are equivalent. They represent the same number for all replacements of a and b.

Example 7 illustrates a useful shortcut worth remembering:

The opposite of $a - b$ is $-a + b$,   or $b - a$.

$$-(a - b) = b - a$$

EXAMPLE 8   Simplify: $9x - 5y - (5x + y - 7)$.

SOLUTION

$$9x - 5y - (5x + y - 7) = 9x - 5y - 5x - y + 7$$ Using the distributive law

$$= 4x - 6y + 7$$ Combining like terms

TRY EXERCISE 49

For many equations, before we use the addition and multiplication principles to *solve*, we must first multiply and combine like terms to *simplify* the expressions within the equation.

EXAMPLE 9   Solve: $5x - 2(x - 5) = 7x - 2$.

SOLUTION

$$5x - 2(x - 5) = 7x - 2$$

$$5x - 2x + 10 = 7x - 2$$ Using the distributive law

$$3x + 10 = 7x - 2$$ Combining like terms

$$3x + 10 - 3x = 7x - 2 - 3x$$ Using the addition principle; adding $-3x$, the opposite of $3x$, to both sides

$$10 = 4x - 2$$ Combining like terms

$$10 + 2 = 4x - 2 + 2$$ Using the addition principle

$$12 = 4x$$ Simplifying

$$\frac{1}{4} \cdot 12 = \frac{1}{4} \cdot 4x$$ Using the multiplication principle; multiplying both sides by $\frac{1}{4}$, the reciprocal of 4

$$3 = x$$ Using the law of reciprocals; simplifying

Check:

$$\begin{array}{c|c}
\multicolumn{2}{c}{5x - 2(x - 5) = 7x - 2} \\
\hline
5 \cdot 3 - 2(3 - 5) & 7 \cdot 3 - 2 \\
15 - 2(-2) & 21 - 2 \\
15 + 4 & 19 \\
19 \stackrel{?}{=} 19 & \text{TRUE}
\end{array}$$

The solution is 3.

▶ TRY EXERCISE 79

Types of Equations

In Examples 3, 4, and 9, we solved *linear equations*. A **linear equation** in one variable—say, x—is an equation equivalent to one of the form $ax = b$ with a and b constants and $a \neq 0$. Since $x = x^1$, the variable in a linear equation is always raised to the first power.

Every equation falls into one of three categories. An **identity** is an equation, like $x + 5 = 3 + x + 2$, that is true for all replacements. A **contradiction** is an equation, like $n = n + 1$, that is *never* true. A **conditional equation**, like $2x + 5 = 17$, is sometimes true and sometimes false, depending on what the replacement of x is. Most of the equations examined in this text are conditional.

EXAMPLE **10**   Solve each of the following equations and classify the equation as an identity, a contradiction, or a conditional equation.

a) $2x + 7 = 7(x + 1) - 5x$ **b)** $3x - 5 = 3(x - 2) + 4$

c) $3 - 8x = 5 - 7x$

SOLUTION

a) $2x + 7 = 7(x + 1) - 5x$

$\quad 2x + 7 = 7x + 7 - 5x$ Using the distributive law

$\quad 2x + 7 = 2x + 7$ Combining like terms

The equation $2x + 7 = 2x + 7$ is true regardless of what x is replaced with, so all real numbers are solutions. Note that $2x + 7 = 2x + 7$ is equivalent to $2x = 2x$, $7 = 7$, or $0 = 0$. All real numbers are solutions and the equation is an identity.

STUDENT NOTES

Write out the steps you use to solve an equation, rather than working only in your head. Work neatly, keeping in mind that the steps in your solution may differ from those in another student's solution. What is important is that each step produce an equivalent equation.

b)

$\quad\quad 3x - 5 = 3(x - 2) + 4$

$\quad\quad\quad 3x - 5 = 3x - 6 + 4$ Using the distributive law

$\quad\quad\quad 3x - 5 = 3x - 2$ Combining like terms

$-3x + 3x - 5 = -3x + 3x - 2$ Using the addition principle

$\quad\quad\quad\quad -5 = -2$

Since the original equation is equivalent to $-5 = -2$, which is false regardless of the choice of x, the original equation has no solution. There is no solution of $3x - 5 = 3(x - 2) + 4$. The equation is a contradiction.

c)

$\quad\quad\quad 3 - 8x = 5 - 7x$

$\quad 3 - 8x + 7x = 5 - 7x + 7x$ Using the addition principle

$\quad\quad\quad 3 - x = 5$ Simplifying

$-3 + 3 - x = -3 + 5$ Using the addition principle

$\quad\quad\quad -x = 2$ Simplifying

$\quad\quad\quad\quad x = \dfrac{2}{-1}, \text{ or } -2$ Dividing both sides by -1 or multiplying both sides by $\dfrac{1}{-1}$, or -1

There is one solution, −2. For other choices of x, the equation is false. This equation is conditional since it can be true or false, depending on the replacement for x.

TRY EXERCISE ▸ 85

We will sometimes refer to the set of solutions, or **solution set**, of a particular equation. Thus the solution set for Example 10(c) is {−2}. The solution set for Example 10(a) is simply ℝ, the set of all real numbers, and the solution set for Example 10(b) is the **empty set**, denoted ∅ or { }. As its name suggests, the empty set is the set containing no elements.

1.3   EXERCISE SET

↶ *Concept Reinforcement*   *Classify each of the following as either a pair of equivalent equations or a pair of equivalent expressions.*

1. $2(x + 7)$, $2x + 14$

2. $2(x + 7) = 11$, $2x + 14 = 11$

3. $3(t − 5) = 18$, $3t − 15 = 18$

4. $3(t − 5)$, $3t − 15$

5. $4x − 9 = 7$, $4x = 16$

6. $4x − 9$, $5x − 9 − x$

7. $8t + 5 − 2t + 1$, $6t + 6$

8. $5t − 2 + t = 8$, $6t = 10$

9. $6x − 3 = 10x + 5$, $−8 = 4x$

10. $9x − 2$, $13x − 6 − 4x + 4$

Determine whether the two equations in each pair are equivalent.

11. $3t = 21$ and $t + 4 = 11$

12. $3t = 27$ and $t − 3 = 5$

13. $12 − x = 3$ and $2x = 20$

14. $3x − 4 = 8$ and $3x = 12$

15. $5x = 2x$ and $\dfrac{4}{x} = 0$

16. $6 = 2x$ and $5 = \dfrac{2}{3 − x}$

Solve. Be sure to check.

17. $x − 2.9 = 13.4$

18. $y + 4.3 = 11.2$

19. $8t = 72$

20. $9t = 63$

21. $\frac{2}{3}x = 30$

22. $\frac{5}{4}x = −80$

23. $4a + 25 = 9$

24. $5a − 11 = 24$

25. $2y − 8 = 9$

26. $3y + 4 = 2$

Simplify to form an equivalent expression by combining like terms. Use the distributive law as needed.

27. $3x + 7x$

28. $9x − 4x$

29. $9t^2 + t^2$

30. $7a^2 + a^2$

31. $16a − a$

32. $11t − t$

33. $n − 8n$

34. $p − 3p$

35. $5x − 3x + 8x$

36. $3x − 11x + 2x$

37. $4x − 2x^2 + 3x$

38. $9a − 5a^2 + 4a$

39. $18p − 12 + 3p + 8$

40. $14y + 6 − 9y + 7$

41. $−7t^2 + 3t + 5t^3 − t^3 + 2t^2 − t$

42. $−9n + 8n^2 + n^3 − 2n^2 − 3n + 4n^3$

43. $2x + 3(5x − 7)$

44. $5x + 4(x + 11)$

45. $7a − (2a + 5)$

46. $x − (5x + 9)$

47. $m − (6m − 2)$

48. $5a − (4a − 3)$

49. $3d − 7 − (5 − 2d)$

50. $8x − 9 − (7 − 5x)$

51. $2(x − 3) + 4(7 − x)$

52. $3(y + 6) + 5(2 − 4y)$

53. $3p − 4 − 2(p + 6)$

54. $8c − 1 − 3(2c + 1)$

55. $−2(a − 5) − [7 − 3(2a − 5)]$

56. $−3(b + 2) − [9 − 5(8b − 1)]$

57. $5\{−2x + 3[2 − 4(5x + 1)]\}$

58. $7\{-7x + 8[5 - 3(4x + 6)]\}$

59. $8y - \{6[2(3y - 4) - (7y + 1)] + 12\}$

60. $2y + \{7[3(2y - 5) - (8y + 7)] + 9\}$

Solve. Be sure to check.

61. $4x + 5x = 63$ **62.** $3x - 7x = 60$

63. $\frac{1}{4}y - \frac{2}{3}y = 5$ **64.** $\frac{3}{5}t - \frac{1}{2}t = 3$

65. $4(t - 3) - t = 6$ **66.** $2(t + 5) + t = 4$

67. $3(x + 4) = 7x$ **68.** $3(y + 5) = 8y$

69. $70 = 10(3t - 2)$ **70.** $27 = 9(5y - 2)$

71. $1.8(2 - n) = 9$ **72.** $2.1(3 - x) = 8.4$

73. $5y - (2y - 10) = 25$

74. $8x - (3x - 5) = 40$

75. $\frac{9}{10}y - \frac{7}{10} = \frac{21}{5}$

76. $\frac{4}{5}t - \frac{3}{10} = \frac{2}{5}$

77. $7r - 2 + 5r = 6r + 6 - 4r$

78. $9m - 15 - 2m = 6m - 1 - m$

79. $\frac{2}{3}(x - 2) - 1 = \frac{1}{4}(x - 3)$

80. $\frac{1}{4}(6t + 48) - 20 = -\frac{1}{3}(4t - 72)$

81. $2(t - 5) - 3(2t - 7) = 12 - 5(3t + 1)$

82. $4t + 8 - 6(2t - 1) = 3(4t - 3) - 7(t - 2)$

83. $3[2 - 4(x - 1)] = 3 - 4(x + 2)$

84. $5 + 2(x - 3) = 2[5 - 4(x + 2)]$

Find each solution set. Then classify each equation as a conditional equation, an identity, or a contradiction.

85. $7x - 2 - 3x = 4x$

86. $3t + 5 + t = 5 + 4t$

87. $2 + 9x = 3(4x + 1) - 1$

88. $4 + 7x = 7(x + 1)$

Aha! **89.** $-9t + 2 = -9t - 7(6 \div 2(49) + 8)$

90. $-9t + 2 = 2 - 9t - 5(8 \div 4(1 + 3^4))$

91. $2\{9 - 3[-2x - 4]\} = 12x + 42$

92. $3\{7 - 2[7x - 4]\} = -40x + 45$

93. Explain the difference between the statements "The equation has no solution." and "The solution of the equation is zero."

94. As the first step in solving
$$2x + 5 = -3,$$
Pat multiplies both sides by $\frac{1}{2}$. Is this incorrect? Why or why not?

Skill Review

To prepare for Section 1.4, review translating phrases to algebraic expressions (Section 1.1).

Translate to an algebraic expression. [1.1]

95. Nine less than twice a number

96. The sum of five and half of a number

Synthesis

97. Explain how the distributive and commutative laws can be used to rewrite $3x + 6y + 4x + 2y$ as $7x + 8y$.

98. Explain the difference between equivalent expressions and equivalent equations.

Solve and check. The symbol ▦ indicates an exercise designed to be solved with a calculator.

▦ **99.** $-0.00458y + 1.7787 = 13.002y - 1.005$

▦ **100.** $4.23x - 17.898 = -1.65x - 42.454$

101. $6x - \{5x - [7x - (4x - (3x + 1))]\} = 6x + 5$

102. $8x - \{3x - [2x - (5x - (7x - 1))]\} = 8x + 7$

103. $23 - 2\{4 + 3[x - 1]\} + 5\{x - 2(x + 3)\}$
$$= 7\{x - 2[5 - (2x + 3)]\}$$

104. $17 - 3\{5 + 2[x - 2]\} + 4\{x - 3(x + 7)\}$
$$= 9\{x + 3[2 + 3(4 - x)]\}$$

105. Create an equation for which it is preferable to use the multiplication principle *before* using the addition principle. Explain why it is best to solve the equation in this manner.

CONNECTING the CONCEPTS

It is important to distinguish between *equivalent expressions* and *equivalent equations*. We *simplify* an expression by writing equivalent expressions; we *solve* an equation by writing equivalent equations.

We use properties such as the commutative, associative, and distributive laws to write equivalent expressions. Equivalent expressions take on the same value when the variables are replaced with numbers. As we use the various laws to remove parentheses and combine like terms, we "simplify" the expression.

As we solve an equation, we write a sequence of equivalent equations using principles such as the addition and multiplication principles for equations. Each equation in the sequence has the same solution.

These ideas are often combined since it is possible to form equivalent equations by replacing part of an equation with an equivalent expression.

Equivalent Expressions

$3x - 2(x - 1)$

$3x - 2x + 2$

$x + 2$

Equivalent Equations

$$3x - 2(x - 1) = 6x$$
$$3x - 2x + 2 = 6x$$
$$x + 2 = 6x$$
$$x + 2 - x = 6x - x$$
$$2 = 5x$$
$$\tfrac{1}{5}(2) = \tfrac{1}{5}(5x)$$
$$\tfrac{2}{5} = x$$

Each line above is an expression that is equivalent to those written above or below. We can write equals signs to indicate that the expressions are equivalent. Because there is no "equation" to solve, the simplified result is an expression—in this case, $x + 2$.

Each line above is a complete equation. Because they are equivalent, all seven equations share the same solution. The solution of the equation is a number—in this case, $\tfrac{2}{5}$.

MIXED REVIEW

Indicate whether each of the following is an expression or an equation. Then either simplify the expression or solve the equation.

1. $3x - 5 - x + 12$

2. $3x - 5 - x + 12 = 1$

3. $4t - (3t - 1)$

4. $5 - (t - 2) = 6$

5. $n - 7n - 2n = 3$

6. $3(x - 2) + 5(x + 7)$

7. $8x + 2[x - (x - 1)]$

8. $9t = 3t - 4 - t$

9. $2x - 6 = 3x + 5$

10. $6(y - 1) - 2(y + 1) = 3(y - 2)$

11. $-(p - 4) - [3 - (9 - 2p)] + p$

12. $3\{-2[4 - (2a - 1) - a] + 1\}$

13. $3(x - 1) - 2(2x + 1) = 5(x - 1)$

14. $\tfrac{1}{2}(y - 3) - \tfrac{3}{2}(5y + 1)$

15. $4n - (5 - n) - 2 = 3n + n + 8$

16. $-(x - 7) - 3[2 - x + 5]$

17. $t - 3t - 2(t + 3 - 2t) + 6 + 4t$

18. $\tfrac{1}{3}(4t - 8) - \tfrac{2}{3}(2t - 7) = 6t$

19. $x = 2 - \{x - 2[3 - 2(x - 7) + 1]\}$

20. $4y - 5 - \{2 - 3[y - (2y + 1)] + 5\}$

1.4   Introduction to Problem Solving

The Five-Step Strategy   ▪   Problem Solving

We now begin to study and practice the "art" of problem solving. Although we are interested mainly in using algebra to solve problems, much of the strategy discussed applies to solving all kinds of problems.

In this text, we do not restrict the use of the word "problem" to computations involving arithmetic or algebra, such as $589 + 437 = a$ or $3x + 5x = 9$. In this text, a problem is simply a question to which we wish to find an answer. Perhaps this can best be illustrated with some sample problems:

1. If I exercise twice a week and eat 3000 calories a day, will I lose weight?
2. Do I have enough time to take 4 courses while working 20 hours a week?
3. My boat travels 12 km/h in still water. How long will it take me to cruise 25 km upstream if the river's current is 3 km/h?

Although these problems differ, there is a strategy that can be applied to all of them.

The Five-Step Strategy

Since you have already studied some algebra, you have some experience with problem solving. The following steps describe a strategy that you may have used already; they form a sound strategy for problem solving in general.

Five Steps for Problem Solving with Algebra

1. *Familiarize* yourself with the problem.
2. *Translate* to mathematical language.
3. *Carry out* some mathematical manipulation.
4. *Check* your possible answer in the original problem.
5. *State* the answer clearly.

Of the five steps, probably the most important is the first: becoming familiar with the problem situation. Here are some ways in which this can be done.

The First Step in Problem Solving with Algebra

To familiarize yourself with the problem:

1. If the problem is written, read it carefully. Then read it again, perhaps aloud. Verbalize the problem to yourself.
2. List the information given and restate the question being asked. Select a variable or variables to represent any unknown(s) and clearly state what each variable represents. Be descriptive! For example, let $t =$ the flight time, in hours; let $p =$ Paul's weight, in pounds; and so on.
3. Find additional information. Look up formulas or definitions with which you are not familiar. Geometric formulas appear at the very end of this text; important words appear in the index. Consult an expert in the field or a reference librarian.
4. Create a table, using variables, in which both known and unknown information are listed. Look for possible patterns.
5. Make and label a drawing.
6. Estimate an answer and check to see whether it is correct.

EXAMPLE **1** How might you familiarize yourself with the situation of Problem 1: "If I exercise twice a week and eat 3000 calories a day, will I lose weight?"

SOLUTION Clearly more information is needed to solve this problem. You might:

a) Research the calorie deficit necessary to lose a pound.

b) Estimate the number of calories used in each exercise session.

c) Visit a personal trainer to find out how many calories per day you burn without exercise.

When enough information is known, it might be wise to make a chart or table to help you reach an answer.

EXAMPLE **2** How might you familiarize yourself with Problem 3: "How long will it take the boat to cruise 25 km upstream?"

SOLUTION First read the question *very* carefully. This may even involve speaking aloud. You may need to reread the problem several times to fully understand what information is given and what information is required. A sketch or table is often helpful.

Current

Boat

Distance to be Traveled	25 km
Speed of Boat in Still Water	12 km/h
Speed of Current	3 km/h
Speed of Boat Upstream	?
Time Required	?

To gain more familiarity with the problem, we should determine, possibly with the aid of outside references, what relationships exist among the various quantities in the problem. With some effort it can be learned that the current's speed should be subtracted from the boat's speed in still water to determine the boat's speed going upstream. We also need to find or recall an extremely important formula:

$$\textbf{Distance} = \textbf{Speed} \times \textbf{Time.}$$ It is important to remember this equation.

We rewrite part of the table, letting t = the number of hours required for the boat to cruise 25 km upstream.

Distance to be Traveled	25 km
Speed of Boat Upstream	$12 - 3 = 9$ km/h
Time Required	t

At this point we might try a guess. Suppose the boat traveled upstream for 2 hr. The boat would have then traveled

$$9\,\frac{\text{km}}{\text{hr}} \times 2\,\text{hr} = 18\,\text{km.}$$ Note that $\frac{\text{km}}{\text{hr}} \cdot \text{hr} = \text{km.}$

$$\text{Speed} \times \text{Time} = \text{Distance}$$

Since $18 \neq 25$, our guess is wrong. Still, examining how we checked our guess sheds light on how to translate the problem to an equation. A better guess, when multiplied by 9, would yield a number closer to 25.

The second step in problem solving is to translate the situation to mathematical language. In algebra, this often means forming an equation. In the third step of our process, we work with the results of the first two steps.

The Second and Third Steps in Problem Solving with Algebra

Translate the problem to mathematical language. This is sometimes done by writing an algebraic expression, but most often in this text it is done by translating to an equation.

Carry out some mathematical manipulation. If you have translated to an equation, this means to solve the equation.

To complete the problem-solving process, we should always check our solution and then state the solution in a clear and precise manner.

The Fourth and Fifth Steps in Problem Solving with Algebra

Check your possible answer in the original problem. Make sure that the answer is reasonable and that all the conditions of the original problem have been satisfied.

State the answer clearly. Write a complete English sentence stating the solution.

The five steps are listed again below. Try to apply them regularly in your work.

> ## Five Steps for Problem Solving with Algebra
> 1. *Familiarize* yourself with the problem.
> 2. *Translate* to mathematical language.
> 3. *Carry out* some mathematical manipulation.
> 4. *Check* your possible answer in the original problem.
> 5. *State* the answer clearly.

Problem Solving

At this point, our study of algebra has just begun. Thus we have few algebraic tools with which to work problems. As the number of tools in our algebraic "toolbox" increases, so will the difficulty of the problems we can solve. For now our problems may seem simple; however, to gain practice with the problem-solving process, you should try to use all five steps. Later some steps may be shortened or combined.

EXAMPLE 3

Purchasing. Renata pays $157.94 for a digital camera. If the price paid includes a 6% sales tax, what is the price of the camera itself?

SOLUTION

1. **Familiarize.** First, we familiarize ourselves with the problem. Note that tax is calculated from, and then added to, the item's price. Let's guess that the camera's price is $140. To check the guess, we calculate the amount of tax, $(0.06)(\$140) = \$8.40$, and add it to $140:

   $$\$140 + (0.06)(\$140) = \$140 + \$8.40$$
   $$= \$148.40. \qquad \$148.40 \neq \$157.94$$

   Our guess was too low, but the manner in which we checked the guess will guide us in the next step. We let

   c = the camera's price, in dollars.

2. **Translate.** Our guess leads us to the following translation:

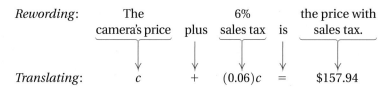

Rewording:	The camera's price	plus	6% sales tax	is	the price with sales tax.
Translating:	c	$+$	$(0.06)c$	$=$	$\$157.94$

3. **Carry out.** Next, we carry out some mathematical manipulation:

   $$c + (0.06)c = 157.94$$
   $$1.06c = 157.94 \qquad \text{Combining like terms:}$$
   $$\qquad\qquad\qquad\qquad 1c + 0.06c = (1 + 0.06)c$$
   $$\frac{1}{1.06} \cdot 1.06c = \frac{1}{1.06} \cdot 157.94 \qquad \text{Using the multiplication principle}$$
   $$c = 149.$$

4. **Check.** To check the answer in the original problem, note that the tax on a camera costing $149 would be $(0.06)(\$149) = \$8.94$. When this is added to $149, we have

$149 + \$8.94$, or $157.94.

Thus, $149 checks in the original problem.

5. **State.** We clearly state the answer: The camera itself costs $149.

> **TRY EXERCISE** 21

EXAMPLE **4**

Home maintenance. In an effort to make their home more energy-efficient, Jess and Drew purchased 200 in. of 3M Press-In-Place™ window glazing. This will be just enough to outline their two square skylights. If the length of the sides of the larger skylight is $1\frac{1}{2}$ times the length of the sides of the smaller one, how should the glazing be cut?

SOLUTION

1. **Familiarize.** Note that the *perimeter* of (distance around) each square is four times the length of a side. Furthermore, if s represents the length of a side of the smaller square, then $\left(1\frac{1}{2}\right)s$ represents the length of a side of the larger square. We make a drawing and note that the two perimeters must add up to 200 in.

Perimeter of a square $= 4 \cdot$ *length of a side*

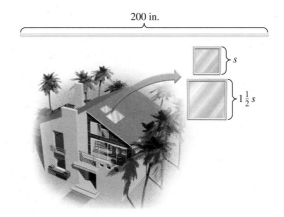

2. **Translate.** Rewording the problem can help us translate:

Rewording: The perimeter   the perimeter
                 of one square   plus   of the other   is   200 in.

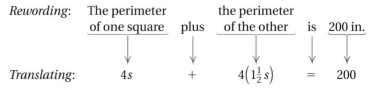

Translating: $4s$ $+$ $4\left(1\frac{1}{2}s\right)$ $=$ 200

3. **Carry out.** We solve the equation:

$$4s + 4\left(1\tfrac{1}{2}s\right) = 200$$

$4s + 6s = 200$ Simplifying; $4\left(1\frac{1}{2}s\right) = 4\left(\frac{3}{2}s\right) = 6s$

$10s = 200$ Combining like terms

$s = \dfrac{1}{10} \cdot 200$ Multiplying both sides by $\frac{1}{10}$

$s = 20.$ Simplifying

4. **Check.** If 20 is the length of the smaller side, then $\left(1\frac{1}{2}\right)(20) = 30$ is the length of the larger side. The two perimeters would then be

$$4 \cdot 20 \text{ in.} = 80 \text{ in.} \quad \text{and} \quad 4 \cdot 30 \text{ in.} = 120 \text{ in.}$$

Since $80 \text{ in.} + 120 \text{ in.} = 200 \text{ in.}$, our answer checks.

5. **State.** The glazing should be cut into two pieces, one 80 in. long and the other 120 in. long.

> TRY EXERCISE ▶ 27

We cannot stress too greatly the importance of labeling the variables in your problem. In Example 4, solving for s is not enough: We need to find $4s$ and $4\left(1\frac{1}{2}s\right)$ to determine the numbers we are after.

EXAMPLE   5   Three numbers are such that the second is 6 less than three times the first and the third is 2 more than two-thirds the first. The sum of the three numbers is 150. Find the largest of the three numbers.

SOLUTION   We proceed according to the five-step process.

1. **Familiarize.** We need to find the largest of three numbers. We list the information given in a table in which x represents the first number.

First Number	x
Second Number	6 less than 3 times the first
Third Number	2 more than $\frac{2}{3}$ the first

$$\text{First} + \text{Second} + \text{Third} = 150$$

Try to check a guess at this point. We will proceed to the next step.

2. **Translate.** Because we wish to write an equation in just one variable, we need to express the second and third numbers using x ("in terms of x"). To do so, we expand the table:

First Number	x	x
Second Number	6 less than 3 times the first	$3x - 6$
Third Number	2 more than $\frac{2}{3}$ the first	$\frac{2}{3}x + 2$

We know that the sum is 150. Substituting, we obtain an equation:

$$\underbrace{\text{First}}_{x} + \underbrace{\text{second}}_{(3x - 6)} + \underbrace{\text{third}}_{\left(\frac{2}{3}x + 2\right)} = \underset{150}{150.}$$

3. Carry out. We solve the equation:

$$x + 3x - 6 + \tfrac{2}{3}x + 2 = 150 \qquad \text{Leaving off unnecessary parentheses}$$

$$\left(4 + \tfrac{2}{3}\right)x - 4 = 150 \qquad \text{Combining like terms}$$

$$\tfrac{14}{3}x - 4 = 150 \qquad \text{Adding within parentheses; } 4\tfrac{2}{3} = \tfrac{14}{3}$$

$$\tfrac{14}{3}x = 154 \qquad \text{Adding 4 to both sides}$$

$$x = \tfrac{3}{14} \cdot 154 \qquad \text{Multiplying both sides by } \tfrac{3}{14}$$

$$x = 33. \qquad \text{Remember, } x \text{ represents the first number.}$$

Going back to the table, we can find the other two numbers:

Second: $3x - 6 = 3 \cdot 33 - 6 = 93$;

Third: $\tfrac{2}{3}x + 2 = \tfrac{2}{3} \cdot 33 + 2 = 24$.

4. Check. We return to the original problem. There are three numbers: 33, 93, and 24. Is the second number 6 less than three times the first?

$$3 \times 33 - 6 = 99 - 6 = 93$$

The answer is *yes*.
Is the third number 2 more than two-thirds the first?

$$\tfrac{2}{3} \times 33 + 2 = 22 + 2 = 24$$

The answer is *yes*.
Is the sum of the three numbers 150?

$$33 + 93 + 24 = 150$$

The answer is *yes*. The numbers do check.

5. State. The problem asks us to find the largest number, so the answer is: "The largest of the three numbers is 93."

TRY EXERCISE ▶ 33

CAUTION! In Example 5, although the equation $x = 33$ enables us to find the largest number, 93, the number 33 is *not* the solution of the problem. By clearly labeling our variable in the first step, we can avoid thinking that the variable always represents the solution of the problem.

Translating for Success

1. *Consecutive integers.* The sum of two consecutive even integers is 102. Find the integers.

2. *Dimensions of a triangle.* One angle of a triangle is twice the measure of a second angle. The third angle measures 102° more than the second angle. Find the measures of the angles.

3. *Salary increase.* After Susanna earned a 5% raise, her new salary was $25,750. What was her former salary?

4. *Dimensions of a rectangle.* The length of a rectangle is 6 in. more than the width. The perimeter of the rectangle is 102 in. Find the length and the width.

5. *Population.* The population of Doddville is decreasing at a rate of 5% per year. The current population is 25,750. What was the population the previous year?

Translate each word problem to an equation or an inequality and select a correct translation from A–O.

A. $0.05(25,750) = x$

B. $x + 2x = 102$

C. $2x + 2(x + 6) = 102$

D. $2x + x + x + 102 = 180$

E. $x - 0.05x = 25,750$

F. $x + (x + 2) = 102$

G. $6x - 102 = 180 - 5x$

H. $x + 5x = 150$

I. $x + 0.05x = 25,750$

J. $x + (2x + 6) = 102$

K. $x + (x + 1) = 102$

L. $102 + x = 180$

M. $0.05x = 25,750$

N. $x + 2x = x + 102$

O. $x + (x + 6) = 102$

Answers on page A–2

An additional, animated version of this activity appears in MyMathLab. To use MyMathLab, you need a course ID and a student access code. Contact your instructor for more information.

6. *Numerical relationship.* One number is 6 more than twice another. The sum of the numbers is 102. Find the numbers.

7. *DVD collections.* Together Ella and Ken have 102 DVDs. If Ken has 6 more DVDs than Ella, how many does each have?

8. *Sales commissions.* Will earns a commission of 5% on his sales. One year he earned commissions totaling $25,750. What were his total sales for the year?

9. *Fencing.* Brian has 102 ft of fencing that he plans to use to enclose two dog runs. The perimeter of one run is to be twice the perimeter of the other. Into what lengths should the fencing be cut?

10. *Quiz scores.* Lupe has a total of 102 points on the first 6 quizzes in her sociology class. How many total points must she earn on the 5 remaining quizzes in order to have 180 points for the semester?

1.4   EXERCISE SET

For each problem, familiarize yourself with the situation. Then translate to mathematical language. You need not actually solve the problem; just carry out the first two steps of the five-step strategy. You will be asked to complete some of the solutions as Exercises 31–38.

1. The sum of two numbers is 91. One of the numbers is 9 more than the other. What are the numbers?

2. The sum of two numbers is 88. One of the numbers is 6 more than the other. What are the numbers?

3. *Kayaking.*   Sea kayaking in Puget Sound is a popular sport in the Seattle area. Stella can maintain a speed of 3.5 mph in a calm sea. The club to which she belongs plans to paddle 6 mi into a 1.9-mph current. How long will it take Stella to make the trip?
   Source: Based on information from the University Kayak Club and the University of Washington

4. *Aviation.*   A Cessna airplane traveling 390 km/h in still air encounters a 65-km/h headwind. How long will it take the plane to travel 725 km into the wind?

5. *Angles in a triangle.*   The degree measures of the angles in a triangle are three consecutive integers. Find the measures of the angles.

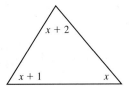

6. *Pricing.*   Becker Lumber gives contractors a 15% discount on all orders. After the discount, a contractor's order cost $272. What was the original cost of the order?

7. *Escalators.*   A 230-ft long escalator at the Wheaton Station stop of the Washington, D.C., Metro is the longest single-span, uninterrupted escalator in the Western Hemisphere. In a rush, Dominik walks up the escalator at 100 ft/min while the escalator is moving up at 90 ft/min. How long will it take him to reach the top of the escalator?
   Source: Washington Metropolitan Area Transit Authority

100 ft/min

90 ft/min

8. *Moving sidewalks.*   A moving sidewalk in Pearson Airport, Ontario, is 912 ft long and moves at a rate of 6 ft/sec. If Alida walks at a rate of 4 ft/sec, how long will it take her to walk the length of the moving sidewalk?
   Source: *The Wall Street Journal*, 8/16/07

9. *Pricing.*   Quick Storage prices flash drives by raising the wholesale price 50% and adding $1.50. What must a drive's wholesale price be if it is being sold for $22.50?

10. *Pricing.*   Miller Oil offers a 5% discount to customers who pay promptly for an oil delivery. The Blancos promptly paid $142.50 for their December oil bill. What would the cost have been had they not promptly paid?

11. *Cruising altitude.*   A Boeing 747 has been instructed to climb from its present altitude of 8000 ft to a cruising altitude of 29,000 ft. If the plane ascends at a rate of 3500 ft/min, how long will it take to reach the cruising altitude?

12. A piece of wire 10 m long is to be cut into two pieces, one of them $\frac{2}{3}$ as long as the other. How should the wire be cut?

13. *Angles in a triangle.*   One angle of a triangle is four times the measure of a second angle. The third angle measures 5° more than twice the second angle. Find the measures of the angles.

14. *Angles in a triangle.*   One angle of a triangle is three times the measure of a second angle. The third angle measures 12° less than twice the second angle. Find the measures of the angles.

15. Find three consecutive odd integers such that the sum of the first, twice the second, and three times the third is 70.

16. Find two consecutive even integers such that two times the first plus three times the second is 76.

17. A steel rod 90 cm long is to be cut into two pieces, each to be bent to make an equilateral triangle. The length of a side of one triangle is to be twice the length of a side of the other. How should the rod be cut?

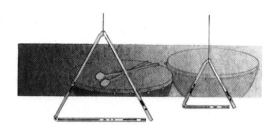

18. A piece of wire 100 cm long is to be cut into two pieces, each to be bent to make a square. The area of one square is to be 144 cm² greater than that of the other. How should the wire be cut? (*Remember:* Do not solve.)

19. *Rescue calls.*   Rescue crews working for Stockton Rescue average 3 calls per shift. After his first four shifts, Cody had received 5, 2, 1, and 3 calls. How many calls will Cody need on his next shift if he is to average 3 calls per shift?

20. *Test scores.*   Olivia's scores on five tests are 93, 89, 72, 80, and 96. What must the score be on her next test so that the average will be 88?

Solve each problem. Use all five problem-solving steps.

21. *Pricing.*   The price that Tess paid for her graphing calculator, $84, is less than what Tony paid by $13. How much did Tony pay for his graphing calculator?

22. *Class size.*   The number of students in Damonte's class, 35, is greater than the number in Rose's class by 12. How many students are in Rose's class?

23. *Day care.*   A family living in Boston, Massachusetts, pays $1089 per month for full-time day care for their toddler. This is $\frac{11}{4}$ of what they paid for comparable day care when they lived in Billings, Montana. How much did the child care cost in Billings?
Source: Based on data from Runzheimer International

24. *Self-employment.*   In 2006, a total of 2,144,000 Americans aged 55 to 64 were self-employed. This was approximately $\frac{4}{3}$ of the number of self-employed persons in that age bracket in 2005. How many Americans aged 55 to 64 were self-employed in 2005?
Source: U.S. Bureau of Labor Statistics

25. *Photography.*   Robbin took graduation pictures for 8 fewer seniors than Michelle did. Together they photographed 40 seniors. How many seniors did Robbin photograph?

26. *Nursing.*   One Friday, Vance gave 11 more flu shots than Mike did. Together they gave 53 flu shots. How many flu shots did Vance give?

27. The length of a rectangular mirror is three times its width and its perimeter is 120 cm. Find the length and the width of the mirror.

28. The length of a rectangular tile is twice its width and its perimeter is 21 cm. Find the length and the width of the tile.

29. The width of a rectangular greenhouse is one-fourth its length and its perimeter is 130 m. Find the length and the width of the greenhouse.

30. The width of a rectangular garden is one-third its length and its perimeter is 32 m. Find the dimensions of the garden.

31. Solve the problem of Exercise 3.

32. Solve the problem of Exercise 6.

33. Solve the problem of Exercise 13.

34. Solve the problem of Exercise 14.

35. Solve the problem of Exercise 10.

36. Solve the problem of Exercise 8.

37. Solve the problem of Exercise 9.

38. Solve the problem of Exercise 16.

39. Write a problem for a classmate to solve for which fractions must be multiplied in order to get the answer.

40. Write a problem for a classmate to solve for which fractions must be divided in order to get the answer.

Skill Review

To prepare for Section 1.5, review solving equations (Section 1.3).

Solve. [1.3]

41. $30 = 5(6x)$

42. $4x = 9x - 7$

43. $3 = \dfrac{x}{-2}$

44. $3 = \frac{2}{3}(x + 4)$

Synthesis

45. How can a guess or estimate help prepare you for the *Translate* step when solving problems?

46. Why is it important to check the solution from the *Carry out* step in the original wording of the problem being solved?

47. *Test scores.* Tico's scores on four tests are 83, 91, 78, and 81. How many points above his current average must Tico score on the next test in order to raise his average 2 points?

48. *Geometry.* The height and sides of a triangle are four consecutive integers. The height is the first integer, and the base is the third integer. The perimeter of the triangle is 42 in. Find the area of the triangle.

49. *Home prices.* Panduski's real estate prices increased 6% from 2004 to 2005 and 2% from 2005 to 2006. From 2006 to 2008, prices dropped 1%. If a townhouse sold for $117,743 in 2008, what was its worth in 2004? (Round to the nearest dollar.)

50. *Adjusted wages.* Emma's salary is reduced *n*% during a period of financial difficulty. By what number should her salary be multiplied in order to bring it back to where it was before the reduction?

COLLABORATIVE CORNER

Who Pays What?

Focus: Problem solving
Time: 15 minutes
Group size: 5

Suppose that two of the five members in each group are celebrating birthdays and the entire group goes out to lunch. Suppose further that each member whose birthday it is gets treated to his or her lunch by the other *four* members. Finally, suppose that all meals cost the same amount and that the total bill is $40.00.*

———————————
*This activity was inspired by "The Birthday-Lunch Problem," *Mathematics Teaching in the Middle School*, vol. 2, no. 1, September–October 1996, pp. 40–42.

ACTIVITY

1. Determine, as a group, how much each group member should pay for the lunch described above. Then explain how this determination was made.

2. Compare the results and methods used for part (1) with those of the other groups in the class.

3. If the total bill is $65, how much should each group member pay? Again compare results with those of other groups.

4. If time permits, generalize the results of parts 1–3 for a total bill of *x* dollars.

1.5   Formulas, Models, and Geometry

Solving Formulas   ●   Mathematical Models

A **formula** is an equation that uses letters to represent a relationship between two or more quantities. For example, in Section 1.4, we made use of the formula $P = 4s$, where P represents the perimeter of a square and s the length of a side. Other important geometric formulas are $A = \pi r^2$ (for the area A of a circle of radius r), $C = \pi d$ (for the circumference C of a circle of diameter d), and $A = b \cdot h$ (for the area A of a parallelogram of height h and base length b).* A more complete list of geometric formulas appears at the very end of this text.

  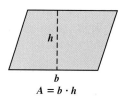

$$A = \pi r^2 \qquad\qquad C = \pi d \qquad\qquad A = b \cdot h$$

Solving Formulas

Suppose we know the floor area and the width of a rectangular room and want to find the length. To do so, we could "solve" the formula $A = l \cdot w$ (Area = Length · Width) for l, using the same principles that we use for solving equations.

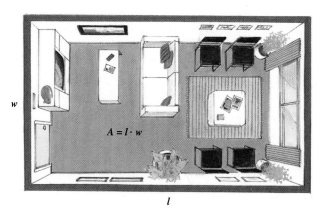

$A = l \cdot w$

l

EXAMPLE 1   *Area of a rectangle.*   Solve the formula $A = l \cdot w$ for l.

SOLUTION

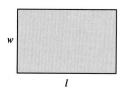

$$A = l \cdot w \qquad \text{We want this letter alone.}$$

$$\dfrac{A}{w} = \dfrac{l \cdot w}{w} \qquad \begin{array}{l} \text{Dividing both sides by } w\text{, or} \\ \text{multiplying both sides by } 1/w \end{array}$$

$$\dfrac{A}{w} = l \cdot \dfrac{w}{w}$$

$$\dfrac{A}{w} = l \qquad\qquad \text{Simplifying by removing a factor equal to 1: } \dfrac{w}{w} = 1$$

TRY EXERCISE ▶ 9

---

*The Greek letter π, read "pi," is *approximately* 3.14159265358979323846264. Often 3.14 or 22/7 is used to approximate π when a calculator with a π key is unavailable.

Thus to find the length of a rectangular room, we can divide the area of the floor by its width. Were we to do this calculation for a variety of rectangular rooms, the formula $l = A/w$ would be more convenient than repeatedly substituting into $A = l \cdot w$ and then dividing.

EXAMPLE 2   *Simple interest.*   The formula $I = Prt$ is used to determine the simple interest I earned when a principal of P dollars is invested for t years at an interest rate r. Solve this formula for t.

SOLUTION

$$I = Prt \qquad \text{We want this letter alone.}$$

$$\frac{I}{Pr} = \frac{Prt}{Pr} \qquad \text{Dividing both sides by } Pr, \text{ or multiplying both sides by } \frac{1}{Pr}$$

$$\frac{I}{Pr} = t \qquad \text{Simplifying by removing a factor equal to 1: } \frac{Pr}{Pr} = 1$$

TRY EXERCISE 13

EXAMPLE 3   *Area of a trapezoid.*   A trapezoid is a geometric shape with four sides, exactly two of which, the bases, are parallel to each other. The formula for calculating the area A of a trapezoid with bases b_1 and b_2 (read "b sub one" and "b sub two") and height h is given by

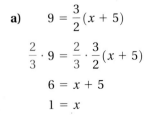

$$A = \frac{h}{2}(b_1 + b_2), \qquad \begin{array}{l}\text{A derivation of this formula is outlined in}\\ \text{Exercise 89 of this section.}\end{array}$$

where the *subscripts* 1 and 2 distinguish one base from the other. Solve for b_1.

SOLUTION   There are several ways to "remove" the parentheses. We could distribute $h/2$, but an easier approach is to multiply both sides by the reciprocal of $h/2$.

$$A = \frac{h}{2}(b_1 + b_2)$$

$$\frac{2}{h} \cdot A = \frac{2}{h} \cdot \frac{h}{2}(b_1 + b_2) \qquad \begin{array}{l}\text{Multiplying both sides by}\\ \frac{2}{h}\left(\text{or dividing by } \frac{h}{2}\right)\end{array}$$

$$\frac{2A}{h} = b_1 + b_2 \qquad \begin{array}{l}\text{Simplifying. The right side is "cleared" of}\\ \text{fractions, since } (2h)/(h2) = 1.\end{array}$$

$$\frac{2A}{h} - b_2 = b_1 \qquad \text{Adding } -b_2 \text{ on both sides}$$

TRY EXERCISE 25

The similarities between solving formulas and solving equations can be seen below. In (a), we solve as we did before; in (b), we do not carry out all calculations; and in (c), we cannot carry out all calculations because the numbers are unknown. The same steps are used each time.

a)
$$9 = \frac{3}{2}(x + 5)$$
$$\frac{2}{3} \cdot 9 = \frac{2}{3} \cdot \frac{3}{2}(x + 5)$$
$$6 = x + 5$$
$$1 = x$$

b)
$$9 = \frac{3}{2}(x + 5)$$
$$\frac{2}{3} \cdot 9 = \frac{2}{3} \cdot \frac{3}{2}(x + 5)$$
$$\frac{2 \cdot 9}{3} = x + 5$$
$$\frac{2 \cdot 9}{3} - 5 = x$$

c)
$$A = \frac{h}{2}(b_1 + b_2)$$
$$\frac{2}{h} \cdot A = \frac{2}{h} \cdot \frac{h}{2}(b_1 + b_2)$$
$$\frac{2A}{h} = b_1 + b_2$$
$$\frac{2A}{h} - b_2 = b_1$$

EXAMPLE 4

STUDENT NOTES

As is often the case, the material in this section builds on the material in the preceding sections. If you experience difficulty in this section, before seeking help make certain that you have a thorough understanding of the material in Sections 1.4 and (especially) 1.3.

Accumulated simple interest.   The formula $A = P + Prt$ gives the amount A that a principal of P dollars will be worth in t years when invested at simple interest rate r. Solve the formula for P.

SOLUTION

$A = P + Prt$	We want this letter alone.
$A = P(1 + rt)$	Factoring (using the distributive law) to write P just once, as a factor
$\dfrac{A}{1 + rt} = \dfrac{P(1 + rt)}{1 + rt}$	Dividing both sides by $1 + rt$, or multiplying both sides by $\dfrac{1}{1 + rt}$
$\dfrac{A}{1 + rt} = P$	Simplifying

This last equation can be used to determine how much should be invested at simple interest rate r in order to have A dollars t years later.

TRY EXERCISE 29

Note in Example 4 that the factoring enabled us to write P once rather than twice. This is comparable to combining like terms when solving an equation like $16 = x + 7x$.

You may find the following summary useful.

> **To Solve a Formula for a Specified Letter**
>
> 1. Get all terms with the specified letter on one side of the equation and all other terms on the other side, using the addition principle. To do this may require removing parentheses.
>   - To remove parentheses, either divide both sides by the multiplier in front of the parentheses or use the distributive law.
> 2. When all terms with the specified letter are on the same side, factor (if necessary) so that the variable is written only once.
> 3. Solve for the letter in question by dividing both sides by the multiplier of that letter.

Mathematical Models

The above formulas from geometry and economics are examples of *mathematical models*. A **mathematical model** can be a formula, or a set of formulas, developed to represent a real-world situation. In problem solving, a mathematical model is formed in the *Translate* step.

EXAMPLE 5

Body mass index.   Peyton Manning, quarterback for the Indianapolis Colts, is 6 ft 5 in. tall and has a body mass index of approximately 27.3. What is his weight?

SOLUTION

1. **Familiarize.** From an outside source, we find that body mass index I depends on a person's height and weight and is found using the formula

$$I = \frac{704.5W}{H^2},$$

where W is the weight, in pounds, and H is the height, in inches.

Source: National Center for Health Statistics

2. Translate. Because we are interested in finding Manning's weight, we solve for W:

$$I = \frac{704.5W}{H^2}$$ We want this letter alone.

$$I \cdot H^2 = \frac{704.5W}{H^2} \cdot H^2$$ Multiplying both sides by H^2 to clear the fraction

$$IH^2 = 704.5W$$ Simplifying

$$\frac{IH^2}{704.5} = \frac{704.5W}{704.5}$$ Dividing by 704.5

$$\frac{IH^2}{704.5} = W.$$

3. Carry out. The model

$$W = \frac{IH^2}{704.5}$$

can be used to calculate the weight of someone whose body mass index and height are known. Using the information given in the problem, we have

$$W = \frac{27.3 \cdot 77^2}{704.5}$$ 6 ft 5 in. is 77 in.

$$\approx 230.$$ Using a calculator

4. Check. We could repeat the calculations or substitute in the original formula and then solve for W. The check is left to the student.

5. State. Peyton Manning weighs about 230 lb.   `TRY EXERCISE` 49

EXAMPLE 6

Density. A collector suspects that a silver coin is not solid silver. The density of silver is 10.5 grams per cubic centimeter (g/cm^3) and the coin is 0.2 cm thick with a radius of 2 cm. If the coin is really silver, how much should it weigh?

SOLUTION

1. Familiarize. From an outside reference, we find that density depends on mass and volume and that, in this setting, mass means weight. The formula for the volume of a right circular cylinder appears at the very end of this text, as well as on p. 45.

2. Translate. We need to use two formulas,

$$D = \frac{m}{V} \quad \text{and} \quad V = \pi r^2 h,$$

where D is the density, m the mass, V the volume, r the length of the radius, and h the height of a right circular cylinder. Since we need a model relating mass to the measurements of the coin, we solve for m and then substitute for V:

$$D = \frac{m}{V}$$

$$V \cdot D = V \cdot \frac{m}{V}$$ Multiplying by V

$$V \cdot D = m$$ Simplifying

$$\pi r^2 h \cdot D = m.$$ Substituting

3. **Carry out.** The model $m = \pi r^2 h D$ can be used to find the mass of any right circular cylinder for which the dimensions and the density are known:

$$m = \pi r^2 h D$$
$$= \pi(2)^2(0.2)(10.5) \qquad \text{Substituting}$$
$$\approx 26.3894. \qquad \text{Using a calculator with a } \pi \text{ key}$$

4. **Check.** To check, we could repeat the calculations. We might also check the model by examining the units:

$$\pi r^2 h \cdot D = \text{cm}^2 \cdot \text{cm} \cdot \frac{\text{g}}{\text{cm}^3} = \text{cm}^3 \cdot \frac{\text{g}}{\text{cm}^3} = \text{g}. \qquad \pi \text{ has no unit.}$$

Since g (grams) is the appropriate unit of mass, we have at least a partial check.

5. **State.** The coin, if it is indeed silver, should weigh about 26 grams.

 TRY EXERCISE 51

The following important formulas will be used throughout the study of algebra.

Area of a rectangle:	$A = lw$
Area of a square:	$A = s^2$
Area of a parallelogram:	$A = bh$
Area of a trapezoid:	$A = \dfrac{h}{2}(b_1 + b_2)$
Area of a triangle:	$A = \frac{1}{2}bh$
Area of a circle:	$A = \pi r^2$
Circumference of a circle:	$C = \pi d$
Volume of a cube:	$V = s^3$
Volume of a right circular cylinder:	$V = \pi r^2 h$
Perimeter of a square:	$P = 4s$
Distance traveled:	$d = rt$
Simple interest:	$I = Prt$

1.5   EXERCISE SET

Concept Reinforcement   *Complete each of the following statements.*

1. A formula is a(n) _____ that uses letters to represent a relationship between two or more quantities.

2. The formula $A = \pi r^2$ is used to calculate the _____ of a circle.

3. The formula $C = \pi d$ is used to calculate the _____ of a circle.

4. The formula _____ is used to calculate the perimeter of a rectangle of length l and width w.

5. The formula _____ is used to calculate the area of a parallelogram of height h and base length b.

6. The formula $l = A/w$ can be used to determine the _____ of a rectangle, given its area and width.

7. In the formula for the area of a trapezoid, $A = \dfrac{h}{2}(b_1 + b_2)$, the numbers 1 and 2 are referred to as _____.

8. When two or more terms on the same side of a formula contain the letter for which we are solving, we can _____ so that the letter is only written once.

Solve.

9. $E = wA$, for A (a nursing formula)

10. $F = ma$, for a (a physics formula)

11. $d = rt$, for r (a distance formula)

12. $P = EI$, for E (an electricity formula)

13. $V = lwh$, for h (a volume formula)

14. $I = Prt$, for r (a formula for interest)

15. $L = \dfrac{k}{d^2}$, for k
    (a formula for intensity of sound or light)

16. $F = \dfrac{mv^2}{r}$, for m (a physics formula)

17. $G = w + 150n$, for n
    (a formula for the gross weight of a bus)

18. $P = b + 0.5t$, for t (a formula for parking prices)

19. $2w + 2h + l = p$, for l
    (a formula used when shipping boxes)

20. $2w + 2h + l = p$, for w

21. $2x + 3y = 4$, for y

22. $3x - 7y = 2$, for y

23. $Ax + By = C$, for y (a formula for graphing lines)

24. $P = 2l + 2w$, for l (a perimeter formula)

25. $C = \dfrac{5}{9}(F - 32)$, for F (a temperature formula)

26. $T = \dfrac{3}{10}(I - 12{,}000)$, for I (a tax formula)

27. $V = \dfrac{4}{3}\pi r^3$, for r^3
    (a formula for the volume of a sphere)

28. $V = \dfrac{4}{3}\pi r^3$, for π

29. $np + nm = t$, for n

30. $ab + ac = d$, for a

31. $uv + wv = x$, for v

32. $st + rt = n$, for t

33. $A = \dfrac{q_1 + q_2 + q_3}{n}$, for n (a formula for averaging)
    (*Hint*: Multiply by n to "clear" fractions.)

34. $g = \dfrac{km_1m_2}{d^2}$, for d^2 (Newton's law of gravitation)

35. $v = \dfrac{d_2 - d_1}{t}$, for t (a physics formula)

36. $v = \dfrac{s_2 - s_1}{m}$, for m

37. $v = \dfrac{d_2 - d_1}{t}$, for d_1

38. $v = \dfrac{s_2 - s_1}{m}$, for s_1

39. $bd = c + ba$, for b

40. $st = n + sm$, for s

41. $v - w = uvw$, for w

42. $p - q = qrs$, for q

43. $n - mk = mt^2$, for m

44. $d - ct = ca^3$, for c

45. *Investing.* Eliana has $2600 to invest for 6 months. If she needs the money to earn $104 in that time, at what rate of simple interest must Eliana invest?

46. *Banking.* Chuma plans to buy a two-year certificate of deposit (CD) that earns 4% simple interest. If he needs the CD to earn $150, how much should Chuma invest?

47. *Geometry.* The area of a parallelogram is 96 cm^2. The base of the figure is 6 cm. What is the height?

48. *Geometry.* The area of a parallelogram is 84 cm^2. The height of the figure is 7 cm. How long is the base?

For Exercises 49–56, make use of the formulas given in Examples 1–6.

49. *Body mass index.* Arnold Schwarzenegger, the current governor of California and a former bodybuilder, is 6 ft 2 in. tall and has a body mass index of 30.2. How much does he weigh?

50. *Body mass index.* Actress Halle Berry has a body mass index of 18 and a height of 5 ft 6 in. What is her weight?

51. *Weight of salt.* The density of salt is 2.16 g/cm^3 (grams per cubic centimeter). An empty cardboard salt canister weighs 28 g, is 13.6 cm tall, and has a 4-cm radius. How much will a filled canister weigh?

52. *Weight of a coin.* The density of gold is 19.3 g/cm^3. If the coin in Example 6 were made of gold instead of silver, how much more would it weigh?

53. *Gardening.* A garden is being constructed in the shape of a trapezoid. The dimensions are as shown in the figure. The unknown dimension is to be such that the area of the garden is 90 ft^2. Find that unknown dimension.

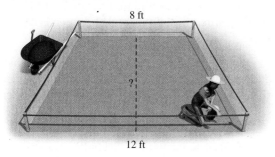

8 ft

?

12 ft

54. *Pet care.* A rectangular kennel is being constructed, and 76 ft of fencing is available. The width of the kennel is to be 13 ft. What should the length be, in order to use just 76 ft of fence?

Aha! **55.** *Investing.* Do Xuan Nam is going to invest $1000 at simple interest at 4%. How long will it take for the investment to be worth $1040?

56. Holli is going to invest $950 at simple interest at 3%. How long will it take for her investment to be worth $1178?

57. *Nursing.* The formula $E = w \cdot A$ is used to find the estimated blood volume E of a patient with weight w, in kilograms, and average blood volume A, in milliliters per kilogram. Find the estimated blood volume of a toddler who weighs 10.2 kg and has an average blood volume of 80 mL/kg.
Source: www.manuelsweb.com

58. *Nursing.* The allowable blood loss L is the amount of blood a patient can lose before a transfusion is necessary. This can be estimated by

$$L = \frac{E(H_i - H_f)}{H_i},$$

where E is the estimated blood volume of the patient, H_i is the initial hemoglobin level, and H_f is the lowest acceptable final hemoglobin level. What is the allowable blood loss for a patient with an estimated blood volume of 3200 mL, an initial hemoglobin of 13 g/dL, and a lowest final hemoglobin of 7 g/dL?
Source: Cecil B. Drain, *Perianesthesia Nursing: A Critical Care Approach.* Saunders, 2003.

Chess ratings. *The formula*

$$R = r + \frac{400(W - L)}{N}$$

is used to establish a chess player's rating R, after he or she has played N games, where W is the number of wins, L is the number of losses, and r is the average rating of the opponents.
Source: U.S. Chess Federation

59. Ulana's rating is 1305 after winning 5 games and losing 3 in tournament play. What was the average rating of her opponents? (Assume there were no draws.)

60. Vladimir's rating fell to 1050 after winning twice and losing 5 times in tournament play. What was the average rating of his opponents? (Assume there were no draws.)

Female caloric needs.   *The number of calories K needed each day by a moderately active woman who weighs w pounds, is h inches tall, and is a years old can be estimated by the formula*

$$K = 917 + 6(w + h - a).$$

Source: M. Parker, *She Does Math*. Mathematical Association of America, p. 96

61. Julie is moderately active, weighs 120 lb, and is 23 years old. If Julie needs 1901 calories per day to maintain her weight, how tall is she?

62. Tawana is moderately active, 31 years old, and 5 ft 4 in. tall. If Tawana needs 1901 calories per day to maintain her weight, how much does she weigh?

Male lean body mass.   *The lean body mass m, in kilograms, of a man over the age of 16 who has a total body mass of w kilograms and is h centimeters tall can be estimated by the formula*

$$m = 0.32810w + 0.33929h - 29.5336.$$

Source: www.medal.org

63. One of the authors of this text, Marv Bittinger, has a lean body mass of 62 kg and is 185 cm tall. What is his total body mass?

64. One of the authors of this text, David Ellenbogen, has a lean body mass of 60 kg and a total body mass of 82 kg. How tall is he?

Blogging.   *A business owner's blog can be an effective marketing and advertising tool. The return on investment r of a blog can be estimated by*

$$r = \frac{tmap}{hs},$$

where t is the average number of visits to the blog each day, m is the percentage of blog visitors who purchase merchandise, a is the average order size, p is the percentage of the average order that is profit, h is the number of hours spent blogging each day, and s is the hourly salary of the blogger. A return on investment less than 1 indicates that the blog is costing the company money.
Source: www.minethatdata.blogspot.com

65. Tomas earns $30 an hour writing a blog for his company. It takes him 4 hr a day to write the blog, and 5% of the blog visitors buy merchandise, with an average order size of $100 and a profit percentage of 15%. He calculates the return on investment to be 3.2. What is his average daily blog traffic?

66. Elyse earns $35 an hour writing a blog for her company. On average, 1200 people visit her blog daily, and 4% of them buy merchandise, with an average order size of $150 and a profit percentage of 14%. She calculates the return on investment to be 4.8. How long does it take her each day to write the blog?

Waiting time.   *In an effort to minimize waiting time for patients at a doctor's office without increasing a physician's idle time, Michael Goiten of Massachusetts General Hospital has developed a model. Goiten suggests that the interval time I, in minutes, between scheduled appointments be related to the total number of minutes T that a physician spends with patients in a day and the number of scheduled appointments N according to the formula I = 1.08(T/N).[*]*

67. Dr. Cruz determines that she has a total of 8 hr a day to see patients. If she insists on an interval time of 15 min, according to Goiten's model, how many appointments should she make in one day?

68. A doctor insists on an interval time of 20 min and must be able to schedule 25 appointments a day. According to Goiten's model, how many hours a day should the doctor be prepared to spend with patients?

Projected birth weight.   *Ultrasonic images of 29-week-old fetuses can be used to predict weight. One model, developed by Thurnau,[†] is P = 9.337da − 299; a second model, developed by Weiner,[‡] is P = 94.593c + 34.227a − 2134.616. For both formulas, P represents the estimated fetal weight in grams, d the diameter of the fetal head in centimeters, c the circumference of the fetal head in centimeters, and a the circumference of the fetal abdomen in centimeters.*

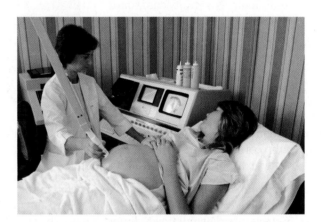

69. Solve Thurnau's model for *d* and use that equation to estimate the diameter of a fetus' head at 29 weeks when the estimated weight is 1614 g and the circumference of the fetal abdomen is 24.1 cm.

70. Solve Weiner's model for *c* and use that equation to estimate the circumference of a fetus' head at 29 weeks when the estimated weight is 1277 g and the circumference of the fetal abdomen is 23.4 cm.

---

[*]*New England Journal of Medicine*, 30 August 1990, pp. 604–608.
[†]Thurnau, G. R., R. K. Tamura, R. E. Sabbagha, et al., *Am. J. Obstet Gynecol* 1983; **145**: 557.
[‡]Weiner, C. P., R. E. Sabbagha, N. Vaisrub, et al., *Obstet Gynecol* 1985; **65**: 812.

71. Is every rectangle a trapezoid? Why or why not?

72. Predictions made using the models of Exercises 69 and 70 are often off by as much as 10%. Does this mean the models should be discarded? Why or why not?

Skill Review

Review simplifying expressions (Section 1.3).

Simplify. [1.3]

73. $2(c - 1) - 5[3 - (c - 5)]$

74. $3 - 2\{3[2t - (6 - t)]\}$

Synthesis

75. Which would you expect to have the greater density, and why: cork or steel?

76. Both of the models used in Exercises 69 and 70 have P alone on one side of the equation. Why?

77. The density of platinum is 21.5 g/cm³. If the ring shown in the figure below is crafted out of platinum, how much will it weigh?

78. The density of a penny is 8.93 g/cm³. The mass of a roll of pennies is 177.6 g. If the diameter of a penny is 1.85 cm, how tall is a roll of pennies?

79. *Baseball.* Baseball analysts use the formula $r = 0.3b - 0.6c$ to estimate the number of runs r due to stolen bases for a runner who stole b bases and was caught stealing c times. In June 2004, Alex Sanchez stole 11 bases and was credited with -2.7 stolen base runs. How many times was he caught stealing?

80. See Exercises 59 and 60. Suppose Heidi plays in a tournament in which all of her opponents have the same rating. Under what circumstances will playing to a draw help or hurt her rating?

Solve.

81. $s = v_i t + \frac{1}{2} a t^2$, for a (a physics formula)

82. $A = 4lw + w^2$, for l

83. $b = \dfrac{h + w + p}{a + w + p + f}$, for w (a baseball formula)

84. $\dfrac{P_1 V_1}{T_1} = \dfrac{P_2 V_2}{T_2}$, for T_2 (a chemistry formula)

85. $\dfrac{b}{a - b} = c$, for b

86. $m = \dfrac{(d/e)}{(e/f)}$, for d

Aha! **87.** $s + \dfrac{s + t}{s - t} = \dfrac{1}{t} + \dfrac{s + t}{s - t}$, for t

88. $\dfrac{a}{a + b} = c$, for a

89. To derive the formula for the area of a trapezoid, consider the area of two trapezoids, one of which is upside down.

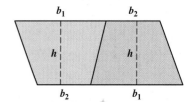

Explain why the total area of the two trapezoids is given by $h(b_1 + b_2)$. Then explain why the area of a trapezoid is given by $\dfrac{h}{2}(b_1 + b_2)$.

1.6   Properties of Exponents

The Product and Quotient Rules • The Zero Exponent • Negative Integers as Exponents •
Simplifying $(a^m)^n$ • Raising a Product or a Quotient to a Power

STUDY SKILLS

Seeking Help Off Campus

Are you aware of all the supplements that exist for this textbook? See the preface for a description of each supplement: the *Student Solutions Manual*, a complete set of videos on DVD, MathXL tutorial exercises on CD, and www.MyMathLab.com. Through this website, you can find additional practice resources, helpful ideas, and links to other learning resources.

In Section 1.1, we discussed how whole-number exponents are used. We now develop rules for manipulating exponents and determine what zero and negative integers will mean as exponents.

The Product and Quotient Rules

The expression $x^3 \cdot x^4$ can be rewritten as follows:

$$x^3 \cdot x^4 = \underbrace{x \cdot x \cdot x}_{3 \text{ factors}} \cdot \underbrace{x \cdot x \cdot x \cdot x}_{4 \text{ factors}}$$

$$= \underbrace{x \cdot x \cdot x \cdot x \cdot x \cdot x \cdot x}_{7 \text{ factors}}$$

$$= x^7.$$

This result is generalized in the *product rule*.

Multiplying with Like Bases: The Product Rule

For any number a and any positive integers m and n,

$$a^m \cdot a^n = a^{m+n}.$$

(When multiplying, if the bases are the same, keep the base and add the exponents.)

EXAMPLE 1   Multiply and simplify: **(a)** $m^5 \cdot m^7$; **(b)** $(5a^2b^3)(3a^4b^5)$.

SOLUTION

a) $m^5 \cdot m^7 = m^{5+7} = m^{12}$ Multiplying powers by adding exponents

b) $(5a^2b^3)(3a^4b^5) = 5 \cdot 3 \cdot a^2 \cdot a^4 \cdot b^3 \cdot b^5$ Using the associative and commutative laws

$$= 15a^{2+4}b^{3+5}$$ Multiplying coefficients; using the product rule

$$= 15a^6b^8$$ TRY EXERCISE 11

STUDENT NOTES

Be careful to distinguish between how coefficients and exponents are handled. For instance, in Example 1(b), the product of the coefficients 5 and 3 is 15, whereas the product of b^3 and b^5 is b^8.

CAUTION!

$$5^8 \cdot 5^6 = 5^{14} \quad \begin{cases} 5^8 \cdot 5^6 \neq 25^{14} & \text{Do not multiply the bases!} \\ 5^8 \cdot 5^6 \neq 5^{48} & \text{Do not multiply the exponents!} \end{cases}$$

Next, we simplify a quotient:

$$\frac{x^8}{x^3} = \frac{x \cdot x \cdot x \cdot x \cdot x \cdot x \cdot x \cdot x}{x \cdot x \cdot x} \qquad \longleftarrow \text{8 factors}$$
$$\longleftarrow \text{3 factors}$$

$$= \frac{x \cdot x \cdot x}{x \cdot x \cdot x} \cdot x \cdot x \cdot x \cdot x \cdot x \qquad \text{Note that } x^3/x^3 \text{ is 1.}$$

$$= x \cdot x \cdot x \cdot x \cdot x \qquad \longleftarrow \text{5 factors}$$

$$= x^5.$$

The generalization of this result is the *quotient rule*.

Dividing with Like Bases: The Quotient Rule

For any nonzero number a and any positive integers m and n, $m > n$,

$$\frac{a^m}{a^n} = a^{m-n}.$$

(When dividing, if the bases are the same, keep the base and subtract the exponent of the denominator from the exponent of the numerator.)

EXAMPLE **2**   Divide and simplify: **(a)** $\dfrac{r^9}{r^3}$; **(b)** $\dfrac{10x^{11}y^5}{2x^4y^3}$.

SOLUTION

a) $\dfrac{r^9}{r^3} = r^{9-3} = r^6$ Using the quotient rule

b) $\dfrac{10x^{11}y^5}{2x^4y^3} = 5 \cdot x^{11-4} \cdot y^{5-3}$ Dividing; using the quotient rule

$$= 5x^7y^2$$

> **TRY EXERCISE** 21

CAUTION!

$$\frac{7^8}{7^2} = 7^6 \qquad \begin{cases} \dfrac{7^8}{7^2} \neq 1^6 & \text{Do not divide the bases!} \\[2mm] \dfrac{7^8}{7^2} \neq 7^4 & \text{Do not divide the exponents!} \end{cases}$$

The Zero Exponent

Suppose now that the bases in the numerator and the denominator are identical and are both raised to the same power. On the one hand, any (nonzero) expression divided by itself is equal to 1. For example,

$$\frac{t^5}{t^5} = 1 \quad \text{and} \quad \frac{6^4}{6^4} = 1.$$

On the other hand, if we continue to subtract exponents when dividing powers with the same base, we have

$$\frac{t^5}{t^5} = t^{5-5} = t^0 \quad \text{and} \quad \frac{6^4}{6^4} = 6^{4-4} = 6^0.$$

This suggests that t^5/t^5 equals both 1 *and* t^0. It also suggests that $6^4/6^4$ equals both 1 *and* 6^0. This leads to the following definition.

> ### The Zero Exponent
>
> For any nonzero real number a,
>
> $$a^0 = 1.$$
>
> (Any nonzero number raised to the zero power is 1. The expression 0^0 is undefined.)

EXAMPLE **3** Evaluate each of the following for $x = 2.9$: **(a)** x^0; **(b)** $-x^0$; **(c)** $(-x)^0$.

SOLUTION

a) $x^0 = 2.9^0 = 1$ Using the definition of 0 as an exponent

b) $-x^0 = -2.9^0 = -1$ The exponent 0 pertains only to the 2.9.

c) $(-x)^0 = (-2.9)^0 = 1$ Because of the parentheses, the base here is -2.9.

TRY EXERCISE 31

Parts (b) and (c) of Example 3 illustrate an important result:

Since $-a^n$ means $-1 \cdot a^n$, in general, $-a^n \neq (-a)^n$.*

Negative Integers as Exponents

Later in this text we will explain what numbers like $\frac{2}{9}$ or $\sqrt{2}$ mean as exponents. Here we develop a definition for negative integer exponents. We can simplify $5^3/5^7$ two ways. First we proceed as in arithmetic:

$$\frac{5^3}{5^7} = \frac{5 \cdot 5 \cdot 5}{5 \cdot 5 \cdot 5 \cdot 5 \cdot 5 \cdot 5 \cdot 5}$$
$$= \frac{5 \cdot 5 \cdot 5 \cdot 1}{5 \cdot 5 \cdot 5 \cdot 5 \cdot 5 \cdot 5 \cdot 5}$$
$$= \frac{5 \cdot 5 \cdot 5}{5 \cdot 5 \cdot 5} \cdot \frac{1}{5 \cdot 5 \cdot 5 \cdot 5}$$
$$= \frac{1}{5^4}.$$

Were we to apply the quotient rule, we would have

$$\frac{5^3}{5^7} = 5^{3-7} = 5^{-4}.$$

These two expressions for $5^3/5^7$ suggest that

$$5^{-4} = \frac{1}{5^4}.$$

This leads to the definition of integer exponents, which includes negative exponents.

---

*When n is odd, it *does* follow that $-a^n = (-a)^n$. However, when n is even, we always have $-a^n \neq (-a)^n$, since $-a^n$ is always negative and $(-a)^n$ is always positive. We assume $a \neq 0$.

> ### Integer Exponents
>
> For any real number a that is nonzero and any integer n,
>
> $$a^{-n} = \frac{1}{a^n}.$$
>
> (The numbers a^{-n} and a^n are reciprocals of each other.)

The definitions above preserve the following pattern:

$$4^3 = 4 \cdot 4 \cdot 4,$$

$$4^2 = 4 \cdot 4, \qquad \text{Dividing both sides by 4}$$

$$4^1 = 4, \qquad \text{Dividing both sides by 4}$$

$$4^0 = 1, \qquad \text{Dividing both sides by 4}$$

$$4^{-1} = \frac{1}{4}, \qquad \text{Dividing both sides by 4}$$

$$4^{-2} = \frac{1}{4 \cdot 4} = \frac{1}{4^2}. \qquad \text{Dividing both sides by 4}$$

CAUTION!   A negative exponent does not, in itself, indicate that an expression is negative. As shown above,

$$4^{-2} \neq 4(-2).$$

EXAMPLE   4   Express each of the following using positive exponents and, if possible, simplify:
(a) 7^{-2}; **(b)** -7^{-2}; **(c)** $(-7)^{-2}$.

SOLUTION

a) $7^{-2} = \dfrac{1}{7^2} = \dfrac{1}{49}$ The base is 7. We use the definition of integer exponents.

b) $-7^{-2} = -\dfrac{1}{7^2} = -\dfrac{1}{49}$ The base is 7.

$$-7^{-2} = -1 \cdot 7^{-2} = -1 \cdot \frac{1}{7^2} = -1 \cdot \frac{1}{49} = -\frac{1}{49}.$$

c) $(-7)^{-2} = \dfrac{1}{(-7)^2} = \dfrac{1}{49}$ The base is -7. We use the definition of integer exponents.

Note that $-7^{-2} \neq (-7)^{-2}$.

▶ TRY EXERCISE 35

EXAMPLE   5   Express each of the following using positive exponents and, if possible, simplify:
(a) $5x^{-4}y^3$; **(b)** $\dfrac{1}{6^{-2}}$.

SOLUTION

a) $5x^{-4}y^3 = 5\left(\dfrac{1}{x^4}\right)y^3 = \dfrac{5y^3}{x^4}$

b) Since $\dfrac{1}{a^n} = a^{-n}$, we have

$$\frac{1}{6^{-2}} = 6^{-(-2)} = 6^2, \text{ or } 36. \qquad \text{Remember: } n \text{ can be a negative integer.}$$

▶ TRY EXERCISE 45

The result from Example 5(b) can be generalized.

Factors and Negative Exponents

For any nonzero real numbers a and b and any integers m and n,

$$\frac{a^{-n}}{b^{-m}} = \frac{b^m}{a^n}.$$

(A factor can be moved to the other side of the fraction bar if the sign of its exponent is changed.)

EXAMPLE 6 Write an equivalent expression without negative exponents:

$$\frac{vx^{-2}y^{-5}}{z^{-4}w^{-3}}.$$

SOLUTION We can move each factor to the other side of the fraction bar if we change the sign of each exponent:

$$\frac{vx^{-2}y^{-5}}{z^{-4}w^{-3}} = \frac{vz^4w^3}{x^2y^5}.$$

> **TRY EXERCISE** 51

The product and quotient rules apply for all integer exponents.

EXAMPLE 7 Simplify: (a) $9^{-3} \cdot 9^8$; (b) $\dfrac{y^{-5}}{y^{-4}}$.

SOLUTION

Using the product rule; adding exponents

a) $9^{-3} \cdot 9^8 = 9^{-3+8}$

$\phantom{9^{-3} \cdot 9^8} = 9^5$

Using the quotient rule; subtracting exponents

b) $\dfrac{y^{-5}}{y^{-4}} = y^{-5-(-4)} = y^{-1}$

$\phantom{\dfrac{y^{-5}}{y^{-4}}} = \dfrac{1}{y}$ Writing the answer without a negative exponent

> **TRY EXERCISE** 71

Example 7(b) can also be simplified as follows:

$$\frac{y^{-5}}{y^{-4}} = \frac{y^4}{y^5} = y^{4-5} = y^{-1} = \frac{1}{y}.$$

Simplifying $(a^m)^n$

Next, consider an expression like $(3^4)^2$:

$$(3^4)^2 = (3^4)(3^4) \qquad \text{We are raising } 3^4 \text{ to the second power.}$$

$$ = (3 \cdot 3 \cdot 3 \cdot 3)(3 \cdot 3 \cdot 3 \cdot 3)$$

$$ = 3 \cdot 3 \cdot 3 \cdot 3 \cdot 3 \cdot 3 \cdot 3 \cdot 3 \qquad \text{Using the associative law}$$

$$ = 3^8.$$

TECHNOLOGY CONNECTION

Most calculators have an exponentiation key, often labeled $\boxed{x^y}$ or $\boxed{\frown}$. To enter 4^7 on most scientific calculators, we press

$\boxed{4}$ $\boxed{x^y}$ $\boxed{7}$ $\boxed{=}$

On most graphing calculators, we press

$\boxed{4}$ $\boxed{\wedge}$ $\boxed{7}$ $\boxed{\text{ENTER}}$

1. List keystrokes that could be used to simplify 2^{-5} on a scientific or graphing calculator.
2. How could 2^{-5} be simplified on a calculator lacking an exponentiation key?

Note that in this case, we could have multiplied the exponents:

$$(3^4)^2 = 3^{4 \cdot 2} = 3^8.$$

Likewise, $(y^8)^3 = (y^8)(y^8)(y^8) = y^{24}$. Once again, we get the same result if we multiply the exponents:

$$(y^8)^3 = y^{8 \cdot 3} = y^{24}.$$

The Power Rule

For any real number a and any integers m and n for which a^m and $(a^m)^n$ exist,

$$(a^m)^n = a^{mn}.$$

(To raise a power to a power, multiply the exponents.)

EXAMPLE   8   Simplify: **(a)** $(3^5)^4$; **(b)** $(y^{-5})^7$; **(c)** $(a^{-3})^{-7}$.

SOLUTION

a) $(3^5)^4 = 3^{5 \cdot 4} = 3^{20}$

b) $(y^{-5})^7 = y^{-5 \cdot 7} = y^{-35} = \dfrac{1}{y^{35}}$

c) $(a^{-3})^{-7} = a^{(-3)(-7)} = a^{21}$

> TRY EXERCISE ▸ 97

Raising a Product or a Quotient to a Power

When an expression inside parentheses is raised to a power, the inside expression is the base. Let's compare $2a^3$ and $(2a)^3$.

$$2a^3 = 2 \cdot a \cdot a \cdot a; \qquad (2a)^3 = (2a)(2a)(2a)$$
$$= 2 \cdot 2 \cdot 2 \cdot a \cdot a \cdot a$$
$$= 2^3 a^3 = 8a^3$$

We see that $2a^3$ and $(2a)^3$ are *not* equivalent. Note also that to simplify $(2a)^3$ we can raise each factor to the power 3. This leads to the following rule.

Raising a Product to a Power

For any integer n, and any real numbers a and b for which $(ab)^n$ exists,

$$(ab)^n = a^n b^n.$$

(To raise a product to a power, raise each factor to that power.)

EXAMPLE   9   Simplify: **(a)** $(-2x)^3$; **(b)** $(-3x^5 y^{-1})^{-4}$.

SOLUTION

a) $(-2x)^3 = (-2)^3 \cdot x^3$ Raising each factor to the third power

$\qquad\quad = -8x^3$

b) $(-3x^5y^{-1})^{-4} = (-3)^{-4}(x^5)^{-4}(y^{-1})^{-4}$ Raising each factor to the negative fourth power

$$= \frac{1}{(-3)^4} \cdot x^{-20}y^4$$ Multiplying exponents; writing $(-3)^{-4}$ as $\dfrac{1}{(-3)^4}$

$$= \frac{1}{81} \cdot \frac{1}{x^{20}} \cdot y^4$$

$$= \frac{y^4}{81x^{20}}$$ **TRY EXERCISE** 101

There is a similar rule for raising a quotient to a power.

Raising a Quotient to a Power

For any integer n, and any real numbers a and b for which a/b, a^n, and b^n exist,

$$\left(\frac{a}{b}\right)^n = \frac{a^n}{b^n}.$$

(To raise a quotient to a power, raise both the numerator and the denominator to that power.)

EXAMPLE 10   Simplify: **(a)** $\left(\dfrac{x^2}{2}\right)^4$; **(b)** $\left(\dfrac{y^2z^3}{5}\right)^{-3}$.

SOLUTION

a) $\left(\dfrac{x^2}{2}\right)^4 = \dfrac{(x^2)^4}{2^4} = \dfrac{x^8}{16} \begin{array}{l} \leftarrow 2 \cdot 4 = 8 \\ \leftarrow 2^4 = 16 \end{array}$

b) $\left(\dfrac{y^2z^3}{5}\right)^{-3} = \dfrac{(y^2z^3)^{-3}}{5^{-3}}$

$$= \frac{5^3}{(y^2z^3)^3}$$ Moving factors to the other side of the fraction bar and reversing the sign of those exponents

$$= \frac{125}{y^6z^9}$$ **TRY EXERCISE** 105

The rule for raising a quotient to a power allows us to derive a useful result for manipulating negative exponents:

$$\left(\frac{a}{b}\right)^{-n} = \frac{a^{-n}}{b^{-n}} = \frac{b^n}{a^n} = \left(\frac{b}{a}\right)^n.$$

Using this result, we can simplify Example 10(b) as follows:

$$\left(\frac{y^2z^3}{5}\right)^{-3} = \left(\frac{5}{y^2z^3}\right)^3$$ Taking the reciprocal of the base and changing the exponent's sign

$$= \frac{5^3}{(y^2z^3)^3}$$

$$= \frac{125}{y^6z^9}.$$

> **Definitions and Properties of Exponents**
>
> The following summary assumes that no denominators are 0 and that 0^0 is not considered, and is true for any integers m and n.
>
> | | |
> |---|---|
> | 1 as an exponent: | $a^1 = a$ |
> | 0 as an exponent: | $a^0 = 1$ |
> | Negative exponents: | $a^{-n} = \dfrac{1}{a^n}$ |
> | | $\dfrac{a^{-n}}{b^{-m}} = \dfrac{b^m}{a^n}$ |
> | | $\left(\dfrac{a}{b}\right)^{-n} = \left(\dfrac{b}{a}\right)^n$ |
> | The Product Rule: | $a^m \cdot a^n = a^{m+n}$ |
> | The Quotient Rule: | $\dfrac{a^m}{a^n} = a^{m-n}$ |
> | The Power Rule: | $(a^m)^n = a^{mn}$ |
> | Raising a product to a power: | $(ab)^n = a^n b^n$ |
> | Raising a quotient to a power: | $\left(\dfrac{a}{b}\right)^n = \dfrac{a^n}{b^n}$ |

1.6   EXERCISE SET

For Extra Help
MyMathLab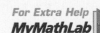
Math XL   PRACTICE   WATCH   DOWNLOAD

Concept Reinforcement *In each of Exercises 1–10, state whether the equation is an example of the product rule, the quotient rule, the power rule, raising a product to a power, or raising a quotient to a power.*

1. $(a^6)^4 = a^{24}$

2. $\left(\dfrac{5}{7}\right)^4 = \dfrac{5^4}{7^4}$

3. $(5x)^7 = 5^7 x^7$

4. $\dfrac{m^9}{m^3} = m^6$

5. $m^6 \cdot m^4 = m^{10}$

6. $(5^2)^7 = 5^{14}$

7. $\left(\dfrac{a}{4}\right)^7 = \dfrac{a^7}{4^7}$

8. $(ab)^{10} = a^{10} b^{10}$

9. $\dfrac{x^{10}}{x^2} = x^8$

10. $r^5 \cdot r^7 = r^{12}$

Multiply and simplify. Leave the answer in exponential notation.

11. $6^4 \cdot 6^7$

12. $3^8 \cdot 3^9$

13. $m^0 \cdot m^8$

14. $t^6 \cdot t^0$

15. $5x^4 \cdot 4x^3$

16. $3a^5 \cdot 2a^4$

17. $(-3a^2)(-8a^6)$

18. $(-4m^7)(6m^2)$

19. $(m^5 n^2)(m^3 n p^0)$

20. $(x^6 y^3)(xy^4 z^0)$

Divide and simplify.

21. $\dfrac{t^8}{t^3}$

22. $\dfrac{a^{11}}{a^8}$

23. $\dfrac{15a^7}{3a^2}$

24. $\dfrac{24t^9}{8t^3}$

25. $\dfrac{m^7 n^9}{m^2 n^5}$

26. $\dfrac{m^{12} n^9}{m^4 n^6}$

27. $\dfrac{32x^8 y^5}{8x^2 y}$

28. $\dfrac{35x^7 y^8}{7xy^2}$

29. $\dfrac{28x^{10} y^9 z^8}{-7x^2 y^3 z^2}$

30. $\dfrac{-20x^8 y^5 z^3}{-4x^2 y^2 z}$

Evaluate each of the following for $x = -2$.

31. $-x^0$

32. $(-x)^0$

33. $(4x)^0$

34. $4x^0$

Write an equivalent expression without negative exponents and, if possible, simplify.

35. t^{-9} **36.** m^{-2} **37.** 6^{-2}

38. 5^{-3} **39.** $(-3)^{-2}$ **40.** $(-2)^{-4}$

41. -3^{-2} **42.** -2^{-4} **43.** -1^{-10}

44. -10^{-2} **45.** $\dfrac{1}{10^{-3}}$ **46.** $\dfrac{1}{2^{-4}}$

47. $6x^{-1}$ **48.** $9x^{-4}$ **49.** $3a^8b^{-6}$

50. $5a^{-7}b^4$ **51.** $\dfrac{2z^{-3}}{x^5}$ **52.** $\dfrac{5a^{-1}}{b}$

53. $\dfrac{3y^2}{z^{-4}}$ **54.** $\dfrac{t^{-6}}{7s^2}$

55. $\dfrac{ab^{-1}}{c^{-1}}$ **56.** $\dfrac{x^{-3}y^4}{z^{-5}}$

57. $\dfrac{pq^{-2}r^{-3}}{2u^5v^{-4}}$ **58.** $\dfrac{5a^{-3}bc^{-1}}{d^{-6}f^2}$

Write an equivalent expression with negative exponents.

59. $\dfrac{1}{x^3}$ **60.** $\dfrac{1}{n^4}$ **61.** $\dfrac{1}{(-10)^3}$

62. $\dfrac{1}{12^5}$ **63.** 8^{10} **64.** $(-6)^4$

65. $4x^2$ **66.** $-4y^5$ **67.** $\dfrac{1}{(5y)^3}$

68. $\dfrac{1}{(5x)^5}$ **69.** $\dfrac{1}{3y^4}$ **70.** $\dfrac{1}{4b^3}$

Simplify. Should negative exponents appear in the answer, write a second answer using only positive exponents.

71. $6^{-3} \cdot 6^{-5}$ **72.** $4^{-2} \cdot 4^{-1}$

73. $a \cdot a^{-8}$ **74.** $b^5 \cdot b^{-2}$

75. $x^{-7} \cdot x^2 \cdot x^5$ **76.** $a^4 \cdot a^2 \cdot a^{-5}$

77. $(4mn^3)(-2m^3n^2)$ **78.** $(6x^6y^{-2})(-3x^2y^3)$

79. $(-7x^4y^{-5})(-5x^{-6}y^8)$

80. $(-4u^{-6}v^8)(-6u^{-4}v^{-2})$

81. $(5a^{-2}b^{-3})(2a^{-4}b)$ **82.** $(3a^{-5}b^{-7})(2ab^{-2})$

83. $\dfrac{10^{-3}}{10^6}$ **84.** $\dfrac{12^{-4}}{12^8}$

85. $\dfrac{2^{-7}}{2^{-5}}$ **86.** $\dfrac{9^{-4}}{9^{-6}}$

87. $\dfrac{y^4}{y^{-5}}$ **88.** $\dfrac{a^3}{a^{-2}}$

89. $\dfrac{24a^5b^3}{-8a^4b}$ **90.** $\dfrac{-12m^4}{-4mn^5}$

91. $\dfrac{15m^5n^3}{10m^{10}n^{-4}}$ **92.** $\dfrac{-24x^6y^7}{18x^{-3}y^9}$

93. $\dfrac{-6x^{-2}y^4z^8}{-24x^{-5}y^6z^{-3}}$ **94.** $\dfrac{8a^6b^{-4}c^8}{32a^{-4}b^5c^9}$

95. $(x^4)^3$ **96.** $(a^3)^2$

97. $(9^3)^{-4}$ **98.** $(8^4)^{-3}$

99. $(t^{-8})^{-5}$ **100.** $(x^{-4})^{-3}$

101. $(-5xy)^2$ **102.** $(-5ab)^3$

103. $(-2a^{-2}b)^{-3}$ **104.** $(-4x^6y^{-2})^{-2}$

105. $\left(\dfrac{m^2n^{-1}}{4}\right)^3$ **106.** $\left(\dfrac{3x^5}{y^{-4}}\right)^2$

107. $\dfrac{(2a^3)^34a^{-3}}{(a^2)^5}$ **108.** $\dfrac{(3x^2)^32x^{-4}}{(x^4)^2}$

Aha! 109. $(8x^{-3}y^2)^{-4}(8x^{-3}y^2)^4$

110. $(2a^{-1}b^3)^{-2}(2a^{-1}b^3)^{-2}$

111. $\dfrac{(5a^3b)^2}{10a^2b}$ **112.** $\dfrac{(3x^3y^4)^3}{6xy^3}$

113. $\left(\dfrac{2x^3y^{-2}}{3y^{-3}}\right)^3$ **114.** $\left(\dfrac{-4x^4y^{-2}}{5x^{-1}y^4}\right)^{-4}$

Aha! 115. $\left(\dfrac{21x^5y^{-7}}{14x^{-2}y^{-6}}\right)^0$ **116.** $\left(\dfrac{6a^{-2}b^6}{8a^{-4}b^0}\right)^{-2}$

117. $\left(\dfrac{5x^0y^{-7}}{2x^{-2}y^4}\right)^{-2}$ **118.** $\left(\dfrac{4a^3b^{-9}}{6a^{-2}b^5}\right)^0$

119. Explain why $(-1)^n = 1$ for any even number n.

120. Explain why $(-17)^{-8}$ is positive.

Skill Review

Review evaluating expressions (Sections 1.1 and 1.2).

Evaluate. [1.1], [1.2]

121. $4.9t^2 + 3t$, for $t = -3$

122. $16t^2 + 10t$, for $t = -2$

Synthesis

123. Explain the different uses and meanings of the "−" sign in the expression $3 - (-2)^{-1}$.

124. Is the following true or false, and why?

$$5^{-6} > 4^{-9}$$

Simplify. Assume that all variables represent nonzero integers.

125. $\dfrac{8a^{x-2}}{2a^{2x+2}}$

126. $[7y(7 - 8)^{-4} - 8y(8 - 7)^{-2}](-2)^2$

127. $\dfrac{(2^{-2})^a \cdot (2^b)^{-a}}{(2^{-2})^{-b}(2^b)^{-2a}}$

128. $\{[(8^{-a})^{-2}]^b\}^{-c} \cdot [(8^0)^a]^c$

129. $(3^{a+2})^a$

130. $\dfrac{-28x^{b+5}y^{4+c}}{7x^{b-5}y^{c-4}}$

131. $\dfrac{4x^{2a+3}y^{2b-1}}{2x^{a+1}y^{b+1}}$

132. $(7^{3-a})^{2b}$

133. $\dfrac{3^{q+3} - 3^2(3^q)}{3(3^{q+4})}$

134. $\dfrac{25x^{a+b}y^{b-a}}{-5x^{a-b}y^{b+a}}$

135. $\left[\left(\dfrac{a^{-2c}}{b^{7c}}\right)^{-3}\left(\dfrac{a^{4c}}{b^{-3c}}\right)^2\right]^{-a}$

1.7   Scientific Notation

Conversions ▪ Significant Digits and Rounding ▪ Scientific Notation in Problem Solving

We write numbers using a variety of symbolism, or *notation*, such as fraction notation, decimal notation, and percent notation. We now study **scientific notation**, so named because of its usefulness in work with the very large and very small numbers that occur in science.

The following are examples of scientific notation:

7.2×10^5 means   720,000;

3.4×10^{-6} means   0.0000034;

4.89×10^{-3} means   0.00489.

> ### Scientific Notation
> *Scientific notation* for a number is an expression of the form $N \times 10^m$, where N is in decimal notation, $1 \le N < 10$, and m is an integer.

Conversions

Note that $10^b/10^b = 10^b \cdot 10^{-b} = 1$. To convert a number to scientific notation, we can multiply by 1, writing 1 in the form $10^b/10^b$ or $10^b \cdot 10^{-b}$.

EXAMPLE 1

Computer algorithms. Scientists at the University of Alberta have proved that the computer program Chinook, designed to play the game of checkers, cannot ever lose. Checkers is the most complex game that has been solved with a computer program, with about 500,000,000,000,000,000,000 possible board positions. Write scientific notation for this number.

Source: *The New York Times, 7/20/07*

SOLUTION   To write 500,000,000,000,000,000,000 as 5×10^m for some integer m, we must move the decimal point in the number 20 places to the left. This can be accomplished by dividing and then multiplying by 10^{20}:

$$500,000,000,000,000,000,000 = \frac{500,000,000,000,000,000,000}{10^{20}} \times 10^{20} \qquad \text{Multiplying by 1: } \frac{10^{20}}{10^{20}} = 1$$

$$= 5 \times 10^{20}. \qquad \text{This is scientific notation.} \qquad \boxed{\text{TRY EXERCISE}} \; 7$$

EXAMPLE   2   Write scientific notation for the mass of a grain of sand:

0.0648 gram (g).

SOLUTION   To write 0.0648 as 6.48×10^m for some integer m, we must move the decimal point 2 places to the right. To do this, we multiply and then divide by 10^2:

$$0.0648 = 0.0648 \times 10^2 / 10^2 \qquad \text{Multiplying by 1: } 10^2/10^2 = 1$$

$$= \frac{6.48}{10^2}$$

$$= 6.48 \times 10^{-2} \text{ g.} \qquad \text{Writing scientific notation}$$

$$\boxed{\text{TRY EXERCISE}} \; 11$$

Try to make conversions to and from scientific notation mentally if possible. In doing so, remember that negative powers of 10 are used when representing small numbers and positive powers of 10 are used when representing large numbers.

EXAMPLE   3   Convert mentally to decimal notation: **(a)** 4.371×10^7; **(b)** 1.73×10^{-5}.

SOLUTION

a) $4.371 \times 10^7 = 43,710,000$ Moving the decimal point 7 places to the right

b) $1.73 \times 10^{-5} = 0.0000173$ Moving the decimal point 5 places to the left

$$\boxed{\text{TRY EXERCISE}} \; 21$$

EXAMPLE   4   Convert mentally to scientific notation: **(a)** 82,500,000; **(b)** 0.0000091.

SOLUTION

a) $82,500,000 = 8.25 \times 10^7$ *Check*: Multiplying 8.25 by 10^7 moves the decimal point 7 places to the right.

b) $0.0000091 = 9.1 \times 10^{-6}$ *Check*: Multiplying 9.1 by 10^{-6} moves the decimal point 6 places to the left.

$$\boxed{\text{TRY EXERCISE}} \; 9$$

Significant Digits and Rounding

In the world of science, it is important to know just how accurate a measurement is. For example, the measurement 5.72×10^4 km is more precise than the measurement 5.7×10^4 km. We say that 5.72×10^4 has three **significant digits** whereas 5.7×10^4 has only two significant digits. If 5.7×10^4, or 57,000, includes no rounding in the hundreds column, we would indicate that by writing 5.70×10^4.

When two or more measurements written in scientific notation are multiplied or divided, the result should be rounded so that it has the same number of significant digits as the measurement with the fewest significant digits. Rounding should be performed at the *end* of the calculation.

Thus,

$$(\underbrace{3.1}_{\text{2 digits}} \times 10^{-3} \text{ mm})(\underbrace{2.45}_{\text{3 digits}} \times 10^{-4} \text{ mm}) = 7.595 \times 10^{-7} \text{ mm}^2$$

should be rounded to

$$\underbrace{7.6}_{\text{2 digits}} \times 10^{-7} \text{ mm}^2.$$

When two or more measurements written in scientific notation are added or subtracted, the result should be rounded so that it has as many decimal places as the measurement with the fewest decimal places.

For example,

$$\underbrace{1.6354}_{\substack{\text{4 decimal} \\ \text{places}}} \times 10^4 \text{ km} + \underbrace{2.078}_{\substack{\text{3 decimal} \\ \text{places}}} \times 10^4 \text{ km} = 3.7134 \times 10^4 \text{ km}$$

should be rounded to

$$\underbrace{3.713}_{\substack{\text{3 decimal} \\ \text{places}}} \times 10^4 \text{ km}.$$

EXAMPLE 5   Multiply and write scientific notation for the answer:

$$(7.2 \times 10^5)(4.3 \times 10^9).$$

SOLUTION   We have

$$(7.2 \times 10^5)(4.3 \times 10^9) = (7.2 \times 4.3)(10^5 \times 10^9) \qquad \text{Using the commutative and associative laws}$$

$$= 30.96 \times 10^{14} \qquad \text{Adding exponents}$$

Since 30.96 is not between 1 and 10, this is not in scientific notation. To find scientific notation for this result, we convert 30.96 to scientific notation and simplify:

$$30.96 \times 10^{14} = (3.096 \times 10^1) \times 10^{14}$$

$$= 3.096 \times 10^{15} \qquad \text{Using the associative law}$$

$$\approx 3.1 \times 10^{15}. \qquad \text{Rounding to 2 significant digits}$$

TRY EXERCISE ▶ 35

TECHNOLOGY CONNECTION

Both graphing and scientific calculators allow expressions to be entered using scientific notation. To do so, a key normally labeled (EE) or EXP is used. Often this is a secondary function and a key labeled 2ND or SHIFT must be pressed first. To check Example 5, we press 7.2 (EE) 5 (×) 4.3 (EE) 9. When we then press ENTER or =, the result 3.096E15 or 3.096   15 appears. We must interpret this result as 3.096×10^{15}.

EXAMPLE **6** Divide and write scientific notation for the answer:

$$\frac{3.48 \times 10^{-7}}{4.64 \times 10^{6}}.$$

SOLUTION

$$\frac{3.48 \times 10^{-7}}{4.64 \times 10^{6}} = \frac{3.48}{4.64} \times \frac{10^{-7}}{10^{6}}$$ Separating factors. Our answer must have 3 significant digits.

$$= 0.75 \times 10^{-13}$$ Subtracting exponents; simplifying

$$= (7.5 \times 10^{-1}) \times 10^{-13}$$ Converting 0.75 to scientific notation

$$= 7.50 \times 10^{-14}$$ Adding exponents. We write 7.50 to indicate 3 significant digits.

TRY EXERCISE 47

Scientific Notation in Problem Solving

Scientific notation can be useful in problem solving.

EXAMPLE **7** *Information technology.* In 2006, about 161 exabytes, or 1.61×10^{14} megabytes, of digital information were generated worldwide by an online population of 694 million people. Find the average amount of information generated by each person who was online in 2006.

Sources: IDC and comScore World Metrix

SOLUTION

1. **Familiarize.** We let $a =$ the average number of megabytes of information generated, per person, in 2006. The amount of digital information is given in scientific notation. To write the online population in scientific notation, note that 1 million is 1,000,000. Thus, 694 million is 694,000,000, or 6.94×10^{8}.

2. **Translate.** To find the average, we need to divide the amount of digital information by the online population:

$$a = \frac{1.61 \times 10^{14} \text{ megabytes}}{6.94 \times 10^{8} \text{ people}}.$$

3. **Carry out.** We calculate and write scientific notation for the result:

$$a = \frac{1.61 \times 10^{14} \text{ megabytes}}{6.94 \times 10^{8} \text{ people}}$$

$$= \frac{1.61}{6.94} \times \frac{10^{14}}{10^{8}} \frac{\text{megabytes}}{\text{people}}$$

$$\approx 0.232 \times 10^{6} \text{ megabytes/person}$$ Rounding to 3 significant digits

$$\approx 2.32 \times 10^{-1} \times 10^{6} \text{ megabytes/person}$$
$$\approx 2.32 \times 10^{5} \text{ megabytes/person.}$$ Writing scientific notation

4. **Check.** To check, we multiply our answer, the average number of megabytes per person, by the worldwide online population:

$$\underbrace{(2.32 \times 10^5 \text{ megabytes/person})}_{\text{Average amount of information generated}}\underbrace{(6.94 \times 10^8 \text{ people})}_{\text{Worldwide online population}}$$

$$= (2.32 \times 6.94)(10^5 \times 10^8)\frac{\text{megabytes}}{\text{person}} \cdot \text{people}$$

$$= 16.1008 \times 10^{13} \text{ megabytes}$$

$$\approx 1.61 \times 10^1 \times 10^{13} \text{ megabytes} \left.\vphantom{\begin{array}{c}a\\b\end{array}}\right\} \quad \begin{array}{l}\text{Rounding to 3 significant digits}\\\text{and writing scientific notation}\end{array}$$

$$\approx 1.61 \times 10^{14} \text{ megabytes}$$

With rounding, our answer checks.

5. **State.** An average of 2.32×10^5 megabytes of information was generated by each person who was online in 2006.

TRY EXERCISE 57

EXAMPLE 8   *Telecommunications.*   A fiber-optic cable will be used for 125 km of transmission line. The cable has a diameter of 0.6 cm. What is the volume of cable needed for the line?

SOLUTION

1. **Familiarize.** Making a drawing, we see that we have a cylinder (a very *long* one). Its length is 125 km and the base has a diameter of 0.6 cm.

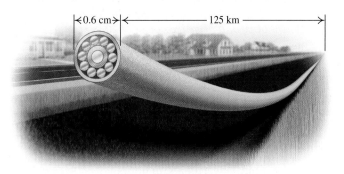

Recall that the formula for the volume of a cylinder is

$$V = \pi r^2 h,$$

where *r* is the radius and *h* is the height (in this case, the length of the cable).

2. **Translate.** Before we use the volume formula, we must make the units consistent. Let's express everything in meters:

Length: 125 km = 125,000 m, or 1.25×10^5 m;

Diameter: 0.6 cm = 0.006 m, or 6×10^{-3} m.

The radius, which we will need in the formula, is half the diameter:

Radius: 3×10^{-3} m.

We now substitute into the above formula:

$$V = \pi(3 \times 10^{-3} \text{ m})^2(1.25 \times 10^5 \text{ m}).$$

3. Carry out. We do the calculations:

$$V = \pi \times (3 \times 10^{-3}\,\text{m})^2(1.25 \times 10^5\,\text{m})$$

$$= \pi \times 3^2 \times 10^{-6}\,\text{m}^2 \times 1.25 \times 10^5\,\text{m} \quad \text{Using the properties of exponents}$$

$$= (\pi \times 3^2 \times 1.25) \times (10^{-6} \times 10^5)\,\text{m}^3$$

$$\approx 35.325 \times 10^{-1}\,\text{m}^3 \quad \text{Using 3.14 for } \pi$$

$$\approx 3.5\,\text{m}^3. \quad \text{Rounding 3.5325 to 2 significant digits because of the 0.6}$$

4. Check. About all we can do here is recheck the translation and calculations. Note that m^3 is a unit of volume, as expected.

5. State. The volume of the cable is about 3.5 m^3 (cubic meters).

TRY EXERCISE 59

1.7 EXERCISE SET

For Extra Help
MyMathLab
Math XL
PRACTICE
WATCH
DOWNLOAD

🪝 *Concept Reinforcement* *State whether scientific notation for each of the following numbers would include either a positive power of 10 or a negative power of 10.*

1. The length of an Olympic marathon, in centimeters

2. The thickness of a cat's whisker, in meters

3. The mass of a hydrogen atom, in grams

4. The mass of a pickup truck, in grams

5. The time between leap years, in seconds

6. The time between a bird's heartbeats, in hours

Convert to scientific notation.

7. 64,000,000,000

8. 3,700,000

9. 1,091,000,000

10. 803,000,000,000,000

11. 0.0000013

12. 0.000078

13. 0.00009

14. 0.00000006

15. 803,000,000,000

16. 3,090,000,000,000

17. 0.000000904

18. 0.00000000802

19. 431,700,000,000

20. 953,400,000,000

Convert to decimal notation.

21. 4×10^5

22. 3×10^{-6}

23. 1.2×10^{-4}

24. 8.6×10^8

25. 3.76×10^{-9}

26. 4.27×10^{-2}

27. 8.056×10^{12}

28. 5.002×10^{10}

29. 7.001×10^{-5}

30. 2.049×10^{-3}

31. 9.06×10^9

32. 1.08×10^6

Simplify and write scientific notation for the answer. Use the correct number of significant digits.

33. $(3.4 \times 10^{-8})(2.6 \times 10^{15})$

34. $(1.8 \times 10^{20})(4.7 \times 10^{-12})$

35. $(2.36 \times 10^6)(1.4 \times 10^{-11})$

36. $(4.26 \times 10^{-6})(8.2 \times 10^{-6})$

37. $(5.2 \times 10^6)(2.6 \times 10^4)$

38. $(6.11 \times 10^3)(1.01 \times 10^{13})$

39. $(7.01 \times 10^{-5})(6.5 \times 10^{-7})$

40. $(4.08 \times 10^{-10})(7.7 \times 10^5)$

Aha! 41. $(2.0 \times 10^6)(3.02 \times 10^{-6})$

42. $(7.04 \times 10^{-9})(9.01 \times 10^{-7})$

43. $\dfrac{6.5 \times 10^{15}}{2.6 \times 10^4}$

44. $\dfrac{8.5 \times 10^{18}}{3.4 \times 10^5}$

45. $\dfrac{9.4 \times 10^{-9}}{4.7 \times 10^{-2}}$

46. $\dfrac{4.0 \times 10^{-6}}{8.0 \times 10^{-3}}$

47. $\dfrac{3.2 \times 10^{-7}}{8.0 \times 10^8}$

48. $\dfrac{1.26 \times 10^9}{4.2 \times 10^{-3}}$

49. $\dfrac{9.36 \times 10^{-11}}{3.12 \times 10^{11}}$

50. $\dfrac{2.42 \times 10^5}{1.21 \times 10^{-5}}$

51. $\dfrac{6.12 \times 10^{19}}{3.06 \times 10^{-7}}$

52. $\dfrac{4.7 \times 10^{-9}}{2.0 \times 10^{-9}}$

53. $4.6 \times 10^{-9} + 3.2 \times 10^{-9}$

54. $2.9 \times 10^{15} + 4.6 \times 10^{15}$

55. $5.9 \times 10^{23} + 6.3 \times 10^{23}$

56. $7.8 \times 10^{-34} + 5.4 \times 10^{-34}$

Solve.

57. *Municipal waste.*   In 2006, the worldwide population of 6.5 billion generated 2.02×10^{12} kg of municipal solid waste. On average, how much municipal solid waste was generated by each person?
Source: Global Waste Management Assessment 2007

58. *Agriculture.*   One of the shrimp ponds on Ana's farm contains 2.5×10^{11} mL of water. If the average number of suspended particles in the pond is 5.41×10^4 particles/mL, how many suspended particles are in the pond?

59. *High-tech fibers.*   A carbon nanotube is a thin cylinder of carbon atoms that, pound for pound, is stronger than steel and may one day be used in clothing. With a diameter of about 4.0×10^{-10} in., a fiber can be made 100 yd long. Find the volume of such a fiber.
Source: The Indianapolis Star, 6/15/03

60. *Home maintenance.*   The thickness of a sheet of plastic is measured in *mils*, where 1 mil = $\frac{1}{1000}$ in. To help conserve heat, the foundation of a 24-ft by 32-ft rectangular home is covered with a 4-ft high sheet of 8-mil plastic. Find the volume of plastic used.

61. *Office supplies.*   A ream of copier paper weighs 2.25 kg. How much does a sheet of copier paper weigh?

62. *Printing and engraving.*   A ton of five-dollar bills is worth $4,540,000. How many pounds does a five-dollar bill weigh?

For Exercises 63 and 64, use the fact that 1 *light year* = 5.88×10^{12} *miles.*

Aha! **63.** *Astronomy.*   The diameter of the Milky Way galaxy is approximately 5.88×10^{17} mi. How many light years is it from one end of the galaxy to the other?

64. *Astronomy.*   The brightest star in the night sky, Sirius, is about 4.704×10^{13} mi from the earth. How many light years is it from the earth to Sirius?

Named in tribute to Anders Ångström, a Swedish physicist who measured light waves, 1 Å *(read "one Angstrom") equals* 10^{-10} *meter. One parsec is about* 3.26 *light years, and one light year equals* 9.46×10^{15} *meters.*

65. How many Angstroms are in one parsec?

66. How many kilometers are in one parsec?

For Exercises 67 and 68, use the approximate average distance from the earth to the sun of 1.50×10^{11} *meters.*

67. Determine the volume of a cylindrical sunbeam that is 3 Å in diameter.

68. Determine the volume of a cylindrical sunbeam that is 5 Å in diameter.

69. *Biology.*   An average of 4.55×10^{11} bacteria live in each pound of U.S. mud. There are 60.0 drops in one teaspoon and 6.0 teaspoons in an ounce. How many bacteria live in a drop of U.S. mud?
Source: Harper's Magazine, April 1996, p. 13

70. *Astronomy.*   If a star 5.9×10^{14} mi from the earth were to explode today, its light would not reach us for 100 years. How far does light travel in 13 weeks?

71. *Astronomy.*   The diameter of Jupiter is about 1.43×10^5 km. A day on Jupiter lasts about 10 hr. At what speed is Jupiter's equator spinning?

72. *Astronomy.*   The average distance of the earth from the sun is about 9.3×10^7 mi. About how far does the earth travel in a yearly orbit about the sun? (Assume a circular orbit.)

73. Write a problem for a classmate to solve. Design the problem so the solution is "The volume of the laser's light beam is 3.14×10^5 mm^3."

74. List two advantages of using scientific notation. Answers may vary.

Skill Review

Review evaluating expressions (Sections 1.1 and 1.2).

Evaluate. [1.1], [1.2]

75. $3x - y \div 2z$, for $x = -2$, $y = -12$, and $z = 3$

76. $x^2 + y^2z$, for $x = -1$, $y = -2$, and $z = -3$

Synthesis

77. A criminal claims to be carrying \$5 million in twenty-dollar bills in a briefcase. Is this possible? Why or why not? (*Hint*: See Exercise 62.)

78. When a calculator indicates that $5^{17} = 7.629394531 \times 10^{11}$, an approximation is being made. How can you tell? (*Hint*: Examine the ones digit.)

79. *Density of the earth.* The volume of the earth is approximately 1.08×10^{12} km^3 and the mass of the earth is about 5.976×10^{24} kg. What is the average density of the earth, in grams per cubic centimeter?

80. The Sartorius Microbalance Model 4108 can weigh objects to an accuracy of 3.5×10^{-10} oz. A chemical compound weighing 1.2×10^{-9} oz is split in half and weighed on the microbalance. Give a weight range for the actual weight of each half.
Source: *Guinness Book of World Records*

81. Compare $8 \cdot 10^{-90}$ and $9 \cdot 10^{-91}$. Which is the larger value? How much larger is it? Write scientific notation for the difference.

82. Write the reciprocal of 8.00×10^{-23} in scientific notation.

83. Evaluate: $(4096)^{0.05}(4096)^{0.2}$.

84. What is the ones digit in 513^{128}?

85. A grain of sand is placed on the first square of a chessboard, two grains on the second square, four grains on the third, eight on the fourth, and so on. Without a calculator, use scientific notation to approximate the number of grains of sand required for the 64th square. (*Hint*: Use the fact that $2^{10} \approx 10^3$.)

CORNER

Paired Problem Solving

Focus: Problem solving, scientific notation, and unit conversion

Time: 15–25 minutes

Group size: 3

ACTIVITY

Given that the earth's average distance from the sun is 1.5×10^{11} meters, determine the earth's orbital speed around the sun in miles per hour. Assume a circular orbit and use the following guidelines.

1. Each group should spend about 10 minutes attempting to solve this problem. Do not worry if the solution is not found.

2. Group members should each describe the interactions that took place, answering these three questions:

   a) What successful strategies were used?
   b) What unsuccessful strategies were used?
   c) What recommendations can you make for students working together to solve a problem?

3. Each group should then share their observations with each other and report to the class as a whole what they feel are their most significant observations.

Study Summary

KEY TERMS AND CONCEPTS	EXAMPLES

SECTION 1.1: SOME BASICS OF ALGEBRA

An **algebraic expression** consists of **variables**, numbers or **constants**, and operation signs.

Phrase	*Translation*
The difference of two numbers	$x - y$
Twelve less than some number	$n - 12$
Nine more than the product of two numbers	$pq + 9$

An algebraic expression can be **evaluated** by **substituting** specific numbers for the variables(s) and carrying out the calculations, following the rules for order of operations.

Evaluate $3 + 4x \div 6y^2$ *for* $x = 12$ *and* $y = -2$.

$$3 + 4x \div 6y^2 = 3 + 4(12) \div 6(-2)^2 \qquad \text{Substituting 12 for } x \text{ and } -2 \text{ for } y$$

$$= 3 + 4(12) \div 6 \cdot 4 \qquad \text{Squaring } -2$$
$$= 3 + 48 \div 6 \cdot 4 \qquad \text{Multiplying}$$
$$= 3 + 8 \cdot 4 \qquad \text{Dividing}$$
$$= 3 + 32 \qquad \text{Multiplying}$$
$$= 35 \qquad \text{Adding}$$

SECTION 1.2: OPERATIONS AND PROPERTIES OF REAL NUMBERS

Absolute Value

$$|x| = \begin{cases} x, & \text{if } x \geq 0, \\ -x, & \text{if } x < 0 \end{cases}$$

$|-15| = 15,$
$|4.8| = 4.8,$
$|0| = 0$

Addition of Real Numbers

1. If the numbers have the same sign, add the absolute values. The answer has the same sign as the numbers.

$7 + 12 = 19,$
$-7 + (-12) = -19$

2. If the numbers have different signs, subtract the absolute values. If the absolute values are the same, the answer is 0. Otherwise, the answer has the same sign as the number with the greater absolute value.

$7 + (-7) = 0,$
$7 + (-12) = -5,$
$-7 + 12 = 5$

3. If one number is zero, the sum is the other number.

$-7 + 0 = -7$

Subtraction of Real Numbers

To subtract, change the sign of the number being subtracted and then add.

$8 - 14 = 8 + (-14) = -6,$
$8 - (-14) = 8 + 14 = 22$

Multiplication and Division of Real Numbers

1. Multiply or divide the absolute values of the numbers.

2. If the signs are different, the answer is negative.

3. If the signs are the same, the answer is positive.

$-3(-5) = 15,$
$10(-2) = -20,$
$-100 \div 25 = -4,$
$\left(-\frac{2}{5}\right) \div \left(-\frac{3}{10}\right) = \left(-\frac{2}{5}\right) \cdot \left(-\frac{10}{3}\right) = \frac{20}{15} = \frac{4}{3}$

The Commutative Laws

For addition: $a + b = b + a$
For multiplication: $a \cdot b = b \cdot a$

$5x + 6 = 6 + 5x$ by the commutative law for addition.
$5x + 6 = x \cdot 5 + 6$ by the commutative law for multiplication.

The Associative Laws

For addition: $a + (b + c) = (a + b) + c$
For multiplication: $a \cdot (b \cdot c) = (a \cdot b) \cdot c$

$2x + (y + 3) = (2x + y) + 3$ by the associative law for addition.
$(5x)y = 5(xy)$ by the associative law for multiplication.

The Distributive Law

$$a(b + c) = ab + ac$$

Multiply: $5(x - 2y) = 5x - 10y$.
Factor: $4a - 20b + 12 = 4(a - 5b + 3)$.

SECTION 1.3: SOLVING EQUATIONS

Like terms have variable factors that are exactly the same. We can use the distributive law to **combine like terms**.

$$
\begin{aligned}
n - 9 - 4(n - 1) &= n - 9 - 4n + 4 &&\text{Using the distributive law}\\
&= n - 4n - 9 + 4 \\
&= -3n - 5 &&\text{Combining like terms}
\end{aligned}
$$

The Addition and Multiplication Principles for Equations

$a = b$ is equivalent to $a + c = b + c$.
$a = b$ is equivalent to $a \cdot c = b \cdot c$, if $c \neq 0$.

Solve: $5t - 3(t - 3) = -t$.

$$
\begin{aligned}
5t - 3(t - 3) &= -t \\
5t - 3t + 9 &= -t &&\text{Multiplying to remove parentheses}\\
2t + 9 &= -t &&\text{Combining like terms}\\
2t + 9 + t &= -t + t &&\text{Using the addition principle; adding } t \\
& &&\text{to both sides.}\\
3t + 9 &= 0 &&\text{Simplifying}\\
3t + 9 - 9 &= 0 - 9 &&\text{Using the addition principle; subtract-}\\
& &&\text{ing 9 or adding } -9 \text{ on both sides}\\
3t &= -9 &&\text{Simplifying}\\
\tfrac{1}{3}(3t) &= \tfrac{1}{3}(-9) &&\text{Using the multiplication principle;}\\
& &&\text{multiplying both sides by } \tfrac{1}{3}\\
t &= -3
\end{aligned}
$$

Check:
$$
\begin{array}{c|c}
\multicolumn{2}{c}{5t - 3(t - 3) = -t} \\
\hline
5(-3) - 3(-3 - 3) & -(-3) \\
-15 - 3(-6) & 3 \\
-15 - (-18) & \\
& 3 \overset{?}{=} 3 \quad \text{TRUE}
\end{array}
$$

The solution is -3.

An **identity** is an equation that is true for all replacements.

$x + 2 = 2 + x$ is an identity. The solution set is $\mathbb{R}$, the set of all real numbers.

A **contradiction** is an equation that is never true.

$x + 1 = x + 2$ is a contradiction. The solution set is $\varnothing$, the empty set.

A **conditional equation** is true for some replacements and false for others.

$x + 2 = 5$ is a conditional equation. The solution set is $\{3\}$.

SECTION 1.4: INTRODUCTION TO PROBLEM SOLVING

Five-Step Problem-Solving Strategy

1. *Familiarize* yourself with the problem.
2. *Translate* to mathematical language.
3. *Carry out* some mathematical manipulation.

Lisa is training for a bicycle race. On Monday and Tuesday, she rode a total of 60 mi. She rode 4 mi farther on Tuesday than she did on Monday. How many miles did she ride each day?

  1. **Familiarize.** Let $x =$ the number of miles she rode on Monday. Then the number of miles she rode on Tuesday is $x + 4$.

4. *Check* your possible answer in the original problem.

5. *State* the answer clearly.

2. Translate. We reword the problem statement and translate.

Rewording: Monday's miles plus Tuesday's miles equals total miles.

Translating: $\qquad x \qquad + \qquad (x + 4) \qquad = \qquad 60$

3. Carry out.

$$x + (x + 4) = 60$$
$$2x + 4 = 60$$
$$2x = 56$$
$$x = 28$$

4. Check. If Lisa rode 28 mi on Monday, then she rode $28 + 4 = 32$ mi on Tuesday. Since $28 + 32 = 60$, the numbers check.

5. State. Lisa rode 28 mi on Monday and 32 mi on Tuesday.

SECTION 1.5: FORMULAS, MODELS, AND GEOMETRY

We can solve a **formula** for a specified letter using the same principles used to solve equations.

Solve $a - c = bc + d$ *for c.*

$$a - c = bc + d$$
$$a - d = c + bc \qquad$$ Using the addition principle; adding c and $-d$ to both sides. All terms containing c are on the right side.

$$a - d = c(1 + b) \qquad$$ Factoring

$$\frac{a - d}{1 + b} = c \qquad$$ Multiplying both sides by $\frac{1}{1 + b}$. The formula is solved for c.

SECTION 1.6: PROPERTIES OF EXPONENTS

For $a, b \neq 0$ and n any integer:

$a^0 = 1;$

$a^{-n} = \dfrac{1}{a^n};$

$\dfrac{a^{-n}}{b^{-m}} = \dfrac{b^m}{a^n};$

$\left(\dfrac{a}{b}\right)^{-n} = \left(\dfrac{b}{a}\right)^n.$

$5^0 = 1$

$5^{-2} = \dfrac{1}{5^2} = \dfrac{1}{25}$

$\dfrac{x^{-4}}{5^{-7}} = \dfrac{5^7}{x^4}$

$\left(\dfrac{x^2}{6}\right)^{-3} = \left(\dfrac{6}{x^2}\right)^3$

The Product Rule

$$a^m \cdot a^n = a^{m+n}$$

$2^5 \cdot 2^{10} = 2^{15}$

The Quotient Rule

$$\frac{a^m}{a^n} = a^{m-n}$$

$\dfrac{3^8}{3^7} = 3^1 = 3$

The Power Rule

$$(a^m)^n = a^{mn}$$

$(4^{-2})^{-5} = 4^{10}$

Raising a product to a power

$$(ab)^n = a^n b^n$$

$(2y^3)^4 = 2^4(y^3)^4 = 16y^{12}$

Raising a quotient to a power

$$\left(\frac{a}{b}\right)^n = \frac{a^n}{b^n}$$

$\left(\dfrac{x^4}{5}\right)^2 = \dfrac{(x^4)^2}{5^2} = \dfrac{x^8}{25}$

SECTION 1.7: SCIENTIFIC NOTATION

Scientific Notation	$1.2 \times 10^5 = 120{,}000$,
$N \times 10^m$, where N is in decimal notation, $1 \le N < 10$, and m is an integer	$3.06 \times 10^{-4} = 0.000306$,
	$10{,}450{,}000{,}000{,}000 = 1.045 \times 10^{13}$,
	$0.0000031452 = 3.1452 \times 10^{-6}$

Review Exercises: Chapter 1

The following review exercises are for practice. Answers are at the back of the book. If you need to, restudy the section indicated alongside the answer.

🢌 *Concept Reinforcement* *In each of Exercises 1–10, match the expression or equation with an equivalent expression or equation from the column on the right.*

1. ____ $2x - 1 = 9$ [1.3]

2. ____ $2x - 1$ [1.3]

3. ____ $\frac{3}{4}x = 5$ [1.3]

4. ____ $\frac{3}{4}x - 5$ [1.3]

5. ____ $2(x + 7)$ [1.2]

6. ____ $2(x + 7) = 6$ [1.3]

7. ____ $4x - 3 + 2x = 5$ [1.3]

8. ____ $4x - 3 + 2x$ [1.3]

9. ____ $6 + 2x$ [1.2]

10. ____ $6 = 2x$ [1.3]

a) $2 + \frac{3}{4}x - 7$

b) $2x + 14 = 6$

c) $6x - 3$

d) $2(3 + x)$

e) $2x = 10$

f) $6x - 3 = 5$

g) $5x - 1 - 3x$

h) $3 = x$

i) $2x + 14$

j) $\frac{4}{3} \cdot \frac{3}{4}x = \frac{4}{3} \cdot 5$

11. Translate to an algebraic expression: Eight less than the quotient of two numbers. [1.1]

12. Evaluate
$$7x^2 - 5y \div zx$$
for $x = -2$, $y = 3$, and $z = -5$. [1.1], [1.2]

13. Name the set consisting of the first five odd natural numbers using both roster notation and set-builder notation. [1.1]

14. Find the area of a triangular flag that has a base of 50 cm and a height of 70 cm. [1.1]

Find the absolute value. [1.2]

15. $|-19|$ 16. $|0|$ 17. $|6.08|$

Perform the indicated operation. [1.2]

18. $-2.3 + (-8.7)$ 19. $\left(-\frac{2}{5}\right) + \frac{1}{3}$

20. $-\frac{3}{4} - \left(-\frac{4}{5}\right)$ 21. $-13 - 12$

22. $10 + (-5.6)$ 23. $12.3 - 16.1$

24. $(-12)(-8)$ 25. $\left(-\frac{2}{3}\right)\left(\frac{5}{8}\right)$

26. $(1.2)(-4)$ 27. $\frac{-24}{-6}$

28. $\frac{72.8}{-8}$ 29. $-7 \div \frac{4}{3}$

30. Find $-a$ if $a = -6.28$. [1.2]

Use a commutative law to write an equivalent expression. [1.2]

31. $12 + x$ 32. $7y$

33. $5x + y$

Use an associative law to write an equivalent expression. [1.2]

34. $(4 + a) + b$ 35. $x(yz)$

36. Obtain an expression that is equivalent to $12m + 4n - 8$ by factoring. [1.2]

37. Combine like terms: $3x^3 - 6x^2 + x^3 + 5$. [1.3]

38. Simplify: $7x - 4[2x + 3(5 - 4x)]$. [1.3]

Solve. If the solution set is $\varnothing$ or $\mathbb{R}$, classify the equation as a contradiction or as an identity. [1.3]

39. $3(t + 1) - t = 4$

40. $\frac{2}{3}n - \frac{5}{6} = \frac{8}{3}$

41. $-9x + 4(2x - 3) = 5(2x - 3) + 7$

42. $3(x - 4) + 2 = x + 2(x - 5)$

43. $5t - (7 - t) = 4t + 2(9 + t)$

44. Translate to an equation but do not solve: Fifteen more than twice a number is 21. [1.4]

45. A number is 19 less than another number. The sum of the numbers is 115. Find the smaller number. [1.4]

46. One angle of a triangle measures three times the second angle. The third angle measures twice the second angle. Find the measures of the angles. [1.4]

47. Solve for c: $x = \dfrac{bc}{t}$. [1.5]

48. Solve for x: $c = mx - rx$. [1.5]

49. The volume of a cylindrical candle is 538.51 cm³, and the radius of the candle is 3.5 cm. Determine the height of the candle. Use 3.14 for π. [1.5]

50. Multiply and simplify: $(-4mn^8)(7m^3n^2)$. [1.6]

51. Divide and simplify: $\dfrac{12x^3y^8}{3x^2y^2}$. [1.6]

52. Evaluate a^0, a^2, and $-a^2$ for $a = -8$. [1.6]

Simplify. Do not use negative exponents in the answer. [1.6]

53. $3^{-5} \cdot 3^7$

54. $(2t^4)^3$

55. $(-5a^{-3}b^2)^{-3}$

56. $\left(\dfrac{x^2y^3}{z^4}\right)^{-2}$

57. $\left(\dfrac{3m^{-5}n}{9m^2n^{-2}}\right)^4$

Simplify. [1.2]

58. $\dfrac{4(9 - 2 \cdot 3) - 3^2}{4^2 - 3^2}$

59. $1 - (2 - 5)^2 + 5 \div 10 \cdot 4^2$

60. Convert 0.000307 to scientific notation. [1.7]

61. One *parsec* (a unit that is used in astronomy) is 30,860,000,000,000 km. Write scientific notation for this number. [1.7]

Simplify and write scientific notation for each answer. Use the correct number of significant digits. [1.7]

62. $(8.7 \times 10^{-9}) \times (4.3 \times 10^{15})$

63. $\dfrac{1.2 \times 10^{-12}}{6.1 \times 10^{-7}}$

64. A sheet of plastic shrink wrap has a thickness of 0.00015 mm. The sheet is 1.2 m by 79 m. Use scientific notation to find the volume of the sheet. [1.7]

Synthesis

65. Describe a method that could be used to write equations that have no solution. [1.3]

66. Under what conditions is each of the following positive? Explain. [1.2], [1.6]
(a) $-(-x)$; (b) $-x^2$; (c) $-x^3$; (d) $(-x)^2$; (e) x^{-2}

67. If the smell of gasoline is detectable at 3 parts per billion, what percent of the air is occupied by the gasoline? [1.7]

68. Evaluate $a + b(c - a^2)^0 + (abc)^{-1}$ for $a = 3$, $b = -2$, and $c = -4$. [1.1], [1.6]

69. What's a better deal: a 13-in. diameter pizza for $8 or a 17-in. diameter pizza for $11? Explain. [1.4], [1.5]

70. The surface area of a cube is 486 cm². Find the volume of the cube. [1.5]

71. Solve for z: $m = \dfrac{x}{y - z}$. [1.5]

72. Simplify: $\dfrac{(3^{-2})^a \cdot (3^b)^{-2a}}{(3^{-2})^b \cdot (9^{-b})^{-3a}}$. [1.6]

73. Each of Garry's test scores counts three times as much as a quiz score. If after 4 quizzes Garry's average is 82.5, what score does he need on the first test in order to raise his average to 85? [1.4]

74. Fill in the following blank so as to ensure that the equation is an identity. [1.3]
$$5x - 7(x + 3) - 4 = 2(7 - x) + \underline{\quad\quad}$$

75. Replace the blank with one term to ensure that the equation is a contradiction. [1.3]
$$20 - 7[3(2x + 4) - 10] = 9 - 2(x - 5) + \underline{\quad\quad}$$

76. Use the commutative law for addition once and the distributive law twice to show that
$$a \cdot 2 + cb + cd + ad = a(d + 2) + c(b + d). \quad [1.2]$$

77. Find an irrational number between $\frac{1}{2}$ and $\frac{3}{4}$. [1.1]

Test: Chapter 1

1. Translate to an algebraic expression: Four less than the product of two numbers.

2. Evaluate $a^3 - 5b + b \div ac$ for $a = -2$, $b = 6$, and $c = 3$.

3. A triangular roof garden in Petach Tikva, Israel, has a base of length 7.8 m and a height of 46.5 m. Find its area.
   Source: www.greenroofs.com

Perform the indicated operation.

4. $-15 + (-16)$

5. $-7.5 + 3.8$

6. $\frac{1}{3} + \left(-\frac{1}{2}\right)$

7. $29.5 - 43.7$

8. $-16.8 - 26.4$

9. $-6.4(5.3)$

10. $-\frac{7}{6} - \left(-\frac{5}{4}\right)$

11. $-\frac{2}{7}\left(-\frac{5}{14}\right)$

12. $\frac{-42.6}{-7.1}$

13. $\frac{2}{5} \div \left(-\frac{3}{10}\right)$

14. Simplify: $7 + (1 - 3)^2 - 9 \div 2^2 \cdot 6$.

15. Use a commutative law to write an expression equivalent to $3 + x$.

Combine like terms.

16. $6y - 9y + 2y$

17. $6a^2b - 5ab^2 + ab^2 - 5a^2b + 2$

Solve. If the solution set is $\mathbb{R}$ or $\varnothing$, classify the equation as an identity or a contradiction.

18. $10x - 7 = 38x + 49$

19. $13t - (5 - 2t) = 5(3t - 1)$

20. Solve for p: $2p = sp + t$.

21. Linda's scores on five tests are 84, 80, 76, 96, and 80. What must Linda score on the sixth test so that her average will be 85?

22. Find three consecutive odd integers such that the sum of four times the first, three times the second, and two times the third is 167.

Simplify. Do not use negative exponents in the answer.

23. $3x - 7 - (4 - 5x)$

24. $6b - [7 - 2(9b - 1)]$

25. $(7x^{-4}y^{-7})(-6x^{-6}y)$

26. -6^{-2}

27. $(-5x^{-1}y^3)^3$

28. $\left(\dfrac{2x^3y^{-6}}{-4y^{-2}}\right)^{-2}$

29. $(7x^3y)^0$

Simplify and write scientific notation for the answer. Use the correct number of significant digits.

30. $(9.05 \times 10^{-3})(2.22 \times 10^{-5})$

31. $\dfrac{1.8 \times 10^{-4}}{4.8 \times 10^{-7}}$

Solve.

32. The lightest known particle in the universe, a neutrino has a maximum mass of 1.8×10^{-36} kg. What is the smallest number of neutrinos that could have the same mass as an alpha particle of mass 3.62×10^{-27} kg that results from the decay of radon?
   Source: *Guinness Book of World Records*

Synthesis

Simplify. Do not use negative exponents in the answer.

33. $(2x^{3a}y^{b+1})^{3c}$

34. $\dfrac{-27a^{x+1}}{3a^{x-2}}$

35. $\dfrac{(-16x^{x-1}y^{y-2})(2x^{x+1}y^{y+1})}{(-7x^{x+2}y^{y+2})(8x^{x-2}y^{y-1})}$

36. Solve: $-\dfrac{5x + 2}{x + 10} = 1$.

Graphs, Functions, and Linear Equations

MELANIE CHAMBERS
MATH TEACHER/RECIPIENT OF
THE MILKEN FAMILY
FOUNDATION'S NATIONAL
EDUCATOR AWARD
Cedar Hill, Texas

Being proficient in math not only
has multiplied my career options
but also opens many doors for
those students who choose to
sharpen their math skills. As a
math instructor, I use math to
calculate students' performance
on assessments, disaggregate
data to differentiate instruction
to better meet student learning
needs, and demonstrate daily
how real-world problem solving
often entails mathematical
analysis.

AN APPLICATION

According to the National Assessment
of Educational Progress (NAEP), the
percentage of fourth-graders who are
proficient in math has grown from
21% in 1996 to 24% in 2000 and 39%
in 2007.

Estimate the percentage of fourth-
graders who showed proficiency in
2004 and predict the percentage who
will demonstrate proficiency in 2011.

Source: nationsreportcard.gov

This problem appears as Example 8 in
Section 2.2.

G raphs help us visualize information and allow us to see relationships. In this chapter, we will examine graphs of equations in two variables. A certain kind of relationship between two variables is known as a *function*. In this chapter, we will explain what a function is as well as how it can be used in problem solving.

2.1   Graphs

Points and Ordered Pairs ▪ Quadrants ▪ Solutions of Equations ▪ Nonlinear Equations

It has often been said that a picture is worth a thousand words. In mathematics this is quite literally the case. Graphs are a compact means of displaying information and provide a visual approach to problem solving.

Points and Ordered Pairs

On a number line, each point corresponds to a number. On a plane, each point corresponds to an *ordered pair* of numbers. We use two perpendicular number lines, called **axes** (pronounced ak-sēz; singular, **axis**) to identify points in a plane. The point at which the axes intersect is called the **origin**. The variable x is most often used for the horizontal axis and the variable y for the vertical axis. Thus we refer to graphing on the *x, y-coordinate system*.

To label a point on the *x, y*-coordinate system, we use a pair of numbers in the form (x, y). The numbers in the pair are called **coordinates**. In the pair $(3, 2)$, the *first coordinate*, or *x-coordinate*, is 3 and the *second coordinate*, or *y-coordinate*,* is 2.

When we are graphing a point, the *x*-coordinate corresponds to that number on the *x*-axis, and the *y*-coordinate corresponds to that number on the *y*-axis. (See the dashed lines in the figure below.) Thus, $(2, 3)$ and $(3, 2)$ are different points. Note that the origin has coordinates $(0, 0)$.

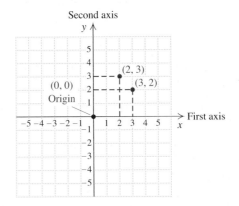

The idea of using axes to identify points in a plane is commonly attributed to the great French mathematician and philosopher René Descartes (1596–1650). In honor of Descartes, this representation is also called the **Cartesian coordinate system**.

---

*The first coordinate is sometimes called the **abscissa** and the second coordinate the **ordinate**.

EXAMPLE **1** Plot the points $(-4, 3)$, $(-5, -3)$, $(0, 4)$, $(4, -5)$, and $(2.5, 0)$.

SOLUTION   To plot $(-4, 3)$, note that the first coordinate, -4, tells us the distance in the first, or horizontal, direction. We go 4 units *left* of the origin. From that location, we go 3 units *up*. The point $(-4, 3)$ is then marked, or "plotted."

The points $(-5, -3)$, $(0, 4)$, $(4, -5)$, and $(2.5, 0)$ are also plotted below.

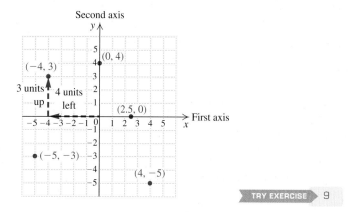

TRY EXERCISE 9

Quadrants

The axes divide the plane into four regions called **quadrants** that are labeled using Roman numerals.

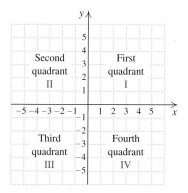

In region I (the *first* quadrant), both coordinates of a point are positive. In region II (the *second* quadrant), the first coordinate is negative and the second coordinate is positive. In the third quadrant, both coordinates are negative, and in the fourth quadrant, the first coordinate is positive while the second coordinate is negative.

Points with one or more 0's as coordinates, such as $(0, -6)$, $(4, 0)$, and $(0, 0)$, are on axes and *not* in quadrants.

Solutions of Equations

The solutions of an equation with two variables are pairs of numbers. When such a solution is written as an ordered pair, the first number listed in the pair generally corresponds to the variable that occurs first alphabetically.

EXAMPLE 2

Determine whether the pairs $(4, 2)$, $(-1, -4)$, and $(2, 5)$ are solutions of the equation $y = 3x - 1$.

SOLUTION   To determine whether each pair is a solution, we replace x with the first coordinate and y with the second coordinate. When the replacements make the equation true, we say that the ordered pair is a solution.

$$
\begin{array}{c|c}
y = 3x - 1 \\
\hline
2 & 3(4) - 1 \\
 & 12 - 1 \\
2 \overset{?}{=} 11
\end{array}
\qquad
\begin{array}{c|c}
y = 3x - 1 \\
\hline
-4 & 3(-1) - 1 \\
 & -3 - 1 \\
-4 \overset{?}{=} -4
\end{array}
\qquad
\begin{array}{c|c}
y = 3x - 1 \\
\hline
5 & 3(2) - 1 \\
 & 6 - 1 \\
5 \overset{?}{=} 5
\end{array}
$$

Since $2 = 11$ is *false*, the pair $(4, 2)$ *is not* a solution.

Since $-4 = -4$ is *true*, the pair $(-1, -4)$ *is* a solution.

Since $5 = 5$ is *true*, the pair $(2, 5)$ *is* a solution.

TRY EXERCISE 21

In fact, there is an infinite number of solutions of $y = 3x - 1$. We can use a graph as a convenient way of representing these solutions. Thus to **graph** an equation means to make a drawing that represents all of its solutions.

EXAMPLE 3

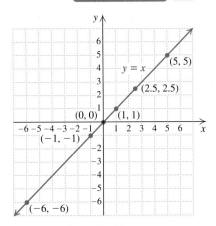

Graph the equation $y = x$.

SOLUTION   We label the horizontal axis as the x-axis and the vertical axis as the y-axis.

Next, we find some ordered pairs that are solutions of the equation. In this case, since $y = x$, no calculations are necessary. Here are a few pairs that satisfy the equation $y = x$:

$$(0, 0), \quad (1, 1), \quad (5, 5), \quad (-1, -1), \quad (-6, -6).$$

Plotting these points, we can see that if we were to plot a hundred solutions, the dots would appear to form a line. Observing the pattern, we can draw the line with a ruler. The line is the graph of the equation $y = x$. We label the line $y = x$.

Note that the coordinates of *any* point on the line—for example, $(2.5, 2.5)$—satisfy the equation $y = x$. The line continues indefinitely in both directions, as indicated by the arrowheads on the line.

TRY EXERCISE 35

EXAMPLE 4

Graph the equation $y = 2x - 1$.

SOLUTION   We find some ordered pairs that are solutions. This time we list the pairs in a table. To find an ordered pair, we can choose *any* number for x and then determine y. For example, if we choose 3 for x, then

$$y = 2x - 1$$
$$y = 2(3) - 1 = 5.$$

We choose some negative values for x, as well as some positive ones (generally, we avoid selecting values that are beyond the edge of the graph paper). Next, we plot these points. If we plotted *many* such points, they would appear to make a solid line. We draw the line with a ruler and label it $y = 2x - 1$.

STUDENT NOTES

There is an infinite number of solutions of $y = 2x - 1$. When you choose a value for x and then compute y, you are determining one solution. Your choices for x may be different from those of a classmate. Although your plotted points will then be different, the graph of the line will be the same.

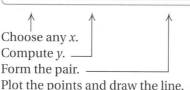

x	y $y = 2x - 1$	(x, y)
0	−1	(0, −1)
1	1	(1, 1)
3	5	(3, 5)
−1	−3	(−1, −3)
−2	−5	(−2, −5)

Choose any x.
Compute y.
Form the pair.
Plot the points and draw the line.

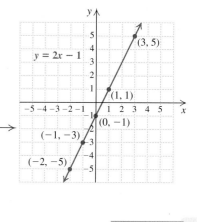

TRY EXERCISE ▶ 37

TECHNOLOGY CONNECTION

The window of a graphing calculator is the rectangular portion of the screen in which a graph appears. Windows are described by four numbers of the form [L, R, B, T], representing the left and right endpoints of the x-axis and the bottom and top endpoints of the y-axis. Below, we have graphed $y = -4x + 3$, using the "standard" $[-10, 10, -10, 10]$ window.

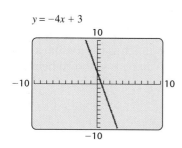

When TRACE is pressed, a cursor can be moved along the graph while its coordinates appear. To find the y-value that is paired with a particular x-value, we simply key in that x-value and press ENTER.

1. Graph $y = -4x + 3$ using a $[-10, 10, -10, 10]$ window. Then TRACE to find coordinates of several points, including the points with the x-values -1.5 and 1.

EXAMPLE 5   Graph the equation $y = -\frac{1}{2}x$.

SOLUTION   Again, we can choose any number for x. In this case, by choosing even integers for x, we can avoid fraction values for y. For example, if we choose 4 for x, we get $y = \left(-\frac{1}{2}\right)(4)$, or -2. When x is -6, we get $y = \left(-\frac{1}{2}\right)(-6)$, or 3. We find several ordered pairs, plot them, and draw the line.

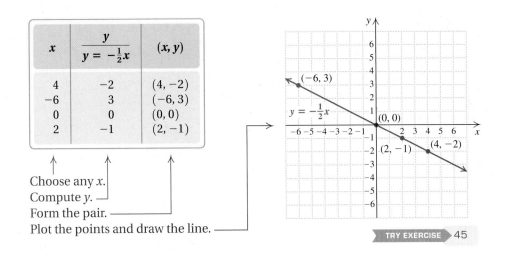

x	$\dfrac{y}{y = -\frac{1}{2}x}$	(x, y)
4	−2	(4, −2)
−6	3	(−6, 3)
0	0	(0, 0)
2	−1	(2, −1)

Choose any x.
Compute y.
Form the pair.
Plot the points and draw the line.

TRY EXERCISE 45

As you can see, the graphs in Examples 3–5 are straight lines. We will refer to any equation whose graph is a straight line as a **linear equation**. To graph a line, be sure to plot at least two points, using a third point as a check. We will develop methods for recognizing and graphing linear equations in Sections 2.3–2.5.

Nonlinear Equations

There are many equations for which the graph is not a straight line. Graphing these **nonlinear equations** often requires plotting many points in order to see the general shape of the graph.

EXAMPLE **6**   Graph: $y = |x|$.

SOLUTION   We select numbers for x and find the corresponding values for y. For example, if we choose −1 for x, we get $y = |-1| = 1$. We list several ordered pairs and plot the points, noting that the absolute value of a positive number is the same as the absolute value of its opposite. Thus the x-values 3 and −3 both are paired with the y-value 3. The graph is V-shaped, as shown below.

x	$\dfrac{y}{y = |x|}$	(x, y)
−3	3	(−3, 3)
−2	2	(−2, 2)
−1	1	(−1, 1)
0	0	(0, 0)
1	1	(1, 1)
2	2	(2, 2)
3	3	(3, 3)

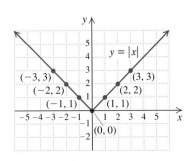

TRY EXERCISE 49

Curves similar to the one in Example 7 below are studied in Chapter 8.

EXAMPLE 7   Graph: $y = x^2 - 5$.

SOLUTION   We select numbers for x and find the corresponding values for y. For example, if we choose -2 for x, we get

$$y = x^2 - 5$$
$$y = (-2)^2 - 5 = 4 - 5 = -1.$$

The table lists several ordered pairs.

x	y $y = x^2 - 5$	(x, y)
0	−5	(0, −5)
−1	−4	(−1, −4)
1	−4	(1, −4)
−2	−1	(−2, −1)
2	−1	(2, −1)
−3	4	(−3, 4)
3	4	(3, 4)

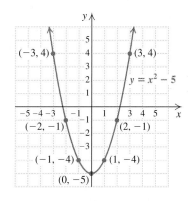

Next, we plot the points. The more points plotted, the clearer the shape of the graph becomes. Since the value of $x^2 - 5$ grows rapidly as x moves away from the origin, the graph rises steeply on either side of the y-axis.

TRY EXERCISE 53

TECHNOLOGY CONNECTION

Most graphing calculators can set up a table of pairs for any equation that is entered. By pressing **2ND** **TBLSET**, we can control the smallest x-value listed using TblStart and the difference between successive x-values using ΔTbl. Setting Indpnt and Depend both to Auto directs the calculator to complete a table automatically. To view the table, we press **2ND** **TABLE**. For the table shown, we used $y_1 = -4x + 3$, with TblStart $= 1.4$ and ΔTbl $= .1$.

TblStart = 1.4 ΔTbl = .1

X	Y1
1.4	−2.6
1.5	−3
1.6	−3.4
1.7	−3.8
1.8	−4.2
1.9	−4.6
2	−5
X = 1.4	

Graph each of the following equations using a $[-10, 10, -10, 10]$ window. Then create a table of ordered pairs in which the x-values start at -1 and are 0.1 unit apart.

1. $y = 5x - 3$
2. $y = x^2 - 4x + 3$
3. $y = (x + 4)^2$
4. $y = \sqrt{x + 2}$
5. $y = |x + 2|$

To the student and the instructor: The [TRY EXERCISES] *for examples are indicated by a shaded block* ▇ *on the exercise number. Complete step-by-step solutions for these exercises appear online at www.pearsonhighered.com/ bittingerellenbogen.*

☛ **Concept Reinforcement**   *Complete each of the following statements.*

1. The two perpendicular number lines that are used for graphing are called _____ .

2. Because the order in which the numbers are listed is important, numbers listed in the form (x, y) are called _____ pairs.

3. In the _____ quadrant, both coordinates of a point are negative.

4. In the fourth quadrant, a point's first coordinate is positive and its second coordinate is _____ .

5. To graph an equation means to make a drawing that represents all _____ of the equation.

6. An equation whose graph is a straight line is said to be a(n) _____ equation.

Give the coordinates of each point.

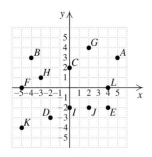

7. A, B, C, D, E, and F

8. G, H, I, J, K, and L

Plot the points. Label each point with the indicated letter.

9. $A(3, 0), B(4, 2), C(5, 4), D(6, 6), E(3, -4), F(3, -3), G(3, -2), H(3, -1)$

10. $A(1, 1), B(2, 3), C(3, 5), D(4, 7), E(-2, 1), F(-2, 2), G(-2, 3), H(-2, 4), J(-2, 5), K(-2, 6)$

11. Plot the points $M(2, 3), N(5, -3)$, and $P(-2, -3)$. Draw $\overline{MN}, \overline{NP}$, and $\overline{MP}$. ($\overline{MN}$ means the line segment from M to N.) What kind of geometric figure is formed? What is its area?

12. Plot the points $Q(-4, 3), R(5, 3), S(2, -1)$, and $T(-7, -1)$. Draw $\overline{QR}, \overline{RS}, \overline{ST}$, and $\overline{TQ}$. What kind of figure is formed? What is its area?

Name the quadrant in which each point is located.

13. $(3, -5)$

14. $(-6, -5)$

15. $(-3, -12)$

16. $(-18, 0.16)$

17. $\left(11, \frac{1}{4}\right)$

18. $(31, 48)$

19. $(-1.2, 46)$

20. $\left(\frac{3}{4}, -\frac{1}{5}\right)$

Determine whether each ordered pair is a solution of the given equation. Remember to use alphabetical order for substitution.

21. $(2, -1); \ y = 3x - 7$

22. $(1, 4); \ y = 5x - 1$

23. $(3, 2); \ 2x - y = 5$

24. $(5, 5); \ 3x - y = 5$

25. $(3, -1); \ a - 5b = 8$

26. $(1, -4); \ 2u - v = -6$

27. $\left(\frac{2}{3}, 0\right); \ 6x + 8y = 4$

28. $\left(0, \frac{3}{5}\right); \ 7a + 10b = 6$

29. $(6, -2); \ r - s = 4$

30. $(4, -3); \ 2x - y = 11$

31. $(2, 1); \ y = 2x^2$

32. $(-2, -1); \ r^2 - s = 5$

33. $(-2, 9); \ x^3 + y = 1$

34. $(3, 2); \ y = x^3 - 5$

Graph.

35. $y = 3x$

36. $y = -x$

37. $y = x + 4$

38. $y = x + 3$

39. $y = x - 4$

40. $y = x - 3$

41. $y = -2x + 3$

42. $y = -3x + 1$

Aha! 43. $y + 2x = 3$

44. $y + 3x = 1$

45. $y = -\frac{3}{2}x$

46. $y = \frac{2}{3}x$

47. $y = \frac{3}{4}x - 1$

48. $y = -\frac{3}{4}x - 1$

49. $y = |x| + 2$

50. $y = |x| + 1$

51. $y = |x| - 2$

52. $y = |x| - 3$

53. $y = x^2 + 2$

54. $y = x^2 + 1$

55. $y = x^2 - 2$ **56.** $y = x^2 - 3$

57. Examine Example 6 and explain why it is unwise to draw a graph after plotting only two points.

58. Points A and B have the same first coordinates and second coordinates that are opposites of each other. How is the location of A related to the location of B?

Skill Review

To prepare for Section 2.2, review evaluating algebraic expressions and solving equations (Sections 1.1, 1.2, and 1.3).

Evaluate.

59. $5t - 7$, for $t = 10$ [1.1]

60. $2r^2 - 7r$, for $r = -1$ [1.2]

Aha! 61. $(3 - x)^2(1 - 2x)^3$, for $x = \frac{1}{2}$ [1.1]

62. $-x$, for $x = -5$ [1.2]

63. $\dfrac{2x + 3}{x - 4}$, for $x = 0$ [1.2]

64. $\dfrac{4 - x}{3x + 1}$, for $x = 4$ [1.1]

Solve. [1.3]

65. $x + 4 = 0$ **66.** $5 - x = 0$

67. $1 - 2x = 0$ **68.** $5x + 3 = 0$

Synthesis

69. Using the same set of axes, graph $y = 2x$, $y = 2x - 3$, and $y = 2x + 3$. Describe the pattern relating each line to the number that is added to $2x$.

70. Graph $y = 6x$, $y = 3x$, $y = \frac{1}{2}x$, $y = -6x$, $y = -3x$, and $y = -\frac{1}{2}x$ using the same set of axes and compare the slants of the lines. Describe the pattern that relates the slant of the line to the multiplier of x.

71. Without making a drawing, how can you tell that the graph of $y = x - 30$ passes through three quadrants?

72. At what point will the line passing through $(a, -1)$ and $(a, 5)$ intersect the line that passes through $(-3, b)$ and $(2, b)$? Why?

73. Match each sentence with the most appropriate of the four graphs shown below.

  a) Austin worked part time until September, full time until December, and overtime until Christmas.

  b) Marlo worked full time until September, half time until December, and full time until Christmas.

  c) Maddie worked overtime until September, full time until December, and overtime until Christmas.

  d) Roberto worked part time until September, half time until December, and full time until Christmas.

I
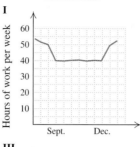

II
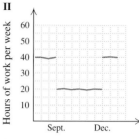

III
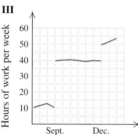

IV
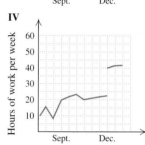

74. Match each sentence with the most appropriate of the four graphs shown below.

  a) Carpooling to work, Jeremy spent 10 min on local streets, then 20 min cruising on the freeway, and then 5 min on local streets to his office.

  b) For her commute to work, Chloe drove 10 min to the train station, rode the express for 20 min, and then walked for 5 min to her office.

  c) For his commute to school, Theo walked 10 min to the bus stop, rode the express for 20 min, and then walked for 5 min to his class.

  d) Coming home from school, Taylor waited 10 min for the school bus, rode the bus for 20 min, and then walked 5 min to her house.

I
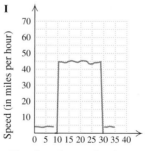
Time from the start (in minutes)

II
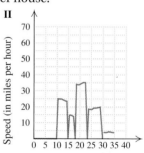
Time from the start (in minutes)

III
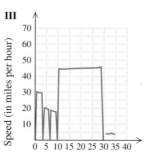
Time from the start (in minutes)

IV
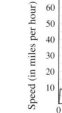
Time from the start (in minutes)

75. Match each program found on an exercise bike with the appropriate graph of speed.

   a) Lakeshore loop
   b) Rocky Mountain monster hill
   c) Interval training
   d) Random mystery ride

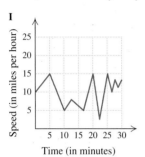

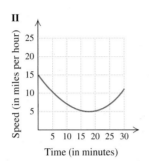

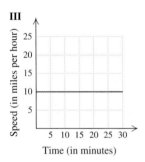

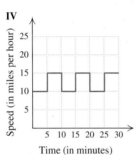

76. Which of the following equations have $\left(-\frac{1}{3}, \frac{1}{4}\right)$ as a solution?

   a) $-\frac{3}{2}x - 3y = -\frac{1}{4}$
   b) $8y - 15x = \frac{7}{2}$
   c) $0.16y = -0.09x + 0.1$
   d) $2(-y + 2) - \frac{1}{4}(3x - 1) = 4$

77. If $(-10, -2)$, $(-3, 4)$, and $(6, 4)$ are the coordinates of three vertices of a parallelogram, determine the coordinates of three different points that could serve as the fourth vertex.

78. If $(2, -3)$ and $(-5, 4)$ are the endpoints of a diagonal of a square, what are the coordinates of the other two vertices? What is the area of the square?

Graph each equation after plotting at least 10 points.

79. $y = 1/x^2$; use values of x from -4 to 4

80. $y = \frac{1}{2}x^2$; use x-values from -4 to 4

81. $y = 1/(x - 2)$; use values of x from -2 to 6

82. $y = 1/x$; use x-values from -4 to 4

83. $y = \sqrt{x} + 1$; use x-values from 0 to 10

84. $y = \sqrt{x}$; use values of x from 0 to 10

85. $y = x^3$; use x-values from -2 to 2

86. $y = x^3 - 5$; use x-values from -2 to 2

87. $y = \dfrac{1}{x} + 3$; use x-values from -4 to 4

Note: Throughout this text, the icon ⌁ *indicates exercises designed for graphing calculators.*

⌁ *In Exercises 88 and 89, use a graphing calculator to draw the graph of each equation. For each equation, select a window that shows the curvature of the graph and create a table of ordered pairs in which x-values extend, by tenths, from 0 to 0.6.*

88. **a)** $y = 2.3x^4 + 3.4x^2 + 1.2x - 4$
     b) $y = -0.25x^2 + 3.7$
     c) $y = 3(x + 2.3)^2 + 2.3$

89. **a)** $y = 0.375x^3$
     b) $y = -3.5x^2 + 6x - 8$
     c) $y = (x - 3.4)^3 + 5.6$

2.2   Functions

Domain and Range ■ Functions and Graphs ■ Function Notation and Equations ■ Applications: Interpolation and Extrapolation

We now develop the idea of a *function*—one of the most important concepts in mathematics.

Domain and Range

A function is a special kind of correspondence between two sets. For example,

To each person in a class	there corresponds	a date of birth.
To each bar code in a store	there corresponds	a price.
To each real number	there corresponds	the cube of that number.

In each example, the first set is called the **domain**. The second set is called the **range**. For any member of the domain, there is *exactly one* member of the range to which it corresponds. This kind of correspondence is called a **function**.

EXAMPLE 1

Determine whether each correspondence is a function.

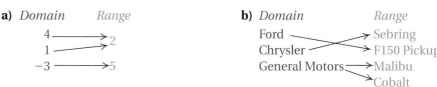

STUDENT NOTES

Note that not all correspondences are functions.

SOLUTION

a) The correspondence *is* a function because each member of the domain corresponds to *exactly one* member of the range.

b) The correspondence *is not* a function because a member of the domain (General Motors) corresponds to more than one member of the range.

TRY EXERCISE 9

> **Function**
>
> A *function* is a correspondence between a first set, called the *domain*, and a second set, called the *range*, such that each member of the domain corresponds to *exactly one* member of the range.

EXAMPLE 2

Determine whether each correspondence is a function.

Domain	Correspondence	Range
a) People in a doctor's waiting room	Each person's weight	A set of positive numbers
b) $\{-2, 0, 1, 2\}$	Each number's square	$\{0, 1, 4\}$
c) Authors of best-selling books	The titles of books written by each author	A set of book titles

SOLUTION

a) The correspondence *is* a function, because each person has *only one* weight.

b) The correspondence *is* a function, because every number has *only one* square.

c) The correspondence *is not* a function, because some authors have written *more than one* book.

TRY EXERCISE 17

A set of ordered pairs is also a correspondence between two sets. The domain is the set of all first coordinates and the range is the set of all second coordinates.

EXAMPLE 3

For the correspondence $\{(-6, 7), (1, 4), (1, -3), (4, -5)\}$, **(a)** write the domain; **(b)** write the range; and **(c)** determine whether the correspondence is a function.

SOLUTION

a) The domain is the set of all first coordinates: $\{-6, 1, 4\}$.

b) The range is the set of all second coordinates: $\{7, 4, -3, -5\}$.

c) We write the correspondence using arrows to determine whether it is a function.

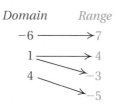

The correspondence *is not* a function because a member of the domain, 1, corresponds to more than one member of the range. We can also see this from the ordered pairs by noting that two of the ordered pairs have the same first coordinate but different second coordinates.

TRY EXERCISE 21

Functions and Graphs

The functions in Examples 1(a) and 2(b) can be expressed as sets of ordered pairs. Example 1(a) can be written $\{(-3, 5), (1, 2), (4, 2)\}$ and Example 2(b) can be written $\{(-2, 4), (0, 0), (1, 1), (2, 4)\}$. We can graph these functions as follows.

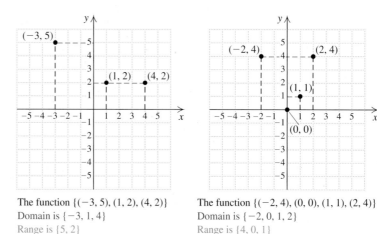

The function $\{(-3, 5), (1, 2), (4, 2)\}$
Domain is $\{-3, 1, 4\}$
Range is $\{5, 2\}$

The function $\{(-2, 4), (0, 0), (1, 1), (2, 4)\}$
Domain is $\{-2, 0, 1, 2\}$
Range is $\{4, 0, 1\}$

We can find the domain and the range of a function directly from its graph. The domain is read from the horizontal axis and the range is read from the vertical axis. Note in the graphs above that if we move along the red dashed lines from the points to the horizontal axis, we find the members, or elements, of the domain. Similarly, if we move along the blue dashed lines from the points to the vertical axis, we find the elements of the range.

Functions are generally named using lowercase or uppercase letters. The function in the following example is named *f*.

EXAMPLE **4** For the function *f* represented below, determine each of the following.

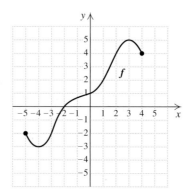

a) The member of the range that is paired with 2

b) The domain of f

c) The member of the domain that is paired with -3

d) The range of f

SOLUTION

a) To determine what member of the range is paired with 2, first note that we are considering 2 in the domain. Thus we locate 2 on the horizontal axis. (See the graph on the left below.) Next, we find the point directly above 2 on the graph of f. From that point, we can look to the vertical axis to find the corresponding y-coordinate, 4. Thus, 4 is the member of the range that is paired with 2.

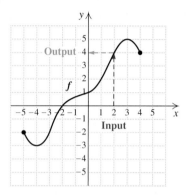 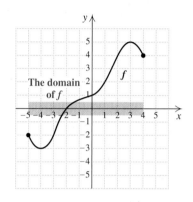

b) The domain of f is the set of all x-values that are used in the points on the curve. (See the graph on the right above.) Because there are no breaks in the graph of f, these extend continuously from -5 to 4 and can be viewed as the curve's shadow, or *projection*, on the x-axis. This is illustrated by the shading on the x-axis. Thus the domain is $\{x \mid -5 \leq x \leq 4\}$.

c) To determine what member of the domain is paired with -3, we note that we are considering -3 in the range. Thus we locate -3 on the vertical axis. (See the graph on the left below.) From there we look at the graph of f to find any points for which -3 is the second coordinate. One such point exists, $(-4, -3)$. We observe that -4 is the only element of the domain paired with -3.

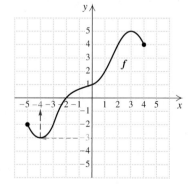 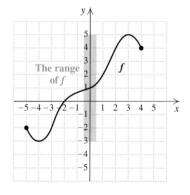

d) The range of f is the set of all y-values that are used in the points on the curve. (See the graph on the right above.) These extend continuously from -3 to 5, and can be viewed as the curve's projection on the y-axis. This is illustrated by the shading on the y-axis. Thus the range is $\{y \mid -3 \leq y \leq 5\}$.

TRY EXERCISE ▶ 27

A closed dot on a graph, such as in Example 4, indicates that the point is part of the function. An open dot indicates that the point is *not* part of the function. (See Exercises 37 and 38 on p. 92.)

The dots in Example 4 also indicate endpoints of the graph. A function may have a domain and/or a range that extends without bound toward positive infinity or negative infinity.

EXAMPLE 5   For the function *g* represented below, determine **(a)** the domain of *g* and **(b)** the range of *g*.

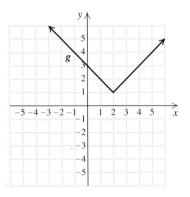

SOLUTION

a) The domain of *g* is the set of all *x*-values that are used in the points on the curve. The arrows on the ends of the graph indicate that it extends both left and right without end. Thus the shadow, or projection, of the graph on the *x*-axis is the entire *x*-axis. (See the graph on the left below.) The domain is $\{x \mid x \text{ is a real number}\}$, or $\mathbb{R}$.

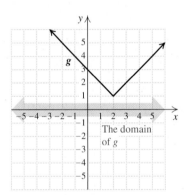
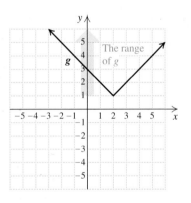

b) The range of *g* is the set of all *y*-values that are used in the points on the curve. The arrows on the ends of the graph indicate that it extends up without end. Thus the projection of the graph on the *y*-axis is the portion of the *y*-axis greater than or equal to 1. (See the graph on the right above.) The range is $\{y \mid y \geq 1\}$.

▸ **TRY EXERCISE** 39

Note that if a graph contains two or more points with the same first coordinate, that graph cannot represent a function (otherwise one member of the domain would correspond to more than one member of the range). This observation is the basis of the *vertical-line test*.

> ### The Vertical-Line Test
>
> If it is possible for a vertical line to cross a graph more than once, then the graph is not the graph of a function.

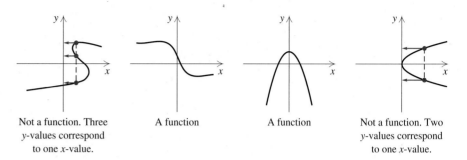

Not a function. Three
y-values correspond
to one *x*-value.

A function

A function

Not a function. Two
y-values correspond
to one *x*-value.

Graphs that do not represent functions still do represent *relations*.

> ### Relation
>
> A *relation* is a correspondence between a first set, called the *domain*, and a second set, called the *range*, such that each member of the domain corresponds to *at least one* member of the range.

Thus, although the correspondences and graphs above are not all functions, they *are* all relations.

Function Notation and Equations

To understand function notation, it helps to imagine a "function machine." Think of putting a member of the domain (an *input*) into the machine. The machine is programmed to produce the appropriate member of the range (the *output*).

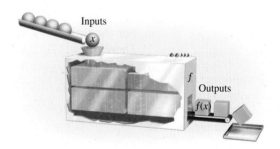

Inputs

Outputs

The function pictured has been named f. Here x represents an input, and $f(x)$ represents the corresponding output. In function notation, "$f(x)$" is read "f of x," or "f at x," or "the value of f at x." In Example 4, we showed that $f(2) = 4$, read "f of 2 equals 4."

CAUTION! $f(x)$ *does not* mean f times x.

Most functions are described by equations. For example, $f(x) = 2x + 3$ describes the function that takes an input x, multiplies it by 2, and then adds 3.

$$\underset{\text{Double}}{\overset{\text{Input}}{f(x) \quad = \quad 2x}} \quad \underset{\text{Add 3}}{+ 3}$$

To calculate the output $f(4)$, we take the input 4, double it, and add 3 to get 11. That is, we substitute 4 into the formula for $f(x)$:

$$f(4) = 2 \cdot 4 + 3$$
$$= 11.$$

Sometimes, in place of $f(x) = 2x + 3$, we write $y = 2x + 3$, where it is understood that the value of y, the *dependent variable*, depends on our choice of x, the *independent variable*. To understand why $f(x)$ notation is so useful, consider two equivalent statements:

a) If $f(x) = 2x + 3$, then $f(4) = 11$.

b) If $y = 2x + 3$, then the value of y is 11 when x is 4.

The notation used in part (a) is far more concise and emphasizes that x is the independent variable.

EXAMPLE **6**   Find each indicated function value.

a) $f(5)$, for $f(x) = 3x + 2$ **b)** $g(-2)$, for $g(r) = 5r^2 + 3r$

c) $h(4)$, for $h(x) = 11$ **d)** $F(a) + 1$, for $F(x) = 2x - 7$

e) $F(a + 1)$, for $F(x) = 2x - 7$

SOLUTION   Finding function values is much like evaluating an algebraic expression.

a) $f(5) = 3(5) + 2 = 17$

b) $g(-2) = 5(-2)^2 + 3(-2)$
$$= 5 \cdot 4 - 6 = 14$$

c) For the function given by $h(x) = 11$, all inputs share the same output, 11. Therefore, $h(4) = 11$. The function h is an example of a *constant function*.

d) $F(a) + 1 = 2(a) - 7 + 1$ The input is a; $F(a) = 2a - 7$
$$= 2a - 6$$

e) $F(a + 1) = 2(a + 1) - 7$ The input is $a + 1$.
$$= 2a + 2 - 7 = 2a - 5$$

TRY EXERCISE 55

STUDENT NOTES ─────

In Example 6(e), it is important to note that the parentheses on the left are for function notation, whereas those on the right indicate multiplication.

Note that whether we write $f(x) = 3x + 2$, or $f(t) = 3t + 2$, or $f(\ ) = 3\ + 2$, we still have $f(5) = 17$. The variable in the parentheses (the independent variable) is the variable used in the algebraic expression. The letter chosen for the independent variable is not as important as the algebraic manipulations to which it is subjected.

When a function is described by an equation, the domain is often unspecified. In such cases, the domain is the set of all numbers for which function values can be calculated. If an x-value is not in the domain of a function, the graph of that function will not include any point above or below that x-value.

EXAMPLE 7

For each equation, determine the domain of f.

a) $f(x) = |x|$ **b)** $f(x) = \dfrac{x}{2x - 6}$

SOLUTION

a) We ask ourselves, "Is there any number x for which we cannot compute $|x|$?" Since we can find the absolute value of *any* number, the answer is no. Thus the domain of f is $\mathbb{R}$, the set of all real numbers.

b) Is there any number x for which $\dfrac{x}{2x - 6}$ cannot be computed? Since $\dfrac{x}{2x - 6}$ cannot be computed when $2x - 6$ is 0, the answer is yes. To determine what x-value would cause $2x - 6$ to be 0, we set up and solve an equation:

$$2x - 6 = 0 \qquad \text{Setting the denominator equal to 0}$$
$$2x = 6 \qquad \text{Adding 6 to both sides}$$
$$x = 3. \qquad \text{Dividing both sides by 2}$$

> *CAUTION!* The denominator cannot be 0, but the numerator can be any number.

Thus, 3 is *not* in the domain of f, whereas all other real numbers are. The domain of f is $\{x \mid x$ is a real number *and* $x \neq 3\}$.

TRY EXERCISE 61

If the domain of a function is not specifically listed, it can be determined from a table, a graph, an equation, or an application.

Domain of a Function

The domain of a function $f(x)$ is the set of all inputs x.

- If the correspondence is listed in a table or as a set of ordered pairs, the domain is the set of all first coordinates.
- If the function is described by a graph, the domain is the set of all x-coordinates of the points on the graph.
- If the function is described by an equation, the domain is the set of all numbers for which the value can be calculated.
- If the function is used in an application, the domain is the set of all numbers that make sense in the problem.

Applications: Interpolation and Extrapolation

Function notation is often used in formulas. For example, to emphasize that the area A of a circle is a function of its radius r, instead of

$$A = \pi r^2,$$

we can write

$$A(r) = \pi r^2.$$

When a function is given as a graph, we can use the graph to estimate an unknown function value using known values. When we estimate the coordinates of an

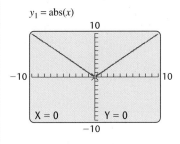
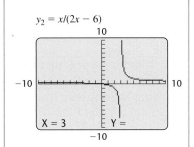

unknown point that lies *between* known points, the process is called **interpolation**. If the unknown point extends *beyond* the known points, the process is called **extrapolation**.

EXAMPLE   8

Elementary school math proficiency.   According to the National Assessment of Educational Progress (NAEP), the percentage of fourth-graders who are proficient in math has grown from 21% in 1996 to 24% in 2000 and 39% in 2007. Estimate the percentage of fourth-graders who showed proficiency in 2004 and predict the percentage who will demonstrate proficiency in 2011.

Source: nationsreportcard.gov

SOLUTION

1., 2. Familiarize., Translate.   The given information enables us to plot and connect three points. We let the horizontal axis represent the year and the vertical axis the percentage of fourth-graders demonstrating mathematical proficiency. We label the function itself P.

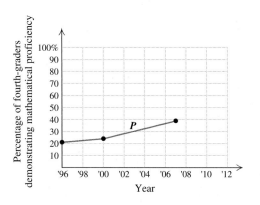

3. Carry out.   To estimate the percentage of fourth-graders showing mathematical proficiency in 2004, we locate the point directly above the year 2004. We then estimate its second coordinate by moving horizontally from that point to the y-axis. Although our result is not exact, we see that $P(2004) \approx 33$.

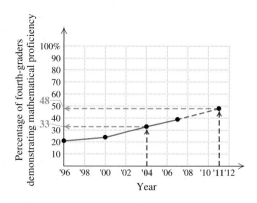

To predict the percentage of fourth-graders showing proficiency in 2011, we extend the graph and extrapolate. It appears that $P(2011) \approx 48$.

4. Check.   A precise check requires consulting an outside information source. Since 33% is between 24% and 39% and 48% is greater than 39%, our estimates seem plausible.

5. State.   In 2004, about 33% of all fourth-graders showed proficiency in math. By 2011, that figure is predicted to grow to 48%.

TRY EXERCISE 87

2.2 EXERCISE SET

Concept Reinforcement *Complete each of the following sentences.*

1. A function is a special kind of _____ between two sets.

2. In any function, each member of the domain is paired with _____ one member of the range.

3. For any function, the set of all inputs, or first values, is called the _____.

4. For any function, the set of all outputs, or second values, is called the _____.

5. When a function is graphed, members of the domain are located on the _____ axis.

6. When a function is graphed, members of the range are located on the _____ axis.

7. The notation $f(3)$ is read _____.

8. The _____-line test can be used to determine whether or not a graph represents a function.

Determine whether each correspondence is a function.

9.

a ⟶ 2
b ⟶
d ⟶ 3
g ⟶ 4
h ⟶

10.
2 ⟶ a
⟶ b
3 ⟶ d
4 ⟶ g
⟶ h

11.
Girl's age (in months)	Average daily weight gain (in grams)
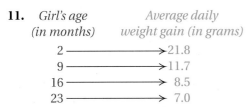
2	21.8
9	11.7
16	8.5
23	7.0

Source: *American Family Physician*, December 1993, p. 1435

12.
Boy's age (in months)	Average daily weight gain (in grams)
2	24.3
9	11.7
16	8.2
23	7.0

Source: *American Family Physician*, December 1993, p. 1435

13.

Birthday	Celebrity

June 9 ⟶ Johnny Depp
⟶ Michael J. Fox
⟶ Amanda Lassiter
October 5 ⟶ Michael Andretti
⟶ Chester A. Arthur
⟶ Kate Winslet

Source: www.leannesbirthdays.com

14.
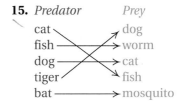
Celebrity	Birthday

Tom Brady ⟶
P. D. James ⟶ August 3
Martha Stewart ⟶
Kim Basinger ⟶
Sinéad O'Connor ⟶ December 8
James Galway ⟶

Source: www.leannesbirthdays.com

15.
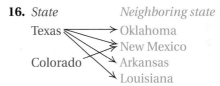
Predator	Prey

cat ⟶ dog
fish ⟶ worm
dog ⟶ cat
tiger ⟶ fish
bat ⟶ mosquito

16.
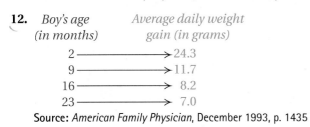
State	Neighboring state

Texas ⟶ Oklahoma
⟶ New Mexico
Colorado ⟶ Arkansas
⟶ Louisiana

Determine whether each of the following is a function. Identify any relations that are not functions.

	Domain	Correspondence	Range
17.	A pile of USB flash drives	The storage capacity of each flash drive	A set of storage capacities
18.	The members of a rock band	An instrument the person can play	A set of instruments
19.	The players on a team	The uniform number of each player	A set of numbers
20.	A set of triangles	The area of each triangle	A set of numbers

For each correspondence, **(a)** *write the domain,* **(b)** *write the range, and* **(c)** *determine whether the correspondence is a function.*

21. $\{(-3, 3), (-2, 5), (0, 9), (4, -10)\}$

22. $\{(0, -1), (1, 3), (2, -1), (5, 3)\}$

23. $\{(1, 1), (2, 1), (3, 1), (4, 1), (5, 1)\}$

24. $\{(1, 1), (1, 2), (1, 3), (1, 4), (1, 5)\}$

25. $\{(4, -2), (-2, 4), (3, -8), (4, 5)\}$

26. $\{(0, 7), (4, 8), (7, 0), (8, 4)\}$

For each graph of a function, determine **(a)** $f(1)$; **(b)** *the domain;* **(c)** *any x-values for which* $f(x) = 2$; *and* **(d)** *the range.*

27.
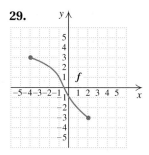

28.
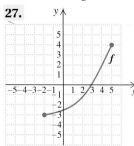

29.
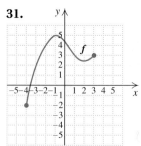

30.
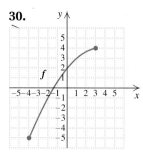

31.
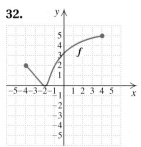

32.
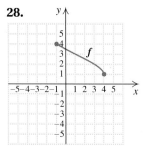

33.
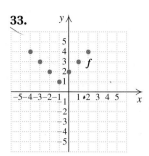

34.
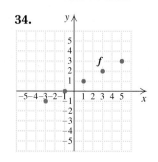

35.
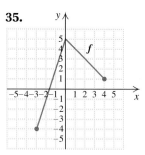

36.
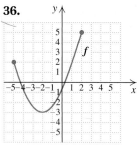

37.
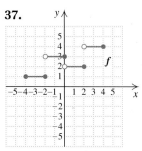

38.
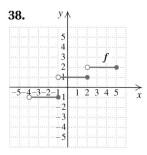

Determine the domain and the range of each function.

39.
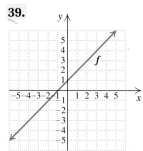

40.
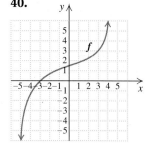

41.
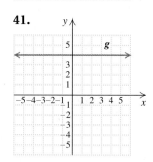

42.
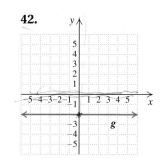

43.
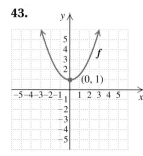

44.
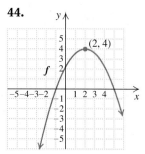

45.

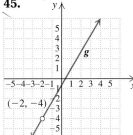

46.

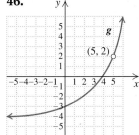

47.

48.

Determine whether each of the following is the graph of a function.

49.

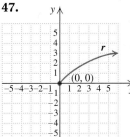

50.

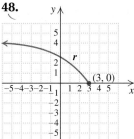

51.

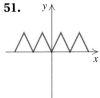

52.

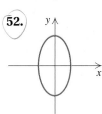

53.

54.

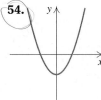

Find the function values. PEMDAS

55. $g(x) = 2x + 5$

  a) $g(0)$ **b)** $g(-4)$ **c)** $g(-7)$
  d) $g(8)$ **e)** $g(a + 2)$ **f)** $g(a) + 2$

56. $h(x) = 5x - 1$

  a) $h(4)$ **b)** $h(8)$ **c)** $h(-3)$
  d) $h(-4)$ **e)** $h(a - 1)$ **f)** $h(a) + 3$

57. $f(n) = 5n^2 + 4n$

  a) $f(0)$ **b)** $f(-1)$ **c)** $f(3)$
  d) $f(t)$ **e)** $f(2a)$ **f)** $f(3) - 9$

58. $g(n) = 3n^2 - 2n$

  a) $g(0)$ **b)** $g(-1)$ **c)** $g(3)$
  d) $g(t)$ **e)** $g(2a)$ **f)** $g(3) - 4$

59. $f(x) = \dfrac{x - 3}{2x - 5}$

  a) $f(0)$ **b)** $f(4)$ **c)** $f(-1)$
  d) $f(3)$ **e)** $f(x + 2)$ **f)** $f(a + h)$

60. $r(x) = \dfrac{3x - 4}{2x + 5}$

  a) $r(0)$ **b)** $r(2)$ **c)** $r\left(\dfrac{4}{3}\right)$
  d) $r(-1)$ **e)** $r(x + 3)$ **f)** $r(a + h)$

Find the domain of each function.

61. $f(x) = \dfrac{5}{x - 3}$

62. $f(x) = \dfrac{7}{6 - x}$

63. $g(x) = 2x + 1$

64. $g(x) = x^2 + 3$

65. $h(x) = |6 - 7x|$

66. $h(x) = |3x - 4|$

67. $f(x) = \dfrac{3}{8 - 5x}$

68. $f(x) = \dfrac{5}{2x + 1}$

69. $h(x) = \dfrac{x}{x + 1}$

70. $h(x) = \dfrac{3x}{x + 7}$

71. $f(x) = \dfrac{3x + 1}{2}$

72. $f(x) = \dfrac{4x - 3}{5}$

73. $g(x) = \dfrac{1}{2x}$

74. $g(x) = \dfrac{1}{2}x$

The function A described by $A(s) = s^2\dfrac{\sqrt{3}}{4}$ gives the area of an equilateral triangle with side s.

75. Find the area when a side measures 4 cm.

76. Find the area when a side measures 6 in.

The function V described by $V(r) = 4\pi r^2$ gives the surface area of a sphere with radius r.

77. Find the surface area when the radius is 3 in.

78. Find the surface area when the radius is 5 cm.

Archaeology. *The function H described by*

$$H(x) = 2.75x + 71.48$$

can be used to predict the height, in centimeters, of a woman whose humerus *(the bone from the elbow to the shoulder) is x cm long. Predict the height of a woman whose humerus is the length given.*

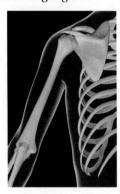

79. 34 cm   **80.** 31 cm

Chemistry. *The function F described by*

$$F(C) = \tfrac{9}{5}C + 32$$

gives the Fahrenheit temperature corresponding to the Celsius temperature C.

81. Find the Fahrenheit temperature equivalent to −5° Celsius.

82. Find the Fahrenheit temperature equivalent to 10° Celsius.

Heart attacks and cholesterol. *For Exercises 83 and 84, use the following graph, which shows the annual heart attack rate per 10,000 men as a function of blood cholesterol level.**

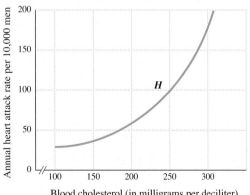

Blood cholesterol (in milligrams per deciliter)

*Copyright 1989, CSPI. Adapted from Nutrition Action Health-letter (1875 Connecticut Avenue, N.W., Suite 300, Washington, DC 20009-5728. $24 for 10 issues).

83. Approximate the annual heart attack rate for those men whose blood cholesterol level is 225 mg/dl. That is, find $H(225)$.

84. Approximate the annual heart attack rate for those men whose blood cholesterol level is 275 mg/dl. That is, find $H(275)$.

Films. *For Exercises 85 and 86, use the following graph, which shows the number of movies released in the United States.*
Source: Nash Information Services

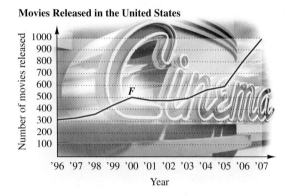

Movies Released in the United States

85. Approximate the number of movies released in 2000. That is, find $F(2000)$.

86. Approximate the number of movies released in 2007. That is, find $F(2007)$.

Energy-saving lightbulbs. *An energy bill signed into law in 2007 requires the United States to phase out standard incandescent lightbulbs. A more efficient replacement is the compact fluorescent (CFL) bulb. The table below lists incandescent wattage and the CFL wattage required to create the same amount of light.*
Source: U.S. Department of Energy

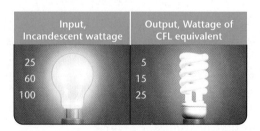

Input, Incandescent wattage	Output, Wattage of CFL equivalent
25	5
60	15
100	25

87. Use the data in the figure above to draw a graph. Estimate the wattage of a CFL bulb that creates light equivalent to a 75-watt incandescent bulb. Then predict the wattage of a CFL bulb that creates light equivalent to a 120-watt incandescent bulb.

88. Use the graph from Exercise 87 to estimate the wattage of a CFL bulb that creates light equivalent to a 40-watt incandescent bulb. Then predict the wattage of a CFL bulb that creates light equivalent to a 150-watt incandescent bulb.

Blood alcohol level.   *The following table can be used to predict the number of drinks required for a person of a specified weight to be considered legally intoxicated (blood alcohol level of 0.08 or above). One 12-oz glass of beer, a 5-oz glass of wine, or a cocktail containing 1 oz of a distilled liquor all count as one drink. Assume that all drinks are consumed within one hour.*

12 oz   5 oz   1 oz

Input, Body Weight (in pounds)	Output, Number of Drinks
100	2.5
160	4
180	4.5
200	5

89. Use the data in the table above to draw a graph and to estimate the number of drinks that a 140-lb person would have to drink to be considered intoxicated. Then predict the number of drinks it would take for a 230-lb person to be considered intoxicated.

90. Use the graph from Exercise 89 to estimate the number of drinks that a 120-lb person would have to drink to be considered intoxicated. Then predict the number of drinks it would take for a 250-lb person to be considered intoxicated.

91. *Retailing.* Mountain View Gifts is experiencing constant growth. They recorded a total of $250,000 in sales in 2003 and $285,000 in 2008. Use a graph that displays the store's total sales as a function of time to estimate sales for 2004 and for 2011.

92. Use the graph in Exercise 91 to estimate sales for 2006 and for 2012.

Researchers at Yale University have suggested that the following graphs* may represent three different aspects of love.

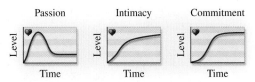

93. In what unit would you measure time if the horizontal length of each graph were ten units? Why?

94. Do you agree with the researchers that these graphs should be shaped as they are? Why or why not?

Skill Review

To prepare for Section 2.3, review simplifying expressions and solving for a variable (Sections 1.2 and 1.5).

Simplify. [1.2]

95. $\dfrac{6 - 3}{-2 - 7}$

96. $\dfrac{-2 - (-4)}{5 - 8}$

97. $\dfrac{-5 - (-5)}{3 - (-10)}$

98. $\dfrac{2 - (-3)}{-3 - 2}$

Solve for y. [1.5]

99. $2x - y = 8$

100. $5x + 5y = 10$

101. $2x + 3y = 6$

102. $5x - 4y = 8$

Synthesis

103. Jaylan is asked to write a function relating the number of fish in an aquarium to the amount of food needed for the fish. Which quantity should he choose as the independent variable? Why?

104. Which would you trust more and why: estimates made using interpolation or those made using extrapolation?

For Exercises 105 and 106, let $f(x) = 3x^2 - 1$ and $g(x) = 2x + 5$.

105. Find $f(g(-4))$ and $g(f(-4))$.

106. Find $f(g(-1))$ and $g(f(-1))$.

107. If f represents the function in Exercise 15, find $f(f(f(f(\text{tiger}))))$.

*From "A Triangular Theory of Love," by R. J. Sternberg, 1986, *Psychological Review,* **93**(2), 119–135. Copyright 1986 by the American Psychological Association, Inc. Reprinted by permission.

Pregnancy. For Exercises 108–111, use the following graph of a woman's "stress test." This graph shows the size of a pregnant woman's contractions as a function of time.

Stress test

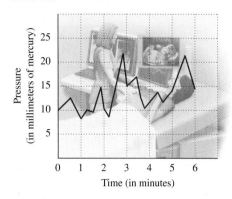

108. How large is the largest contraction that occurred during the test?

109. At what time during the test did the largest contraction occur?

110. On the basis of the information provided, how large a contraction would you expect 60 seconds after the end of the test? Why?

111. What is the frequency of the largest contraction?

112. Suppose that a function g is such that $g(-1) = -7$ and $g(3) = 8$. Find a formula for g if $g(x)$ is of the form $g(x) = mx + b$, where m and b are constants.

113. The *greatest integer function* $f(x) = [\![x]\!]$ is defined as follows: $[\![x]\!]$ is the greatest integer that is less than or equal to x. For example, if $x = 3.74$, then $[\![x]\!] = 3$; and if $x = -0.98$, then $[\![x]\!] = -1$. Graph the greatest integer function for $-5 \le x \le 5$. (The notation $f(x) = \text{INT}(x)$ is used in many graphing calculators and computer programs.)

114. *Energy expenditure.* On the basis of the information given below, what burns more energy: walking $4\frac{1}{2}$ mph for two hours or bicycling 14 mph for one hour?

Approximate Energy Expenditure by a
150-Pound Person in Various Activities

Activity	Calories per Hour
Walking, $2\frac{1}{2}$ mph	210
Bicycling, $5\frac{1}{2}$ mph	210
Walking, $3\frac{3}{4}$ mph	300
Bicycling, 13 mph	660

Source: Based on material prepared by Robert E. Johnson, M.D., Ph.D., and colleagues, University of Illinois.

2.3   Linear Functions: Slope, Graphs, and Models

Slope–Intercept Form of an Equation   ▪   Applications

The functions and real-life models that we examine in this section have graphs that are straight lines. Such functions and their graphs are called *linear* and can be written in the form $f(x) = mx + b$, where m and b are constants.

Slope–Intercept Form of an Equation

Examples 3 and 5 in Section 2.1 suggest that for any number m, the graph of $y = mx$ is a straight line passing through the origin. What happens when we add a number b on the right side and graph the equation $y = mx + b$?

EXAMPLE 1 Graph $y = 2x$ and $y = 2x + 3$, using the same set of axes.

SOLUTION   We first make a table of solutions of both equations.

x	y $y = 2x$	y $y = 2x + 3$
0	0	3
1	2	5
−1	−2	1
2	4	7
−2	−4	−1
3	6	9

We then plot these points. Drawing a blue line for $y = 2x + 3$ and a red line for $y = 2x$, we observe that the graph of $y = 2x + 3$ is simply the graph of $y = 2x$ shifted, or *translated*, 3 units up. The lines are parallel.

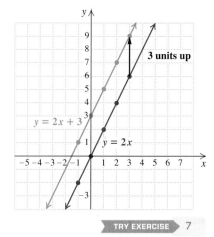

TRY EXERCISE ▸ 7

EXAMPLE 2 Graph $f(x) = \frac{1}{3}x$ and $g(x) = \frac{1}{3}x - 2$, using the same set of axes.

SOLUTION   We first make a table of solutions of both equations. By choos-ing multiples of 3, we can avoid fractions.

x	$f(x)$ $f(x) = \frac{1}{3}x$	$g(x)$ $g(x) = \frac{1}{3}x - 2$
0	0	−2
3	1	−1
−3	−1	−3
6	2	0

We then plot these points. Drawing a blue line for $g(x) = \frac{1}{3}x - 2$ and a red line for $f(x) = \frac{1}{3}x$, we see that the graph of $g(x) = \frac{1}{3}x - 2$ is sim-ply the graph of $f(x) = \frac{1}{3}x$ shifted, or translated, 2 units down.

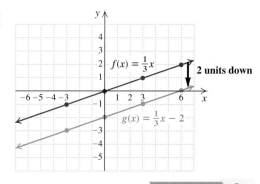

TRY EXERCISE ▸ 9

Note that the graph of $y = 2x + 3$ passes through the point $(0, 3)$ and the graph of $g(x) = \frac{1}{3}x - 2$ passes through the point $(0, -2)$. In general:

The graph of any line written in the form $y = mx + b$ passes through the point $(0, b)$. The point $(0, b)$ is called the y-intercept.

EXAMPLE 3   For each equation, find the y-intercept.

a) $y = -5x + 7$ **b)** $f(x) = 5.3x - 12$

SOLUTION

a) The y-intercept is $(0, 7)$.

b) The y-intercept is $(0, -12)$.   **TRY EXERCISE** 13

A y-intercept $(0, b)$ is plotted by locating b on the y-axis. For this reason, we sometimes refer to the number b as the y-intercept.

In examining the graphs in Examples 1 and 2, note that, in each example, the slant of the red lines seems to match the slant of the blue lines. In general, the graph of a line $y = mx + b$ is parallel to the graph of $y = mx$. Note too that the slant of the lines in Example 1 differs from the slant of the lines in Example 2. This leads us to suspect that it is the number m, in the equation $y = mx + b$, that is responsible for the slant of the line. The following definition enables us to visualize this slant, or *slope*, as a ratio of two lengths.

Slope

The *slope* of the line passing through (x_1, y_1) and (x_2, y_2) is given by

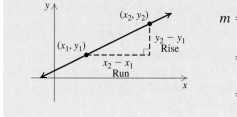

$$m = \frac{\text{rise}}{\text{run}} = \frac{\text{vertical change}}{\text{horizontal change}}$$

$$= \frac{\text{the difference in } y}{\text{the difference in } x}$$

$$= \frac{y_2 - y_1}{x_2 - x_1} = \frac{y_1 - y_2}{x_1 - x_2}.$$

In the definition above, (x_1, y_1) and (x_2, y_2)—read "x sub-one, y sub-one and x sub-two, y sub-two"—represent two different points on a line. It does not matter which point is considered (x_1, y_1) and which is considered (x_2, y_2) so long as coordinates are subtracted in the same order in both the numerator and the denominator.

The letter m is traditionally used for slope. This usage has its roots in the French verb *monter*, meaning "to climb."

EXAMPLE 4   Find the slope of the lines drawn in Examples 1 and 2.

SOLUTION   To find the slope of a line, we can use the coordinates of any two points on that line. We use $(1, 5)$ and $(2, 7)$ to find the slope of the blue line in Example 1:

$$\text{Slope} = \frac{\text{rise}}{\text{run}} = \frac{\text{difference in } y}{\text{difference in } x}$$

$$= \frac{y_2 - y_1}{x_2 - x_1} = \frac{7 - 5}{2 - 1} = 2. \qquad \begin{array}{l} \textit{Any} \text{ pair of points on the line will} \\ \text{give the same slope.} \end{array}$$

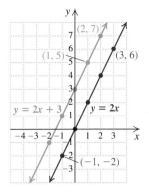

From Example 1

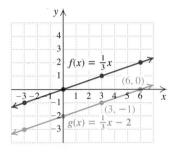

From Example 2

TECHNOLOGY CONNECTION

To use a graphing calculator to examine the effect of m when graphing $y = mx + b$, first graph $y_1 = x + 1$. Then on the same set of axes, graph $y_2 = 2x + 1$, $y_3 = 3x + 1$, and $y_4 = \frac{3}{4}x + 1$. Describe how a positive multiplier of x affects the graph.

To see the effect of a negative value for m, graph the equations $y_1 = x + 1$ and $y_2 = -x + 1$ on the same set of axes. Then graph $y_3 = 2x + 1$ and $y_4 = -2x + 1$ on those same axes.

If your calculator has a Transfrm application, use it to enter and graph $y_1 = Ax + B$. Choose a value for B and enter various positive and negative values for A. Uninstall Transfrm when you are finished.

Describe how the sign of m affects the graph of $y = mx + b$.

To find the slope of the red line in Example 1, we use $(-1, -2)$ and $(3, 6)$:

$$\text{Slope} = \frac{\text{rise}}{\text{run}} = \frac{\text{difference in } y}{\text{difference in } x}$$

$$= \frac{6 - (-2)}{3 - (-1)} = \frac{8}{4} = 2. \qquad \textit{Any pair of points on the line will give the same slope.}$$

To find the slope of the blue line in Example 2, we use $(3, -1)$ and $(6, 0)$:

$$\text{Slope} = \frac{\text{rise}}{\text{run}} = \frac{\text{difference in } y}{\text{difference in } x}$$

$$= \frac{0 - (-1)}{6 - 3} = \frac{1}{3}. \qquad \textit{Any pair of points on the line will give the same slope.}$$

You should confirm that the red line in Example 2 also has a slope of $\frac{1}{3}$.

> **TRY EXERCISE** ▶ 23

In Example 4, we found that the lines given by $y = 2x + 3$, $y = 2x$, $g(x) = \frac{1}{3}x - 2$, and $f(x) = \frac{1}{3}x$ have slopes 2, 2, $\frac{1}{3}$, and $\frac{1}{3}$, respectively. This supports (but does not prove) the following:

The slope of any line written in the form $y = mx + b$ is m.

A proof of this result is outlined in Exercise 107 on p. 108.

Determine the slope of the line given by $y = \frac{2}{3}x + 4$, and graph the line.

SOLUTION Here $m = \frac{2}{3}$, so the slope is $\frac{2}{3}$. This means that from *any* point on the graph, we can locate a second point by simply going *up* 2 units (the *rise*) and *to the right* 3 units (the *run*). Where do we start? Because the y-intercept, $(0, 4)$, is known to be on the graph, we start there. Then $(0 + 3, 4 + 2)$, or $(3, 6)$, is also on the graph. Knowing two points, we can draw the line.

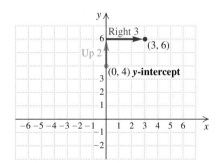 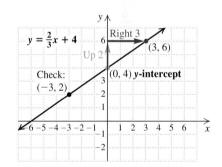

Important! To check the graph, we use some other value for x, say -3, and determine y (in this case, 2). We plot that point and see that it *is* on the line. Were it not, we would know that some error had been made.

Slope–Intercept Form

Any equation of the form $y = mx + b$ is said to be written in *slope–intercept* form and has a graph that is a straight line.

The slope of the line is m.

The y-intercept of the line is $(0, b)$.

EXAMPLE **6**   Determine the slope and the *y*-intercept of the line given by $y = -\frac{1}{3}x + 2$.

SOLUTION   The equation $y = -\frac{1}{3}x + 2$ is written in the form $y = mx + b$:

$$y = \quad mx + b$$

$$y = -\frac{1}{3}x + 2.$$

Since $m = -\frac{1}{3}$, the slope is $-\frac{1}{3}$. Since $b = 2$, the *y*-intercept is $(0, 2)$.

EXAMPLE **7**   Find a linear function whose graph has slope 3 and *y*-intercept $(0, -1)$.

SOLUTION   We use the slope–intercept form, $f(x) = mx + b$:

$$f(x) = 3x - 1. \qquad \text{Substituting 3 for } m \text{ and } -1 \text{ for } b$$

> **TRY EXERCISE** 53

To graph $f(x) = 3x - 1$, we regard the slope of 3 as $\frac{3}{1}$. Then, beginning at the *y*-intercept $(0, -1)$, we count *up* 3 units and *to the right* 1 unit, as shown below.

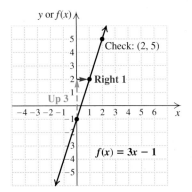

EXAMPLE **8**   Determine the slope and the *y*-intercept of the line given by $f(x) = -\frac{1}{2}x + 5$. Then draw the graph.

SOLUTION   The *y*-intercept is $(0, 5)$. The slope is $-\frac{1}{2}$, or $\frac{-1}{2}$. From the *y*-intercept, we go *down* 1 unit and *to the right* 2 units. That gives us the point $(2, 4)$. We can now draw the graph.

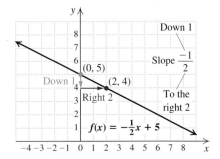

As a new type of check, we rewrite the slope in another form and find another point: $-\frac{1}{2} = \frac{1}{-2}$. Thus we can go *up* 1 unit and then *to the left* 2 units. This gives the point $(-2, 6)$. Since $(-2, 6)$ is on the line, we have a check.

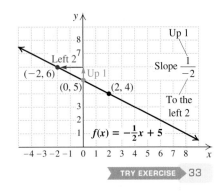

> **TRY EXERCISE** 33

In Examples 1, 2, and 5, the lines slant upward from left to right. In Example 8, the line slants downward from left to right.

Lines with positive slopes slant upward from left to right.

Lines with negative slopes slant downward from left to right.

Often the easiest way to graph an equation is to rewrite it in slope–intercept form and then proceed as in Examples 5 and 8 above.

EXAMPLE **9**   Determine the slope and the y-intercept for the equation $5x - 4y = 8$. Then graph.

SOLUTION   We first convert to slope–intercept form:

$$5x - 4y = 8$$
$$-4y = -5x + 8 \qquad \text{Adding } -5x \text{ to both sides}$$
$$y = -\tfrac{1}{4}(-5x + 8) \qquad \text{Multiplying both sides by } -\tfrac{1}{4}$$
$$y = \tfrac{5}{4}x - 2. \qquad \text{Using the distributive law}$$

Because we have an equation of the form $y = mx + b$, we know that the slope is $\frac{5}{4}$ and the y-intercept is $(0, -2)$. We plot $(0, -2)$, and from there we go *up* 5 units and *to the right* 4 units, and plot a second point at $(4, 3)$. We then draw the graph, as shown at left.

To check that the line is drawn correctly, we calculate the coordinates of another point on the line. For $x = 2$, we have

$$5 \cdot 2 - 4y = 8$$
$$10 - 4y = 8$$
$$-4y = -2$$
$$y = \tfrac{1}{2}.$$

Thus, $(2, \tfrac{1}{2})$ should appear on the graph. Since it *does* appear to be on the line, we have a check.

TRY EXERCISE 39

Applications

Because slope is a ratio that indicates how a change in the vertical direction corresponds to a change in the horizontal direction, it has many real-world applications. Foremost is the use of slope to represent a *rate of change*.

EXAMPLE **10**   *Telephone lines.*   As more people use cell phones as their primary phone line, the number of residential land lines in the United States has been decreasing. Use the following graph to find the rate at which this number is changing. Note that the jagged "break" on the vertical axis is used to avoid including a large portion of unused grid.

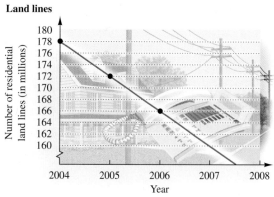

Source: Federal Communications Commission

SOLUTION   Since the graph is linear, we can use any pair of points to determine the rate of change. We choose the points (2004, 178 million) and (2005, 172 million), which gives us

$$\text{Rate of change} = \frac{172 \text{ million} - 178 \text{ million}}{2005 - 2004}$$

$$= \frac{-6 \text{ million}}{1 \text{ year}} = -6 \text{ million per year.}$$

The number of land lines in the United States is changing at the rate of −6 million lines per year.

TRY EXERCISE ▶ 59

EXAMPLE 11   *Construction.*   In 1994, new privately owned single-family homes in the United States had a median floor area of 1951 square feet. By 2006, the median floor area had risen to 2248 square feet. At what rate did the median floor area change?

Source: U. S. Census Bureau

SOLUTION   The rate at which the median floor area changed is given by

$$\text{Rate of change} = \frac{\text{change in median floor area}}{\text{change in time}}$$

$$= \frac{2248 \text{ square feet} - 1951 \text{ square feet}}{2006 - 1994}$$

$$= \frac{297 \text{ square feet}}{12 \text{ years}} = 24.75 \text{ square feet per year.}$$

Between 1994 and 2006, the median floor area of new privately owned single-family homes grew at a rate of 24.75 square feet per year.

TRY EXERCISE ▶ 69

EXAMPLE 12   *Running speed.*   Stephanie runs 10 km during each workout. For the first 7 km, her pace is twice as fast as it is for the last 3 km. Which of the following graphs best describes Stephanie's workout?

A.

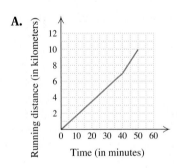

B.

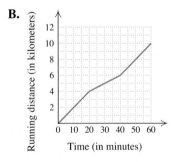

C.

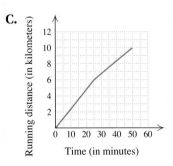

D.

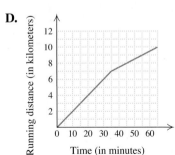

SOLUTION   The slopes in graph A increase as we move to the right. This would indicate that Stephanie ran faster for the *last* part of her workout. Thus graph A is not the correct one.

The slopes in graph B indicate that Stephanie slowed down in the middle of her run and then resumed her original speed. Thus graph B does not correctly model the situation either.

According to graph C, Stephanie slowed down not at the 7-km mark, but at the 6-km mark. Thus graph C is also incorrect.

Graph D indicates that Stephanie ran the first 7 km in 35 min, a rate of 0.2 km/min. It also indicates that she ran the final 3 km in 30 min, a rate of 0.1 km/min. This means that Stephanie's rate was twice as fast for the first 7 km, so graph D provides a correct description of her workout.   **TRY EXERCISE** 67

Linear functions arise continually in today's world. As with the problems in Examples 10–12, it is critical to use proper units in all answers.

EXAMPLE 13   *Salvage value.*   Island Bike Rentals uses the function $S(t) = -125t + 750$ to determine the *salvage value* $S(t)$, in dollars, of a mountain bike t years after its purchase.

a) What do the numbers -125 and 750 signify?

b) How long will it take a bicycle to *depreciate* completely?

c) What is the domain of S?

SOLUTION   Drawing, or at least visualizing, a graph can be useful here.

a) At time $t = 0$, we have $S(0) = -125 \cdot 0 + 750 = 750$. Thus the number 750 signifies the original cost of the mountain bike, in dollars.

This function is written in slope–intercept form. Since the output is measured in dollars and the input in years, the number -125 signifies that the value of the bike is decreasing at a rate of \$125 per year.

b) The bike will have depreciated completely when its value drops to 0. To learn when this occurs, we determine when $S(t) = 0$:

$$S(t) = 0$$
$$-125t + 750 = 0 \qquad \text{Substituting } -125t + 750 \text{ for } S(t)$$
$$-125t = -750 \qquad \text{Subtracting 750 from both sides}$$
$$t = 6. \qquad \text{Dividing both sides by } -125$$

The bike will have depreciated completely in 6 yr.

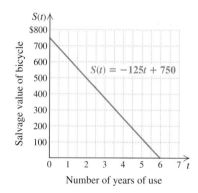

c) Neither the number of years of service nor the salvage value can be negative. In part (b), we found that after 6 yr the salvage value will have dropped to 0. Thus the domain of S is $\{t \mid 0 \le t \le 6\}$. The graph at left serves as a visual check of this result.   **TRY EXERCISE** 85

2.3 EXERCISE SET

For Extra Help
MyMathLab
Math XL
PRACTICE   WATCH   DOWNLOAD

Concept Reinforcement *In each of Exercises 1–6, match the word with the most appropriate choice from the column on the right.*

1. ___ *y*-intercept
2. ___ Slope
3. ___ Rise
4. ___ Run
5. ___ Slope–intercept form
6. ___ Translated

a) $y = mx + b$
b) Shifted
c) $\dfrac{\text{Difference in } y}{\text{Difference in } x}$
d) Difference in x
e) Difference in y
f) $(0, b)$

Graph.

7. $f(x) = 2x - 1$
8. $g(x) = 3x + 4$
9. $g(x) = -\frac{1}{3}x + 2$
10. $f(x) = -\frac{1}{2}x - 5$
11. $h(x) = \frac{2}{5}x - 4$
12. $h(x) = \frac{4}{5}x + 2$

Determine the y-intercept.

13. $y = 5x + 3$
14. $y = 2x - 11$
15. $g(x) = -x - 1$
16. $g(x) = -4x + 5$
17. $y = -\frac{3}{8}x - 4.5$
18. $y = \frac{15}{7}x + 2.2$
19. $f(x) = 1.3x - \frac{1}{4}$
20. $f(x) = -1.2x + \frac{1}{5}$
21. $y = 17x + 138$
22. $y = -52x - 260$

For each pair of points, find the slope of the line containing them.

23. $(10, 11)$ and $(8, 3)$
24. $(2, 9)$ and $(12, 4)$
25. $(-4, -5)$ and $(-8, 3)$
26. $(2, -3)$ and $(6, -2)$
27. $(13, 4)$ and $(-20, -7)$
28. $(-5, -11)$ and $(-8, -21)$
29. $\left(\frac{1}{2}, -\frac{2}{3}\right)$ and $\left(\frac{1}{6}, \frac{1}{6}\right)$
30. $\left(\frac{3}{4}, -\frac{2}{5}\right)$ and $\left(\frac{1}{3}, -\frac{1}{4}\right)$
31. $(-9.7, 43.6)$ and $(4.5, 43.6)$
32. $(-2.8, -3.1)$ and $(-1.8, -2.6)$

Determine the slope and the y-intercept. Then draw a graph. Be sure to check as in Example 5 or Example 8.

33. $y = \frac{5}{2}x - 3$
34. $y = \frac{2}{5}x - 4$
35. $f(x) = -\frac{5}{2}x + 2$
36. $f(x) = -\frac{2}{5}x + 3$
37. $F(x) = 2x + 1$
38. $g(x) = 3x - 2$
39. $4x + y = 3$
40. $4x - y = 1$
41. $6y + x = 6$
42. $4y + 20 = x$
Aha! **43.** $g(x) = -0.25x$
44. $F(x) = 1.5x$
45. $4x - 5y = 10$
46. $5x + 4y = 4$
47. $2x + 3y = 6$
48. $3x - 2y = 8$
49. $5 - y = 3x$
50. $3 + y = 2x$
Aha! **51.** $g(x) = 4.5$
52. $g(x) = \frac{1}{2}$

Find a linear function whose graph has the given slope and y-intercept.

53. Slope 2, *y*-intercept $(0, 5)$
54. Slope -4, *y*-intercept $(0, 1)$
55. Slope $-\frac{2}{3}$, *y*-intercept $(0, -2)$
56. Slope $-\frac{3}{4}$, *y*-intercept $(0, -5)$
57. Slope -7, *y*-intercept $\left(0, \frac{1}{3}\right)$
58. Slope 8, *y*-intercept $\left(0, -\frac{1}{4}\right)$

For each graph, find the rate of change. Remember to use appropriate units. See Example 10.

59.

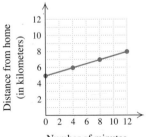

Number of minutes
spent running

60.

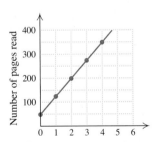

Number of days spent reading

61.

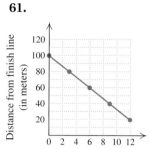

Number of seconds spent running

62.

63.

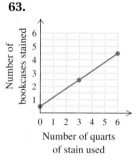

Number of quarts of stain used

64.

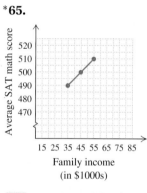

Number of hours spent walking

III

IV

Time of day

68. *Running rate.*   An ultra-marathoner passes the 15-mi point of a race after 2 hr and reaches the 22-mi point 56 min later. Assuming a constant rate, find the speed of the marathoner.

69. *Skiing rate.*   A cross-country skier reaches the 3-km mark of a race in 15 min and the 12-km mark 45 min later. Assuming a constant rate, find the speed of the skier.

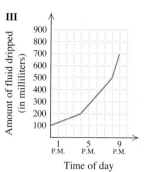

*65.

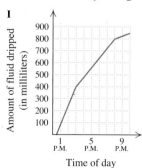

Family income (in $1000s)

*66.

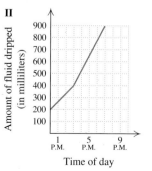

Family income (in $1000s)

70. *Recycling.*   The number of Freecycle recycling groups grew from 2936 at the end of August, 2005, to 4224 at the beginning of January, 2008. Determine the rate at which the number of Freecycle groups were being established.
Source: www.freecycle.org

71. *Work rate.*   As a painter begins work, one-fourth of a house has already been painted. Eight hours later, the house is two-thirds done. Calculate the painter's work rate.

72. *Rate of descent.*   A plane descends to sea level from 12,000 ft after being airborne for $1\frac{1}{2}$ hr. The entire flight time is 2 hr 10 min. Determine the average rate of descent of the plane.

67. *Nursing.*   Match each sentence with the most appropriate of the four graphs shown.

   a) The rate at which fluids were given intravenously was doubled after 3 hr.

   b) The rate at which fluids were given intravenously was gradually reduced to 0.

   c) The rate at which fluids were given intravenously remained constant for 5 hr.

   d) The rate at which fluids were given intravenously was gradually increased.

I

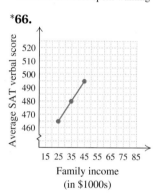

Time of day

II

Time of day

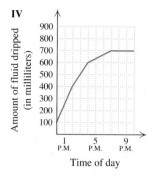

*Based on data from the College Board Online.

73. *Rate of computer hits.*   At the beginning of 2007, Starfarm.com had already received 80,000 hits at their website. At the beginning of 2009, that number had climbed to 430,000. Calculate the rate at which the number of hits is increasing.

74. *Market research.*   Match each sentence with the most appropriate of the four graphs shown.
   a) After January 1, daily sales continued to rise, but at a slower rate.
   b) After January 1, sales decreased faster than they ever grew.
   c) The rate of growth in daily sales doubled after January 1.
   d) After January 1, daily sales decreased at half the rate that they grew in December.

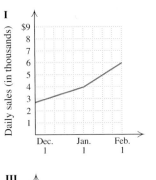

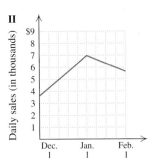

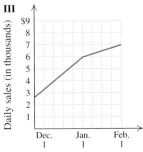

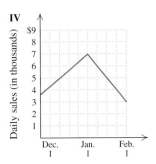

In Exercises 75–84, each model is of the form
$f(x) = mx + b$. *In each case, determine what m and b signify.*

75. *Cost of renting a truck.*   The cost, in dollars, of a one-day truck rental is given by $C(d) = 0.75d + 30$, where d is the number of miles driven.

76. *Weekly pay.*   Each salesperson at Super Electronics is paid $P(x)$ dollars, where $P(x) = 0.05x + 200$ and x is the value of the salesperson's sales for the week.

77. *Hair growth.*   After Lauren donated her hair to Locks of Love, the length $L(t)$ of her hair, in inches, was given by $L(t) = \frac{1}{2}t + 5$, where t is the number of months after she had the haircut.

78. *Electricity demand.*   The demand, in billions of kilowatt-hours, of electricity is estimated by $D(t) = \frac{191}{5}t + 3439$, where t is the number of years after 2000.
Source: Based on data from the U.S. Energy Information Administration

79. *Life expectancy of American women.*   The life expectancy of American women t years after 1970 is given by $A(t) = \frac{1}{8}t + 75.5$.
Source: Based on data from the National Center for Health Statistics

80. *Landscaping.*   After being cut, the length $G(t)$ of the lawn, in inches, at Great Harrington Community College is given by $G(t) = \frac{1}{8}t + 2$, where t is the number of days since the lawn was cut.

81. *Cost of a sports ticket.*   The average price $P(t)$, in dollars, of a major-league baseball ticket is given by $P(t) = 0.89t + 16.63$, where t is the number of years since 2000.
Source: Based on data from Team Marketing Report

82. *Cost of a taxi ride.*   The cost, in dollars, of a taxi ride in New York City is given by $C(d) = 2d + 2.5$,* where d is the number of miles traveled.

83. *Organic cotton.*   The function given by $c(t) = 849t + 5960$ can be used to estimate the number of U.S. acres planted with organic cotton, where t is the number of years since 2006.
Source: Based on data from the Organic Trade Association

84. *Catering.*   When catering a party for x people, Chrissie's Catering uses the formula $C(x) = 25x + 75$, where $C(x)$ is the cost of the food, in dollars.

---

*Rates are higher between 4 P.M. and 8 P.M. (*Source:* Based on data from New York City Taxi and Limousine Commission, 2007)

85. *Salvage value.* Green Glass Recycling uses the function given by $F(t) = -5000t + 90,000$ to determine the salvage value $F(t)$, in dollars, of a waste removal truck t years after it has been put into use.
  a) What do the numbers -5000 and $90,000$ signify?
  b) How long will it take the truck to depreciate completely?
  c) What is the domain of F?

86. *Salvage value.* Consolidated Shirt Works uses the function given by $V(t) = -2000t + 15,000$ to determine the salvage value $V(t)$, in dollars, of a color separator t years after it has been put into use.
  a) What do the numbers -2000 and $15,000$ signify?
  b) How long will it take the machine to depreciate completely?
  c) What is the domain of V?

87. *Trade-in value.* The trade-in value of a Jamis Dakar mountain bike can be determined using the function given by $v(n) = -200n + 1800$. Here $v(n)$ is the trade-in value, in dollars, after n years of use.
  a) What do the numbers -200 and 1800 signify?
  b) When will the trade-in value of the mountain bike be $600?
  c) What is the domain of v?

88. *Trade-in value.* The trade-in value of a John Deere riding lawnmower can be determined using the function given by $T(x) = -300x + 2400$. Here $T(x)$ is the trade-in value, in dollars, after x summers of use.
  a) What do the numbers -300 and 2400 signify?
  b) When will the value of the mower be $1200?
  c) What is the domain of T?

89. *Economics.* In January 2008, the federal debt could be modeled using $D(t) = mt + 9.2$, where $D(t)$ is in trillions of dollars t months after January and m is some constant. If you were president of the United States, would you want m to be positive or negative? Why?
  Source: U.S. Department of the Treasury

90. *Economics.* Examine the function given in Exercise 89. What units of measure must be used for m? Why?

Skill Review

To prepare for Section 2.4, review working with 0.

Simplify. [1.2]

91. $\dfrac{-8 - (-8)}{6 - (-6)}$

92. $\dfrac{-2 - 2}{-3 - (-3)}$

Solve. [1.3]

93. $3 \cdot 0 - 2y = 9$

94. $4x - 7 \cdot 0 = 3$

95. If $f(x) = 2x - 7$, find $f(0)$. [2.2]

96. If $f(x) = 2x - 7$, find any x-values for which $f(x) = 0$. [2.2]

Synthesis

97. The population of Valley Heights is decreasing at a rate of 10% per year. Can this be modeled using a linear function? Why or why not?

98. Janis Hope claims that her firm's profits continue to go up, but the rate of increase is going down.
  a) Sketch a graph that might represent her firm's profits as a function of time.
  b) Explain why the graph can go up while the rate of increase goes down.

99. Match each sentence with the most appropriate of the four graphs shown.
  a) Ellie drove 2 mi to a lake, swam 1 mi, and then drove 3 mi to a store.
  b) During a preseason workout, Rico biked 2 mi, ran for 1 mi, and then walked 3 mi.
  c) Luis bicycled 2 mi to a park, hiked 1 mi over the notch, and then took a 3-mi bus ride back to the park.
  d) After hiking 2 mi, Marcy ran for 1 mi before catching a bus for the 3-mi ride into town.

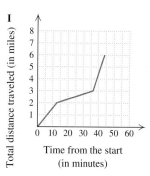

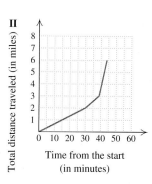

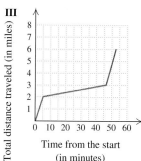

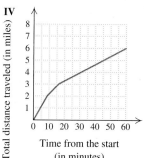

The following graph shows the elevation of each section of a bicycle tour from Sienna to Florence in Italy. Use the graph for Exercises 100–104.

Source: greve-in-chianti.com

Bicycle route elevation

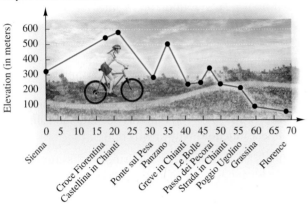

Distance from Sienna (in kilometers)

100. What part of the trip is the steepest?

101. What part of the trip has the longest uphill climb?

A road's grade is the ratio, given as a percent, of the road's change in elevation to its change in horizontal distance and is, by convention, always positive.

102. During one day's ride, Brittany rode uphill and then downhill at about the same grade. She then rode downhill at $\frac{1}{10}$ of the grade of the first two sections. Where did Brittany begin her ride?

103. During one day's ride, Scott biked two downhill sections and one uphill section, all at about the same grade. Where did Scott begin his ride?

104. Calculate the grade of the steepest section of the tour.

In Exercises 105 and 106, assume that r, p, and s are constants and that x and y are variables. Determine the slope and the y-intercept.

105. $rx + py = s - ry$ **106.** $rx + py = s$

107. Let (x_1, y_1) and (x_2, y_2) be two distinct points on the graph of $y = mx + b$. Use the fact that both pairs are solutions of the equation to prove that m is the slope of the line given by $y = mx + b$. (*Hint*: Use the slope formula.)

Given that $f(x) = mx + b$, classify each of the following as either true or false.

108. $f(cd) = f(c)f(d)$

109. $f(c + d) = f(c) + f(d)$

110. $f(c - d) = f(c) - f(d)$

111. $f(kx) = kf(x)$

112. Find k such that the line containing $(-3, k)$ and $(4, 8)$ is parallel to the line containing $(5, 3)$ and $(1, -6)$.

113. Find the slope of the line that contains the given pair of points.
  a) $(5b, -6c), (b, -c)$
  b) $(b, d), (b, d + e)$
  c) $(c + f, a + d), (c - f, -a - d)$

114. *Cost of a speeding ticket.*   The penalty schedule shown below is used to determine the cost of a speeding ticket in certain states. Use this schedule to graph the cost of a speeding ticket as a function of the number of miles per hour over the limit that a driver is going.

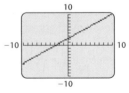

STATE POLICE
SPEEDING VIOLATION
FINES

1–10 mph over limit: $5.00/mph plus $17.50 surcharge
11–20 mph over limit: $6.00/mph plus $17.50 surcharge
21–30 mph over limit: $7.00/mph plus $17.50 surcharge
31+ mph over limit: $8.00/mph plus $17.50 surcharge

Officer will enter mph over limit in line 5a on the front of this document.

115. Graph the equations
$$y_1 = 1.4x + 2,   y_2 = 0.6x + 2,$$
$$y_3 = 1.4x + 5, \text{and} y_4 = 0.6x + 5$$
using a graphing calculator. If possible, use the SIMULTANEOUS mode so that you cannot tell which equation is being graphed first. Then decide which line corresponds to each equation.

116. A student makes a mistake when using a graphing calculator to draw $4x + 5y = 12$ and the following screen appears. Use algebra to show that a mistake has been made. What do you think the mistake was?

117. A student makes a mistake when using a graphing calculator to draw $5x - 2y = 3$ and the following screen appears. Use algebra to show that a mistake has been made. What do you think the mistake was?

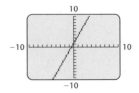

2.4   Another Look at Linear Graphs

Graphing Horizontal and Vertical Lines • Graphing Using Intercepts •
Solving Equations Graphically • Recognizing Linear Equations

In Section 2.3, we graphed linear equations using slopes and y-intercepts. We now graph lines that have a slope of zero or that have an undefined slope. We also graph lines using both x- and y-intercepts and learn how to use graphs to solve certain equations.

Graphing Horizontal and Vertical Lines

To find the slope of a line, we consider two points on the line. Since any two points on a horizontal line have the same y-coordinate, we can label the points (x_1, y_1) and (x_2, y_1). This gives us

$$m = \frac{y_1 - y_1}{x_2 - x_1} = \frac{0}{x_2 - x_1} = 0.$$

Thus the slope of a horizontal line is 0.

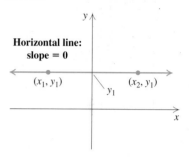

EXAMPLE 1   Use slope–intercept form to graph $f(x) = 3$.

SOLUTION   Recall from Example 6(c) in Section 2.2 that a function of this type is called a *constant function*. Writing $f(x)$ in slope–intercept form,

$$f(x) = 0 \cdot x + 3,$$

we see that the y-intercept is $(0, 3)$ and the slope is 0. Thus we can graph f by plotting $(0, 3)$ and, from there, counting off a slope of 0. Because $0 = 0/2$ (any nonzero number could be used in place of 2), we can draw the graph by going up 0 units and to the right 2 units. As a check, we also find some ordered pairs. Note that for any choice of x-value, $f(x)$ must be 3.

x	$f(x)$
-1	3
0	3
2	3

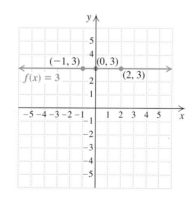

TRY EXERCISE 29

Horizontal Lines

The slope of a horizontal line is 0.

The graph of any function of the form $f(x) = b$ or $y = b$ is a horizontal line that crosses the y-axis at $(0, b)$.

Suppose that two different points are on a vertical line. They then have the same first coordinate. In this case, when we calculate the slope, we have

$$m = \frac{y_2 - y_1}{x_1 - x_1} = \frac{y_2 - y_1}{0}.$$

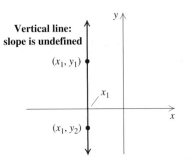

Vertical line:
slope is undefined

(x_1, y_1)

x_1

(x_1, y_2)

Since we cannot divide by 0, this is undefined. Note that when we say that $(y_2 - y_1)/0$ is undefined, it means that we have agreed to not attach any meaning to that expression. Thus the slope of a vertical line is undefined.

EXAMPLE **2**   Graph: $x = -2$.

SOLUTION   With y missing in the equation, no matter which value of y is chosen, x must be -2. Thus the pairs $(-2, 3)$, $(-2, 0)$, and $(-2, -4)$ all satisfy the equation. The graph is a line parallel to the y-axis. Note that since y is missing, this equation cannot be written in slope–intercept form.

x	y
-2	3
-2	0
-2	-4

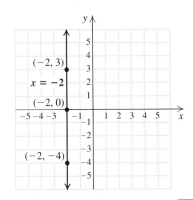

TRY EXERCISE 31

Vertical Lines

The slope of a vertical line is undefined.

The graph of any equation of the form $x = a$ is a vertical line that crosses the x-axis at $(a, 0)$.

EXAMPLE **3**   Find the slope of each given line. If the slope is undefined, state this.

a) $3y + 2 = 14$ **b)** $2x = 10$

STUDENT NOTES

The slope of a horizontal line is 0 and the slope of a vertical line is undefined. Avoid the use of the ambiguous phrase "no slope."

SOLUTION

a) We solve for y:

$3y + 2 = 14$

$3y = 12$ Subtracting 2 from both sides

$y = 4.$ Dividing both sides by 3

The graph of $y = 4$ is a horizontal line. Since $3y + 2 = 14$ is equivalent to $y = 4$, the slope of the line $3y + 2 = 14$ is 0.

b) When *y* does not appear, we solve for *x*:

$$2x = 10$$
$$x = 5. \quad \text{Dividing both sides by 2}$$

The graph of $x = 5$ is a vertical line. Since $2x = 10$ is equivalent to $x = 5$, the slope of the line $2x = 10$ is undefined.   TRY EXERCISE ▸ 11

Graphing Using Intercepts

Any line that is neither horizontal nor vertical will cross both the *x*- and *y*-axes. We have already seen that the point at which a line crosses the *y*-axis is called the *y-intercept*. Similarly, the point at which a line crosses the *x*-axis is called the *x-intercept*. Recall that to find the *y*-intercept, we replace *x* with 0 and solve for *y*. To find the *x*-intercept, we replace *y* with 0 and solve for *x*.

> ### To Determine Intercepts
> The *x*-intercept is of the form $(a, 0)$. To find *a*, let $y = 0$ and solve for *x*.
> The *y*-intercept is of the form $(0, b)$. To find *b*, let $x = 0$ and solve for *y*.

EXAMPLE 4   Graph the equation $3x + 2y = 12$ by using intercepts.

SOLUTION   To find the *y*-intercept, we let $x = 0$ and solve for *y*:

Let $x = 0$. $3 \cdot 0 + 2y = 12$ For points on the *y*-axis, $x = 0$.
$$2y = 12$$
Solve for *y*. $y = 6$.

The *y*-intercept is $(0, 6)$.
To find the *x*-intercept, we let $y = 0$ and solve for *x*:

Let $y = 0$. $3x + 2 \cdot 0 = 12$ For points on the *x*-axis, $y = 0$.
$$3x = 12$$
Solve for *x*. $x = 4$.

The *x*-intercept is $(4, 0)$.
We plot the two intercepts and draw the line. A third point could be calculated and used as a check.

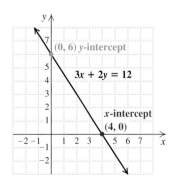

TRY EXERCISE ▸ 39

When the x- and y-intercepts are not both $(0, 0)$, the intercepts can be used to draw the graph of a line.

EXAMPLE 5

Graph $f(x) = 2x + 5$ by using intercepts.

SOLUTION   Because the function is in slope–intercept form, we know that the y-intercept is $(0, 5)$.
To find the x-intercept, we replace $f(x)$ with 0 and solve for x:

$$0 = 2x + 5$$
$$-5 = 2x$$
$$-\tfrac{5}{2} = x.$$

The x-intercept is $\left(-\tfrac{5}{2}, 0\right)$.
We plot the intercepts $(0, 5)$ and $\left(-\tfrac{5}{2}, 0\right)$ and draw the line. As a check, we can calculate the slope:

$$m = \frac{5 - 0}{0 - \left(-\tfrac{5}{2}\right)}$$
$$= \frac{5}{\tfrac{5}{2}}$$
$$= 5 \cdot \frac{2}{5}$$
$$= 2.$$

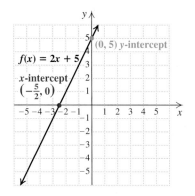

The slope is 2, as expected.

TRY EXERCISE 41

TECHNOLOGY CONNECTION

After graphing a function, we can use **2ND** **CALC** to find the intercepts. By selecting the VALUE option of the menu and entering 0 for x, we can find the y-intercept.
A *zero* or *root* of a function is a value for which $f(x) = 0$. The ZERO option of the menu allows us to find the x-intercept. To find a zero, we enter a value less than the x-intercept and then a value greater than the x-intercept as left and right bounds, respectively. We next enter a guess and the calculator then finds the value of the intercept.

Solving Equations Graphically

In Example 5, the x-intercept, $-\tfrac{5}{2}$, is the solution of $2x + 5 = 0$. Visually, $-\tfrac{5}{2}$ is the x-coordinate of the point at which the graphs of $f(x) = 2x + 5$ and $h(x) = 0$ intersect. Similarly, we can solve $2x + 5 = -3$ by finding the x-coordinate of the point at which the graphs of $f(x) = 2x + 5$ and $g(x) = -3$ intersect. From the graph, it appears that -4 is that x-value. To check, note that $f(-4) = 2(-4) + 5 = -3$.

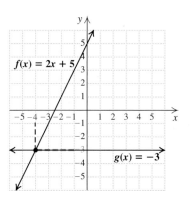

EXAMPLE 6

Solve graphically: $\tfrac{1}{2}x + 3 = 2$.

SOLUTION   To find the x-value for which $\tfrac{1}{2}x + 3$ will equal 2, we graph $f(x) = \tfrac{1}{2}x + 3$ and $g(x) = 2$ on the same set of axes. Since the intersection appears to be $(-2, 2)$, the solution is apparently -2.

STUDENT NOTES ————————

Remember that it is only the first coordinate of the point of intersection that is the solution of the equation.

Check:

$$\frac{1}{2}x + 3 = 2$$

$$\begin{array}{c|c} \frac{1}{2}(-2) + 3 & 2 \\ -1 + 3 & \\ & 2 \overset{?}{=} 2 \quad \text{TRUE} \end{array}$$

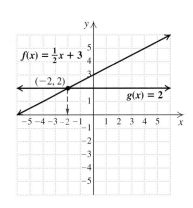

The solution is -2.

TRY EXERCISE ▶ 55

EXAMPLE   **7**

Cell-phone costs.   In 2008, an Apple iPhone cost $400. AT&T offered a plan including 450 daytime minutes a month for $60 per month. Write and graph a mathematical model for the cost. Then use the model to estimate the number of months required for the total cost to reach $700.

Source: www.apple.com

SOLUTION

1. **Familiarize.** For this plan, a monthly fee is charged after an initial purchase has been made. After 1 month of service, the total cost will be $400 + $60 = $460. After 2 months, the total cost will be $400 + $60 · 2 = $520. We can write a general model if we let $C(t)$ represent the total cost, in dollars, for t months of service.

2. **Translate.** We reword and translate as follows:

Rewording:	The total cost	is	the cost of the phone	plus	$60 per month.
Translating:	$C(t)$	$=$	400	$+$	$60 \cdot t$

where $t \geq 0$ (since there cannot be a negative number of months).

3. **Carry out.** Before graphing, we rewrite the model in slope–intercept form: $C(t) = 60t + 400$. We see that the vertical intercept is $(0, 400)$ and the slope, or rate, is $60 per month. Since we want to estimate the time required for the total cost to reach $700, we choose a scale for the vertical axis that will include 700.

We plot $(0, 400)$ and, from there, count up \$60 and to the right 1 month. This takes us to $(1, 460)$. We then draw a line passing through both points.

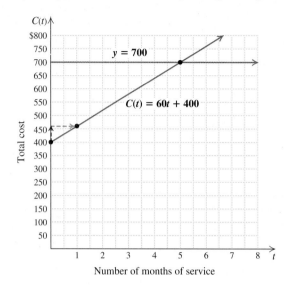

Number of months of service

To estimate the time required for the total cost to reach \$700, we are estimating the solution of

$$700 = 60t + 400. \qquad \text{Replacing } C(t) \text{ with } 700$$

We do this by graphing $y = 700$ and looking for the point of intersection. This point appears to be $(5, 700)$. Thus we estimate that it takes 5 months for the total cost to reach \$700.

4. Check. We evaluate:

$$C(5) = 60 \cdot 5 + 400 = 300 + 400 = 700.$$

Our estimate turns out to be precise.

5. State. It takes 5 months for the total cost to reach \$700.   **TRY EXERCISE** 65

There are limitations to solving equations graphically, as the next example illustrates.

EXAMPLE 8   Solve graphically: $-\frac{3}{4}x + 6 = 2x - 1$.

SOLUTION   We graph $f(x) = -\frac{3}{4}x + 6$ and $g(x) = 2x - 1$ on the same set of axes. It *appears* that the lines intersect at $(2.5, 4)$. This would mean that $f(2.5)$ and $g(2.5)$ are identical. Let's check.

Check:

$$\begin{array}{c|c}
-\frac{3}{4}x + 6 = 2x - 1 & \\
\hline
-\frac{3}{4}(2.5) + 6 & 2(2.5) - 1 \\
-1.875 + 6 & 5 - 1 \\
4.125 \overset{?}{=} 4 & \qquad \text{FALSE}
\end{array}$$

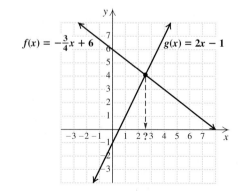

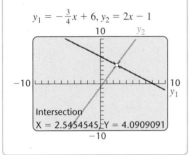

$$y_1 = -\tfrac{3}{4}x + 6, \ y_2 = 2x - 1$$

Intersection
X = 2.5454545 Y = 4.0909091

Our check shows that 2.5 is *not* the solution, although it may not be off by much. To find the exact solution, we need either a more precise way of determining the point of intersection (see the Technology Connection) or an algebraic (non-graphical) approach:

$$-\tfrac{3}{4}x + 6 = 2x - 1$$
$$-\tfrac{3}{4}x + 7 = 2x \qquad \text{Adding 1 to both sides}$$
$$7 = \tfrac{11}{4}x \qquad \text{Adding } \tfrac{3}{4}x \text{ to both sides}$$
$$\tfrac{28}{11} = x. \qquad \text{Multiplying both sides by } \tfrac{4}{11}$$

The solution is $\tfrac{28}{11}$, or about 2.55. A check of this answer is left to the student.

> **CAUTION!** When you are using a graph to solve an equation, it is important to use graph paper and to work as neatly as possible.

Recognizing Linear Equations

Is every equation of the form $Ax + By = C$ linear? To find out, suppose that A and B are nonzero and solve for y:

$$Ax + By = C \qquad A, B, \text{ and } C \text{ are constants.}$$
$$By = -Ax + C \qquad \text{Adding } -Ax \text{ to both sides}$$
$$y = -\frac{A}{B}x + \frac{C}{B}. \qquad \text{Dividing both sides by } B$$

Since the last equation is a slope–intercept equation, we see that $Ax + By = C$ is a linear equation when $A \neq 0$ and $B \neq 0$.

But what if A or B (but not both) is 0? If A is 0, then $By = C$ and $y = C/B$. If B is 0, then $Ax = C$ and $x = C/A$. In the first case, the graph is a horizontal line; in the second case, the line is vertical. In either case, $Ax + By = C$ is a linear equation when A or B (but not both) is 0. We have now justified the following result.

> ### Standard Form of a Linear Equation
> Any equation of the form $Ax + By = C$, where A, B, and C are real numbers and A and B are not both 0, is a linear equation in *standard form* and has a graph that is a straight line.

EXAMPLE 9

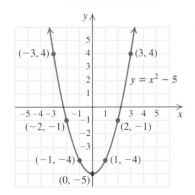

Determine whether the equation $y = x^2 - 5$ is linear.

SOLUTION We attempt to put the equation in standard form:

$$y = x^2 - 5$$
$$-x^2 + y = -5. \qquad \text{Adding } -x^2 \text{ to both sides}$$

This last equation is not linear because it has an x^2-term. We graphed this equation as Example 7 in Section 2.1.

> **TRY EXERCISE** 73

Only linear equations have graphs that are straight lines. Also, only linear graphs have a constant slope. Were you to try to calculate the slope between several pairs of points in Example 9, you would find that the slopes vary.

2.4   EXERCISE SET

↪ *Concept Reinforcement*   *Complete each of the following statements.*

1. Every _____ line has a slope of 0.

2. The graph of any function of the form $f(x) = b$ is a horizontal line that crosses the _____ at $(0, b)$.

3. The slope of a vertical line is _____.

4. The graph of any equation of the form $x = a$ is a(n) _____ line that crosses the x-axis at $(a, 0)$.

5. To find the x-intercept, we let $y =$ _____ and solve the original equation for _____.

6. To find the y-intercept, we let $x =$ _____ and solve the original equation for _____.

7. To solve $3x - 5 = 7$, we can graph $f(x) = 3x - 5$ and $g(x) = 7$ and find the x-value at the point of _____.

8. An equation like $4x + 3y = 8$ is said to be written in _____ form.

9. Only _____ equations have graphs that are straight lines.

10. Linear graphs have a constant _____.

For each equation, find the slope. If the slope is undefined, state this.

11. $y - 2 = 6$

12. $x + 3 = 11$

13. $8x = 6$

14. $y - 3 = 5$

15. $3y = 28$

16. $19 = -6y$

17. $5 - x = 12$

18. $-5x = 13$

19. $2x - 4 = 3$

20. $3 - 2y = 16$

21. $5y - 4 = 35$

22. $2x - 17 = 3$

23. $4y - 3x = 9 - 3x$

24. $x - 4y = 12 - 4y$

25. $5x - 2 = 2x - 7$

26. $5y + 3 = y + 9$

Aha! 27. $y = -\frac{2}{3}x + 5$

28. $y = -\frac{3}{2}x + 4$

Graph.

29. $y = 4$

30. $x = -1$

31. $x = 3$

32. $y = 2$

33. $f(x) = -2$

34. $g(x) = -3$

35. $3x = -15$

36. $2x = 10$

37. $3 \cdot g(x) = 15$

38. $3 - f(x) = 2$

Find the intercepts. Then graph by using the intercepts, if possible, and a third point as a check.

39. $x + y = 4$

40. $x + y = 5$

41. $f(x) = 2x - 6$

42. $f(x) = 3x + 12$

43. $3x + 5y = -15$

44. $5x - 4y = 20$

45. $2x - 3y = 18$

46. $3x + 2y = -18$

47. $3y = -12x$

48. $5y = 15x$

49. $f(x) = 3x - 7$

50. $g(x) = 2x - 9$

51. $5y - x = 5$

52. $y - 3x = 3$

53. $0.2y - 1.1x = 6.6$

54. $\frac{1}{3}x + \frac{1}{2}y = 1$

Solve each equation graphically. Then check your answer by solving the same equation algebraically.

55. $x + 2 = 3$

56. $x - 1 = 2$

57. $2x + 5 = 1$

58. $3x + 7 = 4$

59. $\frac{1}{2}x + 3 = 5$

60. $\frac{1}{3}x - 2 = 1$

61. $x - 8 = 3x - 5$

62. $x + 3 = 5 - x$

63. $4x + 1 = -x + 11$

64. $x + 4 = 3x + 5$

Use a graph to estimate the solution in each of the following. Be sure to use graph paper and a straightedge.

65. *Fitness centers.*   To become a member at Keeping Fit Club, it costs $75 plus a monthly fee of $35. Estimate how many months Kerry has been a member if he has paid a total of $215.

66. *Seminar costs.*   Efficiency Experts charges a $250 booking fee plus $150 per person for a one-day seminar. Estimate how many people attended a seminar if the total cost was $1600.

67. *Healthcare costs.*   Under one particular Aetna student health-insurance plan, an individual pays the first $250 of an emergency room visit plus $\frac{1}{5}$ of all charges in excess of $250. By approximately how much did Gerry's hospital bill exceed $250 if an emergency room visit cost him a total of $520?
Source: Indiana University Student Health Insurance Information (from the Chickering Group, an Aetna company)

68. *Text messaging.*   Recently, AT&T charged $15 per month for up to 1500 text messages plus 10¢ for each message in excess of 1500. One month Whitney's bill for text messages was $18.50. By approximately how many messages did her total exceed 1500?
Source: AT&T

69. *Parking fees.*   Cal's Parking charges $5.00 to park plus 50¢ for each 15-min unit of time. Estimate how long someone can park for $9.50.*

*More precise, nonlinear models of Exercises 69 and 70 appear in Exercises 107 and 106, respectively.

70. *Cost of a road call.*   Kay's Auto Village charges $50 for a road call plus $15 for each 15-min unit of time. Estimate the time required for a road call that cost $140.*

71. *Cost of a FedEx delivery.*   In 2007, for same-day delivery of packages weighing from 26 to 70 lb, FedEx charged $173 plus $1.25 for each pound in excess of 25 lb. Estimate the weight of a package that cost $223 to ship.
Source: FedEx Service Guide

72. *Printing costs.*   FedEx Kinko's charges $9.95 to resize art to fit a banner and $8 per square foot to print the banner. Estimate the area of a banner that cost a total of $153.95.
Source: FedEx Kinko's Customer Service

Determine whether each equation is linear. Find the slope of any nonvertical lines.

73. $5x - 3y = 15$

74. $3x + 5y + 15 = 0$

75. $8x + 40 = 0$

76. $2y - 30 = 0$

77. $4g(x) = 6x^2$

78. $2x + 4f(x) = 8$

79. $3y = 7(2x - 4)$

80. $y(3 - x) = 2$

81. $f(x) - \dfrac{5}{x} = 0$

82. $g(x) - x^3 = 0$

83. $\dfrac{y}{3} = x$

84. $\frac{1}{2}(x - 4) = y$

85. $xy = 10$

86. $y = \dfrac{10}{x}$

87. *Engineering.*   Wind friction, or *air resistance*, increases with speed. Following are some measurements made in a wind tunnel. Plot the data and explain why a linear function does or does not give an approximate fit.

Velocity (in kilometers per hour)	Force of Resistance (in newtons)
10	3
21	4.2
34	6.2
40	7.1
45	15.1
52	29.0

88. *Meteorology.*   Wind chill is a measure of how cold the wind makes you feel. Below are some measurements of wind chill for a 15-mph breeze. How can you tell from the data that a linear function will give an approximate fit?

Temperature	15-mph Wind Chill
30°F	19°F
25°F	13°F
20°F	6°F
15°F	0°F
10°F	−7°F
5°F	−13°F
0°F	−19°F

Source: National Oceanic & Atmospheric Administration, as reported in USA TODAY.com, 2004

Skill Review

To prepare for Section 2.5, review multiplying fractions and simplifying expressions (Sections 1.2 and 1.3).

Simplify.

89. $-\frac{3}{10}\left(\frac{10}{3}\right)$ [1.2]

90. $2\left(-\frac{1}{2}\right)$ [1.2]

91. $-3[x - (-1)]$ [1.3]

92. $-10[x - (-7)]$ [1.3]

93. $\frac{2}{3}\left[x - \left(-\frac{1}{2}\right)\right] - 1$ [1.3]

94. $-\frac{3}{2}\left(x - \frac{2}{5}\right) - 3$ [1.3]

Synthesis

95. Jim tries to avoid fractions as often as possible. Under what conditions will graphing using intercepts allow him to avoid fractions? Why?

96. Under what condition(s) will the *x*- and *y*-intercepts of a line coincide? What would the equation for such a line look like?

97. Give an equation, in standard form, for the line whose *x*-intercept is 5 and whose *y*-intercept is −4.

98. Find the *x*-intercept of $y = mx + b$, assuming that $m \neq 0$.

In Exercises 99–102, assume that r, p, and s are nonzero constants and that x and y are variables. Determine whether each equation is linear.

99. $rx + 3y = p^2 - s$ **100.** $py = sx - r^2y - 9$

101. $r^2x = py + 5$ **102.** $\frac{x}{r} - py = 17$

103. Suppose that two linear equations have the same *y*-intercept but that equation A has an *x*-intercept that is half the *x*-intercept of equation B. How do the slopes compare?

Consider the linear equation

$$ax + 3y = 5x - by + 8.$$

104. Find *a* and *b* if the graph is a horizontal line passing through (0, 4).

105. Find *a* and *b* if the graph is a vertical line passing through (4, 0).

106. (Refer to Exercise 70.) A 32-min road call with Kay's costs the same as a 44-min road call. Thus a linear graph drawn for the solution of Exercise 70 is not a precise representation of the situation. Draw a graph with a series of "steps" that more accurately reflects the situation.

107. (Refer to Exercise 69.) It costs as much to park at Cal's for 16 min as it does for 29 min. Thus a linear graph drawn for the solution of Exercise 69 is not a precise representation of the situation. Draw a graph with a series of "steps" that more accurately reflects the situation.

Solve graphically and then check by solving algebraically.

108. $5x + 3 = 7 - 2x$ **109.** $4x - 1 = 3 - 2x$

110. $3x - 2 = 5x - 9$ **111.** $8 - 7x = -2x - 5$

Solve using a graphing calculator.

112. Weekly pay at The Furniture Gallery is $250 plus a 3.5% sales commission. If a salesperson's pay was $401.03, what did that salesperson's sales total?

113. It costs Gert's Shirts $38 plus $4.25 a shirt to print tee shirts for a day camp. Camp Weehawken paid Gert's $671.25 for shirts. How many shirts were printed?

2.5   Other Equations of Lines

Point–Slope Form   ▪   Parallel and Perpendicular Lines

If we know the slope of a line and a point through which the line passes, then we can draw the line. With this information, we can also write an *equation* of the line.

Point–Slope Form

Suppose that a line of slope m passes through the point (x_1, y_1). For any other point (x, y) to lie on this line, we must have

$$\frac{y - y_1}{x - x_1} = m.$$

Note that for the point (x_1, y_1), the denominator will be 0. To address this concern, we multiply both sides by $x - x_1$:

$$(x - x_1)\frac{y - y_1}{x - x_1} = m(x - x_1)$$

$$y - y_1 = m(x - x_1).\qquad \text{This equation \emph{is} true for } (x_1, y_1).$$

Every point on the line is a solution of this equation. This is the **point–slope** form of a linear equation.

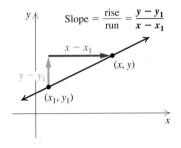

Slope $= \dfrac{\text{rise}}{\text{run}} = \dfrac{\boldsymbol{y - y_1}}{\boldsymbol{x - x_1}}$

> ### Point-Slope Form
>
> Any equation of the form $y - y_1 = m(x - x_1)$ is said to be written in *point–slope* form and has a graph that is a straight line.
>
> The slope of the line is m.
> The line passes through (x_1, y_1).

EXAMPLE 1   Find and graph an equation of the line passing through $(3, 4)$ with slope $-\frac{1}{2}$.

SOLUTION   We substitute in the point–slope equation:

$$y - y_1 = m(x - x_1)$$
$$y - 4 = -\tfrac{1}{2}(x - 3).\qquad \text{Substituting}$$

To graph this point–slope equation, we count off a slope of $-\frac{1}{2}$, starting at $(3, 4)$. Then we draw the line.

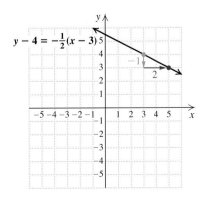

$y - 4 = -\frac{1}{2}(x - 3)$

STUDENT NOTES

To help remember point–slope form, many students begin by writing the equation for slope and then clearing fractions. The equation in Example 1 would then be found as

$$\frac{y - 4}{x - 3} = -\tfrac{1}{2},$$

and

$$y - 4 = -\tfrac{1}{2}(x - 3).$$

TRY EXERCISE   11

EXAMPLE **2**   Find a linear function that has a graph passing through the points $(-1, -5)$ and $(3, -2)$.

SOLUTION   We first determine the slope of the line and then write an equation in point–slope form. Note that

> Find the slope.

$$m = \frac{-5 - (-2)}{-1 - 3} = \frac{-3}{-4} = \frac{3}{4}.$$

Since the line passes through $(3, -2)$, we have

> Substitute the point and the slope in the point–slope form.

$$y - (-2) = \tfrac{3}{4}(x - 3) \qquad \text{Substituting into } y - y_1 = m(x - x_1)$$
$$y + 2 = \tfrac{3}{4}x - \tfrac{9}{4}. \qquad \text{Using the distributive law}$$

Before using function notation, we isolate y:

> Write in slope–intercept form.

$$y = \tfrac{3}{4}x - \tfrac{9}{4} - 2 \qquad \text{Subtracting 2 from both sides}$$
$$y = \tfrac{3}{4}x - \tfrac{17}{4} \qquad -\tfrac{9}{4} - \tfrac{8}{4} = -\tfrac{17}{4}$$
$$f(x) = \tfrac{3}{4}x - \tfrac{17}{4}. \qquad \text{Using function notation}$$

You can check that using $(-1, -5)$ as (x_1, y_1) in $y - y_1 = \tfrac{3}{4}(x - x_1)$ will yield the same expression for $f(x)$.

TRY EXERCISE ▶ 37

EXAMPLE **3**   *Fossil-fuel emissions.* Worldwide carbon-dioxide emissions in 2004 were 27 billion metric tons. If no changes are made to current practices, this amount is expected to grow to 31 billion metric tons in 2010. Assuming constant growth since 2000, what will worldwide carbon-dioxide emissions be in 2015?

Source: U.S. Energy Information Administration

SOLUTION

1. **Familiarize.** Constant growth indicates a constant rate of change, so we can assume a linear relationship. If we let c represent the amount of carbon-dioxide emissions, in billions of metric tons, and t the number of years after 2000, we can form the pairs $(4, 27)$ and $(10, 31)$. After choosing suitable scales on the two axes, we draw the graph.

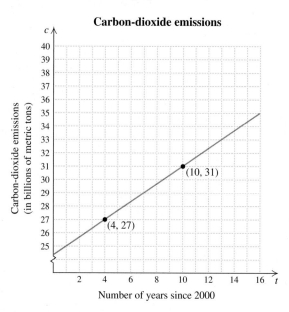

Carbon-dioxide emissions

2. **Translate.** To find an equation relating c and t, we first find the slope of the line. This corresponds to the *growth rate*:

$$m = \frac{31 \text{ billion metric tons} - 27 \text{ billion metric tons}}{10 \text{ years} - 4 \text{ years}}$$

$$= \frac{4 \text{ billion metric tons}}{6 \text{ years}}$$

$$= \frac{2}{3} \text{ billion metric ton per year.}$$

Next, we write point–slope form and solve for c:

$$c - 27 = \tfrac{2}{3}(t - 4) \qquad \text{Using } (4, 27) \text{ to write point–slope form}$$
$$c - 27 = \tfrac{2}{3}t - \tfrac{8}{3} \qquad \text{Using the distributive law}$$
$$c = \tfrac{2}{3}t + 24\tfrac{1}{3}. \qquad -\tfrac{8}{3} + 27 = 27 - \tfrac{8}{3} = 27 - 2\tfrac{2}{3} = 24\tfrac{1}{3}$$

3. **Carry out.** Using function notation, we have

$$c(t) = \tfrac{2}{3}t + 24\tfrac{1}{3}.$$

To predict the amount of carbon-dioxide emissions in 2015, we find

$$c(15) = \tfrac{2}{3} \cdot 15 + 24\tfrac{1}{3} \qquad \text{2015 is 15 yr after 2000.}$$
$$= 34\tfrac{1}{3}. \qquad \tfrac{2}{3} \cdot 15 = 10$$

4. **Check.** To check, we can repeat our calculations. We could also extend the graph to see whether $(15, 34\tfrac{1}{3})$ appears to be on the line.

5. **State.** Assuming constant growth, worldwide carbon-dioxide emissions will reach $34\tfrac{1}{3}$ billion metric tons by 2015. `TRY EXERCISE` ▸ 45

Parallel and Perpendicular Lines

Two lines are parallel if they lie in the same plane and do not intersect no matter how far they are extended. If two lines are vertical, they are parallel. How can we tell if nonvertical lines are parallel? The answer is simple: We look at their slopes (see Examples 1 and 2 in Section 2.3).

> **Slope and Parallel Lines**
>
> Two lines are parallel if they have the same slope or if both lines are vertical.

EXAMPLE 4   Determine whether the line given by $f(x) = -3x + 4.2$ is parallel to the line given by $6x + 2y = 1$.

SOLUTION   We find the slope of each line. If the slopes are the same, the lines are parallel.

The slope of $f(x) = -3x + 4.2$ is -3.

To find the slope of $6x + 2y = 1$, we write the equation in slope–intercept form:

$$6x + 2y = 1$$
$$2y = -6x + 1 \qquad \text{Subtracting } 6x \text{ from both sides}$$
$$y = -3x + \tfrac{1}{2}. \qquad \text{Dividing both sides by 2}$$

The slope of the second line is -3. Since the slopes are equal, the lines are parallel.

`TRY EXERCISE` ▸ 57

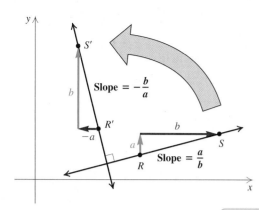

Two lines are perpendicular if they intersect at a right angle. If one line is vertical and another is horizontal, they are perpendicular. There are other instances in which two lines are perpendicular.

Consider a line $\overleftrightarrow{RS}$ as shown at left, with slope a/b. Then think of rotating the figure 90° to get a line $\overleftrightarrow{R'S'}$ perpendicular to $\overleftrightarrow{RS}$. For the new line, the rise and the run are interchanged, but the run is now negative. Thus the slope of the new line is $-b/a$. Let's multiply the slopes:

$$\frac{a}{b}\left(-\frac{b}{a}\right) = -1.$$

This can help us determine which lines are perpendicular.

> ### Slope and Perpendicular Lines
>
> Two lines are perpendicular if the product of their slopes is -1 or if one line is vertical and the other line is horizontal.

Thus, if one line has slope m ($m \neq 0$), the slope of a line perpendicular to it is $-1/m$. That is, we take the reciprocal of m ($m \neq 0$) and change the sign.

EXAMPLE **5** Consider the line given by the equation $8y = 7x - 24$.

a) Find an equation for a parallel line passing through $(-1, 2)$.

b) Find an equation for a perpendicular line passing through $(-1, 2)$.

SOLUTION Both parts (a) and (b) require us to find the slope of the line given by $8y = 7x - 24$. To do so, we solve for y to find slope–intercept form:

$$8y = 7x - 24$$
$$y = \tfrac{7}{8}x - 3. \qquad \text{Multiplying both sides by } \tfrac{1}{8}$$
$$\text{The slope is } \tfrac{7}{8}.$$

a) The slope of any parallel line will be $\tfrac{7}{8}$. The point–slope equation yields

$$y - 2 = \tfrac{7}{8}[x - (-1)] \qquad \text{Substituting } \tfrac{7}{8} \text{ for the slope and} \\ (-1, 2) \text{ for the point}$$
$$y - 2 = \tfrac{7}{8}[x + 1]$$
$$y = \tfrac{7}{8}x + \tfrac{7}{8} + 2 \qquad \text{Using the distributive law and} \\ \text{adding 2 to both sides}$$
$$y = \tfrac{7}{8}x + \tfrac{23}{8}.$$

b) The slope of a perpendicular line is given by the opposite of the reciprocal of $\tfrac{7}{8}$, or $-\tfrac{8}{7}$. The point–slope equation yields

$$y - 2 = -\tfrac{8}{7}[x - (-1)] \qquad \text{Substituting } -\tfrac{8}{7} \text{ for the slope and} \\ (-1, 2) \text{ for the point}$$
$$y - 2 = -\tfrac{8}{7}[x + 1]$$
$$y = -\tfrac{8}{7}x - \tfrac{8}{7} + 2 \qquad \text{Using the distributive law and} \\ \text{adding 2 to both sides}$$
$$y = -\tfrac{8}{7}x + \tfrac{6}{7}.$$

> TRY EXERCISE 63

TECHNOLOGY CONNECTION

To check that the graphs of $y = \tfrac{7}{8}x - 3$ and $y = -\tfrac{8}{7}x + \tfrac{6}{7}$ are perpendicular, we use the ZSQUARE option of the ZOOM menu to create a "squared" window. This corrects for distortion that would result from the units on the axes being of different lengths.

1. Show that the graphs of $y = \tfrac{3}{4}x + 2$ and $y = -\tfrac{4}{3}x - 1$ appear to be perpendicular.

2. Show that the graphs of $y = -\tfrac{2}{5}x - 4$ and $y = \tfrac{5}{2}x + 3$ appear to be perpendicular.

3. To see that this type of check is not foolproof, graph

$$y = \tfrac{31}{40}x + 2$$

and $y = -\tfrac{40}{30}x - 1.$

Are the lines perpendicular? Why or why not?

Visualizing for Success

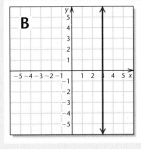

A

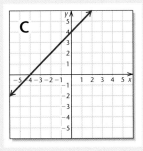

B

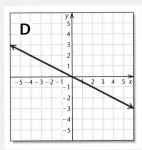

C

D

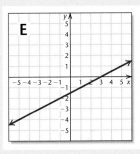

E

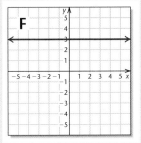

F

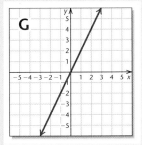

G

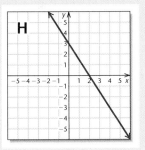

H

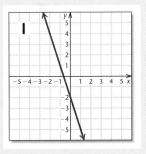

I

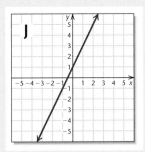

J

Match each equation or function with its graph.

1. $y = x + 4$

2. $y = 2x$

3. $y = 3$

4. $x = 3$

5. $f(x) = -\frac{1}{2}x$

6. $2x - 3y = 6$

7. $f(x) = -3x - 2$

8. $3x + 2y = 6$

9. $y - 3 = 2(x - 1)$

10. $y + 2 = \frac{1}{2}(x + 1)$

Answers on page A-8

An alternate, animated version of this activity appears in MyMathLab. To use MyMathLab, you need a course ID and a student access code. Contact your instructor for more information.

2.5 **EXERCISE SET**

✋ *Concept Reinforcement* *Classify each statement as either true or false.*

1. The equation $y = -3x - 1$ is written in point–slope form.

2. The equation $y - 4 = -3(x - 1)$ is written in point–slope form.

3. Knowing the coordinates of just one point on a line is enough to write an equation of the line.

4. Knowing the slope and the coordinates of just one point on a line is enough to write an equation of the line.

5. Knowing the coordinates of just two points on a line is enough to write an equation of the line.

6. If two lines are perpendicular, then they have the same slope.

7. If the product of the slopes of two lines is 1, then the lines are perpendicular.

8. If two nonvertical lines are parallel, then they have the same slope.

9. If two nonvertical lines are perpendicular, then the product of their slopes is -1.

10. Point–slope form can be used with either point that is used to calculate the slope of that line.

Find an equation in point–slope form of the line having the specified slope and containing the point indicated. Then graph the line.

11. $m = 3$, $(5, 2)$

12. $m = 2$, $(3, 4)$

13. $m = -4$, $(1, 2)$

14. $m = -5$, $(1, 4)$

15. $m = \frac{1}{2}$, $(-2, -4)$

16. $m = 1$, $(-5, -7)$

17. $m = -1$, $(8, 0)$

18. $m = -3$, $(-2, 0)$

For each point–slope equation listed, state the slope and a point on the graph.

19. $y - 3 = \frac{1}{4}(x - 5)$

20. $y - 5 = 6(x - 1)$

21. $y + 1 = -7(x - 2)$

22. $y - 4 = -\frac{2}{3}(x + 8)$

23. $y - 6 = -\frac{10}{3}(x + 4)$

24. $y + 1 = -9(x - 7)$

Aha! 25. $y = 5x$

26. $y = \frac{4}{5}x$

Find an equation of the line having the specified slope and containing the indicated point. Write your final answer as a linear function in slope–intercept form. Then graph the line.

27. $m = 2$, $(1, -4)$

28. $m = -4$, $(-1, 5)$

29. $m = -\frac{3}{5}$, $(-4, 8)$

30. $m = -\frac{1}{5}$, $(-2, 1)$

31. $m = -0.6$, $(-3, -4)$

32. $m = 2.3$, $(4, -5)$

Aha! 33. $m = \frac{2}{7}$, $(0, -6)$

34. $m = \frac{1}{4}$, $(0, 3)$

35. $m = \frac{3}{5}$, $(-4, 6)$

36. $m = -\frac{2}{7}$, $(6, -5)$

Find an equation of the line containing each pair of points. Write your final answer as a linear function in slope–intercept form.

37. $(2, 3)$ and $(3, 7)$

38. $(3, 8)$ and $(1, 4)$

39. $(1.2, -4)$ and $(3.2, 5)$

40. $(-1, -2.5)$ and $(4, 8.5)$

Aha! 41. $(2, -5)$ and $(0, -1)$

42. $(-2, 0)$ and $(0, -7)$

43. $(-6, -10)$ and $(-3, -5)$

44. $(-1, -3)$ and $(-4, -9)$

In Exercises 45–56, assume that a constant rate of change exists for each model formed.

45. *Automobile production.* As demand has grown, worldwide production of small cars rose from 14.5 million in 2002 to 19 million in 2007. Let $a(t)$ represent the number of small cars produced t years after 2000.
Source: *The Wall Street Journal*, 10/22/07

   a) Find a linear function that fits the data.

   b) Use the function from part (a) to predict the number of small cars produced in 2013.

   c) In what year will 25 million small cars be produced?

46. *Convention attendees.* In recent years, Las Vegas has become a popular location for conventions. The number of convention attendees in Las Vegas rose from 4.6 million in 2002 to 6.1 million in 2006. Let $v(t)$ represent the number of convention attendees in Las Vegas t years after 2000.
Source: Las Vegas Convention and Visitors Authority

a) Find a linear function that fits the data.

b) Use the function from part (a) to predict the number of convention attendees in Las Vegas in 2011.

c) In what year will there be 8 million convention attendees in Las Vegas?

47. *Life expectancy of females in the United States.* In 1994, the life expectancy of females was 79.0 yr. In 2004, it was 80.4 yr. Let $E(t)$ represent life expectancy and t the number of years since 1990.
Source: *Statistical Abstract of the United States*, 2007

a) Find a linear function that fits the data.

Aha! **b)** Use the function of part (a) to predict the life expectancy of females in 2012.

48. *Life expectancy of males in the United States.* In 1994, the life expectancy of males was 72.4 yr. In 2004, it was 75.2 yr. Let $E(t)$ represent life expectancy and t the number of years since 1990.
Source: *Statistical Abstract of the United States*, 2007

a) Find a linear function that fits the data.

b) Use the function of part (a) to predict the life expectancy of males in 2012.

49. *PAC contributions.* In 2002, Political Action Committees (PACs) contributed $282 million to federal candidates. In 2006, the figure rose to $372.1 million. Let $A(t)$ represent the amount of PAC contributions, in millions, and t the number of years since 2000.
Source: Federal Election Commission

PAC contributions to federal candidates

2002 $282 million

2006 $372.1 million

a) Find a linear function that fits the data.

b) Use the function of part (a) to predict the amount of PAC contributions in 2010.

50. *Consumer demand.* Suppose that 6.5 million lb of coffee are sold when the price is $8 per pound, and 4.0 million lb are sold when it is $9 per pound.

a) Find a linear function that expresses the amount of coffee sold as a function of the price per pound.

b) Use the function of part (a) to predict how much consumers would be willing to buy at a price of $6 per pound.

51. *Recycling.* In 2000, Americans recycled 52.7 million tons of solid waste. In 2005, the figure grew to 58.4 million tons. Let $N(t)$ represent the number of tons recycled, in millions, and t the number of years since 2000.
Sources: U.S. EPA; Franklin Associates, Ltd.

a) Find a linear function that fits the data.

b) Use the function of part (a) to predict the amount recycled in 2012.

52. *Seller's supply.* Suppose that suppliers are willing to sell 5.0 million lb of coffee at a price of $8 per pound and 7.0 million lb at $9 per pound.

a) Find a linear function that expresses the amount suppliers are willing to sell as a function of the price per pound.

b) Use the function of part (a) to predict how much suppliers would be willing to sell at a price of $6 per pound.

53. *Online banking.* In 2000, about 16 million Americans conducted at least some of their banking online. By 2005, that number had risen to about 63 million. Let $N(t)$ represent the number of Americans using online banking, in millions, t years after 2000.

a) Find a linear function that fits the data.

Aha! **b)** Use the function of part (a) to predict the number of Americans who will use online banking in 2010.

c) In what year will 157 million Americans use online banking?

54. *Records in the 100-meter run.* In 1999, the record for the 100-m run was 9.79 sec. In 2007, it was 9.77 sec. Let $R(t)$ represent the record in the 100-m run and t the number of years since 1999.
Sources: International Association of Athletics Federation; *Guinness World Records*

a) Find a linear function that fits the data.

b) Use the function of part (a) to predict the record in 2015 and in 2030.

c) When will the record be 9.6 sec?

55. *National Park land.*   In 1994, the National Park system consisted of about 74.9 million acres. By 2005, the figure had grown to 79 million acres. Let $A(t)$ represent the amount of land in the National Park system, in millions of acres, t years after 1990.

Source: *Statistical Abstract of the United States*, 2007

 a) Find a linear function that fits the data.
 b) Use the function of part (a) to predict the amount of land in the National Park system in 2010.

56. *Pressure at sea depth.*   The pressure 100 ft beneath the ocean's surface is approximately 4 atm (atmospheres), whereas at a depth of 200 ft, the pressure is about 7 atm.

 a) Find a linear function that expresses pressure as a function of depth.
 b) Use the function of part (a) to determine the pressure at a depth of 690 ft.

Without graphing, tell whether the graphs of each pair of equations are parallel.

57. $x + 2 = y,$
   $y - x = -2$

58. $2x - 1 = y,$
   $2y - 4x = 7$

59. $y + 9 = 3x,$
   $3x - y = -2$

60. $y + 8 = -6x,$
   $-2x + y = 5$

61. $f(x) = 3x + 9,$
   $2y = 8x - 2$

62. $f(x) = -7x - 9,$
   $-3y = 21x + 7$

Write an equation of the line containing the specified point and parallel to the indicated line.

63. $(2, 5),\ x - 2y = 3$

64. $(1, 4),\ 3x + y = 5$

65. $(-3, 2),\ x + y = 7$

66. $(-1, -6),\ x - 5y = 1$

Aha! **67.** $(0, -5),\ y = 4x + 3$

68. $(0, 2),\ y = x - 11$

69. $(-2, -3),\ 2x + 3y = -7$

70. $(3, -4),\ 5x - 6y = 4$

71. $(-6, 2),\ 3x - 9y = 2$

72. $(-7, 0),\ 5x + 2y = 6$

73. $(5, -4),\ x = 2$

74. $(-3, 6),\ y = 7$

Without graphing, tell whether the graphs of each pair of equations are perpendicular.

75. $x - 2y = 3,$
   $4x + 2y = 1$

76. $2x - 5y = -3,$
   $2x + 5y = 4$

77. $f(x) = 3x + 1,$
   $6x + 2y = 5$

78. $y = -x + 7,$
   $f(x) = x + 3$

Write an equation of the line containing the specified point and perpendicular to the indicated line.

79. $(3, 1),\ 2x - 3y = 4$

80. $(6, 0),\ 5x + 4y = 1$

81. $(-4, 2),\ x + y = 6$

82. $(-2, -5),\ x - 2y = 3$

83. $(1, -3),\ 3x - y = 2$

84. $(-5, 6),\ 4x - y = 3$

85. $(-4, -7),\ 3x - 5y = 6$

86. $(-4, 5),\ 7x - 2y = 1$

Aha! **87.** $(0, 6),\ 2x - 5 = y$

88. $(0, -7),\ 4x + 3 = y$

89. $(-3, 7),\ y = 5$

90. $(4, -2),\ x = 1$

91. Suppose that you are given the coordinates of two points on a line, and one of those points is the *y*-intercept. What method would you use to find an equation for the line? Explain the reasoning behind your choice.

92. If two lines are perpendicular, does it follow that the lines have slopes that are negative reciprocals of each other? Why or why not?

Skill Review

To prepare for Section 2.6, review simplifying expressions and finding the domain of a function (Sections 1.3 and 2.2).

Simplify. [1.3]

93. $(2x^2 - x) + (3x - 5)$

94. $(4t + 3) - (6t + 7)$

95. $(2t - 1) - (t - 3)$

96. $(5x^2 - 4) - (9x^2 - 7x)$

Find the domain of each function. [2.2]

97. $f(x) = \dfrac{x}{x - 3}$

98. $g(x) = x^2 - 1$

99. $g(x) = |6x + 11|$

100. $f(x) = \dfrac{x - 7}{2x}$

Synthesis

101. In 2004, Political Action Committees contributed $310.5 million to federal candidates. Does this information make your answer to Exercise 49(b) seem too low or too high? Why?

102. On the basis of your answers to Exercises 47 and 48, would you predict that at some point in the future the life expectancy of males will exceed that of females? Why or why not?

For Exercises 103–107, assume that a linear equation models each situation.

103. *Temperature conversion.*   Water freezes at 32° Fahrenheit and at 0° Celsius. Water boils at 212°F and at 100°C. What Celsius temperature corresponds to a room temperature of 70°F?

104. *Depreciation of a computer.*   After 6 mos of use, the value of Don's computer had dropped to $900. After 8 mos, the value had gone down to $750. How much did the computer cost originally?

105. *Cell-phone charges.*   The total cost of Tam's cell phone was $410 after 5 mos of service and $690 after 9 mos. What costs had Tam already incurred when her service just began? Assume that Tam's monthly charge is constant.

106. *Operating expenses.*   The total cost for operating Ming's Wings was $7500 after 4 mos and $9250 after 7 mos. Predict the total cost after 10 mos.

107. Based on the information given in Exercises 50 and 52, at what price will the supply equal the demand?

108. Specify the domain of your answer to Exercise 50(a).

109. Specify the domain of your answer to Exercise 52(a).

110. For a linear function g, $g(3) = -5$ and $g(7) = -1$.
 a) Find an equation for g.
 b) Find $g(-2)$.
 c) Find a such that $g(a) = 75$.

111. Find the value of k such that the graph of $5y - kx = 7$ and the line containing the points $(7, -3)$ and $(-2, 5)$ are parallel.

112. Find the value of k such that the graph of $7y - kx = 9$ and the line containing the points $(2, -1)$ and $(-4, 5)$ are perpendicular.

113. When several data points are available and they appear to be nearly collinear, a procedure known as *linear regression* can be used to find an equation for the line that most closely fits the data.
 a) Use a graphing calculator with a LINEAR REGRESSION option and the table that follows to find a linear

function that predicts the wattage of a CFL (compact fluorescent) lightbulb as a function of the wattage of a standard incandescent bulb of equivalent brightness. Round coefficients to the nearest thousandth.

Energy Conservation

Incandescent Wattage	CFL Equivalent
25 W	5 W
50 W	9 W
60 W	15 W
100 W	25 W
120 W	28 W

Source: U.S. Department of Energy

 b) Use the function from part (a) to estimate the CFL wattage that is equivalent to a 75-watt incandescent bulb. Then compare your answer with the corresponding answer to Exercise 87 in Section 2.2. Which answer seems more reliable? Why?

114. Use linear regression (see Exercise 113) to find a linear function that predicts a woman's life expectancy as a function of the year in which she was born. Round coefficients to the nearest thousandth. Then use the function to predict the life expectancy in 2012 and compare this with the corresponding answer to Exercise 47 of this exercise set. Which answer seems more reliable? Why?

Life Expectancy of Women

Year, x	Life Expectancy, y (in years)
1920	54.6
1930	61.6
1940	65.2
1950	71.1
1960	73.1
1970	74.7
1980	77.5
1990	78.8
2000	79.5

Source: Statistical Abstract of the United States

115. Use a graphing calculator with a *squared* window to check your answers to Exercises 75–88.

CONNECTING the CONCEPTS

Any line can be described by a number of equivalent equations. For example, all four of the equations below describe the given line.

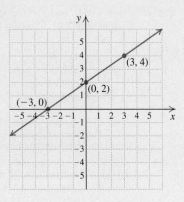

$$y = \tfrac{2}{3}x + 2,$$
$$2x - 3y = -6,$$
$$y - 4 = \tfrac{2}{3}(x - 3),$$
$$2x + 6 = 3y$$

Form of a Linear Equation	Example	Uses
Slope–intercept form: $$y = mx + b \quad \text{or}$$ $$f(x) = mx + b$$	$f(x) = \tfrac{1}{2}x + 6$	Finding slope and y-intercept Graphing using slope and y-intercept Writing an equation given slope and y-intercept Writing linear functions
Standard form: $$Ax + By = C$$	$5x - 3y = 7$	Finding x- and y-intercepts Graphing using intercepts Solving systems of equations (see Chapter 3)
Point–slope form: $$y - y_1 = m(x - x_1)$$	$y - 2 = \tfrac{4}{5}(x - 1)$	Finding slope and a point on the line Graphing using slope and a point on the line Writing an equation given slope and a point on the line or given two points on the line Working with curves and tangents in calculus

MIXED REVIEW

State whether each equation is in slope–intercept form, standard form, point–slope form, or none of these.

1. $2x + 5y = 8$

2. $y = \tfrac{2}{3}x - \tfrac{11}{3}$

3. $x - 13 = 5y$

4. $y - 2 = \tfrac{1}{3}(x - 6)$

5. $x - y = 1$

6. $y = -18x + 3.6$

Write each equation in standard form.

7. $y = \tfrac{2}{5}x + 1$

8. $y - 1 = -2(x - 6)$

Write each equation in slope–intercept form.

9. $3x - 5y = 10$

10. $y + 2 = \tfrac{1}{2}(x - 3)$

Graph.

11. $y = 2x - 1$ **12.** $3x + y = 6$

13. $y - 2 = \frac{1}{2}(x - 1)$

14. $f(x) = 4$

15. $f(x) = -\frac{3}{4}x + 5$

16. Determine the slope and the y-intercept of the line given by $x - 3y = 1$.

17. Find a linear function whose graph has slope -3 and y-intercept 7.

18. Find an equation in point–slope form of the line with slope 5 that contains the point $(-3, 7)$.

19. Find an equation in slope–intercept form of the line with slope $\frac{2}{3}$ that contains the point $(0, -8)$.

20. Find an equation in slope–intercept form of the line containing the points $(4, -1)$ and $(-2, -5)$.

2.6   The Algebra of Functions

The Sum, Difference, Product, or Quotient of Two Functions • Domains and Graphs

We now examine four ways in which functions can be combined.

The Sum, Difference, Product, or Quotient of Two Functions

Suppose that a is in the domain of two functions, f and g. The input a is paired with $f(a)$ by f and with $g(a)$ by g. The outputs can then be added to get $f(a) + g(a)$.

EXAMPLE 1   Let $f(x) = x + 4$ and $g(x) = x^2 + 1$. Find $f(2) + g(2)$.

SOLUTION   We visualize two function machines. Because 2 is in the domain of each function, we can compute $f(2)$ and $g(2)$.

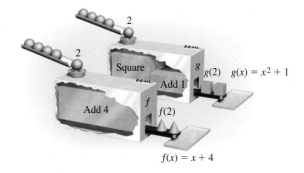

Since

$$f(2) = 2 + 4 = 6 \quad \text{and} \quad g(2) = 2^2 + 1 = 5,$$

we have

$$f(2) + g(2) = 6 + 5 = 11.$$

TRY EXERCISE 7

In Example 1, suppose that we were to write $f(x) + g(x)$ as $(x + 4) + (x^2 + 1)$, or $f(x) + g(x) = x^2 + x + 5$. This could then be regarded as a "new" function. The notation $(f + g)(x)$ is generally used to denote a function formed in this manner. Similar notations exist for subtraction, multiplication, and division of functions.

The Algebra of Functions

If f and g are functions and x is in the domain of both functions, then:

1. $(f + g)(x) = f(x) + g(x)$;
2. $(f - g)(x) = f(x) - g(x)$;
3. $(f \cdot g)(x) = f(x) \cdot g(x)$;
4. $(f/g)(x) = f(x)/g(x)$, provided $g(x) \neq 0$.

EXAMPLE 2   For $f(x) = x^2 - x$ and $g(x) = x + 2$, find the following.

a) $(f + g)(4)$

b) $(f - g)(x)$ and $(f - g)(-1)$

c) $(f/g)(x)$ and $(f/g)(-4)$

d) $(f \cdot g)(4)$

SOLUTION

a) Since $f(4) = 4^2 - 4 = 12$ and $g(4) = 4 + 2 = 6$, we have

$$(f + g)(4) = f(4) + g(4)$$
$$= 12 + 6 \quad \text{Substituting}$$
$$= 18.$$

Alternatively, we could first find $(f + g)(x)$:

$$(f + g)(x) = f(x) + g(x)$$
$$= x^2 - x + x + 2$$
$$= x^2 + 2. \quad \text{Combining like terms}$$

Thus,

$$(f + g)(4) = 4^2 + 2 = 18. \quad \text{Our results match.}$$

b) We have

$$(f - g)(x) = f(x) - g(x)$$
$$= x^2 - x - (x + 2) \quad \text{Substituting}$$
$$= x^2 - 2x - 2. \quad \begin{array}{l}\text{Removing parentheses and}\\\text{combining like terms}\end{array}$$

Thus,

$$(f - g)(-1) = (-1)^2 - 2(-1) - 2 \quad \begin{array}{l}\text{Using } (f - g)(x) \text{ is faster than}\\\text{using } f(x) - g(x).\end{array}$$

$$= 1. \quad \text{Simplifying}$$

c) We have

$$(f/g)(x) = f(x)/g(x)$$
$$= \frac{x^2 - x}{x + 2}. \quad \text{We assume that } x \neq -2.$$

Thus,

$$(f/g)(-4) = \frac{(-4)^2 - (-4)}{-4 + 2}$$ Substituting

$$= \frac{20}{-2} = -10.$$

d) Using our work in part (a), we have

$$(f \cdot g)(4) = f(4) \cdot g(4)$$
$$= 12 \cdot 6$$
$$= 72.$$

It is also possible to compute $(f \cdot g)(4)$ by first multiplying $x^2 - x$ and $x + 2$ using methods we will discuss in Chapter 5.

> TRY EXERCISE > 17

Domains and Graphs

Although applications involving products and quotients of functions rarely appear in newspapers, situations involving sums or differences of functions often do appear in print. For example, the following graphs are similar to those published by the California Department of Education to promote breakfast programs in which students eat a balanced meal of fruit or juice, toast or cereal, and 2% or whole milk. The combination of carbohydrate, protein, and fat gives a sustained release of energy, delaying the onset of hunger for several hours.

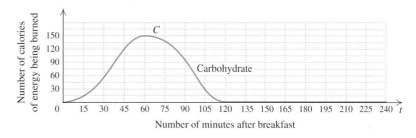

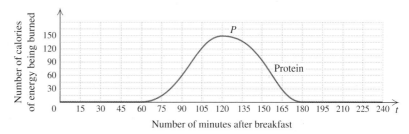

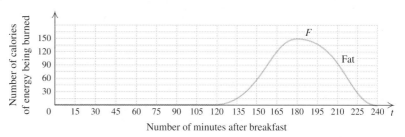

When the three graphs are superimposed, and the calorie expenditures added, it becomes clear that a balanced meal results in a steady, sustained supply of energy.

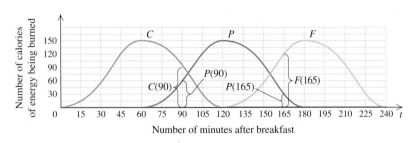

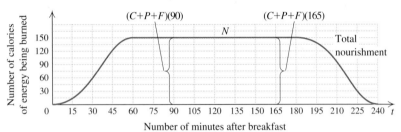

For any point $(t, N(t))$, we have

$$N(t) = (C + P + F)(t) = C(t) + P(t) + F(t).$$

To find $(f + g)(a)$, $(f - g)(a)$, $(f \cdot g)(a)$, or $(f/g)(a)$, we must know that $f(a)$ and $g(a)$ exist. This means a must be in the domain of both f and g.

EXAMPLE 3   Let

$$f(x) = \frac{5}{x} \quad \text{and} \quad g(x) = \frac{2x - 6}{x + 1}.$$

Find the domain of $f + g$, the domain of $f - g$, and the domain of $f \cdot g$.

SOLUTION   Note that because division by 0 is undefined, we have

Domain of $f = \{x \,|\, x \text{ is a real number } and \ x \neq 0\}$

and

Domain of $g = \{x \,|\, x \text{ is a real number } and \ x \neq -1\}$.

In order to find $f(a) + g(a)$, $f(a) - g(a)$, or $f(a) \cdot g(a)$, we must know that a is in *both* of the above domains. Thus,

Domain of $f + g$ = Domain of $f - g$ = Domain of $f \cdot g$

$$= \{x \,|\, x \text{ is a real number } and \ x \neq 0 \ and \ x \neq -1\}.$$

▶ TRY EXERCISE 43

Suppose that for $f(x) = x^2 - x$ and $g(x) = x + 2$, we want to find $(f/g)(-2)$. Finding $f(-2)$ and $g(-2)$ poses no problem:

$$f(-2) = 6 \quad \text{and} \quad g(-2) = 0;$$

but then

$$(f/g)(-2) = f(-2)/g(-2)$$
$$= 6/0. \quad \text{Division by 0 is undefined.}$$

Thus, although -2 is in the domain of both f and g, it is not in the domain of f/g.

We can also see this by writing $(f/g)(x)$:

$$(f/g)(x) = \frac{f(x)}{g(x)} = \frac{x^2 - x}{x + 2}.$$

Since $x + 2 = 0$ when $x = -2$, the domain of f/g must exclude -2.

Determining the Domain

The domain of $f + g$, $f - g$, or $f \cdot g$ is the set of all values common to the domains of f and g.

The domain of f/g is the set of all values common to the domains of f and g, excluding any values for which $g(x)$ is 0.

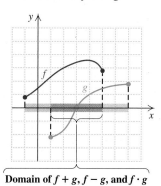

Domain of $f + g$, $f - g$, and $f \cdot g$

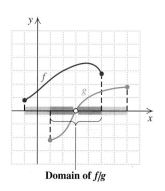

Domain of f/g

EXAMPLE   4

Given $f(x) = 1/x$ and $g(x) = 2x - 7$, find the domains of $f + g$, $f - g$, $f \cdot g$, and f/g.

SOLUTION   We first find the domain of f and the domain of g:

The domain of f is $\{x \,|\, x$ is a real number *and* $x \neq 0\}$.

The domain of g is $\mathbb{R}$.

The domains of $f + g$, $f - g$, and $f \cdot g$ are the set of all elements common to the domains of f and g. This consists of all real numbers except 0.

The domain of $f + g =$ the domain of $f - g =$ the domain of $f \cdot g$
$$= \{x \,|\, x \text{ is a real number } and \ x \neq 0\}.$$

Because we cannot divide by 0, the domain of f/g must also exclude any values of x for which $g(x)$ is 0. We determine those values by solving $g(x) = 0$:

$$g(x) = 0$$
$$2x - 7 = 0 \qquad \text{Replacing } g(x) \text{ with } 2x - 7$$
$$2x = 7$$
$$x = \tfrac{7}{2}.$$

The domain of f/g is the domain of the sum, difference, and product of f and g, found above, excluding $\tfrac{7}{2}$.

The domain of $f/g = \left\{x \,\middle|\, x \text{ is a real number } and \ x \neq 0 \ and \ x \neq \tfrac{7}{2}\right\}$.

TRY EXERCISE   55

STUDENT NOTES

The concern over a denominator being 0 arises throughout this course. Try to develop the habit of checking for any possible input values that would create a denominator of 0 whenever you work with functions.

TECHNOLOGY CONNECTION

A partial check of Example 4 can be performed by setting up a table so the TBLSTART is 0 and the increment of change (ΔTbl) is 0.7. (Other choices, like 0.1, will also work.) Next, we let $y_1 = 1/x$ and $y_2 = 2x - 7$. Using Y-VARS to write $y_3 = y_1 + y_2$ and $y_4 = y_1/y_2$, we can create the table of values shown here. Note that when x is 3.5, a value for y_3 can be found, but y_4 is undefined. If we "de-select" y_1 and y_2 as we enter them, the columns for y_3 and y_4 appear without scrolling through the table.

X	Y3	Y4
0	ERROR	ERROR
.7	−4.171	−.2551
1.4	−3.486	−.1701
2.1	−2.324	−.1701
2.8	−1.043	−.2551
3.5	.28571	ERROR
4.2	1.6381	.17007
X = 0		

Use a similar approach to partially check Example 3.

Division by 0 is not the only condition that can force restrictions on the domain of a function. In Chapter 7, we will examine functions similar to that given by $f(x) = \sqrt{x}$, for which the concern is taking the square root of a negative number.

2.6   EXERCISE SET

Concept Reinforcement   *Make each of the following sentences true by selecting the correct word for each blank.*

1. If f and g are functions, then $(f + g)(x)$ is the _____ of the functions.
   sum/difference

2. One way to compute $(f - g)(2)$ is to _____ $g(2)$ from $f(2)$.
   erase/subtract

3. One way to compute $(f - g)(2)$ is to simplify $f(x) - g(x)$ and then _____ the result
   evaluate/substitute
   for $x = 2$.

4. The domain of $f + g$, $f - g$, and $f \cdot g$ is the set of all values common to the _____ of f and g.
   domains/ranges

5. The domain of f/g is the set of all values common to the domains of f and g, _____ any
   including/excluding
   values for which $g(x)$ is 0.

6. The height of $(f + g)(a)$ on a graph is the _____ of the heights of $f(a)$ and $g(a)$.
   product/sum

Let $f(x) = -2x + 3$ and $g(x) = x^2 - 5$. Find each of the following.

7. $f(3) + g(3)$ 8. $f(4) + g(4)$

9. $f(1) - g(1)$ 10. $f(2) - g(2)$

11. $f(-2) \cdot g(-2)$ 12. $f(-1) \cdot g(-1)$

13. $f(-4)/g(-4)$ 14. $f(3)/g(3)$

15. $g(1) - f(1)$ 16. $g(-3)/f(-3)$

17. $(f + g)(x)$ 18. $(g - f)(x)$

Let $F(x) = x^2 - 2$ and $G(x) = 5 - x$. Find each of the following.

19. $(F + G)(x)$ 20. $(F + G)(a)$

21. $(F - G)(3)$ 22. $(F - G)(2)$

23. $(F \cdot G)(-3)$ 24. $(F \cdot G)(-4)$

25. $(F/G)(x)$ 26. $(G - F)(x)$

27. $(G/F)(-2)$ 28. $(F/G)(-1)$

29. $(F + F)(1)$ 30. $(G \cdot G)(6)$

The following graph shows the number of births in the United States, in millions, from 1970–2004. Here $C(t)$ represents the number of Caesarean section births, $B(t)$ the number of non-Caesarean section births, and $N(t)$ the total number of births in year t.

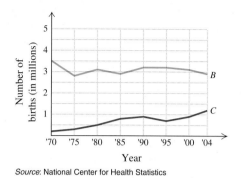

Source: National Center for Health Statistics

31. Use estimates of $C(2004)$ and $B(2004)$ to estimate $N(2004)$.

32. Use estimates of $C(1985)$ and $B(1985)$ to estimate $N(1985)$.

In 2004, a study comparing high doses of the cholesterol-lowering drugs Lipitor and Pravachol indicated that patients taking Lipitor were significantly less likely to have heart attacks or require angioplasty or surgery.

In the graph below, $L(t)$ is the percentage of patients on Lipitor (80 mg) and $P(t)$ is the percentage of patients on Pravachol (40 mg) who suffered heart problems or death t years after beginning to take the medication.
Source: New York Times, March 9, 2004

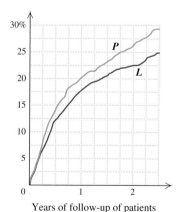

Years of follow-up of patients

Source: New England Journal of Medicine

33. Use estimates of $P(2)$ and $L(2)$ to estimate $(P - L)(2)$.

34. Use estimates of $P(1)$ and $L(1)$ to estimate $(P - L)(1)$.

Often function addition is represented by stacking the individual functions directly on top of each other. The graph below indicates how U.S. municipal solid waste has been managed. The braces indicate the values of the individual functions.

Talking trash

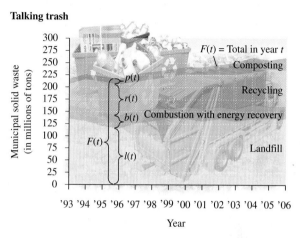

Source: Environmental Protection Agency

35. Estimate $(p + r)('05)$. What does it represent?

36. Estimate $(p + r + b)('05)$. What does it represent?

37. Estimate $F('96)$. What does it represent?

38. Estimate $F('06)$. What does it represent?

39. Estimate $(F - p)('04)$. What does it represent?

40. Estimate $(F - l)('03)$. What does it represent?

For each pair of functions f and g, determine the domain of the sum, difference, and product of the two functions.

41. $f(x) = x^2,$
    $g(x) = 7x - 4$

42. $f(x) = 5x - 1,$
    $g(x) = 2x^2$

43. $f(x) = \dfrac{1}{x + 5},$
    $g(x) = 4x^3$

44. $f(x) = 3x^2,$
    $g(x) = \dfrac{1}{x - 9}$

45. $f(x) = \dfrac{2}{x},$
    $g(x) = x^2 - 4$

46. $f(x) = x^3 + 1,$
    $g(x) = \dfrac{5}{x}$

47. $f(x) = x + \dfrac{2}{x - 1},$
    $g(x) = 3x^3$

48. $f(x) = 9 - x^2,$
    $g(x) = \dfrac{3}{x + 6} + 2x$

49. $f(x) = \dfrac{3}{2x + 9}$,

$g(x) = \dfrac{5}{1 - x}$

50. $f(x) = \dfrac{5}{3 - x}$,

$g(x) = \dfrac{1}{4x - 1}$

For each pair of functions f and g, determine the domain of f/g.

51. $f(x) = x^4$,
$g(x) = x - 3$

52. $f(x) = 2x^3$,
$g(x) = 5 - x$

53. $f(x) = 3x - 2$,
$g(x) = 2x + 8$

54. $f(x) = 5 + x$,
$g(x) = 6 - 2x$

55. $f(x) = \dfrac{3}{x - 4}$,
$g(x) = 5 - x$

56. $f(x) = \dfrac{1}{2 - x}$,
$g(x) = 7 + x$

57. $f(x) = \dfrac{2x}{x + 1}$,
$g(x) = 2x + 5$

58. $f(x) = \dfrac{7x}{x - 2}$,
$g(x) = 3x + 7$

For Exercises 59–66, consider the functions F and G as shown.

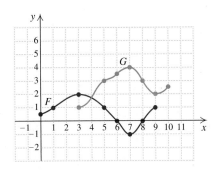

59. Determine $(F + G)(5)$ and $(F + G)(7)$.

60. Determine $(F \cdot G)(6)$ and $(F \cdot G)(9)$.

61. Determine $(G - F)(7)$ and $(G - F)(3)$.

62. Determine $(F/G)(3)$ and $(F/G)(7)$.

63. Find the domains of F, G, $F + G$, and F/G.

64. Find the domains of $F - G$, $F \cdot G$, and G/F.

65. Graph $F + G$.

66. Graph $G - F$.

In the following graph, S(t) represents the number of gallons of carbonated soft drinks consumed by the average American in year t, M(t) the number of gallons of milk, J(t) the number of gallons of fruit juice, and W(t) the number of gallons of bottled water.

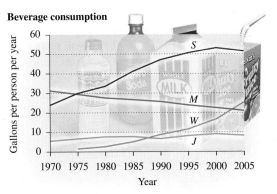

Beverage consumption

Source: Economic Research Service, U. S. Department of Agriculture

67. Between what years did the average American drink more soft drinks than juice, bottled water, and milk combined? Explain how you determined this.

68. Examine the graphs before Exercises 31 and 32. Did the total number of births increase or decrease from 1970 to 2004? Did the percent of births by Caesarean section increase or decrease from 1970 to 2004? Explain how you determined your answers.

Skill Review

To prepare for Chapter 3, review solving an equation for y and translating phrases to algebraic expressions (Sections 1.4 and 1.5).

Solve. [1.5]

69. $x - 6y = 3$, for y

70. $3x - 8y = 5$, for y

71. $5x + 2y = -3$, for y

72. $x + 8y = 4$, for y

Translate each of the following. Do not solve. [1.4]

73. Five more than twice a number is 49.

74. Three less than half of some number is 57.

75. The sum of two consecutive integers is 145.

76. The difference between a number and its opposite is 20.

Synthesis

77. Examine the graphs following Example 2 and explain how they might be modified to represent the absorption of 200 mg of Advil® taken four times a day.

78. If $f(x) = c$, where c is some positive constant, describe how the graphs of $y = g(x)$ and $y = (f + g)(x)$ will differ.

79. Find the domain of F/G, if

$$F(x) = \frac{1}{x - 4} \quad \text{and} \quad G(x) = \frac{x^2 - 4}{x - 3}.$$

80. Find the domain of f/g, if

$$f(x) = \frac{3x}{2x + 5} \quad \text{and} \quad g(x) = \frac{x^4 - 1}{3x + 9}.$$

81. Sketch the graph of two functions f and g such that the domain of f/g is

$$\{x \,|\, {-2} \le x \le 3 \text{ and } x \ne 1\}.$$

82. Find the domains of $f + g$, $f - g$, $f \cdot g$, and f/g, if
$f = \{(-2, 1), (-1, 2), (0, 3), (1, 4), (2, 5)\}$
and
$g = \{(-4, 4), (-3, 3), (-2, 4), (-1, 0), (0, 5), (1, 6)\}.$

83. Find the domain of m/n, if

$$m(x) = 3x \quad \text{for } {-1} < x < 5$$

and

$$n(x) = 2x - 3.$$

84. For f and g as defined in Exercise 82, find $(f + g)(-2)$, $(f \cdot g)(0)$, and $(f/g)(1)$.

85. Write equations for two functions f and g such that the domain of $f + g$ is

$$\{x \,|\, x \text{ is a real number } \text{and } x \ne -2 \text{ and } x \ne 5\}.$$

86. Let $y_1 = 2.5x + 1.5$, $y_2 = x - 3$, and $y_3 = y_1/y_2$. Depending on whether the CONNECTED or DOT mode is used, the graph of y_3 appears as follows. Use algebra to determine which graph more accurately represents y_3.

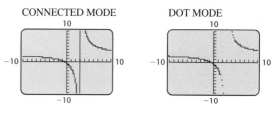

87. Using the window $[-5, 5, -1, 9]$, graph $y_1 = 5$, $y_2 = x + 2$, and $y_3 = \sqrt{x}$. Then predict what shape the graphs of $y_1 + y_2$, $y_1 + y_3$, and $y_2 + y_3$ will take. Use a graphing calculator to check each prediction.

88. Use the TABLE feature on a graphing calculator to check your answers to Exercises 45, 47, 55, and 57. (See the Technology Connection on p. 134.)

COLLABORATIVE CORNER

Time On Your Hands

Focus: The algebra of functions

Time: 10–15 minutes

Group size: 2–3

The graph and the data at right chart the average retirement age $R(x)$ and life expectancy $E(x)$ of U.S. citizens in year x.

ACTIVITY

1. Working as a team, perform the appropriate calculations and then graph $E - R$.

2. What does $(E - R)(x)$ represent? In what fields of study or business might the function $E - R$ prove useful?

3. Should E and R really be calculated separately for men and women? Why or why not?

4. What advice would you give to someone considering early retirement?

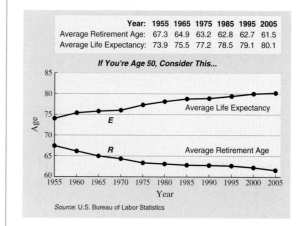

Year:	1955	1965	1975	1985	1995	2005
Average Retirement Age:	67.3	64.9	63.2	62.8	62.7	61.5
Average Life Expectancy:	73.9	75.5	77.2	78.5	79.1	80.1

If You're Age 50, Consider This...

Source: U.S. Bureau of Labor Statistics

Study Summary

KEY TERMS AND CONCEPTS	EXAMPLES

SECTION 2.1: GRAPHS

We can **graph** an equation by selecting values for one variable and finding the corresponding values for the other variable. We plot the resulting ordered pairs and draw the graph.

Graph: $y = x^2 - 3$.

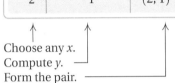

x	y $y = x^2 - 3$	(x, y)
0	−3	$(0, -3)$
−1	−2	$(-1, -2)$
1	−2	$(1, -2)$
−2	1	$(-2, 1)$
2	1	$(2, 1)$

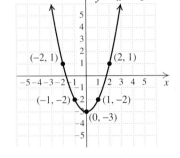

Choose any x.
Compute y.
Form the pair.
Plot the points and draw the graph.

SECTION 2.2: FUNCTIONS

A **function** is a correspondence between a first set, called the **domain**, and a second set, called the **range**, such that each member of the domain corresponds to *exactly one* member of the range.

Consider the function given by
$f(x) = |x| - 3$.

$\quad f(-2) = |-2| - 3 = 2 - 3 = -1$

The input -2 corresponds to the output -1.

The function contains the ordered pair $(-2, -1)$.

The domain of the function is $\mathbb{R}$.

The range of the function is $\{y \,|\, y \geq -3\}$.

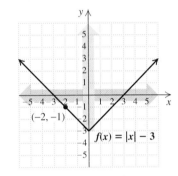

The Vertical-Line Test

If it is possible for a vertical line to cross a graph more than once, then the graph is not the graph of a function.

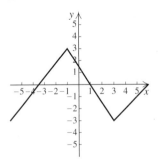   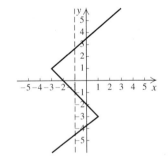

This *is* the graph of a function.   This *is not* the graph of a function.

Unless otherwise stated, the domain of a function is the set of all numbers for which function values can be calculated.

Find the domain of the function given by $f(x) = \dfrac{x + 2}{x - 7}$.

Function values cannot be calculated when the denominator is 0. Since $x - 7 = 0$ when $x = 7$, the domain of f is

$$\{x \,|\, x \text{ is a real number } and \; x \neq 7\}.$$

SECTION 2.3: LINEAR FUNCTIONS: SLOPE, GRAPHS, AND MODELS

Slope

$$m = \frac{\text{rise}}{\text{run}} = \frac{y_2 - y_1}{x_2 - x_1} = \frac{y_1 - y_2}{x_1 - x_2}$$

The slope of the line containing the points $(-1, -4)$ and $(2, -6)$ is

$$m = \frac{y_2 - y_1}{x_2 - x_1} = \frac{-6 - (-4)}{2 - (-1)} = \frac{-2}{3} = -\frac{2}{3}.$$

Slope–Intercept Form

$$y = mx + b$$

The slope of the line is m.
The y-intercept of the line is $(0, b)$.

Determine the slope and the y-intercept of the line given by $y = -\frac{2}{3}x + 7$.

For $y = -\frac{2}{3}x + 7$, the slope is $-\frac{2}{3}$ and the y-intercept is $(0, 7)$.

Find a linear function whose graph has slope 5 *and y-intercept* $(0, -4)$.

$$f(x) = mx + b$$
$$f(x) = 5x + (-4) \qquad \text{Substituting 5 for } m \text{ and } -4 \text{ for } b$$
$$f(x) = 5x - 4$$

To graph an equation written in slope–intercept form, we graph the y-intercept and count off the slope from that point.

Graph: $f(x) = \frac{1}{2}x - 3$.

We plot the y-intercept, $(0, -3)$. From there, we count off a slope of $\frac{1}{2}$: We go up 1 unit and to the right 2 units to the point $(2, -2)$. We then draw the graph.

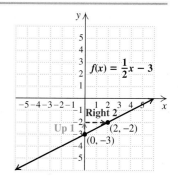

SECTION 2.4: ANOTHER LOOK AT LINEAR GRAPHS

Horizontal Lines

The slope of a horizontal line is 0.
The graph of $f(x) = b$ or $y = b$ is a horizontal line with y-intercept $(0, b)$.

Vertical Lines

The slope of a vertical line is undefined.
The graph of $x = a$ is a vertical line with x-intercept $(0, a)$.

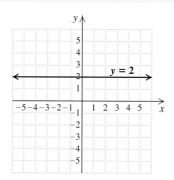

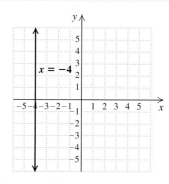

Horizontal line; slope is 0.

Vertical line; slope is undefined.

Intercepts

An x-intercept is of the form $(a, 0)$.
To find a, let $y = 0$ and solve for x.

Graph using intercepts: $x - 2y = 6$.

Let $y = 0$:

$$x - 2 \cdot 0 = 6$$
$$x = 6.$$

The x-intercept is $(6, 0)$.

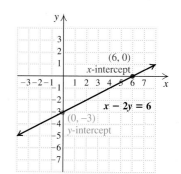

A y-intercept is of the form $(0, b)$.
To find b, let $x = 0$ and solve for y.

Let $x = 0$:

$$0 - 2y = 6$$
$$y = -3.$$

The y-intercept is $(0, -3)$.

SECTION 2.5: OTHER EQUATIONS OF LINES

Point–Slope Form

$$y - y_1 = m(x - x_1)$$

The slope of the line is m.

The line passes through (x_1, y_1).

Find a linear function that has a graph passing through the points $(-2, 8)$ and $(-3, 5)$.

$$m = \frac{y_2 - y_1}{x_2 - x_1} = \frac{5 - 8}{-3 - (-2)} = \frac{-3}{-1} = 3; \quad \text{Finding the slope}$$

$$y - y_1 = m(x - x_1) \qquad \text{Using point–slope form}$$
$$y - 8 = 3(x - (-2)) \qquad \text{Using } (-2, 8) \text{ as } (x_1, y_1)$$
$$y - 8 = 3(x + 2)$$
$$y - 8 = 3x + 6$$
$$y = 3x + 14 \qquad \text{This is slope–intercept form.}$$
$$f(x) = 3x + 14 \qquad \text{Using function notation}$$

Parallel and Perpendicular Lines

Two lines are parallel if they have the same slope or if both are vertical.

Find an equation in slope–intercept form for the line that passes through $(3, -4)$ and is parallel to the line given by $y = \frac{1}{2}x - 5$.

The slope of $y = \frac{1}{2}x - 5$ is $\frac{1}{2}$, so the slope of a parallel line is also $\frac{1}{2}$.

$$y - y_1 = m(x - x_1)$$
$$y - (-4) = \frac{1}{2}(x - 3) \qquad m = \frac{1}{2} \text{ and } (x_1, y_1) = (3, -4)$$
$$y + 4 = \frac{1}{2}x - \frac{3}{2}$$
$$y = \frac{1}{2}x - \frac{11}{2}. \qquad -\frac{3}{2} - 4 = -\frac{3}{2} - \frac{8}{2} = -\frac{11}{2}$$

Two lines are perpendicular if the product of their slopes is -1 or if one line is vertical and the other line is horizontal.

Find an equation in slope–intercept form for the line that passes through $(3, -4)$ and is perpendicular to the line given by $y = \frac{1}{2}x - 5$.

The slope of $y = \frac{1}{2}x - 5$ is $\frac{1}{2}$, so the slope of a perpendicular line is -2, since $\left(\frac{1}{2}\right)(-2) = -1$.

$$y - y_1 = m(x - x_1)$$
$$y - (-4) = -2(x - 3) \qquad m = -2 \text{ and } (x_1, y_1) = (3, -4)$$
$$y + 4 = -2x + 6$$
$$y = -2x + 2$$

SECTION 2.6: THE ALGEBRA OF FUNCTIONS

For $f(x) = x^2 + 3x$ and $g(x) = x - 5$:

$$(f + g)(x) = f(x) + g(x)$$

$$(f + g)(x) = f(x) + g(x)$$
$$= x^2 + 3x + x - 5 = x^2 + 4x - 5;$$

$$(f - g)(x) = f(x) - g(x)$$

$$(f - g)(x) = f(x) - g(x)$$
$$= x^2 + 3x - (x - 5) = x^2 + 2x + 5;$$

$$(f \cdot g)(x) = f(x) \cdot g(x)$$

$$(f \cdot g)(x) = f(x) \cdot g(x)$$
$$= (x^2 + 3x)(x - 5);$$

$$(f/g)(x) = f(x)/g(x), \text{ provided } g(x) \neq 0$$

$$(f/g)(x) = f(x)/g(x), \text{ provided } g(x) \neq 0$$
$$= \frac{x^2 + 3x}{x - 5}, \text{ provided } x \neq 5. \qquad g(5) = 0$$

The domain of $f + g, f - g$, and $f \cdot g$ is the set of all values common to the domains of f and g.

For $f(x) = \dfrac{1}{x}$ and $g(x) = x + 1$:

The domain of f is $\{x \mid x$ is a real number *and* $x \neq 0\}$.

The domain of g is the set of all real numbers $\mathbb{R}$.

The domain of $f + g, f - g$, and $f \cdot g$ is

$\{x \mid x$ is a real number *and* $x \neq 0\}$.

The domain of f/g is the set of all values common to the domains of f and g, excluding any values for which $g(x)$ is 0.

The domain of f/g is

$\{x \mid x$ is a real number *and* $x \neq 0$ *and* $x \neq -1\}$.

Since $g(-1) = 0$, we exclude -1 from the domain.

Review Exercises: Chapter 2

👉 *Concept Reinforcement*   *Classify each statement as either true or false.*

1. No member of a function's range can be used in two different ordered pairs. [2.2]

2. The horizontal-line test is a quick way to determine whether a graph represents a function. [2.2]

3. A line's slope is a measure of how the line is slanted or tilted. [2.3]

4. Every line has a y-intercept. [2.4]

5. Every vertical line has an x-intercept. [2.4]

6. The slope of a vertical line is undefined. [2.4]

7. The slope of the graph of a constant function is 0. [2.4]

8. Extrapolation is done to predict future values. [2.2]

9. If two lines are perpendicular, the slope of one line is the opposite of the slope of the second line. [2.5]

10. In order for $(f/g)(a)$ to exist, we must have $g(a) \neq 0$. [2.6]

Determine whether the ordered pair is a solution of the given equation. [2.1]

11. $(-2, 8),\ x = 2y + 12$

12. $\left(0, -\frac{1}{2}\right),\ 3a - 4b = 2$

13. Name the quadrant in which $(-3, 5)$ is located. [2.1]

14. Graph: $y = -x^2 + 1$. [2.1]

15. Find the rate of change for the graph shown. Be sure to use appropriate units. [2.3]

16. In 2007, there were 214,000 new homes sold in the United States by the end of March and 577,000 sold by the end of August. Calculate the rate at which new homes were being sold. [2.3]
Source: www.census.gov

Find the slope of each line. If the slope is undefined, state this.

17. Containing the points $(4, 5)$ and $(-3, 1)$ [2.3]

18. Containing the points $(-16.4, 2.8)$ and $(-16.4, 3.5)$ [2.4]

19. Containing the points $(-1, -2)$ and $(-5, -1)$ [2.3]

20. Containing the points $\left(\frac{1}{3}, \frac{1}{2}\right)$ and $\left(\frac{1}{6}, \frac{1}{2}\right)$ [2.3]

Find the slope and the y-intercept. [2.3]

21. $g(x) = -5x - 11$

22. $-6y + 5x = 10$

23. The number of calories consumed each day by the average American woman t years after 1971 can be estimated by $C(t) = 11t + 1542$. What do the numbers 11 and 1542 signify? [2.3]
Source: Based on data from Centers for Disease Control and Prevention.

For each equation, find the slope. If the slope is undefined, state this. [2.4]

24. $y + 3 = 7$ 25. $-2x = 9$

26. Find the intercepts of the line given by $3x - 2y = 8$. [2.4]

Graph.

27. $y = -3x + 2$ [2.3]

28. $-2x + 4y = 8$ [2.4]

29. $y = 6$ [2.4]

30. $y + 1 = \frac{3}{4}(x - 5)$ [2.5]

31. $8x + 32 = 0$ [2.4]

32. $g(x) = 15 - x$ [2.3]

33. $f(x) = \frac{1}{2}x - 3$ [2.3]

34. $f(x) = 0$ [2.4]

35. Solve $2 - x = 5 + 2x$ graphically. Then check your answer by solving the equation algebraically. [2.4]

36. Tee Prints charges $120 to print 5 custom-designed tee shirts. Each additional tee shirt costs $8. Use a graph to estimate the number of tee shirts printed if the total cost of the order was $200. [2.4]

Determine whether each pair of lines is parallel, perpendicular, or neither. [2.5]

37. $y + 5 = -x$,
    $x - y = 2$

38. $3x - 5 = 7y$,
    $7y - 3x = 7$

39. Find a linear function whose graph has slope $\frac{2}{9}$ and y-intercept $(0, -4)$. [2.3]

40. Find an equation in point–slope form of the line with slope -5 and containing $(1, 10)$. [2.5]

41. Using function notation, write a slope–intercept equation for the line containing $(2, 5)$ and $(-2, 6)$. [2.5]

Find an equation of the line. [2.5]

42. Containing the point $(2, -5)$ and parallel to the line $3x - 5y = 9$

43. Containing the point $(2, -5)$ and perpendicular to the line $3x - 5y = 9$

The following table shows heating oil prices for various years.

Input, Year	Output, Cost per Gallon of Heating Oil
2002	$0.60
2005	$2.00
2007	$2.25

Source: Bankrate.com

44. Use the data in the table to draw a graph and to estimate the cost per gallon of heating oil in 2004. [2.2]

45. Use the graph from Exercise 44 to estimate the cost per gallon of heating oil in 2008. [2.2]

46. *Records in the 200-meter run.* In 1983, the record for the 200-m run was 19.75 sec.* In 2007, it was 19.32 sec. Let $R(t)$ represent the record in the 200-m run and t the number of years since 1980. [2.5]
Source: International Association of Athletics Federation

   a) Find a linear function that fits the data.
   b) Use the function of part (a) to predict the record in 2013 and in 2020.

Determine whether each of these is a linear equation. [2.4]

47. $2x - 7 = 0$ 48. $3x - \frac{y}{8} = 7$

49. $2x^3 - 7y = 5$ 50. $\frac{2}{x} = y$

*Records are for elevations less than 1000 m.

51. For the following graph of f, determine **(a)** $f(2)$; **(b)** the domain of f; **(c)** any x-values for which $f(x) = 2$; and **(d)** the range of f. [2.2]

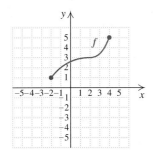

52. Determine the domain and the range of the function g represented below. [2.2]

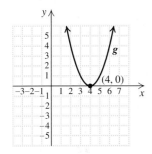

Determine whether each of the following is the graph of a function. [2.2]

53.   **54.**

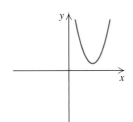 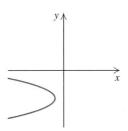

Let $g(x) = 3x - 6$ and $h(x) = x^2 + 1$. Find the following.

55. $g(0)$ [2.2]   **56.** $h(-5)$ [2.2]

57. $g(a + 5)$ [2.2]   **58.** $(g \cdot h)(4)$ [2.6]

59. $(g/h)(-1)$ [2.6]   **60.** $(g + h)(x)$ [2.6]

61. The domain of g [2.2]

62. The domain of $g + h$ [2.6]

63. The domain of h/g [2.6]

Synthesis

64. Explain why every function is a relation, but not every relation is a function. [2.2]

65. Explain why the slope of a vertical line is undefined whereas the slope of a horizontal line is 0. [2.4]

66. Find the y-intercept of the function given by
$$f(x) + 3 = 0.17x^2 + (5 - 2x)^x - 7.$$
[1.6], [2.4]

67. Determine the value of a such that the lines
$$3x - 4y = 12 \quad \text{and} \quad ax + 6y = -9$$
are parallel. [2.5]

68. Treasure Tea charges \$7.99 for each package of loose tea. Shipping charges are \$2.95 per package plus \$20 per order for overnight delivery. Find a linear function for determining the cost of one order of x packages of tea, including shipping and overnight delivery. [2.5]

69. Match each sentence with the most appropriate of the four graphs below. [2.3]

a) Joni walks for 10 min to the train station, rides the train for 15 min, and then walks 5 min to the office.

b) During a workout, Carter bikes for 10 min, runs for 15 min, and then walks for 5 min.

c) Andrew pilots his motorboat for 10 min to the middle of the lake, fishes for 15 min, and then motors for another 5 min to another spot.

d) Patti waits 10 min for her train, rides the train for 15 min, and then runs for 5 min to her job.

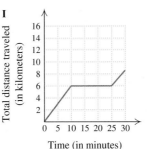

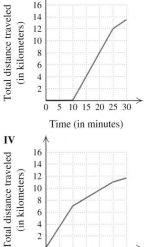

Test: Chapter 2

1. Determine whether the ordered pair is a solution of the given equation.

   $(12, -3), x + 4y = -20$

2. Graph: $f(x) = x^2 + 3$.

3. Find the rate of change for the graph below. Use appropriate units.

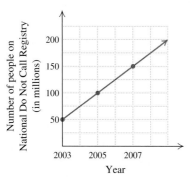

Source: Federal Trade Commission

Find the slope of the line containing the following points. If the slope is undefined, state this.

4. $(-2, -2)$ and $(6, 3)$

5. $(-3.1, 5.2)$ and $(-4.4, 5.2)$

Find the slope and the y-intercept.

6. $f(x) = -\frac{3}{5}x + 12$

7. $-5y - 2x = 7$

Find the slope. If the slope is undefined, state this.

8. $f(x) = -3$ 9. $x - 5 = 11$

10. Find the intercepts of the line given by $5x - y = 15$.

Graph.

11. $f(x) = -3x + 4$

12. $y - 1 = -\frac{1}{2}(x + 4)$

13. $-2x + 5y = 20$

14. $3 - x = 9$

15. Solve $x + 3 = 2x$ graphically. Then check your answer by solving the equation algebraically.

16. There were 41.9 million international visitors to the United States in 2002 and 51.0 million visitors in 2006. Draw a graph and estimate the number of international visitors in 2005.
    Sources: U.S. Department of Commerce, ITA, Office of Travel and Tourism Industries; Global Insight, Inc.

17. Which of these are linear equations?

    a) $8x - 7 = 0$
    b) $4x - 9y^2 = 12$
    c) $2x - 5y = 3$

Determine without graphing whether each pair of lines is parallel, perpendicular, or neither.

18. $4y + 2 = 3x,$
    $-3x + 4y = -12$

19. $y = -2x + 5,$
    $2y - x = 6$

20. Find a linear function whose graph has slope -5 and y-intercept $(0, -1)$.

21. Find an equation in point–slope form of the line with slope 4 and containing $(-2, -4)$.

22. Using function notation, write a slope–intercept equation for the line containing $(3, -1)$ and $(4, -2)$.

Find an equation of the line.

23. Containing $(-3, 2)$ and parallel to the line $2x - 5y = 8$

24. Containing $(-3, 2)$ and perpendicular to the line $2x - 5y = 8$

25. If you rent a truck for one day and drive it 250 mi, the cost is \$100. If you rent it for one day and drive it 300 mi, the cost is \$115. Let $C(m)$ represent the cost, in dollars, of driving m miles.

    a) Find a linear function that fits the data.
    b) Use the function to determine how much it will cost to rent the truck for one day and drive it 500 mi.

26. For the following graph of f, determine **(a)** $f(-2)$; **(b)** the domain of f; **(c)** any x-value for which $f(x) = \frac{1}{2}$; and **(d)** the range of f.

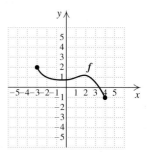

Find the following, given that $g(x) = \dfrac{1}{x}$ and $h(x) = 2x + 1$.

27. $h(-5)$ 28. $(g + h)(x)$

29. The domain of g

30. The domain of $g + h$

31. The domain of g/h

Synthesis

32. The function $f(t) = 5 + 15t$ can be used to determine a bicycle racer's location, in miles from the starting line, measured t hours after passing the 5-mi mark.

    a) How far from the start will the racer be 1 hr and 40 min after passing the 5-mi mark?

    b) Assuming a constant rate, how fast is the racer traveling?

33. The graph of the function $f(x) = mx + b$ contains the points $(r, 3)$ and $(7, s)$. Express s in terms of r if the graph is parallel to the line $3x - 2y = 7$.

34. Given that $f(x) = 5x^2 + 1$ and $g(x) = 4x - 3$, find an expression for $h(x)$ so that the domain of $f/g/h$ is $\left\{x \mid x \text{ is a real number } and \ x \neq \frac{3}{4} \ and \ x \neq \frac{2}{7}\right\}$. Answers may vary.

Cumulative Review: Chapters 1–2

Perform the indicated operation. [1.2]

1. $-3 - (-10)$ **2.** $-\frac{1}{3} + \frac{5}{6}$

3. $-6.1(0.3)$ **4.** $12 \div (-4)$

5. $-\frac{6}{5} \div \left(-\frac{2}{15}\right)$

6. Simplify: $3x - 5(x - 7) + 2$. [1.3]

Solve. If the solution set is $\mathbb{R}$ or $\varnothing$, classify the equation as an identity or a contradiction. [1.3]

7. $2(3t + 1) - t = 5(t + 6)$

8. $-2(4 - x) + 3 = 8 - 6x$

9. Solve for y: $8x - 3y = 12$. [1.5]

Simplify. Do not use negative exponents in the answer. [1.6]

10. $(3x^2y^{-1})^{-1}$ **11.** -2^{-3}

12. $(10a^2b)^0$ **13.** $\left(\dfrac{3x^2y^{-2}}{15x^{-1}y^{-1}}\right)^2$

14. Find the slope of the line containing $(2, 5)$ and $(1, 10)$. [2.3]

15. Find the slope of the line given by $f(x) = 8x + 3$. [2.3]

16. Find the slope of the line given by $y + 6 = -4$. [2.4]

17. Find a linear function whose graph has slope -1 and y-intercept $\left(0, \frac{1}{5}\right)$. [2.3]

18. Find a linear function whose graph contains $(-1, 3)$ and $(-3, -5)$. [2.5]

19. Find an equation of the line containing $(5, -2)$ and perpendicular to the line given by $x - y = 5$. [2.5]

Find the following, given that $f(x) = x + 5$ and $g(x) = x^2 - 1$.

20. $g(-10)$ [2.2]   **21.** $(f \cdot g)(-5)$ [2.6]

22. $(g/f)(x)$ [2.6]

23. Find the domain of f if $f(x) = \dfrac{x}{x + 6}$. [2.2]

24. Determine the domain and the range of the function f represented below. [2.2]

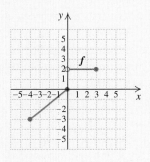

The graph below indicates how the three major airports servicing New York City have been utilized.

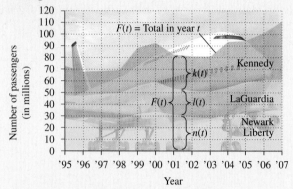

Packing them in

Source: Port Authority of New York and New Jersey

25. Estimate $(n + l)('05)$. What does it represent?   [2.6]

26. Estimate $F('07)$. What does it represent?   [2.6]

27. Estimate $(F - k)('06)$. What does it represent?   [2.6]

Graph.

28. $f(x) = 5$ [2.4]

29. $y - 2 = \frac{1}{3}(x + 1)$ [2.5]

30. $x = -3$ [2.4]

31. $y = |x| + 1$ [2.1]

32. $y = -x - 3$ [2.3]

33. $10x + y = -20$ [2.4]

34. *Enrollment.* In 2005, there were 8.5 million undergraduates enrolled in U.S. four-year colleges. The number of full-time students was four times the number of part-time students. How many full-time students and how many part-time students were enrolled in four-year colleges in 2005?   [1.4]
Source: National Center for Education Statistics, U.S. Department of Education

35. *Tuition.* The average yearly tuition at four-year public colleges was $5188 in 2004. In 2006, the average yearly tuition was $5495. Let $c(t)$ represent the average yearly tuition at a four-year public college t years after 2000.   [2.5]
Source: National Center for Education Statistics, U.S. Department of Education

a) Find a linear function that fits the data.

b) Use the function from part (a) to predict the average yearly tuition at a four-year public college in 2012.

c) In what year will the average tuition at a four-year public college be $7500?

36. *Disaster relief.* Six months after Hurricane Katrina struck the Gulf Coast in 2005, $2.18 billion of relief money had been distributed to disaster victims. This was $\frac{2}{3}$ of the amount raised by charity for disaster relief. How much money was still to be distributed?   [1.4]
Source: www.washingtonpost.com

37. *Money transfer.* The number of Social Security payments, in millions, delivered by paper check is given by $P(t) = -8.3t + 180$, where t is the number of years after 2001.

a) Find the number of Social Security payments delivered by paper check in 2007.   [2.2]

b) What do the numbers -8.3 and 180 signify?   [2.3]

38. *Information.* Approximately 183 billion e-mails were sent each day worldwide in 2006. If each e-mail averages 5.9×10^4 bytes, how many bytes of information were sent in e-mails in 2006? Use scientific notation for your answer.   [1.7]
Source: The Radicati Group

39. *Photo books.* An Everyday Photo Book costs $12 for the first 20 pages plus $0.75 for each additional page. Use a graph to estimate how many pages are in a book that cost $18.   [2.4]
Source: snapfish.com

40. The length of a rectangular painting is three times its width and its perimeter is 52 in. Find the length and the width of the painting.   [1.4]

Synthesis

Translate to an algebraic expression.   [1.1]

41. The difference of two squares

42. The product of the sum and the difference of the same two numbers

43. Simplify: $(3x^a y^{c+a})^{2a}$.   [1.6]

44. Determine the value of k such that the lines
$$2x + y = 5 \quad \text{and} \quad 4x + ky = 3$$
are parallel.   [2.5]

45. Find an equation for a linear function f if $f(2) = 4$ and $f(0) = 3$.   [2.3]

Systems of Linear Equations and Problem Solving

JUDITH L. BRONSTEIN
NATURALIST/ECOLOGIST
Tucson, Arizona

As an ecologist, I often use equations both to estimate numbers of organisms and to make predictions about how those numbers will change under different conditions. In this way, math helps us to answer questions like these: How quickly is the human population growing, and will food production be able to keep up? How many polar bears are left in the wild, how will their numbers be affected as the climate continues to change, and what approaches should be most successful for preserving them?

AN APPLICATION

The number of plant species listed as threatened or endangered has more than tripled in the past 20 years. In 2008, there were 746 species of plants in the United States that were considered threatened or endangered. The number of species considered threatened was 4 less than one-fourth of the number considered endangered. How many U.S. plant species were considered endangered and how many were considered threatened in 2008?

Source: U.S. Fish and Wildlife Service

This problem appears as Example 1 in Section 3.1 and as Example 1 in Section 3.3.

T he most difficult part of problem solving is almost always translating the problem situation to mathematical language. In this chapter, we study *systems of equations* and how to solve them using graphing, substitution, elimination, and matrices. Systems of equations often provide the easiest way to model real-world situations in fields such as psychology, sociology, business, education, engineering, and science.

3.1   Systems of Equations in Two Variables

Translating ■ Identifying Solutions ■ Solving Systems Graphically

Translating

Problems involving two unknown quantities are often translated most easily using two equations in two unknowns. Together these equations form a **system of equations**. We look for a solution to the problem by attempting to find a pair of numbers for which *both* equations are true.

EXAMPLE 1   *Endangered species.*   The number of plant species listed as threatened (likely to become endangered) or endangered (in danger of becoming extinct) has more than tripled in the past 20 years. In 2008, there were 746 species of plants in the United States that were considered threatened or endangered. The number of species considered threatened was 4 less than one-fourth of the number considered endangered. How many U.S. plant species were considered endangered and how many were considered threatened in 2008?

Source: U.S. Fish and Wildlife Service

SOLUTION

1. **Familiarize.** Often statements of problems contain information that has no bearing on the question asked. In this case, the fact that the number of threatened or endangered species has tripled in the past 20 years does not help us solve the problem. Instead, we focus on the number of endangered species and the number of threatened species in 2008. We let t represent the number of threatened plant species and d represent the number of endangered plant species in 2008.

2. **Translate.** There are two statements to translate. First, we look at the total number of endangered or threatened species of plants:

    Rewording: The number of   the number of
                       threatened species plus endangered species was 746.

    Translating: t $+$ d $=$ 746

    The second statement compares the two amounts, d and t:

    Rewording: The number of   4 less than one-fourth of the
                       threatened species was number of endangered species.

    Translating: t $=$ $\frac{1}{4}d - 4$

We have now translated the problem to a pair, or **system**, **of equations**:

$$t + d = 746,$$

$$t = \frac{1}{4}d - 4.$$

We complete the solution of this problem in Section 3.3.

> TRY EXERCISE 41

System of Equations

A *system of equations* is a set of two or more equations, in two or more variables, for which a common solution is sought.

Problems like Example 1 *can* be solved using one variable; however, as problems become complicated, you will find that using more than one variable (and more than one equation) is often the preferable approach.

EXAMPLE 2

Jewelry design.   A jewelry designer purchased 80 beads for a total of $39 (excluding tax) to make a necklace. Some of the beads were sterling silver beads that cost 40¢ each and the rest were gemstone beads that cost 65¢ each. How many of each type did the designer buy?

SOLUTION

1. **Familiarize.** To familiarize ourselves with this problem, let's guess that the designer bought 20 beads at 40¢ each and 60 beads at 65¢ each. The total cost would then be

   $$20 \cdot 40¢ + 60 \cdot 65¢ = 800¢ + 3900¢, \quad \text{or} \quad 4700¢.$$

   Since 4700¢ = $47 and $47 ≠ $39, our guess is incorrect. Rather than guess again, let's see how algebra can be used to translate the problem.

2. **Translate.** We let s = the number of silver beads and g = the number of gemstone beads. Since the cost of each bead is given in cents and the total cost is in dollars, we must choose one of the units to use throughout the problem. We choose to work in cents, so the total cost is 3900¢. The information can be organized in a table, which will help with the translating.

Type of Bead	Silver	Gemstone	Total	
Number Bought	s	g	80	→ $s + g = 80$
Price	40¢	65¢		
Amount	$40s$¢	$65g$¢	3900¢	→ $40s + 65g = 3900$

The first row of the table and the first sentence of the problem indicate that a total of 80 beads were bought:

$$s + g = 80.$$

Since each silver bead cost 40¢ and s beads were bought, $40s$ represents the amount paid, in cents, for the silver beads. Similarly, $65g$ represents the amount paid, in cents, for the gemstone beads. This leads to a second equation:

$$40s + 65g = 3900.$$

We now have the following system of equations as the translation:

$$s + g = 80,$$
$$40s + 65g = 3900.$$

We will complete the solution of this problem in Section 3.3.

TRY EXERCISE ▶ 49

Identifying Solutions

A *solution* of a system of two equations in two variables is an ordered pair of numbers that makes *both* equations true.

EXAMPLE **3**   Determine whether $(-4, 7)$ is a solution of the system

$$x + y = 3,$$
$$5x - y = -27.$$

SOLUTION   As discussed in Chapter 2, unless stated otherwise, we use alphabetical order of the variables. Thus we replace x with -4 and y with 7:

$x + y = 3$	$5x - y = -27$
$-4 + 7 \mid 3$	$5(-4) - 7 \mid -27$
$3 \overset{?}{=} 3$ TRUE	$-20 - 7 \mid$
	$-27 \overset{?}{=} -27$ TRUE

> *CAUTION!*   Be sure to check the ordered pair in *both* equations.

The pair $(-4, 7)$ makes both equations true, so it is a solution of the system. We can also describe the solution by writing $x = -4$ and $y = 7$. Set notation can also be used to list the solution set $\{(-4, 7)\}$.

TRY EXERCISE ▶ 9

Solving Systems Graphically

Recall that the graph of an equation is a drawing that represents its solution set. If we graph the equations in Example 3, we find that $(-4, 7)$ is the only point common to both lines. Thus one way to solve a system of two equations is to graph both equations and identify any points of intersection. **The coordinates of each point of intersection represent a solution of that system.**

$$x + y = 3,$$
$$5x - y = -27$$

The point of intersection of the graphs is $(-4, 7)$.

The solution of the system is $(-4, 7)$.

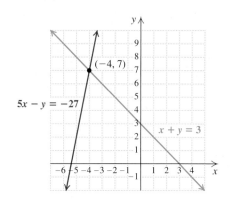

Most pairs of lines have exactly one point in common. We will soon see, however, that this is not always the case.

EXAMPLE **4**   Solve each system graphically.

a) $y - x = 1,$ **b)** $y = -3x + 5,$ **c)** $3y - 2x = 6,$
    $y + x = 3$ $y = -3x - 2$ $-12y + 8x = -24$

SOLUTION

a) We graph each equation using any method studied in Chapter 2. All ordered pairs from line L_1 are solutions of the first equation. All ordered pairs from line L_2 are solutions of the second equation. The point of intersection has coordinates that make *both* equations true. Apparently, $(1, 2)$ is the solution. Graphs are not always accurate, so solving by graphing may yield approximate answers. Our check below shows that $(1, 2)$ is indeed the solution.

Graph both equations.

Look for any points in common.

$$y - x = 1,$$
$$y + x = 3$$

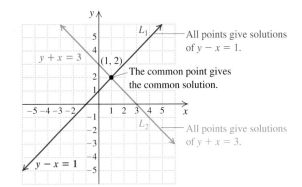

Check.

Check:

$$\frac{y - x = 1}{2 - 1 \;\vert\; 1}$$
$$1 \stackrel{?}{=} 1 \quad \text{TRUE}$$

$$\frac{y + x = 3}{2 + 1 \;\vert\; 3}$$
$$3 \stackrel{?}{=} 3 \quad \text{TRUE}$$

b) We graph the equations. The lines have the same slope, -3, and different y-intercepts, so they are parallel. There is no point at which they cross, so the system has no solution.

$$y = -3x + 5,$$
$$y = -3x - 2$$

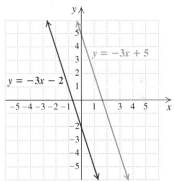

STUDENT NOTES

Although the system in Example 4(c) is true for an infinite number of ordered pairs, those pairs must be of a certain form. Only pairs that are solutions of $3y - 2x = 6$ or $-12y + 8x = -24$ are solutions of the system. It is incorrect to think that *all* ordered pairs are solutions.

c) We graph the equations and find that the same line is drawn twice. Thus any solution of one equation is a solution of the other. Each equation has an infinite number of solutions, so the system itself has an infinite number of solutions. We check one solution, $(0, 2)$, which is the y-intercept of each equation.

$$3y - 2x = 6,$$
$$-12y + 8x = -24$$

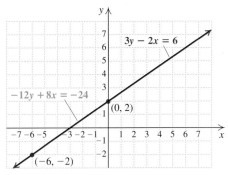

Check:

$$3y - 2x = 6$$
$$\frac{3(2) - 2(0) \mid 6}{}$$
$$6 - 0 \mid$$
$$6 \overset{?}{=} 6 \quad \text{TRUE}$$

$$-12y + 8x = -24$$
$$\frac{-12(2) + 8(0) \mid -24}{}$$
$$-24 + 0 \mid$$
$$-24 \overset{?}{=} -24 \quad \text{TRUE}$$

You can check that $(-6, -2)$ is another solution of both equations. In fact, any pair that is a solution of one equation is a solution of the other equation as well. Thus the solution set is

$$\{(x, y) \mid 3y - 2x = 6\}$$

or, in words, "the set of all pairs (x, y) for which $3y - 2x = 6$." Since the two equations are equivalent, we could have written instead $\{(x, y) \mid -12y + 8x = -24\}$.

TRY EXERCISE ▸ 17

When we graph a system of two linear equations in two variables, one of the following three outcomes will occur.

1. The lines have one point in common, and that point is the only solution of the system (see Example 4a). Any system that has *at least one solution* is said to be **consistent**.

2. The lines are parallel, with no point in common, and the system has *no solution* (see Example 4b). This type of system is called **inconsistent**.

3. The lines coincide, sharing the same graph. Because every solution of one equation is a solution of the other, the system has an infinite number of solutions (see Example 4c). Since it has at least one solution, this type of system is also consistent.

TECHNOLOGY CONNECTION

On most graphing calculators, an INTERSECT option allows us to find the coordinates of the intersection directly.

To illustrate, consider the following system:

$$3.45x + 4.21y = 8.39,$$
$$7.12x - 5.43y = 6.18.$$

After solving for y in each equation, we obtain the graph below. Using INTERSECT, we see that, to the nearest hundredth, the coordinates of the intersection are $(1.47, 0.79)$.

$y_1 = (8.39 - 3.45x)/4.21,$
$y_2 = (6.18 - 7.12x)/(-5.43)$

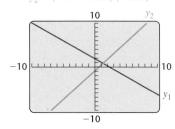

Use a graphing calculator to solve each of the following systems. Round all x- and y-coordinates to the nearest hundredth.

1. $y = -5.43x + 10.89,$
   $y = 6.29x - 7.04$
2. $y = 123.52x + 89.32,$
   $y = -89.22x + 33.76$
3. $2.18x + 7.81y = 13.78,$
   $5.79x - 3.45y = 8.94$
4. $-9.25x - 12.94y = -3.88,$
   $21.83x + 16.33y = 13.69$

When one equation in a system can be obtained by multiplying both sides of another equation by a constant, the two equations are said to be **dependent**. Thus the equations in Example 4(c) are dependent, but those in Examples 4(a) and 4(b) are **independent**. For systems of three or more equations, the definitions of dependent and independent will be slightly modified.

ALGEBRAIC–GRAPHICAL CONNECTION

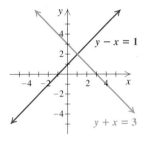

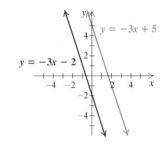

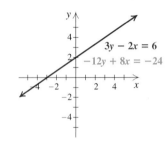

Graphs intersect at one point.	**Graphs are parallel.**	**Equations have the same graph.**

The system

$$y - x = 1,$$
$$y + x = 3$$

is *consistent* and has one solution.

Since neither equation is a multiple of the other, the equations are *independent*.

The system

$$y = -3x - 2,$$
$$y = -3x + 5$$

is *inconsistent* because there is no solution.

Since neither equation is a multiple of the other, the equations are *independent*.

The system

$$3y - 2x = 6,$$
$$-12y + 8x = -24$$

is *consistent* and has an infinite number of solutions.

Since one equation is a multiple of the other, the equations are *dependent*.

Graphing is helpful when solving systems because it allows us to "see" the solution. It can also be used on systems of nonlinear equations, and in many applications, it provides a satisfactory answer. However, graphing often lacks precision, especially when fraction or decimal solutions are involved. In Section 3.2, we will develop two algebraic methods of solving systems. Both methods produce exact answers.

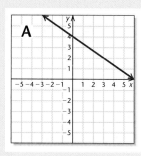

A

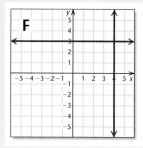

F

Visualizing for Success

Match each equation or system of equations with its graph.

1. $x + y = 2,$
   $x - y = 2$

2. $y = \frac{1}{3}x - 5$

3. $4x - 2y = -8$

4. $2x + y = 1,$
   $x + 2y = 1$

5. $8y + 32 = 0$

6. $f(x) = -x + 4$

7. $\frac{2}{3}x + y = 4$

8. $x = 4,$
   $y = 3$

9. $y = \frac{1}{2}x + 3,$
   $2y - x = 6$

10. $y = -x + 5,$
    $y = 3 - x$

Answers on page A-12

An additional, animated version of this activity appears in MyMathLab. To use MyMathLab, you need a course ID and a student access code. Contact your instructor for more information.

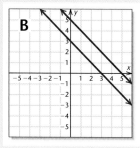

B

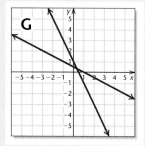

G

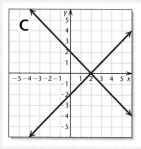

C

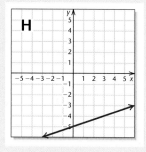

H

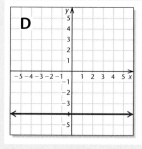

D

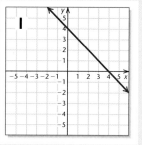

I

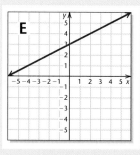

E

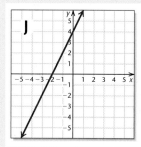

J

3.1 EXERCISE SET

🔖 *Concept Reinforcement*   *Classify each statement as either true or false.*

1. Every system of equations has at least one solution.

2. It is possible for a system of equations to have an infinite number of solutions.

3. Every point of intersection of the graphs of the equations in a system corresponds to a solution of the system.

4. The graphs of the equations in a system of two equations may coincide.

5. The graphs of the equations in a system of two equations could be parallel lines.

6. Any system of equations that has at most one solution is said to be consistent.

7. Any system of equations that has more than one solution is said to be inconsistent.

8. The equations $x + y = 5$ and $2(x + y) = 2(5)$ are dependent.

Determine whether the ordered pair is a solution of the given system of equations. Remember to use alphabetical order of variables.

9. $(2, 3)$; $\begin{aligned} 2x - y &= 1, \\ 5x - 3y &= 1 \end{aligned}$

10. $(4, 0)$; $\begin{aligned} 2x + 7y &= 8, \\ x - 9y &= 4 \end{aligned}$

11. $(-5, 1)$; $\begin{aligned} x + 5y &= 0, \\ y &= 2x + 9 \end{aligned}$

12. $(-1, -2)$; $\begin{aligned} x + 3y &= -7, \\ 3x - 2y &= 12 \end{aligned}$

13. $(0, -5)$; $\begin{aligned} x - y &= 5, \\ y &= 3x - 5 \end{aligned}$

14. $(5, 2)$; $\begin{aligned} a + b &= 7, \\ 2a - 8 &= b \end{aligned}$

Aha! 15. $(3, -1)$; $\begin{aligned} 3x - 4y &= 13, \\ 6x - 8y &= 26 \end{aligned}$

16. $(4, -2)$; $\begin{aligned} -3x - 2y &= -8, \\ 8 &= 3x + 2y \end{aligned}$

Solve each system graphically. Be sure to check your solution. If a system has an infinite number of solutions, use set-builder notation to write the solution set. If a system has no solution, state this.

17. $\begin{aligned} x - y &= 1, \\ x + y &= 5 \end{aligned}$

18. $\begin{aligned} x + y &= 6, \\ x - y &= 4 \end{aligned}$

19. $\begin{aligned} 3x + y &= 5, \\ x - 2y &= 4 \end{aligned}$

20. $\begin{aligned} 2x - y &= 4, \\ 5x - y &= 13 \end{aligned}$

21. $\begin{aligned} 2y &= 3x + 5, \\ x &= y - 3 \end{aligned}$

22. $\begin{aligned} 4x - y &= 9, \\ x - 3y &= 16 \end{aligned}$

23. $\begin{aligned} x &= y - 1, \\ 2x &= 3y \end{aligned}$

24. $\begin{aligned} a &= 1 + b, \\ b &= 5 - 2a \end{aligned}$

25. $\begin{aligned} y &= -1, \\ x &= 3 \end{aligned}$

26. $\begin{aligned} y &= 2, \\ x &= -4 \end{aligned}$

27. $\begin{aligned} t + 2s &= -1, \\ s &= t + 10 \end{aligned}$

28. $\begin{aligned} b + 2a &= 2, \\ a &= -3 - b \end{aligned}$

29. $\begin{aligned} 2b + a &= 11, \\ a - b &= 5 \end{aligned}$

30. $\begin{aligned} y &= -\tfrac{1}{3}x - 1, \\ 4x - 3y &= 18 \end{aligned}$

31. $\begin{aligned} y &= -\tfrac{1}{4}x + 1, \\ 2y &= x - 4 \end{aligned}$

32. $\begin{aligned} 6x - 2y &= 2, \\ 9x - 3y &= 1 \end{aligned}$

33. $\begin{aligned} y - x &= 5, \\ 2x - 2y &= 10 \end{aligned}$

34. $\begin{aligned} y &= x + 2, \\ 3y - 2x &= 4 \end{aligned}$

35. $\begin{aligned} y &= 3 - x, \\ 2x + 2y &= 6 \end{aligned}$

36. $\begin{aligned} 2x - 3y &= 6, \\ 3y - 2x &= -6 \end{aligned}$

37. For the systems in the odd-numbered exercises 17–35, which are consistent?

38. For the systems in the even-numbered exercises 18–36, which are consistent?

39. For the systems in the odd-numbered exercises 17–35, which contain dependent equations?

40. For the systems in the even-numbered exercises 18–36, which contain dependent equations?

Translate each problem situation to a system of equations. Do not attempt to solve, but save for later use.

41. The sum of two numbers is 10. The first number is $\frac{2}{3}$ of the second number. What are the numbers?

42. The sum of two numbers is 30. The first number is twice the second number. What are the numbers?

43. *e-mail usage.*   In 2007, the average e-mail user sent 578 personal and business e-mails each week. The number of business e-mails was 30 more than the number of personal e-mails. How many of each type were sent each week?
Source: *JupiterResearch*

44. *Nontoxic furniture polish.*   A nontoxic wood furniture polish can be made by mixing mineral (or olive) oil with vinegar. To make a 16-oz batch for a squirt bottle, Jazmun uses an amount of mineral oil that is 4 oz more than twice the amount of vinegar. How much of each ingredient is required?
Sources: Based on information from Chittenden Solid Waste District and *Clean House, Clean Planet* by Karen Logan

45. *Geometry.*   Two angles are supplementary.* One angle is 3° less than twice the other. Find the measures of the angles.

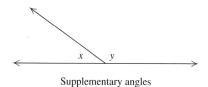

Supplementary angles

46. *Geometry.*   Two angles are complementary.† The sum of the measures of the first angle and half the second angle is 64°. Find the measures of the angles.

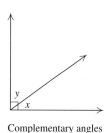

Complementary angles

47. *Basketball scoring.*   Wilt Chamberlain once scored 100 points, setting a record for points scored in an NBA game. Chamberlain took only two-point shots and (one-point) foul shots and made a total of 64 shots. How many shots of each type did he make?

48. *Basketball scoring.*   The Fenton College Cougars made 40 field goals in a recent basketball game, some 2-pointers and the rest 3-pointers. Altogether the 40 baskets counted for 89 points. How many of each type of field goal was made?

---

*The sum of the measures of two supplementary angles is 180°.
†The sum of the measures of two complementary angles is 90°.

49. *Retail sales.*   Simply Souvenirs sold 45 hats and tee shirts. The hats sold for $14.50 each and the tee shirts for $19.50 each. In all, $697.50 was taken in for the souvenirs. How many of each type of souvenir were sold?

50. *Retail sales.*   Cool Treats sold 60 ice cream cones. Single-dip cones sold for $2.50 each and double-dip cones for $4.15 each. In all, $179.70 was taken in for the cones. How many of each size cone were sold?

51. *Sales of pharmaceuticals.*   In 2008, the Diabetic Express charged $83.29 for a 10-mL vial of Humalog insulin and $76.76 for a 10-mL vial of Lantus insulin. If a total of $3981.66 was collected for 50 vials of insulin, how many vials of each type were sold?

52. *Fundraising.*   The Buck Creek Fire Department served 250 dinners. A child's plate cost $5.50 and an adult's plate cost $9.00. A total of $1935 was collected. How many of each type of plate was served?

53. *Lacrosse.*   The perimeter of an NCAA men's lacrosse field is 340 yd. The length is 50 yd longer than the width. Find the dimensions.

P = 340 yd

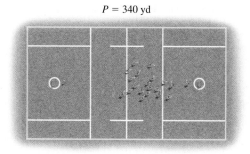

54. *Tennis.*   The perimeter of a standard tennis court used for doubles is 228 ft. The width is 42 ft less than the length. Find the dimensions.

55. Write a problem for a classmate to solve that requires writing a system of two equations. Devise the problem so that the solution is "The Fever made 6 three-point baskets and 31 two-point baskets."

56. Write a problem for a classmate to solve that can be translated into a system of two equations. Devise the problem so that the solution is "In 2009, Diana took five 3-credit classes and two 4-credit classes."

Skill Review

To prepare for Section 3.2, review solving equations and formulas (Sections 1.3 and 1.5).

Solve. [1.3]

57. $3x + 2(5x - 1) = 6$

58. $4(3y + 2) - 7y = 3$

59. $9y = 5 - (y + 6)$

60. $2x - (x - 7) = 18$

Solve. [1.5]

61. $3x - y = 4$, for y

62. $5y - 2x = 7$, for x

Synthesis

Advertising media.   For Exercises 63 and 64, consider the following graph showing the U.S. market share for various advertising media.

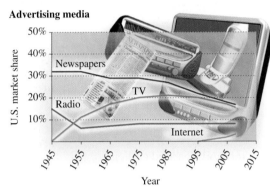

Source: The Wall Street Journal, 12/30/07

63. In what year did no one medium have a higher advertising market share than the others? Explain.

64. Will the Internet advertising market share ever exceed that of radio? TV? newspapers? If so, when? Explain your answers.

65. For each of the following conditions, write a system of equations.

   a) $(5, 1)$ is a solution.
   b) There is no solution.
   c) There is an infinite number of solutions.

66. A system of linear equations has $(1, -1)$ and $(-2, 3)$ as solutions. Determine:

   a) a third point that is a solution, and
   b) how many solutions there are.

67. The solution of the following system is $(4, -5)$. Find A and B.

$$Ax - 6y = 13,$$
$$x - By = -8.$$

Translate to a system of equations. Do not solve.

68. *Ages.*   Tyler is twice as old as his son. Ten years ago, Tyler was three times as old as his son. How old are they now?

69. *Work experience.*   Dell and Juanita are mathematics professors at a state university. Together, they have 46 years of service. Two years ago, Dell had taught 2.5 times as many years as Juanita. How long has each taught at the university?

70. *Design.*   A piece of posterboard has a perimeter of 156 in. If you cut 6 in. off the width, the length becomes four times the width. What are the dimensions of the original piece of posterboard?

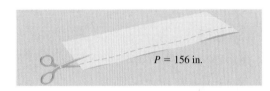

71. *Nontoxic scouring powder.*   A nontoxic scouring powder is made up of 4 parts baking soda and 1 part vinegar. How much of each ingredient is needed for a 16-oz mixture?

72. Solve Exercise 41 graphically.

73. Solve Exercise 44 graphically.

Solve graphically.

74. $y = |x|,$ **75.** $x - y = 0,$
       $3y - x = 8$ $y = x^2$

In Exercises 76–79, use a graphing calculator to solve each system of linear equations for x and y. Round all coordinates to the nearest hundredth.

76. $y = 8.23x + 2.11,$ **77.** $y = -3.44x - 7.72,$
    $y = -9.11x - 4.66$ $y = 4.19x - 8.22$

78. $14.12x + 7.32y = 2.98,$
    $21.88x - 6.45y = -7.22$

79. $5.22x - 8.21y = -10.21,$
   $-12.67x + 10.34y = 12.84$

3.2 # Solving by Substitution or Elimination

The Substitution Method ■ The Elimination Method

The Substitution Method

Algebraic (nongraphical) methods for solving systems are often superior to graphing, especially when fractions are involved. One algebraic method, the *substitution method*, relies on having a variable isolated.

EXAMPLE 1

Solve the system

$$x + y = 4, \quad (1)$$
$$x = y + 1. \quad (2)$$

For easy reference, we have numbered the equations.

SOLUTION   Equation (2) says that x and $y + 1$ name the same number. Thus we can substitute $y + 1$ for x in equation (1):

$$x + y = 4 \qquad \text{Equation (1)}$$
$$(y + 1) + y = 4. \qquad \text{Substituting } y + 1 \text{ for } x$$

We solve this last equation, using methods learned earlier:

$$(y + 1) + y = 4$$
$$2y + 1 = 4 \qquad \text{Removing parentheses and combining like terms}$$
$$2y = 3 \qquad \text{Subtracting 1 from both sides}$$
$$y = \tfrac{3}{2}. \qquad \text{Dividing both sides by 2}$$

We now return to the original pair of equations and substitute $\tfrac{3}{2}$ for y in either equation so that we can solve for x. For this problem, calculations are slightly easier if we use equation (2):

$$x = y + 1 \qquad \text{Equation (2)}$$
$$= \tfrac{3}{2} + 1 \qquad \text{Substituting } \tfrac{3}{2} \text{ for } y$$
$$= \tfrac{3}{2} + \tfrac{2}{2} = \tfrac{5}{2}.$$

We obtain the ordered pair $\left(\tfrac{5}{2}, \tfrac{3}{2}\right)$. A check ensures that it is a solution.

A visualization of Example 1. Note that the coordinates of the intersection are not obvious.

Check:

$x + y = 4$
$\tfrac{5}{2} + \tfrac{3}{2}$ | 4
$\tfrac{8}{2}$
$4 \overset{?}{=} 4$ TRUE

$x = y + 1$
$\tfrac{5}{2}$ | $\tfrac{3}{2} + 1$
$\tfrac{3}{2} + \tfrac{2}{2}$
$\tfrac{5}{2} \overset{?}{=} \tfrac{5}{2}$ TRUE

Since $\left(\tfrac{5}{2}, \tfrac{3}{2}\right)$ checks, it is the solution.

TRY EXERCISE ▶ 7

The exact solution to Example 1 is difficult to find graphically because it involves fractions. The graph shown serves as a partial check and provides a visualization of the problem.

If neither equation in a system has a variable alone on one side, we first isolate a variable in one equation and then substitute.

EXAMPLE 2

Solve the system

$$2x + y = 6, \quad (1)$$
$$3x + 4y = 4. \quad (2)$$

SOLUTION   First, we select an equation and solve for one variable. We can isolate y by subtracting $2x$ from both sides of equation (1):

$$2x + y = 6 \qquad (1)$$
$$y = 6 - 2x. \qquad (3) \qquad \text{Subtracting } 2x \text{ from both sides}$$

Next, we proceed as in Example 1, by substituting:

$$3x + 4(6 - 2x) = 4 \qquad \text{Substituting } 6 - 2x \text{ for } y \text{ in equation (2).}$$
$$\text{Use parentheses!}$$
$$3x + 24 - 8x = 4 \qquad \text{Distributing to remove parentheses}$$
$$3x - 8x = 4 - 24 \qquad \text{Subtracting } 24 \text{ from both sides}$$
$$-5x = -20$$
$$x = 4. \qquad \text{Dividing both sides by } -5$$

Next, we substitute 4 for x in either equation (1), (2), or (3). It is easiest to use equation (3) because it has already been solved for y:

$$y = 6 - 2x$$
$$= 6 - 2(4)$$
$$= 6 - 8 = -2.$$

The pair $(4, -2)$ appears to be the solution. We check in equations (1) and (2).

Check:

$2x + y = 6$		$3x + 4y = 4$	
$2(4) + (-2)$	6	$3(4) + 4(-2)$	4
$8 - 2$		$12 - 8$	
	$6 \overset{?}{=} 6$ TRUE		$4 \overset{?}{=} 4$ TRUE

A visualization of Example 2

Since $(4, -2)$ checks, it is the solution.

TRY EXERCISE ▶ 11

Some systems have no solution, as we saw graphically in Section 3.1. How do we recognize such systems if we are solving by an algebraic method?

EXAMPLE 3   Solve the system

$$y = -3x + 5, \qquad (1)$$
$$y = -3x - 2. \qquad (2)$$

SOLUTION   We solved this system graphically in Example 4(b) of Section 3.1, and found that the lines are parallel and the system has no solution. Let's now try to solve the system by substitution. Proceeding as in Example 1, we substitute $-3x - 2$ for y in the first equation:

$$-3x - 2 = -3x + 5 \qquad \text{Substituting } -3x - 2 \text{ for } y \text{ in equation (1)}$$
$$-2 = 5. \qquad \text{Adding } 3x \text{ to both sides; } -2 = 5 \text{ is a contradiction. The equation is always false.}$$

A visualization of Example 3

Since there is no solution of $-2 = 5$, there is no solution of the system. We state that there is no solution.

TRY EXERCISE ▶ 21

When solving a system algebraically yields a contradiction, the system has no solution.

As we will see in Example 7, when solving a system of two equations algebraically yields an identity, the system has an infinite number of solutions.

The Elimination Method

The *elimination method* for solving systems of equations makes use of the *addition principle*: If $a = b$, then $a + c = b + c$. Consider the following system:

$$2x - 3y = 0,$$
$$-4x + 3y = -1.$$

Note that the $-3y$ in one equation and the $3y$ in the other are opposites. If we add all terms on the left side of the equations, the sum of $-3y$ and $3y$ is 0, so in effect, the variable y is "eliminated."

EXAMPLE 4   Solve the system

$$2x - 3y = 0, \qquad (1)$$
$$-4x + 3y = -1. \qquad (2)$$

SOLUTION   Note that according to equation (2), $-4x + 3y$ and -1 are the same number. Thus we can use the addition principle to work vertically and add $-4x + 3y$ to the left side of equation (1) and -1 to the right side:

$$
\begin{array}{ll}
2x - 3y = 0 & (1) \\
\underline{-4x + 3y = -1} & (2) \\
-2x + 0y = -1. & \text{Adding}
\end{array}
$$

This eliminates the variable y, and leaves an equation with just one variable, x, for which we solve:

$$-2x = -1$$
$$x = \tfrac{1}{2}.$$

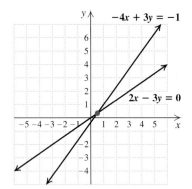

A visualization of Example 4

Next, we substitute $\tfrac{1}{2}$ for x in equation (1) and solve for y:

$$2 \cdot \tfrac{1}{2} - 3y = 0 \qquad \text{Substituting. We also could have used equation (2).}$$
$$1 - 3y = 0$$
$$-3y = -1, \text{ so } y = \tfrac{1}{3}.$$

Check:

$$
\begin{array}{c|c}
2x - 3y = 0 & \\
\hline
2\left(\tfrac{1}{2}\right) - 3\left(\tfrac{1}{3}\right) & 0 \\
1 - 1 & \\
0 \stackrel{?}{=} 0 & \text{TRUE}
\end{array}
\qquad
\begin{array}{c|c}
-4x + 3y = -1 & \\
\hline
-4\left(\tfrac{1}{2}\right) + 3\left(\tfrac{1}{3}\right) & -1 \\
-2 + 1 & \\
-1 \stackrel{?}{=} -1 & \text{TRUE}
\end{array}
$$

Since $\left(\tfrac{1}{2}, \tfrac{1}{3}\right)$ checks, it is the solution. See also the graph at left.

TRY EXERCISE 23

To eliminate a variable, we must sometimes multiply before adding.

EXAMPLE 5   Solve the system

$$5x + 4y = 22, \qquad (1)$$
$$-3x + 8y = 18. \qquad (2)$$

SOLUTION   If we add the left sides of the two equations, we will not eliminate a variable. However, if the $4y$ in equation (1) were changed to $-8y$, we would. To accomplish this change, we multiply both sides of equation (1) by -2:

$$
\begin{array}{ll}
-10x - 8y = -44 & \text{Multiplying both sides of equation (1) by } -2 \\
\underline{-3x + 8y = 18} & \\
-13x + 0 = -26 & \text{Adding} \\
x = 2. & \text{Solving for } x
\end{array}
$$

Then

$$
\begin{array}{ll}
-3 \cdot 2 + 8y = 18 & \text{Substituting 2 for } x \text{ in equation (2)} \\
-6 + 8y = 18 & \\
\left.\begin{array}{l} 8y = 24 \\ y = 3. \end{array}\right\} & \text{Solving for } y
\end{array}
$$

We obtain $(2, 3)$, or $x = 2$, $y = 3$. We leave it to the student to confirm that this checks and is the solution.

TRY EXERCISE 29

Sometimes we must multiply twice in order to make two terms become opposites.

EXAMPLE 6   Solve the system

$$
\begin{array}{ll}
2x + 3y = 17, & (1) \\
5x + 7y = 29. & (2)
\end{array}
$$

SOLUTION   We multiply so that the x-terms will be eliminated when we add.

Eliminate x.

Solve for y.

$$
\begin{array}{l}
2x + 3y = 17, \xrightarrow{\text{Multiplying both sides by 5}} 10x + 15y = 85 \\
5x + 7y = 29 \xrightarrow{\text{Multiplying both sides by } -2} \underline{-10x - 14y = -58} \\
\phantom{5x + 7y = 29 \xrightarrow{\text{Multiplying both}}} 0 + y = 27 \quad \text{Adding} \\
\phantom{5x + 7y = 29 \xrightarrow{\text{Multiplying both}}} y = 27
\end{array}
$$

Next, we substitute to find x:

$$
\begin{array}{ll}
2x + 3 \cdot 27 = 17 & \text{Substituting 27 for } y \text{ in equation (1)} \\
2x + 81 = 17 & \\
\left.\begin{array}{l} 2x = -64 \\ x = -32. \end{array}\right\} & \text{Solving for } x
\end{array}
$$

Solve for x.

Check.

Check:

$$
\begin{array}{c|c}
2x + 3y = 17 & \\
\hline
2(-32) + 3(27) & 17 \\
-64 + 81 & \\
17 \stackrel{?}{=} 17 & \text{TRUE}
\end{array}
\qquad
\begin{array}{c|c}
5x + 7y = 29 & \\
\hline
5(-32) + 7(27) & 29 \\
-160 + 189 & \\
29 \stackrel{?}{=} 29 & \text{TRUE}
\end{array}
$$

We obtain $(-32, 27)$, or $x = -32$, $y = 27$, as the solution.

TRY EXERCISE 31

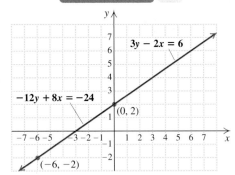

A visualization of Example 7

EXAMPLE 7

Solve the system

$$3y - 2x = 6, \qquad (1)$$
$$-12y + 8x = -24. \qquad (2)$$

SOLUTION We graphed this system in Example 4(c) of Section 3.1, and found that the lines coincide and the system has an infinite number of solutions. Suppose we were to solve this system using the elimination method:

$$12y - 8x = 24 \qquad \text{Multiplying both sides of equation (1) by 4}$$
$$\underline{-12y + 8x = -24}$$
$$0 = 0. \qquad \text{We obtain an identity; } 0 = 0 \text{ is always true.}$$

Note that both variables have been eliminated and what remains is an identity—that is, an equation that is always true. Any pair that is a solution of equation (1) is also a solution of equation (2). The equations are dependent and the solution set is infinite:

$$\{(x, y) \mid 3y - 2x = 6\}, \text{ or equivalently, } \{(x, y) \mid -12y + 8x = -24\}.$$

TRY EXERCISE 47

Example 3 and Example 7 illustrate how to tell algebraically whether a system of two equations is inconsistent or whether the equations are dependent.

Rules for Special Cases

When solving a system of two linear equations in two variables:

1. If we obtain an identity such as $0 = 0$, then the system has an infinite number of solutions. The equations are dependent and, since a solution exists, the system is consistent.*
2. If we obtain a contradiction such as $0 = 7$, then the system has no solution. The system is inconsistent.

Should decimals or fractions appear, it often helps to *clear* before solving.

EXAMPLE 8

Solve the system

$$0.2x + 0.3y = 1.7,$$
$$\tfrac{1}{7}x + \tfrac{1}{5}y = \tfrac{29}{35}.$$

SOLUTION We have

$$0.2x + 0.3y = 1.7, \longrightarrow \text{Multiplying both sides by 10} \longrightarrow 2x + 3y = 17$$
$$\tfrac{1}{7}x + \tfrac{1}{5}y = \tfrac{29}{35} \longrightarrow \text{Multiplying both sides by 35} \longrightarrow 5x + 7y = 29.$$

We multiplied both sides of the first equation by 10 to clear the decimals. Multiplication by 35, the least common denominator, clears the fractions in the second equation. The problem now happens to be identical to Example 6. The solution is $(-32, 27)$, or $x = -32, y = 27$.

TRY EXERCISE 35

The steps for each algebraic method for solving systems of two equations are given below. Note that in both methods, we find the value of one variable and then substitute to find the corresponding value of the other variable.

*Consistent systems and dependent equations are discussed in greater detail in Section 3.4.

> **To Solve a System Using Substitution**
> 1. Isolate a variable in one of the equations (unless one is already isolated).
> 2. Substitute for that variable in the other equation, using parentheses.
> 3. Solve for the remaining variable.
> 4. Substitute the value of the second variable in any of the equations, and solve for the first variable.
> 5. Form an ordered pair and check in the original equations.

> **To Solve a System Using Elimination**
> 1. Write both equations in standard form.
> 2. Multiply both sides of one or both equations by a constant, if necessary, so that the coefficients of one of the variables are opposites.
> 3. Add the left sides and the right sides of the resulting equations. One variable should be eliminated in the sum.
> 4. Solve for the remaining variable.
> 5. Substitute the value of the second variable in any of the equations, and solve for the first variable.
> 6. Form an ordered pair and check in the original equations.

3.2   EXERCISE SET

🦯 *Concept Reinforcement   In each of Exercises 1–6, match the system listed with the choice from the column on the right that would be a subsequent step in solving the system.*

1. _____ $3x - 4y = 6,$
   $5x + 4y = 1$

2. _____ $2x - y = 8,$
   $y = 5x + 3$

3. _____ $x - 2y = 3,$
   $5x + 3y = 4$

4. _____ $8x + 6y = -15,$
   $5x - 3y = 8$

5. _____ $y = 4x - 7,$
   $6x + 3y = 19$

6. _____ $y = 4x - 1,$
   $y = -\frac{2}{3}x - 1$

a) $-5x + 10y = -15,$
   $5x + 3y = 4$

b) The lines intersect at $(0, -1)$.

c) $6x + 3(4x - 7) = 19$

d) $8x = 7$

e) $2x - (5x + 3) = 8$

f) $8x + 6y = -15,$
   $10x - 6y = 16$

For Exercises 7–54, if a system has an infinite number of solutions, use set-builder notation to write the solution set. If a system has no solution, state this.

Solve using the substitution method.

7. $y = 3 - 2x,$
   $3x + y = 5$

8. $3y + x = 4,$
   $x = 2y - 1$

9. $3x + 5y = 3,$
   $x = 8 - 4y$

10. $9x - 2y = 3,$
    $3x - 6 = y$

11. $3s - 4t = 14,$
    $5s + t = 8$

12. $m - 2n = 16,$
    $4m + n = 1$

13. $4x - 2y = 6,$
    $2x - 3 = y$

14. $t = 4 - 2s,$
    $t + 2s = 6$

15. $-5s + t = 11,$
    $4s + 12t = 4$

16. $5x + 6y = 14,$
    $-3y + x = 7$

17. $2x + 2y = 2,$
    $3x - y = 1$

18. $4p - 2q = 16,$
    $5p + 7q = 1$

19. $2a + 6b = 4,$
    $3a - b = 6$

20. $3x - 4y = 5,$
    $2x - y = 1$

21. $2x - 3 = y,$
    $y - 2x = 1$

22. $a - 2b = 3,$
    $3a = 6b + 9$

Solve using the elimination method.

23. $x + 3y = 7,$
    $-x + 4y = 7$

24. $2x + y = 6,$
    $x - y = 3$

25. $x - 2y = 11,$
    $3x + 2y = 17$

26. $5x - 3y = 8,$
    $-5x + y = 4$

27. $9x + 3y = -3,$
    $2x - 3y = -8$

28. $6x - 3y = 18,$
    $6x + 3y = -12$

29. $5x + 3y = 19,$
$\quad x - 6y = 11$

30. $3x + 2y = 3,$
$\quad 9x - 8y = -2$

31. $5r - 3s = 24,$
$\quad 3r + 5s = 28$

32. $5x - 7y = -16,$
$\quad 2x + 8y = 26$

33. $6s + 9t = 12,$
$\quad 4s + 6t = 5$

34. $10a + 6b = 8,$
$\quad 5a + 3b = 2$

35. $\frac{1}{2}x - \frac{1}{6}y = 10,$
$\quad \frac{2}{5}x + \frac{1}{2}y = 8$

36. $\frac{1}{3}x + \frac{1}{5}y = 7,$
$\quad \frac{1}{6}x - \frac{2}{5}y = -4$

37. $\frac{x}{2} + \frac{y}{3} = \frac{7}{6},$
$\quad \frac{2x}{3} + \frac{3y}{4} = \frac{5}{4}$

38. $\frac{2x}{3} + \frac{3y}{4} = \frac{11}{12},$
$\quad \frac{x}{3} + \frac{7y}{18} = \frac{1}{2}$

Aha! **39.** $12x - 6y = -15,$
$\quad -4x + 2y = 5$

40. $8s + 12t = 16,$
$\quad 6s + 9t = 12$

41. $0.3x + 0.2y = 0.3,$
$\quad 0.5x + 0.4y = 0.4$

42. $0.3x + 0.2y = 5,$
$\quad 0.5x + 0.4y = 11$

Solve using any appropriate method.

43. $a - 2b = 16,$
$\quad b + 3 = 3a$

44. $5x - 9y = 7,$
$\quad 7y - 3x = -5$

45. $10x + y = 306,$
$\quad 10y + x = 90$

46. $3(a - b) = 15,$
$\quad 4a = b + 1$

47. $6x - 3y = 3,$
$\quad 4x - 2y = 2$

48. $x + 2y = 8,$
$\quad x = 4 - 2y$

49. $3s - 7t = 5,$
$\quad 7t - 3s = 8$

50. $\quad 2s - 13t = 120,$
$\quad -14s + 91t = -840$

51. $0.05x + 0.25y = 22,$
$\quad 0.15x + 0.05y = 24$

52. $\quad 2.1x - 0.9y = 15,$
$\quad -1.4x + 0.6y = 10$

53. $13a - 7b = 9,$
$\quad 2a - 8b = 6$

54. $3a - 12b = 9,$
$\quad 4a - 5b = 3$

55. Describe a procedure that can be used to write an inconsistent system of equations.

56. Describe a procedure that can be used to write a system that has an infinite number of solutions.

Skill Review

To prepare for Section 3.3, review solving problems using the five-step problem-solving strategy (Section 1.4).

Solve. [1.4]

57. *Energy consumption.*   With average use, a toaster oven and a convection oven together consume 15 kilowatt hours (kWh) of electricity each month. A convection oven uses four times as much electricity as a toaster oven. How much does each use per month?
Source: Lee County Electric Cooperative

58. *Test scores.*   Ellia needs to average 80 on her tests in order to earn a B in her math class. Her average after 4 tests is 77.5. What score is needed on the fifth test in order to raise the average to 80?

59. *Real estate.*   After her house had been on the market for 6 months, Gina reduced the price to $94,500. This was $\frac{9}{10}$ of the original asking price. How much did Gina originally ask for her house?

60. *Car rentals.*   National Car Rental rents minivans to a university for $69 a day plus 30¢ per mile. An English professor rented a minivan for 2 days to take a group of students to a seminar. The bill was $225. How far did the professor drive the van?
Source: www.nationalcar.com

61. *Carpentry.*   Anazi cuts a 96-in. piece of wood trim into three pieces. The second piece is twice as long as the first. The third piece is one-tenth as long as the second. How long is each piece?

62. *Telephone calls.*   Terri's voice over the Internet (VoIP) phone service charges $0.36 for the first minute of each call and $0.06 for each additional $\frac{1}{2}$ minute. One month she was charged $28.20 for 35 calls. How many minutes did she use?

Synthesis

63. Some systems are more easily solved by substitution and some are more easily solved by elimination. What guidelines could be used to help someone determine which method to use?

64. Explain how it is possible to solve Exercise 39 mentally.

65. If $(1, 2)$ and $(-3, 4)$ are two solutions of $f(x) = mx + b$, find m and b.

66. If $(0, -3)$ and $\left(-\frac{3}{2}, 6\right)$ are two solutions of $px - qy = -1$, find p and q.

67. Determine a and b for which $(-4, -3)$ is a solution of the system

$$ax + by = -26,$$
$$bx - ay = 7.$$

68. Solve for x and y in terms of a and b:

$$5x + 2y = a,$$
$$x - y = b.$$

Solve.

69. $\dfrac{x + y}{2} - \dfrac{x - y}{5} = 1,$

$\dfrac{x - y}{2} + \dfrac{x + y}{6} = -2$

70. $3.5x - 2.1y = 106.2,$

$4.1x + 16.7y = -106.28$

Each of the following is a system of nonlinear equations. However, each is reducible to linear, since an appropriate substitution (say, u for 1/x and v for 1/y) yields a linear system. Make such a substitution, solve for the new variables, and then solve for the original variables.

71. $\dfrac{2}{x} + \dfrac{1}{y} = 0,$

$\dfrac{5}{x} + \dfrac{2}{y} = -5$

72. $\dfrac{1}{x} - \dfrac{3}{y} = 2,$

$\dfrac{6}{x} + \dfrac{5}{y} = -34$

73. A student solving the system

$$17x + 19y = 102,$$
$$136x + 152y = 826$$

graphs both equations on a graphing calculator and gets the following screen. The student then (incorrectly) concludes that the equations are dependent and the solution set is infinite. How can algebra be used to convince the student that a mistake has been made?

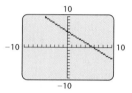

![up arrow] **CONNECTING** ![down arrow] **the CONCEPTS**

We now have three different methods for solving systems of equations. Each method has certain strengths and weaknesses, as outlined below.

Method	Strengths	Weaknesses
Graphical	Solutions are displayed graphically. Can be used with any system that can be graphed.	For some systems, only approximate solutions can be found graphically. The graph drawn may not be large enough to show the solution.
Substitution	Yields exact solutions. Easy to use when a variable has a coefficient of 1.	Introduces extensive computations with fractions when solving more complicated systems. Solutions are not displayed graphically.
Elimination	Yields exact solutions. Easy to use when fractions or decimals appear in the system. The preferred method for systems of 3 or more equations in 3 or more variables (see Section 3.4).	Solutions are not displayed graphically.

(continued)

When selecting a method to use, consider the strengths and weaknesses listed above. If possible, begin solving the system mentally to help discover the method that seems best suited for that particular system.

MIXED REVIEW

Solve using the best method.

1. $x = y,$
$x + y = 2$

2. $x + y = 10,$
$x - y = 8$

3. $y = \frac{1}{2}x + 1,$
$y = 2x - 5$

4. $y = 2x - 3,$
$x + y = 12$

5. $x = 5,$
$y = 10$

6. $3x + 5y = 8,$
$3x - 5y = 4$

7. $2x - y = 1,$
$2y - 4x = 3$

8. $x = 2 - y,$
$3x + 3y = 6$

9. $x + 2y = 3,$
$3x = 4 - y$

10. $9x + 8y = 0,$
$11x - 7y = 0$

11. $10x + 20y = 40,$
$x - \quad y = 7$

12. $y = \frac{5}{3}x + 7,$
$y = \frac{5}{3}x - 8$

13. $2x - 5y = 1,$
$3x + 2y = 11$

14. $\dfrac{x}{2} + \dfrac{y}{3} = \dfrac{2}{3},$
$\dfrac{x}{5} + \dfrac{5y}{2} = \dfrac{1}{4}$

15. $1.1x - 0.3y = 0.8,$
$2.3x + 0.3y = 2.6$

16. $y = -3,$
$x = 11$

17. $x - 2y = 5,$
$3x - 15 = 6y$

18. $12x - 19y = 13,$
$8x + 19y = 7$

19. $0.2x + 0.7y = 1.2,$
$0.3x - 0.1y = 2.7$

20. $\frac{1}{4}x = \frac{1}{3}y,$
$\frac{1}{2}x - \frac{1}{15}y = 2$

3.3   Solving Applications: Systems of Two Equations

Total-Value and Mixture Problems   ●   Motion Problems

You are in a much better position to solve problems now that you know how systems of equations can be used. Using systems often makes the translating step easier.

 EXAMPLE   1

Patch of Brighamia Insignis with flower

Endangered species.   The number of plant species listed as threatened (likely to become endangered) or endangered (in danger of becoming extinct) has more than tripled in the past 20 years. In 2008, there were 746 species of plants in the United States that were considered threatened or endangered. The number considered threatened was 4 less than one-fourth of the number considered endangered. How many U.S. plant species were considered endangered and how many were considered threatened in 2008?

Source: U.S. Fish and Wildlife Service

SOLUTION   The *Familiarize* and *Translate* steps were completed in Example 1 of Section 3.1. The resulting system of equations is

$$t + d = 746,$$
$$t = \tfrac{1}{4}d - 4,$$

where d is the number of endangered plant species and t is the number of threatened plant species in the United States in 2008.

3. Carry out. We solve the system of equations. Since one equation already has a variable isolated, let's use the substitution method:

$$t + d = 746$$
$$\tfrac{1}{4}d - 4 + d = 746 \qquad \text{Substituting } \tfrac{1}{4}d - 4 \text{ for } t$$
$$\tfrac{5}{4}d - 4 = 746 \qquad \text{Combining like terms}$$
$$\tfrac{5}{4}d = 750 \qquad \text{Adding 4 to both sides}$$
$$d = \tfrac{4}{5} \cdot 750 \qquad \text{Multiplying both sides by } \tfrac{4}{5} \colon \ \tfrac{4}{5} \cdot \tfrac{5}{4} = 1$$
$$d = 600. \qquad \text{Simplifying}$$

Next, using either of the original equations, we substitute and solve for t:

$$t = \tfrac{1}{4} \cdot 600 - 4 = 150 - 4 = 146.$$

4. Check. The sum of 600 and 146 is 746, so the total number of species is correct. Since 4 less than one-fourth of 600 is $150 - 4$, or 146, the numbers check.

5. State. In 2008, there were 600 endangered plant species and 146 threatened plant species in the United States.

> TRY EXERCISE ▶ 45

Total-Value and Mixture Problems

Jewelry design. In order to make a necklace, a jewelry designer purchased 80 beads for a total of $39 (excluding tax). Some of the beads were sterling silver beads that cost 40¢ each and the rest were gemstone beads that cost 65¢ each. How many of each type did the designer buy?

SOLUTION   The *Familiarize* and *Translate* steps were completed in Example 2 of Section 3.1.

3. Carry out. We are to solve the system of equations

$$s + g = 80, \qquad (1)$$
$$40s + 65g = 3900, \qquad (2) \qquad \text{Working in cents rather than dollars}$$

where s is the number of silver beads bought and g is the number of gemstone beads bought. Because both equations are in the form $Ax + By = C$, let's use the elimination method to solve the system. We can eliminate s by multiplying both sides of equation (1) by -40 and adding them to the corresponding sides of equation (2):

$$
\begin{array}{rl}
-40s - 40g = -3200 & \text{Multiplying both sides of equation (1) by } -40 \\
\underline{40s + 65g = 3900} & \\
25g = 700 & \text{Adding} \\
g = 28. & \text{Solving for } g
\end{array}
$$

To find s, we substitute 28 for g in equation (1) and then solve for s:

$$s + g = 80 \qquad \text{Equation (1)}$$
$$s + 28 = 80 \qquad \text{Substituting 28 for } g$$
$$s = 52. \qquad \text{Solving for } s$$

We obtain (28, 52), or $g = 28$ and $s = 52$.

STUDENT NOTES

It is very important that you clearly label precisely what each variable represents. Not only will this assist you in writing equations, but it will help you to identify and state solutions.

EXAMPLE 2

4. **Check.** We check in the original problem. Recall that *g* is the number of gemstone beads and *s* the number of silver beads.

Number of beads: $g + s = 28 + 52 = 80$
Cost of gemstone beads: $65g = 65 \times 28 = 1820¢$
Cost of silver beads: $40s = 40 \times 52 = 2080¢$
                                        Total $= 3900¢$

The numbers check.

5. **State.** The designer bought 28 gemstone beads and 52 silver beads.

TRY EXERCISE ▶ 15

Example 2 involved two types of items (silver beads and gemstone beads), the quantity of each type bought, and the total value of the items. We refer to this type of problem as a *total-value problem*.

EXAMPLE 3

Blending teas. Teapots n Treasures sells loose Oolong tea for $2.15 an ounce. Donna mixed Oolong tea with shaved almonds that sell for $0.95 an ounce to create the Market Street Oolong blend that sells for $1.85 an ounce. One week, she made 300 oz of Market Street Oolong. How much tea and how much shaved almonds did Donna use?

SOLUTION

1. **Familiarize.** This problem is similar to Example 2. Rather than silver beads and gemstone beads, we have ounces of tea and ounces of almonds. Instead of a different price for each type of bead, we have a different price per ounce for each ingredient. Finally, rather than knowing the total cost of the beads, we know the weight and the price per ounce of the mixture. Thus we can find the total value of the blend by multiplying 300 ounces times $1.85 per ounce. We let $l =$ the number of ounces of Oolong tea and $a =$ the number of ounces of shaved almonds.

2. **Translate.** Since a 300-oz batch was made, we must have

$l + a = 300.$

To find a second equation, note that the total value of the 300-oz blend must match the combined value of the separate ingredients:

Rewording: The value of the Oolong tea   plus   the value of the almonds   is   the value of the Market Street blend.

Translating: $l \cdot \$2.15$ $+$ $a \cdot \$0.95$ $=$ $300 \cdot \$1.85$

These equations can also be obtained from a table.

	Oolong Tea	Almonds	Market Street Blend	
Number of Ounces	l	a	300	→ $l + a = 300$
Price per Ounce	$2.15	$0.95	$1.85	
Value of Tea	$2.15l	$0.95a	300 · $1.85, or $555	→ $2.15l + 0.95a = 555$

Clearing decimals in the second equation, we have $215l + 95a = 55{,}500$. We have translated to a system of equations:

$$l + \quad a = 300, \qquad (1)$$
$$215l + 95a = 55{,}500. \qquad (2)$$

3. **Carry out.** We can solve using substitution. When equation (1) is solved for l, we have $l = 300 - a$. Substituting $300 - a$ for l in equation (2), we find a:

$215(300 - a) + 95a = 55{,}500$	Substituting
$64{,}500 - 215a + 95a = 55{,}500$	Using the distributive law
$-120a = -9000$	Combining like terms; subtracting $64{,}500$ from both sides
$a = 75.$	Dividing both sides by -120

We have $a = 75$ and, from equation (1) above, $l + a = 300$. Thus, $l = 225$.

4. **Check.** Combining 225 oz of Oolong tea and 75 oz of almonds will give a 300-oz blend. The value of 225 oz of Oolong is $225(\$2.15)$, or \$483.75. The value of 75 oz of almonds is $75(\$0.95)$, or \$71.25. Thus the combined value of the blend is $\$483.75 + \$71.25$, or \$555. A 300-oz blend priced at \$1.85 an ounce would also be worth \$555, so our answer checks.

5. **State.** The Market Street blend was made by combining 225 oz of Oolong tea and 75 oz of almonds.

TRY EXERCISE 23

EXAMPLE 4

Student loans. Rani's student loans totaled \$9600. Part was a PLUS loan made at 8.5% interest and the rest was a Stafford loan made at 6.8% interest. After one year, Rani's loans accumulated \$729.30 in interest. What was the original amount of each loan?

SOLUTION

1. **Familiarize.** We begin with a guess. If \$3000 was borrowed at 8.5% and \$6600 was borrowed at 6.8%, the two loans would total \$9600. The interest would then be $0.085(\$3000)$, or \$255, and $0.068(\$6600)$, or \$448.80, for a total of only \$703.80 in interest. Our guess was wrong, but checking the guess familiarized us with the problem. More than \$3000 was borrowed at the higher rate.

2. **Translate.** We let $p =$ the amount of the PLUS loan and $s =$ the amount of the Stafford loan. Next, we organize a table in which the entries in each column come from the formula for simple interest:

$$Principal \cdot Rate \cdot Time = Interest.$$

	PLUS Loan	Stafford Loan	Total	
Principal	p	s	\$9600	$\rightarrow p + s = 9600$
Rate of Interest	8.5%	6.8%		
Time	1 yr	1 yr		
Interest	$0.085p$	$0.068s$	\$729.30	$\rightarrow 0.085p + 0.068s = 729.30$

The total amount borrowed is found in the first row of the table:

$$p + s = 9600.$$

A second equation, representing the accumulated interest, can be found in the last row:

$$0.085p + 0.068s = 729.30, \quad \text{or} \quad 85p + 68s = 729,300. \qquad \text{Clearing decimals}$$

3. **Carry out.** The system can be solved by elimination:

$$\begin{array}{lll} p + s = 9600, & \longrightarrow \text{Multiplying both} \longrightarrow & -85p - 85s = -816,000 \\ 85p + 68s = 729,300. & \quad\ \text{sides by } -85 & \underline{85p + 68s = 729,300} \\ & & -17s = -86,700 \end{array}$$

$$p + s = 9600 \longleftarrow\!\!\!\longleftarrow s = 5100$$
$$p + 5100 = 9600$$
$$p = 4500.$$

We find that $p = 4500$ and $s = 5100$.

4. **Check.** The total amount borrowed is $4500 + $5100, or $9600. The interest on $4500 at 8.5% for 1 yr is 0.085($4500), or $382.50. The interest on $5100 at 6.8% for 1 yr is 0.068($5100), or $346.80. The total amount of interest is $382.50 + $346.80, or $729.30, so the numbers check.

5. **State.** The PLUS loan was for $4500 and the Stafford loan was for $5100.

Before proceeding to Example 5, briefly scan Examples 2–4 for similarities. Note that in each case, one of the equations in the system is a simple sum while the other equation represents a sum of products. Example 5 continues this pattern with what is commonly called a *mixture problem*.

> **Problem-Solving Tip**
>
> When solving a problem, see if it is patterned or modeled after a problem that you have already solved.

EXAMPLE 5   *Mixing fertilizers.*   Nature's Green Gardening, Inc., carries two brands of fertilizer containing nitrogen and water. "Gentle Grow" is 3% nitrogen and "Sun Saver" is 8% nitrogen. Nature's Green needs to combine the two types of solutions into a 90-L mixture that is 6% nitrogen. How much of each brand should be used?

SOLUTION

1. **Familiarize.** We make a drawing and note that we must consider not only the size of the mixture, but also its strength. Let's make a guess to gain familiarity with the problem.

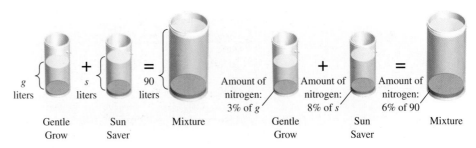

Suppose that 40 L of Gentle Grow and 50 L of Sun Saver are mixed. The resulting mixture will be the right size, 90 L, but will it be the right strength? To find out, note that 40 L of Gentle Grow would contribute $0.03(40) = 1.2$ L of nitrogen to the mixture while 50 L of Sun Saver would contribute $0.08(50) = 4$ L of nitrogen to the mixture. The total amount of nitrogen in the mixture would then be $1.2 + 4$, or 5.2 L. But we want 6% of 90, or 5.4 L, to be nitrogen. Our guess of 40 L and 50 L is close but incorrect. Checking our guess has familiarized us with the problem.

2. **Translate.** Let g = the number of liters of Gentle Grow and s = the number of liters of Sun Saver. The information can be organized in a table.

	Gentle Grow	Sun Saver	Mixture	
Number of Liters	g	s	90	→ $g + s = 90$
Percent of Nitrogen	3%	8%	6%	
Amount of Nitrogen	$0.03g$	$0.08s$	0.06×90, or 5.4 liters	→ $0.03g + 0.08s = 5.4$

If we add g and s in the first row, we get one equation. It represents the total amount of mixture: $g + s = 90$.

If we add the amounts of nitrogen listed in the third row, we get a second equation. This equation represents the amount of nitrogen in the mixture: $0.03g + 0.08s = 5.4$.

After clearing decimals, we have translated the problem to the system

$$\begin{aligned} g + s &= 90, \quad &(1) \\ 3g + 8s &= 540. \quad &(2) \end{aligned}$$

3. **Carry out.** We use the elimination method to solve the system:

$$\begin{aligned} -3g - 3s &= -270 \quad &\text{Multiplying both sides of equation (1) by } -3 \\ \underline{3g + 8s} &= \underline{540} \\ 5s &= 270 \quad &\text{Adding} \\ s &= 54; \quad &\text{Solving for } s \\ g + 54 &= 90 \quad &\text{Substituting into equation (1)} \\ g &= 36. \quad &\text{Solving for } g \end{aligned}$$

4. **Check.** Remember, g is the number of liters of Gentle Grow and s is the number of liters of Sun Saver.

Total amount of mixture: $g + s = 36 + 54 = 90$

Total amount of nitrogen: 3% of 36 + 8% of 54 = $1.08 + 4.32 = 5.4$

Percentage of nitrogen in mixture: $\dfrac{\text{Total amount of nitrogen}}{\text{Total amount of mixture}} = \dfrac{5.4}{90} = 6\%$

The numbers check in the original problem.

5. **State.** Nature's Green Gardening should mix 36 L of Gentle Grow with 54 L of Sun Saver.

TRY EXERCISE 25

Motion Problems

When a problem deals with distance, speed (rate), and time, recall the following.

Distance, Rate, and Time Equations

If r represents rate, t represents time, and d represents distance, then:

$$d = rt, \quad r = \frac{d}{t}, \quad \text{and} \quad t = \frac{d}{r}.$$

Be sure to remember at least one of these equations. The others can be obtained by multiplying or dividing on both sides as needed.

EXAMPLE 6

Train travel. A Vermont Railways freight train, loaded with logs, leaves Boston, heading to Washington D.C., at a speed of 60 km/h. Two hours later, an Amtrak® Metroliner leaves Boston, bound for Washington D.C., on a parallel track at 90 km/h. At what point will the Metroliner catch up to the freight train?

SOLUTION

1. **Familiarize.** Let's make a guess and check to see if it is correct. Suppose the trains meet after traveling 180 km. We can then calculate the time for each train.

	Distance	Rate	Time
Freight Train	180 km	60 km/h	$\frac{180}{60} = 3$ hr
Metroliner	180 km	90 km/h	$\frac{180}{90} = 2$ hr

We see that the distance cannot be 180 km, since the difference in travel times for the trains is *not* 2 hr. Although our guess is wrong, we can use a similar chart to organize the information in this problem.

The distance at which the trains meet is unknown, but we do know that the trains will have traveled the same distance when they meet. We let $d =$ this distance.

The time that the trains are running is also unknown, but we do know that the freight train has a 2-hr head start. Thus if we let $t =$ the number of hours that the freight train is running before they meet, then $t - 2$ is the number of hours that the Metroliner runs before catching up to the freight train.

60 km/h
d kilometers
t hours

90 km/h
d kilometers
$t - 2$ hours

Trains meet here

2. **Translate.** We can organize the information in a chart. Each row is determined by the formula *Distance = Rate · Time*.

	Distance	Rate	Time	
Freight Train	d	60	t	$d = 60t$
Metroliner	d	90	$t - 2$	$d = 90(t - 2)$

Using *Distance* = *Rate* · *Time* twice, we get two equations:

$$d = 60t, \quad (1)$$
$$d = 90(t - 2). \quad (2)$$

3. Carry out. We solve the system using substitution:

$$60t = 90(t - 2) \qquad \text{Substituting } 60t \text{ for } d \text{ in equation (2)}$$
$$60t = 90t - 180$$
$$-30t = -180$$
$$t = 6.$$

STUDENT NOTES ——————

Always be careful to answer the question asked in the problem. In Example 6, the problem asks for distance, not time. Answering "6 hr" would be incorrect.

The time for the freight train is 6 hr, which means that the time for the Metro-liner is 6 − 2, or 4 hr. Remember that it is distance, not time, that the problem asked for. Thus for $t = 6$, we have $d = 60 \cdot 6 = 360$ km.

4. Check. At 60 km/h, the freight train will travel 60 · 6, or 360 km, in 6 hr. At 90 km/h, the Metroliner will travel 90 · (6 − 2) = 360 km in 4 hr. The numbers check.

5. State. The freight train will catch up to the Metroliner at a point 360 km from Boston.

TRY EXERCISE 37

EXAMPLE 7

Jet travel.   A Boeing 747-400 jet flies 4 hr west with a 60-mph tailwind. Returning *against* the wind takes 5 hr. Find the speed of the jet with no wind.

SOLUTION

1. Familiarize. We imagine the situation and make a drawing. Note that the wind *speeds up* the jet on the outbound flight but *slows down* the jet on the return flight.

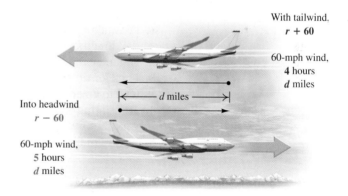

Let's make a guess of the jet's speed if there were no wind. Note that the distances traveled each way must be the same.

Speed with no wind:	400 mph
Speed with the wind:	400 + 60 = 460 mph
Speed against the wind:	400 − 60 = 340 mph
Distance with the wind:	460 · 4 = 1840 mi
Distance against the wind:	340 · 5 = 1700 mi

Since the distances are not the same, our guess of 400 mph is incorrect.

We let r = the speed, in miles per hour, of the jet in still air. Then $r + 60$ = the jet's speed with the wind and $r - 60$ = the jet's speed against the wind. We also let d = the distance traveled, in miles.

2. Translate. The information can be organized in a chart. The distances traveled are the same, so we use *Distance* = *Rate* (or *Speed*) · *Time*. Each row of the chart gives an equation.

	Distance	Rate	Time
With Wind	d	$r + 60$	4
Against Wind	d	$r - 60$	5

$\longrightarrow d = (r + 60)4$

$\longrightarrow d = (r - 60)5$

The two equations constitute a system:

$$d = (r + 60)4, \quad (1)$$
$$d = (r - 60)5. \quad (2)$$

3. Carry out. We solve the system using substitution:

$$(r - 60)5 = (r + 60)4 \qquad \text{Substituting } (r - 60)5 \text{ for } d \text{ in equation (1)}$$
$$5r - 300 = 4r + 240 \qquad \text{Using the distributive law}$$
$$r = 540. \qquad \text{Solving for } r$$

4. Check. When $r = 540$, the speed with the wind is $540 + 60 = 600$ mph, and the speed against the wind is $540 - 60 = 480$ mph. The distance with the wind, $600 \cdot 4 = 2400$ mi, matches the distance into the wind, $480 \cdot 5 = 2400$ mi, so we have a check.

5. State. The speed of the jet with no wind is 540 mph.   ▶ TRY EXERCISE 39

Tips for Solving Motion Problems

1. Draw a diagram using an arrow or arrows to represent distance and the direction of each object in motion.
2. Organize the information in a chart.
3. Look for times, distances, or rates that are the same. These often can lead to an equation.
4. Translating to a system of equations allows for the use of two variables.
5. Always make sure that you have answered the question asked.

3.3 EXERCISE SET

For Extra Help   **MyMathLab**   **Math XL** PRACTICE   WATCH   DOWNLOAD

1.–14. For Exercises 1–14, solve Exercises 41–54 from pp. 155–156.

15. *Recycled paper.*   Staples® recently charged $3.79 per ream (package of 500 sheets) of regular paper and $5.49 per ream of paper made of recycled fibers. Last semester, Valley College spent $582.44 for 116 reams of paper. How many of each type were purchased?

16. *Photocopying.*   Quick Copy recently charged 49¢ per page for color copies and 7¢ per page for black-and-white copies. If Shirlee's bill for 90 copies was $11.34, how many copies of each type were made?

17. *Lighting.* Lowe's Home Improvement recently sold 13-watt Feit Electric Ecobulbs® for $5 each and 18-watt Ecobulbs® for $6 each. If River County Hospital purchased 200 such bulbs for a total of $1140, how many of each type did they purchase?

18. *Office supplies.* Staples® recently charged $17.99 per box of Pilot Precise® rollerball pens and $7.49 per box for Bic® Matic Grip mechanical pencils. If Kelling Community College purchased 120 such boxes for a total of $1234.80, how many boxes of each type did they purchase?

19. *Sales.* Recently, officedepot.com sold a black HP C7115A Laser Jet print cartridge for $64.99 and a color Apple computer M3908GA ink cartridge for $58.99. During a promotion offering free shipping, a total of 450 of these cartridges was purchased for a total of $27,625.50. How many of each type were purchased?

20. *Sales.* Office Max® recently advertised a three-subject notebook for $2.49 and a five-subject notebook for $3.79. At the start of a recent spring semester, a combination of 50 of these notebooks was sold for a total of $166.10. How many of each type were sold?

Aha! **21.** *Blending coffees.* The Roasted Bean charges $13.00 per pound for Fair Trade Organic Mexican coffee and $11.00 per pound for Fair Trade Organic Peruvian coffee. How much of each type should be used to make a 28-lb blend that sells for $12.00 per pound?

22. *Mixed nuts.* Oh Nuts! sells pistachio kernels for $6.50 per pound and almonds for $8.00 per pound. How much of each type should be used to make a 50-lb mixture that sells for $7.40 per pound?

23. *Event planning.* As part of the refreshments for Yvette's 25th birthday party, Kim plans to provide a bowl of M&M candies. She wants to mix custom-printed M&Ms costing 60¢ per ounce with bulk M&Ms costing 25¢ per ounce to create 20 lb of a mixture costing 32¢ per ounce. How much of each type of M&M should she use?
Source: www.mymms.com

24. *Blending spices.* Spice of Life sells ground sumac for $1.35 an ounce and ground thyme for $1.85 an ounce. Aman wants to make a 20-oz Zahtar seasoning blend using the two spices that sells for $1.65 an ounce. How much of each spice should Aman use?

25. *Catering.* Cati's Catering is planning an office reception. The office administrator has requested a candy mixture that is 25% chocolate. Cati has available mixtures that are either 50% chocolate or 10% chocolate. How much of each type should be mixed to get a 20-lb mixture that is 25% chocolate?

26. *Ink remover.* Etch Clean Graphics uses one cleanser that is 25% acid and a second that is 50% acid. How many liters of each should be mixed to get 30 L of a solution that is 40% acid?

27. *Blending granola.* Deep Thought Granola is 25% nuts and dried fruit. Oat Dream Granola is 10% nuts and dried fruit. How much of Deep Thought and how much of Oat Dream should be mixed to form a 20-lb batch of granola that is 19% nuts and dried fruit?

28. *Livestock feed.* Soybean meal is 16% protein and corn meal is 9% protein. How many pounds of each should be mixed to get a 350-lb mixture that is 12% protein?

29. *Student loans.* Stacey's two student loans totaled $12,000. One of her loans was at 6.5% simple interest and the other at 7.2%. After one year, Stacey owed $811.50 in interest. What was the amount of each loan?

30. *Investments.* A self-employed contractor nearing retirement made two investments totaling $15,000. In one year, these investments yielded $1023 in simple interest. Part of the money was invested at 6% and the rest at 7.5%. How much was invested at each rate?

31. *Automotive maintenance.* "Steady State" antifreeze is 18% alcohol and "Even Flow" is 10% alcohol. How many liters of each should be mixed to get 20 L of a mixture that is 15% alcohol?

32. *Chemistry.* E-Chem Testing has a solution that is 80% base and another that is 30% base. A technician needs 150 L of a solution that is 62% base. The 150 L will be prepared by mixing the two solutions on hand. How much of each should be used?

33. *Octane ratings.* The octane rating of a gasoline is a measure of the amount of isooctane in the gas. Manufacturers recommend using 93-octane gasoline on retuned motors. How much 87-octane gas and 95-octane gas should Yousef mix in order to make 10 gal of 93-octane gas for his retuned Ford F-150?
Source: Champlain Electric and Petroleum Equipment

34. *Octane ratings.* The octane rating of a gasoline is a measure of the amount of isooctane in the gas. Subaru recommends 91-octane gasoline for the 2008 Legacy 3.0 R. How much 87-octane gas and 93-octane gas should Kelsey mix in order to make 12 gal of 91-octane gas for her Legacy?
Sources: Champlain Electric and Petroleum Equipment: Dean Team Ballwin

35. *Food science.* The following bar graph shows the milk fat percentages in three dairy products. How many pounds each of whole milk and cream should be mixed to form 200 lb of milk for cream cheese?

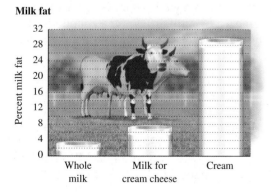

36. *Food science.* How much lowfat milk (1% fat) and how much whole milk (4% fat) should be mixed to make 5 gal of reduced fat milk (2% fat)?

37. *Train travel.* A train leaves Danville Union and travels north at a speed of 75 km/h. Two hours later, an express train leaves on a parallel track and travels north at 125 km/h. How far from the station will they meet?

38. *Car travel.* Two cars leave Salt Lake City, traveling in opposite directions. One car travels at a speed of 80 km/h and the other at 96 km/h. In how many hours will they be 528 km apart?

39. *Canoeing.* Kahla paddled for 4 hr with a 6-km/h current to reach a campsite. The return trip against the same current took 10 hr. Find the speed of Kahla's canoe in still water.

40. *Boating.* Cody's motorboat took 3 hr to make a trip downstream with a 6-mph current. The return trip against the same current took 5 hr. Find the speed of the boat in still water.

41. *Point of no return.* A plane flying the 3458-mi trip from New York City to London has a 50-mph tailwind. The flight's *point of no return* is the point at which the flight time required to return to New York is the same as the time required to continue to London. If the speed of the plane in still air is 360 mph, how far is New York from the point of no return?

42. *Point of no return.* A plane is flying the 2553-mi trip from Los Angeles to Honolulu into a 60-mph headwind. If the speed of the plane in still air is 310 mph, how far from Los Angeles is the plane's point of no return? (See Exercise 41.)

43. *Architecture.* The rectangular ground floor of the John Hancock building has a perimeter of 860 ft. The length is 100 ft more than the width. Find the length and the width.

44. *Real estate.* The perimeter of a rectangular ocean-front lot is 190 m. The width is one-fourth of the length. Find the dimensions.

45. In 2007, Nintendo Co. sold three times as many Wii game machines in Japan as Sony Corp. sold PlayStation 3 consoles. Together, they sold 4.84 million game machines in Japan. How many of each were sold?
Source: Bloomberg.com

46. *Hockey rankings.* Hockey teams receive 2 points for a win and 1 point for a tie. The Wildcats once won a championship with 60 points. They won 9 more games than they tied. How many wins and how many ties did the Wildcats have?

47. *Video rentals.* At one time, Netflix offered an unlimited 1 DVD at-a-time rental plan for $8.99 per month and a rental plan with a limit of 2 DVDs per month for $4.99. During one week, 250 new subscribers paid $1975.50 for these plans. How many of each type of plan were purchased?

48. *Radio airplay.* Roscoe must play 12 commercials during his 1-hr radio show. Each commercial is either 30 sec or 60 sec long. If the total commercial time during that hour is 10 min, how many commercials of each type does Roscoe play?

49. *Making change.* Monica makes a $9.25 purchase at the bookstore with a $20 bill. The store has no bills and gives her the change in quarters and fifty-cent pieces. There are 30 coins in all. How many of each kind are there?

50. *Teller work.* Sabina goes to a bank and gets change for a $50 bill consisting of all $5 bills and $1 bills. There are 22 bills in all. How many of each kind are there?

51. In what ways are Examples 3 and 4 similar? In what sense are their systems of equations similar?

52. Write at least three study tips of your own for someone beginning this exercise set.

Skill Review

To prepare for Section 3.4, review evaluating expressions with three variables (Sections 1.1 and 1.2).

Evaluate. [1.1], [1.2]

53. $2x - 3y - z$, for $x = 5$, $y = 2$, and $z = 3$

54. $4x + y - 6z$, for $x = \frac{1}{2}$, $y = \frac{1}{2}$, and $z = \frac{1}{3}$

55. $x + y + 2z$, for $x = 1$, $y = -4$, and $z = -5$

56. $3a - b + 2c$, for $a = 1$, $b = -6$, and $c = 4$

57. $a - 2b - 3c$, for $a = -2$, $b = 3$, and $c = -5$

58. $2a - 5b - c$, for $a = \frac{1}{4}$, $b = -\frac{1}{4}$, and $c = -\frac{3}{2}$

Synthesis

59. Suppose that in Example 3 you are asked only for the amount of almonds needed for the Market Street blend. Would the method of solving the problem change? Why or why not?

60. Write a problem similar to Example 2 for a classmate to solve. Design the problem so that the solution is "The bakery sold 24 loaves of bread and 18 packages of sandwich rolls."

61. *Recycled paper.* Unable to purchase 60 reams of paper that contains 20% post-consumer fiber, the Naylor School bought paper that was either 0% post-consumer fiber or 30% post-consumer fiber. How many reams of each should be purchased in order to use the same amount of post-consumer fiber as if the 20% post-consumer fiber paper were available?

62. *Automotive maintenance.* The radiator in Natalie's car contains 6.3 L of antifreeze and water. This mixture is 30% antifreeze. How much of this mixture should she drain and replace with pure antifreeze so that there will be a mixture of 50% antifreeze?

63. *Metal alloys.* In order for a metal to be labeled "sterling silver," the silver alloy must contain at least 92.5% pure silver. Nicole has 32 oz of coin silver, which is 90% pure silver. How much pure silver must she add to the coin silver in order to have a sterling-silver alloy?
Source: *The Jewelry Repair Manual*, R. Allen Hardy, Courier Dover Publications, 1996, p. 271.

64. *Exercise.* Elyse jogs and walks to school each day. She averages 4 km/h walking and 8 km/h jogging. From home to school is 6 km and Elyse makes the trip in 1 hr. How far does she jog in a trip?

65. *Book sales.* *American Economic History* can be purchased as a three-volume set for $88 or each volume can be purchased separately for $39. An economics class spent $1641 for 51 volumes. How many three-volume sets were ordered?
Source: National History Day, www.nhd.org

66. The tens digit of a two-digit positive integer is 2 more than three times the units digit. If the digits are interchanged, the new number is 13 less than half the given number. Find the given integer. (*Hint:* Let x = the tens-place digit and y = the units-place digit; then $10x + y$ is the number.)

67. *Wood stains.* Williams' Custom Flooring has 0.5 gal of stain that is 20% brown and 80% neutral. A customer orders 1.5 gal of a stain that is 60% brown and 40% neutral. How much pure brown stain and how much neutral stain should be added to the original 0.5 gal in order to make up the order?*

68. *Train travel.* A train leaves Union Station for Central Station, 216 km away, at 9 A.M. One hour later, a train leaves Central Station for Union Station. They meet at noon. If the second train had started at 9 A.M. and the first train at 10:30 A.M., they would still have met at noon. Find the speed of each train.

69. *Fuel economy.* Grady's station wagon gets 18 miles per gallon (mpg) in city driving and 24 mpg in highway driving. The car is driven 465 mi on 23 gal of gasoline. How many miles were driven in the city and how many were driven on the highway?

70. *Biochemistry.* Industrial biochemists routinely use a machine to mix a buffer of 10% acetone by adding 100% acetone to water. One day, instead of adding 5 L of acetone to create a vat of buffer, a machine added 10 L. How much additional water was needed to bring the concentration down to 10%?

71. See Exercise 67 above. Let x = the amount of pure brown stain added to the original 0.5 gal. Find a function $P(x)$ that can be used to determine the percentage of brown stain in the 1.5-gal mixture. On a graphing calculator, draw the graph of P and use INTERSECT to confirm the answer to Exercise 67.

72. *Gender.* Phil and Phyllis are twins. Phyllis has twice as many brothers as she has sisters. Phil has the same number of brothers as sisters. How many girls and how many boys are in the family?

---

*This problem was suggested by Professor Chris Burditt of Yountville, California.

COLLABORATIVE CORNER

How Many Two's? How Many Three's?

Focus: Systems of linear equations
Time: 20 minutes
Group size: 3

The box score at right, from the 2008 NBA All-Star game, contains information on how many field goals (worth either 2 or 3 points) and free throws (worth 1 point) each player attempted and made. For example, the line "Allen 10-14 3-5 28" means that the East's Ray Allen made 10 field goals out of 14 attempts and 3 free throws out of 5 attempts, for a total of 28 points.

ACTIVITY

1. Work as a group to develop a system of two equations in two unknowns that can be used to determine how many 2-pointers and how many 3-pointers were made by the West.

2. Each group member should solve the system from part (1) in a different way: one person algebraically, one person by making a table and methodically checking all combinations of 2- and 3-pointers, and one person by

guesswork. Compare answers when this has been completed.

3. Determine, as a group, how many 2- and 3-pointers the East made.

East (134)
James 12–22 1–1 27, Bosh 7–15 0–2 14, Howard 7–7 2–3 16, Wade 7–12 0–2 14, Kidd 1–2 0–0 2, Hamilton 4–9 0–0 9, Wallace 1–5 0–0 3, Billups 3–10 0–1 6, Jamison 1–3 0–0 2, Pierce 5–9 0–0 10, Johnson 1–2 0–0 3, Allen 10–14 3–5 28
Totals 59–110 6–14 134

West (128)
Anthony 8–17 2–3 18, Duncan 2–7 0–0 4, Ming 2–5 2–2 6, Bryant 0–0 0–0 0, Iverson 3–7 1–2 7, Nash 4–8 0–0 8, Stoudemire 8–11 1–3 18, Nowitzki 5–14 2–2 13, Paul 7–14 0–0 16, West 3–6 0–0 6, Roy 8–10 0–0 18, Boozer 7–15 0–2 14
Totals 57–114 8–14 128

| East | 34 | 40 | 32 | 28 — 134 |
| West | 28 | 37 | 28 | 35 — 128 |

3.4   Systems of Equations in Three Variables

Identifying Solutions • Solving Systems in Three Variables • Dependency, Inconsistency, and Geometric Considerations

Some problems translate directly to two equations. Others more naturally call for a translation to three or more equations. In this section, we learn how to solve systems of three linear equations. Later, we will use such systems in problem-solving situations.

Identifying Solutions

A **linear equation in three variables** is an equation equivalent to one in the form $Ax + By + Cz = D$, where A, B, C, and D are real numbers. We refer to the form $Ax + By + Cz = D$ as *standard form* for a linear equation in three variables.

A solution of a system of three equations in three variables is an ordered triple (x, y, z) that makes *all three* equations true. The numbers in an ordered triple correspond to the variables in alphabetical order unless otherwise indicated.

EXAMPLE **1**   Determine whether $\left(\frac{3}{2}, -4, 3\right)$ is a solution of the system

$$4x - 2y - 3z = 5,$$
$$-8x - y + z = -5,$$
$$2x + y + 2z = 5.$$

SOLUTION   We substitute $\left(\frac{3}{2}, -4, 3\right)$ into the three equations, using alphabetical order:

$$
\begin{array}{c|c}
4x - 2y - 3z = 5 \\
\hline
4 \cdot \frac{3}{2} - 2(-4) - 3 \cdot 3 & 5 \\
6 + 8 - 9 & \\
& 5 \overset{?}{=} 5 \quad \text{TRUE}
\end{array}
\qquad
\begin{array}{c|c}
-8x - y + z = -5 \\
\hline
-8 \cdot \frac{3}{2} - (-4) + 3 & -5 \\
-12 + 4 + 3 & \\
& -5 \overset{?}{=} -5 \quad \text{TRUE}
\end{array}
$$

$$
\begin{array}{c|c}
2x + y + 2z = 5 \\
\hline
2 \cdot \frac{3}{2} + (-4) + 2 \cdot 3 & 5 \\
3 - 4 + 6 & \\
& 5 \overset{?}{=} 5 \quad \text{TRUE}
\end{array}
$$

The triple makes all three equations true, so it is a solution.

TRY EXERCISE 7

Solving Systems in Three Variables

The graph of a linear equation in three variables is a plane. Because a three-dimensional coordinate system is required, solving systems in three variables graphically is difficult. The substitution method *can* be used but becomes cumbersome unless one or more of the equations has only two variables. Fortunately, the elimination method works well for a system of three equations in three variables. We first eliminate one variable to form a system of two equations in two variables. Once that simpler system has been solved, we substitute into one of the three original equations and solve for the third variable.

EXAMPLE **2**   Solve the following system of equations:

$$
\begin{aligned}
x + y + z &= 4, &&(1) \\
x - 2y - z &= 1, &&(2) \\
2x - y - 2z &= -1. &&(3)
\end{aligned}
$$

SOLUTION   We select *any* two of the three equations and work to get an equation in two variables. Let's add equations (1) and (2):

$$
\begin{array}{ll}
x + y + z = 4 & (1) \\
\underline{x - 2y - z = 1} & (2) \\
2x - y = 5. & (4) \qquad \text{Adding to eliminate } z
\end{array}
$$

> **CAUTION!** Be sure to eliminate the same variable in both pairs of equations.

Next, we select a different pair of equations and eliminate the *same variable* that we did above. Let's use equations (1) and (3) to again eliminate z. Be careful! A common error is to eliminate a different variable in this step.

$$
\begin{array}{l}
x + y + z = 4, \\
2x - y - 2z = -1
\end{array}
\xrightarrow[\text{of equation (1) by 2}]{\text{Multiplying both sides}}
\begin{array}{l}
2x + 2y + 2z = 8 \\
\underline{2x - y - 2z = -1} \\
4x + y = 7 \quad (5)
\end{array}
$$

Now we solve the resulting system of equations (4) and (5). That solution will give us two of the numbers in the solution of the original system.

$$2x - y = 5 \qquad (4)$$
$$4x + y = 7 \qquad (5)$$
$$\overline{\quad6x \quad= 12} \quad \text{Adding}$$
$$x = 2$$

Note that we now have two equations in two variables. Had we not eliminated the *same* variable in both of the above steps, this would not be the case.

We can use either equation (4) or (5) to find y. We choose equation (5):

$$4x + y = 7 \qquad (5)$$
$$4 \cdot 2 + y = 7 \qquad \text{Substituting 2 for } x \text{ in equation (5)}$$
$$8 + y = 7$$
$$y = -1.$$

We now have $x = 2$ and $y = -1$. To find the value for z, we use any of the original three equations and substitute to find the third number, z. Let's use equation (1) and substitute our two numbers in it:

$$x + y + z = 4 \qquad (1)$$
$$2 + (-1) + z = 4 \qquad \text{Substituting 2 for } x \text{ and } -1 \text{ for } y$$
$$1 + z = 4$$
$$z = 3.$$

We have obtained the triple $(2, -1, 3)$. It should check in *all three* equations:

$$\frac{x + y + z = 4}{2 + (-1) + 3 \,\big|\, 4}$$
$$4 \overset{?}{=} 4 \quad \text{TRUE}$$

$$\frac{x - 2y - z = 1}{2 - 2(-1) - 3 \,\big|\, 1}$$
$$1 \overset{?}{=} 1 \quad \text{TRUE}$$

$$\frac{2x - y - 2z = -1}{2 \cdot 2 - (-1) - 2 \cdot 3 \,\big|\, -1}$$
$$-1 \overset{?}{=} -1 \quad \text{TRUE}$$

The solution is $(2, -1, 3)$.

TRY EXERCISE 9

Solving Systems of Three Linear Equations

To use the elimination method to solve systems of three linear equations:

1. Write all equations in the standard form $Ax + By + Cz = D$.
2. Clear any decimals or fractions.
3. Choose a variable to eliminate. Then select two of the three equations and work to get one equation in which the selected variable is eliminated.
4. Next, use a different pair of equations and eliminate the same variable that you did in step (3).
5. Solve the system of equations that resulted from steps (3) and (4).
6. Substitute the solution from step (5) into one of the original three equations and solve for the third variable. Then check.

EXAMPLE 3 Solve the system

$$4x - 2y - 3z = 5, \qquad (1)$$
$$-8x - y + z = -5, \qquad (2)$$
$$2x + y + 2z = 5. \qquad (3)$$

SOLUTION

Write in standard form.

1., 2. The equations are already in standard form with no fractions or decimals.

3. Next, select a variable to eliminate. We decide on y because the y-terms are opposites of each other in equations (2) and (3). We add:

Eliminate a variable. (We choose y.)

$$
\begin{array}{lr}
-8x - y + z = -5 & (2) \\
2x + y + 2z = 5 & (3) \\
\hline
-6x + 3z = 0. & (4) \qquad \text{Adding}
\end{array}
$$

4. We use another pair of equations to create a second equation in x and z. That is, we eliminate the same variable, y, as in step (3). We use equations (1) and (3):

Eliminate the same variable using a different pair of equations.

$$
\begin{array}{l}
4x - 2y - 3z = 5, \\
2x + y + 2z = 5
\end{array}
\quad
\underrightarrow{\begin{array}{c}\text{Multiplying both sides} \\ \text{of equation (3) by 2}\end{array}}
\quad
\begin{array}{lr}
4x - 2y - 3z = 5 \\
4x + 2y + 4z = 10 \\
\hline
8x + z = 15. & (5)
\end{array}
$$

5. Now we solve the resulting system of equations (4) and (5). That allows us to find two parts of the ordered triple.

Solve the system of two equations in two variables.

$$
\begin{array}{l}
-6x + 3z = 0, \\
8x + z = 15
\end{array}
\quad
\underrightarrow{\begin{array}{c}\text{Multiplying both sides} \\ \text{of equation (5) by } -3\end{array}}
\quad
\begin{array}{l}
-6x + 3z = 0 \\
-24x - 3z = -45 \\
\hline
-30x = -45 \\
\\
x = \frac{-45}{-30} = \frac{3}{2}
\end{array}
$$

We use equation (5) to find z:

$$
\begin{array}{ll}
8x + z = 15 & \\
8 \cdot \frac{3}{2} + z = 15 & \text{Substituting } \frac{3}{2} \text{ for } x \\
12 + z = 15 & \\
z = 3. &
\end{array}
$$

6. Finally, we use any of the original equations and substitute to find the third number, y. We choose equation (3):

Solve for the remaining variable.

$$
\begin{array}{ll}
2x + y + 2z = 5 & (3) \\
2 \cdot \frac{3}{2} + y + 2 \cdot 3 = 5 & \text{Substituting } \frac{3}{2} \text{ for } x \text{ and 3 for } z \\
3 + y + 6 = 5 & \\
y + 9 = 5 & \\
y = -4. &
\end{array}
$$

Check.

The solution is $\left(\frac{3}{2}, -4, 3\right)$. The check was performed as Example 1.

TRY EXERCISE 23

Sometimes, certain variables are missing at the outset.

EXAMPLE 4   Solve the system

$$
\begin{array}{lr}
x + y + z = 180, & (1) \\
x - z = -70, & (2) \\
2y - z = 0. & (3)
\end{array}
$$

SOLUTION

1., 2. The equations appear in standard form with no fractions or decimals.

3., 4. Note that there is no y in equation (2). Thus, at the outset, we already have y eliminated from one equation. We need another equation with y eliminated, so we work with equations (1) and (3):

$$x + y + z = 180, \quad \xrightarrow[\text{of equation (1) by } -2]{\text{Multiplying both sides}} \quad \begin{aligned} -2x - 2y - 2z &= -360 \\ 2y - z &= 0 \\ \hline -2x \qquad\;\; - 3z &= -360. \end{aligned} \quad (4)$$

5., 6. Now we solve the resulting system of equations (2) and (4):

$$\begin{aligned} x - z &= -70, \\ -2x - 3z &= -360 \end{aligned} \quad \xrightarrow[\text{of equation (2) by 2}]{\text{Multiplying both sides}} \quad \begin{aligned} 2x - 2z &= -140 \\ -2x - 3z &= -360 \\ \hline -5z &= -500 \\ z &= 100. \end{aligned}$$

Continuing as in Examples 2 and 3, we get the solution $(30, 50, 100)$. The check is left to the student.

> **TRY EXERCISE** 27

Dependency, Inconsistency, and Geometric Considerations

Each equation in Examples 2, 3, and 4 has a graph that is a plane in three dimensions. The solutions are points common to the planes of each system. Since three planes can have an infinite number of points in common or no points at all in common, we need to generalize the concept of *consistency*.

Planes intersect at one point. System is *consistent* and has one solution.

Planes intersect along a common line. System is *consistent* and has an infinite number of solutions.

Three parallel planes. System is *inconsistent;* it has no solution.

Planes intersect two at a time, with no point common to all three. System is *inconsistent;* it has no solution.

Consistency

A system of equations that has at least one solution is said to be **consistent**.

A system of equations that has no solution is said to be **inconsistent**.

EXAMPLE 5   Solve:

$$\begin{aligned} y + 3z &= 4, & (1) \\ -x - y + 2z &= 0, & (2) \\ x + 2y + z &= 1. & (3) \end{aligned}$$

SOLUTION   The variable x is missing in equation (1). By adding equations (2) and (3), we can find a second equation in which x is missing:

$$\begin{aligned} -x - y + 2z &= 0 & (2) \\ x + 2y + z &= 1 & (3) \\ \hline y + 3z &= 1. & (4) \qquad \text{Adding} \end{aligned}$$

Equations (1) and (4) form a system in y and z. We solve as before:

$$y + 3z = 4, \quad \xrightarrow[\text{of equation (1) by } -1]{\text{Multiplying both sides}} \quad \begin{array}{r} -y - 3z = -4 \\ y + 3z = 1 \end{array}$$

$$y + 3z = 1$$

This is a contradiction. $\xrightarrow{} \begin{array}{r} 0 = -3. \end{array}$ Adding

Since we end up with a *false* equation, or contradiction, we state that the system has no solution. It is *inconsistent*.

TRY EXERCISE ▶ 15

The notion of *dependency* from Section 3.1 can also be extended.

EXAMPLE 6   Solve:

$$
\begin{array}{rcl}
2x + y + z &=& 3, \quad (1) \\
x - 2y - z &=& 1, \quad (2) \\
3x + 4y + 3z &=& 5. \quad (3)
\end{array}
$$

SOLUTION   Our plan is to first use equations (1) and (2) to eliminate z. Then we will select another pair of equations and again eliminate z:

$$
\begin{array}{rcl}
2x + y + z &=& 3 \\
x - 2y - z &=& 1 \\
\hline
3x - y \phantom{{}+ z} &=& 4. \quad (4)
\end{array}
$$

Next, we use equations (2) and (3) to eliminate z again:

$$x - 2y - z = 1, \quad \xrightarrow[\text{of equation (2) by } 3]{\text{Multiplying both sides}} \quad \begin{array}{rcl} 3x - 6y - 3z &=& 3 \\ 3x + 4y + 3z &=& 5 \\ \hline 6x - 2y \phantom{{}+ 3z} &=& 8. \quad (5) \end{array}$$

$$3x + 4y + 3z = 5$$

We now try to solve the resulting system of equations (4) and (5):

$$3x - y = 4, \quad \xrightarrow[\text{of equation (4) by } -2]{\text{Multiplying both sides}} \quad \begin{array}{rcl} -6x + 2y &=& -8 \\ 6x - 2y &=& 8 \\ \hline 0 &=& 0. \quad (6) \end{array}$$

$$6x - 2y = 8$$

Equation (6), which is an identity, indicates that equations (1), (2), and (3) are *dependent*. This means that the original system of three equations is equivalent to a system of two equations. One way to see this is to observe that two times equation (1), minus equation (2), is equation (3). Thus removing equation (3) from the system does not affect the solution of the system.* In writing an answer to this problem, we simply state that "the equations are dependent."

TRY EXERCISE ▶ 21

Recall that when dependent equations appeared in Section 3.1, the solution sets were always infinite in size and were written in set-builder notation. There, all systems of dependent equations were *consistent*. This is not always the case for

---

*A set of equations is dependent if at least one equation can be expressed as a sum of multiples of other equations in that set.

systems of three or more equations. The following figures illustrate some possibilities geometrically.

The planes intersect along a common line. The equations are *dependent* and the system is *consistent*. There is an infinite number of solutions.

The planes coincide. The equations are *dependent* and the system is *consistent*. There is an infinite number of solutions.

Two planes coincide. The third plane is parallel. The equations are *dependent* and the system is *inconsistent*. There is no solution.

3.4 EXERCISE SET

🐾 *Concept Reinforcement* *Classify each statement as either true or false.*

1. $3x + 5y + 4z = 7$ is a linear equation in three variables.

2. Every system of three equations in three unknowns has at least one solution.

3. It is not difficult to solve a system of three equations in three unknowns by graphing.

4. If, when we are solving a system of three equations, a false equation results from adding a multiple of one equation to another, the system is inconsistent.

5. If, when we are solving a system of three equations, an identity results from adding a multiple of one equation to another, the equations are dependent.

6. Whenever a system of three equations contains dependent equations, there is an infinite number of solutions.

7. Determine whether $(2, -1, -2)$ is a solution of the system
$$x + y - 2z = 5,$$
$$2x - y - z = 7,$$
$$-x - 2y - 3z = 6.$$

8. Determine whether $(-1, -3, 2)$ is a solution of the system
$$x - y + z = 4,$$
$$x - 2y - z = 3,$$
$$3x + 2y - z = 1.$$

Solve each system. If a system's equations are dependent or if there is no solution, state this.

9. $x - y - z = 0,$
$2x - 3y + 2z = 7,$
$-x + 2y + z = 1$

10. $x + y - z = 0,$
$2x - y + z = 3,$
$-x + 5y - 3z = 2$

11. $x - y - z = 1,$
$2x + y + 2z = 4,$
$x + y + 3z = 5$

12. $x + y - 3z = 4,$
$2x + 3y + z = 6,$
$2x - y + z = -14$

13. $3x + 4y - 3z = 4,$
$5x - y + 2z = 3,$
$x + 2y - z = -2$

14. $2x - 3y + z = 5,$
$x + 3y + 8z = 22,$
$3x - y + 2z = 12$

15. $x + y + z = 0,$
$2x + 3y + 2z = -3,$
$-x - 2y - z = 1$

16. $3a - 2b + 7c = 13,$
$a + 8b - 6c = -47,$
$7a - 9b - 9c = -3$

17. $2x - 3y - z = -9,$
$2x + 5y + z = 1,$
$x - y + z = 3$

18. $4x + y + z = 17,$
$x - 3y + 2z = -8,$
$5x - 2y + 3z = 5$

Aha! 19. $a + b + c = 5,$
$2a + 3b - c = 2,$
$2a + 3b - 2c = 4$

20. $u - v + 6w = 8,$
$3u - v + 6w = 14,$
$-u - 2v - 3w = 7$

21. $-2x + 8y + 2z = 4,$
$x + 6y + 3z = 4,$
$3x - 2y + z = 0$

22. $x - y + z = 4,$
$5x + 2y - 3z = 2,$
$4x + 3y - 4z = -2$

23. $2u - 4v - w = 8,$
$3u + 2v + w = 6,$
$5u - 2v + 3w = 2$

24. $4p + q + r = 3,$
$2p - q + r = 6,$
$2p + 2q - r = -9$

25. $r + \frac{3}{2}s + 6t = 2,$
$2r - 3s + 3t = 0.5,$
$r + s + t = 1$

26. $5x + 3y + \frac{1}{2}z = \frac{7}{2},$
$0.5x - 0.9y - 0.2z = 0.3,$
$3x - 2.4y + 0.4z = -1$

27. $4a + 9b = 8,$
$8a + 6c = -1,$
$6b + 6c = -1$

28. $3p + 2r = 11,$
$q - 7r = 4,$
$p - 6q = 1$

29. $x + y + z = 57,$
$-2x + y = 3,$
$x - z = 6$

30. $x + y + z = 105,$
$10y - z = 11,$
$2x - 3y = 7$

31. $a - 3c = 6,$
$b + 2c = 2,$
$7a - 3b - 5c = 14$

32. $2a - 3b = 2,$
$7a + 4c = \frac{3}{4},$
$2c - 3b = 1$

Aha! **33.** $x + y + z = 83,$
$y = 2x + 3,$
$z = 40 + x$

34. $l + m = 7,$
$3m + 2n = 9,$
$4l + n = 5$

35. $x + z = 0,$
$x + y + 2z = 3,$
$y + z = 2$

36. $x + y = 0,$
$x + z = 1,$
$2x + y + z = 2$

37. $x + y + z = 1,$
$-x + 2y + z = 2,$
$2x - y = -1$

38. $y + z = 1,$
$x + y + z = 1,$
$x + 2y + 2z = 2$

📝 **39.** Rondel always begins solving systems of three equations in three variables by using the first two equations to eliminate x. Is this a good approach? Why or why not?

📝 **40.** Describe a method for writing an inconsistent system of three equations in three variables.

Skill Review

To prepare for Section 3.5, review translating sentences to equations (Section 1.1).

Translate each sentence to an equation. [1.1]

41. One number is half another.

42. The difference of two numbers is twice the first number.

43. The sum of three consecutive numbers is 100.

44. The sum of three numbers is 100.

45. The product of two numbers is five times a third number.

46. The product of two numbers is twice their sum.

Synthesis

📝 **47.** Is it possible for a system of three linear equations to have exactly two ordered triples in its solution set? Why or why not?

📝 **48.** Describe a procedure that could be used to solve a system of four equations in four variables.

Solve.

49. $\dfrac{x + 2}{3} - \dfrac{y + 4}{2} + \dfrac{z + 1}{6} = 0,$
$\dfrac{x - 4}{3} + \dfrac{y + 1}{4} - \dfrac{z - 2}{2} = -1,$
$\dfrac{x + 1}{2} + \dfrac{y}{2} + \dfrac{z - 1}{4} = \dfrac{3}{4}$

50. $w + x - y + z = 0,$
$w - 2x - 2y - z = -5,$
$w - 3x - y + z = 4,$
$2w - x - y + 3z = 7$

51. $w + x + y + z = 2,$
$w + 2x + 2y + 4z = 1,$
$w - x + y + z = 6,$
$w - 3x - y + z = 2$

For Exercises 52 and 53, let u represent $1/x$, v represent $1/y$, and w represent $1/z$. Solve for u, v, and w, and then solve for x, y, and z.

52. $\dfrac{2}{x} + \dfrac{2}{y} - \dfrac{3}{z} = 3,$
$\dfrac{1}{x} - \dfrac{2}{y} - \dfrac{3}{z} = 9,$
$\dfrac{7}{x} - \dfrac{2}{y} + \dfrac{9}{z} = -39$

53. $\dfrac{2}{x} - \dfrac{1}{y} - \dfrac{3}{z} = -1,$
$\dfrac{2}{x} - \dfrac{1}{y} + \dfrac{1}{z} = -9,$
$\dfrac{1}{x} + \dfrac{2}{y} - \dfrac{4}{z} = 17$

Determine k so that each system is dependent.

54. $x - 3y + 2z = 1,$
$2x + y - z = 3,$
$9x - 6y + 3z = k$

55. $5x - 6y + kz = -5,$
$x + 3y - 2z = 2,$
$2x - y + 4z = -1$

In each case, three solutions of an equation in x, y, and z are given. Find the equation.

56. $Ax + By + Cz = 12;$
$\left(1, \frac{3}{4}, 3\right), \left(\frac{4}{3}, 1, 2\right),$ and $(2, 1, 1)$

57. $z = b - mx - ny;$
$(1, 1, 2), (3, 2, -6),$ and $\left(\frac{3}{2}, 1, 1\right)$

58. Write an inconsistent system of equations that contains dependent equations.

📝 **59.** Kadi and Ahmed both correctly solve the system

$x + 2y - z = 1,$
$-x - 2y + z = 3,$
$2x + 4y - 2z = 2.$

Kadi states "the equations are dependent" while Ahmed states "there is no solution." How did each person reach the conclusion?

CORNER

Finding the Preferred Approach

Focus: Systems of three linear equations

Time: 10–15 minutes

Group size: 3

Consider the six steps outlined on p. 181 along with the following system:

$$2x + 4y = 3 - 5z,$$
$$0.3x = 0.2y + 0.7z + 1.4,$$
$$0.04x + 0.03y = 0.07 + 0.04z.$$

ACTIVITY

1. Working independently, each group member should solve the system above. One person should begin by eliminating x, one should

first eliminate y, and one should first eliminate z. Write neatly so that others can follow your steps.

2. Once all group members have solved the system, compare your answers. If the answers do not check, exchange notebooks and check each other's work. If a mistake is detected, allow the person who made the mistake to make the repair.

3. Decide as a group which of the three approaches above (if any) ranks as easiest and which (if any) ranks as most difficult. Then compare your rankings with the other groups in the class.

3.5	## Solving Applications: Systems of Three Equations

Applications of Three Equations in Three Unknowns

Solving systems of three or more equations is important in many applications. Such systems arise in the natural and social sciences, business, and engineering. To begin, let's first look at a purely numerical application.

EXAMPLE 1

The sum of three numbers is 4. The first number minus twice the second, minus the third is 1. Twice the first number minus the second, minus twice the third is -1. Find the numbers.

SOLUTION

1. **Familiarize.** There are three statements involving the same three numbers. Let's label these numbers x, y, and z.

2. **Translate.** We can translate directly as follows.

The sum of the three numbers is 4.

$$x + y + z = 4$$

The first number minus twice the second minus the third is 1.

$$x - 2y - z = 1$$

Twice the first number minus the second minus twice the third is -1.

$$2x - y - 2z = -1$$

We now have a system of three equations:

$$x + y + z = 4,$$
$$x - 2y - z = 1,$$
$$2x - y - 2z = -1.$$

3. **Carry out.** We need to solve the system of equations. Note that we found the solution, $(2, -1, 3)$, in Example 2 of Section 3.4.

4. **Check.** The first statement of the problem says that the sum of the three numbers is 4. That checks, because $2 + (-1) + 3 = 4$. The second statement says that the first number minus twice the second, minus the third is 1: $2 - 2(-1) - 3 = 1$. That checks. The check of the third statement is left to the student.

5. **State.** The three numbers are 2, -1, and 3.

TRY EXERCISE ▶ 1

EXAMPLE 2

Architecture. In a triangular cross section of a roof, the largest angle is 70° greater than the smallest angle. The largest angle is twice as large as the remaining angle. Find the measure of each angle.

SOLUTION

1. **Familiarize.** The first thing we do is make a drawing, or a sketch.

Since we don't know the size of any angle, we use x, y, and z to represent the three measures, from smallest to largest. Recall that the measures of the angles in any triangle add up to 180°.

2. **Translate.** This geometric fact about triangles gives us one equation:

$$x + y + z = 180.$$

Two of the statements can be translated almost directly.

The largest angle ‾‾‾‾‾‾ is ‾ 70° greater than the smallest angle. ‾‾‾‾‾‾‾‾‾‾‾‾

↓ ↓ ↓

$$z \qquad = \qquad x + 70$$

The largest angle ‾‾‾‾‾‾ is ‾ twice as large as the remaining angle. ‾‾‾‾‾‾‾‾‾‾‾‾

↓ ↓ ↓

$$z \qquad = \qquad 2y$$

We now have a system of three equations:

$$
\begin{array}{llll}
x + y + z = 180, & & x + y + z = 180, & \\
x + 70 = z, & \text{or} & x \quad - z = -70, & \text{Rewriting in} \\
2y = z; & & 2y - z = 0. & \text{standard form}
\end{array}
$$

3. **Carry out.** The system was solved in Example 4 of Section 3.4. The solution is $(30, 50, 100)$.

4. **Check.** The sum of the numbers is 180, so that checks. The measure of the largest angle, $100°$, is $70°$ greater than the measure of the smallest angle, $30°$, so that checks. The measure of the largest angle is also twice the measure of the remaining angle, $50°$. Thus we have a check.

5. **State.** The angles in the triangle measure $30°$, $50°$, and $100°$.

▶ TRY EXERCISE ▶ 5

EXAMPLE 3   *Downloads.* Kaya frequently downloads music, TV shows, and iPod games. In January, she downloaded 5 songs, 10 TV shows, and 3 games for a total of $40. In February, she spent a total of $135 for 25 songs, 25 TV shows, and 12 games. In March, she spent a total of $56 for 15 songs, 8 TV shows, and 5 games. Assuming each song is the same price, each TV show is the same price, and each iPod game is the same price, how much does each cost?

Source: www.iTunes.com

SOLUTION

1. **Familiarize.** We let s = the cost, in dollars, per song, t = the cost, in dollars, per TV show, and g = the cost, in dollars, per game. Then in January, Kaya spent $5 \cdot s$ for songs, $10 \cdot t$ for TV shows, and $3 \cdot g$ for iPod games. The sum of these amounts was $40. Each month's downloads will translate to an equation.

2. **Translate.** We can organize the information in a table.

	Cost of Songs	Cost of TV Shows	Cost of iPod Games	Total Cost	
January	$5s$	$10t$	$3g$	40	$\longrightarrow 5s + 10t + 3g = 40$
February	$25s$	$25t$	$12g$	135	$\longrightarrow 25s + 25t + 12g = 135$
March	$15s$	$8t$	$5g$	56	$\longrightarrow 15s + 8t + 5g = 56$

We now have a system of three equations:

$$5s + 10t + 3g = 40, \quad (1)$$
$$25s + 25t + 12g = 135, \quad (2)$$
$$15s + 8t + 5g = 56. \quad (3)$$

3. **Carry out.** We begin by using equations (1) and (2) to eliminate s.

$$\begin{array}{l} 5s + 10t + 3g = 40, \\ 25s + 25t + 12g = 135 \end{array} \xrightarrow[\text{of equation (1) by } -5]{\text{Multiplying both sides}} \begin{array}{r} -25s - 50t - 15g = -200 \\ 25s + 25t + 12g = 135 \\ \hline -25t - 3g = -65 \quad (4) \end{array}$$

We then use equations (1) and (3) to again eliminate s.

$$\begin{array}{l} 5s + 10t + 3g = 40, \\ 15s + 8t + 5g = 56 \end{array} \xrightarrow[\text{of equation (1) by } -3]{\text{Multiplying both sides}} \begin{array}{r} -15s - 30t - 9g = -120 \\ 15s + 8t + 5g = 56 \\ \hline -22t - 4g = -64 \quad (5) \end{array}$$

Now we solve the resulting system of equations (4) and (5).

$-25t - 3g = -65$ $\xrightarrow[\text{of equation (4) by } -4]{\text{Multiplying both sides}}$ $100t + 12g = 260$

$-22t - 4g = -64$ $\xrightarrow[\text{of equation (5) by } 3]{\text{Multiplying both sides}}$ $\dfrac{-66t - 12g = -192}{34t = 68}$

$$t = 2$$

To find g, we use equation (4):

$$-25t - 3g = -65$$
$$-25 \cdot 2 - 3g = -65 \qquad \text{Substituting 2 for } t$$
$$-50 - 3g = -65$$
$$-3g = -15$$
$$g = 5.$$

Finally, we use equation (1) to find s:

$$5s + 10t + 3g = 40$$
$$5s + 10 \cdot 2 + 3 \cdot 5 = 40 \qquad \text{Substituting 2 for } t \text{ and 5 for } g$$
$$5s + 20 + 15 = 40$$
$$5s + 35 = 40$$
$$5s = 5$$
$$s = 1.$$

4. **Check.** If a song costs \$1, a TV show costs \$2, and an iPod game costs \$5, then the total cost for each month's downloads is as follows:

January: $5 \cdot \$1 + 10 \cdot \$2 + 3 \cdot \$5 = \$5 + \$20 + \$15 = \$40;$

February: $25 \cdot \$1 + 25 \cdot \$2 + 12 \cdot \$5 = \$25 + \$50 + \$60 = \$135;$

March: $15 \cdot \$1 + 8 \cdot \$2 + 5 \cdot \$5 = \$15 + \$16 + \$25 = \$56.$

This checks with the information given in the problem.

5. **State.** A song costs \$1, a TV show costs \$2, and an iPod game costs \$5.

TRY EXERCISE ▶ 19

3.5 EXERCISE SET

Solve.

1. The sum of three numbers is 85. The second is 7 more than the first. The third is 2 more than four times the second. Find the numbers.

2. The sum of three numbers is 5. The first number minus the second plus the third is 1. The first minus the third is 3 more than the second. Find the numbers.

3. The sum of three numbers is 26. Twice the first minus the second is 2 less than the third. The third is the second minus three times the first. Find the numbers.

4. The sum of three numbers is 105. The third is 11 less than ten times the second. Twice the first is 7 more than three times the second. Find the numbers.

5. *Geometry.*   In triangle *ABC*, the measure of angle *B* is three times that of angle *A*. The measure of angle *C* is 20° more than that of angle *A*. Find the angle measures.

6. *Geometry.*   In triangle *ABC*, the measure of angle *B* is twice the measure of angle *A*. The measure of angle *C* is 80° more than that of angle *A*. Find the angle measures.

7. *Scholastic Aptitude Test.*   Many high-school students take the Scholastic Aptitude Test (SAT). Beginning in March 2005, students taking the SAT received three scores: a critical reading score, a mathematics score, and a writing score. The average total score of 2007 high-school seniors who took the SAT was 1511. The average mathematics score exceeded the reading score by 13 points and the average writing score was 8 points less than the reading score. What was the average score for each category?
Source: College Entrance Examination Board

8. *Advertising.*   In 2006, U.S. companies spent a total of $123.4 billion on newspaper, television, and magazine ads. The total amount spent on television ads was $7.4 billion more than the amount spent on newspaper and magazine ads together. The amount spent on magazine ads was $2 billion more than the amount spent on newspaper ads. How much was spent on each form of advertising?
Source: TNS Media Intelligence

9. *Nutrition.*   Most nutritionists now agree that a healthy adult diet includes 25–35 g of fiber each day. A breakfast of 2 bran muffins, 1 banana, and a 1-cup serving of Wheaties® contains 9 g of fiber; a breakfast of 1 bran muffin, 2 bananas, and a 1-cup serving of Wheaties® contains 10.5 g of fiber; and a breakfast of 2 bran muffins and a 1-cup serving of Wheaties® contains 6 g of fiber. How much fiber is in each of these foods?
Sources: usda.gov and InteliHealth.com

10. *Nutrition.*   Refer to Exercise 9. A breakfast consisting of 2 pancakes and a 1-cup serving of strawberries contains 4.5 g of fiber, whereas a breakfast of 2 pancakes and a 1-cup serving of Cheerios® contains 4 g of fiber. When a meal consists of 1 pancake, a 1-cup serving of Cheerios®, and a 1-cup serving of strawberries, it contains 7 g of fiber. How much fiber is in each of these foods?
Source: InteliHealth.com

Aha! **11.** *Automobile pricing.*   The basic model of a 2008 Jeep Grand Cherokee Rocky Mountain (2WD) with a tow package costs $30,815. When equipped with a tow package and a rear backup camera, the vehicle's price rose to $31,565. The cost of the basic model with a rear camera was $31,360. Find the basic price, the cost of a tow package, and the cost of a rear camera.
Source: www.jeep.com

12. *Telemarketing.*   Sven, Tina, and Laurie can process 740 telephone orders per day. Sven and Tina together can process 470 orders, while Tina and Laurie together can process 520 orders per day. How many orders can each person process alone?

13. *Coffee prices.*   Reba works at a Starbucks® coffee shop where a 12-oz cup of coffee costs $1.65, a 16-oz cup costs $1.85, and a 20-oz cup costs $1.95. During one busy period, Reba served 55 cups of coffee, emptying six 144-oz "brewers" while collecting a total of $99.65. How many cups of each size did Reba fill?

| 12 oz | 16 oz | 20 oz |
| $1.65 | $1.85 | $1.95 |

14. *Restaurant management.*   Chick-fil-A® recently sold small lemonades for $1.29, medium lemonades for $1.49, and large lemonades for $1.85. During a lunch-time rush, Chris sold 40 lemonades for a total of $59.40. The number of small and large drinks, combined, was 10 fewer than the number of medium drinks. How many drinks of each size were sold?

15. *Small-business loans.*   Chelsea took out three loans for a total of $120,000 to start an organic orchard. Her bank loan was at an interest rate of 8%, the small-business loan was at an interest rate of 5%, and the mortgage on her house was at an interest rate of 4%. The total simple interest due on the loans in one year was $5750. The annual simple interest on the mortgage was $1600 more than the interest on the bank loan. How much did she borrow from each source?

16. *Investments.*   A business class divided an imaginary investment of $80,000 among three mutual funds. The first fund grew by 10%, the second by 6%, and the third by 15%. Total earnings were $8850. The earnings from the first fund were $750 more than the earnings from the third. How much was invested in each fund?

17. *Gold alloys.*   Gold used to make jewelry is often a blend of gold, silver, and copper. The relative amounts of the metals determine the color of the alloy. Red gold is 75% gold, 5% silver, and 20% copper. Yellow gold is 75% gold, 12.5% silver, and 12.5% copper. White gold is 37.5% gold and 62.5% silver. If 100 g of red gold costs $2265.40, 100 g of yellow gold costs $2287.75, and 100 g of white gold costs $1312.50, how much do gold, silver, and copper cost?
Source: World Gold Council

18. *Blending teas.*   Verity has recently created three custom tea blends. A 5-oz package of Southern Sandalwood sells for $13.15 and contains 2 oz of Keemun tea, 2 oz of Assam tea, and 1 oz of a berry blend. A 4-oz package of Golden Sunshine sells for $12.50 and contains 3 oz of Assam tea and 1 oz of the berry blend. A 6-oz package of Mountain Morning sells for $12.50 and contains 2 oz of the berry blend, 3 oz of Keemun tea, and 1 oz of Assam tea. What is the price per ounce of Keemun tea, Assam tea, and the berry blend?

19. *Nutrition.*   A dietician in a hospital prepares meals under the guidance of a physician. Suppose that for a particular patient a physician prescribes a meal to have 800 calories, 55 g of protein, and 220 mg of vitamin C. The dietician prepares a meal of roast beef, baked potatoes, and broccoli according to the data in the following table.

Serving Size	Calories	Protein (in grams)	Vitamin C (in milligrams)
Roast Beef, 3 oz	300	20	0
Baked Potato, 1	100	5	20
Broccoli, 156 g	50	5	100

How many servings of each food are needed in order to satisfy the doctor's orders?

20. *Nutrition.*   Repeat Exercise 19 but replace the broccoli with asparagus, for which a 180-g serving contains 50 calories, 5 g of protein, and 44 mg of vitamin C. Which meal would you prefer eating?

21. Students in a Listening Responses class bought 40 tickets for a piano concert. The number of tickets purchased for seats in either the first mezzanine or the main floor was the same as the number purchased for seats in the second mezzanine. First mezzanine seats cost $52, main floor seats cost $38, and second mezzanine seats cost $28. The total cost of the tickets was $1432. How many of each type of ticket were purchased?

22. *Basketball scoring.*   The New York Knicks recently scored a total of 92 points on a combination of 2-point field goals, 3-point field goals, and 1-point foul shots. Altogether, the Knicks made 50 baskets and 19 more 2-pointers than foul shots. How many shots of each kind were made?

23. *World population growth.*   The world population is projected to be 9.4 billion in 2050. At that time, there is expected to be approximately 3.5 billion more people in Asia than in Africa. The population for the rest of the world will be approximately 0.3 billion less than two-fifths the population of Asia. Find the projected populations of Asia, Africa, and the rest of the world in 2050.
Source: U.S. Census Bureau

24. *History.*   Find the year in which the first U.S. transcontinental railroad was completed. The following are some facts about the number. The sum of the digits in the year is 24. The ones digit is 1 more than the hundreds digit. Both the tens and the ones digits are multiples of 3.

25. Problems like Exercises 13 and 14 could be classified as total-value problems. How do these problems differ from the total-value problems of Section 3.3?

26. Write a problem for a classmate to solve. Design the problem so that it translates to a system of three equations in three variables.

Skill Review

To prepare for Section 3.6, review simplifying expressions (Section 1.2).

Simplify. [1.2]

27. $-2(2x - 3y)$

28. $-(x - 6y)$

29. $-6(x - 2y) + (6x - 5y)$

30. $3(2a + 4b) + (5a - 12b)$

31. $-(2a - b - 6c)$

32. $-10(5a + 3b - c)$

33. $-2(3x - y + z) + 3(-2x + y - 2z)$

34. $(8x - 10y + 7z) + 5(3x + 2y - 4z)$

Synthesis

35. Consider Exercise 22. Suppose there were no foul shots made. Would there still be a solution? Why or why not?

36. Consider Exercise 13. Suppose Reba collected $50. Could the problem still be solved? Why or why not?

37. *Health insurance.*   In 2008, UNICARE® health insurance for a 35-year-old and his or her spouse cost $174/month. That rate increased to $221/month if a child were included and $263/month if two children were included. The rate dropped to $134/month for just the applicant and one child. Find the separate costs for insuring the applicant, the spouse, the first child, and the second child.
Source: UNICARE Life and Health Insurance Company® through www.ehealth.com

38. Find a three-digit positive integer such that the sum of all three digits is 14, the tens digit is 2 more than the ones digit, and if the digits are reversed, the number is unchanged.

39. *Ages.*   Tammy's age is the sum of the ages of Carmen and Dennis. Carmen's age is 2 more than the sum of the ages of Dennis and Mark. Dennis's age is four times Mark's age. The sum of all four ages is 42. How old is Tammy?

40. *Ticket revenue.*   A magic show's audience of 100 people consists of adults, students, and children. The ticket prices are $10 for adults, $3 for students, and 50¢ for children. The total amount of money taken in is $100. How many adults, students, and children are in attendance? Does there seem to be some information missing? Do some more careful reasoning.

41. *Sharing raffle tickets.*   Hal gives Tom as many raffle tickets as Tom first had and Gary as many as Gary first had. In like manner, Tom then gives Hal and Gary as many tickets as each then has. Similarly, Gary gives Hal and Tom as many tickets as each then has. If each finally has 40 tickets, with how many tickets does Tom begin?

42. Find the sum of the angle measures at the tips of the star in this figure.

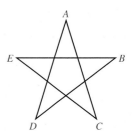

3.6 Elimination Using Matrices

Matrices and Systems ▪ Row-Equivalent Operations

In solving systems of equations, we perform computations with the constants. If we agree to keep all like terms in the same column, we can simplify writing a system by omitting the variables. For example, the system

$$3x + 4y = 5,$$
$$x - 2y = 1$$

simplifies to

3	4	5
1	−2	1

if we do not write the variables, the operation of addition, and the equals signs.

Matrices and Systems

In the example above, we have written a rectangular array of numbers. Such an array is called a **matrix** (plural, **matrices**). We ordinarily write brackets around matrices. The following are matrices:

$$\begin{bmatrix} -3 & 1 \\ 0 & 5 \end{bmatrix}, \begin{bmatrix} 2 & 0 & -1 & 3 \\ -5 & 2 & 7 & -1 \\ 4 & 5 & 3 & 0 \end{bmatrix}, \begin{bmatrix} 2 & 3 \\ 7 & 15 \\ -2 & 23 \\ 4 & 1 \end{bmatrix}$$

The individual numbers are called *elements* or *entries*.

The **rows** of a matrix are horizontal, and the **columns** are vertical.

$$\begin{bmatrix} 5 & -2 & 2 \\ 1 & 0 & 1 \\ 0 & 1 & 2 \end{bmatrix}$$

→ row 1
→ row 2
→ row 3

column 1   column 2   column 3

Let's see how matrices can be used to solve a system.

EXAMPLE 1

Solve the system

$$5x - 4y = -1,$$
$$-2x + 3y = 2.$$

As an aid for understanding, we list the corresponding system in the margin.

$$5x - 4y = -1,$$
$$-2x + 3y = 2$$

SOLUTION We write a matrix using only coefficients and constants, listing *x*-coefficients in the first column and *y*-coefficients in the second. A dashed line separates the coefficients from the constants:

$$\begin{bmatrix} 5 & -4 & \vdots & -1 \\ -2 & 3 & \vdots & 2 \end{bmatrix}.$$

Consult the notes in the margin for further information.

Our goal is to transform

$$\begin{bmatrix} 5 & -4 & \vdots & -1 \\ -2 & 3 & \vdots & 2 \end{bmatrix}$$

into the form

$$\begin{bmatrix} a & b & \vdots & c \\ 0 & d & \vdots & e \end{bmatrix}.$$

We can then reinsert the variables *x* and *y*, form equations, and complete the solution.

Our calculations are similar to those that we would do if we wrote the entire equations. The first step is to multiply and/or interchange the rows so that each number in the first column below the first number is a multiple of that number. Here that means multiplying Row 2 by 5. This corresponds to multiplying both sides of the second equation by 5.

$$5x - 4y = -1,$$
$$-10x + 15y = 10$$

$$\begin{bmatrix} 5 & -4 & \vdots & -1 \\ -10 & 15 & \vdots & 10 \end{bmatrix}$$

New Row 2 = 5(Row 2 from the step above)
$$= 5(-2 \quad 3 \quad \vdots \quad 2) = (-10 \quad 15 \quad \vdots \quad 10)$$

Next, we multiply the first row by 2, add this to Row 2, and write that result as the "new" Row 2. This corresponds to multiplying the first equation by 2 and adding the result to the second equation in order to eliminate a variable. Write out these computations as necessary.

$$5x - 4y = -1,$$
$$7y = 8$$

$$\begin{bmatrix} 5 & -4 & \vdots & -1 \\ 0 & 7 & \vdots & 8 \end{bmatrix}$$

2(Row 1) $= 2(5 \quad -4 \quad \vdots \quad -1) = (10 \quad -8 \quad \vdots \quad -2)$
New Row 2 $= (10 \quad -8 \quad \vdots \quad -2) + (-10 \quad 15 \quad \vdots \quad 10)$
$= (0 \quad 7 \quad \vdots \quad 8)$

If we now reinsert the variables, we have

$$5x - 4y = -1, \quad (1) \qquad \text{From Row 1}$$
$$7y = 8. \quad (2) \qquad \text{From Row 2}$$

Solving equation (2) for y gives us

$$7y = 8 \qquad (2)$$
$$y = \tfrac{8}{7}.$$

Next, we substitute $\tfrac{8}{7}$ for y in equation (1):

$$5x - 4y = -1 \qquad (1)$$
$$5x - 4 \cdot \tfrac{8}{7} = -1 \qquad \text{Substituting } \tfrac{8}{7} \text{ for } y \text{ in equation (1)}$$
$$x = \tfrac{5}{7}. \qquad \text{Solving for } x$$

The solution is $\left(\tfrac{5}{7}, \tfrac{8}{7}\right)$. The check is left to the student.   **TRY EXERCISE** 7

EXAMPLE 2   Solve the system

$$2x - y + 4z = -3,$$
$$x \qquad - 4z = 5,$$
$$6x - y + 2z = 10.$$

SOLUTION   We first write a matrix, using only the constants. Where there are missing terms, we must write 0's:

$$2x - y + 4z = -3,$$
$$x \qquad - 4z = 5,$$
$$6x - y + 2z = 10$$

$$\begin{bmatrix} 2 & -1 & 4 & \vdots & -3 \\ 1 & 0 & -4 & \vdots & 5 \\ 6 & -1 & 2 & \vdots & 10 \end{bmatrix}.$$

Our goal is to transform the matrix to one of the form

$$ax + by + cz = d,$$
$$ey + fz = g,$$
$$hz = i$$

$$\begin{bmatrix} a & b & c & \vdots & d \\ 0 & e & f & \vdots & g \\ 0 & 0 & h & \vdots & i \end{bmatrix}.$$ This matrix is in row-echelon form.

A matrix of this form can be rewritten as a system of equations that is equivalent to the original system, and from which a solution can be easily found.

The first step is to multiply and/or interchange the rows so that each number in the first column is a multiple of the first number in the first row. In this case, we begin by interchanging Rows 1 and 2:

$$x \qquad - 4z = 5,$$
$$2x - y + 4z = -3,$$
$$6x - y + 2z = 10$$

$$\begin{bmatrix} 1 & 0 & -4 & \vdots & 5 \\ 2 & -1 & 4 & \vdots & -3 \\ 6 & -1 & 2 & \vdots & 10 \end{bmatrix}.$$

This corresponds to interchanging the first two equations.

Next, we multiply the first row by -2, add it to the second row, and replace Row 2 with the result:

$$x \qquad - 4z = 5,$$
$$\quad -y + 12z = -13,$$
$$6x - y + 2z = 10$$

$$\begin{bmatrix} 1 & 0 & -4 & \vdots & 5 \\ 0 & -1 & 12 & \vdots & -13 \\ 6 & -1 & 2 & \vdots & 10 \end{bmatrix}.$$

$$-2(1 \quad 0 \quad -4 \;\vdots\; 5) = (-2 \quad 0 \quad 8 \;\vdots\; -10) \text{ and}$$
$$(-2 \quad 0 \quad 8 \;\vdots\; -10) + (2 \quad -1 \quad 4 \;\vdots\; -3) =$$
$$(0 \quad -1 \quad 12 \;\vdots\; -13)$$

Now we multiply the first row by -6, add it to the third row, and replace Row 3 with the result:

$$x \qquad - 4z = 5,$$
$$\quad -y + 12z = -13,$$
$$\quad -y + 26z = -20$$

$$\begin{bmatrix} 1 & 0 & -4 & \vdots & 5 \\ 0 & -1 & 12 & \vdots & -13 \\ 0 & -1 & 26 & \vdots & -20 \end{bmatrix}.$$

$$-6(1 \quad 0 \quad -4 \;\vdots\; 5) = (-6 \quad 0 \quad 24 \;\vdots\; -30) \text{ and}$$
$$(-6 \quad 0 \quad 24 \;\vdots\; -30) + (6 \quad -1 \quad 2 \;\vdots\; 10) =$$
$$(0 \quad -1 \quad 26 \;\vdots\; -20)$$

Next, we multiply Row 2 by -1, add it to the third row, and replace Row 3 with the result:

$$x \qquad - 4z = 5,$$
$$\quad -y + 12z = -13,$$
$$\qquad \quad 14z = -7$$

$$\begin{bmatrix} 1 & 0 & -4 & \vdots & 5 \\ 0 & -1 & 12 & \vdots & -13 \\ 0 & 0 & 14 & \vdots & -7 \end{bmatrix}.$$

$$-1(0 \quad -1 \quad 12 \;\vdots\; -13) = (0 \quad 1 \quad -12 \;\vdots\; 13) \text{ and}$$
$$(0 \quad 1 \quad -12 \;\vdots\; 13) + (0 \quad -1 \quad 26 \;\vdots\; -20) =$$
$$(0 \quad 0 \quad 14 \;\vdots\; -7)$$

Reinserting the variables gives us

$$x \qquad -4z \;=\; 5,$$
$$\quad -y + 12z \;=\; -13,$$
$$\qquad \quad 14z \;=\; -7.$$

We now solve this last equation for z and get $z = -\frac{1}{2}$. Next, we substitute $-\frac{1}{2}$ for z in the preceding equation and solve for y: $-y + 12\left(-\frac{1}{2}\right) = -13$, so $y = 7$. Since there is no y-term in the first equation of this last system, we need only substitute $-\frac{1}{2}$ for z to solve for x: $x - 4\left(-\frac{1}{2}\right) = 5$, so $x = 3$. The solution is $\left(3, 7, -\frac{1}{2}\right)$. The check is left to the student.

TRY EXERCISE 13

The operations used in the preceding example correspond to those used to produce equivalent systems of equations, that is, systems of equations that have the same solution. We call the matrices **row-equivalent** and the operations that produce them **row-equivalent operations.**

Row-Equivalent Operations

Row-Equivalent Operations

Each of the following row-equivalent operations produces a row-equivalent matrix:

a) Interchanging any two rows.
b) Multiplying all elements of a row by a nonzero constant.
c) Replacing a row with the sum of that row and a multiple of another row.

STUDENT NOTES

Note that row-equivalent matrices are not *equal*. It is the solutions of the corresponding systems that are the same.

The best overall method for solving systems of equations is by row-equivalent matrices; even computers are programmed to use them. Matrices are part of a branch of mathematics known as linear algebra. They are also studied in many courses in finite mathematics.

TECHNOLOGY CONNECTION

Row-equivalent operations can be performed on a graphing calculator. For example, to interchange the first and second rows of the matrix, as in step (1) of Example 2 above, we enter the matrix as matrix **A** and select "rowSwap" from the MATRIX MATH menu. Some graphing calculators will not automatically store the matrix produced using a row-equivalent operation, so when several operations are to be performed in succession, it is helpful to store the result of each operation as it is produced. In the window at right, we see both the matrix produced by the rowSwap operation and the indication that this matrix is stored, using STO·, as matrix **B**.

```
rowSwap([A],1,2)→[B]
[[1   0 −4   5]
 [2 −1   4 −3]
 [6 −1   2 10]]
```

1. Use a graphing calculator to proceed through all the steps in Example 2.

<table>
<tr><td>3.6</td><td>EXERCISE SET</td><td>For Extra Help
MyMathLab</td><td></td><td></td><td></td><td></td></tr>
</table>

🖐 *Concept Reinforcement* *Complete each of the following statements.*

1. A(n) _____ is a rectangular array of numbers.

2. The rows of a matrix are _____ and the _____ are vertical.

3. Each number in a matrix is called a(n) _____ or element.

4. The plural of the word matrix is _____ .

5. As part of solving a system using matrices, we can interchange any two _____ .

6. Before we reinsert the variables, the leftmost column in the matrix has zeros in all rows except the _____ one.

Solve using matrices.

7. $x + 2y = 11,$
$3x - y = 5$

8. $x + 3y = 16,$
$6x + y = 11$

9. $3x + y = -1,$
$6x + 5y = 13$

10. $2x - y = 6,$
$8x + 2y = 0$

11. $6x - 2y = 4,$
$7x + y = 13$

12. $3x + 4y = 7,$
$-5x + 2y = 10$

13. $3x + 2y + 2z = 3,$
$x + 2y - z = 5,$
$2x - 4y + z = 0$

14. $4x - y - 3z = 19,$
$8x + y - z = 11,$
$2x + y + 2z = -7$

15. $p - 2q - 3r = 3,$
$2p - q - 2r = 4,$
$4p + 5q + 6r = 4$

16. $x + 2y - 3z = 9,$
$2x - y + 2z = -8,$
$3x - y - 4z = 3$

17. $3p + 2r = 11,$
$q - 7r = 4,$
$p - 6q = 1$

18. $4a + 9b = 8,$
$8a + 6c = -1,$
$6b + 6c = -1$

19. $2x + 2y - 2z - 2w = -10,$
$w + y + z + x = -5,$
$x - y + 4z + 3w = -2,$
$w - 2y + 2z + 3x = -6$

20. $-w - 3y + z + 2x = -8,$
$x + y - z - w = -4,$
$w + y + z + x = 22,$
$x - y - z - w = -14$

Solve using matrices.

21. *Coin value.*   A collection of 42 coins consists of dimes and nickels. The total value is $3.00. How many dimes and how many nickels are there?

22. *Coin value.*   A collection of 43 coins consists of dimes and quarters. The total value is $7.60. How many dimes and how many quarters are there?

23. *Snack mix.*   Bree sells a dried-fruit mixture for $5.80 per pound and Hawaiian macadamia nuts for $14.75 per pound. She wants to blend the two to get a 15-lb mixture that she will sell for $9.38 per pound. How much of each should she use?

24. *Mixing paint.*   Higher quality paint typically contains more solids. Alex has available paint that contains 45% solids and paint that contains 25% solids. How much of each should he use to create 20 gal of paint that contains 39% solids?

25. *Investments.*   Elena receives $212 per year in simple interest from three investments totaling $2500. Part is invested at 7%, part at 8%, and part at 9%. There is $1100 more invested at 9% than at 8%. Find the amount invested at each rate.

26. *Investments.*   Miguel receives $306 per year in simple interest from three investments totaling $3200. Part is invested at 8%, part at 9%, and part at 10%. There is $1900 more invested at 10% than at 9%. Find the amount invested at each rate.

27. Explain how you can recognize dependent equations when solving with matrices.

28. Explain how you can recognize an inconsistent system when solving with matrices.

Skill Review

To prepare for Section 3.7, review order of operations (Section 1.2).

Simplify. [1.2]

29. $3(-1) - (-4)(5)$

30. $7(-5) - 2(-8)$

31. $-2(5 \cdot 3 - 4 \cdot 6) - 3(2 \cdot 7 - 15) + 4(3 \cdot 8 - 5 \cdot 4)$

32. $6(2 \cdot 7 - 3(-4)) - 4(3(-8) - 10) + 5(4 \cdot 3 - (-2)7)$

Synthesis

33. If the matrices

$$\begin{bmatrix} a_1 & b_1 & | & c_1 \\ d_1 & e_1 & | & f_1 \end{bmatrix} \text{ and } \begin{bmatrix} a_2 & b_2 & | & c_2 \\ d_2 & e_2 & | & f_2 \end{bmatrix}$$

share the same solution, does it follow that the corresponding entries are all equal to each other ($a_1 = a_2$, $b_1 = b_2$, etc.)? Why or why not?

34. Explain how the row-equivalent operations make use of the addition, multiplication, and distributive properties.

35. The sum of the digits in a four-digit number is 10. Twice the sum of the thousands digit and the tens digit is 1 less than the sum of the other two digits. The tens digit is twice the thousands digit. The ones digit equals the sum of the thousands digit and the hundreds digit. Find the four-digit number.

36. Solve for x and y:
$$ax + by = c,$$
$$dx + ey = f.$$

3.7   Determinants and Cramer's Rule

Determinants of 2 × 2 Matrices   •   Cramer's Rule: 2 × 2 Systems   •   Cramer's Rule: 3 × 3 Systems

Determinants of 2 × 2 Matrices

When a matrix has m rows and n columns, it is called an "m by n" matrix. Thus its *dimensions* are denoted by $m \times n$. If a matrix has the same number of rows and columns, it is called a **square matrix**. Associated with every square matrix is a number called its **determinant**, defined as follows for 2 × 2 matrices.

2 × 2 Determinants

The determinant of a two-by-two matrix $\begin{bmatrix} a & c \\ b & d \end{bmatrix}$ is denoted $\begin{vmatrix} a & c \\ b & d \end{vmatrix}$ and is defined as follows:

$$\begin{vmatrix} a & c \\ b & d \end{vmatrix} = ad - bc.$$

EXAMPLE 1   Evaluate: $\begin{vmatrix} 2 & -5 \\ 6 & 7 \end{vmatrix}$.

SOLUTION   We multiply and subtract as follows:

$$\begin{vmatrix} 2 & -5 \\ 6 & 7 \end{vmatrix} = 2 \cdot 7 - 6 \cdot (-5) = 14 + 30 = 44.$$

TRY EXERCISE 7

STUDY SKILLS

Find the Highlights

If you do not already own one, consider purchasing a highlighter to use as you read this text and work on the exercises. Often the best time to highlight an important sentence or step in an example is after you have read through the section the first time.

Cramer's Rule: 2 × 2 Systems

One of the many uses for determinants is in solving systems of linear equations in which the number of variables is the same as the number of equations and the constants are not all 0. Let's consider a system of two equations:

$$a_1 x + b_1 y = c_1,$$
$$a_2 x + b_2 y = c_2.$$

If we use the elimination method, a series of steps can show that

$$x = \frac{c_1 b_2 - c_2 b_1}{a_1 b_2 - a_2 b_1} \quad \text{and} \quad y = \frac{a_1 c_2 - a_2 c_1}{a_1 b_2 - a_2 b_1}.$$

These fractions can be rewritten using determinants.

Cramer's Rule: 2 × 2 Systems

The solution of the system

$$a_1 x + b_1 y = c_1,$$
$$a_2 x + b_2 y = c_2,$$

if it is unique, is given by

$$x = \frac{\begin{vmatrix} c_1 & b_1 \\ c_2 & b_2 \end{vmatrix}}{\begin{vmatrix} a_1 & b_1 \\ a_2 & b_2 \end{vmatrix}}, \quad y = \frac{\begin{vmatrix} a_1 & c_1 \\ a_2 & c_2 \end{vmatrix}}{\begin{vmatrix} a_1 & b_1 \\ a_2 & b_2 \end{vmatrix}}.$$

These formulas apply only if the denominator is not 0. If the denominator *is* 0, then one of two things happens:

1. If the denominator is 0 and the numerators are also 0, then the equations in the system are dependent.
2. If the denominator is 0 and at least one numerator is not 0, then the system is inconsistent.

To use Cramer's rule, we find the determinants and compute x and y as shown above. Note that the denominators are identical and the coefficients of x and y appear in the same position as in the original equations. In the numerator of x, the constants c_1 and c_2 replace a_1 and a_2. In the numerator of y, the constants c_1 and c_2 replace b_1 and b_2.

EXAMPLE **2** Solve using Cramer's rule:

$$2x + 5y = 7,$$
$$5x - 2y = -3.$$

SOLUTION We have

$$x = \frac{\begin{vmatrix} 7 & 5 \\ -3 & -2 \end{vmatrix}}{\begin{vmatrix} 2 & 5 \\ 5 & -2 \end{vmatrix}}$$

$\leftarrow$ The constants $\begin{matrix} 7 \\ -3 \end{matrix}$ form the first column.

$\leftarrow$ The columns are the coefficients of the variables.

$$= \frac{7(-2) - (-3)5}{2(-2) - 5 \cdot 5} = \frac{1}{-29} = -\frac{1}{29}$$

and

$$y = \frac{\begin{vmatrix} 2 & 7 \\ 5 & -3 \end{vmatrix}}{\begin{vmatrix} 2 & 5 \\ 5 & -2 \end{vmatrix}}$$

$\leftarrow$ The constants $\begin{matrix} 7 \\ -3 \end{matrix}$ form the second column.

$\leftarrow$ The denominator is the same as in the expression for x.

$$= \frac{2(-3) - 5 \cdot 7}{-29} = \frac{-41}{-29} = \frac{41}{29}.$$

The solution is $\left(-\frac{1}{29}, \frac{41}{29}\right)$. The check is left to the student.   **TRY EXERCISE** 17

Cramer's Rule: 3 × 3 Systems

Cramer's rule can be extended for systems of three linear equations. However, before doing so, we must define what a 3 × 3 determinant is.

3 × 3 Determinants

The determinant of a three-by-three matrix can be defined as follows:

$$\begin{vmatrix} a_1 & b_1 & c_1 \\ a_2 & b_2 & c_2 \\ a_3 & b_3 & c_3 \end{vmatrix} = a_1 \begin{vmatrix} b_2 & c_2 \\ b_3 & c_3 \end{vmatrix} \overset{\text{Subtract.}}{-} a_2 \begin{vmatrix} b_1 & c_1 \\ b_3 & c_3 \end{vmatrix} \overset{\text{Add.}}{+} a_3 \begin{vmatrix} b_1 & c_1 \\ b_2 & c_2 \end{vmatrix}$$

STUDENT NOTES.

Cramer's rule and the evaluation of determinants rely on patterns. Recognizing and remembering the patterns will help you understand and use the definitions.

Note that the a's come from the first column. Note too that the 2 × 2 determinants above can be obtained by crossing out the row and the column in which the a occurs.

For a_1:

$$\begin{vmatrix} a_1 & b_1 & c_1 \\ a_2 & b_2 & c_2 \\ a_3 & b_3 & c_3 \end{vmatrix}$$

For a_2:

$$\begin{vmatrix} a_1 & b_1 & c_1 \\ a_2 & b_2 & c_2 \\ a_3 & b_3 & c_3 \end{vmatrix}$$

For a_3:

$$\begin{vmatrix} a_1 & b_1 & c_1 \\ a_2 & b_2 & c_2 \\ a_3 & b_3 & c_3 \end{vmatrix}$$

EXAMPLE 3   Evaluate:

$$\begin{vmatrix} -1 & 0 & 1 \\ -5 & 1 & -1 \\ 4 & 8 & 1 \end{vmatrix}.$$

SOLUTION   We have

Subtract.   Add.

$$\begin{vmatrix} -1 & 0 & 1 \\ -5 & 1 & -1 \\ 4 & 8 & 1 \end{vmatrix} = -1\begin{vmatrix} 1 & -1 \\ 8 & 1 \end{vmatrix} - (-5)\begin{vmatrix} 0 & 1 \\ 8 & 1 \end{vmatrix} + 4\begin{vmatrix} 0 & 1 \\ 1 & -1 \end{vmatrix}$$

$$= -1(1 + 8) + 5(0 - 8) + 4(0 - 1) \qquad \text{Evaluating the three determinants}$$

$$= -9 - 40 - 4 = -53.$$

TRY EXERCISE 11

— Technology Connection sidebar:

TECHNOLOGY CONNECTION

Determinants can be evaluated on most graphing calculators using **2ND** **MATRIX**. After entering a matrix, we select the determinant operation from the MATRIX MATH menu and enter the name of the matrix. The graphing calculator will return the value of the determinant of the matrix. For example, if

$$\mathbf{A} = \begin{bmatrix} 1 & 6 & -1 \\ -3 & -5 & 3 \\ 0 & 4 & 2 \end{bmatrix},$$

we have

```
det([A])
             26
```

1. Confirm the calculations in Example 4.

Cramer's Rule: 3 × 3 Systems

The solution of the system

$$a_1x + b_1y + c_1z = d_1,$$
$$a_2x + b_2y + c_2z = d_2,$$
$$a_3x + b_3y + c_3z = d_3$$

can be found using the following determinants:

$$D = \begin{vmatrix} a_1 & b_1 & c_1 \\ a_2 & b_2 & c_2 \\ a_3 & b_3 & c_3 \end{vmatrix}, \qquad D_x = \begin{vmatrix} d_1 & b_1 & c_1 \\ d_2 & b_2 & c_2 \\ d_3 & b_3 & c_3 \end{vmatrix},$$

D contains only coefficients.
In D_x the d's replace the a's.

$$D_y = \begin{vmatrix} a_1 & d_1 & c_1 \\ a_2 & d_2 & c_2 \\ a_3 & d_3 & c_3 \end{vmatrix}, \qquad D_z = \begin{vmatrix} a_1 & b_1 & d_1 \\ a_2 & b_2 & d_2 \\ a_3 & b_3 & d_3 \end{vmatrix}.$$

In D_y, the d's replace the b's.
In D_z, the d's replace the c's.

If a unique solution exists, it is given by

$$x = \frac{D_x}{D}, \qquad y = \frac{D_y}{D}, \qquad z = \frac{D_z}{D}.$$

EXAMPLE 4   Solve using Cramer's rule:

$$x - 3y + 7z = 13,$$
$$x + y + z = 1,$$
$$x - 2y + 3z = 4.$$

SOLUTION We compute D, D_x, D_y, and D_z:

$$D = \begin{vmatrix} 1 & -3 & 7 \\ 1 & 1 & 1 \\ 1 & -2 & 3 \end{vmatrix} = -10; \qquad D_x = \begin{vmatrix} 13 & -3 & 7 \\ 1 & 1 & 1 \\ 4 & -2 & 3 \end{vmatrix} = 20;$$

$$D_y = \begin{vmatrix} 1 & 13 & 7 \\ 1 & 1 & 1 \\ 1 & 4 & 3 \end{vmatrix} = -6; \qquad D_z = \begin{vmatrix} 1 & -3 & 13 \\ 1 & 1 & 1 \\ 1 & -2 & 4 \end{vmatrix} = -24.$$

Then

$$x = \frac{D_x}{D} = \frac{20}{-10} = -2;$$

$$y = \frac{D_y}{D} = \frac{-6}{-10} = \frac{3}{5};$$

$$z = \frac{D_z}{D} = \frac{-24}{-10} = \frac{12}{5}.$$

The solution is $\left(-2, \frac{3}{5}, \frac{12}{5}\right)$. The check is left to the student.   **TRY EXERCISE** 21

In Example 4, we need not have evaluated D_z. Once x and y were found, we could have substituted them into one of the equations to find z.

To use Cramer's rule, we divide by D, provided $D \neq 0$. If $D = 0$ and at least one of the other determinants is not 0, then the system is inconsistent. If *all* the determinants are 0, then the equations in the system are dependent.

3.7 EXERCISE SET

👈 *Concept Reinforcement* *Classify each statement as either true or false.*

1. A square matrix has the same number of rows and columns.

2. A 3×4 matrix has 3 rows and 4 columns.

3. A determinant is a number.

4. Cramer's rule exists only for 2×2 systems.

5. Whenever Cramer's rule yields a denominator that is 0, the system has no solution.

6. Whenever Cramer's rule yields a numerator that is 0, the equations are dependent.

Evaluate.

7. $\begin{vmatrix} 3 & 5 \\ 4 & 8 \end{vmatrix}$

8. $\begin{vmatrix} 3 & 2 \\ 2 & -3 \end{vmatrix}$

9. $\begin{vmatrix} 10 & 8 \\ -5 & -9 \end{vmatrix}$

10. $\begin{vmatrix} 3 & 2 \\ -7 & 11 \end{vmatrix}$

11. $\begin{vmatrix} 1 & 4 & 0 \\ 0 & -1 & 2 \\ 3 & -2 & 1 \end{vmatrix}$

12. $\begin{vmatrix} 2 & 4 & -2 \\ 1 & 0 & 2 \\ 0 & 1 & 3 \end{vmatrix}$

13. $\begin{vmatrix} -1 & -2 & -3 \\ 3 & 4 & 2 \\ 0 & 1 & 2 \end{vmatrix}$

14. $\begin{vmatrix} 5 & 2 & 2 \\ 0 & 1 & -1 \\ 3 & 3 & 1 \end{vmatrix}$

15. $\begin{vmatrix} -4 & -2 & 3 \\ -3 & 1 & 2 \\ 3 & 4 & -2 \end{vmatrix}$

16. $\begin{vmatrix} 2 & -1 & 1 \\ 1 & 2 & -1 \\ 3 & 4 & -3 \end{vmatrix}$

Solve using Cramer's rule.

17. $5x + 8y = 1,$
    $3x + 7y = 5$

18. $3x - 4y = 6,$
    $5x + 9y = 10$

19. $5x - 4y = -3,$
    $7x + 2y = 6$

20. $-2x + 4y = 3,$
    $3x - 7y = 1$

21. $3x - y + 2z = 1,$
$\quad x - y + 2z = 3,$
$\quad -2x + 3y + z = 1$

22. $3x + 2y - z = 4,$
$\quad 3x - 2y + z = 5,$
$\quad 4x - 5y - z = -1$

23. $2x - 3y + 5z = 27,$
$\quad x + 2y - z = -4,$
$\quad 5x - y + 4z = 27$

24. $\quad x - y + 2z = -3,$
$\quad x + 2y + 3z = 4,$
$\quad 2x + y + z = -3$

25. $\quad r - 2s + 3t = 6,$
$\quad 2r - s - t = -3,$
$\quad r + s + t = 6$

26. $\quad a \quad\quad - 3c = 6,$
$\quad\quad b + 2c = 2,$
$\quad 7a - 3b - 5c = 14$

27. Describe at least one of the patterns that you see in Cramer's rule.

28. Which version of Cramer's rule do you find more useful: the version for 2×2 systems or the version for 3×3 systems? Why?

Skill Review

To prepare for Section 3.8, review functions (Sections 2.2 and 2.6).

Find each of the following, given $f(x) = 80x + 2500$ and $g(x) = 150x$.

29. $f(90)$ [2.2]

30. $(g - f)(x)$ [2.6]

31. $(g - f)(10)$ [2.6]

32. $(g - f)(100)$ [2.6]

33. All values of x for which $f(x) = g(x)$ [1.3], [2.2]

34. All values of x for which $(g - f)(x) = 0$ [1.3], [2.6]

Synthesis

35. Cramer's rule states that if $a_1 x + b_1 y = c_1$ and $a_2 x + b_2 y = c_2$ are dependent, then
$$\begin{vmatrix} a_1 & b_1 \\ a_2 & b_2 \end{vmatrix} = 0.$$
Explain why this will always happen.

36. Under what conditions can a 3×3 system of linear equations be consistent but unable to be solved using Cramer's rule?

Solve.

37. $\begin{vmatrix} y & -2 \\ 4 & 3 \end{vmatrix} = 44$

38. $\begin{vmatrix} 2 & x & -1 \\ -1 & 3 & 2 \\ -2 & 1 & 1 \end{vmatrix} = -12$

39. $\begin{vmatrix} m+1 & -2 \\ m-2 & 1 \end{vmatrix} = 27$

40. Show that an equation of the line through (x_1, y_1) and (x_2, y_2) can be written
$$\begin{vmatrix} x & y & 1 \\ x_1 & y_1 & 1 \\ x_2 & y_2 & 1 \end{vmatrix} = 0.$$

3.8   Business and Economics Applications

Break-Even Analysis ■ Supply and Demand

Break-Even Analysis

The money that a business spends to manufacture a product is its *cost*. The **total cost** of production can be thought of as a function C, where $C(x)$ is the cost of producing x units. When the company sells the product, it takes in money. This is *revenue* and can be thought of as a function R, where $R(x)$ is the **total revenue** from the sale of x units. **Total profit** is the money taken in less the money spent, or total revenue minus total cost. Total profit from the production and sale of x units is a function P given by

$$\textbf{Profit = Revenue} - \textbf{Cost,} \quad \text{or} \quad P(x) = R(x) - C(x).$$

If $R(x)$ is greater than $C(x)$, there is a gain and $P(x)$ is positive. If $C(x)$ is greater than $R(x)$, there is a loss and $P(x)$ is negative. When $R(x) = C(x)$, the company breaks even.

There are two kinds of costs. First, there are costs like rent, insurance, machinery, and so on. These costs, which must be paid regardless of how many items are produced, are called *fixed costs*. Second, costs for labor, materials, marketing, and so on are called *variable costs*, because they vary according to the amount being produced. The sum of the fixed cost and the variable cost gives the **total cost**.

> *CAUTION!*   Do not confuse "cost" with "price." When we discuss the *cost* of an item, we are referring to what it costs to produce the item. The *price* of an item is what a consumer pays to purchase the item and is used when calculating revenue.

EXAMPLE   1   *Manufacturing chairs.*   Renewable Designs is planning to make a new chair. Fixed costs will be $90,000, and it will cost $25 to produce each chair (variable costs). Each chair sells for $48.

a) Find the total cost $C(x)$ of producing x chairs.

b) Find the total revenue $R(x)$ from the sale of x chairs.

c) Find the total profit $P(x)$ from the production and sale of x chairs.

d) What profit will the company realize from the production and sale of 3000 chairs? of 8000 chairs?

e) Graph the total-cost, total-revenue, and total-profit functions using the same set of axes. Determine the break-even point.

SOLUTION

a) Total cost, in dollars, is given by

$$C(x) = \text{(Fixed costs) plus (Variable costs)},$$
$$\text{or}\quad C(x) = \quad 90{,}000 \quad + \quad 25x$$

where x is the number of chairs produced.

b) Total revenue, in dollars, is given by

$$R(x) = 48x.$$ $48 times the number of chairs sold. We assume that every chair produced is sold.

c) Total profit, in dollars, is given by

$$P(x) = R(x) - C(x)$$ Profit is revenue minus cost.
$$= 48x - (90{,}000 + 25x)$$
$$= 23x - 90{,}000.$$

d) Profits will be

$$P(3000) = 23 \cdot 3000 - 90{,}000 = -\$21{,}000$$

when 3000 chairs are produced and sold, and

$$P(8000) = 23 \cdot 8000 - 90{,}000 = \$94{,}000$$

when 8000 chairs are produced and sold. Thus the company loses money if only 3000 chairs are sold, but makes money if 8000 are sold.

e) The graphs of each of the three functions are shown below:

$C(x) = 90{,}000 + 25x,$ This represents the cost function.

$R(x) = 48x,$ This represents the revenue function.

$P(x) = 23x - 90{,}000.$ This represents the profit function.

$C(x)$, $R(x)$, and $P(x)$ are all in dollars.

The revenue function has a graph that goes through the origin and has a slope of 48. The cost function has an intercept on the $-axis of 90,000 and has a slope of 25. The profit function has an intercept on the $-axis of $-90{,}000$ and has a slope of 23. It is shown by the red and black dashed line. The red portion of the dashed line shows a "negative" profit, which is a loss. (That is what is known as "being in the red.") The black portion of the dashed line shows a "positive" profit, or gain. (That is what is known as "being in the black.")

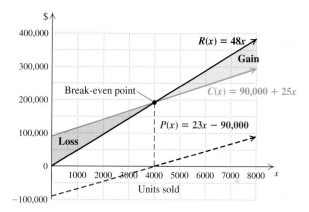

STUDENT NOTES

If you plan to study business or economics, you may want to consult the material in this section when these topics arise in your other courses.

Gains occur where revenue exceeds cost. Losses occur where revenue is less than cost. The **break-even point** occurs where the graphs of R and C cross. Thus to find the break-even point, we solve a system:

$R(x) = 48x,$

$C(x) = 90{,}000 + 25x.$

Since both revenue and cost are in *dollars* and they are equal at the break-even point, the system can be rewritten as

$d = 48x,$ (1)

$d = 90{,}000 + 25x$ (2)

and solved using substitution:

$48x = 90{,}000 + 25x$ Substituting $48x$ for d in equation (2)

$23x = 90{,}000$

$x \approx 3913.04.$

The firm will break even if it produces and sells about 3913 chairs (3913 will yield a tiny loss and 3914 a tiny gain), and takes in a total of $R(3913) = 48 \cdot 3913 = \$187{,}824$ in revenue. Note that the x-coordinate of the break-even point can also be found by solving $P(x) = 0$. The break-even point is (3913 chairs, $187,824).

TRY EXERCISE 9

Supply and Demand

As the price of coffee varies, so too does the amount sold. The table and graph below show that *consumers will demand less as the price goes up.*

Demand Function, *D*

Price, p, per Kilogram	Quantity, $D(p)$ (in millions of kilograms)
$ 8.00	25
9.00	20
10.00	15
11.00	10
12.00	5

As the price of coffee varies, the amount made available varies as well. The table and graph below show that *sellers will supply more as the price goes up.*

Supply Function, *S*

Price, p, per Kilogram	Quantity, $S(p)$ (in millions of kilograms)
$ 9.00	5
9.50	10
10.00	15
10.50	20
11.00	25

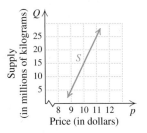

Let's look at the above graphs together. We see that as price increases, demand decreases. As price increases, supply increases. The point of intersection is called the **equilibrium point**. At that price, the amount that the seller will supply is the same amount that the consumer will buy. The situation is similar to a buyer and a seller negotiating the price of an item. The equilibrium point is the price and quantity that they finally agree on.

Any ordered pair of coordinates from the graph is (price, quantity), because the horizontal axis is the price axis and the vertical axis is the quantity axis. If D is a demand function and S is a supply function, then the equilibrium point is where demand equals supply:

$$D(p) = S(p).$$

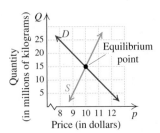

EXAMPLE   2   Find the equilibrium point for the demand and supply functions given:

$$D(p) = 1000 - 60p, \quad (1)$$
$$S(p) = 200 + 4p. \quad (2)$$

SOLUTION   Since both demand and supply are *quantities* and they are equal at the equilibrium point, we rewrite the system as

$$q = 1000 - 60p, \quad (1)$$
$$q = 200 + 4p. \quad (2)$$

We substitute $200 + 4p$ for q in equation (1) and solve:

$$200 + 4p = 1000 - 60p \quad \text{Substituting } 200 + 4p \text{ for } q \text{ in equation (1)}$$
$$200 + 64p = 1000 \quad \text{Adding } 60p \text{ to both sides}$$
$$64p = 800 \quad \text{Adding } -200 \text{ to both sides}$$
$$p = \frac{800}{64} = 12.5.$$

Thus the equilibrium price is \$12.50 per unit.

To find the equilibrium quantity, we substitute \$12.50 into either $D(p)$ or $S(p)$. We use $S(p)$:

$$S(12.5) = 200 + 4(12.5) = 200 + 50 = 250.$$

Therefore, the equilibrium quantity is 250 units, and the equilibrium point is (\$12.50, 250).

TRY EXERCISE ▶ 19

3.8   EXERCISE SET

For Extra Help
MyMathLab   Math XL

PRACTICE   WATCH   DOWNLOAD

🖋 **Concept Reinforcement**   *In each of Exercises 1–8, match the word or phrase with the most appropriate choice from the column on the right.*

1. ____ Total cost

2. ____ Fixed costs

3. ____ Variable costs

4. ____ Total revenue

5. ____ Total profit

6. ____ Price

7. ____ Break-even point

8. ____ Equilibrium point

a) The amount of money that a company takes in

b) The sum of fixed costs and variable costs

c) The point at which total revenue equals total cost

d) What consumers pay per item

e) The difference between total revenue and total cost

f) What companies spend whether or not a product is produced

g) The point at which supply equals demand

h) The costs that vary according to the number of items produced

For each of the following pairs of total-cost and total-revenue functions, find **(a)** *the total-profit function and* **(b)** *the break-even point.*

9. $C(x) = 35x + 200{,}000$,
   $R(x) = 55x$

10. $C(x) = 20x + 500{,}000$,
    $R(x) = 70x$

11. $C(x) = 15x + 3100$,
    $R(x) = 40x$

12. $C(x) = 30x + 49{,}500$,
    $R(x) = 85x$

13. $C(x) = 40x + 22{,}500$,
    $R(x) = 85x$

14. $C(x) = 20x + 10{,}000,$
$R(x) = 100x$

15. $C(x) = 24x + 50{,}000,$
$R(x) = 40x$

16. $C(x) = 40x + 8010,$
$R(x) = 58x$

Aha! **17.** $C(x) = 75x + 100{,}000,$
$R(x) = 125x$

18. $C(x) = 20x + 120{,}000,$
$R(x) = 50x$

Find the equilibrium point for each of the following pairs of demand and supply functions.

19. $D(p) = 2000 - 15p,$
$S(p) = 740 + 6p$

20. $D(p) = 1000 - 8p,$
$S(p) = 350 + 5p$

21. $D(p) = 760 - 13p,$
$S(p) = 430 + 2p$

22. $D(p) = 800 - 43p,$
$S(p) = 210 + 16p$

23. $D(p) = 7500 - 25p,$
$S(p) = 6000 + 5p$

24. $D(p) = 8800 - 30p,$
$S(p) = 7000 + 15p$

25. $D(p) = 1600 - 53p,$
$S(p) = 320 + 75p$

26. $D(p) = 5500 - 40p,$
$S(p) = 1000 + 85p$

Solve.

27. *Manufacturing MP3 players.*   SoundGen, Inc., is planning to manufacture a new type of MP3 player/cell phone. The fixed costs for production are $45,000. The variable costs for producing each unit are estimated to be $40. The revenue from each unit is to be $130. Find the following.

   a) The total cost $C(x)$ of producing x MP3/cell phones

   b) The total revenue $R(x)$ from the sale of x MP3/cell phones

   c) The total profit $P(x)$ from the production and sale of x MP3/cell phones

   d) The profit or loss from the production and sale of 3000 MP3/cell phones; of 400 MP3/cell phones

   e) The break-even point

28. *Computer manufacturing.*   Current Electronics is planning to introduce a new laptop computer. The fixed costs for production are $125,300. The variable costs for producing each computer are $450. The revenue from each computer is $800. Find the following.

   a) The total cost $C(x)$ of producing x computers

   b) The total revenue $R(x)$ from the sale of x computers

   c) The total profit $P(x)$ from the production and sale of x computers

   d) The profit or loss from the production and sale of 100 computers; of 400 computers

   e) The break-even point

29. *Pet safety.*   Ava designed and is now producing a pet car seat. The fixed costs for setting up production are $10,000. The variable costs for producing each seat are $30. The revenue from each seat is to be $80. Find the following.

   a) The total cost $C(x)$ of producing x seats

   b) The total revenue $R(x)$ from the sale of x seats

   c) The total profit $P(x)$ from the production and sale of x seats

   d) The profit or loss from the production and sale of 2000 seats; of 50 seats

   e) The break-even point

30. *Manufacturing caps.*   Martina's Custom Printing is planning on adding painter's caps to its product line. For the first year, the fixed costs for setting up production are $16,404. The variable costs for producing a dozen caps are $6.00. The revenue on each dozen caps will be $18.00. Find the following.

   a) The total cost $C(x)$ of producing x dozen caps

   b) The total revenue $R(x)$ from the sale of x dozen caps

   c) The total profit $P(x)$ from the production and sale of x dozen caps

   d) The profit or loss from the production and sale of 3000 dozen caps; of 1000 dozen caps

   e) The break-even point

31. In Example 1, the slope of the line representing Revenue is the sum of the slopes of the other two lines. This is not a coincidence. Explain why.

32. Variable costs and fixed costs are often compared to the slope and the *y*-intercept, respectively, of an equation for a line. Explain why you feel this analogy is or is not valid.

Skill Review

To prepare for Chapter 4, review solving equations using the addition and multiplication principles (Section 1.3).

Solve. [1.3]

33. $4x - 3 = 21$

34. $5 - x = 7$

35. $3x - 5 = 12x + 6$

36. $x - 4 = 9x - 10$

37. $3 - (x + 2) = 7$

38. $1 - 3(2x + 1) = 3 - 5x$

Synthesis

39. Rosie claims that since her fixed costs are $3000, she need sell only 10 custom birdbaths at $300 each in order to break even. Does this sound plausible? Why or why not?

40. In this section, we examined supply and demand functions for coffee. Does it seem realistic to you for the graph of *D* to have a constant slope? Why or why not?

41. *Yo-yo production.*   Bing Boing Hobbies is willing to produce 100 yo-yo's at $2.00 each and 500 yo-yo's at $8.00 each. Research indicates that the public will buy 500 yo-yo's at $1.00 each and 100 yo-yo's at $9.00 each. Find the equilibrium point.

42. *Loudspeaker production.*   Sonority Speakers, Inc., has fixed costs of $15,400 and variable costs of $100 for each pair of speakers produced. If the speakers sell for $250 per pair, how many pairs of speakers must be produced (and sold) in order to have enough profit to cover the fixed costs of two additional facilities? Assume that all fixed costs are identical.

Use a graphing calculator to solve.

43. *Dog food production.*   Puppy Love, Inc., will soon begin producing a new line of puppy food. The marketing department predicts that the demand function will be $D(p) = -14.97p + 987.35$ and the supply function will be $S(p) = 98.55p - 5.13$.

a) To the nearest cent, what price per unit should be charged in order to have equilibrium between supply and demand?

b) The production of the puppy food involves $87,985 in fixed costs and $5.15 per unit in variable costs. If the price per unit is the value you found in part (a), how many units must be sold in order to break even?

44. *Computer production.*   Brushstroke Computers, Inc., is planning a new line of computers, each of which will sell for $970. The fixed costs in setting up production are $1,235,580 and the variable costs for each computer are $697.

a) What is the break-even point?

b) The marketing department at Brushstroke is not sure that $970 is the best price. Their demand function for the new computers is given by $D(p) = -304.5p + 374,580$ and their supply function is given by $S(p) = 788.7p - 576,504$. To the nearest dollar, what price *p* would result in equilibrium between supply and demand?

c) If the computers are sold for the equilibrium price found in part (b), what is the break-even point?

Study Summary

KEY TERMS AND CONCEPTS	EXAMPLES

SECTION 3.1: SYSTEMS OF EQUATIONS IN TWO VARIABLES

A **system of equations** is a set of two or more equations, in two or more variables. A solution of a system of equations must make all the equations true.

A system is **consistent** if it has at least one solution. Otherwise it is **inconsistent**.

The equations in a system are **dependent** if one of them can be written as a multiple and/or a sum of the other equation(s). Otherwise, they are **independent**.

Systems of two equations in two unknowns can be solved graphically.

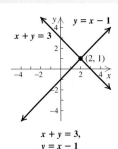

$x + y = 3$,
$y = x - 1$

The graphs intersect at (2, 1).
The solution is (2, 1).
The system is consistent.
The equations are independent.

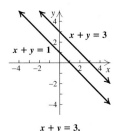

$x + y = 3$,
$x + y = 1$

The graphs do not intersect.
There is no solution.
The system is inconsistent.
The equations are independent.

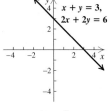

$x + y = 3$,
$2x + 2y = 6$

The graphs are the same.
The solution set is
$\{(x, y) \mid x + y = 3\}$.
The system is consistent.
The equations are dependent.

SECTION 3.2: SOLVING BY SUBSTITUTION OR ELIMINATION

Systems of equations can be solved using substitution.

Solve:

$2x + 3y = 8$,
$x = y + 1$.

Substitute and solve for y:

$2(y + 1) + 3y = 8$
$2y + 2 + 3y = 8$
$y = \frac{6}{5}$.

Substitute and solve for x:

$x = y + 1$
$x = \frac{6}{5} + 1$
$x = \frac{11}{5}$.

The solution is $\left(\frac{11}{5}, \frac{6}{5}\right)$.

Systems of equations can be solved using elimination.

Solve:

$4x - 2y = 6$,
$3x + y = 7$.

Eliminate y and solve for x:

$4x - 2y = 6$
$\underline{6x + 2y = 14}$
$10x \qquad = 20$
$x = 2$.

Substitute and solve for y:

$3x + y = 7$
$3 \cdot 2 + y = 7$
$y = 1$.

The solution is (2, 1).

SECTION 3.3: SOLVING APPLICATIONS: SYSTEMS OF TWO EQUATIONS

Total-value, **mixture**, and **motion problems** often translate directly to systems of equations.

Total Value

A jewelry designer purchased 80 beads for a total of $39 to make a necklace. Some of the beads were sterling silver beads that cost 40¢ each and the rest were gemstone beads that cost 65¢ each. How many of each type were bought? (See Example 2 on pp. 167–168 for a solution.)

Mixture

Nature's Green Gardening, Inc., carries two brands of fertilizer containing nitrogen and water. "Gentle Grow" is 3% nitrogen and "Sun Saver" is 8% nitrogen. Nature's Green needs to combine the two types of solutions into a 90-L mixture that is 6% nitrogen. How much of each brand should be used? (See Example 5 on pp. 170–171 for a solution.)

Motion

A Boeing 747-400 jet flies 4 hr west with a 60-mph tailwind. Returning against the wind takes 5 hr. Find the speed of the jet with no wind. (See Example 7 on pp. 173–174 for a solution.)

SECTION 3.4: SYSTEMS OF EQUATIONS IN THREE VARIABLES

Systems of three equations in three variables are usually easiest to solve using elimination.

Solve:

$$x + y - z = 3, \quad (1)$$
$$-x + y + 2z = -5, \quad (2)$$
$$2x - y - 3z = 9 \quad (3)$$

Eliminate x using two equations:

$$x + y - z = 3 \quad (1)$$
$$-x + y + 2z = -5 \quad (2)$$
$$\overline{2y + z = -2.}$$

Eliminate x again using two different equations:

$$-2x - 2y + 2z = -6 \quad (1)$$
$$2x - y - 3z = 9 \quad (3)$$
$$\overline{-3y - z = 3.}$$

Solve the system of two equations for y and z:

$$2y + z = -2$$
$$-3y - z = 3$$
$$\overline{-y = 1}$$
$$y = -1$$

$$2(-1) + z = -2$$
$$z = 0.$$

Substitute and solve for x:

$$x + y - z = 3$$
$$x + (-1) - 0 = 3$$
$$x = 4.$$

The solution is $(4, -1, 0)$.

SECTION 3.5: SOLVING APPLICATIONS: SYSTEMS OF THREE EQUATIONS

Many problems with three unknowns can be solved after translating to a system of three equations.

In a triangular cross section of a roof, the largest angle is 70° greater than the smallest angle. The largest angle is twice as large as the remaining angle. Find the measure of each angle. (See Example 2 on pp. 188–189 for a solution.)

SECTION 3.6: ELIMINATION USING MATRICES

A **matrix** (plural, **matrices**) is a rectangular array of numbers. The individual numbers are called **entries** or **elements**.

$$\begin{bmatrix} 1 & 3 & -4 \\ -2 & 5 & 11 \end{bmatrix} \begin{matrix} \longrightarrow \text{row 1} \\ \longrightarrow \text{row 2} \end{matrix}$$

column 1   column 2   column 3

The *dimensions* of this matrix are 2×3, read "two by three."

By using **row-equivalent** operations, we can solve systems of equations using matrices.

Solve: $\quad x + 4y = 1,$
$$2x - y = 3.$$

Write as a matrix in row-echelon form:
$$\begin{bmatrix} 1 & 4 & \vdots & 1 \\ 2 & -1 & \vdots & 3 \end{bmatrix} \longrightarrow \begin{bmatrix} 1 & 4 & \vdots & 1 \\ 0 & -9 & \vdots & 1 \end{bmatrix}.$$

Rewrite as equations and solve:
$$-9y = 1 \longrightarrow x + 4\left(-\tfrac{1}{9}\right) = 1$$
$$y = -\tfrac{1}{9} \qquad\qquad x = \tfrac{13}{9}.$$

The solution is $\left(\tfrac{13}{9}, -\tfrac{1}{9}\right)$.

SECTION 3.7: DETERMINANTS AND CRAMER'S RULE

A **determinant** is a number associated with a square matrix.

Determinant of a 2 × 2 Matrix

$$\begin{vmatrix} a & c \\ b & d \end{vmatrix} = ad - bc$$

$$\begin{vmatrix} 2 & 3 \\ -1 & 5 \end{vmatrix} = 2 \cdot 5 - (-1)(3) = 13$$

Determinant of a 3 × 3 Matrix

$$\begin{vmatrix} a_1 & b_1 & c_1 \\ a_2 & b_2 & c_2 \\ a_3 & b_3 & c_3 \end{vmatrix} =$$

$$a_1 \begin{vmatrix} b_2 & c_2 \\ b_3 & c_3 \end{vmatrix} - a_2 \begin{vmatrix} b_1 & c_1 \\ b_3 & c_3 \end{vmatrix} + a_3 \begin{vmatrix} b_1 & c_1 \\ b_2 & c_2 \end{vmatrix}$$

$$\begin{vmatrix} 2 & 3 & 2 \\ 0 & 1 & 0 \\ -1 & 5 & -4 \end{vmatrix} = 2 \begin{vmatrix} 1 & 0 \\ 5 & -4 \end{vmatrix} - 0 \begin{vmatrix} 3 & 2 \\ 5 & -4 \end{vmatrix} + (-1) \begin{vmatrix} 3 & 2 \\ 1 & 0 \end{vmatrix}$$

$$= 2(-4 - 0) - 0 - 1(0 - 2)$$

$$= -8 + 2 = -6$$

We can use matrices and **Cramer's rule** to solve systems of equations.

Cramer's rule for 2 × 2 matrices is given on p. 199.

Cramer's rule for 3 × 3 matrices is given on p. 201.

Solve:

$$x - 3y = 7,$$
$$2x + 5y = 4.$$

$$x = \frac{\begin{vmatrix} 7 & -3 \\ 4 & 5 \end{vmatrix}}{\begin{vmatrix} 1 & -3 \\ 2 & 5 \end{vmatrix}}; \qquad y = \frac{\begin{vmatrix} 1 & 7 \\ 2 & 4 \end{vmatrix}}{\begin{vmatrix} 1 & -3 \\ 2 & 5 \end{vmatrix}}$$

$$x = \frac{47}{11} \qquad\qquad y = \frac{-10}{11}$$

The solution is $\left(\frac{47}{11}, -\frac{10}{11}\right)$.

SECTION 3.8: BUSINESS AND ECONOMICS APPLICATIONS

The **break-even point** occurs where the **revenue** equals the **cost**, or where **profit** is 0.

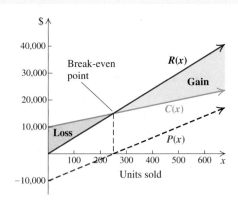

An **equilibrium point** occurs where the **supply** equals the **demand.**

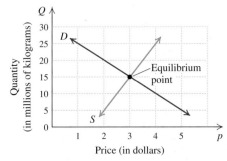

Review Exercises: Chapter 3

Concept Reinforcement *Complete each of the following sentences.*

1. The system

$$5x + 3y = 7,$$
$$y = 2x + 1$$

 is most easily solved using the _____ method. [3.2]

2. The system

$$-2x + 3y = 8,$$
$$2x + 2y = 7$$

 is most easily solved using the _____ method. [3.2]

3. Of the methods used to solve systems of equations, the _____ method may yield only approximate solutions. [3.1], [3.2]

4. When one equation in a system is a multiple of another equation in that system, the equations are said to be _____. [3.1]

5. A system for which there is no solution is said to be _____. [3.1]

6. When we are using an algebraic method to solve a system of equations, obtaining a _____ tells us that the system is inconsistent. [3.2]

7. When we are graphing to solve a system of two equations, if there is no solution, the lines will be _____. [3.1]

8. When a matrix has the same number of rows and columns, it is said to be _____. [3.7]

9. Cramer's rule is a formula in which the numerator and the denominator of each fraction is a(n) _____. [3.7]

10. At the break-even point, the value of the profit function is _____. [3.8]

For Exercises 11–19, if a system has an infinite number of solutions, use set-builder notation to write the solution set. If a system has no solution, state this.

Solve graphically. [3.1]

11. $y = x - 3,$
    $y = \frac{1}{4}x$

12. $2x - 3y = 12,$
    $4x + y = 10$

Solve using the substitution method. [3.2]

13. $5x - 2y = 4,$
    $x = y - 2$

14. $y = x + 2,$
    $y - x = 8$

15. $x - 3y = -2,$
    $7y - 4x = 6$

Solve using the elimination method. [3.2]

16. $2x + 5y = 8,$
    $6x - 5y = 10$

17. $4x - 7y = 18,$
    $9x + 14y = 40$

18. $3x - 5y = 9,$
    $5x - 3y = -1$

19. $1.5x - 3 = -2y,$
    $3x + 4y = 6$

Solve. [3.3]

20. Ana bought two melons and one pineapple for $8.96. If she had purchased one melon and two pineapples, she would have spent $1.49 more. What is the price of a melon? of a pineapple?

21. A freight train leaves Houston at midnight traveling north at a speed of 44 mph. One hour later, a passenger train, going 55 mph, travels north from Houston on a parallel track. How many hours will the passenger train travel before it overtakes the freight train?

22. D'Andre wants 14 L of fruit punch that is 10% juice. At the store, he finds only punch that is 15% juice or punch that is 8% juice. How much of each should he purchase?

Solve. If a system's equations are dependent or if there is no solution, state this. [3.4]

23. $x + 4y + 3z = 2,$
    $2x + y + z = 10,$
    $-x + y + 2z = 8$

24. $4x + 2y - 6z = 34,$
    $2x + y + 3z = 3,$
    $6x + 3y - 3z = 37$

25. $2x - 5y - 2z = -4,$
    $7x + 2y - 5z = -6,$
    $-2x + 3y + 2z = 4$

26. $3x + y = 2,$
    $x + 3y + z = 0,$
    $x + z = 2$

Solve.

27. In triangle ABC, the measure of angle A is four times the measure of angle C, and the measure of angle B is 45° more than the measure of angle C. What are the measures of the angles of the triangle? [3.5]

28. A nontoxic floor wax can be made from lemon juice and food-grade linseed oil. The amount of oil should be twice the amount of lemon juice. How much of each ingredient is needed to make 32 oz of floor wax? (The mix should be spread with a rag and buffed when dry.) [3.3]

29. The sum of the average number of times a man, a woman, and a one-year-old child cry each month is 56.7. A woman cries 3.9 more times than a man. The average number of times a one-year-old cries per month is 43.3 more than the average number of times combined that a man and a woman cry. What is the average number of times per month that each cries? [3.5]

Solve using matrices. Show your work. [3.6]

30. $3x + 4y = -13,$
$5x + 6y = 8$

31. $3x - y + z = -1,$
$2x + 3y + z = 4,$
$5x + 4y + 2z = 5$

Evaluate. [3.7]

32. $\begin{vmatrix} -2 & -5 \\ 3 & 10 \end{vmatrix}$

33. $\begin{vmatrix} 2 & 3 & 0 \\ 1 & 4 & -2 \\ 2 & -1 & 5 \end{vmatrix}$

Solve using Cramer's rule. Show your work. [3.7]

34. $2x + 3y = 6,$
$x - 4y = 14$

35. $2x + y + z = -2,$
$2x - y + 3z = 6,$
$3x - 5y + 4z = 7$

36. Find the equilibrium point for the demand and supply functions

$$S(p) = 60 + 7p$$

and

$$D(p) = 120 - 13p. \quad [3.8]$$

37. Danae is beginning to produce organic honey. For the first year, the fixed costs for setting up production are $54,000. The variable costs for producing each pint of honey are $4.75. The revenue from each pint of honey is $9.25. Find the following. [3.8]

  a) The total cost $C(x)$ of producing x pints of honey

  b) The total revenue $R(x)$ from the sale of x pints of honey

  c) The total profit $P(x)$ from the production and sale of x pints of honey

  d) The profit or loss from the production and sale of 5000 pints of honey; of 15,000 pints of honey

  e) The break-even point

Synthesis

38. How would you go about solving a problem that involves four variables? [3.5]

39. Explain how a system of equations can be both dependent and inconsistent. [3.4]

40. Danae is leaving a job that pays $36,000 a year to make honey (see Exercise 37). How many pints of honey must she produce and sell in order to make as much money as she earned at her previous job? [3.8]

41. Recently, Staples® charged $5.99 for a 2-count pack of Bic® Round Stic Grip mechanical pencils and $7.49 for a 12-count pack of Bic® Matic Grip mechanical pencils. Wiese Accounting purchased 138 of these two types of mechanical pencils for a total of $157.26. How many packs of each did they buy? [3.3]

42. Solve graphically:

$$y = x + 2,$$
$$y = x^2 + 2. \quad [3.1]$$

43. The graph of $f(x) = ax^2 + bx + c$ contains the points $(-2, 3)$, $(1, 1)$, and $(0, 3)$. Find a, b, and c and give a formula for the function. [3.5]

Test: Chapter 3

CHAPTER **Test Prep** VIDEO CD

Step-by-step test solutions are found on the video CD in the front of this book.

For Exercises 1–6, if a system has an infinite number of solutions, use set-builder notation to write the solution set. If a system has no solution, state this.

1. Solve graphically:

$$2x + y = 8,$$
$$y - x = 2.$$

Solve using the substitution method.

2. $x + 3y = -8,$
$4x - 3y = 23$

3. $2x - 4y = -6,$
$x = 2y - 3$

Solve using the elimination method.

4. $3x - y = 7,$
$x + y = 1$

5. $4y + 2x = 18,$
$3x + 6y = 26$

6. $4x - 6y = 3,$
$6x - 4y = -3$

7. The perimeter of a standard basketball court is 288 ft. The length is 44 ft longer than the width. Find the dimensions.

$P = 288$ ft

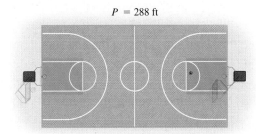

8. Pepperidge Farm® Goldfish is a snack food for which 40% of its calories come from fat. Rold Gold® Pretzels receive 9% of their calories from fat. How many grams of each would be needed to make 620 g of a snack mix for which 15% of the calories are from fat?

9. A truck leaves Gaston at noon traveling 55 mph. An hour later, a car leaves Gaston following the same route as the truck but traveling 65 mph. In how many hours will the car catch up to the truck?

Solve. If a system's equations are dependent or if there is no solution, state this.

10. $-3x + y - 2z = 8,$
$-x + 2y - z = 5,$
$2x + y + z = -3$

11. $6x + 2y - 4z = 15,$
$-3x - 4y + 2z = -6,$
$4x - 6y + 3z = 8$

12. $2x + 2y = 0,$
$4x + 4z = 4,$
$2x + y + z = 2$

13. $3x + 3z = 0,$
$2x + 2y = 2,$
$3y + 3z = 3$

Solve using matrices.

14. $4x + y = 12,$
$3x + 2y = 2$

15. $x + 3y - 3z = 12,$
$3x - y + 4z = 0,$
$-x + 2y - z = 1$

Evaluate.

16. $\begin{vmatrix} 4 & -2 \\ 3 & -5 \end{vmatrix}$

17. $\begin{vmatrix} 3 & 4 & 2 \\ -2 & -5 & 4 \\ 0 & 5 & -3 \end{vmatrix}$

18. Solve using Cramer's rule:
$3x + 4y = -1,$
$5x - 2y = 4.$

19. An electrician, a carpenter, and a plumber are hired to work on a house. The electrician earns $30 per hour, the carpenter $28.50 per hour, and the plumber $34 per hour. The first day on the job, they worked a total of 21.5 hr and earned a total of $673.00. If the plumber worked 2 more hours than the carpenter did, how many hours did each work?

20. Find the equilibrium point for the demand and supply functions
$$D(p) = 79 - 8p \quad \text{and} \quad S(p) = 37 + 6p,$$
where p is the price, in dollars, $D(p)$ is the number of units demanded, and $S(p)$ is the number of units supplied.

21. Kick Back, Inc., is producing a new hammock. For the first year, the fixed costs for setting up production are $44,000. The variable costs for producing each hammock are $25. The revenue from each hammock is $80. Find the following.
a) The total cost $C(x)$ of producing x hammocks
b) The total revenue $R(x)$ from the sale of x hammocks
c) The total profit $P(x)$ from the production and sale of x hammocks
d) The profit or loss from the production and sale of 300 hammocks; of 900 hammocks
e) The break-even point

Synthesis

22. The graph of the function $f(x) = mx + b$ contains the points $(-1, 3)$ and $(-2, -4)$. Find m and b.

23. Some of the world's best and most expensive coffee is Hawaii's Kona coffee. In order for coffee to be labeled "Kona Blend," it must contain at least 30% Kona beans. Bean Town Roasters has 40 lb of Mexican coffee. How much Kona coffee must they add if they wish to market it as Kona Blend?

Simplify. Do not leave negative exponents in your answers.

1. $x^4 \cdot x^{-6} \cdot x^{13}$ [1.6]

2. $(6x^2y^3)^2(-2x^0y^4)^{-3}$ [1.6]

3. $\dfrac{-10a^7b^{-11}}{25a^{-4}b^{22}}$ [1.6]

4. $\left(\dfrac{3x^4y^{-2}}{4x^{-5}}\right)^4$ [1.6]

5. $(1.95 \times 10^{-3})(5.73 \times 10^8)$ [1.7]

6. $\dfrac{2.42 \times 10^5}{6.05 \times 10^{-2}}$ [1.7]

7. Solve $A = \frac{1}{2}h(b + t)$ for b. [1.5]

8. Determine whether $(-3, 4)$ is a solution of $5a - 2b = -23$. [2.1]

Solve.

9. $x + 9.4 = -12.6$ [1.3]

10. $-2.4x = -48$ [1.3]

11. $\frac{3}{8}x + 7 = -14$ [1.3]

12. $-3 + 5x = 2x + 15$ [1.3]

13. $3n - (4n - 2) = 7$ [1.3]

14. $6y - 5(3y - 4) = 10$ [1.3]

15. $9c - [3 - 4(2 - c)] = 10$ [1.3]

16. $3x + y = 4,$
    $y = 6x - 5$ [3.2]

17. $4x + 4y = 4,$
    $5x + 4y = 2$ [3.2]

18. $\quad 6x - 10y = -22,$
    $-11x - 15y = 27$ [3.2]

19. $\quad x + \;\; y + \;\; z = -5,$
    $2x + 3y - 2z = 8,$
    $\quad x - \;\; y + 4z = -21$ [3.4]

20. $\quad 2x + 5y - 3z = -11,$
    $-5x + 3y - 2z = -7,$
    $\quad 3x - 2y + 5z = 12$ [3.4]

Graph.

21. $f(x) = -2x + 8$ [2.3]

22. $y = x^2 - 1$ [2.1]

23. $4x + 16 = 0$ [2.4]

24. $-3x + 2y = 6$ [2.3]

25. Find the slope and the y-intercept of the line with equation $-4y + 9x = 12$. [2.3]

26. Find the slope, if it exists, of the line containing the points $(2, 7)$ and $(-1, 3)$. [2.3]

27. Find an equation in slope–intercept form of the line with slope 4 and containing the point $(2, -11)$. [2.5]

28. Find an equation in slope–intercept form of the line containing the points $(-6, 3)$ and $(4, 2)$. [2.5]

29. Determine whether the lines given by the following equations are parallel, perpendicular, or neither:

    $2x = 4y + 7,$

    $x - 2y = 5.$ [2.5]

30. Find an equation of the line containing the point $(2, 1)$ and perpendicular to the line $x - 2y = 5$. [2.5]

31. For the graph of f shown, determine the domain, the range, $f(-3)$, and any value of x for which $f(x) = 5$. [2.2]

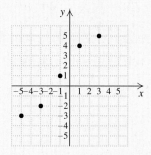

32. Determine the domain of the function given by

    $$f(x) = \frac{7}{x + 10}.\ \text{[2.2]}$$

Given $g(x) = 4x - 3$ and $h(x) = -2x^2 + 1$, find the following function values.

33. $h(4)$ [2.2]

34. $-g(0)$ [2.2]

35. $(g \cdot h)(-1)$ [2.6]

36. $(g - h)(a)$ [2.6]

Evaluate. [3.7]

37. $\begin{vmatrix} 2 & -3 \\ 4 & 1 \end{vmatrix}$

38. $\begin{vmatrix} 1 & 0 & 1 \\ -1 & 2 & 1 \\ 2 & 1 & 3 \end{vmatrix}$

Solve.

39. *ATM fees.* The average out-of-network ATM (automated teller machine) convenience fee charged by banks to customers from other banks is given by $f(t) = 0.15t + 1.03$, where t is the number of years after 2003.
Source: Based on data in *The Wall Street Journal*, 1/26/08

  a) Find the average out-of-network ATM fee in 2008. [2.2]

  b) What do the numbers 0.15 and 1.03 signify? [2.3]

40. *ATMs.* The number of ATMs in the United States rose from 139,000 in 1996 to 400,000 in 2008. Let $A(t)$ represent the number of ATMs, in thousands, in the United States t years after 1992.
Source: *Wall Street Journal*, 1/26/08

  a) Find a linear function that fits the data. [2.5]

  b) Use the function from part (a) to predict the number of ATMs in the United States in 2014. [2.5]

  c) In what year will there be 500,000 ATMs in the United States? [2.5]

41. *Professional memberships.* A one-year professional membership in Investigative Reporters and Editors, Inc. (IRE), costs $60. A one-year student membership costs $25. If, in May, 150 members joined IRE for a total of $6130 in membership dues, how many professionals and how many students joined that month? [3.3]
Source: Investigative Reporters and Editors, Inc.

42. *Saline solutions.* "Sea Spray" is 25% salt and the rest water. "Ocean Mist" is 65% salt and the rest water. How many ounces of each would be needed to obtain 120 oz of a mixture that is 50% salt? [3.3]

43. Find three consecutive odd numbers such that the difference of five times the third number and three times the first number is 54. [1.4]

44. *Perimeter of a hockey field.* The perimeter of the playing field for field hockey is 320 yd. The length is 40 yd longer than the width. What are the dimensions of the field? [3.3]
Source: USA Field Hockey

45. *Tickets.* At a county fair, an adult's ticket sold for $5.50, a senior citizen's ticket for $4.00, and a child's ticket for $1.50. On opening day, the number of adults' and senior citizens' tickets sold was 30 more than the number of children's tickets sold. The number of adults' tickets sold was 6 more than four times the number of senior citizen's tickets sold. Total receipts from the ticket sales were $11,219.50. How many of each type of ticket were sold? [3.5]

46. *Test scores.* Franco's scores on four tests are 93, 85, 100, and 86. What must the score be on the fifth test so that his average will be 90? [1.4]

Synthesis

47. Simplify: $(6x^{a+2}y^{b+2})(-2x^{a-2}y^{y+1})$. [1.6]

Aha! **48.** Chaney Chevrolet discovers that when $1000 is spent on radio advertising, weekly sales increase by $101,000. When $1250 is spent on radio advertising, weekly sales increase by $126,000. Assuming that sales increase according to a linear equation, by what amount would sales increase when $1500 is spent on radio advertising? [2.5]

49. Given that $f(x) = mx + b$ and that $f(5) = -3$ when $f(-4) = 2$, find m and b. [2.5], [3.3]

Inequalities and Problem Solving

ANNE SAMUELS
REGISTERED NURSE/
NURSE MANAGEMENT
Springfield, Massachusetts

As a nurse, I use math every
day to calculate dosages of
medication. If an order prescribes
a dosage of 2.5 mg/kg twice a
day, I need to know the patient's
weight, change that weight from
pounds to kilograms, and then
calculate the dosage. As a nurse
manager, I use math to determine
the number of nurses needed
to care for patients and to work
out a budget for salaries
and equipment.

AN APPLICATION

The number of registered nurses $R(t)$
employed in the United States, in
millions, t years after 2000, can be
approximated by

$$R(t) = 0.05t + 2.2.$$

Determine (using an inequality) those
years for which more than 3 million
registered nurses will be employed in
the United States.

Source: Based on data from the U.S. Department
of Health and Human Services and the Bureau of
Labor Statistics

This problem appears as Example 8 in
Section 4.1.

1 nequalities are mathematical sentences containing symbols such as < (is less than). We solve inequalities using principles similar to those used to solve equations. In this chapter, we solve a variety of inequalities, systems of inequalities, and real-world applications.

4.1 Inequalities and Applications

Solutions of Inequalities ▪ Interval Notation ▪ The Addition Principle for Inequalities ▪
The Multiplication Principle for Inequalities ▪ Using the Principles Together ▪ Problem Solving

Solutions of Inequalities

We now modify our equation-solving skills for the solving of *inequalities*. An **inequality** is any sentence containing $<, >, \leq, \geq,$ or $\neq$ (see Section 1.2)—for example,

$$-2 < a, \qquad x > 4, \qquad x + 3 \leq 6, \qquad 6 - 7y \geq 10y - 4, \quad \text{and} \quad 5x \neq 10.$$

Any value for the variable that makes an inequality true is called a **solution**. The set of all solutions is called the **solution set**. When all solutions of an inequality are found, we say that we have **solved** the inequality.

EXAMPLE **1**   Determine whether the given number is a solution of the inequality.

a) $x + 3 < 6$; 5   **b)** $-3 > -9 - 2x$; -1

SOLUTION

a) We substitute to get $5 + 3 < 6$, or $8 < 6$, a false sentence. Thus, 5 *is not* a solution.

b) We substitute to get $-3 > -9 - 2(-1)$, or $-3 > -7$, a true sentence. Thus, -1 *is* a solution.

▶ TRY EXERCISE 11

The *graph* of an inequality is a visual representation of the inequality's solution set. An inequality in one variable can be graphed on the number line. Inequalities in two variables are graphed on a coordinate plane, and appear later in this chapter.

The solution set of an inequality is often an infinite set. For example, the solution set of $x < 3$ is the set containing all numbers less than 3. To graph this set, we shade the number line to the left of 3. To indicate that 3 is not in the solution set, we use a parenthesis. If 3 were included in the solution set, we would use a bracket.

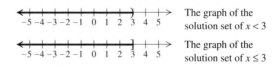

The graph of the
solution set of $x < 3$

The graph of the
solution set of $x \leq 3$

Interval Notation

To write the solution set of $x < 3$, we can use *set-builder notation* (see Section 1.1):

$$\{x | x < 3\}.$$

This is read "The set of all x such that x is less than 3."

Another way to write solutions of an inequality in one variable is to use **interval notation**. Interval notation uses parentheses, ( ), and brackets, [ ].

If a and b are real numbers with $a < b$, we define the **open interval (a, b)** as the set of all numbers x for which $a < x < b$. Using set-builder notation, we write

$$(a, b) = \{x | a < x < b\}. \qquad \text{Parentheses are used to exclude endpoints.}$$

Its graph excludes the endpoints:

The **closed interval $[a, b]$** is defined as the set of all numbers x for which $a \leq x \leq b$. Thus,

$$[a, b] = \{x | a \leq x \leq b\}. \qquad \text{Brackets are used to include endpoints.}$$

Its graph includes the endpoints:*

There are two kinds of **half-open intervals**, defined as follows:

1. $(a, b] = \{x | a < x \leq b\}$. This is open on the left. Its graph is as follows:

2. $[a, b) = \{x | a \leq x < b\}$. This is open on the right. Its graph is as follows:

We use the symbols ∞ and $-\infty$ to represent positive infinity and negative infinity, respectively. Thus the notation (a, ∞) represents the set of all real numbers greater than a, and $(-\infty, a)$ represents the set of all real numbers less than a.

The notation $[a, \infty)$ or $(-\infty, a]$ is used when we want to include the endpoint a.

STUDENT NOTES

You may have noticed which inequality signs in set-builder notation correspond to brackets and which correspond to parentheses. The relationship could be written informally as

$$\leq \quad \geq \quad [\,]$$
$$< \quad > \quad (\,).$$

---

*Some books use the graphs and instead of, respectively, and .

EXAMPLE 2   Graph $y \geq -2$ on a number line and write the solution set using both set-builder and interval notations.

SOLUTION   Using set-builder notation, we write the solution set as $\{y | y \geq -2\}$.
Using interval notation, we write $[-2, \infty)$.*
To graph the solution, we shade all numbers to the right of -2 and use a bracket to indicate that -2 is also a solution.

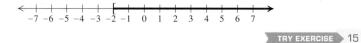

TRY EXERCISE   15

The Addition Principle for Inequalities

Two inequalities are *equivalent* if they have the same solution set. For example, the inequalities $x > 4$ and $4 < x$ are equivalent. Just as the addition principle for equations produces equivalent equations, the addition principle for inequalities produces equivalent inequalities.

> ### The Addition Principle for Inequalities
>
> For any real numbers a, b, and c:
>
> $$a < b \text{ is equivalent to } a + c < b + c;$$
> $$a > b \text{ is equivalent to } a + c > b + c.$$
>
> Similar statements hold for $\leq$ and $\geq$.

As with equations, we try to get the variable alone on one side in order to determine solutions easily.

EXAMPLE 3   Solve and graph: **(a)** $t + 5 > 1$; **(b)** $4x - 1 \geq 5x - 2$.

SOLUTION

a) $t + 5 > 1$

$\qquad t + 5 - 5 > 1 - 5$ Using the addition principle to add -5 to both sides

$\qquad\qquad t > -4$

When an inequality—like this last one—has an infinite number of solutions, we cannot possibly check them all. Instead, we can perform a partial check by substituting one member of the solution set (here we use -2) into the original inequality: $t + 5 = -2 + 5 = 3$ and $3 > 1$, so -2 is a solution.

Using set-builder notation, the solution is $\{t | t > -4\}$.

Using interval notation, the solution is $(-4, \infty)$.

----

*Any letter can be used in place of y. For example, $\{t | t \geq -2\}$ and $\{x | x \geq -2\}$ are both equal to $\{y | y \geq -2\}$. Each set describes the interval $[-2, \infty)$.

On most calculators, Example 3(b) can be checked by graphing $y_1 = 4x - 1 \geq 5x - 2$ (≥ is often found by pressing **2ND MATH**). The solution set is then displayed as an interval (shown by a horizontal line 1 unit above the x-axis).

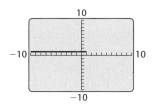

A check can also be made by graphing $y_1 = 4x - 1$ and $y_2 = 5x - 2$ and identifying those x-values for which $y_1 \geq y_2$.

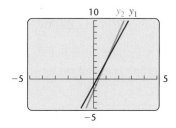

The INTERSECT option helps us find that $y_1 = y_2$ when $x = 1$. Note that $y_1 \geq y_2$ for x-values in the interval $(-\infty, 1]$.

The graph is as follows:

b)
$$4x - 1 \geq 5x - 2$$
$$4x - 1 + 2 \geq 5x - 2 + 2 \qquad \text{Adding 2 to both sides}$$
$$4x + 1 \geq 5x \qquad \text{Simplifying}$$
$$4x + 1 - 4x \geq 5x - 4x \qquad \text{Adding } -4x \text{ to both sides}$$
$$1 \geq x \qquad \text{Simplifying}$$

We know that $1 \geq x$ has the same meaning as $x \leq 1$. You can check that any number less than or equal to 1 is a solution.

Using set-builder notation, the solution is $\{x | 1 \geq x\}$, or $\{x | x \leq 1\}$.

Using interval notation, the solution is $(-\infty, 1]$.

The graph is as follows:

TRY EXERCISE 23

The Multiplication Principle for Inequalities

The multiplication principle for inequalities differs from the multiplication principle for equations.

Consider this true inequality: $4 < 9$. If we multiply both sides of $4 < 9$ by 2, we get another true inequality:

$$4 \cdot 2 < 9 \cdot 2, \quad \text{or} \quad 8 < 18.$$

If we multiply both sides of $4 < 9$ by -2, we get a false inequality:

FALSE ⟶ $4(-2) < 9(-2)$, or $-8 < -18$. ⟵ FALSE

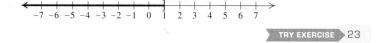

Multiplication (or division) by a negative number changes the sign of the number being multiplied (or divided). When the signs of both numbers in an inequality are changed, the position of the numbers with respect to each other is reversed.

$-8 > -18$. ⟵ TRUE

⤷ The $<$ symbol has been reversed!

> ### The Multiplication Principle for Inequalities
> For any real numbers a and b, and for any *positive* number c,
>
> $a < b$ is equivalent to $ac < bc$;
> $a > b$ is equivalent to $ac > bc$.
>
> For any real numbers a and b, and for any *negative* number c,
>
> $a < b$ is equivalent to $ac > bc$;
> $a > b$ is equivalent to $ac < bc$.
>
> Similar statements hold for $\leq$ and $\geq$.

Since division by *c* is the same as multiplication by $1/c$, there is no need for a separate division principle.

> *CAUTION!* Remember that whenever we multiply or divide both sides of an inequality by a negative number, we must reverse the inequality symbol.

EXAMPLE **4** Solve and graph: **(a)** $3y < \frac{3}{4}$; **(b)** $-5x \geq -80$.

SOLUTION

a) $3y < \frac{3}{4}$ ⟵ The symbol stays the same.

$\frac{1}{3} \cdot 3y < \frac{1}{3} \cdot \frac{3}{4}$ Multiplying both sides by $\frac{1}{3}$ or dividing both sides by 3

$y < \frac{1}{4}$

Any number less than $\frac{1}{4}$ is a solution. The solution set is $\left\{y \mid y < \frac{1}{4}\right\}$, or $\left(-\infty, \frac{1}{4}\right)$. The graph is as follows:

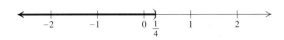

b) $-5x \geq -80$ ⟵ The symbol must be reversed.

$\frac{-5x}{-5} \leq \frac{-80}{-5}$ Dividing both sides by -5 or multiplying both sides by $-\frac{1}{5}$

$x \leq 16$

The solution set is $\{x \mid x \leq 16\}$, or $(-\infty, 16]$. The graph is as follows:

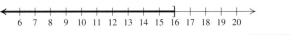

TRY EXERCISE 31

STUDENT NOTES

Try to remember to reverse the inequality symbol as soon as both sides are multiplied or divided by a negative number. Don't wait until after the multiplication or division has been carried out to reverse the symbol.

Using the Principles Together

We use the addition and multiplication principles together in solving inequalities in much the same way as in solving equations.

EXAMPLE **5** Solve: $16 - 7y \geq 10y - 4$.

SOLUTION   We have

$$16 - 7y \geq 10y - 4$$

$$-16 + 16 - 7y \geq -16 + 10y - 4 \qquad \text{Adding } -16 \text{ to both sides}$$

$$-7y \geq 10y - 20$$

$$-10y + (-7y) \geq -10y + 10y - 20 \qquad \text{Adding } -10y \text{ to both sides}$$

$$-17y \geq -20$$

⟵ The symbol must be reversed.

$$-\tfrac{1}{17} \cdot (-17y) \leq -\tfrac{1}{17} \cdot (-20) \qquad \begin{array}{l}\text{Multiplying both sides by } -\tfrac{1}{17} \text{ or} \\ \text{dividing both sides by } -17\end{array}$$

$$y \leq \tfrac{20}{17}.$$

The solution set is $\left\{y \mid y \leq \frac{20}{17}\right\}$, or $\left(-\infty, \frac{20}{17}\right]$.

TRY EXERCISE 53

EXAMPLE 6   Let $f(x) = -3(x + 8) - 5x$ and $g(x) = 4x - 9$. Find all x for which $f(x) > g(x)$.

SOLUTION   We have

$$f(x) > g(x)$$
$$-3(x + 8) - 5x > 4x - 9 \qquad \text{Substituting for } f(x) \text{ and } g(x)$$
$$-3x - 24 - 5x > 4x - 9 \qquad \text{Using the distributive law}$$
$$-24 - 8x > 4x - 9$$
$$-24 - 8x + 8x > 4x - 9 + 8x \qquad \text{Adding } 8x \text{ to both sides}$$
$$-24 > 12x - 9$$
$$-24 + 9 > 12x - 9 + 9 \qquad \text{Adding 9 to both sides}$$
$$-15 > 12x$$
$$\qquad\qquad\qquad \text{The symbol stays the same.}$$
$$-\tfrac{5}{4} > x. \qquad\qquad \text{Dividing by 12 and simplifying}$$

The solution set is $\left\{x \mid -\frac{5}{4} > x\right\}$, or $\left\{x \mid x < -\frac{5}{4}\right\}$, or $\left(-\infty, -\frac{5}{4}\right)$.

TRY EXERCISE ▶ 47

Although radical notation is not discussed in detail until Chapter 7, we know that only nonnegative numbers have square roots that are real numbers. Thus finding the domain of a radical function often involves solving an inequality.

EXAMPLE 7   Find the domain of f if $f(x) = \sqrt{7 - x}$.

SOLUTION   In order for $\sqrt{7 - x}$ to exist as a real number, $7 - x$ must be nonnegative. Thus we solve $7 - x \geq 0$:

$$7 - x \geq 0 \qquad 7 - x \text{ must be nonnegative.}$$
$$-x \geq -7 \qquad \text{Subtracting 7 from both sides}$$
$$\qquad\qquad \text{The symbol must be reversed.}$$
$$x \leq 7. \qquad \text{Multiplying both sides by } -1$$

When $x \leq 7$, the expression $7 - x$ is nonnegative. Thus the domain of f is $\{x \mid x \leq 7\}$, or $(-\infty, 7]$.

TRY EXERCISE ▶ 65

STUDENT NOTES

Note that "is less than" translates to an inequality. This is different from "less than," which translates to subtraction. For example, "Five is less than x" translates to $5 < x$, and "five less than x" translates to $x - 5$.

Problem Solving

Many problem-solving situations translate to inequalities. In addition to "is less than" and "is more than," other phrases are commonly used.

Important Words	Sample Sentence	Translation
is at least	Lia's grade point average is at least 3.4.	$g \geq 3.4$
are at most	There are at most 6 people in the car.	$n \leq 6$
cannot exceed	Total weight in the elevator cannot exceed 2000 pounds.	$w \leq 2000$
must exceed	The height must exceed 56 inches.	$h > 56$
is between	Heather's income is between $23,000 and $35,000.	$23{,}000 < h < 35{,}000$
maximum	The maximum speed is 65 mph.	$s \leq 65$
minimum	Jay must have a minimum of 95 credit hours.	$h \geq 95$
no more than	The child may weigh no more than 50 pounds.	$w \leq 50$
no less than	Ty would accept no less than $5000 for his used car.	$t \geq 5000$

EXAMPLE 8

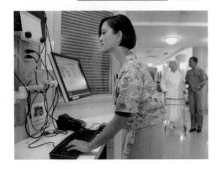

Registered nurses. The number of registered nurses $R(t)$ employed in the United States, in millions, t years after 2000 can be approximated by

$$R(t) = 0.05t + 2.2.$$

Determine (using an inequality) those years for which more than 3 million registered nurses will be employed in the United States.

Source: Based on data from the U.S. Department of Health and Human Services and the Bureau of Labor Statistics

SOLUTION

1. **Familiarize.** We already have a formula. The number 0.05 tells us that employment of registered nurses is growing at a rate of 0.05 million (or 50,000) per year. The number 2.2 tells us that in 2000, there were approximately 2.2 million registered nurses employed in the United States.

2. **Translate.** We are asked to find the years for which *more than* 3 million registered nurses will be employed in the United States. Thus we have

$$R(t) > 3$$
$$0.05t + 2.2 > 3. \qquad \text{Substituting } 0.05t + 2.2 \text{ for } R(t)$$

3. **Carry out.** We solve the inequality:

$$0.05t + 2.2 > 3$$
$$0.05t > 0.8 \qquad \text{Subtracting 2.2 from both sides}$$
$$t > 16. \qquad \text{Dividing both sides by 0.05}$$

Note that this corresponds to years after 2016.

4. **Check.** We can partially check our answer by finding $R(t)$ for a value of t greater than 16. For example,

$$R(20) = 0.05 \cdot 20 + 2.2 = 3.2, \text{ and } 3.2 > 3.$$

5. **State.** More than 3 million registered nurses will be employed in the United States for years after 2016.

> TRY EXERCISE 73

EXAMPLE   **9**

Job offers.   After graduation, Rose had two job offers in sales:

Uptown Fashions: A salary of $600 per month, plus a commission of 4% of sales;

Ergo Designs: A salary of $800 per month, plus a commission of 6% of sales in excess of $10,000.

If sales always exceed $10,000, for what amount of sales would Uptown Fashions provide higher pay?

SOLUTION

1. **Familiarize.** Listing the given information in a table will be helpful.

Uptown Fashions Monthly Income	Ergo Designs Monthly Income
$600 salary 4% of sales *Total*: $600 + 4% of sales	$800 salary 6% of sales over $10,000 *Total*: $800 + 6% of sales over $10,000

Next, suppose that Rose sold a certain amount—say, $12,000—in one month. Which offer would be better? Working for Uptown, she would earn $600 plus 4% of $12,000, or $600 + 0.04(12,000) = \$1080$. Since with Ergo Designs commissions are paid only on sales in excess of $10,000, Rose would earn $800 plus 6% of ($12,000 − $10,000), or $800 + 0.06(2000) = \$920$.

For monthly sales of $12,000, Uptown pays better. Similar calculations will show that for sales of $30,000 a month, Ergo pays better. To determine *all* values for which Uptown pays more money, we must solve an inequality that is based on the calculations above.

2. **Translate.** We let $S =$ the amount of monthly sales, in dollars, and will assume $S > 10,000$ so that both plans will pay a commission. Examining the calculations in the *Familiarize* step, we see that monthly income from Uptown is $600 + 0.04S$ and from Ergo is $800 + 0.06(S − 10,000)$. We want to find all values of S for which

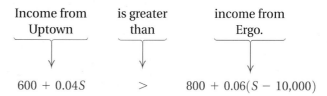

Income from Uptown	is greater than	income from Ergo.
$600 + 0.04S$	$>$	$800 + 0.06(S − 10,000)$

3. **Carry out.** We solve the inequality:

$$600 + 0.04S > 800 + 0.06(S − 10,000)$$
$$600 + 0.04S > 800 + 0.06S − 600 \qquad \text{Using the distributive law}$$
$$600 + 0.04S > 200 + 0.06S \qquad \text{Combining like terms}$$
$$400 > 0.02S \qquad \text{Subtracting 200 and } 0.04S \text{ from both sides}$$
$$20,000 > S, \text{ or } S < 20,000. \qquad \text{Dividing both sides by 0.02}$$

4. **Check.** The above steps indicate that income from Uptown Fashions is higher than income from Ergo Designs for sales less than $20,000. In the *Familiarize* step, we saw that for sales of $12,000, Uptown pays more. Since $12,000 < 20,000$, this is a partial check.

5. **State.** When monthly sales are less than $20,000, Uptown Fashions provides the higher pay.

TRY EXERCISE 81

4.1 EXERCISE SET

Concept Reinforcement *Classify each of the following as equivalent inequalities, equivalent equations, equivalent expressions, or not equivalent.*

1. $5x + 7 = 6 - 3x$, $8x + 7 = 6$

2. $2(4x + 1)$, $8x + 2$

3. $x - 7 > -2$, $x > 5$

4. $t + 3 < 1$, $t < 2$

5. $-4t \leq 12$, $t \leq -3$

6. $\frac{3}{5}a + \frac{1}{5} = 2$, $3a + 1 = 10$

7. $6a + 9$, $3(2a + 3)$

8. $-4x \geq -8$, $x \geq 2$

9. $-\frac{1}{2}x < 7$, $x > 14$

10. $-\frac{1}{3}t \leq -5$, $t \geq 15$

Determine whether the given numbers are solutions of the inequality.

11. $x - 4 \geq 1$
   - a) -4
   - b) 4
   - c) 5
   - d) 8

12. $3x + 1 \leq -5$
   - a) -5
   - b) -2
   - c) 0
   - d) 3

13. $2y + 3 < 6 - y$
   - a) 0
   - b) 1
   - c) -1
   - d) 4

14. $5t - 6 > 1 - 2t$
   - a) 6
   - b) 0
   - c) -3
   - d) 1

Graph each inequality, and write the solution set using both set-builder notation and interval notation.

15. $y < 6$ 16. $x > 4$

17. $x \geq -4$ 18. $t \leq 6$

19. $t > -3$ 20. $y < -3$

21. $x \leq -7$ 22. $x \geq -6$

Solve. Then graph. Write the solution set using both set-builder notation and interval notation.

23. $x + 2 > 1$ 24. $x + 9 > 6$

25. $t - 6 \leq 4$ 26. $t - 1 \geq 5$

27. $x - 12 \geq -11$ 28. $x - 11 \leq -2$

29. $9t < -81$ 30. $8x \geq 24$

31. $-0.3x > -15$ 32. $-0.5x < -30$

33. $-9x \geq 8.1$ 34. $-8y \leq 3.2$

35. $\frac{3}{4}y \geq -\frac{5}{8}$ 36. $\frac{5}{6}x \leq -\frac{3}{4}$

37. $3x + 1 < 7$ 38. $2x - 5 \geq 9$

39. $3 - x \geq 12$ 40. $8 - x < 15$

41. $\frac{2x + 7}{5} < -9$ 42. $\frac{5y + 13}{4} > -2$

43. $\frac{3t - 7}{-4} \leq 5$ 44. $\frac{2t - 9}{-3} \geq 7$

45. $\frac{9 - x}{-2} \geq -6$ 46. $\frac{3 - x}{-5} < -2$

47. Let $f(x) = 7 - 3x$ and $g(x) = 2x - 3$. Find all values of x for which $f(x) \leq g(x)$.

48. Let $f(x) = 8x - 9$ and $g(x) = 3x - 11$. Find all values of x for which $f(x) \leq g(x)$.

49. Let $f(x) = 2x - 7$ and $g(x) = 5x - 9$. Find all values of x for which $f(x) < g(x)$.

50. Let $f(x) = 0.4x + 5$ and $g(x) = 1.2x - 4$. Find all values of x for which $g(x) \geq f(x)$.

51. Let $y_1 = \frac{3}{8} + 2x$ and $y_2 = 3x - \frac{1}{8}$. Find all values of x for which $y_2 \geq y_1$.

52. Let $y_1 = 2x + 1$ and $y_2 = -\frac{1}{2}x + 6$. Find all values of x for which $y_1 < y_2$.

Solve. Write the solution set using both set-builder notation and interval notation.

53. $3 - 8y \geq 9 - 4y$

54. $4m + 7 \geq 9m - 3$

55. $5(t - 3) + 4t < 2(7 + 2t)$

56. $2(4 + 2x) > 2x + 3(2 - 5x)$

57. $5[3m - (m + 4)] > -2(m - 4)$

58. $8x - 3(3x + 2) - 5 \geq 3(x + 4) - 2x$

59. $19 - (2x + 3) \leq 2(x + 3) + x$

60. $13 - (2c + 2) \geq 2(c + 2) + 3c$

61. $\frac{1}{4}(8y + 4) - 17 < -\frac{1}{2}(4y - 8)$

62. $\frac{1}{3}(6x + 24) - 20 > -\frac{1}{4}(12x - 72)$

63. $2[8 - 4(3 - x)] - 2 \geq 8[2(4x - 3) + 7] - 50$

64. $5[3(7 - t) - 4(8 + 2t)] - 20 \leq -6[2(6 + 3t) - 4]$

Find the domain of each function.

65. $f(x) = \sqrt{x - 10}$

66. $f(x) = \sqrt{x + 2}$

67. $f(x) = \sqrt{3 - x}$

68. $f(x) = \sqrt{11 - x}$

69. $f(x) = \sqrt{2x + 7}$

70. $f(x) = \sqrt{8 - 5x}$

71. $f(x) = \sqrt{8 - 2x}$

72. $f(x) = \sqrt{2x - 10}$

Solve.

73. *Photography.*   Eli will photograph a wedding for a flat fee of $900 or for an hourly rate of $120. For what lengths of time would the hourly rate be less expensive?

74. *Truck rentals.*   Jenn can rent a moving truck for either $99 with unlimited mileage or $49 plus 80¢ per mile. For what mileages would the unlimited mileage plan save money?

75. *Exam scores.*   There are 80 questions on a college entrance examination. Two points are awarded for each correct answer, and one half point is deducted for each incorrect answer. How many questions does Tami need to answer correctly in order to score at least 100 on the test? Assume that Tami answers every question.

76. *Insurance claims.*   After a serious automobile accident, most insurance companies will replace the damaged car with a new one if repair costs exceed 80% of the NADA, or "blue-book," value of

the car. Lorenzo's car recently sustained $9200 worth of damage but was not replaced. What was the blue-book value of his car?

Phone rates.   The Waitsfield and Champlain Valley Telecom recently charged customers $13.40 for monthly service plus 1¢ per minute for local phone calls between 9 A.M. and 9 P.M. weekdays. The charge for off-peak local calls was 0.5¢ per minute. Calls were free after the total monthly charges reached $41.40.

77. Assume that only peak local calls were made. For how long must a customer speak on the phone if the $41.40 maximum charge is to apply?

78. Assume that only off-peak calls were made. For how long must a customer speak on the phone if the $41.40 maximum charge is to apply?

79. *ATM rates.*   The Intercity Bank offers two account plans. Their Local plan charges a $5 monthly service fee plus $3.00 per out-of-network ATM (automated teller machine) transaction. Their Anywhere plan charges a $15 monthly service fee plus $1.75 per out-of-network ATM transaction. For what number of out-of-network ATM transactions per month will the Anywhere plan cost less?

80. *Legal fees.*   Bridgewater Legal Offices charges a $250 retainer fee for real estate transactions plus $180 an hour. Dockside Legal charges a $100 retainer fee plus $230 an hour. For what number of hours does Bridgewater charge more?

81. *Wages.*   Toni can be paid in one of two ways:

> *Plan A*:   A salary of $400 per month, plus a commission of 8% of gross sales;
>
> *Plan B*:   A salary of $610 per month, plus a commission of 5% of gross sales.

For what amount of gross sales should Toni select plan A?

82. *Wages.*   Eric can be paid for his masonry work in one of two ways:

> *Plan A*: $300 plus $9.00 per hour;
>
> *Plan B*:   Straight $12.50 per hour.

Suppose that the job takes *n* hours. For what values of *n* is plan B better for Eric?

83. *Insurance benefits.*   Under the "Green Badge" medical insurance plan, Carlee would pay the first $2000 of her medical bills and 30% of all remaining bills. Under the "Blue Seal" plan, Carlee would pay the first $2500 of bills, but only 20% of the rest. For what amount of medical bills will the "Blue Seal" plan save Carlee money? (Assume that her bills will exceed $2500.)

84. *Checking accounts.* North Bank charges $10 per month for a student checking account. The first 8 checks are free, and each additional check costs $0.75. South Bank offers a student checking account with no monthly charge. The first 8 checks are free, and each additional check costs $3. For what numbers of checks is the South Bank plan more expensive? (Assume that the student will always write more than 8 checks.)

85. *Crude-oil production.* The yearly U.S. production of crude oil $C(t)$, in millions of barrels, t years after 2000 can be approximated by

$$C(t) = -40.5t + 2159.$$

Determine (using an inequality) those years for which domestic production will drop below 1750 million barrels.

Source: U.S. Energy Information Administration

86. *HDTVs.* The percentage of U.S. households $p(t)$ with an HDTV t years after 2005 can be approximated by

$$p(t) = 8t + 12.5.$$

Determine (using an inequality) those years for which more than half of all U.S. households will have an HDTV.

Source: Based on data from Consumer Electronics Association

87. *Body fat percentage.* The function given by

$$F(d) = (4.95/d - 4.50) \times 100$$

can be used to estimate the body fat percentage $F(d)$ of a person with an average body density d, in kilograms per liter.

a) A man is considered obese if his body fat percentage is at least 25%. Find the body densities of an obese man.

b) A woman is considered obese if her body fat percentage is at least 32%. Find the body densities of an obese woman.

88. *Temperature conversion.* The function

$$C(F) = \tfrac{5}{9}(F - 32)$$

can be used to find the Celsius temperature $C(F)$ that corresponds to $F°$ Fahrenheit.

a) Gold is solid at Celsius temperatures less than 1063°C. Find the Fahrenheit temperatures for which gold is solid.

b) Silver is solid at Celsius temperatures less than 960.8°C. Find the Fahrenheit temperatures for which silver is solid.

89. *Manufacturing.* Bright Ideas is planning to make a new kind of lamp. Fixed costs will be $90,000, and variable costs will be $25 for the production of each lamp. The total-cost function for x lamps is

$$C(x) = 90,000 + 25x.$$

The company makes $48 in revenue for each lamp sold. The total-revenue function for x lamps is

$$R(x) = 48x.$$

(See Section 3.8.)

a) When $R(x) < C(x)$, the company loses money. Find the values of x for which the company loses money.

b) When $R(x) > C(x)$, the company makes a profit. Find the values of x for which the company makes a profit.

90. *Publishing.* The demand and supply functions for a locally produced poetry book are approximated by

$$D(p) = 2000 - 60p \quad \text{and}$$
$$S(p) = 460 + 94p,$$

where p is the price, in dollars (see Section 3.8).

a) Find those values of p for which demand exceeds supply.

b) Find those values of p for which demand is less than supply.

91. How is the solution of $x + 3 = 8$ related to the solution sets of

$$x + 3 > 8 \quad \text{and} \quad x + 3 < 8?$$

92. Why isn't roster notation used to write solutions of inequalities?

Skill Review

To prepare for Section 4.2, review finding domains of functions (Section 2.2).

Find the domain of f. [2.2]

93. $f(x) = \dfrac{5}{x}$

94. $f(x) = \dfrac{3}{x - 6}$

95. $f(x) = \dfrac{x - 2}{2x + 1}$

96. $f(x) = \dfrac{x + 3}{5x - 7}$

97. $f(x) = \dfrac{x + 10}{8}$

98. $f(x) = \dfrac{3}{x} + 5$

Synthesis

99. The percentage of the U.S. population that owns an HDTV cannot exceed 100%. How does this affect the answer to Exercise 86?

100. Explain how the addition principle can be used to avoid ever needing to multiply or divide both sides of an inequality by a negative number.

Solve for x and y. Assume that a, b, c, d, and m are positive constants.

101. $3ax + 2x \geq 5ax - 4$; assume $a > 1$

102. $6by - 4y \leq 7by + 10$

103. $a(by - 2) \geq b(2y + 5)$; assume $a > 2$

104. $c(6x - 4) < d(3 + 2x)$; assume $3c > d$

105. $c(2 - 5x) + dx > m(4 + 2x)$; assume $5c + 2m < d$

106. $a(3 - 4x) + cx < d(5x + 2)$; assume $c > 4a + 5d$

Determine whether each statement is true or false. If false, give an example that shows this.

107. For any real numbers a, b, c, and d, if $a < b$ and $c < d$, then $a - c < b - d$.

108. For all real numbers x and y, if $x < y$, then $x^2 < y^2$.

109. Are the inequalities

$$x < 3 \quad \text{and} \quad x + \frac{1}{x} < 3 + \frac{1}{x}$$

equivalent? Why or why not?

110. Are the inequalities

$$x < 3 \quad \text{and} \quad 0 \cdot x < 0 \cdot 3$$

equivalent? Why or why not?

Solve. Then graph.

111. $x + 5 \leq 5 + x$

112. $x + 8 < 3 + x$

113. $x^2 > 0$

114. Assume that the graphs of $y_1 = -\frac{1}{2}x + 5$, $y_2 = x - 1$, and $y_3 = 2x - 3$ are as shown below. Solve each inequality, referring only to the figure.

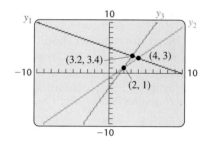

a) $-\frac{1}{2}x + 5 > x - 1$
b) $x - 1 \leq 2x - 3$
c) $2x - 3 \geq -\frac{1}{2}x + 5$

115. Using an approach similar to that in the Technology Connection on p. 223, use a graphing calculator to check your answers to Exercises 23, 47, and 63.

116. Use a graphing calculator to confirm the domains of the functions in Exercises 65, 67, and 71.

CORNER

Saving on Shipping Costs

Focus: Inequalities and problem solving

Time: 20–30 minutes

Group size: 2–3

For overnight delivery packages weighing up to 10 lb sent by Express Mail, the United States Postal Service charges (as of May 2008) $19.00 for up to one pound delivered to Zone 3 and, on average, $2.12 for each pound or part of a pound after the first. UPS Next Day Air charges $22.05 for a one-pound delivery to Zone 3 and each additional pound or part of a pound costs $1.45.*

ACTIVITY

1. One group member should determine the function p, where $p(x)$ represents the

   cost, in dollars, of mailing x pounds using Express Mail.

2. One member should determine the function r, where $r(x)$ represents the cost, in dollars, of shipping x pounds using UPS Next Day Air.

3. A third member should graph p and r on the same set of axes.

4. Finally, working together, use the graph to determine those weights for which Express Mail is less expensive than UPS Next Day Air shipping. Express your answer in both set-builder notation and interval notation.

*This activity is based on an article by Michael Contino in *Mathematics Teacher*, May 1995.

4.2   Intersections, Unions, and Compound Inequalities

Intersections of Sets and Conjunctions of Sentences • Unions of Sets and Disjunctions of Sentences • Interval Notation and Domains

Two inequalities joined by the word "and" or the word "or" are called **compound inequalities**. Thus, "$x < -3 \ or \ x > 0$" and "$x < 5 \ and \ x > 3$" are two examples of compound inequalities. To discuss how to solve compound inequalities, we must first study ways in which sets can be combined.

Intersections of Sets and Conjunctions of Sentences

The **intersection** of two sets A and B is the set of all elements that are common to both A and B. We denote the intersection of sets A and B as

$$A \cap B.$$

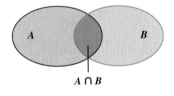

$A \cap B$

The intersection of two sets is represented by the purple region shown in the figure at left. For example, if $A = \{$all students who are taking a math class$\}$ and $B = \{$all students who are taking a history class$\}$, then $A \cap B = \{$all students who are taking a math class *and* a history class$\}$.

EXAMPLE 1   Find the intersection: $\{1, 2, 3, 4, 5\} \cap \{-2, -1, 0, 1, 2, 3\}$.

SOLUTION   The numbers 1, 2, and 3 are common to both sets, so the intersection is $\{1, 2, 3\}$.

TRY EXERCISE 11

When two or more sentences are joined by the word *and* to make a compound sentence, the new sentence is called a **conjunction** of the sentences. The following is a conjunction of inequalities:

$$-2 < x \quad and \quad x < 1.$$

A number is a solution of a conjunction if it is a solution of *both* of the separate parts. For example, -1 is a solution because it is a solution of $-2 < x$ as well as $x < 1$; that is, -1 is *both* greater than -2 *and* less than 1.

> *The solution set of a conjunction is the intersection of the solution sets of the individual sentences.*

EXAMPLE 2   Graph and write interval notation for the conjunction

$$-2 < x \quad and \quad x < 1.$$

SOLUTION   We first graph $-2 < x$, then $x < 1$, and finally the conjunction $-2 < x$ *and* $x < 1$.

$\{x \mid -2 < x\}$

$(-2, \infty)$

$\{x \mid x < 1\}$

$(-\infty, 1)$

$\{x \mid -2 < x\} \cap \{x \mid x < 1\}$
$= \{x \mid -2 < x \text{ and } x < 1\}$

$(-2, 1)$

Because there are numbers that are both greater than -2 and less than 1, the solution set of the conjunction $-2 < x$ *and* $x < 1$ is the interval $(-2, 1)$. In set-builder notation, this is written $\{x \mid -2 < x < 1\}$, the set of all numbers that are *simultaneously* greater than -2 *and* less than 1.

TRY EXERCISE 33

For $a < b$,

$$a < x \quad and \quad x < b \quad \textbf{can be abbreviated} \quad a < x < b;$$

and, equivalently,

$$b > x \quad and \quad x > a \quad \textbf{can be abbreviated} \quad b > x > a.$$

Mathematical Use of the Word "and"

The word "and" corresponds to "intersection" and to the symbol "$\cap$". Any solution of a conjunction must make each part of the conjunction true.

EXAMPLE 3   Solve and graph: $-1 \leq 2x + 5 < 13$.

SOLUTION   This inequality is an abbreviation for the conjunction

$$-1 \leq 2x + 5 \quad and \quad 2x + 5 < 13.$$

The word *and* corresponds to set *intersection*. To solve the conjunction, we solve each of the two inequalities separately and then find the intersection of the solution sets:

$$-1 \leq 2x + 5 \quad and \quad 2x + 5 < 13$$
$$-6 \leq 2x \qquad and \qquad 2x < 8 \qquad \text{Subtracting 5 from both sides of each inequality}$$
$$-3 \leq x \qquad and \qquad x < 4. \qquad \text{Dividing both sides of each inequality by 2}$$

The solution of the conjunction is the intersection of the two separate solution sets.

$\{x \mid -3 \leq x\}$ $[-3, \infty)$

$\{x \mid x < 4\}$ $(-\infty, 4)$

$\{x \mid -3 \leq x\} \cap \{x \mid x < 4\}$
$= \{x \mid -3 \leq x < 4\}$ $[-3, 4)$

We can abbreviate the answer as $-3 \leq x < 4$. The solution set is $\{x \mid -3 \leq x < 4\}$, or, in interval notation, $[-3, 4)$. ▶ TRY EXERCISE 45

The steps in Example 3 are often combined as follows:

$$-1 \leq 2x + 5 < 13$$
$$-1 - 5 \leq 2x + 5 - 5 < 13 - 5 \qquad \text{Subtracting 5 from all three regions}$$
$$-6 \leq 2x < 8$$
$$-3 \leq x < 4. \qquad \text{Dividing by 2 in all three regions}$$

Such an approach saves some writing and will prove useful in Section 4.3.

> **CAUTION!** The abbreviated form of a conjunction, like $-3 \leq x < 4$, can be written only if both inequality symbols point in the same direction. It is *not acceptable* to write a sentence like $-1 > x < 5$ since doing so does not indicate if *both* $-1 > x$ and $x < 5$ must be true or if it is enough for one of the separate inequalities to be true.

STUDY SKILLS

Two Books Are Better Than One

Many students find it helpful to use a second book as a reference when studying. Perhaps you or a friend own a text from a previous math course that can serve as a resource. Often professors have older texts that they will happily give away. Library book sales and thrift shops can also be excellent sources for extra books. Saving your text when you finish a math course can provide you with an excellent aid for your next course.

EXAMPLE 4 Solve and graph: $2x - 5 \geq -3 \ and \ 5x + 2 \geq 17$.

SOLUTION   We first solve each inequality, retaining the word *and*:

$$2x - 5 \geq -3 \quad and \quad 5x + 2 \geq 17$$
$$2x \geq 2 \qquad and \qquad 5x \geq 15$$
$$x \geq 1 \qquad and \qquad x \geq 3.$$
Keep the word "and."

Next, we find the intersection of the two separate solution sets.

$\{x \mid x \geq 1\}$ $[1, \infty)$

$\{x \mid x \geq 3\}$ $[3, \infty)$

$\{x \mid x \geq 1\} \cap \{x \mid x \geq 3\}$
$= \{x \mid x \geq 3\}$ $[3, \infty)$

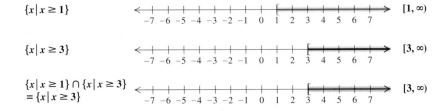

The numbers common to both sets are those greater than or equal to 3. Thus the solution set is $\{x \mid x \geq 3\}$, or, in interval notation, $[3, \infty)$. You should check that any number in $[3, \infty)$ satisfies the conjunction whereas numbers outside $[3, \infty)$ do not.

TRY EXERCISE 67

Sometimes there is no way to solve both parts of a conjunction at once.

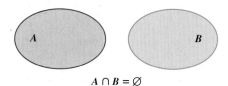

When $A \cap B = \varnothing$,
A and B are said
to be *disjoint*.

$A \cap B = \varnothing$

EXAMPLE 5

Solve and graph: $2x - 3 > 1$ *and* $3x - 1 < 2$.

SOLUTION   We solve each inequality separately:

$$2x - 3 > 1 \quad and \quad 3x - 1 < 2$$
$$2x > 4 \quad and \quad 3x < 3$$
$$x > 2 \quad and \quad x < 1.$$

The solution set is the intersection of the individual inequalities.

$\{x \mid x > 2\}$ — $(2, \infty)$

$\{x \mid x < 1\}$ — $(-\infty, 1)$

$\{x \mid x > 2\} \cap \{x \mid x < 1\}$
$= \{x \mid x > 2 \ and \ x < 1\} = \varnothing$ — $\varnothing$

Since no number is both greater than 2 and less than 1, the solution set is the empty set, $\varnothing$.

TRY EXERCISE 69

Unions of Sets and Disjunctions of Sentences

The **union** of two sets A and B is the collection of elements belonging to A and/or B. We denote the union of A and B by

$A \cup B$.

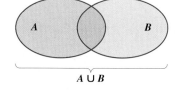

$A \cup B$

The union of two sets is often pictured as shown at left. For example, if $A = \{$all students who are taking a math class$\}$ and $B = \{$all students who are taking a history class$\}$, then $A \cup B = \{$all students who are taking a math class *or* a history class$\}$. Note that this set includes students who are taking a math class *and* a history class.

EXAMPLE 6

Find the union: $\{2, 3, 4\} \cup \{3, 5, 7\}$.

SOLUTION   The numbers in either or both sets are 2, 3, 4, 5, and 7, so the union is $\{2, 3, 4, 5, 7\}$.

TRY EXERCISE 13

When two or more sentences are joined by the word *or* to make a compound sentence, the new sentence is called a **disjunction** of the sentences. Here is an example:

$$x < -3 \quad or \quad x > 3.$$

A number is a solution of a disjunction if it is a solution of at least one of the separate parts. For example, -5 is a solution of this disjunction since -5 is a solution of $x < -3$.

> *The solution set of a disjunction is the union of the solution sets of the individual sentences.*

EXAMPLE **7** Graph and write interval notation for the disjunction

$$x < -3 \quad or \quad x > 3.$$

SOLUTION We first graph $x < -3$, then $x > 3$, and finally the disjunction $x < -3$ or $x > 3$.

$\{x \mid x < -3\}$ $(-\infty, -3)$

$\{x \mid x > 3\}$ $(3, \infty)$

$\{x \mid x < -3\} \cup \{x \mid x > 3\}$
$= \{x \mid x < -3 \text{ or } x > 3\}$ $(-\infty, -3) \cup (3, \infty)$

The solution set of $x < -3$ or $x > 3$ is $\{x \mid x < -3 \text{ or } x > 3\}$, or, in interval notation, $(-\infty, -3) \cup (3, \infty)$. There is no simpler way to write the solution.

TRY EXERCISE 27

Mathematical Use of the Word "or"

The word "or" corresponds to "union" and to the symbol "$\cup$". For a number to be a solution of a disjunction, it must be in *at least one* of the solution sets of the individual sentences.

EXAMPLE **8** Solve and graph: $7 + 2x < -1 \text{ or } 13 - 5x \le 3$.

SOLUTION We solve each inequality separately, retaining the word *or*:

$$7 + 2x < -1 \quad or \quad 13 - 5x \le 3$$
$$2x < -8 \quad or \quad -5x \le -10$$

Dividing by a negative and reversing the symbol

$$x < -4 \quad or \quad x \ge 2.$$

To find the solution set of the disjunction, we consider the individual graphs. We graph $x < -4$ and then $x \ge 2$. Then we take the union of the graphs.

$\{x \mid x < -4\}$ $(-\infty, -4)$

$\{x \mid x \ge 2\}$ $[2, \infty)$

$\{x \mid x < -4\} \cup \{x \mid x \ge 2\}$
$= \{x \mid x < -4 \text{ or } x \ge 2\}$ $(-\infty, -4) \cup [2, \infty)$

The solution set is $\{x \mid x < -4 \text{ or } x \ge 2\}$, or $(-\infty, -4) \cup [2, \infty)$.

TRY EXERCISE 63

CAUTION!   A compound inequality like

$$x < -4 \quad or \quad x \geq 2,$$

as in Example 8, *cannot* be expressed as $2 \leq x < -4$ because to do so would be to say that x is *simultaneously* less than -4 and greater than or equal to 2. No number is both less than -4 *and* greater than 2, but many are less than -4 *or* greater than 2.

EXAMPLE   9

Solve: $-2x - 5 < -2 \ or \ x - 3 < -10$.

SOLUTION   We solve the individual inequalities, retaining the word *or*:

$$-2x - 5 < -2 \quad or \quad x - 3 < -10$$
$$-2x < 3 \quad or \quad x < -7$$

Dividing by a negative and reversing the symbol ———   ——— Keep the word "or."

$$x > -\tfrac{3}{2} \quad or \quad x < -7.$$

The solution set is $\left\{x \mid x < -7 \ or \ x > -\tfrac{3}{2}\right\}$, or $(-\infty, -7) \cup \left(-\tfrac{3}{2}, \infty\right)$.

TRY EXERCISE   65

EXAMPLE   10

Solve: $3x - 11 < 4 \ or \ 4x + 9 \geq 1$.

SOLUTION   We solve the individual inequalities separately, retaining the word *or*:

$$3x - 11 < 4 \quad or \quad 4x + 9 \geq 1$$
$$3x < 15 \quad or \quad 4x \geq -8$$
$$x < 5 \quad or \quad x \geq -2.$$

———   Keep the word "or."

To find the solution set, we first look at the individual graphs.

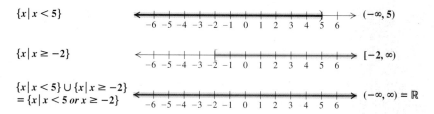

Since *all* numbers are less than 5 or greater than or equal to -2, the two sets fill the entire number line. Thus the solution set is $\mathbb{R}$, the set of all real numbers.

TRY EXERCISE   51

Interval Notation and Domains

In Section 2.2, we saw that if $g(x) = \dfrac{5x - 2}{x - 3}$, then the number 3 is not in the domain of g. We can represent the domain of g using set-builder notation or interval notation.

EXAMPLE **11** Use interval notation to write the domain of g if $g(x) = \dfrac{5x - 2}{x - 3}$.

SOLUTION The expression $\dfrac{5x - 2}{x - 3}$ is not defined when the denominator is 0. We set $x - 3$ equal to 0 and solve:

$$x - 3 = 0$$
$$x = 3. \qquad \text{The number 3 is \textit{not} in the domain.}$$

We have the domain of $g = \{x \mid x \text{ is a real number } and\ x \neq 3\}$. If we graph this set, we see that the domain can be written as a union of two intervals.

Thus the domain of $g = (-\infty, 3) \cup (3, \infty)$.

TRY EXERCISE 73

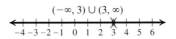

4.2 **EXERCISE SET**

Concept Reinforcement In each of Exercises 1–10, match the set with the most appropriate choice from the column on the right.

1. ____ $\{x \mid x < -2 \text{ or } x > 2\}$

2. ____ $\{x \mid x < -2 \text{ and } x > 2\}$

3. ____ $\{x \mid x > -2\} \cap \{x \mid x < 2\}$

4. ____ $\{x \mid x \leq -2\} \cup \{x \mid x \geq 2\}$

5. ____ $\{x \mid x \leq -2\} \cup \{x \mid x \leq 2\}$

6. ____ $\{x \mid x \leq -2\} \cap \{x \mid x \leq 2\}$

7. ____ $\{x \mid x \geq -2\} \cap \{x \mid x \geq 2\}$

8. ____ $\{x \mid x \geq -2\} \cup \{x \mid x \geq 2\}$

9. ____ $\{x \mid x \leq 2\} \text{ and } \{x \mid x \geq -2\}$

10. ____ $\{x \mid x \leq 2\} \text{ or } \{x \mid x \geq -2\}$

a)
b)
c)
d)
e)
f)
g)
h)
i) $\mathbb{R}$
j) $\varnothing$

Find each indicated intersection or union.

11. $\{2, 4, 16\} \cap \{4, 16, 256\}$

12. $\{1, 2, 4\} \cup \{4, 6, 8\}$

13. $\{0, 5, 10, 15\} \cup \{5, 15, 20\}$

14. $\{2, 5, 9, 13\} \cap \{5, 8, 10\}$

15. $\{a, b, c, d, e, f\} \cap \{b, d, f\}$

16. $\{u, v, w\} \cup \{u, w\}$

17. $\{x, y, z\} \cup \{u, v, x, y, z\}$

18. $\{m, n, o, p\} \cap \{m, o, p\}$

19. $\{3, 6, 9, 12\} \cap \{5, 10, 15\}$

20. $\{1, 5, 9\} \cup \{4, 6, 8\}$

21. $\{1, 3, 5\} \cup \varnothing$

22. $\{1, 3, 5\} \cap \varnothing$

Graph and write interval notation for each compound inequality.

23. $1 < x < 3$

24. $0 \le y \le 5$

25. $-6 \le y \le 0$

26. $-8 < x \le -2$

27. $x \le -1 \ or \ x > 4$

28. $x < -5 \ or \ x > 1$

29. $x \le -2 \ or \ x > 1$

30. $x \le -5 \ or \ x > 2$

31. $-4 \le -x < 2$

32. $x > -7 \ and \ x < -2$

33. $x > -2 \ and \ x < 4$

34. $3 > -x \ge -1$

35. $5 > a \ or \ a > 7$

36. $t \ge 2 \ or \ -3 > t$

37. $x \ge 5 \ or \ -x \ge 4$

38. $-x < 3 \ or \ x < -6$

39. $7 > y \ and \ y \ge -3$

40. $6 > -x \ge 0$

41. $-x < 7 \ and \ -x \ge 0$

42. $x \ge -3 \ and \ x < 3$

Aha! **43.** $t < 2 \ or \ t < 5$

44. $t > 4 \ or \ t > -1$

Solve and graph each solution set.

45. $-3 \le x + 2 < 9$

46. $-1 < x - 3 < 5$

47. $0 < t - 4 \ and \ t - 1 \le 7$

48. $-6 \le t + 1 \ and \ t + 8 < 2$

49. $-7 \le 2a - 3 \ and \ 3a + 1 < 7$

50. $-4 \le 3n + 5 \ and \ 2n - 3 \le 7$

Aha! **51.** $x + 3 \le -1 \ or \ x + 3 > -2$

52. $x + 5 < -3 \ or \ x + 5 \ge 4$

53. $-10 \le 3x - 1 \le 5$

54. $-18 \le 4x + 2 \le 30$

55. $5 > \dfrac{x - 3}{4} > 1$

56. $3 \ge \dfrac{x - 1}{2} \ge -4$

57. $-2 \le \dfrac{x + 2}{-5} \le 6$

58. $-10 \le \dfrac{x + 6}{-3} \le -8$

59. $2 \le f(x) \le 8$, where $f(x) = 3x - 1$

60. $7 \ge g(x) \ge -2$, where $g(x) = 3x - 5$

61. $-21 \le f(x) < 0$, where $f(x) = -2x - 7$

62. $4 > g(t) \ge 2$, where $g(t) = -3t - 8$

63. $f(t) < 3 \ or \ f(t) > 8$, where $f(t) = 5t + 3$

64. $g(x) \le -2 \ or \ g(x) \ge 10$, where $g(x) = 3x - 5$

65. $6 > 2a - 1 \ or \ -4 \le -3a + 2$

66. $3a - 7 > -10 \ or \ 5a + 2 \le 22$

67. $a + 3 < -2 \ and \ 3a - 4 < 8$

68. $1 - a < -2 \ and \ 2a + 1 > 9$

69. $3x + 2 < 2 \ and \ 3 - x < 1$

70. $2x - 1 > 5 \ and \ 2 - 3x > 11$

71. $2t - 7 \le 5 \ or \ 5 - 2t > 3$

72. $5 - 3a \le 8 \ or \ 2a + 1 > 7$

For $f(x)$ as given, use interval notation to write the domain of f.

73. $f(x) = \dfrac{9}{x + 6}$

74. $f(x) = \dfrac{2}{x - 5}$

75. $f(x) = \dfrac{1}{x}$

76. $f(x) = -\dfrac{6}{x}$

77. $f(x) = \dfrac{x + 3}{2x - 8}$

78. $f(x) = \dfrac{x - 1}{3x + 6}$

79. Why can the conjunction $2 < x \ and \ x < 5$ be rewritten as $2 < x < 5$, but the disjunction $2 < x \ or \ x < 5$ cannot be rewritten as $2 < x < 5$?

80. Can the solution set of a disjunction be empty? Why or why not?

Skill Review

To prepare for Section 4.3, review graphing and solving equations by graphing (Sections 2.1 and 2.4).

Graph. [2.1]

81. $g(x) = 2x$

82. $f(x) = 4$

83. $g(x) = -3$

84. $f(x) = |x|$

Solve by graphing. [2.4]

85. $x + 4 = 3$

86. $x - 1 = -5$

Synthesis

87. What can you conclude about a, b, c, and d, if $[a, b] \cup [c, d] = [a, d]$? Why?

88. What can you conclude about a, b, c, and d, if $[a, b] \cap [c, d] = [a, b]$? Why?

89. Use the accompanying graph of $f(x) = 2x - 5$ to solve $-7 < 2x - 5 < 7$.

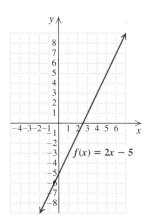

90. Use the accompanying graph of $g(x) = 4 - x$ to solve $4 - x < -2$ *or* $4 - x > 7$.

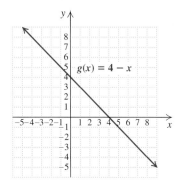

91. *Counseling.* The function given by
$$s(t) = 500t + 16,500$$
can be used to estimate the number of student visits to the Cornell University's counseling center *t* years after 2000. For what years is the number of student visits between 18,000 and 21,000?
Source: Based on data from Cornell University

92. *Pressure at sea depth.* The function given by
$$P(d) = 1 + \frac{d}{33}$$
gives the pressure, in atmospheres (atm), at a depth of *d* feet in the sea. For what depths *d* is the pressure at least 1 atm and at most 7 atm?

93. *Converting dress sizes.* The function given by
$$f(x) = 2(x + 10)$$
can be used to convert dress sizes *x* in the United States to dress sizes *f(x)* in Italy. For what dress sizes in the United States will dress sizes in Italy be between 32 and 46?

94. *Solid-waste generation.* The function given by
$$w(t) = 0.0125t + 4.525$$
can be used to estimate the number of pounds of solid waste, *w(t)*, produced daily, on average, by each person in the United States, *t* years after 2000. For what years will waste production range from 4.6 to 4.8 lb per person per day?

95. *Body fat percentage.* The function given by
$$F(d) = (4.95/d - 4.50) \times 100$$
can be used to estimate the body fat percentage *F(d)* of a person with an average body density *d*, in kilograms per liter. A woman's body fat percentage is considered acceptable if $25 \le F(d) \le 31$. What body densities are considered acceptable for a woman?

96. *Temperatures of liquids.* The formula
$$C = \frac{5}{9}(F - 32)$$
can be used to convert Fahrenheit temperatures *F* to Celsius temperatures *C*.
   a) Gold is liquid for Celsius temperatures *C* such that $1063° \le C < 2660°$. Find a comparable inequality for Fahrenheit temperatures.
   b) Silver is liquid for Celsius temperatures *C* such that $960.8° \le C < 2180°$. Find a comparable inequality for Fahrenheit temperatures.

97. *Minimizing tolls.* A $6.00 toll is charged to cross the bridge to Sanibel Island from mainland Florida. A six-month reduced-fare pass costs $50 and reduces the toll to $2.00. A six-month unlimited-trip pass costs $300 and allows for free crossings. How many crossings in six months does it take for the reduced-fare pass to be the more economical choice?
Source: www.leewayinfo.com

Solve and graph.

98. $4a - 2 \le a + 1 \le 3a + 4$

99. $4m - 8 > 6m + 5$ *or* $5m - 8 < -2$

100. $x - 10 < 5x + 6 \le x + 10$

101. $3x < 4 - 5x < 5 + 3x$

Determine whether each sentence is true or false for all real numbers a, b, and c.

102. If $-b < -a$, then $a < b$.

103. If $a \le c$ and $c \le b$, then $b > a$.

104. If $a < c$ and $b < c$, then $a < b$.

105. If $-a < c$ and $-c > b$, then $a > b$.

For f(x) as given, use interval notation to write the domain of f.

106. $f(x) = \dfrac{\sqrt{5 + 2x}}{x - 1}$

107. $f(x) = \dfrac{\sqrt{3 - 4x}}{x + 7}$

108. For $f(x) = \sqrt{x - 5}$ and $g(x) = \sqrt{9 - x}$, use interval notation to write the domain of $f + g$.

109. Let $y_1 = -1$, $y_2 = 2x + 5$, and $y_3 = 13$. Then use the graphs of y_1, y_2, and y_3 to check the solution to Example 3.

110. Let $y_1 = -2x - 5$, $y_2 = -2$, $y_3 = x - 3$, and $y_4 = -10$. Then use the graphs of y_1, y_2, y_3, and y_4 to check the solution to Example 9.

111. Use a graphing calculator to check your answers to Exercises 45–48 and Exercises 63–66.

112. On many graphing calculators, the TEST key provides access to inequality symbols, while the LOGIC option of that same key accesses the conjunction *and* and the disjunction *or*. Thus, if $y_1 = x > -2$ and $y_2 = x < 4$, Exercise 33 can be checked by forming the expression $y_3 = y_1$ *and* y_2. The interval(s) in the solution set appears as a horizontal line 1 unit above the *x*-axis. (Be careful to "deselect" y_1 and y_2 so that only y_3 is drawn.) Use the TEST key to check Exercises 35, 39, 41, and 43.

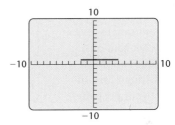

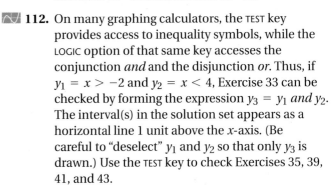

COLLABORATIVE CORNER

Reduce, Reuse, and Recycle

Focus: Compound inequalities

Time: 15–20 minutes

Group size: 2

In the United States, the amount of solid waste (rubbish) being recovered is slowly catching up to the amount being generated. In 2002, each person generated, on average, 4.55 lb of solid waste every day, of which 1.34 lb was recovered. In 2006, each person generated, on average, 4.60 lb of solid waste, of which 1.50 lb was recovered.

Source: U.S. Environmental Protection Agency

ACTIVITY

Assume that the amount of solid waste being generated and the amount being recovered are both increasing linearly. One group member should find a linear function *w* for which $w(t)$ represents the number of pounds of waste generated per person per day *t* years after 2000. The other group member should find a linear function *r* for which $r(t)$ represents the number of pounds recovered per person per day *t* years after 2000. Finally, working together, the group should determine those years for which the amount recovered will be more than $\frac{1}{3}$ of but less than $\frac{1}{2}$ of the amount generated.

4.3   Absolute-Value Equations and Inequalities

Equations with Absolute Value   •   Inequalities with Absolute Value

Equations with Absolute Value

Recall from Section 1.2 the definition of absolute value.

> ### Absolute Value
>
> The absolute value of x, denoted $|x|$, is defined as
> $$|x| = \begin{cases} x, & \text{if } x \geq 0, \\ -x, & \text{if } x < 0. \end{cases}$$
>
> (When x is nonnegative, the absolute value of x is x. When x is negative, the absolute value of x is the opposite of x.)

To better understand this definition, suppose x is -5. Then $|x| = |-5| = 5$, and 5 is the opposite of -5. This shows that when x represents a negative number, the absolute value of x is the opposite of x (which is positive).

Since distance is always nonnegative, we can think of a number's absolute value as its distance from zero on the number line.

EXAMPLE   **1**   Find the solution set: **(a)** $|x| = 4$;   **(b)** $|x| = 0$;   **(c)** $|x| = -7$.

SOLUTION

a) We interpret $|x| = 4$ to mean that the number x is 4 units from zero on the number line. There are two such numbers, 4 and -4. Thus the solution set is $\{-4, 4\}$.

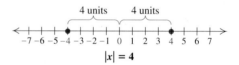

$|x| = 4$

b) We interpret $|x| = 0$ to mean that x is 0 units from zero on the number line. The only number that satisfies this is 0 itself. Thus the solution set is $\{0\}$.

c) Since distance is always nonnegative, it doesn't make sense to talk about a number that is -7 units from zero. Remember: The absolute value of a number is nonnegative. Thus, $|x| = -7$ has no solution; the solution set is $\varnothing$.

TRY EXERCISE 15

Example 1 leads us to the following principle for solving equations.

> ### The Absolute-Value Principle for Equations
>
> For any positive number p and any algebraic expression X:
>
> **a)** The solutions of $|X| = p$ are those numbers that satisfy
>
> $$X = -p \quad or \quad X = p.$$
>
> **b)** The equation $|X| = 0$ is equivalent to the equation $X = 0$.
> **c)** The equation $|X| = -p$ has no solution.

EXAMPLE   2   Find the solution set: **(a)** $|2x + 5| = 13$; **(b)** $|4 - 7x| = -8$.

SOLUTION

a) We use the absolute-value principle, knowing that $2x + 5$ must be either 13 or -13:

$$|X| = p$$
$$|2x + 5| = 13 \quad \text{Substituting}$$
$$2x + 5 = -13 \quad or \quad 2x + 5 = 13$$
$$2x = -18 \quad or \qquad 2x = 8$$
$$x = -9 \quad or \qquad x = 4.$$

Check: For -9:

$$\begin{array}{c|c} |2x + 5| = 13 & \\ \hline |2(-9) + 5| & 13 \\ |-18 + 5| & \\ |-13| & \\ & 13 \stackrel{?}{=} 13 \quad \text{TRUE} \end{array}$$

For 4:

$$\begin{array}{c|c} |2x + 5| = 13 & \\ \hline |2 \cdot 4 + 5| & 13 \\ |8 + 5| & \\ |13| & \\ & 13 \stackrel{?}{=} 13 \quad \text{TRUE} \end{array}$$

The number $2x + 5$ is 13 units from zero if x is replaced with -9 or 4. The solution set is $\{-9, 4\}$.

b) The absolute-value principle reminds us that absolute value is always nonnegative. The equation $|4 - 7x| = -8$ has no solution. The solution set is $\varnothing$.

TRY EXERCISE   21

To use the absolute-value principle, we must be sure that the absolute-value expression is alone on one side of the equation.

EXAMPLE   3   Given that $f(x) = 2|x + 3| + 1$, find all x for which $f(x) = 15$.

SOLUTION   Since we are looking for $f(x) = 15$, we substitute:

$$f(x) = 15$$
$$2|x + 3| + 1 = 15 \quad \text{Replacing } f(x) \text{ with } 2|x + 3| + 1$$
$$2|x + 3| = 14 \quad \text{Subtracting 1 from both sides}$$
$$|x + 3| = 7 \quad \text{Dividing both sides by 2}$$
$$x + 3 = -7 \quad or \quad x + 3 = 7 \quad \begin{array}{l} \text{Using the absolute-value principle} \\ \text{for equations} \end{array}$$
$$x = -10 \quad or \qquad x = 4.$$

The student should check that $f(-10) = f(4) = 15$. The solution set is $\{-10, 4\}$.

TRY EXERCISE   43

EXAMPLE   4   Solve: $|x - 2| = 3$.

SOLUTION   Because this equation is of the form $|a - b| = c$, it can be solved in two different ways.

CAUTION! There are two solutions of $|x - 2| = 3$. Simply solving $x - 2 = 3$ will yield only one of those solutions.

Method 1. We interpret $|x - 2| = 3$ as stating that the number $x - 2$ is 3 units from zero. Using the absolute-value principle, we replace X with $x - 2$ and p with 3:

$$|X| = p$$
$$|x - 2| = 3$$
$$x - 2 = -3 \quad or \quad x - 2 = 3 \qquad \text{Using the absolute-value principle}$$
$$x = -1 \quad or \qquad x = 5.$$

Method 2. This approach is helpful in calculus. The expressions $|a - b|$ and $|b - a|$ can be used to represent the *distance between a and b* on the number line. For example, the distance between 7 and 8 is given by $|8 - 7|$ or $|7 - 8|$. From this viewpoint, the equation $|x - 2| = 3$ states that the distance between x and 2 is 3 units. We draw a number line and locate all numbers that are 3 units from 2.

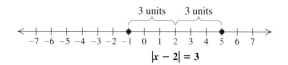

The solutions of $|x - 2| = 3$ are -1 and 5.

Check: The check consists of observing that both methods give the same solutions. The solution set is $\{-1, 5\}$.

⟩ TRY EXERCISE ⟩ 25

Sometimes an equation has two absolute-value expressions. Consider $|a| = |b|$. This means that a and b are the same distance from zero.

If a and b are the same distance from zero, then either they are the same number or they are opposites.

EXAMPLE 5

Solve: $|2x - 3| = |x + 5|$.

SOLUTION   The given equation tells us that $2x - 3$ and $x + 5$ are the same distance from zero. This means that they are either the same number or opposites:

This assumes the two numbers are the same.   This assumes the two numbers are opposites.

$$2x - 3 = x + 5 \quad or \quad 2x - 3 = -(x + 5)$$
$$x - 3 = 5 \qquad or \quad 2x - 3 = -x - 5$$
$$x = 8 \qquad or \quad 3x - 3 = -5$$
$$3x = -2$$
$$x = -\tfrac{2}{3}.$$

The check is left to the student. The solutions are 8 and $-\tfrac{2}{3}$ and the solution set is $\left\{-\tfrac{2}{3}, 8\right\}$.

⟩ TRY EXERCISE ⟩ 47

Inequalities with Absolute Value

Our methods for solving equations with absolute value can be adapted for solving inequalities. Inequalities of this sort arise regularly in more advanced courses.

EXAMPLE 6

Solve $|x| < 4$. Then graph.

SOLUTION   The solutions of $|x| < 4$ are all numbers whose *distance from zero is less than* 4. By substituting or by looking at the number line, we can see that numbers like $-3, -2, -1, -\frac{1}{2}, -\frac{1}{4}, 0, \frac{1}{4}, \frac{1}{2}, 1, 2$, and 3 are all solutions. In fact, the solutions are all the numbers between -4 and 4. The solution set is $\{x \mid -4 < x < 4\}$, or, in interval notation, $(-4, 4)$. The graph is as follows:

$|x| < 4$

TRY EXERCISE 57

EXAMPLE 7

Solve $|x| \geq 4$. Then graph.

SOLUTION   The solutions of $|x| \geq 4$ are all numbers that are at least 4 units from zero—in other words, those numbers x for which $x \leq -4$ or $4 \leq x$. The solution set is $\{x \mid x \leq -4 \text{ or } x \geq 4\}$. In interval notation, the solution set is $(-\infty, -4] \cup [4, \infty)$. We can check mentally with numbers like $-4.1, -5, 4.1$, and 5. The graph is as follows:

$|x| \geq 4$

TRY EXERCISE 59

Examples 1, 6, and 7 illustrate three types of problems in which absolute-value symbols appear. The general principle for solving such problems follows.

Principles for Solving Absolute-Value Problems

For any positive number p and any expression X:

a) The solutions of $|X| = p$ are those numbers that satisfy

$$X = -p \quad \text{or} \quad X = p.$$

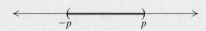

b) The solutions of $|X| < p$ are those numbers that satisfy

$$-p < X < p.$$

c) The solutions of $|X| > p$ are those numbers that satisfy

$$X < -p \quad \text{or} \quad p < X.$$

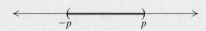

The above principles are true for any positive number p.

If p is negative, any value of X will satisfy the inequality $|X| > p$ because absolute value is never negative. Thus, $|2x - 7| > -3$ is true for any real number x, and the solution set is $\mathbb{R}$.

If p is not positive, the inequality $|X| < p$ has no solution. Thus, $|2x - 7| < -3$ has no solution, and the solution set is $\varnothing$.

EXAMPLE   8   Solve $|3x - 2| < 4$. Then graph.

SOLUTION   The number $3x - 2$ must be less than 4 units from zero. This is of the form $|X| < p$, so part (b) of the principles listed above applies:

$$|X| < p \qquad \text{This corresponds to } -p < X < p.$$
$$|3x - 2| < 4 \qquad \text{Replacing } X \text{ with } 3x - 2 \text{ and } p \text{ with } 4$$
$$-4 < 3x - 2 < 4 \qquad \text{The number } 3x - 2 \text{ must be within 4 units of zero.}$$
$$-2 < \quad 3x \quad < 6 \qquad \text{Adding 2}$$
$$-\tfrac{2}{3} < \quad x \quad < 2. \qquad \text{Multiplying by } \tfrac{1}{3}$$

The solution set is $\{x \mid -\tfrac{2}{3} < x < 2\}$. In interval notation, the solution is $\left(-\tfrac{2}{3}, 2\right)$. The graph is as follows:

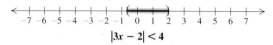

$$|3x - 2| < 4$$

TRY EXERCISE   61

EXAMPLE   9   Given that $f(x) = |4x + 2|$, find all x for which $f(x) \geq 6$.

SOLUTION   We have

$$f(x) \geq 6,$$
$$\text{or} \qquad |4x + 2| \geq 6. \qquad \text{Substituting}$$

To solve, we use part (c) of the principles listed above. In this case, X is $4x + 2$ and p is 6:

$$|X| \geq p \qquad \text{This corresponds to } X < -p \text{ or } p < X.$$
$$|4x + 2| \geq 6 \qquad \text{Replacing } X \text{ with } 4x + 2 \text{ and } p \text{ with } 6$$
$$4x + 2 \leq -6 \quad \text{or} \quad 6 \leq 4x + 2 \qquad \text{The number } 4x + 2 \text{ must be at least 6 units from zero.}$$
$$4x \leq -8 \quad \text{or} \quad 4 \leq 4x \qquad \text{Adding } -2$$
$$x \leq -2 \quad \text{or} \quad 1 \leq x. \qquad \text{Multiplying by } \tfrac{1}{4}$$

The solution set is $\{x \mid x \leq -2 \text{ or } x \geq 1\}$. In interval notation, the solution is $(-\infty, -2] \cup [1, \infty)$. The graph is as follows:

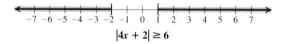

$$|4x + 2| \geq 6$$

TRY EXERCISE   87

ALGEBRAIC–GRAPHICAL CONNECTION

We can visualize Examples 1(a), 6, and 7 by graphing $f(x) = |x|$ and $g(x) = 4$.

Solve: $|x| = 4$.

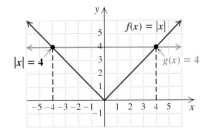

The graphs intersect at $(-4, 4)$ and $(4, 4)$.

$|x| = 4$ when $x = -4$ or $x = 4$.

The solution set of $|x| = 4$ is $\{-4, 4\}$.

Solve: $|x| < 4$.

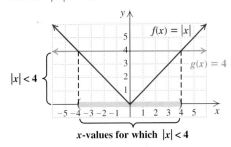

x-values for which $|x| < 4$

The graphs intersect at $(-4, 4)$ and $(4, 4)$.

$|x| < 4$ when $-4 < x < 4$.

The solution set of $|x| < 4$ is $(-4, 4)$.

Solve: $|x| \geq 4$.

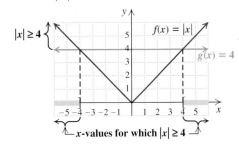

x-values for which $|x| \geq 4$

The graphs intersect at $(-4, 4)$ and $(4, 4)$.

$|x| \geq 4$ when $x \leq -4$ or $x \geq 4$.

The solution set of $|x| \geq 4$ is $(-\infty, -4] \cup [4, \infty)$.

TECHNOLOGY CONNECTION

To enter an absolute-value function on a graphing calculator, we press **MATH** and use the abs(option in the NUM menu. To solve $|4x + 2| = 6$, we graph $y_1 = \text{abs}(4x + 2)$ and $y_2 = 6$.

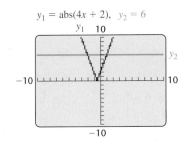

$y_1 = \text{abs}(4x + 2), \quad y_2 = 6$

Using the INTERSECT option of the CALC menu, we find that the graphs intersect at $(-2, 6)$ and $(1, 6)$. The x-coordinates -2 and 1 are the solutions. To solve $|4x + 2| \geq 6$, note where the graph of y_1 is *on or above* the line $y = 6$. The corresponding x-values are the solutions of the inequality.

1. How can the same graph be used to solve $|4x + 2| < 6$?
2. Solve Example 8.
3. Use a graphing calculator to show that $|4x + 2| = -6$ has no solution.

4.3   EXERCISE SET

For Extra Help   PRACTICE WATCH DOWNLOAD

Concept Reinforcement   Classify each statement as either true or false.

1. $|x|$ is never negative.

2. $|x|$ is always positive.

3. If x is negative, then $|x| = -x$.

4. The number a is $|a|$ units from 0.

5. The distance between a and b can be expressed as $|a - b|$.

6. There are two solutions of $|3x - 8| = 17$.

7. There is no solution of $|4x + 9| > -5$.

8. All real numbers are solutions of $|2x - 7| < -3$.

Match each equation or inequality with an equivalent statement from the column on the right. Letters may be used more than once or not at all.

9. $|x - 3| = 5$ a) The solution set is $\varnothing$.

10. $|x - 3| < 5$ b) The solution set is $\mathbb{R}$.

11. $|x - 3| > 5$ c) $x - 3 > 5$

12. $|x - 3| < -5$ d) $x - 3 < -5 \text{ or } x - 3 > 5$

13. $|x - 3| = -5$ e) $x - 3 = 5$

14. $|x - 3| > -5$ f) $x - 3 < 5$

     g) $x - 3 = -5 \text{ or } x - 3 = 5$

     h) $-5 < x - 3 < 5$

Solve.

15. $|x| = 10$

16. $|x| = 5$

Aha! 17. $|x| = -1$

18. $|x| = -8$

19. $|p| = 0$

20. $|y| = 7.3$

21. $|2x - 3| = 4$

22. $|5x + 2| = 7$

23. $|3x + 5| = -8$

24. $|7x - 2| = -9$

25. $|x - 2| = 6$

26. $|x - 3| = 11$

27. $|x - 7| = 1$

28. $|x - 4| = 5$

29. $|t| + 1.1 = 6.6$

30. $|m| + 3 = 3$

31. $|5x| - 3 = 37$

32. $|2y| - 5 = 13$

33. $7|q| + 2 = 9$

34. $5|z| + 2 = 17$

35. $\left|\dfrac{2x - 1}{3}\right| = 4$

36. $\left|\dfrac{4 - 5x}{6}\right| = 3$

37. $|5 - m| + 9 = 16$

38. $|t - 7| + 1 = 4$

39. $5 - 2|3x - 4| = -5$

40. $3|2x - 5| - 7 = -1$

41. Let $f(x) = |2x + 6|$. Find all x for which $f(x) = 8$.

42. Let $f(x) = |2x - 4|$. Find all x for which $f(x) = 10$.

43. Let $f(x) = |x| - 3$. Find all x for which $f(x) = 5.7$.

44. Let $f(x) = |x| + 7$. Find all x for which $f(x) = 18$.

45. Let $f(x) = \left|\dfrac{1 - 2x}{5}\right|$. Find all x for which $f(x) = 2$.

46. Let $f(x) = \left|\dfrac{3x + 4}{3}\right|$. Find all x for which $f(x) = 1$.

Solve.

47. $|x - 7| = |2x + 1|$

48. $|3x + 2| = |x - 6|$

49. $|x + 4| = |x - 3|$

50. $|x - 9| = |x + 6|$

51. $|3a - 1| = |2a + 4|$

52. $|5t + 7| = |4t + 3|$

Aha! 53. $|n - 3| = |3 - n|$

54. $|y - 2| = |2 - y|$

55. $|7 - 4a| = |4a + 5|$

56. $|6 - 5t| = |5t + 8|$

Solve and graph.

57. $|a| \le 3$

58. $|x| < 5$

59. $|t| > 0$

60. $|t| \ge 1$

61. $|x - 1| < 4$

62. $|x - 1| < 3$

63. $|n + 2| \le 6$

64. $|a + 4| \le 0$

65. $|x - 3| + 2 > 7$

66. $|x - 4| + 5 > 10$

Aha! 67. $|2y - 9| > -5$

68. $|3y - 4| > -8$

69. $|3a + 4| + 2 \ge 8$

70. $|2a + 5| + 1 \ge 9$

71. $|y - 3| < 12$

72. $|p - 2| < 3$

73. $9 - |x + 4| \leq 5$

74. $12 - |x - 5| \leq 9$

75. $6 + |3 - 2x| > 10$

76. $|7 - 2y| < -8$

Aha! **77.** $|5 - 4x| < -6$

78. $7 + |4a - 5| \leq 26$

79. $\left| \dfrac{1 + 3x}{5} \right| > \dfrac{7}{8}$

80. $\left| \dfrac{2 - 5x}{4} \right| \geq \dfrac{2}{3}$

81. $|m + 3| + 8 \leq 14$

82. $|t - 7| + 3 \geq 4$

83. $25 - 2|a + 3| > 19$

84. $30 - 4|a + 2| > 12$

85. Let $f(x) = |2x - 3|$. Find all x for which $f(x) \leq 4$.

86. Let $f(x) = |5x + 2|$. Find all x for which $f(x) \leq 3$.

87. Let $f(x) = 5 + |3x - 4|$. Find all x for which $f(x) \geq 16$.

88. Let $f(x) = |2 - 9x|$. Find all x for which $f(x) \geq 25$.

89. Let $f(x) = 7 + |2x - 1|$. Find all x for which $f(x) < 16$.

90. Let $f(x) = 5 + |3x + 2|$. Find all x for which $f(x) < 19$.

91. Explain in your own words why -7 is not a solution of $|x| < 5$.

92. Explain in your own words why $[6, \infty)$ is only part of the solution of $|x| \geq 6$.

Skill Review

To prepare for Section 4.4, review graphing equations and solving systems of equations (Sections 2.3, 2.4, and 3.2).

Graph.

93. $3x - y = 6$ [2.4]

94. $y = \frac{1}{2}x - 1$ [2.3]

95. $x = -2$ [2.4]

96. $y = 4$ [2.4]

Solve using substitution or elimination. [3.2]

97. $x - 3y = 8,$
$2x + 3y = 4$

98. $x - 2y = 3,$
$x = y + 4$

99. $y = 1 - 5x,$
$2x - y = 4$

100. $3x - 2y = 4,$
$5x - 3y = 5$

Synthesis

101. Describe a procedure that could be used to solve any equation of the form $g(x) < c$ graphically.

102. Explain why the inequality $|x + 5| \geq 2$ can be interpreted as "the number x is at least 2 units from -5."

103. From the definition of absolute value, $|x| = x$ only when $x \geq 0$. Solve $|3t - 5| = 3t - 5$ using this same reasoning.

Solve.

104. $|3x - 5| = x$

105. $|x + 2| > x$

106. $2 \leq |x - 1| \leq 5$

107. $|5t - 3| = 2t + 4$

108. $t - 2 \leq |t - 3|$

Find an equivalent inequality with absolute value.

109. $-3 < x < 3$

110. $-5 \leq y \leq 5$

111. $x \leq -6 \; or \; 6 \leq x$

112. $x < -4 \; or \; 4 < x$

113. $x < -8 \; or \; 2 < x$

114. $-5 < x < 1$

115. x is less than 2 units from 7.

116. x is less than 1 unit from 5.

Write an absolute-value inequality for which the interval shown is the solution.

117. ![number line from -7 to 7, interval from -1 to 7] $-7\ -6\ -5\ -4\ -3\ -2\ -1\ \ 0\ \ 1\ \ 2\ \ 3\ \ 4\ \ 5\ \ 6\ \ 7$

118. ![number line from -5 to 9, interval from -4 to 8] $-5\ -4\ -3\ -2\ -1\ \ 0\ \ 1\ \ 2\ \ 3\ \ 4\ \ 5\ \ 6\ \ 7\ \ 8\ \ 9$

119. ![number line from -7 to 7, interval from -7 to -1] $-7\ -6\ -5\ -4\ -3\ -2\ -1\ \ 0\ \ 1\ \ 2\ \ 3\ \ 4\ \ 5\ \ 6\ \ 7$

120. ![number line from 0 to 14, interval from 2 to 12] $0\ \ 1\ \ 2\ \ 3\ \ 4\ \ 5\ \ 6\ \ 7\ \ 8\ \ 9\ \ 10\ 11\ 12\ 13\ 14$

121. *Bungee jumping.* A bungee jumper is bouncing up and down so that her distance d above a river satisfies the inequality $|d - 60 \text{ ft}| \leq 10 \text{ ft}$ (see the figure below). If the bridge from which she jumped is 150 ft above the river, how far is the bungee jumper from the bridge at any given time?

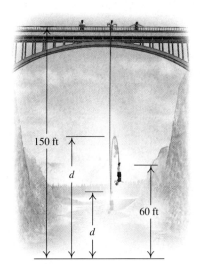

122. *Water level.* Depending on how dry or wet the weather has been, water in a well will rise and fall. The distance d, in feet, that a well's water level is below the ground satisfies the inequality $|d - 15| \leq 2.5$ (see the figure below).

a) Solve for d.

b) How tall a column of water is in the well at any given time?

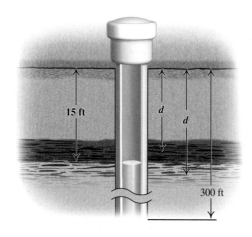

123. Use this graph of $f(x) = |2x - 6|$ to solve $|2x - 6| \leq 4$.

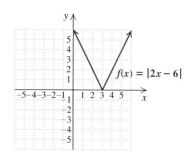

124. Is it possible for an equation in x of the form $|ax + b| = c$ to have exactly one solution? Why or why not?

125. Isabel is using the following graph to solve $|x - 3| < 4$. How can you tell that a mistake has been made in entering $y = \text{abs}(x - 3)$?

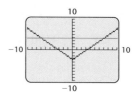

CONNECTING the CONCEPTS

In Chapters 1 and 4, we have learned to solve a variety of equations and inequalities. As we continue our study of algebra, we will learn to solve additional types of equations and inequalities, some of which will require new principles for solving. Following is a list of the principles we have used so far to solve equations and inequalities. Unless otherwise stated, a, b, and c can represent any real number.

The Addition and Multiplication Principles for Equations

$a = b$ is equivalent to $a + c = b + c$.

$a = b$ is equivalent to $ac = bc$, provided $c \neq 0$.

The Addition Principle for Inequalities

$a < b$ is equivalent to $a + c < b + c$.

$a > b$ is equivalent to $a + c > b + c$.

The Multiplication Principle for Inequalities

For any *positive* number c,

$a < b$ is equivalent to $ac < bc$;

$a > b$ is equivalent to $ac > bc$.

For any *negative* number c,

$a < b$ is equivalent to $ac > bc$;

$a > b$ is equivalent to $ac < bc$.

The Absolute-Value Principles for Equations and Inequalities

For any positive number p and any algebraic expression X:

The solutions of $|X| = p$ are those numbers that satisfy

$$X = -p \ \text{ or } \ X = p.$$

The solutions of $|X| < p$ are those numbers that satisfy

$$-p < X < p.$$

The solutions of $|X| > p$ are those numbers that satisfy

$$X < -p \ \text{ or } \ p < X.$$

MIXED REVIEW

Solve.

1. $2x + 3 = 7$

2. $3x - 1 > 8$

3. $3(t - 5) = 4 - (t + 1)$

4. $|2x + 1| = 7$

5. $-x \leq 6$

6. $5|t| < 20$

7. $2(3n + 6) - n = 4 - 3(n + 1)$

8. $3(2a + 9) = 5(3a - 7) - 6a$

9. $2 + |3x| = 10$

10. $|x - 3| \leq 10$

11. $\frac{1}{2}x - 7 = \frac{3}{4} + \frac{1}{4}x$

12. $|t| < 0$

13. $|2x + 5| + 1 \geq 13$

14. $2(x - 3) - x = 5x + 7 - 4x$

15. $|m + 6| - 8 < 10$

16. $\left|\dfrac{x + 2}{5}\right| = 8$

17. $4 - |7 - t| \leq 1$

18. $0.3x + 0.7 = 0.5x$

19. $8 - 5|a + 6| > 3$

20. $|5x + 7| + 9 \geq 4$

4.4 Inequalities in Two Variables

Graphs of Linear Inequalities ■ Systems of Linear Inequalities

In Section 4.1, we graphed inequalities in one variable on a number line. Now we graph inequalities in two variables on a plane.

Graphs of Linear Inequalities

When the equals sign in a linear equation is replaced with an inequality sign, a **linear inequality** is formed. Solutions of linear inequalities are ordered pairs.

EXAMPLE 1

STUDENT NOTES

Pay careful attention to the inequality symbol when determining whether an ordered pair is a solution of an inequality. Writing the symbol at the end of the check, as in Example 1, will help you compare the numbers correctly.

Determine whether $(-3, 2)$ and $(6, -7)$ are solutions of $5x - 4y > 13$.

SOLUTION   Below, on the left, we replace x with -3 and y with 2. On the right, we replace x with 6 and y with -7.

$$\begin{array}{c|c} \multicolumn{2}{c}{5x - 4y > 13} \\ \hline 5(-3) - 4 \cdot 2 & 13 \\ -15 - 8 & \\ -23 \overset{?}{>} 13 & \text{FALSE} \end{array}$$

$$\begin{array}{c|c} \multicolumn{2}{c}{5x - 4y > 13} \\ \hline 5(6) - 4(-7) & 13 \\ 30 + 28 & \\ 58 \overset{?}{>} 13 & \text{TRUE} \end{array}$$

Since $-23 > 13$ is false, $(-3, 2)$ *is not* a solution.

Since $58 > 13$ is true, $(6, -7)$ *is* a solution.

TRY EXERCISE 7

The graph of a linear equation is a straight line. The graph of a linear inequality is a **half-plane**, with a **boundary** that is a straight line. To find the equation of the boundary, we replace the inequality sign with an equals sign.

EXAMPLE   2

Graph: $y \leq x$.

SOLUTION   We first graph the equation of the boundary, $y = x$. Every solution of $y = x$ is an ordered pair, like $(3, 3)$, in which both coordinates are the same. The graph of $y = x$ is shown on the left below. Since the inequality symbol is $\leq$, the line is drawn solid and is part of the graph of $y \leq x$.

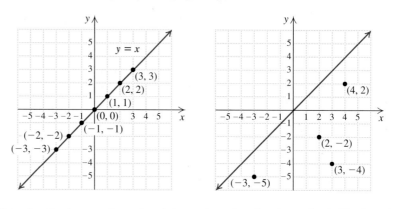

Note that in the graph on the right each ordered pair on the half-plane below $y = x$ contains a y-coordinate that is less than the x-coordinate. All these pairs represent solutions of $y \leq x$. We check one pair, $(4, 2)$, as follows:

$$\frac{y \leq x}{2 \overset{?}{\leq} 4} \quad \text{TRUE}$$

It turns out that *any* point on the same side of $y = x$ as $(4, 2)$ is also a solution. Thus, if one point in a half-plane is a solution, then *all* points in that half-plane are solutions.

We finish drawing the solution set by shading the half-plane below $y = x$. The complete solution set consists of the shaded half-plane as well as the boundary line itself.

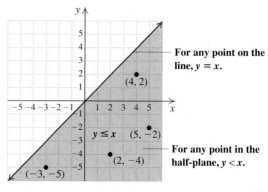

For any point on the line, $y = x$.

For any point in the half-plane, $y < x$.

TRY EXERCISE ▶ 11

From Example 2, we see that for any inequality of the form $y \leq f(x)$ or $y < f(x)$, we shade *below* the graph of $y = f(x)$.

EXAMPLE 3   Graph: $8x + 3y > 24$.

SOLUTION   First, we sketch the graph of $8x + 3y = 24$. Since the inequality sign is $>$, points on this line do not represent solutions of the inequality, and the line is drawn dashed. Points representing solutions of $8x + 3y > 24$ are in either the half-plane above the line or the half-plane below the line. To determine which, we select a point that is not on the line and check whether it is a solution of $8x + 3y > 24$. Let's use $(1, 1)$ as this *test point*:

$$\begin{array}{c|c} \multicolumn{2}{c}{8x + 3y > 24} \\ \hline 8(1) + 3(1) & 24 \\ 8 + 3 & \\ 11 \overset{?}{>} 24 & \text{FALSE} \end{array}$$

Since $11 > 24$ is *false*, $(1, 1)$ is not a solution. Thus no point in the half-plane containing $(1, 1)$ is a solution. The points in the other half-plane *are* solutions, so we shade that half-plane and obtain the graph shown below.

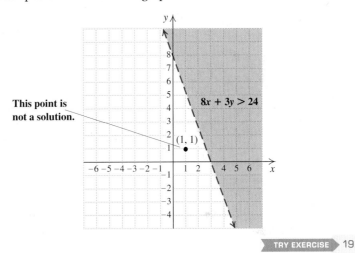

This point is not a solution.

$8x + 3y > 24$

$(1, 1)$

TRY EXERCISE   19

Steps for Graphing Linear Inequalities

1. Replace the inequality sign with an equals sign and graph this line as the boundary. If the inequality symbol is $<$ or $>$, draw the line dashed. If the symbol is $\leq$ or $\geq$, draw the line solid.

2. The graph of the inequality consists of a half-plane on one side of the line and, if the line is solid, the line as well.

   a) If the inequality is of the form $y < mx + b$ or $y \leq mx + b$, shade *below* the line.
      If the inequality is of the form $y > mx + b$ or $y \geq mx + b$, shade *above* the line.

   b) If y is not isolated, either solve for y and graph as in part (a) or simply graph the boundary and use a test point not on the line (as in Example 3). If the test point *is* a solution, shade the half-plane containing the point. If it is not a solution, shade the other half-plane.

EXAMPLE **4**

Graph: $6x - 2y < 12$.

SOLUTION We could graph $6x - 2y = 12$ and use a test point, as in Example 3. Instead, let's solve $6x - 2y < 12$ for y:

$$6x - 2y < 12$$
$$-2y < -6x + 12 \qquad \text{Adding } -6x \text{ to both sides}$$
$$y > 3x - 6. \qquad \text{Dividing both sides by } -2 \text{ and reversing the } < \text{ symbol}$$

The graph consists of the half plane above the dashed boundary line $y = 3x - 6$ (see the graph below).

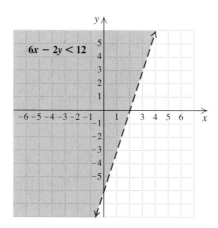

TRY EXERCISE 21

EXAMPLE **5**

Graph $x > -3$ on a plane.

SOLUTION There is only one variable in this inequality. If we graph the inequality on a line, its graph is as follows:

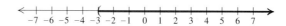

However, we can also write this inequality as $x + 0y > -3$ and graph it on a plane. We can use the same technique as in the examples above. First, we graph the boundary $x = -3$ in the plane, using a dashed line. Then we test some point, say, $(2, 5)$:

$$\frac{x + 0y > -3}{2 + 0 \cdot 5 \mid -3}$$
$$2 \overset{?}{>} -3 \quad \text{TRUE}$$

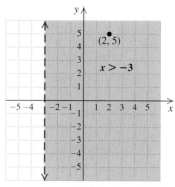

Since $(2, 5)$ is a solution, all points in the half-plane containing $(2, 5)$ are solutions. We shade that half-plane. Another approach is to simply note that the solutions of $x > -3$ are all pairs with first coordinates greater than -3.

TRY EXERCISE 25

TECHNOLOGY CONNECTION

On most graphing calculators, an inequality like $y < \frac{6}{5}x + 3.49$ can be drawn by entering $(6/5)x + 3.49$ as y_1, moving the cursor to the GraphStyle icon just to the left of y_1, pressing **ENTER** until ◣ appears, and then pressing **GRAPH**.

Many calculators have an INEQUALZ program that is accessed using the **APPS** key. Running this program allows us to write inequalities at the **Y=** screen by pressing **ALPHA** and then one of the five keys just below the screen.

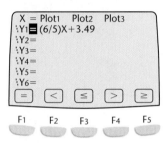

Although the graphs should be identical regardless of the method used, when we are using INEQUALZ, the boundary line appears dashed when < or > is selected.

$$y_1 < (6/5)x + 3.49, \text{ or}$$
$$◣ \, y_1 = (6/5)x + 3.49$$

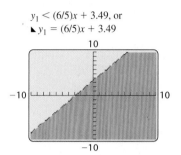

Graph each of the following. Solve for y first if necessary.

1. $y > x + 3.5$
2. $7y \leq 2x + 5$
3. $8x - 2y < 11$
4. $11x + 13y + 4 \geq 0$

EXAMPLE 6   Graph $y \leq 4$ on a plane.

SOLUTION   The inequality is of the form $y \leq mx + b$ (with $m = 0$), so we shade below the solid horizontal line representing $y = 4$.

This inequality can also be graphed by drawing $y = 4$ and testing a point above or below the line. The student should check that this results in a graph identical to the one at right.

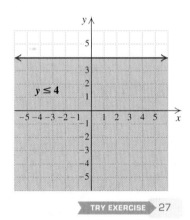

TRY EXERCISE 27

Systems of Linear Inequalities

To graph a system of equations, we graph the individual equations and then find the intersection of the graphs. We do the same thing for a system of inequalities: We graph each inequality and find the intersection of the graphs.

EXAMPLE 7   Graph the system

$$x + y \leq 4,$$
$$x - y < 4.$$

SOLUTION   To graph $x + y \leq 4$, we graph $x + y = 4$ using a solid line. Since the test point $(0, 0)$ *is* a solution and $(0, 0)$ is below the line, we shade the half-plane below the graph red. The arrows near the ends of the line are another way of indicating the half-plane containing solutions.

Next, we graph $x - y < 4$. We graph $x - y = 4$ using a dashed line and consider $(0, 0)$ as a test point. Again, $(0, 0)$ is a solution, so we shade that side of the line blue. The solution set of the system is the region that is shaded purple (both red and blue) and part of the line $x + y = 4$.

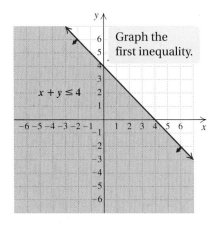

Graph the first inequality.

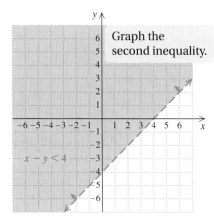

Graph the second inequality.

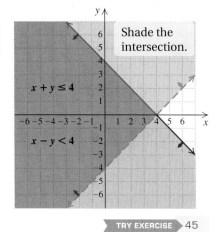

Shade the intersection.

TRY EXERCISE 45

EXAMPLE   8

STUDENT NOTES ——————

If you don't use differently colored pencils or pens to shade different regions, consider using a pencil to make slashes that tilt in different directions in each region. You may also find it useful to attach arrows to the lines, as in the graphs shown.

Graph: $-2 < x \leq 3$.

SOLUTION   This is a system of inequalities:

$$-2 < x,$$
$$x \leq 3.$$

We graph the equation $-2 = x$, and see that the graph of the first inequality is the half-plane to the right of the boundary $-2 = x$. It is shaded red.

We graph the second inequality, starting with the boundary line $x = 3$. The inequality's graph is the line and the half-plane to its left. It is shaded blue.

The solution set of the system is the region that is the intersection of the individual graphs. Since it is shaded both blue and red, it appears to be purple. All points in this region have x-coordinates that are greater than -2 but do not exceed 3.

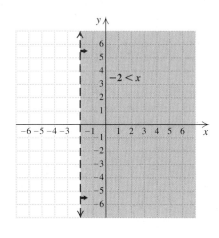

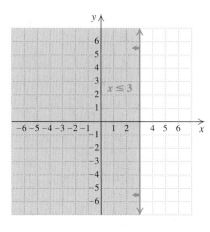

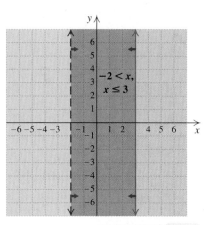

TRY EXERCISE 31

A system of inequalities may have a graph that consists of a polygon and its interior. In Section 4.5, we will have use for the corners, or *vertices* (singular, *vertex*), of such a graph.

EXAMPLE 9   Graph the system of inequalities. Find the coordinates of any vertices formed.

$$6x - 2y \leq 12, \quad (1)$$
$$y - 3 \leq 0, \quad (2)$$
$$x + y \geq 0 \quad (3)$$

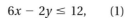

SOLUTION   We graph the boundaries

$$6x - 2y = 12,$$
$$y - 3 = 0,$$
and $$x + y = 0$$

using solid lines. The regions for each inequality are indicated by the arrows near the ends of the lines. We note where the regions overlap and shade the region of solutions purple.

To find the vertices, we solve three different systems of two equations. The system of boundary equations from inequalities (1) and (2) is

$$6x - 2y = 12, \qquad \text{The student can use graphing, substitution, or}$$
$$y - 3 = 0. \qquad \text{elimination to solve these systems.}$$

Solving, we obtain the vertex $(3, 3)$.

The system of boundary equations from inequalities (1) and (3) is

$$6x - 2y = 12,$$
$$x + y = 0.$$

Solving, we obtain the vertex $\left(\frac{3}{2}, -\frac{3}{2}\right)$.

The system of boundary equations from inequalities (2) and (3) is

$$y - 3 = 0,$$
$$x + y = 0.$$

Solving, we obtain the vertex $(-3, 3)$.

▶ **TRY EXERCISE** 49

TECHNOLOGY CONNECTION

Systems of inequalities can be graphed by solving for y and then graphing each inequality as in the Technology Connection on p. 255. To graph systems directly using the INEQUALZ application, enter the correct inequalities, press ⌜GRAPH⌟, and then press **ALPHA** and Shades (⌜F1⌟ or ⌜F2⌟). At the SHADES menu, select Ineq Intersection to see the final graph. To find the vertices, or points of intersection, select PoI-Trace from the graph menu.

$$y_1 \geq 3x - 6, \quad y_2 \leq 3, \quad y_3 \geq -x$$

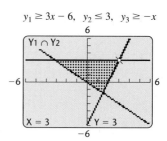

1. Use a graphing calculator to check the solution of Example 7.

Visualizing for Success

A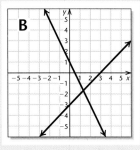

B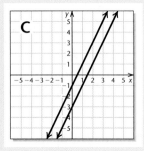

C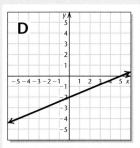

D

E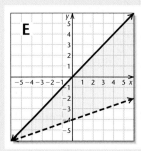

F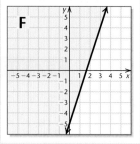

G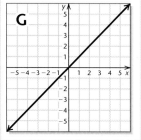

H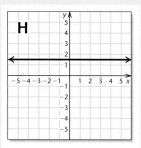

I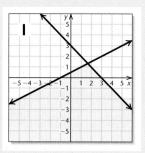

J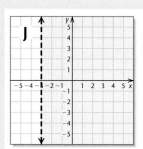

Match each equation, inequality, or system of equations or inequalities with its graph.

1. $x - y = 3,$
$2x + y = 1$

2. $3x - y \leq 5$

3. $x > -3$

4. $y = \dfrac{1}{3}x - 4$

5. $y > \dfrac{1}{3}x - 4,$
$y \leq x$

6. $x = y$

7. $y = 2x - 1,$
$y = 2x - 3$

8. $2x - 5y = 10$

9. $x + y \leq 3,$
$2y \leq x + 1$

10. $y = \dfrac{3}{2}$

Answers on page A-17

An additional, animated version of this activity appears in MyMathLab. To use MyMathLab, you need a course ID and a student access code. Contact your instructor for more information.

4.4 EXERCISE SET

👁 *Concept Reinforcement* In each of Exercises 1–6, match the phrase with the most appropriate choice from the column on the right.

1. ____ A solution of a linear inequality

2. ____ The graph of a linear inequality

3. ____ The graph of a system of linear inequalities

4. ____ Often a convenient test point

5. ____ The name for the corners of a graph of a system of linear inequalities

6. ____ A dashed line

a) $(0, 0)$

b) Vertices

c) A half-plane

d) The intersection of two or more half-planes

e) An ordered pair that satisfies the inequality

f) Indicates the line is not part of the solution

Determine whether each ordered pair is a solution of the given inequality.

7. $(-2, 3)$; $2x - y > -4$

8. $(1, -6)$; $3x + y \geq -3$

9. $(5, 8)$; $3y - 5x \leq 0$

10. $(6, 20)$; $5y - 8x < 40$

Graph on a plane.

11. $y \geq \frac{1}{2}x$

12. $y \leq 3x$

13. $y > x - 3$

14. $y < x + 3$

15. $y \leq x + 2$

16. $y \geq x - 5$

17. $x - y \leq 4$

18. $x + y < 4$

19. $2x + 3y < 6$

20. $3x + 4y \leq 12$

21. $2y - x \leq 4$

22. $2y - 3x > 6$

23. $2x - 2y \geq 8 + 2y$

24. $3x - 2 \leq 5x + y$

25. $x > -2$

26. $x \geq 3$

27. $y \leq 6$

28. $y < -1$

29. $-2 < y < 7$

30. $-4 < y < -1$

31. $-5 \leq x < 4$

32. $-2 < y \leq 1$

33. $0 \leq y \leq 3$

34. $0 \leq x \leq 6$

Graph each system.

35. $y > x,$
    $y < -x + 3$

36. $y < x,$
    $y > -x + 1$

37. $y \leq x,$
    $y \leq 2x - 5$

38. $y \geq x,$
    $y \leq -x + 4$

39. $y \leq -3,$
    $x \geq -1$

40. $y \geq -3,$
    $x \geq 1$

41. $x > -4,$
    $y < -2x + 3$

42. $x < 3,$
    $y > -3x + 2$

43. $y \leq 5,$
    $y \geq -x + 4$

44. $y \geq -2,$
    $y \geq x + 3$

45. $x + y \leq 6,$
    $x - y \leq 4$

46. $x + y < 1,$
    $x - y < 2$

47. $y + 3x > 0,$
    $y + 3x < 2$

48. $y - 2x \geq 1,$
    $y - 2x \leq 3$

Graph each system of inequalities. Find the coordinates of any vertices formed.

49. $y \leq 2x - 3,$
    $y \geq -2x + 1,$
    $x \leq 5$

50. $2y - x \leq 2,$
    $y - 3x \geq -4,$
    $y \geq -1$

51. $x + 2y \leq 12,$
    $2x + y \leq 12,$
    $x \geq 0,$
    $y \geq 0$

52. $x - y \leq 2,$
    $x + 2y \geq 8,$
    $y \leq 4$

53. $8x + 5y \leq 40,$
    $x + 2y \leq 8,$
    $x \geq 0,$
    $y \geq 0$

54. $4y - 3x \geq -12,$
    $4y + 3x \geq -36,$
    $y \leq 0,$
    $x \leq 0$

55. $y - x \geq 2,$
    $y - x \leq 4,$
    $2 \leq x \leq 5$

56. $3x + 4y \geq 12,$
    $5x + 6y \leq 30,$
    $1 \leq x \leq 3$

57. Explain in your own words why the boundary line is drawn dashed for the symbols $<$ and $>$ and why it is drawn solid for the symbols $\leq$ and $\geq$.

58. When graphing linear inequalities, Ron makes a habit of always shading above the line when the symbol $\geq$ is used. Is this wise? Why or why not?

Skill Review

To prepare for Section 4.5, review solving applications using the five-step problem-solving strategy (Sections 1.4 and 3.3).

Solve.

59. *Interest rate.* What rate of interest is required in order for a principal of $1560 to earn $25.35 in half a year? [1.4]

60. *Interest.* Luke invested $5000 in two accounts. He put $2200 in an account paying 4% simple interest and the rest in an account paying 5% simple interest. How much interest did he earn in one year from both accounts? [1.4]

61. *Investments.* Gina invested $10,000 in two accounts, one paying 3% simple interest and one paying 5% simple interest. After one year, she had earned $428 from both accounts. How much did she invest in each? [3.3]

62. *Catering.* Janice provided 20 lb of fresh vegetables for a reception. Carrots were $1.50 per pound and broccoli was $2.50 per pound. If she spent $38, how much of each vegetable did she buy? [3.3]

63. *Admissions.* There were 170 tickets sold for a high school basketball game. Tickets were $1 each for students and $3 each for adults. The total amount of money collected was $386. How many of each type of ticket were sold? [3.3]

64. *Agriculture.* Josh planted 400 acres in corn and soybeans. He planted 80 more acres in corn than he did in soybeans. How many acres of each did he plant? [3.3]

Synthesis

65. Explain how a system of linear inequalities could have a solution set containing exactly one pair.

66. In Example 7 on pp. 255–256, is the point $(4, 0)$ part of the solution set? Why or why not?

Graph.

67. $x + y > 8,$
$\quad x + y \leq -2$

68. $\quad x + y \geq 1,$
$\quad -x + y \geq 2,$
$\qquad x \geq -2,$
$\qquad y \geq 2,$
$\qquad y \leq 4,$
$\qquad x \leq 2$

69. $\quad x - 2y \leq 0,$
$\quad -2x + y \leq 2,$
$\qquad\quad x \leq 2,$
$\qquad\quad y \leq 2,$
$\quad x + y \leq 4$

70. Write four systems of four inequalities that describe a 2-unit by 2-unit square that has $(0, 0)$ as one of the vertices.

71. *Luggage size.* Unless an additional fee is paid, most major airlines will not check any luggage for which the sum of the item's length, width, and height exceeds 62 in. The U.S. Postal Service will ship a package only if the sum of the package's length and girth (distance around its midsection) does not exceed 130 in. Video Promotions is ordering several 30-in. long cases that will be both mailed and checked as luggage. Using w and h for width and height (in inches), respectively, write and graph an inequality that represents all acceptable combinations of width and height.
Sources: U.S. Postal Service; www.case2go.com

72. *Hockey wins and losses.* The Skating Stars believe they need at least 60 points for the season in order to make the playoffs. A win is worth 2 points and a tie is worth 1 point. Graph a system of inequalities that describes the situation. (*Hint*: Let w = the number of wins and t = the number of ties.)

73. *Graduate-school admissions.* Students entering the Master of Science program in Computer Science and Engineering at University of Texas Arlington must meet minimum score requirements on the Graduate Records Examination (GRE). The GRE Quantitative score must be at least 700 and the GRE Verbal score must be at least 400. The sum of the GRE Quantitative and Verbal scores must be at least 1150. Both scores have a maximum of 800. Using q for the quantitative score and v for the verbal score, write and graph a system of inequalities that represents all combinations that meet the requirements for entrance into the program.
Source: University of Texas Arlington

74. *Widths of a basketball floor.* Sizes of basketball floors vary due to building sizes and other constraints such as cost. The length L is to be at most 94 ft and the width W is to be at most 50 ft. Graph a system of inequalities that describes the possible dimensions of a basketball floor.

75. *Elevators.* Many elevators have a capacity of 1 metric ton (1000 kg). Suppose that c children, each weighing 35 kg, and a adults, each 75 kg, are on an elevator. Graph a system of inequalities that indicates when the elevator is overloaded.

76. *Age of marriage.* The following rule of thumb for determining an appropriate difference in age between a bride and a groom appears in many Internet blogs: *The younger spouse's age should be at least seven more than half the age of the older spouse.* Let b = the age of the bride, in years, and g = the age of the groom, in years. Write and graph a system of inequalities that represents all combinations of ages that follow this rule of thumb. Should a minimum or maximum age for marriage exist? How would the graph of the system of inequalities change with such a requirement?

77. *Waterfalls.* In order for a waterfall to be classified as a classical waterfall, its height must be less than twice its crest width, and its crest width cannot exceed one-and-a-half times its height. The tallest waterfall in the world is about 3200 ft high. Let h represent a waterfall's height, in feet, and w the crest width, in feet. Write and graph a system of inequalities that represents all possible combinations of heights and crest widths of classical waterfalls.

78. Use a graphing calculator to check your answers to Exercises 35–48. Then use INTERSECT to determine any point(s) of intersection.

79. Use a graphing calculator to graph each inequality.
  a) $3x + 6y > 2$
  b) $x - 5y \le 10$
  c) $13x - 25y + 10 \le 0$
  d) $2x + 5y > 0$

CONNECTING the CONCEPTS

We have now solved a variety of equations, inequalities, systems of equations, and systems of inequalities. Below is a list of the different types of problems we have solved, illustrations of each type, and descriptions of the solutions. Note that a solution set may be empty.

Type	Example	Solution	Graph
Linear equations in one variable	$2x - 8 = 3(x + 5)$	A number	
Linear inequalities in one variable	$-3x + 5 > 2$	A set of numbers; an interval	
Linear equations in two variables	$2x + y = 7$	A set of ordered pairs; a line	
Linear inequalities in two variables	$x + y \geq 4$	A set of ordered pairs; a half-plane	
System of equations in two variables	$x + y = 3,$ $5x - y = -27$	An ordered pair or a set of ordered pairs	
System of inequalities in two variables	$6x - 2y \leq 12,$ $y - 3 \leq 0,$ $x + y \geq 0$	A set of ordered pairs; a region of a plane	

MIXED REVIEW

Graph each solution on a number line.

1. $x + 2 = 7$

2. $x + 2 > 7$

3. $x + 2 \leq 7$

4. $3(x - 7) - 2 = 5 - (2 - x)$

5. $6 - 2x \geq 8$

6. $7 > 5 - x$

Graph on a plane.

7. $x + y = 2$

8. $x + y < 2$

9. $x + y \geq 2$

10. $y = 3x - 3$

11. $x = 4$

12. $2x - 5y = -10$

13. $x + y = 1,$
    $x - y = 1$

14. $y \geq 1 - x,$
    $y \leq x - 3,$
    $y \leq 2$

15. $2x + y < 6$

16. $x > 6y - 6$

17. $4x = 3y$

18. $y = 2x - 3,$
    $y = -\frac{1}{2}x + 1$

19. $x - y \leq 3,$
    $y \geq 2x,$
    $2y - x \leq 2$

20. $3y = 8$

4.5   Applications Using Linear Programming

Objective Functions and Constraints   •   Linear Programming

There are many real-world situations in which we need to find a greatest value (a maximum) or a least value (a minimum). For example, most businesses like to know how to make the *most* profit with the *least* expense possible. Some such problems can be solved using systems of inequalities.

Objective Functions and Constraints

Often a quantity we wish to maximize depends on two or more other quantities. For example, a gardener's profits P might depend on the number of shrubs s and the number of trees t that are planted. If the gardener makes a $10 profit from each shrub and an $18 profit from each tree, the total profit, in dollars, is given by the **objective function**

$$P = 10s + 18t.$$

Thus the gardener might be tempted to simply plant lots of trees since they yield the greater profit. This would be a good idea were it not for the fact that the number of trees and shrubs planted—and thus the total profit—is subject to the demands, or **constraints**, of the situation. For example, to improve drainage, the gardener might be required to plant at least 3 shrubs. Thus the objective function would be subject to the *constraint*

$$s \geq 3.$$

Because of limited space, the gardener might also be required to plant no more than 10 plants. This would subject the objective function to a *second* constraint:

$$s + t \leq 10.$$

Finally, the gardener might be told to spend no more than $700 on the plants. If the shrubs cost $40 each and the trees cost $100 each, the objective function is subject to a *third* constraint:

The cost of the shrubs   plus   the cost of the trees   cannot exceed $700.

$$40s \qquad + \qquad 100t \qquad\qquad \leq \qquad 700$$

In short, the gardener wishes to maximize the objective function

$$P = 10s + 18t,$$

subject to the constraints

$$s \geq 3,$$
$$s + t \leq 10,$$
$$40s + 100t \leq 700,$$
$$s \geq 0,$$
$$t \geq 0.$$

Because the number of trees and shrubs cannot be negative

These constraints form a system of linear inequalities that can be graphed.

Linear Programming

The gardener's problem is "How many shrubs and trees should be planted, subject to the constraints listed, in order to maximize profit?" To solve such a problem, we use a result from a branch of mathematics known as **linear programming**.

The Corner Principle

Suppose an objective function $F = ax + by + c$ depends on x and y (with a, b, and c constant). Suppose also that F is subject to constraints on x and y, which form a system of linear inequalities. If F has a minimum or a maximum value, then it can be found as follows:

1. Graph the system of inequalities and find the vertices.
2. Find the value of the objective function at each vertex. The greatest and the least of those values are the maximum and the minimum of the function, respectively.
3. The ordered pair at which the maximum or minimum occurs indicates the choice of (x, y) for which that maximum or minimum occurs.

This result was proven during World War II, when linear programming was developed to help allocate troops and supplies bound for Europe.

EXAMPLE   1   Solve the gardener's problem discussed above.

SOLUTION   We are asked to maximize $P = 10s + 18t$, subject to the constraints

$$s \geq 3,$$
$$s + t \leq 10,$$
$$40s + 100t \leq 700,$$
$$s \geq 0,$$
$$t \geq 0.$$

We graph the system, using the techniques of Section 4.4. The portion of the graph that is shaded represents all pairs that satisfy the constraints. It is sometimes called the *feasible region*.

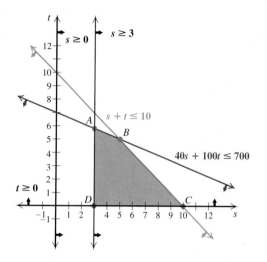

According to the corner principle, *P* is maximized at one of the vertices of the shaded region. To determine the coordinates of the vertices, we solve the following systems:

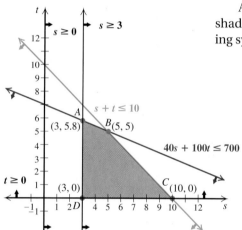

A: $\left.\begin{array}{r} 40s + 100t = 700, \\ s = 3; \end{array}\right\}$ The student can verify that the solution of this system is $(3, 5.8)$. The coordinates of point A are $(3, 5.8)$.

B: $\left.\begin{array}{r} s + t = 10, \\ 40s + 100t = 700; \end{array}\right\}$ The student can verify that the solution of this system is $(5, 5)$. The coordinates of point B are $(5, 5)$.

C: $\left.\begin{array}{r} s + t = 10, \\ t = 0; \end{array}\right\}$ The solution of this system is $(10, 0)$. The coordinates of point C are $(10, 0)$.

D: $\left.\begin{array}{r} t = 0, \\ s = 3. \end{array}\right\}$ The solution of this system is $(3, 0)$. The coordinates of point D are $(3, 0)$.

We now find the value of *P* at each vertex.

Vertex (s, t)	Profit $P = 10s + 18t$	
A $(3, 5.8)$	$10(3) + 18(5.8) = 134.4$	
B $(5, 5)$	$10(5) + 18(5) = 140$	⟵ Maximum
C $(10, 0)$	$10(10) + 18(0) = 100$	
D $(3, 0)$	$10(3) + 18(0) = 30$	⟵ Minimum

The greatest value of *P* occurs at $(5, 5)$. Thus profit is maximized at \$140 if the gardener plants 5 shrubs and 5 trees. Incidentally, we have also shown that profit is minimized at \$30 if 3 shrubs and 0 trees are planted.

TRY EXERCISE 13

EXAMPLE 2   *Grading.*   For his history grade, Cy can write book summaries for 70 points each or research papers for 80 points each. He estimates that each book summary will take 9 hr and each research paper will take 15 hr and that he will have at most 120 hr to spend. He may turn in a total of no more than 12 summaries or papers. How many of each should he write in order to receive the highest score?

SOLUTION

1. **Familiarize.** Since we are looking for the number of book summaries and the number of research papers, we let $b =$ the number of book summaries and $r =$ the number of research papers. Cy is limited by the number of hours he can spend and by the number of summaries and papers he can turn in. These two limits are the constraints.

2. **Translate.** We organize the information in a table.

Type	Number of Points for Each	Time Required for Each	Number Written	Total Time for Each Type	Total Points for Each Type
Book summary	70	9 hr	b	$9b$	$70b$
Research paper	80	15 hr	r	$15r$	$80r$
Total			$b + r \leq 12$	$9b + 15r \leq 120$	$70b + 80r$

Because no more than 12 may be turned in   Because the time cannot exceed 120 hr   We wish to maximize the total score.

Let T represent the total score. We see from the table that

$$T = 70b + 80r.$$

We wish to maximize T subject to the number and time constraints:

$$b + r \leq 12,$$
$$9b + 15r \leq 120,$$
$$b \geq 0, \}$$
$$r \geq 0. \}$$

We include this because the number of summaries and papers cannot be negative.

3. **Carry out.** We graph the system and evaluate T at each vertex. The graph is as follows:

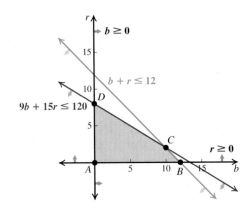

We find the coordinates of each vertex by solving a system of two linear equations. The coordinates of point A are obviously $(0, 0)$. To find the coordinates of point C, we solve the system

$$b + \quad r = 12, \qquad (1)$$
$$9b + 15r = 120. \qquad (2)$$

We multiply both sides of equation (1) by -9 and add:

$$-9b - \quad 9r = -108$$
$$\underline{\quad 9b + 15r = \quad 120}$$
$$6r = \quad 12$$
$$r = 2.$$

Substituting, we find that $b = 10$. Thus the coordinates of C are $(10, 2)$. Point B is the intersection of $b + r = 12$ and $r = 0$, so B is $(12, 0)$. Point D is the intersection of $9b + 15r = 120$ and $b = 0$, so D is $(0, 8)$. Computing the score for each ordered pair, we obtain the table at left.

The greatest score in the table is 860, obtained when Cy writes 10 book summaries and 2 research papers.

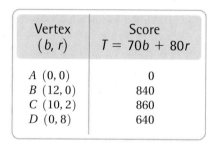

Vertex (b, r)	Score $T = 70b + 80r$
A $(0, 0)$	0
B $(12, 0)$	840
C $(10, 2)$	860
D $(0, 8)$	640

4. **Check.** We can check that $T \leq 860$ for any other pair in the shaded region. This is left to the student.

5. **State.** In order to maximize his score, Cy should write 10 book summaries and 2 research papers.

TRY EXERCISE 19

4.5   EXERCISE SET

For Extra Help MyMathLab Math XP PRACTICE WATCH DOWNLOAD

Concept Reinforcement   *Complete each of the following sentences.*

1. In linear programming, the quantity we wish to maximize or minimize is given by the _____ function.

2. In linear programming, the demands arising from the given situation are known as _____.

3. To solve a linear programming problem, we make use of the _____ principle.

4. The shaded portion of a graph that represents all points that satisfy a problem's constraints is known as the _____ region.

5. In linear programming, the corners of the shaded portion of the graph are referred to as _____.

6. If it exists, the maximum value of an objective function occurs at a _____ of the feasible region.

Find the maximum and the minimum values of each objective function and the values of x and y at which they occur.

7. $F = 2x + 14y$,
   subject to
   $5x + 3y \leq 34$,
   $3x + 5y \leq 30$,
   $x \geq 0$,
   $y \geq 0$

8. $G = 7x + 8y$,
   subject to
   $3x + 2y \leq 12$,
   $2y - x \leq 4$,
   $x \geq 0$,
   $y \geq 0$

9. $P = 8x - y + 20$,
   subject to
   $6x + 8y \leq 48$,
   $0 \leq y \leq 4$,
   $0 \leq x \leq 7$

10. $Q = 24x - 3y + 52$,
    subject to
    $5x + 4y \leq 20$,
    $0 \leq y \leq 4$,
    $0 \leq x \leq 3$

11. $F = 2y - 3x$,
    subject to
    $y \leq 2x + 1$,
    $y \geq -2x + 3$,
    $x \leq 3$

12. $G = 5x + 2y + 4$,
    subject to
    $y \leq 2x + 1$,
    $y \geq -x + 3$,
    $x \leq 5$

13. *Lunch-time profits.*   Art sells gumbo and sandwiches. To stay in business, Art must sell at least 10 orders of gumbo and 30 sandwiches each day. Because of limited space, no more than 40 orders of gumbo or 70 sandwiches can be made. The total number of orders cannot exceed 90. If profit is $1.65 per gumbo order and $1.05 per sandwich, how many of each item should Art sell in order to maximize profit?

14. *Gas mileage.*   Caroline owns a car and a moped. She has at most 12 gal of gasoline to be used between the car and the moped. The car's tank holds at most 18 gal and the moped's 3 gal. The mileage for the car is 20 mpg and for the moped is 100 mpg. How many gallons of gasoline should each vehicle use if Caroline wants to travel as far as possible? What is the maximum number of miles?

15. *Photo albums.*   Photo Perfect prints pages of photographs for albums. A page containing 4 photos costs $3 and a page containing 6 photos costs $5. Ann can spend no more than $90 for photo pages of her recent vacation, and can use no more than 20 pages in her album. What combination of 4-photo pages and 6-photo pages will maximize the number of photos she can display? What is the maximum number of photos that she can display?

16. *Milling.*   Picture Rocks Lumber can convert logs into either lumber or plywood. In a given week, the mill can turn out 400 units of production, of which 100 units of lumber and 150 units of plywood are required by regular customers. The profit on a unit of lumber is $20 and on a unit of plywood is $30. How many units of each type should the mill produce in order to maximize profit?

Aha! 17. *Investing.*   Rosa is planning to invest up to $40,000 in corporate or municipal bonds, or both. She must invest from $6000 to $22,000 in corporate bonds, and she refuses to invest more than $30,000 in municipal bonds. The interest on corporate bonds is 8% and on municipal bonds is $7\frac{1}{2}$%. This is simple interest for one year. How much should Rosa invest in each type of bond in order to earn the most interest? What is the maximum interest?

18. *Investing.*   Jamaal is planning to invest up to $22,000 in City Bank or the Southwick Credit Union, or both. He wants to invest at least $2000 but no more than $14,000 in City Bank. Because of insurance limitations, he will invest no more than $15,000 in the Southwick Credit Union. The interest in City Bank is 6% and in the credit union is $6\frac{1}{2}$%. This is simple interest for one year. How much should Jamaal invest in each bank in order to earn the most interest? What is the maximum interest?

19. *Test scores.*   Corinna is taking a test in which short-answer questions are worth 10 points each and essay questions are worth 15 points each. She estimates that it will take 3 min to answer each short-answer question and 6 min to answer each essay question. The total time allowed is 60 min, and no more than 16 questions can be answered. Assuming that all her answers are correct, how many questions of each type should Corinna answer to get the best score?

20. *Test scores.*   Edy is about to take a test that contains short-answer questions worth 4 points each and word problems worth 7 points each. Edy must do at least 5 short-answer questions, but time restricts doing more than 10. She must do at least 3 word problems, but time restricts doing more than 10. Edy can do no more than 18 questions in total. How many of each type of question must Edy do in order to maximize her score? What is this maximum score?

21. *Grape growing.*   Auggie's vineyard consists of 240 acres upon which he wishes to plant Merlot and Cabernet grapes. Profit per acre of Merlot is $400 and profit per acre of Cabernet is $300. Furthermore, the total number of hours of labor available during the harvest season is 3200. Each acre of Merlot requires 20 hr of labor and each acre of Cabernet requires 10 hr of labor. Determine how the land should be divided between Merlot and Cabernet in order to maximize profit.

22. *Coffee blending.*   The Coffee Peddler has 1440 lb of Sumatran coffee and 700 lb of Kona coffee. A batch of Hawaiian Blend requires 8 lb of Kona and 12 lb of Sumatran, and yields a profit of $90. A batch of Classic Blend requires 4 lb of Kona and 16 lb of Sumatran, and yields a $55 profit. How many batches of each kind should be made in order to maximize profit? What is the maximum profit? (*Hint:* Organize the information in a table.)

23. *Nutrition.*   Becca is supposed to have at least 15 mg but no more than 45 mg of iron each day. She should also have at least 1500 mg but no more than 2500 mg of calcium per day. One serving of goat cheese contains 1 mg of iron, 500 mg of calcium, and 264 calories. One serving of hazelnuts contains 5 mg of iron, 100 mg of calcium, and 628 calories. How many servings of goat cheese and how many servings of hazelnuts should Becca eat in order to meet the daily requirements of iron and calcium but minimize the total number of calories?

24. *Textile production.*   It takes Cosmic Stitching 2 hr of cutting and 4 hr of sewing to make a knit suit. To make a worsted suit, it takes 4 hr of cutting and 2 hr of sewing. At most 20 hr per day are available for cutting and at most 16 hr per day are available for sewing. The profit on a knit suit is $68 and on a worsted suit is $62. How many of each kind of suit should be made in order to maximize profit?

25. Before a student begins work in this section, what three sections of the text would you suggest he or she study? Why?

26. What does the use of the word "constraint" in this section have in common with the use of the word in everyday speech?

Skill Review

To prepare for Section 5.1, review evaluating and simplifying algebraic expressions (Sections 1.1 and 1.2).

Evaluate.

27. $3x^3 - 5x^2 - 8x + 7$, for $x = -1$ [1.1], [1.2]

28. $t^3 + 6t^2 - 10$, for $t = 2$ [1.1]

Simplify. [1.2]

29. $3(2t - 7) + 5(3t + 1)$

30. $6(5x + 1) + 8(3 - x)$

31. $(8t + 6) - (7t + 6)$

32. $(9x - 5) - (10 - 3x)$

Synthesis

33. Explain how Exercises 17 and 18 can be answered by logical reasoning without linear programming.

34. Write a linear programming problem for a classmate to solve. Devise the problem so that profit must be maximized subject to at least two (nontrivial) constraints.

35. *Airplane production.*   Alpha Tours has two types of airplanes, the T3 and the S5, and contracts requiring accommodations for a minimum of 2000 first-class, 1500 tourist-class, and 2400 economy-class passengers. The T3 costs $30 per mile to operate and can accommodate 40 first-class, 40 tourist-class, and 120 economy-class passengers, whereas the S5 costs $25 per mile to operate and can accommodate 80 first-class, 30 tourist-class, and 40 economy-class passengers. How many of each type of airplane should be used in order to minimize the operating cost?

36. *Airplane production.*   A new airplane, the T4, is now available, having an operating cost of $37.50 per mile and accommodating 40 first-class, 40 tourist-class, and 80 economy-class passengers. If the T3 of Exercise 35 were replaced with the T4, how many S5's and how many T4's would be needed in order to minimize the operating cost?

37. *Furniture production.*   P. J. Edward Furniture Design produces chairs and sofas. The chairs require 20 ft of wood, 1 lb of foam rubber, and 2 sq yd of fabric. The sofas require 100 ft of wood, 50 lb of foam rubber, and 20 sq yd of fabric. The company has 1900 ft of wood, 500 lb of foam rubber, and 240 sq yd of fabric. The chairs can be sold for $80 each and the sofas for $1200 each. How many of each should be produced in order to maximize income?

Study Summary

KEY TERMS AND CONCEPTS

EXAMPLES

SECTION 4.1: INEQUALITIES AND APPLICATIONS

An **inequality** is any sentence containing $<, >, \leq, \geq$, or $\neq$. Solution sets of inequalities can be **graphed** and written in **set-builder notation** or **interval notation**.

Interval Notation	Set-builder Notation	Graph
(a, b)	$\{x \mid a < x < b\}$	
$[a, b]$	$\{x \mid a \leq x \leq b\}$	
$[a, b)$	$\{x \mid a \leq x < b\}$	
$(a, b]$	$\{x \mid a < x \leq b\}$	
(a, ∞)	$\{x \mid a < x\}$	
$(-\infty, a)$	$\{x \mid x < a\}$	

The Addition Principle for Inequalities

For any real numbers a, b, and c:

$a < b$ is equivalent to $a + c < b + c$;
$a > b$ is equivalent to $a + c > b + c$.

Similar statements hold for $\leq$ and $\geq$.

$$x + 3 \leq 5 \quad \text{is equivalent to}$$
$$x + 3 - 3 \leq 5 - 3$$
$$x \leq 2.$$

The solution set is $\{x \mid x \leq 2\}$, or $(-\infty, 2]$.

The Multiplication Principle for Inequalities

For any real numbers a and b, and for any *positive* number c,

$a < b$ is equivalent to $ac < bc$;
$a > b$ is equivalent to $ac > bc$.

$$3x > 9 \quad \text{is equivalent to}$$
$$\frac{1}{3} \cdot 3x > \frac{1}{3} \cdot 9 \qquad \text{The inequality symbol does not change, because } \tfrac{1}{3} \text{ is positive.}$$
$$x > 3$$

The solution set is $\{x \mid x > 3\}$, or $(3, \infty)$.

For any real numbers a and b, and for any *negative* number c.

$a < b$ is equivalent to $ac > bc$;
$a > b$ is equivalent to $ac < bc$.

Similar statements hold for $\leq$ and $\geq$.

$$-4x \geq 20 \quad \text{is equal to}$$
$$-\frac{1}{4} \cdot (-4x) \leq -\frac{1}{4} \cdot 20 \qquad \text{The inequality symbol is reversed, because } -\tfrac{1}{4} \text{ is negative.}$$
$$x \leq -5$$

The solution set is $\{x \mid x \leq -5\}$, or $(-\infty, -5]$.

Many problem-solving situations translate to inequalities. See p. 226 for a list of common phrases and their translations.

Phrase	Translation
The essay must contain *no more than* 500 words.	$w \leq 500$
Cal must make *no less than* \$35.	$p \geq 35$

SECTION 4.2: INTERSECTIONS, UNIONS, AND COMPOUND INEQUALITIES

A **conjunction** consists of two or more sentences joined by the word *and*. The solution set of the conjunction is the **intersection** of the solution sets of the individual sentences.

$$-4 \leq x - 1 \leq 5$$

$$-4 \leq x - 1 \quad and \quad x - 1 \leq 5$$
$$-3 \leq x \qquad and \qquad x \leq 6$$

The solution set is $\{x | -3 \leq x \leq 6\}$, or $[-3, 6]$.

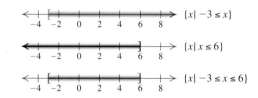

A **disjunction** consists of two or more sentences joined by the word *or*. The solution set of the disjunction is the **union** of the solution sets of the individual sentences.

$$2x + 9 < 1 \qquad or \quad 5x - 2 \geq 3$$
$$2x < -8 \quad or \qquad 5x \geq 5$$
$$x < -4 \quad or \qquad x \geq 1$$

The solution set is $\{x | x < -4 \text{ or } x \geq 1\}$, or $(-\infty, -4) \cup [1, \infty)$.

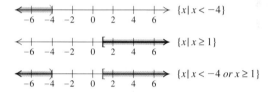

SECTION 4.3: ABSOLUTE-VALUE EQUATIONS AND INEQUALITIES

The Absolute-Value Principles for Equations and Inequalities

For any positive number p and any algebraic expression X:

a) The solutions of $|X| = p$ are those numbers that satisfy

$$X = -p \quad or \quad X = p.$$

b) The solutions of $|X| < p$ are those numbers that satisfy

$$-p < X < p.$$

c) The solutions of $|X| > p$ are those numbers that satisfy

$$X < -p \quad or \quad p < X.$$

If $|X| = 0$, then $X = 0$. If p is negative, then $|X| = p$ and $|X| < p$ have no solution, and any value of X will satisfy $|X| > p$.

$$|x + 3| = 4$$
$$x + 3 = 4 \quad or \quad x + 3 = -4 \qquad \text{Using part (a)}$$
$$x = 1 \quad or \qquad x = -7$$

The solution set is $\{-7, 1\}$.

$$|x + 3| < 4$$
$$-4 < x + 3 < 4 \qquad \text{Using part (b)}$$
$$-7 < x < 1$$

The solution set is $\{x | -7 < x < 1\}$, or $(-7, 1)$.

$$|x + 3| \geq 4$$
$$x + 3 \leq -4 \quad or \quad 4 \leq x + 3 \qquad \text{Using part (c)}$$
$$x \leq -7 \quad or \quad 1 \leq x$$

The solution set is $\{x | x \leq -7 \text{ or } x \geq 1\}$, or $(-\infty, -7] \cup [1, \infty)$.

SECTION 4.4: INEQUALITIES IN TWO VARIABLES

To graph a linear inequality:

1. Graph the **boundary line**. Draw a dashed line if the inequality symbol is $<$ or $>$, and draw a solid line if the inequality symbol is $\leq$ or $\geq$.

2. Determine which side of the boundary line contains the solution set, and shade that **half-plane**.

Graph: $x + y < -1$.

1. Graph $x + y = -1$ using a dashed line.

2. Choose a test point not on the line: $(0, 0)$.

$$\frac{x + y < -1}{0 + 0 \;\big|\; -1}$$
$$0 \overset{?}{<} -1 \quad \text{FALSE}$$

Since $0 < -1$ is false, shade the half-plane that does *not* contain $(0, 0)$.

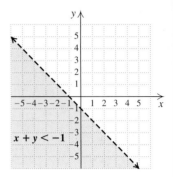

SECTION 4.5: APPLICATIONS USING LINEAR PROGRAMMING

The Corner Principle

The maximum or minimum value of an **objective function** over a *feasible region* is the maximum or minimum value of the function at a **vertex** of that region.

Maximize $F = x + 2y$ subject to

$$x + y \leq 5,$$
$$x \geq 0,$$
$$y \geq 1.$$

1. Graph the feasible region.
2. Find the value of F at the vertices.

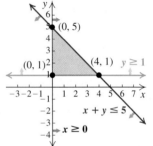

Vertex	$F = x + 2y$
$(0, 1)$	2
$(0, 5)$	10
$(4, 1)$	6

⟵ The maximum value of F is 10.

Review Exercises: Chapter 4

🖐 *Concept Reinforcement*   *Classify each statement as either true or false.*

1. If x cannot exceed 10, then $x \leq 10$. [4.1]

2. It is always true that if $a > b$, then $ac > bc$. [4.1]

3. The solution of $|3x - 5| \leq 8$ is a closed interval. [4.3]

4. The inequality $2 < 5x + 1 < 9$ is equivalent to $2 < 5x + 1 \text{ or } 5x + 1 < 9$. [4.2]

5. The solution set of a disjunction is the union of two solution sets. [4.2]

6. The equation $|x| = r$ has no solution when r is negative. [4.3]

7. $|f(x)| > 3$ is equivalent to $f(x) < -3 \text{ or } f(x) > 3$. [4.3]

8. A test point is used to determine whether the line in a linear inequality is drawn solid or dashed. [4.4]

9. The graph of a system of linear inequalities is always a half-plane. [4.4]

10. The corner principle states that every objective function has a maximum or minimum value. [4.5]

Graph each inequality and write the solution set using both set-builder notation and interval notation. [4.1]

11. $x \le -1$

12. $a + 3 \le 7$

13. $4y > -15$

14. $-0.2y < 6$

15. $-6x - 5 < 4$

16. $-\frac{1}{2}x - \frac{1}{4} > \frac{1}{2} - \frac{1}{4}x$

17. $0.3y - 7 < 2.6y + 15$

18. $-2(x - 5) \ge 6(x + 7) - 12$

19. Let $f(x) = 3x + 2$ and $g(x) = 10 - x$. Find all values of x for which $f(x) \le g(x)$. [4.1]

Solve. [4.1]

20. Mariah has two offers for a summer job. She can work in a sandwich shop for $8.40 an hour, or she can do carpentry work for $16 an hour. In order to do the carpentry work, she must spend $950 for tools. For how many hours must Mariah work in order for carpentry to be more profitable than the sandwich shop?

21. Clay is going to invest $9000, part at 3% and the rest at 3.5%. What is the most he can invest at 3% and still be guaranteed $300 in interest each year?

22. Find the intersection:
$$\{a, b, c, d\} \cap \{a, c, e, f, g\}.\ [4.2]$$

23. Find the union:
$$\{a, b, c, d\} \cup \{a, c, e, f, g\}.\ [4.2]$$

Graph and write interval notation. [4.2]

24. $x \le 2$ *and* $x > -3$

25. $x \le 3$ *or* $x > -5$

Solve and graph each solution set. [4.2]

26. $-3 < x + 5 \le 5$

27. $-15 < -4x - 5 < 0$

28. $3x < -9$ *or* $-5x < -5$

29. $2x + 5 < -17$ *or* $-4x + 10 \le 34$

30. $2x + 7 \le -5$ *or* $x + 7 \ge 15$

31. $f(x) < -5$ *or* $f(x) > 5$, where $f(x) = 3 - 5x$

For $f(x)$ as given, use interval notation to write the domain of f.

32. $f(x) = \dfrac{2x}{x + 3}$ [4.2]

33. $f(x) = \sqrt{5x - 10}$ [4.1]

34. $f(x) = \sqrt{1 - 4x}$ [4.1]

Solve. [4.3]

35. $|x| = 11$

36. $|t| \ge 21$

37. $|x - 8| = 3$

38. $|4a + 3| < 11$

39. $|3x - 4| \ge 15$

40. $|2x + 5| = |x - 9|$

41. $|5n + 6| = -11$

42. $\left| \dfrac{x + 4}{6} \right| \le 2$

43. $2|x - 5| - 7 > 3$

44. $19 - 3|x + 1| \ge 4$

45. Let $f(x) = |8x - 3|$. Find all x for which $f(x) < 0$. [4.3]

46. Graph $x - 2y \ge 6$ on a plane. [4.4]

Graph each system of inequalities. Find the coordinates of any vertices formed. [4.4]

47. $x + 3y > -1,$
  $x + 3y < 4$

48. $x - 3y \le 3,$
  $x + 3y \ge 9,$
  $y \le 6$

49. Find the maximum and the minimum values of
$$F = 3x + y + 4$$
  subject to
  $$y \le 2x + 1,$$
  $$x \le 7,$$
  $$y \ge 3. \qquad [4.5]$$

50. Custom Computers has two manufacturing plants. The Oregon plant cannot produce more than 60 computers per week, while the Ohio plant cannot produce more than 120 computers per week. The Electronics Outpost sells at least 160 Custom computers each week. It costs $40 to ship a computer to The Electronics Outpost from the Oregon plant and $25 to ship from the Ohio plant. How many computers should be shipped from each plant in order to minimize cost? [4.5]

Synthesis

51. Explain in your own words why $|X| = p$ has two solutions when p is positive and no solution when p is negative. [4.3]

52. Explain why the graph of the solution of a system of linear inequalities is the intersection, not the union, of the individual graphs. [4.4]

53. Solve: $|2x + 5| \le |x + 3|$. [4.3]

54. Classify as true or false: If $x < 3$, then $x^2 < 9$. If false, give an example showing why. [4.1]

55. Super Lock manufactures brass doorknobs with a 2.5-in. diameter and a ± 0.003-in. manufacturing tolerance, or allowable variation in diameter. Write the tolerance as an inequality with absolute value. [4.3]

Test: Chapter 4

Graph each inequality and write the solution set using both set-builder notation and interval notation.

1. $x - 3 < 8$

2. $-\frac{1}{2}t < 12$

3. $-4y - 3 \geq 5$

4. $3a - 5 \leq -2a + 6$

5. $3(7 - x) < 2x + 5$

6. $-2(3x - 1) - 5 \geq 6x - 4(3 - x)$

7. Let $f(x) = -5x - 1$ and $g(x) = -9x + 3$. Find all values of x for which $f(x) > g(x)$.

8. Dani can rent a van for either $80 with unlimited mileage or $45 with 100 free miles and an extra charge of 40¢ for each mile over 100. For what numbers of miles traveled would the unlimited mileage plan save Dani money?

9. A refrigeration repair company charges $80 for the first half-hour of work and $60 for each additional hour. Blue Mountain Camp has budgeted $200 to repair its walk-in cooler. For what lengths of a service call will the budget not be exceeded?

10. Find the intersection:
$$\{a, e, i, o, u\} \cap \{a, b, c, d, e\}.$$

11. Find the union:
$$\{a, e, i, o, u\} \cup \{a, b, c, d, e\}.$$

For $f(x)$ as given, use interval notation to write the domain of f.

12. $f(x) = \sqrt{6 - 3x}$

13. $f(x) = \dfrac{x}{x - 7}$

Solve and graph each solution set.

14. $-5 < 4x + 1 \leq 3$

15. $3x - 2 < 7 \; or \; x - 2 > 4$

16. $-3x > 12 \; or \; 4x \geq -10$

17. $1 \leq 3 - 2x \leq 9$

18. $|n| = 15$

19. $|a| > 5$

20. $|3x - 1| < 7$

21. $|-5t - 3| \geq 10$

22. $|2 - 5x| = -12$

23. $g(x) < -3 \; or \; g(x) > 3$, where $g(x) = 4 - 2x$

24. Let $f(x) = |2x - 1|$ and $g(x) = |2x + 7|$. Find all values of x for which $f(x) = g(x)$.

25. Graph $y \leq 2x + 1$ on a plane.

Graph the system of inequalities. Find the coordinates of any vertices formed.

26. $x + y \geq 3,$
   $x - y \geq 5$

27. $2y - x \geq -7,$
   $2y + 3x \leq 15,$
   $y \leq 0,$
   $x \leq 0$

28. Find the maximum and the minimum values of
$$F = 5x + 3y$$
   subject to
   $x + y \leq 15,$
   $1 \leq x \leq 6,$
   $0 \leq y \leq 12.$

29. Swift Cuts makes $12 on each manicure and $18 on each haircut. A manicure takes 30 min and a haircut takes 50 min, and there are 5 stylists who each work 6 hr a day. If the salon can schedule 50 appointments a day, how many should be manicures and how many haircuts in order to maximize profit? What is the maximum profit?

Synthesis

Solve. Write the solution set using interval notation.

30. $|2x - 5| \leq 7 \; and \; |x - 2| \geq 2$

31. $7x < 8 - 3x < 6 + 7x$

32. Write an absolute-value inequality for which the interval shown is the solution.

Simplify. Do not leave negative exponents in your answers.

1. $-\frac{1}{9} - \left(-\frac{2}{3}\right)$ [1.2]

2. $3 + 24 \div 2^2 \cdot 3 - (6 - 7)$ [1.2]

3. $3c - [8 - 2(1 - c)]$ [1.3]

4. -10^{-2} [1.6]

5. $(3xy^{-4})(-2x^3y)$ [1.6]

6. $\left(\dfrac{18a^2b^{-1}}{12a^{-1}b}\right)^2$ [1.6]

Solve.

7. $2x - 4 = 8x$ [1.3]

8. $3(x - 2) = 14 - x$ [1.3]

9. $x - 2 < 6 \text{ or } 2x + 1 > 5$ [4.2]

10. $2x + 5y = 2,$
    $3x - y = 4$ [3.2]

11. $y = \frac{1}{2}x - 7,$
    $2x - 4y = 3$ [3.2]

12. $x + 3y = 8,$
    $2x - 3y = 7$ [3.2]

13. $|2x - 1| = 8$ [4.3]

14. $9(x - 3) - 4x < 2 - (3 - x)$ [4.1]

15. $|4t| > 12$ [4.3]

16. $|3x - 2| \leq 8$ [4.3]

Graph on a plane.

17. $y = \frac{2}{3}x - 4$ [2.3]

18. $x = -3$ [2.4]

19. $3x - y = 3$ [2.4]

20. $x + y \geq -2$ [4.4]

21. $f(x) = -x + 1$ [2.3]

22. $x - 2y > 4,$
    $x + 2y \geq -2$ [4.4]

23. Find the slope and the y-intercept of the line given by $4x - 9y = 18$. [2.3]

24. Using function notation, write a slope–intercept equation for the line with slope -7 and containing the point $(-3, -4)$. [2.5]

25. Find an equation of the line with y-intercept $(0, 4)$ and perpendicular to the line given by $3x + 2y = 1$. [2.5]

26. For the graph of f shown, determine the domain and the range of f. [2.2], [4.2]

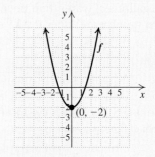

27. Determine the domain of the function given by $f(x) = \dfrac{3}{2x + 5}$. [2.2], [4.2]

28. Find $g(-2)$ if $g(x) = 3x^2 - 5x$. [2.2]

29. Find $(f - g)(x)$ if $f(x) = x^2 + 3x$ and $g(x) = 9 - 3x$. [2.6]

30. Graph the solution set of $-3 \leq f(x) \leq 2$, where $f(x) = 1 - x$. [4.2]

31. Find the domain of h/g if $h(x) = \dfrac{1}{x}$ and $g(x) = 3x - 1$. [2.6]

32. Solve for t: $at - dt = c$. [1.5]

33. *Water usage.* On average, it takes about 750,000 gal of water to create an acre of machine-made snow. Resorts in the Alps make about 60,000 acres of machine-made snow each year. Using scientific notation, find the amount of water used each year to make machine-made snow in the Alps. [1.7]

Sources: Swiss Federal Institute for Snow and Avalanche Research; www.telegraph.co.uk

34. *Water usage.*   In dry climates, it takes about 11,600 gal of water to produce a pound of beef and a pound of wheat. The pound of beef requires 7000 more gallons of water than the pound of wheat. How much water does it take to produce each? [3.3]
Source: *The Wall Street Journal*, 1/28/08

35. *Book sales.*   U.S. sales of books and maps were $34.6 billion in 2001 and $41.4 billion in 2004. Let $b(t)$ represent U.S. book sales t years after 2001. [2.5]
Source: Bureau of Economic Analysis

  a) Find a linear function that fits the data.
  b) Use the function from part (a) to predict U.S. book sales in 2010.
  c) In what year will U.S. book sales be $50 billion?

36. *Fundraising.*   Michelle is planning a fundraising dinner for Happy Hollow Children's Camp. The banquet facility charges a rental fee of $1500, but will waive the rental fee if more than $6000 is spent for catering. Michelle knows that 150 people will attend the dinner. [4.1]

  a) How much should each dinner cost in order for the rental fee to be waived?
  b) For what costs per person will the total cost (including the rental fee) exceed $6000?

37. *Perimeter of a rectangle.*   The perimeter of a rectangle is 32 cm. If five times the width equals three times the length, what are the dimensions of the rectangle? [3.3]

38. *Utility bills.*   One month Lori and Tony spent $920 for electricity, rent, and cell phone. The electric bill was $\frac{1}{4}$ of the rent, and the phone bill was $40 less than the electric bill. How much was the rent? [3.5]

39. *Banking.*   Banks charge a fee to a customer whose checking account does not contain enough money to pay for a debit-card purchase or a written check. These insufficient-funds fees totaled $35 billion in the United States in a recent year. This was 70% of the total fee income of banks. What was the total fee income of banks in that year? [1.4]

40. *Catering.*   Dan charges $35 per person for a vegetarian meal and $40 per person for a steak dinner. For one event, he served 28 dinners for a total cost of $1060. How many dinners were vegetarian and how many were steak? [3.3]

Synthesis

41. If $(2, 6)$ and $(-1, 5)$ are two solutions of $f(x) = mx + b$, find m and b. [3.2]

42. Find k such that the line containing $(-2, k)$ and $(3, 8)$ is parallel to the line containing $(1, 6)$ and $(4, -2)$. [2.5]

43. Use interval notation to write the domain of the function given by
$$f(x) = \frac{\sqrt{x + 4}}{x}. \quad [4.2]$$

44. Simplify: $\dfrac{2^{a-1} \cdot 2^{4a}}{2^{3(-2a+5)}}$. [1.6]

Polynomials and Polynomial Functions

BEN GIVENS
WIND FARM OPERATIONS MANAGER
Trent, Texas

The operations and maintenance of a wind farm depend heavily on data. We use many complicated mathematical algorithms to analyze the performance of the turbines. Production data are used to determine royalty payments and revenue for the project on a monthly basis. Without math, there would be no way to perform these tasks.

AN APPLICATION

The number of watts of power P generated by a particular turbine at a wind speed of x miles per hour can be approximated by the polynomial function given by

$$P(x) = 0.0157x^3 + 0.1163x^2 - 1.3396x + 3.7063.$$

Use the graph of the function to estimate the power generated by a 10-mph wind.

Source: Based on data from *QST*, November 2006

This exercise appears as Exercise 41 in Section 5.1.

I n many ways, polynomials, like $2x + 5$ or $x^2 - 3$, are central to the study of algebra. We have already used many polynomials in this text. Here in Chapter 5 we will clearly define what polynomials are and discuss how to manipulate them. We will then use polynomials and polynomial functions in problem solving.

5.1   Introduction to Polynomials and Polynomial Functions

Algebraic Expressions and Polynomials   •   Polynomial Functions   •   Adding Polynomials   •
Opposites and Subtraction

In this section, we define a type of algebraic expression known as a *polynomial*. After developing some vocabulary, we study addition and subtraction of polynomials, and evaluate *polynomial functions*.

Algebraic Expressions and Polynomials

In Chapter 1, we introduced algebraic expressions like

$$5x^2, \quad \frac{3}{x^2 + 5}, \quad 9a^3b^4, \quad 3x^{-2}, \quad 6x^2 + 3x + 1, \quad -9, \quad \text{and} \quad 5 - 2x.$$

Of the expressions listed, $5x^2$, $9a^3b^4$, $3x^{-2}$, and -9 are examples of *terms*. A **term** is simply a number or a variable raised to a power or a product of numbers and/or variables raised to powers. The power used may be 1 and thus not written. Note that a term may be a product but not a sum or a difference.

When all variables in a term are raised to whole-number powers, the term is a **monomial**. Of the terms listed above, $5x^2$, $9a^3b^4$, and -9 are monomials. Since -2 is not a whole number, $3x^{-2}$ is *not* a monomial. The **degree** of a monomial is the sum of the exponents of the variables. Thus, $5x^2$ has degree 2 and $9a^3b^4$ has degree 7. Nonzero constant terms, like -9, can be written $-9x^0$ and therefore have degree 0. The term 0 itself is said to have no degree.

The number 5 is said to be the **coefficient** of $5x^2$. Thus the coefficient of $9a^3b^4$ is 9 and the coefficient of $-2x$ is -2. The coefficient of a constant term is just that constant.

A **polynomial** is a monomial or a sum of monomials. Of the expressions listed, $5x^2$, $9a^3b^4$, $6x^2 + 3x + 1$, -9, and $5 - 2x$ are polynomials. In fact, with the exception of $9a^3b^4$, these are all polynomials *in one variable*. The expression $9a^3b^4$ is a *polynomial in two variables*. Note that $5 - 2x$ is the sum of 5 and $-2x$. Thus, 5 and $-2x$ are the terms in the polynomial $5 - 2x$.

The **leading term** of a polynomial is the term of highest degree. Its coefficient is called the **leading coefficient**. The **degree of a polynomial** is the same as the degree of its leading term.

STUDY SKILLS

Keep Your Focus

When studying with someone else, it can be tempting to talk about subjects that have little to do with mathematics. If you see that this may happen, explain to your partner(s) that you enjoy the nonmathematical conversation, but would enjoy it more later—after the math work has been completed.

EXAMPLE   1   For each polynomial given, find the degree of each term, the degree of the polynomial, the leading term, and the leading coefficient.

a) $2x^3 + 8x^2 - 17x - 3$

b) $6x^2 + 8x^2y^3 - 17xy - 24xy^2z^4 + 2y + 3$

SOLUTION

a) $2x^3 + 8x^2 - 17x - 3$ b) $6x^2 + 8x^2y^3 - 17xy - 24xy^2z^4 + 2y + 3$

Term	$2x^3$	$8x^2$	$-17x$	-3	$6x^2$	$8x^2y^3$	$-17xy$	$-24xy^2z^4$	$2y$	3
Degree of Each Term	3	2	1	0	2	5	2	7	1	0
Leading Term	$2x^3$				$-24xy^2z^4$					
Leading Coefficient	2				-24					
Degree of Polynomial	3				7					

TRY EXERCISE 11

A polynomial of degree 0 or 1 is called **linear**. A polynomial in one variable is said to be **quadratic** if it is of degree 2, **cubic** if it is of degree 3, and **quartic** if it is of degree 4.

The following are some names for certain kinds of polynomials.

Type	Definition	Examples				
Monomial	A polynomial of one term	4	$-3p$	$5x^2$	$-7a^2b^3$	0 xyz
Binomial	A polynomial of two terms	$2x + 7$		$a - 3b$	$5x^2 + 7y^3$	
Trinomial	A polynomial of three terms	$x^2 - 7x + 12$		$4a^2 + 2ab + b^2$		

We generally arrange polynomials in one variable so that the exponents *decrease* from left to right. This is called **descending order**. Some polynomials may be written with exponents *increasing* from left to right, which is **ascending order**. Generally, if an exercise is written in one kind of order, the answer is written in that same order.

EXAMPLE 2   Arrange in ascending order: $12 + 2x^3 - 7x + x^2$.

SOLUTION

$$12 + 2x^3 - 7x + x^2 = 12 - 7x + x^2 + 2x^3$$

Polynomials in several variables can be arranged with respect to the powers of one of the variables.

EXAMPLE 3   Arrange in descending powers of x: $y^4 + 2 - 5x^2 + 3x^3y + 7xy^2$.

SOLUTION

$$y^4 + 2 - 5x^2 + 3x^3y + 7xy^2 = 3x^3y - 5x^2 + 7xy^2 + y^4 + 2$$

TRY EXERCISES 19 and 23

Polynomial Functions

A *polynomial function* is a function in which ordered pairs are determined by evaluating a polynomial. For example, the function P given by

$$P(x) = 5x^7 + 3x^5 - 4x^2 - 5$$

is an example of a polynomial function. To evaluate a polynomial function, we substitute a number for the variable just as in Chapter 2. In this text, we limit ourselves to polynomial functions in one variable.

EXAMPLE 4   For the polynomial function $P(x) = -x^2 + 4x - 1$, find the following: **(a)** $P(5)$; **(b)** $P(-5)$.

SOLUTION

a) $P(5) = -5^2 + 4(5) - 1$ We square each input before taking its opposite.

$ = -25 + 20 - 1 = -6$

b) $P(-5) = -(-5)^2 + 4(-5) - 1$ Use parentheses when an input is negative.

$ = -25 - 20 - 1 = -46$ **TRY EXERCISE** 29

EXAMPLE 5   *Medicine.*   Ibuprofen is a medication used to relieve pain. The polynomial function

$$M(t) = 0.5t^4 + 3.45t^3 - 96.65t^2 + 347.7t, \quad 0 \le t \le 6$$

can be used to estimate the number of milligrams of ibuprofen in the bloodstream t hours after 400 mg of the medication has been swallowed.

Source: Based on data from Dr. P. Carey, Burlington, VT

a) How many milligrams of ibuprofen are in the bloodstream 2 hr after 400 mg has been swallowed?

b) Use the graph below to estimate $M(4)$.

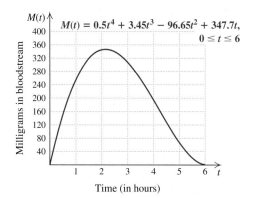

Time (in hours)

SOLUTION

a) We evaluate the function for $t = 2$:

$$M(2) = 0.5(2)^4 + 3.45(2)^3 - 96.65(2)^2 + 347.7(2)$$

We carry out the calculation using the rules for order of operations.

$$= 0.5(16) + 3.45(8) - 96.65(4) + 695.4$$

$$= 8 + 27.6 - 386.6 + 695.4$$

$$= 344.4.$$

Approximately 344 mg of ibuprofen is in the bloodstream 2 hr after 400 mg has been swallowed.

b) To estimate $M(4)$, the amount in the bloodstream after 4 hr, we locate 4 on the horizontal axis. From there we move vertically to the graph of the function and

One way to evaluate a function is to enter and graph it as y_1 and then select TRACE. We can then enter any x-value that appears in that window and the corresponding y-value will appear. We use this approach below to check Example 5(a).

$$y_1 = 0.5x^4 + 3.45x^3 - 96.65x^2 + 347.7x$$

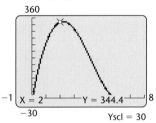

Use this approach to check Examples 5(b) and 4.

then horizontally to the $M(t)$-axis, as shown below. This locates a value for $M(4)$ of about 190.

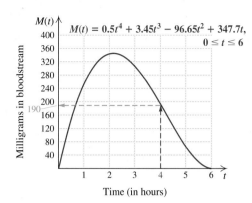

Time (in hours)

After 4 hr, approximately 190 mg of ibuprofen is still in a person's bloodstream, assuming an original dosage of 400 mg.   TRY EXERCISE ▶ 41

Adding Polynomials

Recall from Section 1.3 that when two terms have the same variable(s) raised to the same power(s), they are **similar**, or **like, terms** and can be "combined" or "collected."

EXAMPLE 6   Combine like terms.

a) $3x^2 - 4y + 2x^2$ **b)** $4t^3 - 6t - 8t^2 + t^3 + 9t^2$

c) $3x^2y + 5xy^2 - 3x^2y - xy^2$

SOLUTION

a) $3x^2 - 4y + 2x^2 = 3x^2 + 2x^2 - 4y$ Rearranging terms using the commutative law for addition

$\qquad\qquad\qquad = (3 + 2)x^2 - 4y$ Using the distributive law

$\qquad\qquad\qquad = 5x^2 - 4y$

b) $4t^3 - 6t - 8t^2 + t^3 + 9t^2 = 5t^3 + t^2 - 6t$ We usually perform the middle steps mentally and write just the answer.

c) $3x^2y + 5xy^2 - 3x^2y - xy^2 = 4xy^2$ TRY EXERCISE ▶ 57

We add polynomials by combining like terms.

EXAMPLE 7   Add: $(-3x^3 + 2x - 4) + (4x^3 + 3x^2 + 2)$.

SOLUTION

$$(-3x^3 + 2x - 4) + (4x^3 + 3x^2 + 2) = x^3 + 3x^2 + 2x - 2$$

TRY EXERCISE ▶ 63

Using columns is sometimes helpful. To do so, we write the polynomials one under the other, listing like terms under one another and leaving spaces for any missing terms.

EXAMPLE   8   Add: $4ax^2 + 4bx - 5$ and $-6ax^2 + 8$.

SOLUTION

$$
\begin{array}{l}
4ax^2 + 4bx - 5 \\
\underline{-6ax^2 \qquad\quad + 8} \qquad \text{Leaving space for the missing } bx\text{-term} \\
-2ax^2 + 4bx + 3 \qquad \text{Combining like terms}
\end{array}
$$

▶ TRY EXERCISE ▷ 65

EXAMPLE   9   Add: $13x^3y + 3x^2y - 5y$ and $x^3y + 4x^2y - 3xy$.

SOLUTION

$$(13x^3y + 3x^2y - 5y) + (x^3y + 4x^2y - 3xy) = 14x^3y + 7x^2y - 3xy - 5y$$

▶ TRY EXERCISE ▷ 67

TECHNOLOGY CONNECTION

By pressing **2ND** **TBLSET** and selecting AUTO, we can use a Table to check that polynomials in one variable have been added or subtracted correctly. To check Example 7, we enter $y_1 = (-3x^3 + 2x - 4) + (4x^3 + 3x^2 + 2)$ and $y_2 = x^3 + 3x^2 + 2x - 2$. If the addition is correct, the values of y_1 and y_2 will match, regardless of the x-values used.

We can also check by using graphs, as discussed on p. 294.

X	Y₁	Y₂
-2	-2	-2
-1	-2	-2
0	-2	-2
1	4	4
2	22	22
3	58	58
4	118	118

X = -2

Use a table to determine whether each sum or difference is correct.

1. $(x^3 - 2x^2 + 3x - 7) + (3x^2 - 4x + 5) = x^3 + x^2 - x - 2$

2. $(2x^2 + 3x - 6) + (5x^2 - 7x + 4) = 7x^2 + 4x - 2$

3. $(x^4 + 2x^2 + x) - (3x^4 - 5x + 1) = -2x^4 + 2x^2 + 6x - 1$

4. $(3x^4 - 2x^2 - 1) - (2x^4 - 3x^2 - 4) = x^4 + x^2 - 5$

Opposites and Subtraction

If the sum of two polynomials is 0, the polynomials are *opposites,* or *additive inverses,* of each other. For example,

$$(3x^2 - 5x + 2) + (-3x^2 + 5x - 2) = 0,$$

so the opposite of $(3x^2 - 5x + 2)$ must be $(-3x^2 + 5x - 2)$. We can say the same thing using algebraic symbolism, as follows:

The opposite of $\underbrace{(3x^2 - 5x + 2)}$ is $\underbrace{(-3x^2 + 5x - 2)}$.

$$-\qquad (3x^2 - 5x + 2) \quad = \quad -3x^2 + 5x - 2$$

To form the opposite of a polynomial, we can think of distributing the "−" sign, or multiplying each term of the polynomial by −1, and removing the parentheses. The effect is to change the sign of each term in the polynomial.

> ### The Opposite of a Polynomial
>
> The *opposite* of a polynomial P can be written as $-P$ or, equivalently, by replacing each term with its opposite.

EXAMPLE 10 Write two equivalent expressions for the opposite of

$$7xy^2 - 6xy - 4y + 3.$$

SOLUTION

a) The opposite of $7xy^2 - 6xy - 4y + 3$ can be written with parentheses as

$$-(7xy^2 - 6xy - 4y + 3). \qquad \text{Writing the opposite of } P \text{ as } -P$$

b) The opposite of $7xy^2 - 6xy - 4y + 3$ can be written without parentheses as

$$-7xy^2 + 6xy + 4y - 3. \qquad \text{Multiplying each term by } -1$$

> TRY EXERCISE 73

To subtract a polynomial, we add its opposite.

EXAMPLE 11 Subtract: $(-3x^2 + 4xy) - (2x^2 - 5xy + 7y^2)$.

SOLUTION

$$
\begin{aligned}
&(-3x^2 + 4xy) - (2x^2 - 5xy + 7y^2) \\
&= (-3x^2 + 4xy) + (-2x^2 + 5xy - 7y^2) \qquad &&\text{Adding the opposite} \\
&= -3x^2 + 4xy - 2x^2 + 5xy - 7y^2 \qquad &&\text{Try to go directly to this step.} \\
&= -5x^2 + 9xy - 7y^2 \qquad &&\text{Combining like terms}
\end{aligned}
$$

> TRY EXERCISE 83

With practice, you may find that you can skip some steps, by mentally taking the opposite of each term being subtracted and then combining like terms.

To use columns for subtraction, we mentally change the signs of the terms being subtracted.

EXAMPLE 12 Subtract: $(3x^4 - 2x^3 + 6x - 1) - (3x^4 - 9x^3 - x^2 + 7)$.

SOLUTION

Write: (Subtract)   *Think*: (Add)

$$
\begin{array}{l}
3x^4 - 2x^3 + 6x - 1 \\
-(3x^4 - 9x^3 - x^2 + 7) \\
\hline
\end{array}
\qquad
\begin{array}{l}
3x^4 - 2x^3 + 6x - 1 \\
-3x^4 + 9x^3 + x^2 - 7 \\
\hline
7x^3 + x^2 + 6x - 8
\end{array}
$$

Take the opposite of each term mentally and add.

> TRY EXERCISE 85

Concept Reinforcement   *In each of Exercises 1–10, match the item with the best example of that item from the column on the right.*

1. ____ A binomial

2. ____ A trinomial

3. ____ A monomial

4. ____ A sixth-degree polynomial

5. ____ A polynomial written in ascending powers of t

6. ____ A term that is not a monomial

7. ____ A polynomial with a leading term of degree 5

8. ____ A polynomial with a leading coefficient of 5

9. ____ A cubic polynomial

10. ____ A polynomial containing similar terms

a) $9a^7$

b) $6s^2 - 2t + 4st^2 - st^3$

c) $4t^{-2}$

d) $t^4 - st + s^3$

e) $7t^3 - 13 + 5t^4 - 2t$

f) $4t^3 + 12t^2 + 9t - 7$

g) $5 + a$

h) $4t^6 + 7t - 8t^2 + 5$

i) $8st^3 - 6s^2t + 4st^3 - 2s + 7$

j) $7s^3t^2 - 4s^2t + 3st^2 + 1$

In Exercises 11–14, for each polynomial given, answer the following questions.

a) *How many terms are there?*
b) *What is the degree of each term?*
c) *What is the degree of the polynomial?*
d) *What is the leading term?*
e) *What is the leading coefficient?*

11. $-5x^6 + x^4 + 7x^3 - 2x - 10$

12. $t^3 + 5t^2 - t + 9$

13. $7a^4 + a^3b^2 - 5a^2b + 3$

14. $-uv + 8v^4 + 9u^2v^5 - 6u^2 - 1$

Determine the degree of each polynomial.

15. $8y^2 + y^5 - 9 - 2y + 3y^4$

16. $3x^2 - 5x + 8x^4 + 12$

17. $3p^4 - 5pq + 2p^3q^3 + 8pq^2 - 7$

18. $2xy^3 + 9y^2 - 8x^3 + 7x^2y^2 + y^7$

Arrange in descending order. Then find the leading term and the leading coefficient.

19. $4 - 8t + 5t^2 + 2t^3 - 15t^4$

20. $4 - 7y^2 + 6y^4 - 2y - y^5$

21. $3x + 6x^5 - 5 - x^6 + 7x^2$

22. $a - a^2 + 12a^7 + 3a^4 - 15$

Arrange in ascending powers of x.

23. $4x + 5x^3 - x^6 - 9$

24. $7 - x + 3x^6 + 2x^4$

25. $2x^2y + 5xy^3 - x^3 + 8y$

26. $2ax - 9ab + 4x^5 - 7bx^2$

Find g(3) for each polynomial function.

27. $g(x) = x - 5x^2 + 4$

28. $g(x) = 2 - x + 4x^2$

Find f(−1) for each polynomial function.

29. $f(x) = -3x^4 + 5x^3 + 6x - 2$

30. $f(x) = -5x^3 + 4x^2 - 7x + 9$

Find the specified function values.

31. Find $F(2)$ and $F(5)$: $F(x) = 2x^2 - 6x - 9$.

32. Find $P(4)$ and $P(0)$: $P(x) = 3x^2 - 2x + 7$.

33. Find $Q(-3)$ and $Q(0)$:
$$Q(y) = -8y^3 + 7y^2 - 4y - 9.$$

34. Find $G(3)$ and $G(-1)$:
$$G(x) = -6x^2 - 5x + x^3 - 1.$$

35. *Skateboarding.* The distance $s(t)$, in feet, traveled by an object falling freely from rest in t seconds is approximated by

$$s(t) = 16t^2.$$

In 2006, Danny Way set a record for a free-fall skateboard drop. Way was airborne for about 1.32 sec. How far did he drop?

Source: Based on data from *Guinness Book of World Records*

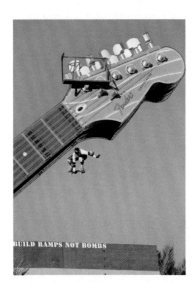

36. A paintbrush falls from a scaffold and takes 3 sec to hit the ground (see Exercise 35). How high is the scaffold?

Electing officers. *For a club consisting of p people, the number of ways N in which a president, vice president, and treasurer can be elected can be determined using the function given by*

$$N(p) = p^3 - 3p^2 + 2p.$$

37. The Southside Rugby Club has 20 members. In how many ways can they elect a president, vice president, and treasurer?

38. The Stage Right drama club has 12 members. In how many ways can a president, vice president, and treasurer be elected?

Horsepower. *The amount of horsepower needed to overcome air resistance by a car traveling v miles per hour can be approximated by the polynomial function given by*

$$h(v) = \frac{0.354}{8250}v^3.$$

Source: "The Physics of Racing," Brian Beckman, www.miata.net

39. How much horsepower does a race car traveling 180 mph need to overcome air resistance?

40. How much horsepower does a car traveling 65 mph need to overcome air resistance?

Wind energy. *The number of watts of power P generated by a particular turbine at a wind speed of x miles per hour can be approximated by the polynomial function given by*

$$P(x) = 0.0157x^3 + 0.1163x^2 - 1.3396x + 3.7063.$$

Use the following graph for Exercises 41– 44.

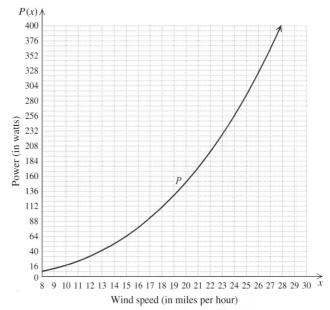

Source: Based on data from *QST*, November 2006

41. Estimate the power, in watts, generated by a 10-mph wind.

42. Estimate the power, in watts, generated by a 25-mph wind.

43. Approximate $P(20)$.

44. Approximate $P(15)$.

45. *Stacking spheres.* In 2004, the journal *Annals of Mathematics* accepted a proof of the so-called Kepler Conjecture: that the most efficient way to pack spheres is in the shape of a square pyramid.

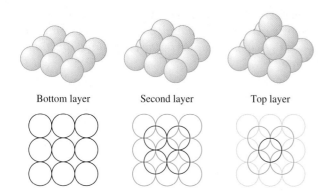

Bottom layer   Second layer   Top layer

The number *N* of balls in the stack is given by the polynomial function

$$N(x) = \tfrac{1}{3}x^3 + \tfrac{1}{2}x^2 + \tfrac{1}{6}x,$$

where *x* is the number of layers. Use both the function and the figure to find $N(3)$. Then calculate the number of oranges in a pyramid with 5 layers.
Source: *The New York Times* 4/6/04

46. *Stacking cannonballs.* The function in Exercise 45 was discovered by Thomas Harriot, assistant to Sir Walter Raleigh, when preparing for an expedition at sea. How many cannonballs did they pack if there were 10 layers to their pyramid?
Source: *The New York Times* 4/7/04

Veterinary science. *Gentamicin is an antibiotic frequently used by veterinarians. The concentration, in micrograms per milliliter (mcg/ mL), of Gentamicin in a horse's bloodstream t hours after injection can be approximated by the polynomial function*

$$C(t) = -0.005t^4 + 0.003t^3 + 0.35t^2 + 0.5t.$$

Use the following graph for Exercises 47–50.
Source: Michele Tulis, DVM, telephone interview

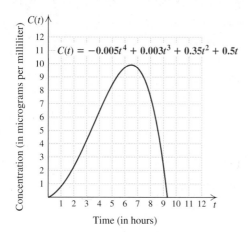

47. Estimate the concentration, in mcg/mL, of Gentamicin in the bloodstream 2 hr after injection.

48. Estimate the concentration, in mcg/mL, of Gentamicin in the bloodstream 4 hr after injection.

49. Approximate $C(2)$.

50. Approximate $C(4)$.

Surface area of a right circular cylinder. *The surface area of a right circular cylinder is given by the polynomial*

$$2\pi rh + 2\pi r^2,$$

where h is the height, r is the radius of the base, and h and r are given in the same units.

51. A 16-oz beverage can has height 6.3 in. and radius 1.2 in. Find the surface area of the can. (Use a calculator with a $\boxed{\pi}$ key or use 3.141592654 for π.)

52. A 12-oz beverage can has height 4.7 in. and radius 1.2 in. Find the surface area of the can. (Use a calculator with a $\boxed{\pi}$ key or use 3.141592654 for π.)

Total revenue. *An electronics firm is marketing a Blu-Ray disc player. The firm determines that when it sells x Blu-Ray players, its total revenue is*

$$R(x) = 280x - 0.4x^2 \text{ dollars.}$$

53. What is the total revenue from the sale of 75 Blu-Ray players?

54. What is the total revenue from the sale of 100 Blu-Ray players?

Total cost. *The electronics firm determines that the total cost, in dollars, of producing x Blu-Ray players is given by*

$$C(x) = 5000 + 0.6x^2.$$

55. What is the total cost of producing 75 Blu-Ray players?

56. What is the total cost of producing 100 Blu-Ray players?

Combine like terms to write an equivalent expression.

57. $8x + 2 - 5x + 3x^3 - 4x - 1$

58. $2a + 11 - 8a + 5a + 7a^2 + 9$

59. $3a^2b + 4b^2 - 9a^2b - 7b^2$

60. $5x^2y^2 + 4x^3 - 8x^2y^2 - 12x^3$

61. $9x^2 - 3xy + 12y^2 + x^2 - y^2 + 5xy + 4y^2$

62. $a^2 - 2ab + b^2 + 9a^2 + 5ab - 4b^2 + a^2$

Add.

63. $(5t^4 - 2t^3 + t) + (-t^4 - t^3 + 6t^2)$

64. $(3x^3 - 2x - 4) + (-5x^3 + x^2 - 10)$

65. $(x^2 + 2x - 3xy - 7) + (-3x^2 - x + 2y^2 + 6)$

66. $(3a^2 - 2b + a + 6) + (-a^2 + 5b - 5ab - 5)$

67. $(8x^2y - 3xy^2 + 4xy) + (-2x^2y - xy^2 + xy)$

68. $(9ab - 3ac + 5bc) + (13ab - 15ac - 8bc)$

69. $(2r^2 + 12r - 11) + (6r^2 - 2r + 4) + (r^2 - r - 2)$

70. $(5x^2 + 19x - 23) + (7x^2 - 2x + 1) + (-x^2 - 9x + 8)$

71. $\left(\frac{1}{8}xy - \frac{3}{5}x^3y^2 + 4.3y^3\right) + \left(-\frac{1}{3}xy - \frac{3}{4}x^3y^2 - 2.9y^3\right)$

72. $\left(\frac{2}{3}xy + \frac{5}{6}xy^2 + 5.1x^2y\right) + \left(-\frac{4}{5}xy + \frac{3}{4}xy^2 - 3.4x^2y\right)$

Write two expressions, one with parentheses and one without, for the opposite of each polynomial.

73. $3t^4 + 8t^2 - 7t - 1$

74. $-4x^5 - 3x^2 - x + 11$

75. $-12y^5 + 4ay^4 - 7by^2$

76. $7ax^3y^2 - 8by^4 - 7abx - 12ay$

Subtract.

77. $(4x - 6) - (-3x + 2)$

78. $(7y + 11) - (-7y - 4)$

79. $(-3x^2 + 2x + 9) - (x^2 + 5x - 4)$

80. $(-7y^2 + 5y + 6) - (4y^2 + 3y - 2)$

81. $(8a - 3b + c) - (2a + 3b - 4c)$

82. $(9r - 5s - t) - (7r - 5s + 3t)$

83. $(6a^2 + 5ab - 4b^2) - (8a^2 - 7ab + 3b^2)$

84. $(4y^2 - 13yz - 9z^2) - (9y^2 - 6yz + 3z^2)$

85. $(6ab - 4a^2b + 6ab^2) - (3ab^2 - 10ab - 12a^2b)$

86. $(10xy - 4x^2y^2 - 3y^3) - (-9x^2y^2 + 4y^3 - 7xy)$

87. $\left(\frac{5}{8}x^4 - \frac{1}{4}x^2 - \frac{1}{2}\right) - \left(-\frac{3}{8}x^4 + \frac{3}{4}x^2 + \frac{1}{2}\right)$

88. $\left(\frac{5}{6}y^4 - \frac{1}{2}y^2 - 7.8y\right) - \left(-\frac{3}{8}y^4 + \frac{3}{4}y^2 + 3.4y\right)$

Perform the indicated operations.

89. $(6t^2 + 7) - (2t^2 + 3) + (t^2 + t)$

90. $(9x^2 + 1) - (x^2 + 7) + (4x^2 - 3x)$

91. $(8r^2 - 6r) - (2r - 6) + (5r^2 - 7)$

92. $(7s^2 - 5s) - (4s - 1) + (3s^2 - 5)$

Aha! **93.** $(x^2 - 4x + 7) + (3x^2 - 9) - (x^2 - 4x + 7)$

94. $(t^2 - 5t + 6) + (5t - 8) - (t^2 + 3t - 4)$

Total profit. Total profit is defined as total revenue minus total cost. In Exercises 95 and 96, let $R(x)$ and $C(x)$ represent the revenue and the cost in dollars, respectively, from the sale of x bookcases.

95. If $R(x) = 280x - 0.4x^2$ and $C(x) = 5000 + 0.6x^2$, find the profit from the sale of 70 bookcases.

96. If $R(x) = 280x - 0.7x^2$ and $C(x) = 8000 + 0.5x^2$, find the profit from the sale of 100 bookcases.

97. Is the sum of two binomials always a binomial? Why or why not?

98. Ani claims that she can add any two polynomials but finds subtraction difficult. What advice would you offer her?

Skill Review

To prepare for Section 5.2, review working with exponents (Section 1.6).

Simplify. [1.6]

99. $x^5 \cdot x^3$

100. $y^2 \cdot y^8$

101. $(a^2b^3)(a^4b)$

102. $(t^4)^2$

103. $(5y^3)^2$

104. $(2x^5y)^2$

Synthesis

105. For $P(x)$ as given in Exercises 41–44, calculate

$$\frac{P(20) - P(10)}{20 - 10}.$$

Explain what this number represents graphically and what meaning it has in the application.

106. For $C(t)$ as given in Exercises 47–50, calculate

$$\frac{C(9) - C(6)}{9 - 6}.$$

Explain what this number represents graphically and what meaning it has in the application.

For $P(x)$ and $Q(x)$ as given, find the following.

$P(x) = 13x^5 - 22x^4 - 36x^3 + 40x^2 - 16x + 75,$
$Q(x) = 42x^5 - 37x^4 + 50x^3 - 28x^2 + 34x + 100$

107. $2[P(x)] + Q(x)$

108. $3[P(x)] - Q(x)$

109. $2[Q(x)] - 3[P(x)]$

110. $4[P(x)] + 3[Q(x)]$

111. *Volume of a display.* The number of spheres in a triangular pyramid with x layers is given by the function

$$N(x) = \tfrac{1}{6}x^3 + \tfrac{1}{2}x^2 + \tfrac{1}{3}x.$$

The volume of a sphere of radius r is given by the function

$$V(r) = \tfrac{4}{3}\pi r^3,$$

where π can be approximated as 3.14.

Greta's Chocolate has a window display of truffles piled in a triangular pyramid formation 5 layers deep. If the diameter of each truffle is 3 cm, find the volume of chocolate in the display.

112. If one large truffle were to have the same volume as the display of truffles in Exercise 111, what would be its diameter?

113. Find a polynomial function that gives the outside surface area of the box shown, with an open top and dimensions as shown.

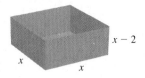

114. Develop a formula for the surface area of a right circular cylinder in which h is the height, in *centimeters*, and r is the radius, in *meters*. (See Exercises 51 and 52.)

Perform the indicated operation. Assume that the exponents are natural numbers.

115. $(2x^{2a} + 4x^a + 3) + (6x^{2a} + 3x^a + 4)$

116. $(3x^{6a} - 5x^{5a} + 4x^{3a} + 8) - (2x^{6a} + 4x^{4a} + 3x^{3a} + 2x^{2a})$

117. $(2x^{5b} + 4x^{4b} + 3x^{3b} + 8) - (x^{5b} + 2x^{3b} + 6x^{2b} + 9x^b + 8)$

118. Use a graphing calculator to check your answers to Exercises 29, 37, and 57.

119. Use a graphing calculator to check your answers to Exercises 32, 38, and 58.

120. A student who is trying to graph

$$p(x) = 0.05x^4 - x^2 + 5$$

gets the following screen. How can the student tell at a glance that a mistake has been made?

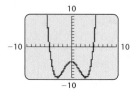

COLLABORATIVE

CORNER

How Many Handshakes?

Focus: Polynomial functions
Time: 20 minutes
Group size: 5

ACTIVITY

1. All group members should shake hands with each other. Without "double counting," determine how many handshakes occurred.
2. Complete the table in the next column.
3. Join another group to determine the number of handshakes for a group of size 10.
4. Try to find a function of the form $H(n) = an^2 + bn$, for which $H(n)$ is the number of different handshakes that are possible in a group of n people. Make sure

Group Size	Number of Handshakes
1	
2	
3	
4	
5	

that $H(n)$ produces all of the values in the table above. (*Hint*: Use the table to twice select n and $H(n)$. Then solve the resulting system of equations for a and b.)

5.2   Multiplication of Polynomials

Multiplying Monomials • Multiplying Monomials and Binomials • Multiplying Any Two Polynomials • The Product of Two Binomials: FOIL • Squares of Binomials • Products of Sums and Differences • Function Notation

Just like numbers, polynomials can be multiplied. We begin by finding products of monomials.

Multiplying Monomials

To multiply two monomials, we multiply their coefficients and we multiply their variables. To do so, we use the rules for exponents and the commutative and associative laws. With practice, we can work mentally, writing only the answer.

EXAMPLE 1

Multiply and simplify: **(a)** $(-8x^4y^7)(5x^3y^2)$; **(b)** $(-3a^5bc^6)(-4a^2b^5c^8)$.

SOLUTION

a) $(-8x^4y^7)(5x^3y^2) = -8 \cdot 5 \cdot x^4 \cdot x^3 \cdot y^7 \cdot y^2$ Using the associative and commutative laws

$= -40x^{4+3}y^{7+2}$ Multiplying coefficients; adding exponents

$= -40x^7y^9$

b) $(-3a^5bc^6)(-4a^2b^5c^8) = (-3)(-4) \cdot a^5 \cdot a^2 \cdot b \cdot b^5 \cdot c^6 \cdot c^8$

$= 12a^7b^6c^{14}$ Multiplying coefficients; adding exponents

STUDENT NOTES

If the meaning of a word is unclear to you, take time to look it up before continuing your reading. In Example 1, the word "coefficient" appears. This was defined in Section 5.1. The *coefficient* of $-8x^4y^7$ is -8.

TRY EXERCISE 11

Multiplying Monomials and Binomials

The distributive law is required when we are multiplying polynomials other than two monomials.

EXAMPLE **2**   Multiply: **(a)** $2t(3t - 5)$; **(b)** $3a^2b(a^2 - b^2)$.

SOLUTION

a) $2t(3t - 5) = 2t \cdot 3t - 2t \cdot 5$ Using the distributive law

$\qquad = 6t^2 - 10t$ Multiplying monomials

b) $3a^2b(a^2 - b^2) = 3a^2b \cdot a^2 - 3a^2b \cdot b^2$ Using the distributive law

$\qquad = 3a^4b - 3a^2b^3$

TRY EXERCISE 15

The distributive law is also used for multiplying two binomials. In this case, however, we begin by distributing a *binomial* rather than a monomial. With practice, some of the following steps can be combined.

EXAMPLE **3**   Multiply: $(y^3 - 5)(2y^3 + 4)$.

SOLUTION

$(y^3 - 5)\,(2y^3 + 4) = (y^3 - 5)\,2y^3 + (y^3 - 5)\,4$ "Distributing" the $y^3 - 5$

$\qquad = 2y^3(y^3 - 5) + 4(y^3 - 5)$ Using the commutative law for multiplication. Try to do this step mentally.

$\qquad = 2y^3 \cdot y^3 - 2y^3 \cdot 5 + 4 \cdot y^3 - 4 \cdot 5$ Using the distributive law (twice)

$\qquad = 2y^6 - 10y^3 + 4y^3 - 20$ Multiplying the monomials

$\qquad = 2y^6 - 6y^3 - 20$ Combining like terms

TRY EXERCISE 21

Multiplying Any Two Polynomials

Repeated use of the distributive law enables us to multiply *any* two polynomials, regardless of how many terms are in each.

EXAMPLE **4**   Multiply: $(p + 2)(p^4 - 2p^3 + 3)$.

SOLUTION   We can use the distributive law from right to left if we wish:

$(p + 2)\,(p^4 - 2p^3 + 3) = p\,(p^4 - 2p^3 + 3) + 2\,(p^4 - 2p^3 + 3)$

$\qquad = p \cdot p^4 - p \cdot 2p^3 + p \cdot 3 + 2 \cdot p^4 - 2 \cdot 2p^3 + 2 \cdot 3$

$\qquad = p^5 - 2p^4 + 3p + 2p^4 - 4p^3 + 6$

$\qquad = p^5 - 4p^3 + 3p + 6.$ Combining like terms

TRY EXERCISE 23

The Product of Two Polynomials

The *product* of two polynomials P and Q is found by multiplying each term of P by every term of Q and then combining like terms.

It is also possible to stack the polynomials, multiplying each term at the top by every term below, keeping like terms in columns, and leaving spaces for missing terms. Then we add just as we do in long multiplication with numbers.

EXAMPLE 5   Multiply: $(5x^3 + x - 4)(-2x^2 + 3x + 6)$.

SOLUTION

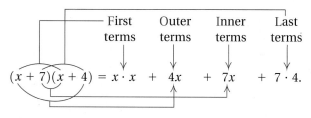

$$
\begin{array}{r}
5x^3 + x - 4 \\
-2x^2 + 3x + 6 \\
\hline
30x^3 + 6x - 24 \\
15x^4 + 3x^2 - 12x \\
-10x^5 - 2x^3 + 8x^2 \\
\hline
-10x^5 + 15x^4 + 28x^3 + 11x^2 - 6x - 24
\end{array}
$$

Multiplying by 6
Multiplying by $3x$
Multiplying by $-2x^2$
Adding

TRY EXERCISE   27

The Product of Two Binomials: FOIL

We now consider what are called *special products*. These products of polynomials occur often and can be simplified using shortcuts that we now develop.

To find a special-product rule for the product of any two binomials, consider $(x + 7)(x + 4)$. We multiply each term of $(x + 7)$ by each term of $(x + 4)$:

$$(x + 7)(x + 4) = x \cdot x + x \cdot 4 + 7 \cdot x + 7 \cdot 4.$$

This multiplication illustrates a pattern that occurs anytime two binomials are multiplied:

A visualization of $(x + 7)(x + 4)$ using areas

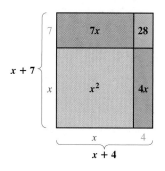

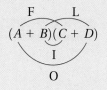

$$
\underset{\substack{\text{First}\\\text{terms}}}{} \quad \underset{\substack{\text{Outer}\\\text{terms}}}{} \quad \underset{\substack{\text{Inner}\\\text{terms}}}{} \quad \underset{\substack{\text{Last}\\\text{terms}}}{}
$$

$$(x + 7)(x + 4) = x \cdot x \;+\; 4x \;+\; 7x \;+\; 7 \cdot 4.$$

We use the mnemonic device FOIL to remember this method for multiplying.

The FOIL Method

To multiply two binomials $A + B$ and $C + D$, multiply the First terms AC, the Outer terms AD, the Inner terms BC, and then the Last terms BD. Then combine like terms, if possible.

$$(A + B)(C + D) = AC + AD + BC + BD$$

1. Multiply First terms: AC.
2. Multiply Outer terms: AD.
3. Multiply Inner terms: BC.
4. Multiply Last terms: BD.

$$\downarrow$$
FOIL

$$
\begin{array}{c}
\text{F} \qquad\quad \text{L} \\
(A + B)(C + D) \\
\text{I} \\
\text{O}
\end{array}
$$

EXAMPLE 6   Multiply.

a) $(x + 5)(x - 8)$ **b)** $(2x + 3y)(x - 4y)$ **c)** $(t + 2)(t - 4)(t + 5)$

SOLUTION

$$\quad\quad\quad\quad\quad \text{F} \quad\quad \text{O} \quad\quad \text{I} \quad\quad \text{L}$$

a) $(x + 5)(x - 8) = x^2 - 8x + 5x - 40$

$\quad\quad\quad\quad\quad\quad\quad\quad = x^2 - 3x - 40 \quad$ Combining like terms

b) $(2x + 3y)(x - 4y) = 2x^2 - 8xy + 3xy - 12y^2 \quad$ Using FOIL

$\quad\quad\quad\quad\quad\quad\quad\quad\quad = 2x^2 - 5xy - 12y^2 \quad\quad\quad$ Combining like terms

c) $(t + 2)(t - 4)(t + 5) = (t^2 - 4t + 2t - 8)(t + 5) \quad$ Using FOIL

$\quad\quad\quad\quad\quad\quad\quad\quad = (t^2 - 2t - 8)(t + 5)$

$\quad\quad\quad\quad\quad\quad\quad\quad = (t^2 - 2t - 8) \cdot t + (t^2 - 2t - 8) \cdot 5 \quad$ Using the distributive law

$\quad\quad\quad\quad\quad\quad\quad\quad = t^3 - 2t^2 - 8t + 5t^2 - 10t - 40$

$\quad\quad\quad\quad\quad\quad\quad\quad = t^3 + 3t^2 - 18t - 40 \quad$ Combining like terms

TRY EXERCISE 33

Squares of Binomials

A fast method for squaring any binomial can be developed using FOIL:

$$(A + B)^2 = (A + B)(A + B)$$

$\quad\quad\quad\quad = A^2 + AB + AB + B^2 \quad$ Note that AB occurs twice.

$\quad\quad\quad\quad = A^2 + 2AB + B^2;$

$$(A - B)^2 = (A - B)(A - B)$$

$\quad\quad\quad\quad = A^2 - AB - AB + B^2 \quad$ Note that $-AB$ occurs twice.

$\quad\quad\quad\quad = A^2 - 2AB + B^2.$

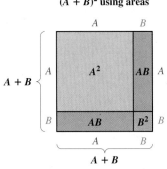

A visualization of
$(A + B)^2$ **using areas**

Squaring a Binomial

$$(A + B)^2 = A^2 + 2AB + B^2;$$
$$(A - B)^2 = A^2 - 2AB + B^2$$

The square of a binomial is the square of the first term, plus twice the product of the two terms, plus the square of the last term.

Trinomials that can be written in the form $A^2 + 2AB + B^2$ or $A^2 - 2AB + B^2$ are called *perfect-square trinomials.*

It can help to remember the words of the rules and say them while multiplying.

EXAMPLE 7   Multiply: **(a)** $(y - 5)^2$; **(b)** $(2x + 3y)^2$; **(c)** $\left(\frac{1}{2}x - 3y^4\right)^2$.

SOLUTION

$$(A - B)^2 = A^2 - 2 \cdot A \cdot B + B^2$$

$$\quad\quad\quad\downarrow\ \downarrow\quad\quad\downarrow\quad\ \downarrow\ \downarrow\ \downarrow\quad\quad\downarrow$$

a) $(y - 5)^2 = y^2 - 2 \cdot y \cdot 5 + 5^2 \quad$ Note that $-2 \cdot y \cdot 5$ is twice the product of y and -5.

$\quad\quad\quad\quad = y^2 - 10y + 25 \quad$ The square of a binomial is always a trinomial.

b) $(2x + 3y)^2 = (2x)^2 + 2 \cdot 2x \cdot 3y + (3y)^2$

$\qquad = 4x^2 + 12xy + 9y^2$ Raising a product to a power

c) $\left(\frac{1}{2}x - 3y^4\right)^2 = \left(\frac{1}{2}x\right)^2 - 2 \cdot \frac{1}{2}x \cdot 3y^4 + (3y^4)^2$ $2 \cdot \frac{1}{2}x \cdot (-3y^4) = -2 \cdot \frac{1}{2}x \cdot 3y^4$

$\qquad = \frac{1}{4}x^2 - 3xy^4 + 9y^8$ Raising a product to a power; multiplying exponents

TRY EXERCISE 43

> ***CAUTION!*** Note that $(y - 5)^2 \neq y^2 - 5^2$. (To see this, replace y with 6 and note that $(6 - 5)^2 = 1^2 = 1$ and $6^2 - 5^2 = 36 - 25 = 11$.) More generally,
>
> $$(A + B)^2 \neq A^2 + B^2 \quad \text{and} \quad (A - B)^2 \neq A^2 - B^2.$$

Products of Sums and Differences

Another pattern emerges when we are multiplying a sum and a difference of the same two terms. Note the following:

$$\overset{F \quad\ O \quad\ I \quad\ L}{(A + B)(A - B) = A^2 - AB + AB - B^2}$$

$$= A^2 - B^2. \quad -AB + AB = 0$$

The Product of a Sum and a Difference

$$(A + B)(A - B) = A^2 - B^2 \quad \text{This is called a } \textit{difference of two squares.}$$

The product of the sum and the difference of the same two terms is the square of the first term minus the square of the second term.

EXAMPLE 8   Multiply.

a) $(t + 5)(t - 5)$ 　　　　　　　　**b)** $(2xy^2 + 3x)(2xy^2 - 3x)$

c) $(0.2t - 1.4m)(0.2t + 1.4m)$ 　　**d)** $\left(\frac{2}{3}n - m^3\right)\left(\frac{2}{3}n + m^3\right)$

SOLUTION

$$(A + B)(A - B) = A^2 - B^2$$

a) $(t + 5)(t - 5) = t^2 - 5^2$ Replacing A with t and B with 5

$\qquad\qquad\quad = t^2 - 25$ Try to do problems like this in one step.

b) $(2xy^2 + 3x)(2xy^2 - 3x) = (2xy^2)^2 - (3x)^2$

$\qquad\qquad\qquad\qquad\qquad = 4x^2y^4 - 9x^2$ Raising a product to a power

c) $(0.2t - 1.4m)(0.2t + 1.4m) = (0.2t)^2 - (1.4m)^2$

$\qquad\qquad\qquad\qquad\qquad = 0.04t^2 - 1.96m^2$

d) $\left(\frac{2}{3}n - m^3\right)\left(\frac{2}{3}n + m^3\right) = \left(\frac{2}{3}n\right)^2 - (m^3)^2$

$\qquad\qquad\qquad\qquad\qquad = \frac{4}{9}n^2 - m^6$

TRY EXERCISE 59

TECHNOLOGY CONNECTION

One way to check problems like Example 8(a) is to use a Table, much as in the Technology Connection on p. 282. Another check is to note that if the multiplication is correct, then $(t + 5)(t - 5) = t^2 - 25$ is an identity and $t^2 - 25 - (t + 5)(t - 5)$ must be 0. In the window below, we set the MODE to G-T so that we can view both a graph and a table. We use a heavy line to distinguish the graph from the x-axis.

$y_1 = x^2 - 25 - (x + 5)(x - 5)$

X	Y1
-2	0
-1	0
0	0
1	0
2	0
3	0
4	0
X = -2	

Had we found $y_1 \neq 0$, we would have known that a mistake had been made.

1. Use this procedure to show that $(x - 3)(x + 3) = x^2 - 9$.
2. Use this procedure to show that $(t - 4)^2 = t^2 - 8t + 16$.
3. Show that the graphs of $y_1 = x^2 - 4$ and $y_2 = (x + 2)(x - 2)$ coincide, using the Sequential MODE with a heavier-weight line for y_2. Then, use the Y-VARS option of the VARS key to key in $y_3 = y_2 - y_1$. What do you expect the graph of y_3 to look like?

EXAMPLE 9   Multiply and simplify.

a) $(5y + 4 + 3x)(5y + 4 - 3x)$

b) $(3xy^2 + 4y)(-3xy^2 + 4y)$

c) $(2t + 3)^2 - (t - 1)(t + 1)$

SOLUTION

a) By far the easiest way to multiply $(5y + 4 + 3x)(5y + 4 - 3x)$ is to note that it is in the form $(A + B)(A - B)$:

$$(5y + 4 + 3x)(5y + 4 - 3x) = (5y + 4)^2 - (3x)^2 \qquad \text{Try to be alert for situations like this.}$$

$$= 25y^2 + 40y + 16 - 9x^2$$

We can also multiply $(5y + 4 + 3x)(5y + 4 - 3x)$ using columns, but not as quickly.

b) $(3xy^2 + 4y)(-3xy^2 + 4y) = (4y + 3xy^2)(4y - 3xy^2) \qquad \text{Using a commutative law}$

$$= (4y)^2 - (3xy^2)^2$$

$$= 16y^2 - 9x^2y^4$$

c) $(2t + 3)^2 - (t - 1)(t + 1) = 4t^2 + 12t + 9 - (t^2 - 1) \qquad \begin{array}{l}\text{Squaring a} \\ \text{binomial;} \\ \text{simplifying} \\ (t - 1)(t + 1)\end{array}$

$$= 4t^2 + 12t + 9 - t^2 + 1 \qquad \text{Subtracting}$$

$$= 3t^2 + 12t + 10 \qquad \begin{array}{l}\text{Combining like} \\ \text{terms}\end{array}$$

TRY EXERCISES 69 and 75

STUDENT NOTES

To remember the special products, look for differences between the rules. When we are squaring a binomial, after combining like terms, we have a trinomial.

In the product of a sum and a difference, the binomials are not the same; one is a sum and the other a difference. After combining like terms, we have a binomial.

$(A + B)(A + B) = A^2 + 2AB + B^2;$

$(A - B)(A - B) = A^2 - 2AB + B^2;$

$(A + B)(A - B) = A^2 - B^2$

Function Notation

Let's stop for a moment and look back at what we have done in this section. We have shown, for example, that

$$(x - 2)(x + 2) = x^2 - 4,$$

that is, $x^2 - 4$ and $(x - 2)(x + 2)$ are equivalent expressions.

From the viewpoint of functions, if we have

$$f(x) = x^2 - 4$$

and

$$g(x) = (x - 2)(x + 2),$$

then for any given input x, the outputs $f(x)$ and $g(x)$ are identical. Thus the graphs of these functions are identical and we say that f and g represent the same function. Functions like these are graphed in detail in Chapter 8.

x	$f(x)$	$g(x)$
3	5	5
2	0	0
1	-3	-3
0	-4	-4
-1	-3	-3
-2	0	0
-3	5	5

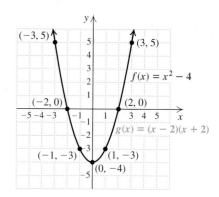

Our work with multiplying can be used when evaluating functions.

EXAMPLE 10   Given $f(x) = x^2 - 4x + 5$, find and simplify each of the following.

a) $f(a) + 3$ **b)** $f(a + 3)$ **c)** $f(a + h) - f(a)$

SOLUTION

a) To find $f(a) + 3$, we replace x with a to find $f(a)$. Then we add 3 to the result:

$$f(a) + 3 = a^2 - 4a + 5 + 3 \quad \text{Evaluating } f(a)$$
$$= a^2 - 4a + 8.$$

b) To find $f(a + 3)$, we replace x with $a + 3$. Then we simplify:

$$f(a + 3) = (a + 3)^2 - 4(a + 3) + 5$$
$$= a^2 + 6a + 9 - 4a - 12 + 5$$
$$= a^2 + 2a + 2.$$

> *CAUTION!*   Note from parts (a) and (b) that, in general,
> $$f(a + 3) \neq f(a) + 3.$$

c) To find $f(a + h)$ and $f(a)$, we replace x with $a + h$ and a, respectively:

$$f(a + h) - f(a) = [(a + h)^2 - 4(a + h) + 5] - [a^2 - 4a + 5]$$
$$= [a^2 + 2ah + h^2 - 4a - 4h + 5] - [a^2 - 4a + 5]$$
$$= a^2 + 2ah + h^2 - 4a - 4h + 5 - a^2 + 4a - 5$$
$$= 2ah + h^2 - 4h.$$

TRY EXERCISE 79

5.2 **EXERCISE SET**

Concept Reinforcement *Classify each statement as either true or false.*

1. The coefficient of $3x^5$ is 5.

2. The product of two monomials is a monomial.

3. The product of a monomial and a binomial is found using the distributive law.

4. To simplify the product of two binomials, we often need to combine like terms.

5. FOIL can be used whenever two monomials are multiplied.

6. The square of a binomial is a difference of two squares.

7. The product of the sum and the difference of the same two terms is a binomial.

8. In general, $f(a + 5) \neq f(a) + 5$.

Multiply.

9. $3x^4 \cdot 5x$

10. $-2x^3 \cdot 4x$

11. $6a^2(-8ab^2)$

12. $-3uv^2(5u^2v^2)$

13. $(-4x^3y^2)(-9x^2y^4)$

14. $(-7a^2bc^4)(-8ab^3c^2)$

15. $7x(3 - x)$

16. $3a(a^2 - 4a)$

17. $5cd(4c^2d - 5cd^2)$

18. $a^2(2a^2 - 5a^3)$

19. $(x + 3)(x + 5)$

20. $(t - 1)(t - 4)$

21. $(2a + 3)(4a - 1)$

22. $(3r - 4)(2r + 1)$

23. $(x + 2)(x^2 - 3x + 1)$

24. $(a + 3)(a^2 - 4a + 2)$

25. $(t - 5)(t^2 + 2t - 3)$

26. $(x - 4)(x^2 + x - 7)$

27. $(a^2 + a - 1)(a^2 + 4a - 5)$

28. $(x^2 - 2x + 1)(x^2 + x + 2)$

29. $(x + 3)(x^2 - 3x + 9)$

30. $(y + 4)(y^2 - 4y + 16)$

31. $(a - b)(a^2 + ab + b^2)$

32. $(x - y)(x^2 + xy + y^2)$

33. $(t - 3)(t + 2)$

34. $(x + 6)(x - 1)$

35. $(5x + 2y)(4x + y)$

36. $(3t + 2)(2t + 7)$

37. $\left(t - \frac{1}{3}\right)\left(t - \frac{1}{4}\right)$

38. $\left(x - \frac{1}{2}\right)\left(x - \frac{1}{5}\right)$

39. $(1.2t + 3s)(2.5t - 5s)$

40. $(30a - 0.5b)(0.2a + 10b)$

41. $(r + 3)(r + 2)(r - 1)$

42. $(t + 4)(t + 1)(t - 2)$

43. $(x + 5)^2$

44. $(t + 6)^2$

45. $(2y - 7)^2$

46. $(3x - 4)^2$

47. $(5c - 2d)^2$

48. $(8x - 3y)^2$

49. $(3a^3 - 10b^2)^2$

50. $(3s^2 + 4t^3)^2$

51. $(x^3y^4 + 5)^2$

52. $(a^4b^2 - 3)^2$

53. Let $P(x) = 3x^2 - 5$ and $Q(x) = 4x^2 - 7x + 1$. Find $P(x) \cdot Q(x)$.

54. Let $P(x) = x^2 - x + 1$ and $Q(x) = x^3 + x^2 + 5$. Find $P(x) \cdot Q(x)$.

55. Let $P(x) = 5x - 2$. Find $P(x) \cdot P(x)$.

56. Let $Q(x) = 3x^2 + 1$. Find $Q(x) \cdot Q(x)$.

57. Let $F(x) = 2x - \frac{1}{3}$. Find $[F(x)]^2$.

58. Let $G(x) = 5x - \frac{1}{2}$. Find $[G(x)]^2$.

Multiply and, if possible, simplify.

59. $(c + 7)(c - 7)$

60. $(x - 3)(x + 3)$

61. $(1 - 4x)(1 + 4x)$

62. $(5 + 2y)(5 - 2y)$

63. $\left(3m - \frac{1}{2}n\right)\left(3m + \frac{1}{2}n\right)$

64. $(0.4c - 0.5d)(0.4c + 0.5d)$

65. $(x^3 + yz)(x^3 - yz)$

66. $(2a^4 + ab)(2a^4 - ab)$

67. $(-mn + 3m^2)(mn + 3m^2)$

68. $(-6u + v^2)(6u + v^2)$

69. $(x + 7)^2 - (x + 3)(x - 3)$

70. $(t + 5)^2 - (t - 4)(t + 4)$

71. $(2m - n)(2m + n) - (m - 2n)^2$

72. $(3x + y)(3x - y) - (2x + y)^2$

Aha! **73.** $(a + b + 1)(a + b - 1)$

74. $(m + n + 2)(m + n - 2)$

75. $(2x + 3y + 4)(2x + 3y - 4)$

76. $(3a - 2b + c)(3a - 2b - c)$

77. *Compounding interest.* Suppose that P dollars is invested in a savings account at interest rate r, compounded annually, for 2 yr. The amount A in the account after 2 yr is given by

$$A = P(1 + r)^2,$$

where r is in decimal form. Find an equivalent expression for A.

78. *Compounding interest.* Suppose that P dollars is invested in a savings account at interest rate r, compounded semiannually, for 1 yr. The amount A in the account after 1 yr is given by

$$A = P\left(1 + \frac{r}{2}\right)^2,$$

where r is in decimal form. Find an equivalent expression for A.

79. Given $f(x) = x^2 + 5$, find and simplify.
   a) $f(t - 1)$
   b) $f(a + h) - f(a)$
   c) $f(a) - f(a - h)$

80. Given $f(x) = x^2 + 7$, find and simplify.
   a) $f(p + 1)$
   b) $f(a + h) - f(a)$
   c) $f(a) - f(a - h)$

81. Given $f(x) = x^2 + x$, find and simplify each of the following.
   a) $f(a) + f(-a)$
   b) $f(a + h)$
   c) $f(a + h) - f(a)$

82. Given $f(x) = x^2 - x$, find and simplify each of the following.
   a) $f(a) - f(-a)$
   b) $f(a + h)$
   c) $f(a + h) - f(a)$

83. Find two binomials whose product is $x^2 - 25$ and explain how you decided on those two binomials.

84. Find two binomials whose product is $x^2 - 6x + 9$ and explain how you decided on those two binomials.

Skill Review

To prepare for Section 5.3, review factoring using the distributive law (Section 1.2).

Find an equivalent expression by factoring. [1.2]

85. $5x + 15y - 5$

86. $8x + 8 - 40y$

87. $16t - 64$

88. $18a + 30b$

89. $ax + bx - cx$

90. $bx + by - b$

Synthesis

91. We have seen that $(a - b)(a + b) = a^2 - b^2$. Explain how this result can be used to develop a fast way of calculating $95 \cdot 105$.

92. A student incorrectly claims that since $2x^2 \cdot 2x^2 = 4x^4$, it follows that $5x^5 \cdot 5x^5 = 25x^{25}$. How could you convince the student that a mistake has been made?

Multiply. Assume that variables in exponents represent natural numbers.

93. $(x^2 + y^n)(x^2 - y^n)$

94. $(a^n + b^n)^2$

95. $x^2 y^3 (5x^n + 4y^n)$

96. $a^n b^m (7a^2 - 3b^3)$

97. $(x^n - 4)(x^{2n} + 3x^n - 2)$

98. $(t^n + 3)(t^{2n} - 2t^n + 1)$

Aha! **99.** $(a - b + c - d)(a + b + c + d)$

100. $[(a + b)(a - b)][5 - (a + b)][5 + (a + b)]$

101. $(x^2 - 3x + 5)(x^2 + 3x + 5)$

102. $\left(\frac{2}{3}x + \frac{1}{3}y + 1\right)\left(\frac{2}{3}x - \frac{1}{3}y - 1\right)$

103. $(x - 1)(x^2 + x + 1)(x^3 + 1)$

104. $(x^a + y^b)(x^a - y^b)(x^{2a} + y^{2b})$

105. $(x^{a-b})^{a+b}$

Aha! **106.** $(M^{x+y})^{x+y}$

107. $(x - a)(x - b)(x - c) \cdots (x - z)$

108. Given $f(x) = x^2 + 7$, find and simplify
$$\frac{f(a + h) - f(a)}{h}.$$

109. Given $g(x) = x^2 - 9$, find and simplify

$$\frac{g(a + h) - g(a)}{h}.$$

110. Draw rectangles similar to those on p. 291 to show that $(x + 2)(x + 5) = x^2 + 7x + 10$.

111. Draw rectangles similar to those on p. 292 to show that $(A - B)^2 = A^2 - 2AB + B^2$.

112. Use a graphing calculator to check your answers to Exercises 15, 33, and 57.

113. Use a graphing calculator to determine which of the following is an identity.

a) $(x - 1)^2 = x^2 - 1$
b) $(x - 2)(x + 3) = x^2 + x - 6$
c) $(x - 1)^3 = x^3 - 3x^2 + 3x - 1$
d) $(x + 1)^4 = x^4 + 1$
e) $(x + 1)^4 = x^4 + 4x^3 + 8x^2 + 4x + 1$

COLLABORATIVE CORNER

Algebra and Number Tricks

Focus: Polynomial multiplication

Time: 15–20 minutes

Group size: 2

Consider the following dialogue:

Jinny: Cal, let me do a number trick with you. Think of a number between 1 and 7. I'll have you perform some manipulations to this number, you'll tell me the result, and I'll tell you your number.

Cal: Okay. I've thought of a number.

Jinny: Good. Write it down so I can't see it, double it, and then subtract x from the result.

Cal: Hey, this is algebra!

Jinny: I know. Now square your binomial and subtract x^2.

Cal: How did you know I had an x^2? Is this rigged?

Jinny: It is. Now, divide by 4 and tell me either your constant term or your x-term. I'll tell you the other term and the number you chose.

Cal: Okay. The constant term is 16.

Jinny: Then the other term is $-4x$ and the number you chose is 4.

Cal: You're right! How did you do it?

ACTIVITY

1. Each group member should follow Jinny's instructions. Then determine how Jinny determined Cal's number and the other term.

2. Suppose that, at the end, Cal told Jinny the x-term. How would Jinny have determined Cal's number and the other term?

3. Would Jinny's "trick" work with *any* real number? Why do you think she specified numbers between 1 and 7?

4. Each group member should create a new number "trick" and perform it on the other group member. Be sure to include a variable to provide practice with polynomials.

<table>
<tr><td>**5.3**</td><td colspan="2">## Common Factors and Factoring by Grouping</td></tr>
<tr><td></td><td>Terms with Common Factors</td><td>•   Factoring by Grouping</td></tr>
</table>

Factoring is the reverse of multiplication. To **factor** an expression means to write an equivalent expression that is a product. We will use factoring when working with polynomial functions and solving polynomial equations later in this chapter.

Terms with Common Factors

When factoring a polynomial, we always look for a factor common to every term. If one exists, we then use the distributive law.

EXAMPLE   1   Factor out a common factor: $6y^2 - 18$.

SOLUTION   We have

$$6y^2 - 18 = 6 \cdot y^2 - 6 \cdot 3 \qquad \text{Noting that 6 is a common factor}$$
$$= 6(y^2 - 3). \qquad \text{Using the distributive law}$$

Check: $6(y^2 - 3) = 6y^2 - 18$.

TRY EXERCISE 9

Suppose in Example 1 that the common factor 2 were used:

$$6y^2 - 18 = 2 \cdot 3y^2 - 2 \cdot 9 \qquad \text{2 is a common factor.}$$
$$= 2(3y^2 - 9). \qquad \text{Using the distributive law}$$

Note that $3y^2 - 9$ itself has a common factor, 3. It is standard practice to factor out the *largest*, or *greatest, common factor*, so that the resulting polynomial factor cannot be factored any further. Thus, by now factoring out 3, we can complete the factorization:

$$6y^2 - 18 = 2(3y^2 - 9)$$
$$= 2 \cdot 3(y^2 - 3) = 6(y^2 - 3). \qquad \begin{array}{l}\text{Remember to multiply}\\\text{the two common factors:}\\2 \cdot 3 = 6.\end{array}$$

To determine the greatest common factor of a polynomial, we multiply the greatest common factor of the coefficients by the greatest common factor of the variables appearing in every term. Thus, to find the greatest common factor of $30x^4 + 20x^5$, we multiply the greatest common factor of 30 and 20, which is 10, by the greatest common factor of x^4 and x^5, which is x^4:

$$30x^4 + 20x^5 = 10 \cdot 3 \cdot x^4 + 10 \cdot 2 \cdot x^4 \cdot x$$
$$= 10x^4(3 + 2x). \qquad \text{The greatest common factor is } 10x^4.$$

EXAMPLE   2   Write an expression equivalent to $8p^6q^2 - 4p^5q^3 + 10p^4q^4$ by factoring out the greatest common factor.

SOLUTION   First, we look for the greatest positive common factor of the coefficients
of $8p^6q^2 - 4p^5q^3 + 10p^4q^4$:

$$8, \ -4, \ 10 \ \longrightarrow \ \text{Greatest common factor} = 2.$$

Second, we look for the greatest common factor of the powers of p:

$$p^6, \ p^5, \ p^4 \ \longrightarrow \ \text{Greatest common factor} = p^4.$$

Third, we look for the greatest common factor of the powers of q:

$$q^2, \ q^3, \ q^4 \ \longrightarrow \ \text{Greatest common factor} = q^2.$$

Thus, $2p^4q^2$ is the greatest common factor of the given polynomial. Then

$$8p^6q^2 - 4p^5q^3 + 10p^4q^4 = 2p^4q^2 \cdot 4p^2 - 2p^4q^2 \cdot 2pq + 2p^4q^2 \cdot 5q^2$$
$$= 2p^4q^2(4p^2 - 2pq + 5q^2).$$

As a final check, note that

$$2p^4q^2(4p^2 - 2pq + 5q^2) = 2p^4q^2 \cdot 4p^2 - 2p^4q^2 \cdot 2pq + 2p^4q^2 \cdot 5q^2$$
$$= 8p^6q^2 - 4p^5q^3 + 10p^4q^4,$$

which is the original polynomial. Since $4p^2 - 2pq + 5q^2$ has no common factor,
we know that $2p^4q^2$ is the greatest common factor.   **TRY EXERCISE** ▸ 21

    The polynomials in Examples 1 and 2 have been **factored completely**. They
cannot be factored further. The factors in the resulting factorizations are said to
be **prime polynomials**.
    When the leading coefficient is a negative number, we generally factor out a
common factor with a negative coefficient.

EXAMPLE   **3**   Write an equivalent expression by factoring out a common factor with a negative
coefficient.

a) $-4x - 24$ **b)** $-2x^3 + 6x^2 - 2x$

SOLUTION

a) $-4x - 24 = -4(x + 6)$

b) $-2x^3 + 6x^2 - 2x = -2x(x^2 - 3x + 1)$ The 1 is essential—without it, the
                                                factorization does not check.

   TRY EXERCISE ▸ 25

EXAMPLE   **4**   *Height of a thrown object.*   Suppose that a baseball is thrown upward with an
initial velocity of 64 ft/sec. Its height in feet, $h(t)$, after t seconds is given by

$$h(t) = -16t^2 + 64t.$$

Find an equivalent expression for $h(t)$ by factoring out a common factor.

$h(t) = -16t^2 + 64t$

To check Example 4 with a table, let $y_1 = -16x^2 + 64x$ and $y_2 = -16x(x - 4)$. Then compare values of y_1 and y_2.

ΔTBL = 1

X	Y₁	Y₂
0	0	0
1	48	48
2	64	64
3	48	48
4	0	0
5	−80	−80
6	−192	−192

X = 0

1. How can $y_3 = y_2 - y_1$ and a table be used as a check?

SOLUTION   We factor out $-16t$ as follows:

$$h(t) = -16t^2 + 64t = -16t(t - 4). \quad \textit{Check:} \; -16t \cdot t = -16t^2 \text{ and } -16t(-4) = 64t.$$

Note that we can obtain function values using either expression for $h(t)$, since factoring forms equivalent expressions. For example,

$$h(1) = -16 \cdot 1^2 + 64 \cdot 1 = 48$$

and

$$h(1) = -16 \cdot 1(1 - 4) = 48. \quad \text{Using the factorization}$$

> TRY EXERCISE ▸ 57

In Example 4, we could have evaluated $-16t^2 + 64t$ and $-16t(t - 4)$ using any value for t. The results should always match. Thus a quick partial check of any factorization is to evaluate the factorization and the original polynomial for one or two convenient replacements. The check in Example 4 becomes foolproof if three replacements are used. In general, an nth-degree factorization is correct if it checks for $n + 1$ different replacements. The proof of this useful result is beyond the scope of this text.

Factoring by Grouping

The largest common factor is sometimes a binomial.

EXAMPLE   5

Write an equivalent expression by factoring:

$$(a - b)(x + 5) + (a - b)(x - y^2).$$

SOLUTION   Here the largest common factor is the binomial $a - b$:

$$(a - b)(x + 5) + (a - b)(x - y^2) = (a - b)[(x + 5) + (x - y^2)]$$
$$= (a - b)[2x + 5 - y^2].$$

> TRY EXERCISE ▸ 41

Often, in order to identify a common binomial factor, we must regroup into two groups of two terms each.

EXAMPLE   6

Write an equivalent expression by factoring.

a) $y^3 + 3y^2 + 4y + 12$ **b)** $4x^3 - 15 + 20x^2 - 3x$

SOLUTION

a)
$$y^3 + 3y^2 + 4y + 12 = (y^3 + 3y^2) + (4y + 12) \qquad \text{Each grouping has a common factor.}$$
$$= y^2(y + 3) + 4(y + 3) \qquad \text{Factoring out a common factor from each binomial}$$
$$= (y + 3)(y^2 + 4) \qquad \text{Factoring out } y + 3$$

b) When we try grouping $4x^3 - 15 + 20x^2 - 3x$ as

$$(4x^3 - 15) + (20x^2 - 3x),$$

we are unable to factor $4x^3 - 15$. When this happens, we can rearrange the polynomial and try a different grouping:

$$
\begin{aligned}
4x^3 - 15 + 20x^2 - 3x &= 4x^3 + 20x^2 - 3x - 15 && \text{Using the commutative law} \\
&= 4x^2(x + 5) - 3(x + 5) && \text{By factoring out } -3 \text{ instead} \\
&= (x + 5)(4x^2 - 3). && \text{of 3, we see that } x + 5 \text{ is a} \\
& && \text{common factor.}
\end{aligned}
$$

> **TRY EXERCISE** 45

In Example 7 of Section 1.3 (see p. 25), we saw that

$$b - a, \quad -a + b, \quad -(a - b), \quad \text{and} \quad -1(a - b)$$

are all equivalent. Remembering this can help whenever we wish to reverse the order in subtraction (see the third step below).

EXAMPLE 7   Write an equivalent expression by factoring: $ax - bx + by - ay$.

SOLUTION   We have

$$
\begin{aligned}
ax - bx + by - ay &= (ax - bx) + (by - ay) && \text{Grouping} \\
&= x(a - b) + y(b - a) && \text{Factoring each} \\
& && \text{binomial} \\
&= x(a - b) + y(-1)(a - b) && \text{Factoring out } -1 \text{ to} \\
& && \text{reverse } b - a \\
&= x(a - b) - y(a - b) && \text{Simplifying} \\
&= (a - b)(x - y). && \text{Factoring out } a - b
\end{aligned}
$$

Check: To check, note that $a - b$ and $x - y$ are both prime and

$$(a - b)(x - y) = ax - ay - bx + by = ax - bx + by - ay.$$

> **TRY EXERCISE** 49

STUDENT NOTES

In Example 7, make certain that you understand why -1 or $-y$ must be factored from $by - ay$.

Some polynomials with four terms, like $x^3 + x^2 + 3x - 3$, are prime. Not only is there no common monomial factor, but no matter how we group terms, there is no common binomial factor:

$$
\begin{aligned}
x^3 + x^2 + 3x - 3 &= x^2(x + 1) + 3(x - 1); && \text{No common factor} \\
x^3 + 3x + x^2 - 3 &= x(x^2 + 3) + (x^2 - 3); && \text{No common factor} \\
x^3 - 3 + x^2 + 3x &= (x^3 - 3) + x(x + 3). && \text{No common factor}
\end{aligned}
$$

5.3   EXERCISE SET

🔖 *Concept Reinforcement*   *Classify each statement as either true or false.*

1. It is possible for a polynomial to contain several different common factors.

2. The largest common factor of $10x^4 + 15x^2$ is $5x$.

3. When the leading coefficient of a polynomial is negative, we generally factor out a common factor with a negative coefficient.

4. A polynomial $3x + 40$ is prime.

5. A binomial can be a common factor.

6. Every polynomial with four terms can be factored by grouping.

7. The expressions $b - a$, $-(a - b)$, and $-1(a - b)$ are all equivalent.

8. The complete factorization of $12x^3 - 20x^2$ is $4x(3x^2 - 5x)$.

Write an equivalent expression by factoring out the greatest common factor.

9. $10x^2 + 35$

10. $8y^2 + 20$

11. $2y^2 - 18y$

12. $6t^2 - 12t$

13. $5t^3 - 15t + 5$

14. $9x^2 - 3x + 3$

15. $a^6 + 2a^4 - a^3$

16. $3y^7 - y^6 - y^2$

17. $12x^4 - 30x^3 + 42x$

18. $16t^8 + 40t^6 - 24t$

19. $6a^2b - 2ab - 9b$

20. $4x^2y + 10xy + 5y$

21. $15m^4n + 30m^5n^2 + 25m^3n^3$

22. $24s^2t^4 - 18st^3 - 42s^4t^5$

23. $9x^3y^6z^2 - 12x^4y^4z^4 + 15x^2y^5z^3$

24. $14a^4b^3c^5 + 21a^3b^5c^4 - 35a^4b^4c^3$

Write an equivalent expression by factoring out a factor with a negative coefficient.

25. $-5x - 40$

26. $-5x - 35$

27. $-16t^2 + 96$

28. $-16t^2 + 128$

29. $-2x^2 + 12x + 40$

30. $-2x^2 + 4x - 12$

31. $5 - 10y$

32. $7 - 35t$

33. $8d^2 - 12cd$

34. $12q^2 - 21pq$

35. $-m^3 + 8$

36. $-x^2 + 100$

37. $-p^3 - 2p^2 - 5p + 2$

38. $-a^5 - 5a^4 - 11a + 10$

Write an equivalent expression by factoring.

39. $a(b - 5) + c(b - 5)$

40. $r(t - 3) - s(t - 3)$

41. $(x + 7)(x - 1) + (x + 7)(x - 2)$

42. $(a + 5)(a - 2) + (a + 5)(a + 1)$

43. $a^2(x - y) + 5(y - x)$

44. $5x^2(x - 6) + 2(6 - x)$

45. $xy + xz + wy + wz$

46. $ac + ad + bc + bd$

47. $y^3 - y^2 + 3y - 3$

48. $b^3 - b^2 + 2b - 2$

49. $t^3 + 6t^2 - 2t - 12$

50. $a^3 - 3a^2 + 6 - 2a$

51. $12a^4 - 21a^3 - 9a^2$

52. $72x^3 - 36x^2 + 24x$

53. $y^8 - 1 - y^7 + y$

54. $t^6 - 1 - t^5 + t$

55. $2xy + 3x - x^2y - 6$

56. $2y^5 + 15 - 6y^4 - 5y$

57. *Height of a baseball.* A baseball is popped up with an upward velocity of 72 ft/sec. Its height in feet, $h(t)$, after t seconds is given by
$$h(t) = -16t^2 + 72t.$$
a) Find an equivalent expression for $h(t)$ by factoring out a common factor with a negative coefficient.
b) Perform a partial check of part (a) by evaluating both expressions for $h(t)$ at $t = 1$.

58. *Height of a rocket.* A water rocket is launched upward with an initial velocity of 96 ft/sec. Its height in feet, $h(t)$, after t seconds is given by
$$h(t) = -16t^2 + 96t.$$
a) Find an equivalent expression for $h(t)$ by factoring out a common factor with a negative coefficient.
b) Check your factoring by evaluating both expressions for $h(t)$ at $t = 1$.

59. *Surface area of a silo.* A silo is a right circular cylinder with a half sphere on top. The surface area of a silo of height h and radius r (including the area of the base) is given by the polynomial $2\pi rh + \pi r^2$. Find an equivalent expression by factoring out a common factor.

60. *Airline routes.* When an airline links n cities so that from any one city it is possible to fly directly to each of the other cities, the total number of direct routes is given by
$$R(n) = n^2 - n.$$
Find an equivalent expression for $R(n)$ by factoring out a common factor.

61. *Total profit.* After t weeks of production, Pedal Up, Inc., is making a profit of $P(t) = t^2 - 5t$ from sales of their bicycles. Find an equivalent expression by factoring out a common factor.

62. *Total profit.* When x hundred cameras are sold, Digital Electronics collects a profit of $P(x)$, where

$$P(x) = x^2 - 3x,$$

and $P(x)$ is in thousands of dollars. Find an equivalent expression by factoring out a common factor.

63. *Total revenue.* Urban Sounds is marketing a new MP3 player. The firm determines that when it sells x units, the total revenue $R(x)$, in dollars, is given by the polynomial function

$$R(x) = 280x - 0.4x^2.$$

Find an equivalent expression for $R(x)$ by factoring out $0.4x$.

64. *Total cost.* Urban Sounds determines that the total cost $C(x)$, in dollars, of producing x MP3 players is given by the polynomial function

$$C(x) = 0.18x + 0.6x^2.$$

Find an equivalent expression for $C(x)$ by factoring out $0.6x$.

65. *Number of diagonals.* The number of diagonals of a polygon having n sides is given by the polynomial function

$$P(n) = \tfrac{1}{2}n^2 - \tfrac{3}{2}n.$$

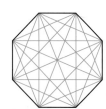

Find an equivalent expression for $P(n)$ by factoring out $\tfrac{1}{2}$.

66. *Number of games in a league.* If there are n teams in a league and each team plays every other team once, we can find the total number of games played by using the polynomial function $f(n) = \tfrac{1}{2}n^2 - \tfrac{1}{2}n$. Find an equivalent expression by factoring out $\tfrac{1}{2}$.

67. *Counting spheres in a pile.* The number N of spheres in a triangular pile like the one shown here is a polynomial function given by

$$N(x) = \tfrac{1}{6}x^3 + \tfrac{1}{2}x^2 + \tfrac{1}{3}x,$$

where x is the number of layers and $N(x)$ is the number of spheres. Find an equivalent expression for $N(x)$ by factoring out $\tfrac{1}{6}$.

68. *High-fives.* When a team of n players all give each other high-fives, a total of $H(n)$ hand slaps occurs, where

$$H(n) = \tfrac{1}{2}n^2 - \tfrac{1}{2}n.$$

Find an equivalent expression by factoring out $\tfrac{1}{2}n$.

69. What is the *prime factorization* of a polynomial? How does it correspond to the prime factorization of a number?

70. Explain in your own words why $-(a - b) = b - a$.

Skill Review

Review graphing linear equations (Sections 2.3, 2.4, and 2.5).

Graph.

71. $f(x) = -\tfrac{1}{2}x + 3$ [2.3]

72. $3x - y = 9$ [2.4]

73. $y - 1 = 2(x + 3)$ [2.5]

74. $y = -3$ [2.4]

75. $6x = 3$ [2.4]

76. $3x = 2y - 4$ [2.3]

Synthesis

77. Under what conditions would it be easier to evaluate a polynomial *after* it has been factored?

78. Following Example 4, we stated that checking the factorization of a second-degree polynomial by making a single replacement is only a *partial* check. Write an *incorrect* factorization and explain how evaluating both the polynomial and the factorization might not catch the mistake.

Complete each of the following.

79. $x^5y^4 + \underline{\hspace{1cm}} = x^3y(\underline{\hspace{1cm}} + xy^5)$

80. $a^3b^7 - \underline{\hspace{1cm}} = \underline{\hspace{1cm}}(ab^4 - c^2)$

Write an equivalent expression by factoring.

81. $rx^2 - rx + 5r + sx^2 - sx + 5s$

82. $3a^2 + 6a + 30 + 7a^2b + 14ab + 70b$

83. $a^4x^4 + a^4x^2 + 5a^4 + a^2x^4 + a^2x^2 + 5a^2 + 5x^4 + 5x^2 + 25$
(*Hint:* Use three groups of three.)

Write an equivalent expression by factoring out the smallest power of x in each of the following.

84. $x^{-8} + x^{-4} + x^{-6}$

85. $x^{-6} + x^{-9} + x^{-3}$

86. $x^{3/4} + x^{1/2} - x^{1/4}$

87. $x^{1/3} - 5x^{1/2} + 3x^{3/4}$

88. $x^{-3/2} + x^{-1/2}$

89. $x^{-5/2} + x^{-3/2}$

90. $x^{-3/4} - x^{-5/4} + x^{-1/2}$

91. $x^{-4/5} - x^{-7/5} + x^{-1/3}$

Write an equivalent expression by factoring. Assume that all exponents are natural numbers.

92. $2x^{3a} + 8x^a + 4x^{2a}$

93. $3a^{n+1} + 6a^n - 15a^{n+2}$

94. $4x^{a+b} + 7x^{a-b}$

95. $7y^{2a+b} - 5y^{a+b} + 3y^{a+2b}$

96. Use the TABLE feature of a graphing calculator to check your answers to Exercises 25, 35, and 41.

97. Use a graphing calculator to show that
$$(x^2 - 3x + 2)^4 = x^8 + 81x^4 + 16$$
is *not* an identity.

CONNECTING the CONCEPTS

When we add, subtract, multiply, or factor polynomials, we are forming equivalent expressions.

Operation	Procedure	Example
Add	Combine like terms.	$(2x^2 - 5x) + (-x^2 + 6) = 2x^2 + (-x^2) - 5x + 6$ $= x^2 - 5x + 6$
Subtract	Add the opposite of the polynomial being subtracted. Combine like terms.	$(4x^2 - x) - (7x^2 - 6x) = 4x^2 - x - 7x^2 + 6x$ $= -3x^2 + 5x$
Multiply	Use the distributive law to multiply each term of one polynomial by every term in the other polynomial. Combine like terms.	$(x - 3)(x + 4) = x^2 + 4x - 3x - 12$ $= x^2 + x - 12$
Factor	Write as a product.	$6x^2y - 9xy^2 + 3xy = 3xy(2x - 3y + 1)$

MIXED REVIEW

Perform the indicated operation.

1. Add: $(4t^3 - 2t + 6) + (8t^2 - 11t - 7)$.

2. Subtract: $(3a + 4b - c) - (a + 2b - 3c)$.

3. Multiply: $4x^2y(3xy - 2x^3 + 6y^2)$.

4. Factor: $30x^2 - 40$.

5. Subtract: $(8n^2 + 5n - 2) - (-n^2 + 6n - 2)$.

6. Factor: $16t^3 - 40t^2 + 8t$.

7. Add: $(2x^2 - 5xy + y^2) + (-x^2 + 5xy - 9y^2)$.

8. Factor: $2x + 2y + ax + ay$.

9. Multiply: $(x + 1)(x + 7)$.

10. Multiply: $(2x - 3)(5x - 1)$.

11. Factor: $p^3 - p^2 + 7p - 7$.

12. Add: $\left(\frac{1}{2}x^2 + \frac{1}{3}x - \frac{3}{2}\right) + \left(\frac{2}{3}x^2 - \frac{1}{2}x - \frac{1}{3}\right)$.

13. Multiply: $(3m - 10)^2$.

14. Subtract: $(1.2x^2 - 3.7x) - (2.8x^2 - x + 1.4)$.

15. Multiply: $(a + 2)(a^2 - a - 6)$.

16. Factor: $3t^3 - 3t^2 - 1 + t$.

17. Multiply: $(c + 9)(c - 9)$.

18. Multiply: $(-s + 6t)(s + 6t)$.

19. Factor: $8x^2y^3z + 12x^3y^2 - 16x^2yz^3$.

20. Factor: $x^3 - 3x^2 + 15 - 5x$.

5.4 Factoring Trinomials

Factoring Trinomials of the Type $x^2 + bx + c$ ▪ Factoring Trinomials of the Type $ax^2 + bx + c, a \neq 1$

Our study of how to factor trinomials begins with trinomials of the type $x^2 + bx + c$. We then move on to the type $ax^2 + bx + c$, where $a \neq 1$.

Factoring Trinomials of the Type $x^2 + bx + c$

When trying to factor trinomials of the type $x^2 + bx + c$, we can use a trial-and-error procedure.

CONSTANT TERM POSITIVE

Recall the FOIL method of multiplying two binomials:

$$
\begin{array}{cccc}
& \text{F} & \text{O} \quad \text{I} & \text{L} \\
(x + 3)(x + 5) = & x^2 + & \underline{5x + 3x} + & 15 \\
& & \downarrow & \\
= & x^2 + & 8x & + 15.
\end{array}
$$

Because the leading coefficient in each binomial is 1, the leading coefficient in the product is also 1. To factor $x^2 + 8x + 15$, we think of FOIL: The term x^2 is the product of the First terms in each of two binomial factors, so the first term in each binomial must be x. The challenge is to find two numbers p and q such that

$$
x^2 + 8x + 15 = (x + p)(x + q)
$$
$$
= x^2 + qx + px + pq.
$$

Note that the Outer and Inner products, qx and px, are like terms and can be combined as $(p + q)(x)$. The Last product, pq, is a constant. Thus the numbers p and q must be selected so that their product is 15 and their sum is 8. In this case, we know from above that these numbers are 3 and 5. The factorization is

$$(x + 3)(x + 5), \quad \text{or} \quad (x + 5)(x + 3). \qquad \text{Using a commutative law}$$

In general, to factor $x^2 + (p + q)x + pq$, we use FOIL in reverse:

$$x^2 + (p + q)x + pq = (x + p)(x + q).$$

When the constant term of a trinomial is positive, the product pq must be positive. Thus the constant terms in the binomial factors must be either both positive or both negative. The sign used is that of the trinomial's middle term.

EXAMPLE 1 Write an equivalent expression by factoring: $x^2 + 9x + 8$.

SOLUTION We think of FOIL in reverse. The first term of each factor is x. We look for numbers p and q such that

$$x^2 + 9x + 8 = (x + p)(x + q) = x^2 + (p + q)x + pq.$$

We list pairs of factors of 8 and choose the pair whose sum is 9.

Pair of Factors	Sum of Factors
2, 4	6
1, 8	9

Both factors are positive.

The numbers we need are 1 and 8, forming the factorization $(x + 1)(x + 8)$.

Check: $(x + 1)(x + 8) = x^2 + 8x + x + 8 = x^2 + 9x + 8.$

The factorization is $(x + 1)(x + 8)$.

 TRY EXERCISE   9

Note in Example 1 that we found the factorization by listing all pairs of factors of 8 along with their sums. If, instead, you form binomial factors without calculating sums, you must carefully check that possible factorization. For example, if we attempt the factorization

$$x^2 + 9x + 8 \overset{?}{=} (x + 2)(x + 4),$$

a check reveals that

$$(x + 2)(x + 4) = x^2 + 6x + 8 \neq x^2 + 9x + 8.$$

This type of trial-and-error procedure becomes easier with time. As you gain experience, you will find that many trials can be performed mentally.

EXAMPLE   2

Factor: $y^2 - 9y + 20.$

SOLUTION   Since the constant term is positive and the coefficient of the middle term is negative, we look for a factorization of 20 in which both factors are negative. Their sum must be -9.

Pair of Factors	Sum of Factors
−1, −20	−21
−2, −10	−12
−4, −5	−9

Both factors are negative.

The numbers we need are −4 and −5.

Check: $(y - 4)(y - 5) = y^2 - 5y - 4y + 20 = y^2 - 9y + 20.$

The factorization of $y^2 - 9y + 20$ is $(y - 4)(y - 5)$.

TRY EXERCISE   11

CONSTANT TERM NEGATIVE

When the constant term of a trinomial is negative, we look for one negative factor and one positive factor. The sum of the factors must still be the coefficient of the middle term.

EXAMPLE **3**

Factor: $x^3 - x^2 - 30x$.

SOLUTION   *Always* look first for a common factor! This time there is one, x. We factor it out:

$$x^3 - x^2 - 30x = x(x^2 - x - 30).$$

STUDENT NOTES

Factoring a polynomial may require more than one step. *Always* look first for a common factor. If one exists, factor out the *greatest* common factor. Then concentrate on trying to factor the polynomial in the parentheses.

Now we consider $x^2 - x - 30$. We need a factorization of -30 for which the sum of the factors is -1. Since both the product and the sum are to be negative, we need one positive factor and one negative factor, and the negative factor must have the greater absolute value.

Pair of Factors	Sum of Factors
1, −30	−29
2, −15	−13
3, −10	−7
5, −6	−1

Each pair of factors gives a negative product and a negative sum.

The numbers we need are 5 and −6.

The factorization of $x^2 - x - 30$ is $(x + 5)(x - 6)$. *Don't forget the factor that was factored out earlier!* We check $x(x + 5)(x - 6)$.

Check: $x(x + 5)(x - 6) = x[x^2 - 6x + 5x - 30]$
$= x[x^2 - x - 30]$
$= x^3 - x^2 - 30x.$

The factorization of $x^3 - x^2 - 30x$ is $x(x + 5)(x - 6)$.   **TRY EXERCISE** 21

EXAMPLE **4**

Factor: $2x^2 + 34x - 220$.

SOLUTION   *Always* look first for a common factor! This time we can factor out 2:

$$2x^2 + 34x - 220 = 2(x^2 + 17x - 110).$$

TECHNOLOGY CONNECTION

The method described in the Technology Connection on p. 294 can be used to check Example 4: Let

$y_1 = 2x^2 + 34x - 220$,

$y_2 = 2(x - 5)(x + 22)$, and

$y_3 = y_2 - y_1$.

1. How should the graphs of y_1 and y_2 compare?
2. What should the graph of y_3 look like?
3. Check Example 3 with a graphing calculator.
4. Use graphs to show that $(2x + 5)(x - 3)$ is *not* a factorization of $2x^2 + x - 15$.

We next look for a factorization of -110 for which the sum of the factors is 17. Since the product is to be negative and the sum positive, we need one positive factor and one negative factor, and the positive factor must have the larger absolute value.

Pair of Factors	Sum of Factors
−1, 110	109
−2, 55	53
−5, 22	17

Each pair of factors gives a negative product and a positive sum.

The numbers we need are −5 and 22. We stop listing pairs of factors when we have found the correct sum.

Thus, $x^2 + 17x - 110 = (x - 5)(x + 22)$. The factorization of the original trinomial, $2x^2 + 34x - 220$, is $2(x - 5)(x + 22)$. The check is left to the student.

TRY EXERCISE 17

Some polynomials are not factorable using integers.

EXAMPLE 5 Factor: $x^2 - x - 7$.

SOLUTION The only pairs of factors of -7 are $-1, 7$ and $1, -7$. Neither pair gives a sum of -1. This trinomial is *not* factorable into binomials with integer coefficients. Although $x^2 - x - 7$ can be factored using noninteger coefficients and more advanced techniques, for our purposes the polynomial is *prime*.

TRY EXERCISE 35

Tips for Factoring $x^2 + bx + c$

1. If necessary, rewrite the trinomial in descending order.
2. Find a pair of factors that have c as their product and b as their sum. Remember the following:

   - If c is positive, its factors will have the same sign as b.

   - If c is negative, one factor will be positive and the other will be negative. Select the factors such that the factor with the larger absolute value is the factor with the same sign as b.

   - If the sum of the two factors is the opposite of b, changing the signs of both factors will give the desired factors whose sum is b (see Example 7).

3. Check the result by multiplying the binomials.

These tips still apply when a trinomial has more than one variable.

EXAMPLE 6 Factor: $x^2 - 2xy - 48y^2$.

SOLUTION We look for numbers p and q such that

$$x^2 - 2xy - 48y^2 = (x + py)(x + qy).$$ The x's and y's can be written in the binomials first:
$$(x + \ y)(x + \ y).$$

Our thinking is much the same as if we were factoring $x^2 - 2x - 48$. We look for factors of -48 whose sum is -2. Those factors are 6 and -8. Thus,

$$x^2 - 2xy - 48y^2 = (x + 6y)(x - 8y).$$

The check is left to the student.

TRY EXERCISE 37

Factoring Trinomials of the Type $ax^2 + bx + c, a \neq 1$

Now we look at trinomials in which the leading coefficient is not 1. We consider two methods. Use what works best for you or what your instructor chooses.

METHOD 1: REVERSING FOIL

We first consider the **FOIL method** for factoring trinomials of the type

$$ax^2 + bx + c, \quad \text{where } a \neq 1.$$

Consider the following multiplication.

$$
\begin{array}{cccc}
\text{F} & \text{O} & \text{I} & \text{L} \\
\downarrow & \downarrow & \downarrow & \downarrow
\end{array}
$$

$$(3x + 2)(4x + 5) = 12x^2 + \underbrace{15x + 8x} + 10$$

$$= 12x^2 + 23x + 10$$

To factor $12x^2 + 23x + 10$, we must reverse what we just did. We look for two binomials whose product is this trinomial. The product of the First terms must be $12x^2$. The product of the Outer terms plus the product of the Inner terms must be $23x$. The product of the Last terms must be 10. We know from the preceding discussion that the factorization is

$$(3x + 2)(4x + 5).$$

In general, however, finding such a factorization involves trial and error. We use the following method.

To Factor $ax^2 + bx + c$ by Reversing FOIL

1. Factor out the largest common factor, if one exists. Here we assume none does.
2. List possible **First** terms whose product is ax^2:

$$(\,x + \;)(\,x + \;) = ax^2 + bx + c.$$
$$\underbrace{}_{\text{FOIL}}$$

3. List possible **Last** terms whose product is c:

$$(\;x + \;)(\;x + \;) = ax^2 + bx + c.$$
$$\underbrace{}_{\text{FOIL}}$$

4. Using the possibilities from steps (2) and (3), find a combination for which the sum of the **O**uter and **I**nner products is bx:

$$(\,x + \;)(\,x + \;) = ax^2 + bx + c.$$
$$\underset{\text{O}}{\underbrace{}} \quad \text{FOIL}$$

EXAMPLE 7   Factor: $3x^2 + 10x - 8$.

SOLUTION

1. First, observe that there is no common factor (other than 1 or -1).
2. Next, factor the first term, $3x^2$. The only possibility for factors is $3x \cdot x$. Thus, if a factorization exists, it must be of the form

$$(3x + \;)(x + \;).$$

   We need to find the right numbers for the blanks.

3. The constant term, -8, can be factored as

$$
\begin{array}{lll}
(-8)(1), & (1)(-8), & \text{When } a \neq 1, \text{ the} \\
(8)(-1), & (-1)(8), & \text{order of the factors} \\
(-2)(4), \ \text{as well as} & (4)(-2), & \text{can affect the} \\
(2)(-4), & (-4)(2). & \text{middle term.}
\end{array}
$$

4. Find binomial factors for which the sum of the Outer and Inner products is the middle term, $10x$. Check each possibility by multiplying:

$$(3x - 8)(x + 1) = 3x^2 - 5x - 8. \qquad O + I = 3x + (-8x) = -5x$$

This gives a middle term with a negative coefficient. We try again:

$$(3x + 8)(x - 1) = 3x^2 + 5x - 8. \qquad O + I = -3x + 8x = 5x$$

Note that changing the signs of the two constant terms changes only the sign of the middle term. We try another possibility:

$$(3x - 2)(x + 4) = 3x^2 + 10x - 8. \qquad \text{This is what we wanted.}$$

Thus the desired factorization of $3x^2 + 10x - 8$ is $(3x - 2)(x + 4)$.

TRY EXERCISE ▶ 45

EXAMPLE 8 Factor: $6x^6 - 19x^5 + 10x^4$.

SOLUTION

1. First, factor out the greatest common factor x^4:

$$x^4(6x^2 - 19x + 10).$$

2. Note that $6x^2 = 6x \cdot x$ and $6x^2 = 3x \cdot 2x$. Thus, $6x^2 - 19x + 10$ may factor into

$$(3x + \quad)(2x + \quad) \quad \text{or} \quad (6x + \quad)(x + \quad).$$

3. We factor the last term, 10. The possibilities are

$(10)(1)$,		$(1)(10)$,
$(-10)(-1)$,	as well as	$(-1)(-10)$,
$(5)(2)$,		$(2)(5)$,
$(-5)(-2)$.		$(-2)(-5)$.

4. There are 8 possibilities for *each* factorization in step (2). The sum of the "outer" and "inner" parts of FOIL must be the middle term, $-19x$. Since the x-coefficient is negative, we consider pairs of negative factors. We check each possible factorization by multiplying:

$$(3x - 10)(2x - 1) = 6x^2 - 23x + 10, \quad \longleftarrow \text{Wrong middle term}$$
$$(3x - 5)(2x - 2) = 6x^2 - 16x + 10. \quad \longleftarrow \text{Wrong middle term}$$

Actually this last attempt could have been rejected by noting that $2x - 2$ has a common factor, 2. Since the *largest* common factor was removed in step (1), no other common factors can exist. We try again, reversing the -5 and -2:

$$(3x - 2)(2x - 5) = 6x^2 - 19x + 10. \qquad \text{This is what we wanted.}$$

The factorization of $6x^2 - 19x + 10$ is $(3x - 2)(2x - 5)$. *But do not forget the common factor!* We must include it to get the complete factorization of the original trinomial:

$$6x^6 - 19x^5 + 10x^4 = x^4(3x - 2)(2x - 5).$$

TRY EXERCISE ▶ 59

STUDENT NOTES

Keep your work organized so that you can see what you have already considered. When factoring $6x^2 - 19x + 10$, we can list all possibilities and cross out those in which a common factor appears:

$(3x - 10)(2x - 1)$,
$\cancel{(3x - 1)(2x - 10)}$,
$\cancel{(3x - 5)(2x - 2)}$,
$(3x - 2)(2x - 5)$,
$\cancel{(6x - 10)(x - 1)}$,
$(6x - 1)(x - 10)$,
$(6x - 5)(x - 2)$,
$\cancel{(6x - 2)(x - 5)}$.

By being organized and not erasing, we can see that there are only four possible factorizations.

Tips for Factoring with FOIL

1. If the largest common factor has been factored out of the original trinomial, then no binomial factor can have a common factor (other than 1 or −1).
2. If a and c are both positive, then the signs in the factors will be the same as the sign of b.
3. When a possible factoring produces the opposite of the desired middle term, reverse the signs of the constants in the factors.
4. Be systematic about your trials. Keep track of those possibilities that you have tried and those that you have not.

Keep in mind that this method of factoring involves trial and error. With practice, you will find yourself making fewer and better guesses.

METHOD 2: THE GROUPING METHOD

The second method for factoring trinomials of the type $ax^2 + bx + c, a \neq 1$, is known as the *grouping method*, or the *ac-method*. It involves not only trial and error and FOIL but also factoring by grouping. To see how it works, let's factor $x^2 + 7x + 10$ by grouping:

$$x^2 + 7x + 10 = x^2 + 2x + 5x + 10 \qquad \text{We "split" } 7x \text{ as } 2x + 5x.$$
$$= x(x + 2) + 5(x + 2)$$
$$= (x + 2)(x + 5).$$

If the leading coefficient is not 1, we need two more steps to split the middle term.* Consider $6x^2 + 23x + 20$. First, multiply the leading coefficient, 6, and the constant, 20, to get 120. Then find a factorization of 120 in which the sum of the factors is the coefficient of the middle term: 23. Split the middle term into a sum or difference using these factors.

$$6x^2 + 23x + 20$$

(1) Multiply 6 and 20: $6 \cdot 20 = 120$.
(2) Factor 120: $120 = 8 \cdot 15$, and $8 + 15 = 23$.
(3) Split the middle term: $23x = 8x + 15x$.
(4) Factor by grouping.

We factor by grouping as follows:

$$6x^2 + 23x + 20 = 6x^2 + 8x + 15x + 20$$
$$= 2x(3x + 4) + 5(3x + 4) \qquad \text{Factoring by grouping}$$
$$= (3x + 4)(2x + 5).$$

To Factor $ax^2 + bx + c$ Using Grouping

1. Make sure that any common factors have been factored out.
2. Multiply the leading coefficient a and the constant c.
3. Find a pair of factors, p and q, so that $pq = ac$ and $p + q = b$.
4. Rewrite the trinomial's middle term, bx, as $px + qx$.
5. Factor by grouping.

---

*The rationale behind these steps is outlined in Exercise 111.

EXAMPLE **9**   Factor: $3x^2 + 10x - 8$.

SOLUTION

1. First, look for a common factor. There is none (other than 1 or -1).
2. Multiply the leading coefficient and the constant, 3 and -8:

$$3(-8) = -24.$$

3. Try to factor -24 so that the sum of the factors is 10:

$$-24 = 12(-2) \quad \text{and} \quad 12 + (-2) = 10.$$

4. Split $10x$ using the results of step (3):

$$10x = 12x - 2x.$$

5. Finally, factor by grouping:

$$3x^2 + 10x - 8 = 3x^2 + 12x - 2x - 8 \qquad \text{Substituting } 12x - 2x \text{ for } 10x$$

$$= 3x(x + 4) - 2(x + 4) \rbrace$$
$$= (x + 4)(3x - 2). \qquad \text{Factoring by grouping}$$

The check is left to the student.

TRY EXERCISE 47

EXAMPLE **10**   Factor: $6x^4 - 116x^3 - 80x^2$.

SOLUTION

Pair of Factors	Sum of Factors
1, -120	-119
2, -60	-58
3, -40	-37
4, -30	-26
5, -24	-19
6, -20	-14
8, -15	-7
10, -12	-2

1. First, factor out the greatest common factor, if any. Here $2x^2$ is common to all three terms: $2x^2(3x^2 - 58x - 40)$.
2. To factor $3x^2 - 58x - 40$, we first multiply the leading coefficient, 3, and the constant, -40: $3(-40) = -120$.
3. Next, try to factor -120 so that the sum of the factors is -58. Since -58 is negative, the negative factor of -120 must have the larger absolute value. We see from the table at left that the factors we need are 2 and -60.
4. Split the middle term, $-58x$, using the results of step (3): $-58x = 2x - 60x$.
5. Factor by grouping:

$$3x^2 - 58x - 40 = 3x^2 + 2x - 60x - 40 \qquad \text{Substituting } 2x - 60x \text{ for } -58x$$

$$= x(3x + 2) - 20(3x + 2) \rbrace$$
$$= (3x + 2)(x - 20). \qquad \text{Factoring by grouping}$$

The factorization of $3x^2 - 58x - 40$ is $(3x + 2)(x - 20)$. *But don't forget the common factor!* We must include it to factor the original trinomial:

$$6x^4 - 116x^3 - 80x^2 = 2x^2(3x + 2)(x - 20).$$

Check: $2x^2(3x + 2)(x - 20) = 2x^2(3x^2 - 58x - 40)$
$$= 6x^4 - 116x^3 - 80x^2.$$

The complete factorization is $2x^2(3x + 2)(x - 20)$.

TRY EXERCISE 67

5.4   EXERCISE SET

Concept Reinforcement *Classify each statement as either true or false.*

1. The first step in factoring any polynomial is to look for a common factor.

2. When factoring a trinomial, we look for two binomial factors.

3. If c is prime, $x^2 + bx + c$ cannot be factored.

4. If a trinomial contains a common factor, it cannot be factored using binomials.

5. Whenever the product of two numbers is negative, the factors have the same sign.

6. If $p + q = -17$, then $-p + (-q) = 17$.

7. If a trinomial has no common factor, then neither of its binomial factors can have a common factor.

8. Trinomials in more than one variable cannot be factored.

Factor. If a polynomial is prime, state this.

9. $x^2 + 5x + 4$

10. $x^2 + 7x + 12$

11. $y^2 - 12y + 27$

12. $t^2 - 8t + 15$

13. $t^2 - 2t - 8$

14. $y^2 - 3y - 10$

15. $a^2 + a - 2$

16. $n^2 + n - 20$

17. $2x^2 + 6x - 108$

18. $3p^2 - 9p - 120$

19. $14a + a^2 + 45$

20. $11y + y^2 + 24$

21. $p^3 - p^2 - 72p$

22. $x^3 + 2x^2 - 63x$

23. $a^2 - 11a + 28$

24. $t^2 - 14t + 45$

25. $x + x^2 - 6$

26. $3x + x^2 - 10$

27. $5y^2 + 40y + 35$

28. $3x^2 + 15x + 18$

29. $32 + 4y - y^2$

30. $56 + x - x^2$

31. $56x + x^2 - x^3$

32. $32y + 4y^2 - y^3$

33. $y^4 + 5y^3 - 84y^2$

34. $x^4 + 11x^3 - 80x^2$

35. $x^2 - 3x + 5$

36. $x^2 + 12x + 13$

37. $x^2 + 12xy + 27y^2$

38. $p^2 - 5pq - 24q^2$

39. $x^2 - 14xy + 49y^2$

40. $y^2 + 8yz + 16z^2$

41. $n^5 - 80n^4 + 79n^3$

42. $t^5 - 50t^4 + 49t^3$

43. $x^6 + 2x^5 - 63x^4$

44. $x^6 + 7x^5 - 18x^4$

45. $3x^2 - 4x - 4$

46. $2x^2 - x - 10$

47. $6t^2 + t - 15$

48. $10y^2 + 7y - 12$

49. $6p^2 - 20p + 16$

50. $24a^2 - 14a + 2$

51. $9a^2 + 18a + 8$

52. $35y^2 + 34y + 8$

53. $8y^2 + 30y^3 - 6y$

54. $4t^2 + 10t^3 - 6t$

55. $18x^2 - 24 - 6x$

56. $8x^2 - 16 - 28x$

57. $t^8 + 5t^7 - 14t^6$

58. $a^6 + a^5 - 6a^4$

59. $70x^4 - 68x^3 + 16x^2$

60. $14x^4 - 19x^3 - 3x^2$

61. $18y^2 - 9y - 20$

62. $20x^2 + x - 30$

63. $16x^2 + 24x + 5$

64. $2y^2 + 9y + 9$

Aha! 65. $5x^2 + 24x + 16$
(*Hint:* See Exercise 63.)

66. $9y^2 + 9y + 2$

67. $-8t^2 - 8t + 30$

68. $-36a^2 + 21a - 3$

69. $18xy^3 + 3xy^2 - 10xy$

70. $3x^3y^2 - 5x^2y^2 - 2xy^2$

71. $24x^2 - 2 - 47x$

72. $15y^2 - 10 - 47y$

73. $63x^3 + 111x^2 + 36x$

74. $50y^3 + 115y^2 + 60y$

75. $48x^4 + 4x^3 - 30x^2$

76. $40y^4 + 4y^3 - 12y^2$

77. $12a^2 - 17ab + 6b^2$

78. $20p^2 - 23pq + 6q^2$

79. $2x^2 + xy - 6y^2$

80. $8m^2 - 6mn - 9n^2$

81. $6x^2 - 29xy + 28y^2$

82. $10p^2 + 7pq - 12q^2$

83. $9x^2 - 30xy + 25y^2$

84. $4p^2 + 12pq + 9q^2$

85. $9x^2y^2 + 5xy - 4$

86. $7a^2b^2 + 13ab + 6$

87. How can one conclude that $x^2 + 5x + 200$ is a prime polynomial without performing any trials?

88. How can one conclude that $x^2 - 59x + 6$ is a prime polynomial without performing any trials?

Skill Review

To prepare for Section 5.5, review the product and power rules for exponents and special polynomial products (Sections 1.6 and 5.2).

Simplify. [1.6]

89. $(5a)^2$

90. $(3x^4)^2$

Multiply. [5.2]

91. $(x + 3)^2$

92. $(2t - 5)^2$

93. $(y + 1)(y - 1)$

94. $(4x^2 + 3y)(4x^2 - 3y)$

Synthesis

95. Describe in your own words an approach that can be used to factor any trinomial of the form $ax^2 + bx + c$ that is not prime.

96. Suppose $(rx + p)(sx - q) = ax^2 - bx + c$ is true. Explain how this can be used to factor $ax^2 + bx + c$.

Factor. Assume that variables in exponents represent positive integers.

97. $60x^8y^6 + 35x^4y^3 + 5$

98. $x^2 + \frac{3}{5}x - \frac{4}{25}$

99. $y^2 - \frac{8}{49} + \frac{2}{7}y$

100. $y^2 + 0.4y - 0.05$

101. $20a^3b^6 - 3a^2b^4 - 2ab^2$

102. $4x^{2a} - 4x^a - 3$

103. $x^{2a} + 5x^a - 24$

104. $bdx^2 + adx + bcx + ac$

105. $2ar^2 + 4asr + as^2 - asr$

106. $a^2p^{2a} + a^2p^a - 2a^2$

Aha! **107.** $(x + 3)^2 - 2(x + 3) - 35$

108. $6(x - 7)^2 + 13(x - 7) - 5$

109. Find all integers m for which $x^2 + mx + 75$ can be factored.

110. Find all integers q for which $x^2 + qx - 32$ can be factored.

111. To better understand factoring $ax^2 + bx + c$ by grouping, suppose that
$$ax^2 + bx + c = (mx + r)(nx + s).$$
Show that if $P = ms$ and $Q = rn$, then $P + Q = b$ and $PQ = ac$.

112. One factor of $x^2 - 345x - 7300$ is $x + 20$. Find the other factor.

113. Use the TABLE feature to check your answers to Exercises 15, 57, and 99.

114. Let $y_1 = 3x^2 + 10x - 8$, $y_2 = (x + 4)(3x - 2)$, and $y_3 = y_2 - y_1$ to check Example 9 graphically.

115. Explain how the following graph of
$$y = x^2 + 3x - 2 - (x - 2)(x + 1)$$
can be used to show that
$$x^2 + 3x - 2 \neq (x - 2)(x + 1).$$

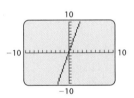

| 5.5 | **Factoring Perfect-Square Trinomials and Differences of Squares** |

Perfect-Square Trinomials ■ Differences of Squares ■ More Factoring by Grouping

STUDENT NOTES

If you're not already quick to recognize that $1^2 = 1$, $2^2 = 4$, $3^2 = 9$, $4^2 = 16$, $5^2 = 25$, $6^2 = 36$, $7^2 = 49$, $8^2 = 64$, $9^2 = 81$, $10^2 = 100$, $11^2 = 121$, and $12^2 = 144$, this is a good time to familiarize yourself with these numbers.

We now introduce a faster way to factor trinomials that are squares of binomials. A method for factoring differences of squares is also developed.

Perfect-Square Trinomials

Consider the trinomial
$$x^2 + 6x + 9.$$

To factor it, we can proceed as in Section 5.4 and look for factors of 9 that add to 6. These factors are 3 and 3 and the factorization is
$$x^2 + 6x + 9 = (x + 3)(x + 3) = (x + 3)^2.$$

Note that the result is the square of a binomial. Because of this, we call $x^2 + 6x + 9$ a **perfect-square trinomial**. Although trial and error can be used to factor a perfect-square trinomial, once recognized, a perfect-square trinomial can be quickly factored.

To Recognize a Perfect-Square Trinomial

- Two terms must be squares, such as A^2 and B^2. Both of these terms will be positive.
- The remaining term must be $2AB$ or its opposite, $-2AB$.

EXAMPLE **1** Determine whether each polynomial is a perfect-square trinomial.

a) $x^2 + 10x + 25$

b) $4x + 16 + 3x^2$

c) $100y^2 + 81 - 180y$

SOLUTION

a) • Two of the terms in $x^2 + 10x + 25$ are squares: x^2 and 25.

   • Twice the product of the square roots is $2 \cdot x \cdot 5$, or $10x$. This is the remaining term.

   Thus, $x^2 + 10x + 25$ *is* a perfect square.

b) In $4x + 16 + 3x^2$, only one term, 16, is a square ($3x^2$ is not a square because 3 is not a perfect square; $4x$ is not a square because x is not a square).

   Thus, $4x + 16 + 3x^2$ *is not* a perfect square.

c) It can help to first write the polynomial in descending order:

$$100y^2 - 180y + 81.$$

   • Two of the terms, $100y^2$ and 81, are squares.

   • Twice the product of the square roots is $2(10y)(9)$, or $180y$. The remaining term is the opposite of $180y$.

   Thus, $100y^2 + 81 - 180y$ *is* a perfect-square trinomial.

To factor a perfect-square trinomial, we reuse the patterns that we learned in Section 5.2.

Factoring a Perfect-Square Trinomial

$A^2 + 2AB + B^2 = (A + B)^2;$

$A^2 - 2AB + B^2 = (A - B)^2$

EXAMPLE **2** Factor.

a) $x^2 - 10x + 25$

b) $16y^2 + 49 + 56y$

c) $-20xy + 4y^2 + 25x^2$

SOLUTION

a) $x^2 - 10x + 25 = (x - 5)^2$.   We find the square terms and write
                                       the square roots with a minus sign
                                       between them.

Note the sign!

Check: $(x - 5)^2 = (x - 5)(x - 5)$
                  $= x^2 - 5x - 5x + 25$
                  $= x^2 - 10x + 25$.

The factorization is $(x - 5)^2$.

b) $16y^2 + 49 + 56y = 16y^2 + 56y + 49$ Using a commutative law

                       $= (4y + 7)^2$ We find the square terms and
                                          write the square roots with a
                                          plus sign between them.

The check is left to the student.

c) $-20xy + 4y^2 + 25x^2 = 4y^2 - 20xy + 25x^2$ Writing descending order
                                                      with respect to y

                          $= (2y - 5x)^2$

This square can also be expressed as

$$25x^2 - 20xy + 4y^2 = (5x - 2y)^2.$$

The student should confirm that both factorizations check.

TRY EXERCISE 11

When factoring, always look first for a factor common to all the terms.

EXAMPLE 3   Factor: **(a)** $2x^2 - 12xy + 18y^2$; **(b)** $-4y^2 - 144y^8 + 48y^5$.

SOLUTION

a) We first look for a common factor. This time, there is a common factor, 2.

$2x^2 - 12xy + 18y^2 = 2(x^2 - 6xy + 9y^2)$ Factoring out the 2
                    $= 2(x - 3y)^2$ Factoring the perfect-
                                        square trinomial

The check is left to the student.

b) $-4y^2 - 144y^8 + 48y^5 = -4y^2(1 + 36y^6 - 12y^3)$ Factoring out the
                                                            common factor

                            $= -4y^2(36y^6 - 12y^3 + 1)$ Changing order. Note
                                                            that $(y^3)^2 = y^6$.

                            $= -4y^2(6y^3 - 1)^2$ Factoring the perfect-
                                                     square trinomial

Check: $-4y^2(6y^3 - 1)^2 = -4y^2(6y^3 - 1)(6y^3 - 1)$
                          $= -4y^2(36y^6 - 12y^3 + 1)$
                          $= -144y^8 + 48y^5 - 4y^2$
                          $= -4y^2 - 144y^8 + 48y^5$

The factorization $-4y^2(6y^3 - 1)^2$ checks.

TRY EXERCISE 33

Differences of Squares

An expression of the form $A^2 - B^2$ is a **difference of squares**. Note that, unlike a perfect-square trinomial, $A^2 - B^2$ has only two terms and one of these must be negative.

When an expression like $x^2 - 9$ is recognized as a difference of two squares, we can reverse another pattern first seen in Section 5.2.

> **Factoring a Difference of Two Squares**
> $$A^2 - B^2 = (A + B)(A - B)$$
>
> To factor a difference of two squares, write the product of the sum and the difference of the quantities being squared.

EXAMPLE **4**   Factor: **(a)** $x^2 - 9$; **(b)** $25y^6 - 49x^2$.

SOLUTION

a) $x^2 - 9 = x^2 - 3^2 = (x + 3)(x - 3)$

$$A^2 \quad - \quad B^2 \quad = (A \quad + B)(A \quad - B)$$
$$\downarrow \qquad \downarrow \qquad \downarrow \quad \downarrow \quad \downarrow \qquad \downarrow$$

b) $25y^6 - 49x^2 = (5y^3)^2 - (7x)^2 = (5y^3 + 7x)(5y^3 - 7x)$

> TRY EXERCISE 35

As always, the first step in factoring is to look for common factors.

EXAMPLE **5**   Factor: **(a)** $5 - 5p^2q^6$; **(b)** $16x^4y - 81y$.

SOLUTION

a) $5 - 5p^2q^6 = 5(1 - p^2q^6)$ 　　　　　Factoring out the common factor
$\qquad\qquad\quad = 5[1^2 - (pq^3)^2]$ 　　　Rewriting p^2q^6 as a quantity squared
$\qquad\qquad\quad = 5(1 + pq^3)(1 - pq^3)$ 　Factoring the difference of squares

Check: $5(1 + pq^3)(1 - pq^3) = 5(1 - pq^3 + pq^3 - p^2q^6)$
$$= 5(1 - p^2q^6) = 5 - 5p^2q^6$$

The factorization $5(1 + pq^3)(1 - pq^3)$ checks.

b) $16x^4y - 81y = y(16x^4 - 81)$ 　　　　　　　Factoring out the common factor
$\qquad\qquad\quad = y[(4x^2)^2 - 9^2]$
$\qquad\qquad\quad = y(4x^2 + 9)(4x^2 - 9)$ 　　　Factoring the difference of squares
$\qquad\qquad\quad = y(4x^2 + 9)(2x + 3)(2x - 3)$ 　Factoring $4x^2 - 9$, which is itself a difference of squares

The check is left to the student. 　　　　　　　　　　　TRY EXERCISE 47

Note in Example 5(b) that $4x^2 - 9$ *could* be factored further. Whenever a factor itself can be factored, do so. We say that we have factored completely when none of the factors can be factored further.

More Factoring by Grouping

Sometimes, when factoring a polynomial with four terms, we may be able to factor further.

EXAMPLE **6**   Factor: $x^3 + 3x^2 - 4x - 12$.

SOLUTION

$$
\begin{aligned}
x^3 + 3x^2 - 4x - 12 &= x^2(x + 3) - 4(x + 3) && \text{Factoring by grouping} \\
&= (x + 3)(x^2 - 4) && \text{Factoring out } x + 3 \\
&= (x + 3)(x + 2)(x - 2) && \text{Factoring } x^2 - 4
\end{aligned}
$$

TRY EXERCISE ▸ 61

A difference of squares can have four or more terms. For example, one of the squares may be a trinomial. In this case, a new type of grouping can be used.

EXAMPLE **7**   Factor: **(a)** $x^2 + 6x + 9 - y^2$; **(b)** $a^2 - b^2 + 8b - 16$.

SOLUTION

a)
$$
\begin{aligned}
x^2 + 6x + 9 - y^2 &= (x^2 + 6x + 9) - y^2 && \text{Grouping as a perfect-square} \\
&&& \text{trinomial minus } y^2 \text{ to show a} \\
&&& \text{difference of squares} \\
&= (x + 3)^2 - y^2 \\
&= (x + 3 + y)(x + 3 - y)
\end{aligned}
$$

b) Grouping $a^2 - b^2 + 8b - 16$ into two groups of two terms does not yield a common binomial factor, so we look for a perfect-square trinomial. In this case, the perfect-square trinomial is being subtracted from a^2:

$$
\begin{aligned}
a^2 - b^2 + 8b - 16 &= a^2 - (b^2 - 8b + 16) && \text{Factoring out } -1 \text{ and} \\
&&& \text{rewriting as subtraction} \\
&= a^2 - (b - 4)^2 && \text{Factoring the perfect-square} \\
&&& \text{trinomial} \\
&= (a + (b - 4))(a - (b - 4)) && \text{Factoring a} \\
&&& \text{difference of} \\
&&& \text{squares} \\
&= (a + b - 4)(a - b + 4). && \text{Removing} \\
&&& \text{parentheses}
\end{aligned}
$$

TRY EXERCISE ▸ 59

5.5   EXERCISE SET

For Extra Help

MyMathLab   PRACTICE   WATCH   DOWNLOAD

🖐 *Concept Reinforcement*   *Classify each of the following as a perfect-square trinomial, a difference of two squares, a polynomial having a common factor, or none of these.*

1. $x^2 - 100$

2. $t^2 - 18t + 81$

3. $36x^2 - 12x + 1$

4. $36a^2 - 25$

5. $4r^2 + 8r + 9$

6. $9x^2 - 12$

7. $4x^2 + 8x + 10$

8. $t^2 - 6t + 8$

9. $4t^2 + 9s^2 + 12st$

10. $9rt^2 - 5rt + 6r$

Factor completely.

11. $x^2 + 20x + 100$

12. $x^2 + 6x + 9$

13. $t^2 - 2t + 1$

14. $t^2 - 4t + 4$

15. $4a^2 - 24a + 36$

16. $9a^2 + 18a + 9$

17. $y^2 + 36 + 12y$

18. $y^2 + 36 - 12y$

19. $-18y^2 + y^3 + 81y$

20. $24a^2 + a^3 + 144a$

21. $2x^2 - 40x + 200$

22. $32x^2 + 48x + 18$

23. $1 - 8d + 16d^2$

24. $64 + 25y^2 - 80y$

25. $-y^3 - 8y^2 - 16y$

26. $-a^3 + 10a^2 - 25a$

27. $0.25x^2 + 0.30x + 0.09$

28. $0.04x^2 - 0.28x + 0.49$

29. $p^2 - 2pq + q^2$

30. $m^2 + 2mn + n^2$

31. $25a^2 + 30ab + 9b^2$

32. $49p^2 - 84pq + 36q^2$

33. $5a^2 + 10ab + 5b^2$ **34.** $4t^2 - 8tr + 4r^2$

35. $x^2 - 25$ **36.** $x^2 - 16$

37. $m^2 - 64$ **38.** $p^2 - 49$

39. $4a^2 - 81$ **40.** $100c^2 - 1$

41. $12c^2 - 12d^2$ **42.** $6x^2 - 6y^2$

43. $7xy^4 - 7xz^4$ **44.** $25ab^4 - 25az^4$

45. $4a^3 - 49a$ **46.** $9x^4 - 25x^2$

47. $3x^8 - 3y^8$ **48.** $9a^4 - a^2b^2$

49. $p^2q^2 - 100$ **50.** $a^2b^2 - 121$

51. $9a^4 - 25a^2b^4$ **52.** $16x^6 - 81x^2y^4$

53. $y^2 - \frac{1}{4}$ **54.** $x^2 - \frac{1}{9}$

55. $\frac{1}{100} - x^2$ **56.** $\frac{1}{16} - y^2$

57. $(a + b)^2 - 36$ **58.** $(p + q)^2 - 64$

59. $x^2 - 6x + 9 - y^2$ **60.** $a^2 - 8a + 16 - b^2$

61. $t^3 + 8t^2 - t - 8$

62. $x^3 - 7x^2 - 4x + 28$

63. $r^3 - 3r^2 - 9r + 27$ **64.** $t^3 + 2t^2 - 4t - 8$

65. $m^2 - 2mn + n^2 - 25$ **66.** $x^2 + 2xy + y^2 - 9$

67. $81 - (x + y)^2$ **68.** $49 - (a + b)^2$

69. $r^2 - 2r + 1 - 4s^2$

70. $c^2 + 4cd + 4d^2 - 9p^2$

Aha! **71.** $16 - a^2 - 2ab - b^2$

72. $100 - x^2 - 2xy - y^2$

73. $x^3 + 5x^2 - 4x - 20$

74. $t^3 + 6t^2 - 9t - 54$

75. $a^3 - ab^2 - 2a^2 + 2b^2$

76. $p^2q - 25q + 3p^2 - 75$

77. Describe a procedure that could be used to determine whether a polynomial is a difference of squares.

78. Are the product and power rules for exponents (see Section 1.6) important when factoring differences of squares? Why or why not?

Skill Review

To prepare for Section 5.6, review the product and power rules for exponents and multiplication of polynomials (Sections 1.6 and 5.2).

Simplify. [1.6]

79. $(2x^2y^4)^3$ **80.** $(3ab^6)^3$

81. $(-10x^{10})^3$ **82.** $(-5x^2y)^3$

Multiply. [5.2]

83. $(x + 1)(x + 1)(x + 1)$ **84.** $(x - 1)^3$

85. $(p + q)^3$ **86.** $(p - q)^3$

Synthesis

87. Gretchen plans to use FOIL to factor polynomials rather than looking for perfect-square trinomials or differences of squares. How might you convince her that it is worthwhile to learn the factoring techniques of this section?

88. Without finding the entire factorization, determine the number of factors of $x^{256} - 1$. Explain how you arrived at your answer.

Factor completely. Assume that variables in exponents represent positive integers.

89. $-\frac{8}{27}r^2 - \frac{10}{9}rs - \frac{1}{6}s^2 + \frac{2}{3}rs$

90. $\frac{1}{36}x^8 + \frac{2}{9}x^4 + \frac{4}{9}$

91. $0.09x^8 + 0.48x^4 + 0.64$

92. $a^2 + 2ab + b^2 - c^2 + 6c - 9$

93. $r^2 - 8r - 25 - s^2 - 10s + 16$

94. $x^{2a} - y^2$

95. $x^{4a} - y^{2b}$

96. $4y^{4a} + 20y^{2a} + 20y^{2a} + 100$

97. $25y^{2a} - (x^{2b} - 2x^b + 1)$

98. $(a - 3)^2 - 8(a - 3) + 16$

99. $3(x + 1)^2 + 12(x + 1) + 12$

100. $m^2 + 4mn + 4n^2 + 5m + 10n$

101. $s^2 - 4st + 4t^2 + 4s - 8t + 4$

102. $5c^{100} - 80d^{100}$

103. $9x^{2n} - 6x^n + 1$

104. $c^{2w+1} + 2c^{w+1} + c$

105. If $P(x) = x^2$, use factoring to simplify

$$P(a + h) - P(a).$$

106. If $P(x) = x^4$, use factoring to simplify

$$P(a + h) - P(a).$$

107. *Volume of carpeting.*   The volume of a carpet that is rolled up can be estimated by the polynomial $\pi R^2 h - \pi r^2 h$.

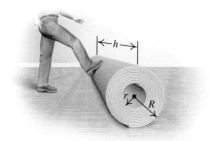

a) Factor the polynomial.
b) Use both the original and the factored forms to find the volume of a roll for which $R = 50$ cm, $r = 10$ cm, and $h = 4$ m. Use 3.14 for π.

108. Use a graphing calculator to check your answers to Exercises 11, 35, and 45 graphically by examining $y_1 =$ the original polynomial, $y_2 =$ the factored polynomial, and $y_3 = y_2 - y_1$.

109. Check your answers to Exercises 11, 35, and 45 by using tables of values (see Exercise 108).

5.6   Factoring Sums or Differences of Cubes

Formulas for Factoring Sums or Differences of Cubes   •   Using the Formulas

Formulas for Factoring Sums or Differences of Cubes

We have seen that a difference of two squares can always be factored, but a *sum* of two squares is usually prime. The situation is different with cubes: The difference *or sum* of two cubes can always be factored. To see this, consider the following products:

$$\begin{aligned}(A + B)(A^2 - AB + B^2) &= A(A^2 - AB + B^2) + B(A^2 - AB + B^2) \\ &= A^3 - A^2 B + AB^2 + A^2 B - AB^2 + B^3 \\ &= A^3 + B^3 \quad \text{Combining like terms}\end{aligned}$$

and

$$\begin{aligned}(A - B)(A^2 + AB + B^2) &= A(A^2 + AB + B^2) - B(A^2 + AB + B^2) \\ &= A^3 + A^2 B + AB^2 - A^2 B - AB^2 - B^3 \\ &= A^3 - B^3. \quad \text{Combining like terms}\end{aligned}$$

These products allow us to factor a sum or a difference of two cubes. Observe how the location of the + and − signs changes.

> **Factoring a Sum or a Difference of Two Cubes**
> $$A^3 + B^3 = (A + B)(A^2 - AB + B^2);$$
> $$A^3 - B^3 = (A - B)(A^2 + AB + B^2)$$

Using the Formulas

When factoring a sum or a difference of cubes, it can be helpful to remember that $2^3 = 8$, $3^3 = 27$, $4^3 = 64$, $5^3 = 125$, $6^3 = 216$, and so on. We say that 2 is the *cube root* of 8, that 3 is the cube root of 27, and so on.

EXAMPLE **1** Write an equivalent expression by factoring: $x^3 + 27$.

SOLUTION   We first observe that

$$x^3 + 27 = x^3 + 3^3. \qquad \text{This is a sum of cubes.}$$

Next, in one set of parentheses, we write the first cube root, x, plus the second cube root, 3:

$$(x + 3)(\qquad).$$

To get the other factor, we think of $x + 3$ and do the following:

Square the first term: x^2.
Multiply the terms and then change the sign: $-3x$.
Square the second term: 3^2, or 9.

$$(x + 3)(x^2 - 3x + 9).$$

Check: $(x + 3)(x^2 - 3x + 9) = x^3 - 3x^2 + 9x + 3x^2 - 9x + 27$
$$= x^3 + 27. \qquad \text{Combining like terms}$$

Thus, $x^3 + 27 = (x + 3)(x^2 - 3x + 9)$.   **TRY EXERCISE** 13

In Example 2, you will see that the pattern used to write the trinomial factor in Example 1 can be used when factoring a *difference* of two cubes as well.

EXAMPLE **2** Factor.

a) $125x^3 - y^3$ **b)** $m^6 + 64$
c) $128y^7 - 250x^6y$ **d)** $r^6 - s^6$

SOLUTION

a) We have

$$125x^3 - y^3 = (5x)^3 - y^3. \qquad \text{This is a difference of cubes.}$$

In one set of parentheses, we write the cube root of the first term, $5x$, minus the cube root of the second term, y:

$$(5x - y)(\qquad). \qquad \text{This can be regarded as } 5x \text{ plus the cube root of } (-y)^3, \text{ since } -y^3 = (-y)^3.$$

STUDENT NOTES —————

If you think of $A^3 - B^3$ as $A^3 + (-B)^3$, it is then sufficient to remember only the pattern for factoring a sum of two cubes. Be sure to simplify your result if you do this.

To get the other factor, we think of $5x + y$ and do the following:

Square the first term: $(5x)^2$, or $25x^2$.
Multiply the terms and then change the sign: $5xy$.
Square the second term: $(-y)^2 = y^2$.

$$(5x - y)(25x^2 + 5xy + y^2).$$

Check:
$(5x - y)(25x^2 + 5xy + y^2) = 125x^3 + 25x^2y + 5xy^2 - 25x^2y - 5xy^2 - y^3$
$$= 125x^3 - y^3. \qquad \text{Combining like terms}$$

Thus, $125x^3 - y^3 = (5x - y)(25x^2 + 5xy + y^2)$.

b) We have

$$m^6 + 64 = (m^2)^3 + 4^3. \qquad \text{Rewriting as a sum of quantities cubed}$$

Next, we reuse the pattern used in Example 1:

$$A^3 \quad + \quad B^3 = (A + B)(\quad A^2 \quad - \quad A \cdot B + B^2)$$
$$\downarrow \qquad \downarrow$$
$$(m^2)^3 + 4^3 = (m^2 + 4)((m^2)^2 - m^2 \cdot 4 + 4^2)$$
$$= (m^2 + 4)(m^4 - 4m^2 + 16). \qquad \text{The check is left to the student.}$$

c) We have

$$128y^7 - 250x^6y = 2y(64y^6 - 125x^6) \qquad \text{Remember: } \textit{Always} \text{ look for a common factor.}$$

$$= 2y[(4y^2)^3 - (5x^2)^3]. \qquad \text{Rewriting as a difference of quantities cubed}$$

To factor $(4y^2)^3 - (5x^2)^3$, we reuse the pattern in part (a) above:

$$A^3 \quad - \quad B^3 \quad = (A \quad - \quad B)(\quad A^2 \quad + \quad A \cdot B \quad + \quad B^2)$$
$$\downarrow \qquad \qquad \downarrow$$
$$(4y^2)^3 - (5x^2)^3 = (4y^2 - 5x^2)((4y^2)^2 + 4y^2 \cdot 5x^2 + (5x^2)^2)$$
$$= (4y^2 - 5x^2)(16y^4 + 20x^2y^2 + 25x^4).$$

The check is left to the student. We have

$$128y^7 - 250x^6y = 2y(4y^2 - 5x^2)(16y^4 + 20x^2y^2 + 25x^4).$$

d) We have

$$r^6 - s^6 = (r^3)^2 - (s^3)^2$$
$$= (r^3 + s^3)(r^3 - s^3) \qquad \text{Factoring a difference of two } \textit{squares}$$
$$= (r + s)(r^2 - rs + s^2)(r - s)(r^2 + rs + s^2). \qquad \text{Factoring the sum and the difference of two cubes}$$

To check, read the steps in reverse order and inspect the multiplication.

> TRY EXERCISES ▷ 31 and 41

In Example 2(d), suppose we first factored $r^6 - s^6$ as a difference of two cubes:

$$(r^2)^3 - (s^2)^3 = (r^2 - s^2)(r^4 + r^2s^2 + s^4)$$
$$= (r + s)(r - s)(r^4 + r^2s^2 + s^4).$$

In this case, we might have missed some factors; $r^4 + r^2s^2 + s^4$ can be factored as $(r^2 - rs + s^2)(r^2 + rs + s^2)$, but we probably would never have suspected that such a factorization exists. Given a choice, it is generally better to factor as a difference of squares before factoring as a sum or a difference of cubes.

Useful Factoring Facts

Sum of cubes: $\qquad A^3 + B^3 = (A + B)(A^2 - AB + B^2)$

Difference of cubes: $\quad A^3 - B^3 = (A - B)(A^2 + AB + B^2)$

Difference of squares: $\; A^2 - B^2 = (A + B)(A - B)$

There is no formula for factoring a sum of two squares.

5.6   EXERCISE SET

For Extra Help

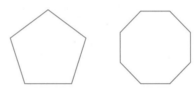

Concept Reinforcement   *Classify each binomial as either a sum of cubes, a difference of cubes, a difference of squares, or none of these.*

1. $x^3 - 1$

2. $8 + t^3$

3. $9x^4 - 25$

4. $9x^2 + 25$

5. $1000t^3 + 1$

6. $x^3y^3 - 27z^3$

7. $25x^2 + 8x$

8. $100y^8 - 25x^4$

9. $s^{21} - t^{15}$

10. $14x^3 - 2x$

Factor completely.

11. $x^3 - 64$

12. $t^3 - 27$

13. $z^3 + 1$

14. $x^3 + 8$

15. $t^3 - 1000$

16. $m^3 + 125$

17. $27x^3 + 1$

18. $8a^3 + 1$

19. $64 - 125x^3$

20. $27 - 8t^3$

21. $x^3 - y^3$

22. $y^3 - z^3$

23. $a^3 + \frac{1}{8}$

24. $x^3 + \frac{1}{27}$

25. $8t^3 - 8$

26. $2y^3 - 128$

27. $54x^3 + 2$

28. $8a^3 + 1000$

29. $rs^4 + 64rs$

30. $ab^5 + 1000ab^2$

31. $5x^3 - 40z^3$

32. $2y^3 - 54z^3$

33. $y^3 - \frac{1}{1000}$

34. $x^3 - \frac{1}{8}$

35. $x^3 + 0.001$

36. $y^3 + 0.125$

37. $64x^6 - 8t^6$

38. $125c^6 - 8d^6$

39. $54y^4 - 128y$

40. $3z^5 - 3z^2$

41. $z^6 - 1$

42. $t^6 + 1$

43. $t^6 + 64y^6$

44. $p^6 - q^6$

45. $x^{12} - y^3z^{12}$

46. $a^9 + b^{12}c^{15}$

47. How could you use factoring to convince someone that $x^3 + y^3 \neq (x + y)^3$?

48. Is the following statement true or false and why? If A^3 and B^3 have a common factor, then A and B have a common factor.

Skill Review

Review solving applications using the five-step problem-solving strategy (see Section 1.4).

Solve.

49. *Geometry.*   The perimeter of a triangle is 108 cm. The lengths of the three sides are consecutive numbers. What are the lengths of the sides? [1.4]

50. *Geometry.*   A regular pentagon (all five sides the same length) has the same perimeter as a regular octagon (all eight sides the same length). One side of the regular pentagon is 1 cm less than twice the length of one side of the regular octagon. Find the perimeter of each shape. [3.3]

51. *Value of coins.*   There are 50 dimes in a roll of dimes, 40 nickels in a roll of nickels, and 40 quarters in a roll of quarters. Jenna has 10 rolls of coins, which have a total value of $77. There are twice as many rolls of quarters as there are rolls of dimes. How many of each type of roll does she have? [3.5]

52. *Study time.*   In order to keep up with his history course, Kyle needs to average at least 45 min a day reading his text. One week, he read 30 min on Monday, nothing on Tuesday, 50 min on Wednesday, and 80 min on Thursday. How long must he read on Friday in order to average at least 45 min a day for the week? [4.1]

53. *Conservation.*   Using helicopters, Ken and Kathy counted alligator nests in a 285-mi^2 area. Ken found 8 more nests than Kathy did, and together they counted 100 nests. How many nests did each count? [1.4]

54. *Wages.*   Mihkel worked a total of 12 days last month at bicycle races. He earned $100 a day during the week and $140 a day during the weekend. During the month, Mihkel earned $1520. How many weekdays did he work? [3.3]

Synthesis

55. Explain how the geometric model below can be used to verify the formula for factoring $a^3 - b^3$.

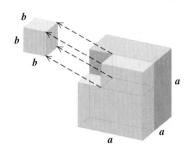

56. Explain how someone could construct a binomial that is both a difference of two cubes and a difference of two squares.

Factor.

57. $x^{6a} - y^{3b}$

58. $2x^{3a} + 16y^{3b}$

Aha! **59.** $(x + 5)^3 + (x - 5)^3$

60. $\frac{1}{16}x^{3a} + \frac{1}{2}y^{6a}z^{9b}$

61. $5x^3y^6 - \frac{5}{8}$

62. $x^3 - (x + y)^3$

63. $x^{6a} - (x^{2a} + 1)^3$

64. $(x^{2a} - 1)^3 - x^{6a}$

65. $t^4 - 8t^3 - t + 8$

66. If $P(x) = x^3$, use factoring to simplify
$$P(a + h) - P(a).$$

67. If $Q(x) = x^6$, use factoring to simplify
$$Q(a + h) - Q(a).$$

68. Using one set of axes, graph the following.
a) $f(x) = x^3$
b) $g(x) = x^3 - 8$
c) $h(x) = (x - 2)^3$

69. Use a graphing calculator to check Example 1: Let $y_1 = x^3 + 27$, $y_2 = (x + 3)(x^2 - 3x + 9)$, and $y_3 = y_1 - y_2$.

70. Use the approach of Exercise 69 to check your answers to Exercises 15, 25, and 39.

5.7 Factoring: A General Strategy

Mixed Factoring Problems

Factoring is an important algebraic skill. The following strategy for factoring emphasizes recognizing the type of expression being factored.

A Strategy for Factoring

A. Always factor out the greatest common factor, if possible.

B. Once the greatest common factor has been factored out, *count the number of terms* in the other factor:

Two terms: Try factoring as a difference of squares first. Next, try factoring as a sum or a difference of cubes.

Three terms: Try factoring as a perfect-square trinomial. Next, try trial and error, either reversing the FOIL method or using the grouping method.

Four or more terms: Try factoring by grouping and factoring out a common binomial factor. Next, try grouping into a difference of squares, one of which is a trinomial.

C. Always *factor completely*. If a factor with more than one term can itself be factored further, do so.

D. *Check* the factorization by multiplying.

EXAMPLE **1**   Write an equivalent expression by factoring: $10a^2x - 40b^2x$.

SOLUTION

A. Look first for a common factor:

$$10a^2x - 40b^2x = 10x(a^2 - 4b^2).$$ Factoring out the largest common factor

STUDENT NOTES —————

Try to remember that anytime a new set of parentheses is created while factoring, the expression within it must be checked to see if it can be factored further.

B. The factor $a^2 - 4b^2$ has two terms. It is a difference of squares. We factor it, keeping the common factor:

$$10a^2x - 40b^2x = 10x(a + 2b)(a - 2b).$$

C. Have we factored completely? Yes, because no factor with more than one term can be factored further.

D. *Check:* $10x(a + 2b)(a - 2b) = 10x(a^2 - 4b^2) = 10a^2x - 40b^2x.$

> TRY EXERCISE 17

EXAMPLE **2**   Factor: $x^6 - 64$.

SOLUTION

A. Look for a common factor. There is none (other than 1 or -1).

B. There are two terms, a difference of squares: $(x^3)^2 - (8)^2$. We factor it:

$$x^6 - 64 = (x^3 + 8)(x^3 - 8).$$ Note that $x^6 = (x^3)^2$.

C. One factor is a sum of two cubes, and the other factor is a difference of two cubes. We factor both:

$$x^6 - 64 = (x + 2)(x^2 - 2x + 4)(x - 2)(x^2 + 2x + 4).$$

The factorization is complete because no factor can be factored further.

D. The check is left to the student.

> TRY EXERCISE 29

EXAMPLE **3**   Factor: $7x^6 + 35y^2$.

SOLUTION

A. Factor out the largest common factor:

$$7x^6 + 35y^2 = 7(x^6 + 5y^2).$$

B. The binomial $x^6 + 5y^2$ is not a difference of squares, a difference of cubes, or a sum of cubes. It cannot be factored.

C. We cannot factor further.

D. *Check:* $7(x^6 + 5y^2) = 7x^6 + 35y^2.$

> TRY EXERCISE 63

EXAMPLE **4**   Factor: $2x^2 + 50a^2 - 20ax$.

SOLUTION

A. Factor out the largest common factor: $2(x^2 + 25a^2 - 10ax)$.

B. Next, we rearrange the trinomial in descending powers of x: $2(x^2 - 10ax + 25a^2)$. The trinomial is a perfect-square trinomial:

$$2x^2 + 50a^2 - 20ax = 2(x^2 - 10ax + 25a^2) = 2(x - 5a)^2.$$

C. We cannot factor further. Had we used descending powers of a, we would have discovered an equivalent factorization, $2(5a - x)^2$.

D. *Check:* $2(x - 5a)^2 = 2(x^2 - 10ax + 25a^2)$
$$= 2x^2 - 20ax + 50a^2$$
$$= 2x^2 + 50a^2 - 20ax.$$

TRY EXERCISE 21

EXAMPLE 5   Factor: $12x^2 - 40x - 32$.

SOLUTION

A. Factor out the largest common factor: $4(3x^2 - 10x - 8)$.

B. The trinomial factor is not a square. We factor using trial and error:
$$12x^2 - 40x - 32 = 4(x - 4)(3x + 2).$$

C. We cannot factor further.

D. *Check:* $4(x - 4)(3x + 2) = 4(3x^2 + 2x - 12x - 8)$
$$= 4(3x^2 - 10x - 8)$$
$$= 12x^2 - 40x - 32.$$

TRY EXERCISE 23

EXAMPLE 6   Factor: $3x + 12 + ax^2 + 4ax$.

SOLUTION

A. There is no common factor (other than 1 or -1).

B. There are four terms. We try grouping to find a common binomial factor:

$$3x + 12 + ax^2 + 4ax = 3(x + 4) + ax(x + 4) \qquad \text{Factoring two grouped binomials}$$

$$= (x + 4)(3 + ax). \qquad \text{Removing the common binomial factor}$$

C. We cannot factor further.

D. *Check:* $(x + 4)(3 + ax) = 3x + ax^2 + 12 + 4ax = 3x + 12 + ax^2 + 4ax$.

TRY EXERCISE 41

EXAMPLE 7   Factor: $y^2 - 9a^2 + 12y + 36$.

SOLUTION

A. There is no common factor (other than 1 or -1).

B. There are four terms. We try grouping to remove a common binomial factor, but find none. Next, we try grouping as a difference of squares:

$$(y^2 + 12y + 36) - 9a^2 \qquad \text{Grouping}$$
$$= (y + 6)^2 - (3a)^2 \qquad \text{Rewriting as a difference of squares}$$
$$= (y + 6 + 3a)(y + 6 - 3a). \qquad \text{Factoring the difference of squares}$$

C. No factor with more than one term can be factored further.

D. The check is left to the student.

TRY EXERCISE 31

EXAMPLE **8**   Factor: $x^3 - xy^2 + x^2y - y^3$.

SOLUTION

A. There is no common factor (other than 1 or -1).

B. There are four terms. We try grouping to remove a common binomial factor:

$$x^3 - xy^2 + x^2y - y^3$$
$$= x(x^2 - y^2) + y(x^2 - y^2) \qquad \text{Factoring two grouped binomials}$$
$$= (x^2 - y^2)(x + y). \qquad \text{Removing the common binomial factor}$$

C. The factor $x^2 - y^2$ can be factored further:

$$x^3 - xy^2 + x^2y - y^3 = (x + y)(x - y)(x + y), \text{ or } (x + y)^2(x - y).$$

No factor can be factored further, so we have factored completely.

D. The check is left to the student.

TRY EXERCISE 49

5.7   **EXERCISE SET**

🔖 *Concept Reinforcement*   *Choose the item from the column on the right that corresponds to the type of polynomial. Items may be used more than once.*

1. ____ $25y^2 - 49$

2. ____ $36x^2y - 9xy$

3. ____ $9y^6 + 16x^8$

4. ____ $8a^3 - b^6c^9$

5. ____ $c^{12} + 1$

6. ____ $4t^2 - 12t + 9$

7. ____ $4a^2 + 8a + 16$

8. ____ $9x^2 + 24x - 16$

a) Polynomial with a common factor

b) Difference of two squares

c) Sum of two cubes

d) Difference of two cubes

e) Perfect-square trinomial

f) None of these

For each polynomial, tell what type of factoring is needed. Then give the factorization of the polynomial. For example, for $3x^2 - 300$, write "Factor out a common factor; factor a sum of squares; $3(x + 10)(x - 10)$."

9. $x^2 - 3x - 4$

10. $x^6 - 1$

11. $2x^3 - 5x^2 - 2x + 5$

12. $t^2 + 100 - 20t$

13. $24a^2 - 4a - 8$

14. $6x^2 + 6x + 12$

Factor completely.

15. $x^2 - 81$

16. $a^2 - 4$

17. $9m^4 - 900$

18. $t^5 - 49t$

19. $2x^3 + 12x^2 + 16x$

20. $10x^2 - 40x + 40$

21. $a^2 + 25 + 10a$

22. $8a^3 - 18a^2 - 5a$

23. $2y^2 - 11y + 12$

24. $6y^2 - 13y - 5$

25. $3x^2 + 15x - 252$

26. $2y^2 + 10y - 132$

27. $25x^2 - 9y^2$

28. $16a^2 - 81b^2$

29. $t^6 + 1$

30. $64t^6 - 1$

31. $x^2 + 6x - y^2 + 9$

32. $t^2 + 10t - p^2 + 25$

33. $128a^3 + 250b^3$

34. $343x^3 + 27y^3$

35. $7x^3 - 14x^2 - 105x$

36. $2t^3 + 20t^2 - 48t$

37. $-9t^2 + 16t^4$

38. $-24x^6 + 6x^4$

39. $8m^3 + m^6 - 20$

40. $-37x^2 + x^4 + 36$

41. $ac + cd - ab - bd$

42. $xw - yw + xz - yz$

43. $4c^2 - 4cd + d^2$

44. $70b^2 - 3ab - a^2$

45. $40x^2 + 3xy - y^2$

46. $p^2 - 10pq + 25q^2$

47. $4a - 5a^2 - 10 + 2a^3$

48. $24 + 3t^3 - 9t^2 - 8t$

49. $2x^3 + 6x^2 - 8x - 24$

50. $3x^3 + 6x^2 - 27x - 54$

51. $54a^3 - 16b^3$

52. $54x^3 - 250y^3$

53. $36y^2 - 35 + 12y$

54. $2b - 28a^2b + 10ab$

55. $4m^4 - 64n^4$

56. $2x^4 - 32$

57. $a^5b - 16ab^5$

58. $x^3y - 25xy^3$

59. $34t^3 - 6t$

60. $13t^3 - 26t$

Aha! **61.** $(a - 3)(a + 7) + (a - 3)(a - 1)$

62. $x^2(x + 3) - 4(x + 3)$

63. $7a^4 - 14a^3 + 21a^2 - 7a$

64. $a^3 - ab^2 + a^2b - b^3$

65. $42ab + 27a^2b^2 + 8$

66. $-23xy + 20x^2y^2 + 6$

67. $-10t^3 + 15t$

68. $-9x^3 + 12x$

69. $-6x^4 + 8x^3 - 12x$

70. $-15t^4 + 10t$

71. $p - 64p^4$

72. $125a - 8a^4$

Aha! **73.** $a^2 - b^2 - 6b - 9$

74. $m^2 - n^2 - 8n - 16$

75. Emily has factored a polynomial as $(a - b)(x - y)$, while Jorge has factored the same polynomial as $(b - a)(y - x)$. Can they both be correct? Why or why not?

76. In your own words, outline a procedure that can be used to factor any polynomial.

Skill Review

To prepare for Section 5.8, review solving equations and finding the domain of a function (Sections 1.3, 2.2, and 4.2).

Solve. [1.3]

77. $x + 2 = 0$

78. $2x - 5 = 0$

79. $4x = 0$

80. $x = 0$

Find the domain of each function. [2.2], [4.2]

81. $f(x) = \dfrac{2x}{3x - 2}$

82. $f(x) = \dfrac{x + 5}{2x + 1}$

Synthesis

83. Explain how one could construct a polynomial that is a difference of squares that contains a sum of two cubes and a difference of two cubes as factors.

84. Explain how one could construct a polynomial with four terms that can be factored by grouping three terms together.

Factor completely.

85. $28a^3 - 25a^2bc + 3ab^2c^2$

86. $-16 + 17(5 - y^2) - (5 - y^2)^2$

Aha! **87.** $(x - p)^2 - p^2$

88. $a^4 - 50a^2b^2 + 49b^4$

89. $(y - 1)^4 - (y - 1)^2$

90. $x^6 - 2x^5 + x^4 - x^2 + 2x - 1$

91. $4x^2 + 4xy + y^2 - r^2 + 6rs - 9s^2$

92. $(1 - x)^3 - (x - 1)^6$

93. $\dfrac{x^{27}}{1000} - 1$

94. $a - by^8 + b - ay^8$

95. $3(x + 1)^2 - 9(x + 1) - 12$

96. $3a^2 + 3b^2 - 3c^2 - 3d^2 + 6ab - 6cd$

97. $3(a + 2)^2 + 30(a + 2) + 75$

98. $(m - 1)^3 - (m + 1)^3$

99. $27x^{6s} + 64y^{3t}$

100. $24t^{2a} - 6$

101. $a^{2w+1} + 2a^{w+1} + a$

102. If $\left(x + \dfrac{2}{x}\right)^2 = 6$, find $x^3 + \dfrac{8}{x^3}$.

CONNECTING the CONCEPTS

The strategy for factoring on p. 325 provides a helpful approach when factoring a mixed set of polynomials.

MIXED REVIEW

Factor completely.

1. $t^2 - 2t + 1$
2. $2x^2 - 16x + 30$
3. $x^3 - 64x$
4. $6a^2 - a - 1$
5. $5t^3 + 500t$
6. $x^3 - 64$
7. $4x^3 + 100x + 40x^2$
8. $2x^3 - 3 - 6x^2 + x$
9. $12y^3 + y^2 - 6y$
10. $24 + n^6 - 10n^3$
11. $7t^3 + 7$
12. $6m^4 + 96m^3 + 384m^2$
13. $x^3 + 3x^2 - x - 3$
14. $a^2 - \frac{1}{9}$
15. $0.25 - y^2$
16. $3n^2 - 21n$
17. $x^4 + 4 - 5x^2$

18. $-10c^3 + 25c^2$
19. $1 - 64t^6$
20. $6x^5 - 15x^4 + 18x^2 + 9x$
21. $2x^5 + 6x^4 + 3x^2 + 9x$
22. $yz - 2tx - ty + 2xz$
23. $50a^2b^2 - 32c^4$
24. $x^2 + 10x + 25 - y^2$
25. $2x^2 - 12xy - 32y^2$
26. $4x^2y^6 + 20xy^3 + 25$
27. $m^2 - n^2 + 12n - 36$
28. $6a^2b - 9ab - 60b$
29. $p^2 + 121q^2 - 22pq$
30. $8a^2b^3c - 40ab^2c^3 + 4ab^2c$

5.8   Applications of Polynomial Equations

The Principle of Zero Products   ▪   Problem Solving

We now turn our focus to solving a new type of equation in which factoring plays an important role.

Whenever two polynomials are set equal to each other, we have a **polynomial equation**. Some examples of polynomial equations are

$$4x^3 + x^2 + 5x = 6x - 3, \qquad x^2 - x = 6, \quad \text{and} \quad 3y^4 + 2y^2 + 2 = 0.$$

The *degree of a polynomial equation* is the same as the highest degree of any term in the equation. Thus, from left to right, the degree of each equation listed above is 3, 2, and 4. A second-degree polynomial equation in one variable is called a **quadratic equation**. Of the equations listed above, only $x^2 - x = 6$ is a quadratic equation.

Polynomial equations occur frequently in applications, so the ability to solve them is an important skill. One way of solving certain polynomial equations involves factoring.

The Principle of Zero Products

When we multiply two or more numbers, the product is 0 if any one of those numbers (factors) is 0. Conversely, if a product is 0, then at least one of the factors must be 0. This property of 0 gives us a new principle for solving equations.

The Principle of Zero Products

For any real numbers a and b:

If $ab = 0$, then $a = 0$ or $b = 0$. If $a = 0$ or $b = 0$, then $ab = 0$.

Thus, if $(t - 7)(2t + 5) = 0$, then $t - 7 = 0$ or $2t + 5 = 0$. To solve a quadratic equation using the principle of zero products, we first write it in *standard form*: with 0 on one side of the equation and the leading coefficient positive. We then factor and determine when each factor is 0.

EXAMPLE **1**   Solve: $x^2 - x = 6$.

SOLUTION   To apply the principle of zero products, we need 0 on one side of the equation. Thus we subtract 6 from both sides:

Get 0 on one side.

$$x^2 - x - 6 = 0.   \text{Getting 0 on one side}$$

To express the polynomial as a product, we factor:

Factor.

$$(x - 3)(x + 2) = 0.   \text{Factoring}$$

The principle of zero products says that since $(x - 3)(x + 2)$ is 0, then

Set each factor equal to 0.

$$x - 3 = 0 \quad or \quad x + 2 = 0.   \text{Using the principle of zero products}$$

Each of these linear equations is then solved separately:

Solve.

$$x = 3 \quad or \quad x = -2.$$

We check as follows:

Check.   *Check:*

$$\begin{array}{c|c} x^2 - x = 6 \\ \hline 3^2 - 3 & 6 \\ 9 - 3 & \\ & 6 \overset{?}{=} 6 \quad \text{TRUE} \end{array} \qquad \begin{array}{c|c} x^2 - x = 6 \\ \hline (-2)^2 - (-2) & 6 \\ 4 + 2 & \\ & 6 \overset{?}{=} 6 \quad \text{TRUE} \end{array}$$

Both 3 and -2 are solutions. The solution set is $\{-2, 3\}$.

TRY EXERCISE 27

To Use the Principle of Zero Products

1. Write an equivalent equation with 0 on one side, using the addition principle.
2. Factor the nonzero side of the equation.
3. Set each factor that is not a constant equal to 0.
4. Solve the resulting equations.

CAUTION!   When we are using the principle of zero products, there must be a 0 on one side of the equation. If neither side of the equation is 0, the procedure will not work.

To see this, consider $x^2 - x = 6$ in Example 1 as

$$x(x - 1) = 6.$$

Knowing that the product of two numbers is 6 tells us nothing about either number. The numbers may be $2 \cdot 3$ or $6 \cdot 1$ or $-12 \cdot \left(-\frac{1}{2}\right)$, and so on.

Suppose we *incorrectly* set each factor equal to 6:

$$x = 6 \quad or \quad x - 1 = 6 \longleftarrow \text{This is wrong!}$$
$$x = 7.$$

Neither 6 nor 7 checks, as shown below:

$x^2 - x = 6$		$x^2 - x = 6$	
$6^2 - 6$	6	$7^2 - 7$	6
$36 - 6$		$49 - 7$	
$30 \overset{?}{=} 6$	FALSE	$42 \overset{?}{=} 6$	FALSE

EXAMPLE 2   Solve: **(a)** $5b^2 = 10b$; **(b)** $x^2 - 6x + 9 = 0$.

SOLUTION

a) We have

$$5b^2 = 10b$$
$$5b^2 - 10b = 0 \qquad \text{Getting 0 on one side}$$
$$5b(b - 2) = 0 \qquad \text{Factoring}$$
$$5b = 0 \quad or \quad b - 2 = 0 \qquad \text{Using the principle of zero products}$$
$$b = 0 \quad or \quad b = 2. \qquad \text{The checks are left to the student.}$$

The solutions are 0 and 2. The solution set is $\{0, 2\}$.

b) We have

$$x^2 - 6x + 9 = 0$$
$$(x - 3)(x - 3) = 0 \qquad \text{Factoring}$$
$$x - 3 = 0 \quad or \quad x - 3 = 0 \qquad \text{Using the principle of zero products}$$
$$x = 3 \quad or \quad x = 3. \qquad \textit{Check:}$$
$$3^2 - 6 \cdot 3 + 9 = 9 - 18 + 9 = 0.$$

There is only one solution, 3. The solution set is $\{3\}$.   ▶ **TRY EXERCISE** ▶ 23

EXAMPLE 3   Given that $f(x) = 3x^2 - 4x$, find all values of a for which $f(a) = 4$.

SOLUTION   We want all numbers a for which $f(a) = 4$. Since $f(a) = 3a^2 - 4a$, we must have

$$3a^2 - 4a = 4 \qquad \text{Setting } f(a) \text{ equal to 4}$$
$$3a^2 - 4a - 4 = 0 \qquad \text{Getting 0 on one side}$$
$$(3a + 2)(a - 2) = 0 \qquad \text{Factoring}$$
$$3a + 2 = 0 \quad or \quad a - 2 = 0$$
$$a = -\tfrac{2}{3} \quad or \quad a = 2.$$

Check: $f\left(-\frac{2}{3}\right) = 3\left(-\frac{2}{3}\right)^2 - 4\left(-\frac{2}{3}\right) = 3 \cdot \frac{4}{9} + \frac{8}{3} = \frac{4}{3} + \frac{8}{3} = \frac{12}{3} = 4.$

$f(2) = 3(2)^2 - 4(2) = 3 \cdot 4 - 8 = 12 - 8 = 4.$

To have $f(a) = 4$, we must have $a = -\frac{2}{3}$ or $a = 2$.   **TRY EXERCISE** 55

EXAMPLE 4

Let $f(x) = 3x^3 - 30x$ and $g(x) = 9x^2$. Find all x-values for which $f(x) = g(x)$.

SOLUTION   We substitute the polynomial expressions for $f(x)$ and $g(x)$ and solve the resulting equation:

$$f(x) = g(x)$$

$3x^3 - 30x = 9x^2$ Substituting

$3x^3 - 9x^2 - 30x = 0$ Getting 0 on one side and writing in descending order

$3x(x^2 - 3x - 10) = 0$ Factoring out a common factor

$3x(x + 2)(x - 5) = 0$ Factoring the trinomial

$3x = 0 \quad or \quad x + 2 = 0 \quad or \quad x - 5 = 0$ Using the principle of zero products

$x = 0 \quad or \quad\quad x = -2 \quad or \quad\quad x = 5.$

Check: $f(0) = 3 \cdot 0^3 - 30 \cdot 0 = 3 \cdot 0 - 0 = 0$, and

$g(0) = 9 \cdot 0^2 = 9 \cdot 0 = 0;$

$f(-2) = 3(-2)^3 - 30(-2) = 3(-8) - (-60) = -24 + 60 = 36$, and

$g(-2) = 9(-2)^2 = 9 \cdot 4 = 36;$

$f(5) = 3 \cdot 5^3 - 30 \cdot 5 = 3 \cdot 125 - 150 = 375 - 150 = 225$, and

$g(5) = 9 \cdot 5^2 = 9 \cdot 25 = 225.$

For $x = 0, -2$, and 5, we have $f(x) = g(x)$.   **TRY EXERCISE** 63

EXAMPLE 5

Find the domain of F if $F(x) = \dfrac{x - 2}{x^2 + 2x - 15}$.

SOLUTION   The domain of F is the set of all values for which $F(x)$ is a real number. Since division by 0 is undefined, $F(x)$ cannot be calculated for any x-value for which the denominator, $x^2 + 2x - 15$, is 0. To make sure these values are *excluded*, we solve:

$x^2 + 2x - 15 = 0$ Setting the denominator equal to 0

$(x - 3)(x + 5) = 0$ Factoring

$x - 3 = 0 \quad or \quad x + 5 = 0$

$x = 3 \quad or \quad\quad x = -5.$ These are the values to *exclude.*

The domain of F is $\{x \mid x$ is a real number *and* $x \neq -5$ *and* $x \neq 3\}$. In interval notation, the domain is $(-\infty, -5) \cup (-5, 3) \cup (3, \infty)$.   **TRY EXERCISE** 65

ALGEBRAIC–GRAPHICAL CONNECTION

Let's return for a moment to the equation in Example 1, $x^2 - x = 6$. One way to begin solving this equation is to either graph by hand or use a graphing calculator to draw the graph of the function given by $f(x) = x^2 - x$. We then look for any x-value that is paired with 6, as shown on the left below. In Chapter 8, we develop methods for quickly generating these graphs by hand.

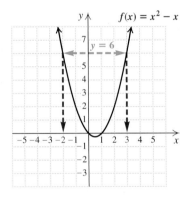
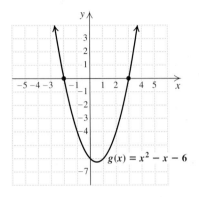

Equivalently, we could graph the function given by $g(x) = x^2 - x - 6$ and look for values of x for which $g(x) = 0$. Using this method, we can visualize what we call the *roots*, or *zeros*, of a polynomial function.

It appears from the graph that $f(x) = 6$ and $g(x) = 0$ when $x \approx -2$ or $x \approx 3$. Although making a graph is not the fastest or most precise method of solving this equation, it gives us a visualization and is useful with problems that are more difficult to solve algebraically. In some cases, the x-intercepts of a graph can be used to help find a factorization.

TECHNOLOGY CONNECTION

To use the INTERSECT option (see p. 115 in Section 2.4) to check Example 1, let $y_1 = x^2 - x$ and $y_2 = 6$. One intersection occurs at $(-2, 6)$. You should confirm that the other occurs at $(3, 6)$.

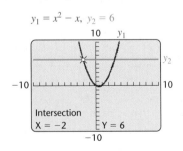

Another approach is to find where the graph of $y_3 = x^2 - x - 6$ crosses the x-axis, using the ZERO option of the CALC menu (see p. 112 in Section 2.4). One zero is -2.

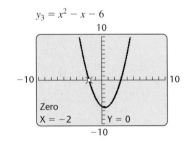

The student should confirm that the other zero is 3.

(continued)

To visualize Example 4, we let $y_1 = 3x^3 - 30x$ and $y_2 = 9x^2$ and use a viewing window of $[-4, 6, -110, 360]$.

1. Use the INTERSECT option of the CALC menu to confirm all three solutions of Example 4.
2. As a second check, show that $y_3 = 3x^3 - 9x^2 - 30x$ has zeros at 0, −2, and 5.

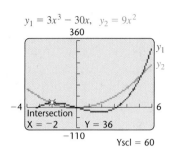

$y_1 = 3x^3 - 30x, \quad y_2 = 9x^2$

Intersection X = −2   Y = 36

Yscl = 60

Problem Solving

Some problems can be translated to quadratic equations, which we can now solve. The problem-solving process is the same as for other kinds of problems.

EXAMPLE 6

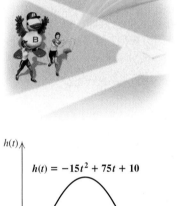

Prize tee shirts.   During intermission at sporting events, it has become common for team mascots to use a powerful slingshot to launch tightly rolled tee shirts into the stands. The height $h(t)$, in feet, of an airborne tee shirt t seconds after being launched can be approximated by

$$h(t) = -15t^2 + 75t + 10.$$

After peaking, a rolled-up tee shirt is caught by a fan 70 ft above ground level. How long was the tee shirt in the air?

SOLUTION

1. **Familiarize.** We make a drawing and label it, using the information provided (see the figure at left). If we wanted to, we could evaluate $h(t)$ for a few values of t. Note that t cannot be negative, since it represents time from launch.

2. **Translate.** The relevant function has been provided. Since we are asked to determine how long it will take for the shirt to reach someone 70 ft above ground level, we are interested in the value of t for which $h(t) = 70$:

$$-15t^2 + 75t + 10 = 70.$$

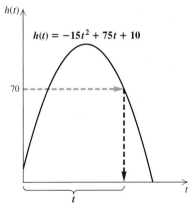

$h(t) = -15t^2 + 75t + 10$

3. **Carry out.** We solve the quadratic equation:

$$-15t^2 + 75t + 10 = 70$$
$$-15t^2 + 75t - 60 = 0 \qquad \text{Subtracting 70 from both sides}$$
$$\left.\begin{array}{l} -15(t^2 - 5t + 4) = 0 \\ -15(t - 4)(t - 1) = 0 \end{array}\right\} \quad \text{Factoring}$$
$$t - 4 = 0 \quad or \quad t - 1 = 0$$
$$t = 4 \quad or \quad t = 1.$$

The solutions appear to be 4 and 1.

4. **Check.** We have

$$h(4) = -15 \cdot 4^2 + 75 \cdot 4 + 10 = -240 + 300 + 10 = 70 \text{ ft};$$
$$h(1) = -15 \cdot 1^2 + 75 \cdot 1 + 10 = -15 + 75 + 10 = 70 \text{ ft}.$$

Both 4 and 1 check. However, the problem states that the tee shirt is caught after peaking. Thus we reject 1 since that would indicate when the height of the tee shirt was 70 ft on the way *up*.

5. **State.** The tee shirt was in the air for 4 sec before being caught 70 ft above ground level.

TRY EXERCISE 93

The following problem involves the **Pythagorean theorem**, which relates the lengths of the sides of a right triangle. A **right triangle** has a 90°, or right, angle, which is indicated in the triangle by the symbol ⌐ or ⌐. The longest side, called the **hypotenuse**, is opposite the 90° angle. The other sides, called **legs**, form the two sides of the right angle.

The Pythagorean Theorem

In any right triangle, if a and b are the lengths of the legs and c is the length of the hypotenuse, then

$$a^2 + b^2 = c^2.$$

The symbol ⌐ denotes a 90° angle.

EXAMPLE 7

Carpentry. In order to build a deck at a right angle to their house, Lucinda and Felipe decide to plant a stake in the ground a precise distance from the back wall of their house. This stake will combine with two marks on the house to form a right triangle. From a course in geometry, Lucinda remembers that there are three consecutive integers that can work as sides of a right triangle. Find the measurements of that triangle.

SOLUTION

1. **Familiarize.** Recall that x, $x + 1$, and $x + 2$ can be used to represent three unknown consecutive integers. Since $x + 2$ is the largest number, it must represent the hypotenuse. The legs serve as the sides of the right angle, so one leg must be formed by the marks on the house. We make a drawing in which

$$x = \text{the distance between the marks on the house,}$$
$$x + 1 = \text{the length of the other leg, and}$$
$$x + 2 = \text{the length of the hypotenuse.}$$

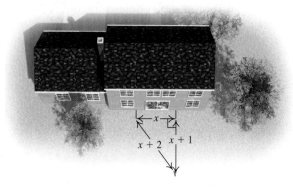

2. **Translate.** Applying the Pythagorean theorem, we translate as follows:

$$a^2 + b^2 = c^2$$
$$x^2 + (x + 1)^2 = (x + 2)^2.$$

3. **Carry out.** We solve the equation as follows:

$x^2 + (x^2 + 2x + 1) = x^2 + 4x + 4$	Squaring the binomials
$2x^2 + 2x + 1 = x^2 + 4x + 4$	Combining like terms
$x^2 - 2x - 3 = 0$	Subtracting $x^2 + 4x + 4$ from both sides
$(x - 3)(x + 1) = 0$	Factoring
$x - 3 = 0 \quad or \quad x + 1 = 0$	Using the principle of zero products
$x = 3 \quad or \qquad x = -1.$	

4. **Check.** The integer -1 cannot be a length of a side because it is negative. For $x = 3$, we have $x + 1 = 4$, and $x + 2 = 5$. Since $3^2 + 4^2 = 5^2$, the lengths 3, 4, and 5 determine a right triangle. Thus, 3, 4, and 5 check.

5. **State.** Lucinda and Felipe should use a triangle with sides having a ratio of 3:4:5. Thus, if the marks on the house are 3 yd apart, they should locate the stake at the point in the yard that is precisely 4 yd from one mark and 5 yd from the other mark.

> **TRY EXERCISE** 85

EXAMPLE 8

Display of a sports card. A valuable sports card is 4 cm wide and 5 cm long. The card is to be sandwiched by two pieces of Lucite, each of which is $5\frac{1}{2}$ times the area of the card. Determine the dimensions of the Lucite that will ensure a uniform border around the card.

SOLUTION

1. **Familiarize.** We make a drawing and label it, using x to represent the width of the border, in centimeters. Since the border extends uniformly around the entire card, the length of the Lucite must be $5 + 2x$ and the width must be $4 + 2x$.

2. **Translate.** We rephrase the information given and translate as follows:

Area of Lucite   is $5\frac{1}{2}$ times   area of card.

$$(5 + 2x)(4 + 2x) = 5\frac{1}{2} \cdot 5 \cdot 4$$

3. **Carry out.** We solve the equation:

$$(5 + 2x)(4 + 2x) = 5\frac{1}{2} \cdot 5 \cdot 4$$

$$20 + 10x + 8x + 4x^2 = 110 \qquad \text{Multiplying}$$

$$4x^2 + 18x - 90 = 0 \qquad \text{Finding standard form}$$

$$\left.\begin{array}{r} 2(2x^2 + 9x - 45) = 0 \\ 2(2x + 15)(x - 3) = 0 \end{array}\right\} \qquad \text{Factoring}$$

$$2x + 15 = 0 \quad or \quad x - 3 = 0 \qquad \text{Principle of zero products}$$

$$x = -7\frac{1}{2} \quad or \qquad x = 3.$$

4. **Check.** We check 3 in the original problem. (Note that $-7\frac{1}{2}$ is not a solution because measurements cannot be negative.) If the border is 3 cm wide, the Lucite will have a length of $5 + 2 \cdot 3$, or 11 cm, and a width of $4 + 2 \cdot 3$, or 10 cm. The area of the Lucite is thus $11 \cdot 10$, or 110 cm². Since the area of the card is 20 cm² and 110 cm² is $5\frac{1}{2}$ times 20 cm², the number 3 checks.

5. **State.** Each piece of Lucite should be 11 cm long and 10 cm wide.

> **TRY EXERCISE** 77

Visualizing for Success

A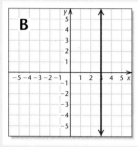

F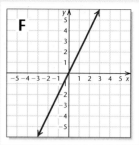

Match each equation or function with its graph.

1. $f(x) = x$

B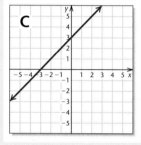

2. $f(x) = |x|$

G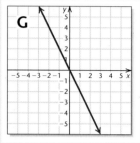

3. $f(x) = x^2$

4. $f(x) = 3$

C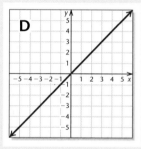

5. $x = 3$

H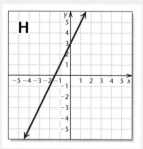

6. $y = x + 3$

7. $y = x - 3$

D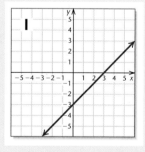

8. $y = 2x$

I

9. $y = -2x$

10. $y = 2x + 3$

E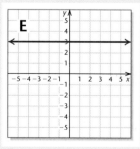

Answers on page A-24

J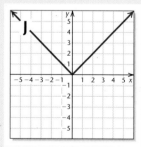

An alternate, animated version of this activity appears in MyMathLab. To use MyMathLab, you need a course ID and a student access code. Contact your instructor for more information.

5.8 EXERCISE SET

Concept Reinforcement *Classify each statement as either true or false.*

1. Not every polynomial equation is quadratic.

2. The equations $3x^2 - 5x + 2 = 0$, $4t^2 = t + 2$, and $9n^2 = 4$ are all examples of quadratic equations.

3. Every quadratic equation has two solutions.

4. To use the principle of zero products, we must have an equation with 0 on one side.

5. Every triangle has a hypotenuse.

6. When we are solving an applied problem, a solution of the translated equation may not be a solution of the problem.

Solve.

7. $(x - 2)(x - 5) = 0$

8. $(x + 3)(x + 7) = 0$

9. $x^2 + 8x + 7 = 0$

10. $x^2 - 12x + 20 = 0$

11. $9t(2t + 1) = 0$

12. $3t(2t - 5) = 0$

13. $15t^2 - 12t = 0$

14. $24t^2 + 8t = 0$

15. $(2t + 5)(t - 7) = 0$

16. $(2t - 7)(t + 2) = 0$

17. $x^2 - 3x - 18 = 0$

18. $x^2 + 2x - 35 = 0$

19. $t^2 - 10t = 0$

20. $t^2 - 8t = 0$

21. $(3x - 1)(4x - 5) = 0$

22. $(5x + 3)(2x - 7) = 0$

23. $4a^2 = 10a$

24. $6a^2 = 8a$

25. $t^2 - 6t - 16 = 0$

26. $t^2 - 3t - 18 = 0$

27. $t^2 - 3t = 28$

28. $x^2 - 4x = 45$

29. $r^2 + 16 = 8r$

30. $a^2 + 1 = 2a$

31. $a^2 + 20a + 100 = 0$

32. $z^2 + 6z + 9 = 0$

33. $8y + y^2 + 15 = 0$

34. $9x + x^2 + 20 = 0$

Aha! 35. $n^2 - 81 = 0$

36. $p^2 - 144 = 0$

37. $x^3 - 2x^2 = 63x$

38. $a^3 - 3a^2 = 40a$

39. $t^2 = 25$

40. $r^2 = 4$

41. $(a - 4)(a + 4) = 20$

42. $(t - 6)(t + 6) = 45$

43. $-9x^2 + 15x - 4 = 0$

44. $3x^2 - 8x + 4 = 0$

45. $-8y^3 - 10y^2 - 3y = 0$

46. $-4t^3 - 11t^2 - 6t = 0$

47. $(z + 4)(z - 2) = -5$

48. $(y - 3)(y + 2) = 14$

49. $x(5 + 12x) = 28$

50. $a(1 + 21a) = 10$

51. $a^2 - \frac{1}{100} = 0$

52. $x^2 - \frac{1}{64} = 0$

53. $t^4 - 26t^2 + 25 = 0$

54. $t^4 - 13t^2 + 36 = 0$

55. Let $f(x) = x^2 + 12x + 40$. Find a such that $f(a) = 8$.

56. Let $f(x) = x^2 + 14x + 50$. Find a such that $f(a) = 5$.

57. Let $g(x) = 2x^2 + 5x$. Find a such that $g(a) = 12$.

58. Let $g(x) = 2x^2 - 15x$. Find a such that $g(a) = -7$.

59. Let $h(x) = 12x + x^2$. Find a such that $h(a) = -27$.

60. Let $h(x) = 4x - x^2$. Find a such that $h(a) = -32$.

61. If $f(x) = 12x^2 - 15x$ and $g(x) = 8x - 5$, find all x-values for which $f(x) = g(x)$.

62. If $f(x) = 10x^2 + 20$ and $g(x) = 43x - 8$, find all x-values for which $f(x) = g(x)$.

63. If $f(x) = 2x^3 - 5x$ and $g(x) = 10x - 7x^2$, find all x-values for which $f(x) = g(x)$.

64. If $f(x) = 3x^3 - 4x$ and $g(x) = 8x^2 + 12x$, find all x-values for which $f(x) = g(x)$.

Find the domain of the function f given by each of the following.

65. $f(x) = \dfrac{3}{x^2 - 3x - 4}$

66. $f(x) = \dfrac{2}{x^2 - 7x + 10}$

67. $f(x) = \dfrac{x}{6x^2 - 54}$

68. $f(x) = \dfrac{2x}{5x^2 - 20}$

69. $f(x) = \dfrac{x - 5}{9x - 18x^2}$

70. $f(x) = \dfrac{1 + x}{3x - 15x^2}$

71. $f(x) = \dfrac{7}{5x^3 - 35x^2 + 50x}$

72. $f(x) = \dfrac{3}{2x^3 - 2x^2 - 12x}$

Solve.

73. *Postage rates.*   The maximum size envelope that can be mailed with a Large Envelope rate is 3 in. longer than it is wide. The area is $180\,\text{in}^2$. Find the length and the width.
Source: USPS

74. *Photo size.*   A photo is 3 cm longer than it is wide. Find the length and the width if the area is $108\,\text{cm}^2$.

75. *Geometry.*   If each of the sides of a square is lengthened by 4 m, the area becomes $49\,\text{m}^2$. Find the length of a side of the original square.

76. *Geometry.*   If each of the sides of a square is lengthened by 6 cm, the area becomes $144\,\text{cm}^2$. Find the length of a side of the original square.

77. *Framing a picture.*   A picture frame measures 12 cm by 20 cm, and $84\,\text{cm}^2$ of picture shows. Find the width of the frame.

78. *Framing a picture.*   A picture frame measures 14 cm by 20 cm, and $160\,\text{cm}^2$ of picture shows. Find the width of the frame.

79. *Catering.*   A rectangular table is 60 in. long and 40 in. wide. A tablecloth that is twice the area of the table will be centered on the table. How far will the tablecloth hang down on each side?

80. *Landscaping.*   A rectangular garden is 30 ft by 40 ft. Part of the garden is removed in order to install a walkway of uniform width around it. The area of the new garden is one-half the area of the old garden. How wide is the walkway?

81. Three consecutive even integers are such that the square of the third is 76 more than the square of the second. Find the three integers.

82. Three consecutive even integers are such that the square of the first plus the square of the third is 136. Find the three integers.

83. *Furniture.*   The base of a triangular tabletop is 20 in. longer than the height. The area is $750\,\text{in}^2$. Find the height and the base.

84. *Tent design.*   The triangular entrance to a tent is 2 ft taller than it is wide. The area of the entrance is $12\,\text{ft}^2$. Find the height and the base.

85. *Building lots.*   A lot for sale in New York City is in the shape of a right triangle, as shown below. The side not bordering a street is 25 ft long. One of the other sides is 5 ft longer than the remaining side. Find the lengths of the sides.

86. *Antenna wires.*   A wire is stretched from the ground to the top of an antenna tower, as shown. The wire is 20 m long. The height of the tower is 4 m greater than the distance d from the tower's base to the bottom of the wire. Find the distance *d* and the height of the tower.

20 m

d

87. *Ladder location.*   The foot of an extension ladder is 9 ft from a wall. The height that the ladder reaches on the wall and the length of the ladder are consecutive integers. How long is the ladder?

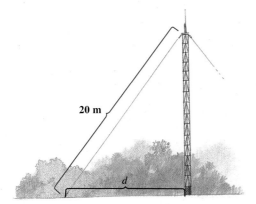

9 ft

88. *Ladder location.*   The foot of an extension ladder is 10 ft from a wall. The ladder is 2 ft longer than the height that it reaches on the wall. How far up the wall does the ladder reach?

89. *Garden design.*   Ignacio is planning a garden that is 25 m longer than it is wide. The garden will have an area of 7500 m². What will its dimensions be?

90. *Garden design.*   A flower bed is to be 3 m longer than it is wide. The flower bed will have an area of 108 m². What will its dimensions be?

91. *Cabinet making.*   Dovetail Woodworking determines that the revenue *R*, in thousands of dollars, from the sale of *x* sets of cabinets is given by $R(x) = 2x^2 + x$. If the cost *C*, in thousands of dollars, of producing *x* sets of cabinets is given by $C(x) = x^2 - 2x + 10$, how many sets must be produced and sold in order for the company to break even?

92. *Camera production.*   Suppose that the cost of making *x* cameras is $C(x) = \frac{1}{9}x^2 + 2x + 1$, where $C(x)$ is in thousands of dollars. If the revenue from the sale of *x* cameras is given by $R(x) = \frac{5}{36}x^2 + 2x$, where $R(x)$ is in thousands of dollars, how many

cameras must be sold in order for the firm to break even?

93. *Prize tee shirts.*   Using the model in Example 6, determine how long a tee shirt has been airborne if it is caught on the way *up* by a fan 100 ft above ground level.

94. *Prize tee shirts.*   Using the model in Example 6, determine how long a tee shirt has been airborne if it is caught on the way *down* by a fan 10 ft above ground level.

95. *Fireworks displays.*   Fireworks are typically launched from a mortar with an upward velocity (initial speed) of about 64 ft/sec. The height $h(t)$, in feet, of a "weeping willow" display, *t* seconds after having been launched from an 80-ft high rooftop, is given by

$$h(t) = -16t^2 + 64t + 80.$$

How long will it take the cardboard shell from the fireworks to reach the ground?

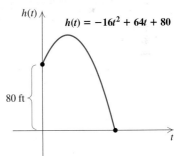

$h(t)$

$h(t) = -16t^2 + 64t + 80$

80 ft

t

96. *Safety flares.*   Suppose that a flare is launched upward with an initial velocity of 80 ft/sec from a height of 224 ft. Its height in feet, $h(t)$, after *t* seconds is given by

$$h(t) = -16t^2 + 80t + 224.$$

How long will it take the flare to reach the ground?

97. *Advertising.*   The amount spent by U.S. companies for online advertising can be approximated by

$$a(t) = \tfrac{1}{2}t^2 - \tfrac{19}{10}t + 8,$$

where $a(t)$ is in billions of dollars and *t* is the number of years after 2000. In what year after 2000 did U.S. companies spend about $11 billion in online advertising?

Source: eMarketer

98. *Advertising.* The amount spent in political advertising in U.S. newspapers can be approximated by

$$p(t) = \tfrac{1}{5}(12t^2 - 5t + 130),$$

where $p(t)$ is in billions of dollars and t is the number of years after 2000. In what year after 2000 was $81 billion spent in political advertising in U.S. newspapers?
Source: PQMedia

99. Suppose you are given a detailed graph of $y = p(x)$, where $p(x)$ is some polynomial in x. How could the graph be used to help solve the equation $p(x) = 0$?

100. Can the number of solutions of a quadratic equation exceed two? Why or why not?

Skill Review

To prepare for Section 6.1, review multiplying, dividing, and simplifying fractions (Section 1.2).

Find the reciprocal. [1.2]

101. $\tfrac{2}{3}$

102. -2

Simplify. [1.2]

103. $\dfrac{5}{12} \cdot \left(-\dfrac{45}{8}\right)$

104. $\dfrac{5}{12} \div \left(-\dfrac{45}{8}\right)$

105. $\dfrac{6 - 4 \cdot 8}{-2 + 3 \cdot 5}$

106. $\dfrac{2 \cdot 3 - 5 \cdot 6}{7 - 5^2}$

107. $\dfrac{240}{280}$

108. $-\dfrac{462}{252}$

Synthesis

109. Explain how one could write a quadratic equation that has -3 and 5 as solutions.

110. If the graph of $f(x) = ax^2 + bx + c$ has no x-intercepts, what can you conclude about the equation $ax^2 + bx + c = 0$?

Solve.

111. $(8x + 11)(12x^2 - 5x - 2) = 0$

112. $(x + 1)^3 = (x - 1)^3 + 26$

113. Use the following graph of $g(x) = -x^2 - 2x + 3$ to solve $-x^2 - 2x + 3 = 0$ and to solve $-x^2 - 2x + 3 \geq -5$.

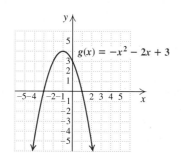

114. Use the following graph of $f(x) = x^2 - 2x - 3$ to solve $x^2 - 2x - 3 = 0$ and to solve $x^2 - 2x - 3 < 5$.

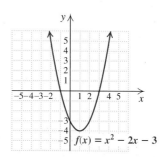

115. Find a polynomial function f for which $f(2) = 0, f(-1) = 0, f(3) = 0,$ and $f(0) = 30$.

116. Find a polynomial function g for which $g(-3) = 0, g(1) = 0, g(5) = 0,$ and $g(0) = 45$.

117. *Box construction.* A rectangular piece of tin is twice as long as it is wide. Squares 2 cm on a side are cut out of each corner, and the ends are turned up to make a box whose volume is 480 cm³. What are the dimensions of the piece of tin?

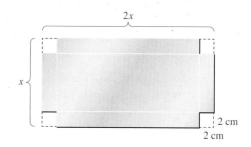

118. *Navigation.* A tugboat and a freighter leave the same port at the same time at right angles. The freighter travels 7 km/h slower than the tugboat. After 4 hr, they are 68 km apart. Find the speed of each boat.

119. *Skydiving.* During the first 13 sec of a jump, a skydiver falls approximately $11.12t^2$ feet in t seconds. A small heavy object (with less wind resistance) falls about $15.4t^2$ feet in t seconds. Suppose that a skydiver jumps from 30,000 ft, and 1 sec later a camera falls out of the airplane. How long will it take the camera to catch up to the skydiver?

120. Use the TABLE feature of a graphing calculator to check that -5 and 3 are not in the domain of F as shown in Example 5.

121. Use the TABLE feature of a graphing calculator to check your answers to Exercises 67, 69, and 71.

In Exercises 122–125, use a graphing calculator to find any real-number solutions that exist accurate to two decimal places.

122. $x^2 - 2x - 8 = 0$ (Check by factoring.)

123. $-x^2 + 13.80x = 47.61$

124. $-x^2 + 3.63x + 34.34 = x^2$

125. $x^3 - 3.48x^2 + x = 3.48$

126. Mary Louise is attempting to solve $x^3 + 20x^2 + 4x + 80 = 0$ with a graphing calculator. Unfortunately, when she graphs $y_1 = x^3 + 20x^2 + 4x + 80$ in a standard $[-10, 10, -10, 10]$ window, she sees no graph at all, let alone any x-intercept. Can this problem be solved graphically? If so, how? If not, why not?

CONNECTING the CONCEPTS

Be careful not to confuse a polynomial expression with a polynomial equation. We can form equivalent polynomial expressions by combining like terms, adding, subtracting, multiplying, and factoring. An expression cannot be "solved." Compare the following.

Factor: $x^2 - x - 12$.

$$x^2 - x - 12 = (x - 4)(x + 3)$$

The factorization is $(x - 4)(x + 3)$.

The expressions $x^2 - x - 12$ and $(x - 4)(x + 3)$ are equivalent.

Solve: $x^2 - x - 12 = 0$.

$$x^2 - x - 12 = 0$$
$$(x - 4)(x + 3) = 0 \qquad \text{Factoring}$$
$$x - 4 = 0 \quad or \quad x + 3 = 0 \qquad \text{Using the principle of zero products}$$
$$x = 4 \quad or \qquad x = -3$$

The solution set is $\{-3, 4\}$.

The numbers -3 and 4 are solutions.

MIXED REVIEW

1. Factor: $x^2 + 5x + 6$.

2. Solve: $x^2 + 5x + 6 = 0$.

3. Solve: $x^2 + 6 = 5x$.

4. Combine like terms: $3x^2 - x + x^2 - 5$.

5. Subtract: $(3x^2 - x) - (x^2 - 5)$.

6. Factor: $a^2 - 1$.

7. Multiply: $(a + 1)(a - 1)$.

8. Solve: $(a + 1)(a - 1) = 24$.

9. Solve: $x^2 = 19x$.

10. Add: $(3x^4 + x^2 - 2) + (4x^2 - 7x + 1)$.

11. Combine like terms: $3x^2 - 14 + x^2 + 3x$.

12. Solve: $3x^2 - 14 = x^2 + 3x$.

13. Factor: $t^3 + 1$.

14. Factor: $10m^2 - 29m - 21$.

15. Multiply: $(5n - 6)^2$.

16. Solve: $81t^2 + 36t + 4 = 0$.

17. Factor: $2x^2 + 8x - 42$.

18. Multiply: $(5y - 1)(8y + 3)$.

19. Subtract: $(x^2 - 5x - 7) - (4x^2 - 5x + 2)$.

20. Solve: $x^2 = 64$.

Study Summary

KEY TERMS AND CONCEPTS	EXAMPLES

SECTION 5.1: INTRODUCTION TO POLYNOMIALS AND POLYNOMIAL FUNCTIONS

A **polynomial** is a monomial or a sum of monomials.

$3x^6 - 5x^2 + x - 8$ is a polynomial.
It is arranged in **descending order**.
$3x^6$ is the **leading term**.
3 is the **leading coefficient**.
6 is the **degree** of the polynomial.

Polynomials with one, two, or three terms have special names.

Monomial (one term): $4x^3$
Binomial (two terms): $x^2 - 5$
Trinomial (three terms): $3t^3 + 2t - 10$

Polynomials of degree 4 or less have special names.

Constant (degree 0): -9
Linear (degree 1): $\frac{1}{3}x + 8$
Quadratic (degree 2): $x^2 + 2x + 3$
Cubic (degree 3): $-x^3 - 5x$
Quartic (degree 4): $15x^4 + x^2 + 1$

To add polynomials, we combine like terms.

$(2x^2 - 5x + 7) + (x^3 + 3x - 10) = x^3 + 2x^2 - 2x - 3$

To subtract polynomials, we add the opposite of the polynomial being subtracted.

$(4x^3 - x^2 - 5x) - (7x^3 - x^2 - 11x)$
$= 4x^3 - x^2 - 5x - 7x^3 + x^2 + 11x$
$= -3x^3 + 6x$

SECTION 5.2: MULTIPLICATION OF POLYNOMIALS

To multiply polynomials, we multiply each term of one polynomial by each term of the other.

$(x + 2)(x^2 - x - 1) = x \cdot x^2 - x \cdot x - x \cdot 1 + 2 \cdot x^2 - 2 \cdot x - 2 \cdot 1$
$= x^3 - x^2 - x + 2x^2 - 2x - 2$
$= x^3 + x^2 - 3x - 2$

Special Products

$(A + B)(C + D) = AC + AD + BC + BD$
$(A + B)^2 = A^2 + 2AB + B^2$
$(A - B)^2 = A^2 - 2AB + B^2$
$(A + B)(A - B) = A^2 - B^2$

$(y^3 - 2)(y + 4) = y^4 + 4y^3 - 2y - 8$
$(t + 6)^2 = t^2 + 2 \cdot t \cdot 6 + 36 = t^2 + 12t + 36$
$(c - 5d)^2 = c^2 - 2 \cdot c \cdot 5d + (5d)^2 = c^2 - 10cd + 25d^2$
$(x^2 + 3)(x^2 - 3) = (x^2)^2 - (3)^2 = x^4 - 9$

SECTION 5.3: COMMON FACTORS AND FACTORING BY GROUPING

To **factor** a polynomial means to write it as a product. We begin by factoring out the **largest common factor**.

$12x^4 - 30x^3 = 6x^3(2x - 5)$ $6x^3$ is the largest common factor.

Some polynomials with four terms can be **factored by grouping**.

$3x^3 - x^2 - 6x + 2 = x^2(3x - 1) - 2(3x - 1)$
$= (3x - 1)(x^2 - 2)$

SECTION 5.4: FACTORING TRINOMIALS

Some trinomials can be written as the product of two binomials.

$$x^2 - 11x + 18 = (x - 2)(x - 9)$$
$$6x^2 - 5x - 6 = (3x + 2)(2x - 3)$$ Using FOIL or the grouping method

SECTION 5.5: FACTORING PERFECT-SQUARE TRINOMIALS AND DIFFERENCES OF SQUARES

Factoring a Perfect-Square Trinomial
$$A^2 + 2AB + B^2 = (A + B)^2$$
$$A^2 - 2AB + B^2 = (A - B)^2$$

$$y^2 + 20y + 100 = (y + 10)^2$$
$$m^2 - 14mn + 49n^2 = (m - 7n)^2$$

Factoring a Difference of Squares
$$A^2 - B^2 = (A + B)(A - B)$$

$$9t^2 - 1 = (3t + 1)(3t - 1)$$

SECTION 5.6: FACTORING SUMS OR DIFFERENCES OF CUBES

Factoring a Sum or a Difference of Cubes
$$A^3 + B^3 = (A + B)(A^2 - AB + B^2)$$
$$A^3 - B^3 = (A - B)(A^2 + AB + B^2)$$

$$x^3 + 1000 = (x + 10)(x^2 - 10x + 100)$$
$$z^6 - 8w^3 = (z^2 - 2w)(z^4 + 2wz^2 + 4w^2)$$

SECTION 5.7: FACTORING: A GENERAL STRATEGY

To factor a polynomial:

A. Factor out the largest common factor.

B. Look at the number of terms.

Two terms: Try to factor as a difference of squares, a sum of cubes, or a difference of cubes.

Three terms: Try to factor as a trinomial square. Then try FOIL or grouping.

Four terms: Try factoring by grouping.

C. Factor completely.

D. Check by multiplying.

$$5x^5 - 80x = 5x(x^4 - 16)$$ Factoring out a common factor

$$= 5x(x^2 + 4)(x^2 - 4)$$ Factoring a difference of squares

$$= 5x(x^2 + 4)(x + 2)(x - 2)$$ Factoring a difference of squares

Check: $5x(x^2 + 4)(x + 2)(x - 2) = 5x(x^2 + 4)(x^2 - 4)$
$$= 5x(x^4 - 16) = 5x^5 - 80x$$

$$-x^2y^3 - 3xy^2 + 10y = -y(x^2y^2 + 3xy - 10)$$ Factoring out a common factor

$$= -y(xy + 5)(xy - 2)$$ Factoring a trinomial

Check: $-y(xy + 5)(xy - 2) = -y(x^2y^2 + 3xy - 10)$
$$= -x^2y^3 - 3xy^2 + 10y$$

SECTION 5.8: APPLICATIONS OF POLYNOMIAL EQUATIONS

The Principle of Zero Products

For any real numbers a and b:

If $ab = 0$, then $a = 0$ or $b = 0$.
If $a = 0$ or $b = 0$, then $ab = 0$.

Solve: $x^2 + 7x = 30$.

$$x^2 + 7x = 30$$
$$x^2 + 7x - 30 = 0$$ Getting 0 on one side
$$(x + 10)(x - 3) = 0$$ Factoring
$$x + 10 = 0 \quad or \quad x - 3 = 0$$ Using the principle of zero products
$$x = -10 \quad or \quad x = 3$$

The solution set is $\{-10, 3\}$.

The Pythagorean Theorem

In any right triangle, if a and b are the lengths of the legs and c is the length of the hypotenuse, then $a^2 + b^2 = c^2$.

For the right triangle shown,

$$x^2 + (x + 1)^2 = 5^2.$$

Review Exercises: Chapter 5

Concept Reinforcement *In each of Exercises 1–10, match the item with the most appropriate choice from the column on the right.*

1. ____ A polynomial with four terms [5.1]

2. ____ A term that is not a monomial [5.1]

3. ____ A polynomial written in ascending order [5.1]

4. ____ A polynomial that cannot be factored [5.7]

5. ____ A difference of two squares [5.5]

6. ____ A perfect-square trinomial [5.5]

7. ____ A difference of two cubes [5.6]

8. ____ The principle of zero products [5.8]

9. ____ A quadratic equation [5.8]

10. ____ The longest side in any right triangle [5.8]

a) $5x + 2x^2 - 4x^3$

b) $3x^{-1}$

c) If $a \cdot b = 0$, then $a = 0$ or $b = 0$.

d) Prime

e) $t^2 - 9$

f) Hypotenuse

g) $8x^3 - 4x^2 + 12x + 14$

h) $t^3 - 27$

i) $2x^2 - 4x = 7$

j) $4a^2 - 12a + 9$

11. Determine the degree of $2xy^6 - 7x^8y^3 + 2x^3 + 9$. [5.1]

12. Arrange $3x - 5x^3 + 2x^2 + 9$ in descending order and determine the leading term and the leading coefficient. [5.1]

13. Arrange in ascending powers of x:
$8x^6y - 7x^8y^3 + 2x^3 - 3x^2$. [5.1]

14. Find $P(0)$ and $P(-1)$:
$P(x) = x^3 - x^2 + 4x$. [5.1]

15. Given $P(x) = x^2 + 10x$, find and simplify
$P(a + h) - P(a)$. [5.2]

Combine like terms. [5.1]

16. $6 - 4a + a^2 - 2a^3 - 10 + a$

17. $4x^2y - 3xy^2 - 5x^2y + xy^2$

Add. [5.1]

18. $(-7x^3 - 4x^2 + 3x + 2) + (5x^3 + 2x + 6x^2 + 1)$

19. $(4n^3 + 2n^2 - 12n + 7) + (-6n^3 + 9n + 4 + n)$

20. $(-9xy^2 - xy - 6x^2y) + (-5x^2y - xy + 4xy^2)$

Subtract. [5.1]

21. $(8x - 5) - (-6x + 2)$

22. $(4a - b - 3c) - (6a - 7b - 3c)$

23. $(8x^2 - 4xy + y^2) - (2x^2 - 3y^2 - 9y)$

Simplify as indicated. [5.2]

24. $(3x^2y)(-6xy^3)$

25. $(x^4 - 2x^2 + 3)(x^4 + x^2 - 1)$

26. $(4ab + 3c)(2ab - c)$

27. $(7t + 1)(7t - 1)$

28. $(3x - 4y)^2$

29. $(x + 3)(2x - 1)$

30. $(x^2 + 4y^3)^2$

31. $(3t - 5)^2 - (2t + 3)^2$

32. $\left(x - \frac{1}{3}\right)\left(x - \frac{1}{6}\right)$

Factor.

33. $7x^2 + 6x$ [5.3]

34. $-3y^4 - 9y^2 + 12y$ [5.3]

35. $15x^4 - 18x^3 + 21x^2 - 3x$ [5.3]

36. $a^2 - 12a + 27$ [5.4]

37. $3m^2 - 10m - 8$ [5.4]

38. $25x^2 + 20x + 4$ [5.5]

39. $4y^2 - 16$ [5.5]

40. $5x^2 + x^3 - 14x$ [5.4]

41. $ax + 2bx - ay - 2by$ [5.3]

42. $3y^3 + 6y^2 - 5y - 10$ [5.3]

43. $a^4 - 81$ [5.5]

44. $4x^4 + 4x^2 + 20$ [5.3]

45. $27x^3 + 8$ [5.6]

46. $\frac{1}{125}b^3 - \frac{1}{8}c^6$ [5.6]

47. $490t^2 - 640t^4$ [5.5]

48. $n^7 + n^5$ [5.3]

49. $0.01x^4 - 1.44y^6$ [5.5]

50. $4x^2y + 100y - 40xy$ [5.5]

51. $6t^2 + 17pt + 5p^2$ [5.4]

52. $x^3 + 2x^2 - 9x - 18$ [5.5]

53. $a^2 - 2ab + b^2 - 4t^2$ [5.5]

Solve. [5.8]

54. $x^2 - 12x + 36 = 0$

55. $6b^2 + 6 = 13b$

56. $8y^2 = 14y$

57. $3r^2 = 12$

58. $a^3 = 4a^2 + 21a$

59. $(y - 1)(y - 4) = 10$

60. Let $f(x) = x^2 - 7x - 40$. Find a such that $f(a) = 4$. [5.8]

61. Find the domain of the function f given by
$$f(x) = \frac{x - 5}{x^2 - x - 56}. \quad [5.8]$$

Solve. [5.8]

62. A triangular sail is 9 ft taller than it is wide. The area is 56 ft². Find the height and the base of the sail.

Area = 56 ft²

63. The sum of the squares of three consecutive odd numbers is 155. Find the numbers.

64. A photograph is 3 in. longer than it is wide. When a $1\frac{1}{2}$-in. border is placed around the photograph, the total area of the photograph and the border is 88 in². Find the dimensions of the photograph.

65. Hassan is designing a garden in the shape of a right triangle. One leg of the triangle is 8 ft long. The other leg and the hypotenuse are consecutive odd integers. How long are the other two sides of the garden?

66. The number of mortgage foreclosure filings in the United States can be approximated by the polynomial function given by $m(t) = 620t^2 - 1700t + 1920$, where $m(t)$ is the number of foreclosures, in thousands, t years after 2005. In what year were there 1,000,000 foreclosure filings?
Source: Based on information from RealtyTrac

Synthesis

67. Explain the difference between multiplying two polynomials and factoring a polynomial. [5.2], [5.7]

68. Explain in your own words why there must be a 0 on one side of an equation before you can use the principle of zero products. [5.8]

Factor. [5.6]

69. $128x^6 - 2y^6$

70. $(x - 1)^3 - (x + 1)^3$

Multiply.

71. $[a - (b - 1)][(b - 1)^2 + a(b - 1) + a^2]$ [5.2], [5.6]

72. $\left(z^{n^2}\right)^{n^3}\left(z^{4n^3}\right)^{n^2}$ [5.2]

73. Solve: $(x + 1)^3 = x^2(x + 1)$. [5.8]

Test: Chapter 5

 **Test Prep** *Step-by-step test solutions are found on the video CD in the front of this book.*

Given the polynomial $8xy^3 - 14x^2y + 5x^5y^4 - 9x^4y$.

1. Determine the degree of the polynomial.

2. Arrange in descending powers of x.

3. Determine the leading term of the polynomial $7a - 12 + a^2 - 5a^3$.

4. Given $P(x) = 2x^3 + 3x^2 - x + 4$, find $P(0)$ and $P(-2)$.

5. Given $P(x) = x^2 - 3x$, find and simplify
$$P(a + h) - P(a).$$

6. Combine like terms:
$$6xy - 2xy^2 - 2xy + 5xy^2.$$

Add.

7. $(-4y^3 + 6y^2 - y) + (3y^3 - 9y - 7)$

8. $(2m^3 - 4m^2n - 5n^2) + (8m^3 - 3mn^2 + 6n^2)$

Subtract.

9. $(8a - 4b) - (3a + 4b)$

10. $(9y^2 - 2y - 5y^3) - (4y^2 - 2y - 6y^3)$

Multiply.

11. $(-4x^2y^3)(-16xy^5)$ **12.** $(6a - 5b)(2a + b)$

13. $(x - y)(x^2 - xy - y^2)$ **14.** $(4t - 3)^2$

15. $(5a^3 + 9)^2$ **16.** $(x - 2y)(x + 2y)$

Factor.

17. $45x^2 + 5x^4$ **18.** $y^3 + 5y^2 - 4y - 20$

19. $p^2 - 12p - 28$ **20.** $12m^2 + 20m + 3$

21. $9y^2 - 25$ **22.** $3r^3 - 3$

23. $9x^2 + 25 - 30x$ **24.** $x^8 - y^8$

25. $y^2 + 8y + 16 - 100t^2$ **26.** $20a^2 - 5b^2$

27. $24x^2 - 46x + 10$ **28.** $16a^7b + 54ab^7$

29. $4y^4x + 36yx^2 + 8y^2x^3 + 4xy$

Solve.

30. $x^2 - 40 - 6x = 0$ **31.** $5y^2 = 125$

32. $2x^3 + 21x = -17x^2$ **33.** $12r^2 + 6r = 0$

34. Let $f(x) = 3x^2 - 15x + 11$. Find a such that $f(a) = 11$.

35. Find the domain of the function f given by
$$f(x) = \frac{8 - x}{x^2 + 2x + 1}.$$

36. A framed painting is 20 in. long and 18 in. wide, and 195 in^2 of the painting shows. Find the width of the frame.

37. The hypotenuse of a right triangle is 26 cm long. One leg of the triangle is 4 cm longer than twice the length of the other leg. Find the lengths of the legs of the triangle.

38. Fireworks are launched over a lake from a dam 36 ft above the water. The height of a display, t seconds after it has been launched, is given by
$$h(t) = -16t^2 + 64t + 36$$
How long will it take for a "dud" firework to reach the water?

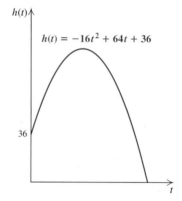

Synthesis

39. a) Multiply: $(x^2 + x + 1)(x^3 - x^2 + 1)$.
b) Factor: $x^5 + x + 1$.

40. Factor: $6x^{2n} - 7x^n - 20$.

Cumulative Review: Chapters 1–5

Perform the indicated operations and simplify.

1. $-120 \div 2 \cdot 5 - 3 \cdot (-1)^3$ [1.2]

2. $(-4x^2 - 6x + x^3) + (5x^2 - 2x - 7)$ [5.1]

3. $(2a^2b - b^2 - 3a^2) - (5a^2 - 4a^2b - 4b^2)$ [5.1]

4. $(2x + 9)(2x - 9)$ [5.2]

5. $(2x + 9y)^2$ [5.2]

6. $(5m^3 + n)(2m^2 - n)$ [5.2]

Factor.

7. $3t^2 - 48$ [5.5]

8. $a^2 - 14a + 49$ [5.5]

9. $36x^3y^2 - 27x^4y + 45x^2y^3$ [5.3]

10. $125a^3 + 64b^3$ [5.6]

11. $m^2n - 4mn - 21n$ [5.4]

12. $x^8 - x^2$ [5.6]

13. $12y^2 + 7y - 10$ [5.4]

14. $d^2 - a^2 + 2ab - b^2$ [5.5]

Solve.

15. $2x - 3(x + 2) = 6 - (x - 1)$ [1.3]

16. $x - 2y = 7,$
$\quad y = \frac{1}{4}x$ [3.2]

17. $\quad x - 2y + 3z = 16,$
$\quad 3x + y + z = -5,$
$\quad x - y - z = -3$ [3.4]

18. $3x - 7(x + 1) \geq 4x - 5$ [4.1]

19. $-2 < x + 6 < 0$ [4.2]

20. $|2x - 1| \leq 5$ [4.3]

21. $x^2 = 10x + 24$ [5.8]

22. $5n^2 = 30n$ [5.8]

23. Solve $3a - 5b = 6 + b$ for b. [1.5]

24. Determine the slope and the y-intercept for the line given by $x = -2y + 8$. [2.3]

25. Find the slope of the line containing $(2, -4)$ and $(-1, -5)$. [2.4]

26. Find a linear function whose graph has slope $\frac{1}{3}$ and y-intercept $\left(0, -\frac{1}{4}\right)$. [2.3]

27. Find a linear function whose graph contains $(-2, 5)$ and $(-1, -4)$. [2.5]

Graph on a plane.

28. $y = \frac{2}{3}x - 1$ [2.3]

29. $4y = -2$ [2.4]

30. $x = y + 3$ [2.3]

31. $4x - 3y \leq 12$ [4.4]

32. Find $P(-1)$ if $P(x) = 3x^4 - 2x^3 - 5x^2 + 6x$. [5.1]

33. Find the domain of f if
$$f(x) = \frac{x + 1}{x^2 - 3x + 2}. \quad [5.8]$$

34. Write the domain of f using interval notation if $f(x) = \sqrt{x + 3}$. [4.1]

35. If $f(x) = x^2 + 1$ and $g(x) = x - 2$, find $(f \cdot g)(x)$. [2.6], [5.2]

36. *Cooling costs.* The installation of a rooftop garden on Chicago's City Hall has dramatically reduced summer cooling costs for the building. The annual power bill fell $10,000, or 11%. What was the annual power bill before the garden was installed? [1.4]
Source: *The Wall Street Journal*, 2/11/08

37. *Utility bills.* In the summer, Georgia Power charges residential customers approximately 4.6¢ per kilowatt-hour (kWh) for the first 650 kWh of electricity and 7.7¢ per kWh for usage over 650 kWh but under 1000 kWh. Eileen never uses more than 1000 kWh a month, and her bill is always between $45.30 and $53 each month. How much electricity does Eileen use each month? [4.1], [4.2]
Source: *The Wall Street Journal*, 2/11/08

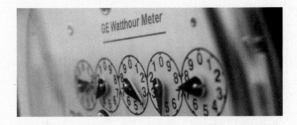

38. *Scrapbooking.* A photo is cut so that its length is 2 cm longer than its width. It is then centered on a background to form a 3-cm border around the photo. The area of the background paper is 168 cm². What are the dimensions of the photo? [5.8]

39. *Scrapbooking.* A photo is cropped so that its length is 2 cm longer than its width. It is then placed on a page and a narrow ribbon is glued around the perimeter of the photo. If the length of the ribbon is 50 cm, what are the dimensions of the cropped photo? [1.4]

40. *Ordering pizza.* The Westville Marching Band ordered 48 pizzas for their annual party. Single-topping pizzas were $16.95 each, and two-topping pizzas were $19.95 each. If the total cost of the pizzas was $870.60, how many of each type did the band order? [3.3]

41. *Payday loans.* The amount of payday loans in the United States has grown from approximately $13 billion in 1999 to $41 billion in 2006. Let $p(t)$ represent the amount of payday loans, in billions of dollars, t years after 1999. [2.5]
Source: Stephens Inc. research-reports

  a) Find a linear function that fits the data.
  b) Use the function from part (a) to predict the amount of payday loans in 2010.
  c) In what year will there be $70 billion in payday loans?

42. *Landscaping.* A traffic median in the shape of a triangle has a hypotenuse that is 20 ft long. One leg is 4 ft longer than the other leg. What is the length of each leg? [5.8]

43. *Cleaning solutions.* Carlee uses vinegar in a variety of cleaning solutions. Her window-cleaning solution is 50% vinegar and the rest water, and her electric-razor-cleaning solution is 5% vinegar. How much of each of the two solutions should she mix in order to make 90 oz of a floor-cleaning solution that is 10% vinegar? [3.3]
Source: Based on recipes from www.apple-cider-vinegar-benefits.com

Synthesis

44. Solve $\dfrac{c + 2d}{c - d} = t$ for d. [1.5]

45. Determine the value of k such that the lines
$$2x - 5y = 8 \quad \text{and} \quad 3x + ky = -4$$
are parallel. [2.5]

46. Write an absolute-value inequality for which the interval shown is the solution. [4.3]

47. Factor: $4x^{6n} - 8x^{3n} - 5$. Assume that n is a positive integer. [5.4]

Rational Expressions, Equations, and Functions

ELLA JOHNSON
KITEBOARDING INSTRUCTOR
San Francisco, California

As a competing kiteboarder and as an instructor, I find numbers to be part of my everyday schedule. Between calculating the wind speed or finding a kite suitable to a student's weight, I must use mathematics in order to ensure the safety of those around me: Too large a kite could potentially lift a student dozens of feet into the air; too small a kite could cause someone to drift down the river or ocean.

AN APPLICATION

The speed v of a train of ocean waves varies directly as the swell period t, or time between successive waves. Waves with a swell period of 12 sec are traveling 21 mph. How fast are waves traveling that have a swell period of 20 sec?

Source: www.rodntube.com

This problem appears as Example 5 in Section 6.8.

A rational expression is an expression, similar to a fraction in arithmetic, that indicates division. In this chapter, we add, subtract, multiply, and divide rational expressions, and use them in equations and functions. We then use rational expressions to solve problems that we could not have solved before.

6.1 Rational Expressions and Functions: Multiplying and Dividing

Rational Functions   ▪   Multiplying   ▪   Simplifying Rational Expressions and Functions   ▪   Dividing and Simplifying

An expression that consists of a polynomial divided by a nonzero polynomial is called a **rational expression**. The following are examples of rational expressions:

$$\frac{3}{4}, \quad \frac{x}{y}, \quad \frac{9}{a+b}, \quad \frac{x^2+7xy-4}{x^3-y^3}, \quad \frac{1+z^3}{1-z^6}.$$

Rational Functions

Like polynomials, certain rational expressions are used to describe functions. Such functions are called **rational functions**.

EXAMPLE 1   The function given by

$$H(t) = \frac{t^2+5t}{2t+5}$$

gives the time, in hours, for two machines, working together, to complete a job that the first machine could do alone in t hours and the second machine could do in $t + 5$ hours. How long will the two machines, working together, require for the job if the first machine alone would take **(a)** 1 hour? **(b)** 6 hours?

SOLUTION

a) $H(1) = \dfrac{1^2+5\cdot 1}{2\cdot 1+5} = \dfrac{1+5}{2+5} = \dfrac{6}{7}\,\text{hr}$

b) $H(6) = \dfrac{6^2+5\cdot 6}{2\cdot 6+5} = \dfrac{36+30}{12+5} = \dfrac{66}{17}\ \text{or}\ 3\dfrac{15}{17}\,\text{hr}$

TRY EXERCISE ▶ 15

Since division by 0 is undefined, the domain of a rational function must exclude any numbers for which the denominator is 0. For the function H above, the denominator is 0 when t is $-\frac{5}{2}$, so

the domain of H is $\left(-\infty, -\frac{5}{2}\right) \cup \left(-\frac{5}{2}, \infty\right)$.

Although graphing rational functions is beyond the scope of this text, it is educational to examine a computer-generated graph of the above function. Note that the graph consists of two unconnected "branches." Since $-\frac{5}{2}$ is not in the domain of H, the graph of H does not touch a vertical line passing through $-\frac{5}{2}$.

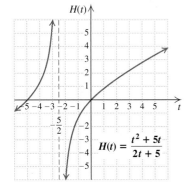

$$H(t) = \frac{t^2+5t}{2t+5}$$

Multiplying

The calculations that are performed with rational expressions resemble those performed with fractions in arithmetic.

> ### Products of Rational Expressions
> To multiply two rational expressions, multiply numerators and multiply denominators:
> $$\frac{A}{B} \cdot \frac{C}{D} = \frac{AC}{BD}, \quad \text{where } B \neq 0, D \neq 0.$$

EXAMPLE 2   Multiply: $\dfrac{x+1}{y-3} \cdot \dfrac{x^2}{y+1}$.

SOLUTION

$$\frac{x+1}{y-3} \cdot \frac{x^2}{y+1} = \frac{(x+1)x^2}{(y-3)(y+1)}$$ Multiplying the numerators and multiplying the denominators

TRY EXERCISE 17

Recall from arithmetic that multiplication by 1 can be used to find equivalent expressions:

$$\frac{3}{5} = \frac{3}{5} \cdot \frac{2}{2}$$ Multiplying by $\frac{2}{2}$, which is 1

$$= \frac{6}{10}.$$ $\frac{3}{5}$ and $\frac{6}{10}$ represent the same number.

Similarly, multiplication by 1 can be used to find equivalent rational expressions:

$$\begin{aligned} \frac{x-5}{x+2} &= \frac{x-5}{x+2} \cdot \frac{x+3}{x+3} \\ &= \frac{(x-5)(x+3)}{(x+2)(x+3)}. \end{aligned}$$ Multiplying by $\dfrac{x+3}{x+3}$, which is 1, provided $x \neq -3$

So long as x is replaced with a number other than -2 or -3, the expressions

$$\frac{x-5}{x+2} \quad \text{and} \quad \frac{(x-5)(x+3)}{(x+2)(x+3)}$$

represent the same number. For example, if $x = 4$, then

$$\frac{x-5}{x+2} = \frac{4-5}{4+2} = \frac{-1}{6}$$

and

$$\frac{(x-5)(x+3)}{(x+2)(x+3)} = \frac{(4-5)(4+3)}{(4+2)(4+3)} = \frac{-1 \cdot 7}{6 \cdot 7} = \frac{-7}{42} = \frac{-1}{6}.$$

Simplifying Rational Expressions and Functions

As in arithmetic, rational expressions are *simplified* by "removing" a factor equal to 1. This reverses the process shown above:

$$\frac{6}{10} = \frac{3 \cdot 2}{5 \cdot 2} = \frac{3}{5} \cdot \frac{2}{2} = \frac{3}{5}.$$ We "removed" the factor that equals 1: $\frac{2}{2} = 1$.

Similarly,

$$\frac{(x-5)(x+3)}{(x+2)(x+3)} = \frac{x-5}{x+2} \cdot \frac{x+3}{x+3} = \frac{x-5}{x+2}.$$

We "removed" the factor that equals 1: $\frac{x+3}{x+3} = 1$.

It is important that a rational function's domain not be changed as a result of simplifying. For example, above we wrote

$$\frac{(x-5)(x+3)}{(x+2)(x+3)} = \frac{x-5}{x+2}.$$

There is a serious problem with stating that the functions

$$F(x) = \frac{(x-5)(x+3)}{(x+2)(x+3)} \quad \text{and} \quad G(x) = \frac{x-5}{x+2}$$

represent the same function. The difficulty arises from the fact that the domain of each function is assumed to be all real numbers for which the denominator is nonzero. Thus, unless we specify otherwise,

Domain of $F = \{x \mid x \neq -2, x \neq -3\}$, and

Domain of $G = \{x \mid x \neq -2\}$.*

Thus, as presently written, the domain of G includes -3, but the domain of F does not. This problem is easily addressed by specifying

$$F(x) = \frac{(x-5)(x+3)}{(x+2)(x+3)} = \frac{x-5}{x+2} \quad \text{with } x \neq -3.$$

EXAMPLE 3 Write the function given by

$$f(t) = \frac{7t^2 + 21t}{14t}$$

in simplified form.

SOLUTION   We first factor the numerator and the denominator, looking for the largest factor common to both. Once the greatest common factor is found, we use it to write 1 and simplify:

$$f(t) = \frac{7t^2 + 21t}{14t}$$ Note that the domain of $f = \{t \mid t \neq 0\}$.

$$= \frac{7t(t+3)}{7 \cdot 2 \cdot t}$$ Factoring. The greatest common factor is $7t$.

$$= \frac{7t}{7t} \cdot \frac{t+3}{2}$$ Rewriting as a product of two rational expressions

$$= 1 \cdot \frac{t+3}{2}$$ For $t \neq 0$, we have $\frac{7t}{7t} = 1$. Try to do this step mentally.

$$= \frac{t+3}{2}, \ t \neq 0.$$ Removing the factor 1. To keep the same domain, we specify that $t \neq 0$.

Thus simplified form is $f(t) = \frac{t+3}{2}$, with $t \neq 0$.   TRY EXERCISE 33

*This use of set-builder notation assumes that, apart from the restrictions listed, all other real numbers are in the domain.

A rational expression is said to be **simplified** when no factors equal to 1 can be removed. Often, this requires two or more steps. For example, suppose we remove $7/7$ instead of $(7t)/(7t)$ in Example 3. We would then have

$$\frac{7t^2 + 21t}{14t} = \frac{7(t^2 + 3t)}{7 \cdot 2t}$$

$$= \frac{t^2 + 3t}{2t}.$$

Removing a factor equal to 1: $\frac{7}{7} = 1$.
Note that $t \neq 0$.

Here, since another common factor remains, we need to simplify further:

$$\frac{t^2 + 3t}{2t} = \frac{t(t + 3)}{t \cdot 2}$$

$$= \frac{t + 3}{2}, \ t \neq 0.$$

Removing another factor equal to 1: $t/t = 1$. The rational expression is now simplified. We must still specify $t \neq 0$.

EXAMPLE 4   Write the function given by

$$g(x) = \frac{x^2 + 3x - 10}{2x^2 - 3x - 2}$$

in simplified form, and list all restrictions on the domain.

SOLUTION   We have

$$g(x) = \frac{x^2 + 3x - 10}{2x^2 - 3x - 2}$$

$$= \frac{(x - 2)(x + 5)}{(2x + 1)(x - 2)}$$

Factoring the numerator and the denominator. Note that $x \neq -\frac{1}{2}$ and $x \neq 2$.

$$= \frac{x - 2}{x - 2} \cdot \frac{x + 5}{2x + 1}$$

Rewriting as a product of two rational expressions

$$= \frac{x + 5}{2x + 1}, x \neq -\frac{1}{2}, 2.$$

Removing a factor equal to 1: $\frac{x - 2}{x - 2} = 1$. We list both restrictions.

Thus, $g(x) = \frac{x + 5}{2x + 1}, x \neq -\frac{1}{2}, 2.$

TRY EXERCISE 37

We generally list restrictions only when using function notation.

STUDENT NOTES

Note that factoring the numerator and the denominator is the first step in simplifying a rational expression. All operations with rational expressions involve polynomials. If you are not comfortable with factoring polynomials, it will be worth your time to review Sections 5.3–5.7.

EXAMPLE 5   Simplify: **(a)** $\dfrac{9x^2 + 6xy - 3y^2}{12x^2 - 12y^2}$; **(b)** $\dfrac{4 - t}{3t - 12}$.

SOLUTION

a) $\dfrac{9x^2 + 6xy - 3y^2}{12x^2 - 12y^2} = \dfrac{3(x + y)(3x - y)}{12(x + y)(x - y)}$

Factoring the numerator and the denominator

$$= \frac{3(x + y)}{3(x + y)} \cdot \frac{3x - y}{4(x - y)}$$

Rewriting as a product of two rational expressions

$$= \frac{3x - y}{4(x - y)}$$

Removing a factor equal to 1: $\dfrac{3(x + y)}{3(x + y)} = 1$

For purposes of later work, we usually do not multiply out the numerator and the denominator of the simplified expression.

b) $\dfrac{4 - t}{3t - 12} = \dfrac{4 - t}{3(t - 4)}$ Factoring

Since $4 - t$ is the opposite of $t - 4$, we factor out -1 to reverse the subtraction:

$$\dfrac{4 - t}{3t - 12} = \dfrac{-1(t - 4)}{3(t - 4)} \qquad 4 - t = -1(-4 + t) = -1(t - 4)$$

$$= \dfrac{-1}{3} \cdot \dfrac{t - 4}{t - 4}$$

$$= -\dfrac{1}{3}. \qquad \text{Removing a factor equal to 1:}$$
$$\qquad\qquad\qquad (t - 4)/(t - 4) = 1$$

TRY EXERCISE ▶ 35

TECHNOLOGY CONNECTION

To check that a simplified expression is equivalent to the original expression, we can let y_1 = the original expression, y_2 = the simplified expression, and $y_3 = y_1 - y_2$ (or $y_2 - y_1$). If y_1 and y_2 are indeed equivalent, (TABLE) or (TRACE) can be used to show that, except when y_1 or y_2 is undefined, we have $y_1 = y_2$ and $y_3 = 0$.

1. Use a graphing calculator to check Example 3. Be sure to use parentheses as needed.
2. Use a graphing calculator to show that $\dfrac{x + 3}{x} \neq 3$.
   (See the Caution! box concerning factoring.)

CANCELING

"Canceling" is a shortcut often used for removing a factor equal to 1 when working with fractions. With caution, we mention it as a possible way to speed up your work. Canceling removes factors equal to 1 in products. It *cannot* be done in sums or when adding expressions together. If your instructor permits canceling (not all do), it must be done with care and understanding. Example 5(a) might have been done with less writing as follows:

$$\dfrac{9x^2 + 6xy - 3y^2}{12x^2 - 12y^2} = \dfrac{\cancel{3}\,\cancel{(x + y)}(3x - y)}{\cancel{3} \cdot 4\,\cancel{(x + y)}(x - y)} \qquad \begin{array}{l}\text{When a factor that equals}\\ \text{1 is found, it is "canceled"}\\ \text{as shown.}\end{array}$$

$$= \dfrac{3x - y}{4(x - y)}. \qquad \begin{array}{l}\text{Removing a factor equal to 1:}\\ \dfrac{3(x + y)}{3(x + y)} = 1\end{array}$$

CAUTION!   Canceling is often performed incorrectly:

$$\dfrac{\cancel{x} + 3}{\cancel{x}} = 3, \qquad \dfrac{\cancel{4}x + 3}{\cancel{2}} = 2x + 3, \qquad \dfrac{\cancel{5}}{\cancel{5} + x} = \dfrac{1}{x} \qquad \begin{array}{l}\text{To check that}\\ \text{these are not}\\ \text{equivalent,}\\ \text{substitute a}\\ \text{number for } x.\end{array}$$

Incorrect!   Incorrect!   Incorrect!

In each incorrect situation, the expressions canceled are *not* both factors. Factors are parts of products. If it's not a factor, it can't be canceled! If in doubt, don't cancel!

If, after multiplying two rational expressions, it is possible to simplify, it is common practice to do so.

EXAMPLE   6   Multiply. Then simplify by removing a factor equal to 1.

a) $\dfrac{x + 2}{x - 3} \cdot \dfrac{x^2 - 4}{x^2 + x - 2}$

b) $\dfrac{1 - a^3}{a^2} \cdot \dfrac{a^5}{a^2 - 1}$

SOLUTION

a) $\dfrac{x+2}{x-3} \cdot \dfrac{x^2-4}{x^2+x-2} = \dfrac{(x+2)(x^2-4)}{(x-3)(x^2+x-2)}$ Multiplying the numerators and also the denominators

Factor. $= \dfrac{(x+2)(x-2)(x+2)}{(x-3)(x+2)(x-1)}$ Factoring the numerator and the denominator and finding common factors

Remove a factor equal to 1. $= \dfrac{(x+2)\cancel{(x+2)}(x-2)}{(x-3)\cancel{(x+2)}(x-1)}$ Removing a factor equal to 1: $\dfrac{x+2}{x+2} = 1$

$= \dfrac{(x+2)(x-2)}{(x-3)(x-1)}$ Simplifying

b) $\dfrac{1-a^3}{a^2} \cdot \dfrac{a^5}{a^2-1} = \dfrac{(1-a^3)a^5}{a^2(a^2-1)}$

$= \dfrac{(1-a)(1+a+a^2)a^5}{a^2(a-1)(a+1)}$ Factoring a difference of cubes and a difference of squares

Factor. $= \dfrac{-1(a-1)(1+a+a^2)a^5}{a^2(a-1)(a+1)}$ *Important!* Factoring out -1 reverses the subtraction.

Remove a factor equal to 1. $= \dfrac{\cancel{(a-1)}a^2 \cdot a^3(-1)(1+a+a^2)}{\cancel{(a-1)}a^2(a+1)}$ Rewriting a^5 as $a^2 \cdot a^3$; removing a factor equal to 1: $\dfrac{(a-1)a^2}{(a-1)a^2} = 1$

$= \dfrac{-a^3(1+a+a^2)}{a+1}$ Simplifying

As in Example 5, there is no need for us to multiply out the numerator or the denominator of the final result.

TRY EXERCISE ▸ 57

Dividing and Simplifying

Two expressions are reciprocals of each other if their product is 1. As in arithmetic, to find the reciprocal of a rational expression, we interchange numerator and denominator.

The reciprocal of $\dfrac{x}{x^2+3}$ is $\dfrac{x^2+3}{x}$.

The reciprocal of $y-8$ is $\dfrac{1}{y-8}$.

Quotients of Rational Expressions

For any rational expressions A/B and C/D, with $B, C, D \neq 0$,

$$\frac{A}{B} \div \frac{C}{D} = \frac{A}{B} \cdot \frac{D}{C}.$$

(To divide two rational expressions, multiply by the reciprocal of the divisor. We often say that we "*invert* and multiply.")

EXAMPLE **7** Divide. Simplify by removing a factor equal to 1 if possible.

$$\frac{x-2}{x+1} \div \frac{x+5}{x-3}$$

SOLUTION

$$\frac{x-2}{x+1} \div \frac{x+5}{x-3} = \frac{x-2}{x+1} \cdot \frac{x-3}{x+5} \qquad \text{Multiplying by the reciprocal of the divisor}$$

$$= \frac{(x-2)(x-3)}{(x+1)(x+5)} \qquad \text{Multiplying the numerators and the denominators}$$

> **TRY EXERCISE** 63

EXAMPLE **8** Write the function given by

$$g(a) = \frac{a^2 - 2a + 1}{a + 5} \div \frac{a^3 - a}{a - 2}$$

in simplified form, and list all restrictions on the domain.

SOLUTION A number is not in the domain of a rational function if it makes a divisor zero. There are three divisors in this rational function:

$$a + 5, \quad a - 2, \quad \text{and} \quad \frac{a^3 - a}{a - 2}.$$

None of these can be zero:

$$a + 5 = 0 \text{ when } a = -5,$$
$$a - 2 = 0 \text{ when } a = 2, \text{ and}$$
$$\frac{a^3 - a}{a - 2} = 0 \text{ when } a^3 - a = 0.$$

We solve $a^3 - a = 0$:

$$a^3 - a = 0$$
$$a(a^2 - 1) = 0$$
$$a(a + 1)(a - 1) = 0$$
$$a = 0 \quad or \quad a + 1 = 0 \quad or \quad a - 1 = 0$$
$$a = 0 \quad or \quad a = -1 \quad or \quad a = 1.$$

Thus the domain of *g* does not contain $-5, -1, 0, 1,$ or 2:

$$g(a) = \frac{a^2 - 2a + 1}{a + 5} \div \frac{a^3 - a}{a - 2} = \frac{a^2 - 2a + 1}{a + 5} \cdot \frac{a - 2}{a^3 - a} \qquad \begin{array}{l}\text{Multiplying by}\\\text{the reciprocal of}\\\text{the divisor}\end{array}$$

$$= \frac{(a^2 - 2a + 1)(a - 2)}{(a + 5)(a^3 - a)} \qquad \begin{array}{l}\text{Multiplying the}\\\text{numerators and}\\\text{the denominators}\end{array}$$

$$= \frac{(a - 1)(a - 1)(a - 2)}{(a + 5)a(a + 1)(a - 1)} \qquad \begin{array}{l}\text{Factoring the}\\\text{numerator and}\\\text{the denominator}\end{array}$$

$$= \frac{(a - 1)\cancel{(a - 1)}(a - 2)}{(a + 5)a(a + 1)\cancel{(a - 1)}} \qquad \begin{array}{l}\text{Removing a}\\\text{factor equal}\\\text{to } 1: \dfrac{a - 1}{a - 1} = 1\end{array}$$

$$= \frac{(a - 1)(a - 2)}{a(a + 5)(a + 1)}, \quad a \neq -5, -1, 0, 1, 2 \qquad \text{Simplifying}$$

> **TRY EXERCISE** 77

Visualizing for Success

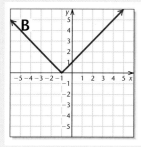

A

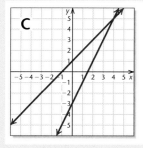

B

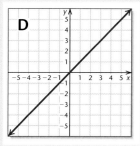

C

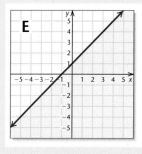

D

E

Match each equation, inequality, set of equations, or set of inequalities with its graph.

1. $y = -2$

2. $y = x$

3. $y = x^2$

4. $y = \dfrac{1}{x}$

5. $y = x + 1$

6. $y = |x + 1|$

7. $y \le x + 1$

8. $y > 2x - 3$

9. $y = x + 1,$
$y = 2x - 3$

10. $y \le x + 1,$
$y > 2x - 3$

Answers on page A-26

An alternate, animated version of this activity appears in MyMathLab. To use MyMathLab, you need a course ID and a student access code. Contact your instructor for more information.

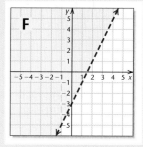

F

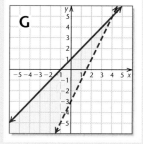

G

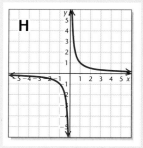

H

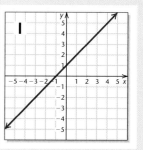

I

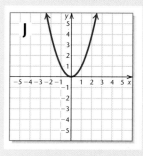

J

6.1 EXERCISE SET

For Extra Help

MyMathLab   PRACTICE   WATCH   DOWNLOAD

🐦 *Concept Reinforcement* *In each of Exercises 1–10, match the function described with the appropriate domain from the column on the right.*

1. ____ $f(x) = \dfrac{2 - x}{x - 5}$

2. ____ $g(x) = \dfrac{x + 2}{x + 5}$

3. ____ $h(x) = \dfrac{x - 5}{x - 2}$

4. ____ $f(x) = \dfrac{x + 5}{x + 2}$

5. ____ $g(x) = \dfrac{x - 3}{(x - 2)(x - 5)}$

6. ____ $h(x) = \dfrac{x - 3}{(x + 2)(x + 5)}$

7. ____ $f(x) = \dfrac{x + 3}{(x - 2)(x + 5)}$

8. ____ $g(x) = \dfrac{x + 3}{(x + 2)(x - 5)}$

9. ____ $h(x) = \dfrac{(x - 2)(x - 3)}{x + 3}$

10. ____ $f(x) = \dfrac{(x + 2)(x + 3)}{x - 3}$

a) $\{x \mid x \neq -5, x \neq 2\}$

b) $\{x \mid x \neq 3\}$

c) $\{x \mid x \neq -2\}$

d) $\{x \mid x \neq -3\}$

e) $\{x \mid x \neq 5\}$

f) $\{x \mid x \neq -2, x \neq 5\}$

g) $\{x \mid x \neq 2\}$

h) $\{x \mid x \neq -5, x \neq -2\}$

i) $\{x \mid x \neq 2, x \neq 5\}$

j) $\{x \mid x \neq -5\}$

For each rational function, find the function values indicated, provided the value exists.

11. $f(x) = \dfrac{2x^2 - x - 5}{x - 1}$; **(a)** $f(0)$; **(b)** $f(-1)$; **(c)** $f(3)$

12. $v(t) = \dfrac{t^2 + 5t - 9}{t + 4}$; **(a)** $v(0)$; **(b)** $v(-3)$; **(c)** $v(6)$

13. $r(t) = \dfrac{t^2 - 8t - 9}{t^2 - 4}$; **(a)** $r(0)$; **(b)** $r(2)$; **(c)** $r(-1)$

14. $g(x) = \dfrac{2x^3 - x}{x^2 - 6x + 9}$; **(a)** $g(0)$; **(b)** $g(-2)$; **(c)** $g(3)$

Restaurant management. *Gregory usually takes 1 hr more than Alayna does to prepare the day's soups for Roux Palace. If Alayna takes t hr to prepare the day's soups, the function given by*

$$H(t) = \dfrac{t^2 + t}{2t + 1}$$

can be used to determine how long it would take if they worked together.

15. How long will it take them, working together, to prepare the soups if Alayna can prepare them alone in 5 hr?

16. How long will it take them, working together, to prepare the soups if Alayna can prepare them alone in 2 hr?

Multiply to obtain equivalent expressions. Do not simplify. Assume that all denominators are nonzero.

17. $\dfrac{9x}{9x} \cdot \dfrac{x + 2}{x - 5}$

18. $\dfrac{3 - a^2}{a - 7} \cdot \dfrac{-1}{-1}$

19. $\dfrac{t - 2}{t + 3} \cdot \dfrac{-1}{-1}$

20. $\dfrac{x - 4}{x + 5} \cdot \dfrac{x - 5}{x - 5}$

Simplify by removing a factor equal to 1.

21. $\dfrac{8t^4}{40t}$

22. $\dfrac{35n}{5n^2}$

23. $\dfrac{24x^3y}{30x^5y^8}$

24. $\dfrac{10yz^4}{40y^2z^9}$

25. $\dfrac{2a - 10}{2}$

26. $\dfrac{3a + 12}{3}$

27. $\dfrac{5}{25y - 30}$

28. $\dfrac{21}{6x - 9}$

29. $\dfrac{3x - 12}{3x + 15}$

30. $\dfrac{4y - 20}{4y + 12}$

Write simplified form for each of the following. Be sure to list all restrictions on the domain, as in Example 4.

31. $f(x) = \dfrac{5x + 30}{x^2 + 6x}$

32. $f(x) = \dfrac{3x + 30}{x^2 + 10x}$

33. $g(x) = \dfrac{x^2 - 9}{5x + 15}$

34. $g(x) = \dfrac{8x - 16}{x^2 - 4}$

35. $h(x) = \dfrac{2 - x}{7x - 14}$

36. $h(x) = \dfrac{4 - x}{12x - 48}$

37. $f(t) = \dfrac{t^2 - 16}{t^2 - 8t + 16}$

38. $f(t) = \dfrac{t^2 - 25}{t^2 + 10t + 25}$

39. $g(t) = \dfrac{21 - 7t}{3t - 9}$

40. $g(t) = \dfrac{12 - 6t}{5t - 10}$

41. $h(t) = \dfrac{t^2 + 5t + 4}{t^2 - 8t - 9}$

42. $h(t) = \dfrac{t^2 - 3t - 4}{t^2 + 9t + 8}$

43. $f(x) = \dfrac{9x^2 - 4}{3x - 2}$

44. $f(x) = \dfrac{4x^2 - 1}{2x - 1}$

45. $g(t) = \dfrac{16 - t^2}{t^2 - 8t + 16}$

46. $g(p) = \dfrac{25 - p^2}{p^2 + 10p + 25}$

Multiply and, if possible, simplify.

47. $\dfrac{3y^3}{5z} \cdot \dfrac{10z^4}{7y^6}$

48. $\dfrac{20y}{9z^7} \cdot \dfrac{6z^4}{5y^2}$

49. $\dfrac{8x - 16}{5x} \cdot \dfrac{x^3}{5x - 10}$

50. $\dfrac{5t^3}{4t - 8} \cdot \dfrac{6t - 12}{10t}$

51. $\dfrac{y^2 - 9}{y^2} \cdot \dfrac{y^2 - 3y}{y^2 - y - 6}$

52. $\dfrac{y^2 + 10y + 25}{y^2 - 9} \cdot \dfrac{y^2 + 3y}{y + 5}$

53. $\dfrac{7a - 14}{4 - a^2} \cdot \dfrac{5a^2 + 6a + 1}{35a + 7}$

54. $\dfrac{a^2 - 1}{2 - 5a} \cdot \dfrac{15a - 6}{a^2 + 5a - 6}$

Aha! **55.** $\dfrac{t^3 - 4t}{t - t^4} \cdot \dfrac{t^4 - t}{4t - t^3}$

56. $\dfrac{x^2 - 6x + 9}{12 - 4x} \cdot \dfrac{x^6 - 9x^4}{x^3 - 3x^2}$

57. $\dfrac{c^3 + 8}{c^5 - 4c^3} \cdot \dfrac{c^6 - 4c^5 + 4c^4}{c^2 - 2c + 4}$

58. $\dfrac{t^3 - 27}{t^4 - 9t^2} \cdot \dfrac{t^5 - 6t^4 + 9t^3}{t^2 + 3t + 9}$

59. $\dfrac{a^3 - b^3}{3a^2 + 9ab + 6b^2} \cdot \dfrac{a^2 + 2ab + b^2}{a^2 - b^2}$

60. $\dfrac{x^3 + y^3}{x^2 + 2xy - 3y^2} \cdot \dfrac{x^2 - y^2}{3x^2 + 6xy + 3y^2}$

Divide and, if possible, simplify.

61. $\dfrac{12a^3}{5b^2} \div \dfrac{4a^2}{15b}$

62. $\dfrac{9x^7}{8y} \div \dfrac{15x^2}{4y}$

63. $\dfrac{5x + 20}{x^6} \div \dfrac{x + 4}{x^2}$

64. $\dfrac{3a + 15}{a^9} \div \dfrac{a + 5}{a^8}$

65. $\dfrac{25x^2 - 4}{x^2 - 9} \div \dfrac{2 - 5x}{x + 3}$

66. $\dfrac{4a^2 - 1}{a^2 - 4} \div \dfrac{2a - 1}{2 - a}$

67. $\dfrac{5y - 5x}{15y^3} \div \dfrac{x^2 - y^2}{3x + 3y}$

68. $\dfrac{x^2 - y^2}{4x + 4y} \div \dfrac{3y - 3x}{12x^2}$

69. $\dfrac{y^2 - 36}{y^2 - 8y + 16} \div \dfrac{3y - 18}{y^2 - y - 12}$

70. $\dfrac{x^2 - 16}{x^2 - 10x + 25} \div \dfrac{3x - 12}{x^2 - 3x - 10}$

71. $\dfrac{x^3 - 64}{x^3 + 64} \div \dfrac{x^2 - 16}{x^2 - 4x + 16}$

72. $\dfrac{8y^3 - 27}{64y^3 - 1} \div \dfrac{4y^2 - 9}{16y^2 + 4y + 1}$

Write simplified form for each of the following. Be sure to list all restrictions on the domain.

73. $f(t) = \dfrac{t^2 - 100}{5t + 20} \cdot \dfrac{t + 4}{t - 10}$

74. $g(n) = \dfrac{n+5}{n-5} \cdot \dfrac{n^2-25}{2n+2}$

75. $g(x) = \dfrac{x^2-2x-35}{2x^3-3x^2} \cdot \dfrac{4x^3-9x}{7x-49}$

76. $h(t) = \dfrac{t^2-10t+9}{t^2-1} \cdot \dfrac{1-t^2}{t^2-5t-36}$

77. $f(x) = \dfrac{x^2-4}{x^3} \div \dfrac{x^5-2x^4}{x+4}$

78. $g(x) = \dfrac{x^2-9}{x^2} \div \dfrac{x^5+3x^4}{x+2}$

79. $h(n) = \dfrac{n^3+3n}{n^2-9} \div \dfrac{n^2+5n-14}{n^2+4n-21}$

80. $f(x) = \dfrac{x^3+4x}{x^2-16} \div \dfrac{x^2+8x+15}{x^2+x-20}$

Perform the indicated operations and, if possible, simplify. Recall that multiplications and divisions are performed in order from left to right.

81. $\dfrac{4x^2-9y^2}{8x^3-27y^3} \div \dfrac{4x+6y}{3x-9y} \cdot \dfrac{4x^2+6xy+9y^2}{4x^2-8xy+3y^2}$

82. $\dfrac{5x^2-5y^2}{27x^3+8y^3} \div \dfrac{x^2-2xy+y^2}{9x^2-6xy+4y^2} \cdot \dfrac{6x+4y}{10x-15y}$

83. $\dfrac{a^3-ab^2}{2a^2+3ab+b^2} \cdot \dfrac{4a^2-b^2}{a^2-2ab+b^2} \div \dfrac{a^2+a}{a-1}$

84. $\dfrac{2x+4y}{2x^2+5xy+2y^2} \cdot \dfrac{4x^2-y^2}{8x^2-8} \div \dfrac{x^2+4xy+4y^2}{x^2-6xy+9y^2}$

85. Nancy *incorrectly* simplifies $\dfrac{x+2}{x}$ as

$$\dfrac{x+2}{x} = \dfrac{\cancel{x}+2}{\cancel{x}} = 1+2 = 3.$$

She insists this is correct because it checks when x is replaced with 1. Explain her misconception.

86. Give a step-by-step procedure that a classmate could use to divide rational expressions.

Skill Review

To prepare for Section 6.2, review subtraction of fractions and polynomials (Sections 1.2 and 5.1).

Simplify.

87. $\dfrac{7}{12} - \dfrac{2}{15}$ [1.2]

88. $\dfrac{7}{10} - \dfrac{1}{6}$ [1.2]

89. $\dfrac{1}{5} \cdot \dfrac{3}{4} - \dfrac{7}{10} \cdot \dfrac{3}{5}$ [1.2]

90. $\dfrac{2}{7} \cdot \dfrac{3}{5} - \dfrac{1}{2} \cdot \dfrac{5}{7}$ [1.2]

91. $(5x^2-6x+1)-(x^2-6x+3)$ [5.1]

92. $(t^2+8t-9)-(2t^2-t-7)$ [5.1]

93. $(y+1)(y-2)-(y+3)(y-5)$ [5.1]

94. $(2x+1)(x+3)-(x-7)(3x-1)$ [5.1]

Synthesis

95. Explain why the graphs of $f(x) = 5x$ and $g(x) = \dfrac{5x^2}{x}$ differ.

96. Todd *incorrectly* argues that since

$$\dfrac{a^2-4}{a-2} = \dfrac{a^2}{a} + \dfrac{-4}{-2} = a+2$$

is correct, it follows that

$$\dfrac{x^2+9}{x+1} = \dfrac{x^2}{x} + \dfrac{9}{1} = x+9.$$

Explain his misconception.

97. Calculate the slope of the line passing through $(a, f(a))$ and $(a+h, f(a+h))$ for the function f given by $f(x) = x^2 + 5$. Be sure your answer is simplified.

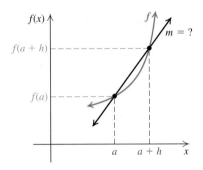

98. Calculate the slope of the line passing through the points $(a, f(a))$ and $(a+h, f(a+h))$ for the function f given by $f(x) = 3x^2$. Be sure your answer is simplified.

99. Graph the function given by

$$f(x) = \dfrac{x^2-9}{x-3}.$$

(*Hint*: Determine the domain of f and simplify.)

Perform the indicated operations and simplify.

100. $\dfrac{r^2-4s^2}{r+2s} \div (r+2s)^2 \left(\dfrac{2s}{r-2s}\right)^2$

101. $\dfrac{d^2-d}{d^2-6d+8} \cdot \dfrac{d-2}{d^2+5d} \div \left(\dfrac{5d^2}{d^2-9d+20}\right)^2$

Aha! 102. $\dfrac{6t^2 - 26t + 30}{8t^2 - 15t - 21} \cdot \dfrac{5t^2 - 9t - 15}{6t^2 - 14t - 20}$

$\div \dfrac{5t^2 - 9t - 15}{6t^2 - 14t - 20}$

Simplify.

103. $\dfrac{m^2 - t^2}{m^2 + t^2 + m + t + 2mt}$

104. $\dfrac{a^3 - 2a^2 + 2a - 4}{a^3 - 2a^2 - 3a + 6}$

105. $\dfrac{x^3 + x^2 - y^3 - y^2}{x^2 - 2xy + y^2}$

106. $\dfrac{u^6 + v^6 + 2u^3v^3}{u^3 - v^3 + u^2v - uv^2}$

107. $\dfrac{x^5 - x^3 + x^2 - 1 - (x^3 - 1)(x + 1)^2}{(x^2 - 1)^2}$

108. Let

$$g(x) = \dfrac{2x + 3}{4x - 1}.$$

Determine each of the following.

a) $g(x + h)$
b) $g(2x - 2) \cdot g(x)$
c) $g\left(\tfrac{1}{2}x + 1\right) \cdot g(x)$

109. Let

$$f(x) = \dfrac{4}{x^2 - 1} \quad \text{and} \quad g(x) = \dfrac{4x^2 + 8x + 4}{x^3 - 1}.$$

Find each of the following.

a) $(f \cdot g)(x)$
b) $(f/g)(x)$
c) $(g/f)(x)$

110. Use a graphing calculator to check Example 4. Use the method described on p. 356.

111. Use a graphing calculator to check your answers to Exercises 37, 43, and 71. Use the method described on p. 356.

112. Use a graphing calculator to show that

$$\dfrac{x^2 - 16}{x + 2} \neq x - 8.$$

113. To check Example 3, Kara lets

$$y_1 = \dfrac{7x^2 + 21x}{14x} \quad \text{and} \quad y_2 = \dfrac{x + 3}{2}.$$

Since the graphs of y_1 and y_2 appear to be identical, Kara believes that the domains of the functions described by y_1 and y_2 are the same, $\mathbb{R}$. How could you convince Kara otherwise?

6.2 Rational Expressions and Functions: Adding and Subtracting

When Denominators Are the Same ▪ When Denominators Are Different

Rational expressions are added and subtracted in much the same way as the fractions of arithmetic.

When Denominators Are the Same

Recall that $\dfrac{2}{7} + \dfrac{3}{7} = \dfrac{2 + 3}{7} = \dfrac{5}{7}$.

Addition and Subtraction with Like Denominators

To add or subtract when denominators are the same, add or subtract the numerators and keep the same denominator.

$$\dfrac{A}{C} + \dfrac{B}{C} = \dfrac{A + B}{C} \quad \text{and} \quad \dfrac{A}{C} - \dfrac{B}{C} = \dfrac{A - B}{C}, \quad \text{where } C \neq 0.$$

EXAMPLE **1** Add: $\dfrac{3 + x}{x} + \dfrac{4}{x}$.

SOLUTION

$$\dfrac{3 + x}{x} + \dfrac{4}{x} = \dfrac{3 + x + 4}{x} = \dfrac{x + 7}{x}$$

> *CAUTION!* Because x is a *term* in the numerator and not a *factor*, $\dfrac{x + 7}{x}$ cannot be simplified.

TRY EXERCISE 9

EXAMPLE **2** Add: $\dfrac{4x^2 - 5xy}{x^2 - y^2} + \dfrac{2xy - y^2}{x^2 - y^2}$.

SOLUTION

Add the numerators. Keep the denominators.

$$\dfrac{4x^2 - 5xy}{x^2 - y^2} + \dfrac{2xy - y^2}{x^2 - y^2} = \dfrac{4x^2 - 3xy - y^2}{x^2 - y^2}$$

Adding the numerators and combining like terms. The denominator is unchanged.

Factor.

$$= \dfrac{(x - y)(4x + y)}{(x - y)(x + y)}$$

Factoring the numerator and the denominator and looking for common factors

Remove a factor equal to 1.

$$= \dfrac{\cancel{(x - y)}(4x + y)}{\cancel{(x - y)}(x + y)}$$

Removing a factor equal to 1: $\dfrac{x - y}{x - y} = 1$

$$= \dfrac{4x + y}{x + y}$$

Simplifying

TRY EXERCISE 17

Recall that a fraction bar is a grouping symbol. The next example shows that when a numerator is subtracted, care must be taken to subtract, or change the sign of, *each* term in that polynomial.

EXAMPLE **3** If

$$f(x) = \dfrac{4x + 5}{x + 3} - \dfrac{x - 2}{x + 3},$$

find a simplified form for $f(x)$ and list all restrictions on the domain.

SOLUTION

$$f(x) = \dfrac{4x + 5}{x + 3} - \dfrac{x - 2}{x + 3}$$ Note that $x \neq -3$.

$$= \dfrac{4x + 5 - (x - 2)}{x + 3}$$ The parentheses remind us to subtract *both* terms.

$$= \dfrac{4x + 5 - x + 2}{x + 3}$$

$$= \dfrac{3x + 7}{x + 3}, \quad x \neq -3$$

TRY EXERCISE 27

TECHNOLOGY CONNECTION

Example 3 can be checked by comparing the graphs of

$$y_1 = \dfrac{4x + 5}{x + 3} - \dfrac{x - 2}{x + 3}$$

and

$$y_2 = \dfrac{3x + 7}{x + 3}$$

on the same set of axes. Since the equations are equivalent, one curve (it has two branches) should appear. Equivalently, you can show that $y_3 = y_2 - y_1$ is 0 for all x not equal to -3. The TABLE or TRACE feature can assist in either type of check. Use parentheses around each numerator and each denominator in the expressions.

When Denominators Are Different

Recall that when adding fractions with different denominators, we first find common denominators:

$$\frac{1}{6} + \frac{4}{15} = \frac{1}{6} \cdot \frac{5}{5} + \frac{4}{15} \cdot \frac{2}{2} = \frac{5}{30} + \frac{8}{30} = \frac{13}{30}.$$

Our work is easier when we use the *least common multiple* (LCM) of the denominators.

Least Common Multiple

To find the least common multiple (LCM) of two or more expressions:

1. Find the prime factorization of each expression.
2. Form a product that contains each factor the greatest number of times that it occurs in any one prime factorization.

EXAMPLE 4   Find the least common multiple of each pair of polynomials.

a) $21x$ and $3x^2$

b) $x^2 + x - 12$ and $x^2 - 16$

SOLUTION

a) We write the prime factorizations of $21x$ and $3x^2$:

$$21x = 3 \cdot 7 \cdot x \quad \text{and} \quad 3x^2 = 3 \cdot x \cdot x.$$

The factors 3, 7, and x must appear in the LCM if $21x$ is to be a factor of the LCM. The factors 3, x, and x must appear in the LCM if $3x^2$ is to be a factor of the LCM. These do not all appear in $3 \cdot 7 \cdot x$. However, if $3 \cdot 7 \cdot x$ is multiplied by another factor of x, a product is formed that contains both $21x$ and $3x^2$ as factors:

$$\text{LCM} = 3 \cdot 7 \cdot x \cdot x = 21x^2.$$

$21x$ is a factor.

$3x^2$ is a factor.

Note that each factor (3, 7, and x) is used the greatest number of times that it occurs as a factor of either $21x$ or $3x^2$. The LCM is $3 \cdot 7 \cdot x \cdot x$, or $21x^2$.

b) We factor both expressions:

$$x^2 + x - 12 = (x - 3)(x + 4),$$
$$x^2 - 16 = (x + 4)(x - 4).$$

The LCM must contain each polynomial as a factor. By multiplying the factors of $x^2 + x - 12$ by $x - 4$, we form a product that contains both $x^2 + x - 12$ and $x^2 - 16$ as factors:

$x^2 + x - 12$ is a factor.

$$\text{LCM} = (x - 3)(x + 4)(x - 4). \qquad \text{There is no need to multiply this out.}$$

$x^2 - 16$ is a factor.

TRY EXERCISE 21

> **To Add or Subtract Rational Expressions**
>
> 1. Determine the *least common denominator* (LCD) by finding the least common multiple of the denominators.
> 2. Rewrite each of the original rational expressions, as needed, in an equivalent form that has the LCD.
> 3. Add or subtract the resulting rational expressions, as indicated.
> 4. Simplify the result, if possible, and list any restrictions on the domain of functions.

EXAMPLE 5

Add: $\dfrac{2}{21x} + \dfrac{5}{3x^2}$.

SOLUTION In Example 4(a), we found that the LCD is $3 \cdot 7 \cdot x \cdot x$, or $21x^2$. We now multiply each rational expression by 1, using expressions for 1 that give us the LCD in each expression. To determine what to use, ask "$21x$ times what is $21x^2$?" and "$3x^2$ times what is $21x^2$?" The answers are x and 7, respectively, so we multiply by x/x and $7/7$:

$$\frac{2}{21x} \cdot \frac{x}{x} + \frac{5}{3x^2} \cdot \frac{7}{7} = \frac{2x}{21x^2} + \frac{35}{21x^2} \qquad \text{We now have a common denominator.}$$

$$= \frac{2x + 35}{21x^2}. \qquad \text{This expression cannot be simplified.}$$

> **TRY EXERCISE** 29

EXAMPLE 6

Add: $\dfrac{x^2}{x^2 + 2xy + y^2} + \dfrac{2x - 2y}{x^2 - y^2}$.

SOLUTION To find the LCD, we first factor the denominators:

$$\frac{x^2}{x^2 + 2xy + y^2} + \frac{2x - 2y}{x^2 - y^2} = \frac{x^2}{(x + y)(x + y)} + \frac{2x - 2y}{(x + y)(x - y)}.$$

STUDENT NOTES

When working with rational expressions, it is helpful to always begin by factoring all numerators and denominators. For addition and subtraction, this will allow you to identify any expressions that can be simplified as well as identify the LCD.

Although the numerators need not always be factored, doing so may enable us to simplify. In this case, the rightmost rational expression can be simplified:

$$\frac{x^2}{x^2 + 2xy + y^2} + \frac{2x - 2y}{x^2 - y^2} = \frac{x^2}{(x + y)(x + y)} + \frac{2(x - y)}{(x + y)(x - y)} \qquad \text{Factoring}$$

$$= \frac{x^2}{(x + y)(x + y)} + \frac{2}{x + y}. \qquad \text{Removing a factor equal to 1: } \frac{x - y}{x - y} = 1$$

Note that the LCM of $(x + y)(x + y)$ and $(x + y)$ is $(x + y)(x + y)$. To get the LCD in the second expression, we multiply by 1, using $(x + y)/(x + y)$. Then we add and, if possible, simplify.

$$\frac{x^2}{(x + y)(x + y)} + \frac{2}{x + y} = \frac{x^2}{(x + y)(x + y)} + \frac{2}{x + y} \cdot \frac{x + y}{x + y}$$

$$= \frac{x^2}{(x + y)(x + y)} + \frac{2x + 2y}{(x + y)(x + y)} \qquad \text{We have the LCD.}$$

$$= \frac{x^2 + 2x + 2y}{(x + y)(x + y)} \qquad \text{Since the numerator cannot be factored, we cannot simplify further.}$$

> **TRY EXERCISE** 33

EXAMPLE **7**   Subtract: $\dfrac{2y + 1}{y^2 - 7y + 6} - \dfrac{y + 3}{y^2 - 5y - 6}$.

SOLUTION

$$\dfrac{2y + 1}{y^2 - 7y + 6} - \dfrac{y + 3}{y^2 - 5y - 6}$$

Determine the LCD.

$$= \dfrac{2y + 1}{(y - 6)(y - 1)} - \dfrac{y + 3}{(y - 6)(y + 1)}$$ The LCD is $(y - 6)(y - 1)(y + 1)$.

Multiply by 1 so that both expressions have the LCD.

$$= \dfrac{2y + 1}{(y - 6)(y - 1)} \cdot \dfrac{y + 1}{y + 1} - \dfrac{y + 3}{(y - 6)(y + 1)} \cdot \dfrac{y - 1}{y - 1}$$

Multiplying by 1 to get the LCD in each expression

Subtract the numerators.

$$= \dfrac{(2y + 1)(y + 1) - (y + 3)(y - 1)}{(y - 6)(y - 1)(y + 1)}$$

$$= \dfrac{2y^2 + 3y + 1 - (y^2 + 2y - 3)}{(y - 6)(y - 1)(y + 1)}$$ Performing the multiplications in the numerator. The parentheses are important.

$$= \dfrac{2y^2 + 3y + 1 - y^2 - 2y + 3}{(y - 6)(y - 1)(y + 1)}$$

Factor and, if possible, simplify.

$$= \dfrac{y^2 + y + 4}{(y - 6)(y - 1)(y + 1)}$$ The numerator cannot be factored. We leave the denominator in factored form.

TRY EXERCISE 39

EXAMPLE **8**   Add: $\dfrac{3}{8a} + \dfrac{1}{-8a}$.

SOLUTION

$$\dfrac{3}{8a} + \dfrac{1}{-8a} = \dfrac{3}{8a} + \dfrac{-1}{-1} \cdot \dfrac{1}{-8a}$$

> When denominators are opposites, we multiply one rational expression by $-1/-1$ to get the LCD.

$$= \dfrac{3}{8a} + \dfrac{-1}{8a} = \dfrac{2}{8a}$$

$$= \dfrac{\cancel{2} \cdot 1}{\cancel{2} \cdot 4a} = \dfrac{1}{4a}$$ Simplifying by removing a factor equal to 1: $\dfrac{2}{2} = 1$

TRY EXERCISE 41

EXAMPLE **9**   Subtract: $\dfrac{5x}{x - 2y} - \dfrac{3y - 7}{2y - x}$.

SOLUTION

$$\dfrac{5x}{x - 2y} - \dfrac{3y - 7}{2y - x} = \dfrac{5x}{x - 2y} - \dfrac{-1}{-1} \cdot \dfrac{3y - 7}{2y - x}$$ Note that $x - 2y$ and $2y - x$ are opposites.

$$= \dfrac{5x}{x - 2y} - \dfrac{7 - 3y}{x - 2y}$$ Performing the multiplication. *Note*: $-1(2y - x) = -2y + x$
$\phantom{= \dfrac{5x}{x - 2y} - \dfrac{7 - 3y}{x - 2y}}$ $= x - 2y.$

$$= \dfrac{5x - (7 - 3y)}{x - 2y}$$

$$= \dfrac{5x - 7 + 3y}{x - 2y}$$ Subtracting. The parentheses are important.

TRY EXERCISE 43

In Example 9, you may have noticed that when $3y - 7$ is multiplied by -1 and subtracted, the result is $-7 + 3y$, which is equivalent to the original $3y - 7$. Thus, instead of multiplying the numerator by -1 and then subtracting, we could have simply *added* $3y - 7$ to $5x$, as in the following:

$$\frac{5x}{x - 2y} - \frac{3y - 7}{2y - x} = \frac{5x}{x - 2y} + (-1) \cdot \frac{3y - 7}{2y - x} \qquad \text{Rewriting subtraction as addition}$$

$$= \frac{5x}{x - 2y} + \frac{1}{-1} \cdot \frac{3y - 7}{2y - x} \qquad \text{Writing } -1 \text{ as } \frac{1}{-1}$$

$$= \frac{5x}{x - 2y} + \frac{3y - 7}{x - 2y} \qquad \begin{array}{l} \text{The opposite of } 2y - x \\ \text{is } x - 2y. \end{array}$$

$$= \frac{5x + 3y - 7}{x - 2y}. \qquad \begin{array}{l} \text{This checks with the} \\ \text{answer to Example 9.} \end{array}$$

EXAMPLE 10   Find simplified form for the function given by

$$f(x) = \frac{2x}{x^2 - 4} + \frac{5}{2 - x} - \frac{1}{2 + x}$$

and list all restrictions on the domain.

SOLUTION   We have

$$\frac{2x}{x^2 - 4} + \frac{5}{2 - x} - \frac{1}{2 + x} = \frac{2x}{(x - 2)(x + 2)} + \frac{5}{2 - x} - \frac{1}{2 + x} \qquad \begin{array}{l} \text{Factoring.} \\ \text{Note that} \\ x \neq -2, 2. \end{array}$$

$$= \frac{2x}{(x - 2)(x + 2)} + \frac{-1}{-1} \cdot \frac{5}{(2 - x)} - \frac{1}{x + 2} \qquad \begin{array}{l} \text{Multiplying by } -1/-1 \text{ since} \\ 2 - x \text{ is the opposite of} \\ x - 2 \end{array}$$

$$= \frac{2x}{(x - 2)(x + 2)} + \frac{-5}{x - 2} - \frac{1}{x + 2} \qquad \text{The LCD is } (x - 2)(x + 2).$$

$$= \frac{2x}{(x - 2)(x + 2)} + \frac{-5}{x - 2} \cdot \frac{x + 2}{x + 2} - \frac{1}{x + 2} \cdot \frac{x - 2}{x - 2} \qquad \begin{array}{l} \text{Multiplying by 1 to} \\ \text{get the LCD} \end{array}$$

$$= \frac{2x - 5(x + 2) - (x - 2)}{(x - 2)(x + 2)} = \frac{2x - 5x - 10 - x + 2}{(x - 2)(x + 2)}$$

$$= \frac{-4x - 8}{(x - 2)(x + 2)} = \frac{-4(x + 2)}{(x - 2)(x + 2)}$$

$$= \frac{-4\cancel{(x + 2)}}{(x - 2)\cancel{(x + 2)}} \qquad \begin{array}{l} \text{Removing a factor equal to 1:} \\ \dfrac{x + 2}{x + 2} = 1, x \neq -2 \end{array}$$

$$= \frac{-4}{x - 2}, \text{ or } -\frac{4}{x - 2}, x \neq -2, 2.$$

STUDENT NOTES

It is usually best to double-check each step of your work as you proceed. Waiting until the end of the problem to check your work is generally a less efficient use of your time.

An equivalent answer is $\dfrac{4}{2 - x}$. It is found by writing $-\dfrac{4}{x - 2}$ as $\dfrac{4}{-(x - 2)}$ and then using the distributive law to remove parentheses. **TRY EXERCISE** ▶ 67

Our work in Example 10 indicates that for

$$f(x) = \frac{2x}{x^2 - 4} + \frac{5}{2 - x} - \frac{1}{2 + x} \quad \text{and} \quad g(x) = \frac{-4}{x - 2},$$

with $x \neq -2$ and $x \neq 2$, we have $f = g$. Note that whereas the domain of f includes all real numbers except -2 or 2, the domain of g excludes only 2. This is illustrated in the following graphs. Methods for drawing such graphs by hand are discussed in more advanced courses. The graphs are for visualization only.

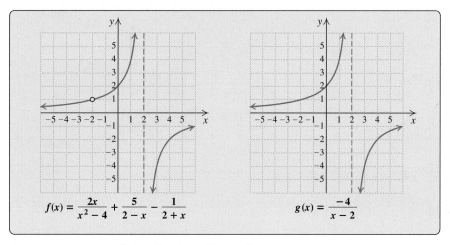

$$f(x) = \frac{2x}{x^2 - 4} + \frac{5}{2 - x} - \frac{1}{2 + x}$$

$$g(x) = \frac{-4}{x - 2}$$

A computer-generated visualization of Example 10

A quick, partial check of any simplification is to evaluate both the original and the simplified expressions for a convenient choice of *x*. For instance, to check Example 10, if $x = 1$, we have

$$f(1) = \frac{2 \cdot 1}{1^2 - 4} + \frac{5}{2 - 1} - \frac{1}{2 + 1} = \frac{2}{-3} + \frac{5}{1} - \frac{1}{3} = 5 - \frac{3}{3} = 4$$

and

$$g(1) = \frac{-4}{1 - 2} = \frac{-4}{-1} = 4.$$

Since both functions include the pair $(1, 4)$, our algebra was *probably* correct. Although this is only a partial check (on rare occasions, an incorrect answer might "check"), because it is so easy to perform, it is quite useful. Further evaluation provides a more definitive check.

6.2 EXERCISE SET

  Concept Reinforcement *Classify each statement as either true or false.*

1. A common denominator is required in order to add or subtract rational expressions.

2. To find the least common denominator, we find the least common multiple of the denominators.

3. It is rarely necessary to factor in order to find a common denominator.

4. The sum of two rational expressions is the sum of the numerators over the sum of the denominators.

5. The least common multiple of two expressions is always the product of those two expressions.

6. To add two rational expressions, it is often necessary to multiply at least one of those expressions by a form of 1.

7. After two rational expressions are added, it is unnecessary to simplify the result.

8. Parentheses are particularly important when we are subtracting rational expressions.

Perform the indicated operations. Simplify when possible.

9. $\dfrac{4}{3a} + \dfrac{11}{3a}$

10. $\dfrac{2}{5n} + \dfrac{8}{5n}$

11. $\dfrac{5}{3m^2n^2} - \dfrac{4}{3m^2n^2}$

12. $\dfrac{1}{4a^2b} - \dfrac{5}{4a^2b}$

13. $\dfrac{x - 3y}{x + y} + \dfrac{x + 5y}{x + y}$

14. $\dfrac{a - 5b}{a + b} + \dfrac{a + 7b}{a + b}$

15. $\dfrac{3t + 2}{t - 4} - \dfrac{t - 2}{t - 4}$

16. $\dfrac{4y + 2}{y - 2} - \dfrac{y - 3}{y - 2}$

17. $\dfrac{5 - 7x}{x^2 - 3x - 10} + \dfrac{8x - 3}{x^2 - 3x - 10}$

18. $\dfrac{4 - 2x}{x^2 - 9} + \dfrac{3x - 1}{x^2 - 9}$

19. $\dfrac{a - 2}{a^2 - 25} - \dfrac{2a - 7}{a^2 - 25}$

20. $\dfrac{5a - 4}{a^2 - 6a - 7} - \dfrac{6a - 11}{a^2 - 6a - 7}$

Find the least common multiple of each pair of polynomials.

21. $8x^2,\ 12x^5$

22. $15y,\ 18y^3$

23. $x^2 - 9,\ x^2 - 6x + 9$

24. $x^2 - x - 12,\ x^2 - 16$

Find simplified form for $f(x)$ and list all restrictions on the domain.

25. $f(x) = \dfrac{2x + 1}{x^2 + 6x + 5} + \dfrac{x - 2}{x^2 + 6x + 5}$

26. $f(x) = \dfrac{x - 6}{x^2 - 4x + 3} + \dfrac{5x - 1}{x^2 - 4x + 3}$

27. $f(x)\quad \dfrac{x - 4}{x^2 - 1}\ \ \dfrac{2x + 1}{x^2 - 1}$

28. $f(x) = \dfrac{3x + 11}{x^2 - 4} - \dfrac{2x - 8}{x^2 - 4}$

Perform the indicated operations. Simplify when possible.

29. $\dfrac{2}{15x^2} + \dfrac{3}{5x}$

30. $\dfrac{8}{9y} - \dfrac{5}{18y^2}$

31. $\dfrac{y + 1}{y - 2} - \dfrac{y - 1}{2y - 4}$

32. $\dfrac{x - 3}{2x + 6} + \dfrac{x + 2}{x + 3}$

33. $\dfrac{4xy}{x^2 - y^2} + \dfrac{x - y}{x + y}$

34. $\dfrac{5ab}{a^2 - b^2} + \dfrac{a + b}{a - b}$

35. $\dfrac{8}{2x^2 - 7x + 5} + \dfrac{3x + 2}{2x^2 - x - 10}$

36. $\dfrac{3y + 2}{y^2 + 5y - 24} + \dfrac{7}{y^2 + 4y - 32}$

37. $\dfrac{5ab}{a^2 - b^2} - \dfrac{a - b}{a + b}$

38. $\dfrac{6xy}{x^2 - y^2} - \dfrac{x + y}{x - y}$

39. $\dfrac{x}{x^2 + 9x + 20} - \dfrac{4}{x^2 + 7x + 12}$

40. $\dfrac{x}{x^2 + 11x + 30} - \dfrac{5}{x^2 + 9x + 20}$

41. $\dfrac{3}{t} - \dfrac{6}{-t}$

42. $\dfrac{8}{p} - \dfrac{7}{-p}$

43. $\dfrac{s^2}{r - s} + \dfrac{r^2}{s - r}$

44. $\dfrac{a^2}{a - b} + \dfrac{b^2}{b - a}$

45. $\dfrac{a + 2}{a - 4} + \dfrac{a - 2}{a + 3}$

46. $\dfrac{a + 3}{a - 5} + \dfrac{a - 2}{a + 4}$

47. $4 + \dfrac{x - 3}{x + 1}$

48. $3 + \dfrac{y + 2}{y - 5}$

49. $\dfrac{x + 6}{5x + 10} - \dfrac{x - 2}{4x + 8}$

50. $\dfrac{a + 3}{5a + 25} - \dfrac{a - 1}{3a + 15}$

51. $\dfrac{4}{x + 1} + \dfrac{x + 2}{x^2 - 1} + \dfrac{3}{x - 1}$

52. $\dfrac{-2}{y + 2} + \dfrac{5}{y - 2} + \dfrac{y + 3}{y^2 - 4}$

53. $\dfrac{y - 4}{y^2 - 25} - \dfrac{9 - 2y}{25 - y^2}$

54. $\dfrac{x - 7}{x^2 - 16} - \dfrac{x - 1}{16 - x^2}$

55. $\dfrac{y^2 - 5}{y^4 - 81} + \dfrac{4}{81 - y^4}$

56. $\dfrac{t^2 + 3}{t^4 - 16} + \dfrac{7}{16 - t^4}$

57. $\dfrac{r - 6s}{r^3 - s^3} - \dfrac{5s}{s^3 - r^3}$

58. $\dfrac{m - 3n}{m^3 - n^3} - \dfrac{2n}{n^3 - m^3}$

59. $\dfrac{3y}{y^2 - 7y + 10} - \dfrac{2y}{y^2 - 8y + 15}$

60. $\dfrac{5x}{x^2 - 6x + 8} - \dfrac{3x}{x^2 - x - 12}$

61. $\dfrac{2x + 1}{x - y} + \dfrac{5x^2 - 5xy}{x^2 - 2xy + y^2}$

62. $\dfrac{2 - 3a}{a - b} + \dfrac{3a^2 + 3ab}{a^2 - b^2}$

63. $\dfrac{2y - 6}{y^2 - 9} - \dfrac{y}{y - 1} + \dfrac{y^2 + 2}{y^2 + 2y - 3}$

64. $\dfrac{x - 1}{x^2 - 1} - \dfrac{x}{x - 2} + \dfrac{x^2 + 2}{x^2 - x - 2}$

Aha! **65.** $\dfrac{5y}{1 - 4y^2} - \dfrac{2y}{2y + 1} + \dfrac{5y}{4y^2 - 1}$

66. $\dfrac{4x}{x^2 - 1} + \dfrac{3x}{1 - x} - \dfrac{4}{x - 1}$

Find simplified form for $f(x)$ and list all restrictions on the domain.

67. $f(x) = 2 + \dfrac{x}{x - 3} - \dfrac{18}{x^2 - 9}$

68. $f(x) = 5 + \dfrac{x}{x + 2} - \dfrac{8}{x^2 - 4}$

69. $f(x) = \dfrac{3x - 1}{x^2 + 2x - 3} - \dfrac{x + 4}{x^2 - 16}$

70. $f(x) = \dfrac{3x - 2}{x^2 + 2x - 24} - \dfrac{x - 3}{x^2 - 9}$

71. $f(x) = \dfrac{1}{x^2 + 5x + 6} - \dfrac{2}{x^2 + 3x + 2} - \dfrac{1}{x^2 + 5x + 6}$

72. $f(x) = \dfrac{2}{x^2 - 5x + 6} - \dfrac{4}{x^2 - 2x - 3} + \dfrac{2}{x^2 + 4x + 3}$

73. Badar found that the sum of two rational expressions was $(3 - x)/(x - 5)$. The answer given at the back of the book is $(x - 3)/(5 - x)$. Is Badar's answer incorrect? Why or why not?

74. When two rational expressions are added or subtracted, should the numerator of the result be factored? Why or why not?

Skill Review

To prepare for Section 6.3, review negative exponents and multiplying using the distributive law (Sections 1.6 and 1.2).

Simplify. Use only positive exponents in your answer. [1.6]

75. $2x^{-1}$

76. $4x^{-2}$

77. $ab(a + b)^{-2}$

78. $3p^2(3 - p)^{-1}$

Multiply and simplify. [1.2], [1.6]

79. $9x^3\left(\dfrac{1}{x^2} - \dfrac{2}{3x^3}\right)$

80. $8a^2b^5\left(\dfrac{3}{8ab^2} + \dfrac{a}{4b^5}\right)$

Synthesis

81. Many students make the mistake of always multiplying denominators when looking for a common denominator. Use Example 7 to explain why this approach can yield results that are more difficult to simplify.

82. Is the sum of two rational expressions always a rational expression? Why or why not?

83. *Prescription drugs.*   After visiting her doctor, Jinney went to the pharmacy for a two-week supply of Zyrtec®, a 20-day supply of Albuterol, and a 30-day supply of Pepcid®. Jinney refills each prescription as soon as her supply runs out. How long will it be until she can refill all three prescriptions on the same day?

84. *Astronomy.*   Earth, Jupiter, Saturn, and Uranus all revolve around the sun. Earth takes 1 yr, Jupiter 12 yr, Saturn 30 yr, and Uranus 84 yr. How frequently do these four planets line up with each other?

85. *Music.*   To duplicate a common African poly-rhythm, a drummer needs to play sextuplets (6 beats per measure) on a tom-tom while simultaneously playing quarter notes (4 beats per measure) on a bass drum. Into how many equally sized parts must a measure be divided, in order to precisely execute this rhythm?

86. *Home appliances.*   Electric ranges last an average of 16 yr, water heaters about 12 yr, and refrigerators about 14 yr. In 1992, Golden Years Apartments installed new refrigerators in each of its units. In 2000, the apartments all received new ranges and water heaters. Predict the year in which the apartment complex will need to replace all three types of appliances at once.

Source: U.S. Department of Energy

Find the LCM.

87. $x^8 - x^4,\ x^5 - x^2,\ x^5 - x^3,\ x^5 + x^2$

88. $2a^3 + 2a^2b + 2ab^2,\ a^6 - b^6,$
$2b^2 + ab - 3a^2,\ 2a^2b + 4ab^2 + 2b^3$

89. The LCM of two expressions is $8a^4b^7$. One of the expressions is $2a^3b^7$. List all the possibilities for the other expression.

90. Determine the domain and the range of the function graphed below.

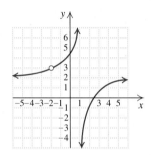

If

$$f(x) = \frac{x^3}{x^2 - 4} \quad and \quad g(x) = \frac{x^2}{x^2 + 3x - 10},$$

find each of the following.

91. $(f + g)(x)$ **92.** $(f - g)(x)$

93. $(f \cdot g)(x)$ **94.** $(f/g)(x)$

Perform the indicated operations and simplify.

95. $x^{-2} + 2x^{-1}$

96. $a^{-3}b - ab^{-3}$

97. $5(x - 3)^{-1} + 4(x + 3)^{-1} - 2(x + 3)^{-2}$

98. $4(y - 1)(2y - 5)^{-1} + 5(2y + 3)(5 - 2y)^{-1} +$
$(y - 4)(2y - 5)^{-1}$

99. $\dfrac{x + 4}{6x^2 - 20x} \cdot \left(\dfrac{x}{x^2 - x - 20} + \dfrac{2}{x + 4} \right)$

100. $\dfrac{x^2 - 7x + 12}{x^2 - x - 29/3} \cdot \left(\dfrac{3x + 2}{x^2 + 5x - 24} + \dfrac{7}{x^2 + 4x - 32} \right)$

101. $\dfrac{8t^5}{2t^2 - 10t + 12} \div \left(\dfrac{2t}{t^2 - 8t + 15} - \dfrac{3t}{t^2 - 7t + 10} \right)$

102. $\dfrac{9t^3}{3t^3 - 12t^2 + 9t} \div \left(\dfrac{t + 4}{t^2 - 9} - \dfrac{3t - 1}{t^2 + 2t - 3} \right)$

103. Use a graphing calculator to check your answers to Exercises 27, 53, and 63. Use the method discussed on p. 364.

104. Let

$$f(x) = 2 + \frac{x - 3}{x + 1}.$$

Use algebra, together with a graphing calculator, to determine the domain and the range of *f*.

6.3   Complex Rational Expressions

Multiplying by 1   ▪   Dividing Two Rational Expressions

A **complex rational expression** is a rational expression that contains rational expressions within its numerator and/or its denominator. Here are some examples:

$$\frac{x + \dfrac{5}{x}}{4x}, \quad \frac{\dfrac{x - y}{x + y}}{\dfrac{2x - y}{3x + y}}, \quad \frac{\dfrac{7x}{3} - \dfrac{4}{x}}{\dfrac{5x}{6} + \dfrac{8}{3}}, \quad \frac{\dfrac{r}{6} + \dfrac{r^2}{144}}{\dfrac{r}{12}}.$$

The rational expressions within each complex rational expression are red.

Complex rational expressions arise in a variety of real-world applications. For example, the last complex rational expression in the list above is used when calculating the size of certain loan payments.

Two methods are used to simplify complex rational expressions. Determining restrictions on variables may now require the solution of equations not yet studied. *Thus for this section we will not state restrictions on variables.*

Method 1: Multiplying by 1

One method of simplifying a complex rational expression is to multiply the entire expression by 1. To write 1, we use the LCD of the rational expressions within the complex rational expression.

EXAMPLE 1   Simplify:

$$\frac{\dfrac{1}{a^3b} + \dfrac{1}{b}}{\dfrac{1}{a^2b^2} - \dfrac{1}{b^2}}.$$

SOLUTION   The denominators within the complex rational expression are a^3b, b, a^2b^2, and b^2. Thus the LCD is a^3b^2. We multiply by 1, using $(a^3b^2)/(a^3b^2)$:

Find the LCD.

Using the LCD, multiply by 1.

$$\frac{\dfrac{1}{a^3b} + \dfrac{1}{b}}{\dfrac{1}{a^2b^2} - \dfrac{1}{b^2}} = \frac{\dfrac{1}{a^3b} + \dfrac{1}{b}}{\dfrac{1}{a^2b^2} - \dfrac{1}{b^2}} \cdot \frac{a^3b^2}{a^3b^2}$$

Multiplying by 1, using the LCD

$$= \frac{\left(\dfrac{1}{a^3b} + \dfrac{1}{b}\right)a^3b^2}{\left(\dfrac{1}{a^2b^2} - \dfrac{1}{b^2}\right)a^3b^2}$$

Multiplying in the numerator and in the denominator. Remember to use parentheses.

Distribute and simplify.

$$= \frac{\dfrac{1}{a^3b} \cdot a^3b^2 + \dfrac{1}{b} \cdot a^3b^2}{\dfrac{1}{a^2b^2} \cdot a^3b^2 - \dfrac{1}{b^2} \cdot a^3b^2}$$

Using the distributive law to carry out the multiplications

$$= \frac{\dfrac{a^3b}{a^3b} \cdot b + \dfrac{b}{b} \cdot a^3b}{\dfrac{a^2b^2}{a^2b^2} \cdot a - \dfrac{b^2}{b^2} \cdot a^3}$$

Removing factors that equal 1. Study this carefully.

$$= \frac{b + a^3b}{a - a^3}$$

Simplifying

$$= \frac{b(1 + a^3)}{a(1 - a^2)}$$

Factoring

Factor and simplify.

$$= \frac{b(1 + a)(1 - a + a^2)}{a(1 + a)(1 - a)}$$

Factoring further and identifying a factor that equals 1

$$= \frac{b(1 - a + a^2)}{a(1 - a)}.$$

Simplifying

TRY EXERCISE 15

> **Using Multiplication by 1 to Simplify a Complex Rational Expression**
>
> **1.** Find the LCD of all rational expressions *within* the complex rational expression.
> **2.** Multiply the complex rational expression by 1, writing 1 as the LCD divided by itself.
> **3.** Distribute and simplify so that the numerator and the denominator of the complex rational expression are polynomials.
> **4.** Factor and, if possible, simplify.

Note that use of the LCD, when multiplying by a form of 1, clears the numerator and the denominator of the complex rational expression of all rational expressions.

EXAMPLE **2** Simplify:

$$\dfrac{\dfrac{3}{2x-2}-\dfrac{1}{x+1}}{\dfrac{1}{x-1}+\dfrac{x}{x^2-1}}.$$

STUDENT NOTES

In Example 1, we found the LCD for four rational expressions. To find this LCD, you may prefer to find the LCD for the expressions in the numerator, then find the LCD for the expressions in the denominator, and, finally, find the LCM of those two LCDs.

SOLUTION In this case, to find the LCD, we have to factor first:

$$\dfrac{\dfrac{3}{2x-2}-\dfrac{1}{x+1}}{\dfrac{1}{x-1}+\dfrac{x}{x^2-1}}=\dfrac{\dfrac{3}{2(x-1)}-\dfrac{1}{x+1}}{\dfrac{1}{x-1}+\dfrac{x}{(x-1)(x+1)}} \qquad \text{The LCD is } 2(x-1)(x+1).$$

$$=\dfrac{\dfrac{3}{2(x-1)}-\dfrac{1}{x+1}}{\dfrac{1}{x-1}+\dfrac{x}{(x-1)(x+1)}}\cdot\dfrac{2(x-1)(x+1)}{2(x-1)(x+1)} \qquad \begin{array}{l}\text{Multiplying by 1,}\\\text{using the LCD}\end{array}$$

$$=\dfrac{\dfrac{3}{2(x-1)}\cdot 2(x-1)(x+1)-\dfrac{1}{x+1}\cdot 2(x-1)(x+1)}{\dfrac{1}{x-1}\cdot 2(x-1)(x+1)+\dfrac{x}{(x-1)(x+1)}\cdot 2(x-1)(x+1)} \qquad \text{Using the distributive law}$$

$$=\dfrac{\dfrac{2(x-1)}{2(x-1)}\cdot 3(x+1)-\dfrac{x+1}{x+1}\cdot 2(x-1)}{\dfrac{x-1}{x-1}\cdot 2(x+1)+\dfrac{(x-1)(x+1)}{(x-1)(x+1)}\cdot 2x} \qquad \begin{array}{l}\text{Removing factors}\\\text{that equal 1}\end{array}$$

$$=\dfrac{3(x+1)-2(x-1)}{2(x+1)+2x} \qquad \text{Simplifying}$$

$$=\dfrac{3x+3-2x+2}{2x+2+2x} \qquad \text{Using the distributive law}$$

$$=\dfrac{x+5}{4x+2}. \qquad \text{Combining like terms}$$

▶ **TRY EXERCISE** 27

Method 2: Dividing Two Rational Expressions

Another method for simplifying complex rational expressions involves first adding or subtracting, as necessary, to obtain one rational expression in the numerator and one rational expression in the denominator. The problem is then regarded as one rational expression divided by another.

EXAMPLE 3   Simplify:

$$\dfrac{\dfrac{3}{x} - \dfrac{2}{x^2}}{\dfrac{3}{x-2} + \dfrac{1}{x^2}}.$$

SOLUTION

$$\dfrac{\dfrac{3}{x} - \dfrac{2}{x^2}}{\dfrac{3}{x-2} + \dfrac{1}{x^2}} = \dfrac{\dfrac{3}{x} \cdot \dfrac{x}{x} - \dfrac{2}{x^2}}{\dfrac{3}{x-2} \cdot \dfrac{x^2}{x^2} + \dfrac{1}{x^2} \cdot \dfrac{x-2}{x-2}}$$

Multiplying $3/x$ by 1 to obtain x^2 as a common denominator

Multiplying by 1, twice, to obtain $x^2(x-2)$ as a common denominator

$$= \dfrac{\dfrac{3x}{x^2} - \dfrac{2}{x^2}}{\dfrac{3x^2}{(x-2)x^2} + \dfrac{x-2}{x^2(x-2)}}$$

The common denominator is x^2.

The common denominator is $(x-2)x^2$.

> We now have one rational expression divided by another rational expression.

$$= \dfrac{\dfrac{3x-2}{x^2}}{\dfrac{3x^2 + x - 2}{(x-2)x^2}}$$

Subtracting in the numerator

Adding in the denominator

$$= \dfrac{3x-2}{x^2} \div \dfrac{3x^2 + x - 2}{(x-2)x^2}$$

Rewriting with a division symbol

$$= \dfrac{3x-2}{x^2} \cdot \dfrac{(x-2)x^2}{3x^2 + x - 2}$$

To divide, multiply by the reciprocal of the divisor.

$$= \dfrac{\cancel{(3x-2)}(x-2)\cancel{x^2}}{\cancel{x^2}\cancel{(3x-2)}(x+1)}$$

Factoring and removing a factor equal to 1: $\dfrac{x^2(3x-2)}{x^2(3x-2)} = 1$

$$= \dfrac{x-2}{x+1}$$

TRY EXERCISE 45

> **Using Division to Simplify a Complex Rational Expression**
>
> 1. Add or subtract, as necessary, to get one rational expression in the numerator.
> 2. Add or subtract, as necessary, to get one rational expression in the denominator.
> 3. Perform the indicated division (invert the divisor and multiply).
> 4. Simplify, if possible, by removing any factors that equal 1.

EXAMPLE **4**

Simplify:

$$\frac{1 + \dfrac{2}{x}}{1 - \dfrac{4}{x^2}}.$$

SOLUTION   We have

$$\frac{1 + \dfrac{2}{x}}{1 - \dfrac{4}{x^2}} = \frac{\dfrac{x}{x} + \dfrac{2}{x}}{\dfrac{x^2}{x^2} - \dfrac{4}{x^2}} \quad \left.\begin{array}{l}\\ \end{array}\right\} \text{ Finding a common denominator}$$
$$\left.\begin{array}{l}\\ \end{array}\right\} \text{ Finding a common denominator}$$

$$= \frac{\dfrac{x + 2}{x}}{\dfrac{x^2 - 4}{x^2}} \quad \left.\begin{array}{l}\\ \end{array}\right\} \text{ Adding in the numerator}$$
$$\left.\begin{array}{l}\\ \end{array}\right\} \text{ Subtracting in the denominator}$$

$$= \frac{x + 2}{x} \cdot \frac{x^2}{x^2 - 4} \qquad \begin{array}{l}\text{Multiplying by the reciprocal of}\\ \text{the divisor}\end{array}$$

$$= \frac{(x + 2) \cdot x^2}{x(x + 2)(x - 2)} \qquad \begin{array}{l}\text{Factoring. Remember to simplify}\\ \text{when possible.}\end{array}$$

$$= \frac{\cancel{(x + 2)}x \cdot x}{x\cancel{(x + 2)}(x - 2)} \qquad \begin{array}{l}\text{Removing a factor equal to 1:}\\ \dfrac{(x + 2)x}{(x + 2)x} = 1\end{array}$$

$$= \frac{x}{x - 2}. \qquad \text{Simplifying}$$

As a quick partial check, we select a convenient value for *x*—say, 1:

$$\frac{1 + \dfrac{2}{1}}{1 - \dfrac{4}{1^2}} = \frac{1 + 2}{1 - 4} = \frac{3}{-3} = -1 \qquad \begin{array}{l}\text{We evaluated the original expression}\\ \text{for } x = 1.\end{array}$$

and

$$\frac{1}{1 - 2} = \frac{1}{-1} = -1. \qquad \begin{array}{l}\text{We evaluated the simplified}\\ \text{expression for } x = 1.\end{array}$$

Since both expressions yield the same result, our simplification is probably correct. More evaluation would provide a more definitive check.

TRY EXERCISE ▶ 29

To check Example 4, we can show that the graphs of

$$y_1 = \frac{1 + \dfrac{2}{x}}{1 - \dfrac{4}{x^2}}$$

and

$$y_2 = \frac{x}{x - 2}$$

coincide, or we can show that (except for $x = -2$ or 0) their tables of values are identical. We can also check by showing that (except for $x = -2$ or 0 or 2) $y_2 - y_1 = 0$.

1. Use a graphing calculator to check Example 3. What values, if any, can *x* *not* equal?

If negative exponents occur, we first find an equivalent expression using positive exponents and then simplify as in the preceding examples.

EXAMPLE   5   Simplify:

$$\frac{a^{-1} + b^{-1}}{a^{-3} + b^{-3}}.$$

SOLUTION

$$\frac{a^{-1} + b^{-1}}{a^{-3} + b^{-3}} = \frac{\dfrac{1}{a} + \dfrac{1}{b}}{\dfrac{1}{a^3} + \dfrac{1}{b^3}}$$ Rewriting with positive exponents.

$$= \frac{\dfrac{1}{a} \cdot \dfrac{b}{b} + \dfrac{1}{b} \cdot \dfrac{a}{a}}{\dfrac{1}{a^3} \cdot \dfrac{b^3}{b^3} + \dfrac{1}{b^3} \cdot \dfrac{a^3}{a^3}}$$ Finding a common denominator

Finding a common denominator

$$= \frac{\dfrac{b}{ab} + \dfrac{a}{ab}}{\dfrac{b^3}{a^3b^3} + \dfrac{a^3}{a^3b^3}}$$

Add or subtract in the numerator.

Add or subtract in the denominator.

$$= \frac{\dfrac{b + a}{ab}}{\dfrac{b^3 + a^3}{a^3b^3}}$$ Adding in the numerator

Adding in the denominator

Invert the divisor and multiply.

$$= \frac{b + a}{ab} \cdot \frac{a^3b^3}{b^3 + a^3}$$ Multiplying by the reciprocal of the divisor

$$= \frac{(b + a) \cdot ab \cdot a^2b^2}{ab(b + a)(b^2 - ab + a^2)}$$ Factoring and looking for common factors

Factor and simplify.

$$= \frac{\cancel{(b + a)} \cdot \cancel{ab} \cdot a^2b^2}{\cancel{ab}\,\cancel{(b + a)}(b^2 - ab + a^2)}$$ Removing a factor equal to 1: $\dfrac{(b + a)ab}{ab(b + a)} = 1$

$$= \frac{a^2b^2}{b^2 - ab + a^2}$$

TRY EXERCISE 31

There is no one method that is best to use. For expressions like

$$\frac{\dfrac{3x + 1}{x - 5}}{\dfrac{2 - x}{x + 3}} \quad \text{or} \quad \frac{\dfrac{3}{x} - \dfrac{2}{x}}{\dfrac{1}{x + 1} + \dfrac{5}{x + 1}},$$

the second method is probably easier to use since it is little or no work to write the expression as a quotient of two rational expressions.

On the other hand, expressions like

$$\frac{\dfrac{3}{a^2b} - \dfrac{4}{bc^3}}{\dfrac{1}{b^3c} + \dfrac{2}{ac^4}} \quad \text{or} \quad \frac{\dfrac{5}{a^2 - b^2} + \dfrac{2}{a^2 + 2ab + b^2}}{\dfrac{1}{a - b} + \dfrac{4}{a + b}}$$

require fewer steps if we use the first method. Either method can be used to simplify any complex rational expression.

6.3   EXERCISE SET

For Extra Help
MyMathLab   Math XL   PRACTICE   WATCH   DOWNLOAD

🪝 *Concept Reinforcement   In each of Exercises 1–6, match the expression(s) from the column on the right to the description with reference to the complex rational expression*

$$\frac{\dfrac{x-6}{x^2}+\dfrac{2}{5x}}{\dfrac{x}{x+1}-\dfrac{x}{x-1}}.$$

1. _____ The rational expressions within the complex rational expression

2. _____ The denominator of the complex rational expression

3. _____ The denominators within the complex rational expression

4. _____ The LCD of the rational expressions in the numerator

5. _____ The LCD of the rational expressions in the denominator

6. _____ The LCD of all the rational expressions within the complex rational expression

a) $\dfrac{x}{x+1}-\dfrac{x}{x-1}$

b) $\dfrac{x-6}{x^2},\ \dfrac{2}{5x},\ \dfrac{x}{x+1},\ \dfrac{x}{x-1}$

c) $5x^2$

d) $(x+1)(x-1)$

e) $5x^2(x+1)(x-1)$

f) $x^2,\ 5x,\ x+1,\ x-1$

Simplify. If possible, use a second method or evaluation as a check.

7. $\dfrac{\dfrac{x+5}{x-3}}{\dfrac{x-2}{x+1}}$

8. $\dfrac{\dfrac{x-3}{x+4}}{\dfrac{x+6}{x-1}}$

9. $\dfrac{\dfrac{3}{x}+\dfrac{2}{x^3}}{\dfrac{5}{x}-\dfrac{3}{x^2}}$

10. $\dfrac{\dfrac{5}{y^2}+\dfrac{3}{y^4}}{\dfrac{3}{y}-\dfrac{2}{y^3}}$

11. $\dfrac{\dfrac{3}{m-n}}{\dfrac{5}{m+n}}$

12. $\dfrac{\dfrac{5}{a+b}}{\dfrac{3}{a-b}}$

13. $\dfrac{\dfrac{6}{r}-\dfrac{1}{s}}{\dfrac{2}{r}+\dfrac{3}{s}}$

14. $\dfrac{\dfrac{9}{a}-\dfrac{5}{b}}{\dfrac{4}{a}+\dfrac{1}{b}}$

15. $\dfrac{\dfrac{3}{z^2}+\dfrac{2}{yz}}{\dfrac{4}{zy^2}-\dfrac{1}{y}}$

16. $\dfrac{\dfrac{6}{x^3}+\dfrac{7}{y}}{\dfrac{7}{xy^2}-\dfrac{6}{x^2y^2}}$

17. $\dfrac{\dfrac{a^2-b^2}{ab}}{\dfrac{a-b}{b}}$

18. $\dfrac{\dfrac{x^2-y^2}{xy}}{\dfrac{x+y}{y}}$

19. $\dfrac{1-\dfrac{2}{3x}}{x-\dfrac{4}{9x}}$

20. $\dfrac{\dfrac{3x}{y}-x}{2y-\dfrac{y}{x}}$

21. $\dfrac{y^{-1}-x^{-1}}{\dfrac{x^2-y^2}{xy}}$

22. $\dfrac{a^{-1}+b^{-1}}{\dfrac{a^2-b^2}{ab}}$

23. $\dfrac{\dfrac{1}{x+h}-\dfrac{1}{x}}{h}$

24. $\dfrac{\dfrac{1}{a-h}-\dfrac{1}{a}}{h}$

25. $\dfrac{\dfrac{a^2 - 4}{a^2 + 3a + 2}}{\dfrac{a^2 - 5a - 6}{a^2 - 6a - 7}}$

26. $\dfrac{\dfrac{x^2 - x - 12}{x^2 - 2x - 15}}{\dfrac{x^2 + 8x + 12}{x^2 - 5x - 14}}$

27. $\dfrac{\dfrac{x}{x^2 + 3x - 4} - \dfrac{1}{x^2 + 3x - 4}}{\dfrac{x}{x^2 + 6x + 8} + \dfrac{3}{x^2 + 6x + 8}}$

28. $\dfrac{\dfrac{x}{x^2 + 5x - 6} + \dfrac{6}{x^2 + 5x - 6}}{\dfrac{x}{x^2 - 5x + 4} - \dfrac{2}{x^2 - 5x + 4}}$

29. $\dfrac{\dfrac{1}{y} + 2}{\dfrac{1}{y} - 3}$

30. $\dfrac{7 + \dfrac{1}{a}}{\dfrac{1}{a} - 3}$

31. $\dfrac{y + y^{-2}}{y - y^{-2}}$

32. $\dfrac{x - x^{-2}}{x + x^{-2}}$

33. $\dfrac{\dfrac{1}{x - 2} + \dfrac{3}{x - 1}}{\dfrac{2}{x - 1} + \dfrac{5}{x - 2}}$

34. $\dfrac{\dfrac{2}{y - 3} + \dfrac{1}{y + 1}}{\dfrac{3}{y + 1} + \dfrac{4}{y - 3}}$

35. $\dfrac{a(a + 3)^{-1} - 2(a - 1)^{-1}}{a(a + 3)^{-1} - (a - 1)^{-1}}$

36. $\dfrac{a(a + 2)^{-1} - 3(a - 3)^{-1}}{a(a + 2)^{-1} - (a - 3)^{-1}}$

37. $\dfrac{\dfrac{2}{a^2 - 1} + \dfrac{1}{a + 1}}{\dfrac{3}{a^2 - 1} + \dfrac{2}{a - 1}}$

38. $\dfrac{\dfrac{3}{a^2 - 9} + \dfrac{2}{a + 3}}{\dfrac{4}{a^2 - 9} + \dfrac{1}{a + 3}}$

39. $\dfrac{\dfrac{5}{x^2 - 4} - \dfrac{3}{x - 2}}{\dfrac{4}{x^2 - 4} - \dfrac{2}{x + 2}}$

40. $\dfrac{\dfrac{4}{x^2 - 1} - \dfrac{3}{x + 1}}{\dfrac{5}{x^2 - 1} - \dfrac{2}{x - 1}}$

41. $\dfrac{\dfrac{y^3}{y^2 - 4} + \dfrac{125}{4 - y^2}}{\dfrac{y}{y^2 - 4} + \dfrac{5}{4 - y^2}}$

42. $\dfrac{\dfrac{y}{y^2 - 1} - \dfrac{3}{1 - y^2}}{\dfrac{y^3}{y^2 - 1} - \dfrac{27}{1 - y^2}}$

43. $\dfrac{\dfrac{y^2}{y^2 - 25} - \dfrac{y}{y - 5}}{\dfrac{y}{y^2 - 25} - \dfrac{1}{y + 5}}$

44. $\dfrac{\dfrac{y^2}{y^2 - 9} - \dfrac{y}{y + 3}}{\dfrac{y}{y^2 - 9} - \dfrac{1}{y - 3}}$

45. $\dfrac{\dfrac{a}{a + 2} + \dfrac{5}{a}}{\dfrac{a}{2a + 4} + \dfrac{1}{3a}}$

46. $\dfrac{\dfrac{a}{a + 3} + \dfrac{4}{5a}}{\dfrac{a}{2a + 6} + \dfrac{3}{a}}$

47. $\dfrac{\dfrac{1}{x^2 - 3x + 2} + \dfrac{1}{x^2 - 4}}{\dfrac{1}{x^2 + 4x + 4} + \dfrac{1}{x^2 - 4}}$

48. $\dfrac{\dfrac{1}{x^2 + 3x + 2} + \dfrac{1}{x^2 - 1}}{\dfrac{1}{x^2 - 1} + \dfrac{1}{x^2 - 4x + 3}}$

49. $\dfrac{\dfrac{3}{a^2 - 4a + 3} + \dfrac{3}{a^2 - 5a + 6}}{\dfrac{3}{a^2 - 3a + 2} + \dfrac{3}{a^2 + 3a - 10}}$

50. $\dfrac{\dfrac{1}{a^2 + 7a + 10} - \dfrac{2}{a^2 - 7a + 12}}{\dfrac{2}{a^2 - a - 6} - \dfrac{1}{a^2 + a - 20}}$

Aha! **51.** $\dfrac{\dfrac{y}{y^2 - 4} - \dfrac{2y}{y^2 + y - 6}}{\dfrac{2y}{y^2 + y - 6} - \dfrac{y}{y^2 - 4}}$

52. $\dfrac{\dfrac{y}{y^2 - 1} - \dfrac{3y}{y^2 + 5y + 4}}{\dfrac{3y}{y^2 - 1} - \dfrac{y}{y^2 - 4y + 3}}$

53. $\dfrac{\dfrac{3}{x^2 + 2x - 3} - \dfrac{1}{x^2 - 3x - 10}}{\dfrac{3}{x^2 - 6x + 5} - \dfrac{1}{x^2 + 5x + 6}}$

54. $\dfrac{\dfrac{1}{a^2 + 7a + 12} + \dfrac{1}{a^2 + a - 6}}{\dfrac{1}{a^2 + 2a - 8} + \dfrac{1}{a^2 + 5a + 4}}$

55. Michael *incorrectly* simplifies

$$\dfrac{a + b^{-1}}{a + c^{-1}} \quad \text{as} \quad \dfrac{a + c}{a + b}.$$

What mistake is he probably making and how could you convince him that this is incorrect?

56. To simplify a complex rational expression in which the sum of two fractions is divided by the difference of the same two fractions, which method is easier? Why?

Skill Review

To prepare for Section 6.4, review solving equations (Sections 1.3 and 5.8).

Solve.

57. $2(y + 3) - 5(y - 1) = 10y$ [1.3]

58. $6(x - 3) + (x + 7) = 12$ [1.3]

59. $x^2 = 25$ [5.8]   **60.** $a^2 + 8 = 6a$ [5.8]

61. $\frac{1}{3}x - \frac{1}{4} = \frac{1}{6} - \frac{1}{2}x$ [1.3]   **62.** $\frac{y}{8} - \frac{1}{3} = \frac{y}{12} + \frac{5}{6}$ [1.3]

Synthesis

63. In arithmetic, we are taught that

$$\frac{a}{b} \div \frac{c}{d} = \frac{a}{b} \cdot \frac{d}{c}$$

(to divide by a fraction, we invert the divisor and multiply). Use method 1 to explain *why* this is the correct approach.

64. An LCD is used in both method 1 and method 2. Explain how the use of the LCD differs in these methods.

Simplify.

65. $\dfrac{5x^{-2} + 10x^{-1}y^{-1} + 5y^{-2}}{3x^{-2} - 3y^{-2}}$

66. $(a^2 - ab + b^2)^{-1}(a^2b^{-1} + b^2a^{-1}) \times$
$(a^{-2} - b^{-2})(a^{-2} + 2a^{-1}b^{-1} + b^{-2})^{-1}$

67. *Astronomy.*   When two galaxies are moving in opposite directions at velocities v_1 and v_2, an observer in one of the galaxies would see the other galaxy receding at speed

$$\frac{v_1 + v_2}{1 + \dfrac{v_1 v_2}{c^2}},$$

where c is the speed of light. Determine the observed speed if v_1 and v_2 are both one-fourth the speed of light.

Find and simplify

$$\frac{f(x + h) - f(x)}{h}$$

for each rational function f in Exercises 68–71.

68. $f(x) = \dfrac{2}{x^2}$ **69.** $f(x) = \dfrac{3}{x}$

70. $f(x) = \dfrac{x}{1 - x}$ **71.** $f(x) = \dfrac{2x}{1 + x}$

72. If

$$F(x) = \frac{3 + \dfrac{1}{x}}{2 - \dfrac{8}{x^2}},$$

find the domain of F.

73. If

$$G(x) = \frac{x - \dfrac{1}{x^2 - 1}}{\dfrac{1}{9} - \dfrac{1}{x^2 - 16}},$$

find the domain of G.

74. Find the reciprocal of y if

$$y = x^2 + x + 1 + \frac{1}{x} + \frac{1}{x^2}.$$

75. For $f(x) = \dfrac{2}{2 + x}$, find $f(f(a))$.

76. For $g(x) = \dfrac{x + 3}{x - 1}$, find $g(g(a))$.

77. Let

$$f(x) = \left[\frac{\dfrac{x + 3}{x - 3} + 1}{\dfrac{x + 3}{x - 3} - 1} \right]^4.$$

Find a simplified form of $f(x)$ and specify the domain of f.

78. Use a graphing calculator to check your answers to Exercises 29, 33, 45, and 72.

79. *Financial planning.*   Austin wishes to invest a portion of each month's pay in an account that pays 7.5% interest. If he wants to have $30,000 in the account after 10 yr, the amount invested each month is given by

$$\frac{30{,}000 \cdot \dfrac{0.075}{12}}{\left(1 + \dfrac{0.075}{12}\right)^{120} - 1}.$$

Find the amount of Austin's monthly investment.

80. Use algebra to determine the domain of the function given by

$$f(x) = \frac{\dfrac{1}{x - 2}}{\dfrac{x}{x - 2} - \dfrac{5}{x - 2}}.$$

Then explain how a graphing calculator could be used to check your answer.

CORNER

Which Method Is Better?

Focus: Complex rational expressions
Time: 10–15 minutes
Group size: 3–4

ACTIVITY

Consider the steps in Examples 2 and 3 for simplifying a complex rational expression by each of the two methods. Then, work as a group to simplify

$$\frac{\dfrac{5}{x+1} - \dfrac{1}{x}}{\dfrac{2}{x^2} + \dfrac{4}{x}}$$

using the following procedure.

1. The group should predict which method will more easily simplify this expression.

2. Using the method selected in part (1), one group member should perform the first step in the simplification and then pass the problem on to another member of the group. That person then checks the work, performs the next step, and passes the problem on to another group member. If a mistake is found, the problem should be passed to the person who made the mistake for repair. This process continues until, eventually, the simplification is complete.

3. At the same time that part (2) is being performed, another group member should perform the first step of the solution using the method not selected in part (1). He or she should then pass the problem to another group member and so on, just as in part (2).

4. What method *was* easier? Why? Compare your responses with those of other groups.

6.4 Rational Equations

Solving Rational Equations

In Sections 6.1–6.3, we learned how to *simplify expressions*. We now learn to *solve* a new type of *equation*. A **rational equation** is an equation that contains one or more rational expressions. Here are some examples:

$$\frac{2}{3} - \frac{5}{6} = \frac{1}{t}, \qquad \frac{a-1}{a-5} = \frac{4}{a^2-25}, \qquad x^3 + \frac{6}{x} = 5.$$

Equations of this type occur frequently in applications.

Solving Rational Equations

Recall that one way to *clear fractions* from an equation is to multiply both sides of the equation by the LCD. For example, we have cleared fractions in equations like

$$\frac{x}{2} - \frac{1}{4} = \frac{1}{3}$$

by multiplying both sides of the equation by the LCD, 12:

$$12\left(\frac{x}{2} - \frac{1}{4}\right) = 12 \cdot \frac{1}{3}$$

$$\frac{12x}{2} - \frac{12}{4} = \frac{12}{3}$$

$$6x - 3 = 4 \qquad \text{The fractions are } \textit{cleared.}$$

$$6x = 7$$

$$x = \frac{7}{6}.$$

Most of the rational equations that we will encounter contain a variable in at least one denominator. Since division by 0 is undefined, any replacement for the variable that makes a denominator 0 cannot be a solution of the equation. We can rule out these numbers before we even attempt to find a solution. After we have solved the equation, we must check that no possible solution makes a denominator 0.

> **To Solve a Rational Equation**
>
> 1. List any numbers that will make a denominator 0. State that the variable *cannot* equal these numbers.
> 2. Clear fractions by multiplying both sides of the equation by the LCD. Use the distributive law as needed.
> 3. Solve the equation.
> 4. Check possible solutions against the list of numbers that cannot be solutions and in the original, rational equation.

EXAMPLE **1**   Solve: $\dfrac{x + 4}{3x} + \dfrac{x + 8}{5x} = 2$.

SOLUTION

1. Because the left side of this equation is undefined when x is 0, we state at the outset that $x \neq 0$.

2. We multiply both sides of the equation by the LCD, $3 \cdot 5 \cdot x$, or $15x$:

$$15x\left(\frac{x + 4}{3x} + \frac{x + 8}{5x}\right) = 15x \cdot 2 \qquad \begin{array}{l}\text{Multiplying by the LCD}\\ \text{to clear fractions}\end{array}$$

$$15x \cdot \frac{x + 4}{3x} + 15x \cdot \frac{x + 8}{5x} = 15x \cdot 2 \qquad \begin{array}{l}\text{Using the distributive}\\ \text{law}\end{array}$$

$$\frac{5 \cdot 3x \cdot (x + 4)}{3x} + \frac{3 \cdot 5x \cdot (x + 8)}{5x} = 30x \qquad \begin{array}{l}\text{Locating factors}\\ \text{equal to 1}\end{array}$$

$$5(x + 4) + 3(x + 8) = 30x. \qquad \begin{array}{l}\text{Removing factors}\\ \text{equal to 1:}\\ \frac{3x}{3x} = 1; \frac{5x}{5x} = 1\end{array}$$

| We have solved equations like this before. |

3. We solve the equation:

$$5x + 20 + 3x + 24 = 30x \qquad \text{Using the distributive law}$$

$$8x + 44 = 30x$$

$$44 = 22x$$

$$2 = x.$$

4. We stated that $x \neq 0$, so 2 should check.

Check:
$$\frac{x+4}{3x} + \frac{x+8}{5x} = 2$$

$$\frac{2+4}{3 \cdot 2} + \frac{2+8}{5 \cdot 2} \,\bigg|\, 2$$

$$\frac{6}{6} + \frac{10}{10} \,\bigg|$$

$$2 \overset{?}{=} 2 \quad \text{TRUE}$$

The number 2 is the solution.

[TRY EXERCISE] 17

When we clear fractions, all denominators are eliminated. This produces an equation without rational expressions, which we already know how to solve. In the remaining examples, we follow, but do not list, the above steps for solving rational equations.

[EXAMPLE] 2

Solve: $\dfrac{x-1}{x-5} = \dfrac{4}{x-5}$.

STUDENT NOTES ———

When we are solving rational equations, as in Examples 1–5, the LCD is never actually used as a denominator. Rather, it is used as a multiplier to eliminate the denominators in the equation. This differs from how the LCD is used in Section 6.2.

SOLUTION   To ensure that neither denominator is 0, we state that $x \neq 5$. Then we multiply both sides by the LCD, $x - 5$:

$$(x-5) \cdot \frac{x-1}{x-5} = (x-5) \cdot \frac{4}{x-5} \qquad \text{Multiplying to clear fractions}$$

$$x - 1 = 4$$

$$x = 5. \qquad \text{But recall that } x \neq 5.$$

Because of the restriction above, 5 cannot be a solution. A check confirms the necessity of that restriction.

Check:
$$\frac{x-1}{x-5} = \frac{4}{x-5}$$

$$\frac{5-1}{5-5} \,\bigg|\, \frac{4}{5-5}$$

$$\frac{4}{0} \overset{?}{=} \frac{4}{0} \qquad \text{Division by 0 is undefined.}$$

This equation has no solution.

[TRY EXERCISE] 29

To see why 5 is not a solution of Example 2, note that the multiplication principle for equations requires that we multiply both sides by a *nonzero* number. When both sides of an equation are multiplied by an expression containing variables, certain replacements may make that expression 0. Thus it is safe to say that *if* a solution of

$$\frac{x-1}{x-5} = \frac{4}{x-5}$$

exists, then it is also a solution of $x - 1 = 4$. We *cannot* conclude that every solution of $x - 1 = 4$ is a solution of the original equation.

> *CAUTION!*   When solving rational equations, be sure to list any restrictions as your first step. Refer to the restriction(s) as you proceed, and check all possible solutions in the original equation.

EXAMPLE **3**   Solve: $\dfrac{x^2}{x-3} = \dfrac{9}{x-3}$.

SOLUTION   Note that $x \neq 3$. Since the LCD is $x - 3$, we multiply both sides by $x - 3$:

$$(x-3) \cdot \dfrac{x^2}{x-3} = (x-3) \cdot \dfrac{9}{x-3} \qquad \text{Multiplying to clear fractions}$$

$$x^2 = 9 \qquad \begin{array}{l}\text{Simplifying. The fractions}\\\text{are cleared.}\end{array}$$

$$x^2 - 9 = 0 \qquad \text{Getting 0 on one side}$$

$$(x-3)(x+3) = 0 \qquad \text{Factoring}$$

$$x = 3 \quad or \quad x = -3. \qquad \begin{array}{l}\text{Using the principle of}\\\text{zero products}\end{array}$$

Although 3 is a solution of $x^2 = 9$, it must be rejected as a solution of the rational equation because of the restriction stated in red above. The number -3 *does* check in the original equation. The solution is -3.

`TRY EXERCISE` 35

EXAMPLE **4**   Solve: $\dfrac{2}{x+5} + \dfrac{1}{x-5} = \dfrac{16}{x^2 - 25}$.

SOLUTION   To find all restrictions and to assist in finding the LCD, we factor:

$$\dfrac{2}{x+5} + \dfrac{1}{x-5} = \dfrac{16}{(x+5)(x-5)}. \qquad \text{Factoring } x^2 - 25$$

Note that $x \neq -5$ and $x \neq 5$. We multiply by the LCD, $(x+5)(x-5)$, and then use the distributive law:

$$(x+5)(x-5)\left(\dfrac{2}{x+5} + \dfrac{1}{x-5}\right) = (x+5)(x-5) \cdot \dfrac{16}{(x+5)(x-5)}$$

$$(x+5)(x-5)\dfrac{2}{x+5} + (x+5)(x-5)\dfrac{1}{x-5} = \dfrac{(x+5)(x-5)16}{(x+5)(x-5)}$$

$$2(x-5) + (x+5) = 16 \qquad \text{The fractions are cleared.}$$

$$2x - 10 + x + 5 = 16$$

$$3x - 5 = 16$$

$$3x = 21$$

$$x = 7. \qquad \text{7 should check.}$$

A check will confirm that the solution is 7.

`TRY EXERCISE` 41

Rational equations often appear when we are working with functions.

EXAMPLE **5**   Let $f(x) = x + \dfrac{6}{x}$. Find all values of a for which $f(a) = 5$.

SOLUTION   Since $f(a) = a + \dfrac{6}{a}$, the problem asks that we find all values of a for which

$$a + \dfrac{6}{a} = 5.$$

There are several ways in which Example 5 can be checked. One way is to confirm that the graphs of $y_1 = x + 6/x$ and $y_2 = 5$ intersect at $x = 2$ and $x = 3$. You can also use a table to check that $y_1 = y_2$ when x is 2 and again when x is 3.

Use a graphing calculator to check Examples 1–3.

First note that $a \neq 0$. We multiply both sides of the equation by the LCD, a:

$$a\left(a + \frac{6}{a}\right) = 5 \cdot a \qquad \text{Multiplying to clear fractions}$$

$$a \cdot a + a \cdot \frac{6}{a} = 5a \qquad \text{Using the distributive law}$$

$$a^2 + 6 = 5a \qquad \text{Simplifying. The fractions are cleared.}$$

$$a^2 - 5a + 6 = 0 \qquad \text{Getting 0 on one side}$$

$$(a - 3)(a - 2) = 0 \qquad \text{Factoring}$$

$$a = 3 \quad or \quad a = 2. \qquad \text{Using the principle of zero products}$$

Check: $f(3) = 3 + \dfrac{6}{3} = 3 + 2 = 5;$

$f(2) = 2 + \dfrac{6}{2} = 2 + 3 = 5.$

The solutions are 2 and 3. For $a = 2$ or $a = 3$, we have $f(a) = 5$.

TRY EXERCISE 55

ALGEBRAIC–GRAPHICAL CONNECTION

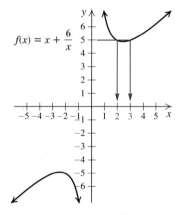

$f(x) = x + \dfrac{6}{x}$

A computer-generated visualization of Example 5

One way to visualize the solution to Example 5 is to make a graph. This can be done by graphing $f(x) = x + 6/x$ with a calculator, with a computer, or by hand. We then inspect the graph for any x-values that are paired with 5. (Note that no y-value is paired with 0, since 0 is not in the domain of f.) It appears from the graph that $f(x) = 5$ when $x \approx 2$ or $x \approx 3$. Although making a graph is not the fastest or most precise method of solving a rational equation, it provides visualization and is useful when problems are too difficult to solve algebraically.

6.4   EXERCISE SET

Concept Reinforcement   *Classify each of the following as either an expression or an equation.*

1. $\dfrac{2}{3} = \dfrac{1}{x}$

2. $\dfrac{2}{x + 1} - \dfrac{x - 5}{x^2 - 2x}$

3. $\dfrac{4}{t^2 - 1} + \dfrac{3}{t + 1}$

4. $\dfrac{2}{t^2 - 1} = \dfrac{3}{t + 1}$

5. $\dfrac{2}{x + 7} + \dfrac{6}{5x} = 4$

6. $\dfrac{5}{2x} - \dfrac{3}{x^2} = 7$

7. $\dfrac{t + 3}{t - 4} = \dfrac{t - 5}{t - 7}$

8. $\dfrac{7t}{2t - 3} \div \dfrac{3t}{2t + 3}$

9. $\dfrac{5x}{x^2 - 4} \cdot \dfrac{7}{x^2 - 5x + 4}$

10. $\dfrac{7x}{2 - x} = \dfrac{3}{4 - x}$

Solve. If no solution exists, state this.

11. $\dfrac{t}{10} + \dfrac{t}{15} = 1$

12. $\dfrac{t}{45} + \dfrac{t}{30} = 1$

13. $\dfrac{3}{4} - \dfrac{1}{x} = \dfrac{7}{8}$

14. $\dfrac{2}{3} - \dfrac{1}{y} = \dfrac{5}{6}$

15. $\dfrac{a+1}{3} + \dfrac{a-4}{5} = \dfrac{2a}{9}$

16. $\dfrac{x-5}{4} + \dfrac{x+6}{6} = \dfrac{5x}{8}$

17. $\dfrac{1}{3t} + \dfrac{1}{t} = \dfrac{1}{2}$

18. $\dfrac{1}{t} + \dfrac{1}{2t} = \dfrac{1}{5}$

19. $\dfrac{3}{x-1} + \dfrac{3}{10} = \dfrac{5}{2x-2}$

20. $\dfrac{3}{2n+10} + \dfrac{5}{4} = \dfrac{7}{n+5}$

Aha! **21.** $\dfrac{2}{6} + \dfrac{1}{2x} = \dfrac{1}{3}$

22. $\dfrac{12}{15} - \dfrac{1}{3x} = \dfrac{4}{5}$

23. $y + \dfrac{4}{y} = -5$

24. $t + \dfrac{6}{t} = -5$

25. $x - \dfrac{12}{x} = 4$

26. $y - \dfrac{14}{y} = 5$

27. $\dfrac{9}{10} = \dfrac{1}{y}$

28. $-\dfrac{5}{6} = \dfrac{1}{x}$

29. $\dfrac{t-1}{t-3} = \dfrac{2}{t-3}$

30. $\dfrac{x-2}{x-4} = \dfrac{2}{x-4}$

31. $\dfrac{x}{x-5} = \dfrac{25}{x^2-5x}$

32. $\dfrac{t}{t-6} = \dfrac{36}{t^2-6t}$

33. $\dfrac{5}{4t} = \dfrac{7}{5t-2}$

34. $\dfrac{3}{x-2} = \dfrac{5}{x+4}$

Aha! **35.** $\dfrac{x^2+4}{x-1} = \dfrac{5}{x-1}$

36. $\dfrac{x^2-1}{x+2} = \dfrac{3}{x+2}$

37. $\dfrac{6}{a+1} = \dfrac{a}{a-1}$

38. $\dfrac{4}{a-7} = \dfrac{-2a}{a+3}$

39. $\dfrac{60}{t-5} - \dfrac{18}{t} = \dfrac{40}{t}$

40. $\dfrac{50}{t-2} - \dfrac{16}{t} = \dfrac{30}{t}$

41. $\dfrac{4}{y^2+y-12} = \dfrac{1}{y+4} - \dfrac{2}{y-3}$

42. $\dfrac{3}{a^2-7a+10} = \dfrac{2}{a-2} + \dfrac{1}{a-5}$

43. $\dfrac{3}{x-3} + \dfrac{5}{x+2} = \dfrac{5x}{x^2-x-6}$

44. $\dfrac{2}{x-2} + \dfrac{1}{x+4} = \dfrac{x}{x^2+2x-8}$

45. $\dfrac{3}{x} + \dfrac{x}{x+2} = \dfrac{4}{x^2+2x}$

46. $\dfrac{x}{x+1} + \dfrac{5}{x} = \dfrac{1}{x^2+x}$

47. $\dfrac{2}{t-4} + \dfrac{1}{t} = \dfrac{t}{4-t}$

48. $\dfrac{2t}{3-t} - \dfrac{4}{t} = \dfrac{1}{t-3}$

49. $\dfrac{5}{x+2} - \dfrac{3}{x-2} = \dfrac{2x}{4-x^2}$

50. $\dfrac{y+3}{y+2} - \dfrac{y}{y^2-4} = \dfrac{y}{y-2}$

51. $\dfrac{1}{x^2+2x+1} = \dfrac{x-1}{3x+3} + \dfrac{x+2}{5x+5}$

52. $\dfrac{3}{x^2-6x+9} + \dfrac{x-2}{3x-9} = \dfrac{x}{2x-6}$

53. $\dfrac{3-2y}{y+1} - \dfrac{10}{y^2-1} = \dfrac{2y+3}{1-y}$

54. $\dfrac{1-2x}{x+2} + \dfrac{20}{x^2-4} = \dfrac{2x+1}{2-x}$

In Exercises 55–60, a rational function f is given. Find all values of a for which f(a) is the indicated value.

55. $f(x) = 2x - \dfrac{15}{x};\ f(a) = 7$

56. $f(x) = 2x - \dfrac{6}{x};\ f(a) = 1$

57. $f(x) = \dfrac{x-5}{x+1};\ f(a) = \dfrac{3}{5}$

58. $f(x) = \dfrac{x-3}{x+2};\ f(a) = \dfrac{1}{5}$

59. $f(x) = \dfrac{12}{x} - \dfrac{12}{2x};\ f(a) = 8$

60. $f(x) = \dfrac{6}{x} - \dfrac{6}{2x};\ f(a) = 5$

For each pair of functions f and g, find all values of a for which f(a) = g(a).

61. $f(x) = \dfrac{3x-1}{x^2-7x+10}$,

$g(x) = \dfrac{x-1}{x^2-4} + \dfrac{2x+1}{x^2-3x-10}$

62. $f(x) = \dfrac{2x+5}{x^2+4x+3}$,

$g(x) = \dfrac{x+2}{x^2-9} + \dfrac{x-1}{x^2-2x-3}$

63. $f(x) = \dfrac{2}{x^2-8x+7}$,

$g(x) = \dfrac{3}{x^2-2x-3} - \dfrac{1}{x^2-1}$

64. $f(x) = \dfrac{4}{x^2 + 3x - 10}$,

$g(x) = \dfrac{3}{x^2 - x - 12} + \dfrac{1}{x^2 + x - 6}$

65. Below are unlabeled graphs of $f(x) = x + 2$ and $g(x) = (x^2 - 4)/(x - 2)$. How can you tell which graph represents f and which graph represents g?

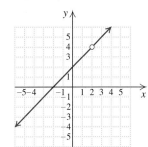 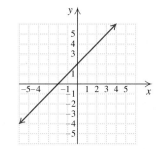

66. Explain how one can easily produce rational equations for which no solution exists. (*Hint*: Examine Example 2.)

Skill Review

To prepare for Section 6.5, review solving applications using the five-step problem-solving strategy (Sections 1.4 and 5.8).

Solve.

67. *Aviation.*   Gail's Cessna travels 135 mph in still air. With a tailwind of 15 mph, how long will it take Gail to travel 200 mi? [1.4]

68. *Canoeing.*   Pe'Rez paddles 55 m per minute in still water. If he paddles 135 m upstream in 3 min, what is the speed of the current? [1.4]

69. *Recycling.*   Kylie sorted twice as many pounds of plastic as Brenton. Together, they sorted 123 lb of plastic. How much did each sort? [1.4]

70. *Carpentry.*   The area of a wooden shutter is 900 in². The shutter is four times as long as it is wide. Find the length and the width of the shutter. [5.8]

71. *Picture framing.*   A photo is placed under a mat that is 10 in. wide and 13 in. long. The area of the photo that shows is 70 in². How wide is the mat? [5.8]

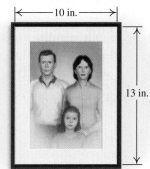

72. Find two consecutive even integers whose product is 120. [5.8]

Synthesis

73. Is the following statement true or false: "For any real numbers a, b, and c, if $ac = bc$, then $a = b$"? Explain why you answered as you did.

74. When checking a possible solution of a rational equation, is it sufficient to check that the "solution" does not make any denominator equal to 0? Why or why not?

For each pair of functions f and g, find all values of a for which $f(a) = g(a)$.

75. $f(x) = \dfrac{x - \dfrac{2}{3}}{x + \dfrac{1}{2}}$, $g(x) = \dfrac{x + \dfrac{2}{3}}{x - \dfrac{3}{2}}$

76. $f(x) = \dfrac{2 - \dfrac{x}{4}}{2}$, $g(x) = \dfrac{\dfrac{x}{4} - 2}{\dfrac{x}{2} + 2}$

77. $f(x) = \dfrac{x + 3}{x + 2} - \dfrac{x + 4}{x + 3}$, $g(x) = \dfrac{x + 5}{x + 4} - \dfrac{x + 6}{x + 5}$

78. $f(x) = \dfrac{1}{1 + x} + \dfrac{x}{1 - x}$, $g(x) = \dfrac{1}{1 - x} - \dfrac{x}{1 + x}$

79. $f(x) = \dfrac{0.793}{x} + 18.15$, $g(x) = \dfrac{6.034}{x} - 43.17$

80. $f(x) = \dfrac{2.315}{x} - \dfrac{12.6}{17.4}$, $g(x) = \dfrac{6.71}{x} + 0.763$

Recall that identities are true for any possible replacement of the variable(s). Determine whether each of the following equations is an identity.

81. $\dfrac{x^2 + 6x - 16}{x - 2} = x + 8$, $x \neq 2$

82. $\dfrac{x^3 + 8}{x^2 - 4} = \dfrac{x^2 - 2x + 4}{x - 2}$, $x \neq -2$, $x \neq 2$

83. Use a graphing calculator to check your answers to Exercises 11, 47, and 57.

84. Use a graphing calculator to check your answers to Exercises 12, 36, and 58.

85. Use a graphing calculator with a TABLE feature to show that 2 is not in the domain of g, if $g(x) = (x^2 - 4)/(x - 2)$. (See Exercise 65.)

86. Can Exercise 85 be answered on a graphing calculator using only graphs? Why or why not?

CONNECTING the CONCEPTS

Simplifying an expression is different from solving an equation. An equation contains an equals sign; an expression does not. When expressions are simplified, the result is an equivalent expression. When equations are solved, the result is a solution. Compare the following.

Simplify: $\dfrac{x}{x+1} + \dfrac{2}{3}$.   There is no equals sign.

SOLUTION

$$\dfrac{x}{x+1} + \dfrac{2}{3} = \dfrac{x}{x+1} \cdot \dfrac{3}{3} + \dfrac{2}{3} \cdot \dfrac{x+1}{x+1}$$

$$= \dfrac{3x}{3(x+1)} + \dfrac{2x+2}{3(x+1)} \qquad \text{Writing with the LCD, } 3(x+1)$$

$$= \dfrac{5x+2}{3(x+1)} \qquad \boxed{\text{The equals signs indicate that all the expressions are equivalent.}}$$

The result is an expression equivalent to $\dfrac{x}{x+1} + \dfrac{2}{3}$.

Solve: $\dfrac{x}{x+1} = \dfrac{2}{3}$.   There is an equals sign.

SOLUTION   Note that $x \neq -1$.

$$\dfrac{x}{x+1} = \dfrac{2}{3}$$

$$3(x+1) \cdot \dfrac{x}{x+1} = 3(x+1) \cdot \dfrac{2}{3} \qquad \text{Multiplying by the LCD, } 3(x+1)$$

$$3x = 2(x+1)$$

$$3x = 2x+2 \qquad \boxed{\text{Each line is an equivalent equation to be solved.}}$$

$$x = 2$$

The result is a solution; 2 is the solution of $\dfrac{x}{x+1} = \dfrac{2}{3}$.

MIXED REVIEW

Determine whether each of the following is an expression or an equation. Then simplify the expression or solve the equation.

1. Simplify: $\dfrac{5x^2 - 10x}{5x^2 + 5x}$.

2. Add and, if possible, simplify: $\dfrac{5}{3t} + \dfrac{1}{2t-1}$.

3. Solve: $\dfrac{t}{2} + \dfrac{t}{3} = 5$.

4. Solve: $\dfrac{1}{y} - \dfrac{1}{2} = \dfrac{5}{6y}$.

5. Simplify: $\dfrac{\dfrac{1}{z} + 1}{\dfrac{1}{z^2} - 1}$.

6. Multiply and, if possible, simplify: $\dfrac{a+1}{6a} \cdot \dfrac{8a^2}{a^2-1}$.

7. Solve: $\dfrac{5}{x+3} = \dfrac{3}{x+2}$.

8. Simplify: $\dfrac{a^{-1} + b^{-1}}{ab^{-1} - ba^{-1}}$.

9. Divide and, if possible, simplify: $\dfrac{27a^2}{8} \div \dfrac{12}{5a}$.

10. Solve: $\dfrac{18}{d} = \dfrac{d}{2}$.

11. Subtract and, if possible, simplify: $\dfrac{2n-1}{n-2} - \dfrac{n-3}{n+1}$.

12. Divide and, if possible, simplify: $\dfrac{n^3+1}{15n} \div \dfrac{n^2+n}{25}$.

13. Multiply and, if possible, simplify: $\dfrac{8t+8}{2t^2+t-1} \cdot \dfrac{t^2-1}{t^2-2t+1}$.

14. Solve: $\dfrac{15}{x} - \dfrac{15}{x + 2} = 2$.

15. Solve: $\dfrac{5}{t} = \dfrac{4}{3}$.

16. Subtract and, if possible, simplify:

$$\dfrac{2a}{a + 1} - \dfrac{4a}{1 - a^2}.$$

17. Solve: $\dfrac{1}{6x} = \dfrac{x}{x + 1}$.

18. Solve: $\dfrac{z}{z - 3} - \dfrac{3z}{z + 2} = \dfrac{5z}{z^2 - z - 6}$.

19. Add and, if possible, simplify:

$$\dfrac{4}{x^2 - 6x - 16} + \dfrac{x}{x^2 - x - 6}.$$

20. Solve: $\dfrac{2}{1 - x} = \dfrac{-4}{x^2 - 1}$.

6.5 Solving Applications Using Rational Equations

Problems Involving Work • Problems Involving Motion

Now that we are able to solve rational equations, we can also solve new types of applications. The five problem-solving steps remain the same.

Problems Involving Work

EXAMPLE 1

The roof of Finn and Paige's townhouse needs to be reshingled. Finn can do the job alone in 8 hr and Paige can do the job alone in 10 hr. How long will it take the two of them, working together, to reshingle the roof?

SOLUTION

1. **Familiarize.** This *work problem* is a type of problem we have not yet encountered. Work problems are often *incorrectly* translated to mathematical language in several ways.

   a) Add the times together: $8\,\text{hr} + 10\,\text{hr} = 18\,\text{hr}$. ← Incorrect

   This cannot be the correct approach since Finn and Paige working together should not take longer than either of them working alone.

   b) Average the times: $(8\,\text{hr} + 10\,\text{hr})/2 = 9\,\text{hr}$. ← Incorrect

   Again, this is longer than it would take Finn to do the job alone.

   c) Assume that each person does half the job. ← Incorrect

   Finn would reshingle $\frac{1}{2}$ the roof in $\frac{1}{2}(8\,\text{hr})$, or 4 hr, and Paige would reshingle $\frac{1}{2}$ the roof in $\frac{1}{2}(10\,\text{hr})$, or 5 hr, so Finn would finish an hour before Paige. The problem assumes that the two are working together, so Finn will help Paige after completing his half. This tells us that the job will take between 4 and 5 hr.

   Each incorrect approach started with the time it took each worker to do the job. The correct approach instead focuses on the *rate* of work, or the amount of the job that each person completes in 1 hr.

Since Finn takes 8 hr to reshingle the entire roof, in 1 hr he reshingles $\frac{1}{8}$ of the roof. Since Paige takes 10 hr to reshingle the entire roof, in 1 hr she reshingles $\frac{1}{10}$ of the roof. Thus Finn works at a rate of $\frac{1}{8}$ roof per hour, and Paige works at a rate of $\frac{1}{10}$ roof per hour.

Working together, Finn and Paige reshingle $\frac{1}{8} + \frac{1}{10}$ roof in 1 hr, so their rate, as a team, is $\frac{1}{8} + \frac{1}{10} = \frac{5}{40} + \frac{4}{40} = \frac{9}{40}$ roof per hour.

We are looking for the time required to reshingle 1 entire roof, not just a part of it. Setting up a table helps us organize the information.

Time	Fraction of the Roof Reshingled		
	By Finn	**By Paige**	**Together**
1 hr	$\frac{1}{8}$	$\frac{1}{10}$	$\frac{1}{8} + \frac{1}{10}$, or $\frac{9}{40}$
2 hr	$\frac{1}{8} \cdot 2$	$\frac{1}{10} \cdot 2$	$\left(\frac{1}{8} + \frac{1}{10}\right)2$, or $\frac{9}{40} \cdot 2$, or $\frac{9}{20}$
3 hr	$\frac{1}{8} \cdot 3$	$\frac{1}{10} \cdot 3$	$\left(\frac{1}{8} + \frac{1}{10}\right)3$, or $\frac{9}{40} \cdot 3$, or $\frac{27}{40}$
t hr	$\frac{1}{8} \cdot t$	$\frac{1}{10} \cdot t$	$\left(\frac{1}{8} + \frac{1}{10}\right)t$, or $\frac{9}{40} \cdot t$

2. Translate. From the table, we see that t must be some number for which

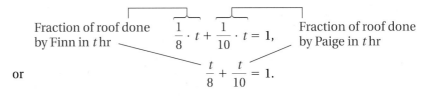

Fraction of roof done by Finn in t hr ⟶ $\frac{1}{8} \cdot t + \frac{1}{10} \cdot t = 1,$ ⟵ Fraction of roof done by Paige in t hr

or

$$\frac{t}{8} + \frac{t}{10} = 1.$$

3. Carry out. We solve the equation:

$$\frac{t}{8} + \frac{t}{10} = 1 \qquad \text{The LCD is 40.}$$

$$40\left(\frac{t}{8} + \frac{t}{10}\right) = 40 \cdot 1 \qquad \text{Multiplying to clear fractions}$$

$$\frac{40t}{8} + \frac{40t}{10} = 40$$

$$5t + 4t = 40 \qquad \text{Simplifying}$$

$$9t = 40$$

$$t = \frac{40}{9}, \text{ or } 4\frac{4}{9}.$$

4. Check. In $\frac{40}{9}$ hr, Finn reshingles $\frac{1}{8} \cdot \frac{40}{9}$, or $\frac{5}{9}$, of the roof and Paige reshingles $\frac{1}{10} \cdot \frac{40}{9}$, or $\frac{4}{9}$, of the roof. Together, they reshingle $\frac{5}{9} + \frac{4}{9}$, or 1 roof. The fact that our solution is between 4 and 5 hr (see step 1 above) is also a check.

5. State. It will take $4\frac{4}{9}$ hr for Finn and Paige, working together, to reshingle the roof.

TRY EXERCISE ▸ 13

EXAMPLE 2

It takes Manuel 9 hr longer than Zoe to rebuild an engine. Working together, they can do the job in 20 hr. How long would it take each, working alone, to rebuild an engine?

SOLUTION

1. **Familiarize.** Unlike Example 1, this problem does not provide us with the times required by the individuals to do the job alone. Let's have z = the number of hours it would take Zoe working alone and $z + 9$ = the number of hours it would take Manuel working alone.

2. **Translate.** Using the same reasoning as in Example 1, we see that Zoe completes $\frac{1}{z}$ of the job in 1 hr and Manuel completes $\frac{1}{z + 9}$ of the job in 1 hr. Thus, in 20 hr,

   Zoe completes $\frac{1}{z} \cdot 20$ of the job and Manuel completes $\frac{1}{z + 9} \cdot 20$ of the job.

   We know that, together, Zoe and Manuel can complete the entire job in 20 hr. This gives the following:

   Fraction of job done by Zoe in 20 hr $\overbrace{\quad\frac{1}{z} \cdot 20 + \frac{1}{z + 9} \cdot 20 = 1,\quad}$ Fraction of job done by Manuel in 20 hr

   or $\dfrac{20}{z} + \dfrac{20}{z + 9} = 1.$

3. **Carry out.** We solve the equation:

$$\frac{20}{z} + \frac{20}{z + 9} = 1 \qquad \text{The LCD is } z(z + 9).$$

$$z(z + 9)\left(\frac{20}{z} + \frac{20}{z + 9}\right) = z(z + 9)1 \qquad \text{Multiplying to clear fractions}$$

$$(z + 9)20 + z \cdot 20 = z(z + 9) \qquad \text{Distributing and simplifying}$$

$$40z + 180 = z^2 + 9z$$

$$0 = z^2 - 31z - 180 \qquad \text{Getting 0 on one side}$$

$$0 = (z - 36)(z + 5) \qquad \text{Factoring}$$

$$z - 36 = 0 \quad or \quad z + 5 = 0 \qquad \text{Principle of zero products}$$

$$z = 36 \quad or \qquad z = -5.$$

4. **Check.** Since negative time has no meaning in the problem, −5 is not a solution to the original problem. The number 36 checks since, if Zoe takes 36 hr alone and Manuel takes 36 + 9 = 45 hr alone, in 20 hr they would have finished

$$\frac{20}{36} + \frac{20}{45} = \frac{5}{9} + \frac{4}{9} = 1 \text{ complete rebuild.}$$

5. **State.** It would take Zoe 36 hr to rebuild an engine alone, and Manuel 45 hr.

 TRY EXERCISE ▶ 25

The equations used in Examples 1 and 2 can be generalized as follows.

> ### Modeling Work Problems
> If
>
> a = the time needed for A to complete the work alone,
>
> b = the time needed for B to complete the work alone, and
>
> t = the time needed for A and B to complete the work together,
>
> then
>
> $$\frac{t}{a} + \frac{t}{b} = 1.$$
>
> The following are equivalent equations that can also be used:
>
> $$\frac{1}{a} \cdot t + \frac{1}{b} \cdot t = 1 \quad \text{and} \quad \frac{1}{a} + \frac{1}{b} = \frac{1}{t}.$$

Problems Involving Motion

Problems dealing with distance, rate (or speed), and time are called **motion problems**. To translate them, we use either the basic motion formula, $d = rt$, or the formulas $r = d/t$ or $t = d/r$, which can be derived from $d = rt$.

EXAMPLE **3**

On her road bike, Olivia bikes 15 km/h faster than Jason does on his mountain bike. In the time it takes Olivia to travel 80 km, Jason travels 50 km. Find the speed of each bicyclist.

SOLUTION

1. **Familiarize.** Let's guess that Jason is going 10 km/h. Then the following would be true.

    Olivia's speed would be $10 + 15$, or 25 km/h.

    Jason would cover 50 km in $50/10 = 5$ hr.

    Olivia would cover 80 km in $80/25 = 3.2$ hr.

Since the times are not the same, our guess is wrong, but we can make some observations.

   If r = the rate, in kilometers per hour, of Jason's bike, then the rate of Olivia's bike = $r + 15$.

   Jason's travel time is the same as Olivia's travel time.

We make a drawing and construct a table, listing the information we know.

STUDENT NOTES

You need remember only the motion formula $d = rt$. Then you can divide both sides by t to get $r = d/t$, or you can divide both sides by r to get $t = d/r$.

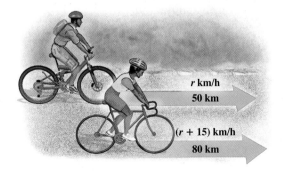

	Distance	Speed	Time
Jason's Mountain Bike	50	r	
Olivia's Road Bike	80	$r + 15$	

2. **Translate.** By looking at how we checked our guess, we see that we can fill in the **Time** column of the table using the formula *Time = Distance/Rate*, as follows.

	Distance	Speed	Time
Jason's Mountain Bike	50	r	$50/r$
Olivia's Road Bike	80	$r + 15$	$80/(r + 15)$

Since we know that the times are the same, we can write an equation:

$$\frac{50}{r} = \frac{80}{r + 15}.$$

3. **Carry out.** We solve the equation:

$$\frac{50}{r} = \frac{80}{r + 15} \qquad \text{The LCD is } r(r + 15).$$

$$r(r + 15)\frac{50}{r} = r(r + 15)\frac{80}{r + 15} \qquad \text{Multiplying to clear fractions}$$

$$50r + 750 = 80r \qquad \text{Simplifying}$$

$$750 = 30r$$

$$25 = r.$$

4. **Check.** If our answer checks, Jason's mountain bike is going 25 km/h and Olivia's road bike is going $25 + 15 = 40$ km/h.

   Traveling 80 km at 40 km/h, Olivia is riding for $\frac{80}{40} = 2$ hr. Traveling 50 km at 25 km/h, Jason is riding for $\frac{50}{25} = 2$ hr. Our answer checks since the two times are the same.

5. **State.** Olivia's speed is 40 km/h, and Jason's speed is 25 km/h.

> TRY EXERCISE 33

In the next example, although distance is the same in both directions, the key to the translation lies in an additional piece of given information.

EXAMPLE 4   A Hudson River tugboat goes 10 mph in still water. It travels 24 mi upstream and 24 mi back in a total time of 5 hr. What is the speed of the current?

Sources: Based on information from the Department of the Interior, U.S. Geological Survey, and *The Tugboat Captain*, Montgomery County Community College

SOLUTION

1. **Familiarize.** Let's guess that the speed of the current is 4 mph. Then the following would be true.

   The tugboat would move $10 - 4 = 6$ mph upstream.

   The tugboat would move $10 + 4 = 14$ mph downstream.

   To travel 24 mi upstream would require $\frac{24}{6} = 4$ hr.

   To travel 24 mi downstream would require $\frac{24}{14} = 1\frac{5}{7}$ hr.

Since the total time, $4 + 1\frac{5}{7} = 5\frac{5}{7}$ hr, is not the 5 hr mentioned in the problem, our guess is wrong, but we can make some observations.

If c = the current's rate, in miles per hour, we have the following.

The tugboat's speed upstream is $(10 - c)$ mph.

The tugboat's speed downstream is $(10 + c)$ mph.

The total travel time is 5 hr.

We make a sketch and construct a table, listing the information we know.

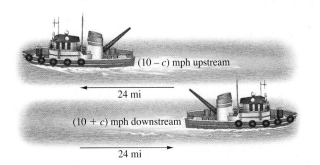

	Distance	Speed	Time
Upstream	24	$10 - c$	
Downstream	24	$10 + c$	

2. **Translate.** From examining our guess, we see that the time traveled can be represented using the formula *Time = Distance/Rate*:

	Distance	Speed	Time
Upstream	24	$10 - c$	$24/(10 - c)$
Downstream	24	$10 + c$	$24/(10 + c)$

Since the total time upstream and back is 5 hr, we use the last column of the table to form an equation:

$$\frac{24}{10 - c} + \frac{24}{10 + c} = 5.$$

3. **Carry out.** We solve the equation:

$$\frac{24}{10 - c} + \frac{24}{10 + c} = 5 \qquad \text{The LCD is } (10 - c)(10 + c).$$

$$(10 - c)(10 + c)\left[\frac{24}{10 - c} + \frac{24}{10 + c}\right] = (10 - c)(10 + c)5 \qquad \text{Multiplying to clear fractions}$$

$$24(10 + c) + 24(10 - c) = (100 - c^2)5$$

$$480 = 500 - 5c^2 \qquad \text{Simplifying}$$

$$5c^2 - 20 = 0$$

$$5(c^2 - 4) = 0$$

$$5(c - 2)(c + 2) = 0$$

$$c = 2 \quad or \quad c = -2.$$

4. **Check.** Since speed cannot be negative in this problem, -2 cannot be a solution. You should confirm that 2 checks in the original problem.

5. **State.** The speed of the current is 2 mph.

TRY EXERCISE ▶ 37

6.5 EXERCISE SET

 Concept Reinforcement   *Find each rate.*

1. If Sandy can decorate a cake in 2 hr, what is her rate?

2. If Eric can decorate a cake in 3 hr, what is his rate?

3. If Sandy can decorate a cake in 2 hr and Eric can decorate the same cake in 3 hr, what is their rate, working together?

4. If Lisa and Mark can mow a lawn together in 1 hr, what is their rate?

5. If Lisa can mow a lawn by herself in 3 hr, what is her rate?

6. If Lisa and Mark can mow a lawn together in 1 hr, and Lisa can mow the same lawn by herself in 3 hr, what is Mark's rate, working alone?

Solve.

7. The reciprocal of 3, plus the reciprocal of 6, is the reciprocal of what number?

8. The reciprocal of 10, plus the reciprocal of 15, is the reciprocal of what number?

9. The sum of a number and 6 times its reciprocal is -5. Find the number.

10. The sum of a number and 21 times its reciprocal is -10. Find the number.

11. The reciprocal of the product of two consecutive integers is $\frac{1}{90}$. Find the two integers.

12. The reciprocal of the product of two consecutive integers is $\frac{1}{30}$. Find the two integers.

13. *Home restoration.*   Bryan can refinish the floor of an apartment in 8 hr. Caroline can refinish the floor in 6 hr. How long will it take them, working together, to refinish the floor?

14. *Custom embroidery.*   Chandra can embroider logos on a team's sweatshirts in 6 hr. Traci, a new employee, needs 9 hr to complete the same job. Working together, how long will it take them to do the job?

15. *Filling a pool.*   The San Paulo community swimming pool can be filled in 12 hr if water enters through a pipe alone or in 30 hr if water enters through a hose alone. If water is entering through both the pipe and the hose, how long will it take to fill the pool?

16. *Filling a tank.*   A community water tank can be filled in 18 hr by the town office well alone and in 22 hr by the high school well alone. How long will it take to fill the tank if both wells are working?

17. *Pumping water.*   A $\frac{1}{2}$ HP Wayne stainless-steel sump pump can remove water from Carmen's flooded basement in 42 min. The $\frac{1}{2}$ HP Craftsman Professional sump pump can complete the same job in 35 min. How long would it take the two pumps together to pump out the basement?
Source: Based on data from manufacturers

18. *Hotel management.*   The Honeywell Enviracaire Silent Comfort air cleaner can clean the air in an 11-ft by 17-ft conference room in 10 min. The Blueair 201 Air Purifier can clean the air in a room of the same size in 12 min. How long would it take the two machines together to clean the air in such a room?
Source: Based on information from manufacturers' and retailers' websites

19. *Photocopiers.*   The HP Officejet H470 Mobile Printer takes twice the time required by the HP Officejet Pro K5400 color printer to photocopy brochures for the New Bretton Arts Council annual arts fair. If, working together, the two machines can complete the job in 45 min, how long would it take each machine, working alone, to copy the brochures?
Source: www.shoppinghp.com

20. *Cutting firewood.*   Kent can cut and split a cord of wood twice as fast as Brent can. When they work together, it takes them 4 hr. How long would it take each of them to do the job alone?

21. *Hotel management.*   The Austin Healthmate 400 can purify the air in a conference hall in 15 fewer minutes than it takes the Airgle 750 Air Purifier to do the same job. Together the two machines can purify the air in the conference hall in 10 min. How long would it take each machine, working alone, to purify the air in the room?
Source: Based on information from manufacturers' and retailers' websites

22. *Photo printing.*   It takes the Canon PIXMA iP6310D 15 min longer to print a set of photo proofs than it takes the HP Officejet H470b Mobile Printer. Together it would take them $\frac{180}{7}$, or $25\frac{5}{7}$ min to print the photos. How long would it take each machine, working alone, to print the photos?

Sources: www.shoppinghp.com; www.staples.com

23. *Newspaper delivery.*   Elliot can deliver papers three times as fast as Sara can. If they work together, it takes them 1 hr. How long would it take each to deliver the papers alone?

24. *Forest fires.*   The Erickson Air-Crane helicopter can scoop water and douse a certain forest fire four times as fast as an S-58T helicopter. Working together, the two helicopters can douse the fire in 8 hr. How long would it take each helicopter, working alone, to douse the fire?

Sources: Based on information from www.emergency.com and www.arishelicopters.com

25. *Painting.*   Zeno takes 3 hr longer to paint a floor than it takes Lia. When they work together, it takes them 2 hr. How long would each take to do the job alone?

26. *Waxing a car.*   It takes Monica 48 min longer to wax the family car than it takes Gretchen. When they work together, they can wax the car in 45 min. How long would it take Gretchen, working by herself, to wax the car?

27. *Sorting recyclables.*   Together, it takes Kim and Chris 2 hr 55 min to sort recyclables. Alone, Kim would require 2 hr more than Chris. How long would it take Chris to do the job alone? (*Hint:* Convert minutes to hours or hours to minutes.)

28. *Paving.*   Together, Steve and Bill require 4 hr 48 min to pave a driveway. Alone, Steve would require 4 hr more than Bill. How long would it take Bill to do the job alone? (*Hint:* Convert minutes to hours.)

29. *Kayaking.* The speed of the current in Catamount Creek is 3 mph. Sean can kayak 4 mi upstream in the same time it takes him to kayak 10 mi downstream. What is the speed of Sean's kayak in still water?

30. *Boating.*   The current in the Lazy River moves at a rate of 4 mph. Nicole's dinghy motors 6 mi upstream in the same time it takes to motor 12 mi downstream. What is the speed of the dinghy in still water?

31. *Moving sidewalks.*   The moving sidewalk at O'Hare Airport in Chicago moves 1.8 ft/sec. Walking on the moving sidewalk, Roslyn travels 105 ft forward in the time it takes to travel 51 ft in the opposite direction. How fast does Roslyn walk on a nonmoving sidewalk?

32. *Moving sidewalks.*   Newark Airport's moving sidewalk moves at a speed of 1.7 ft/sec. Walking on the moving sidewalk, Drew can travel 120 ft forward in the same time it takes to travel 52 ft in the opposite direction. What is Drew's walking speed on a nonmoving sidewalk?

33. *Train speed.*   The speed of the A&M freight train is 14 mph less than the speed of the A&M passenger train. The passenger train travels 400 mi in the same time that the freight train travels 330 mi. Find the speed of each train.

34. *Walking.*   Courtney walks 2 mph slower than Brittany. In the time it takes Brittany to walk 8 mi, Courtney walks 5 mi. Find the speed of each person.

Aha! 35. *Bus travel.*   A local bus travels 7 mph slower than the express. The express travels 45 mi in the time it takes the local to travel 38 mi. Find the speed of each bus.

36. *Train speed.*   The A train goes 12 mph slower than the E train. The A train travels 230 mi in the same time that the E train travels 290 mi. Find the speed of each train.

37. *Boating.*   LeBron's Mercruiser travels 15 km/h in still water. He motors 140 km downstream in the same time it takes to travel 35 km upstream. What is the speed of the river?

38. *Boating.*   Annette's paddleboat travels 2 km/h in still water. The boat is paddled 4 km downstream in the same time it takes to go 1 km upstream. What is the speed of the river?

39. *Shipping.*   A barge moves 7 km/h in still water. It travels 45 km upriver and 45 km downriver in a total time of 14 hr. What is the speed of the current?

40. *Moped speed.*   Cameron's moped travels 8 km/h faster than Ellia's. Cameron travels 69 km in the same time that Ellia travels 45 km. Find the speed of each person's moped.

41. *Aviation.*   A Citation CV jet travels 460 mph in still air and flies 525 mi into the wind and 525 mi with the wind in a total of 2.3 hr. Find the wind speed.

Source: Blue Star Jets, Inc.

42. *Canoeing.*   Chad paddles 55 m/min in still water. He paddles 150 m upstream and 150 m downstream in a total time of 5.5 min. What is the speed of the current?

43. *Train travel.* A freight train covered 120 mi at a certain speed. Had the train been able to travel 10 mph faster, the trip would have been 2 hr shorter. How fast did the train go?

44. *Boating.* Fiona's Boston Whaler cruised 45 mi upstream and 45 mi back in a total of 8 hr. The speed of the river is 3 mph. Find the speed of the boat in still water.

45. Two steamrollers are paving a parking lot. Working together, will the two steamrollers take less than half as long as the slower steamroller would working alone? Why or why not?

46. Two fuel lines are filling a freighter with oil. Will the faster fuel line take more or less than twice as long to fill the freighter by itself? Why?

Skill Review

To prepare for Section 6.6, review the quotient rule for exponents and subtraction of polynomials (Sections 1.6 and 5.1).

Simplify.

47. $\dfrac{42x^8y^9}{7x^2y}$ [1.6]

48. $\dfrac{-20a^4b^3}{4a^3b^3}$ [1.6]

49. $\dfrac{4x^2y}{-xy^2}$ [1.6]

50. $3x^4 - x^2 - 6x - (x^4 + x^3 - x^2)$ [5.1]

51. $\begin{array}{r} 4x^3 - 3x^2 \quad\quad - 7 \\ -(4x^3 - 8x^2 + 4x) \\ \hline \end{array}$
[5.1]

52. $\begin{array}{r} -3x^2 - 2x + 1 \\ -(-3x^2 - x + 6) \\ \hline \end{array}$
[5.1]

Synthesis

53. Write a work problem for a classmate to solve. Devise the problem so that the solution is "Beth and Leanne will take 4 hr to complete the job, working together."

54. Write a work problem for a classmate to solve. Devise the problem so that the solution is "Rosa takes 5 hr and Romano takes 6 hr to complete the job alone."

55. *Filling a bog.* The Norwich cranberry bog can be filled in 9 hr and drained in 11 hr. How long will it take to fill the bog if the drainage gate is left open?

56. *Filling a tub.* Jillian's hot tub can be filled in 10 min and drained in 8 min. How long will it take to empty a full tub if the water is left on?

57. Refer to Exercise 29. How long will it take Sean to kayak 5 mi downstream?

58. Refer to Exercise 30. How long will it take Nicole to motor 3 mi downstream?

59. *Escalators.* Together, a 100-cm wide escalator and a 60-cm wide escalator can empty a 1575-person auditorium in 14 min. The wider escalator moves twice as many people as the narrower one. How many people per hour does the 60-cm wide escalator move?
Source: *McGraw-Hill Encyclopedia of Science and Technology*

60. *Aviation.* A Coast Guard plane has enough fuel to fly for 6 hr, and its speed in still air is 240 mph. The plane departs with a 40-mph tailwind and returns to the same airport flying into the same wind. How far can the plane travel under these conditions?

61. *Boating.* Shoreline Travel operates a 3-hr paddleboat cruise on the Missouri River. If the speed of the boat in still water is 12 mph, how far upriver can the pilot travel against a 5-mph current before it is time to turn around?

62. *Travel by car.* Angenita drives to work at 50 mph and arrives 1 min late. She drives to work at 60 mph and arrives 5 min early. How far does Angenita live from work?

63. *Photocopying.* The printer in an admissions office can print a 500-page document in 50 min, while the printer in the business office can print the same document in 40 min. If the two printers work together to print the document, with the faster machine starting on page 1 and the slower machine working backwards from page 500, at what page will the two machines meet to complete the job?

64. At what time after 4:00 will the minute hand and the hour hand of a clock first be in the same position?

65. At what time after 10:30 will the hands of a clock first be perpendicular?

Average speed is defined as total distance divided by total time.

66. Ferdaws drove 200 km. For the first 100 km of the trip, she drove at a speed of 40 km/h. For the second half of the trip, she traveled at a speed of 60 km/h. What was the average speed of the entire trip? (It was *not* 50 km/h.)

67. For the first 50 mi of a 100-mi trip, Garry drove 40 mph. What speed would he have to travel for the last half of the trip so that the average speed for the entire trip would be 45 mph?

CORNER

Can We Count on the Model?

Focus: Testing a mathematical model

Time: 10–15 minutes

Group size: 2–3

Materials: Two stopwatches, a blackboard or a whiteboard, chalk or markers

Two students who are willing to write on the board should each take a piece of chalk or a marker and stand at the chalkboard or whiteboard. Two other students, each with a stopwatch, should be ready to time each of the students at the board.

ACTIVITY

1. The "job" is to write the numbers from 1 to 100 on the board. One student at the board should work carefully and neatly and at a steady pace. The other should work as quickly as possible. They should begin at the same time, and the time it takes each to complete the job should be recorded.

2. Using the times found in part (1), each group should use algebra to predict how long it will take the students, working together, to write the numbers from 1 to 100.

3. In order to do the job together, one student will write numbers counting up from 1 and the other will count down from 100 until they meet. Each group should predict the number at which the students will meet.

4. The students at the board should again write the numbers from 1 to 100. This time, the one working quickly should start at 1 and the one working more carefully should start at 100 and count down. Again, they should start at the same time, and the time should be recorded at which the numbers meet.

5. How accurate were your predictions? What may have caused the results to differ from your predictions?

6.6 Division of Polynomials

Dividing by a Monomial ▪ Dividing by a Polynomial

A rational expression indicates division. Division of polynomials, like division of real numbers, relies on our multiplication and subtraction skills.

Dividing by a Monomial

To divide a monomial by a monomial, we divide coefficients and, if the bases are the same, subtract exponents (see Section 1.6).

$$\text{Dividend} \longrightarrow \frac{45x^{10}}{3x^4} = 15x^{10-4} = 15x^6, \qquad \frac{8a^2b^5}{-2ab^2} = -4a^{2-1}b^{5-2} = -4ab^3.$$

Divisor $\longrightarrow$

$\uparrow$ Quotient

To divide a polynomial by a monomial, we regard the division as a sum of quotients of monomials. This uses the fact that since

$$\frac{A}{C} + \frac{B}{C} = \frac{A+B}{C}, \quad \text{we know that} \quad \frac{A+B}{C} = \frac{A}{C} + \frac{B}{C}.$$

EXAMPLE 1   Divide $12x^3 + 8x^2 + x + 4$ by $4x$.

SOLUTION

$$(12x^3 + 8x^2 + x + 4) \div (4x) = \frac{12x^3 + 8x^2 + x + 4}{4x}$$ Writing a rational expression

$$= \frac{12x^3}{4x} + \frac{8x^2}{4x} + \frac{x}{4x} + \frac{4}{4x}$$ Writing as a sum of quotients

$$= 3x^2 + 2x + \frac{1}{4} + \frac{1}{x}$$ Performing the four indicated divisions

TRY EXERCISE 7

EXAMPLE 2   Divide: $(8x^4y^5 - 3x^3y^4 + 5x^2y^3) \div (-x^2y^3)$.

SOLUTION

$$\frac{8x^4y^5 - 3x^3y^4 + 5x^2y^3}{-x^2y^3} = \frac{8x^4y^5}{-x^2y^3} - \frac{3x^3y^4}{-x^2y^3} + \frac{5x^2y^3}{-x^2y^3}$$ Try to perform this step mentally.

$$= -8x^2y^2 + 3xy - 5$$ TRY EXERCISE 11

Division by a Monomial

To divide a polynomial by a monomial, divide each term of the polynomial by the monomial.

Dividing by a Polynomial

When the divisor has more than one term, we use a procedure very similar to long division in arithmetic.

EXAMPLE 3   Divide $2x^2 - 7x - 15$ by $x - 5$.

SOLUTION   We have

$$
\begin{array}{r}
2x \\
x - 5 \overline{) 2x^2 - 7x - 15} \\
-(2x^2 - 10x) \\
\hline
3x - 15
\end{array}
$$

Divide $2x^2$ by x: $2x^2/x = 2x$.
Multiply $x - 5$ by $2x$.
Subtract by mentally changing signs and adding: $-7x - (-10x) = -7x + 10x = 3x$.

We next divide the leading term of this remainder, $3x$, by the leading term of the divisor, x.

$$
\begin{array}{r}
2x + 3 \\
x - 5 \overline{) 2x^2 - 7x - 15} \\
2x^2 - 10x \\
\hline
3x - 15 \\
-(3x - 15) \\
\hline
0
\end{array}
$$

Divide $3x$ by x: $3x/x = 3$.
Multiply $x - 5$ by 3.
Subtract.   Our remainder is now 0.

Check: $(x - 5)(2x + 3) = 2x^2 - 7x - 15$. The answer checks.

The quotient is $2x + 3$.

TRY EXERCISE 19

To understand why we perform long division as we do, note that Example 3 amounts to "filling in" an unknown polynomial:

$$(x - 5)(\quad ? \quad) = 2x^2 - 7x - 15.$$

We see that $2x$ must be in the unknown polynomial if we are to get the first term, $2x^2$, from the multiplication. To see what else is needed, note that

$$(x - 5)(2x \qquad) = 2x^2 - 10x \neq 2x^2 - 7x - 15.$$

The $2x$ can be regarded as a (poor) approximation of the quotient that we are seeking. To see how far off the approximation is, we subtract:

$$\begin{array}{r} 2x^2 - 7x - 15 \\ -(2x^2 - 10x) \\ \hline 3x - 15 \end{array}$$

Note where this appeared in the long division above.

$\longleftarrow$ This is the first remainder.

To get the needed terms, $3x - 15$, we need another term in the unknown polynomial. We use 3 because $(x - 5) \cdot 3$ is $3x - 15$:

$$(x - 5)(2x + 3) = 2x^2 - 10x + 3x - 15$$
$$= 2x^2 - 7x - 15.$$

Now when we subtract the product $(x - 5)(2x + 3)$ from $2x^2 - 7x - 15$, the remainder is 0.

If a nonzero remainder occurs, when do we stop dividing? We continue until the degree of the remainder is less than the degree of the divisor.

EXAMPLE 4

Divide $x^2 + 5x + 8$ by $x + 3$.

SOLUTION We have

$$\begin{array}{r} x \\ x + 3 \overline{)\,x^2 + 5x + 8} \\ \underline{x^2 + 3x} \\ 2x + 8 \end{array}$$

Divide the first term of the dividend by the first term of the divisor: $x^2/x = x$.

$\longleftarrow$ Multiply x above by $x + 3$.

Subtract: $(x^2 + 5x) - (x^2 + 3x) = 2x$.

Note that

$$x^2 + 5x + 8 - (x^2 + 3x) = x^2 + 5x + 8 - x^2 - 3x.$$

Remember: To subtract, add the opposite (change the sign of every term, then add).

We now focus on the current remainder, $2x + 8$, and repeat the process:

$$\begin{array}{r} x + 2 \\ x + 3 \overline{)\,x^2 + 5x + 8} \\ \underline{x^2 + 3x} \\ 2x + 8 \\ \underline{2x + 6} \\ 2 \end{array}$$

Divide the first term by the first term: $2x/x = 2$.

$2x + 8$ is the first remainder.

$\longleftarrow$ Multiply 2 by $x + 3$.

Subtract: $(2x + 8) - (2x + 6)$.

The quotient is $x + 2$, with remainder 2. Note that the degree of the remainder is 0 and the degree of the divisor, $x + 3$, is 1. Since $0 < 1$, the process stops.

Check: $(x + 3)(x + 2) + 2 = x^2 + 5x + 6 + 2$ Add the remainder to the product.

$$= x^2 + 5x + 8$$

See if, after studying Examples 3 and 4, you can explain why long division of numbers is performed as it is.

We write our answer as $x + 2$, R 2, or as

$$\text{Quotient} + \frac{\text{Remainder}}{\text{Divisor}}$$

$$x + 2 \quad + \quad \frac{2}{x + 3}.$$

This is how answers are listed at the back of the book.

TRY EXERCISE 21

The last answer in Example 4 can also be checked by multiplying:

$$(x + 3)\left[(x + 2) + \frac{2}{x + 3}\right] = (x + 3)(x + 2) + (x + 3)\frac{2}{x + 3}$$

Using the distributive law

$$= x^2 + 5x + 6 + 2$$

$$= x^2 + 5x + 8.$$ This was the dividend in Example 4.

As shown in Example 4, the quicker check is to multiply the divisor by the quotient and then add the remainder.

You may have noticed that it is helpful to have all polynomials written in descending order.

Dividing Polynomials

1. Arrange polynomials in descending order.
2. If there are missing terms in the dividend, either write them with 0 coefficients or leave space for them.
3. Perform the long division process until the degree of the remainder is less than the degree of the divisor.

EXAMPLE 5

Divide: $(9a^2 + a^3 - 5) \div (a^2 - 1)$.

SOLUTION   We rewrite the problem in descending order:

$$(a^3 + 9a^2 - 5) \div (a^2 - 1).$$

We write $0a$ for the missing term in the dividend.

$$
\require{enclose}
\begin{array}{r}
a + 9 \\
a^2 - 1 \enclose{longdiv}{a^3 + 9a^2 + 0a - 5} \\
\underline{a^3 - a } \\
9a^2 + a - 5 \\
\underline{9a^2 - 9} \\
a + 4
\end{array}
$$

When there is a missing term in the dividend, we can write it in, as shown here, or leave space, as in Example 6 below.

Subtracting: $0a - (-a) = a$.

The degree of the remainder is less than the degree of the divisor, so we are finished.

The answer is $a + 9 + \dfrac{a + 4}{a^2 - 1}$.

TRY EXERCISE 35

EXAMPLE **6** Let $f(x) = 125x^3 - 8$ and $g(x) = 5x - 2$. If $F(x) = (f/g)(x)$, find a simplified expression for $F(x)$ and list all restrictions on the domain.

SOLUTION Recall that $(f/g)(x) = f(x)/g(x)$. Thus,

$$F(x) = \frac{125x^3 - 8}{5x - 2}$$

and

$$
\begin{array}{r}
25x^2 + 10x + 4 \\
5x - 2\overline{)125x^3 \qquad\qquad - 8} \\
\underline{125x^3 - 50x^2} \\
50x^2 \qquad - 8 \\
\underline{50x^2 - 20x} \\
20x - 8 \\
\underline{20x - 8} \\
0
\end{array}
$$

Leaving space for the missing terms

Subtracting: $125x^3 - (125x^3 - 50x^2) = 50x^2$

Subtracting

Note that, because $F(x) = f(x)/g(x)$, it follows that $g(x)$ cannot be 0. Since $g(x)$ is 0 for $x = \frac{2}{5}$ (check this), we have

$$F(x) = 25x^2 + 10x + 4, \quad \text{provided } x \neq \tfrac{2}{5}.$$

TRY EXERCISE 39

6.6 EXERCISE SET

For Extra Help

Concept Reinforcement *Fill in each blank with reference to the following division.*

$$
\begin{array}{r}
x + 2 \\
x - 3\overline{)x^2 - \quad x - 1} \\
\underline{x^2 - 3x} \\
2x - 1 \\
\underline{2x - 6} \\
5
\end{array}
$$

1. The divisor is _____.

2. The dividend is _____.

3. The quotient is _____.

4. The remainder is _____.

5. The degree of the divisor is _____.

6. The degree of the remainder is _____.

Divide and check.

7. $\dfrac{36x^6 + 18x^5 - 27x^2}{9x^2}$

8. $\dfrac{30y^8 - 15y^6 + 40y^4}{5y^4}$

9. $\dfrac{21a^3 + 7a^2 - 3a - 14}{-7a}$

10. $\dfrac{-25x^3 + 20x^2 - 3x + 7}{-5x}$

11. $\dfrac{16y^4z^2 - 8y^6z^4 + 12y^8z^3}{-4y^4z}$

12. $\dfrac{6p^2q^2 - 9p^2q + 12pq^2}{-3pq}$

13. $(16y^3 - 9y^2 - 8y) \div (2y^2)$

14. $(6a^4 + 9a^2 - 8) \div (2a)$

15. $(15x^7 - 21x^4 - 3x^2) \div (-3x^2)$

16. $(36y^6 - 18y^4 - 12y^2) \div (-6y)$

17. $(a^2b - a^3b^3 - a^5b^5) \div (a^2b)$

18. $(x^3y^2 - x^3y^3 - x^4y^2) \div (x^2y^2)$

19. $(x^2 + 10x + 21) \div (x + 7)$

20. $(y^2 - 8y + 16) \div (y - 4)$

21. $(y^2 - 10y - 25) \div (y - 5)$

22. $(a^2 - 8a - 16) \div (a + 4)$

23. $(x^2 - 9x + 21) \div (x - 4)$

24. $(y^2 - 11y + 33) \div (y - 6)$

25. $(y^2 - 25) \div (y + 5)$

26. $(a^2 - 81) \div (a - 9)$

27. $(y^3 - 4y^2 + 3y - 6) \div (y - 2)$

28. $(x^3 - 2x^2 + 4x - 5) \div (x + 3)$

29. $(2x^3 + 3x^2 - x - 3) \div (x + 2)$

30. $(3x^3 - 5x^2 - 3x - 2) \div (x - 2)$

31. $(a^3 - 10a + 24) \div (a + 4)$

32. $(x^3 - x + 6) \div (x + 2)$

33. $(10y^3 + 6y^2 - 9y + 10) \div (5y - 2)$

34. $(6x^3 - 11x^2 + 11x - 2) \div (2x - 3)$

35. $(3x^4 + x^3 - 8x^2 - 3x - 3) \div (x^2 - 3)$

36. $(2x^4 - 2x^3 + 3x^2 - 2x + 1) \div (x^2 + 1)$

37. $(2x^4 - x^3 - 5x^2 + x - 6) \div (x^2 + 2)$

38. $(3x^4 + 2x^3 - 11x^2 - 2x + 5) \div (x^2 - 2)$

For Exercises 39–46, $f(x)$ and $g(x)$ are as given. Find a simplified expression for $F(x)$ if $F(x) = (f/g)(x)$. (See Example 6.) Be sure to list all restrictions on the domain of $F(x)$.

39. $f(x) = 6x^2 - 11x - 10, \ g(x) = 3x + 2$

40. $f(x) = 8x^2 - 22x - 21, \ g(x) = 2x - 7$

41. $f(x) = 8x^3 - 27, \ g(x) = 2x - 3$

42. $f(x) = 64x^3 + 8, \ g(x) = 4x + 2$

43. $f(x) = x^4 - 24x^2 - 25, \ g(x) = x^2 - 25$

44. $f(x) = x^4 - 3x^2 - 54, \ g(x) = x^2 - 9$

45. $f(x) = 8x^2 - 3x^4 - 2x^3 + 2x^5 - 5, \ g(x) = x^2 - 1$

46. $f(x) = 4x - x^3 - 10x^2 + 3x^4 - 8, \ g(x) = x^2 - 4$

47. Explain how factoring could be used to solve Example 6.

48. Explain how to construct a polynomial of degree 4 that has a remainder of 3 when divided by $x + 1$.

Skill Review

Review graphing equations and inequalities (Sections 2.3, 2.4, and 4.4).

Graph on a plane.

49. $3x - y = 9$ [2.4]

50. $5y = -15$ [2.4]

51. $y < \frac{5}{2}x$ [4.4]

52. $x + y \geq 3$ [4.4]

53. $y = -\frac{3}{4}x + 1$ [2.3]

54. $x \geq -1$ [4.4]

Synthesis

55. Explain how to construct a polynomial of degree 4 that has a remainder of 2 when divided by $x + c$.

56. Do addition, subtraction, and multiplication of polynomials always result in a polynomial? Does division? Why or why not?

Divide.

57. $(4a^3b + 5a^2b^2 + a^4 + 2ab^3) \div (a^2 + 2b^2 + 3ab)$

58. $(x^4 - x^3y + x^2y^2 + 2x^2y - 2xy^2 + 2y^3) \div (x^2 - xy + y^2)$

59. $(a^7 + b^7) \div (a + b)$

60. Find k such that when $x^3 - kx^2 + 3x + 7k$ is divided by $x + 2$, the remainder is 0.

61. When $x^2 - 3x + 2k$ is divided by $x + 2$, the remainder is 7. Find k.

62. Let
$$f(x) = \frac{3x + 7}{x + 2}.$$

a) Use division to find an expression equivalent to $f(x)$. Then graph f.

b) On the same set of axes, sketch both $g(x) = 1/(x + 2)$ and $h(x) = 1/x$.

c) How do the graphs of f, g, and h compare?

63. D'Andre incorrectly states that
$$(x^3 + 9x^2 - 6) \div (x^2 - 1) = x + 9 + \frac{x + 4}{x^2 - 1}.$$

Without performing any long division, how could you show D'Andre that his division cannot possibly be correct?

64. Use a graphing calculator to check Example 3 by setting $y_1 = (2x^2 - 7x - 15)/(x - 5)$ and $y_2 = 2x + 3$. Then press either $\boxed{\text{TRACE}}$ (after selecting the ZOOM ZINTEGER option) or $\boxed{\text{TABLE}}$ (with TblMin = 0 and ΔTbl = 1) to show that $y_1 \neq y_2$ for $x = 5$.

65. Use a graphing calculator to check Example 5. Perform the check using
$y_1 = (9x^2 + x^3 - 5)/(x^2 - 1)$,
$y_2 = x + 9 + (x + 4)/(x^2 - 1)$, and $y_3 = y_2 - y_1$.

6.7 Synthetic Division

Streamlining Long Division • The Remainder Theorem

Streamlining Long Division

To divide a polynomial by a binomial of the type $x - a$, we can streamline the usual procedure to develop a process called *synthetic division*.

Compare the following. In each stage, we attempt to write less than in the previous stage, while retaining enough essentials to solve the problem.

STAGE 1

When a polynomial is written in descending order, the coefficients provide the essential information:

$$
\begin{array}{r}
4x^2 + 5x + 11 \\
x - 2\overline{)4x^3 - 3x^2 + x + 7} \\
\underline{4x^3 - 8x^2} \\
5x^2 + x \\
\underline{5x^2 - 10x} \\
11x + 7 \\
\underline{11x - 22} \\
29
\end{array}
\qquad
\begin{array}{r}
4 + 5 + 11 \\
1 - 2\overline{)4 - 3 + 1 + 7} \\
\underline{4 - 8} \\
5 + 1 \\
\underline{5 - 10} \\
11 + 7 \\
\underline{11 - 22} \\
29
\end{array}
$$

Because the leading coefficient in $x - 2$ is 1, each time we multiply it by a term in the answer, the leading coefficient of that product duplicates a coefficient in the answer. In the next stage, rather than duplicate these numbers we focus on where -2 is used and drop the 1 from the divisor.

STAGE 2

$$
\begin{array}{r}
4x^2 + 5x + 11 \\
x - 2\overline{)4x^3 - 3x^2 + x + 7} \\
\underline{4x^3 - 8x^2} \\
5x^2 + x \\
\underline{5x^2 - 10x} \\
11x + 7 \\
\underline{11x - 22} \\
29
\end{array}
$$

$$
\begin{array}{r}
4 + 5 + 11 \\
-2\overline{)4 - 3 + 1 + 7} \\
-8 \longleftarrow \text{Multiply: } -2 \cdot 4 = -8. \\
5 + 1 \quad \searrow \text{Subtract: } -3 - (-8) = 5. \\
-10 \longleftarrow \text{Multiply: } -2 \cdot 5 = -10. \\
11 + 7 \quad \searrow \text{Subtract: } 1 - (-10) = 11. \\
-22 \longleftarrow \text{Multiply: } -2 \cdot 11 = -22. \\
29 \longleftarrow \text{Subtract: } 7 - (-22) = 29.
\end{array}
$$

To simplify further, we now reverse the sign of the -2 in the divisor and, in exchange, *add* at each step in the long division.

STAGE 3

$$
x - 2 \overline{)\begin{array}{c} 4x^2 + 5x + 11 \\ 4x^3 - 3x^2 + x + 7 \end{array}}
$$

$$
\begin{array}{r}
4x^3 - 8x^2 \\ \hline
5x^2 + x \\
5x^2 - 10x \\ \hline
11x + 7 \\
11x - 22 \\ \hline
29
\end{array}
$$

$$
2 \overline{)\begin{array}{rrrr} 4 + 5 + 11 \\ 4 - 3 + 1 + 7 \end{array}}
$$ Replace the -2 with 2.

$8 \longleftarrow$ Multiply: $2 \cdot 4 = 8$.
$5 + 1$ Add: $-3 + 8 = 5$.
$10 \longleftarrow$ Multiply: $2 \cdot 5 = 10$.
$11 + 7$ Add: $1 + 10 = 11$.
$22 \longleftarrow$ Multiply: $2 \cdot 11 = 22$.
$29 \longleftarrow$ Add: $7 + 22 = 29$.

The blue numbers can be eliminated if we look at the red numbers instead.

STAGE 4

$$
x - 2 \overline{)\begin{array}{c} 4x^2 + 5x + 11 \\ 4x^3 - 3x^2 + x + 7 \end{array}}
$$

$$
\begin{array}{r}
4x^3 - 8x^2 \\ \hline
5x^2 + x \\
5x^2 - 10x \\ \hline
11x + 7 \\
11x - 22 \\ \hline
29
\end{array}
$$

$$
\begin{array}{c|rrrr}
 & 4 & 5 & 11 & \\
2 & 4 & -3 & 1 & 7 \\
 & & 8 & 10 & 22 \\ \hline
 & 5 & 11 & 29 &
\end{array}
$$

STUDENT NOTES ────

You will not need to write out all five stages when performing synthetic division on your own. We show the steps to help you understand the reasoning behind the method.

Note that the 5 and the 11 preceding the remainder 29 coincide with the 5 and the 11 following the 4 on the top line. By writing a 4 to the left of 5 on the bottom line, we can eliminate the top line in stage 4 and read our answer from the bottom line. This final stage is commonly called **synthetic division**.

STAGE 5

$$
\begin{array}{c|rrrr}
 & 4 & 5 & 11 & \\
2 & 4 & -3 & 1 & 7 \\
 & & 8 & 10 & 22 \\ \hline
 & 5 & 11 & 29 &
\end{array}
$$

$$
2 \,\big|\; \begin{array}{rrrr} 4 & -3 & 1 & 7 \\ & 8 & 10 & 22 \end{array}
$$
$$
\begin{array}{rrr|r} 4 & 5 & 11 & 29 \end{array} \longleftarrow \text{This is the remainder.}
$$

This is the zero-degree coefficient.
This is the first-degree coefficient.
This is the second-degree coefficient.

The quotient is $4x^2 + 5x + 11$ with a remainder of 29.

> Remember that for this method to work, the divisor must be of the form $x - a$, that is, a variable minus a constant.

EXAMPLE **1**   Use synthetic division to divide: $(x^3 + 6x^2 - x - 30) \div (x - 2)$.

SOLUTION

$$\begin{array}{r|rrr} 2 & 1 & 6 & -1 & -30 \\ \hline & 1 \end{array}$$ Write the 2 of $x - 2$ and the coefficients of the dividend.

Bring down the first coefficient.

$$\begin{array}{r|rrr} 2 & 1 & 6 & -1 & -30 \\ & & 2 \\ \hline & 1 & 8 \end{array}$$ Multiply 1 by 2 to get 2.

Add 6 and 2.

$$\begin{array}{r|rrr} 2 & 1 & 6 & -1 & -30 \\ & & 2 & 16 \\ \hline & 1 & 8 & 15 \end{array}$$ Multiply 8 by 2.

Add -1 and 16.

$$\begin{array}{r|rrr} 2 & 1 & 6 & -1 & -30 \\ & & 2 & 16 & 30 \\ \hline & 1 & 8 & 15 & 0 \end{array}$$ Multiply 15 by 2 and add.

The answer is $x^2 + 8x + 15$ with R 0, or just $x^2 + 8x + 15$.

TRY EXERCISE 7

EXAMPLE **2**   Use synthetic division to divide.

a) $(2x^3 + 7x^2 - 5) \div (x + 3)$
b) $(10x^2 - 13x + 3x^3 - 20) \div (4 + x)$

SOLUTION

a) $(2x^3 + 7x^2 - 5) \div (x + 3)$

The dividend has no x-term, so we need to write 0 as the coefficient of x. Note that $x + 3 = x - (-3)$, so we write -3 inside the ⌞.

$$\begin{array}{r|rrrr} -3 & 2 & 7 & 0 & -5 \\ & & -6 & -3 & 9 \\ \hline & 2 & 1 & -3 & \vert\; 4 \end{array}$$

The answer is $2x^2 + x - 3$, with R 4, or $2x^2 + x - 3 + \dfrac{4}{x + 3}$.

b) We first rewrite $(10x^2 - 13x + 3x^3 - 20) \div (4 + x)$ in descending order:

$$(3x^3 + 10x^2 - 13x - 20) \div (x + 4).$$

Next, we use synthetic division. Note that $x + 4 = x - (-4)$.

$$\begin{array}{r|rrrr} -4 & 3 & 10 & -13 & -20 \\ & & -12 & 8 & 20 \\ \hline & 3 & -2 & -5 & \vert\; 0 \end{array}$$

The answer is $3x^2 - 2x - 5$.

TRY EXERCISE 15

The Remainder Theorem

Because the remainder is 0, Example 1 shows that $x - 2$ is a factor of $x^3 + 6x^2 - x - 30$ and that $x^3 + 6x^2 - x - 30 = (x - 2)(x^2 + 8x + 15)$. Thus if $f(x) = x^3 + 6x^2 - x - 30$, then $f(2) = 0$ (since $x - 2$ is a factor of $f(x)$). Similarly, from Example 2(b), we know that if $g(x) = 10x^2 - 13x + 3x^3 - 20$, then $x + 4$ is a factor of $g(x)$ and $g(-4) = 0$. In both examples, the remainder from the division, 0, can serve as a function value. Remarkably, this pattern extends to nonzero remainders. For example, the remainder in Example 2(a) is 4, and if $f(x) = 2x^3 + 7x^2 - 5$, then $f(-3)$ is also 4 (you should check this). The fact that the remainder and the function value coincide is predicted by the remainder theorem.

> **The Remainder Theorem**
>
> The remainder obtained by dividing $P(x)$ by $x - r$ is $P(r)$.

A proof of this result is outlined in Exercise 37.

EXAMPLE 3 Let $f(x) = 8x^5 - 6x^3 + x - 8$. Use synthetic division to find $f(2)$.

SOLUTION The remainder theorem tells us that $f(2)$ is the remainder when $f(x)$ is divided by $x - 2$. We use synthetic division to find that remainder:

$$
\begin{array}{r|rrrrrr}
2 & 8 & 0 & -6 & 0 & 1 & -8 \\
  &   & 16 & 32 & 52 & 104 & 210 \\
\hline
  & 8 & 16 & 26 & 52 & 105 & 202
\end{array}
$$

Although the bottom line can be used to find the quotient for the division $(8x^5 - 6x^3 + x - 8) \div (x - 2)$, what we are really interested in is the remainder. It tells us that $f(2) = 202$.   **TRY EXERCISE** 21

The remainder theorem is often used to check division. Thus Example 2(a) can be checked by computing $P(-3) = 2(-3)^3 + 7(-3)^2 - 5$. Since $P(-3) = 4$ and the remainder in Example 2(a) is also 4, our division was probably correct.

6.7 **EXERCISE SET**

🖐 *Concept Reinforcement* *Classify each statement as either true or false.*

1. If $x - 2$ is a factor of some polynomial $P(x)$, then $P(2) = 0$.

2. If $p(3) = 0$ for some polynomial $p(x)$, then $x - 3$ is a factor of $p(x)$.

3. If $P(-5) = 39$ and $P(x) = x^3 + 7x^2 + 3x + 4$, then

$$
\begin{array}{r|rrrr}
-5 & 1 & 7 & 3 & 4 \\
   &   & -5 & -10 & 35 \\
\hline
   & 1 & 2 & -7 & 39
\end{array}
$$

4. In order for $f(x)/g(x)$ to exist, $g(x)$ must be 0.

5. In order to use synthetic division, we must be sure that the divisor is of the form $x - a$.

6. Synthetic division can be used in problems in which long division could not be used.

Use synthetic division to divide.

7. $(x^3 - 4x^2 - 2x + 5) \div (x - 1)$

8. $(x^3 - 4x^2 + 5x - 6) \div (x - 3)$

9. $(a^2 + 8a + 11) \div (a + 3)$

10. $(a^2 + 8a + 11) \div (a + 5)$

11. $(2x^3 - x^2 - 7x + 14) \div (x + 2)$

12. $(3x^3 - 10x^2 - 9x + 15) \div (x - 4)$

13. $(a^3 - 10a + 12) \div (a - 2)$

14. $(a^3 - 14a + 15) \div (a - 3)$

15. $(3y^3 - 7y^2 - 20) \div (y - 3)$

16. $(2x^3 - 3x^2 + 8) \div (x + 2)$

17. $(x^5 - 32) \div (x - 2)$

18. $(y^5 - 1) \div (y - 1)$

19. $(3x^3 + 1 - x + 7x^2) \div \left(x + \frac{1}{3}\right)$

20. $(8x^3 - 1 + 7x - 6x^2) \div \left(x - \frac{1}{2}\right)$

Use synthetic division to find the indicated function value.

21. $f(x) = 5x^4 + 12x^3 + 28x + 9;\ f(-3)$

22. $g(x) = 3x^4 - 25x^2 - 18;\ g(3)$

23. $P(x) = 2x^4 - x^3 - 7x^2 + x + 2;\ P(-3)$

24. $F(x) = 3x^4 + 8x^3 + 2x^2 - 7x - 4;\ F(-2)$

25. $f(x) = x^4 - 6x^3 + 11x^2 - 17x + 20;\ f(4)$

26. $p(x) = x^4 + 7x^3 + 11x^2 - 7x - 12;\ p(2)$

27. Why is it that we *add* when performing synthetic division, but *subtract* when performing long division?

28. Explain how synthetic division could be useful when attempting to factor a polynomial.

Skill Review

To prepare for Section 6.8, review solving a formula for a variable (Section 1.5).

Solve. [1.5]

29. $ac = b$, for c

30. $x - wz = y$, for w

31. $pq - rq = st$, for q

32. $ab = d - cb$, for b

33. $ab - cd = 3b + d$, for b

34. $ab - cd = 3b + d$, for d

Synthesis

35. Let $Q(x)$ be a polynomial function with $p(x)$ a factor of $Q(x)$. If $p(3) = 0$, does it follow that $Q(3) = 0$? Why or why not? If $Q(3) = 0$, does it follow that $p(3) = 0$? Why or why not?

36. What adjustments must be made if synthetic division is to be used to divide a polynomial by a binomial of the form $ax + b$, with $a > 1$?

37. To prove the remainder theorem, note that any polynomial $P(x)$ can be rewritten as $(x - r) \cdot Q(x) + R$, where $Q(x)$ is the quotient polynomial that arises when $P(x)$ is divided by $x - r$, and R is some constant (the remainder).

  a) How do we know that R must be a constant?
  b) Show that $P(r) = R$ (this says that $P(r)$ is the remainder when $P(x)$ is divided by $x - r$).

38. Let $f(x) = 6x^3 - 13x^2 - 79x + 140$. Find $f(4)$ and then solve the equation $f(x) = 0$.

39. Let $f(x) = 4x^3 + 16x^2 - 3x - 45$. Find $f(-3)$ and then solve the equation $f(x) = 0$.

40. Use the TRACE feature on a graphing calculator to check your answer to Exercise 38.

41. Use the TRACE feature on a graphing calculator to check your answer to Exercise 39.

Nested evaluation. One way to evaluate a polynomial function like $P(x) = 3x^4 - 5x^3 + 4x^2 - 1$ is to successively factor out x as shown:

$$P(x) = x(x(x(3x - 5) + 4) + 0) - 1.$$

Computations are then performed using this "nested" form of $P(x)$.

42. Use nested evaluation to find $f(4)$ in Exercise 38. Note the similarities to the calculations performed with synthetic division.

43. Use nested evaluation to find $f(-3)$ in Exercise 39. Note the similarities to the calculations performed with synthetic division.

6.8 Formulas, Applications, and Variation

Formulas ▪ Direct Variation ▪ Inverse Variation ▪ Joint Variation and Combined Variation

Formulas

Formulas occur frequently as mathematical models. Many formulas contain rational expressions, and to solve such formulas for a specified letter, we proceed as when solving rational equations.

EXAMPLE **1**   *Electronics.* The formula

$$\frac{1}{R} = \frac{1}{r_1} + \frac{1}{r_2}$$

is used by electricians to determine the resistance R of two resistors r_1 and r_2 connected in parallel.* Solve for r_1.

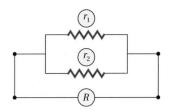

SOLUTION   We use the same approach as in Section 6.4:

$$Rr_1r_2 \cdot \frac{1}{R} = Rr_1r_2 \cdot \left(\frac{1}{r_1} + \frac{1}{r_2} \right)$$ Multiplying both sides by the LCD to clear fractions

$$Rr_1r_2 \cdot \frac{1}{R} = Rr_1r_2 \cdot \frac{1}{r_1} + Rr_1r_2 \cdot \frac{1}{r_2}$$ Multiplying to remove parentheses

$$r_1r_2 = Rr_2 + Rr_1.$$ Simplifying by removing factors equal to 1: $\frac{R}{R} = 1; \frac{r_1}{r_1} = 1; \frac{r_2}{r_2} = 1$

At this point it is tempting to multiply by $1/r_2$ to get r_1 alone on the left, *but* note that there is an r_1 on the right. We must get all the terms involving r_1 on the *same side* of the equation.

$$r_1r_2 - Rr_1 = Rr_2$$ Subtracting Rr_1 from both sides

$$r_1(r_2 - R) = Rr_2$$ Factoring out r_1 in order to combine like terms

$$r_1 = \frac{Rr_2}{r_2 - R}$$ Dividing both sides by $r_2 - R$ to get r_1 alone

This formula can be used to calculate r_1 whenever R and r_2 are known.

TRY EXERCISE ▶ 17

EXAMPLE **2**   *Astronomy.* The formula

$$\frac{V^2}{R^2} = \frac{2g}{R + h}$$

is used to find a satellite's *escape velocity V*, where R is a planet's radius, h is the satellite's height above the planet, and g is the planet's gravitational constant. Solve for h.

SOLUTION   We first multiply by the LCD, $R^2(R + h)$, to clear fractions:

$$\frac{V^2}{R^2} = \frac{2g}{R + h}$$

$$R^2(R + h)\frac{V^2}{R^2} = R^2(R + h)\frac{2g}{R + h}$$ Multiplying to clear fractions

$$\frac{R^2(R + h)V^2}{R^2} = \frac{R^2(R + h)2g}{R + h}$$

$$(R + h)V^2 = R^2 \cdot 2g.$$ Removing factors equal to 1: $\frac{R^2}{R^2} = 1$ and $\frac{R + h}{R + h} = 1$

---

*Recall that the subscripts 1 and 2 merely indicate that r_1 and r_2 are different variables representing similar quantities.

Remember: We are solving for *h*. Although we *could* distribute V^2, since *h* appears only within the factor $R + h$, it is easier to divide both sides by V^2:

$$\frac{(R + h)V^2}{V^2} = \frac{2R^2g}{V^2} \qquad \text{Dividing both sides by } V^2$$

$$R + h = \frac{2R^2g}{V^2} \qquad \text{Removing a factor equal to 1: } \frac{V^2}{V^2} = 1$$

$$h = \frac{2R^2g}{V^2} - R. \qquad \text{Subtracting } R \text{ from both sides}$$

The last equation can be used to determine the height of a satellite above a planet when the planet's radius and gravitational constant, along with the satellite's escape velocity, are known.

TRY EXERCISE ▶ 29

EXAMPLE 3

Acoustics (the Doppler Effect).   The formula

$$f = \frac{sg}{s + v}$$

is used to determine the frequency *f* of a sound that is moving at velocity *v* toward a listener who hears the sound as frequency *g*. Here *s* is the speed of sound in a particular medium. Solve for *s*.

STUDENT NOTES

The steps used to solve equations are precisely the same steps used to solve formulas. If you feel "rusty" in this regard, study the earlier section in which this type of equation first appeared. Then make sure that you can consistently solve those equations before returning to the work with formulas.

SOLUTION   We first clear fractions by multiplying by the LCD, $s + v$:

$$f \cdot (s + v) = \frac{sg}{s + v}(s + v)$$

$$fs + fv = sg. \qquad \begin{array}{l}\text{The variable for which we are solving, } s, \\ \text{appears on both sides, forcing us to} \\ \text{distribute on the left side.}\end{array}$$

Next, we must get all terms containing *s* on one side:

$$fv = sg - fs \qquad \text{Subtracting } fs \text{ from both sides}$$

$$fv = s(g - f) \qquad \text{Factoring out } s. \text{ This is like combining like terms.}$$

$$\frac{fv}{g - f} = s. \qquad \text{Dividing both sides by } g - f$$

Since *s* is isolated on one side, we have solved for *s*. This last equation can be used to determine the speed of sound whenever *f*, *v*, and *g* are known.

TRY EXERCISE ▶ 19

> **To Solve a Rational Equation for a Specified Variable**
> 1. Multiply both sides by the LCD to clear fractions, if necessary.
> 2. Multiply to remove parentheses, if necessary.
> 3. Get all terms with the specified variable alone on one side.
> 4. Factor out the specified variable if it is in more than one term.
> 5. Multiply or divide on both sides to isolate the specified variable.

Variation

To extend our study of formulas and functions, we now examine three real-world situations: direct variation, inverse variation, and combined variation.

DIRECT VARIATION

A computer technician earns $22 per hour. In 1 hr, $22 is earned. In 2 hr, $44 is earned. In 3 hr, $66 is earned, and so on. This gives rise to a set of ordered pairs:

$$(1, 22), (2, 44), (3, 66), (4, 88), \quad \text{and so on.}$$

Note that the ratio of earnings E to time t is $\frac{22}{1}$ in every case.

If a situation is modeled by pairs for which the ratio is constant, we say there is **direct variation**. Here earnings *vary directly* as the time:

We have $\dfrac{E}{t} = 22$, so $E = 22t$ or, using function notation, $E(t) = 22t$.

> **Direct Variation**
>
> When a situation is modeled by a linear function of the form $f(x) = kx$, or $y = kx$, where k is a nonzero constant, we say that there is *direct variation*, that *y varies directly* as x, or that *y is proportional to x*. The number k is called the *variation constant*, or *constant of proportionality*.

Note that for $k > 0$, any equation of the form $y = kx$ indicates that as x increases, y increases as well.

EXAMPLE 4   Find the variation constant and an equation of variation if y varies directly as x, and $y = 32$ when $x = 2$.

SOLUTION   We know that $(2, 32)$ is a solution of $y = kx$. Therefore,

$$32 = k \cdot 2 \qquad \text{Substituting}$$

$$\frac{32}{2} = k, \quad \text{or} \quad k = 16. \qquad \text{Solving for } k$$

The variation constant is 16. The equation of variation is $y = 16x$. The notation $y(x) = 16x$ or $f(x) = 16x$ is also used.

> TRY EXERCISE ▶ 43

EXAMPLE 5

Ocean waves. The speed v of a train of ocean waves varies directly as the swell period t, or time between successive waves. Waves with a swell period of 12 sec are traveling 21 mph. How fast are waves traveling that have a swell period of 20 sec?

Source: www.rodntube.com

SOLUTION

1. **Familiarize.** Because of the phrase "v . . . varies directly as . . . t," we express the speed of the wave v, in miles per hour, as a function of the swell period t, in seconds. Thus, $v(t) = kt$, where k is the variation constant. Because we are using ratios, we can use the units "seconds" and "miles per hour" without converting sec to hr or hr to sec. Knowing that waves with a swell period of 12 sec are traveling 21 mph, we have $v(12) = 21$.

2. **Translate.** We find the variation constant using the data and then use it to write the equation of variation:

$$v(t) = kt$$
$$v(12) = k \cdot 12 \qquad \text{Replacing } t \text{ with 12}$$
$$21 = k \cdot 12 \qquad \text{Replacing } v(12) \text{ with 21}$$
$$\frac{21}{12} = k \qquad \text{Solving for } k$$
$$1.75 = k. \qquad \text{This is the variation constant.}$$

The equation of variation is $v(t) = 1.75t$. This is the translation.

3. **Carry out.** To find the speed of waves with a swell period of 20 sec, we compute $v(20)$:

$$v(t) = 1.75t$$
$$v(20) = 1.75(20) \qquad \text{Substituting 20 for } t$$
$$= 35.$$

4. **Check.** To check, we could reexamine all our calculations. Note that our answer seems reasonable since the ratios 21/12 and 35/20 are both 1.75.

5. **State.** Waves with a swell period of 20 sec are traveling 35 mph.

> TRY EXERCISE 55

INVERSE VARIATION

Suppose a bus travels 20 mi. At 20 mph, the trip takes 1 hr. At 40 mph, it takes $\frac{1}{2}$ hr. At 60 mph, it takes $\frac{1}{3}$ hr, and so on. This gives pairs of numbers, all having the same product:

$$(20, 1), \left(40, \tfrac{1}{2}\right), \left(60, \tfrac{1}{3}\right), \left(80, \tfrac{1}{4}\right), \quad \text{and so on.}$$

Note that the product of each pair is 20. When a situation is modeled by pairs for which the product is constant, we say that there is **inverse variation**. Since $r \cdot t = 20$, we have

$$t = \frac{20}{r} \quad \text{or, using function notation,} \quad t(r) = \frac{20}{r}.$$

> ### Inverse Variation
>
> When a situation is modeled by a rational function of the form $f(x) = k/x$, or $y = k/x$, where k is a nonzero constant, we say that there is *inverse variation,* that *y varies inversely as x*, or that *y is inversely proportional to x*. The number k is called the *variation constant,* or *constant of proportionality.*

Note that for $k > 0$, any equation of the form $y = k/x$ indicates that as x increases, y decreases.

EXAMPLE 6   Find the variation constant and an equation of variation if y varies inversely as x, and $y = 32$ when $x = 0.2$.

SOLUTION   We know that $(0.2, 32)$ is a solution of

$$y = \frac{k}{x}.$$

Therefore,

$$32 = \frac{k}{0.2} \qquad \text{Substituting}$$

$$(0.2)32 = k$$

$$6.4 = k. \qquad \text{Solving for } k$$

The variation constant is 6.4. The equation of variation is

$$y = \frac{6.4}{x}.$$

> TRY EXERCISE   49

There are many real-life quantities that vary inversely.

EXAMPLE 7   *Movie downloads.*   The time t that it takes to download a movie file varies inversely as the transfer speed s of the Internet connection. A typical full-length movie file will transfer in 48 min at a transfer speed of 256 KB/s (kilobytes per second). How long will it take to transfer the same movie file at a transfer speed of 32 KB/s?

Source: www.xsvidmovies.com

SOLUTION

1. **Familiarize.** Because of the phrase ". . . varies inversely as the transfer speed," we express the download time t, in minutes, as a function of the transfer speed s, in kilobytes per second. Thus, $t(s) = k/s$.

2. **Translate.** We use the given information to solve for k. We will then use that result to write the equation of variation.

$$t(s) = \frac{k}{s}$$

$$t(256) = \frac{k}{256} \qquad \text{Replacing } s \text{ with } 256$$

$$48 = \frac{k}{256} \qquad \text{Replacing } t(256) \text{ with } 48$$

$$12{,}288 = k.$$

The equation of variation is $t(s) = 12{,}288/s$. This is the translation.

3. **Carry out.** To find the download time at a transfer speed of 32 KB/s, we calculate $t(32)$:

$$t(32) = \frac{12{,}288}{32} = 384.$$

4. **Check.** Note that, as expected, as the transfer speed goes *down*, the download time goes *up*. Also, the products $48 \cdot 256$ and $32 \cdot 384$ are both 12,288.

5. **State.** At a transfer speed of 32 KB/s, it will take 384 min, or 6 hr 24 min, to download the movie file.   `TRY EXERCISE` 57

JOINT VARIATION AND COMBINED VARIATION

When a variable varies directly with more than one other variable, we say that there is *joint variation*. For example, in the formula for the volume of a right circular cylinder, $V = \pi r^2 h$, we say that V varies *jointly* as h and the square of r.

> ### Joint Variation
> y varies *jointly* as x and z if, for some nonzero constant k, $y = kxz$.

EXAMPLE **8**   Find an equation of variation if y varies jointly as x and z, and $y = 30$ when $x = 2$ and $z = 3$.

SOLUTION   We have

$$y = kxz,$$

so

$$30 = k \cdot 2 \cdot 3$$
$$k = 5. \qquad \text{The variation constant is 5.}$$

The equation of variation is $y = 5xz$.   `TRY EXERCISE` 73

Joint variation is one form of *combined variation*. In general, when a variable varies directly and/or inversely, at the same time, with more than one other variable, there is **combined variation**. Examples 8 and 9 are both examples of combined variation.

EXAMPLE **9**   Find an equation of variation if y varies jointly as x and z and inversely as the square of w, and $y = 105$ when $x = 3$, $z = 20$, and $w = 2$.

SOLUTION   The equation of variation is of the form

$$y = k \cdot \frac{xz}{w^2},$$

so, substituting, we have

$$105 = k \cdot \frac{3 \cdot 20}{2^2}$$
$$105 = k \cdot 15$$
$$k = 7.$$

Thus,

$$y = 7 \cdot \frac{xz}{w^2}.$$ `TRY EXERCISE` 75

6.8 EXERCISE SET

For Extra Help  PRACTICE WATCH DOWNLOAD

Concept Reinforcement *Match each statement with the correct term that completes it from the list on the right.*

1. To clear fractions, we can multiply both sides of an equation by the ____.

2. With direct variation, pairs of numbers have a constant ____.

3. With inverse variation, pairs of numbers have a constant ____.

4. If $y = k/x$, then y varies ____ as x.

5. If $y = kx$, then y varies ____ as x.

6. If $y = kxz$, then y varies ____ as x and z.

a) Directly

b) Inversely

c) Jointly

d) LCD

e) Product

f) Ratio

Determine whether each situation represents direct variation or inverse variation.

7. Two painters can scrape a house in 9 hr, whereas three painters can scrape the house in 6 hr.

8. Andres planted 5 bulbs in 20 min and 7 bulbs in 28 min.

9. Salma swam 2 laps in 7 min and 6 laps in 21 min.

10. It took 2 band members 80 min to set up for a show; with 4 members working, it took 40 min.

11. It took 3 hr for 4 volunteers to wrap the campus' collection of Toys for Tots, but only 1.5 hr with 8 volunteers working.

12. Ayana's air conditioner cooled off 1000 ft^3 in 10 min and 3000 ft^3 in 30 min.

Solve each formula for the specified variable.

13. $f = \dfrac{L}{d}$; d

14. $\dfrac{W_1}{W_2} = \dfrac{d_1}{d_2}$; W_1

15. $s = \dfrac{(v_1 + v_2)t}{2}$; v_1

16. $s = \dfrac{(v_1 + v_2)t}{2}$; t

17. $\dfrac{t}{a} + \dfrac{t}{b} = 1$; b

18. $\dfrac{1}{R} = \dfrac{1}{r_1} + \dfrac{1}{r_2}$; R

19. $R = \dfrac{gs}{g + s}$; g

20. $K = \dfrac{rt}{r - t}$; t

21. $I = \dfrac{nE}{R + nr}$; n

22. $I = \dfrac{nE}{R + nr}$; r

23. $\dfrac{1}{p} + \dfrac{1}{q} = \dfrac{1}{f}$; q

24. $\dfrac{1}{p} + \dfrac{1}{q} = \dfrac{1}{f}$; p

25. $S = \dfrac{H}{m(t_1 - t_2)}$; t_1

26. $S = \dfrac{H}{m(t_1 - t_2)}$; H

27. $\dfrac{E}{e} = \dfrac{R + r}{r}$; r

28. $\dfrac{E}{e} = \dfrac{R + r}{R}$; R

29. $S = \dfrac{a}{1 - r}$; r

30. $S = \dfrac{a - ar^n}{1 - r}$; a

Aha! 31. $c = \dfrac{f}{(a + b)c}$; $a + b$

32. $d = \dfrac{g}{d(c + f)}$; $c + f$

33. *Interest.* The formula

$$P = \dfrac{A}{1 + r}$$

is used to determine what principal P should be invested for one year at $(100 \cdot r)\%$ simple interest in order to have A dollars after a year. Solve for r.

34. *Taxable interest.* The formula

$$I_t = \dfrac{I_f}{1 - T}$$

gives the *taxable interest rate* I_t equivalent to the *tax-free interest rate* I_f for a person in the $(100 \cdot T)\%$ tax bracket. Solve for T.

35. *Average speed.* The formula

$$v = \frac{d_2 - d_1}{t_2 - t_1}$$

gives an object's average speed v when that object has traveled d_1 miles in t_1 hours and d_2 miles in t_2 hours. Solve for t_1.

36. *Average acceleration.* The formula

$$a = \frac{v_2 - v_1}{t_2 - t_1}$$

gives a vehicle's *average acceleration* when its velocity changes from v_1 at time t_1 to v_2 at time t_2. Solve for t_2.

37. *Work rate.* The formula

$$\frac{1}{t} = \frac{1}{a} + \frac{1}{b}$$

gives the total time t required for two workers to complete a job, if the workers' individual times are a and b. Solve for t.

38. *Planetary orbits.* The formula

$$\frac{x^2}{a^2} + \frac{y^2}{b^2} = 1$$

can be used to plot a planet's elliptical orbit of width $2a$ and length $2b$ (see p. 665 in Section 10.2). Solve for b^2.

39. *Semester average.* The formula

$$A = \frac{2Tt + Qq}{2T + Q}$$

gives a student's average A after T tests and Q quizzes, where each test counts as 2 quizzes, t is the test average, and q is the quiz average. Solve for Q.

40. *Astronomy.* The formula

$$L = \frac{dR}{D - d},$$

where D is the diameter of the sun, d is the diameter of the earth, R is the earth's distance from the sun, and L is some fixed distance, is used in calculating when lunar eclipses occur. Solve for D.

41. *Body-fat percentage.* The YMCA calculates men's body-fat percentage p using the formula

$$p = \frac{-98.42 + 4.15c - 0.082w}{w},$$

where c is the waist measurement, in inches, and w is the weight, in pounds. Solve for w.
Source: YMCA guide to Physical Fitness Assessment

42. *Preferred viewing distance.* Researchers model the distance D from which an observer prefers to watch television in "picture heights"—that is, mul-

tiples of the height of the viewing screen. The preferred viewing distance is given by

$$D = \frac{3.55H + 0.9}{H},$$

where D is in picture heights and H is in meters. Solve for H.
Source: www.tid.es, Telefonica Investigación y Desarrollo, S.A. Unipersonal

Find the variation constant and an equation of variation if y varies directly as x and the following conditions apply.

43. $y = 30$ when $x = 5$

44. $y = 80$ when $x = 16$

45. $y = 3.4$ when $x = 2$

46. $y = 2$ when $x = 5$

47. $y = 2$ when $x = \frac{1}{5}$

48. $y = 0.9$ when $x = 0.5$

Find the variation constant and an equation of variation in which y varies inversely as x, and the following conditions exist.

49. $y = 5$ when $x = 20$

50. $y = 40$ when $x = 8$

51. $y = 11$ when $x = 4$

52. $y = 9$ when $x = 10$

53. $y = 27$ when $x = \frac{1}{3}$

54. $y = 81$ when $x = \frac{1}{9}$

55. *Hooke's law.* Hooke's law states that the distance d that a spring is stretched by a hanging object varies directly as the mass m of the object. If the distance is 20 cm when the mass is 3 kg, what is the distance when the mass is 5 kg?

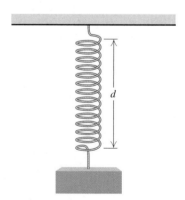

56. *Ohm's law.* The electric current I, in amperes, in a circuit varies directly as the voltage V. When 15 volts are applied, the current is 5 amperes. What is the current when 18 volts are applied?

57. *Work rate.*   The time T required to do a job varies inversely as the number of people P working. It takes 5 hr for 7 volunteers to pick up rubbish from 1 mi of roadway. How long would it take 10 volunteers to complete the job?

58. *Pumping rate.*   The time t required to empty a tank varies inversely as the rate r of pumping. If a Briggs and Stratton pump can empty a tank in 45 min at the rate of 600 kL/min, how long will it take the pump to empty the tank at 1000 kL/min?

59. *Water from melting snow.*   The number of centimeters W of water produced from melting snow varies directly as the number of centimeters S of snow. Meteorologists know that under certain conditions, 150 cm of snow will melt to 16.8 cm of water. The average annual snowfall in Alta, Utah, is 500 in. Assuming the above conditions, how much water will replace the 500 in. of snow?

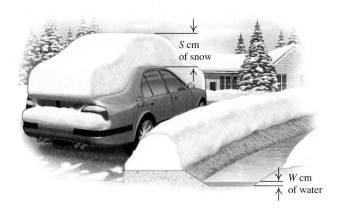

60. *Gardening.*   The number of calories burned by a gardener is directly proportional to the time spent gardening. It takes 30 min to burn 180 calories. How long would it take to burn 240 calories when gardening?
Source: www.healthstatus.com

Aha! **61.** *Mass of water in a human.*   The number of kilograms W of water in a human body varies directly as the mass of the body. A 96-kg person contains 64 kg of water. How many kilograms of water are in a 48-kg person?

62. *Weight on Mars.*   The weight M of an object on Mars varies directly as its weight E on Earth. A person who weighs 95 lb on Earth weighs 38 lb on Mars. How much would a 100-lb person weigh on Mars?

63. *String length and frequency.*   The frequency of a string is inversely proportional to its length. A violin string that is 33 cm long vibrates with a frequency of 260 Hz. What is the frequency when the string is shortened to 30 cm?

64. *Wavelength and frequency.*   The wavelength W of a radio wave varies inversely as its frequency F. A wave with a frequency of 1200 kilohertz has a length of 300 meters. What is the length of a wave with a frequency of 800 kilohertz?

65. *Ultraviolet index.*   At an ultraviolet, or UV, rating of 4, those people who are less sensitive to the sun will burn in 75 min. Given that the number of minutes it takes to burn, t, varies inversely with the UV rating, u, how long will it take less sensitive people to burn when the UV rating is 14?
Source: *The Electronic Textbook of Dermatology* at www.telemedicine.org

66. *Current and resistance.*   The current I in an electrical conductor varies inversely as the resistance R of the conductor. If the current is $\frac{1}{2}$ ampere when the resistance is 240 ohms, what is the current when the resistance is 540 ohms?

67. *Air pollution.*   The average U.S. household of 2.6 people released 0.94 tons of carbon monoxide into the environment in a recent year. How many tons were released nationally? Use 305,000,000 as the U.S. population.
Sources: Based on data from the U.S. Environmental Protection Agency and the U.S. Census Bureau

68. *Relative aperture.*   The relative aperture, or f-stop, of a 23.5-mm lens is directly proportional to the focal length F of the lens. If a lens with a 150-mm focal length has an f-stop of 6.3, find the f-stop of a 23.5-mm lens with a focal length of 80 mm.

Find an equation of variation in which:

69. y varies directly as the square of x, and $y = 50$ when $x = 10$.

70. y varies directly as the square of x, and $y = 0.15$ when $x = 0.1$.

71. y varies inversely as the square of x, and $y = 50$ when $x = 10$.

72. y varies inversely as the square of x, and $y = 0.15$ when $x = 0.1$.

73. y varies jointly as x and z, and $y = 105$ when $x = 14$ and $z = 5$.

74. y varies jointly as x and z, and $y = \frac{3}{2}$ when $x = 2$ and $z = 10$.

75. y varies jointly as w and the square of x and inversely as z, and $y = 49$ when $w = 3$, $x = 7$, and $z = 12$.

76. y varies directly as x and inversely as w and the square of z, and $y = 4.5$ when $x = 15$, $w = 5$, and $z = 2$.

77. *Stopping distance of a car.* The stopping distance d of a car after the brakes have been applied varies directly as the square of the speed r. Once the brakes are applied, a car traveling 60 mph can stop in 138 ft. What stopping distance corresponds to a speed of 40 mph?
Source: Based on data from Edmunds.com

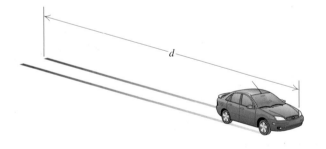

78. *Reverberation time.* A sound's reverberation time T is the time it takes for the sound level to decrease by 60 dB (decibels) after the sound has been turned off. Reverberation time varies directly as the volume V of a room and inversely as the sound absorption A of the room. A given sound has a reverberation time of 1.5 sec in a room with a volume of 90 m^3 and a sound absorption of 9.6. What is the reverberation time of the same sound in a room with a volume of 84 m^3 and a sound absorption of 10.5?
Source: www.isover.co.uk

79. *Volume of a gas.* The volume V of a given mass of a gas varies directly as the temperature T and inversely as the pressure P. If $V = 231$ cm^3 when $T = 300°$K (Kelvin) and $P = 20$ lb/cm^2, what is the volume when $T = 320°$K and $P = 16$ lb/cm^2?

80. *Intensity of a signal.* The intensity I of a television signal varies inversely as the square of the distance d from the transmitter. If the intensity is 25 W/m^2 at a distance of 2 km, what is the intensity 6.25 km from the transmitter?

81. *Atmospheric drag.* Wind resistance, or atmospheric drag, tends to slow down moving objects. Atmospheric drag W varies jointly as an object's surface area A and velocity v. If a car traveling at a speed of 40 mph with a surface area of 37.8 ft^2 experiences a drag of 222 N (Newtons), how fast must a car with 51 ft^2 of surface area travel in order to experience a drag force of 430 N?

82. *Drag force.* The drag force F on a boat varies jointly as the wetted surface area A and the square of the velocity of the boat. If a boat traveling 6.5 mph experiences a drag force of 86 N when the wetted surface area is 41.2 ft^2, find the wetted surface area of a boat traveling 8.2 mph with a drag force of 94 N.

83. If y varies directly as x, does doubling x cause y to be doubled as well? Why or why not?

84. Which exercise did you find easier to work: Exercise 15 or Exercise 19? Why?

Skill Review

Review function notation and domains of functions (Sections 2.2, 4.1, and 5.8).

85. If $f(x) = 4x - 7$, find $f(a) + h$. [2.2]

86. If $f(x) = 4x - 7$, find $f(a + h)$. [2.2]

Find the domain of f.

87. $f(x) = \dfrac{x - 5}{2x + 1}$ [2.2], [4.2]

88. $f(x) = \dfrac{3x}{x^2 + 1}$ [2.2]

89. $f(x) = \sqrt{2x + 8}$ [4.1]

90. $f(x) = \dfrac{3x}{x^2 - 1}$ [5.8]

Synthesis

91. Suppose that the number of customer complaints is inversely proportional to the number of employees hired. Will a firm reduce the number of complaints more by expanding from 5 to 10 employees, or from 20 to 25? Explain. Consider using a graph to help justify your answer.

92. Why do you think subscripts are used in Exercises 15 and 25 but not in Exercises 27 and 28?

93. *Escape velocity.*   A satellite's escape velocity is 6.5 mi/sec, the radius of the earth is 3960 mi, and the earth's gravitational constant is 32.2 ft/sec^2. How far is the satellite from the surface of the earth? (See Example 2.)

94. The *harmonic mean* of two numbers a and b is a number M such that the reciprocal of M is the average of the reciprocals of a and b. Find a formula for the harmonic mean.

95. *Health-care.*   Young's rule for determining the size of a particular child's medicine dosage c is

$$c = \frac{a}{a + 12} \cdot d,$$

where a is the child's age and d is the typical adult dosage. If a child's age is doubled, the dosage increases. Find the ratio of the larger dosage to the smaller dosage. By what percent does the dosage increase?
Source: Olsen, June Looby, Leon J. Ablon, and Anthony Patrick Giangrasso, *Medical Dosage Calculations*, 6th ed.

96. Solve for x:

$$x^2\left(1 - \frac{2pq}{x}\right) = \frac{2p^2q^3 - pq^2x}{-q}.$$

97. *Average acceleration.*   The formula

$$a = \frac{\dfrac{d_4 - d_3}{t_4 - t_3} - \dfrac{d_2 - d_1}{t_2 - t_1}}{t_4 - t_2}$$

can be used to approximate average acceleration, where the d's are distances and the t's are the corresponding times. Solve for t_1.

98. If y varies inversely as the cube of x and x is multiplied by 0.5, what is the effect on y?

99. *Intensity of light.*   The intensity I of light from a bulb varies directly as the wattage of the bulb and inversely as the square of the distance d from the bulb. If the wattage of a light source and its distance from reading matter are both doubled, how does the intensity change?

100. Describe in words the variation represented by $W = \dfrac{km_1M_1}{d^2}$. Assume k is a constant.

101. *Tension of a musical string.*   The tension T on a string in a musical instrument varies jointly as the string's mass per unit length m, the square of its length l, and the square of its fundamental frequency f. A 2-m long string of mass 5 gm/m with a fundamental frequency of 80 has a tension of 100 N (Newtons). How long should the same string be if its tension is going to be changed to 72 N?

102. *Volume and cost.*   A peanut butter jar in the shape of a right circular cylinder is 4 in. high and 3 in. in diameter and sells for $1.20. If we assume that cost is proportional to volume, how much should a jar 6 in. high and 6 in. in diameter cost?

103. *Golf distance finder.*   A device used in golf to estimate the distance d to a hole measures the size s that the 7-ft pin *appears* to be in a viewfinder. The viewfinder uses the principle, diagrammed here, that s gets bigger when d gets smaller. If $s = 0.56$ in. when $d = 50$ yd, find an equation of variation that expresses d as a function of s. What is d when $s = 0.40$ in.?

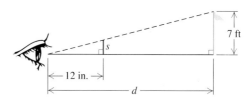

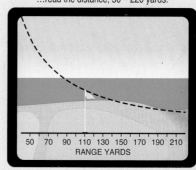

HOW IT WORKS:

Just sight the flagstick through the viewfinder…
fit flag between top dashed line and the solid line below…
…read the distance, 50 – 220 yards.

Nothing to focus.

Gives you exact distance that your ball lies from the flagstick.

Choose proper club on every approach shot.

Figure new pin placement instantly.

Train your naked eye for formal and tournament play.

Eliminate the need to remember every stake, tree, and bush on the course.

Study Summary

SECTION 6.1: RATIONAL EXPRESSIONS AND FUNCTIONS: MULTIPLYING AND DIVIDING

A **rational expression** can be written as a quotient of two polynomials, and is undefined when the denominator is zero.

To simplify rational expressions, we remove a factor equal to 1.

We list any restrictions when simplifying functions.

Simplify: $f(x) = \dfrac{x^2 - 3x - 4}{x^2 - 1}.$

$$f(x) = \frac{x^2 - 3x - 4}{x^2 - 1} = \frac{(x+1)(x-4)}{(x+1)(x-1)}$$

Factoring. We list the restrictions $x \neq -1, x \neq 1.$

$$f(x) = \frac{x - 4}{x - 1}, x \neq -1, x \neq 1$$

Removing a factor equal to 1:
$$\frac{(x+1)}{(x+1)} = 1$$

Multiplication

$$\frac{A}{B} \cdot \frac{C}{D} = \frac{AC}{BD}$$

$$\frac{5v + 5}{v - 2} \cdot \frac{2v^2 - 8v + 8}{v^2 - 1}$$

$$= \frac{5(v+1) \cdot 2(v-2)(v-2)}{(v-2)(v+1)(v-1)}$$

Multiplying and factoring

$$= \frac{10(v - 2)}{v - 1}$$

$$\frac{(v+1)(v-2)}{(v+1)(v-2)} = 1$$

Division

$$\frac{A}{B} \div \frac{C}{D} = \frac{A}{B} \cdot \frac{D}{C} = \frac{AD}{BC}$$

$$(x^2 - 5x - 6) \div \frac{x^2 - 1}{x + 6} = \frac{x^2 - 5x - 6}{1} \cdot \frac{x + 6}{x^2 - 1}$$

Multiplying by the reciprocal

$$= \frac{(x - 6)(x+1)(x + 6)}{(x+1)(x - 1)}$$

Multiplying and factoring

$$= \frac{(x - 6)(x + 6)}{x - 1}$$

$$\frac{x + 1}{x + 1} = 1$$

SECTION 6.2: RATIONAL EXPRESSIONS AND FUNCTIONS: ADDING AND SUBTRACTING

Addition and Subtraction with Like Denominators

$$\frac{A}{B} + \frac{C}{B} = \frac{A + C}{B}$$

$$\frac{A}{B} - \frac{C}{B} = \frac{A - C}{B}$$

$$\frac{7x - 6}{2x + 1} - \frac{x - 9}{2x + 1} = \frac{7x - 6 - (x - 9)}{2x + 1}$$

Subtracting numerators; keeping the denominator

$$= \frac{7x - 6 - x + 9}{2x + 1}$$

$$= \frac{6x + 3}{2x + 1}$$

$$= \frac{3(2x+1)}{1(2x+1)}$$

Factoring

$$= 3$$

$$(2x + 1)/(2x + 1) = 1$$

Addition and Subtraction with Unlike Denominators

1. Determine the least common denominator (LCD).

2. Rewrite each expression with the LCD.

3. Add or subtract, as indicated.

4. Simplify, if possible.

$$\frac{2x}{x^2 - 16} + \frac{x}{x - 4} = \frac{2x}{(x + 4)(x - 4)} + \frac{x}{x - 4}$$

The LCD is $(x + 4)(x - 4)$.

$$= \frac{2x}{(x + 4)(x - 4)} + \frac{x}{x - 4} \cdot \frac{x + 4}{x + 4}$$

Writing with the LCD

$$= \frac{2x}{(x + 4)(x - 4)} + \frac{x^2 + 4x}{(x + 4)(x - 4)}$$

$$= \frac{x^2 + 6x}{(x + 4)(x - 4)}$$

SECTION 6.3: COMPLEX RATIONAL EXPRESSIONS

Complex rational expressions contain one or more rational expressions within the numerator and/or the denominator. They can be simplified either by using division or by multiplying by a form of 1 to clear the fractions.

Using division to simplify:

$$\frac{\dfrac{1}{6} - \dfrac{1}{x}}{\dfrac{6-x}{6}} = \frac{\dfrac{1}{6} \cdot \dfrac{x}{x} - \dfrac{1}{x} \cdot \dfrac{6}{6}}{\dfrac{6-x}{6}} = \frac{\dfrac{x-6}{6x}}{\dfrac{6-x}{6}}$$

Subtracting to form a single rational expression in the numerator

$$= \frac{x-6}{6x} \div \frac{6-x}{6} = \frac{x-6}{6x} \cdot \frac{6}{6-x}$$

Dividing

$$= \frac{6(x-6)}{6x(-1)(x-6)} = \frac{1}{-x} = -\frac{1}{x}$$

$\dfrac{6(x-6)}{6(x-6)} = 1$

Multiplying by 1:

$$\frac{\dfrac{4}{x}}{\dfrac{3}{x} + \dfrac{2}{x^2}} = \frac{\dfrac{4}{x}}{\dfrac{3}{x} + \dfrac{2}{x^2}} \cdot \frac{x^2}{x^2} = \frac{\dfrac{4}{x} \cdot \dfrac{x^2}{1}}{\left(\dfrac{3}{x} + \dfrac{2}{x^2}\right) \cdot \dfrac{x^2}{1}}$$

Multiplying by $\dfrac{x^2}{x^2}$, a form of 1

$$= \frac{\dfrac{4 \cdot x \cdot x}{x}}{\dfrac{3 \cdot x \cdot x}{x} + \dfrac{2 \cdot x^2}{x^2}} = \frac{4x}{3x+2}$$

SECTION 6.4: RATIONAL EQUATIONS

To Solve a Rational Equation

1. List any numbers that will make a denominator zero.
2. Clear fractions by multiplying by the LCD.
3. Solve the equation.
4. Check possible solutions.

Solve: $\dfrac{x+1}{x+3} + 1 = \dfrac{x+1}{x+3}.$

$$\frac{x+1}{x+3} + 1 = \frac{x+1}{x+3} \qquad x \neq -3$$

$$(x+3) \cdot \frac{x+1}{x+3} + (x+3) \cdot 1 = \frac{x+1}{x+3} \cdot (x+3)$$

Multiplying to clear fractions

$$x + 1 + x + 3 = x + 1$$
$$2x + 4 = x + 1$$
$$x = -3$$

Since we stated that $x \neq -3$, this equation has no solution.

SECTION 6.5: SOLVING APPLICATIONS USING RATIONAL EQUATIONS

Modeling Work Problems

If

a = the time needed for A to complete the work alone,

b = the time needed for B to complete the work alone,

and

t = the time needed for A and B to complete the work together, then:

$$\frac{t}{a} + \frac{t}{b} = 1; \quad \frac{1}{a} \cdot t + \frac{1}{b} \cdot t = 1;$$

$$\frac{1}{a} + \frac{1}{b} = \frac{1}{t}.$$

It takes Manuel 9 hr longer than Zoe to rebuild an engine. Working together, they can do the job in 20 hr. How long would it take each, working alone, to rebuild an engine?

We model the situation using the work principle, with $a = z$, $b = z + 9$, and $t = 20$:

$$\frac{20}{z} + \frac{20}{z+9} = 1.$$

See Example 2 in Section 6.5 for the complete solution.

Modeling Motion Problems

Organize the information in a chart and use one of the forms of the motion formula:

$$d = r \cdot t; \quad r = \frac{d}{t}; \quad t = \frac{d}{r}.$$

On her road bike, Olivia bikes 15 km/h faster than Jason does on his mountain bike. In the time it takes Olivia to travel 80 km, Jason travels 50 km. Find the speed of each bicyclist.

Jason's speed: r km/h
Olivia's speed: $(r + 15)$ km/h
Jason's time: $50/r$ hr
Olivia's time: $80/(r + 15)$ hr

The times are equal:

$$\frac{50}{r} = \frac{80}{r + 15}.$$

See Example 3 in Section 6.5 for the complete solution.

SECTION 6.6: DIVISION OF POLYNOMIALS

To divide a polynomial by a monomial, divide each term by the monomial. Divide coefficients and subtract exponents.

$$\frac{18x^2y^3 - 9xy^2 - 3x^2y}{9xy} = 2xy^2 - y - \frac{1}{3}x$$

To divide a polynomial by a binomial, use long division.

$$
\begin{array}{r}
x + 11 \\
x - 5 \overline{\smash{)}\, x^2 + 6x - 8} \\
\underline{x^2 - 5x} \\
11x - 8 \\
\underline{11x - 55} \\
47
\end{array}
$$

$$(x^2 + 6x - 8) \div (x - 5) = x + 11 + \frac{47}{x - 5}$$

SECTION 6.7: SYNTHETIC DIVISION

The Remainder Theorem

The remainder obtained by dividing a polynomial $P(x)$ by $x - r$ is $P(r)$.

The remainder can be found using **synthetic division**.

$$
\begin{array}{r|rrrr}
2 & 3 & -4 & 0 & 6 \\
  &   & 6 & 4 & 8 \\
\hline
  & 3 & 2 & 4 & 14
\end{array}
$$

$$(3x^3 - 4x^2 + 6) \div (x - 2)$$
$$= 3x^2 + 2x + 4 + \frac{14}{x - 2}$$

For $P(x) = 3x^3 - 4x^2 + 6$, we have $P(2) = 14$.

SECTION 6.8: FORMULAS, APPLICATIONS, AND VARIATION

Direct Variation

$$y = kx$$

If y varies directly as x and y = 45 when x = 0.15, find the equation of variation.

$$y = kx$$
$$45 = k(0.15)$$
$$300 = k$$

The equation of variation is $y = 300x$.

Inverse Variation

$$y = \frac{k}{x}$$

If y varies inversely as x and y = 45 when x = 0.15, find the equation of variation.

$$y = \frac{k}{x}$$
$$45 = \frac{k}{0.15}$$
$$6.75 = k$$

The equation of variation is $y = \dfrac{6.75}{x}$.

Joint Variation

$$y = kxz$$

If y varies jointly as x and z and y = 40 when x = 5 and z = 4, find the equation of variation.

$$y = kxz$$
$$40 = k \cdot 5 \cdot 4$$
$$2 = k$$

The equation of variation is $y = 2xz$.

Review Exercises: Chapter 6

✏ *Concept Reinforcement* *Classify each statement as either true or false.*

1. If $f(x) = \dfrac{x-3}{x^2-4}$, the domain of f is assumed to be $\{x \mid x \neq -2, x \neq 2\}$. [6.1]

2. We write numerators and denominators in factored form before we simplify rational expressions. [6.1]

3. The LCM of $x - 3$ and $3 - x$ is $(x-3)(3-x)$. [6.2]

4. Checking the solution of a rational equation is no more important than checking the solution of any other equation. [6.4]

5. If Camden can do a job alone in t_1 hr and Jacob can do the same job in t_2 hr, then working together it will take them $(t_1 + t_2)/2$ hr. [6.5]

6. If Skye swims 5 km/h in still water and heads into a current of 2 km/h, her speed will change to 3 km/h. [6.5]

7. The fomulas $d = rt$, $r = d/t$, and $t = d/r$ are equivalent. [6.5]

8. A remainder of 0 indicates that the divisor is a factor of the quotient. [6.7]

9. To divide two polynomials using synthetic division, we must make sure that the divisor is of the form $x - a$. [6.7]

10. If x varies inversely as y, then there exists some constant k for which $x = k/y$. [6.8]

---

11. If
$$f(t) = \frac{t^2 - 3t + 2}{t^2 - 9},$$
find the following function values. [6.1]
   a) $f(0)$
   b) $f(-1)$
   c) $f(1)$

Find the LCM of the polynomials. [6.2]

12. $20x^3$, $24x^2$

13. $x^2 + 8x - 20$, $x^2 + 7x - 30$

Perform the indicated operations and, if possible, simplify.

14. $\dfrac{x^2}{x-8} - \dfrac{64}{x-8}$ [6.2]

15. $\dfrac{12a^2b^3}{5c^3d^2} \cdot \dfrac{25c^9d^4}{9a^7b}$ [6.1]

16. $\dfrac{5}{6m^2n^3p} + \dfrac{7}{9mn^4p^2}$ [6.2]

17. $\dfrac{x^3 - 8}{x^2 - 25} \cdot \dfrac{x^2 + 10x + 25}{x^2 + 2x + 4}$ [6.1]

18. $\dfrac{x^2 - 4x - 12}{x^2 - 6x + 8} \div \dfrac{x^2 - 4}{x^3 - 64}$ [6.1]

19. $\dfrac{x}{x^2 + 5x + 6} - \dfrac{2}{x^2 + 3x + 2}$ [6.2]

20. $\dfrac{-4xy}{x^2 - y^2} + \dfrac{x+y}{x-y}$ [6.2]

21. $\dfrac{5a^2}{a-b} + \dfrac{5b^2}{b-a}$ [6.2]

22. $\dfrac{3}{y+4} - \dfrac{y}{y-1} + \dfrac{y^2 + 3}{y^2 + 3y - 4}$ [6.2]

Find simplified form for $f(x)$ and list all restrictions on the domain.

23. $f(x) = \dfrac{4x - 2}{x^2 - 5x + 4} - \dfrac{3x + 2}{x^2 - 5x + 4}$ [6.2]

24. $f(x) = \dfrac{x+8}{x+5} \cdot \dfrac{2x+10}{x^2 - 64}$ [6.1]

25. $f(x) = \dfrac{9x^2 - 1}{x^2 - 9} \div \dfrac{3x + 1}{x + 3}$ [6.1]

Simplify. [6.3]

26. $\dfrac{\dfrac{4}{x} - 4}{\dfrac{9}{x} - 9}$

27. $\dfrac{\dfrac{3}{a} + \dfrac{3}{b}}{\dfrac{6}{a^3} + \dfrac{6}{b^3}}$

28. $\dfrac{\dfrac{y^2 + 4y - 77}{y^2 - 10y + 25}}{\dfrac{y^2 - 5y - 14}{y^2 - 25}}$

29. $\dfrac{\dfrac{5}{x^2 - 9} - \dfrac{3}{x+3}}{\dfrac{4}{x^2 + 6x + 9} + \dfrac{2}{x-3}}$

Solve. [6.4]

30. $\dfrac{3}{x} + \dfrac{7}{x} = 5$

31. $\dfrac{5}{3x+2} = \dfrac{3}{2x}$

32. $\dfrac{4x}{x+1} + \dfrac{4}{x} + 9 = \dfrac{4}{x^2 + x}$

33. $\dfrac{x+6}{x^2+x-6} + \dfrac{x}{x^2+4x+3} = \dfrac{x+2}{x^2-x-2}$

34. $\dfrac{x}{x-3} - \dfrac{3x}{x+2} = \dfrac{5}{x^2-x-6}$

35. If

$$f(x) = \frac{2}{x-1} + \frac{2}{x+2},$$

find all a for which $f(a) = 1$. [6.4]

Solve. [6.5]

36. Meg can arrange the books for a book sale in 9 hr. Kelly can set up for the same book sale in 12 hr. How long would it take them, working together, to set up for the book sale?

37. A research company uses employees' computers to process data while the employee is not using the computer. An Intel Core 2 Quad processor can process a data file in 15 sec less time than an Intel Core 2 Duo processor. Working together, the computers can process the file in 18 sec. How long does it take each computer to process the file?

38. The Black River's current is 6 mph. A boat travels 50 mi downstream in the same time that it takes to travel 30 mi upstream. What is the speed of the boat in still water?

39. A car and a motorcycle leave a rest area at the same time, with the car traveling 8 mph faster than the motorcycle. The car then travels 105 mi in the time it takes the motorcycle to travel 93 mi. Find the speed of each vehicle.

Divide. [6.6]

40. $(30r^2s^3 + 25r^2s^2 - 20r^3s^3) \div (10r^2s)$

41. $(y^3 + 8) \div (y + 2)$

42. $(4x^3 + 3x^2 - 5x - 2) \div (x^2 + 1)$

43. Divide using synthetic division:
$(x^3 + 3x^2 + 2x - 6) \div (x - 3)$. [6.7]

44. If $f(x) = 4x^3 - 6x^2 - 9$, use synthetic division to find $f(5)$. [6.7]

Solve. [6.8]

45. $I = \dfrac{2V}{R+2r}$, for r

46. $S = \dfrac{H}{m(t_1 - t_2)}$, for m

47. $\dfrac{1}{ac} = \dfrac{2}{ab} - \dfrac{3}{bc}$, for c

48. $T = \dfrac{A}{v(t_2 - t_1)}$, for t_1

49. For those people with highly sensitive skin, an ultraviolet, or UV, rating of 6 will cause sunburn after 10 min. Given that the number of minutes it takes to burn t varies inversely as the UV rating u, how long will it take a highly sensitive person to burn on a day with a UV rating of 4?
Source: *The Electronic Textbook of Dermatology* found at www.telemedicine.org

50. The amount of waste generated by a family varies directly as the number of people in the family. The average U.S. family has 3.2 people and generates 14.4 lb of waste daily. How many pounds of waste would be generated daily by a family of 5?
Sources: Based on data from the U.S. Census Bureau and the U.S. Statistical Abstract 2007

51. *Electrical safety.* The amount of time t needed for an electrical shock to stop a 150-lb person's heart varies inversely as the square of the current flowing through the body. It is known that a 0.089-amp current is deadly to a 150-lb person after 3.4 sec. How long would it take a 0.096-amp current to be deadly?
Source: Safety Consulting Services

Synthesis

52. Discuss at least three different uses of the LCD studied in this chapter. [6.2], [6.3], [6.4]

53. Explain the difference between a rational expression and a rational equation. [6.1], [6.4]

Solve.

54. $\dfrac{5}{x-13} - \dfrac{5}{x} = \dfrac{65}{x^2-13x}$ [6.4]

55. $\dfrac{\dfrac{x}{x^2-25} + \dfrac{2}{x-5}}{\dfrac{3}{x-5} - \dfrac{4}{x^2-10x+25}} = 1$ [6.3], [6.4]

56. An Intel Core 2 Extreme processor can process the data file in Exercise 37 in 20 sec. How long would it take the Core 2 Extreme working together with the Core 2 Duo and the Core 2 Quad processors to process the file? [6.5]

Test: Chapter 6

Step-by-step test solutions are found on the video CD in the front of this book.

Simplify.

1. $\dfrac{t+1}{t+3} \cdot \dfrac{5t+15}{4t^2-4}$

2. $\dfrac{x^3+27}{x^2-16} \div \dfrac{x^2+8x+15}{x^2+x-20}$

Perform the indicated operation and simplify when possible.

3. $\dfrac{25x}{x+5} + \dfrac{x^3}{x+5}$

4. $\dfrac{3a^2}{a-b} - \dfrac{3b^2-6ab}{b-a}$

5. $\dfrac{4ab}{a^2-b^2} + \dfrac{a^2+b^2}{a+b}$

6. $\dfrac{6}{x^3-64} - \dfrac{4}{x^2-16}$

Find simplified form for $f(x)$ and list all restrictions on the domain.

7. $f(x) = \dfrac{4}{x+3} - \dfrac{x}{x-2} + \dfrac{x^2+4}{x^2+x-6}$

8. $f(x) = \dfrac{x^2-1}{x+2} \div \dfrac{x^2-2x}{x^2+x-2}$

Simplify.

9. $\dfrac{\dfrac{2}{a}+\dfrac{3}{b}}{\dfrac{5}{ab}+\dfrac{1}{a^2}}$

10. $\dfrac{\dfrac{x^2-5x-36}{x^2-36}}{\dfrac{x^2+x-12}{x^2-12x+36}}$

11. $\dfrac{\dfrac{2}{x+3}-\dfrac{1}{x^2-3x+2}}{\dfrac{3}{x-2}+\dfrac{4}{x^2+2x-3}}$

Solve.

12. $\dfrac{4}{5x-2} = \dfrac{6}{5x}$

13. $\dfrac{t+11}{t^2-t-12} + \dfrac{1}{t-4} = \dfrac{4}{t+3}$

14. $\dfrac{x-1}{x+2} - \dfrac{1}{x^2-4} = \dfrac{x-4}{2-x}$

For Exercises 15 and 16, let $f(x) = \dfrac{x+5}{x-1}$.

15. Find $f(0)$ and $f(-3)$.

16. Find all a for which $f(a) = 10$.

17. Ella can install a countertop in 5 hr. Sari can perform the same job in 4 hr. How long will it take them, working together, to install the countertop?

Divide.

18. $(16a^4b^3c - 10a^5b^2c^2 + 12a^2b^2c) \div (4a^2b)$

19. $(y^2 - 20y + 64) \div (y-6)$

20. $(6x^4 + 3x^2 + 5x + 4) \div (x^2+2)$

21. Divide using synthetic division:
$$(x^3 + 5x^2 + 4x - 7) \div (x-2).$$

22. If $f(x) = 3x^4 - 5x^3 + 2x - 7$, use synthetic division to find $f(4)$.

23. Solve $R = \dfrac{gs}{g+s}$ for s.

24. The product of the reciprocals of two consecutive integers is $\frac{1}{110}$. Find the integers.

25. Terrel bicycles 12 mph with no wind. Against the wind, he bikes 8 mi in the same time that it takes to bike 14 mi with the wind. What is the speed of the wind?

26. The number of workers n needed to clean a stadium after a game varies inversely as the amount of time t allowed for the cleanup. If it takes 25 workers to clean the stadium when there are 6 hr allowed for the job, how many workers are needed if the stadium must be cleaned in 5 hr?

27. The surface area of a balloon varies directly as the square of its radius. The area is 325 in² when the radius is 5 in. What is the area when the radius is 7 in.?

Synthesis

28. Let
$$f(x) = \dfrac{1}{x+3} + \dfrac{5}{x-2}.$$
Find all a for which $f(a) = f(a+5)$.

29. Solve: $\dfrac{6}{x-15} - \dfrac{6}{x} = \dfrac{90}{x^2-15x}$.

30. Find the x- and y-intercepts for the function given by
$$f(x) = \dfrac{\dfrac{5}{x+4}-\dfrac{3}{x-2}}{\dfrac{2}{x-3}+\dfrac{1}{x+4}}.$$

31. One summer, Alex mowed 4 lawns for every 3 lawns mowed by his brother Ryan. Together, they mowed 98 lawns. How many lawns did each mow?

Cumulative Review: Chapters 1–6

1. Evaluate
$$\frac{2x - y^2}{x + y}$$
for $x = 3$ and $y = -4$. [1.1], [1.2]

2. Convert to scientific notation: 391,000,000. [1.7]

3. Determine the slope and the y-intercept for the line given by $7x - 4y = 12$. [2.3]

4. Find an equation for the line that passes through the points $(-1, 7)$ and $(4, -3)$. [2.5]

5. If
$$f(x) = \frac{x - 3}{x^2 - 11x + 30},$$
find (a) $f(3)$ and (b) the domain of f. [2.2], [5.8]

6. Write the domain of f using interval notation if $f(x) = \sqrt{x - 9}$. [4.1]

Graph on a plane.

7. $5x = y$ [2.3]

8. $8y + 2x = 16$ [2.4]

9. $4x \geq 5y + 12$ [4.4]

10. $y = \frac{1}{3}x - 2$ [2.3]

Perform the indicated operations and simplify.

11. $(8x^3y^2)(-3xy^2)$ [5.2]

12. $(5x^2 - 2x + 1)(3x^2 + x - 2)$ [5.2]

13. $(3x^2 + y)^2$ [5.2]

14. $(2x^2 - 9)(2x^2 + 9)$ [5.2]

15. $(-5m^3n^2 - 3mn^3) + (-4m^2n^2 + 4m^3n^2) - (2mn^3 - 3m^2n^2)$ [5.1]

16. $\dfrac{y^2 - 36}{2y + 8} \cdot \dfrac{y + 4}{y + 6}$ [6.1]

17. $\dfrac{x^4 - 1}{x^2 - x - 2} \div \dfrac{x^2 + 1}{x - 2}$ [6.1]

18. $\dfrac{5ab}{a^2 - b^2} + \dfrac{a + b}{a - b}$ [6.2]

19. $\dfrac{2}{m + 1} + \dfrac{3}{m - 5} - \dfrac{m^2 - 1}{m^2 - 4m - 5}$ [6.2]

20. $y - \dfrac{2}{3y}$ [6.2]

21. Simplify: $\dfrac{\dfrac{1}{x} - \dfrac{1}{y}}{x + y}$. [6.3]

22. Divide: $(9x^3 + 5x^2 + 2) \div (x + 2)$. [6.6]

Factor.

23. $4x^3 + 400x$ [5.3]

24. $x^2 + 8x - 84$ [5.4]

25. $16y^2 - 25$ [5.5]

26. $64x^3 + 8$ [5.6]

27. $t^2 - 16t + 64$ [5.5]

28. $x^6 - x^2$ [5.5]

29. $\frac{1}{8}b^3 - c^3$ [5.6]

30. $3t^2 + 17t - 28$ [5.4]

31. $x^5 - x^3y + x^2y - y^2$ [5.3]

Solve.

32. $8x = 1 + 16x^2$ [5.8]

33. $288 = 2y^2$ [5.8]

34. $\frac{1}{3}x - \frac{1}{5} \geq \frac{1}{5}x - \frac{1}{3}$ [4.1]

35. $-13 < 3x + 2 < -1$ [4.2]

36. $3x - 2 < -6 \text{ or } x + 3 > 9$ [4.2]

37. $|x| > 6.4$ [4.3]

38. $|3x - 2| \leq 14$ [4.3]

39. $\dfrac{6}{x - 5} = \dfrac{2}{2x}$ [6.4]

40. $\dfrac{3x}{x - 2} - \dfrac{6}{x + 2} = \dfrac{24}{x^2 - 4}$ [6.4]

41. $5x - 2y = -23,$
    $3x + 4y = 7$ [3.2]

42. $-3x + 4y + z = -5,$
    $x - 3y - z = 6,$
    $2x + 3y + 5z = -8$ [3.4]

43. $P = \dfrac{4a}{a + b}$, for a [6.8]

44. *Broadway revenue.* Gross revenue from Broadway shows has grown from $20 million in 1986–1987 to $939 million in 2006–2007. Let $r(t)$ represent gross revenue, in millions of dollars, from Broadway shows t seasons after the 1986–1987 season. [2.5]
Source: The League of American Theatres and Producers

a) Find a linear function that fits the data.
b) Use the function from part (a) to predict the gross revenue from Broadway shows in 2009–2010.
c) In what season will the gross revenue from Broadway shows reach $1.4 billion?

45. *Broadway performances.* In January 2006, *The Phantom of the Opera* became the longest-running Broadway show with 7486 performances. By January 2008, the show had played 8302 times. Calculate the rate at which the number of performances was rising. [2.3]

46. *Trail mix.* Kenny mixes Himalayan Diamonds trail mix, which contains 40% nuts, with Alpine Gold trail mix, which contains 25% nuts, to create 20 lb of a mixture that is 30% nuts. How much of each type of trail mix does he use? [3.3]

47. *Quilting.* A rectangular quilted wall hanging is 4 in. longer than it is wide. The area of the quilt is 320 in^2. Find the perimeter of the quilt. [5.8]

48. *Hotel management.* The IQAir HealthPro Plus air purifier can clean the air in a 20-ft by 25-ft meeting room in 5 fewer minutes than it takes the Austin Healthmate HM400 to do the same job. Together the two machines can purify the air in the room in 6 min. How long would it take each machine, working alone, to purify the air in the room? [6.5]
Source: Manufacturers' and retailers' websites

49. *Driving delays.* According to the National Surface Transportation Policy and Revenue Study Commission, the best-case scenario for driving delays due to road work in 2055 will be 250% of the delays in 2005. If the commission predicts 30 billion hr of driving delays in 2055, how many hours of driving delays were there in 2005? [1.4]

50. *Driving time.* The time t that it takes for Johann to drive to work varies inversely as his speed. On a day when Johann averages 45 mph, it takes him 20 min to drive to work. How long will it take him to drive to work when he averages only 40 mph? [6.8]

51. *Amusement parks.* Together, Magic Kingdom, Disneyland, and California Adventure have 133 rides and attractions. Magic Kingdom has 5 fewer rides and attractions than half the total at California Adventure and Disneyland. Disneyland has three-fourths as many rides and attractions as the total number at Magic Kingdom and California Adventure. How many rides and attractions does each amusement park have? [3.5]
Source: The Walt Disney Company

Synthesis

52. Multiply: $(x - 4)^3$. [5.2]

53. Find all roots for $f(x) = x^4 - 34x^2 + 225$. [5.8]

Solve.

54. $4 \leq |3 - x| \leq 6$ [4.2], [4.3]

55. $\dfrac{18}{x - 9} + \dfrac{10}{x + 5} = \dfrac{28x}{x^2 - 4x - 45}$ [6.4]

56. $16x^3 = x$ [5.8]

Exponents and Radicals

MICHAEL MANOLAKIS
FIREFIGHTER
Boston, Massachusetts

Firefighters use math every day. Hosing down a fire may seem simple, but there are considerations to take into account when getting water from its source to the end of the hose where the "pipeman" is attacking the fire. For example, the safest force or psi (pounds per square inch) needed at the tip of a hose is determined by its size. The safest force needed at the tip of a $1\frac{3}{4}$-in. line is 120 psi. To achieve this pressure, we must consider the distance and height of the hose line from the source of water. These two factors are necessary to accurately calculate and deliver the appropriate psi at the tip.

AN APPLICATION

The velocity of water flow, in feet per second, from a nozzle is given by

$$v(p) = 12.1\sqrt{p},$$

where p is the nozzle pressure, in pounds per square inch. Find the nozzle pressure if the water flow is 100 feet per second.

Source: Houston Fire Department Continuing Education

This problem appears as Exercise 65 in Section 7.6.

n this chapter, we learn about square roots, cube roots, fourth roots, and so on. These roots can be expressed in radical notation and appear in both radical expressions and radical equations. Exponents that are fractions are also studied and will ease some of our work with radicals. The chapter closes with an introduction to the complex-number system.

7.1   Radical Expressions and Functions

Square Roots and Square-Root Functions • Expressions of the Form $\sqrt{a^2}$ • Cube Roots • Odd and Even *n*th Roots

In this section, we consider roots, such as square roots and cube roots. We look at the symbolism that is used and ways in which symbols can be manipulated to get equivalent expressions. All of this will be important in problem solving.

Square Roots and Square-Root Functions

When a number is multiplied by itself, we say that the number is squared. Often we need to know what number was squared in order to produce some value *a*. If such a number can be found, we call that number a *square root* of *a*.

> **Square Root**
>
> The number *c* is a *square root* of *a* if $c^2 = a$.

For example,

9 has -3 and 3 as square roots because $(-3)^2 = 9$ and $3^2 = 9$.

25 has -5 and 5 as square roots because $(-5)^2 = 25$ and $5^2 = 25$.

-4 does not have a real-number square root because there is no real number *c* for which $c^2 = -4$.

Note that every positive number has two square roots, whereas 0 has only itself as a square root. Negative numbers do not have real-number square roots, although later in this chapter we introduce the *complex-number* system in which such square roots do exist.

EXAMPLE 1   Find the two square roots of 36.

SOLUTION   The square roots are 6 and -6, because $6^2 = 36$ and $(-6)^2 = 36$.

> TRY EXERCISE   9

Whenever we refer to *the* square root of a number, we mean the nonnegative square root of that number. This is often referred to as the *principal square root* of the number.

> **Principal Square Root**
>
> The *principal square root* of a nonnegative number is its nonnegative square root. The symbol $\sqrt{}$ is called a *radical sign* and is used to indicate the principal square root of the number over which it appears.

STUDENT NOTES

It is important to remember the difference between *the* square root of 9 and *a* square root of 9. *A* square root of 9 means either 3 or -3, whereas *the* square root of 9, denoted $\sqrt{9}$, means the principal square root of 9, or 3.

EXAMPLE 2   Simplify each of the following.

a) $\sqrt{25}$ **b)** $\sqrt{\dfrac{25}{64}}$ **c)** $-\sqrt{64}$ **d)** $\sqrt{0.0049}$

SOLUTION

a) $\sqrt{25} = 5$ $\sqrt{}$ indicates the principal square root. Note that $\sqrt{25} \neq -5$.

b) $\sqrt{\dfrac{25}{64}} = \dfrac{5}{8}$ Since $\left(\dfrac{5}{8}\right)^2 = \dfrac{25}{64}$

c) $-\sqrt{64} = -8$ Since $\sqrt{64} = 8, -\sqrt{64} = -8$.

d) $\sqrt{0.0049} = 0.07$ $(0.07)(0.07) = 0.0049$. Note too that
$$\sqrt{0.0049} = \sqrt{\dfrac{49}{10,000}} = \dfrac{7}{100}.$$

TRY EXERCISE 19

In addition to being read as "the principal square root of *a*," $\sqrt{a}$ is also read as "the square root of *a*," "root *a*," or "radical *a*." Any expression in which a radical sign appears is called a *radical expression*. The following are radical expressions:

$$\sqrt{5}, \quad \sqrt{a}, \quad -\sqrt{3x}, \quad \sqrt{\dfrac{y^2 + 7}{y}}, \quad \sqrt{x} + 8.$$

The expression under the radical sign is called the **radicand**. In the expressions above, the radicands are 5, *a*, 3*x*, $(y^2 + 7)/y$, and *x*, respectively.

Values for square roots found on calculators are, for the most part, approximations. For example, a calculator will show a number like

2.23606798

for $\sqrt{5}$. The exact value of $\sqrt{5}$ is not given by any repeating or terminating decimal. In general, for any whole number *a* that is not a perfect square, $\sqrt{a}$ is a non-terminating, nonrepeating decimal or an *irrational number*.

The square-root function, given by

$$f(x) = \sqrt{x},$$

has $[0, \infty)$ as its domain and $[0, \infty)$ as its range. We can draw its graph by selecting convenient values for *x* and calculating the corresponding outputs. Once these ordered pairs have been graphed, a smooth curve can be drawn.

$$f(x) = \sqrt{x}$$

x	$\sqrt{x}$	$(x, f(x))$
0	0	$(0, 0)$
1	1	$(1, 1)$
4	2	$(4, 2)$
9	3	$(9, 3)$

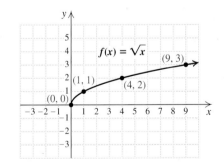

EXAMPLE 3   For each function, find the indicated function value.

a) $f(x) = \sqrt{3x - 2};\ f(1)$ **b)** $g(z) = -\sqrt{6z + 4};\ g(3)$

SOLUTION

a) $f(1) = \sqrt{3 \cdot 1 - 2}$ Substituting
     $= \sqrt{1} = 1$ Simplifying

b) $g(3) = -\sqrt{6 \cdot 3 + 4}$ Substituting

$\qquad\quad = -\sqrt{22}$ Simplifying. This answer is exact.

$\qquad\quad \approx -4.69041576$ Using a calculator to write an approximation

TRY EXERCISE ▶ 35

Expressions of the Form $\sqrt{a^2}$

As the next example shows, $\sqrt{a^2}$ does not always simplify to a.

EXAMPLE **4**   Evaluate $\sqrt{a^2}$ for the following values: **(a)** 5; **(b)** 0; **(c)** −5.

SOLUTION

a) $\sqrt{5^2} = \sqrt{25} = 5$

         └──────────┴────── Same

b) $\sqrt{0^2} = \sqrt{0} = 0$

         └──────────┴────── Same

c) $\sqrt{(-5)^2} = \sqrt{25} = 5$

         └──────────┴────── Opposites   Note that $\sqrt{(-5)^2} \neq -5$.

You may have noticed that evaluating $\sqrt{a^2}$ is just like evaluating $|a|$.

Simplifying $\sqrt{a^2}$

For any real number a,

$$\sqrt{a^2} = |a|.$$

(The principal square root of a^2 is the absolute value of a.)

When a radicand is the square of a variable expression, like $(x + 5)^2$ or $36t^2$, absolute-value signs are needed when simplifying. We use absolute-value signs unless we know that the expression being squared is nonnegative. This ensures that our result is never negative.

EXAMPLE **5**   Simplify each expression. Assume that the variable can represent any real number.

a) $\sqrt{(x + 1)^2}$ **b)** $\sqrt{x^2 - 8x + 16}$

c) $\sqrt{a^8}$ **d)** $\sqrt{t^6}$

SOLUTION

a) $\sqrt{(x + 1)^2} = |x + 1|$ Since $x + 1$ can be negative (for example, if $x = -3$), absolute-value notation is required.

b) $\sqrt{x^2 - 8x + 16} = \sqrt{(x - 4)^2} = |x - 4|$ Since $x - 4$ can be negative, absolute-value notation is required.

c) Note that $(a^4)^2 = a^8$ and that a^4 is never negative. Thus,

$$\sqrt{a^8} = a^4.$$ Absolute-value notation is unnecessary here.

TECHNOLOGY CONNECTION

To see the necessity of absolute-value signs, let y_1 represent the left side and y_2 the right side of each of the following equations. Then use a graph or table to determine whether these equations are true.

1. $\sqrt{x^2} \overset{?}{=} x$

2. $\sqrt{x^2} \overset{?}{=} |x|$

3. $x \overset{?}{=} |x|$

d) Note that $(t^3)^2 = t^6$. Thus,

$$\sqrt{t^6} = |t^3|. \quad \text{Since } t^3 \text{ can be negative, absolute-value notation is required.}$$

TRY EXERCISE 43

If we assume that the expression being squared is nonnegative, then absolute-value notation is not necessary.

EXAMPLE 6   Simplify each expression. Assume that no radicands were formed by squaring negative quantities.

a) $\sqrt{y^2}$ **b)** $\sqrt{a^{10}}$ **c)** $\sqrt{9x^2 - 6x + 1}$

SOLUTION

a) $\sqrt{y^2} = y$ We assume that y is nonnegative, so no absolute-value notation is necessary. When y is negative, $\sqrt{y^2} \neq y$.

b) $\sqrt{a^{10}} = a^5$ Assuming that a^5 is nonnegative. Note that $(a^5)^2 = a^{10}$.

c) $\sqrt{9x^2 - 6x + 1} = \sqrt{(3x - 1)^2} = 3x - 1$ Assuming that $3x - 1$ is nonnegative

TRY EXERCISE 69

Cube Roots

We often need to know what number cubed produces a certain value. When such a number is found, we say that we have found a *cube root*. For example,

2 is the cube root of 8 because $2^3 = 2 \cdot 2 \cdot 2 = 8$;

-4 is the cube root of -64 because $(-4)^3 = (-4)(-4)(-4) = -64$.

> ### Cube Root
>
> The number c is the *cube root* of a if $c^3 = a$. In symbols, we write $\sqrt[3]{a}$ to denote the cube root of a.

Each real number has only one real-number cube root. The cube-root function, given by

$$f(x) = \sqrt[3]{x},$$

has $\mathbb{R}$ as its domain and $\mathbb{R}$ as its range. To draw its graph, we select convenient values for x and calculate the corresponding outputs. Once these ordered pairs have been graphed, a smooth curve is drawn. Note that the cube root of a positive number is positive, and the cube root of a negative number is negative.

$$f(x) = \sqrt[3]{x}$$

x	$\sqrt[3]{x}$	$(x, f(x))$
0	0	$(0, 0)$
1	1	$(1, 1)$
8	2	$(8, 2)$
-1	-1	$(-1, -1)$
-8	-2	$(-8, -2)$

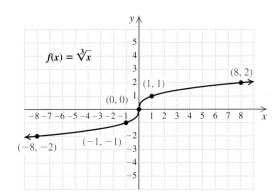

EXAMPLE **7** For each function, find the indicated function value.

a) $f(y) = \sqrt[3]{y}$; $f(125)$
b) $g(x) = \sqrt[3]{x - 1}$; $g(-26)$

SOLUTION

a) $f(125) = \sqrt[3]{125} = 5$ Since $5 \cdot 5 \cdot 5 = 125$

b) $g(-26) = \sqrt[3]{-26 - 1}$
$$= \sqrt[3]{-27}$$
$$= -3 \quad \text{Since } (-3)(-3)(-3) = -27$$

TRY EXERCISE 89

EXAMPLE **8** Simplify: $\sqrt[3]{-8y^3}$.

SOLUTION

$$\sqrt[3]{-8y^3} = -2y \quad \text{Since } (-2y)(-2y)(-2y) = -8y^3$$

TRY EXERCISE 83

Odd and Even *n*th Roots

The fourth root of a number a is the number c for which $c^4 = a$. There are also 5th roots, 6th roots, and so on. We write $\sqrt[n]{a}$ for the principal *n*th root. The number n is called the *index* (plural, *indices*). When the index is 2, we do not write it.

When the index n is odd, we are taking an *odd root*. Note that every number has exactly one real root when n is odd. Odd roots of positive numbers are positive and odd roots of negative numbers are negative. Absolute-value signs are not used when finding odd roots.

EXAMPLE **9** Simplify each expression.

a) $\sqrt[5]{32}$ b) $\sqrt[5]{-32}$ c) $-\sqrt[5]{32}$

d) $-\sqrt[5]{-32}$ e) $\sqrt[7]{x^7}$ f) $\sqrt[9]{(t - 1)^9}$

SOLUTION

a) $\sqrt[5]{32} = 2$ Since $2^5 = 32$

b) $\sqrt[5]{-32} = -2$ Since $(-2)^5 = -32$

c) $-\sqrt[5]{32} = -2$ Taking the opposite of $\sqrt[5]{32}$

d) $-\sqrt[5]{-32} = -(-2) = 2$ Taking the opposite of $\sqrt[5]{-32}$

e) $\sqrt[7]{x^7} = x$ No absolute-value signs are needed.

f) $\sqrt[9]{(t - 1)^9} = t - 1$

TRY EXERCISE 81

When the index n is even, we are taking an *even root*. Every positive real number has two real *n*th roots when n is even—one positive and one negative. For example, the fourth roots of 16 are -2 and 2. Negative numbers do not have real *n*th roots when n is even.

When n is even, the notation $\sqrt[n]{a}$ indicates the nonnegative *n*th root. Thus, when we simplify even *n*th roots, absolute-value signs are often required.

Compare the following.

Odd Root	Even Root
$\sqrt[3]{8} = 2$	$\sqrt[4]{16} = 2$
$\sqrt[3]{-8} = -2$	$\sqrt[4]{-16}$ is not a real number.
$\sqrt[3]{x^3} = x$	$\sqrt[4]{x^4} = \lvert x \rvert$

EXAMPLE 10 Simplify each expression, if possible. Assume that variables can represent any real number.

a) $\sqrt[4]{81}$ **b)** $-\sqrt[4]{81}$ **c)** $\sqrt[4]{-81}$
d) $\sqrt[4]{81x^4}$ **e)** $\sqrt[6]{(y+7)^6}$

SOLUTION

a) $\sqrt[4]{81} = 3$ Since $3^4 = 81$
b) $-\sqrt[4]{81} = -3$ Taking the opposite of $\sqrt[4]{81}$
c) $\sqrt[4]{-81}$ cannot be simplified. $\sqrt[4]{-81}$ is not a real number.
d) $\sqrt[4]{81x^4} = \lvert 3x \rvert$, or $3\lvert x \rvert$ Use absolute-value notation since x could represent a negative number.
e) $\sqrt[6]{(y+7)^6} = \lvert y + 7 \rvert$ Use absolute-value notation since $y + 7$ is negative for $y < -7$.

TRY EXERCISE 59

We summarize as follows.

Simplifying nth Roots

n	a	$\sqrt[n]{a}$	$\sqrt[n]{a^n}$
Even	Positive	Positive	$\lvert a \rvert$
	Negative	Not a real number	
Odd	Positive	Positive	a
	Negative	Negative	

EXAMPLE 11 Determine the domain of g if $g(x) = \sqrt[6]{7 - 3x}$.

SOLUTION Since the index is even, the radicand, $7 - 3x$, must be nonnegative. We solve the inequality:

$7 - 3x \geq 0$ We cannot find the 6th root of a negative number.
$-3x \geq -7$
$x \leq \frac{7}{3}$. Multiplying both sides by $-\frac{1}{3}$ and reversing the inequality

Thus,

Domain of $g = \left\{ x \mid x \leq \frac{7}{3} \right\}$
$= \left(-\infty, \frac{7}{3} \right]$.

TRY EXERCISE 95

TECHNOLOGY CONNECTION

To enter cube or higher roots on a graphing calculator, select options 4 or 5 of the **MATH** menu. The characters $6\sqrt[x]{}$ indicate the sixth root.

1. Use a **TABLE** or **GRAPH** and **TRACE** to check the solution of Example 11.

Visualizing for Success

A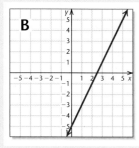

F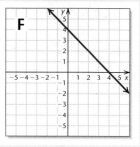

Match each function with its graph.

1. $f(x) = 2x - 5$

B

2. $f(x) = x^2 - 1$

G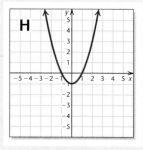

3. $f(x) = \sqrt{x}$

C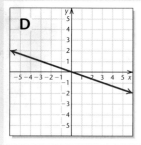

4. $f(x) = x - 2$

H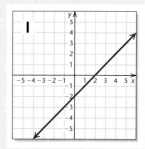

5. $f(x) = -\frac{1}{3}x$

D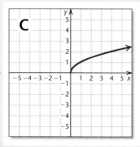

6. $f(x) = 2x$

7. $f(x) = 4 - x$

I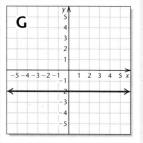

8. $f(x) = |2x - 5|$

9. $f(x) = -2$

E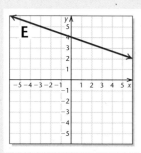

10. $f(x) = -\frac{1}{3}x + 4$

Answers on page A-30

J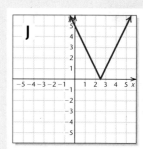

An additional, animated version of this activity appears in MyMathLab. To use MyMathLab, you need a course ID and a student access code. Consult your instructor for more information.

7.1 EXERCISE SET

🌢 *Concept Reinforcement* *Select the appropriate word to complete each of the following.*

1. Every positive number has _____ square root(s).
one/two

2. The principal square root is never _____.
negative/positive

3. For any _____ number a, we have
negative/positive
$\sqrt{a^2} = a$.

4. For any _____ number a, we have
negative/positive
$\sqrt{a^2} = -a$.

5. If a is a whole number that is not a perfect square, then $\sqrt{a}$ is a(n) _____ number.
irrational/rational

6. The domain of the function f given by $f(x) = \sqrt[3]{x}$ is all _____ numbers.
whole/real/positive

7. If $\sqrt[4]{x}$ is a real number, then x must be
_____.
negative/positive/nonnegative

8. If $\sqrt[3]{x}$ is negative, then x must be _____.
negative/positive

For each number, find all of its square roots.

9. 64

10. 81

11. 100

12. 121

13. 400

14. 2500

15. 625

16. 225

Simplify.

17. $\sqrt{49}$

18. $\sqrt{144}$

19. $-\sqrt{16}$

20. $-\sqrt{100}$

21. $\sqrt{\dfrac{36}{49}}$

22. $\sqrt{\dfrac{4}{9}}$

23. $-\sqrt{169}$

24. $-\sqrt{196}$

25. $-\sqrt{\dfrac{16}{81}}$

26. $-\sqrt{\dfrac{81}{144}}$

27. $\sqrt{0.04}$

28. $\sqrt{0.36}$

29. $\sqrt{0.0081}$

30. $\sqrt{0.0016}$

Identify the radicand and the index for each expression.

31. $5\sqrt{p^2 + 4}$

32. $-7\sqrt{y^2 - 8}$

33. $x^2 y^3 \sqrt[5]{\dfrac{x}{y + 4}}$

34. $\dfrac{a^2}{b} \sqrt[6]{a(a + b)}$

For each function, find the specified function value, if it exists.

35. $f(t) = \sqrt{5t - 10}$; $f(3), f(2), f(1), f(-1)$

36. $g(x) = \sqrt{x^2 - 25}$; $g(-6), g(3), g(6), g(13)$

37. $t(x) = -\sqrt{2x^2 - 1}$; $t(5), t(0), t(-1), t\left(-\frac{1}{2}\right)$

38. $p(z) = \sqrt{2z - 20}$; $p(4), p(10), p(12), p(0)$

39. $f(t) = \sqrt{t^2 + 1}$; $f(0), f(-1), f(-10)$

40. $g(x) = -\sqrt{(x + 1)^2}$; $g(-3), g(4), g(-5)$

Simplify. Remember to use absolute-value notation when necessary. If a root cannot be simplified, state this.

41. $\sqrt{100x^2}$

42. $\sqrt{16t^2}$

43. $\sqrt{(8 - t)^2}$

44. $\sqrt{(a + 3)^2}$

45. $\sqrt{y^2 + 16y + 64}$

46. $\sqrt{x^2 - 4x + 4}$

47. $\sqrt{4x^2 + 28x + 49}$

48. $\sqrt{9x^2 - 30x + 25}$

49. $-\sqrt[4]{256}$

50. $-\sqrt[4]{625}$

51. $\sqrt[3]{-1}$

52. $-\sqrt[3]{-1000}$

53. $-\sqrt[5]{-\dfrac{32}{243}}$

54. $\sqrt[5]{-\dfrac{1}{32}}$

55. $\sqrt[6]{x^6}$

56. $\sqrt[8]{y^8}$

57. $\sqrt[9]{t^9}$

58. $\sqrt[5]{a^5}$

59. $\sqrt[4]{(6a)^4}$

60. $\sqrt[4]{(7b)^4}$

61. $\sqrt[10]{(-6)^{10}}$

62. $\sqrt[12]{(-10)^{12}}$

63. $\sqrt[414]{(a + b)^{414}}$

64. $\sqrt[1976]{(2a + b)^{1976}}$

65. $\sqrt{a^{22}}$

66. $\sqrt{x^{10}}$

67. $\sqrt{-25}$

68. $\sqrt{-16}$

Simplify. Assume that no radicands were formed by raising negative quantities to even powers.

69. $\sqrt{16x^2}$

70. $\sqrt{25t^2}$

71. $-\sqrt{(3t)^2}$

72. $-\sqrt{(7c)^2}$

73. $\sqrt{(-5b)^2}$

74. $\sqrt{(-10a)^2}$

75. $\sqrt{a^2 + 2a + 1}$

76. $\sqrt{9 - 6y + y^2}$

77. $\sqrt[3]{27}$

78. $-\sqrt[3]{64}$

79. $\sqrt[4]{16x^4}$

80. $\sqrt[4]{81x^4}$

81. $\sqrt[5]{(x-1)^5}$

82. $-\sqrt[5]{(7y)^5}$

83. $-\sqrt[3]{-125y^3}$

84. $\sqrt[3]{-64x^3}$

85. $\sqrt{t^{18}}$

86. $\sqrt{a^{14}}$

87. $\sqrt{(x-2)^8}$

88. $\sqrt{(x+3)^{10}}$

For each function, find the specified function value, if it exists.

89. $f(x) = \sqrt[3]{x+1};\ f(7), f(26), f(-9), f(-65)$

90. $g(x) = -\sqrt[3]{2x-1};\ g(0), g(-62), g(-13), g(63)$

91. $g(t) = \sqrt[4]{t-3};\ g(19), g(-13), g(1), g(84)$

92. $f(t) = \sqrt[5]{t+1};\ f(0), f(15), f(-82), f(80)$

Determine the domain of each function described.

93. $f(x) = \sqrt{x-6}$

94. $g(x) = \sqrt{x+8}$

95. $g(t) = \sqrt[4]{t+8}$

96. $f(x) = \sqrt[4]{x-9}$

97. $g(x) = \sqrt[4]{10-2x}$

98. $g(t) = \sqrt[3]{2t-6}$

99. $f(t) = \sqrt[5]{2t+7}$

100. $f(t) = \sqrt[6]{4+3t}$

101. $h(z) = -\sqrt[6]{5z+2}$

102. $d(x) = -\sqrt[4]{5-7x}$

Aha! **103.** $f(t) = 7 + \sqrt[8]{t^8}$

104. $g(t) = 9 + \sqrt[6]{t^6}$

105. Explain how to write the negative square root of a number using radical notation.

106. Does the square root of a number's absolute value always exist? Why or why not?

Skill Review

To prepare for Section 7.2, review exponents (Section 1.6).

Simplify. Do not use negative exponents in your answer. [1.6]

107. $(a^2b)(a^4b)$

108. $(3xy^8)(5x^2y)$

109. $(5x^2y^{-3})^3$

110. $(2a^{-1}b^2c)^{-3}$

111. $\left(\dfrac{10x^{-1}y^5}{5x^2y^{-1}}\right)^{-1}$

112. $\left(\dfrac{8x^3y^{-2}}{2xz^4}\right)^{-2}$

Synthesis

113. Under what conditions does the nth root of x^3 exist? Explain your reasoning.

114. Under what conditions does the nth root of x^2 exist? Explain your reasoning.

115. *Biology.* The number of species S of plants in Guyana in an area of A hectares can be estimated using the formula

$$S = 88.63\sqrt[4]{A}.$$

The Kaieteur National Park in Guyana has an area of 63,000 hectares. How many species of plants are in the park?

Source: Hans ter Steege, "A Perspective on Guyana and its Plant Richness," as found on www.bio.uu.nl

116. *Spaces in a parking lot.* A parking lot has attendants to park the cars. The number N of stalls needed for waiting cars before attendants can get to them is given by the formula $N = 2.5\sqrt{A}$, where A is the number of arrivals in peak hours. Find the number of spaces needed for the given number of arrivals in peak hours: **(a)** 25; **(b)** 36; **(c)** 49; **(d)** 64.

Determine the domain of each function described. Then draw the graph of each function.

117. $f(x) = \sqrt{x+5}$

118. $g(x) = \sqrt{x} + 5$

119. $g(x) = \sqrt{x} - 2$

120. $f(x) = \sqrt{x-2}$

121. Find the domain of f if

$$f(x) = \frac{\sqrt{x+3}}{\sqrt[4]{2-x}}.$$

122. Find the domain of g if

$$g(x) = \frac{\sqrt[4]{5-x}}{\sqrt[6]{x+4}}.$$

123. Find the domain of F if $F(x) = \dfrac{x}{\sqrt{x^2-5x-6}}.$

124. Use a graphing calculator to check your answers to Exercises 41, 45, and 59. On some graphing calculators, a MATH key is needed to enter higher roots.

125. Use a graphing calculator to check your answers to Exercises 117 and 118. (See Exercise 124.)

7.2   Rational Numbers as Exponents

Rational Exponents ▪ Negative Rational Exponents ▪ Laws of Exponents ▪
Simplifying Radical Expressions

In Chapter 1, we first considered natural-number exponents and then integer exponents. We now expand the study of exponents further to include all rational numbers. This will give meaning to expressions like $7^{1/3}$ and $(2x)^{-4/5}$. Such notation will help us simplify certain radical expressions.

Rational Exponents

When defining rational exponents, we want the rules for exponents to hold for rational exponents just as they do for integer exponents. In particular, we still want to add exponents when multiplying.

If $a^{1/2} \cdot a^{1/2} = a^{1/2+1/2} = a^1$,   then $a^{1/2}$ should mean $\sqrt{a}$.

If $a^{1/3} \cdot a^{1/3} \cdot a^{1/3} = a^{1/3+1/3+1/3} = a^1$, then $a^{1/3}$ should mean $\sqrt[3]{a}$.

$$a^{1/n} = \sqrt[n]{a}$$

$a^{1/n}$ means $\sqrt[n]{a}$. When a is nonnegative, n can be any natural number greater than 1. When a is negative, n can be any odd natural number greater than 1.

Thus, $a^{1/5} = \sqrt[5]{a}$ and $a^{1/10} = \sqrt[10]{a}$. Note that the denominator of the exponent becomes the index and the base becomes the radicand.

EXAMPLE 1   Write an equivalent expression using radical notation and, if possible, simplify.

a) $16^{1/2}$ **b)** $(-8)^{1/3}$ **c)** $(abc)^{1/5}$ **d)** $(25x^{16})^{1/2}$

SOLUTION

a) $16^{1/2} = \sqrt{16} = 4$ ⎫   The denominator of the exponent becomes
b) $(-8)^{1/3} = \sqrt[3]{-8} = -2$ ⎬   the index. The base becomes the radicand.
c) $(abc)^{1/5} = \sqrt[5]{abc}$ ⎭   Recall that for square roots, the index 2 is understood without being written.

d) $(25x^{16})^{1/2} = 25^{1/2}x^8 = \sqrt{25} \cdot x^8 = 5x^8$ **TRY EXERCISE** 11

EXAMPLE 2   Write an equivalent expression using exponential notation.

a) $\sqrt[5]{9ab}$ **b)** $\sqrt[7]{\dfrac{x^3y}{4}}$ **c)** $\sqrt{5x}$

SOLUTION   Parentheses are required to indicate the base.

a) $\sqrt[5]{9ab} = (9ab)^{1/5}$ ⎫
                                          The index becomes the denominator of the
b) $\sqrt[7]{\dfrac{x^3y}{4}} = \left(\dfrac{x^3y}{4}\right)^{1/7}$ ⎬   exponent. The radicand becomes the base.

c) $\sqrt{5x} = (5x)^{1/2}$ The index 2 is understood without being written.
                                  We assume $x \geq 0$.   **TRY EXERCISE** 31

How shall we define $a^{2/3}$? If the property for multiplying exponents is to hold, we must have $a^{2/3} = (a^{1/3})^2$ and $a^{2/3} = (a^2)^{1/3}$. This would suggest that $a^{2/3} = (\sqrt[3]{a})^2$ and $a^{2/3} = \sqrt[3]{a^2}$. We make our definition accordingly.

Positive Rational Exponents

For any natural numbers m and n ($n \neq 1$) and any real number a for which $\sqrt[n]{a}$ exists,

$$a^{m/n} \quad \text{means} \quad (\sqrt[n]{a})^m, \quad \text{or} \quad \sqrt[n]{a^m}.$$

EXAMPLE **3**   Write an equivalent expression using radical notation and simplify.

a) $27^{2/3}$ **b)** $25^{3/2}$

SOLUTION

a) $27^{2/3}$ means $(\sqrt[3]{27})^2$ or, equivalently, $\sqrt[3]{27^2}$. Let's see which is easier to simplify:

$$(\sqrt[3]{27})^2 = 3^2 \qquad \sqrt[3]{27^2} = \sqrt[3]{729}$$
$$= 9; \qquad\qquad\qquad = 9.$$

The simplification on the left is probably easier for most people.

b) $25^{3/2}$ means $(\sqrt[2]{25})^3$ or, equivalently, $\sqrt[2]{25^3}$ (the index 2 is normally omitted). Since $\sqrt{25}$ is more commonly known than $\sqrt{25^3}$, we use that form:

$$25^{3/2} = (\sqrt{25})^3 = 5^3 = 125.$$

TRY EXERCISE 23

STUDENT NOTES

It is important to remember both meanings of $a^{m/n}$. When the root of the base a is known, $(\sqrt[n]{a})^m$ is generally easier to work with. When it is not known, $\sqrt[n]{a^m}$ is often more convenient.

EXAMPLE **4**   Write an equivalent expression using exponential notation.

a) $\sqrt[3]{9^4}$ **b)** $(\sqrt[4]{7xy})^5$

SOLUTION

a) $\sqrt[3]{9^4} = 9^{4/3}$

b) $(\sqrt[4]{7xy})^5 = (7xy)^{5/4}$ The index becomes the denominator of the fraction that is the exponent.

TRY EXERCISE 37

Negative Rational Exponents

Recall from Section 1.6 that $x^{-2} = 1/x^2$. Negative rational exponents behave similarly.

Negative Rational Exponents

For any rational number m/n and any nonzero real number a for which $a^{m/n}$ exists,

$$a^{-m/n} \quad \text{means} \quad \frac{1}{a^{m/n}}.$$

CAUTION! A negative exponent does not indicate that the expression in which it appears is negative: $a^{-1} \neq -a$.

TECHNOLOGY CONNECTION

To approximate $7^{2/3}$, we enter 7 ⌃ (2/3).

1. Why are the parentheses needed above?
2. Compare the graphs of $y_1 = x^{1/2}$, $y_2 = x$, and $y_3 = x^{3/2}$ and determine those x-values for which $y_1 > y_3$.

EXAMPLE   5   Write an equivalent expression with positive exponents and, if possible, simplify.

a) $9^{-1/2}$ **b)** $(5xy)^{-4/5}$ **c)** $64^{-2/3}$

d) $4x^{-2/3}y^{1/5}$ **e)** $\left(\dfrac{3r}{7s}\right)^{-5/2}$

SOLUTION

a) $9^{-1/2} = \dfrac{1}{9^{1/2}}$ $9^{-1/2}$ is the reciprocal of $9^{1/2}$.

Since $9^{1/2} = \sqrt{9} = 3$, the answer simplifies to $\dfrac{1}{3}$.

b) $(5xy)^{-4/5} = \dfrac{1}{(5xy)^{4/5}}$ $(5xy)^{-4/5}$ is the reciprocal of $(5xy)^{4/5}$.

c) $64^{-2/3} = \dfrac{1}{64^{2/3}}$ $64^{-2/3}$ is the reciprocal of $64^{2/3}$.

Since $64^{2/3} = \left(\sqrt[3]{64}\right)^2 = 4^2 = 16$, the answer simplifies to $\dfrac{1}{16}$.

d) $4x^{-2/3}y^{1/5} = 4 \cdot \dfrac{1}{x^{2/3}} \cdot y^{1/5} = \dfrac{4y^{1/5}}{x^{2/3}}$

e) In Section 1.6, we found that $(a/b)^{-n} = (b/a)^n$. This property holds for *any* negative exponent:

$$\left(\frac{3r}{7s}\right)^{-5/2} = \left(\frac{7s}{3r}\right)^{5/2}.$$ Writing the reciprocal of the base and changing the sign of the exponent

TRY EXERCISE 53

Laws of Exponents

The same laws hold for rational exponents as for integer exponents.

> ### Laws of Exponents
>
> For any real numbers a and b and any rational exponents m and n for which a^m, a^n, and b^m are defined:
>
> **1.** $a^m \cdot a^n = a^{m+n}$ When multiplying, add exponents if the bases are the same.
>
> **2.** $\dfrac{a^m}{a^n} = a^{m-n}$ When dividing, subtract exponents if the bases are the same. (Assume $a \neq 0$.)
>
> **3.** $(a^m)^n = a^{m \cdot n}$ To raise a power to a power, multiply the exponents.
>
> **4.** $(ab)^m = a^m b^m$ To raise a product to a power, raise each factor to the power and multiply.

EXAMPLE   6   Use the laws of exponents to simplify.

a) $3^{1/5} \cdot 3^{3/5}$ **b)** $\dfrac{a^{1/4}}{a^{1/2}}$

c) $(7.2^{2/3})^{3/4}$ **d)** $(a^{-1/3}b^{2/5})^{1/2}$

SOLUTION

a) $3^{1/5} \cdot 3^{3/5} = 3^{1/5+3/5} = 3^{4/5}$ Adding exponents

b) $\dfrac{a^{1/4}}{a^{1/2}} = a^{1/4-1/2} = a^{1/4-2/4}$ Subtracting exponents after finding a common denominator

$= a^{-1/4}$, or $\dfrac{1}{a^{1/4}}$ $a^{-1/4}$ is the reciprocal of $a^{1/4}$.

c) $(7.2^{2/3})^{3/4} = 7.2^{(2/3)(3/4)} = 7.2^{6/12}$ Multiplying exponents

$= 7.2^{1/2}$ Using arithmetic to simplify the exponent

d) $(a^{-1/3}b^{2/5})^{1/2} = a^{(-1/3)(1/2)} \cdot b^{(2/5)(1/2)}$ Raising a product to a power and multiplying exponents

$= a^{-1/6}b^{1/5}$, or $\dfrac{b^{1/5}}{a^{1/6}}$ **TRY EXERCISE** 69

Simplifying Radical Expressions

Many radical expressions contain radicands or factors of radicands that are powers. When these powers and the index share a common factor, rational exponents can be used to simplify the expression.

To Simplify Radical Expressions
1. Convert radical expressions to exponential expressions.
2. Use arithmetic and the laws of exponents to simplify.
3. Convert back to radical notation as needed.

EXAMPLE 7 Use rational exponents to simplify. Do not use exponents that are fractions in the final answer.

a) $\sqrt[6]{(5x)^3}$ **b)** $\sqrt[5]{t^{20}}$

c) $\left(\sqrt[3]{ab^2c}\right)^{12}$ **d)** $\sqrt{\sqrt[3]{x}}$

SOLUTION

a) $\sqrt[6]{(5x)^3} = (5x)^{3/6}$ Converting to exponential notation

$= (5x)^{1/2}$ Simplifying the exponent

$= \sqrt{5x}$ Returning to radical notation

b) $\sqrt[5]{t^{20}} = t^{20/5}$ Converting to exponential notation

$= t^4$ Simplifying the exponent

c) $\left(\sqrt[3]{ab^2c}\right)^{12} = (ab^2c)^{12/3}$ Converting to exponential notation

$= (ab^2c)^4$ Simplifying the exponent

$= a^4b^8c^4$ Using the laws of exponents

d) $\sqrt{\sqrt[3]{x}} = \sqrt{x^{1/3}}$ Converting the radicand to exponential notation

$= (x^{1/3})^{1/2}$ Try to go directly to this step.

$= x^{1/6}$ Using the laws of exponents

$= \sqrt[6]{x}$ Returning to radical notation

TRY EXERCISE 87

| **7.2** | **EXERCISE SET** |

Concept Reinforcement   *In each of Exercises 1–8, match the expression with the equivalent expression from the column on the right.*

1. ___ $x^{2/5}$ a) $x^{3/5}$

2. ___ $x^{5/2}$ b) $\left(\sqrt[5]{x}\right)^4$

3. ___ $x^{-5/2}$ c) $\sqrt{x^5}$

4. ___ $x^{-2/5}$ d) $x^{1/2}$

5. ___ $x^{1/5} \cdot x^{2/5}$

6. ___ $(x^{1/5})^{5/2}$ e) $\dfrac{1}{\left(\sqrt{x}\right)^5}$

7. ___ $\sqrt[5]{x^4}$ f) $\sqrt[4]{x^5}$

8. ___ $\left(\sqrt[4]{x}\right)^5$ g) $\sqrt[5]{x^2}$

        h) $\dfrac{1}{\left(\sqrt[5]{x}\right)^2}$

Note: Assume for all exercises that all variables are nonnegative and that all denominators are nonzero.

Write an equivalent expression using radical notation and, if possible, simplify.

9. $y^{1/3}$ 10. $t^{1/4}$

11. $36^{1/2}$ 12. $125^{1/3}$

13. $32^{1/5}$ 14. $81^{1/4}$

15. $64^{1/2}$ 16. $100^{1/2}$

17. $(xyz)^{1/2}$ 18. $(ab)^{1/4}$

19. $(a^2b^2)^{1/5}$ 20. $(x^3y^3)^{1/4}$

21. $t^{5/6}$ 22. $a^{3/2}$

23. $16^{3/4}$ 24. $4^{7/2}$

25. $125^{4/3}$ 26. $9^{5/2}$

27. $(81x)^{3/4}$ 28. $(125a)^{2/3}$

29. $(25x^4)^{3/2}$ 30. $(9y^6)^{3/2}$

Write an equivalent expression using exponential notation.

31. $\sqrt[3]{18}$ 32. $\sqrt[4]{10}$

33. $\sqrt{30}$ 34. $\sqrt{22}$

35. $\sqrt{x^7}$ 36. $\sqrt{a^3}$

37. $\sqrt[5]{m^2}$ 38. $\sqrt[5]{n^4}$

39. $\sqrt[4]{pq}$ 40. $\sqrt[3]{cd}$

41. $\sqrt[5]{xy^2z}$ 42. $\sqrt[7]{x^3y^2z^2}$

43. $\left(\sqrt{3mn}\right)^3$ 44. $\left(\sqrt[3]{7xy}\right)^4$

45. $\left(\sqrt[7]{8x^2y}\right)^5$ 46. $\left(\sqrt[6]{2a^5b}\right)^7$

47. $\dfrac{2x}{\sqrt[3]{z^2}}$ 48. $\dfrac{3a}{\sqrt[5]{c^2}}$

Write an equivalent expression with positive exponents and, if possible, simplify.

49. $a^{-1/4}$ 50. $m^{-1/3}$

51. $(2rs)^{-3/4}$ 52. $(5xy)^{-5/6}$

53. $\left(\dfrac{1}{16}\right)^{-3/4}$ 54. $\left(\dfrac{1}{8}\right)^{-2/3}$

55. $\dfrac{8c}{a^{-3/5}}$ 56. $\dfrac{3b}{a^{-5/7}}$

57. $2a^{3/4}b^{-1/2}c^{2/3}$ 58. $5x^{-2/3}y^{4/5}z$

59. $3^{-5/2}a^3b^{-7/3}$ 60. $2^{-1/3}x^4y^{-2/7}$

61. $\left(\dfrac{2ab}{3c}\right)^{-5/6}$ 62. $\left(\dfrac{7x}{8yz}\right)^{-3/5}$

63. $\dfrac{6a}{\sqrt[4]{b}}$ 64. $\dfrac{5y}{\sqrt[3]{z}}$

Use the laws of exponents to simplify. Do not use negative exponents in any answers.

65. $11^{1/2} \cdot 11^{1/3}$ 66. $5^{1/4} \cdot 5^{1/8}$

67. $\dfrac{3^{5/8}}{3^{-1/8}}$ 68. $\dfrac{8^{7/11}}{8^{-2/11}}$

69. $\dfrac{4.3^{-1/5}}{4.3^{-7/10}}$ 70. $\dfrac{2.7^{-11/12}}{2.7^{-1/6}}$

71. $(10^{3/5})^{2/5}$ 72. $(5^{5/4})^{3/7}$

73. $a^{2/3} \cdot a^{5/4}$ 74. $x^{3/4} \cdot x^{1/3}$

Aha! 75. $(64^{3/4})^{4/3}$ 76. $(27^{-2/3})^{3/2}$

77. $(m^{2/3}n^{-1/4})^{1/2}$ 78. $(x^{-1/3}y^{2/5})^{1/4}$

Use rational exponents to simplify. Do not use fraction exponents in the final answer.

79. $\sqrt[9]{x^3}$ 80. $\sqrt[12]{a^3}$

81. $\sqrt[3]{y^{15}}$ 82. $\sqrt[4]{y^{40}}$

83. $\sqrt[12]{a^6}$ 84. $\sqrt[30]{x^5}$

85. $\left(\sqrt[7]{xy}\right)^{14}$ 86. $\left(\sqrt[3]{ab}\right)^{15}$

87. $\sqrt[4]{(7a)^2}$

88. $\sqrt[8]{(3x)^2}$

89. $\left(\sqrt[8]{2x}\right)^6$

90. $\left(\sqrt[10]{3a}\right)^5$

91. $\sqrt{\sqrt[5]{m}}$

92. $\sqrt[6]{\sqrt{n}}$

93. $\sqrt[4]{(xy)^{12}}$

94. $\sqrt{(ab)^6}$

95. $\left(\sqrt[5]{a^2b^4}\right)^{15}$

96. $\left(\sqrt[3]{x^2y^5}\right)^{12}$

97. $\sqrt[3]{\sqrt[4]{xy}}$

98. $\sqrt[5]{\sqrt[3]{2a}}$

99. If $f(x) = (x + 5)^{1/2}(x + 7)^{-1/2}$, find the domain of f. Explain how you found your answer.

100. Let $f(x) = 5x^{-1/3}$. Under what condition will we have $f(x) > 0$? Why?

Skill Review

To prepare for Section 7.3, review multiplying and factoring polynomials (Sections 5.2–5.6).

Multiply. [5.2]

101. $(x + 5)(x - 5)$

102. $(x - 2)(x^2 + 2x + 4)$

Factor. [5.5]

103. $4x^2 + 20x + 25$

104. $9a^2 - 24a + 16$

105. $5t^2 - 10t + 5$

106. $3n^2 + 12n + 12$

Synthesis

107. Explain why $\sqrt[3]{x^6} = x^2$ for any value of x, whereas $\sqrt[2]{x^6} = x^3$ only when $x \geq 0$.

108. If $g(x) = x^{3/n}$, in what way does the domain of g depend on whether n is odd or even?

Use rational exponents to simplify.

109. $\sqrt{x\sqrt[3]{x^2}}$

110. $\sqrt[4]{\sqrt[3]{8x^3y^6}}$

111. $\sqrt[14]{c^2 - 2cd + d^2}$

Music.   The function given by $f(x) = k2^{x/12}$ can be used to determine the frequency, in cycles per second, of a musical note that is x half-steps above a note with frequency k.*

112. The frequency of concert A for a trumpet is 440 cycles per second. Find the frequency of the A that is two octaves (24 half-steps) above concert A (few trumpeters can reach this note.)

---

*This application was inspired by information provided by Dr. Homer B. Tilton of Pima Community College East.

113. Show that the G that is 7 half-steps (a "perfect fifth") above middle C (262 cycles per second) has a frequency that is about 1.5 times that of middle C.

114. Show that the C sharp that is 4 half-steps (a "major third") above concert A (see Exercise 112) has a frequency that is about 25% greater than that of concert A.

115. *Road pavement messages.*   In a psychological study, it was determined that the proper length L of the letters of a word printed on pavement is given by

$$L = \frac{0.000169d^{2.27}}{h},$$

where d is the distance of a car from the lettering and h is the height of the eye above the surface of the road. All units are in meters. This formula says that from a vantage point h meters above the surface of the road, if a driver is to be able to recognize a message d meters away, that message will be the most recognizable if the length of the letters is L. Find L to the nearest tenth of a meter, given d and h.

a) $h = 1$ m, $d = 60$ m
b) $h = 0.9906$ m, $d = 75$ m
c) $h = 2.4$ m, $d = 80$ m
d) $h = 1.1$ m, $d = 100$ m

116. *Baseball.*   The statistician Bill James has found that a baseball team's winning percentage P can be approximated by

$$P = \frac{r^{1.83}}{r^{1.83} + \sigma^{1.83}},$$

where r is the total number of runs scored by that team and σ (sigma) is the total number of runs scored by their opponents. During a recent season, the San Francisco Giants scored 799 runs and their opponents scored 749 runs. Use James's formula to predict the Giants' winning percentage (the team actually won 55.6% of their games).

Source: M. Bittinger, *One Man's Journey Through Mathematics.* Boston: Addison–Wesley, 2004

117. *Forestry.*   The total wood volume T, in cubic feet, in a California black oak can be estimated using the formula

$$T = 0.936\, d^{1.97} h^{0.85},$$

where d is the diameter of the tree at breast height and h is the total height of the tree. How much wood is in a California black oak that is 3 ft in diameter at breast height and 80 ft high?

Source: Norman H. Pillsbury and Michael L. Kirkley, 1984. Equations for total, wood, and saw-log volume for thirteen California hardwoods, USDA Forest Service PNW Research Note No. 414: 52 p.

118. *Physics.*   The equation $m = m_0(1 - v^2 c^{-2})^{-1/2}$, developed by Albert Einstein, is used to determine the mass m of an object that is moving v meters per second and has mass m_0 before the motion begins. The constant c is the speed of light, approximately 3×10^8 m/sec. Suppose that a particle with mass 8 mg is accelerated to a speed of $\frac{9}{5} \times 10^8$ m/sec. Without using a calculator, find the new mass of the particle.

119. Using a graphing calculator, select **MODE** SIMUL and the FORMAT EXPROFF. Then graph

$$y_1 = x^{1/2}, \qquad y_2 = 3x^{2/5},$$
$$y_3 = x^{4/7}, \quad \text{and} \quad y_4 = \tfrac{1}{5} x^{3/4}.$$

Looking only at coordinates, match each graph with its equation.

COLLABORATIVE CORNER

Are Equivalent Fractions Equivalent Exponents?

Focus: Functions and rational exponents

Time: 10–20 minutes

Group size: 3

Materials: Graph paper

In arithmetic, we have seen that $\frac{1}{3}, \frac{1}{6} \cdot 2$, and $2 \cdot \frac{1}{6}$ all represent the same number. Interestingly,

$$f(x) = x^{1/3},$$
$$g(x) = (x^{1/6})^2, \quad \text{and}$$
$$h(x) = (x^2)^{1/6}$$

represent three *different* functions.

ACTIVITY

1. Selecting a variety of values for x and using the definition of positive rational exponents, one group member should graph f, a second group member should graph g, and a third group member should graph h. Be sure to check whether negative x-values are in the domain of the function.

2. Compare the three graphs and check each other's work. How and why do the graphs differ?

3. Decide as a group which graph, if any, would best represent the graph of $k(x) = x^{2/6}$. Then be prepared to explain your reasoning to the entire class. (*Hint*: Study the definition of $a^{m/n}$ on p. 440 carefully.)

7.3   Multiplying Radical Expressions

Multiplying Radical Expressions   •   Simplifying by Factoring   •   Multiplying and Simplifying

Multiplying Radical Expressions

Note that $\sqrt{4}\sqrt{25} = 2 \cdot 5 = 10$. Also $\sqrt{4 \cdot 25} = \sqrt{100} = 10$. Likewise,

$$\sqrt[3]{27}\,\sqrt[3]{8} = 3 \cdot 2 = 6 \quad \text{and} \quad \sqrt[3]{27 \cdot 8} = \sqrt[3]{216} = 6.$$

These examples suggest the following.

The Product Rule for Radicals

For any real numbers $\sqrt[n]{a}$ and $\sqrt[n]{b}$,

$$\sqrt[n]{a} \cdot \sqrt[n]{b} = \sqrt[n]{a \cdot b}.$$

(The product of two nth roots is the nth root of the product of the two radicands.)

Rational exponents can be used to derive this rule:

$$\sqrt[n]{a} \cdot \sqrt[n]{b} = a^{1/n} \cdot b^{1/n} = (a \cdot b)^{1/n} = \sqrt[n]{a \cdot b}.$$

EXAMPLE   **1**   Multiply.

a) $\sqrt{2} \cdot \sqrt{7}$

b) $\sqrt{x + 3}\,\sqrt{x - 3}$

c) $\sqrt[3]{4} \cdot \sqrt[3]{5}$

d) $\sqrt[4]{\dfrac{y}{5}} \cdot \sqrt[4]{\dfrac{7}{x}}$

SOLUTION

a) When no index is written, roots are understood to be square roots with an unwritten index of two. We apply the product rule:

$$\sqrt{2} \cdot \sqrt{7} = \sqrt{2 \cdot 7}$$
$$= \sqrt{14}.$$

b) $\sqrt{x + 3}\,\sqrt{x - 3} = \sqrt{(x + 3)(x - 3)}$ The product of two square roots is
$$= \sqrt{x^2 - 9}$$ the square root of the product.

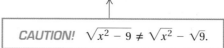

CAUTION! $\sqrt{x^2 - 9} \neq \sqrt{x^2} - \sqrt{9}$.

c) Both $\sqrt[3]{4}$ and $\sqrt[3]{5}$ have indices of three, so to multiply we can use the product rule:

$$\sqrt[3]{4} \cdot \sqrt[3]{5} = \sqrt[3]{4 \cdot 5} = \sqrt[3]{20}.$$

d) $\sqrt[4]{\dfrac{y}{5}} \cdot \sqrt[4]{\dfrac{7}{x}} = \sqrt[4]{\dfrac{y}{5} \cdot \dfrac{7}{x}} = \sqrt[4]{\dfrac{7y}{5x}}$ In Section 7.4, we discuss other ways to write answers like this.

TRY EXERCISE   7

To check Example 1(b), let $y_1 = \sqrt{x+3}\sqrt{x-3}$ and $y_2 = \sqrt{x^2-9}$ and compare:

$y_1 = \sqrt{(x+3)}\sqrt{(x-3)}$

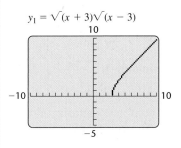

$y_2 = \sqrt{(x^2-9)}$

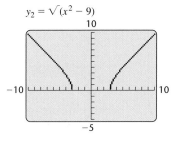

Because $y_1 = y_2$ for all x-values that can be used in *both* y_1 and y_2, Example 1(b) *is* correct.

1. Why do the graphs above differ in appearance? (*Hint*: What are the domains of the two related functions?)

CAUTION!   The product rule for radicals applies only when radicals have the same index:

$$\sqrt[n]{a} \cdot \sqrt[m]{b} \neq \sqrt[nm]{a \cdot b}.$$

Simplifying by Factoring

The number p is a *perfect square* if there exists a rational number q for which $q^2 = p$. We say that p is a *perfect cube* if $q^3 = p$ for some rational number q. In general, p is a *perfect nth power* if $q^n = p$ for some rational number q. Thus, 16 and $\frac{1}{10,000}$ are both perfect 4th powers since $2^4 = 16$ and $\left(\frac{1}{10}\right)^4 = \frac{1}{10,000}$.

The product rule allows us to simplify $\sqrt[n]{ab}$ whenever ab contains a factor that is a perfect nth power.

> **Using the Product Rule to Simplify**
> $$\sqrt[n]{ab} = \sqrt[n]{a} \cdot \sqrt[n]{b}.$$
> $\left(\sqrt[n]{a} \text{ and } \sqrt[n]{b} \text{ must both be real numbers.}\right)$

To illustrate, suppose we wish to simplify $\sqrt{20}$. Since this is a *square* root, we check to see if there is a factor of 20 that is a perfect square. There is one, 4, so we express 20 as $4 \cdot 5$ and use the product rule:

$\sqrt{20} = \sqrt{4 \cdot 5}$ Factoring the radicand (4 is a perfect square)

$\quad\quad = \sqrt{4} \cdot \sqrt{5}$ Factoring into two radicals

$\quad\quad = 2\sqrt{5}.$ Finding the square root of 4

> **To Simplify a Radical Expression with Index *n* by Factoring**
> 1. Express the radicand as a product in which one factor is the largest perfect *n*th power possible.
> 2. Rewrite the expression as the *n*th root of each factor.
> 3. Simplify the expression containing the perfect *n*th power.
> 4. Simplification is complete when no radicand has a factor that is a perfect *n*th power.

It is often safe to assume that a radicand does not represent a negative number raised to an even power. We will henceforth make this assumption—unless functions are involved—and discontinue use of absolute-value notation when taking even roots.

EXAMPLE   2

Simplify by factoring: **(a)** $\sqrt{200}$; **(b)** $\sqrt{18x^2y}$; **(c)** $\sqrt[3]{-72}$; **(d)** $\sqrt[4]{162x^6}$.

SOLUTION

a) $\sqrt{200} = \sqrt{100 \cdot 2}$ 100 is the largest perfect-square factor of 200.

$\quad\quad\quad = \sqrt{100} \cdot \sqrt{2} = 10\sqrt{2}$

Express the radicand as a product.   **b)** $\sqrt{18x^2y} = \sqrt{9 \cdot 2 \cdot x^2 \cdot y}$ $9x^2$ is the largest perfect-square factor of $18x^2y$.

Rewrite as the nth root of each factor.

$$= \sqrt{9x^2} \cdot \sqrt{2y}$$ Factoring into two radicals

Simplify. $$= 3x\sqrt{2y}$$ Taking the square root of $9x^2$

c) $\sqrt[3]{-72} = \sqrt[3]{-8 \cdot 9}$ -8 is a perfect-cube (third-power) factor of -72.

$$= \sqrt[3]{-8} \cdot \sqrt[3]{9} = -2\sqrt[3]{9}$$

d) $\sqrt[4]{162x^6} = \sqrt[4]{81 \cdot 2 \cdot x^4 \cdot x^2}$ $81 \cdot x^4$ is the largest perfect fourth-power factor of $162x^6$.

$$= \sqrt[4]{81x^4} \cdot \sqrt[4]{2x^2}$$ Factoring into two radicals

$$= 3x\sqrt[4]{2x^2}$$ Taking the fourth root of $81x^4$

Let's look at this example another way. We write a complete factorization and look for quadruples of factors. Each quadruple makes a perfect fourth power:

$$\sqrt[4]{162x^6} = \sqrt[4]{\boxed{3 \cdot 3 \cdot 3 \cdot 3} \cdot 2 \cdot \boxed{x \cdot x \cdot x \cdot x} \cdot x \cdot x}$$ $3 \cdot 3 \cdot 3 \cdot 3 = 3^4$ and $x \cdot x \cdot x \cdot x = x^4$

$$= 3 \cdot x \cdot \sqrt[4]{2 \cdot x \cdot x}$$

$$= 3x\sqrt[4]{2x^2}.$$ ▸ TRY EXERCISE 31

EXAMPLE   3   If $f(x) = \sqrt{3x^2 - 6x + 3}$, find a simplified form for $f(x)$. Because we are working with a function, assume that x can be any real number.

SOLUTION

$$f(x) = \sqrt{3x^2 - 6x + 3}$$

$$= \sqrt{3(x^2 - 2x + 1)}$$ Factoring the radicand; $x^2 - 2x + 1$ is a perfect square.

$$= \sqrt{(x-1)^2 \cdot 3}$$

$$= \sqrt{(x-1)^2} \cdot \sqrt{3}$$ Factoring into two radicals

$$= |x - 1|\sqrt{3}$$ Finding the square root of $(x-1)^2$

▸ TRY EXERCISE 43

TECHNOLOGY CONNECTION

To check Example 3, let $y_1 = \sqrt{(3x^2 - 6x + 3)}$, $y_2 = \text{abs}(x - 1)\sqrt{(3)}$, and $y_3 = (x - 1)\sqrt{(3)}$. Do the graphs all coincide? Why or why not?

EXAMPLE   4   Simplify: **(a)** $\sqrt{x^7y^{11}z^9}$; **(b)** $\sqrt[3]{16a^7b^{14}}$.

SOLUTION

a) There are many ways to factor $x^7y^{11}z^9$. Because of the square root (index of 2), we identify the largest exponents that are multiples of 2:

$$\sqrt{x^7y^{11}z^9} = \sqrt{x^6 \cdot x \cdot y^{10} \cdot y \cdot z^8 \cdot z}$$ The largest perfect-square factor is $x^6y^{10}z^8$.

$$= \sqrt{x^6y^{10}z^8}\,\sqrt{xyz}$$ Factoring into two radicals

$$= x^{6/2}y^{10/2}z^{8/2}\sqrt{xyz}$$ Converting to rational exponents. Try to do this mentally.

$$= x^3y^5z^4\sqrt{xyz}.$$ Simplifying

Check: $\left(x^3y^5z^4\sqrt{xyz}\right)^2 = (x^3)^2(y^5)^2(z^4)^2\left(\sqrt{xyz}\right)^2$

$$= x^6 \cdot y^{10} \cdot z^8 \cdot xyz = x^7y^{11}z^9.$$

Our check shows that $x^3y^5z^4\sqrt{xyz}$ is the square root of $x^7y^{11}z^9$.

b) There are many ways to factor $16a^7b^{14}$. Because of the cube root (index of 3), we identify factors with the largest exponents that are multiples of 3:

$$\sqrt[3]{16a^7b^{14}} = \sqrt[3]{8 \cdot 2 \cdot a^6 \cdot a \cdot b^{12} \cdot b^2} \qquad \text{The largest perfect-cube factor is } 8a^6b^{12}.$$

$$= \sqrt[3]{8a^6b^{12}}\sqrt[3]{2ab^2} \qquad \text{Rewriting as a product of cube roots}$$

$$= 2a^2b^4\sqrt[3]{2ab^2}. \qquad \text{Simplifying the expression containing the perfect cube}$$

As a check, let's redo the problem using a complete factorization of the radicand:

$$\sqrt[3]{16a^7b^{14}} = \sqrt[3]{\boxed{2 \cdot 2 \cdot 2} \cdot 2 \cdot \boxed{a \cdot a \cdot a} \cdot \boxed{a \cdot a \cdot a} \cdot a \cdot \boxed{b \cdot b \cdot b} \cdot \boxed{b \cdot b \cdot b} \cdot \boxed{b \cdot b \cdot b} \cdot \boxed{b \cdot b \cdot b} \cdot b \cdot b}$$

Each triple of factors makes a cube.

$$= 2 \cdot a \cdot a \cdot b \cdot b \cdot b \cdot b \cdot \sqrt[3]{2 \cdot a \cdot b \cdot b}$$

$$= 2a^2b^4\sqrt[3]{2ab^2}. \qquad \text{Our answer checks.}$$

> TRY EXERCISE 51

> *Remember*: To simplify an *n*th root, identify factors in the radicand with exponents that are multiples of *n*.

Multiplying and Simplifying

We have used the product rule for radicals to find products and also to simplify radical expressions. For some radical expressions, it is possible to do both: First find a product and then simplify.

EXAMPLE 5   Multiply and simplify.

a) $\sqrt{15}\sqrt{6}$ **b)** $3\sqrt[3]{25} \cdot 2\sqrt[3]{5}$ **c)** $\sqrt[4]{8x^3y^5}\sqrt[4]{4x^2y^3}$

SOLUTION

a) $\sqrt{15}\sqrt{6} = \sqrt{15 \cdot 6}$ Multiplying radicands

$$= \sqrt{90} = \sqrt{9}\sqrt{10} \qquad \text{9 is a perfect square.}$$

$$= 3\sqrt{10}$$

b) $3\sqrt[3]{25} \cdot 2\sqrt[3]{5} = 3 \cdot 2 \cdot \sqrt[3]{25 \cdot 5}$ Using a commutative law; multiplying radicands

$$= 6 \cdot \sqrt[3]{125} \qquad \text{125 is a perfect cube.}$$

$$= 6 \cdot 5, \text{ or } 30$$

c) $\sqrt[4]{8x^3y^5}\sqrt[4]{4x^2y^3} = \sqrt[4]{32x^5y^8}$ Multiplying radicands

$$= \sqrt[4]{16x^4y^8 \cdot 2x} \qquad \text{Identifying the largest perfect fourth-power factor}$$

$$= \sqrt[4]{16x^4y^8}\sqrt[4]{2x} \qquad \text{Factoring into radicals}$$

$$= 2xy^2\sqrt[4]{2x} \qquad \text{Finding the fourth root; assume } x \geq 0.$$

The checks are left to the student.

> TRY EXERCISE 65

7.3 EXERCISE SET

Concept Reinforcement *Classify each statement as either true or false.*

1. For any real numbers $\sqrt[n]{a}$ and $\sqrt[n]{b}$, $\sqrt[n]{a} \cdot \sqrt[n]{b} = \sqrt[n]{ab}$.

2. For any real numbers $\sqrt[n]{a}$ and $\sqrt[n]{b}$, $\sqrt[n]{a} + \sqrt[n]{b} = \sqrt[n]{a + b}$.

3. For any real numbers $\sqrt[n]{a}$ and $\sqrt[m]{b}$, $\sqrt[n]{a} \cdot \sqrt[m]{b} = \sqrt[nm]{ab}$.

4. For $x > 0$, $\sqrt{x^2 - 9} = x - 3$.

5. The expression $\sqrt[3]{X}$ is not simplified if X contains a factor that is a perfect cube.

6. It is often possible to simplify $\sqrt{A \cdot B}$ even though $\sqrt{A}$ and $\sqrt{B}$ cannot be simplified.

Multiply.

7. $\sqrt{3}\,\sqrt{10}$

8. $\sqrt{6}\,\sqrt{5}$

9. $\sqrt[3]{7}\,\sqrt[3]{5}$

10. $\sqrt[3]{2}\,\sqrt[3]{3}$

11. $\sqrt[4]{6}\,\sqrt[4]{9}$

12. $\sqrt[4]{4}\,\sqrt[4]{10}$

13. $\sqrt{2x}\,\sqrt{13y}$

14. $\sqrt{5a}\,\sqrt{6b}$

15. $\sqrt[5]{8y^3}\,\sqrt[5]{10y}$

16. $\sqrt[5]{9t^2}\,\sqrt[5]{2t}$

17. $\sqrt{y - b}\,\sqrt{y + b}$

18. $\sqrt{x - a}\,\sqrt{x + a}$

19. $\sqrt[3]{0.7y}\,\sqrt[3]{0.3y}$

20. $\sqrt[3]{0.5x}\,\sqrt[3]{0.2x}$

21. $\sqrt[5]{x - 2}\,\sqrt[5]{(x - 2)^2}$

22. $\sqrt[4]{x - 1}\,\sqrt[4]{x^2 + x + 1}$

23. $\sqrt{\dfrac{2}{t}}\,\sqrt{\dfrac{3s}{11}}$

24. $\sqrt{\dfrac{7p}{6}}\,\sqrt{\dfrac{5}{q}}$

25. $\sqrt[7]{\dfrac{x - 3}{4}}\,\sqrt[7]{\dfrac{5}{x + 2}}$

26. $\sqrt[6]{\dfrac{a}{b - 2}}\,\sqrt[6]{\dfrac{3}{b + 2}}$

Simplify by factoring.

27. $\sqrt{12}$

28. $\sqrt{300}$

29. $\sqrt{45}$

30. $\sqrt{27}$

31. $\sqrt{8x^9}$

32. $\sqrt{75y^5}$

33. $\sqrt{120}$

34. $\sqrt{350}$

35. $\sqrt{36a^4b}$

36. $\sqrt{175y^8}$

37. $\sqrt[3]{8x^3y^2}$

38. $\sqrt[3]{27ab^6}$

39. $\sqrt[3]{-16x^6}$

40. $\sqrt[3]{-32a^6}$

Find a simplified form of $f(x)$. Assume that x can be any real number.

41. $f(x) = \sqrt[3]{40x^6}$

42. $f(x) = \sqrt[3]{27x^5}$

43. $f(x) = \sqrt{49(x - 3)^2}$

44. $f(x) = \sqrt{81(x - 1)^2}$

45. $f(x) = \sqrt{5x^2 - 10x + 5}$

46. $f(x) = \sqrt{2x^2 + 8x + 8}$

Simplify. Assume that no radicands were formed by raising negative numbers to even powers.

47. $\sqrt{a^{10}b^{11}}$

48. $\sqrt{x^8y^7}$

49. $\sqrt[3]{x^5y^6z^{10}}$

50. $\sqrt[3]{a^6b^7c^{13}}$

51. $\sqrt[3]{16x^5y^{11}}$

52. $\sqrt[5]{-32a^7b^{11}}$

53. $\sqrt[5]{x^{13}y^8z^{17}}$

54. $\sqrt[5]{a^6b^8c^9}$

55. $\sqrt[3]{-80a^{14}}$

56. $\sqrt[4]{810x^9}$

Multiply and simplify. Assume that no radicands were formed by raising negative numbers to even powers.

57. $\sqrt{5}\,\sqrt{10}$

58. $\sqrt{2}\,\sqrt{6}$

59. $\sqrt{6}\,\sqrt{33}$

60. $\sqrt{10}\,\sqrt{35}$

61. $\sqrt[3]{9}\,\sqrt[3]{3}$

62. $\sqrt[3]{2}\,\sqrt[3]{4}$

Aha! 63. $\sqrt{24y^5}\,\sqrt{24y^5}$

64. $\sqrt{120t^9}\,\sqrt{120t^9}$

65. $\sqrt[3]{5a^2}\,\sqrt[3]{2a}$

66. $\sqrt[3]{7x}\,\sqrt[3]{3x^2}$

67. $\sqrt{2x^5}\,\sqrt{10x^2}$

68. $\sqrt{5a^7}\,\sqrt{15a^3}$

69. $\sqrt[3]{s^2t^4}\,\sqrt[3]{s^4t^6}$

70. $\sqrt[3]{x^2y^4}\,\sqrt[3]{x^2y^6}$

71. $\sqrt[3]{(x - y)^2}\,\sqrt[3]{(x - y)^{10}}$

72. $\sqrt[3]{(t + 4)^5}\,\sqrt[3]{(t + 4)}$

73. $\sqrt[4]{20a^3b^7}\,\sqrt[4]{4a^2b^5}$

74. $\sqrt[4]{9x^7y^2}\,\sqrt[4]{9x^2y^9}$

75. $\sqrt[5]{x^3(y + z)^6}\,\sqrt[5]{x^3(y + z)^4}$

76. $\sqrt[5]{a^3(b - c)^4}\,\sqrt[5]{a^7(b - c)^4}$

77. Explain how you could convince a friend that $\sqrt{x^2 - 16} \neq \sqrt{x^2} - \sqrt{16}$.

78. Why is it incorrect to say that, in general, $\sqrt{x^2} = x$?

Skill Review

Review simplifying rational expressions (Sections 6.1, 6.2, and 6.3).

Perform the indicated operation and, if possible, simplify.

79. $\dfrac{15a^2x}{8b} \cdot \dfrac{24b^2x}{5a}$ [6.1]

80. $\dfrac{x^2 - 1}{x^2 - 4} \div \dfrac{x^2 - x - 2}{x^2 + x - 2}$ [6.1]

81. $\dfrac{x - 3}{2x - 10} - \dfrac{3x - 5}{x^2 - 25}$ [6.2]

82. $\dfrac{6x}{25y^2} + \dfrac{3y}{10x}$ [6.2]

83. $\dfrac{a^{-1} + b^{-1}}{ab}$ [6.3]

84. $\dfrac{\dfrac{1}{x + 1} - \dfrac{2}{x}}{\dfrac{3}{x} + \dfrac{1}{x + 1}}$ [6.3]

Synthesis

85. Explain why it is true that $\sqrt[n]{ab} = \sqrt[n]{a} \cdot \sqrt[n]{b}$ for any real numbers $\sqrt[n]{a}$ and $\sqrt[n]{b}$.

86. Is the equation $\sqrt{(2x + 3)^8} = (2x + 3)^4$ always, sometimes, or never true? Why?

87. *Radar range.*   The function given by

$$R(x) = \frac{1}{2} \sqrt[4]{\frac{x \cdot 3.0 \times 10^6}{\pi^2}}$$

can be used to determine the maximum range $R(x)$, in miles, of an ARSR-3 surveillance radar with a peak power of x watts. Determine the maximum radar range when the peak power is 5×10^4 watts.
Source: Introduction to RADAR Techniques, Federal Aviation Administration, 1988

88. *Speed of a skidding car.*   Police can estimate the speed at which a car was traveling by measuring its skid marks. The function given by

$$r(L) = 2\sqrt{5L}$$

can be used, where L is the length of a skid mark, in feet, and $r(L)$ is the speed, in miles per hour. Find

the exact speed and an estimate (to the nearest tenth mile per hour) for the speed of a car that left skid marks **(a)** 20 ft long; **(b)** 70 ft long; **(c)** 90 ft long. See also Exercise 102.

89. *Wind chill temperature.*   When the temperature is T degrees Celsius and the wind speed is v meters per second, the *wind chill temperature*, T_w, is the temperature (with no wind) that it feels like. Here is a formula for finding wind chill temperature:

$$T_w = 33 - \frac{(10.45 + 10\sqrt{v} - v)(33 - T)}{22}.$$

Estimate the wind chill temperature (to the nearest tenth of a degree) for the given actual temperatures and wind speeds.

a) $T = 7°C$, $v = 8$ m/sec
b) $T = 0°C$, $v = 12$ m/sec
c) $T = -5°C$, $v = 14$ m/sec
d) $T = -23°C$, $v = 15$ m/sec

Simplify. Assume that all variables are nonnegative.

90. $\left(\sqrt{r^3t}\right)^7$ **91.** $\left(\sqrt[3]{25x^4}\right)^4$

92. $\left(\sqrt[3]{a^2b^4}\right)^5$ **93.** $\left(\sqrt{a^3b^5}\right)^7$

Draw and compare the graphs of each group of equations.

94. $f(x) = \sqrt{x^2 - 2x + 1}$,
$g(x) = x - 1$,
$h(x) = |x - 1|$

95. $f(x) = \sqrt{x^2 + 2x + 1}$,
$g(x) = x + 1$,
$h(x) = |x + 1|$

96. If $f(t) = \sqrt{t^2 - 3t - 4}$, what is the domain of f?

97. What is the domain of g, if $g(x) = \sqrt{x^2 - 6x + 8}$?

Solve.

98. $\sqrt[3]{5x^{k+1}} \, \sqrt[3]{25x^k} = 5x^7$, for k

99. $\sqrt[5]{4a^{3k+2}} \, \sqrt[5]{8a^{6-k}} = 2a^4$, for k

100. Use a graphing calculator to check your answers to Exercises 21 and 41.

101. Blair is puzzled. When he uses a graphing calculator to graph $y = \sqrt{x} \cdot \sqrt{x}$, he gets the following screen. Explain why Blair did not get the complete line $y = x$.

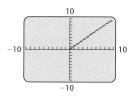

102. Does a car traveling twice as fast as another car leave a skid mark that is twice as long? (See Exercise 88.) Why or why not?

7.4   Dividing Radical Expressions

Dividing and Simplifying   ▪   Rationalizing Denominators or Numerators With One Term

Dividing and Simplifying

Just as the root of a product can be expressed as the product of two roots, the root of a quotient can be expressed as the quotient of two roots. For example,

$$\sqrt[3]{\frac{27}{8}} = \frac{3}{2} \quad \text{and} \quad \frac{\sqrt[3]{27}}{\sqrt[3]{8}} = \frac{3}{2}.$$

This example suggests the following.

> **The Quotient Rule for Radicals**
> For any real numbers $\sqrt[n]{a}$ and $\sqrt[n]{b}$, $b \neq 0$,
>
> $$\sqrt[n]{\frac{a}{b}} = \frac{\sqrt[n]{a}}{\sqrt[n]{b}}.$$

Remember that an nth root is simplified when its radicand has no factors that are perfect nth powers. Unless functions are involved, we assume that no radicands represent negative quantities raised to an even power.

EXAMPLE **1**   Simplify by taking the roots of the numerator and the denominator.

a) $\sqrt[3]{\dfrac{27}{125}}$ **b)** $\sqrt{\dfrac{25}{y^2}}$

SOLUTION

a) $\sqrt[3]{\dfrac{27}{125}} = \dfrac{\sqrt[3]{27}}{\sqrt[3]{125}} = \dfrac{3}{5}$ Taking the cube roots of the numerator and the denominator

b) $\sqrt{\dfrac{25}{y^2}} = \dfrac{\sqrt{25}}{\sqrt{y^2}} = \dfrac{5}{y}$ Taking the square roots of the numerator and the denominator. Assume $y > 0$.   **TRY EXERCISE** ▶ 9

Any radical expressions appearing in the answers should be simplified as much as possible.

EXAMPLE **2**   Simplify: **(a)** $\sqrt{\dfrac{16x^3}{y^8}}$; **(b)** $\sqrt[3]{\dfrac{27y^{14}}{8x^3}}$.

SOLUTION

a) $\sqrt{\dfrac{16x^3}{y^8}} = \dfrac{\sqrt{16x^3}}{\sqrt{y^8}}$

$= \dfrac{\sqrt{16x^2 \cdot x}}{\sqrt{y^8}}$

$= \dfrac{4x\sqrt{x}}{y^4}$ Simplifying the numerator and the denominator

b) $\sqrt[3]{\dfrac{27y^{14}}{8x^3}} = \dfrac{\sqrt[3]{27y^{14}}}{\sqrt[3]{8x^3}}$

$\qquad = \dfrac{\sqrt[3]{27y^{12}y^2}}{\sqrt[3]{8x^3}}$ y^{12} is the largest perfect-cube factor of y^{14}.

$\qquad = \dfrac{\sqrt[3]{27y^{12}}\,\sqrt[3]{y^2}}{\sqrt[3]{8x^3}}$

$\qquad = \dfrac{3y^4\sqrt[3]{y^2}}{2x}$ Simplifying the numerator and the denominator

> **TRY EXERCISE** 17

If we read from right to left, the quotient rule tells us that to divide two radical expressions that have the same index, we can divide the radicands.

EXAMPLE 3   Divide and, if possible, simplify.

a) $\dfrac{\sqrt{80}}{\sqrt{5}}$ 　　　　　　　　　　**b)** $\dfrac{5\sqrt[3]{32}}{\sqrt[3]{2}}$

c) $\dfrac{\sqrt{72xy}}{2\sqrt{2}}$ 　　　　　　　　**d)** $\dfrac{\sqrt[4]{18a^9b^5}}{\sqrt[4]{3b}}$

SOLUTION

STUDENT NOTES

When writing radical signs, pay careful attention to what is included as the radicand. Each of the following represents a *different* number:

$$\sqrt{\dfrac{5\cdot 2}{3}},\quad \dfrac{\sqrt{5\cdot 2}}{3},\quad \dfrac{\sqrt{5}\cdot 2}{3}.$$

a) $\dfrac{\sqrt{80}}{\sqrt{5}} = \sqrt{\dfrac{80}{5}} = \sqrt{16} = 4$

> Because the indices match, we can divide the radicands.

b) $\dfrac{5\sqrt[3]{32}}{\sqrt[3]{2}} = 5\sqrt[3]{\dfrac{32}{2}} = 5\sqrt[3]{16}$

$\qquad = 5\sqrt[3]{8\cdot 2}$ 8 is the largest perfect-cube factor of 16.

$\qquad = 5\sqrt[3]{8}\,\sqrt[3]{2} = 5\cdot 2\sqrt[3]{2}$

$\qquad = 10\sqrt[3]{2}$

c) $\dfrac{\sqrt{72xy}}{2\sqrt{2}} = \dfrac{1}{2}\sqrt{\dfrac{72xy}{2}}$

> Because the indices match, we can divide the radicands.

$\qquad = \dfrac{1}{2}\sqrt{36xy} = \dfrac{1}{2}\cdot 6\sqrt{xy}$

$\qquad = 3\sqrt{xy}$

d) $\dfrac{\sqrt[4]{18a^9b^5}}{\sqrt[4]{3b}} = \sqrt[4]{\dfrac{18a^9b^5}{3b}}$

$\qquad = \sqrt[4]{6a^9b^4} = \sqrt[4]{a^8b^4}\,\sqrt[4]{6a}$ Note that 8 is the largest power less than 9 that is a multiple of the index 4.

$\qquad = a^2b\sqrt[4]{6a}$ *Partial check:* $(a^2b)^4 = a^8b^4$

> **TRY EXERCISE** 27

Rationalizing Denominators or Numerators With One Term*

The expressions

$$\frac{1}{\sqrt{2}} \quad \text{and} \quad \frac{\sqrt{2}}{2}$$

are equivalent, but the second expression does not have a radical expression in the denominator.† We can **rationalize the denominator** of a radical expression if we multiply by 1 in either of two ways.

One way is to multiply by 1 *under* the radical to make the denominator of the radicand a perfect power.

EXAMPLE 4   Rationalize each denominator.

a) $\sqrt{\dfrac{7}{3}}$ **b)** $\sqrt[3]{\dfrac{5}{16}}$

SOLUTION

a) We multiply by 1 under the radical, using $\frac{3}{3}$. We do this so that the denominator of the radicand will be a perfect square:

$$\sqrt{\frac{7}{3}} = \sqrt{\frac{7}{3} \cdot \frac{3}{3}} \qquad \text{Multiplying by 1 under the radical}$$

$$= \sqrt{\frac{21}{9}} \qquad \text{The denominator, 9, is now a perfect square.}$$

$$= \frac{\sqrt{21}}{\sqrt{9}} \qquad \text{Using the quotient rule for radicals}$$

$$= \frac{\sqrt{21}}{3}.$$

b) Note that $16 = 4^2$. Thus, to make the denominator a perfect cube, we multiply under the radical by $\frac{4}{4}$:

$$\sqrt[3]{\frac{5}{16}} = \sqrt[3]{\frac{5}{4 \cdot 4} \cdot \frac{4}{4}} \qquad \text{Since the index is 3, we need 3 identical factors in the denominator.}$$

$$= \sqrt[3]{\frac{20}{4^3}} \qquad \text{The denominator is now a perfect cube.}$$

$$= \frac{\sqrt[3]{20}}{\sqrt[3]{4^3}}$$

$$= \frac{\sqrt[3]{20}}{4}. \qquad \qquad \boxed{\text{TRY EXERCISE} \blacktriangleright 41}$$

Another way to rationalize a denominator is to multiply by 1 *outside* the radical.

EXAMPLE 5   Rationalize each denominator.

a) $\sqrt{\dfrac{4}{5b}}$ **b)** $\dfrac{\sqrt[3]{a}}{\sqrt[3]{25bc^5}}$ **c)** $\dfrac{3x}{\sqrt[5]{2x^2y^3}}$

---

*Denominators and numerators with two terms are rationalized in Section 7.5.
†See Exercise 73 on p. 457.

SOLUTION

a) We rewrite the expression as a quotient of two radicals. Then we simplify and multiply by 1:

$$\sqrt{\frac{4}{5b}} = \frac{\sqrt{4}}{\sqrt{5b}} = \frac{2}{\sqrt{5b}} \qquad \text{We assume } b > 0.$$

$$= \frac{2}{\sqrt{5b}} \cdot \frac{\sqrt{5b}}{\sqrt{5b}} \qquad \text{Multiplying by 1}$$

$$= \frac{2\sqrt{5b}}{\left(\sqrt{5b}\right)^2} \qquad \text{Try to do this step mentally.}$$

$$= \frac{2\sqrt{5b}}{5b}.$$

b) Note that the radicand $25bc^5$ is $5 \cdot 5 \cdot b \cdot c \cdot c \cdot c \cdot c \cdot c$. In order for this to be a cube, we need another factor of 5, two more factors of b, and one more factor of c. Thus we multiply by 1, using $\sqrt[3]{5b^2c}/\sqrt[3]{5b^2c}$:

$$\frac{\sqrt[3]{a}}{\sqrt[3]{25bc^5}} = \frac{\sqrt[3]{a}}{\sqrt[3]{25bc^5}} \cdot \frac{\sqrt[3]{5b^2c}}{\sqrt[3]{5b^2c}} \qquad \text{Multiplying by 1}$$

$$= \frac{\sqrt[3]{5ab^2c}}{\sqrt[3]{125b^3c^6}} \longleftarrow \text{This radicand is now a perfect cube.}$$

$$= \frac{\sqrt[3]{5ab^2c}}{5bc^2}.$$

c) To change the radicand $2x^2y^3$ into a perfect fifth power, we need four more factors of 2, three more factors of x, and two more factors of y. Thus we multiply by 1, using $\sqrt[5]{2^4x^3y^2}/\sqrt[5]{2^4x^3y^2}$, or $\sqrt[5]{16x^3y^2}/\sqrt[5]{16x^3y^2}$:

$$\frac{3x}{\sqrt[5]{2x^2y^3}} = \frac{3x}{\sqrt[5]{2x^2y^3}} \cdot \frac{\sqrt[5]{16x^3y^2}}{\sqrt[5]{16x^3y^2}} \qquad \text{Multiplying by 1}$$

$$= \frac{3x\sqrt[5]{16x^3y^2}}{\sqrt[5]{32x^5y^5}} \longleftarrow \text{This radicand is now a perfect fifth power.}$$

$$= \frac{3x\sqrt[5]{16x^3y^2}}{2xy} = \frac{3\sqrt[5]{16x^3y^2}}{2y}. \qquad \text{Always simplify if possible.}$$

TRY EXERCISE 47

Sometimes in calculus it is necessary to rationalize a numerator. To do so, we multiply by 1 to make the radicand in the *numerator* a perfect power.

EXAMPLE **6**   Rationalize the numerator: $\dfrac{\sqrt[3]{4a^2}}{\sqrt[3]{5b}}$.

SOLUTION

$$\frac{\sqrt[3]{4a^2}}{\sqrt[3]{5b}} = \frac{\sqrt[3]{4a^2}}{\sqrt[3]{5b}} \cdot \frac{\sqrt[3]{2a}}{\sqrt[3]{2a}} \qquad \text{Multiplying by 1}$$

$$= \frac{\sqrt[3]{8a^3}}{\sqrt[3]{10ba}} \longleftarrow \text{This radicand is now a perfect cube.}$$

$$= \frac{2a}{\sqrt[3]{10ab}}$$

TRY EXERCISE 59

In Section 7.5, we will discuss rationalizing denominators and numerators in which two terms appear.

7.4 EXERCISE SET

🦢 *Concept Reinforcement   In each of Exercises 1–8, match the expression with an equivalent expression from the column on the right. Assume a, b > 0.*

1. ___ $\sqrt[4]{\dfrac{16a^6}{a^2}}$

2. ___ $\dfrac{\sqrt[3]{a^6}}{\sqrt[3]{b^9}}$

3. ___ $\sqrt[5]{\dfrac{a^6}{b^4}}$

4. ___ $\sqrt{\dfrac{a}{b^3}}$

5. ___ $\dfrac{\sqrt[5]{a^2}}{\sqrt[5]{b^2}}$

6. ___ $\dfrac{\sqrt{5a^4}}{\sqrt{5a^3}}$

7. ___ $\dfrac{\sqrt[5]{a^2}}{\sqrt[5]{b^3}}$

8. ___ $\sqrt[3]{\dfrac{a^2}{b^6}}$

a) $\dfrac{\sqrt[5]{a^2}\sqrt[5]{b^2}}{\sqrt[5]{b^5}}$

b) $\dfrac{a^2}{b^3}$

c) $\sqrt{\dfrac{a \cdot b}{b^3 \cdot b}}$

d) $\sqrt{a}$

e) $\dfrac{\sqrt[3]{a^2}}{b^2}$

f) $\sqrt[5]{\dfrac{a^6 b}{b^4 \cdot b}}$

g) $2a$

h) $\dfrac{\sqrt[5]{a^2 b^3}}{\sqrt[5]{b^5}}$

Simplify by taking the roots of the numerator and the denominator. Assume all variables represent positive numbers.

9. $\sqrt{\dfrac{49}{100}}$

10. $\sqrt{\dfrac{81}{25}}$

11. $\sqrt[3]{\dfrac{125}{8}}$

12. $\sqrt[3]{\dfrac{1000}{27}}$

13. $\sqrt{\dfrac{121}{t^2}}$

14. $\sqrt{\dfrac{144}{p^2}}$

15. $\sqrt{\dfrac{36y^3}{x^4}}$

16. $\sqrt{\dfrac{25a^5}{b^6}}$

17. $\sqrt[3]{\dfrac{27a^4}{8b^3}}$

18. $\sqrt[3]{\dfrac{64x^7}{216y^6}}$

19. $\sqrt[4]{\dfrac{32a^4}{2b^4 c^8}}$

20. $\sqrt[4]{\dfrac{81x^4}{y^8 z^4}}$

21. $\sqrt[4]{\dfrac{a^5 b^8}{c^{10}}}$

22. $\sqrt[4]{\dfrac{x^9 y^{12}}{z^6}}$

23. $\sqrt[5]{\dfrac{32x^6}{y^{11}}}$

24. $\sqrt[5]{\dfrac{243a^9}{b^{13}}}$

25. $\sqrt[6]{\dfrac{x^6 y^8}{z^{15}}}$

26. $\sqrt[6]{\dfrac{a^9 b^{12}}{c^{13}}}$

Divide and, if possible, simplify. Assume all variables represent positive numbers.

27. $\dfrac{\sqrt{18y}}{\sqrt{2y}}$

28. $\dfrac{\sqrt{700x}}{\sqrt{7x}}$

29. $\dfrac{\sqrt[3]{26}}{\sqrt[3]{13}}$

30. $\dfrac{\sqrt[3]{35}}{\sqrt[3]{5}}$

31. $\dfrac{\sqrt{40xy^3}}{\sqrt{8x}}$

32. $\dfrac{\sqrt{56ab^3}}{\sqrt{7a}}$

33. $\dfrac{\sqrt[3]{96a^4 b^2}}{\sqrt[3]{12a^2 b}}$

34. $\dfrac{\sqrt[3]{189x^5 y^7}}{\sqrt[3]{7x^2 y^2}}$

35. $\dfrac{\sqrt{100ab}}{5\sqrt{2}}$

36. $\dfrac{\sqrt{75ab}}{3\sqrt{3}}$

37. $\dfrac{\sqrt[4]{48x^9 y^{13}}}{\sqrt[4]{3xy^{-2}}}$

38. $\dfrac{\sqrt[5]{64a^{11} b^{28}}}{\sqrt[5]{2ab^{-2}}}$

39. $\dfrac{\sqrt[3]{x^3 - y^3}}{\sqrt[3]{x - y}}$

40. $\dfrac{\sqrt[3]{r^3 + s^3}}{\sqrt[3]{r + s}}$

Hint: Factor and then simplify.

Rationalize each denominator. Assume all variables represent positive numbers.

41. $\sqrt{\dfrac{2}{5}}$

42. $\sqrt{\dfrac{7}{2}}$

43. $\dfrac{2\sqrt{5}}{7\sqrt{3}}$

44. $\dfrac{3\sqrt{5}}{2\sqrt{7}}$

45. $\sqrt[3]{\dfrac{5}{4}}$

46. $\sqrt[3]{\dfrac{2}{9}}$

47. $\dfrac{\sqrt[3]{3a}}{\sqrt[3]{5c}}$

48. $\dfrac{\sqrt[3]{7x}}{\sqrt[3]{3y}}$

49. $\dfrac{\sqrt[4]{5y^6}}{\sqrt[4]{9x}}$

50. $\dfrac{\sqrt[5]{3a^4}}{\sqrt[5]{2b^7}}$

51. $\sqrt[3]{\dfrac{2}{x^2 y}}$

52. $\sqrt[3]{\dfrac{5}{ab^2}}$

53. $\sqrt{\dfrac{7a}{18}}$

54. $\sqrt{\dfrac{3x}{20}}$

55. $\sqrt[5]{\dfrac{9}{32x^5 y}}$

56. $\sqrt[4]{\dfrac{7}{64a^2 b^4}}$ Aha! 57. $\sqrt{\dfrac{10ab^2}{72a^3 b}}$

58. $\sqrt{\dfrac{21x^2 y}{75xy^5}}$

Rationalize each numerator. Assume all variables represent positive numbers.

59. $\sqrt{\dfrac{5}{11}}$ **60.** $\sqrt{\dfrac{2}{3}}$ **61.** $\dfrac{2\sqrt{6}}{5\sqrt{7}}$

62. $\dfrac{3\sqrt{10}}{2\sqrt{3}}$ **63.** $\dfrac{\sqrt{8}}{2\sqrt{3x}}$ **64.** $\dfrac{\sqrt{12}}{\sqrt{5y}}$

65. $\dfrac{\sqrt[3]{7}}{\sqrt[3]{2}}$ **66.** $\dfrac{\sqrt[3]{5}}{\sqrt[3]{4}}$ **67.** $\sqrt{\dfrac{7x}{3y}}$

68. $\sqrt{\dfrac{7a}{6b}}$ **69.** $\sqrt[3]{\dfrac{2a^5}{5b}}$ **70.** $\sqrt[3]{\dfrac{2a^4}{7b}}$

71. $\sqrt{\dfrac{x^3y}{2}}$ **72.** $\sqrt{\dfrac{ab^5}{3}}$

73. Explain why it is easier to approximate

$$\frac{\sqrt{2}}{2} \quad \text{than} \quad \frac{1}{\sqrt{2}}$$

if no calculator is available and $\sqrt{2} \approx 1.414213562$.

74. A student *incorrectly* claims that

$$\frac{5 + \sqrt{2}}{\sqrt{18}} = \frac{5 + \sqrt{1}}{\sqrt{9}} = \frac{5 + 1}{3}.$$

How could you convince the student that a mistake has been made? How would you explain the correct way of rationalizing the denominator?

Skill Review

To prepare for Section 7.5, review factoring expressions and multiplying polynomials (Sections 5.2 and 5.3).

Factor. [5.3]

75. $3x - 8xy + 2xz$ **76.** $4a^2c + 9ac - 3a^3c$

Multiply. [5.2]

77. $(a + b)(a - b)$ **78.** $(a^2 - 2y)(a^2 + 2y)$

79. $(8 + 3x)(7 - 4x)$ **80.** $(2y - x)(3a - c)$

Synthesis

81. Is the quotient of two irrational numbers always an irrational number? Why or why not?

82. Is it possible to understand how to rationalize a denominator without knowing how to multiply rational expressions? Why or why not?

83. *Pendulums.* The *period* of a pendulum is the time it takes to complete one cycle, swinging to and fro. For a pendulum that is L centimeters long, the period T is given by the formula

$$T = 2\pi\sqrt{\frac{L}{980}},$$

where T is in seconds. Find, to the nearest hundredth of a second, the period of a pendulum of length **(a)** 65 cm; **(b)** 98 cm; **(c)** 120 cm. Use a calculator's $\boxed{\pi}$ key if possible.

Perform the indicated operations.

84. $\dfrac{7\sqrt{a^2b}\,\sqrt{25xy}}{5\sqrt{a^{-4}b^{-1}}\sqrt{49x^{-1}y^{-3}}}$ **85.** $\dfrac{\left(\sqrt[3]{81mn^2}\right)^2}{\left(\sqrt[3]{mn}\right)^2}$

86. $\dfrac{\sqrt{44x^2y^9z}\,\sqrt{22y^9z^6}}{\left(\sqrt{11xy^8z^2}\right)^2}$

87. $\sqrt{a^2 - 3} - \dfrac{a^2}{\sqrt{a^2 - 3}}$

88. $5\sqrt{\dfrac{x}{y}} + 4\sqrt{\dfrac{y}{x}} - \dfrac{3}{\sqrt{xy}}$

89. Provide a reason for each step in the following derivation of the quotient rule:

$$\sqrt[n]{\frac{a}{b}} = \left(\frac{a}{b}\right)^{1/n} \quad \rule{1.5cm}{0.4pt}$$

$$= \frac{a^{1/n}}{b^{1/n}} \quad \rule{1.5cm}{0.4pt}$$

$$= \frac{\sqrt[n]{a}}{\sqrt[n]{b}} \quad \rule{1.5cm}{0.4pt}$$

90. Show that $\dfrac{\sqrt[n]{a}}{\sqrt[n]{b}}$ is the nth root of $\dfrac{a}{b}$ by raising it to the nth power and simplifying.

91. Let $f(x) = \sqrt{18x^3}$ and $g(x) = \sqrt{2x}$. Find $(f/g)(x)$ and specify the domain of f/g.

92. Let $f(t) = \sqrt{2t}$ and $g(t) = \sqrt{50t^3}$. Find $(f/g)(t)$ and specify the domain of f/g.

93. Let $f(x) = \sqrt{x^2 - 9}$ and $g(x) = \sqrt{x - 3}$. Find $(f/g)(x)$ and specify the domain of f/g.

7.5 Expressions Containing Several Radical Terms

Adding and Subtracting Radical Expressions ▪ Products and Quotients of Two or More Radical Terms ▪ Rationalizing Denominators or Numerators With Two Terms ▪ Terms with Differing Indices

Radical expressions like $6\sqrt{7} + 4\sqrt{7}$ or $(\sqrt{a} + \sqrt{b})(\sqrt{a} - \sqrt{b})$ contain more than one *radical term* and can sometimes be simplified.

Adding and Subtracting Radical Expressions

When two radical expressions have the same indices and radicands, they are said to be **like radicals**. Like radicals can be combined (added or subtracted) in much the same way that we combine like terms.

EXAMPLE 1

Simplify by combining like radical terms.

a) $6\sqrt{7} + 4\sqrt{7}$

b) $\sqrt[3]{2} - 7x\sqrt[3]{2} + 5\sqrt[3]{2}$

c) $6\sqrt[5]{4x} + 3\sqrt[5]{4x} - \sqrt[3]{4x}$

SOLUTION

a) $6\sqrt{7} + 4\sqrt{7} = (6 + 4)\sqrt{7}$ Using the distributive law (factoring out $\sqrt{7}$)

$= 10\sqrt{7}$ You can think: 6 square roots of 7 plus 4 square roots of 7 is 10 square roots of 7.

b) $\sqrt[3]{2} - 7x\sqrt[3]{2} + 5\sqrt[3]{2} = (1 - 7x + 5)\sqrt[3]{2}$ Factoring out $\sqrt[3]{2}$

$= (6 - 7x)\sqrt[3]{2}$ These parentheses are important!

c) $6\sqrt[5]{4x} + 3\sqrt[5]{4x} - \sqrt[3]{4x} = (6 + 3)\sqrt[5]{4x} - \sqrt[3]{4x}$ Try to do this step mentally.

$= 9\sqrt[5]{4x} - \sqrt[3]{4x}$ The indices are different. We cannot combine these terms.

TRY EXERCISE ▶ 7

Our ability to simplify radical expressions can help us to find like radicals even when, at first, it may appear that there are none.

EXAMPLE 2

Simplify by combining like radical terms, if possible.

a) $3\sqrt{8} - 5\sqrt{2}$

b) $9\sqrt{5} - 4\sqrt{3}$

c) $\sqrt[3]{2x^6y^4} + 7\sqrt[3]{2y}$

SOLUTION

a) $3\sqrt{8} - 5\sqrt{2} = 3\sqrt{4 \cdot 2} - 5\sqrt{2}$

$= 3\sqrt{4} \cdot \sqrt{2} - 5\sqrt{2}$ ⎫

$= 3 \cdot 2 \cdot \sqrt{2} - 5\sqrt{2}$ ⎬ Simplifying $\sqrt{8}$

$= 6\sqrt{2} - 5\sqrt{2}$ ⎭

$= \sqrt{2}$ Combining like radicals

b) $9\sqrt{5} - 4\sqrt{3}$ cannot be simplified.   The radicands are different.

c) $\sqrt[3]{2x^6 y^4} + 7\sqrt[3]{2y} = \sqrt[3]{x^6 y^3 \cdot 2y} + 7\sqrt[3]{2y}$

$\quad\quad\quad\quad = \sqrt[3]{x^6 y^3} \cdot \sqrt[3]{2y} + 7\sqrt[3]{2y}$ Simplifying $\sqrt[3]{2x^6 y^4}$

$\quad\quad\quad\quad = x^2 y \cdot \sqrt[3]{2y} + 7\sqrt[3]{2y}$

$\quad\quad\quad\quad = (x^2 y + 7)\sqrt[3]{2y}$ Factoring to combine like radical terms

TRY EXERCISE 17

Products and Quotients of Two or More Radical Terms

Radical expressions often contain factors that have more than one term. Multiplying such expressions is similar to finding products of polynomials. Some products will yield like radical terms, which we can now combine.

EXAMPLE 3   Multiply.

a) $\sqrt{3}(x - \sqrt{5})$

b) $\sqrt[3]{y}\left(\sqrt[3]{y^2} + \sqrt[3]{2}\right)$

c) $(4\sqrt{3} + \sqrt{2})(\sqrt{3} - 5\sqrt{2})$

d) $(\sqrt{a} + \sqrt{b})(\sqrt{a} - \sqrt{b})$

SOLUTION

a) $\sqrt{3}(x - \sqrt{5}) = \sqrt{3} \cdot x - \sqrt{3} \cdot \sqrt{5}$ Using the distributive law

$\quad\quad\quad\quad = x\sqrt{3} - \sqrt{15}$ Multiplying radicals

b) $\sqrt[3]{y}\left(\sqrt[3]{y^2} + \sqrt[3]{2}\right) = \sqrt[3]{y} \cdot \sqrt[3]{y^2} + \sqrt[3]{y} \cdot \sqrt[3]{2}$ Using the distributive law

$\quad\quad\quad\quad = \sqrt[3]{y^3} + \sqrt[3]{2y}$ Multiplying radicals

$\quad\quad\quad\quad = y + \sqrt[3]{2y}$ Simplifying $\sqrt[3]{y^3}$

$\quad\quad\quad\quad\quad\quad\quad\quad\quad\quad\quad\quad\quad$ F $\quad\quad$ O $\quad\quad$ I $\quad\quad$ L

c) $(4\sqrt{3} + \sqrt{2})(\sqrt{3} - 5\sqrt{2}) = 4(\sqrt{3})^2 - 20\sqrt{3} \cdot \sqrt{2} + \sqrt{2} \cdot \sqrt{3} - 5(\sqrt{2})^2$

$\quad\quad\quad\quad = 4 \cdot 3 - 20\sqrt{6} + \sqrt{6} - 5 \cdot 2$ Multiplying radicals

$\quad\quad\quad\quad = 12 - 20\sqrt{6} + \sqrt{6} - 10$

$\quad\quad\quad\quad = 2 - 19\sqrt{6}$ Combining like terms

d) $(\sqrt{a} + \sqrt{b})(\sqrt{a} - \sqrt{b}) = (\sqrt{a})^2 - \sqrt{a}\sqrt{b} + \sqrt{a}\sqrt{b} - (\sqrt{b})^2$ Using FOIL

$\quad\quad\quad\quad = a - b$ Combining like terms

TRY EXERCISE 41

In Example 3(d) above, you may have noticed that since the outer and inner products in FOIL are opposites, the result, $a - b$, is not itself a radical expression. Pairs of radical expressions like $\sqrt{a} + \sqrt{b}$ and $\sqrt{a} - \sqrt{b}$ are called **conjugates**.

Rationalizing Denominators or Numerators With Two Terms

The use of conjugates allows us to rationalize denominators or numerators that contain two terms.

EXAMPLE **4** Rationalize each denominator: **(a)** $\dfrac{4}{\sqrt{3} + x}$; **(b)** $\dfrac{4 + \sqrt{2}}{\sqrt{5} - \sqrt{2}}$.

SOLUTION

a) $\dfrac{4}{\sqrt{3} + x} = \dfrac{4}{\sqrt{3} + x} \cdot \dfrac{\sqrt{3} - x}{\sqrt{3} - x}$ Multiplying by 1, using the conjugate of $\sqrt{3} + x$, which is $\sqrt{3} - x$

$\qquad = \dfrac{4(\sqrt{3} - x)}{(\sqrt{3} + x)(\sqrt{3} - x)}$ Multiplying numerators and denominators

$\qquad = \dfrac{4(\sqrt{3} - x)}{(\sqrt{3})^2 - x^2}$ Using FOIL in the denominator

$\qquad = \dfrac{4\sqrt{3} - 4x}{3 - x^2}$ Simplifying. No radicals remain in the denominator.

b) $\dfrac{4 + \sqrt{2}}{\sqrt{5} - \sqrt{2}} = \dfrac{4 + \sqrt{2}}{\sqrt{5} - \sqrt{2}} \cdot \dfrac{\sqrt{5} + \sqrt{2}}{\sqrt{5} + \sqrt{2}}$ Multiplying by 1, using the conjugate of $\sqrt{5} - \sqrt{2}$, which is $\sqrt{5} + \sqrt{2}$

$\qquad = \dfrac{(4 + \sqrt{2})(\sqrt{5} + \sqrt{2})}{(\sqrt{5} - \sqrt{2})(\sqrt{5} + \sqrt{2})}$ Multiplying numerators and denominators

$\qquad = \dfrac{4\sqrt{5} + 4\sqrt{2} + \sqrt{2}\sqrt{5} + (\sqrt{2})^2}{(\sqrt{5})^2 - (\sqrt{2})^2}$ Using FOIL

$\qquad = \dfrac{4\sqrt{5} + 4\sqrt{2} + \sqrt{10} + 2}{5 - 2}$ Squaring in the denominator and the numerator

$\qquad = \dfrac{4\sqrt{5} + 4\sqrt{2} + \sqrt{10} + 2}{3}$ No radicals remain in the denominator.

TRY EXERCISE 61

To rationalize a numerator with two terms, we use the conjugate of the numerator.

EXAMPLE **5** Rationalize the numerator: $\dfrac{4 + \sqrt{2}}{\sqrt{5} - \sqrt{2}}$.

SOLUTION

$\qquad \dfrac{4 + \sqrt{2}}{\sqrt{5} - \sqrt{2}} = \dfrac{4 + \sqrt{2}}{\sqrt{5} - \sqrt{2}} \cdot \dfrac{4 - \sqrt{2}}{4 - \sqrt{2}}$ Multiplying by 1, using the conjugate of $4 + \sqrt{2}$, which is $4 - \sqrt{2}$

$\qquad\qquad = \dfrac{16 - (\sqrt{2})^2}{4\sqrt{5} - \sqrt{5}\sqrt{2} - 4\sqrt{2} + (\sqrt{2})^2}$

$\qquad\qquad = \dfrac{14}{4\sqrt{5} - \sqrt{10} - 4\sqrt{2} + 2}$

TRY EXERCISE 71

Terms with Differing Indices

To multiply or divide radical terms with identical radicands but different indices, we can convert to exponential notation, use the rules for exponents, and then convert back to radical notation.

EXAMPLE 6

Divide and, if possible, simplify: $\dfrac{\sqrt[4]{(x + y)^3}}{\sqrt{x + y}}$.

STUDENT NOTES

Expressions similar to the one in Example 6 are most easily simplified by rewriting the expression using exponents in place of radicals. After simplifying, remember to write your final result in radical notation. In general, if a problem is presented in one form, it is expected that the final result be presented in the same form.

SOLUTION

$$\frac{\sqrt[4]{(x + y)^3}}{\sqrt{x + y}} = \frac{(x + y)^{3/4}}{(x + y)^{1/2}}$$ Converting to exponential notation

$$= (x + y)^{3/4 - 1/2}$$ Since the bases are identical, we can subtract exponents: $\frac{3}{4} - \frac{1}{2} = \frac{3}{4} - \frac{2}{4} = \frac{1}{4}$.

$$= (x + y)^{1/4}$$
$$= \sqrt[4]{x + y}$$ Converting back to radical notation

TRY EXERCISE 95

The steps used in Example 6 can be used in a variety of situations.

To Simplify Products or Quotients with Differing Indices

1. Convert all radical expressions to exponential notation.
2. When the bases are identical, subtract exponents to divide and add exponents to multiply. This may require finding a common denominator.
3. Convert back to radical notation and, if possible, simplify.

EXAMPLE 7

Multiply and simplify: $\sqrt{x^3}\sqrt[3]{x}$.

SOLUTION

$$\sqrt{x^3}\sqrt[3]{x} = x^{3/2} \cdot x^{1/3}$$ Converting to exponential notation

$$= x^{11/6}$$ Adding exponents: $\frac{3}{2} + \frac{1}{3} = \frac{9}{6} + \frac{2}{6}$

$$= \sqrt[6]{x^{11}}$$ Converting back to radical notation

$$= \sqrt[6]{x^6}\sqrt[6]{x^5}$$
$$= x\sqrt[6]{x^5}$$ Simplifying

TRY EXERCISE 79

EXAMPLE 8

If $f(x) = \sqrt[3]{x^2}$ and $g(x) = \sqrt{x} + \sqrt[4]{x}$, find $(f \cdot g)(x)$.

SOLUTION   Recall from Section 2.6 that $(f \cdot g)(x) = f(x) \cdot g(x)$. Thus,

$$(f \cdot g)(x) = \sqrt[3]{x^2}\left(\sqrt{x} + \sqrt[4]{x}\right)$$ x is assumed to be nonnegative.

$$= x^{2/3}(x^{1/2} + x^{1/4})$$ Converting to exponential notation

$$= x^{2/3} \cdot x^{1/2} + x^{2/3} \cdot x^{1/4}$$ Using the distributive law

$$= x^{2/3 + 1/2} + x^{2/3 + 1/4}$$ Adding exponents:

$$= x^{7/6} + x^{11/12}$$ $\frac{2}{3} + \frac{1}{2} = \frac{4}{6} + \frac{3}{6}; \frac{2}{3} + \frac{1}{4} = \frac{8}{12} + \frac{3}{12}$

$$= \sqrt[6]{x^7} + \sqrt[12]{x^{11}}$$ Converting back to radical notation

$$= \sqrt[6]{x^6}\sqrt[6]{x} + \sqrt[12]{x^{11}}$$ Simplifying

$$= x\sqrt[6]{x} + \sqrt[12]{x^{11}}.$$

TRY EXERCISE 103

We often can write the final result as a single radical expression by finding a common denominator in the exponents.

EXAMPLE 9   Divide and, if possible, simplify: $\dfrac{\sqrt[3]{a^2 b^4}}{\sqrt{ab}}$.

SOLUTION

$$\frac{\sqrt[3]{a^2 b^4}}{\sqrt{ab}} = \frac{(a^2 b^4)^{1/3}}{(ab)^{1/2}} \qquad \text{Converting to exponential notation}$$

$$= \frac{a^{2/3} b^{4/3}}{a^{1/2} b^{1/2}} \qquad \text{Using the product and power rules}$$

$$= a^{2/3 - 1/2} b^{4/3 - 1/2} \qquad \text{Subtracting exponents}$$

$$= a^{1/6} b^{5/6}$$

$$= \sqrt[6]{a}\,\sqrt[6]{b^5} \qquad \text{Converting to radical notation}$$

$$= \sqrt[6]{ab^5} \qquad \text{Using the product rule for radicals}$$

TRY EXERCISE 91

7.5   EXERCISE SET

For Extra Help

🐦 *Concept Reinforcement*   *For each of Exercises 1–6, fill in the blanks by selecting from the following words (which may be used more than once):*

radicand(s), indices, conjugate(s), base(s), denominator(s), numerator(s).

1. To add radical expressions, the _____ and the _____ must be the same.

2. To multiply radical expressions, the _____ must be the same.

3. To find a product by adding exponents, the _____ must be the same.

4. To add rational expressions, the _____ must be the same.

5. To rationalize the _____ of $\dfrac{\sqrt{c} - \sqrt{a}}{5}$, we multiply by a form of 1, using the _____ of $\sqrt{c} - \sqrt{a}$, or $\sqrt{c} + \sqrt{a}$, to write 1.

6. To find a quotient by subtracting exponents, the _____ must be the same.

Add or subtract. Simplify by combining like radical terms, if possible. Assume that all variables and radicands represent positive real numbers.

7. $4\sqrt{3} + 7\sqrt{3}$

8. $6\sqrt{5} + 2\sqrt{5}$

9. $7\sqrt[3]{4} - 5\sqrt[3]{4}$

10. $14\sqrt[5]{2} - 8\sqrt[5]{2}$

11. $\sqrt[3]{y} + 9\sqrt[3]{y}$

12. $4\sqrt[4]{t} - \sqrt[4]{t}$

13. $8\sqrt{2} - \sqrt{2} + 5\sqrt{2}$

14. $\sqrt{6} + 3\sqrt{6} - 8\sqrt{6}$

15. $9\sqrt[3]{7} - \sqrt{3} + 4\sqrt[3]{7} + 2\sqrt{3}$

16. $5\sqrt{7} - 8\sqrt[4]{11} + \sqrt{7} + 9\sqrt[4]{11}$

17. $4\sqrt{27} - 3\sqrt{3}$

18. $9\sqrt{50} - 4\sqrt{2}$

19. $3\sqrt{45} - 8\sqrt{20}$

20. $5\sqrt{12} + 16\sqrt{27}$

21. $3\sqrt[3]{16} + \sqrt[3]{54}$

22. $\sqrt[3]{27} - 5\sqrt[3]{8}$

23. $\sqrt{a} + 3\sqrt{16a^3}$

24. $2\sqrt{9x^3} - \sqrt{x}$

25. $\sqrt[3]{6x^4} - \sqrt[3]{48x}$

26. $\sqrt[3]{54x} - \sqrt[3]{2x^4}$

27. $\sqrt{4a - 4} + \sqrt{a - 1}$

28. $\sqrt{9y + 27} + \sqrt{y + 3}$

29. $\sqrt{x^3 - x^2} + \sqrt{9x - 9}$

30. $\sqrt{4x - 4} - \sqrt{x^3 - x^2}$

Multiply. Assume all variables represent nonnegative real numbers.

31. $\sqrt{2}(5 + \sqrt{2})$

32. $\sqrt{3}(6 - \sqrt{3})$

33. $3\sqrt{5}(\sqrt{6} - \sqrt{7})$

34. $4\sqrt{2}(\sqrt{3} + \sqrt{5})$

35. $\sqrt{2}(3\sqrt{10} - \sqrt{8})$

36. $\sqrt{3}(2\sqrt{15} - 3\sqrt{4})$

37. $\sqrt[3]{3}(\sqrt[3]{9} - 4\sqrt[3]{21})$

38. $\sqrt[3]{2}(\sqrt[3]{4} - 2\sqrt[3]{32})$

39. $\sqrt[3]{a}\left(\sqrt[3]{a^2} + \sqrt[3]{24a^2}\right)$

40. $\sqrt[3]{x}\left(\sqrt[3]{3x^2} - \sqrt[3]{81x^2}\right)$

41. $(2 + \sqrt{6})(5 - \sqrt{6})$

42. $(4 - \sqrt{5})(2 + \sqrt{5})$

43. $(\sqrt{2} + \sqrt{7})(\sqrt{3} - \sqrt{7})$

44. $(\sqrt{7} - \sqrt{2})(\sqrt{5} + \sqrt{2})$

45. $(2 - \sqrt{3})(2 + \sqrt{3})$

46. $(3 + \sqrt{11})(3 - \sqrt{11})$

47. $(\sqrt{10} - \sqrt{15})(\sqrt{10} + \sqrt{15})$

48. $(\sqrt{12} + \sqrt{5})(\sqrt{12} - \sqrt{5})$

49. $(3\sqrt{7} + 2\sqrt{5})(2\sqrt{7} - 4\sqrt{5})$

50. $(4\sqrt{5} - 3\sqrt{2})(2\sqrt{5} + 4\sqrt{2})$

51. $(4 + \sqrt{7})^2$

52. $(3 + \sqrt{10})^2$

53. $(\sqrt{3} - \sqrt{2})^2$

54. $(\sqrt{5} - \sqrt{3})^2$

55. $(\sqrt{2t} + \sqrt{5})^2$

56. $(\sqrt{3x} - \sqrt{2})^2$

57. $(3 - \sqrt{x + 5})^2$

58. $(4 + \sqrt{x - 3})^2$

59. $\left(2\sqrt[4]{7} - \sqrt[4]{6}\right)\left(3\sqrt[4]{9} + 2\sqrt[4]{5}\right)$

60. $\left(4\sqrt[3]{3} + \sqrt[3]{10}\right)\left(2\sqrt[3]{7} + 5\sqrt[3]{6}\right)$

Rationalize each denominator.

61. $\dfrac{6}{3 - \sqrt{2}}$

62. $\dfrac{5}{4 - \sqrt{5}}$

63. $\dfrac{2 + \sqrt{5}}{6 + \sqrt{3}}$

64. $\dfrac{1 + \sqrt{2}}{3 + \sqrt{5}}$

65. $\dfrac{\sqrt{a}}{\sqrt{a} + \sqrt{b}}$

66. $\dfrac{\sqrt{z}}{\sqrt{x} - \sqrt{z}}$

Aha! **67.** $\dfrac{\sqrt{7} - \sqrt{3}}{\sqrt{3} - \sqrt{7}}$

68. $\dfrac{\sqrt{7} + \sqrt{5}}{\sqrt{5} + \sqrt{2}}$

69. $\dfrac{3\sqrt{2} - \sqrt{7}}{4\sqrt{2} + 2\sqrt{5}}$

70. $\dfrac{5\sqrt{3} - \sqrt{11}}{2\sqrt{3} - 5\sqrt{2}}$

Rationalize each numerator. If possible, simplify your result.

71. $\dfrac{\sqrt{5} + 1}{4}$

72. $\dfrac{\sqrt{15} - 3}{6}$

73. $\dfrac{\sqrt{6} - 2}{\sqrt{3} + 7}$

74. $\dfrac{\sqrt{10} + 4}{\sqrt{2} - 3}$

75. $\dfrac{\sqrt{x} - \sqrt{y}}{\sqrt{x} + \sqrt{y}}$

76. $\dfrac{\sqrt{a} + \sqrt{b}}{\sqrt{a} - \sqrt{b}}$

77. $\dfrac{\sqrt{a + h} - \sqrt{a}}{h}$

78. $\dfrac{\sqrt{x - h} - \sqrt{x}}{h}$

Perform the indicated operation and simplify. Assume all variables represent positive real numbers.

79. $\sqrt[3]{a}\sqrt[6]{a}$

80. $\sqrt[10]{a}\sqrt[5]{a^2}$

81. $\sqrt{b^3}\sqrt[5]{b^4}$

82. $\sqrt[3]{b^4}\sqrt[4]{b^3}$

83. $\sqrt{xy^3}\,\sqrt[3]{x^2y}$

84. $\sqrt[5]{a^3b}\,\sqrt{ab}$

85. $\sqrt[4]{9ab^3}\sqrt{3a^4b}$

86. $\sqrt{2x^3y^3}\,\sqrt[3]{4xy^2}$

87. $\sqrt{a^4b^3c^4}\,\sqrt[3]{ab^2c}$

88. $\sqrt[3]{xy^2z}\sqrt{x^3yz^2}$

89. $\dfrac{\sqrt[3]{a^2}}{\sqrt[4]{a}}$

90. $\dfrac{\sqrt[3]{x^2}}{\sqrt[5]{x}}$

91. $\dfrac{\sqrt[4]{x^2y^3}}{\sqrt[3]{xy}}$

92. $\dfrac{\sqrt[5]{a^4b}}{\sqrt[3]{ab}}$

93. $\dfrac{\sqrt{ab^3}}{\sqrt[5]{a^2b^3}}$

94. $\dfrac{\sqrt[5]{x^3y^4}}{\sqrt{xy}}$

95. $\dfrac{\sqrt{(7 - y)^3}}{\sqrt[3]{(7 - y)^2}}$

96. $\dfrac{\sqrt[5]{(y - 9)^3}}{\sqrt{y - 9}}$

97. $\dfrac{\sqrt[4]{(5 + 3x)^3}}{\sqrt[3]{(5 + 3x)^2}}$

98. $\dfrac{\sqrt[3]{(2x + 1)^2}}{\sqrt[5]{(2x + 1)^2}}$

99. $\sqrt[3]{x^2y}\left(\sqrt{xy} - \sqrt[5]{xy^3}\right)$

100. $\sqrt[4]{a^2b}\left(\sqrt[3]{a^2b} - \sqrt[5]{a^2b^2}\right)$

101. $\left(m + \sqrt[3]{n^2}\right)\left(2m + \sqrt[4]{n}\right)$

102. $\left(r - \sqrt[4]{s^3}\right)\left(3r - \sqrt[5]{s}\right)$

In Exercises 103–106, $f(x)$ and $g(x)$ are as given. Find $(f \cdot g)(x)$. Assume all variables represent nonnegative real numbers.

103. $f(x) = \sqrt[4]{x}, \ g(x) = 2\sqrt{x} - \sqrt[3]{x^2}$

104. $f(x) = \sqrt[4]{2x} + 5\sqrt{2x}, \ g(x) = \sqrt[3]{2x}$

105. $f(x) = x + \sqrt{7}, \ g(x) = x - \sqrt{7}$

106. $f(x) = x - \sqrt{2}, \ g(x) = x + \sqrt{6}$

Let $f(x) = x^2$. Find each of the following.

107. $f(3 - \sqrt{2})$

108. $f(5 - \sqrt{3})$

109. $f(\sqrt{6} + \sqrt{21})$

110. $f(\sqrt{2} + \sqrt{10})$

111. In what way(s) is combining like radical terms similar to combining like terms that are monomials?

112. Why do we need to know how to multiply radical expressions before learning how to add them?

Skill Review

To prepare for Section 7.6, review solving equations (Sections 1.3, 5.8, and 6.4).

Solve.

113. $3x - 1 = 125$ [1.3]

114. $x + 5 - 2x = 3x + 6 - x$ [1.3]

115. $x^2 + 2x + 1 = 22 - 2x$ [5.8]

116. $9x^2 - 6x + 1 = 7 + 5x - x^2$ [5.8]

117. $\dfrac{1}{x} + \dfrac{1}{2} = \dfrac{1}{6}$ [6.4]

118. $\dfrac{x}{x - 4} + \dfrac{2}{x + 4} = \dfrac{x - 2}{x^2 - 16}$ [6.4]

Synthesis

119. Ramon *incorrectly* writes
$$\sqrt[5]{x^2} \cdot \sqrt{x^3} = x^{2/5} \cdot x^{3/2} = \sqrt[5]{x^3}.$$
What mistake do you suspect he is making?

120. After examining the expression $\sqrt[4]{25xy^3}\ \sqrt{5x^4y}$, Dyan (correctly) concludes that x and y are both nonnegative. Explain how she could reach this conclusion.

Find a simplified form for $f(x)$. Assume $x \geq 0$.

121. $f(x) = \sqrt{x^3 - x^2} + \sqrt{9x^3 - 9x^2} - \sqrt{4x^3 - 4x^2}$

122. $f(x) = \sqrt{20x^2 + 4x^3} - 3x\sqrt{45 + 9x} + \sqrt{5x^2 + x^3}$

123. $f(x) = \sqrt[4]{x^5 - x^4} + 3\sqrt[4]{x^9 - x^8}$

124. $f(x) = \sqrt[4]{16x^4 + 16x^5} - 2\sqrt[4]{x^8 + x^9}$

Simplify.

125. $7x\sqrt{(x + y)^3} - 5xy\sqrt{x + y} - 2y\sqrt{(x + y)^3}$

126. $\sqrt{27a^5(b + 1)}\ \sqrt[3]{81a(b + 1)^4}$

127. $\sqrt{8x(y + z)^5}\ \sqrt[3]{4x^2(y + z)^2}$

128. $\frac{1}{2}\sqrt{36a^5bc^4} - \frac{1}{2}\sqrt[3]{64a^4bc^6} + \frac{1}{6}\sqrt{144a^3bc^6}$

129. $\dfrac{\dfrac{1}{\sqrt{w}} - \sqrt{w}}{\dfrac{\sqrt{w} + 1}{\sqrt{w}}}$

130. $\dfrac{1}{4 + \sqrt{3}} + \dfrac{1}{\sqrt{3}} + \dfrac{1}{\sqrt{3} - 4}$

Express each of the following as the product of two radical expressions.

131. $x - 5$ **132.** $y - 7$

133. $x - a$

Multiply.

134. $\sqrt{9 + 3\sqrt{5}}\sqrt{9 - 3\sqrt{5}}$

135. $(\sqrt{x + 2} - \sqrt{x - 2})^2$

136. Use a graphing calculator to check your answers to Exercises 25, 39, and 81.

CONNECTING the CONCEPTS

Many radical expressions can be simplified. It is important to know under which conditions radical expressions can be multiplied and divided and radical terms can be combined.

Multiplication and division: The indices must be the same.

$$\frac{\sqrt{50t^5}}{\sqrt{2t^{11}}} = \sqrt{\frac{50t^5}{2t^{11}}} = \sqrt{\frac{25}{t^6}} = \frac{5}{t^3}; \qquad \sqrt[4]{8x^3} \cdot \sqrt[4]{2x} = \sqrt[4]{16x^4} = 2x$$

Combining like terms: The indices and the radicands must both be the same.

$$\sqrt{75x} + \sqrt{12x} - \sqrt{3x} = 5\sqrt{3x} + 2\sqrt{3x} - \sqrt{3x} = 6\sqrt{3x}$$

Radical expressions with differing indices can sometimes be simplified using rational exponents.

$$\sqrt[3]{x^2}\sqrt{x} = x^{2/3}x^{1/2} = x^{4/6}x^{3/6} = x^{7/6} = \sqrt[6]{x^7} = x\sqrt[6]{x}$$

MIXED REVIEW

Simplify. Assume that all variables represent non-negative numbers. Thus no absolute-value signs are needed in an answer.

1. $\sqrt{(t+5)^2}$

2. $\sqrt[3]{-27a^{12}}$

3. $\sqrt{6x}\sqrt{15x}$

4. $\dfrac{\sqrt{20y}}{\sqrt{45y}}$

5. $\sqrt{15t} + 4\sqrt{15t}$

6. $\sqrt[5]{a^5b^{10}c^{11}}$

7. $\sqrt{6}(\sqrt{10} - \sqrt{33})$

8. $\dfrac{-\sqrt[4]{80a^2b}}{\sqrt[4]{5a^{-1}b^{-6}}}$

9. $\dfrac{\sqrt{t}}{\sqrt[8]{t^3}}$

10. $\sqrt[5]{\dfrac{3a^{12}}{96a^2}}$

11. $2\sqrt{3} - 5\sqrt{12}$

12. $(\sqrt{5}+3)(\sqrt{5}-3)$

13. $(\sqrt{15}+\sqrt{10})^2$

14. $\sqrt{25x-25} - \sqrt{9x-9}$

15. $\sqrt{x^3y}\sqrt[5]{xy^4}$

16. $\sqrt[3]{5000} + \sqrt[3]{625}$

17. $\sqrt{\sqrt[5]{x^2}}$

18. $\sqrt{3x^2+6x+3}$

19. $\left(\sqrt[4]{a^2b^3}\right)^2$

20. $\sqrt[3]{12x^2y^5}\sqrt[3]{18x^7y}$

7.6 Solving Radical Equations

The Principle of Powers • Equations with Two Radical Terms

In Sections 7.1–7.5, we learned how to manipulate radical expressions as well as expressions containing rational exponents. We performed this work to find *equivalent expressions*.

Now that we know how to work with radicals and rational exponents, we can learn how to solve a new type of equation.

The Principle of Powers

A **radical equation** is an equation in which the variable appears in a radicand. Examples are

$$\sqrt[3]{2x} + 1 = 5, \qquad \sqrt{a-2} = 7, \qquad \text{and} \qquad 4 - \sqrt{3x+1} = \sqrt{6-x}.$$

To solve such equations, we need a new principle. Suppose $a = b$ is true. If we square both sides, we get another true equation: $a^2 = b^2$. This can be generalized.

> **The Principle of Powers**
> If $a = b$, then $a^n = b^n$ for any exponent n.

Note that the principle of powers is an "if–then" statement. The statement obtained by interchanging the two parts of the sentence—"if $a^n = b^n$ for some exponent n, then $a = b$"—*is not always true*. For example, "if $x = 3$, then $x^2 = 9$" is true, but the statement "if $x^2 = 9$, then $x = 3$" is *not* true when x is replaced with -3. For this reason, when both sides of an equation are raised to an even exponent, it is essential to check the answer(s) in the *original* equation.

EXAMPLE **1** Solve: $\sqrt{x} - 3 = 4$.

SOLUTION Before using the principle of powers, we need to isolate the radical term:

$$\sqrt{x} - 3 = 4$$
$$\sqrt{x} = 7 \qquad \text{Isolating the radical by adding 3 to both sides}$$
$$(\sqrt{x})^2 = 7^2 \qquad \text{Using the principle of powers}$$
$$x = 49.$$

Check:

$$\begin{array}{c|c} \sqrt{x} - 3 = 4 \\ \hline \sqrt{49} - 3 & 4 \\ 7 - 3 & \\ 4 \overset{?}{=} 4 & \text{TRUE} \end{array}$$

The solution is 49.

TRY EXERCISE 7

EXAMPLE **2** Solve: $\sqrt{x} + 5 = 3$.

SOLUTION

$$\sqrt{x} + 5 = 3$$
$$\sqrt{x} = -2 \qquad \text{Isolating the radical by adding } -5 \text{ to both sides}$$

The equation $\sqrt{x} = -2$ has no solution because the principal square root of a number is never negative. We continue as in Example 1 for comparison.

$$(\sqrt{x})^2 = (-2)^2 \qquad \text{Using the principle of powers}$$
$$x = 4$$

Check:

$$\begin{array}{c|c} \sqrt{x} + 5 = 3 \\ \hline \sqrt{4} + 5 & 3 \\ 2 + 5 & \\ 7 \overset{?}{=} 3 & \text{FALSE} \end{array}$$

The number 4 does not check. Thus, $\sqrt{x} + 5 = 3$ has no solution.

TRY EXERCISE 27

CAUTION! Raising both sides of an equation to an even power may not produce an equivalent equation. In this case, a check is essential.

Note in Example 2 that $x = 4$ has the solution 4, but $\sqrt{x} + 5 = 3$ has *no* solution. Thus the equations $x = 4$ and $\sqrt{x} + 5 = 3$ are *not* equivalent.

To Solve an Equation with a Radical Term

1. Isolate the radical term on one side of the equation.
2. Use the principle of powers and solve the resulting equation.
3. Check any possible solution in the original equation.

EXAMPLE 3

Solve: $x = \sqrt{x + 7} + 5$.

SOLUTION

$$x = \sqrt{x + 7} + 5$$

$$x - 5 = \sqrt{x + 7} \qquad \text{Isolating the radical by subtracting 5 from both sides}$$

$$\left.\begin{array}{l} (x - 5)^2 = (\sqrt{x + 7})^2 \\ x^2 - 10x + 25 = x + 7 \end{array}\right\} \quad \begin{array}{l} \text{Using the principle of powers;} \\ \text{squaring both sides} \end{array}$$

$$x^2 - 11x + 18 = 0 \qquad \begin{array}{l} \text{Adding } -x - 7 \text{ to both sides to} \\ \text{write the quadratic equation in} \\ \text{standard form} \end{array}$$

$$(x - 9)(x - 2) = 0 \qquad \text{Factoring}$$

$$x = 9 \quad or \quad x = 2 \qquad \text{Using the principle of zero products}$$

The possible solutions are 9 and 2. Let's check.

Check: For 9:

$$x = \sqrt{x + 7} + 5$$

$$\begin{array}{c|c} 9 & \sqrt{9 + 7} + 5 \\ \hline & 9 \stackrel{?}{=} 9 \qquad \text{TRUE} \end{array}$$

For 2:

$$x = \sqrt{x + 7} + 5$$

$$\begin{array}{c|c} 2 & \sqrt{2 + 7} + 5 \\ \hline & 2 \stackrel{?}{=} 8 \qquad \text{FALSE} \end{array}$$

Since 9 checks but 2 does not, the solution is 9.

TRY EXERCISE 39

It is important to isolate a radical term before using the principle of powers. Suppose in Example 3 that both sides of the equation were squared *before* isolating the radical. We then would have had the expression $(\sqrt{x + 7} + 5)^2$ or $x + 7 + 10\sqrt{x + 7} + 25$ on the right side, and the radical would have remained in the problem.

TECHNOLOGY CONNECTION

To solve Example 3, we can graph $y_1 = x$ and $y_2 = (x + 7)^{1/2} + 5$ and then use the INTERSECT option of the CALC menu to find the point of intersection. The intersection occurs at $x = 9$. Note that there is no intersection when $x = 2$, as predicted in the check of Example 3.

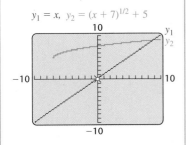

$y_1 = x, \; y_2 = (x + 7)^{1/2} + 5$

1. Use a graphing calculator to solve Examples 1, 2, 4, 5, and 6. Compare your answers with those found using the algebraic methods shown.

EXAMPLE 4

Solve: $(2x + 1)^{1/3} + 5 = 0$.

SOLUTION   We can use exponential notation to solve:

$$(2x + 1)^{1/3} + 5 = 0$$

$$(2x + 1)^{1/3} = -5 \qquad \text{Subtracting 5 from both sides}$$

$$[(2x + 1)^{1/3}]^3 = (-5)^3 \qquad \text{Cubing both sides}$$

$$(2x + 1)^1 = (-5)^3 \qquad \begin{array}{l} \text{Multiplying exponents. Try to do this} \\ \text{mentally.} \end{array}$$

$$2x + 1 = -125$$

$$2x = -126 \qquad \text{Subtracting 1 from both sides}$$

$$x = -63.$$

Because both sides were raised to an *odd* power, a check is not *essential*. It is wise, however, for the student to confirm that -63 checks and is the solution.

TRY EXERCISE 25

Equations with Two Radical Terms

A strategy for solving equations with two or more radical terms is as follows.

> **To Solve an Equation with Two or More Radical Terms**
>
> 1. Isolate one of the radical terms.
> 2. Use the principle of powers.
> 3. If a radical remains, perform steps (1) and (2) again.
> 4. Solve the resulting equation.
> 5. Check possible solutions in the original equation.

EXAMPLE 5   Solve: $\sqrt{2x-5} = 1 + \sqrt{x-3}$.

SOLUTION

$$\sqrt{2x-5} = 1 + \sqrt{x-3}$$

$$(\sqrt{2x-5})^2 = (1 + \sqrt{x-3})^2 \qquad \text{One radical is already isolated. We square both sides.}$$

> This is like squaring a binomial. We square 1, then find twice the product of 1 and $\sqrt{x-3}$, and finally square $\sqrt{x-3}$. Study this carefully.

$$2x - 5 = 1 + 2\sqrt{x-3} + (\sqrt{x-3})^2$$

$$2x - 5 = 1 + 2\sqrt{x-3} + (x-3)$$

$$x - 3 = 2\sqrt{x-3} \qquad \text{Isolating the remaining radical term}$$

$$(x-3)^2 = (2\sqrt{x-3})^2 \qquad \text{Squaring both sides}$$

$$x^2 - 6x + 9 = 4(x-3) \qquad \begin{array}{l}\text{Remember to square both the 2 and} \\ \text{the } \sqrt{x-3} \text{ on the right side.}\end{array}$$

$$x^2 - 6x + 9 = 4x - 12$$

$$x^2 - 10x + 21 = 0$$

$$(x-7)(x-3) = 0 \qquad \text{Factoring}$$

$$x = 7 \quad or \quad x = 3 \qquad \text{Using the principle of zero products}$$

We leave it to the student to show that 7 and 3 both check and are the solutions.

TRY EXERCISE 41

CAUTION!   A common error in solving equations like

$$\sqrt{2x-5} = 1 + \sqrt{x-3}$$

is to obtain $1 + (x-3)$ as the square of the right side. This is wrong because $(A + B)^2 \neq A^2 + B^2$. For example,

$$\left.\begin{array}{c}(1+2)^2 \neq 1^2 + 2^2 \\ 3^2 \neq 1 + 4 \\ 9 \neq 5.\end{array}\right\} \quad \begin{array}{l}\text{See Example 5 for the correct} \\ \text{expansion of } (1 + \sqrt{x-3})^2.\end{array}$$

EXAMPLE 6   Let $f(x) = \sqrt{x + 5} - \sqrt{x - 7}$. Find all x-values for which $f(x) = 2$.

SOLUTION   We must have $f(x) = 2$, or

$$\sqrt{x + 5} - \sqrt{x - 7} = 2. \quad \text{Substituting for } f(x)$$

To solve, we isolate one radical term and square both sides:

Isolate a radical term.	$\sqrt{x + 5} = 2 + \sqrt{x - 7}$	Adding $\sqrt{x - 7}$ to both sides. This isolates one of the radical terms.
Raise both sides to the same power.	$(\sqrt{x + 5})^2 = (2 + \sqrt{x - 7})^2$	Using the principle of powers (squaring both sides)
	$x + 5 = 4 + 4\sqrt{x - 7} + (x - 7)$	Using $(A + B)^2 = A^2 + 2AB + B^2$
	$5 = 4\sqrt{x - 7} - 3$	Adding $-x$ to both sides and combining like terms
Isolate a radical term.	$8 = 4\sqrt{x - 7}$	Isolating the remaining radical term
	$2 = \sqrt{x - 7}$	
Raise both sides to the same power.	$2^2 = (\sqrt{x - 7})^2$	Squaring both sides
	$4 = x - 7$	
Solve.	$11 = x.$	
Check.	*Check:* $f(11) = \sqrt{11 + 5} - \sqrt{11 - 7}$ $= \sqrt{16} - \sqrt{4}$ $= 4 - 2 = 2.$	

We have $f(x) = 2$ when $x = 11$.

TRY EXERCISE 49

7.6   EXERCISE SET

➥ *Concept Reinforcement*   *Classify each statement as either true or false.*

1. If $x^2 = 25$, then $x = 5$.

2. If $t = 7$, then $t^2 = 49$.

3. If $\sqrt{x} = 3$, then $(\sqrt{x})^2 = 3^2$.

4. If $x^2 = 36$, then $x = 6$.

5. $\sqrt{x} - 8 = 7$ is equivalent to $\sqrt{x} = 15$.

6. $\sqrt{t} + 5 = 8$ is equivalent to $\sqrt{t} = 3$.

Solve.

7. $\sqrt{5x + 1} = 4$

8. $\sqrt{7x - 3} = 5$

9. $\sqrt{3x} + 1 = 5$

10. $\sqrt{2x} - 1 = 2$

11. $\sqrt{y + 5} - 4 = 1$

12. $\sqrt{x - 2} - 7 = -4$

13. $\sqrt{8 - x} + 7 = 10$

14. $\sqrt{y + 4} + 6 = 7$

15. $\sqrt[3]{y + 3} = 2$

16. $\sqrt[3]{x - 2} = 3$

17. $\sqrt[4]{t - 10} = 3$

18. $\sqrt[4]{t + 5} = 2$

19. $6\sqrt{x} = x$

20. $7\sqrt{y} = y$

21. $2y^{1/2} - 13 = 7$

22. $3x^{1/2} + 12 = 9$

23. $\sqrt[3]{x} = -5$

24. $\sqrt[3]{y} = -4$

25. $z^{1/4} + 8 = 10$

26. $x^{1/4} - 2 = 1$

Aha! 27. $\sqrt{n} = -2$

28. $\sqrt{a} = -1$

29. $\sqrt[4]{3x + 1} - 4 = -1$

30. $\sqrt[4]{2x + 3} - 5 = -2$

31. $(21x + 55)^{1/3} = 10$

32. $(5y + 31)^{1/4} = 2$

33. $\sqrt[3]{3y + 6} + 7 = 8$

34. $\sqrt[3]{6x + 9} + 5 = 2$

35. $\sqrt{3t + 4} = \sqrt{4t + 3}$

36. $\sqrt{2t - 7} = \sqrt{3t - 12}$

37. $3(4 - t)^{1/4} = 6^{1/4}$

38. $2(1 - x)^{1/3} = 4^{1/3}$

39. $3 + \sqrt{5 - x} = x$

40. $x = \sqrt{x - 1} + 3$

41. $\sqrt{4x - 3} = 2 + \sqrt{2x - 5}$

42. $3 + \sqrt{z - 6} = \sqrt{z + 9}$

43. $\sqrt{20 - x} + 8 = \sqrt{9 - x} + 11$

44. $4 + \sqrt{10 - x} = 6 + \sqrt{4 - x}$

45. $\sqrt{x + 2} + \sqrt{3x + 4} = 2$

46. $\sqrt{6x + 7} - \sqrt{3x + 3} = 1$

47. If $f(x) = \sqrt{x} + \sqrt{x - 9}$, find any x for which $f(x) = 1$.

48. If $g(x) = \sqrt{x} + \sqrt{x - 5}$, find any x for which $g(x) = 5$.

49. If $f(t) = \sqrt{t - 2} - \sqrt{4t + 1}$, find any t for which $f(t) = -3$.

50. If $g(t) = \sqrt{2t + 7} - \sqrt{t + 15}$, find any t for which $g(t) = -1$.

51. If $f(x) = \sqrt{2x - 3}$ and $g(x) = \sqrt{x + 7} - 2$, find any x for which $f(x) = g(x)$.

52. If $f(x) = 2\sqrt{3x + 6}$ and $g(x) = 5 + \sqrt{4x + 9}$, find any x for which $f(x) = g(x)$.

53. If $f(t) = 4 - \sqrt{t - 3}$ and $g(t) = (t + 5)^{1/2}$, find any t for which $f(t) = g(t)$.

54. If $f(t) = 7 + \sqrt{2t - 5}$ and $g(t) = 3(t + 1)^{1/2}$, find any t for which $f(t) = g(t)$.

55. Explain in your own words why it is important to check your answers when using the principle of powers.

56. The principle of powers is an "if–then" statement that becomes false when the sentence parts are interchanged. Give an example of another such if–then statement from everyday life (answers will vary).

Skill Review

To prepare for Section 7.7, review finding dimensions of triangles and rectangles (Sections 1.4 and 5.8).

Solve.

57. *Sign dimensions.* The largest sign in the United States is a rectangle with a perimeter of 430 ft. The length of the rectangle is 5 ft longer than thirteen times the width. Find the dimensions of the sign. [1.4]

Source: Florida Center for Instructional Technology

58. *Sign dimensions.* The base of a triangular sign is 4 in. longer than twice the height. The area of the sign is 255 in². Find the dimensions of the sign. [5.8]

59. *Photograph dimensions.* A rectangular family photo is 4 in. longer than it is wide. The area of the photo is 140 in². Find the dimensions of the photograph. [5.8]

60. *Sidewalk length.* The length of a rectangular lawn between classroom buildings is 2 yd less than twice the width of the lawn. A path that is 34 yd long stretches diagonally across the area. What are the dimensions of the lawn? [5.8]

61. The sides of a right triangle are consecutive even integers. Find the length of each side. [5.8]

62. One leg of a right triangle is 5 cm long. The hypotenuse is 1 cm longer than the other leg. Find the length of the hypotenuse. [5.8]

Synthesis

63. Describe a procedure that could be used to create radical equations that have no solution.

64. Is checking essential when the principle of powers is used with an odd power n? Why or why not?

65. *Firefighting.* The velocity of water flow, in feet per second, from a nozzle is given by

$$v(p) = 12.1\sqrt{p},$$

where p is the nozzle pressure, in pounds per square inch (psi). Find the nozzle pressure if the water flow is 100 feet per second.

Source: Houston Fire Department Continuing Education

66. *Firefighting.* The velocity of water flow, in feet per second, from a water tank that is h feet high is given by

$$v(h) = 8\sqrt{h}.$$

Find the height of a water tank that provides a water flow of 60 feet per second.

Source: Houston Fire Department Continuing Education

67. *Music.*   The frequency of a violin string varies directly with the square root of the tension on the string. A violin string vibrates with a frequency of 260 Hz when the tension on the string is 28 N. What is the frequency when the tension is 32 N?

68. *Music.*   The frequency of a violin string varies inversely with the square root of the density of the string. A nylon violin string with a density of 1200 kg/m^3 vibrates with a frequency of 250 Hz. What is the frequency of a silk violin string with a density of 1300 kg/m^3?
Source: www.speech.kth.se

Steel manufacturing.   In the production of steel and other metals, the temperature of the molten metal is so great that conventional thermometers melt. Instead, sound is transmitted across the surface of the metal to a receiver on the far side and the speed of the sound is measured. The formula

$$S(t) = 1087.7\sqrt{\frac{9t + 2617}{2457}}$$

gives the speed of sound S(t), in feet per second, at a temperature of t degrees Celsius.

69. Find the temperature of a blast furnace where sound travels 1880 ft/sec.

70. Find the temperature of a blast furnace where sound travels 1502.3 ft/sec.

71. Solve the above equation for *t*.

Automotive repair.   For an engine with a displacement of 2.8 L, the function given by

$$d(n) = 0.75\sqrt{2.8n}$$

can be used to determine the diameter size of the carburetor's opening, in millimeters. Here n is the number of rpm's at which the engine achieves peak performance.
Source: macdizzy.com

72. If the diameter of a carburetor's opening is 81 mm, for what number of rpm's will the engine produce peak power?

73. If a carburetor's opening is 84 mm, for what number of rpm's will the engine produce peak power?

Escape velocity.   A formula for the escape velocity v of a satellite is

$$v = \sqrt{2gr}\sqrt{\frac{h}{r + h}},$$

where g is the force of gravity, r is the planet or star's radius, and h is the height of the satellite above the planet or star's surface.

74. Solve for *h*.

75. Solve for *r*.

Solve.

76. $\left(\frac{z}{4} - 5\right)^{2/3} = \frac{1}{25}$

77. $\dfrac{x + \sqrt{x + 1}}{x - \sqrt{x + 1}} = \dfrac{5}{11}$

78. $\sqrt{\sqrt{y} + 49} = 7$

79. $(z^2 + 17)^{3/4} = 27$

80. $x^2 - 5x - \sqrt{x^2 - 5x - 2} = 4$
(*Hint:* Let $u = x^2 - 5x - 2$.)

81. $\sqrt{8 - b} = b\sqrt{8 - b}$

Without graphing, determine the x-intercepts of the graphs given by each of the following.

82. $f(x) = \sqrt{x - 2} - \sqrt{x + 2} + 2$

83. $g(x) = 6x^{1/2} + 6x^{-1/2} - 37$

84. $f(x) = (x^2 + 30x)^{1/2} - x - (5x)^{1/2}$

85. Use a graphing calculator to check your answers to Exercises 9, 15, and 31.

86. Saul is trying to solve Exercise 73 using a graphing calculator. Without resorting to trial and error, how can he determine a suitable viewing window for finding the solution?

87. Use a graphing calculator to check your answers to Exercises 27, 35, and 41.

CORNER

Tailgater Alert

Focus: Radical equations and problem solving
Time: 15–25 minutes
Group size: 2–3
Materials: Calculators or square-root tables

The faster a car is traveling, the more distance it needs to stop. Thus it is important for drivers to allow sufficient space between their vehicle and the vehicle in front of them. Police recommend that for each 10 mph of speed, a driver allow 1 car length. Thus a driver going 30 mph should have at least 3 car lengths between his or her vehicle and the one in front.

In Exercise Set 7.3, the function $r(L) = 2\sqrt{5L}$ was used to find the speed, in miles per hour, that a car was traveling when it left skid marks L feet long.

ACTIVITY

1. Each group member should estimate the length of a car in which he or she frequently travels. (Each should use a different length, if possible.)

2. Using a calculator as needed, each group member should complete the table below. Column 1 gives a car's speed s, and column 2 lists the minimum amount of space between cars traveling s miles per hour, as recommended by police. Column 3 is the speed that a vehicle *could* travel were it forced to stop in the distance listed in column 2, using the above function.

Column 1 s (in miles per hour)	Column 2 $L(s)$ (in feet)	Column 3 $r(L)$ (in miles per hour)
20		
30		
40		
50		
60		
70		

3. Determine whether there are any speeds at which the "1 car length per 10 mph" guideline might not suffice. On what reasoning do you base your answer? Compare tables to determine how car length affects the results. What recommendations would your group make to a new driver?

7.7 The Distance and Midpoint Formulas and Other Applications

Using the Pythagorean Theorem ▪ Two Special Triangles ▪ The Distance and Midpoint Formulas

Using the Pythagorean Theorem

There are many kinds of problems that involve powers and roots. Many also involve right triangles and the Pythagorean theorem, which we studied in Section 5.8 and restate here.

STUDY SKILLS

Making Sketches

One need not be an artist to make highly useful mathematical sketches. That said, it is important to make sure that your sketches are drawn accurately enough to represent the relative sizes within each shape. For example, if one side of a triangle is clearly the longest, make sure your drawing reflects this.

The Pythagorean Theorem*

In any right triangle, if a and b are the lengths of the legs and c is the length of the hypotenuse, then

$$a^2 + b^2 = c^2.$$

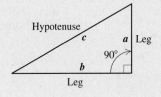

In using the Pythagorean theorem, we often make use of the following principle.

The Principle of Square Roots

For any nonnegative real number n,

If $x^2 = n$, then $x = \sqrt{n}$ or $x = -\sqrt{n}$.

For most real-world applications involving length or distance, $-\sqrt{n}$ is not needed.

EXAMPLE **1**

Baseball. A baseball diamond is actually a square 90 ft on a side. Suppose a catcher fields a ball while standing on the third-base line 10 ft from home plate, as shown in the figure. How far is the catcher's throw to first base? Give an exact answer and an approximation to three decimal places.

SOLUTION We make a drawing and let $d =$ the distance, in feet, to first base. Note that a right triangle is formed in which the leg from home plate to first base measures 90 ft and the leg from home plate to where the catcher fields the ball measures 10 ft.

We substitute these values into the Pythagorean theorem to find d:

$$d^2 = 90^2 + 10^2$$
$$d^2 = 8100 + 100$$
$$d^2 = 8200.$$

We now use the principle of square roots: If $d^2 = 8200$, then $d = \sqrt{8200}$ or $d = -\sqrt{8200}$. Since d represents a length, it follows that d is the positive square root of 8200:

$d = \sqrt{8200}$ ft   This is an exact answer.

$d \approx 90.554$ ft.   Using a calculator for an approximation

TRY EXERCISE 19

---

*The converse of the Pythagorean theorem also holds. That is, if a, b, and c are the lengths of the sides of a triangle and $a^2 + b^2 = c^2$, then the triangle is a right triangle.

EXAMPLE **2**   *Guy wires.*   The base of a 40-ft long guy wire is located 15 ft from the telephone pole that it is anchoring. How high up the pole does the guy wire reach? Give an exact answer and an approximation to three decimal places.

SOLUTION   We make a drawing and let h = the height, in feet, to which the guy wire reaches. A right triangle is formed in which one leg measures 15 ft and the hypotenuse measures 40 ft. Using the Pythagorean theorem, we have

$$h^2 + 15^2 = 40^2$$
$$h^2 + 225 = 1600$$
$$h^2 = 1375$$
$$h = \sqrt{1375}.$$

Exact answer:

$h = \sqrt{1375}$ ft   Using the positive square root

Approximation:

$h \approx 37.081$ ft   Using a calculator

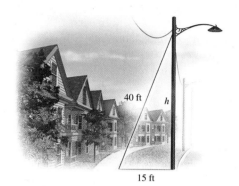

TRY EXERCISE ▶ 23

Two Special Triangles

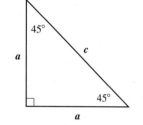

When both legs of a right triangle are the same size, as shown at left, we call the triangle an *isosceles right triangle*. If one leg of an isosceles right triangle has length a, we can find a formula for the length of the hypotenuse as follows:

$$c^2 = a^2 + b^2$$
$$c^2 = a^2 + a^2$$ Because the triangle is isosceles, both legs are the same size: $a = b$.
$$c^2 = 2a^2.$$ Combining like terms

Next, we use the principle of square roots. Because a, b, and c are lengths, there is no need to consider negative square roots or absolute values. Thus,

$$c = \sqrt{2a^2}$$ Using the principle of square roots
$$c = \sqrt{a^2 \cdot 2} = a\sqrt{2}.$$

EXAMPLE **3**   One leg of an isosceles right triangle measures 7 cm. Find the length of the hypotenuse. Give an exact answer and an approximation to three decimal places.

SOLUTION   We substitute:

$$c = a\sqrt{2}$$ This equation is worth remembering.
$$c = 7\sqrt{2}.$$

Exact answer: $c = 7\sqrt{2}$ cm

Approximation: $c \approx 9.899$ cm   Using a calculator

TRY EXERCISE ▶ 29

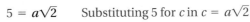

When the hypotenuse of an isosceles right triangle is known, the lengths of the legs can be found.

EXAMPLE 4   The hypotenuse of an isosceles right triangle is 5 ft long. Find the length of a leg. Give an exact answer and an approximation to three decimal places.

SOLUTION   We replace c with 5 and solve for a:

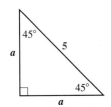

$$5 = a\sqrt{2} \qquad \text{Substituting 5 for } c \text{ in } c = a\sqrt{2}$$

$$\frac{5}{\sqrt{2}} = a \qquad \text{Dividing both sides by } \sqrt{2}$$

$$\frac{5\sqrt{2}}{2} = a. \qquad \text{Rationalize the denominator if desired.}$$

Exact answer: $a = \dfrac{5}{\sqrt{2}}$ ft, or $\dfrac{5\sqrt{2}}{2}$ ft

Approximation: $a \approx 3.536$ ft   Using a calculator

TRY EXERCISE 35

A second special triangle is known as a 30°–60°–90° triangle, so named because of the measures of its angles. Note that in an equilateral triangle, all sides have the same length and all angles are 60°. An altitude, drawn dashed in the figure, bisects, or splits in half, one angle and one side. Two 30°–60°–90° right triangles are thus formed.

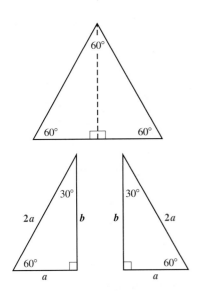

If we let a represent the length of the shorter leg in a 30°–60°–90° triangle, then $2a$ represents the length of the hypotenuse. We have

$$a^2 + b^2 = (2a)^2 \qquad \text{Using the Pythagorean theorem}$$
$$a^2 + b^2 = 4a^2$$
$$b^2 = 3a^2 \qquad \text{Subtracting } a^2 \text{ from both sides}$$
$$b = \sqrt{3a^2} \qquad \text{Considering only the positive square root}$$
$$b = \sqrt{a^2 \cdot 3}$$
$$b = a\sqrt{3}.$$

EXAMPLE 5   The shorter leg of a 30°–60°–90° triangle measures 8 in. Find the lengths of the other sides. Give exact answers and, where appropriate, an approximation to three decimal places.

SOLUTION   The hypotenuse is twice as long as the shorter leg, so we have

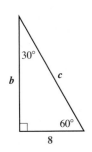

$$c = 2a \qquad \text{This relationship is worth remembering.}$$
$$= 2 \cdot 8 = 16 \text{ in.} \qquad \text{This is the length of the hypotenuse.}$$

The length of the longer leg is the length of the shorter leg times $\sqrt{3}$. This gives us

$$b = a\sqrt{3} \qquad \text{This is also worth remembering.}$$
$$= 8\sqrt{3} \text{ in.} \qquad \text{This is the length of the longer leg.}$$

Exact answer: $c = 16$ in., $b = 8\sqrt{3}$ in.
Approximation: $b \approx 13.856$ in.

TRY EXERCISE 37

EXAMPLE   6   The length of the longer leg of a 30°–60°–90° triangle is 14 cm. Find the length of the hypotenuse. Give an exact answer and an approximation to three decimal places.

SOLUTION   The length of the hypotenuse is twice the length of the shorter leg. We first find a, the length of the shorter leg, by using the length of the longer leg:

$$14 = a\sqrt{3} \qquad \text{Substituting 14 for } b \text{ in } b = a\sqrt{3}$$

$$\frac{14}{\sqrt{3}} = a. \qquad \text{Dividing by } \sqrt{3}$$

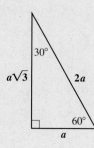

Since the hypotenuse is twice as long as the shorter leg, we have

$$c = 2a$$

$$= 2 \cdot \frac{14}{\sqrt{3}} \qquad \text{Substituting}$$

$$= \frac{28}{\sqrt{3}} \text{ cm.}$$

Exact answer: $c = \dfrac{28}{\sqrt{3}}$ cm, or $\dfrac{28\sqrt{3}}{3}$ cm if the denominator is rationalized.

Approximation: $c \approx 16.166$ cm

TRY EXERCISE ▶ 33

STUDENT NOTES —————

Perhaps the easiest way to remember the important results listed in the adjacent box is to write out, on your own, the derivations shown on pp. 475 and 476.

Lengths Within Isosceles and 30°–60°–90° Right Triangles

The length of the hypotenuse in an isosceles right triangle is the length of a leg times $\sqrt{2}$.

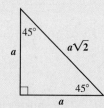

The length of the longer leg in a 30°–60°–90° right triangle is the length of the shorter leg times $\sqrt{3}$. The hypotenuse is twice as long as the shorter leg.

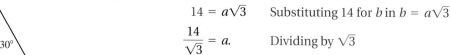

The Distance and Midpoint Formulas

We can use the Pythagorean theorem to find the distance between two points on a plane.

To find the distance between two points on the number line, we subtract. Depending on the order in which we subtract, the difference may be positive or negative. However, if we take the absolute value of the difference, we always obtain a positive value for the distance:

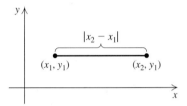

$$|4 - (-3)| = |7| = 7$$
$$|-3 - 4| = |-7| = 7$$

If two points are on a horizontal line, they have the same second coordinate. We can find the distance between them by subtracting their first coordinates and taking the absolute value of that difference.

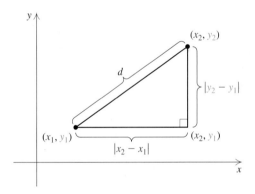

The distance between the points (x_1, y_1) and (x_2, y_1) on a horizontal line is thus $|x_2 - x_1|$. Similarly, the distance between the points (x_2, y_1) and (x_2, y_2) on a vertical line is $|y_2 - y_1|$.

Now consider two points (x_1, y_1) and (x_2, y_2). If $x_1 \neq x_2$ and $y_1 \neq y_2$, these points, along with the point (x_2, y_1), describe a right triangle. The lengths of the legs are $|x_2 - x_1|$ and $|y_2 - y_1|$. We find d, the length of the hypotenuse, by using the Pythagorean theorem:

$$d^2 = |x_2 - x_1|^2 + |y_2 - y_1|^2.$$

Since the square of a number is the same as the square of its opposite, we can replace the absolute-value signs with parentheses:

$$d^2 = (x_2 - x_1)^2 + (y_2 - y_1)^2.$$

Taking the principal square root, we have a formula for distance.

> ## The Distance Formula
> The distance d between any two points (x_1, y_1) and (x_2, y_2) is given by
> $$d = \sqrt{(x_2 - x_1)^2 + (y_2 - y_1)^2}.$$

EXAMPLE **7**

Find the distance between $(5, -1)$ and $(-4, 6)$. Find an exact answer and an approximation to three decimal places.

SOLUTION We substitute into the distance formula:

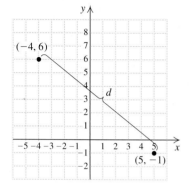

$$d = \sqrt{(-4 - 5)^2 + [6 - (-1)]^2} \quad \text{Substituting. A drawing is optional.}$$
$$= \sqrt{(-9)^2 + 7^2}$$
$$= \sqrt{130} \quad \text{This is exact.}$$
$$\approx 11.402. \quad \text{Using a calculator for an approximation}$$

> TRY EXERCISE 51

The distance formula is needed to verify a formula for the coordinates of the *midpoint* of a segment connecting two points. We state the midpoint formula and leave its proof to the exercises.

STUDENT NOTES

To help remember the formulas correctly, note that the distance formula (a variation on the Pythagorean theorem) involves both subtraction and addition, whereas the midpoint formula does not include any subtraction.

> ## The Midpoint Formula
> If the endpoints of a segment are (x_1, y_1) and (x_2, y_2), then the coordinates of the midpoint are
> $$\left(\frac{x_1 + x_2}{2}, \frac{y_1 + y_2}{2}\right).$$
>
>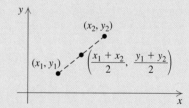
>
> (To locate the midpoint, average the x-coordinates and average the y-coordinates.)

EXAMPLE **8**

Find the midpoint of the segment with endpoints $(-2, 3)$ and $(4, -6)$.

SOLUTION Using the midpoint formula, we obtain

$$\left(\frac{-2 + 4}{2}, \frac{3 + (-6)}{2}\right), \quad \text{or} \quad \left(\frac{2}{2}, \frac{-3}{2}\right), \quad \text{or} \quad \left(1, -\frac{3}{2}\right).$$

> TRY EXERCISE 65

7.7   **EXERCISE SET**

Concept Reinforcement   *Complete each sentence with the best choice from the column on the right.*

1. In any _____ triangle, the square of the length of the hypotenuse is the sum of the squares of the lengths of the legs.

2. The shortest side of a right triangle is always one of the two _____.

3. The principle of _____ states that if $x^2 = n$, then $x = \sqrt{n}$ or $x = -\sqrt{n}$.

4. In a(n) _____ right triangle, both legs have the same length.

5. In a(n) _____ right triangle, the hypotenuse is twice as long as the shorter leg.

6. If both legs in a right triangle have measure a, then the _____ measures $a\sqrt{2}$.

a) Hypotenuse

b) Isosceles

c) Legs

d) Right

e) Square roots

f) 30°–60°–90°

In a right triangle, find the length of the side not given. Give an exact answer and, where appropriate, an approximation to three decimal places.

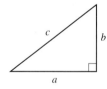

7. $a = 5$, $b = 3$

8. $a = 8$, $b = 10$

Aha! 9. $a = 9$, $b = 9$

10. $a = 10$, $b = 10$

11. $b = 15$, $c = 17$

12. $a = 7$, $c = 25$

In Exercises 13–18, give an exact answer and, where appropriate, an approximation to three decimal places.

13. A right triangle's hypotenuse is 8 m and one leg is $4\sqrt{3}$ m. Find the length of the other leg.

14. A right triangle's hypotenuse is 6 cm and one leg is $\sqrt{5}$ cm. Find the length of the other leg.

15. The hypotenuse of a right triangle is $\sqrt{20}$ in. and one leg measures 1 in. Find the length of the other leg.

16. The hypotenuse of a right triangle is $\sqrt{15}$ ft and one leg measures 2 ft. Find the length of the other leg.

Aha! 17. One leg in a right triangle is 1 m and the hypotenuse measures $\sqrt{2}$ m. Find the length of the other leg.

18. One leg of a right triangle is 1 yd and the hypotenuse measures 2 yd. Find the length of the other leg.

In Exercises 19–28, give an exact answer and, where appropriate, an approximation to three decimal places.

19. *Bicycling.* Clare routinely bicycles across a rectangular parking lot on her way to work. If the lot is 200 ft long and 150 ft wide, how far does Clare travel when she rides across the lot diagonally?

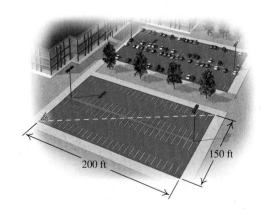

200 ft   150 ft

20. *Guy wire.* How long is a guy wire if it reaches from the top of a 15-ft pole to a point on the ground 10 ft from the pole?

21. *Softball.* A slow-pitch softball diamond is actually a square 65 ft on a side. How far is it from home plate to second base?

22. *Baseball.* Suppose the catcher in Example 1 makes a throw to second base from the same location. How far is that throw?

23. *Television sets.* What does it mean to refer to a 51-in. TV set? Such units refer to the diagonal of the screen. A 51-in. TV set has a width of 45 in. What is its height?

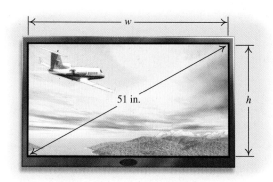

24. *Television sets.* A 53-in. TV set has a screen with a height of 28 in. What is its width? (See Exercise 23.)

25. *Speaker placement.* A stereo receiver is in a corner of a 12-ft by 14-ft room. Wire will run under a rug, diagonally, to a subwoofer in the far corner. If 4 ft of slack is required on each end, how long a piece of wire should be purchased?

26. *Distance over water.* To determine the width of a pond, a surveyor locates two stakes at either end of the pond and uses instrumentation to place a third stake so that the distance across the pond is the length of a hypotenuse. If the third stake is 90 m from one stake and 70 m from the other, what is the distance across the pond?

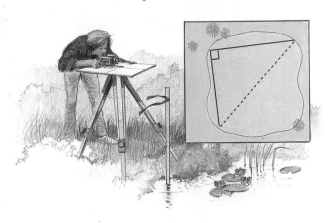

27. *Walking.* Students at Pohlman Community College have worn a path that cuts diagonally across the campus "quad." If the quad is actually a rectangle that Marissa measured to be 70 paces long and 40 paces wide, how many paces will Marissa save by using the diagonal path?

28. *Crosswalks.* The diagonal crosswalk at the intersection of State St. and Main St. is the hypotenuse of a triangle in which the crosswalks across State St. and Main St. are the legs. If State St. is 28 ft wide and Main St. is 40 ft wide, how much shorter is the distance traveled by pedestrians using the diagonal crosswalk?

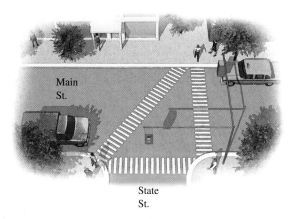

For each triangle, find the missing length(s). Give an exact answer and, where appropriate, an approximation to three decimal places.

29.

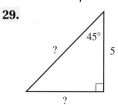

30.

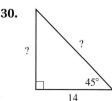

31.

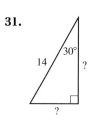

32.

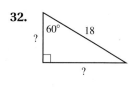

33.

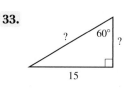

34.

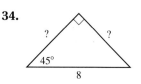

35.

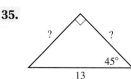

36.

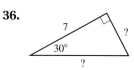

37.

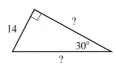

14, ?, 30°, ?

38.

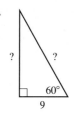

?, ?, 60°, 9

39.

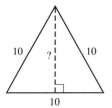

10, ?, 10, 10

40.

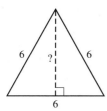

6, ?, 6, 6

41.

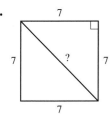

7, ?, 7, 7

42.

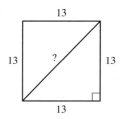

13, ?, 13, 13, 13

43.

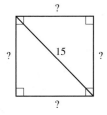

?, ?, 15, ?, ?

44.

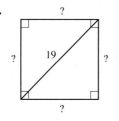

?, ?, 19, ?, ?

In Exercises 45–48, give an exact answer and, where appropriate, an approximation to three decimal places.

45. *Bridge expansion.* During the summer heat, a 2-mi bridge expands 2 ft in length. If we assume that the bulge occurs straight up the middle, how high is the bulge? (The answer may surprise you. Most bridges have expansion spaces to avoid such buckling.)

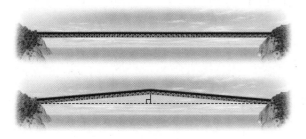

46. Triangle *ABC* has sides of lengths 25 ft, 25 ft, and 30 ft. Triangle *PQR* has sides of lengths 25 ft, 25 ft, and 40 ft. Which triangle, if either, has the greater area and by how much?

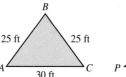
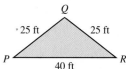

47. *Architecture.* The Rushton Triangular Lodge in Northamptonshire, England, was designed and constructed by Sir Thomas Tresham between 1593 and 1597. The building is in the shape of an equilateral triangle with walls of length 33 ft. How many square feet of land is covered by the lodge?
Source: The Internet Encyclopedia of Science

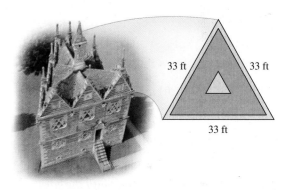

48. *Antenna length.* As part of an emergency radio communication station, Rik sets up an "Inverted-V" antenna. He stretches a copper wire from one point on the ground to a point on a tree and then back down to the ground, forming two 30°–60°–90° triangles. If the wire is fastened to the tree 34 ft above the ground, how long is the copper wire?

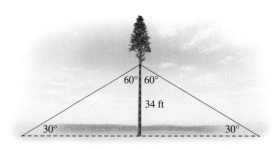

49. Find all points on the *y*-axis of a Cartesian coordinate system that are 5 units from the point (3, 0).

50. Find all points on the *x*-axis of a Cartesian coordinate system that are 5 units from the point (0, 4).

Find the distance between each pair of points. Where appropriate, find an approximation to three decimal places.

51. (4, 5) and (7, 1)

52. (0, 8) and (6, 0)

53. (0, −5) and (1, −2)

54. (−1, −4) and (−3, −5)

55. (−4, 4) and (6, −6)

56. (5, 21) and (−3, 1)

Aha! **57.** (8.6, −3.4) and (−9.2, −3.4)

58. (5.9, 2) and (3.7, −7.7)

59. $\left(\frac{1}{2}, \frac{1}{3}\right)$ and $\left(\frac{5}{6}, -\frac{1}{6}\right)$

60. $\left(\frac{5}{7}, \frac{1}{14}\right)$ and $\left(\frac{1}{7}, \frac{11}{14}\right)$

61. $(-\sqrt{6}, \sqrt{6})$ and (0, 0)

62. $(\sqrt{5}, -\sqrt{3})$ and (0, 0)

63. (−1, −30) and (−2, −40)

64. (0.5, 100) and (1.5, −100)

Find the midpoint of each segment with the given endpoints.

65. (−2, 5) and (8, 3)

66. (1, 4) and (9, −6)

67. (2, −1) and (5, 8)

68. (−1, 2) and (1, −3)

69. (−8, −5) and (6, −1)

70. (8, −2) and (−3, 4)

71. (−3.4, 8.1) and (4.8, −8.1)

72. (4.1, 6.9) and (5.2, −8.9)

73. $\left(\frac{1}{6}, -\frac{3}{4}\right)$ and $\left(-\frac{1}{3}, \frac{5}{6}\right)$

74. $\left(-\frac{4}{5}, -\frac{2}{3}\right)$ and $\left(\frac{1}{8}, \frac{3}{4}\right)$

75. $(\sqrt{2}, -1)$ and $(\sqrt{3}, 4)$

76. $(9, 2\sqrt{3})$ and $(-4, 5\sqrt{3})$

77. Are there any right triangles, other than those with sides measuring 3, 4, and 5, that have consecutive numbers for the lengths of the sides? Why or why not?

78. If a 30°−60°−90° triangle and an isosceles right triangle have the same perimeter, which will have the greater area? Why?

Skill Review

Review graphing (Sections 2.3, 2.4, and 4.4).

Graph on a plane.

79. $y = 2x - 3$ [2.3]

80. $y < x$ [4.4]

81. $8x - 4y = 8$ [2.4]

82. $2y - 1 = 7$ [2.4]

83. $x \geq 1$ [4.4]

84. $x - 5 = 6 - 2y$ [2.3]

Synthesis

85. Describe a procedure that uses the distance formula to determine whether three points, (x_1, y_1), (x_2, y_2), and (x_3, y_3), are vertices of a right triangle.

86. Outline a procedure that uses the distance formula to determine whether three points, (x_1, y_1), (x_2, y_2), and (x_3, y_3), are collinear (lie on the same line).

87. The perimeter of a regular hexagon is 72 cm. Determine the area of the shaded region shown.

88. If the perimeter of a regular hexagon is 120 ft, what is its area? (*Hint*: See Exercise 87.)

89. Each side of a regular octagon has length *s*. Find a formula for the distance *d* between the parallel sides of the octagon.

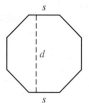

90. *Roofing.*   Kit's home, which is 24 ft wide and 32 ft long, needs a new roof. By counting clapboards that are 4 in. apart, Kit determines that the peak of the roof is 6 ft higher than the sides. A packet of shingles covers 100 ft^2. How many packets will the job require?

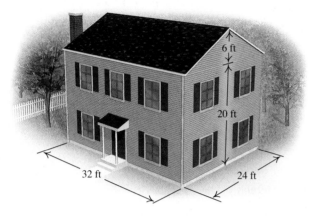

91. *Painting.*   (Refer to Exercise 90.) A gallon of Benjamin Moore® exterior acrylic paint covers 450–500 ft^2. If Kit's house has dimensions as shown above, how many gallons of paint should be bought to paint the house? What assumption(s) is made in your answer?

92. *Contracting.*   Oxford Builders has an extension cord on their generator that permits them to work, with electricity, anywhere in a circular area of 3850 ft^2. Find the dimensions of the largest square room they could work on without having to relocate the generator to reach each corner of the floor plan.

93. *Contracting.*   Cleary Construction has a hose attached to their insulation blower that permits them to reach anywhere in a circular area of 6160 ft^2. Find the dimensions of the largest square room with 12-ft ceilings in which they could reach all corners with the hose while leaving the blower centrally located. Assume that the blower sits on the floor.

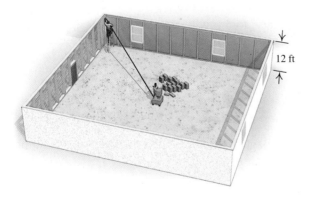

94. The length and the width of a rectangle are given by consecutive integers. The area of the rectangle is 90 cm^2. Find the length of a diagonal of the rectangle.

95. A cube measures 5 cm on each side. How long is the diagonal that connects two opposite corners of the cube? Give an exact answer.

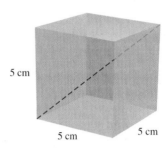

96. Prove the midpoint formula by showing that

   i) the distance from (x_1, y_1) to
   $$\left(\frac{x_1 + x_2}{2}, \frac{y_1 + y_2}{2} \right)$$
   equals the distance from (x_2, y_2) to
   $$\left(\frac{x_1 + x_2}{2}, \frac{y_1 + y_2}{2} \right);$$
   and

   ii) the points
   $$(x_1, y_1), \left(\frac{x_1 + x_2}{2}, \frac{y_1 + y_2}{2} \right),$$
   and
   $$(x_2, y_2)$$
   lie on the same line (see Exercise 86).

7.8 The Complex Numbers

Imaginary and Complex Numbers ■ Addition and Subtraction ■ Multiplication ■
Conjugates and Division ■ Powers of *i*

Imaginary and Complex Numbers

Negative numbers do not have square roots in the real-number system. However, a
larger number system that contains the real-number system is designed so that
negative numbers *do* have square roots. That system is called the **complex-number
system**, and it will allow us to solve equations like $x^2 + 1 = 0$. The complex-
number system makes use of *i*, a number that is, by definition, a square root of -1.

> **The Number *i***
>
> *i* is the unique number for which $i = \sqrt{-1}$ and $i^2 = -1$.

We can now define the square root of a negative number as follows:

$$\sqrt{-p} = \sqrt{-1}\sqrt{p} = i\sqrt{p} \text{ or } \sqrt{p}i, \text{ for any positive number } p.$$

EXAMPLE **1** Express in terms of *i*: **(a)** $\sqrt{-7}$; **(b)** $\sqrt{-16}$; **(c)** $-\sqrt{-13}$; **(d)** $-\sqrt{-50}$.

SOLUTION

a) $\sqrt{-7} = \sqrt{-1 \cdot 7} = \sqrt{-1} \cdot \sqrt{7} = i\sqrt{7}$, or $\sqrt{7}i$ *i* is *not* under the radical.

b) $\sqrt{-16} = \sqrt{-1 \cdot 16} = \sqrt{-1} \cdot \sqrt{16} = i \cdot 4 = 4i$

c) $-\sqrt{-13} = -\sqrt{-1 \cdot 13} = -\sqrt{-1} \cdot \sqrt{13} = -i\sqrt{13}$, or $-\sqrt{13}i$

d) $-\sqrt{-50} = -\sqrt{-1} \cdot \sqrt{25} \cdot \sqrt{2} = -i \cdot 5 \cdot \sqrt{2} = -5i\sqrt{2}$, or $-5\sqrt{2}i$

TRY EXERCISE 9

> **Imaginary Numbers**
>
> An *imaginary number* is a number that can be written in the form
> $a + bi$, where *a* and *b* are real numbers and $b \neq 0$.

Don't let the name "imaginary" fool you. Imaginary numbers appear in fields
such as engineering and the physical sciences. The following are examples of
imaginary numbers:

$5 + 4i$,   Here $a = 5$, $b = 4$.

$\sqrt{3} - \pi i$,   Here $a = \sqrt{3}$, $b = -\pi$.

$\sqrt{7}i$ Here $a = 0$, $b = \sqrt{7}$.

The union of the set of all imaginary numbers and the set of all real numbers
is the set of all **complex numbers**.

> ## Complex Numbers
>
> A *complex number* is any number that can be written in the form $a + bi$, where a and b are real numbers. (Note that a and b both can be 0.)

The following are examples of complex numbers:

$7 + 3i$ (here $a \neq 0$, $b \neq 0$); $\quad\quad$ $4i$ (here $a = 0$, $b \neq 0$);

8 (here $a \neq 0$, $b = 0$); $\quad\quad\quad$ 0 (here $a = 0$, $b = 0$).

Complex numbers like $17i$ or $4i$, in which $a = 0$ and $b \neq 0$, are called *pure imaginary numbers*.

For $b = 0$, we have $a + 0i = a$, so every real number is a complex number. The relationships among various real and complex numbers are shown below.

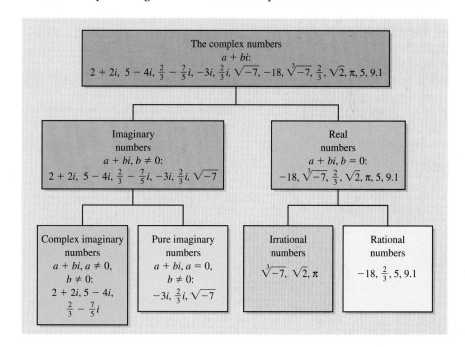

Note that although $\sqrt{-7}$ and $\sqrt[3]{-7}$ are both complex numbers, $\sqrt{-7}$ is imaginary whereas $\sqrt[3]{-7}$ is real.

Addition and Subtraction

The complex numbers obey the commutative, associative, and distributive laws. Thus we can add and subtract them as we do binomials.

EXAMPLE 2   Add or subtract and simplify.

a) $(8 + 6i) + (3 + 2i)$ $\quad\quad\quad\quad\quad\quad$ **b)** $(4 + 5i) - (6 - 3i)$

SOLUTION

a) $(8 + 6i) + (3 + 2i) = (8 + 3) + (6i + 2i)$ $\quad$ Combining the real parts and the imaginary parts

$= 11 + (6 + 2)i = 11 + 8i$

b) $(4 + 5i) - (6 - 3i) = (4 - 6) + [5i - (-3i)]$ Note that the 6 and the $-3i$ are *both* being subtracted.

$$= -2 + 8i$$

TRY EXERCISE 27

STUDENT NOTES

The rule developed in Section 7.3, $\sqrt[n]{a} \cdot \sqrt[n]{b} = \sqrt[n]{a \cdot b}$, does *not* apply when n is 2 and either a or b is negative. Indeed this condition is stated on p. 446 when it is specified that $\sqrt[n]{a}$ and $\sqrt[n]{b}$ are both *real* numbers.

Multiplication

To multiply square roots of negative real numbers, we first express them in terms of i. For example,

$$\sqrt{-2} \cdot \sqrt{-5} = \sqrt{-1} \cdot \sqrt{2} \cdot \sqrt{-1} \cdot \sqrt{5}$$
$$= i \cdot \sqrt{2} \cdot i \cdot \sqrt{5}$$
$$= i^2 \cdot \sqrt{10}$$
$$= -1\sqrt{10} = -\sqrt{10} \text{ is correct!}$$

> *CAUTION!* With complex numbers, simply multiplying radicands is *incorrect* when both radicands are negative: $\sqrt{-2} \cdot \sqrt{-5} \neq \sqrt{10}$.

With this in mind, we can now multiply complex numbers.

EXAMPLE 3   Multiply and simplify. When possible, write answers in the form $a + bi$.

a) $\sqrt{-4}\sqrt{-25}$ **b)** $\sqrt{-5} \cdot \sqrt{-7}$ **c)** $-3i \cdot 8i$

d) $-4i(3 - 5i)$ **e)** $(1 + 2i)(4 + 3i)$

SOLUTION

a) $\sqrt{-4}\sqrt{-25} = \sqrt{-1} \cdot \sqrt{4} \cdot \sqrt{-1} \cdot \sqrt{25}$
$$= i \cdot 2 \cdot i \cdot 5$$
$$= i^2 \cdot 10$$
$$= -1 \cdot 10 \qquad i^2 = -1$$
$$= -10$$

b) $\sqrt{-5} \cdot \sqrt{-7} = \sqrt{-1} \cdot \sqrt{5} \cdot \sqrt{-1} \cdot \sqrt{7}$ Try to do this step mentally.
$$= i \cdot \sqrt{5} \cdot i \cdot \sqrt{7}$$
$$= i^2 \cdot \sqrt{35}$$
$$= -1 \cdot \sqrt{35} \qquad i^2 = -1$$
$$= -\sqrt{35}$$

c) $-3i \cdot 8i = -24 \cdot i^2$
$$= -24 \cdot (-1) \qquad i^2 = -1$$
$$= 24$$

d) $-4i(3 - 5i) = -4i \cdot 3 + (-4i)(-5i)$ Using the distributive law
$$= -12i + 20i^2$$
$$= -12i - 20 \qquad\qquad i^2 = -1$$
$$= -20 - 12i \qquad\qquad \text{Writing in the form } a + bi$$

e) $(1 + 2i)(4 + 3i) = 4 + 3i + 8i + 6i^2$ Multiplying each term of $4 + 3i$ by each term of $1 + 2i$ (FOIL)

$$= 4 + 3i + 8i - 6 \qquad i^2 = -1$$
$$= -2 + 11i \qquad\qquad\quad \text{Combining like terms}$$

TRY EXERCISES 35 and 49

Conjugates and Division

Recall that the conjugate of $4 + \sqrt{2}$ is $4 - \sqrt{2}$.

Conjugates of complex numbers are defined in a similar manner.

Conjugate of a Complex Number

The *conjugate* of a complex number $a + bi$ is $a - bi$, and the *conjugate* of $a - bi$ is $a + bi$.

EXAMPLE   4   Find the conjugate of each number.

a) $-3 - 7i$ **b)** $4i$

SOLUTION

a) $-3 - 7i$ The conjugate is $-3 + 7i$.

b) $4i$ The conjugate is $-4i$. Note that $4i = 0 + 4i$.

The product of a complex number and its conjugate is a real number.

EXAMPLE   5   Multiply: $(5 + 7i)(5 - 7i)$.

SOLUTION

$$
\begin{aligned}
(5 + 7i)(5 - 7i) &= 5^2 - (7i)^2 \qquad \text{Using } (A + B)(A - B) = A^2 - B^2 \\
&= 25 - 49i^2 \\
&= 25 - 49(-1) \qquad i^2 = -1 \\
&= 25 + 49 = 74
\end{aligned}
$$

> TRY EXERCISE 55

Conjugates are used when dividing complex numbers. The procedure is much like that used to rationalize denominators in Section 7.5.

EXAMPLE   6   Divide and simplify to the form $a + bi$.

a) $\dfrac{-2 + 9i}{1 - 3i}$ **b)** $\dfrac{7 + 4i}{5i}$

SOLUTION

a) To divide and simplify $(-2 + 9i)/(1 - 3i)$, we multiply by 1, using the conjugate of the denominator to form 1:

$$
\begin{aligned}
\frac{-2 + 9i}{1 - 3i} &= \frac{-2 + 9i}{1 - 3i} \cdot \frac{1 + 3i}{1 + 3i} \qquad && \text{Multiplying by 1 using the conjugate of the denominator in the symbol for 1} \\[2mm]
&= \frac{(-2 + 9i)(1 + 3i)}{(1 - 3i)(1 + 3i)} && \text{Multiplying numerators; multiplying denominators} \\[2mm]
&= \frac{-2 - 6i + 9i + 27i^2}{1^2 - 9i^2} && \text{Using FOIL} \\[2mm]
&= \frac{-2 + 3i + (-27)}{1 - (-9)} && i^2 = -1 \\[2mm]
&= \frac{-29 + 3i}{10} && \text{Writing in the form } a + bi; \\[2mm]
&= -\frac{29}{10} + \frac{3}{10}i. && \text{note that } \frac{X + Y}{Z} = \frac{X}{Z} + \frac{Y}{Z}
\end{aligned}
$$

b) The conjugate of $5i$ is $-5i$, so we *could* multiply by $-5i/(-5i)$. However, when the denominator is a pure imaginary number, it is easiest if we multiply by i/i:

$$\frac{7 + 4i}{5i} = \frac{7 + 4i}{5i} \cdot \frac{i}{i} \qquad \text{Multiplying by 1 using } i/i. \text{ We can also use the conjugate of } 5i \text{ to write } -5i/(-5i).$$

$$= \frac{7i + 4i^2}{5i^2} \qquad \text{Multiplying}$$

$$= \frac{7i + 4(-1)}{5(-1)} \qquad i^2 = -1$$

$$= \frac{7i - 4}{-5} = \frac{-4}{-5} + \frac{7}{-5} i, \text{ or } \frac{4}{5} - \frac{7}{5} i. \qquad \text{Writing in the form } a + bi$$

TRY EXERCISE ▶ 73

Powers of i

Answers to problems involving complex numbers are generally written in the form $a + bi$. In the following discussion, we show why there is no need to use powers of i (other than 1) when writing answers.

Recall that -1 raised to an *even* power is 1, and -1 raised to an *odd* power is -1. Simplifying powers of i can then be done by using the fact that $i^2 = -1$ and expressing the given power of i in terms of i^2. Consider the following:

$$i^2 = -1, \longleftarrow$$
$$i^3 = i^2 \cdot i = (-1)i = -i,$$
$$i^4 = (i^2)^2 = (-1)^2 = 1,$$
$$i^5 = i^4 \cdot i = (i^2)^2 \cdot i = (-1)^2 \cdot i = i,$$
$$i^6 = (i^2)^3 = (-1)^3 = -1. \longleftarrow \text{The pattern is now repeating.}$$

The powers of i cycle themselves through the values i, -1, $-i$, and 1. Even powers of i are -1 or 1 whereas odd powers of i are i or $-i$.

EXAMPLE 7   Simplify: **(a)** i^{18}; **(b)** i^{24}.

SOLUTION

a) $i^{18} = (i^2)^9 \qquad$ Using the power rule

$\qquad = (-1)^9 = -1 \qquad$ Raising -1 to a power

b) $i^{24} = (i^2)^{12}$

$\qquad = (-1)^{12} = 1$

TRY EXERCISE ▶ 83

To simplify i^n when n is odd, we rewrite i^n as $i^{n-1} \cdot i$.

EXAMPLE 8   Simplify: **(a)** i^{29}; **(b)** i^{75}.

SOLUTION

a) $i^{29} = i^{28} i^1 \qquad$ Using the product rule. This is a key step when i is raised to an odd power.

$\qquad = (i^2)^{14} i \qquad$ Using the power rule

$\qquad = (-1)^{14} i$

$\qquad = 1 \cdot i = i$

b) $i^{75} = i^{74} i^1 \qquad$ Using the product rule

$\qquad = (i^2)^{37} i \qquad$ Using the power rule

$\qquad = (-1)^{37} i$

$\qquad = -1 \cdot i = -i$

TRY EXERCISE ▶ 85

7.8 EXERCISE SET

🖐 **Concept Reinforcement** *Classify each statement as either true or false.*

1. Imaginary numbers are so named because they have no real-world applications.

2. Every real number is imaginary, but not every imaginary number is real.

3. Every imaginary number is a complex number, but not every complex number is imaginary.

4. Every real number is a complex number, but not every complex number is real.

5. We add complex numbers by combining real parts and combining imaginary parts.

6. The product of a complex number and its conjugate is always a real number.

7. The square of a complex number is always a real number.

8. The quotient of two complex numbers is always a complex number.

Express in terms of i.

9. $\sqrt{-100}$

10. $\sqrt{-9}$

11. $\sqrt{-5}$

12. $\sqrt{-7}$

13. $\sqrt{-8}$

14. $\sqrt{-12}$

15. $-\sqrt{-11}$

16. $-\sqrt{-17}$

17. $-\sqrt{-49}$

18. $-\sqrt{-81}$

19. $-\sqrt{-300}$

20. $-\sqrt{-75}$

21. $6 - \sqrt{-84}$

22. $4 - \sqrt{-60}$

23. $-\sqrt{-76} + \sqrt{-125}$

24. $\sqrt{-4} + \sqrt{-12}$

25. $\sqrt{-18} - \sqrt{-64}$

26. $\sqrt{-72} - \sqrt{-25}$

Perform the indicated operation and simplify. Write each answer in the form a + bi.

27. $(3 + 4i) + (2 - 7i)$

28. $(5 - 6i) + (8 + 9i)$

29. $(9 + 5i) - (2 + 3i)$

30. $(8 + 7i) - (2 + 4i)$

31. $(7 - 4i) - (5 - 3i)$

32. $(5 - 3i) - (9 + 2i)$

33. $(-5 - i) - (7 + 4i)$

34. $(-2 + 6i) - (-7 + i)$

35. $5i \cdot 8i$

36. $3i \cdot 9i$

37. $(-4i)(-6i)$

38. $7i \cdot (-8i)$

39. $\sqrt{-36}\sqrt{-9}$

40. $\sqrt{-49}\sqrt{-16}$

41. $\sqrt{-3}\sqrt{-10}$

42. $\sqrt{-6}\sqrt{-7}$

43. $\sqrt{-6}\sqrt{-21}$

44. $\sqrt{-15}\sqrt{-10}$

45. $5i(2 + 6i)$

46. $2i(7 + 3i)$

47. $-7i(3 + 4i)$

48. $-4i(6 - 5i)$

49. $(1 + i)(3 + 2i)$

50. $(4 + i)(2 + 3i)$

51. $(6 - 5i)(3 + 4i)$

52. $(5 - 6i)(2 + 5i)$

53. $(7 - 2i)(2 - 6i)$

54. $(-4 + 5i)(3 - 4i)$

55. $(3 + 8i)(3 - 8i)$

56. $(1 + 2i)(1 - 2i)$

57. $(-7 + i)(-7 - i)$

58. $(-4 + 5i)(-4 - 5i)$

59. $(4 - 2i)^2$

60. $(1 - 2i)^2$

61. $(2 + 3i)^2$

62. $(3 + 2i)^2$

63. $(-2 + 3i)^2$

64. $(-5 - 2i)^2$

65. $\dfrac{10}{3 + i}$

66. $\dfrac{26}{5 + i}$

67. $\dfrac{2}{3 - 2i}$

68. $\dfrac{4}{2 - 3i}$

69. $\dfrac{2i}{5 + 3i}$

70. $\dfrac{3i}{4 + 2i}$

71. $\dfrac{5}{6i}$

72. $\dfrac{4}{7i}$

73. $\dfrac{5 - 3i}{4i}$

74. $\dfrac{2 + 7i}{5i}$

Aha! 75. $\dfrac{7i + 14}{7i}$

76. $\dfrac{6i + 3}{3i}$

77. $\dfrac{4 + 5i}{3 - 7i}$

78. $\dfrac{5 + 3i}{7 - 4i}$

79. $\dfrac{2 + 3i}{2 + 5i}$

80. $\dfrac{3 + 2i}{4 + 3i}$

81. $\dfrac{3 - 2i}{4 + 3i}$

82. $\dfrac{5 - 2i}{3 + 6i}$

Simplify.

83. i^{32} **84.** i^{19}

85. i^{15} **86.** i^{38}

87. i^{42} **88.** i^{64}

89. i^9 **90.** $(-i)^{71}$

91. $(-i)^6$ **92.** $(-i)^4$

93. $(5i)^3$ **94.** $(-3i)^5$

95. $i^2 + i^4$ **96.** $5i^5 + 4i^3$

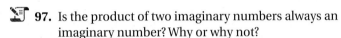 **97.** Is the product of two imaginary numbers always an imaginary number? Why or why not?

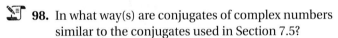 **98.** In what way(s) are conjugates of complex numbers similar to the conjugates used in Section 7.5?

Skill Review

To prepare for Section 8.1, review solving quadratic equations (Section 5.8).

Solve. [5.8]

99. $x^2 - x - 6 = 0$

100. $(x - 5)^2 = 0$

101. $t^2 = 100$

102. $2t^2 - 50 = 0$

103. $15x^2 = 14x + 8$

104. $6x^2 = 5x + 6$

Synthesis

105. Is the set of real numbers a subset of the set of complex numbers? Why or why not?

106. Is the union of the set of imaginary numbers and the set of real numbers the set of complex numbers? Why or why not?

Complex numbers are often graphed on a plane. The horizontal axis is the real axis and the vertical axis is the imaginary axis. A complex number such as $5 - 2i$ then corresponds to 5 on the real axis and -2 on the imaginary axis.

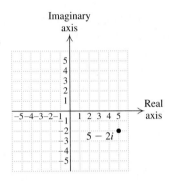

107. Graph each of the following.

   a) $3 + 2i$ **b)** $-1 + 4i$

   c) $3 - i$ **d)** $-5i$

108. Graph each of the following.

   a) $1 - 4i$ **b)** $-2 - 3i$

   c) i **d)** 4

The absolute value of a complex number $a + bi$ is its distance from the origin. Using the distance formula, we have $|a + bi| = \sqrt{a^2 + b^2}$. Find the absolute value of each complex number.

109. $|3 + 4i|$ **110.** $|8 - 6i|$

111. $|-1 + i|$ **112.** $|-3 - i|$

A function g is given by

$$g(z) = \dfrac{z^4 - z^2}{z - 1}.$$

113. Find $g(3i)$.   **114.** Find $g(1 + i)$.

115. Find $g(5i - 1)$.   **116.** Find $g(2 - 3i)$.

117. Evaluate

$$\dfrac{1}{w - w^2} \quad \text{for} \quad w = \dfrac{1 - i}{10}.$$

Simplify.

118. $\dfrac{i^5 + i^6 + i^7 + i^8}{(1 - i)^4}$ **119.** $(1 - i)^3(1 + i)^3$

120. $\dfrac{5 - \sqrt{5}i}{\sqrt{5}i}$ **121.** $\dfrac{6}{1 + \dfrac{3}{i}}$

122. $\left(\dfrac{1}{2} - \dfrac{1}{3}i\right)^2 - \left(\dfrac{1}{2} + \dfrac{1}{3}i\right)^2$

123. $\dfrac{i - i^{38}}{1 + i}$

Study Summary

KEY TERMS AND CONCEPTS	EXAMPLES

SECTION 7.1: RADICAL EXPRESSIONS AND FUNCTIONS

c is a **square root** of a if $c^2 = a$.

c is a **cube root** of a if $c^3 = a$.

$\sqrt{a}$ indicates the **principal** square root of a.

$\sqrt[n]{a}$ indicates the **nth root** of a.

index

$\sqrt[n]{a}$ ——**radicand**

radical symbol

The square roots of 25 are -5 and 5.

The cube root of -8 is -2.

$\sqrt{25} = 5$

$\sqrt[3]{-8} = -2$

For all a,

$$\sqrt[n]{a^n} = |a| \text{ when } n \text{ is even;}$$
$$\sqrt[n]{a^n} = a \text{ when } n \text{ is odd.}$$

If a represents a nonnegative number,

$$\sqrt[n]{a^n} = a.$$

Assume that x can be any real number.

$$\sqrt{(3 + x)^2} = |3 + x|$$

Assume that x represents a nonnegative number.

$$\sqrt{(7x)^2} = 7x$$

SECTION 7.2: RATIONAL NUMBERS AS EXPONENTS

$a^{1/n}$ means $\sqrt[n]{a}$.

$a^{m/n}$ means $\left(\sqrt[n]{a}\right)^m$ or $\sqrt[n]{a^m}$.

$a^{-m/n}$ means $\dfrac{1}{a^{m/n}}$.

$64^{1/2} = \sqrt{64} = 8$

$125^{2/3} = \left(\sqrt[3]{125}\right)^2 = 5^2 = 25$

$8^{-1/3} = \dfrac{1}{8^{1/3}} = \dfrac{1}{2}$

SECTION 7.3: MULTIPLYING RADICAL EXPRESSIONS

The Product Rule for Radicals

For any real numbers $\sqrt[n]{a}$ and $\sqrt[n]{b}$,

$$\sqrt[n]{a} \cdot \sqrt[n]{b} = \sqrt[n]{a \cdot b}.$$

$$\sqrt[3]{4x} \cdot \sqrt[3]{5y} = \sqrt[3]{20xy}$$

Using the Product Rule to Simplify

For any real numbers $\sqrt[n]{a}$ and $\sqrt[n]{b}$,

$$\sqrt[n]{a \cdot b} = \sqrt[n]{a} \cdot \sqrt[n]{b}.$$

$$\begin{aligned}\sqrt{75x^8 y^{11}} &= \sqrt{25 \cdot x^8 \cdot y^{10} \cdot 3 \cdot y} \\ &= \sqrt{25} \cdot \sqrt{x^8} \cdot \sqrt{y^{10}} \cdot \sqrt{3y} \\ &= 5x^4 y^5 \sqrt{3y} \quad \text{Assuming } y \text{ is nonnegative}\end{aligned}$$

SECTION 7.4: DIVIDING RATIONAL EXPRESSIONS

The Quotient Rule for Radicals

For any real numbers $\sqrt[n]{a}$ and $\sqrt[n]{b}$, $b \neq 0$,

$$\sqrt[n]{\frac{a}{b}} = \frac{\sqrt[n]{a}}{\sqrt[n]{b}}.$$

$$\sqrt[3]{\frac{8y^4}{125}} = \frac{\sqrt[3]{8y^4}}{\sqrt[3]{125}} = \frac{2y\sqrt[3]{y}}{5}$$

$$\frac{\sqrt{18a^9}}{\sqrt{2a^3}} = \sqrt{\frac{18a^9}{2a^3}} = \sqrt{9a^6} = 3a^3 \quad \text{Assuming } a \text{ is positive}$$

We can **rationalize a denominator** by multiplying by 1.

$$\frac{\sqrt[3]{5}}{\sqrt[3]{4y}} = \frac{\sqrt[3]{5}}{\sqrt[3]{4y}} \cdot \frac{\sqrt[3]{2y^2}}{\sqrt[3]{2y^2}} = \frac{\sqrt[3]{10y^2}}{\sqrt[3]{8y^3}} = \frac{\sqrt[3]{10y^2}}{2y}$$

SECTION 7.5: EXPRESSIONS CONTAINING SEVERAL RADICAL TERMS

Like radicals have the same indices and radicands and can be combined.

$$\sqrt{12} + 5\sqrt{3} = \sqrt{4 \cdot 3} + 5\sqrt{3} = 2\sqrt{3} + 5\sqrt{3} = 7\sqrt{3}$$

Radical expressions are multiplied in much the same way that polynomials are multiplied.

$$(1 + 5\sqrt{6})(4 - \sqrt{6}) = 1 \cdot 4 - 1\sqrt{6} + 4 \cdot 5\sqrt{6} - 5\sqrt{6} \cdot \sqrt{6}$$
$$= 4 - \sqrt{6} + 20\sqrt{6} - 5 \cdot 6$$
$$= -26 + 19\sqrt{6}$$

To rationalize a denominator containing two terms, we use the **conjugate** of the denominator to write a form of 1.

$$\frac{2}{1 - \sqrt{3}} = \frac{2}{1 - \sqrt{3}} \cdot \frac{1 + \sqrt{3}}{1 + \sqrt{3}} \qquad \text{$1 + \sqrt{3}$ is the conjugate of $1 - \sqrt{3}$.}$$
$$= \frac{2(1 + \sqrt{3})}{-2} = -1 - \sqrt{3}$$

When terms have different indices, we can often use rational exponents to simplify.

$$\sqrt[3]{p} \cdot \sqrt[4]{q^3} = p^{1/3} \cdot q^{3/4}$$
$$= p^{4/12} \cdot q^{9/12} \qquad \text{Finding a common denominator}$$
$$= \sqrt[12]{p^4 q^9}$$

SECTION 7.6: SOLVING RADICAL EQUATIONS

The Principle of Powers

If $a = b$, then $a^n = b^n$.

To solve a radical equation, use the principle of powers and the steps on pp. 466 and 468.

Solutions found using the principle of powers must be checked in the original equation.

$$x - 7 = \sqrt{x - 5}$$
$$(x - 7)^2 = (\sqrt{x - 5})^2$$
$$x^2 - 14x + 49 = x - 5$$
$$x^2 - 15x + 54 = 0$$
$$(x - 6)(x - 9) = 0$$
$$x = 6 \ \ or \ \ x = 9$$

Only 9 checks and is the solution.

$$2 + \sqrt{t} = \sqrt{t + 8}$$
$$(2 + \sqrt{t})^2 = (\sqrt{t + 8})^2$$
$$4 + 4\sqrt{t} + t = t + 8$$
$$4\sqrt{t} = 4$$
$$\sqrt{t} = 1$$
$$(\sqrt{t})^2 = (1)^2$$
$$t = 1$$

1 checks and is the solution.

SECTION 7.7: THE DISTANCE AND MIDPOINT FORMULAS AND OTHER APPLICATIONS

The Pythagorean Theorem

In any right triangle, if a and b are the lengths of the legs and c is the length of the hypotenuse, then
$$a^2 + b^2 = c^2.$$

Find the length of the hypotenuse of a right triangle with legs of lengths 4 and 7. Give an exact answer in radical notation, as well as a decimal approximation to three decimal places.

$$a^2 + b^2 = c^2$$
$$4^2 + 7^2 = c^2 \qquad \text{Substituting}$$
$$16 + 49 = c^2$$
$$65 = c^2$$
$$\sqrt{65} = c \qquad \text{This is exact.}$$
$$8.062 \approx c \qquad \text{This is approximate.}$$

Special Triangles

The length of the hypotenuse in an isosceles right triangle is the length of a leg times $\sqrt{2}$.

Find the missing lengths. Give an exact answer and, where appropriate, an approximation to three decimal places.

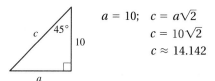

$$a = 10; \quad c = a\sqrt{2}$$
$$c = 10\sqrt{2}$$
$$c \approx 14.142$$

The length of the longer leg in a $30°$–$60°$–$90°$ triangle is the length of the shorter leg times $\sqrt{3}$. The hypotenuse is twice as long as the shorter leg.

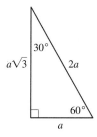

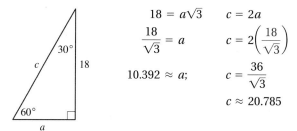

$$18 = a\sqrt{3} \qquad c = 2a$$
$$\frac{18}{\sqrt{3}} = a \qquad c = 2\left(\frac{18}{\sqrt{3}}\right)$$
$$10.392 \approx a; \qquad c = \frac{36}{\sqrt{3}}$$
$$c \approx 20.785$$

The Distance Formula

The distance d between any two points (x_1, y_1) and (x_2, y_2) is given by

$$d = \sqrt{(x_2 - x_1)^2 + (y_2 - y_1)^2}.$$

Find the distance between $(3, -5)$ and $(-1, -2)$.

$$d = \sqrt{(-1 - 3)^2 + (-2 - (-5))^2}$$
$$= \sqrt{(-4)^2 + (3)^2}$$
$$= \sqrt{16 + 9} = \sqrt{25} = 5$$

The Midpoint Formula

If the endpoints of a segment are (x_1, y_1) and (x_2, y_2), then the coordinates of the midpoint are

$$\left(\frac{x_1 + x_2}{2}, \frac{y_1 + y_2}{2}\right).$$

Find the midpoint of the segment with endpoints $(3, -5)$ and $(-1, -2)$.

$$\left(\frac{3 + (-1)}{2}, \frac{-5 + (-2)}{2}\right), \text{ or } \left(1, -\frac{7}{2}\right)$$

SECTION 7.8: THE COMPLEX NUMBERS

A **complex number** is any number that can be written in the form $a + bi$, where a and b are real numbers,

$$i = \sqrt{-1}, \quad \text{and} \quad i^2 = -1.$$

$$(3 + 2i) + (4 - 7i) = 7 - 5i$$

$$(8 + 6i) - (5 + 2i) = 3 + 4i$$

$$(2 + 3i)(4 - i) = 8 - 2i + 12i - 3i^2$$
$$= 8 + 10i - 3(-1) = 11 + 10i$$

$$\frac{1 - 4i}{3 - 2i} = \frac{1 - 4i}{3 - 2i} \cdot \frac{3 + 2i}{3 + 2i} \qquad \text{The conjugate of } 3 - 2i \text{ is } 3 + 2i.$$

$$= \frac{3 + 2i - 12i - 8i^2}{9 + 6i - 6i - 4i^2}$$

$$= \frac{3 - 10i - 8(-1)}{9 - 4(-1)} = \frac{11 - 10i}{13} = \frac{11}{13} - \frac{10}{13}i$$

Review Exercises: Chapter 7

Concept Reinforcement *Classify each statement as either true or false.*

1. $\sqrt{ab} = \sqrt{a} \cdot \sqrt{b}$ for any real numbers $\sqrt{a}$ and $\sqrt{b}$. [7.3]

2. $\sqrt{a + b} = \sqrt{a} + \sqrt{b}$ for any real numbers $\sqrt{a}$ and $\sqrt{b}$. [7.5]

3. $\sqrt{a^2} = a$, for any real number a. [7.1]

4. $\sqrt[3]{a^3} = a$, for any real number a. [7.1]

5. $x^{2/5}$ means $\sqrt[5]{x^2}$ and $(\sqrt[5]{x})^2$. [7.2]

6. The hypotenuse of a right triangle is never shorter than either leg. [7.7]

7. Some radical equations have no solution. [7.6]

8. If $f(x) = \sqrt{x - 5}$, then the domain of f is the set of all nonnegative real numbers. [7.1]

Simplify. [7.1]

9. $\sqrt{\dfrac{100}{121}}$

10. $-\sqrt{0.36}$

Let $f(x) = \sqrt{x + 10}$. Find the following. [7.1]

11. $f(15)$

12. The domain of f

Simplify. Assume that each variable can represent any real number. [7.1]

13. $\sqrt{64t^2}$

14. $\sqrt{(c + 7)^2}$

15. $\sqrt{4x^2 + 4x + 1}$

16. $\sqrt[5]{-32}$

17. Write an equivalent expression using exponential notation: $(\sqrt[3]{5ab})^4$. [7.2]

18. Write an equivalent expression using radical notation: $(16a^6)^{3/4}$. [7.2]

Use rational exponents to simplify. Assume $x, y \geq 0$. [7.2]

19. $\sqrt{x^6 y^{10}}$

20. $(\sqrt[6]{x^2 y})^2$

Simplify. Do not use negative exponents in the answers. [7.2]

21. $(x^{-2/3})^{3/5}$

22. $\dfrac{7^{-1/3}}{7^{-1/2}}$

23. If $f(x) = \sqrt{25(x - 6)^2}$, find a simplified form for $f(x)$. [7.3]

Simplify. Write all answers using radical notation. Assume that all variables represent nonnegative numbers.

24. $\sqrt[4]{16x^{20}y^8}$ [7.3]

25. $\sqrt{250x^3 y^2}$ [7.3]

26. $\sqrt{5a}\sqrt{7b}$ [7.3]

27. $\sqrt[3]{3x^4 b}\sqrt[3]{9xb^2}$ [7.3]

28. $\sqrt[3]{-24x^{10}y^8}\,\sqrt[3]{18x^7 y^4}$ [7.3]

29. $\sqrt[3]{-\dfrac{27y^{12}}{64}}$ [7.4]

30. $\dfrac{\sqrt[3]{60xy^3}}{\sqrt[3]{10x}}$ [7.4]

31. $\dfrac{\sqrt{75x}}{2\sqrt{3}}$ [7.4]

32. $\sqrt[4]{\dfrac{48a^{11}}{c^8}}$ [7.4]

33. $5\sqrt[3]{4y} + 2\sqrt[3]{4y}$ [7.5]

34. $2\sqrt{75} - 9\sqrt{3}$ [7.5]

35. $\sqrt[3]{8x^4} + \sqrt[3]{xy^6}$ [7.5]

36. $\sqrt{50} + 2\sqrt{18} + \sqrt{32}$ [7.5]

37. $(3 + \sqrt{10})(3 - \sqrt{10})$ [7.5]

38. $(\sqrt{3} - 3\sqrt{8})(\sqrt{5} + 2\sqrt{8})$ [7.5]

39. $\sqrt[4]{x}\sqrt{x}$ [7.5]

40. $\dfrac{\sqrt[3]{x^2}}{\sqrt[4]{x}}$ [7.5]

41. If $f(x) = x^2$, find $f(2 - \sqrt{a})$. [7.5]

42. Rationalize the denominator:
$$\dfrac{4\sqrt{5}}{\sqrt{2} + \sqrt{3}}.$$ [7.5]

43. Rationalize the numerator of the expression in Exercise 42. [7.5]

Solve. [7.6]

44. $\sqrt{y + 6} - 2 = 3$

45. $(x + 1)^{1/3} = -5$

46. $1 + \sqrt{x} = \sqrt{3x - 3}$

47. If $f(x) = \sqrt{x + 2} + x$, find a such that $f(a) = 4$. [7.6]

Solve. Give an exact answer and, where appropriate, an approximation to three decimal places. [7.7]

48. The diagonal of a square has length 10 cm. Find the length of a side of the square.

49. A skate-park jump has a ramp that is 6 ft long and is 2 ft high. How long is its base?

6 ft   2 ft
?

50. Find the missing lengths. Give exact answers and, where appropriate, an approximation to three decimal places.

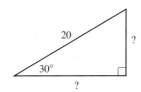

20   ?
30°
?

51. Find the distance between $(-6, 4)$ and $(-1, 5)$. Give an exact answer and an approximation to three decimal places. [7.7]

52. Find the midpoint of the segment with endpoints $(-7, -2)$ and $(3, -1)$. [7.7]

53. Express in terms of i and simplify: $\sqrt{-45}$. [7.8]

54. Add: $(-4 + 3i) + (2 - 12i)$. [7.8]

55. Subtract: $(9 - 7i) - (3 - 8i)$. [7.8]

Simplify. [7.8]

56. $(2 + 5i)(2 - 5i)$

57. i^{34}

58. $(6 - 3i)(2 - i)$

59. Divide. Write the answer in the form $a + b$.
$$\frac{7 - 2i}{3 + 4i}$$ [7.8]

Synthesis

60. What makes some complex numbers real and others imaginary? [7.8]

61. Explain why $\sqrt[n]{x^n} = |x|$ when n is even, but $\sqrt[n]{x^n} = x$ when n is odd. [7.1]

62. Write a quotient of two imaginary numbers that is a real number (answers may vary). [7.8]

63. Solve:
$$\sqrt{11x + \sqrt{6 + x}} = 6.$$ [7.6]

64. Simplify:
$$\frac{2}{1 - 3i} - \frac{3}{4 + 2i}.$$ [7.8]

65. Don's Discount Shoes has two locations. The sign at the original location is shaped like an isosceles right triangle. The sign at the newer location is shaped like a 30°–60°–90° triangle. The hypotenuse of each sign measures 6 ft. Which sign has the greater area and by how much? (Round to three decimal places.) [7.7]

Simplify. Assume that variables can represent any real number.

1. $\sqrt{50}$

2. $\sqrt[3]{-\dfrac{8}{x^6}}$

3. $\sqrt{81a^2}$

4. $\sqrt{x^2 - 8x + 16}$

5. Write an equivalent expression using exponential notation: $\sqrt{7xy}$.

6. Write an equivalent expression using radical notation: $(4a^3b)^{5/6}$.

7. If $f(x) = \sqrt{2x - 10}$, determine the domain of f.

8. If $f(x) = x^2$, find $f(5 + \sqrt{2})$.

Simplify. Write all answers using radical notation. Assume that all variables represent positive numbers.

9. $\sqrt[5]{32x^{16}y^{10}}$

10. $\sqrt[3]{4w}\sqrt[3]{4v^2}$

11. $\sqrt{\dfrac{100a^4}{9b^6}}$

12. $\dfrac{\sqrt[5]{48x^6y^{10}}}{\sqrt[5]{16x^2y^9}}$

13. $\sqrt[4]{x^3}\sqrt{x}$

14. $\dfrac{\sqrt{y}}{\sqrt[10]{y}}$

15. $8\sqrt{2} - 2\sqrt{2}$

16. $\sqrt{x^4y} + \sqrt{9y^3}$

17. $(7 + \sqrt{x})(2 - 3\sqrt{x})$

18. Rationalize the denominator:
$$\dfrac{\sqrt{3}}{5 + \sqrt{2}}.$$

Solve.

19. $6 = \sqrt{x - 3} + 5$

20. $x = \sqrt{3x + 3} - 1$

21. $\sqrt{2x} = \sqrt{x + 1} + 1$

Solve. For Exercises 22–24, give exact answers and approximations to three decimal places.

22. A referee jogs diagonally from one corner of a 50-ft by 90-ft basketball court to the far corner. How far does she jog?

23. The hypotenuse of a 30°–60°–90° triangle is 10 cm long. Find the lengths of the legs.

24. Find the distance between the points $(3, 7)$ and $(-1, 8)$.

25. Find the midpoint of the segment with endpoints $(2, -5)$ and $(1, -7)$.

26. Express in terms of i and simplify: $\sqrt{-50}$.

27. Subtract: $(9 + 8i) - (-3 + 6i)$.

28. Multiply. Write the answer in the form $a + bi$.
$$(4 - i)^2$$

29. Divide. Write the answer in the form $a + bi$.
$$\dfrac{-2 + i}{3 - 5i}$$

30. Simplify: i^{37}.

Synthesis

31. Solve:
$$\sqrt{2x - 2} + \sqrt{7x + 4} = \sqrt{13x + 10}.$$

32. Simplify:
$$\dfrac{1 - 4i}{4i(1 + 4i)^{-1}}.$$

33. The function $D(h) = 1.2\sqrt{h}$ can be used to approximate the distance D, in miles, that a person can see to the horizon from a height h, in feet. How far above sea level must a pilot fly in order to see a horizon that is 180 mi away?

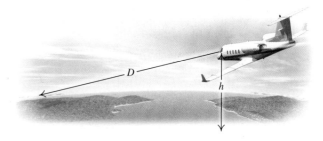

Cumulative Review: Chapters 1–7

Solve.

1. $2(x - 5) - 3 = 3(2x + 5)$ [1.3]

2. $x(x + 2) = 35$ [5.8]

3. $2y^2 = 50$ [5.8]

4. $\dfrac{1}{x} = \dfrac{2}{5}$ [6.4]

5. $\sqrt[3]{t} = -1$ [7.6]

6. $25x^2 - 10x + 1 = 0$ [5.8]

7. $|x - 2| \le 5$ [4.3]

8. $2x + 5 > 6 \ or \ x - 3 \le 9$ [4.2]

9. $\dfrac{2x}{x - 1} + \dfrac{x}{x - 3} = 2$ [6.4]

10. $x = \sqrt{2x - 5} + 4$ [7.6]

11. $3x + y = 5,$
    $x - y = -5$ [3.2]

12. $2x - \ y + \ z = 1,$
    $\ x + 2y + \ z = -3,$
    $5x - \ y + 3z = 0$ [3.4]

Graph on a plane.

13. $3y = -6$ [2.4]

14. $y = -x + 5$ [2.3]

15. $x + y \le 2$ [4.4]

16. $2x = y$ [2.3]

17. Determine the slope and the y-intercept of the line given by $y = -6 - x.$ [2.3]

18. Find an equation for the line parallel to the line given by $y = 7x$ and passing through the point $(0, -11)$. [2.5]

Perform the indicated operations and, if possible, simplify. For radical expressions, assume that all variables represent positive numbers.

19. $18 \div 3 \cdot 2 - 6^2 \div (2 + 4)$ [1.2]

20. $(x^2y - 3x^2 - 4xy^2) - (x^2y - 3x^2 + 4xy^2)$ [5.1]

21. $(2a - 5b)^2$ [5.2]

22. $(c^2 - 3d)(c^2 + 3d)$ [5.2]

23. $\dfrac{1}{x} + \dfrac{1}{x + 1}$ [6.2]

24. $\dfrac{x + 3}{x - 2} - \dfrac{x + 5}{x + 1}$ [6.2]

25. $\dfrac{a^2 - a - 6}{a^2 - 1} \div \dfrac{a^2 - 6a + 9}{2a^2 + 3a + 1}$ [6.1]

26. $\dfrac{\dfrac{1}{x} + \dfrac{1}{x + 1}}{\dfrac{x}{x + 1}}$ [6.3]

27. $\sqrt{200} - 5\sqrt{8}$ [7.5]

28. $(1 + \sqrt{5})(4 - \sqrt{5})$ [7.5]

29. $\sqrt{10a^2b} \cdot \sqrt{15ab^3}$ [7.3]

30. $\sqrt[3]{y}\sqrt[5]{y}$ [7.5]

Factor.

31. $x^2 - 5x - 14$ [5.4]

32. $4y^8 - 4y^5$ [5.6]

33. $100c^2 - 25d^2$ [5.5]

34. $3t^2 - 5t - 8$ [5.4]

35. $3x^2 - 6x - 21$ [5.3]

36. $yt - xt - yz^2 + xz^2$ [5.3]

Find the domain of each function.

37. $f(x) = \dfrac{2x - 3}{x^2 - 6x + 9}$ [5.8]

38. $f(x) = \sqrt{2x - 11}$ [7.1]

Find each of the following, if $f(x) = \sqrt{2x - 3}$ and $g(x) = x^2.$

39. $f(14)$ [7.1]

40. $g(1 - \sqrt{5})$ [7.5]

41. $(f + g)(x)$ [2.6], [7.5]

Solve.

42. *Flood rescue.*   A flood rescue team uses a boat that travels 10 mph in still water. To reach a stranded family, they travel 7 mi against the current and return 7 mi with the current in a total time of $1\frac{2}{3}$ hr. What is the speed of the current?   [6.5]

43. *Emergency shelter.*   The entrance to a tent used by a rescue team is the shape of an equilateral triangle. If the base of the tent is 4 ft wide, how tall is the tent? Give an exact answer and an approximation to three decimal places.   [7.7]

44. *Age at marriage.*   The median age at first marriage for U.S. men has grown from 25.1 in 2001 to 25.5 in 2006. Let $m(t)$ represent the median age of men at first marriage t years after 2000.   [2.5]
Source: U.S. Census Bureau

a) Find a linear function that fits the data.
b) Use the function from part (a) to predict the median age of men at first marriage in 2020.
c) In what year will the median age of men at first marriage be 28?

45. *Salary.*   Neil's annual salary is $38,849. This includes a 6% superior performance raise. What would Neil's salary have been without the performance raise?   [1.4]

46. *Food service.*   Melted Goodness mixes Swiss chocolate and whipping cream to make a dessert fondue. Swiss chocolate costs $1.20 per ounce and whipping cream costs $0.30 per ounce. How much of each does Melted Goodness use to make 65 oz of fondue at a cost of $60.00?   [3.3]

47. *Data storage.*   The size of an average program written for a calculator is 5 kilobytes (5×10^3 bytes). Toni's calculator contains a 2-gigabyte (2×10^9 bytes) memory card. How many programs of average length can Toni store on her calculator?   [1.7]

48. *Food cost.*   The average cost of a Thanksgiving dinner in the United States rose from $34.56 in 2002 to $42.26 in 2007. What was the rate of increase?   [2.4]
Sources: Purdue University; American Farm Bureau Federation

49. *Landscaping.*   A rectangular parking lot is 80 ft by 100 ft. Part of the asphalt is removed in order to install a landscaped border of uniform width around it. The area of the new parking lot is 6300 ft². How wide is the landscaped border?   [5.8]

Synthesis

50. Give an equation in standard form for the line whose x-intercept is $(-3, 0)$ and whose y-intercept is $(0, 5)$.   [2.4]

51. Solve by graphing:
$$y = x - 1,$$
$$y = x^2 - 1. \quad [3.1]$$

Solve.

52. $\dfrac{\dfrac{1}{x} + \dfrac{1}{x+1}}{\dfrac{1}{x} - 1} = 1$ [6.3], [6.4]

53. $3\sqrt{2x - 11} = 2 + \sqrt{5x - 1}$ [7.6]

Quadratic Functions and Equations

JANET FISHER
MUSIC PUBLISHER AND
RECORD LABEL OWNER
Los Angeles, California

As a music publisher and record label owner, I pay our songwriters and artists royalties based on sales and uses of their songs, both physical and digital. In the world of digital downloads, these royalties run from fractions of cents for a "streamed listen," to a set number of cents per download, depending on the site from which the purchase or stream is made. Varying advances are split between publishers and writers when a song is used in a film or TV show. When a writer has a combination of uses for his or her song, you can imagine how important math is in order to pay them properly.

AN APPLICATION

As more listeners download their music purchases, sales of compact discs are decreasing. According to Nielsen SoundScan, sales of music CDs increased from 500 million in 1997 to 700 million in 2001 and then decreased to 450 million in 2007. Find a quadratic function that fits the data, and use the function to estimate the sales of music CDs in 2009.

This problem appears as Example 3 in Section 8.8.

T he mathematical translation of a problem is often a function or an equation containing a second-degree polynomial in one variable. Such functions or equations are said to be *quadratic*. In this chapter, we examine a variety of ways to solve quadratic equations and look at graphs and applications of quadratic functions.

8.1 Quadratic Equations

The Principle of Square Roots ▪ Completing the Square ▪ Problem Solving

The general form of a quadratic function is

$$f(x) = ax^2 + bx + c, \quad \text{with } a \neq 0.$$

The graph of a quadratic function is a *parabola*. Such graphs open up or down and can have 0, 1, or 2 *x*-intercepts. We learn to graph quadratic functions later in this chapter.

ALGEBRAIC–GRAPHICAL CONNECTION

The graphs of the quadratic function $f(x) = x^2 + 6x + 8$ and the linear function $g(x) = 0$ are shown below.

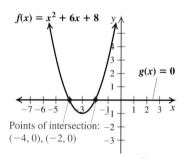

Note that $(-4, 0)$ and $(-2, 0)$ are the points of intersection of the graphs of $f(x) = x^2 + 6x + 8$ and $g(x) = 0$ (the *x*-axis). In Sections 8.6 and 8.7, we will develop efficient ways to graph quadratic functions. For now, the graphs help us visualize solutions.

In Chapter 5, we solved equations like $x^2 + 6x + 8 = 0$ by factoring:

$$x^2 + 6x + 8 = 0$$
$$(x + 4)(x + 2) = 0 \qquad \text{Factoring}$$
$$x + 4 = 0 \quad or \quad x + 2 = 0 \qquad \text{Using the principle of zero products}$$
$$x = -4 \quad or \qquad x = -2.$$

Note that -4 and -2 are the first coordinates of the points of intersection (or the *x*-intercepts) of the graph of $f(x)$ above.

In this section and the next, we develop algebraic methods for solving *any* quadratic equation, whether it is factorable or not.

EXAMPLE   **1**   Solve: $x^2 = 25$.

SOLUTION   We have

$$x^2 = 25$$
$$x^2 - 25 = 0 \qquad \text{Writing in standard form}$$
$$(x - 5)(x + 5) = 0 \qquad \text{Factoring}$$
$$x - 5 = 0 \quad or \quad x + 5 = 0 \qquad \text{Using the principle of zero products}$$
$$x = 5 \quad or \qquad x = -5.$$

The solutions are 5 and -5. A graph in which $f(x) = x^2$ represents the left side of the original equation and $g(x) = 25$ represents the right side provides a check (see the figure at left). Of course, we can also check by substituting 5 and -5 into the original equation.

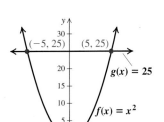

A visualization of Example 1

> TRY EXERCISE ▶ 7

The Principle of Square Roots

Let's reconsider $x^2 = 25$. We know from Chapter 7 that the number 25 has two real-number square roots, 5 and -5, the solutions of the equation in Example 1. Thus we see that square roots provide quick solutions for equations of the type $x^2 = k$.

> ### The Principle of Square Roots
>
> For any real number k, if $x^2 = k$, then
> $$x = \sqrt{k} \quad or \quad x = -\sqrt{k}.$$

EXAMPLE   **2**   Solve: $3x^2 = 6$. Give exact solutions and approximations to three decimal places.

SOLUTION   We have

$$3x^2 = 6$$
$$x^2 = 2 \qquad \text{Isolating } x^2$$
$$x = \sqrt{2} \quad or \quad x = -\sqrt{2}. \qquad \text{Using the principle of square roots}$$

We can use the symbol $\pm\sqrt{2}$ to represent both of the solutions.

> *CAUTION!*   There are *two* solutions: $\sqrt{2}$ and $-\sqrt{2}$. Don't forget the second solution.

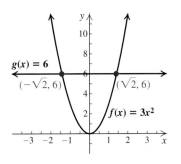

A visualization of Example 2

Check: For $\sqrt{2}$:

$$\begin{array}{c|c} 3x^2 = 6 & \\ \hline 3(\sqrt{2})^2 & 6 \\ 3 \cdot 2 & \\ & 6 \overset{?}{=} 6 \quad \text{TRUE} \end{array}$$

For $-\sqrt{2}$:

$$\begin{array}{c|c} 3x^2 = 6 & \\ \hline 3(-\sqrt{2})^2 & 6 \\ 3 \cdot 2 & \\ & 6 \overset{?}{=} 6 \quad \text{TRUE} \end{array}$$

The solutions are $\sqrt{2}$ and $-\sqrt{2}$, or $\pm\sqrt{2}$, which round to 1.414 and -1.414.

> TRY EXERCISE ▶ 11

EXAMPLE 3

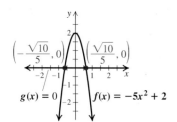

$\left(-\frac{\sqrt{10}}{5}, 0\right)$ $\left(\frac{\sqrt{10}}{5}, 0\right)$

$g(x) = 0$ $f(x) = -5x^2 + 2$

A visualization of Example 3

Solve: $-5x^2 + 2 = 0$.

SOLUTION   We have

$$-5x^2 + 2 = 0$$

$$x^2 = \frac{2}{5} \qquad \text{Isolating } x^2$$

$$x = \sqrt{\frac{2}{5}} \quad \text{or} \quad x = -\sqrt{\frac{2}{5}}. \qquad \text{Using the principle of square roots}$$

The solutions are $\sqrt{\frac{2}{5}}$ and $-\sqrt{\frac{2}{5}}$, or simply $\pm\sqrt{\frac{2}{5}}$. If we rationalize the denomina-

tor, the solutions are written $\pm\dfrac{\sqrt{10}}{5}$. The checks are left to the student.

TRY EXERCISE 15

Sometimes we get solutions that are imaginary numbers.

EXAMPLE 4

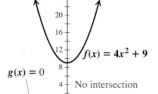

$f(x) = 4x^2 + 9$

$g(x) = 0$

No intersection

A visualization of Example 4

Solve: $4x^2 + 9 = 0$.

SOLUTION   We have

$$4x^2 + 9 = 0$$

$$x^2 = -\frac{9}{4} \qquad \text{Isolating } x^2$$

$$x = \sqrt{-\frac{9}{4}} \quad \text{or} \quad x = -\sqrt{-\frac{9}{4}} \qquad \text{Using the principle of square roots}$$

$$x = \sqrt{\frac{9}{4}}\sqrt{-1} \quad \text{or} \quad x = -\sqrt{\frac{9}{4}}\sqrt{-1}$$

$$x = \frac{3}{2}i \quad \text{or} \quad x = -\frac{3}{2}i. \qquad \text{Recall that } \sqrt{-1} = i.$$

Check: For $\frac{3}{2}i$:

$$\begin{array}{c|c} 4x^2 + 9 = 0 & \\ \hline 4\left(\frac{3}{2}i\right)^2 + 9 & 0 \\ 4 \cdot \frac{9}{4} \cdot i^2 + 9 & \\ 9(-1) + 9 & \\ & 0 \overset{?}{=} 0 \quad \text{TRUE} \end{array}$$

For $-\frac{3}{2}i$:

$$\begin{array}{c|c} 4x^2 + 9 = 0 & \\ \hline 4\left(-\frac{3}{2}i\right)^2 + 9 & 0 \\ 4 \cdot \frac{9}{4} \cdot i^2 + 9 & \\ 9(-1) + 9 & \\ & 0 \overset{?}{=} 0 \quad \text{TRUE} \end{array}$$

The solutions are $\frac{3}{2}i$ and $-\frac{3}{2}i$, or $\pm\frac{3}{2}i$. The graph at left confirms that there are no real-number solutions.

TRY EXERCISE 19

The principle of square roots can be restated in a more general form for any equation in which some algebraic expression squared equals a constant.

The Principle of Square Roots (Generalized Form)

For any real number k and any algebraic expression X:

$$\text{If } X^2 = k, \quad \text{then} \quad X = \sqrt{k} \quad \text{or} \quad X = -\sqrt{k}.$$

EXAMPLE 5

Let $f(x) = (x - 2)^2$. Find all x-values for which $f(x) = 7$.

SOLUTION   We are asked to find all x-values for which

$$f(x) = 7,$$

or

$$(x - 2)^2 = 7. \qquad \text{Substituting } (x - 2)^2 \text{ for } f(x)$$

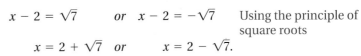

The generalized principle of square roots gives us

$$x - 2 = \sqrt{7} \qquad or \quad x - 2 = -\sqrt{7} \qquad \text{Using the principle of square roots}$$

$$x = 2 + \sqrt{7} \quad or \qquad x = 2 - \sqrt{7}.$$

Check: $f(2 + \sqrt{7}) = (2 + \sqrt{7} - 2)^2 = (\sqrt{7})^2 = 7.$

Similarly,

$$f(2 - \sqrt{7}) = (2 - \sqrt{7} - 2)^2 = (-\sqrt{7})^2 = 7.$$

The solutions are $2 + \sqrt{7}$ and $2 - \sqrt{7}$, or simply $2 \pm \sqrt{7}$.

 35

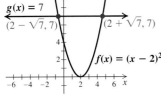

A visualization of Example 5

Example 5 is of the form $(x - a)^2 = c$, where a and c are constants. Sometimes we must factor in order to obtain this form.

EXAMPLE   6

Solve: $x^2 + 6x + 9 = 2.$

SOLUTION   We have

$$x^2 + 6x + 9 = 2 \qquad \text{The left side is the square of a binomial.}$$
$$(x + 3)^2 = 2 \qquad \text{Factoring}$$
$$x + 3 = \sqrt{2} \qquad or \quad x + 3 = -\sqrt{2} \qquad \text{Using the principle of square roots}$$
$$x = -3 + \sqrt{2} \quad or \qquad x = -3 - \sqrt{2}. \qquad \text{Adding } -3 \text{ to both sides}$$

The solutions are $-3 + \sqrt{2}$ and $-3 - \sqrt{2}$, or $-3 \pm \sqrt{2}$. The checks are left to the student.

 29

$f(x) = x^2 + 6x + 9$

A visualization of Example 6

Completing the Square

Not all quadratic equations are in the form $X^2 = k$. By using a method called *completing the square*, we can use the principle of square roots to solve *any* quadratic equation by writing it in this form.

Suppose we want to solve the quadratic equation

$$x^2 + 6x + 4 = 0.$$

The trinomial $x^2 + 6x + 4$ is not a perfect square. We can, however, create an equivalent equation with a perfect-square trinomial on one side:

$$x^2 + 6x + 4 = 0$$
$$x^2 + 6x \quad\ = -4 \qquad \text{Only variable terms are on the left side.}$$
$$x^2 + 6x + 9 = -4 + 9 \qquad \text{Adding 9 to both sides. We explain this shortly.}$$
$$(x + 3)^2 = 5. \qquad \text{We could now use the principle of square roots to solve.}$$

We chose to add 9 to both sides because it creates a perfect-square trinomial on the left side. The 9 was determined by taking half of the coefficient of x and squaring it—that is,

$$\left(\tfrac{1}{2} \cdot 6\right)^2 = 3^2, \quad or \quad 9.$$

To understand why this procedure works, examine the following drawings.

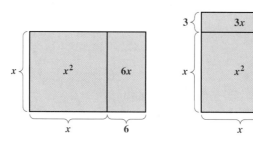

Note that the shaded areas in both figures represent the same area, $x^2 + 6x$. However, only the figure on the right, in which the $6x$ is halved, can be converted into a square with the addition of a constant term. The constant 9 is the "missing" piece that *completes* the square.

To complete the square for $x^2 + bx$, we add $(b/2)^2$.

Example 7, which follows, provides practice in finding numbers that complete the square. We will then use this skill to solve equations.

EXAMPLE 7   Replace the blanks in each equation with constants to form a true equation.

a) $x^2 + 14x + \underline{\hspace{0.4cm}} = (x + \underline{\hspace{0.4cm}})^2$

b) $x^2 - 5x + \underline{\hspace{0.4cm}} = (x - \underline{\hspace{0.4cm}})^2$

c) $x^2 + \frac{3}{4}x + \underline{\hspace{0.4cm}} = (x + \underline{\hspace{0.4cm}})^2$

SOLUTION   We take half of the coefficient of x and square it.

a) Half of 14 is 7, and $7^2 = 49$. Thus, $x^2 + 14x + 49$ is a perfect-square trinomial and is equivalent to $(x + 7)^2$. We have

$$x^2 + 14x + 49 = (x + 7)^2.$$

b) Half of -5 is $-\frac{5}{2}$, and $\left(-\frac{5}{2}\right)^2 = \frac{25}{4}$. Thus, $x^2 - 5x + \frac{25}{4}$ is a perfect-square trinomial and is equivalent to $\left(x - \frac{5}{2}\right)^2$. We have

$$x^2 - 5x + \frac{25}{4} = \left(x - \frac{5}{2}\right)^2.$$

c) Half of $\frac{3}{4}$ is $\frac{3}{8}$, and $\left(\frac{3}{8}\right)^2 = \frac{9}{64}$. Thus, $x^2 + \frac{3}{4}x + \frac{9}{64}$ is a perfect-square trinomial and is equivalent to $\left(x + \frac{3}{8}\right)^2$. We have

$$x^2 + \frac{3}{4}x + \frac{9}{64} = \left(x + \frac{3}{8}\right)^2.$$

TRY EXERCISE 39

STUDENT NOTES

In problems like Examples 7(b) and (c), it is best to avoid decimal notation. Most students have an easier time recognizing $\frac{9}{64}$ as $\left(\frac{3}{8}\right)^2$ than seeing 0.140625 as 0.375^2.

We can now use the method of completing the square to solve equations.

EXAMPLE 8   Solve: $x^2 - 8x - 7 = 0$.

SOLUTION   We begin by adding 7 to both sides:

$$x^2 - 8x - 7 = 0$$

$$x^2 - 8x = 7 \qquad \text{Adding 7 to both sides. We can now complete the square on the left side.}$$

$$x^2 - 8x + 16 = 7 + 16 \qquad \text{Adding 16 to both sides to complete the square: } \frac{1}{2}(-8) = -4, \text{ and } (-4)^2 = 16$$

$$(x - 4)^2 = 23 \qquad \text{Factoring and simplifying}$$

$$x - 4 = \pm\sqrt{23} \qquad \text{Using the principle of square roots}$$

$$x = 4 \pm \sqrt{23}. \qquad \text{Adding 4 to both sides}$$

Check: For $4 + \sqrt{23}$:

$$x^2 - 8x - 7 = 0$$

$$\begin{array}{c|c} (4 + \sqrt{23})^2 - 8(4 + \sqrt{23}) - 7 & 0 \\ 16 + 8\sqrt{23} + 23 - 32 - 8\sqrt{23} - 7 & \\ 16 + 23 - 32 - 7 + 8\sqrt{23} - 8\sqrt{23} & \\ \end{array}$$

$$0 \stackrel{?}{=} 0 \quad \text{TRUE}$$

For $4 - \sqrt{23}$:

$$x^2 - 8x - 7 = 0$$

$$\begin{array}{c|c} (4 - \sqrt{23})^2 - 8(4 - \sqrt{23}) - 7 & 0 \\ 16 - 8\sqrt{23} + 23 - 32 + 8\sqrt{23} - 7 & \\ 16 + 23 - 32 - 7 - 8\sqrt{23} + 8\sqrt{23} & \\ \end{array}$$

$$0 \stackrel{?}{=} 0 \quad \text{TRUE}$$

The solutions are $4 + \sqrt{23}$ and $4 - \sqrt{23}$, or $4 \pm \sqrt{23}$.   **TRY EXERCISE** 53

Recall that the value of $f(x)$ must be 0 at any x-intercept of the graph of f. If $f(a) = 0$, then $(a, 0)$ is an x-intercept of the graph.

EXAMPLE 9   Find the x-intercepts of the graph of $f(x) = x^2 + 5x - 3$.

SOLUTION   We set $f(x)$ equal to 0 and solve:

$$f(x) = 0$$

$$x^2 + 5x - 3 = 0 \qquad \text{Substituting}$$

$$x^2 + 5x = 3 \qquad \text{Adding 3 to both sides}$$

$$x^2 + 5x + \frac{25}{4} = 3 + \frac{25}{4} \qquad \begin{array}{l} \text{Completing the square:} \\ \frac{1}{2} \cdot 5 = \frac{5}{2}, \text{ and } \left(\frac{5}{2}\right)^2 = \frac{25}{4} \end{array}$$

$$\left(x + \frac{5}{2}\right)^2 = \frac{37}{4} \qquad \text{Factoring and simplifying}$$

$$x + \frac{5}{2} = \pm \frac{\sqrt{37}}{2} \qquad \begin{array}{l} \text{Using the principle of square roots} \\ \text{and the quotient rule for radicals} \end{array}$$

$$x = -\frac{5}{2} \pm \frac{\sqrt{37}}{2}, \quad \text{or} \quad \frac{-5 \pm \sqrt{37}}{2}. \qquad \text{Adding } -\frac{5}{2} \text{ to both sides}$$

The x-intercepts are

$$\left(-\frac{5}{2} - \frac{\sqrt{37}}{2}, 0\right) \quad \text{and} \quad \left(-\frac{5}{2} + \frac{\sqrt{37}}{2}, 0\right), \quad \text{or}$$

$$\left(\frac{-5 - \sqrt{37}}{2}, 0\right) \quad \text{and} \quad \left(\frac{-5 + \sqrt{37}}{2}, 0\right).$$

The checks are left to the student.   **TRY EXERCISE** 59

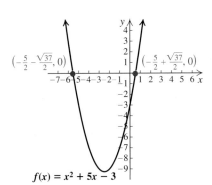

$\left(-\frac{5}{2} - \frac{\sqrt{37}}{2}, 0\right)$ $\left(-\frac{5}{2} + \frac{\sqrt{37}}{2}, 0\right)$

$f(x) = x^2 + 5x - 3$

A visualization of Example 9

Before we complete the square in a quadratic equation, the leading coefficient must be 1. When it is not 1, we divide both sides of the equation by whatever that coefficient may be.

> **To Solve a Quadratic Equation in x by Completing the Square**
>
> 1. Isolate the terms with variables on one side of the equation, and arrange them in descending order.
> 2. Divide both sides by the coefficient of x^2 if that coefficient is not 1.
> 3. Complete the square by taking half of the coefficient of x and adding its square to both sides.
> 4. Express the trinomial as the square of a binomial (factor the trinomial) and simplify the other side.
> 5. Use the principle of square roots (find the square roots of both sides).
> 6. Solve for x by adding or subtracting on both sides.

EXAMPLE 10   Solve: $3x^2 + 7x - 2 = 0$.

SOLUTION   We follow the steps listed above:

$$3x^2 + 7x - 2 = 0$$

Isolate the variable terms.

$$3x^2 + 7x = 2 \qquad \text{Adding 2 to both sides}$$

Divide both sides by the x^2-coefficient.

$$x^2 + \frac{7}{3}x = \frac{2}{3} \qquad \text{Dividing both sides by 3}$$

Complete the square.

$$x^2 + \frac{7}{3}x + \frac{49}{36} = \frac{2}{3} + \frac{49}{36} \qquad \text{Completing the square: } \left(\frac{1}{2} \cdot \frac{7}{3}\right)^2 = \frac{49}{36}$$

Factor the trinomial.

$$\left(x + \frac{7}{6}\right)^2 = \frac{73}{36} \qquad \text{Factoring and simplifying}$$

Use the principle of square roots.

$$x + \frac{7}{6} = \pm\frac{\sqrt{73}}{6} \qquad \text{Using the principle of square roots and the quotient rule for radicals}$$

Solve for x.

$$x = -\frac{7}{6} \pm \frac{\sqrt{73}}{6}, \text{ or } \frac{-7 \pm \sqrt{73}}{6}. \qquad \text{Adding } -\frac{7}{6} \text{ to both sides}$$

The checks are left to the student. The solutions are $-\dfrac{7}{6} \pm \dfrac{\sqrt{73}}{6}$, or $\dfrac{-7 \pm \sqrt{73}}{6}$.

This can be written as

$$-\frac{7}{6} + \frac{\sqrt{73}}{6} \text{ and } -\frac{7}{6} - \frac{\sqrt{73}}{6}, \text{ or } \frac{-7 + \sqrt{73}}{6} \text{ and } \frac{-7 - \sqrt{73}}{6}.$$

TRY EXERCISE 69

Any quadratic equation can be solved by completing the square. The procedure is also useful when graphing quadratic equations and will be used in the next section to develop a formula for solving quadratic equations.

Problem Solving

After one year, an amount of money P, invested at 4% per year, is worth 104% of P, or $P(1.04)$. If that amount continues to earn 4% interest per year, after the second year the investment will be worth 104% of $P(1.04)$, or $P(1.04)^2$. This is called **compounding interest** since after the first time period, interest is earned on both the initial investment *and* the interest from the first time period. Continuing the above pattern, we see that after the third year, the investment will be worth 104% of $P(1.04)^2$. Generalizing, we have the following.

> ### The Compound-Interest Formula
>
> If an amount of money P is invested at interest rate r, compounded annually, then in t years, it will grow to the amount A given by
>
> $$A = P(1 + r)^t.   \text{(r is written in decimal notation.)}$$

We can use quadratic equations to solve certain interest problems.

EXAMPLE 11

Investment growth.   Katia invested $4000 at interest rate r, compounded annually. In 2 yr, it grew to $4410. What was the interest rate?

SOLUTION

1. **Familiarize.** We are already familiar with the compound-interest formula. If we were not, we would need to consult an outside source.

2. **Translate.** The translation consists of substituting into the formula:

$$A = P(1 + r)^t$$
$$4410 = 4000(1 + r)^2.   \text{Substituting}$$

3. **Carry out.** We solve for r:

$$4410 = 4000(1 + r)^2$$
$$\frac{4410}{4000} = (1 + r)^2   \text{Dividing both sides by 4000}$$
$$\frac{441}{400} = (1 + r)^2   \text{Simplifying}$$
$$\pm\sqrt{\frac{441}{400}} = 1 + r   \text{Using the principle of square roots}$$
$$\pm\frac{21}{20} = 1 + r   \text{Simplifying}$$
$$-\frac{20}{20} \pm \frac{21}{20} = r   \text{Adding } -1, \text{ or } -\frac{20}{20}, \text{ to both sides}$$
$$\frac{1}{20} = r   \text{or}   -\frac{41}{20} = r.$$

4. **Check.** Since the interest rate cannot be negative, we need check only $\frac{1}{20}$, or 5%. If $4000 were invested at 5% interest, compounded annually, then in 2 yr it would grow to $4000(1.05)^2$, or $4410. The rate 5% checks.

5. **State.** The interest rate was 5%.

TRY EXERCISE 75

EXAMPLE 12

Free-falling objects.   The formula $s = 16t^2$ is used to approximate the distance s, in feet, that an object falls freely from rest in t seconds. The Grand Canyon Skywalk is 4000 ft above the Colorado River. How long will it take a stone to fall from the Skywalk to the river? Round to the nearest tenth of a second.

Source: www.grandcanyonskywalk.com

SOLUTION

1. **Familiarize.** We agree to disregard air resistance and use the given formula.

2. **Translate.** We substitute into the formula:

$$s = 16t^2$$
$$4000 = 16t^2.$$

3. Carry out. We solve for t:

$$4000 = 16t^2$$

$$250 = t^2$$

$$\sqrt{250} = t \quad \text{Using the principle of square roots; rejecting the negative square root since } t \text{ cannot be negative in this problem}$$

$$15.8 \approx t. \quad \text{Using a calculator and rounding to the nearest tenth}$$

4. Check. Since $16(15.8)^2 = 3994.24 \approx 4000$, our answer checks.

5. State. It takes about 15.8 sec for a stone to fall freely from the Grand Canyon Skywalk to the river.

TRY EXERCISE 79

TECHNOLOGY CONNECTION

As we saw in Section 5.8, a graphing calculator can be used to find approximate solutions of any quadratic equation that has real-number solutions.

To check Example 8, we graph $y = x^2 - 8x - 7$ and use the ZERO or ROOT option of the CALC menu. When asked for a Left and Right Bound, we enter cursor positions to the left of and to the right of the root. A Guess between the bounds is entered and a value for the root then appears.

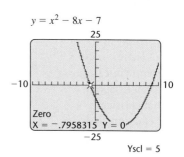

1. Use a graphing calculator to check the second solution of Example 8.
2. Use a graphing calculator to confirm the solutions in Example 9.
3. Can a graphing calculator be used to find *exact* solutions in Example 10? Why or why not?
4. Use a graphing calculator to confirm that there are no real-number solutions of $x^2 - 6x + 11 = 0$.

8.1   EXERCISE SET

For Extra Help MyMathLab MathXL PRACTICE WATCH DOWNLOAD

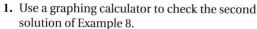 *Concept Reinforcement* *Complete each of the following to form true statements.*

1. The principle of square roots states that if $x^2 = k$, then $x = $ ____ or $x = $ ____.

2. If $(x + 5)^2 = 49$, then $x + 5 = $ ____ or $x + 5 = $ ____.

3. If $t^2 + 6t + 9 = 17$, then (____ $)^2 = 17$ and ____ $= \pm\sqrt{17}$.

4. The equations $x^2 + 8x + $ ____ $= 23$ and $x^2 + 8x = 7$ are equivalent.

5. The expressions $t^2 + 10t + $ ____ and $(t + $ ____ $)^2$ are equivalent.

6. The expressions $x^2 - 6x + $ ____ and $(x - $ ____ $)^2$ are equivalent.

Solve.

7. $x^2 = 100$

8. $t^2 = 144$

9. $p^2 - 50 = 0$

10. $c^2 - 8 = 0$

11. $5y^2 = 30$

12. $4y^2 = 12$

13. $9x^2 - 49 = 0$

14. $36a^2 - 25 = 0$

15. $6t^2 - 5 = 0$

16. $7x^2 - 5 = 0$

17. $a^2 + 1 = 0$

18. $t^2 + 4 = 0$

19. $4d^2 + 81 = 0$

20. $25y^2 + 16 = 0$

21. $(x - 3)^2 = 16$

22. $(x + 1)^2 = 100$

23. $(t + 5)^2 = 12$

24. $(y - 4)^2 = 18$

25. $(x + 1)^2 = -9$

26. $(x - 1)^2 = -49$

27. $\left(y + \frac{3}{4}\right)^2 = \frac{17}{16}$

28. $\left(t + \frac{3}{2}\right)^2 = \frac{7}{2}$

29. $x^2 - 10x + 25 = 64$ **30.** $x^2 - 6x + 9 = 100$

31. Let $f(x) = x^2$. Find x such that $f(x) = 19$.

32. Let $f(x) = x^2$. Find x such that $f(x) = 11$.

33. Let $f(x) = (x - 5)^2$. Find x such that $f(x) = 16$.

34. Let $g(x) = (x - 2)^2$. Find x such that $g(x) = 25$.

35. Let $F(t) = (t + 4)^2$. Find t such that $F(t) = 13$.

36. Let $f(t) = (t + 6)^2$. Find t such that $f(t) = 15$.

Aha! **37.** Let $g(x) = x^2 + 14x + 49$. Find x such that $g(x) = 49$.

38. Let $F(x) = x^2 + 8x + 16$. Find x such that $F(x) = 9$.

Replace the blanks in each equation with constants to complete the square and form a true equation.

39. $x^2 + 16x +$ ___ $= (x +$ ___$)^2$

40. $x^2 + 12x +$ ___ $= (x +$ ___$)^2$

41. $t^2 - 10t +$ ___ $= (t -$ ___$)^2$

42. $t^2 - 6t +$ ___ $= (t -$ ___$)^2$

43. $t^2 - 2t +$ ___ $= (t -$ ___$)^2$

44. $x^2 + 2x +$ ___ $= (x +$ ___$)^2$

45. $x^2 + 3x +$ ___ $= \left(x +$ ___$\right)^2$

46. $t^2 - 9t +$ ___ $= \left(t -$ ___$\right)^2$

47. $x^2 + \frac{2}{5}x +$ ___ $= \left(x +$ ___$\right)^2$

48. $x^2 + \frac{2}{3}x +$ ___ $= \left(x +$ ___$\right)^2$

49. $t^2 - \frac{5}{6}t +$ ___ $= \left(t -$ ___$\right)^2$

50. $t^2 - \frac{5}{3}t +$ ___ $= \left(t -$ ___$\right)^2$

Solve by completing the square. Show your work.

51. $x^2 + 6x = 7$

52. $x^2 + 8x = 9$

53. $t^2 - 10t = -23$

54. $t^2 - 4t = -1$

55. $x^2 + 12x + 32 = 0$

56. $x^2 + 16x + 15 = 0$

57. $t^2 + 8t - 3 = 0$

58. $t^2 + 6t - 5 = 0$

Complete the square to find the x-intercepts of each function given by the equation listed.

59. $f(x) = x^2 + 6x + 7$

60. $f(x) = x^2 + 10x - 2$

61. $g(x) = x^2 + 9x - 25$

62. $g(x) = x^2 + 5x + 2$

63. $f(x) = x^2 - 10x - 22$

64. $f(x) = x^2 - 8x - 10$

Solve by completing the square. Remember to first divide, as in Example 10, to make sure that the coefficient of x^2 is 1.

65. $9x^2 + 18x = -8$ **66.** $4x^2 + 8x = -3$

67. $3x^2 - 5x - 2 = 0$ **68.** $2x^2 - 5x - 3 = 0$

69. $5x^2 + 4x - 3 = 0$ **70.** $4x^2 + 3x - 5 = 0$

71. Find the x-intercepts of the function given by $f(x) = 4x^2 + 2x - 3$.

72. Find the x-intercepts of the function given by $f(x) = 3x^2 + x - 5$.

73. Find the x-intercepts of the function given by $g(x) = 2x^2 - 3x - 1$.

74. Find the x-intercepts of the function given by $g(x) = 3x^2 - 5x - 1$.

Interest.   Use $A = P(1 + r)^t$ to find the interest rate in Exercises 75–78. Refer to Example 11.

75. $2000 grows to $2420 in 2 yr

76. $1000 grows to $1440 in 2 yr

77. $6250 grows to $6760 in 2 yr

78. $6250 grows to $7290 in 2 yr

Free-falling objects.   Use $s = 16t^2$ for Exercises 79–82. Refer to Example 12 and neglect air resistance.

79. At a height of 290 ft, the Rainbow Bridge in Lake Powell National Monument, Utah, is the world's highest natural arch. How long would it take an object to fall freely from the bridge?
Source: *Guinness World Records* 2008

80. The Sears Tower in Chicago is 1454 ft tall. How long would it take an object to fall freely from the top?

81. At 2063 ft, the KVLY-TV tower in North Dakota is the tallest supported tower. How long would it take an object to fall freely from the top?
Source: North Dakota Tourism Division

82. El Capitan in Yosemite National Park is 3593 ft high. How long would it take a carabiner to fall freely from the top?
Source: *Guinness World Records* 2008

83. Explain in your own words a sequence of steps that can be used to solve any quadratic equation in the quickest way.

84. Write an interest-rate problem for a classmate to solve. Devise the problem so that the solution is "The loan was made at 7% interest."

Skill Review

To prepare for Section 8.2, review evaluating expressions and simplifying radical expressions (Sections 1.2, 7.3, and 7.8).

Evaluate. [1.2]

85. $b^2 - 4ac$, for $a = 3$, $b = 2$, and $c = -5$

86. $b^2 - 4ac$, for $a = 1$, $b = -1$, and $c = 4$

Simplify. [7.3], [7.8]

87. $\sqrt{200}$

88. $\sqrt{96}$

89. $\sqrt{-4}$

90. $\sqrt{-25}$

91. $\sqrt{-8}$

92. $\sqrt{-24}$

Synthesis

93. What would be better: to receive 3% interest every 6 months, or to receive 6% interest every 12 months? Why?

94. Write a problem involving a free-falling object for a classmate to solve (see Example 12). Devise the problem so that the solution is "The object takes about 4.5 sec to fall freely from the top of the structure."

Find b such that each trinomial is a square.

95. $x^2 + bx + 81$

96. $x^2 + bx + 49$

97. If $f(x) = 2x^5 - 9x^4 - 66x^3 + 45x^2 + 280x$ and $x^2 - 5$ is a factor of $f(x)$, find all a for which $f(a) = 0$.

98. If $f(x) = \left(x - \frac{1}{3}\right)(x^2 + 6)$ and $g(x) = \left(x - \frac{1}{3}\right)\left(x^2 - \frac{2}{3}\right)$, find all a for which $(f + g)(a) = 0$.

99. *Boating.* A barge and a fishing boat leave a dock at the same time, traveling at a right angle to each other. The barge travels 7 km/h slower than the fishing boat. After 4 hr, the boats are 68 km apart. Find the speed of each boat.

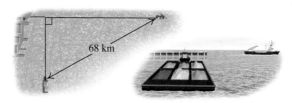

100. Find three consecutive integers such that the square of the first plus the product of the other two is 67.

101. Exercises 29, 33, and 53 can be solved on a graphing calculator without first rewriting in standard form. Simply let y_1 represent the left side of the equation and y_2 the right side. Then use a graphing calculator to determine the x-coordinate of any point of intersection. Use a graphing calculator to solve Exercises 29, 33, and 53 in this manner.

102. Use a graphing calculator to check your answers to Exercises 5, 13, 71, and 73.

103. Example 11 can be solved with a graphing calculator by graphing each side of
$$4410 = 4000(1 + r)^2.$$
How could you determine, from a reading of the problem, a suitable viewing window? What might that window be?

Solving Using the Quadratic Formula   ▪   Approximating Solutions

We can use the process of completing the square to develop a general formula for solving quadratic equations.

Solving Using the Quadratic Formula

Each time we solve by completing the square, the procedure is the same. When a procedure is repeated many times, we can often develop a formula to speed up our work.

We begin with a quadratic equation in standard form,

$$ax^2 + bx + c = 0,$$

with $a > 0$. For $a < 0$, a slightly different derivation is needed (see Exercise 60), but the result is the same. Let's solve by completing the square. As the steps are performed, compare them with Example 10 on p. 506.

$$ax^2 + bx = -c \qquad \text{Adding to both sides}$$

$$x^2 + \frac{b}{a}x = -\frac{c}{a} \qquad \text{Dividing both sides by } a$$

Half of $\dfrac{b}{a}$ is $\dfrac{b}{2a}$ and $\left(\dfrac{b}{2a}\right)^2$ is $\dfrac{b^2}{4a^2}$. We add $\dfrac{b^2}{4a^2}$ to both sides:

$$x^2 + \frac{b}{a}x + \frac{b^2}{4a^2} = -\frac{c}{a} + \frac{b^2}{4a^2} \qquad \text{Adding } \dfrac{b^2}{4a^2} \text{ to complete the square}$$

$$\left(x + \frac{b}{2a}\right)^2 = -\frac{4ac}{4a^2} + \frac{b^2}{4a^2} \qquad \text{Factoring on the left side; finding a common denominator on the right side}$$

$$\left(x + \frac{b}{2a}\right)^2 = \frac{b^2 - 4ac}{4a^2}$$

$$x + \frac{b}{2a} = \pm\frac{\sqrt{b^2 - 4ac}}{2a} \qquad \text{Using the principle of square roots and the quotient rule for radicals. Since } a > 0, \sqrt{4a^2} = 2a.$$

$$x = \frac{-b \pm \sqrt{b^2 - 4ac}}{2a}. \qquad \text{Adding } -\dfrac{b}{2a} \text{ to both sides}$$

It is important to remember the quadratic formula and know how to use it.

The Quadratic Formula

The solutions of $ax^2 + bx + c = 0, a \neq 0$, are given by

$$x = \frac{-b \pm \sqrt{b^2 - 4ac}}{2a}.$$

EXAMPLE   **1**   Solve $5x^2 + 8x = -3$ using the quadratic formula.

SOLUTION   We first find standard form and determine a, b, and c:

$$5x^2 + 8x + 3 = 0;$$ Adding 3 to both sides to get 0 on one side

$$a = 5, \quad b = 8, \quad c = 3.$$

Next, we use the quadratic formula:

$$x = \frac{-b \pm \sqrt{b^2 - 4ac}}{2a}$$ It is important to remember this formula.

$$x = \frac{-8 \pm \sqrt{8^2 - 4 \cdot 5 \cdot 3}}{2 \cdot 5}$$ Substituting

$$x = \frac{-8 \pm \sqrt{64 - 60}}{10}$$ Be sure to write the fraction bar all the way across.

$$x = \frac{-8 \pm \sqrt{4}}{10} = \frac{-8 \pm 2}{10}$$

$$x = \frac{-8 + 2}{10} \quad or \quad x = \frac{-8 - 2}{10}$$ The symbol $\pm$ indicates two solutions.

$$x = \frac{-6}{10} \quad or \quad x = \frac{-10}{10}$$

$$x = -\frac{3}{5} \quad or \quad x = -1.$$

The solutions are $-\frac{3}{5}$ and -1. The checks are left to the student.

 **TRY EXERCISE** 25

Because $5x^2 + 8x + 3$ can be factored, the quadratic formula may not have been the fastest way of solving Example 1. However, because the quadratic formula works for *any* quadratic equation, we need not spend too much time struggling to solve a quadratic equation by factoring.

To Solve a Quadratic Equation

1. If the equation can be easily written in the form $ax^2 = p$ or $(x + k)^2 = d$, use the principle of square roots as in Section 8.1.
2. If step (1) does not apply, write the equation in the form $ax^2 + bx + c = 0$.
3. Try factoring and using the principle of zero products.
4. If factoring seems difficult or impossible, use the quadratic formula. Completing the square can also be used.

The solutions of a quadratic equation can always be found using the quadratic formula. They cannot always be found by factoring.

Recall that a second-degree polynomial in one variable is said to be quadratic. Similarly, a second-degree polynomial function in one variable is said to be a **quadratic function**.

EXAMPLE   2

For the quadratic function given by $f(x) = 3x^2 - 6x - 4$, find all x for which $f(x) = 0$.

SOLUTION   We substitute and solve for x:

$$f(x) = 0$$
$$3x^2 - 6x - 4 = 0. \quad \text{Substituting}$$

Since $3x^2 - 6x - 4$ does not factor, we use the quadratic formula with $a = 3$, $b = -6$, and $c = -4$:

$$x = \frac{-(-6) \pm \sqrt{(-6)^2 - 4 \cdot 3 \cdot (-4)}}{2 \cdot 3}$$

$$= \frac{6 \pm \sqrt{36 + 48}}{6} \qquad (-6)^2 - 4 \cdot 3 \cdot (-4) = 36 - (-48) = 36 + 48$$

$$= \frac{6 \pm \sqrt{84}}{6} \qquad \text{Note that 4 is a perfect-square factor of 84.}$$

$$= \frac{6}{6} \pm \frac{\sqrt{84}}{6} \qquad \text{Writing as two fractions to simplify each separately}$$

$$= 1 \pm \frac{\sqrt{4}\sqrt{21}}{6} \qquad 84 = 4 \cdot 21$$

$$\left. \begin{array}{l} = 1 \pm \dfrac{2\sqrt{21}}{2 \cdot 3} \\[2mm] = 1 \pm \dfrac{\sqrt{21}}{3}. \end{array} \right\} \quad \text{Simplifying by removing a factor of 1: } \dfrac{2}{2} = 1$$

The solutions are $1 - \dfrac{\sqrt{21}}{3}$ and $1 + \dfrac{\sqrt{21}}{3}$. The checks are left to the student.

TRY EXERCISE   39

TECHNOLOGY CONNECTION

To check Example 2 by graphing $y_1 = 3x^2 - 6x - 4$, press ⟨TRACE⟩ and enter $1 + \sqrt{21}/3$. A rational approximation and the y-value 0 should appear.

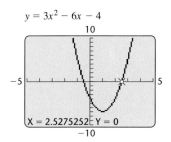

$y = 3x^2 - 6x - 4$

X = 2.5275252   Y = 0

Use this approach to check the other solution of Example 2.

Some quadratic equations have solutions that are imaginary numbers.

EXAMPLE   3

Solve: $x(x + 5) = 2(2x - 1)$.

SOLUTION   We first find standard form:

$$x^2 + 5x = 4x - 2 \qquad \text{Multiplying}$$
$$x^2 + x + 2 = 0. \qquad \text{Subtracting } 4x \text{ and adding 2 to both sides}$$

Since we cannot factor $x^2 + x + 2$, we use the quadratic formula with $a = 1$, $b = 1$, and $c = 2$:

$$x = \frac{-1 \pm \sqrt{1^2 - 4 \cdot 1 \cdot 2}}{2 \cdot 1} \qquad \text{Substituting}$$

$$= \frac{-1 \pm \sqrt{1 - 8}}{2}$$

$$= \frac{-1 \pm \sqrt{-7}}{2}$$

$$= \frac{-1 \pm i\sqrt{7}}{2}, \text{ or } -\frac{1}{2} \pm \frac{\sqrt{7}}{2}i.$$

The solutions are $-\dfrac{1}{2} - \dfrac{\sqrt{7}}{2}i$ and $-\dfrac{1}{2} + \dfrac{\sqrt{7}}{2}i$. The checks are left to the student.

TRY EXERCISE   35

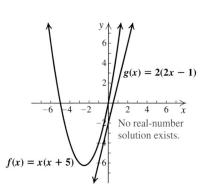

$g(x) = 2(2x - 1)$

No real-number solution exists.

$f(x) = x(x + 5)$

A visualization of Example 3

The quadratic formula can be used to solve certain rational equations.

If $f(t) = 2 + \dfrac{7}{t}$ and $g(t) = \dfrac{4}{t^2}$, find all t for which $f(t) = g(t)$.

SOLUTION   We set $f(t)$ equal to $g(t)$ and solve:

$$f(t) = g(t)$$

$$2 + \frac{7}{t} = \frac{4}{t^2}.\qquad \text{Substituting. Note that } t \neq 0.$$

This is a rational equation similar to those in Section 6.4. To solve, we multiply both sides by the LCD, t^2:

$$t^2\left(2 + \frac{7}{t}\right) = t^2 \cdot \frac{4}{t^2}$$

$$2t^2 + 7t = 4 \qquad \text{Simplifying}$$

$$2t^2 + 7t - 4 = 0. \qquad \text{Subtracting 4 from both sides}$$

We use the quadratic formula with $a = 2$, $b = 7$, and $c = -4$:

$$t = \frac{-7 \pm \sqrt{7^2 - 4 \cdot 2 \cdot (-4)}}{2 \cdot 2}$$

$$= \frac{-7 \pm \sqrt{49 + 32}}{4} \qquad 7^2 - 4 \cdot 2 \cdot (-4) = 49 - (-32) = 49 + 32$$

$$= \frac{-7 \pm \sqrt{81}}{4}$$

$$= \frac{-7 \pm 9}{4} \qquad \text{This means } \frac{-7 + 9}{4} \text{ or } \frac{-7 - 9}{4}.$$

$$t = \frac{2}{4} = \frac{1}{2} \quad \text{or} \quad t = \frac{-16}{4} = -4. \qquad \begin{array}{l}\text{Both answers should check}\\ \text{since } t \neq 0.\end{array}$$

You can confirm that $f\left(\tfrac{1}{2}\right) = g\left(\tfrac{1}{2}\right)$ and $f(-4) = g(-4)$. The solutions are $\tfrac{1}{2}$ and -4.

TRY EXERCISE 41

We saw in Sections 5.8 and 8.1 how graphing calculators can solve quadratic equations. To determine whether quadratic equations are solved more quickly on a graphing calculator or by using the quadratic formula, solve Examples 2 and 4 both ways. Which method is faster? Which method is more precise? Why?

Approximating Solutions

When the solution of an equation is irrational, a rational-number approximation is often useful. This is often the case in real-world applications similar to those found in Section 8.4.

Use a calculator to approximate, to three decimal places, the solutions of Example 2.

SOLUTION   On most calculators, one of the following sequences of keystrokes can be used to approximate $1 + \sqrt{21}/3$:

Similar keystrokes can be used to approximate $1 - \sqrt{21}/3$.

    The solutions are approximately 2.527525232 and -0.5275252317. Rounded to three decimal places, the solutions are approximately 2.528 and -0.528.

TRY EXERCISE 45

It is important that you understand both the rules for order of operations *and* the manner in which your calculator applies those rules.

8.2   EXERCISE SET

 Concept Reinforcement *Classify each statement as either true or false.*

1. The quadratic formula can be used to solve *any* quadratic equation.

2. The steps used to derive the quadratic formula are the same as those used when solving by completing the square.

3. The quadratic formula does not work if solutions are imaginary numbers.

4. Solving by factoring is always slower than using the quadratic formula.

5. A quadratic equation can have as many as four solutions.

6. It is possible for a quadratic equation to have no real-number solutions.

Solve.

7. $2x^2 + 3x - 5 = 0$

8. $3x^2 - 7x + 2 = 0$

9. $u^2 + 2u - 4 = 0$

10. $u^2 - 2u - 2 = 0$

11. $t^2 + 3 = 6t$

12. $t^2 + 4t = 1$

13. $x^2 = 3x + 5$

14. $x^2 + 5x + 3 = 0$

15. $3t(t + 2) = 1$

16. $2t(t + 2) = 1$

17. $\dfrac{1}{x^2} - 3 = \dfrac{8}{x}$

18. $\dfrac{9}{x} - 2 = \dfrac{5}{x^2}$

19. $t^2 + 10 = 6t$

20. $t^2 + 10t + 26 = 0$

21. $p^2 - p + 1 = 0$

22. $p^2 + p + 4 = 0$

23. $x^2 + 4x + 6 = 0$

24. $x^2 + 11 = 6x$

25. $12t^2 + 17t = 40$

26. $15t^2 + 7t = 2$

27. $25x^2 - 20x + 4 = 0$

28. $36x^2 + 84x + 49 = 0$

29. $7x(x + 2) + 5 = 3x(x + 1)$

30. $5x(x - 1) - 7 = 4x(x - 2)$

31. $14(x - 4) - (x + 2) = (x + 2)(x - 4)$

32. $11(x - 2) + (x - 5) = (x + 2)(x - 6)$

33. $51p = 2p^2 + 72$

34. $72 = 3p^2 + 50p$

35. $x(x - 3) = x - 9$

36. $x(x - 1) = 2x - 7$

37. $x^3 - 8 = 0$ (*Hint:* Factor the difference of cubes. Then use the quadratic formula.)

38. $x^3 + 1 = 0$

39. Let $f(x) = 6x^2 - 7x - 20$. Find x such that $f(x) = 0$.

40. Let $g(x) = 4x^2 - 2x - 3$. Find x such that $g(x) = 0$.

41. Let
$$f(x) = \frac{7}{x} + \frac{7}{x + 4}.$$
Find all x for which $f(x) = 1$.

42. Let
$$g(x) = \frac{2}{x} + \frac{2}{x + 3}.$$
Find all x for which $g(x) = 1$.

43. Let
$$F(x) = \frac{3 - x}{4} \quad \text{and} \quad G(x) = \frac{1}{4x}.$$
Find all x for which $F(x) = G(x)$.

44. Let
$$f(x) = x + 5 \quad \text{and} \quad g(x) = \frac{3}{x - 5}.$$
Find all x for which $f(x) = g(x)$.

Solve using the quadratic formula. Then use a calculator to approximate, to three decimal places, the solutions as rational numbers.

45. $x^2 + 4x - 7 = 0$

46. $x^2 + 6x + 4 = 0$

Aha! 47. $x^2 - 6x + 4 = 0$

48. $x^2 - 4x + 1 = 0$

49. $2x^2 - 3x - 7 = 0$

50. $3x^2 - 3x - 2 = 0$

51. Are there any equations that can be solved by the quadratic formula but not by completing the square? Why or why not?

52. Suppose you are solving a quadratic equation with no constant term ($c = 0$). Would you use factoring or the quadratic formula to solve? Why?

Skill Review

To prepare for Section 8.3, review multiplying and simplifying radical and complex-number expressions (Sections 7.5 and 7.8).

Multiply and simplify.

53. $(x - 2i)(x + 2i)$ [7.8]

54. $(x - 6\sqrt{5})(x + 6\sqrt{5})$ [7.5]

55. $(x - (2 - \sqrt{7}))(x - (2 + \sqrt{7}))$ [7.5]

56. $(x - (-3 + 5i))(x - (-3 - 5i))$ [7.8]

Simplify.

57. $\dfrac{-6 \pm \sqrt{(-4)^2 - 4(2)(2)}}{2(2)}$ [7.3]

58. $\dfrac{-(-1) \pm \sqrt{(6)^2 - 4(3)(5)}}{2(3)}$ [7.8]

Synthesis

59. Explain how you could use the quadratic formula to help factor a quadratic polynomial.

60. If $a < 0$ and $ax^2 + bx + c = 0$, then $-a$ is positive and the equivalent equation, $-ax^2 - bx - c = 0$, can be solved using the quadratic formula.

   a) Find this solution, replacing a, b, and c in the formula with $-a$, $-b$, and $-c$ from the equation.

   b) How does the result of part (a) indicate that the quadratic formula "works" regardless of the sign of a?

For Exercises 61–63, let

$$f(x) = \frac{x^2}{x - 2} + 1 \quad and \quad g(x) = \frac{4x - 2}{x - 2} + \frac{x + 4}{2}.$$

61. Find the x-intercepts of the graph of f.

62. Find the x-intercepts of the graph of g.

63. Find all x for which $f(x) = g(x)$.

Solve. Approximate the solutions to three decimal places.

64. $x^2 - 0.75x - 0.5 = 0$

65. $z^2 + 0.84z - 0.4 = 0$

Solve.

66. $(1 + \sqrt{3})x^2 - (3 + 2\sqrt{3})x + 3 = 0$

67. $\sqrt{2}x^2 + 5x + \sqrt{2} = 0$

68. $ix^2 - 2x + 1 = 0$

69. One solution of $kx^2 + 3x - k = 0$ is -2. Find the other.

70. Use a graphing calculator to solve Exercises 9, 27, and 43.

71. Use a graphing calculator to solve Exercises 11, 33, and 41. Use the method of graphing each side of the equation.

72. Can a graphing calculator be used to solve *any* quadratic equation? Why or why not?

8.3   Studying Solutions of Quadratic Equations

The Discriminant   •   Writing Equations from Solutions

The Discriminant

It is sometimes enough to know what *type* of number a solution will be, without actually solving the equation. Suppose we want to know if $4x^2 - 5x - 2 = 0$ has rational solutions (and thus can be solved by factoring). Using the quadratic formula, we would have

$$x = \frac{-b \pm \sqrt{b^2 - 4ac}}{2a}$$

$$= \frac{-(-5) \pm \sqrt{(-5)^2 - 4 \cdot 4(-2)}}{2 \cdot 4}.$$

Since $(-5)^2 - 4 \cdot 4 \cdot (-2) = 25 - 16(-2) = 25 + 32 = 57$ and since 57 is not a perfect square, the solutions of the equation are not rational numbers. This means that $4x^2 - 5x - 2 = 0$ *cannot* be solved by factoring. Note that the radicand, 57, determines what type of number the solutions will be.

The radicand $b^2 - 4ac$ is known as the **discriminant**. If a, b, and c are rational, then we have the following.

- When $b^2 - 4ac$ simplifies to 0, it doesn't matter if we use $+\sqrt{b^2 - 4ac}$ or $-\sqrt{b^2 - 4ac}$; we get the same solution twice. Thus, when the discriminant is 0, there is one *repeated* solution and it is rational.

  Example: $9x^2 + 6x + 1 = 0 \rightarrow b^2 - 4ac = 6^2 - 4 \cdot 9 \cdot 1 = 0$.
  Solving $9x^2 + 6x + 1 = 0$ gives the (repeated) solution $-\frac{1}{3}$.

- When $b^2 - 4ac$ is positive, there are two different real-number solutions: If $b^2 - 4ac$ is a perfect square, these solutions are rational numbers.

  Example: $6x^2 + 5x + 1 = 0 \rightarrow b^2 - 4ac = 5^2 - 4 \cdot 6 \cdot 1 = 1$.
  Solving $6x^2 + 5x + 1 = 0$ gives the solutions $-\frac{1}{3}$ and $-\frac{1}{2}$.

- When $b^2 - 4ac$ is positive but not a perfect square, there are two irrational solutions and they are conjugates of each other (see p. 459).

  Example: $x^2 + 4x + 2 = 0 \rightarrow b^2 - 4ac = 4^2 - 4 \cdot 1 \cdot 2 = 8$.
  Solving $x^2 + 4x + 2 = 0$ gives the solutions $-2 + \sqrt{2}$ and $-2 - \sqrt{2}$.

- When the discriminant is negative, there are two imaginary-number solutions and they are complex conjugates of each other.

  Example: $x^2 + 4x + 5 = 0 \rightarrow b^2 - 4ac = 4^2 - 4 \cdot 1 \cdot 5 = -4$.
  Solving $x^2 + 4x + 5 = 0$ gives the solutions $-2 + i$ and $-2 - i$.

Note that any equation for which $b^2 - 4ac$ is a perfect square can be solved by factoring.

Discriminant $b^2 - 4ac$	Nature of Solutions
0	One solution; a rational number
Positive Perfect square Not a perfect square	Two different real-number solutions Solutions are rational. Solutions are irrational conjugates.
Negative	Two different imaginary-number solutions (complex conjugates)

EXAMPLE 1   For each equation, determine what type of number the solutions are and how many solutions exist.

a) $9x^2 - 12x + 4 = 0$ **b)** $x^2 + 5x + 8 = 0$ **c)** $2x^2 + 7x - 3 = 0$

SOLUTION

a) For $9x^2 - 12x + 4 = 0$, we have

$$a = 9, \quad b = -12, \quad c = 4.$$

We substitute and compute the discriminant:

$$b^2 - 4ac = (-12)^2 - 4 \cdot 9 \cdot 4$$
$$= 144 - 144 = 0.$$

There is exactly one solution, and it is rational. This indicates that $9x^2 - 12x + 4 = 0$ can be solved by factoring.

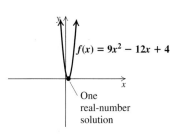

$f(x) = 9x^2 - 12x + 4$

One real-number solution

A visualization of part (a)

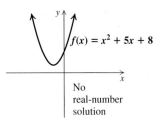

A visualization of part (b)

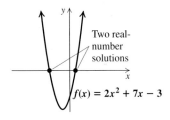

A visualization of part (c)

b) For $x^2 + 5x + 8 = 0$, we have

$$a = 1, \quad b = 5, \quad c = 8.$$

We substitute and compute the discriminant:

$$b^2 - 4ac = 5^2 - 4 \cdot 1 \cdot 8$$
$$= 25 - 32 = -7.$$

Since the discriminant is negative, there are two different imaginary-number solutions that are complex conjugates of each other.

c) For $2x^2 + 7x - 3 = 0$, we have

$$a = 2, \quad b = 7, \quad c = -3.$$

We substitute and compute the discriminant:

$$b^2 - 4ac = 7^2 - 4 \cdot 2(-3)$$
$$= 49 - (-24) = 73.$$

The discriminant is a positive number that is not a perfect square. Thus there are two different irrational solutions that are conjugates of each other.

TRY EXERCISE ▶ 7

Discriminants can also be used to determine the number of *x*-intercepts of the graph of a quadratic function.

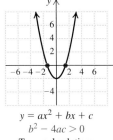

$y = ax^2 + bx + c$
$b^2 - 4ac > 0$
Two real solutions
of $ax^2 + bx + c = 0$
Two *x*-intercepts

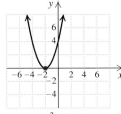

$y = ax^2 + bx + c$
$b^2 - 4ac = 0$
One real solution
of $ax^2 + bx + c = 0$
One *x*-intercept

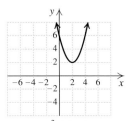

$y = ax^2 + bx + c$
$b^2 - 4ac < 0$
No real solutions
of $ax^2 + bx + c = 0$
No *x*-intercept

Writing Equations from Solutions

We know by the principle of zero products that $(x - 2)(x + 3) = 0$ has solutions 2 and -3. If we wish for two given numbers to be solutions of an equation, we can create such an equation, using the principle in reverse.

EXAMPLE 2   Find an equation for which the given numbers are solutions.

a) 3 and $-\frac{2}{5}$ **b)** $2i$ and $-2i$

c) $5\sqrt{7}$ and $-5\sqrt{7}$ **d)** $-4, 0,$ and 1

SOLUTION

a)

$$x = 3 \quad or \quad x = -\tfrac{2}{5}$$
$$x - 3 = 0 \quad or \quad x + \tfrac{2}{5} = 0 \qquad \text{Getting 0's on one side}$$
$$(x - 3)\left(x + \tfrac{2}{5}\right) = 0 \qquad \text{Using the principle of zero products (multiplying)}$$
$$x^2 + \tfrac{2}{5}x - 3x - 3 \cdot \tfrac{2}{5} = 0 \qquad \text{Multiplying}$$
$$x^2 - \tfrac{13}{5}x - \tfrac{6}{5} = 0 \qquad \text{Combining like terms}$$
$$5x^2 - 13x - 6 = 0 \qquad \text{Multiplying both sides by 5 to clear fractions}$$

Note that multiplying both sides by the LCD, 5, clears the equation of fractions. Had we preferred, we could have multiplied $x + \frac{2}{5} = 0$ by 5, thus clearing fractions *before* using the principle of zero products.

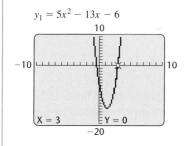

TECHNOLOGY CONNECTION

To check Example 2(a), we can let $y_1 = 5x^2 - 13x - 6$ and verify that the x-intercepts are 3 and $-\frac{2}{5}$. One way to do this is to press $\boxed{\text{TRACE}}$ and then enter 3. The cursor then appears on the curve at $x = 3$. Since this is an x-intercept, we know that $5x^2 - 13x - 6 = 0$ is an equation that has 3 as a solution.

$y_1 = 5x^2 - 13x - 6$

X = 3   Y = 0

1. Confirm that $5x^2 - 13x - 6 = 0$ also has $-\frac{2}{5}$ as a solution.
2. Check Example 2(c).
3. Check Example 2(d).

b)
$$x = 2i \quad or \quad x = -2i$$
$$x - 2i = 0 \quad or \quad x + 2i = 0 \qquad \text{Getting 0's on one side}$$
$$(x - 2i)(x + 2i) = 0 \qquad \text{Using the principle of zero products (multiplying)}$$
$$x^2 - (2i)^2 = 0 \qquad \text{Finding the product of a sum and a difference}$$
$$x^2 - 4i^2 = 0$$
$$x^2 + 4 = 0 \qquad i^2 = -1$$

c)
$$x = 5\sqrt{7} \quad or \quad x = -5\sqrt{7}$$
$$x - 5\sqrt{7} = 0 \quad or \quad x + 5\sqrt{7} = 0 \qquad \text{Getting 0's on one side}$$
$$(x - 5\sqrt{7})(x + 5\sqrt{7}) = 0 \qquad \text{Using the principle of zero products}$$
$$x^2 - (5\sqrt{7})^2 = 0 \qquad \text{Finding the product of a sum and a difference}$$
$$x^2 - 25 \cdot 7 = 0$$
$$x^2 - 175 = 0$$

d)
$$x = -4 \quad or \quad x = 0 \quad or \quad x = 1$$
$$x + 4 = 0 \quad or \quad x = 0 \quad or \quad x - 1 = 0 \qquad \text{Getting 0's on one side}$$
$$(x + 4)x(x - 1) = 0 \qquad \text{Using the principle of zero products}$$
$$x(x^2 + 3x - 4) = 0 \qquad \text{Multiplying}$$
$$x^3 + 3x^2 - 4x = 0$$

TRY EXERCISE ▶ 29

To check any of these equations, we can simply substitute one or more of the given solutions. For example, in Example 2(d) above,
$$(-4)^3 + 3(-4)^2 - 4(-4) = -64 + 3 \cdot 16 + 16$$
$$= -64 + 48 + 16 = 0.$$

The other checks are left to the student.

8.3 EXERCISE SET

For Extra Help

MyMathLab | Math XL PRACTICE | WATCH | DOWNLOAD

🖐 *Concept Reinforcement* Match the nature of the solution(s) with each discriminant. Answers may be used more than once.

1. ___ $b^2 - 4ac = 9$

2. ___ $b^2 - 4ac = 0$

3. ___ $b^2 - 4ac = -1$

4. ___ $b^2 - 4ac = 1$

5. ___ $b^2 - 4ac = 8$

6. ___ $b^2 - 4ac = 12$

a) One rational solution

b) Two different rational solutions

c) Two different irrational solutions

d) Two different imaginary-number solutions

For each equation, determine what type of number the solutions are and how many solutions exist.

7. $x^2 - 7x + 5 = 0$

8. $x^2 - 5x + 3 = 0$

9. $x^2 + 11 = 0$

10. $x^2 + 7 = 0$

11. $x^2 - 11 = 0$

12. $x^2 - 7 = 0$

13. $4x^2 + 8x - 5 = 0$

14. $4x^2 - 12x + 9 = 0$

15. $x^2 + 4x + 6 = 0$

16. $x^2 - 2x + 4 = 0$

17. $9t^2 - 48t + 64 = 0$

18. $10t^2 - t - 2 = 0$

Aha! 19. $9t^2 + 3t = 0$

20. $4m^2 + 7m = 0$

21. $x^2 + 4x = 8$ **22.** $x^2 + 5x = 9$

23. $2a^2 - 3a = -5$ **24.** $3a^2 + 5 = -7a$

25. $7x^2 = 19x$ **26.** $5x^2 = 48x$

27. $y^2 + \frac{9}{4} = 4y$ **28.** $x^2 = \frac{1}{2}x - \frac{3}{5}$

Write a quadratic equation having the given numbers as solutions.

29. $-5,\ 4$

30. $-2,\ 8$

31. 3, only solution (*Hint*: It must be a repeated solution.)

32. -5, only solution

33. $-1,\ -3$

34. $-2,\ -5$

35. $5,\ \frac{3}{4}$

36. $4,\ \frac{2}{3}$

37. $-\frac{1}{4},\ -\frac{1}{2}$

38. $\frac{1}{2},\ \frac{1}{3}$

39. $2.4,\ -0.4$

40. $-0.6,\ 1.4$

41. $-\sqrt{3},\ \sqrt{3}$

42. $-\sqrt{7},\ \sqrt{7}$

43. $2\sqrt{5},\ -2\sqrt{5}$

44. $3\sqrt{2},\ -3\sqrt{2}$

45. $4i,\ -4i$

46. $3i,\ -3i$

47. $2 - 7i,\ 2 + 7i$

48. $5 - 2i,\ 5 + 2i$

49. $3 - \sqrt{14},\ 3 + \sqrt{14}$

50. $2 - \sqrt{10},\ 2 + \sqrt{10}$

51. $1 - \dfrac{\sqrt{21}}{3},\ 1 + \dfrac{\sqrt{21}}{3}$

52. $\dfrac{5}{4} - \dfrac{\sqrt{33}}{4},\ \dfrac{5}{4} + \dfrac{\sqrt{33}}{4}$

Write a third-degree equation having the given numbers as solutions.

53. $-2,\ 1,\ 5$ **54.** $-5,\ 0,\ 2$

55. $-1,\ 0,\ 3$ **56.** $-2,\ 2,\ 3$

57. Explain why there are not two different solutions when the discriminant is 0.

58. Describe a procedure that could be used to write an equation having the first 7 natural numbers as solutions.

Skill Review

To prepare for Section 8.4, review solving formulas and solving motion problems (Sections 3.3, 6.5, and 6.8).

Solve each formula for the specified variable. [6.8]

59. $\dfrac{c}{d} = c + d$, for c

60. $\dfrac{p}{q} = \dfrac{a + b}{b}$, for b

61. $x = \dfrac{3}{1 - y}$, for y

Solve.

62. *Boating.*   Kiara's motorboat took 4 hr to make a trip downstream with a 2-mph current. The return trip against the same current took 6 hr. Find the speed of the boat in still water. [3.3]

63. *Walking.*   Jamal walks 1.5 mph faster than Kade. In the time it takes Jamal to walk 7 mi, Kade walks 4 mi. Find the speed of each person. [6.5]

64. *Aviation.*   Taryn's Cessna travels 120 mph in still air. She flies 140 mi into the wind and 140 mi with the wind in a total of 2.4 hr. Find the wind speed. [6.5]

Synthesis

65. If we assume that a quadratic equation has integers for coefficients, will the product of the solutions always be a real number? Why or why not?

66. Can a fourth-degree equation with rational coefficients have exactly three irrational solutions? Why or why not?

67. The graph of an equation of the form

$$y = ax^2 + bx + c$$

is a curve similar to the one shown below. Determine a, b, and c from the information given.

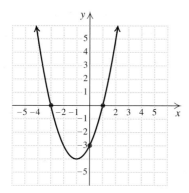

68. Show that the product of the solutions of $ax^2 + bx + c = 0$ is c/a.

For each equation under the given condition, (a) find k and (b) find the other solution.

69. $kx^2 - 2x + k = 0$; one solution is -3

70. $x^2 - kx + 2 = 0$; one solution is $1 + i$

71. $x^2 - (6 + 3i)x + k = 0$; one solution is 3

72. Show that the sum of the solutions of $ax^2 + bx + c = 0$ is $-b/a$.

73. Show that whenever there is just one solution of $ax^2 + bx + c = 0$, that solution is of the form $-b/(2a)$.

74. Find h and k, where $3x^2 - hx + 4k = 0$, the sum of the solutions is -12, and the product of the solutions is 20. (*Hint*: See Exercises 68 and 72.)

75. Suppose that $f(x) = ax^2 + bx + c$, with $f(-3) = 0$, $f\left(\frac{1}{2}\right) = 0$, and $f(0) = -12$. Find a, b, and c.

76. Find an equation for which $2 - \sqrt{3}$, $2 + \sqrt{3}$, $5 - 2i$, and $5 + 2i$ are solutions.

Aha! **77.** Write a quadratic equation with integer coefficients for which $-\sqrt{2}$ is one solution.

78. Write a quadratic equation with integer coefficients for which $10i$ is one solution.

79. Find an equation with integer coefficients for which $1 - \sqrt{5}$ and $3 + 2i$ are two of the solutions.

80. A discriminant that is a perfect square indicates that factoring can be used to solve the quadratic equation. Why?

81. While solving a quadratic equation of the form $ax^2 + bx + c = 0$ with a graphing calculator, Keisha gets the following screen. How could the sign of the discriminant help her check the graph?

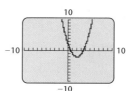

8.4   Applications Involving Quadratic Equations

Solving Problems   •   Solving Formulas

Solving Problems

As we found in Section 6.5, some problems translate to rational equations. The solution of such rational equations can involve quadratic equations.

EXAMPLE 1   *Motorcycle travel.*   Fiona rode her motorcycle 300 mi at a certain average speed. Had she traveled 10 mph faster, the trip would have taken 1 hr less. Find Fiona's average speed.

SOLUTION

1. **Familiarize.** We make a drawing, labeling it with the information provided. As in Section 6.5, we can create a table. We let r represent the rate, in miles per hour, and t the time, in hours, for Fiona's trip.

300 miles

Time t Speed r

300 miles

Time $t - 1$ Speed $r + 10$

Distance	Speed	Time
300	r	t
300	$r + 10$	$t - 1$

$$r = \frac{300}{t}$$

$$r + 10 = \frac{300}{t - 1}$$

Recall that the definition of speed, $r = d/t$, relates the three quantities.

2. **Translate.** From the table, we obtain

$$r = \frac{300}{t} \quad \text{and} \quad r + 10 = \frac{300}{t - 1}.$$

3. **Carry out.** A system of equations has been formed. We substitute for r from the first equation into the second and solve the resulting equation:

$$\frac{300}{t} + 10 = \frac{300}{t - 1} \qquad \text{Substituting } 300/t \text{ for } r$$

$$t(t - 1) \cdot \left[\frac{300}{t} + 10 \right] = t(t - 1) \cdot \frac{300}{t - 1} \qquad \text{Multiplying by the LCD to clear fractions}$$

$$t(t - 1) \cdot \frac{300}{t} + t(t - 1) \cdot 10 = t(t - 1) \cdot \frac{300}{t - 1} \qquad \text{Using the distributive law}$$

$$\frac{\cancel{t}(t - 1)}{1} \cdot \frac{300}{\cancel{t}} + t(t - 1) \cdot 10 = \frac{t\cancel{(t - 1)}}{1} \cdot \frac{300}{\cancel{t - 1}} \qquad \begin{array}{l} \text{Removing factors} \\ \text{that equal 1:} \\ t/t = 1 \text{ and} \\ (t - 1)/(t - 1) = 1 \end{array}$$

$$\left. \begin{array}{l} 300(t - 1) + 10(t^2 - t) = 300t \\ 300t - 300 + 10t^2 - 10t = 300t \\ 10t^2 - 10t - 300 = 0 \end{array} \right\} \qquad \begin{array}{l} \text{Rewriting in} \\ \text{standard form} \end{array}$$

$$t^2 - t - 30 = 0 \qquad \begin{array}{l} \text{Multiplying by } \frac{1}{10} \\ \text{or dividing by 10} \end{array}$$

$$(t - 6)(t + 5) = 0 \qquad \text{Factoring}$$

$$t = 6 \quad \text{or} \quad t = -5. \qquad \begin{array}{l} \text{Principle of zero} \\ \text{products} \end{array}$$

4. Check. Note that we have solved for *t*, not *r* as required. Since negative time has no meaning here, we disregard the −5 and use 6 hr to find *r* :

$$r = \frac{300 \text{ mi}}{6 \text{ hr}} = 50 \text{ mph}.$$

> **CAUTION!**   Always make sure that you find the quantity asked for in the problem.

To see if 50 mph checks, we increase the speed 10 mph to 60 mph and see how long the trip would have taken at that speed:

$$t = \frac{d}{r} = \frac{300 \text{ mi}}{60 \text{ mph}} = 5 \text{ hr}.$$ Note that mi/mph = mi ÷ $\frac{\text{mi}}{\text{hr}}$ =

$$\cancel{\text{mi}} \cdot \frac{\text{hr}}{\cancel{\text{mi}}} = \text{hr}.$$

This is 1 hr less than the trip actually took, so the answer checks.

5. State. Fiona traveled at an average speed of 50 mph.   **TRY EXERCISE** 1

Solving Formulas

Recall that to solve a formula for a certain letter, we use the principles for solving equations to get that letter alone on one side.

EXAMPLE 2

Period of a pendulum.   The time *T* required for a pendulum of length *l* to swing back and forth (complete one period) is given by the formula $T = 2\pi \sqrt{l/g}$, where *g* is the earth's gravitational constant. Solve for *l*.

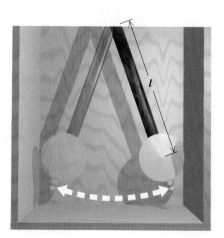

SOLUTION   We have

$$T = 2\pi\sqrt{\frac{l}{g}}$$ This is a radical equation (see Section 7.6).

$$T^2 = \left(2\pi\sqrt{\frac{l}{g}}\right)^2$$ Squaring both sides

$$T^2 = 2^2\pi^2\frac{l}{g}$$

$$gT^2 = 4\pi^2 l$$ Multiplying both sides by *g* to clear fractions

$$\frac{gT^2}{4\pi^2} = l.$$ Dividing both sides by $4\pi^2$

We now have *l* alone on one side and *l* does not appear on the other side, so the formula is solved for *l*.   **TRY EXERCISE** 21

In formulas for which variables represent only nonnegative numbers, there is no need for absolute-value signs when taking square roots.

EXAMPLE **3**

*Hang time.** An athlete's *hang time* is the amount of time that the athlete can remain airborne when jumping. A formula relating an athlete's vertical leap V, in inches, to hang time T, in seconds, is $V = 48T^2$. Solve for T.

SOLUTION   We have

$$48T^2 = V$$

$$T^2 = \frac{V}{48}$$ Dividing by 48 to isolate T^2

$$T = \frac{\sqrt{V}}{\sqrt{48}}$$ Using the principle of square roots and the quotient rule for radicals. We assume $V, T \geq 0$.

$$= \frac{\sqrt{V}}{\sqrt{16}\sqrt{3}}$$

$$= \frac{\sqrt{V}}{4\sqrt{3}}$$

$$\left. \begin{array}{c} = \frac{\sqrt{V}}{4\sqrt{3}} \cdot \frac{\sqrt{3}}{\sqrt{3}} \\[2mm] = \frac{\sqrt{3V}}{12}. \end{array} \right\}$$ Rationalizing the denominator

TRY EXERCISE 15

EXAMPLE **4**

Falling distance.   An object tossed downward with an initial speed (velocity) of v_0 will travel a distance of s meters, where $s = 4.9t^2 + v_0t$ and t is measured in seconds. Solve for t.

SOLUTION   Since t is squared in one term and raised to the first power in the other term, the equation is quadratic in t.

$$4.9t^2 + v_0t = s$$

$$4.9t^2 + v_0t - s = 0$$ Writing standard form

$$a = 4.9, \quad b = v_0, \quad c = -s$$

$$t = \frac{-v_0 \pm \sqrt{(v_0)^2 - 4(4.9)(-s)}}{2(4.9)}$$ Using the quadratic formula

Since the negative square root would yield a negative value for t, we use only the positive root:

$$t = \frac{-v_0 + \sqrt{(v_0)^2 + 19.6s}}{9.8}.$$

TRY EXERCISE 25

STUDENT NOTES ———————

After identifying which numbers to use as a, b, and c, be careful to replace only the *letters* in the quadratic formula.

---

*This formula is taken from an article by Peter Brancazio, "The Mechanics of a Slam Dunk," *Popular Mechanics,* November 1991. Courtesy of Professor Peter Brancazio, Brooklyn College.

The following list of steps should help you when solving formulas for a given letter. Try to remember that when solving a formula, you use the same approach that you would to solve an equation.

To Solve a Formula for a Letter—Say, h

1. Clear fractions and use the principle of powers, as needed. Perform these steps until radicals containing h are gone and h is not in any denominator.
2. Combine all like terms.
3. If the only power of h is h^1, the equation can be solved as in Sections 1.5 and 6.8. (See Example 2.)
4. If h^2 appears but h does not, solve for h^2 and use the principle of square roots to solve for h. (See Example 3.)
5. If there are terms containing both h and h^2, put the equation in standard form and use the quadratic formula. (See Example 4.)

8.4 EXERCISE SET

Solve.

1. *Car trips.* During the first part of a trip, Tara's Honda traveled 120 mi at a certain speed. Tara then drove another 100 mi at a speed that was 10 mph slower. If the total time of Tara's trip was 4 hr, what was her speed on each part of the trip?

2. *Canoeing.* During the first part of a canoe trip, Ken covered 60 km at a certain speed. He then traveled 24 km at a speed that was 4 km/h slower. If the total time for the trip was 8 hr, what was the speed on each part of the trip?

3. *Car trips.* Diane's Dodge travels 200 mi averaging a certain speed. If the car had gone 10 mph faster, the trip would have taken 1 hr less. Find Diane's average speed.

4. *Car trips.* Stuart's Subaru travels 280 mi averaging a certain speed. If the car had gone 5 mph faster, the trip would have taken 1 hr less. Find Stuart's average speed.

5. *Air travel.* A Cessna flies 600 mi at a certain speed. A Beechcraft flies 1000 mi at a speed that is 50 mph faster, but takes 1 hr longer. Find the speed of each plane.

6. *Air travel.* A turbo-jet flies 50 mph faster than a super-prop plane. If a turbo-jet goes 2000 mi in 3 hr less time than it takes the super-prop to go 2800 mi, find the speed of each plane.

7. *Bicycling.* Naoki bikes the 36 mi to Hillsboro averaging a certain speed. The return trip is made at a speed 3 mph slower. Total time for the round trip is 7 hr. Find Naoki's average speed on each part of the trip.

8. *Car speed.* On a sales trip, Mark drives the 600 mi to Richmond averaging a certain speed. The return trip is made at an average speed that is 10 mph slower. Total time for the round trip is 22 hr. Find Mark's average speed on each part of the trip.

9. *Navigation.* The Hudson River flows at a rate of 3 mph. A patrol boat travels 60 mi upriver and returns in a total time of 9 hr. What is the speed of the boat in still water?

■ 10. *Navigation.* The current in a typical Mississippi River shipping route flows at a rate of 4 mph. In order for a barge to travel 24 mi upriver and then return in a total of 5 hr, approximately how fast must the barge be able to travel in still water?

11. *Filling a pool.* A well and a spring are filling a swimming pool. Together, they can fill the pool in 3 hr. The well, working alone, can fill the pool in 8 hr less time than the spring. How long would the spring take, working alone, to fill the pool?

12. *Filling a tank.* Two pipes are connected to the same tank. Working together, they can fill the tank in 4 hr. The larger pipe, working alone, can fill the tank in 6 hr less time than the smaller one. How long would the smaller one take, working alone, to fill the tank?

■ 13. *Paddleboats.* Kofi paddles 1 mi upstream and 1 mi back in a total time of 1 hr. The speed of the river is 2 mph. Find the speed of Kofi's paddleboat in still water.

■ 14. *Rowing.* Abby rows 10 km upstream and 10 km back in a total time of 3 hr. The speed of the river is 5 km/h. Find Abby's speed in still water.

Solve each formula for the indicated letter. Assume that all variables represent nonnegative numbers.

15. $A = 4\pi r^2$, for r
(Surface area of a sphere of radius r)

16. $A = 6s^2$, for s
(Surface area of a cube with sides of length s)

17. $A = 2\pi r^2 + 2\pi rh$, for r
(Surface area of a right cylindrical solid with radius r and height h)

18. $N = \dfrac{k^2 - 3k}{2}$, for k
(Number of diagonals of a polygon with k sides)

19. $F = \dfrac{Gm_1m_2}{r^2}$, for r
(Law of gravity)

20. $N = \dfrac{kQ_1Q_2}{s^2}$, for s
(Number of phone calls between two cities)

21. $c = \sqrt{gH}$, for H
(Velocity of ocean wave)

22. $V = 3.5\sqrt{h}$, for h
(Distance to horizon from a height)

23. $a^2 + b^2 = c^2$, for b
(Pythagorean formula in two dimensions)

24. $a^2 + b^2 + c^2 = d^2$, for c
(Pythagorean formula in three dimensions)

25. $s = v_0t + \dfrac{gt^2}{2}$, for t
(A motion formula)

26. $A = \pi r^2 + \pi rs$, for r
(Surface area of a cone)

27. $N = \frac{1}{2}(n^2 - n)$, for n
(Number of games if n teams play each other once)

28. $A = A_0(1 - r)^2$, for r
(A business formula)

29. $T = 2\pi\sqrt{\dfrac{l}{g}}$, for g
(A pendulum formula)

30. $W = \sqrt{\dfrac{1}{LC}}$, for L
(An electricity formula)

Aha! 31. $at^2 + bt + c = 0$, for t
(An algebraic formula)

32. $A = P_1(1 + r)^2 + P_2(1 + r)$, for r
(Amount in an account when P_1 is invested for 2 yr and P_2 for 1 yr at interest rate r)

Solve.

33. *Falling distance.* (Use $4.9t^2 + v_0t = s$.)

a) A bolt falls off an airplane at an altitude of 500 m. Approximately how long does it take the bolt to reach the ground?

b) A ball is thrown downward at a speed of 30 m/sec from an altitude of 500 m. Approximately how long does it take the ball to reach the ground?

c) Approximately how far will an object fall in 5 sec, when thrown downward at an initial velocity of 30 m/sec from a plane?

34. *Falling distance.* (Use $4.9t^2 + v_0t = s$.)

a) A ring is dropped from a helicopter at an altitude of 75 m. Approximately how long does it take the ring to reach the ground?

b) A coin is tossed downward with an initial velocity of 30 m/sec from an altitude of 75 m. Approximately how long does it take the coin to reach the ground?

c) Approximately how far will an object fall in 2 sec, if thrown downward at an initial velocity of 20 m/sec from a helicopter?

35. *Bungee jumping.* Chad is tied to one end of a 40-m elasticized (bungee) cord. The other end of the cord is tied to the middle of a bridge. If Chad jumps

off the bridge, for how long will he fall before the cord begins to stretch? (Use $4.9t^2 = s$.)

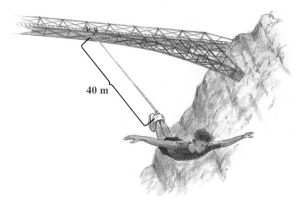

40 m

36. *Bungee jumping.* Chika is tied to a bungee cord (see Exercise 35) and falls for 2.5 sec before her cord begins to stretch. How long is the bungee cord?

37. *Hang time.* The NBA's Dwight Howard has a vertical leap of 38 in. What is his hang time? (Use $V = 48T^2$.)

Source: www.dwighthoward.com

38. *League schedules.* In a bowling league, each team plays each of the other teams once. If a total of 66 games is played, how many teams are in the league? (See Exercise 27.)

For Exercises 39 and 40, use $4.9t^2 + v_0t = s$.

39. *Downward speed.* A stone thrown downward from a 100-m cliff travels 51.6 m in 3 sec. What was the initial velocity of the object?

40. *Downward speed.* A pebble thrown downward from a 200-m cliff travels 91.2 m in 4 sec. What was the initial velocity of the object?

For Exercises 41 and 42, use $A = P_1(1 + r)^2 + P_2(1 + r)$. (See Exercise 32.)

41. *Compound interest.* A firm invests $3200 in a savings account for 2 yr. At the beginning of the second

year, an additional $1800 is invested. If a total of $5375.48 is in the account at the end of the second year, what is the annual interest rate?

42. *Compound interest.* A business invests $10,000 in a savings account for 2 yr. At the beginning of the second year, an additional $3500 is invested. If a total of $14,822.75 is in the account at the end of the second year, what is the annual interest rate?

43. Marti is tied to a bungee cord that is twice as long as the cord tied to Rafe's. Will Marti's fall take twice as long as Rafe's before their cords begin to stretch? Why or why not? (See Exercises 35 and 36.)

44. Under what circumstances would a negative value for t, time, have meaning?

Skill Review

To prepare for Section 8.5, review raising a power to a power and solving rational equations and radical equations (Sections 1.6, 6.4, and 7.6).

Simplify.

45. $(m^{-1})^2$ [1.6]

46. $(t^{1/3})^2$ [7.2]

47. $(y^{1/6})^2$ [7.2]

48. $(z^{1/4})^2$ [7.2]

Solve.

49. $t^{-1} = \dfrac{1}{2}$ [6.4], [7.6]

50. $x^{1/4} = 3$ [7.6]

Synthesis

51. Write a problem for a classmate to solve. Devise the problem so that (a) the solution is found after solving a rational equation and (b) the solution is "The express train travels 90 mph."

52. In what ways do the motion problems of this section (like Example 1) differ from the motion problems in Chapter 6 (see p. 392)?

53. *Biochemistry.* The equation

$$A = 6.5 - \frac{20.4t}{t^2 + 36}$$

is used to calculate the acid level A in a person's blood t minutes after sugar is consumed. Solve for t.

54. *Special relativity.* Einstein found that an object with initial mass m_0 and traveling velocity v has mass

$$m = \frac{m_0}{\sqrt{1 - \dfrac{v^2}{c^2}}},$$

where c is the speed of light. Solve the formula for c.

55. Find a number for which the reciprocal of 1 less than the number is the same as 1 more than the number.

56. *Purchasing.* A discount store bought a quantity of potted plants for $250 and sold all but 15 at a profit of $3.50 per plant. With the total amount received, the manager could buy 4 more than twice as many as were bought before. Find the cost per plant.

57. *Art and aesthetics.* For over 2000 yr, artists, sculptors, and architects have regarded the proportions of a "golden" rectangle as visually appealing. A rectangle of width w and length l is considered "golden" if

$$\frac{w}{l} = \frac{l}{w + l}.$$

Solve for l.

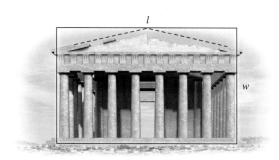

58. *Diagonal of a cube.* Find a formula that expresses the length of the three-dimensional diagonal of a cube as a function of the cube's surface area.

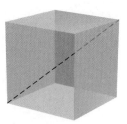

59. Solve for n:

$$mn^4 - r^2pm^3 - r^2n^2 + p = 0.$$

60. *Surface area.* Find a formula that expresses the diameter of a right cylindrical solid as a function of its surface area and its height. (See Exercise 17.)

61. A sphere is inscribed in a cube as shown in the figure below. Express the surface area of the sphere as a function of the surface area S of the cube. (See Exercise 15.)

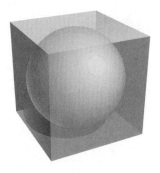

8.5   Equations Reducible to Quadratic

Recognizing Equations in Quadratic Form   ■   Radical Equations and Rational Equations

Recognizing Equations in Quadratic Form

Certain equations that are not really quadratic can be thought of in such a way that they can be solved as quadratic. For example, because the square of x^2 is x^4, the equation $x^4 - 9x^2 + 8 = 0$ is said to be "quadratic in x^2":

$$x^4 - 9x^2 + 8 = 0$$
$$\downarrow \qquad \downarrow \qquad \downarrow \quad \downarrow$$
$$(x^2)^2 - 9(x^2) + 8 = 0 \qquad \text{Thinking of } x^4 \text{ as } (x^2)^2$$
$$\downarrow \qquad \downarrow \qquad \downarrow \quad \downarrow$$
$$u^2 - 9u + 8 = 0. \qquad \text{To make this clearer, write } u \text{ instead of } x^2.$$

The equation $u^2 - 9u + 8 = 0$ can be solved for u by factoring or by the quadratic formula. Then, remembering that $u = x^2$, we can solve for x. Equations that can be solved like this are *reducible to quadratic* and are said to be *in quadratic form*.

EXAMPLE **1**   Solve: $x^4 - 9x^2 + 8 = 0$.

SOLUTION   We begin by letting $u = x^2$ and finding u^2:

$$u = x^2$$
$$u^2 = (x^2)^2 = x^4.$$

Then we solve by substituting u^2 for x^4 and u for x^2:

$$u^2 - 9u + 8 = 0$$

$(u - 8)(u - 1) = 0$	Factoring
$u - 8 = 0$ *or* $u - 1 = 0$	Principle of zero products
$u = 8$ *or* $u = 1$.	

STUDENT NOTES

To identify an equation in quadratic form, look for two variable expressions in the equation. The exponent in one expression is twice the exponent in the other expression.

We replace u with x^2 and solve these equations:

$$x^2 = 8 \quad or \quad x^2 = 1$$
$$x = \pm\sqrt{8} \quad or \quad x = \pm 1$$
$$x = \pm 2\sqrt{2} \quad or \quad x = \pm 1.$$

$\boxed{\text{We are solving for } x.}$

To check, note that for both $x = 2\sqrt{2}$ and $-2\sqrt{2}$, we have $x^2 = 8$ and $x^4 = 64$. Similarly, for both $x = 1$ and -1, we have $x^2 = 1$ and $x^4 = 1$. Thus instead of making four checks, we need make only two.

Check: For $\pm 2\sqrt{2}$:

$$\frac{x^4 - 9x^2 + 8 = 0}{(\pm 2\sqrt{2})^4 - 9(\pm 2\sqrt{2})^2 + 8 \mid 0}$$
$$64 - 9 \cdot 8 + 8$$
$$0 \overset{?}{=} 0 \quad \text{TRUE}$$

For ± 1:

$$\frac{x^4 - 9x^2 + 8 = 0}{(\pm 1)^4 - 9(\pm 1)^2 + 8 \mid 0}$$
$$1 - 9 + 8$$
$$0 \overset{?}{=} 0 \quad \text{TRUE}$$

The solutions are $1, -1, 2\sqrt{2},$ and $-2\sqrt{2}$.

TRY EXERCISE 17

CAUTION!   A common error when working on problems like Example 1 is to solve for u but forget to solve for x. Remember to solve for the *original* variable!

Equations like those in Example 1 can be solved directly by factoring:

$$x^4 - 9x^2 + 8 = 0$$
$$(x^2 - 1)(x^2 - 8) = 0$$
$$x^2 - 1 = 0 \quad or \quad x^2 - 8 = 0$$
$$x^2 = 1 \quad or \quad x^2 = 8$$
$$x = \pm 1 \quad or \quad x = \pm 2\sqrt{2}.$$

However, it often becomes difficult to solve the equation without first making a substitution.

To recognize an equation in quadratic form, inspect all the variable expressions in the equation. For an equation to be written in the form $au^2 + bu + c = 0$, it is necessary to identify one variable expression as u and a second variable expression as u^2.

EXAMPLE 2

Find the x-intercepts of the graph of $f(x) = (x^2 - 1)^2 - (x^2 - 1) - 2$.

SOLUTION   The x-intercepts occur where $f(x) = 0$ so we must have

$$(x^2 - 1)^2 - (x^2 - 1) - 2 = 0. \qquad \text{Setting } f(x) \text{ equal to } 0$$

If we identify $x^2 - 1$ as u, the equation can be written in quadratic form:

$$u = x^2 - 1$$
$$u^2 = (x^2 - 1)^2.$$

Substituting, we have

$$u^2 - u - 2 = 0 \qquad \begin{array}{l} \text{Substituting in} \\ (x^2 - 1)^2 - (x^2 - 1) - 2 = 0 \end{array}$$
$$(u - 2)(u + 1) = 0$$
$$u = 2 \quad or \quad u = -1. \qquad \begin{array}{l} \text{Using the principle of} \\ \text{zero products} \end{array}$$

Next, we replace u with $x^2 - 1$ and solve these equations:

$$x^2 - 1 = 2 \qquad or \quad x^2 - 1 = -1$$
$$x^2 = 3 \qquad or \qquad x^2 = 0 \qquad \text{Adding 1 to both sides}$$
$$x = \pm\sqrt{3} \quad or \qquad x = 0. \qquad \begin{array}{l} \text{Using the principle of} \\ \text{square roots} \end{array}$$

The x-intercepts occur at $(-\sqrt{3}, 0)$, $(0, 0)$, and $(\sqrt{3}, 0)$.   **TRY EXERCISE** 49

Radical Equations and Rational Equations

Sometimes rational equations, radical equations, or equations containing exponents that are fractions are reducible to quadratic. It is especially important that answers to these equations be checked in the original equation.

EXAMPLE 3

Solve: $x - 3\sqrt{x} - 4 = 0$.

SOLUTION   This radical equation could be solved using the method discussed in Section 7.6. However, if we note that the square of $\sqrt{x}$ is x, we can regard the equation as "quadratic in $\sqrt{x}$."
    We determine u and u^2:

$$u = \sqrt{x}$$
$$u^2 = x.$$

Substituting, we have

$$x - 3\sqrt{x} - 4 = 0$$
$$u^2 - 3u - 4 = 0$$
$$(u - 4)(u + 1) = 0$$
$$u = 4 \quad or \quad u = -1. \qquad \begin{array}{l} \text{Using the principle of} \\ \text{zero products} \end{array}$$

Next, we replace u with $\sqrt{x}$ and solve these equations:

$$\sqrt{x} = 4 \quad or \quad \sqrt{x} = -1.$$

Squaring gives us $x = 16$ or $x = 1$ and also makes checking essential.

Check: For 16:

$$\frac{x - 3\sqrt{x} - 4 = 0}{16 - 3\sqrt{16} - 4 \;\big|\; 0}$$

$$16 - 3 \cdot 4 - 4 \;\Big|$$

$$0 \overset{?}{=} 0 \quad \text{TRUE}$$

For 1:

$$\frac{x - 3\sqrt{x} - 4 = 0}{1 - 3\sqrt{1} - 4 \;\big|\; 0}$$

$$1 - 3 \cdot 1 - 4 \;\Big|$$

$$-6 \overset{?}{=} 0 \quad \text{FALSE}$$

The number 16 checks, but 1 does not. Had we noticed that $\sqrt{x} = -1$ has no solution (since principal square roots are never negative), we could have solved only the equation $\sqrt{x} = 4$. The solution is 16.   **TRY EXERCISE** 21

The following tips may prove useful.

> **To Solve an Equation That Is Reducible to Quadratic**
>
> 1. Look for two variable expressions in the equation. One expression should be the square of the other.
> 2. Write down any substitutions that you are making.
> 3. Remember to solve for the variable that is used in the original equation.
> 4. Check possible answers in the original equation.

EXAMPLE 4   Solve: $2m^{-2} + m^{-1} - 15 = 0$.

SOLUTION   Note that the square of m^{-1} is $(m^{-1})^2$, or m^{-2}. We let $u = m^{-1}$:

Determine u and u^2.

$$u = m^{-1}$$
$$u^2 = m^{-2}.$$

Substituting, we have

Substitute.

$$2u^2 + u - 15 = 0 \qquad \text{Substituting in } 2m^{-2} + m^{-1} - 15 = 0$$

$$(2u - 5)(u + 3) = 0$$

$$2u - 5 = 0 \quad or \quad u + 3 = 0 \qquad \text{Using the principle of zero products}$$

$$2u = 5 \quad or \qquad u = -3$$

Solve for u.

$$u = \frac{5}{2} \quad or \qquad u = -3.$$

Now we replace u with m^{-1} and solve:

$$m^{-1} = \frac{5}{2} \quad or \quad m^{-1} = -3$$

$$\frac{1}{m} = \frac{5}{2} \quad or \quad \frac{1}{m} = -3 \qquad \text{Recall that } m^{-1} = \frac{1}{m}.$$

$$1 = \frac{5}{2}m \quad or \quad 1 = -3m \qquad \text{Multiplying both sides by } m$$

Solve for the original variable.

$$\frac{2}{5} = m \quad or \quad -\frac{1}{3} = m. \qquad \text{Solving for } m$$

Check.

Check:

For $\frac{2}{5}$:

$$\begin{array}{c|c} 2m^{-2} + m^{-1} - 15 = 0 & \\ \hline 2\left(\frac{2}{5}\right)^{-2} + \left(\frac{2}{5}\right)^{-1} - 15 & 0 \\ 2\left(\frac{5}{2}\right)^2 + \left(\frac{5}{2}\right) - 15 & \\ 2\left(\frac{25}{4}\right) + \frac{5}{2} - 15 & \\ \frac{25}{2} + \frac{5}{2} - 15 & \\ \frac{30}{2} - 15 & \\ & 0 \stackrel{?}{=} 0 \quad \text{TRUE} \end{array}$$

For $-\frac{1}{3}$:

$$\begin{array}{c|c} 2m^{-2} + m^{-1} - 15 = 0 & \\ \hline 2\left(-\frac{1}{3}\right)^{-2} + \left(-\frac{1}{3}\right)^{-1} - 15 & 0 \\ 2\left(-\frac{3}{1}\right)^2 + \left(-\frac{3}{1}\right) - 15 & \\ 2(9) + (-3) - 15 & \\ 18 - 3 - 15 & \\ & 0 \stackrel{?}{=} 0 \quad \text{TRUE} \end{array}$$

Both numbers check. The solutions are $-\frac{1}{3}$ and $\frac{2}{5}$.

> TRY EXERCISE 29

Note that Example 4 can also be written

$$\frac{2}{m^2} + \frac{1}{m} - 15 = 0.$$

It can then be solved by letting $u = 1/m$ and $u^2 = 1/m^2$ or by clearing fractions as in Section 6.4.

EXAMPLE 5

Solve: $t^{2/5} - t^{1/5} - 2 = 0$.

SOLUTION Note that the square of $t^{1/5}$ is $(t^{1/5})^2$, or $t^{2/5}$. The equation is therefore quadratic in $t^{1/5}$, so we let $u = t^{1/5}$:

$$u = t^{1/5}$$
$$u^2 = t^{2/5}.$$

Substituting, we have

$$u^2 - u - 2 = 0 \qquad \text{Substituting in } t^{2/5} - t^{1/5} - 2 = 0$$
$$(u - 2)(u + 1) = 0$$
$$u = 2 \quad or \quad u = -1. \qquad \text{Using the principle of zero products}$$

Now we replace u with $t^{1/5}$ and solve:

$$t^{1/5} = 2 \quad or \quad t^{1/5} = -1$$
$$t = 32 \quad or \qquad t = -1. \qquad \text{Principle of powers; raising both sides to the 5th power}$$

Check:

For 32:

$$\begin{array}{c|c} t^{2/5} - t^{1/5} - 2 = 0 & \\ \hline 32^{2/5} - 32^{1/5} - 2 & 0 \\ (32^{1/5})^2 - 32^{1/5} - 2 & \\ 2^2 - 2 - 2 & \\ & 0 \stackrel{?}{=} 0 \quad \text{TRUE} \end{array}$$

For -1:

$$\begin{array}{c|c} t^{2/5} - t^{1/5} - 2 = 0 & \\ \hline (-1)^{2/5} - (-1)^{1/5} - 2 & 0 \\ [(-1)^{1/5}]^2 - (-1)^{1/5} - 2 & \\ (-1)^2 - (-1) - 2 & \\ & 0 \stackrel{?}{=} 0 \quad \text{TRUE} \end{array}$$

Both numbers check. The solutions are 32 and -1.

> TRY EXERCISE 33

EXAMPLE **6**   Solve: $(5 + \sqrt{r})^2 + 6(5 + \sqrt{r}) + 2 = 0$.

SOLUTION   We determine u and u^2:

$$u = 5 + \sqrt{r}$$
$$u^2 = (5 + \sqrt{r})^2.$$

Substituting, we have

$$u^2 + 6u + 2 = 0$$

$$u = \frac{-6 \pm \sqrt{6^2 - 4 \cdot 1 \cdot 2}}{2 \cdot 1} \qquad \text{Using the quadratic formula}$$

$$= \frac{-6 \pm \sqrt{28}}{2}$$

$$= \left. \frac{-6}{2} \pm \frac{2\sqrt{7}}{2} \right\} \qquad \text{Simplifying; } \sqrt{28} = \sqrt{4}\sqrt{7}$$

$$= -3 \pm \sqrt{7}.$$

Now we replace u with $5 + \sqrt{r}$ and solve for r:

$$5 + \sqrt{r} = -3 + \sqrt{7} \quad or \quad 5 + \sqrt{r} = -3 - \sqrt{7} \qquad u = -3 + \sqrt{7}$$
$$\sqrt{r} = -8 + \sqrt{7} \quad or \qquad \sqrt{r} = -8 - \sqrt{7}. \qquad or \ u = -3 - \sqrt{7}$$

We could now solve for r and check possible solutions, but first let's examine $-8 + \sqrt{7}$ and $-8 - \sqrt{7}$. Since $\sqrt{7} \approx 2.6$, both $-8 + \sqrt{7}$ and $-8 - \sqrt{7}$ are negative. Since the principal square root of r is never negative, both values of $\sqrt{r}$ must be rejected. Note too that in the original equation, $(5 + \sqrt{r})^2$, $6(5 + \sqrt{r})$, and 2 are all positive. Thus it is impossible for their sum to be 0.

The original equation has no solution.

TRY EXERCISE 25

8.5   EXERCISE SET

⤷ *Concept Reinforcement   In each of Exercises 1–8, match the equation with a substitution from the column on the right that could be used to reduce the equation to quadratic form.*

1. ____ $4x^6 - 2x^3 + 1 = 0$ a) $u = x^{-1/3}$

2. ____ $3x^4 + 4x^2 - 7 = 0$ b) $u = x^{1/3}$

3. ____ $5x^8 + 2x^4 - 3 = 0$ c) $u = x^{-2}$

4. ____ $2x^{2/3} - 5x^{1/3} + 4 = 0$ d) $u = x^2$

5. ____ $3x^{4/3} + 4x^{2/3} - 7 = 0$ e) $u = x^{-2/3}$

6. ____ $2x^{-2/3} + x^{-1/3} + 6 = 0$ f) $u = x^3$

7. ____ $4x^{-4/3} - 2x^{-2/3} + 3 = 0$ g) $u = x^{2/3}$

8. ____ $3x^{-4} + 4x^{-2} - 2 = 0$ h) $u = x^4$

Write the substitution that could be used to make each equation quadratic in u.

9. For $3p - 4\sqrt{p} + 6 = 0$, use $u = $ _____.

10. For $x^{1/2} - x^{1/4} - 2 = 0$, use $u = $ _____.

11. For $(x^2 + 3)^2 + (x^2 + 3) - 7 = 0$, use $u = $ _____.

12. For $t^{-6} + 5t^{-3} - 6 = 0$, use $u = $ _____.

13. For $(1 + t)^4 + (1 + t)^2 + 4 = 0$, use $u = $ _____.

14. For $w^{1/3} - 3w^{1/6} + 8 = 0$, use $w = $ _____.

Solve.

15. $x^4 - 13x^2 + 36 = 0$ 16. $x^4 - 17x^2 + 16 = 0$

17. $t^4 - 7t^2 + 12 = 0$ 18. $t^4 - 11t^2 + 18 = 0$

19. $4x^4 - 9x^2 + 5 = 0$

20. $9x^4 - 38x^2 + 8 = 0$

21. $w + 4\sqrt{w} - 12 = 0$

22. $s + 3\sqrt{s} - 40 = 0$

23. $(x^2 - 7)^2 - 3(x^2 - 7) + 2 = 0$

24. $(x^2 - 2)^2 - 12(x^2 - 2) + 20 = 0$

25. $r - 2\sqrt{r} - 6 = 0$

26. $s - 4\sqrt{s} - 1 = 0$

27. $(1 + \sqrt{x})^2 + 5(1 + \sqrt{x}) + 6 = 0$

28. $(3 + \sqrt{x})^2 + 3(3 + \sqrt{x}) - 10 = 0$

29. $x^{-2} - x^{-1} - 6 = 0$

30. $2x^{-2} - x^{-1} - 1 = 0$

31. $4t^{-2} - 3t^{-1} - 1 = 0$

32. $2m^{-2} + 7m^{-1} - 15 = 0$

33. $t^{2/3} + t^{1/3} - 6 = 0$

34. $w^{2/3} - 2w^{1/3} - 8 = 0$

35. $y^{1/3} - y^{1/6} - 6 = 0$

36. $t^{1/2} + 3t^{1/4} + 2 = 0$

37. $t^{1/3} + 2t^{1/6} = 3$

38. $m^{1/2} + 6 = 5m^{1/4}$

39. $(3 - \sqrt{x})^2 - 10(3 - \sqrt{x}) + 23 = 0$

40. $(5 + \sqrt{x})^2 - 12(5 + \sqrt{x}) + 33 = 0$

41. $16\left(\dfrac{x - 1}{x - 8}\right)^2 + 8\left(\dfrac{x - 1}{x - 8}\right) + 1 = 0$

42. $9\left(\dfrac{x + 2}{x + 3}\right)^2 - 6\left(\dfrac{x + 2}{x + 3}\right) + 1 = 0$

43. $x^4 + 5x^2 - 36 = 0$

44. $x^4 + 5x^2 + 4 = 0$

45. $(n^2 + 6)^2 - 7(n^2 + 6) + 10 = 0$

46. $(m^2 + 7)^2 - 6(m^2 + 7) - 16 = 0$

Find all x-intercepts of the given function f. If none exists, state this.

47. $f(x) = 5x + 13\sqrt{x} - 6$

48. $f(x) = 3x + 10\sqrt{x} - 8$

49. $f(x) = (x^2 - 3x)^2 - 10(x^2 - 3x) + 24$

50. $f(x) = (x^2 - 6x)^2 - 2(x^2 - 6x) - 35$

51. $f(x) = x^{2/5} + x^{1/5} - 6$

52. $f(x) = x^{1/2} - x^{1/4} - 6$

Aha! **53.** $f(x) = \left(\dfrac{x^2 + 2}{x}\right)^4 + 7\left(\dfrac{x^2 + 2}{x}\right)^2 + 5$

54. $f(x) = \left(\dfrac{x^2 + 1}{x}\right)^4 + 4\left(\dfrac{x^2 + 1}{x}\right)^2 + 12$

55. To solve $25x^6 - 10x^3 + 1 = 0$, Jose lets $u = 5x^3$ and Robin lets $u = x^3$. Can they both be correct? Why or why not?

56. Jenn writes that the solutions of $x^4 - 5x^2 + 6 = 0$ are 2 and 3. What mistake is she making?

Skill Review

To prepare for Section 8.6, review graphing functions (Sections 2.1 and 2.3).

Graph.

57. $f(x) = x$ [2.3]

58. $g(x) = x + 2$ [2.3]

59. $h(x) = x - 2$ [2.3]

60. $f(x) = x^2$ [2.1]

61. $g(x) = x^2 + 2$ [2.1]

62. $h(x) = x^2 - 2$ [2.1]

Synthesis

63. Describe a procedure that could be used to solve any equation of the form $ax^4 + bx^2 + c = 0$.

64. Describe a procedure that could be used to write an equation that is quadratic in $3x^2 - 1$. Then explain how the procedure could be adjusted to write equations that are quadratic in $3x^2 - 1$ and have no real-number solution.

Solve.

65. $3x^4 + 5x^2 - 1 = 0$

66. $5x^4 - 7x^2 + 1 = 0$

67. $(x^2 - 5x - 1)^2 - 18(x^2 - 5x - 1) + 65 = 0$

68. $(x^2 - 4x - 2)^2 - 13(x^2 - 4x - 2) + 30 = 0$

69. $\dfrac{x}{x - 1} - 6\sqrt{\dfrac{x}{x - 1}} - 40 = 0$

70. $\left(\sqrt{\dfrac{x}{x - 3}}\right)^2 - 24 = 10\sqrt{\dfrac{x}{x - 3}}$

71. $a^5(a^2 - 25) + 13a^3(25 - a^2) + 36a(a^2 - 25) = 0$

72. $a^3 - 26a^{3/2} - 27 = 0$

73. $x^6 - 28x^3 + 27 = 0$

74. $x^6 + 7x^3 - 8 = 0$

75. Use a graphing calculator to check your answers to Exercises 15, 17, 41, and 53.

76. Use a graphing calculator to solve
$$x^4 - x^3 - 13x^2 + x + 12 = 0.$$

77. While trying to solve $0.05x^4 - 0.8 = 0$ with a graphing calculator, Salam gets the screen at right. Can Salam solve this equation with a graphing calculator? Why or why not?

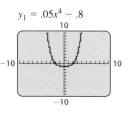
$y_1 = .05x^4 - .8$

CONNECTING the CONCEPTS

We have studied four different ways of solving quadratic equations. Each method has advantages and disadvantages, as outlined below. Note that although the quadratic formula can be used to solve *any* quadratic equation, the other methods are sometimes faster and easier to use. Also note that any of these methods can be used when solving equations that are reducible to quadratic.

Method	Advantages	Disadvantages	Example
Factoring	Can be very fast.	Can be used only on certain equations. Many equations are difficult or impossible to solve by factoring.	$x^2 - x - 6 = 0$ $(x - 3)(x + 2) = 0$ $x = 3 \quad or \quad x = -2$
The principle of square roots	Fastest way to solve equations of the form $X^2 = k$. Can be used to solve *any* quadratic equation.	Can be slow when original equation is not written in the form $X^2 = k$.	$(x - 5)^2 = 2$ $x - 5 = \pm\sqrt{2}$ $x = 5 \pm \sqrt{2}$
Completing the square	Works well on equations of the form $x^2 + bx = -c$, when b is even. Can be used to solve *any* quadratic equation.	Can be complicated when $a \neq 1$ or when b is not even in $x^2 + bx = -c$.	$x^2 + 14x = -2$ $x^2 + 14x + 49 = -2 + 49$ $(x + 7)^2 = 47$ $x + 7 = \pm\sqrt{47}$ $x = -7 \pm \sqrt{47}$
The quadratic formula	Can be used to solve *any* quadratic equation.	Can be slower than factoring or the principle of square roots for certain equations.	$x^2 - 2x - 5 = 0$ $x = \dfrac{-(-2) \pm \sqrt{(-2)^2 - 4(1)(-5)}}{2 \cdot 1}$ $= \dfrac{2 \pm \sqrt{24}}{2}$ $= \dfrac{2}{2} \pm \dfrac{2\sqrt{6}}{2} = 1 \pm \sqrt{6}$

MIXED REVIEW

Solve. Examine each exercise carefully, and try to solve using the easiest method.

1. $x^2 - 3x - 10 = 0$

2. $x^2 = 121$

3. $x^2 + 6x = 10$

4. $x^2 + x - 3 = 0$

5. $(x + 1)^2 = 2$

6. $x^2 - 10x + 25 = 0$

7. $x^2 - x - 1 = 0$

8. $x^2 - 2x = 6$

9. $4t^2 = 11$

(*continued*)

10. $2t^2 + 1 = 3t$

11. $c^2 + c + 1 = 0$

12. $16c^2 = 7c$

13. $6y^2 - 7y - 10 = 0$

14. $y^2 - 2y + 8 = 0$

15. $x^4 - 10x^2 + 9 = 0$

16. $x^4 - 8x^2 - 9 = 0$

17. $t(t - 3) = 2t(t + 1)$

18. $(t + 4)(t - 3) = 18$

19. $(m^2 + 3)^2 - 4(m^2 + 3) - 5 = 0$

20. $m^{-4} - 5m^{-2} + 6 = 0$

8.6   Quadratic Functions and Their Graphs

The Graph of $f(x) = ax^2$ ▪   The Graph of $f(x) = a(x - h)^2$ ▪   The Graph of $f(x) = a(x - h)^2 - k$

We have seen that the graph of any linear function $f(x) = mx + b$ is a straight line. In this section and the next, we will see that the graph of any quadratic function $f(x) = ax^2 + bx + c$ is a *parabola*. We examine the shape of such graphs by first looking at quadratic functions with $b = 0$ and $c = 0$.

The Graph of $f(x) = ax^2$

The most basic quadratic function is $f(x) = x^2$.

EXAMPLE   **1**

Graph: $f(x) = x^2$.

SOLUTION   We choose some values for x and compute $f(x)$ for each. Then we plot the ordered pairs and connect them with a smooth curve.

x	$f(x) = x^2$	$(x, f(x))$
-3	9	$(-3, 9)$
-2	4	$(-2, 4)$
-1	1	$(-1, 1)$
0	0	$(0, 0)$
1	1	$(1, 1)$
2	4	$(2, 4)$
3	9	$(3, 9)$

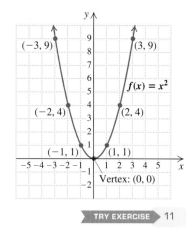

TRY EXERCISE   11

All quadratic functions have graphs similar to the one in Example 1. Such curves are called **parabolas**. They are U-shaped and can open upward, as in Example 1, or downward. The "turning point" of the graph is called the **vertex** of the parabola. The vertex of the graph in Example 1 is (0, 0).

A parabola is symmetric with respect to a line that goes through the center of the parabola and the vertex. This line is known as the parabola's **axis of symmetry**. In Example 1, the y-axis (the vertical line $x = 0$) is the axis of symmetry. Were the paper folded on this line, the two halves of the curve would match.

STUDENT NOTES

By paying attention to the symmetry of each parabola and the location of the vertex, you save yourself considerable work. Note too that when we are graphing ax^2, the x-values 1 unit to the right or left of the vertex are paired with the y-value a units above the vertex. Thus the graph of $y = \frac{3}{2}x^2$ includes the points $\left(-1, \frac{3}{2}\right)$ and $\left(1, \frac{3}{2}\right)$.

The graph of any function of the form $y = ax^2$ has a vertex of $(0, 0)$ and an axis of symmetry $x = 0$. By plotting points, we can compare the graphs of $g(x) = \frac{1}{2}x^2$ and $h(x) = 2x^2$ with the graph of $f(x) = x^2$.

x	$g(x) = \frac{1}{2}x^2$
-3	$\frac{9}{2}$
-2	2
-1	$\frac{1}{2}$
0	0
1	$\frac{1}{2}$
2	2
3	$\frac{9}{2}$

x	$h(x) = 2x^2$
-3	18
-2	8
-1	2
0	0
1	2
2	8
3	18

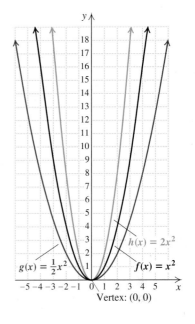

TECHNOLOGY CONNECTION

To explore the effect of a on the graph of $y = ax^2$, let $y_1 = x^2$, $y_2 = 3x^2$, and $y_3 = \frac{1}{3}x^2$. Graph the equations and use (TRACE) to see how the y-values compare, using ⌃ or ⌄ to hop the cursor from one curve to the next.

Many graphing calculators include a Transfrm application. If you run that application and let $y_1 = Ax^2$, the graph becomes interactive. A value for A can be entered while viewing the graph, or the values can be stepped up or down by pressing ◁ or ▷.

1. Compare the graphs of $y_1 = \frac{1}{5}x^2$, $y_2 = x^2$, $y_3 = \frac{5}{2}x^2$, $y_4 = -\frac{1}{5}x^2$, $y_5 = -x^2$, and $y_6 = -\frac{5}{2}x^2$.
2. Describe the effect that A has on each graph.

Note that the graph of $g(x) = \frac{1}{2}x^2$ is "wider" than the graph of $f(x) = x^2$, and the graph of $h(x) = 2x^2$ is "narrower." The vertex and the axis of symmetry, however, remain $(0, 0)$ and the line $x = 0$, respectively.

When we consider the graph of $k(x) = -\frac{1}{2}x^2$, we see that the parabola is the same shape as the graph of $g(x) = \frac{1}{2}x^2$, but opens downward. We say that the graphs of k and g are *reflections* of each other across the x-axis.

x	$k(x) = -\frac{1}{2}x^2$
-3	$-\frac{9}{2}$
-2	-2
-1	$-\frac{1}{2}$
0	0
1	$-\frac{1}{2}$
2	-2
3	$-\frac{9}{2}$

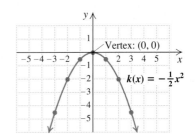

> Graphing $f(x) = ax^2$
>
> The graph of $f(x) = ax^2$ is a parabola with $x = 0$ as its axis of symmetry. Its vertex is the origin.
>
> For $a > 0$, the parabola opens upward. For $a < 0$, the parabola opens downward.
>
> If $|a|$ is greater than 1, the parabola is narrower than $y = x^2$.
>
> If $|a|$ is between 0 and 1, the parabola is wider than $y = x^2$.

The width of a parabola and whether it opens upward or downward are determined by the coefficient a in $f(x) = ax^2 + bx + c$. In the remainder of this section, we graph quadratic functions that are written in a form from which the vertex can be read directly.

The Graph of $f(x) = a(x - h)^2$

EXAMPLE 2   Graph: $f(x) = (x - 3)^2$.

SOLUTION   We choose some values for x and compute $f(x)$. Since $(x - 3)^2 = 1 \cdot (x - 3)^2$, $a = 1$, and the graph opens upward. It is important to note that when an input here is 3 more than an input for Example 1, the outputs match. We plot the points and draw the curve.

x	$f(x) = (x - 3)^2$
-1	16
0	9
1	4
2	1
3	0	← Vertex
4	1
5	4
6	9

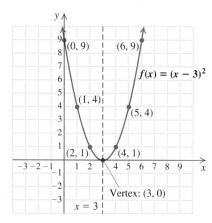

Note that $f(x)$ is smallest when $x - 3$ is 0, that is, for $x = 3$. Thus the line $x = 3$ is now the axis of symmetry and the point $(3, 0)$ is the vertex. Had we recognized earlier that $x = 3$ is the axis of symmetry, we could have computed some values on one side, such as $(4, 1)$, $(5, 4)$, and $(6, 9)$, and then used symmetry to get their mirror images $(2, 1)$, $(1, 4)$, and $(0, 9)$ without further computation.

TRY EXERCISE 19

The result of Example 2 can be generalized:

The vertex of the graph of $f(x) = a(x - h)^2$ is $(h, 0)$.

EXAMPLE 3   Graph: $g(x) = -2(x + 4)^2$.

SOLUTION   We choose some values for x and compute $g(x)$. Since $a = -2$, the graph will open downward. Note that $g(x)$ is greatest when $x + 4$ is 0, that is, for $x = -4$. Thus the line given by $x = -4$ is the axis of symmetry and the point $(-4, 0)$ is the vertex. We plot some points and draw the curve.

To explore the effect of h on the graph of $f(x) = a(x - h)^2$, let $y_1 = 7x^2$ and $y_2 = 7(x - 1)^2$. Graph both y_1 and y_2 and compare y-values, beginning at $x = 1$ and increasing x by one unit at a time. The G-T or HORIZ **MODE** can be used to view a split screen showing both the graph and a table.

Next, let $y_3 = 7(x - 2)^2$ and compare its graph and y-values with those of y_1 and y_2. Then let $y_4 = 7(x + 1)^2$ and $y_5 = 7(x + 2)^2$.

1. Compare graphs and y-values and describe the effect of h on the graph of $f(x) = a(x - h)^2$.
2. If the Transfrm application is available, let $y_1 = A(x - B)^2$ and describe the effect that A and B have on each graph.

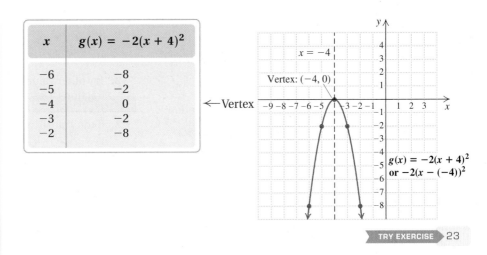

x	$g(x) = -2(x + 4)^2$
-6	-8
-5	-2
-4	0
-3	-2
-2	-8

TRY EXERCISE ▶ 23

In Example 2, the graph of $f(x) = (x - 3)^2$ looks just like the graph of $y = x^2$, except that it is moved, or *translated*, 3 units to the right. In Example 3, the graph of $g(x) = -2(x + 4)^2$ looks like the graph of $y = -2x^2$, except that it is shifted 4 units to the left. These results are generalized as follows.

> **Graphing $f(x) = a(x - h)^2$**
>
> The graph of $f(x) = a(x - h)^2$ has the same shape as the graph of $y = ax^2$.
>
> - If h is positive, the graph of $y = ax^2$ is shifted h units to the right.
> - If h is negative, the graph of $y = ax^2$ is shifted $|h|$ units to the left.
> - The vertex is $(h, 0)$ and the axis of symmetry is $x = h$.

The Graph of $f(x) = a(x - h)^2 + k$

Given a graph of $f(x) = a(x - h)^2$, what happens if we add a constant k? Suppose that we add 2. This increases $f(x)$ by 2, so the curve is moved up. If k is negative, the curve is moved down. The axis of symmetry for the parabola remains $x = h$, but the vertex will be at (h, k), or, equivalently, $(h, f(h))$.

Because of the shape of their graphs, quadratic functions have either a *minimum* value or a *maximum* value. Many real-world applications involve finding that value. For example, a business owner is concerned with minimizing cost and maximizing profit. If a parabola opens upward ($a > 0$), the function value, or y-value, at the vertex is a least, or minimum, value. That is, it is less than the y-value at any other point on the graph. If the parabola opens downward ($a < 0$), the function value at the vertex is a greatest, or maximum, value.

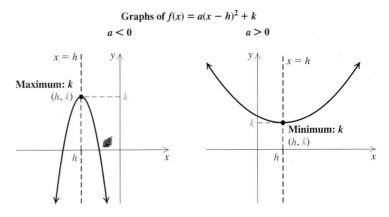
Graphs of $f(x) = a(x - h)^2 + k$

Graphing $f(x) = a(x - h)^2 + k$

The graph of $f(x) = a(x - h)^2 + k$ has the same shape as the graph of $y = a(x - h)^2$.

- If k is positive, the graph of $y = a(x - h)^2$ is shifted k units up.
- If k is negative, the graph of $y = a(x - h)^2$ is shifted $|k|$ units down.
- The vertex is (h, k), and the axis of symmetry is $x = h$.
- For $a > 0$, the minimum function value is k. For $a < 0$, the maximum function value is k.

EXAMPLE   4   Graph $g(x) = (x - 3)^2 - 5$, and find the minimum function value.

SOLUTION   The graph will look like that of $f(x) = (x - 3)^2$ (see Example 2) but shifted 5 units down. You can confirm this by plotting some points. For instance, $g(4) = (4 - 3)^2 - 5 = -4$, whereas in Example 2, $f(4) = (4 - 3)^2 = 1$.
The vertex is now $(3, -5)$, and the minimum function value is -5.

x	$g(x) = (x - 3)^2 - 5$
0	4
1	−1
2	−4
3	−5	← Vertex
4	−4
5	−1
6	4

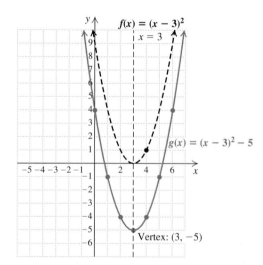

TRY EXERCISE ▶ 39

EXAMPLE **5** Graph $h(x) = \frac{1}{2}(x - 3)^2 + 6$, and find the minimum function value.

SOLUTION The graph looks just like that of $f(x) = \frac{1}{2}x^2$ but moved 3 units to the right and 6 units up. The vertex is $(3, 6)$, and the axis of symmetry is $x = 3$. We draw $f(x) = \frac{1}{2}x^2$ and then shift the curve over and up. The minimum function value is 6. By plotting some points, we have a check.

x	$h(x) = \frac{1}{2}(x - 3)^2 + 6$
0	$10\frac{1}{2}$
1	8
3	6
5	8
6	$10\frac{1}{2}$

←——— Vertex

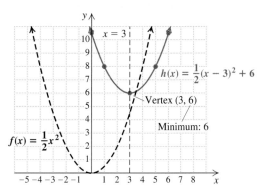

TRY EXERCISE 43

EXAMPLE **6** Graph $y = -2(x + 3)^2 + 5$. Find the vertex, the axis of symmetry, and the maximum or minimum value.

SOLUTION We first express the equation in the equivalent form

$$y = -2[x - (-3)]^2 + 5. \qquad \text{This is in the form } y = a(x - h)^2 + k.$$

The graph looks like that of $y = -2x^2$ translated 3 units to the left and 5 units up. The vertex is $(-3, 5)$, and the axis of symmetry is $x = -3$. Since -2 is negative, the graph opens downward, and we know that 5, the second coordinate of the vertex, is the maximum y-value.

We compute a few points as needed, selecting convenient x-values on either side of the vertex. The graph is shown here.

x	$y = -2(x + 3)^2 + 5$
-4	3
-3	5
-2	3

←——Vertex

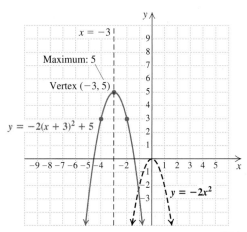

TRY EXERCISE 45

Concept Reinforcement   *In each of Exercises 1–8, match the equation with the corresponding graph from those shown.*

1. ____ $f(x) = 2(x - 1)^2 + 3$

2. ____ $f(x) = -2(x - 1)^2 + 3$

3. ____ $f(x) = 2(x + 1)^2 + 3$

4. ____ $f(x) = 2(x - 1)^2 - 3$

5. ____ $f(x) = -2(x + 1)^2 + 3$

6. ____ $f(x) = -2(x + 1)^2 - 3$

7. ____ $f(x) = 2(x + 1)^2 - 3$

8. ____ $f(x) = -2(x - 1)^2 - 3$

c)
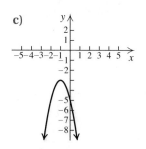

d)
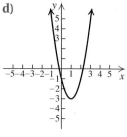

e)
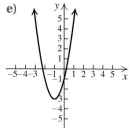

f)
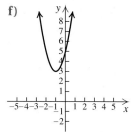

a)
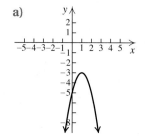

b)
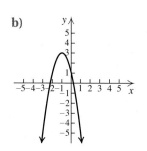

g)
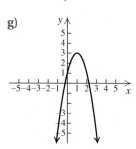

h)
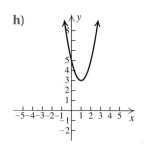

Graph.

9. $f(x) = x^2$

10. $f(x) = -x^2$

11. $f(x) = -2x^2$

12. $f(x) = -3x^2$

13. $g(x) = \frac{1}{3}x^2$

14. $g(x) = \frac{1}{4}x^2$

Aha! 15. $h(x) = -\frac{1}{3}x^2$

16. $h(x) = -\frac{1}{4}x^2$

17. $f(x) = \frac{5}{2}x^2$

18. $f(x) = \frac{3}{2}x^2$

For each of the following, graph the function, label the vertex, and draw the axis of symmetry.

19. $g(x) = (x + 1)^2$

20. $g(x) = (x + 4)^2$

21. $f(x) = (x - 2)^2$

22. $f(x) = (x - 1)^2$

23. $f(x) = -(x + 1)^2$

24. $f(x) = -(x - 1)^2$

25. $g(x) = -(x - 2)^2$

26. $g(x) = -(x + 4)^2$

27. $f(x) = 2(x + 1)^2$

28. $f(x) = 2(x + 4)^2$

29. $g(x) = 3(x - 4)^2$

30. $g(x) = 3(x - 5)^2$

31. $h(x) = -\frac{1}{2}(x - 4)^2$

32. $h(x) = -\frac{3}{2}(x - 2)^2$

33. $f(x) = \frac{1}{2}(x - 1)^2$

34. $f(x) = \frac{1}{3}(x + 2)^2$

35. $f(x) = -2(x + 5)^2$

36. $f(x) = -3(x + 7)^2$

37. $h(x) = -3\left(x - \frac{1}{2}\right)^2$

38. $h(x) = -2\left(x + \frac{1}{2}\right)^2$

For each of the following, graph the function and find the vertex, the axis of symmetry, and the maximum value or the minimum value.

39. $f(x) = (x - 5)^2 + 2$

40. $f(x) = (x + 3)^2 - 2$

41. $f(x) = (x + 1)^2 - 3$

42. $f(x) = (x - 1)^2 + 2$

43. $g(x) = \frac{1}{2}(x + 4)^2 + 1$

44. $g(x) = -(x - 2)^2 - 4$

45. $h(x) = -2(x - 1)^2 - 3$

46. $h(x) = -2(x + 1)^2 + 4$

47. $f(x) = 2(x + 3)^2 + 1$

48. $f(x) = 2(x - 5)^2 - 3$

49. $g(x) = -\frac{3}{2}(x - 2)^2 + 4$

50. $g(x) = \frac{3}{2}(x + 2)^2 - 1$

Without graphing, find the vertex, the axis of symmetry, and the maximum value or the minimum value.

51. $f(x) = 5(x - 3)^2 + 9$

52. $f(x) = 2(x - 1)^2 - 10$

53. $f(x) = -\frac{3}{7}(x + 8)^2 + 2$

54. $f(x) = -\frac{1}{4}(x + 4)^2 - 12$

55. $f(x) = \left(x - \frac{7}{2}\right)^2 - \frac{29}{4}$

56. $f(x) = -\left(x + \frac{3}{4}\right)^2 + \frac{17}{16}$

57. $f(x) = -\sqrt{2}(x + 2.25)^2 - \pi$

58. $f(x) = 2\pi(x - 0.01)^2 + \sqrt{15}$

59. Explain, without plotting points, why the graph of $y = x^2 - 4$ looks like the graph of $y = x^2$ translated 4 units down.

60. Explain, without plotting points, why the graph of $y = (x + 2)^2$ looks like the graph of $y = x^2$ translated 2 units to the left.

Skill Review

To prepare for Section 8.7, review finding intercepts and completing the square (Sections 2.4, 5.8, and 8.1).

Find the x-intercept and the y-intercept. [2.4]

61. $8x - 6y = 24$

62. $3x + 4y = 8$

Find the x-intercepts. [5.8]

63. $f(x) = x^2 + 8x + 15$

64. $g(x) = 2x^2 - x - 3$

Replace the blanks with constants to form a true equation. [8.1]

65. $x^2 - 14x + \underline{\quad} = (x - \underline{\quad})^2$

66. $x^2 + 7x + \underline{\quad} = \left(x + \underline{\quad}\right)^2$

Synthesis

67. Before graphing a quadratic function, Martha always plots five points. First, she calculates and plots the coordinates of the vertex. Then she plots *four* more points after calculating *two* more ordered pairs. How is this possible?

68. If the graphs of $f(x) = a_1(x - h_1)^2 + k_1$ and $g(x) = a_2(x - h_2)^2 + k_2$ have the same shape, what, if anything, can you conclude about the a's, the h's, and the k's? Why?

Write an equation for a function having a graph with the same shape as the graph of $f(x) = \frac{3}{5}x^2$, but with the given point as the vertex.

69. $(1, 3)$

70. $(2, 8)$

71. $(4, -7)$

72. $(9, -6)$

73. $(-2, -5)$

74. $(-4, -2)$

For each of the following, write the equation of the parabola that has the shape of $f(x) = 2x^2$ or $g(x) = -2x^2$ and has a maximum value or a minimum value at the specified point.

75. Minimum: $(2, 0)$

76. Minimum: $(-4, 0)$

77. Maximum: $(0, -5)$

78. Maximum: $(3, 8)$

Use the following graph of $f(x) = a(x - h)^2 + k$ for Exercises 79–82.

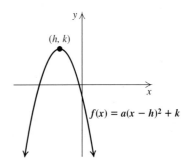

79. Describe what will happen to the graph if h is increased.

80. Describe what will happen to the graph if k is decreased.

81. Describe what will happen to the graph if a is replaced with $-a$.

82. Describe what will happen to the graph if $(x - h)$ is replaced with $(x + h)$.

Find an equation for the quadratic function F that satisfies the following conditions.

83. The graph of F is the same shape as the graph of f, where $f(x) = 3(x + 2)^2 + 7$, and $F(x)$ is a minimum at the same point that $g(x) = -2(x - 5)^2 + 1$ is a maximum.

84. The graph of F is the same shape as the graph of f, where $f(x) = -\frac{1}{3}(x - 2)^2 + 7$, and $F(x)$ is a maximum at the same point that $g(x) = 2(x + 4)^2 - 6$ is a minimum.

Functions other than parabolas can be translated. When calculating $f(x)$, if we replace x with $x - h$, where h is a constant, the graph will be moved horizontally. If we replace $f(x)$ with $f(x) + k$, the graph will be moved vertically. Use the graph below for Exercises 85–90.

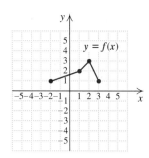

Draw a graph of each of the following.

85. $y = f(x - 1)$ **86.** $y = f(x + 2)$

87. $y = f(x) + 2$ **88.** $y = f(x) - 3$

89. $y = f(x + 3) - 2$ **90.** $y = f(x - 3) + 1$

91. Use the TRACE and/or TABLE features of a graphing calculator to confirm the maximum and minimum values given as answers to Exercises 51, 53, and 55. Be sure to adjust the window appropriately. On many graphing calculators, a maximum or minimum option may be available by using a CALC key.

92. Use a graphing calculator to check your graphs for Exercises 18, 28, and 48.

93. While trying to graph $y = -\frac{1}{2}x^2 + 3x + 1$, Yusef gets the following screen. How can Yusef tell at a glance that a mistake has been made?

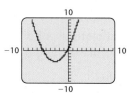

COLLABORATIVE CORNER

Match the Graph

Focus: Graphing quadratic functions
Time: 15–20 minutes
Group size: 6
Materials: Index cards

ACTIVITY

1. On each of six index cards, write one of the following equations:

   $y = \frac{1}{2}(x - 3)^2 + 1$; $y = \frac{1}{2}(x - 1)^2 + 3$;
   $y = \frac{1}{2}(x + 1)^2 - 3$; $y = \frac{1}{2}(x + 3)^2 + 1$;
   $y = \frac{1}{2}(x + 3)^2 - 1$; $y = \frac{1}{2}(x + 1)^2 + 3$.

2. Fold each index card and mix up the six cards in a hat or bag. Then, one by one, each group member should select one of the equations. Do not let anyone see your equation.

3. Each group member should carefully graph the equation selected. Make the graph large enough so that when it is finished, it can be easily viewed by the rest of the group. Be sure to scale the axes and label the vertex, but **do not label the graph with the equation used**.

4. When all group members have drawn a graph, place the graphs in a pile. The group should then match and agree on the correct equation for each graph *with no help from the person who drew the graph*. If a mistake has been made and a graph has no match, determine what its equation *should* be.

5. Compare your group's labeled graphs with those of other groups to reach consensus within the class on the correct label for each graph.

8.7   More About Graphing Quadratic Functions

Completing the Square ▪ Finding Intercepts

Completing the Square

By *completing the square* (see Section 8.1), we can rewrite any polynomial $ax^2 + bx + c$ in the form $a(x - h)^2 + k$. Once that has been done, the procedures discussed in Section 8.6 will enable us to graph any quadratic function.

EXAMPLE 1

Graph: $g(x) = x^2 - 6x + 4$. Label the vertex and the axis of symmetry.

SOLUTION   We have

$$g(x) = x^2 - 6x + 4$$
$$= (x^2 - 6x) + 4.$$

To complete the square inside the parentheses, we take half the x-coefficient, $\frac{1}{2} \cdot (-6) = -3$, and square it to get $(-3)^2 = 9$. Then we add $9 - 9$ inside the parentheses:

$$g(x) = (x^2 - 6x + 9 - 9) + 4 \qquad \text{The effect is of adding 0.}$$
$$= (x^2 - 6x + 9) + (-9 + 4) \qquad \text{Using the associative law of addition to regroup}$$
$$= (x - 3)^2 - 5. \qquad \text{Factoring and simplifying}$$

This equation appeared as Example 4 of Section 8.6. The graph is that of $f(x) = x^2$ translated 3 units right and 5 units down. The vertex is $(3, -5)$, and the axis of symmetry is $x = 3$.

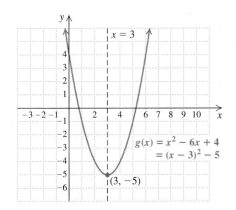

TRY EXERCISE 19

When the leading coefficient is not 1, we factor out that number from the first two terms. Then we complete the square and use the distributive law.

EXAMPLE 2

Graph: $f(x) = 3x^2 + 12x + 13$. Label the vertex and the axis of symmetry.

SOLUTION   Since the coefficient of x^2 is not 1, we need to factor out that number—in this case, 3—from the first two terms. Remember that we want the form $f(x) = a(x - h)^2 + k$:

$$f(x) = 3x^2 + 12x + 13$$
$$= 3(x^2 + 4x) + 13.$$

STUDENT NOTES

In this section, we add and subtract the same number when completing the square instead of adding the same number to both sides of the equation. We do this because we are using function notation. The effect is the same with both approaches: An equivalent equation is formed.

Now we complete the square as before. We take half of the x-coefficient, $\frac{1}{2} \cdot 4 = 2$, and square it: $2^2 = 4$. Then we add $4 - 4$ inside the parentheses:

$$f(x) = 3(x^2 + 4x + 4 - 4) + 13. \qquad \text{Adding } 4 - 4 \text{, or } 0 \text{, inside the parentheses}$$

The distributive law allows us to separate the -4 from the perfect-square trinomial so long as it is multiplied by 3. *This step is critical:*

$$f(x) = 3(x^2 + 4x + 4) + 3(-4) + 13 \qquad \text{This leaves a perfect-square trinomial inside the parentheses.}$$

$$= 3(x + 2)^2 + 1. \qquad \text{Factoring and simplifying}$$

The vertex is $(-2, 1)$, and the axis of symmetry is $x = -2$. The coefficient of x^2 is 3, so the graph is narrow and opens upward. We choose a few x-values on either side of the vertex, compute y-values, and then graph the parabola.

x	$f(x) = 3(x + 2)^2 + 1$
-2	1
-3	4
-1	4

⟵ Vertex

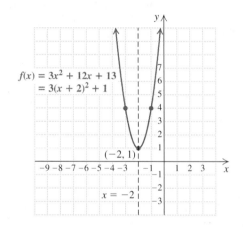

TRY EXERCISE ▸ 23

EXAMPLE **3** Graph $f(x) = -2x^2 + 10x - 7$, and find the maximum or minimum function value.

SOLUTION We first find the vertex by completing the square. To do so, we factor out -2 from the first two terms of the expression. This makes the coefficient of x^2 inside the parentheses 1:

$$f(x) = -2x^2 + 10x - 7$$
$$= -2(x^2 - 5x) - 7.$$

Now we complete the square as before. We take half of the x-coefficient and square it to get $\frac{25}{4}$. Then we add $\frac{25}{4} - \frac{25}{4}$ inside the parentheses:

$$f(x) = -2\left(x^2 - 5x + \frac{25}{4} - \frac{25}{4}\right) - 7$$
$$= -2\left(x^2 - 5x + \frac{25}{4}\right) + (-2)\left(-\frac{25}{4}\right) - 7 \qquad \begin{array}{l}\text{Multiplying by } -2, \\ \text{using the distributive} \\ \text{law, and regrouping}\end{array}$$
$$= -2\left(x - \frac{5}{2}\right)^2 + \frac{11}{2}. \qquad \begin{array}{l}\text{Factoring and} \\ \text{simplifying}\end{array}$$

The vertex is $\left(\frac{5}{2}, \frac{11}{2}\right)$, and the axis of symmetry is $x = \frac{5}{2}$. The coefficient of x^2, -2, is negative, so the graph opens downward and the second coordinate of the vertex, $\frac{11}{2}$, is the maximum function value.

We plot a few points on either side of the vertex, including the y-intercept, $f(0)$, and graph the parabola.

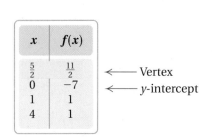

x	$f(x)$	
$\frac{5}{2}$	$\frac{11}{2}$	←— Vertex
0	-7	←— y-intercept
1	1	
4	1	

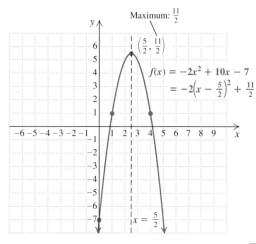

TRY EXERCISE ▶ 39

The method used in Examples 1–3 can be generalized to find a formula for locating the vertex. We complete the square as follows:

$$f(x) = ax^2 + bx + c$$

$$= a\left(x^2 + \frac{b}{a}x\right) + c. \qquad \text{Factoring } a \text{ out of the first two terms. Check by multiplying.}$$

Half of the x-coefficient, $\frac{b}{a}$, is $\frac{b}{2a}$. We square it to get $\frac{b^2}{4a^2}$ and add $\frac{b^2}{4a^2} - \frac{b^2}{4a^2}$ inside the parentheses. Then we distribute the a and regroup terms:

$$f(x) = a\left(x^2 + \frac{b}{a}x + \frac{b^2}{4a^2} - \frac{b^2}{4a^2}\right) + c$$

$$= a\left(x^2 + \frac{b}{a}x + \frac{b^2}{4a^2}\right) + a\left(-\frac{b^2}{4a^2}\right) + c \qquad \text{Using the distributive law}$$

$$= a\left(x + \frac{b}{2a}\right)^2 + \frac{-b^2}{4a} + \frac{4ac}{4a} \qquad \text{Factoring and finding a common denominator}$$

$$= a\left[x - \left(-\frac{b}{2a}\right)\right]^2 + \frac{4ac - b^2}{4a}.$$

Thus we have the following.

The Vertex of a Parabola

The vertex of the parabola given by $f(x) = ax^2 + bx + c$ is

$$\left(-\frac{b}{2a},\ f\left(-\frac{b}{2a}\right)\right), \quad \text{or} \quad \left(-\frac{b}{2a},\ \frac{4ac - b^2}{4a}\right).$$

- The x-coordinate of the vertex is $-b/(2a)$.
- The axis of symmetry is $x = -b/(2a)$.
- The second coordinate of the vertex is most commonly found by computing $f\left(-\frac{b}{2a}\right)$.

Let's reexamine Example 3 to see how we could have found the vertex directly. From the formula above,

$$\text{the } x\text{-coordinate of the vertex is } -\frac{b}{2a} = -\frac{10}{2(-2)} = \frac{5}{2}.$$

Substituting $\frac{5}{2}$ into $f(x) = -2x^2 + 10x - 7$, we find the second coordinate of the vertex:

$$f\left(\tfrac{5}{2}\right) = -2\left(\tfrac{5}{2}\right)^2 + 10\left(\tfrac{5}{2}\right) - 7$$
$$= -2\left(\tfrac{25}{4}\right) + 25 - 7$$
$$= -\tfrac{25}{2} + 18$$
$$= -\tfrac{25}{2} + \tfrac{36}{2} = \tfrac{11}{2}.$$

The vertex is $\left(\frac{5}{2}, \frac{11}{2}\right)$. The axis of symmetry is $x = \frac{5}{2}$.

We have actually developed two methods for finding the vertex. One is by completing the square and the other is by using a formula. You should check to see if your instructor prefers one method over the other or wants you to use both.

Finding Intercepts

All quadratic functions have a y-intercept and 0, 1, or 2 x-intercepts. For $f(x) = ax^2 + bx + c$, the y-intercept is $(0, f(0))$, or $(0, c)$. To find x-intercepts, if any exist, we look for points where $y = 0$ or $f(x) = 0$. Thus, for $f(x) = ax^2 + bx + c$, the x-intercepts occur at those x-values for which

$$ax^2 + bx + c = 0.$$

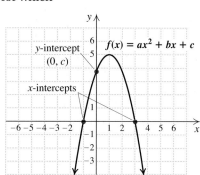

EXAMPLE   **4**   Find any x-intercepts and the y-intercept of the graph of $f(x) = x^2 - 2x - 2$.

SOLUTION   The y-intercept is simply $(0, f(0))$, or $(0, -2)$. To find any x-intercepts, we solve

$$0 = x^2 - 2x - 2.$$

We are unable to factor $x^2 - 2x - 2$, so we use the quadratic formula and get $x = 1 \pm \sqrt{3}$. Thus the x-intercepts are $(1 - \sqrt{3}, 0)$ and $(1 + \sqrt{3}, 0)$.

If graphing, we would approximate, to get $(-0.7, 0)$ and $(2.7, 0)$.

TRY EXERCISE 43

If the solutions of $f(x) = 0$ are imaginary, the graph of f has no x-intercepts.

Visualizing for Success

A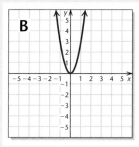

B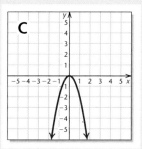

C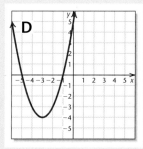

D

E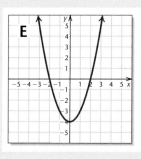

F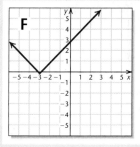

G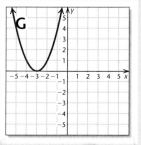

H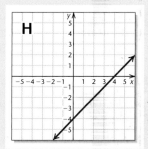

I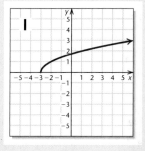

J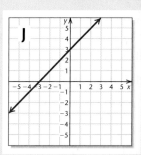

Match each function with its graph.

1. $f(x) = 3x^2$

2. $f(x) = x^2 - 4$

3. $f(x) = (x - 4)^2$

4. $f(x) = x - 4$

5. $f(x) = -2x^2$

6. $f(x) = x + 3$

7. $f(x) = |x + 3|$

8. $f(x) = (x + 3)^2$

9. $f(x) = \sqrt{x + 3}$

10. $f(x) = (x + 3)^2 - 4$

Answers on page 37

An additional, animated version of this activity appears in MyMathLab. To use MyMathLab, you need a course ID and a student access code. Contact your instructor for more information.

8.7	EXERCISE SET

↘ **Concept Reinforcement** *Classify each statement as either true or false.*

1. The graph of $f(x) = 3x^2 - x + 6$ opens upward.

2. The function given by $g(x) = -x^2 + 3x + 1$ has a minimum value.

3. The graph of $f(x) = -2(x - 3)^2 + 7$ has its vertex at $(3, 7)$.

4. The graph of $g(x) = 4(x + 6)^2 - 2$ has its vertex at $(-6, -2)$.

5. The graph of $g(x) = \frac{1}{2}\left(x - \frac{3}{2}\right)^2 + \frac{1}{4}$ has $x = \frac{1}{4}$ as its axis of symmetry.

6. The function given by $f(x) = (x - 2)^2 - 5$ has a minimum value of -5.

7. The y-intercept of the graph of $f(x) = 2x^2 - 6x + 7$ is $(7, 0)$.

8. If the graph of a quadratic function f opens upward and has a vertex of $(1, 5)$, then the graph has no x-intercepts.

Complete the square to write each function in the form $f(x) = a(x - h)^2 + k.$

9. $f(x) = x^2 - 8x + 2$

10. $f(x) = x^2 - 6x - 1$

11. $f(x) = x^2 + 3x - 5$

12. $f(x) = x^2 + 5x + 3$

13. $f(x) = 3x^2 + 6x - 2$

14. $f(x) = 2x^2 - 20x - 3$

15. $f(x) = -x^2 - 4x - 7$

16. $f(x) = -2x^2 - 8x + 4$

17. $f(x) = 2x^2 - 5x + 10$

18. $f(x) = 3x^2 + 7x - 3$

For each quadratic function, **(a)** *find the vertex and the axis of symmetry and* **(b)** *graph the function.*

19. $f(x) = x^2 + 4x + 5$

20. $f(x) = x^2 + 2x - 5$

21. $f(x) = x^2 + 8x + 20$

22. $f(x) = x^2 - 10x + 21$

23. $h(x) = 2x^2 - 16x + 25$

24. $h(x) = 2x^2 + 16x + 23$

25. $f(x) = -x^2 + 2x + 5$

26. $f(x) = -x^2 - 2x + 7$

27. $g(x) = x^2 + 3x - 10$

28. $g(x) = x^2 + 5x + 4$

29. $h(x) = x^2 + 7x$

30. $h(x) = x^2 - 5x$

31. $f(x) = -2x^2 - 4x - 6$

32. $f(x) = -3x^2 + 6x + 2$

For each quadratic function, **(a)** *find the vertex, the axis of symmetry, and the maximum or minimum function value and* **(b)** *graph the function.*

33. $g(x) = x^2 - 6x + 13$

34. $g(x) = x^2 - 4x + 5$

35. $g(x) = 2x^2 - 8x + 3$

36. $g(x) = 2x^2 + 5x - 1$

37. $f(x) = 3x^2 - 24x + 50$

38. $f(x) = 4x^2 + 16x + 13$

39. $f(x) = -3x^2 + 5x - 2$

40. $f(x) = -3x^2 - 7x + 2$

41. $h(x) = \frac{1}{2}x^2 + 4x + \frac{19}{3}$

42. $h(x) = \frac{1}{2}x^2 - 3x + 2$

Find any x-intercepts and the y-intercept. If no x-intercepts exist, state this.

43. $f(x) = x^2 - 6x + 3$

44. $f(x) = x^2 + 5x + 4$

45. $g(x) = -x^2 + 2x + 3$

46. $g(x) = x^2 - 6x + 9$

Aha! 47. $f(x) = x^2 - 9x$

48. $f(x) = x^2 - 7x$

49. $h(x) = -x^2 + 4x - 4$

50. $h(x) = -2x^2 - 20x - 50$

51. $g(x) = x^2 + x - 5$

52. $g(x) = 2x^2 + 3x - 1$

53. $f(x) = 2x^2 - 4x + 6$

54. $f(x) = x^2 - x + 2$

55. The graph of a quadratic function f opens downward and has no x-intercepts. In what quadrant(s) must the vertex lie? Explain your reasoning.

56. Is it possible for the graph of a quadratic function to have only one x-intercept if the vertex is off the x-axis? Why or why not?

Skill Review

To prepare for Section 8.8, review solving systems of three equations in three unknowns (Section 3.4).

Solve. [3.4]

57. $x + y + z = 3,$
$x - y + z = 1,$
$-x - y + z = -1$

58. $x - y + z = -6,$
$2x + y + z = 2,$
$3x + y + z = 0$

59. $z = 8,$
$x + y + z = 23,$
$2x + y - z = 17$

60. $z = -5,$
$2x - y + 3z = -27,$
$x + 2y + 7z = -26$

61. $1.5 = c,$
$52.5 = 25a + 5b + c,$
$7.5 = 4a + 2b + c$

62. $\frac{1}{2} = c,$
$5 = 9a + 6b + 2c,$
$29 = 81a + 9b + c$

Synthesis

63. If the graphs of two quadratic functions have the same x-intercepts, will they also have the same vertex? Why or why not?

64. Suppose that the graph of $f(x) = ax^2 + bx + c$ has $(x_1, 0)$ and $(x_2, 0)$ as x-intercepts. Explain why the graph of $g(x) = -ax^2 - bx - c$ will also have $(x_1, 0)$ and $(x_2, 0)$ as x-intercepts.

For each quadratic function, find **(a)** *the maximum or minimum value and* **(b)** *any x-intercepts and the y-intercept.*

65. $f(x) = 2.31x^2 - 3.135x - 5.89$

66. $f(x) = -18.8x^2 + 7.92x + 6.18$

67. Graph the function
$$f(x) = x^2 - x - 6.$$
Then use the graph to approximate solutions to each of the following equations.
a) $x^2 - x - 6 = 2$
b) $x^2 - x - 6 = -3$

68. Graph the function
$$f(x) = \frac{x^2}{2} + x - \frac{3}{2}.$$
Then use the graph to approximate solutions to each of the following equations.
a) $\frac{x^2}{2} + x - \frac{3}{2} = 0$
b) $\frac{x^2}{2} + x - \frac{3}{2} = 1$
c) $\frac{x^2}{2} + x - \frac{3}{2} = 2$

Find an equivalent equation of the type
$$f(x) = a(x - h)^2 + k.$$

69. $f(x) = mx^2 - nx + p$

70. $f(x) = 3x^2 + mx + m^2$

71. A quadratic function has $(-1, 0)$ as one of its intercepts and $(3, -5)$ as its vertex. Find an equation for the function.

72. A quadratic function has $(4, 0)$ as one of its intercepts and $(-1, 7)$ as its vertex. Find an equation for the function.

Graph.

73. $f(x) = |x^2 - 1|$

74. $f(x) = |x^2 - 3x - 4|$

75. $f(x) = |2(x - 3)^2 - 5|$

76. Use a graphing calculator to check your answers to Exercises 25, 41, 53, 65, and 67.

8.8 Problem Solving and Quadratic Functions

Maximum and Minimum Problems ▪ Fitting Quadratic Functions to Data

Let's look now at some of the many situations in which quadratic functions are used for problem solving.

Maximum and Minimum Problems

We have seen that for any quadratic function f, the value of $f(x)$ at the vertex is either a maximum or a minimum. Thus problems in which a quantity must be maximized or minimized can be solved by finding the coordinates of a vertex, assuming the problem can be modeled with a quadratic function.

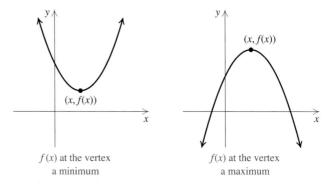

$f(x)$ at the vertex a minimum $f(x)$ at the vertex a maximum

EXAMPLE 1

Museum attendance. After the admission fee was dropped, attendance at the Indianapolis Museum of Art began to rise after several years of decline. The number of museum admissions, in thousands, t years after 2000 can be approximated by $m(t)$, where $m(t) = 32t^2 - 320t + 975$. In what year was the museum attendance the lowest, and how many people went to the museum that year?

Source: Based on information in the *Indianapolis Star*, 9/9/07

SOLUTION

1., 2. Familiarize and **Translate.** We are given the function for museum attendance. Note that it is a quadratic function of the number of years since 2000. The coefficient of the squared term is positive, so the graph opens upward and there is a minimum value. The calculator-generated graph at left confirms this.

3. Carry out. We can either complete the square or use the formula for the vertex. Completing the square, we have

$$m(t) = 32t^2 - 320t + 975$$
$$= 32(t^2 - 10t) + 975$$
$$= 32(t^2 - 10t + 25 - 25) + 975 \qquad \text{Completing the square}$$
$$= 32(t^2 - 10t + 25) - (32)(25) + 975$$
$$= 32(t - 5)^2 + 175. \qquad \text{Factoring and simplifying}$$

There is a minimum value of 175 when $t = 5$.

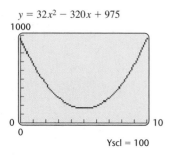

$y = 32x^2 - 320x + 975$

Yscl = 100

A visualization for Example 1

4. **Check.** Using the formula, we have $-b/(2a) = -(-320)/64 = 5$. Then

$$m(5) = 32(5)^2 - 320(5) + 975 = 175.$$

Both approaches give the same minimum, and that minimum is also confirmed by the graph. The answer checks.

5. **State.** The minimum attendance was 175,000. It occurred 5 yr after 2000, or in 2005.

▸ TRY EXERCISE 7

EXAMPLE 2

Swimming area.   A lifeguard has 100 m of roped-together flotation devices with which to cordon off a rectangular swimming area at North Beach. If the shoreline forms one side of the rectangle, what dimensions will maximize the size of the area for swimming?

SOLUTION

1. **Familiarize.** We make a drawing and label it, letting w = the width of the rectangle, in meters, and l = the length of the rectangle, in meters.

Recall that Area = $l \cdot w$ and Perimeter = $2w + 2l$. Since the beach forms one length of the rectangle, the flotation devices comprise three sides. Thus

$$2w + l = 100.$$

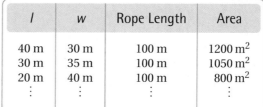

To get a better feel for the problem, we can look at some possible dimensions for a rectangular area that can be enclosed with 100 m of flotation devices. All possibilities are chosen so that $2w + l = 100$.

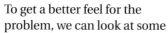

l	w	Rope Length	Area	
40 m	30 m	100 m	1200 m²	⎫ What choice of l and
30 m	35 m	100 m	1050 m²	⎬ w will maximize A?
20 m	40 m	100 m	800 m²	⎭
⋮	⋮	⋮	⋮	

2. **Translate.**   We have two equations: One guarantees that all 100 m of flotation devices are used; the other expresses area in terms of length and width.

$$2w + l = 100,$$
$$A = l \cdot w$$

3. **Carry out.**   We need to express A as a function of l or w but not both. To do so, we solve for l in the first equation to obtain $l = 100 - 2w$. Substituting for l in the second equation, we get a quadratic function:

$A = (100 - 2w)w$ Substituting for l

$\quad = 100w - 2w^2$.   This represents a parabola opening downward, so a maximum exists.

Factoring and completing the square, we get

$A = -2(w^2 - 50w + 625 - 625)$ We could also use the vertex formula.

$\quad = -2(w - 25)^2 + 1250$.

There is a maximum value of 1250 when $w = 25$.

4. **Check.** If $w = 25$ m, then $l = 100 - 2 \cdot 25 = 50$ m. These dimensions give an area of 1250 m². Note that 1250 m² is greater than any of the values for A found in the *Familiarize* step. To be more certain, we could check values other than those used in that step. For example, if $w = 26$ m, then $l = 48$ m, and $A = 26 \cdot 48 = 1248$ m². Since 1250 m² is greater than 1248 m², it appears that we have a maximum.

5. **State.** The largest rectangular area for swimming that can be enclosed is 25 m by 50 m.

> **TRY EXERCISE** 11

Fitting Quadratic Functions to Data

Whenever a certain quadratic function fits a situation, that function can be determined if three inputs and their outputs are known. Each of the given ordered pairs is called a *data point*.

EXAMPLE 3

Music CDs.   As more listeners download their music purchases, sales of compact discs are decreasing. According to Nielsen SoundScan, sales of music CDs increased from 500 million in 1997 to 700 million in 2001 and then decreased to 450 million in 2007. As the graph suggests, sales of music CDs can be modeled by a quadratic function.

Years After 1997	Number of Music CDs Sold in the United States (in millions)
0	500
4	700
10	450

Source: Nielsen SoundScan

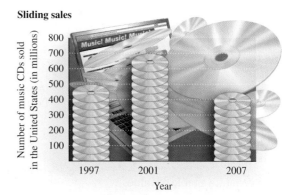

Sliding sales

a) Let t represent the number of years since 1997 and $S(t)$ the total number of CDs sold, in millions. Use the data points (0, 500), (4, 700), and (10, 450) to find a quadratic function that fits the data.

b) Use the function from part (a) to estimate the sales of music CDs in 2009.

SOLUTION

a) We are looking for a function of the form $S(t) = at^2 + bt + c$ given that $S(0) = 500$, $S(4) = 700$, and $S(10) = 450$. Thus,

$$500 = a \cdot 0^2 + b \cdot 0 + c, \qquad \text{Using the data point (0, 500)}$$
$$700 = a \cdot 4^2 + b \cdot 4 + c, \qquad \text{Using the data point (4, 700)}$$
$$450 = a \cdot 10^2 + b \cdot 10 + c. \qquad \text{Using the data point (10, 450)}$$

STUDENT NOTES ————

Try to keep the "big picture" in mind on problems like Example 3. Solving a system of three equations is but one part of the solution.

After simplifying, we see that we need to solve the system

$$500 = c, \qquad \qquad \textbf{(1)}$$
$$700 = 16a + 4b + c, \qquad \textbf{(2)}$$
$$450 = 100a + 10b + c. \quad \textbf{(3)}$$

To use a graphing calculator to fit a quadratic function to the data in Example 3, we first select EDIT in the **STAT** menu and enter the given data.

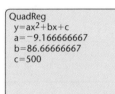

L1	L2	L3	2
0	500	------	
4	700		
10	450		

L2(4) =

To fit a quadratic function to the data, we press **STAT** ▷ **5** **VARS** ▷ **1** **1** **ENTER**. The first three keystrokes select QuadReg from the STAT CALC menu. The keystrokes **VARS** ▷ **1** **1** copy the regression equation to the equation-editor screen as y_1.

QuadReg
y=ax²+bx+c
a=−9.166666667
b=86.66666667
c=500

We see that the regression equation is $y = -9.166666667x^2 + 86.66666667x + 500$. We press **Y=** **▲** **ENTER** to turn on the PLOT feature and **ZOOM** **9** to see the regression equation graphed with the data points.

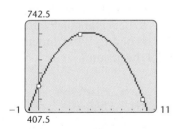

To check Example 3(b), we set Indpnt to Ask in the Table Setup and enter $X = 12$ in the table. A Y1-value of 220 confirms our answer.

1. Use the above approach to estimate the sales of music CDs in 2005.

We know from equation (1) that $c = 500$. Substituting that value into equations (2) and (3), we have

$$700 = 16a + 4b + 500,$$
$$450 = 100a + 10b + 500.$$

Subtracting 500 from both sides of each equation, we have

$$200 = 16a + 4b, \qquad \textbf{(4)}$$
$$-50 = 100a + 10b. \qquad \textbf{(5)}$$

To solve, we multiply equation (4) by 5 and equation (5) by -2. We then add to eliminate b:

$$1000 = 80a + 20b$$
$$\underline{100 = -200a - 20b}$$
$$1100 = -120a$$
$$-\frac{1100}{120} = a, \quad \text{or} \quad a = -\frac{55}{6}. \qquad \text{Simplifying}$$

Next, we solve for b, using equation (5) above:

$$-50 = 100\left(-\frac{55}{6}\right) + 10b$$

$$-50 = -\frac{2750}{3} + 10b$$

$$\frac{2600}{3} = 10b \qquad \text{Adding } \tfrac{2750}{3} \text{ to both sides and simplifying}$$

$$\frac{2600}{30} = b, \quad \text{or} \quad b = \frac{260}{3}. \qquad \text{Dividing both sides by 10 and simplifying}$$

We can now write $S(t) = at^2 + bt + c$ as

$$S(t) = -\frac{55}{6}t^2 + \frac{260}{3}t + 500.$$

b) To find the sales of CDs in 2009, we evaluate the function. Note that 2009 is 12 yr after 1997. Thus,

$$S(12) = -\frac{55}{6} \cdot 12^2 + \frac{260}{3} \cdot 12 + 500$$
$$= 220.$$

In 2009, an estimated 220 million music CDs will be sold.

TRY EXERCISE 35

8.8   EXERCISE SET

Concept Reinforcement   *In each of Exercises 1–6, match the description with the graph that displays that characteristic.*

1. ____ A minimum value of $f(x)$ exists.

2. ____ A maximum value of $f(x)$ exists.

3. ____ No maximum or minimum value of $f(x)$ exists.

4. ____ The data points appear to suggest a linear model for g.

5. ____ The data points appear to suggest that g is a quadratic function with a maximum.

6. ____ The data points appear to suggest that g is a quadratic function with a minimum.

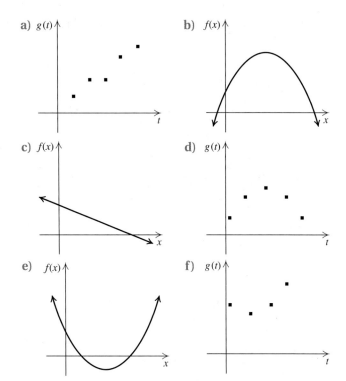

Solve.

7. *Newborn calves.*   The number of pounds of milk per day recommended for a calf that is x weeks old can be approximated by $p(x)$, where

$$p(x) = -0.2x^2 + 1.3x + 6.2.$$

When is a calf's milk consumption greatest and how much milk does it consume at that time?
Source: C. Chaloux, University of Vermont, 1998

8. *Stock prices.*   The value of a share of I. J. Solar can be represented by $V(x) = x^2 - 6x + 13$, where x is the number of months after January 2009. What is the lowest value $V(x)$ will reach, and when did that occur?

9. *Minimizing cost.*   Sweet Harmony Crafts has determined that when x hundred dulcimers are built, the average cost per dulcimer can be estimated by

$$C(x) = 0.1x^2 - 0.7x + 2.425,$$

where $C(x)$ is in hundreds of dollars. What is the minimum average cost per dulcimer and how many dulcimers should be built in order to achieve that minimum?

10. *Maximizing profit.*   Recall that total profit P is the difference between total revenue R and total cost C. Given $R(x) = 1000x - x^2$ and $C(x) = 3000 + 20x$, find the total profit, the maximum value of the total profit, and the value of x at which it occurs.

11. *Architecture.*   An architect is designing an atrium for a hotel. The atrium is to be rectangular with a perimeter of 720 ft of brass piping. What dimensions will maximize the area of the atrium?

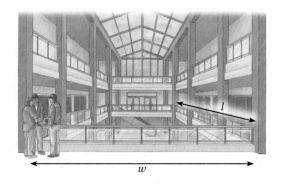

12. *Furniture design.*   A furniture builder is designing a rectangular end table with a perimeter of 128 in. What dimensions will yield the maximum area?

13. *Patio design.*   A stone mason has enough stones to enclose a rectangular patio with 60 ft of perimeter, assuming that the attached house forms one side of the rectangle. What is the maximum area that the mason can enclose? What should the dimensions of the patio be in order to yield this area?

14. *Garden design.*   Ginger is fencing in a rectangular garden, using the side of her house as one side of the rectangle. What is the maximum area that she can enclose with 40 ft of fence? What should the dimensions of the garden be in order to yield this area?

15. *Molding plastics.*   Economite Plastics plans to produce a one-compartment vertical file by bending the long side of an 8-in. by 14-in. sheet of plastic along two lines to form a U shape. How tall should the file be in order to maximize the volume that the file can hold?

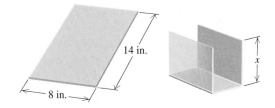

16. *Composting.*   A rectangular compost container is to be formed in a corner of a fenced yard, with 8 ft of chicken wire completing the other two sides of the rectangle. If the chicken wire is 3 ft high, what dimensions of the base will maximize the container's volume?

17. What is the maximum product of two numbers that add to 18? What numbers yield this product?

18. What is the maximum product of two numbers that add to 26? What numbers yield this product?

19. What is the minimum product of two numbers that differ by 8? What are the numbers?

20. What is the minimum product of two numbers that differ by 7? What are the numbers?

Aha! **21.** What is the maximum product of two numbers that add to −10? What numbers yield this product?

22. What is the maximum product of two numbers that add to −12? What numbers yield this product?

Choosing models.   *For the scatterplots and graphs in Exercises 23–34, determine which, if any, of the following functions might be used as a model for the data: Linear, with* $f(x) = mx + b$; *quadratic, with* $f(x) = ax^2 + bx + c$, $a > 0$; *quadratic, with* $f(x) = ax^2 + bx + c$, $a < 0$; *neither quadratic nor linear.*

23.   **Sonoma Sunshine**

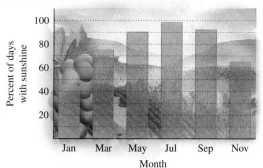

Source: www.city-data.com

24.   **Sonoma Precipitation**

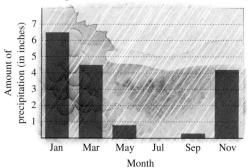

Source: www.city-data.com

25. **Safe sight distance to the left**

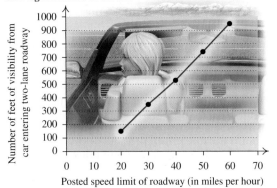

Source: Institute of Traffic Engineers

26. **Safe sight distance to the right**

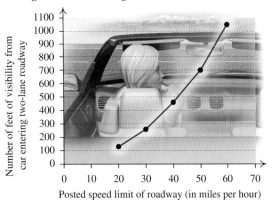

Source: Institute of Traffic Engineers

27. **Winter Olympic volunteers**

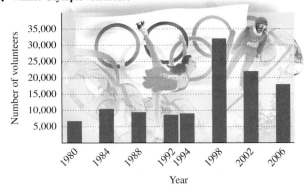

28. **U.S. senior population**

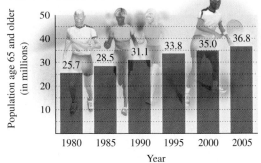

Source: U.S. Bureau of Labor Statistics

29. **Changing work force**

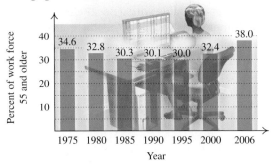

Source: U.S. Department of Labor, Bureau of Labor Statistics

30. **Airline bumping rate**

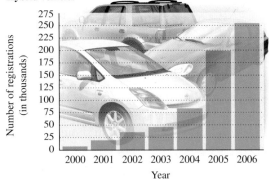

Source: U.S. Department of Transportation

31. **Hybrid vehicles**

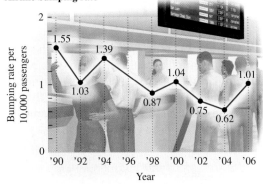

Source: R.L. Polk & Co.

32.

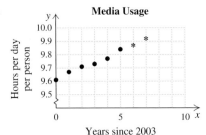

Media Usage

* Projected

Source: Statistical Abstract of the United States

33. **Employee contribution to health insurance premium**

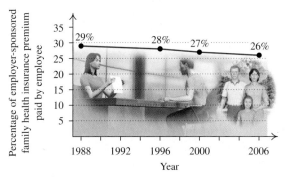

Source: Based on data from Kaiser

34. **Average number of live births per 1000 women, 2005**

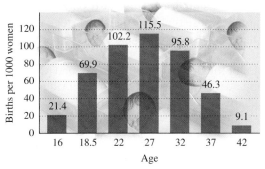

Source: U.S. Centers for Disease Control

Find a quadratic function that fits the set of data points.

35. $(1, 4), (-1, -2), (2, 13)$

36. $(1, 4), (-1, 6), (-2, 16)$

37. $(2, 0), (4, 3), (12, -5)$

38. $(-3, -30), (3, 0), (6, 6)$

39. a) Find a quadratic function that fits the following data.

Travel Speed (in kilometers per hour)	Number of Nighttime Accidents (for every 200 million kilometers driven)
60	400
80	250
100	250

b) Use the function to estimate the number of nighttime accidents that occur at 50 km/h.

40. a) Find a quadratic function that fits the following data.

Travel Speed (in kilometers per hour)	Number of Daytime Accidents (for every 200 million kilometers driven)
60	100
80	130
100	200

b) Use the function to estimate the number of daytime accidents that occur at 50 km/h.

41. *Archery.* The Olympic flame tower at the 1992 Summer Olympics was lit at a height of about 27 m by a flaming arrow that was launched about 63 m from the base of the tower. If the arrow landed about 63 m beyond the tower, find a quadratic function that expresses the height h of the arrow as a function of the distance d that it traveled horizontally.

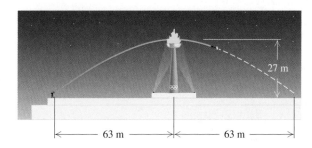

42. *Outsourcing.* The revenue, in billions of dollars, from India's outsourcing industry is shown in the following table.

Year	Outsourcing Revenue (in billions of dollars)
2001	$12
2004	21
2007	48

Source: Nasscom

a) Let t represent the number of years since 2000 and $r(t)$ the revenue, in billions of dollars. Find a quadratic function that fits the data.

b) Use the function to estimate India's outsourcing revenue in 2012.

43. Does every nonlinear function have a minimum or a maximum value? Why or why not?

44. Explain how the leading coefficient of a quadratic function can be used to determine if a maximum or a minimum function value exists.

Skill Review

To prepare for Section 8.9, review solving inequalities and rational expressions and equations (Chapters 4 and 6).

Solve.

45. $2x - 3 > 5$ [4.1]

46. $4 - x \leq 7$ [4.1]

47. $|9 - x| \geq 2$ [4.3]

48. $|4x + 1| < 11$ [4.3]

Find simplified form for $f(x)$ and list all restrictions on the domain. [6.2]

49. $f(x) = \dfrac{x - 3}{x + 4} - 5$

50. $f(x) = \dfrac{x}{x - 1} - 1$

Solve. [6.4]

51. $\dfrac{x - 3}{x + 4} = 5$

52. $\dfrac{x}{x - 1} = 1$

53. $\dfrac{x}{(x - 3)(x + 7)} = 0$

54. $\dfrac{(x + 6)(x - 9)}{x + 5} = 0$

Synthesis

The following graphs can be used to compare the baseball statistics of pitcher Roger Clemens with the 31 other pitchers since 1968 who started at least 10 games in at least 15 seasons and pitched at least 3000 innings. Use the graphs to answer questions 55 and 56.

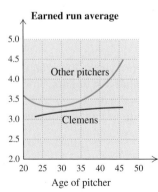

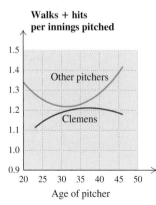

Source: The New York Times, February 10, 2008;
Eric Bradlow, Shane Jensen, Justin Wolfers and Adi Wyner

55. The earned run average describes how many runs a pitcher has allowed per game. The lower the earned run average, the better a pitcher. Compare, in terms of maximums or minimums, the earned run average of Roger Clemens with that of other pitchers. Is there any reason to suspect that the aging process was unusual for Clemens? Explain.

56. The statistic "Walks + hits per innings pitched" is related to how often a pitcher allows a batter to reach a base. The lower this statistic, the better. Compare, in terms of maximums or minimums, the "walks + hits" statistic of Roger Clemens with that of other pitchers.

57. *Bridge design.* The cables supporting a straight-line suspension bridge are nearly parabolic in shape. Suppose that a suspension bridge is being designed with concrete supports 160 ft apart and with vertical cables 30 ft above road level at the midpoint of the bridge and 80 ft above road level at a point 50 ft from the midpoint of the bridge. How long are the longest vertical cables?

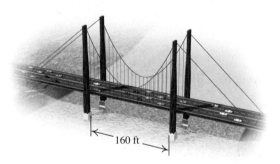

58. *Trajectory of a launched object.* The height above the ground of a launched object is a quadratic function of the time that it is in the air. Suppose that a flare is launched from a cliff 64 ft above sea level. If 3 sec after being launched the flare is again level with the cliff, and if 2 sec after that it lands in the sea, what is the maximum height that the flare will reach?

59. *Cover charges.* When the owner of Sweet Sounds charges a $10 cover charge, an average of 80 people will attend a show. For each 25¢ increase in admission price, the average number attending decreases by 1. What should the owner charge in order to make the most money?

60. *Crop yield.* An orange grower finds that she gets an average yield of 40 bushels (bu) per tree when she plants 20 trees on an acre of ground. Each time she adds one tree per acre, the yield per tree decreases by 1 bu, due to congestion. How many trees per acre should she plant for maximum yield?

61. *Norman window.*   A *Norman window* is a rectangle with a semicircle on top. Big Sky Windows is designing a Norman window that will require 24 ft of trim. What dimensions will allow the maximum amount of light to enter a house?

62. *Minimizing area.*   A 36-in. piece of string is cut into two pieces. One piece is used to form a circle while the other is used to form a square. How should the string be cut so that the sum of the areas is a minimum?

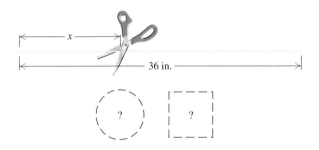

Regression can be used to find the "best"-fitting quadratic function when more than three data points are provided. In Exercises 63 and 64, six data points are given, but the approach used in the Technology Connection on p. 555 still applies.

63. *Hybrid vehicles.*   The number of hybrid vehicles in the United States during several years is shown in the table below.

Year	Number of Vehicles
2001	19,963
2002	35,934
2003	45,943
2004	83,153
2005	199,148
2006	254,545

Source: R. L. Polk & Co.

a) Use regression to find a quadratic function that can be used to estimate the number of hybrid vehicles $h(x)$ in the United States x years after 2000.

b) Use the function found in part (a) to predict the number of hybrid vehicles in the United States in 2010.

64. *Hydrology.*   The drawing below shows the cross section of a river. Typically rivers are deepest in the middle, with the depth decreasing to 0 at the edges. A hydrologist measures the depths D, in feet, of a river at distances x, in feet, from one bank. The results are listed in the table below.

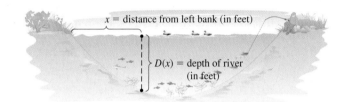

Distance x, from the Left Bank (in feet)	Depth, D, of the River (in feet)
0	0
15	10.2
25	17
50	20
90	7.2
100	0

a) Use regression to find a quadratic function that fits the data.

b) Use the function to estimate the depth of the river 70 ft from the left bank.

CORNER

Parabolic Pizza

Focus: Modeling

Time: 20–30 minutes

Group size: 3

Materials: Graphing calculators are optional.

College Pizza on Chestnut Street in Philadelphia, PA, sells a 10-in.–diameter cheese pizza for $5.00, a 14-in. cheese pizza for $7.50, and an 18-in. cheese pizza for $11.00. Which models better the price of the pizza: a linear function or a quadratic function of the diameter?

Source: Campusfood.com

ACTIVITY

1. As a group, carefully graph the ordered pairs from the data above in the form (diameter, price). Do the data appear to be quadratic or linear?

2. Each group member should choose one of the following to fit a model to the data, where $p(x)$ is the price, in dollars, of an x-inch–diameter pizza. Then, using a different color for each graph, that member should graph the function on the same graph as the ordered pairs.

a) Linear function $p(x) = mx + b$, using the points (10, 5) and (14, 7.5)
b) Linear function $p(x) = mx + b$, using the points (10, 5) and (18, 11)
c) Quadratic function $p(x) = ax^2 + bx + c$, using all three points

3. As a group, determine which function from part (1) appears to be the best fit.

4. One way to tell whether a function is a good fit is to see how well it predicts another known value. College Pizza also sells a 16-in. cheese pizza for $9.00. Each group member should use the function from part (2) to predict the price of a 16-in. cheese pizza. Which function came the closest to predicting the actual value?

5. If a graphing calculator is available, use the LINREG and QUADREG options to fit and graph linear and quadratic functions for the four data points (three pairs from part (1) and one pair from part (4)). Which function appears to give the best fit?

6. Because the area of a circle is given by $A = \pi r^2$, would you expect the price of a cheese pizza to be quadratic or linear?

8.9   Polynomial and Rational Inequalities

Quadratic and Other Polynomial Inequalities   ▪   Rational Inequalities

Quadratic and Other Polynomial Inequalities

Inequalities like the following are called *polynomial inequalities*:

$$x^3 - 5x > x^2 + 7, \qquad 4x - 3 < 9, \qquad 5x^2 - 3x + 2 \geq 0.$$

Second-degree polynomial inequalities in one variable are called *quadratic inequalities*. To solve polynomial inequalities, we often focus attention on where the outputs of a polynomial function are positive and where they are negative.

EXAMPLE   1   Solve: $x^2 + 3x - 10 > 0$.

SOLUTION   Consider the "related" function $f(x) = x^2 + 3x - 10$. We are looking for those x-values for which $f(x) > 0$. Graphically, function values are positive when the graph is above the x-axis.

The graph of f opens upward since the leading coefficient is positive. Thus y-values are positive *outside* the interval formed by the x-intercepts. To find the intercepts, we set the polynomial equal to 0 and solve:

$$x^2 + 3x - 10 = 0$$
$$(x + 5)(x - 2) = 0$$
$$x + 5 = 0 \quad or \quad x - 2 = 0$$
$$x = -5 \quad or \quad x = 2. \qquad \text{The } x\text{-intercepts are } (-5, 0) \text{ and } (2, 0).$$

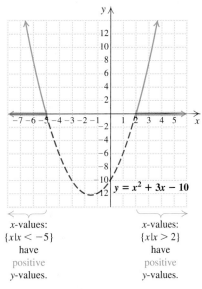

x-values:
$\{x | x < -5\}$
have
positive
y-values.

x-values:
$\{x | x > 2\}$
have
positive
y-values.

Thus the solution set of the inequality is

$$(-\infty, -5) \cup (2, \infty), \text{ or } \{x | x < -5 \text{ or } x > 2\}.$$ TRY EXERCISE ▶ 13

Any inequality with 0 on one side can be solved by considering a graph of the related function and finding intercepts as in Example 1. Sometimes the quadratic formula is needed to find the intercepts.

EXAMPLE **2** Solve: $x^2 - 2x \le 2$.

SOLUTION We first write the quadratic inequality in standard form, with 0 on one side:

$$x^2 - 2x - 2 \le 0. \qquad \text{This is equivalent to the original inequality.}$$

The graph of $f(x) = x^2 - 2x - 2$ is a parabola opening upward. Values of $f(x)$ are negative for x-values between the x-intercepts. We find the x-intercepts by solving $f(x) = 0$:

$$x = \frac{-b \pm \sqrt{b^2 - 4ac}}{2a}$$

$$= \frac{-(-2) \pm \sqrt{(-2)^2 - 4 \cdot 1(-2)}}{2 \cdot 1}$$

$$= \frac{2 \pm \sqrt{12}}{2} = \frac{2}{2} \pm \frac{2\sqrt{3}}{2} = 1 \pm \sqrt{3}.$$

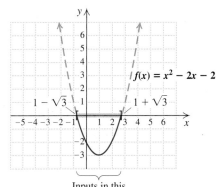

Inputs in this interval have negative or 0 outputs.

At the x-intercepts, $1 - \sqrt{3}$ and $1 + \sqrt{3}$, the value of $f(x)$ is 0. Since the inequality symbol is $\leq$, the solution set will include all values of x for which $f(x)$ is negative *or* $f(x)$ is 0. Thus the solution set of the inequality is

$$[1 - \sqrt{3}, 1 + \sqrt{3}], \quad \text{or} \quad \{x | 1 - \sqrt{3} \leq x \leq 1 + \sqrt{3}\}.$$

> **TRY EXERCISE** 21

In Example 2, it was not essential to draw the graph. The important information came from finding the x-intercepts and the sign of $f(x)$ on each side of those intercepts. We now solve a third-degree polynomial inequality, without graphing, by locating the x-intercepts, or **zeros**, of f and then using *test points* to determine the sign of $f(x)$ over each interval of the x-axis.

EXAMPLE 3

For $f(x) = 5x^3 + 10x^2 - 15x$, find all x-values for which $f(x) > 0$.

SOLUTION   We first solve the related equation:

$$
\begin{aligned}
f(x) &= 0 \\
5x^3 + 10x^2 - 15x &= 0 \qquad \text{Substituting} \\
5x(x^2 + 2x - 3) &= 0 \\
5x(x + 3)(x - 1) &= 0 \\
5x = 0 \quad or \quad x + 3 &= 0 \quad or \quad x - 1 = 0 \\
x = 0 \quad or \qquad x &= -3 \quad or \qquad x = 1.
\end{aligned}
$$

The zeros of f are $-3, 0$, and 1. These zeros divide the number line, or x-axis, into four intervals: A, B, C, and D.

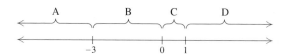

Next, selecting one convenient test value from each interval, we determine the sign of $f(x)$ for that interval. We know that, within each interval, the sign of $f(x)$ cannot change. If it did, there would need to be another zero in that interval. Using the factored form of $f(x)$ eases the computations:

$$f(x) = 5x(x + 3)(x - 1).$$

For interval A,

$$f(-4) = \underbrace{5(-4)}\,\underbrace{((-4) + 3)}\,\underbrace{((-4) - 1)}$$

-4 is a convenient value in interval A.

$$\underbrace{\text{Negative} \cdot \text{Negative} \cdot \text{Negative}}_{\text{Negative}}$$

Only the sign is important. The product of three negative numbers is negative, so $f(-4)$ is negative.

For interval B,

$$f(-1) = \underbrace{5(-1)}\,\underbrace{((-1) + 3)}\,\underbrace{((-1) - 1)}$$

-1 is a convenient value in interval B.

$$\underbrace{\text{Negative} \cdot \text{Positive} \cdot \text{Negative}}_{\text{Positive}}$$

$f(-1)$ is positive.

STUDENT NOTES

When we are evaluating test values, there is often no need to do lengthy computations since all we need to determine is the sign of the result.

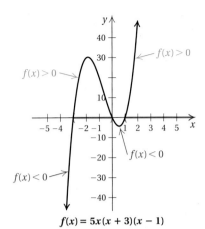

$f(x) = 5x(x + 3)(x − 1)$

A visualization of Example 3

For interval C,

$$f\left(\tfrac{1}{2}\right) = \underbrace{5 \cdot \tfrac{1}{2} \cdot \left(\tfrac{1}{2} + 3\right) \cdot \left(\tfrac{1}{2} − 1\right)}.$$ $\tfrac{1}{2}$ is a convenient value in interval C.

$$\underbrace{\text{Positive} \cdot \text{Positive} \cdot \text{Negative}}$$

$$\text{Negative}$$ $f\left(\tfrac{1}{2}\right)$ is negative.

For interval D,

$$f(2) = \underbrace{5 \cdot 2 \cdot (2 + 3) \cdot (2 − 1)}.$$ 2 is a convenient value in interval D.

$$\underbrace{\text{Positive} \cdot \text{Positive} \cdot \text{Positive}}$$ $f(2)$ is positive.

Recall that we are looking for all x for which $5x^3 + 10x^2 − 15x > 0$. The calculations above indicate that $f(x)$ is positive for any number in intervals B and D. The solution set of the original inequality is

$$(−3, 0) \cup (1, \infty), \quad \text{or} \quad \{x \mid −3 < x < 0 \, or \, x > 1\}.$$

TRY EXERCISE 29

The calculations in Example 3 were made simpler by using a factored form of the polynomial and by focusing on only the *sign* of $f(x)$. By looking at how many positive or negative factors are multiplied, we are able to determine the sign of the polynomial function.

> ### To Solve a Polynomial Inequality Using Factors
>
> 1. Add or subtract to get 0 on one side and solve the related polynomial equation by factoring.
> 2. Use the numbers found in step (1) to divide the number line into intervals.
> 3. Using a test value from each interval, determine the sign of the function over each interval. First find the sign of each factor, and then determine the sign of the product of the factors. Remember that the product of an odd number of negative numbers is negative.
> 4. Select the interval(s) for which the inequality is satisfied and write interval notation or set-builder notation for the solution set. Include endpoints of intervals when $\leq$ or $\geq$ is used.

EXAMPLE 4   For $f(x) = 4x^3 − 4x$, find all x-values for which $f(x) \leq 0$.

SOLUTION   We first solve the related equation:

Solve $f(x) = 0$.

$$f(x) = 0$$
$$4x^3 − 4x = 0$$
$$4x(x^2 − 1) = 0$$
$$4x(x + 1)(x − 1) = 0$$
$$4x = 0 \quad or \quad x + 1 = 0 \quad or \quad x − 1 = 0$$
$$x = 0 \quad or \quad x = −1 \quad or \quad x = 1.$$

Divide the number line into intervals.

The function f has zeros at $−1, 0$, and 1, so we divide the number line into four intervals:

The product $4x(x + 1)(x - 1)$ is positive or negative, depending on the signs of $4x$, $x + 1$, and $x - 1$. This can be determined by making a chart.

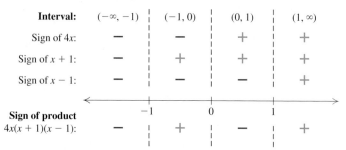

Determine the sign of the function over each interval.

A product is negative when it has an odd number of negative factors. Since the $\leq$ sign allows for equality, the endpoints -1, 0, and 1 are solutions. From the chart, we see that the solution set is

Select the interval(s) for which the inequality is satisfied.

$$(-\infty, -1] \cup [0, 1], \quad \text{or} \quad \{x \mid x \leq -1 \text{ or } 0 \leq x \leq 1\}.$$

> TRY EXERCISE ▶ 31

TECHNOLOGY CONNECTION

To solve $2.3x^2 \leq 9.11 - 2.94x$, we write the inequality in the form $2.3x^2 + 2.94x - 9.11 \leq 0$ and graph the function $f(x) = 2.3x^2 + 2.94x - 9.11$.

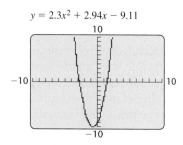

The region in which the graph lies *on or below* the x-axis begins somewhere between -3 and -2, and

continues to somewhere between 1 and 2. Using the ZERO option of CALC and rounding, we find that the endpoints are -2.73 and 1.45. The solution set is approximately $\{x \mid -2.73 \leq x \leq 1.45\}$.

Had the inequality been $2.3x^2 > 9.11 - 2.94x$, we would look for portions of the graph that lie *above* the x-axis. An approximate solution set of this inequality is $\{x \mid x < -2.73 \text{ or } x > 1.45\}$.

Use a graphing calculator to solve each inequality. Round the values of the endpoints to the nearest hundredth.

1. $4.32x^2 - 3.54x - 5.34 \leq 0$
2. $7.34x^2 - 16.55x - 3.89 \geq 0$
3. $10.85x^2 + 4.28x + 4.44$
   $> 7.91x^2 + 7.43x + 13.03$
4. $5.79x^3 - 5.68x^2 + 10.68x$
   $> 2.11x^3 + 16.90x - 11.69$

Rational Inequalities

Inequalities involving rational expressions are called **rational inequalities**. Like polynomial inequalities, rational inequalities can be solved using test values. Unlike polynomials, however, rational expressions often have values for which the expression is undefined. These values must be used when dividing the number line into intervals.

EXAMPLE 5   Solve: $\dfrac{x - 3}{x + 4} \geq 2$.

SOLUTION   We write the related equation by changing the $\geq$ symbol to $=$:

$$\frac{x - 3}{x + 4} = 2. \qquad \text{Note that } x \neq -4.$$

Next, we solve this related equation:

$$(x + 4) \cdot \frac{x - 3}{x + 4} = (x + 4) \cdot 2 \qquad \text{Multiplying both sides by the LCD, } x + 4$$

$$x - 3 = 2x + 8$$

$$-11 = x. \qquad \text{Solving for } x$$

Since -11 is a solution of the related equation, we use -11 when dividing the number line into intervals. Since the rational expression is undefined for $x = -4$, we also use -4:

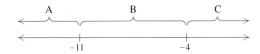

We test a number from each interval to see where the original inequality is satisfied:

$$\frac{x - 3}{x + 4} \geq 2.$$

For Interval A,

$$\text{Test } -15, \quad \frac{-15 - 3}{-15 + 4} = \frac{-18}{-11}$$

$$= \frac{18}{11} \not\geq 2. \qquad \begin{array}{l} -15 \text{ } is \text{ } not \text{ a solution, so interval A is} \\ \text{not part of the solution set.} \end{array}$$

For Interval B,

$$\text{Test } -8, \quad \frac{-8 - 3}{-8 + 4} = \frac{-11}{-4}$$

$$= \frac{11}{4} \geq 2. \qquad \begin{array}{l} -8 \text{ } is \text{ a solution, so interval B is part of} \\ \text{the solution set.} \end{array}$$

For Interval C,

$$\text{Test } 1, \quad \frac{1 - 3}{1 + 4} = \frac{-2}{5}$$

$$= -\frac{2}{5} \not\geq 2. \qquad \begin{array}{l} 1 \text{ } is \text{ } not \text{ a solution, so interval C is not} \\ \text{part of the solution set.} \end{array}$$

The solution set includes interval B. The endpoint -11 is included because the inequality symbol is $\geq$ and -11 is a solution of the related equation. The number -4 is *not* included because $(x - 3)/(x + 4)$ is undefined for $x = -4$. Thus the solution set of the original inequality is

$$[-11, -4), \quad \text{or} \quad \{x | -11 \leq x < -4\}. \qquad \text{▸ TRY EXERCISE} \triangleright 37$$

ALGEBRAIC–GRAPHICAL CONNECTION

To compare the algebraic solution of Example 5 with a graphical solution, we graph $f(x) = (x - 3)/(x + 4)$ and the line $y = 2$. The solutions of $(x - 3)/(x + 4) \geq 2$ are found by locating all x-values for which $f(x) \geq 2$.

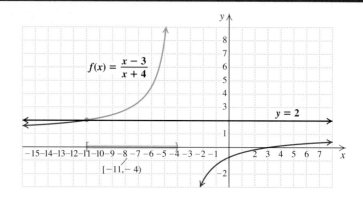

To Solve a Rational Inequality

1. Find any replacements for which any rational expression is undefined.
2. Change the inequality symbol to an equals sign and solve the related equation.
3. Use the numbers found in steps (1) and (2) to divide the number line into intervals.
4. Substitute a test value from each interval into the inequality. If the number is a solution, then the interval to which it belongs is part of the solution set.
5. Select the interval(s) and any endpoints for which the inequality is satisfied and use interval notation or set-builder notation for the solution set. If the inequality symbol is $\leq$ or $\geq$, then the solutions from step (2) are also included in the solution set. All numbers found in step (1) must be excluded from the solution set, even if they are solutions from step (2).

8.9   EXERCISE SET

🖐 *Concept Reinforcement*   *Classify each statement as either true or false.*

1. The solution of $(x - 3)(x + 2) \leq 0$ is $[-2, 3]$.

2. The solution of $(x + 5)(x - 4) \geq 0$ is $[-5, 4]$.

3. The solution of $(x - 1)(x - 6) > 0$ is $\{x | x < 1 \ or \ x > 6\}$.

4. The solution of $(x + 4)(x + 2) < 0$ is $(-4, -2)$.

5. To solve $\dfrac{x + 2}{x - 3} < 0$ using intervals, we divide the number line into the intervals $(-\infty, -2)$ and $(-2, \infty)$.

6. To solve $\dfrac{x - 5}{x + 4} \geq 0$ using intervals, we divide the number line into the intervals $(-\infty, -4)$, $(-4, 5)$, and $(5, \infty)$.

Solve each inequality using the graph provided.

7. $p(x) \le 0$

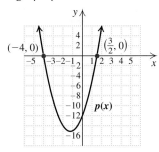

8. $p(x) < 0$

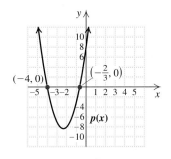

9. $x^4 + 12x > 3x^3 + 4x^2$

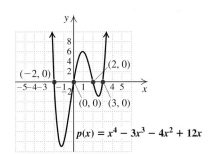

10. $x^4 + x^3 \ge 6x^2$

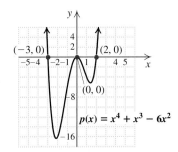

11. $\dfrac{x - 1}{x + 2} < 3$

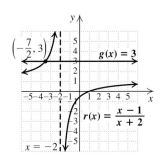

12. $\dfrac{2x - 1}{x - 5} \ge 1$

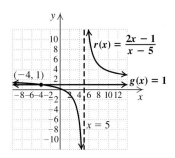

Solve.

13. $(x - 6)(x - 5) < 0$

14. $(x + 8)(x + 10) > 0$

15. $(x + 7)(x - 2) \ge 0$

16. $(x - 1)(x + 4) \le 0$

17. $x^2 - x - 2 > 0$

18. $x^2 + x - 2 < 0$

Aha! **19.** $x^2 + 4x + 4 < 0$

20. $x^2 + 6x + 9 < 0$

21. $x^2 - 4x \le 3$

22. $x^2 + 6x \ge 2$

23. $3x(x + 2)(x - 2) < 0$

24. $5x(x + 1)(x - 1) > 0$

25. $(x - 1)(x + 2)(x - 4) \ge 0$

26. $(x + 3)(x + 2)(x - 1) < 0$

27. For $f(x) = 7 - x^2$, find all x-values for which $f(x) \ge 3$.

28. For $f(x) = 14 - x^2$, find all x-values for which $f(x) > 5$.

29. For $g(x) = (x - 2)(x - 3)(x + 1)$, find all x-values for which $g(x) > 0$.

30. For $g(x) = (x + 3)(x - 2)(x + 1)$, find all x-values for which $g(x) < 0$.

31. For $F(x) = x^3 - 7x^2 + 10x$, find all x-values for which $F(x) \le 0$.

32. For $G(x) = x^3 - 8x^2 + 12x$, find all x-values for which $G(x) \ge 0$.

Solve.

33. $\dfrac{1}{x-5} < 0$

34. $\dfrac{1}{x+4} > 0$

35. $\dfrac{x+1}{x-3} \geq 0$

36. $\dfrac{x-2}{x+4} \leq 0$

37. $\dfrac{x+1}{x+6} \geq 1$

38. $\dfrac{x-1}{x-2} \leq 1$

39. $\dfrac{(x-2)(x+1)}{x-5} \leq 0$

40. $\dfrac{(x+4)(x-1)}{x+3} \geq 0$

41. $\dfrac{x}{x+3} \geq 0$

42. $\dfrac{x-2}{x} \leq 0$

43. $\dfrac{x-5}{x} < 1$

44. $\dfrac{x}{x-1} > 2$

45. $\dfrac{x-1}{(x-3)(x+4)} \leq 0$

46. $\dfrac{x+2}{(x-2)(x+7)} \geq 0$

47. For $f(x) = \dfrac{5-2x}{4x+3}$, find all x-values for which $f(x) \geq 0$.

48. For $g(x) = \dfrac{2+3x}{2x-4}$, find all x-values for which $g(x) \geq 0$.

49. For $G(x) = \dfrac{1}{x-2}$, find all x-values for which $G(x) \leq 1$.

50. For $F(x) = \dfrac{1}{x-3}$, find all x-values for which $F(x) \leq 2$.

51. Explain how any quadratic inequality can be solved by examining a parabola.

52. Describe a method for creating a quadratic inequality for which there is no solution.

Skill Review

To prepare for Section 9.1, review function notation (Section 2.1).

Graph each function. [2.1]

53. $f(x) = x^3 - 2$

54. $g(x) = \dfrac{2}{x}$

55. If $f(x) = x + 7$, find $f\left(\dfrac{1}{a^2}\right)$. [2.1]

56. If $g(x) = x^2 - 3$, find $g(\sqrt{a-5})$. [2.1], [7.1]

57. If $g(x) = x^2 + 2$, find $g(2a+5)$. [2.1], [5.2]

58. If $f(x) = \sqrt{4x+1}$, find $g(3a-5)$. [2.1]

Synthesis

59. Step (5) on p. 568 states that even when the inequality symbol is $\leq$ or $\geq$, the solutions from step (2) may not be part of the solution set. Why?

60. Describe a method that could be used to create a quadratic inequality that has $(-\infty, a] \cup [b, \infty)$ as the solution set. Assume $a < b$.

Find each solution set.

61. $x^2 + 2x < 5$

62. $x^4 + 2x^2 \geq 0$

63. $x^4 + 3x^2 \leq 0$

64. $\left|\dfrac{x+2}{x-1}\right| \leq 3$

65. *Total profit.* Derex, Inc., determines that its total-profit function is given by
$$P(x) = -3x^2 + 630x - 6000.$$
a) Find all values of x for which Derex makes a profit.
b) Find all values of x for which Derex loses money.

66. *Height of a thrown object.* The function
$$S(t) = -16t^2 + 32t + 1920$$
gives the height S, in feet, of an object thrown from a cliff that is 1920 ft high. Here t is the time, in seconds, that the object is in the air.
a) For what times does the height exceed 1920 ft?
b) For what times is the height less than 640 ft?

67. *Number of handshakes.* There are n people in a room. The number N of possible handshakes by the people is given by the function
$$N(n) = \dfrac{n(n-1)}{2}.$$
For what number of people n is $66 \leq N \leq 300$?

68. *Number of diagonals.* A polygon with n sides has D diagonals, where D is given by the function
$$D(n) = \dfrac{n(n-3)}{2}.$$
Find the number of sides n if
$$27 \leq D \leq 230.$$

Use a graphing calculator to graph each function and find solutions of $f(x) = 0$. Then solve the inequalities $f(x) < 0$ and $f(x) > 0$.

69. $f(x) = x^3 - 2x^2 - 5x + 6$

70. $f(x) = \dfrac{1}{3}x^3 - x + \dfrac{2}{3}$

71. $f(x) = x + \dfrac{1}{x}$

72. $f(x) = x - \sqrt{x}, x \geq 0$

73. $f(x) = \dfrac{x^3 - x^2 - 2x}{x^2 + x - 6}$

74. $f(x) = x^4 - 4x^3 - x^2 + 16x - 12$

Find the domain of each function

75. $f(x) = \sqrt{x^2 - 4x - 45}$

76. $f(x) = \sqrt{9 - x^2}$

77. $f(x) = \sqrt{x^2 + 8x}$

78. $f(x) = \sqrt{x^2 + 2x + 1}$

79. Describe a method that could be used to create a rational inequality that has $(-\infty, a] \cup (b, \infty)$ as the solution set. Assume $a < b$.

80. Use a graphing calculator to solve Exercises 43 and 49 by drawing two curves, one for each side of the inequality.

Study Summary

KEY TERMS AND CONCEPTS	EXAMPLES

SECTION 8.1: QUADRATIC EQUATIONS

A **quadratic equation in standard form** is written $ax^2 + bx + c = 0$, with a, b, and c constant and $a \neq 0$.

Some quadratic equations can be solved by factoring.

$$x^2 - 3x - 10 = 0$$
$$(x + 2)(x - 5) = 0$$
$$x + 2 = 0 \quad or \quad x - 5 = 0$$
$$x = -2 \quad or \quad x = 5$$

The Principle of Square Roots

For any real number k, if $X^2 = k$, then $X = \sqrt{k}$ or $X = -\sqrt{k}$.

$$x^2 - 8x + 16 = 25$$
$$(x - 4)^2 = 25$$
$$x - 4 = -5 \quad or \quad x - 4 = 5$$
$$x = -1 \quad or \quad x = 9$$

Any quadratic equation can be solved by **completing the square**.

$$x^2 + 6x = 1$$
$$x^2 + 6x + \left(\tfrac{6}{2}\right)^2 = 1 + \left(\tfrac{6}{2}\right)^2$$
$$x^2 + 6x + 9 = 1 + 9$$
$$(x + 3)^2 = 10$$
$$x + 3 = \pm\sqrt{10}$$
$$x = -3 \pm \sqrt{10}$$

SECTION 8.2: THE QUADRATIC FORMULA

The Quadratic Formula

The solutions of $ax^2 + bx + c = 0$ are given by

$$x = \frac{-b \pm \sqrt{b^2 - 4ac}}{2a}.$$

$3x^2 - 2x - 5 = 0 \qquad a = 3, b = -2, c = -5$

$$x = \frac{-(-2) \pm \sqrt{(-2)^2 - 4 \cdot 3(-5)}}{2 \cdot 3}$$

$$x = \frac{2 \pm \sqrt{4 + 60}}{6}$$

$$x = \frac{2 \pm \sqrt{64}}{6}$$

$$x = \frac{2 \pm 8}{6}$$

$$x = \frac{10}{6} = \frac{5}{3} \quad or \quad x = \frac{-6}{6} = -1$$

SECTION 8.3: STUDYING SOLUTIONS OF QUADRATIC EQUATIONS

The **discriminant** of the quadratic formula is $b^2 - 4ac$.

$b^2 - 4ac = 0 \rightarrow$ One solution; a rational number

For $4x^2 - 12x + 9 = 0$, $b^2 - 4ac = (-12)^2 - 4(4)(9)$

$$= 144 - 144 = 0. \quad \text{The discriminant is zero.}$$

Thus, $4x^2 - 12x + 9 = 0$ has one rational solution.

$b^2 - 4ac > 0 \rightarrow$ Two real solutions; both are rational if $b^2 - 4ac$ is a perfect square.

For $x^2 + 6x - 2 = 0$, $b^2 - 4ac = (6)^2 - 4(1)(-2)$

$$= 36 + 8 = 44. \quad \text{The discriminant is not a perfect square.}$$

Thus, $x^2 + 6x - 2 = 0$ has two irrational real-number solutions.

$b^2 - 4ac < 0 \rightarrow$ Two imaginary-number solutions

For $2x^2 - 3x + 5 = 0$, $b^2 - 4ac = (-3)^2 - 4(2)(5)$

$$= 9 - 40 = -31. \quad \text{The discriminant is negative.}$$

Thus, $2x^2 - 3x + 5 = 0$ has two imaginary-number solutions.

SECTION 8.4: APPLICATIONS INVOLVING QUADRATIC EQUATIONS

To solve a formula for a letter, use the same principles used to solve equations.

Solve $y = pn^2 + dn$ for n.

$pn^2 + dn - y = 0$ Writing standard form of a quadratic equation

$$n = \frac{-d \pm \sqrt{d^2 - 4p(-y)}}{2 \cdot p}$$ Using the quadratic formula; $a = p, b = d, c = -y$

$$n = \frac{-d \pm \sqrt{d^2 + 4py}}{2p}$$

SECTION 8.5: EQUATIONS REDUCIBLE TO QUADRATIC

Equations that are **reducible to quadratic** or in **quadratic form** can be solved by making an appropriate substitution.

$x^4 - 10x^2 + 9 = 0$ Let $u = x^2$. Then $u^2 = x^4$.

$u^2 - 10u + 9 = 0$ Substituting

$(u - 9)(u - 1) = 0$

$u - 9 = 0 \quad or \quad u - 1 = 0$

$u = 9 \quad or \quad u = 1$ Solving for u

$x^2 = 9 \quad or \quad x^2 = 1$

$x = \pm 3 \quad or \quad x = \pm 1$ Solving for x

SECTION 8.6: QUADRATIC FUNCTIONS AND THEIR GRAPHS
SECTION 8.7: MORE ABOUT GRAPHING QUADRATIC FUNCTIONS

The graph of a quadratic function

$$f(x) = ax^2 + bx + c = a(x - h)^2 + k$$

is a **parabola**. The graph opens upward for $a > 0$ and downward for $a < 0$.

The **vertex** is (h, k) and the **axis of symmetry** is $x = h$.

If $a > 0$, the function has a **minimum** value of k, and if $a < 0$, the function has a **maximum** value of k.

The vertex and the axis of symmetry occur at $x = -\dfrac{b}{2a}$.

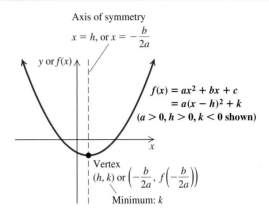

Axis of symmetry

$x = h$, or $x = -\dfrac{b}{2a}$

$f(x) = ax^2 + bx + c$
$= a(x - h)^2 + k$
$(a > 0, h > 0, k < 0$ shown$)$

Vertex
(h, k) or $\left(-\dfrac{b}{2a}, f\left(-\dfrac{b}{2a}\right)\right)$

Minimum: k

SECTION 8.8: PROBLEM SOLVING AND QUADRATIC FUNCTIONS

Some problem situations can be **modeled** using quadratic functions. For those problems, a quantity can often be **maximized** or **minimized** by finding the coordinates of a vertex.

A lifeguard has 100 m of roped-together flotation devices with which to cordon off a rectangular swimming area at North Beach. If the shoreline forms one side of the rectangle, what dimensions will maximize the size of the area for swimming?

This problem and its solution appear as Example 2 on pp. 553–554.

SECTION 8.9: POLYNOMIAL AND RATIONAL INEQUALITIES

The x-intercepts, or **zeros**, of a function are used to divide the x-axis into intervals when solving a **polynomial inequality**. (See p. 565.)

Solve: $x^2 - 2x - 15 > 0$.

$$x^2 - 2x - 15 = 0 \qquad \text{Solving the related equation}$$
$$(x - 5)(x + 3) = 0$$
$$x = 5 \quad or \quad x = -3 \qquad \begin{array}{l}-3 \text{ and } 5 \text{ divide the number line} \\ \text{into three intervals.}\end{array}$$

Since $f(x) = x^2 - 2x - 15 = (x - 5)(x + 3)$,

    $f(x)$ is positive for $x < -3$;

    $f(x)$ is negative for $-3 < x < 5$;

    $f(x)$ is positive for $x > 5$.

Thus, $x^2 - 2x - 15 > 0$ for $(-\infty, -3) \cup (5, \infty)$, or $\{x | x < -3 \ or \ x > 5\}$.

The solutions of a rational equation and any replacements that make a denominator zero are both used to divide the x-axis into intervals when solving a **rational inequality**. (See p. 568.)

Review Exercises: Chapter 8

🔖 *Concept Reinforcement*   *Classify each statement as either true or false.*

1. Every quadratic equation has two different solutions. [8.3]

2. Every quadratic equation has at least one solution. [8.3]

3. If an equation cannot be solved by completing the square, it cannot be solved by the quadratic formula. [8.2]

4. A negative discriminant indicates two imaginary-number solutions of a quadratic equation. [8.3]

5. The graph of $f(x) = 2(x + 3)^2 - 4$ has its vertex at $(3, -4)$. [8.6]

6. The graph of $g(x) = 5x^2$ has $x = 0$ as its axis of symmetry. [8.6]

7. The graph of $f(x) = -2x^2 + 1$ has no minimum value. [8.6]

8. The zeros of $g(x) = x^2 - 9$ are -3 and 3. [8.6]

9. If a quadratic function has two different imaginary-number zeros, the graph of the function has two x-intercepts. [8.7]

10. To solve a polynomial inequality, we often must solve a polynomial equation. [8.9]

Solve.

11. $9x^2 - 2 = 0$ [8.1]

12. $8x^2 + 6x = 0$ [8.1]

13. $x^2 - 12x + 36 = 9$ [8.1]

14. $x^2 - 4x + 8 = 0$ [8.2]

15. $x(3x + 4) = 4x(x - 1) + 15$ [8.2]

16. $x^2 + 9x = 1$ [8.2]

17. $x^2 - 5x - 2 = 0$. Use a calculator to approximate, to three decimal places, the solutions with rational numbers. [8.2]

18. Let $f(x) = 4x^2 - 3x - 1$. Find x such that $f(x) = 0$. [8.2]

Replace the blanks with constants to form a true equation. [8.1]

19. $x^2 - 18x +$ ____ $= (x -$ ____ $)^2$

20. $x^2 + \frac{3}{5}x +$ ____ $= \left(x +$ ____ $\right)^2$

21. Solve by completing the square. Show your work.
$$x^2 - 6x + 1 = 0 \quad [8.1]$$

22. \$2500 grows to \$2704 in 2 yr. Use the formula $A = P(1 + r)^t$ to find the interest rate. [8.1]

23. The U.S. Bank Tower in Los Angeles, California, is 1018 ft tall. Use $s = 16t^2$ to approximate how long it would take an object to fall from the top. [8.1]

For each equation, determine whether the solutions are real or imaginary. If they are real, specify whether they are rational or irrational. [8.3]

24. $x^2 + 3x - 6 = 0$ 25. $x^2 + 2x + 5 = 0$

26. Write a quadratic equation having the solutions $3i$ and $-3i$. [8.3]

27. Write a quadratic equation having -5 as its only solution. [8.3]

Solve. [8.4]

28. Horizons has a manufacturing plant located 300 mi from company headquarters. Their corporate pilot must fly from headquarters to the plant and back in 4 hr. If there is a 20-mph headwind going and a 20-mph tailwind returning, how fast must the plane be able to travel in still air?

29. Working together, Dani and Cheri can reply to a day's worth of customer-service e-mails in 4 hr. Working alone, Dani takes 6 hr longer than Cheri. How long would it take Cheri to reply to the e-mails alone?

30. Find all x-intercepts of the graph of $f(x) = x^4 - 13x^2 + 36$. [8.5]

Solve. [8.5]

31. $15x^{-2} - 2x^{-1} - 1 = 0$

32. $(x^2 - 4)^2 - (x^2 - 4) - 6 = 0$

33. a) Graph: $f(x) = -3(x + 2)^2 + 4$. [8.6]
    b) Label the vertex.
    c) Draw the axis of symmetry.
    d) Find the maximum or the minimum value.

34. For the function given by $f(x) = 2x^2 - 12x + 23$: [8.7]
    a) find the vertex and the axis of symmetry;
    b) graph the function.

35. Find any x-intercepts and the y-intercept of the graph of
$$f(x) = x^2 - 9x + 14. \quad [8.7]$$

36. Solve $N = 3\pi\sqrt{\dfrac{1}{p}}$ for p. [8.4]

37. Solve $2A + T = 3T^2$ for T. [8.4]

State whether each graph appears to represent a quadratic function or a linear function. [8.8]

38. **Increase in health insurance premiums**

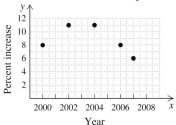

Source: Kaiser/HRET

39. **Health benefits in small firms**

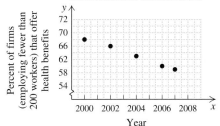

Source: Kaiser/HRET

40. Eastgate Consignments wants to build a rectangular area in a corner for children to play in while their parents shop. They have 30 ft of low fencing. What is the maximum area they can enclose? What dimensions will yield this area? [8.8]

41. The following table lists the percent increase in health insurance premiums x years after 2000. (See Exercise 38.) [8.8]

Years Since 2000	Percent Increase in Health Insurance Premiums
0	8
2	11
6	8

   a) Find the quadratic function that fits the data.
   b) Use the function to estimate the percent increase in health insurance premiums in 2005.

Solve. [8.9]

42. $x^3 - 3x > 2x^2$

43. $\dfrac{x - 5}{x + 3} \le 0$

Synthesis

44. Explain how the x-intercepts of a quadratic function can be used to help find the maximum or minimum value of the function. [8.7], [8.8]

45. Suppose that the quadratic formula is used to solve a quadratic equation. If the discriminant is a perfect square, could factoring have been used to solve the equation? Why or why not? [8.2], [8.3]

46. What is the greatest number of solutions that an equation of the form $ax^4 + bx^2 + c = 0$ can have? Why? [8.5]

47. Discuss two ways in which completing the square was used in this chapter. [8.1], [8.2], [8.7]

48. A quadratic function has x-intercepts at -3 and 5. If the y-intercept is at -7, find an equation for the function. [8.7]

49. Find h and k if, for $3x^2 - hx + 4k = 0$, the sum of the solutions is 20 and the product of the solutions is 80. [8.3]

50. The average of two positive integers is 171. One of the numbers is the square root of the other. Find the integers. [8.5]

Test: Chapter 8

Solve.

1. $25x^2 - 7 = 0$

2. $4x(x - 2) - 3x(x + 1) = -18$

3. $x^2 + 2x + 3 = 0$

4. $2x + 5 = x^2$

5. $x^{-2} - x^{-1} = \frac{3}{4}$

6. $x^2 + 3x = 5$. Use a calculator to approximate, to three decimal places, the solutions with rational numbers.

7. Let $f(x) = 12x^2 - 19x - 21$. Find x such that $f(x) = 0$.

Replace the blanks with constants to form a true equation.

8. $x^2 - 20x + \underline{\quad} = (x - \underline{\quad})^2$

9. $x^2 + \frac{2}{7}x + \underline{\quad} = (x + \underline{\quad})^2$

10. Solve by completing the square. Show your work.
$$x^2 + 10x + 15 = 0$$

11. Determine the type of number that the solutions of $x^2 + 2x + 5 = 0$ will be.

12. Write a quadratic equation having solutions $\sqrt{11}$ and $-\sqrt{11}$.

Solve.

13. The Connecticut River flows at a rate of 4 km/h for the length of a popular scenic route. In order for a cruiser to travel 60 km upriver and then return in a total of 8 hr, how fast must the boat be able to travel in still water?

14. Dal and Kim can assemble a swing set in $1\frac{1}{2}$ hr. Working alone, it takes Kim 4 hr longer than Dal to assemble the swing set. How long would it take Dal, working alone, to assemble the swing set?

15. Find all x-intercepts of the graph of
$$f(x) = x^4 - 15x^2 - 16.$$

16. **a)** Graph: $f(x) = 4(x - 3)^2 + 5$.
    b) Label the vertex.
    c) Draw the axis of symmetry.
    d) Find the maximum or the minimum function value.

17. For the function $f(x) = 2x^2 + 4x - 6$:
    a) find the vertex and the axis of symmetry;
    b) graph the function.

18. Find the x- and y-intercepts of
$$f(x) = x^2 - x - 6.$$

19. Solve $V = \frac{1}{3}\pi(R^2 + r^2)$ for r. Assume all variables are positive.

20. State whether the graph appears to represent a linear function, a quadratic function, or neither.

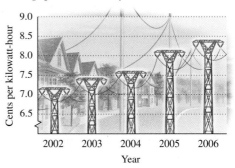

Average price of electricity

Source: Energy Information Administration, U.S. Department of Energy

21. Jay's Metals has determined that when x hundred storage cabinets are built, the average cost per cabinet is given by
$$C(x) = 0.2x^2 - 1.3x + 3.4025,$$
where $C(x)$ is in hundreds of dollars. What is the minimum cost per cabinet and how many cabinets should be built to achieve that minimum?

22. Find the quadratic function that fits the data points $(0, 0)$, $(3, 0)$, and $(5, 2)$.

Solve.

23. $x^2 + 5x < 6$

24. $x - \dfrac{1}{x} \geq 0$

Synthesis

25. One solution of $kx^2 + 3x - k = 0$ is -2. Find the other solution.

26. Find a fourth-degree polynomial equation, with integer coefficients, for which $-\sqrt{3}$ and $2i$ are solutions.

27. Solve: $x^4 - 4x^2 - 1 = 0$.

Cumulative Review: Chapters 1–8

Simplify.

1. $-3 \cdot 8 \div (-2)^3 \cdot 4 - 6(5 - 7)$ [1.2]

2. $\dfrac{18a^5bc^{10}}{24a^{-5}bc^3}$ [1.6]

3. $(5x^2y - 8xy - 6xy^2) - (2xy - 9x^2y + 3xy^2)$ [5.1]

4. $(9p^2q + 8t)(9p^2q - 8t)$ [5.2]

5. $\dfrac{t^2 - 25}{9t^2 + 24t + 16} \div \dfrac{3t^2 - 11t - 20}{t^2 + t}$ [6.1]

6. $\dfrac{1}{4 - x} + \dfrac{8}{x^2 - 16} - \dfrac{2}{x + 4}$ [6.2]

7. $\sqrt[3]{18x^4y} \cdot \sqrt[3]{6x^2y}$ [7.3]

8. $(3\sqrt{2} + i)(2\sqrt{2} - i)$ [7.8]

Factor.

9. $12x^4 - 75y^4$ [5.5]

10. $x^3 - 24x^2 + 80x$ [5.4]

11. $100m^6 - 100$ [5.6]

12. $6t^2 + 35t + 36$ [5.4]

Solve.

13. $2(5x - 3) - 8x = 4 - (3 - x)$ [1.3]

14. $2(5x - 3) - 8x < 4 - (3 - x)$ [4.1]

15. $2x - 6y = 3,$
$-3x + 8y = -5$ [3.2]

16. $x(x - 5) = 66$ [5.8]

17. $\dfrac{2}{t} + \dfrac{1}{t - 1} = 2$ [6.4]

18. $\sqrt{x} = 1 + \sqrt{2x - 7}$ [7.6]

19. $m^2 + 10m + 25 = 2$ [8.1]

20. $3x^2 + 1 = x$ [8.2]

Graph.

21. $9x - 2y = 18$ [2.4]

22. $x < \frac{1}{2}y$ [4.4]

23. $y = 2(x - 3)^2 + 1$ [8.6]

24. $f(x) = x^2 + 4x + 3$ [8.7]

25. Find an equation in slope–intercept form whose graph has slope -5 and y-intercept $\left(0, \frac{1}{2}\right)$. [2.3]

26. Find the slope of the line containing $(8, 3)$ and $(-2, 10)$. [2.3]

27. For the function described by $f(x) = 3x^2 - 8x - 7$, find $f(-2)$. [5.1]

Find the domain of each function.

28. $f(x) = \sqrt{10 - x}$ [4.1]

29. $f(x) = \dfrac{x + 3}{x - 4}$ [2.2], [4.2]

Solve each formula for the specified letter.

30. $b = \dfrac{a + c}{2a}$, for a [6.8]

31. $p = 2\sqrt{\dfrac{r}{3t}}$, for t [8.4]

Solve.

32. *Mobile ad spending.* The amount spent worldwide in advertising on mobile devices can be estimated by $f(x) = 0.4x^2 + 0.01x + 0.9$, where x is the number of years after 2005 and $f(x)$ is in billions of dollars.
Source: Based on data from eMarketer

a) How much was spent worldwide for mobile ads in 2008? [5.1]
b) When will worldwide mobile ad spending reach $41 billion? [5.8], [8.4]

33. *Wi-fi hotspots.* The number of Wi-fi hotspots worldwide grew from 19,000 in 2002 to 118,000 in 2005. Let $h(t)$ represent the number of hotspots, in thousands, t years after 2000. [2.5]
Source: IDC

a) Find a linear function that fits the data.
b) Use the function from part (a) to predict the number of Wi-fi hotspots in 2010.
c) In what year will there be 500,000 Wi-fi hotspots?

34. *Gold prices.*   Annette is selling some of her gold jewelry. She has 4 bracelets and 1 necklace that weigh a total of 3 oz.   [1.4]

a) Annette's jewelry is 58% gold. How many ounces of gold does her jewelry contain?

b) A gold dealer offers Annette $1044 for the jewelry. How much per ounce of gold was she offered?

c) The price of gold at the time of Annette's sale was $800 an ounce. What percent of the gold price was she offered for her jewelry?

35. *Education.*   Sven ordered number tiles at $9 per set and alphabet tiles at $15 per set for his classroom. He ordered a total of 36 sets for $384. How many sets of each did he order?   [3.3]

36. *Minimizing cost.*   Dormitory Furnishings has determined that when x bunk beds are built, the average cost, in dollars, per bunk bed can be estimated by $c(x) = 0.004375x^2 - 3.5x + 825$. What is the minimum average cost per bunk bed and how many bunk beds should be built to achieve that minimum?   [8.8]

37. *Volunteer work.*   It takes Deanna twice as long to set up a fundraising auction as it takes Donna. Together they can set up for the auction in 4 hr. How long would it take each of them to do the job alone?   [6.5]

38. *Canoeing.*   Kent paddled for 2 hr with a 5-km/h current to reach a campsite. The return trip against the same current took 7 hr. Find the speed of Kent's canoe in still water.   [3.3]

39. *Truck rentals.*   Josh and Lindsay plan to rent a moving truck. The truck costs $70 plus 40¢ per mile. They have budgeted $90 for the truck rental. For what mileages will they not exceed their budget?   [4.1]

Synthesis

Solve.

40. $\dfrac{\dfrac{1}{x}}{2 + \dfrac{1}{x-1}} = 3$ [6.3], [8.2]

41. $x^4 + 5x^2 \leq 0$ [8.9]

42. The graph of the function $f(x) = mx + b$ contains the point $(2, 3)$ and is perpendicular to the line containing the points $(-1, 4)$ and $(-2, 5)$. Find the equation of the function.   [2.5]

43. Find the points of intersection of the graphs of $f(x) = x^2 + 8x + 1$ and $g(x) = 10x + 6$.   [8.2]

Exponential and Logarithmic Functions

PAUL WILLIAMS
DATA SYSTEMS SPECIALIST
Indianapolis, Indiana

Because I work with computers and networking every day, everything around me is based on math. Computers and their algorithms are based in the binary number system. We use mathematical tools such as statistics and graphs to study network bottlenecks in computers and transmission pipes and to detect network saturation points.

AN APPLICATION

In 2000, there were approximately 12 million text messages sent each month in the United States. This number has increased exponentially at an average rate of 108% per year. Find the exponential growth function that models the data, and estimate the number of text messages sent each month in 2009.

Source: Based on information from www.cellulist.com

This problem appears as Example 4 in Section 9.7.

T he functions that we consider in this chapter have rich applications in many fields, such as epidemiology (the study of the spread of disease), population growth, and marketing.

The theory centers on functions with variable exponents (*exponential functions*). Results follow from those functions, their properties, and properties of their closely related *inverse* functions.

9.1   Composite and Inverse Functions

Composite Functions ■ Inverses and One-to-One Functions ■ Finding Formulas for Inverses ■ Graphing Functions and Their Inverses ■ Inverse Functions and Composition

Later in this chapter, we introduce two closely related types of functions: exponential and logarithmic functions. In order to properly understand the link between these functions, we must first understand composite and inverse functions.

Composite Functions

In the real world, functions frequently occur in which some quantity depends on a variable that, in turn, depends on another variable. For instance, a firm's profits may depend on the number of items the firm produces, which may in turn depend on the number of employees hired. Functions like this are called **composite functions**.

For example, the function g that gives a correspondence between women's shoe sizes in the United States and those in Britain is given by $g(x) = x - 2$, where x is the U.S. size and $g(x)$ is the British size. Thus a U.S. size 4 corresponds to a shoe size of $g(4) = 4 - 2$, or 2, in Britain.

A second function converts women's shoe sizes in Britain to those in Italy. This particular function is given by $f(x) = 2x + 28$, where x is the British size and $f(x)$ is the corresponding Italian size. Thus a British size 2 corresponds to an Italian size $f(2) = 2 \cdot 2 + 28$, or 32.

It is correct to conclude that a U.S. size 4 corresponds to an Italian size 32 and that some function h describes this correspondence.

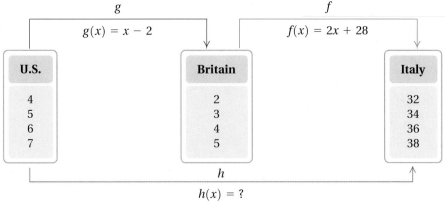

STUDENT NOTES ———

Throughout this chapter, keep in mind that equations such as $g(x) = x - 2$ and $g(t) = t - 2$ describe the same function g. Both equations tell us to find a function value by subtracting 2 from the input.

Size x shoes in the United States correspond to size $g(x)$ shoes in Britain, where

$$g(x) = x - 2.$$

Size n shoes in Britain correspond to size $f(n)$ shoes in Italy. Similarly, size $g(x)$ shoes in Britain correspond to size $f(g(x))$ shoes in Italy. Since the x in the expression $f(g(x))$ represents a U.S. shoe size, we can find the Italian shoe size that corresponds to a U.S. size x as follows:

$$f(g(x)) = f(x - 2) = 2(x - 2) + 28 \qquad \text{Using } g(x) \text{ as an input}$$
$$= 2x - 4 + 28 = 2x + 24.$$

This gives a formula for h: $h(x) = 2x + 24$. Thus U.S. size 4 corresponds to Italian size $h(4) = 2(4) + 24$, or 32. We call h the *composition* of f and g and denote it by $f \circ g$ (read "the composition of f and g," "f composed with g," or "f circle g").

Composition of Functions

The *composite function* $f \circ g$, the *composition* of f and g, is defined as

$$(f \circ g)(x) = f(g(x)).$$

We can visualize the composition of functions as follows.

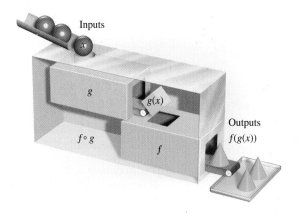

EXAMPLE **1**   Given $f(x) = 3x$ and $g(x) = 1 + x^2$:

a) Find $(f \circ g)(5)$ and $(g \circ f)(5)$.   **b)** Find $(f \circ g)(x)$ and $(g \circ f)(x)$.

SOLUTION   Consider each function separately:

$$f(x) = 3x \qquad \text{This function multiplies each input by 3.}$$

and

$$g(x) = 1 + x^2. \qquad \text{This function adds 1 to the square of each input.}$$

a) To find $(f \circ g)(5)$, we find $g(5)$ and then use that as an input for f:

$$(f \circ g)(5) = f(g(5)) = f(1 + 5^2) \qquad \text{Using } g(x) = 1 + x^2$$
$$= f(26) = 3 \cdot 26 = 78. \qquad \text{Using } f(x) = 3x$$

To find $(g \circ f)(5)$, we find $f(5)$ and then use that as an input for g:

$$(g \circ f)(5) = g(f(5)) = g(3 \cdot 5) \qquad \text{Note that } f(5) = 3 \cdot 5 = 15.$$
$$= g(15) = 1 + 15^2 = 1 + 225 = 226.$$

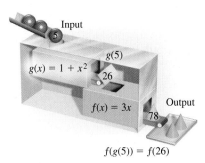

A composition machine for Example 1

b) We find $(f \circ g)(x)$ by substituting $g(x)$ for x in the equation for $f(x)$:

$$(f \circ g)(x) = f(g(x)) = f(1 + x^2) \qquad \text{Using } g(x) = 1 + x^2$$
$$= 3 \cdot (1 + x^2) = 3 + 3x^2. \qquad \text{Using } f(x) = 3x$$

To find $(g \circ f)(x)$, we substitute $f(x)$ for x in the equation for $g(x)$:

$$(g \circ f)(x) = g(f(x)) = g(3x) \qquad \text{Substituting } 3x \text{ for } f(x)$$
$$= 1 + (3x)^2 = 1 + 9x^2.$$

We can now find the function values of part (a) using the functions of part (b):

$$(f \circ g)(5) = 3 + 3(5)^2 = 3 + 3 \cdot 25 = 78;$$
$$(g \circ f)(5) = 1 + 9(5)^2 = 1 + 9 \cdot 25 = 226.$$ **TRY EXERCISE** 9

Example 1 shows that, in general, $(f \circ g)(x) \neq (g \circ f)(x)$.

EXAMPLE   2   Given $f(x) = \sqrt{x}$ and $g(x) = x - 1$, find $(f \circ g)(x)$ and $(g \circ f)(x)$.

SOLUTION

$$(f \circ g)(x) = f(g(x)) = f(x - 1) = \sqrt{x - 1}; \qquad \text{Using } g(x) = x - 1$$
$$(g \circ f)(x) = g(f(x)) = g(\sqrt{x}) = \sqrt{x} - 1 \qquad \text{Using } f(x) = \sqrt{x}$$

TRY EXERCISE 15

In fields ranging from chemistry to geology and economics, one needs to recognize how a function can be regarded as the composition of two "simpler" functions. This is sometimes called *de*composition.

EXAMPLE   3   If $h(x) = (7x + 3)^2$, find f and g such that $h(x) = (f \circ g)(x)$.

SOLUTION   We can think of $h(x)$ as the result of first finding $7x + 3$ and then squaring that. This suggests that $g(x) = 7x + 3$ and $f(x) = x^2$. We check by forming the composition:

$$(f \circ g)(x) = f(g(x))$$
$$= f(7x + 3) = (7x + 3)^2 = h(x), \text{ as desired.}$$

This may be the most "obvious" solution, but there are other less obvious answers. For example, if $f(x) = (x - 1)^2$ and $g(x) = 7x + 4$, then

$$(f \circ g)(x) = f(g(x)) = f(7x + 4)$$
$$= (7x + 4 - 1)^2 = (7x + 3)^2 = h(x).$$ **TRY EXERCISE** 21

TECHNOLOGY CONNECTION

In Example 3, we see that if $g(x) = 7x + 3$ and $f(x) = x^2$, then $f(g(x)) = (7x + 3)^2$. One way to show this is to let $y_1 = 7x + 3$ and $y_2 = x^2$. If we let $y_3 = (7x + 3)^2$ and $y_4 = y_2(y_1)$, we can use graphs or a table to show that $y_3 = y_4$.

1. Check Example 2 by using the above approach.

Inverses and One-to-One Functions

Let's view the following two functions as relations, or correspondences.

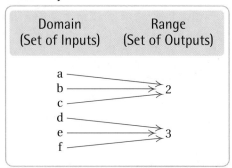

Countries and Their Capitals

Domain (Set of Inputs)	Range (Set of Outputs)
Australia	Canberra
China	Beijing
Germany	Berlin
Madagascar	Antananaviro
Turkey	Ankara
United States	Washington, D.C.

Phone Keys

Domain (Set of Inputs)	Range (Set of Outputs)
a	
b	2
c	
d	
e	3
f	

Suppose we reverse the arrows. We obtain what is called the **inverse relation**. Are these inverse relations functions?

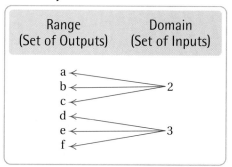

Countries and Their Capitals

Range (Set of Outputs)	Domain (Set of Inputs)
Australia ←	Canberra
China ←	Beijing
Germany ←	Berlin
Madagascar ←	Antananaviro
Turkey ←	Ankara
United States ←	Washington, D.C.

Phone Keys

Range (Set of Outputs)	Domain (Set of Inputs)
a ←	
b ←	2
c ←	
d ←	
e ←	3
f ←	

Recall that for each input, a function has exactly one output. However, it is possible for different inputs to correspond to the same output. Only when this possibility is *excluded* will the inverse be a function. For the functions listed above, this means the inverse of the "Capitals" correspondence is a function, but the inverse of the "Phone Keys" correspondence is not.

In the Capitals function, each input has its own output, so it is a **one-to-one-function**. In the Phone Keys function, a and b are both paired with 2. Thus the Phone Keys function is not a one-to-one function.

One-To-One Function

A function f is *one-to-one* if different inputs have different outputs. That is, if for a and b in the domain of f with $a \neq b$, we have $f(a) \neq f(b)$, then f is one-to-one. If a function is one-to-one, then its inverse correspondence is also a function.

How can we tell graphically whether a function is one-to-one?

EXAMPLE   4   At left is the graph of a function similar to those we will study in Section 9.2. Determine whether the function is one-to-one and thus has an inverse that is a function.

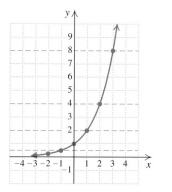

SOLUTION   A function is one-to-one if different inputs have different outputs—that is, if no two x-values have the same y-value. For this function, we cannot find two x-values that have the same y-value. Note that this means that no horizontal line can be drawn so that it crosses the graph more than once. The function is one-to-one so its inverse is a function.

TRY EXERCISE ▶ 31

The graph of every function must pass the vertical-line test. In order for a function to have an inverse that is a function, it must pass the *horizontal-line test* as well.

The Horizontal-Line Test

If it is impossible to draw a horizontal line that intersects a function's graph more than once, then the function is one-to-one. For every one-to-one function, an inverse function exists.

EXAMPLE 5   Determine whether the function $f(x) = x^2$ is one-to-one and thus has an inverse that is a function.

SOLUTION   The graph of $f(x) = x^2$ is shown here. Many horizontal lines cross the graph more than once. For example, the line $y = 4$ crosses where the first coordinates are -2 and 2. Although these are different inputs, they have the same output. That is, $-2 \neq 2$, but

$$f(-2) = (-2)^2 = 4 = 2^2 = f(2).$$

Thus the function is not one-to-one and no inverse function exists.

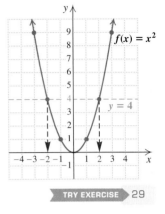

TRY EXERCISE 29

Finding Formulas for Inverses

When the inverse of f is also a function, it is denoted f^{-1} (read "f-inverse").

CAUTION!   The -1 in f^{-1} is *not* an exponent!

Suppose a function is described by a formula. If its inverse is a function, how do we find a formula for that inverse? For any equation in two variables, if we interchange the variables, we form an equation of the inverse correspondence. If it is a function, we proceed as follows to find a formula for f^{-1}.

To Find a Formula for f^{-1}

First make sure that f is one-to-one. Then:

1. Replace $f(x)$ with y.
2. Interchange x and y. (This gives the inverse function.)
3. Solve for y.
4. Replace y with $f^{-1}(x)$. (This is inverse function notation.)

EXAMPLE 6   Determine whether each function is one-to-one and if it is, find a formula for $f^{-1}(x)$.

a) $f(x) = x + 2$ **b)** $f(x) = 2x - 3$

SOLUTION

a) The graph of $f(x) = x + 2$ is shown at left. It passes the horizontal-line test, so it is one-to-one. Thus its inverse is a function.

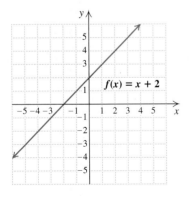

1. Replace $f(x)$ with y: $y = x + 2$.
2. Interchange x and y: $x = y + 2$.   This gives the inverse function.
3. Solve for y: $x - 2 = y$.
4. Replace y with $f^{-1}(x)$: $f^{-1}(x) = x - 2$.   We also "reversed" the equation.

In this case, the function f adds 2 to all inputs. Thus, to "undo" f, the function f^{-1} must subtract 2 from its inputs.

b) The function $f(x) = 2x - 3$ is also linear. Any linear function that is not constant will pass the horizontal-line test. Thus, f is one-to-one.

1. Replace $f(x)$ with y: $y = 2x - 3$.

2. Interchange x and y: $x = 2y - 3$.

3. Solve for y: $x + 3 = 2y$

$$\frac{x + 3}{2} = y.$$

4. Replace y with $f^{-1}(x)$: $f^{-1}(x) = \dfrac{x + 3}{2}$.

In this case, the function f doubles all inputs and then subtracts 3. Thus, to "undo" f, the function f^{-1} adds 3 to each input and then divides by 2.

> TRY EXERCISE 35

Graphing Functions and Their Inverses

How do the graphs of a function and its inverse compare?

EXAMPLE **7**   Graph $f(x) = 2x - 3$ and $f^{-1}(x) = (x + 3)/2$ on the same set of axes. Then compare.

SOLUTION   The graph of each function follows. Note that the graph of f^{-1} can be drawn by reflecting the graph of f across the line $y = x$. That is, if we graph $f(x) = 2x - 3$ in wet ink and fold the paper along the line $y = x$, the graph of $f^{-1}(x) = (x + 3)/2$ will appear as the impression made by f.

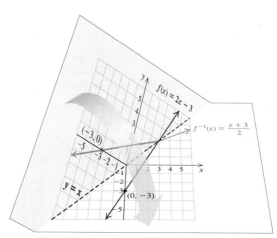

When x and y are interchanged to find a formula for the inverse, we are, in effect, reflecting or flipping the graph of $f(x) = 2x - 3$ across the line $y = x$. For example, when $(0, -3)$, the coordinates of the y-intercept of the graph of f, are reversed, we get $(-3, 0)$, the x-intercept of the graph of f^{-1}.

> TRY EXERCISE 59

Visualizing Inverses

The graph of f^{-1} is a reflection of the graph of f across the line $y = x$.

EXAMPLE **8**   Consider $g(x) = x^3 + 2$.

a) Determine whether the function is one-to-one.

b) If it is one-to-one, find a formula for its inverse.

c) Graph the inverse, if it exists.

SOLUTION

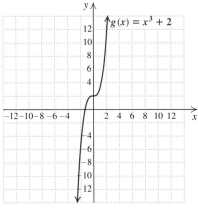

a) The graph of $g(x) = x^3 + 2$ is shown at right. It passes the horizontal-line test and thus is one-to-one and has an inverse that is a function.

b) **1.** Replace $g(x)$ with y: $y = x^3 + 2$.   Using $g(x) = x^3 + 2$

   2. Interchange x and y: $x = y^3 + 2$.

   3. Solve for y: $x - 2 = y^3$

                                 $\sqrt[3]{x - 2} = y$.   Each real number has only one cube root, so we can solve for y.

   4. Replace y with $g^{-1}(x)$: $g^{-1}(x) = \sqrt[3]{x - 2}$.

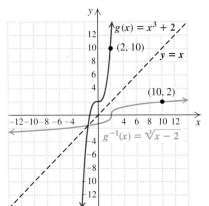

c) To graph g^{-1}, we can reflect the graph of $g(x) = x^3 + 2$ across the line $y = x$, as we did in Example 7. We also could graph $g^{-1}(x) = \sqrt[3]{x - 2}$ by plotting points. Note that $(2, 10)$ is on the graph of g, whereas $(10, 2)$ is on the graph of g^{-1}. The graphs of g and g^{-1} are shown at left. **TRY EXERCISE** 61

Inverse Functions and Composition

Let's consider inverses of functions in terms of function machines. Suppose that a one-to-one function f is programmed into a machine. If the machine is run in reverse, it will perform the inverse function f^{-1}. Inputs then enter at the opposite end, and the entire process is reversed.

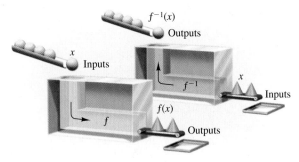

Consider $f(x) = x^3 + 2$ and $f^{-1}(x) = \sqrt[3]{x - 2}$ from Example 8. For the input 3,

$$f(3) = 3^3 + 2 = 27 + 2 = 29.$$

The output is 29. Let's now use 29 for the input in the inverse:

$$f^{-1}(29) = \sqrt[3]{29 - 2} = \sqrt[3]{27} = 3.$$

The function f takes 3 to 29. The inverse function f^{-1} takes the number 29 back to 3.

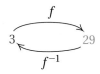

In general, if f is one-to-one, then f^{-1} takes the output $f(x)$ back to x. Similarly, f takes the output $f^{-1}(x)$ back to x.

Composition and Inverses

If a function f is one-to-one, then f^{-1} is the unique function for which

$$(f^{-1} \circ f)(x) = f^{-1}(f(x)) = x \quad \text{and} \quad (f \circ f^{-1})(x) = f(f^{-1}(x)) = x.$$

EXAMPLE 9   Let $f(x) = 2x + 1$. Show that

$$f^{-1}(x) = \frac{x - 1}{2}.$$

SOLUTION   We find $(f^{-1} \circ f)(x)$ and $(f \circ f^{-1})(x)$ and check to see that each is x.

$$(f^{-1} \circ f)(x) = f^{-1}(f(x)) = f^{-1}(2x + 1)$$

$$= \frac{(2x + 1) - 1}{2}$$

$$= \frac{2x}{2} = x \qquad \text{Thus, } (f^{-1} \circ f)(x) = x.$$

$$(f \circ f^{-1})(x) = f(f^{-1}(x)) = f\left(\frac{x - 1}{2}\right)$$

$$= 2 \cdot \frac{x - 1}{2} + 1$$

$$= x - 1 + 1 = x \qquad \text{Thus, } (f \circ f^{-1})(x) = x.$$

TRY EXERCISE ▶ 69

TECHNOLOGY CONNECTION

To determine whether $y_1 = 2x + 6$ and $y_2 = \frac{1}{2}x - 3$ are inverses of each other, we can graph both functions, along with the line $y = x$, on a "squared" set of axes. It *appears* that y_1 and y_2 are inverses of each other. A more precise check is achieved by selecting the DRAWINV option of the (DRAW) menu. The resulting graph of the inverse of y_1 should coincide with y_2.

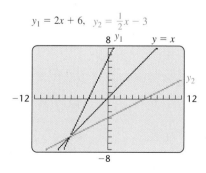

$y_1 = 2x + 6, \quad y_2 = \frac{1}{2}x - 3$

For a more dependable check, examine a TABLE in which $y_1 = 2x + 6$ and $y_2 = \frac{1}{2} \cdot y_1 - 3$. Note that y_2 "undoes" what y_1 does.

TBLSTART $= -3$ ΔTBL $= 1$ $y_2 = \frac{1}{2}y_1 - 3$

X	Y1	Y2
−3	0	−3
−2	2	−2
−1	4	−1
0	6	0
1	8	1
2	10	2
3	12	3

X = 3

1. Use a graphing calculator to check Examples 7, 8, and 9.
2. Will DRAWINV work for *any* choice of y_1? Why or why not?

9.1   **EXERCISE SET**

Concept Reinforcement   *Classify each statement as either true or false.*

1. The composition of two functions f and g is written $f \circ g$.

2. The notation $(f \circ g)(x)$ means $f(g(x))$.

3. If $f(x) = x^2$ and $g(x) = x + 3$, then $(g \circ f)(x) = (x + 3)^2$.

4. For any function h, there is only one way to decompose the function as $h = f \circ g$.

5. The function f is one-to-one if $f(1) = 1$.

6. The -1 in f^{-1} is an exponent.

7. The function f is the inverse of f^{-1}.

8. If g and h are inverses of each other, then $(g \circ h)(x) = x$.

For each pair of functions, find **(a)** $(f \circ g)(1)$;
(b) $(g \circ f)(1)$; **(c)** $(f \circ g)(x)$; **(d)** $(g \circ f)(x)$.

9. $f(x) = x^2 + 1$; $g(x) = x - 3$

10. $f(x) = x + 4$; $g(x) = x^2 - 5$

11. $f(x) = 5x + 1$; $g(x) = 2x^2 - 7$

12. $f(x) = 3x^2 + 4$; $g(x) = 4x - 1$

13. $f(x) = x + 7$; $g(x) = 1/x^2$

14. $f(x) = 1/x^2$; $g(x) = x + 2$

15. $f(x) = \sqrt{x}$; $g(x) = x + 3$

16. $f(x) = 10 - x$; $g(x) = \sqrt{x}$

17. $f(x) = \sqrt{4x}$; $g(x) = 1/x$

18. $f(x) = \sqrt{x + 3}$; $g(x) = 13/x$

19. $f(x) = x^2 + 4$; $g(x) = \sqrt{x - 1}$

20. $f(x) = x^2 + 8$; $g(x) = \sqrt{x + 17}$

Find $f(x)$ and $g(x)$ such that $h(x) = (f \circ g)(x)$. Answers may vary.

21. $h(x) = (3x - 5)^4$

22. $h(x) = (2x + 7)^3$

23. $h(x) = \sqrt{9x + 1}$

24. $h(x) = \sqrt[3]{4x - 5}$

25. $h(x) = \dfrac{6}{5x - 2}$

26. $h(x) = \dfrac{3}{x} + 4$

Determine whether each function is one-to-one.

27. $f(x) = -x$

28. $f(x) = x + 5$

29. $f(x) = x^2 + 3$

30. $f(x) = 3 - x^2$

31.
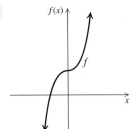

32.
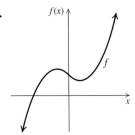

33.
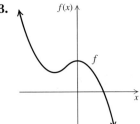

34.
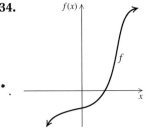

For each function, **(a)** *determine whether it is one-to-one;* **(b)** *if it is one-to-one, find a formula for the inverse.*

35. $f(x) = x + 3$

36. $f(x) = x + 2$

37. $f(x) = 2x$

38. $f(x) = 3x$

39. $g(x) = 3x - 1$

40. $g(x) = 2x - 3$

41. $f(x) = \dfrac{1}{2}x + 1$

42. $f(x) = \dfrac{1}{3}x + 2$

43. $g(x) = x^2 + 5$

44. $g(x) = x^2 - 4$

45. $h(x) = -10 - x$

46. $h(x) = 7 - x$

Aha! 47. $f(x) = \dfrac{1}{x}$

48. $f(x) = \dfrac{3}{x}$

49. $g(x) = 1$

50. $h(x) = 8$

51. $f(x) = \dfrac{2x + 1}{3}$

52. $f(x) = \dfrac{3x + 2}{5}$

53. $f(x) = x^3 + 5$

54. $f(x) = x^3 - 4$

55. $g(x) = (x - 2)^3$

56. $g(x) = (x + 7)^3$

57. $f(x) = \sqrt{x}$

58. $f(x) = \sqrt{x - 1}$

Graph each function and its inverse using the same set of axes.

59. $f(x) = \dfrac{2}{3}x + 4$

60. $g(x) = \dfrac{1}{4}x + 2$

61. $f(x) = x^3 + 1$

62. $f(x) = x^3 - 1$

63. $g(x) = \dfrac{1}{2}x^3$

64. $g(x) = \dfrac{1}{3}x^3$

65. $F(x) = -\sqrt{x}$

66. $f(x) = \sqrt{x}$

67. $f(x) = -x^2, x \geq 0$

68. $f(x) = x^2 - 1, x \leq 0$

69. Let $f(x) = \sqrt[3]{x - 4}$. Show that
$$f^{-1}(x) = x^3 + 4.$$

70. Let $f(x) = 3/(x + 2)$. Show that
$$f^{-1}(x) = \dfrac{3}{x} - 2.$$

71. Let $f(x) = (1 - x)/x$. Show that
$$f^{-1}(x) = \dfrac{1}{x + 1}.$$

72. Let $f(x) = x^3 - 5$. Show that
$$f^{-1}(x) = \sqrt[3]{x + 5}.$$

73. *Dress sizes in the United States and Italy.* A size-6 dress in the United States is size 36 in Italy. A function that converts dress sizes in the United States to those in Italy is
$$f(x) = 2(x + 12).$$
a) Find the dress sizes in Italy that correspond to sizes 8, 10, 14, and 18 in the United States.
b) Determine whether f has an inverse that is a function. If so, find a formula for the inverse.

c) Use the inverse function to find dress sizes in the United States that correspond to sizes 40, 44, 52, and 60 in Italy.

74. *Dress sizes in the United States and France.* A size-6 dress in the United States is size 38 in France. A function that converts dress sizes in the United States to those in France is
$$f(x) = x + 32.$$
a) Find the dress sizes in France that correspond to sizes 8, 10, 14, and 18 in the United States.
b) Determine whether f has an inverse that is a function. If so, find a formula for the inverse.
c) Use the inverse function to find dress sizes in the United States that correspond to sizes 40, 42, 46, and 50 in France.

75. Is there a one-to-one relationship between items in a store and the price of each of those items? Why or why not?

76. Mathematicians usually try to select "logical" words when forming definitions. Does the term "one-to-one" seem logical? Why or why not?

Skill Review

To prepare for Section 9.2, review simplifying exponential expressions and graphing equations (Sections 1.6, 2.1, and 7.2).

Simplify.

77. 2^{-3} [1.6]

78. $5^{(1-3)}$ [1.6]

79. $4^{5/2}$ [7.2]

80. $3^{7/10}$ [7.2]

Graph. [2.1]

81. $y = x^3$

82. $x = y^3$

Synthesis

83. The function $V(t) = 750(1.2)^t$ is used to predict the value $V(t)$ of a certain rare stamp t years from 2008. Do not calculate $V^{-1}(t)$, but explain how V^{-1} could be used.

84. An organization determines that the cost per person $C(x)$, in dollars, of chartering a bus with x passengers is given by

$$C(x) = \frac{100 + 5x}{x}.$$

Determine $C^{-1}(x)$ and explain how this inverse function could be used.

For Exercises 85 and 86, graph the inverse of f.

85.

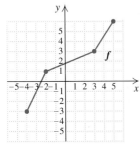

86.

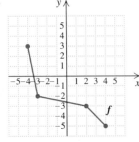

87. *Dress sizes in France and Italy.* Use the information in Exercises 73 and 74 to find a function for the French dress size that corresponds to a size x dress in Italy.

88. *Dress sizes in Italy and France.* Use the information in Exercises 73 and 74 to find a function for the Italian dress size that corresponds to a size x dress in France.

89. What relationship exists between the answers to Exercises 87 and 88? Explain how you determined this.

90. Show that function composition is associative by showing that $((f \circ g) \circ h)(x) = (f \circ (g \circ h))(x)$.

91. Show that if $h(x) = (f \circ g)(x)$, then $h^{-1}(x) = (g^{-1} \circ f^{-1})(x)$. (*Hint*: Use Exercise 90.)

Determine whether or not the given pairs of functions are inverses of each other.

92. $f(x) = 0.75x^2 + 2;\ g(x) = \sqrt{\dfrac{4(x - 2)}{3}}$

93. $f(x) = 1.4x^3 + 3.2;\ g(x) = \sqrt[3]{\dfrac{x - 3.2}{1.4}}$

94. $f(x) = \sqrt{2.5x + 9.25};$
$g(x) = 0.4x^2 - 3.7, x \geq 0$

95. $f(x) = 0.8x^{1/2} + 5.23;$
$g(x) = 1.25(x^2 - 5.23), x \geq 0$

96. $f(x) = 2.5(x^3 - 7.1);$
$g(x) = \sqrt[3]{0.4x + 7.1}$

97. Match each function in Column A with its inverse from Column B.

Column A

(1) $y = 5x^3 + 10$

(2) $y = (5x + 10)^3$

(3) $y = 5(x + 10)^3$

(4) $y = (5x)^3 + 10$

Column B

A. $y = \dfrac{\sqrt[3]{x} - 10}{5}$

B. $y = \sqrt[3]{\dfrac{x}{5}} - 10$

C. $y = \sqrt[3]{\dfrac{x - 10}{5}}$

D. $y = \dfrac{\sqrt[3]{x - 10}}{5}$

98. Examine the following table. Is it possible that f and g are inverses of each other? Why or why not?

x	$f(x)$	$g(x)$
6	6	6
7	6.5	8
8	7	10
9	7.5	12
10	8	14
11	8.5	16
12	9	18

99. The following window appears on a graphing calculator.

X	Y1	Y2
0	1	−2
1	1.5	0
2	2	2
3	2.5	4
4	3	6
5	3.5	8
6	4	10

X = 0

a) What evidence is there that the functions Y1 and Y2 are inverses of each other?

b) Find equations for Y1 and Y2, assuming that both are linear functions.

c) On the basis of your answer to part (b), are Y1 and Y2 inverses of each other?

9.2   Exponential Functions

Graphing Exponential Functions ▪ Equations with *x* and *y* Interchanged ▪ Applications of Exponential Functions

STUDY SKILLS ─────────

Know Your Machine

Whether you use a scientific or a graphing calculator, it is a wise investment of time to study the user's manual. If you cannot find a paper manual to consult, an electronic version can usually be found, online, at the manufacturer's website. Experimenting by pressing various combinations of keystrokes can also be useful.

In this section, we introduce a new type of function, the *exponential function*. These functions and their inverses, called *logarithmic functions*, have applications in many fields.

Consider the graph below. The rapidly rising curve approximates the graph of an *exponential function*.

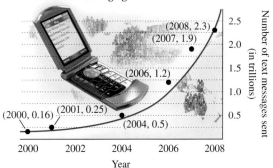

Worldwide text messaging

(2008, 2.3)
(2007, 1.9)
(2006, 1.2)
(2000, 0.16) (2001, 0.25)
(2004, 0.5)

Number of text messages sent (in trillions)

Year

Source: Mobile SMS Marketing, Gartner

Graphing Exponential Functions

In Chapter 7, we studied exponential expressions with rational-number exponents, such as

$$5^{1/4}, \quad 3^{-3/4}, \quad 7^{2.34}, \quad 5^{1.73}.$$

For example, $5^{1.73}$, or $5^{173/100}$, represents the 100th root of 5 raised to the 173rd power. What about expressions with irrational exponents, such as $5^{\sqrt{3}}$ or $7^{-\pi}$? To attach meaning to $5^{\sqrt{3}}$, consider a rational approximation, r, of $\sqrt{3}$. As r gets closer to $\sqrt{3}$, the value of 5^r gets closer to some real number p.

r closes in on $\sqrt{3}$.	5^r closes in on some real number p.
$1.7 < r < 1.8$	$15.426 \approx 5^{1.7} < p < 5^{1.8} \approx 18.119$
$1.73 < r < 1.74$	$16.189 \approx 5^{1.73} < p < 5^{1.74} \approx 16.452$
$1.732 < r < 1.733$	$16.241 \approx 5^{1.732} < p < 5^{1.733} \approx 16.267$

We define $5^{\sqrt{3}}$ to be the number p. To eight decimal places,

$$5^{\sqrt{3}} \approx 16.24245082.$$

Any positive irrational exponent can be interpreted in a similar way. Negative irrational exponents are then defined using reciprocals. Thus, so long as a is positive, a^x has meaning for *any* real number x. All of the laws of exponents still hold, but we will not prove that here. We can now define an *exponential function*.

> **Exponential Function**
>
> The function $f(x) = a^x$, where a is a positive constant, $a \neq 1$, and x is any real number, is called the *exponential function*, base a.

We require the base a to be positive to avoid imaginary numbers that would result from taking even roots of negative numbers. The restriction $a \neq 1$ is made to exclude the constant function $f(x) = 1^x$, or $f(x) = 1$.

The following are examples of exponential functions:

$$f(x) = 2^x, \qquad f(x) = \left(\tfrac{1}{3}\right)^x, \qquad f(x) = 5^{-3x}. \qquad \text{Note that } 5^{-3x} = (5^{-3})^x.$$

Like polynomial functions, the domain of an exponential function is the set of all real numbers. Unlike polynomial functions, exponential functions have a variable exponent. Because of this, graphs of exponential functions either rise or fall dramatically.

EXAMPLE   1   Graph the exponential function given by $y = f(x) = 2^x$.

SOLUTION   We compute some function values, thinking of y as $f(x)$, and list the results in a table. It is a good idea to start by letting $x = 0$.

$$f(0) = 2^0 = 1; \qquad\qquad f(-1) = 2^{-1} = \frac{1}{2^1} = \frac{1}{2};$$
$$f(1) = 2^1 = 2;$$
$$f(2) = 2^2 = 4; \qquad\qquad f(-2) = 2^{-2} = \frac{1}{2^2} = \frac{1}{4};$$
$$f(3) = 2^3 = 8;$$
$$\qquad\qquad\qquad\qquad f(-3) = 2^{-3} = \frac{1}{2^3} = \frac{1}{8}$$

x	y, or $f(x)$
0	1
1	2
2	4
3	8
-1	$\frac{1}{2}$
-2	$\frac{1}{4}$
-3	$\frac{1}{8}$

Next, we plot these points and connect them with a smooth curve.

The curve comes very close to the x-axis, but does not touch or cross it.

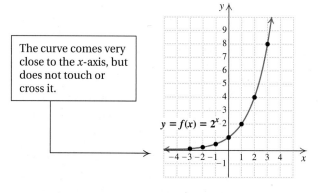

Be sure to plot enough points to determine how steeply the curve rises.

Note that as x increases, the function values increase without bound. As x decreases, the function values decrease, getting closer to 0. The x-axis, or the line $y = 0$, is a horizontal *asymptote*, meaning that the curve gets closer and closer to this line the further we move to the left.

TRY EXERCISE 7

EXAMPLE   2   Graph: $y = f(x) = \left(\tfrac{1}{2}\right)^x$.

SOLUTION   We compute some function values, thinking of y as $f(x)$, and list the results in a table. Before we do this, note that

$$y = f(x) = \left(\tfrac{1}{2}\right)^x = (2^{-1})^x = 2^{-x}.$$

Then we have

$$f(0) = 2^{-0} = 1; \qquad\qquad f(3) = 2^{-3} = \frac{1}{2^3} = \frac{1}{8};$$
$$f(1) = 2^{-1} = \frac{1}{2^1} = \frac{1}{2}; \qquad\qquad f(-1) = 2^{-(-1)} = 2^1 = 2;$$
$$\qquad\qquad\qquad\qquad f(-2) = 2^{-(-2)} = 2^2 = 4;$$
$$f(2) = 2^{-2} = \frac{1}{2^2} = \frac{1}{4}; \qquad\qquad f(-3) = 2^{-(-3)} = 2^3 = 8.$$

x	y, or $f(x)$
0	1
1	$\frac{1}{2}$
2	$\frac{1}{4}$
3	$\frac{1}{8}$
-1	2
-2	4
-3	8

Graphing calculators are helpful when graphing equations like $y = 5000(1.075)^x$. To choose a window, we note that y-values are positive and increase rapidly. One suitable window is $[-10, 10, 0, 15000]$, with a y-scale of 1000.

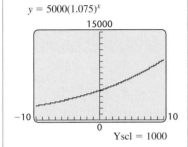

$y = 5000(1.075)^x$

Yscl = 1000

Graph each pair of functions. Select an appropriate window and scale.

1. $y_1 = \left(\frac{5}{2}\right)^x$ and $y_2 = \left(\frac{2}{5}\right)^x$
2. $y_1 = 3.2^x$ and $y_2 = 3.2^{-x}$
3. $y_1 = \left(\frac{3}{7}\right)^x$ and $y_2 = \left(\frac{7}{3}\right)^x$
4. $y_1 = 5000(1.08)^x$ and $y_2 = 5000(1.08)^{x-3}$

Next, we plot these points and connect them with a smooth curve. This curve is a mirror image, or *reflection*, of the graph of $y = 2^x$ (see Example 1) across the y-axis. The line $y = 0$ is again the horizontal asymptote.

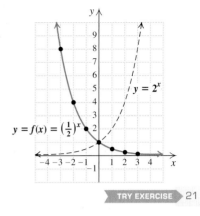

TRY EXERCISE 21

From Examples 1 and 2, we can make the following observations.

- For $a > 1$, the graph of $f(x) = a^x$ increases from left to right. The greater the value of a, the steeper the curve. (See the figure on the left below.)

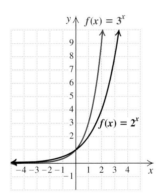
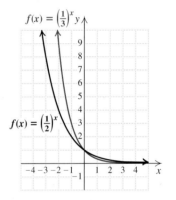

- For $0 < a < 1$, the graph of $f(x) = a^x$ decreases from left to right. For smaller values of a, the curve is steeper. (See the figure on the right above.)
- All graphs of $f(x) = a^x$ go through the y-intercept $(0, 1)$.
- All graphs of $f(x) = a^x$ have the x-axis as the horizontal asymptote.
- If $f(x) = a^x$, with $a > 0, a \neq 1$, the domain of f is all real numbers, and the range of f is all positive real numbers.
- For $a > 0, a \neq 1$, the function given by $f(x) = a^x$ is one-to-one. Its graph passes the horizontal-line test.

EXAMPLE 3

STUDENT NOTES

When using translations, make sure that you are shifting in the correct direction. When in doubt, substitute a value for x and make some calculations.

Graph: $y = f(x) = 2^{x-2}$.

SOLUTION   We construct a table of values. Then we plot the points and connect them with a smooth curve. Here $x - 2$ is the *exponent*.

$f(0) = 2^{0-2} = 2^{-2} = \frac{1}{4}$;
$f(1) = 2^{1-2} = 2^{-1} = \frac{1}{2}$;
$f(2) = 2^{2-2} = 2^0 = 1$;
$f(3) = 2^{3-2} = 2^1 = 2$;
$f(4) = 2^{4-2} = 2^2 = 4$;
$f(-1) = 2^{-1-2} = 2^{-3} = \frac{1}{8}$;
$f(-2) = 2^{-2-2} = 2^{-4} = \frac{1}{16}$

x	y, or $f(x)$
0	$\frac{1}{4}$
1	$\frac{1}{2}$
2	1
3	2
4	4
−1	$\frac{1}{8}$
−2	$\frac{1}{16}$

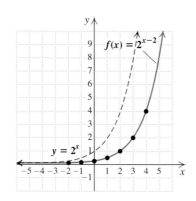

The graph looks just like the graph of $y = 2^x$, but it is translated 2 units to the right. The y-intercept of $y = 2^x$ is $(0, 1)$. The y-intercept of $y = 2^{x-2}$ is $\left(0, \frac{1}{4}\right)$. The line $y = 0$ is again the horizontal asymptote.

▶ **TRY EXERCISE** 17

Equations with x and y Interchanged

It will be helpful in later work to be able to graph an equation in which the x and the y in $y = a^x$ are interchanged.

EXAMPLE 4

Graph: $x = 2^y$.

SOLUTION   Note that x is alone on one side of the equation. To find ordered pairs that are solutions, we choose values for y and then compute values for x.

For $y = 0$, $x = 2^0 = 1$.

For $y = 1$, $x = 2^1 = 2$.

For $y = 2$, $x = 2^2 = 4$.

For $y = 3$, $x = 2^3 = 8$.

For $y = -1$, $x = 2^{-1} = \dfrac{1}{2}$.

For $y = -2$, $x = 2^{-2} = \dfrac{1}{4}$.

For $y = -3$, $x = 2^{-3} = \dfrac{1}{8}$.

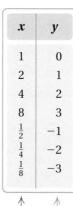

x	y
1	0
2	1
4	2
8	3
$\frac{1}{2}$	−1
$\frac{1}{4}$	−2
$\frac{1}{8}$	−3

— (1) Choose values for y.
— (2) Compute values for x.

We plot the points and connect them with a smooth curve.

This curve does not touch or cross the y-axis, which serves as a vertical asymptote.

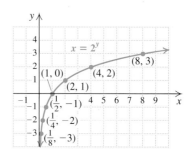

Note too that this curve looks just like the graph of $y = 2^x$, except that it is re-flected across the line $y = x$, as shown here.

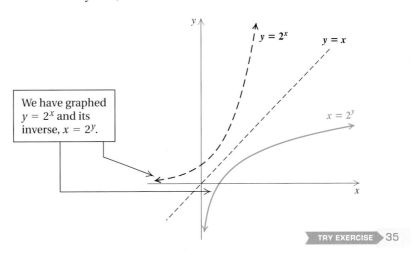

We have graphed $y = 2^x$ and its inverse, $x = 2^y$.

TRY EXERCISE ▶ 35

Applications of Exponential Functions

EXAMPLE **5**

Interest compounded annually. The amount of money A that a principal P will be worth after t years at interest rate i, compounded annually, is given by the formula

$$A = P(1 + i)^t.$$ You might review Example 11 in Section 8.1.

Suppose that $100,000 is invested at 8% interest, compounded annually.

a) Find a function for the amount in the account after t years.

b) Find the amount of money in the account at $t = 0$, $t = 4$, $t = 8$, and $t = 10$.

c) Graph the function.

SOLUTION

a) If $P = \$100,000$ and $i = 8\% = 0.08$, we can substitute these values and form the following function:

$$A(t) = \$100,000(1 + 0.08)^t$$ Using $A = P(1 + i)^t$
$$= \$100,000(1.08)^t.$$

b) To find the function values, a calculator with a power key is helpful.

$$A(0) = \$100,000(1.08)^0 \qquad A(8) = \$100,000(1.08)^8$$
$$= \$100,000(1) \qquad\qquad \approx \$100,000(1.85093021)$$
$$= \$100,000 \qquad\qquad\quad \approx \$185,093.02$$

$$A(4) = \$100,000(1.08)^4 \qquad A(10) = \$100,000(1.08)^{10}$$
$$= \$100,000(1.36048896) \qquad \approx \$100,000(2.158924997)$$
$$\approx \$136,048.90 \qquad\qquad\quad \approx \$215,892.50$$

TECHNOLOGY CONNECTION

Graphing calculators can quickly find many function values. Let $y_1 = 100,000(1.08)^x$. Then use the TABLE feature with INDPNT set to ASK to check Example 5(b).

c) We use the function values computed in part (b), and others if we wish, to draw the graph as follows. Note that the axes are scaled differently because of the large numbers.

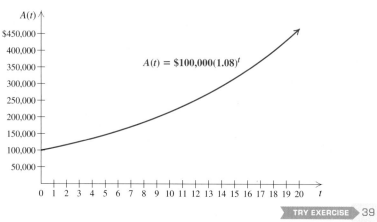

$$A(t) = \$100{,}000(1.08)^t$$

TRY EXERCISE ▶ 39

<table>
<tr><td>**9.2**</td><td>**EXERCISE SET**</td><td>*For Extra Help*
MyMathLab   Math XL
PRACTICE   WATCH   DOWNLOAD</td></tr>
</table>

Concept Reinforcement   *Classify each statement as either true or false.*

1. The graph of $f(x) = a^x$ always passes through the point $(0, 1)$.

2. The graph of $g(x) = \left(\frac{1}{2}\right)^x$ gets closer and closer to the x-axis as x gets larger and larger.

3. The graph of $f(x) = 2^{x-3}$ looks just like the graph of $y = 2^x$, but it is translated 3 units to the right.

4. The graph of $g(x) = 2^x - 3$ looks just like the graph of $y = 2^x$, but it is translated 3 units up.

5. The graph of $y = 3^x$ gets close to, but never touches, the y-axis.

6. The graph of $x = 3^y$ gets close to, but never touches, the y-axis.

Graph.

7. $y = f(x) = 3^x$ **8.** $y = f(x) = 4^x$

9. $y = 6^x$ **10.** $y = 5^x$

11. $y = 2^x + 1$ **12.** $y = 2^x + 3$

13. $y = 3^x - 2$ **14.** $y = 3^x - 1$

15. $y = 2^x - 5$ **16.** $y = 2^x - 4$

17. $y = 2^{x-3}$ **18.** $y = 2^{x-1}$

19. $y = 2^{x+1}$ **20.** $y = 2^{x+3}$

21. $y = \left(\frac{1}{4}\right)^x$ **22.** $y = \left(\frac{1}{5}\right)^x$

23. $y = \left(\frac{1}{3}\right)^x$ **24.** $y = \left(\frac{1}{10}\right)^x$

25. $y = 2^{x+1} - 3$ **26.** $y = 2^{x-3} - 1$

27. $x = 6^y$ **28.** $x = 3^y$

29. $x = 3^{-y}$ **30.** $x = 2^{-y}$

31. $x = 4^y$ **32.** $x = 5^y$

33. $x = \left(\frac{4}{3}\right)^y$ **34.** $x = \left(\frac{3}{2}\right)^y$

Graph each pair of equations on the same set of axes.

35. $y = 3^x,\ x = 3^y$ **36.** $y = 2^x,\ x = 2^y$

37. $y = \left(\frac{1}{2}\right)^x,\ x = \left(\frac{1}{2}\right)^y$ **38.** $y = \left(\frac{1}{4}\right)^x,\ x = \left(\frac{1}{4}\right)^y$

Solve.

39. *Music downloads.* The number $M(t)$ of single tracks downloaded, in billions, t years after 2003 can be approximated by

$$M(t) = 0.353\,(1.244)^t.$$

Source: International Federation of the Phonographic Industry

a) Estimate the number of single tracks downloaded in 2006, in 2008, and in 2012.

b) Graph the function.

40. *Growth of bacteria.* The bacteria *Escherichia coli* are commonly found in the human bladder. Suppose that 3000 of the bacteria are present at time $t = 0$. Then t minutes later, the number of bacteria present can be approximated by

$$N(t) = 3000(2)^{t/20}.$$

a) How many bacteria will be present after 10 min? 20 min? 30 min? 40 min? 60 min?

b) Graph the function.

41. *Smoking cessation.* The percentage of smokers *P* who receive telephone counseling to quit smoking and are still successful *t* months later can be approximated by

$$P(t) = 21.4(0.914)^t.$$

Sources: *New England Journal of Medicine;* data from California's Smokers' Hotline

a) Estimate the percentage of smokers receiving telephone counseling who are successful in quitting for 1 month, 3 months, and 1 year.

b) Graph the function.

42. *Smoking cessation.* The percentage of smokers *P* who, without telephone counseling, have successfully quit smoking for *t* months (see Exercise 41) can be approximated by

$$P(t) = 9.02(0.93)^t.$$

Sources: *New England Journal of Medicine;* data from California's Smokers' Hotline

a) Estimate the percentage of smokers not receiving telephone counseling who are successful in quitting for 1 month, 3 months, and 1 year.

b) Graph the function.

43. *Marine biology.* Due to excessive whaling prior to the mid 1970s, the humpback whale is considered an endangered species. The worldwide population of humpbacks, $P(t)$, in thousands, *t* years after 1900 ($t < 70$) can be approximated by*

$$P(t) = 150(0.960)^t.$$

a) How many humpback whales were alive in 1930? in 1960?

b) Graph the function.

44. *Salvage value.* A laser printer is purchased for $1200. Its value each year is about 80% of the value of the preceding year. Its value, in dollars, after *t* years is given by the exponential function

$$V(t) = 1200(0.8)^t.$$

a) Find the value of the printer after 0 yr, 1 yr, 2 yr, 5 yr, and 10 yr.

b) Graph the function.

45. *Marine biology.* As a result of preservation efforts in most countries in which whaling was common, the humpback whale population has grown since the 1970s. The worldwide population of hump-

backs, $P(t)$, in thousands, *t* years after 1982 can be approximated by*

$$P(t) = 5.5(1.047)^t.$$

a) How many humpback whales were alive in 1992? in 2004?

b) Graph the function.

46. *Recycling aluminum cans.* It is estimated that $\frac{1}{2}$ of all aluminum cans distributed will be recycled each year. A beverage company distributes 250,000 cans. The number still in use after time *t*, in years, is given by the exponential function

$$N(t) = 250{,}000\left(\tfrac{1}{2}\right)^t.$$

Source: The Aluminum Association, Inc., 2005

a) How many cans are still in use after 0 yr? 1 yr? 4 yr? 10 yr?

b) Graph the function.

47. *Spread of zebra mussels.* Beginning in 1988, infestations of zebra mussels started spreading throughout North American waters.[†] These mussels spread with such speed that water treatment facilities, power plants, and entire ecosystems can become threatened. The function

$$A(t) = 10 \cdot 34^t$$

can be used to estimate the number of square centimeters of lake bottom that will be covered with mussels *t* years after an infestation covering 10 cm^2 first occurs.

a) How many square centimeters of lake bottom will be covered with mussels 5 yr after an infestation covering 10 cm^2 first appears? 7 yr after the infestation first appears?

b) Graph the function.

---

*Based on information from the American Cetacean Society, 2001, and the ASK Archive, 1998.

[†]Many thanks to Dr. Gerald Mackie of the Department of Zoology at the University of Guelph in Ontario for the background information for this exercise.

48. *Cell phones.* The number of cell phones in use in the United States is increasing exponentially. The number N, in millions, in use can be estimated by

$$N(t) = 7.12(1.3)^t,$$

where t is the number of years after 1990.

Source: Based on data from CTIA-The Wireless Association

a) Estimate the number of cell phones in use in 1995, in 2005, and in 2010.
b) Graph the function.

49. Without using a calculator, explain why 2^π must be greater than 8 but less than 16.

50. Suppose that $1000 is invested for 5 yr at 7% interest, compounded annually. In what year will the most interest be earned? Why?

Skill Review

Review factoring polynomials (Sections 5.3–5.6).

Factor.

51. $3x^2 - 48$ [5.5]

52. $x^2 - 20x + 100$ [5.5]

53. $6x^2 + x - 12$ [5.4]

54. $8x^6 - 64y^6$ [5.6]

55. $t^2 - y^2 + 2y - 1$ [5.5]

56. $5x^4 - 10x^3 - 3x^2 + 6x$ [5.3]

Synthesis

57. Examine Exercise 48. Do you believe that the equation for the number of cell phones in use in the United States will be accurate 20 yr from now? Why or why not?

58. Explain why the graph of $x = 2^y$ is the graph of $y = 2^x$ reflected across the line $y = x$.

Determine which of the two numbers is larger. Do not use a calculator.

59. $\pi^{1.3}$ or $\pi^{2.4}$

60. $\sqrt{8^3}$ or $8^{\sqrt{3}}$

Graph.

61. $f(x) = 2.5^x$

62. $f(x) = 0.5^x$

63. $y = 2^x + 2^{-x}$

64. $y = \left|\left(\frac{1}{2}\right)^x - 1\right|$

65. $y = |2^x - 2|$

66. $y = 2^{-(x-1)^2}$

67. $y = |2^{x^2} - 1|$

68. $y = 3^x + 3^{-x}$

Graph both equations using the same set of axes.

69. $y = 3^{-(x-1)}, \ x = 3^{-(y-1)}$

70. $y = 1^x, \ x = 1^y$

71. *Navigational devices.* The number of GPS navigational devices in use in the United States has grown from 0.5 million in 2000 to 4 million in 2004 to 50 million in 2008. After pressing **STAT** and entering the data, use the ExpReg option in the STAT CALC menu to find an exponential function that models the number of navigational devices in use t years after 2000. Then use that function to predict the total number of devices in use in 2012.

Source: Telematics Research Group

72. *Keyboarding speed.* Trey is studying keyboarding. After he has studied for t hours, Trey's speed, in words per minute, is given by the exponential function

$$S(t) = 200[1 - (0.99)^t].$$

Use a graph and/or table of values to predict Trey's speed after studying for 10 hr, 40 hr, and 80 hr.

73. The following graph shows growth in the height of ocean waves over time, assuming a steady surface wind.

Source: magicseaweed.com

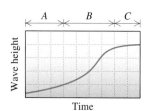

Source: magicseaweed.com

a) Consider the portions of the graph marked A, B, and C. Suppose that each portion can be labeled Exponential Growth, Linear Growth, or Saturation. How would you label each portion?
b) Small vertical movements in wind, surface roughness of water, and gravity are three forces that create waves. How might these forces be related to the shape of the wave-height graph?

74. Consider any exponential function of the form $f(x) = a^x$ with $a > 1$. Will it always follow that $f(3) - f(2) > f(2) - f(1)$, and, in general, $f(n + 2) - f(n + 1) > f(n + 1) - f(n)$? Why or why not? (*Hint:* Think graphically.)

75. On many graphing calculators, it is possible to enter and graph $y_1 = A \wedge (X - B) + C$ after first pressing **APPS** Transfrm. Use this application to graph $f(x) = 2.5^{x-3} + 2, g(x) = 2.5^{x+3} + 2,$ $h(x) = 2.5^{x-3} - 2,$ and $k(x) = 2.5^{x+3} - 2.$

CORNER

The True Cost of a New Car

Focus: Car loans and exponential functions

Time: 30 minutes

Group size: 2

Materials: Calculators with exponentiation keys

The formula

$$M = \frac{Pr}{1 - (1 + r)^{-n}}$$

is used to determine the payment size, M, when a loan of P dollars is to be repaid in n equally sized monthly payments. Here r represents the monthly interest rate. Loans repaid in this fashion are said to be *amortized* (spread out equally) over a period of n months.

ACTIVITY

1. Suppose one group member is selling the other a car for $2600, financed at 1% interest per month for 24 months. What should be the size of each monthly payment?

2. Suppose both group members are shopping for the same model new car. To save time, each group member visits a different dealer. One dealer offers the car for $13,000 at 10.5% interest (0.00875 monthly interest) for 60 months (no down payment). The other dealer offers the same car for $12,000, but at 12% interest (0.01 monthly interest) for 48 months (no down payment).

   a) Determine the monthly payment size for each offer. Then determine the total amount paid for the car under each offer. How much of each total is interest?

   b) Work together to find the annual interest rate for which the total cost of 60 monthly payments for the $13,000 car would equal the total amount paid for the $12,000 car (as found in part a above).

9.3   Logarithmic Functions

Graphs of Logarithmic Functions ▪ Equivalent Equations ▪ Solving Certain Logarithmic Equations

We are now ready to study inverses of exponential functions. These functions have many applications and are called *logarithm,* or *logarithmic, functions.*

STUDY SKILLS

When a Turn Is Trouble

Occasionally a page turn can interrupt your thoughts as you work through a section. You may find it helpful to rewrite (in pencil) the last equation or sentence appearing on one page at the very top of the next page.

Graphs of Logarithmic Functions

Consider the exponential function $f(x) = 2^x$. Like all exponential functions, f is one-to-one. Can a formula for f^{-1} be found? To answer this, we use the method of Section 9.1:

1. Replace $f(x)$ with y: $y = 2^x$.
2. Interchange x and y: $x = 2^y$.

3. Solve for y: $y =$ the exponent to which we raise 2 to get x.

4. Replace y with $f^{-1}(x)$: $f^{-1}(x) =$ the exponent to which we raise 2 to get x.

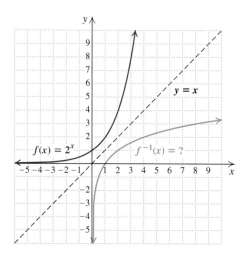

We now define a new symbol to replace the words "the exponent to which we raise 2 to get x":

$\log_2 x$, read "the logarithm, base 2, of x," or "log, base 2, of x," means "the exponent to which we raise 2 to get x."

Thus if $f(x) = 2^x$, then $f^{-1}(x) = \log_2 x$. Note that $f^{-1}(8) = \log_2 8 = 3$, because 3 is *the exponent to which we raise* 2 *to get* 8.

EXAMPLE   1

Simplify: **(a)** $\log_2 32$; **(b)** $\log_2 1$; **(c)** $\log_2 \frac{1}{8}$.

SOLUTION

a) Think of $\log_2 32$ as the exponent to which we raise 2 to get 32. That exponent is 5. Therefore, $\log_2 32 = 5$.

b) We ask ourselves: "To what exponent do we raise 2 in order to get 1?" That exponent is 0 (recall that $2^0 = 1$). Thus, $\log_2 1 = 0$.

c) To what exponent do we raise 2 in order to get $\frac{1}{8}$? Since $2^{-3} = \frac{1}{8}$, we have $\log_2 \frac{1}{8} = -3$.

TRY EXERCISE 9

Although numbers like $\log_2 13$ can be only approximated, we must remember that $\log_2 13$ represents *the exponent to which we raise* 2 *to get* 13. That is, $2^{\log_2 13} = 13$. A calculator indicates that $\log_2 13 \approx 3.7$ and $2^{3.7} \approx 13$.

For any exponential function $f(x) = a^x$, the inverse is called a **logarithmic function, base a**. The graph of the inverse can be drawn by reflecting the graph of $f(x) = a^x$ across the line $y = x$. It will be helpful to remember that the inverse of $f(x) = a^x$ is given by $f^{-1}(x) = \log_a x$.

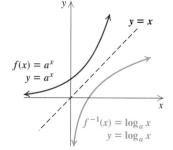

The Meaning of $\log_a x$

For $x > 0$ and a a positive constant other than 1, $\log_a x$ is the exponent to which a must be raised in order to get x. Thus,

$$\log_a x = m \quad \text{means} \quad a^m = x$$

or equivalently,

$$\log_a x \text{ is that unique exponent for which } a^{\log_a x} = x.$$

STUDENT NOTES ————

As an aid in remembering what $\log_a x$ means, note that a is called the *base*, just as it is the base in $a^y = x$.

It is important to remember that *a logarithm is an exponent.* It might help to verbalize: "The logarithm, base *a*, of a number *x* is the exponent to which *a* must be raised in order to get *x*."

EXAMPLE 2

Simplify: $7^{\log_7 85}$.

SOLUTION   Remember that $\log_7 85$ is the exponent to which 7 is raised to get 85. Raising 7 to that exponent, we have

$$7^{\log_7 85} = 85.$$

TRY EXERCISE 35

Because logarithmic and exponential functions are inverses of each other, the result in Example 2 should come as no surprise: If $f(x) = \log_7 x$, then

for $f(x) = \log_7 x$, we have $f^{-1}(x) = 7^x$

and $f^{-1}(f(x)) = f^{-1}(\log_7 x) = 7^{\log_7 x} = x$.

Thus, $f^{-1}(f(85)) = 7^{\log_7 85} = 85$.

The following is a comparison of exponential and logarithmic functions.

Exponential Function	Logarithmic Function
$y = a^x$	$x = a^y$
$f(x) = a^x$	$g(x) = \log_a x$
$a > 0, a \neq 1$	$a > 0, a \neq 1$
The domain is $\mathbb{R}$.	The range is $\mathbb{R}$.
$y > 0$ (Outputs are positive.)	$x > 0$ (Inputs are positive.)
$f^{-1}(x) = \log_a x$	$g^{-1}(x) = a^x$

EXAMPLE 3

Graph: $y = f(x) = \log_5 x$.

SOLUTION   If $y = \log_5 x$, then $5^y = x$. We can find ordered pairs that are solutions by choosing values for y and computing the x-values.

For $y = 0, x = 5^0 = 1$.

For $y = 1, x = 5^1 = 5$.

For $y = 2, x = 5^2 = 25$.

For $y = -1, x = 5^{-1} = \frac{1}{5}$.

For $y = -2, x = 5^{-2} = \frac{1}{25}$.

(1) Select y.

(2) Compute x.

This table shows the following:

$\log_5 1 = 0;$
$\log_5 5 = 1;$
$\log_5 25 = 2;$
$\log_5 \frac{1}{5} = -1;$
$\log_5 \frac{1}{25} = -2.$

These can all be checked using the equations above.

x, or 5^y	y
1	0
5	1
25	2
$\frac{1}{5}$	-1
$\frac{1}{25}$	-2

TECHNOLOGY CONNECTION

To see that $f(x) = 10^x$ and $g(x) = \log_{10} x$ are inverses of each other, let $y_1 = 10^x$ and $y_2 = \log_{10} x = \log x$. Then, using a squared window, compare both graphs. If possible, select DrawInv from the (DRAW) menu and then press (VARS) (▷) (1) (1) (ENTER) to see another representation of f^{-1}. Finally, let $y_3 = y_1(y_2)$ and $y_4 = y_2(y_1)$ to show, using a table or graphs, that, for $x > 0$, $y_3 = y_4 = x$.

We plot the set of ordered pairs and connect the points with a smooth curve. The graphs of $y = 5^x$ and $y = x$ are shown only for reference.

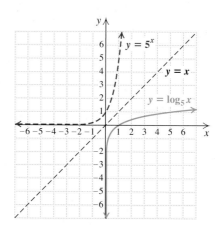

TRY EXERCISE ▶ 37

Equivalent Equations

We use the definition of logarithm to rewrite a *logarithmic equation* as an equivalent *exponential equation* or the other way around:

$$m = \log_a x \quad \text{is equivalent to} \quad a^m = x.$$

> *CAUTION!* **Do not forget this relationship!** It is probably the most important definition in the chapter. Many times this definition will be used to justify a property we are considering.

EXAMPLE 4   Rewrite each as an equivalent exponential equation: **(a)** $y = \log_3 5$; **(b)** $-2 = \log_a 7$; **(c)** $a = \log_b d$.

SOLUTION

a) $y = \log_3 5$ is equivalent to $3^y = 5$ The logarithm is the exponent.
                                                     The base remains the base.

b) $-2 = \log_a 7$ is equivalent to $a^{-2} = 7$

c) $a = \log_b d$ is equivalent to $b^a = d$ TRY EXERCISE ▶ 47

We also use the definition of logarithm to rewrite an exponential equation as an equivalent logarithmic equation.

EXAMPLE 5   Rewrite each as an equivalent logarithmic equation: **(a)** $8 = 2^x$; **(b)** $y^{-1} = 4$; **(c)** $a^b = c$.

SOLUTION

a) $8 = 2^x$ is equivalent to $x = \log_2 8$ The exponent is the logarithm.
                                                     The base remains the base.

b) $y^{-1} = 4$ is equivalent to $-1 = \log_y 4$

c) $a^b = c$ is equivalent to $b = \log_a c$ TRY EXERCISE ▶ 63

Solving Certain Logarithmic Equations

Many logarithmic equations can be solved by rewriting them as equivalent exponential equations.

EXAMPLE 6   Solve: **(a)** $\log_2 x = -3$; **(b)** $\log_x 16 = 2$.

SOLUTION

a) $\log_2 x = -3$

$$2^{-3} = x \qquad \text{Rewriting as an exponential equation}$$
$$\tfrac{1}{8} = x \qquad \text{Computing } 2^{-3}$$

Check: $\log_2 \tfrac{1}{8}$ is the exponent to which 2 is raised to get $\tfrac{1}{8}$. Since that exponent is -3, we have a check. The solution is $\tfrac{1}{8}$.

b) $\log_x 16 = 2$

$$x^2 = 16 \qquad \text{Rewriting as an exponential equation}$$
$$x = 4 \quad or \quad x = -4 \qquad \text{Principle of square roots}$$

Check: $\log_4 16 = 2$ because $4^2 = 16$. Thus, 4 is a solution of $\log_x 16 = 2$. Because all logarithmic bases must be positive, -4 cannot be a solution. Logarithmic bases must be positive because logarithms are defined using exponential functions that require positive bases. The solution is 4.

TRY EXERCISE ▶ 79

One method for solving certain logarithmic and exponential equations relies on the following property, which results from the fact that exponential functions are one-to-one.

The Principle of Exponential Equality

For any real number b, where $b \neq -1, 0$, or 1,

$$b^{x_1} = b^{x_2} \quad \text{is equivalent to} \quad x_1 = x_2.$$

(Powers of the same base are equal if and only if the exponents are equal.)

EXAMPLE 7   Solve: **(a)** $\log_{10} 1000 = x$; **(b)** $\log_4 1 = t$.

SOLUTION

a) We rewrite $\log_{10} 1000 = x$ in exponential form and solve:

$$10^x = 1000 \qquad \text{Rewriting as an exponential equation}$$
$$10^x = 10^3 \qquad \text{Writing 1000 as a power of 10}$$
$$x = 3. \qquad \text{Equating exponents}$$

Check: This equation can also be solved directly by determining the exponent to which we raise 10 in order to get 1000. In both cases we find that $\log_{10} 1000 = 3$, so we have a check. The solution is 3.

b) We rewrite $\log_4 1 = t$ in exponential form and solve:

$$4^t = 1 \qquad \text{Rewriting as an exponential equation}$$
$$4^t = 4^0 \qquad \text{Writing 1 as a power of 4. This can be done mentally.}$$
$$t = 0. \qquad \text{Equating exponents}$$

Check: As in part (a), this equation can be solved directly by determining the exponent to which we raise 4 in order to get 1. In both cases we find that $\log_4 1 = 0$, so we have a check. The solution is 0.

TRY EXERCISE 81

Example 7 illustrates an important property of logarithms.

$\log_a 1$

The logarithm, base a, of 1 is always 0: $\log_a 1 = 0$.

This follows from the fact that $a^0 = 1$ is equivalent to the logarithmic equation $\log_a 1 = 0$. Thus, $\log_{10} 1 = 0$, $\log_7 1 = 0$, and so on.

Another property results from the fact that $a^1 = a$. This is equivalent to the logarithmic equation $\log_a a = 1$.

$\log_a a$

The logarithm, base a, of a is always 1: $\log_a a = 1$.

Thus, $\log_{10} 10 = 1$, $\log_8 8 = 1$, and so on.

9.3 EXERCISE SET

↪ **Concept Reinforcement** *In each of Exercises 1–8, match the expression or equation with an equivalent expression or equation from the column on the right.*

1. ___ $\log_5 25$

2. ___ $2^5 = x$

3. ___ $\log_5 5$

4. ___ $\log_2 1$

5. ___ $\log_5 5^x$

6. ___ $\log_x 27 = 5$

7. ___ $5 = 2^x$

8. ___ $x^{-2} = 5$

a) 1

b) x

c) $x^5 = 27$

d) $\log_2 x = 5$

e) $\log_2 5 = x$

f) $\log_x 5 = -2$

g) 2

h) 0

Simplify.

9. $\log_{10} 1000$

10. $\log_{10} 100$

11. $\log_7 49$

12. $\log_2 8$

13. $\log_3 81$

14. $\log_3 9$

15. $\log_5 \frac{1}{25}$

16. $\log_5 \frac{1}{5}$

17. $\log_8 \frac{1}{8}$

18. $\log_8 \frac{1}{64}$

19. $\log_5 625$

20. $\log_5 125$

21. $\log_7 7$

22. $\log_9 1$

23. $\log_3 1$

24. $\log_3 3$

Aha! 25. $\log_6 6^5$

26. $\log_6 6^9$

27. $\log_{10} 0.01$

28. $\log_{10} 0.1$

29. $\log_{16} 4$ **30.** $\log_{100} 10$

31. $\log_9 27$ **32.** $\log_4 32$

33. $\log_{1000} 100$ **34.** $\log_{16} 8$

35. $3^{\log_3 29}$ **36.** $6^{\log_6 13}$

Graph.

37. $y = \log_{10} x$ **38.** $y = \log_2 x$

39. $y = \log_3 x$ **40.** $y = \log_7 x$

41. $f(x) = \log_6 x$ **42.** $f(x) = \log_4 x$

43. $f(x) = \log_{2.5} x$ **44.** $f(x) = \log_{1/2} x$

Graph both functions using the same set of axes.

45. $f(x) = 3^x,\ f^{-1}(x) = \log_3 x$

46. $f(x) = 4^x,\ f^{-1}(x) = \log_4 x$

Rewrite each of the following as an equivalent exponential equation. Do not solve.

47. $x = \log_{10} 8$ **48.** $y = \log_8 10$

49. $\log_9 9 = 1$ **50.** $\log_6 36 = 2$

51. $\log_{10} 0.1 = -1$ **52.** $\log_{10} 0.01 = -2$

53. $\log_{10} 7 = 0.845$ **54.** $\log_{10} 3 = 0.4771$

55. $\log_c m = 8$ **56.** $\log_b n = 23$

57. $\log_r C = t$ **58.** $\log_m P = a$

59. $\log_e 0.25 = -1.3863$

60. $\log_e 0.989 = -0.0111$

61. $\log_r T = -x$ **62.** $\log_c M = -w$

Rewrite each of the following as an equivalent logarithmic equation. Do not solve.

63. $10^2 = 100$ **64.** $10^4 = 10{,}000$

65. $5^{-3} = \frac{1}{125}$ **66.** $2^{-5} = \frac{1}{32}$

67. $16^{1/4} = 2$ **68.** $8^{1/3} = 2$

69. $10^{0.4771} = 3$ **70.** $10^{0.3010} = 2$

71. $z^m = 6$ **72.** $m^n = r$

73. $p^t = q$ **74.** $y^t = x$

75. $e^3 = 20.0855$ **76.** $e^2 = 7.3891$

77. $e^{-4} = 0.0183$ **78.** $e^{-2} = 0.1353$

Solve.

79. $\log_6 x = 2$ **80.** $\log_4 x = 3$

81. $\log_2 32 = x$ **82.** $\log_5 25 = x$

83. $\log_x 9 = 1$ **84.** $\log_x 12 = 1$

85. $\log_x 7 = \frac{1}{2}$ **86.** $\log_x 9 = \frac{1}{2}$

87. $\log_3 x = -2$ **88.** $\log_2 x = -1$

89. $\log_{32} x = \frac{2}{5}$ **90.** $\log_8 x = \frac{2}{3}$

91. In what way is a logarithm an exponent?

92. Is it easier to find x given $x = \log_9 \frac{1}{3}$ or given $9^x = \frac{1}{3}$? Explain your reasoning.

Skill Review

Review simplifying rational and radical expressions (Chapters 6 and 7).

Simplify.

93. $\sqrt{18a^3 b}\sqrt{50ab^7}$ [7.3]

94. $(2\sqrt{3} + \sqrt{5})(2\sqrt{3} - \sqrt{10})$ [7.5]

95. $\sqrt{192x} - \sqrt{75x}$ [7.5]   **96.** $\sqrt[4]{\sqrt[3]{x}}$ [7.2]

97. $\dfrac{\dfrac{3}{x} - \dfrac{2}{xy}}{\dfrac{2}{x^2} + \dfrac{1}{xy}}$ [6.3]   **98.** $\dfrac{\dfrac{4+x}{x^2 + 2x + 1}}{\dfrac{3}{x+1} - \dfrac{2}{x+2}}$ [6.3]

Synthesis

99. Would a manufacturer be pleased or unhappy if sales of a product grew logarithmically? Why?

100. Explain why the number $\log_2 13$ must be between 3 and 4.

101. Graph both equations using the same set of axes:
$$y = \left(\tfrac{3}{2}\right)^x, \qquad y = \log_{3/2} x.$$

Graph.

102. $y = \log_2 (x - 1)$ **103.** $y = \log_3 |x + 1|$

Solve.

104. $|\log_3 x| = 2$

105. $\log_4 (3x - 2) = 2$

106. $\log_8 (2x + 1) = -1$

107. $\log_{10} (x^2 + 21x) = 2$

Simplify.

108. $\log_{1/4} \frac{1}{64}$ **109.** $\log_{1/5} 25$

110. $\log_{81} 3 \cdot \log_3 81$

111. $\log_{10} (\log_4 (\log_3 81))$

112. $\log_2 (\log_2 (\log_4 256))$

113. Show that $b^{x_1} = b^{x_2}$ is *not* equivalent to $x_1 = x_2$ for $b = 0$ or $b = 1$.

114. If $\log_b a = x$, does it follow that $\log_a b = 1/x$? Why or why not?

9.4 Properties of Logarithmic Functions

Logarithms of Products ■ Logarithms of Powers ■ Logarithms of Quotients ■
Using the Properties Together

Logarithmic functions are important in many applications and in more advanced mathematics. We now establish some basic properties that are useful in manipulating expressions involving logarithms. As their proofs reveal, the properties of logarithms are related to the properties of exponents.

Logarithms of Products

The first property we discuss is related to the product rule for exponents: $a^m \cdot a^n = a^{m+n}$. Its proof appears immediately after Example 2.

> **The Product Rule for Logarithms**
>
> For any positive numbers M, N, and a ($a \neq 1$),
>
> $$\log_a (MN) = \log_a M + \log_a N.$$
>
> (The logarithm of a product is the sum of the logarithms of the factors.)

EXAMPLE 1   Express as an equivalent expression that is a sum of logarithms: $\log_2 (4 \cdot 16)$.

SOLUTION   We have

$$\log_2 (4 \cdot 16) = \log_2 4 + \log_2 16. \qquad \text{Using the product rule for logarithms}$$

As a check, note that

$$\log_2 (4 \cdot 16) = \log_2 64 = 6 \qquad 2^6 = 64$$

and that

$$\log_2 4 + \log_2 16 = 2 + 4 = 6. \qquad 2^2 = 4 \text{ and } 2^4 = 16$$

TRY EXERCISE 7

EXAMPLE 2   Express as an equivalent expression that is a single logarithm: $\log_b 7 + \log_b 5$.

SOLUTION   We have

$$\log_b 7 + \log_b 5 = \log_b (7 \cdot 5) \qquad \text{Using the product rule for logarithms}$$
$$= \log_b 35.$$

TRY EXERCISE 13

A Proof of the Product Rule. Let $\log_a M = x$ and $\log_a N = y$. Converting to exponential equations, we have $a^x = M$ and $a^y = N$.

Now we multiply the left side of the first exponential equation by the left side of the second equation and similarly multiply the right sides to obtain

$$MN = a^x \cdot a^y, \quad \text{or} \quad MN = a^{x+y}.$$

Converting back to a logarithmic equation, we get

$$\log_a (MN) = x + y.$$

Recalling what x and y represent, we have

$$\log_a(MN) = \log_a M + \log_a N.$$

Logarithms of Powers

The second basic property is related to the power rule for exponents: $(a^m)^n = a^{mn}$. Its proof follows Example 3.

The Power Rule for Logarithms

For any positive numbers M and a ($a \neq 1$), and any real number p,

$$\log_a M^p = p \cdot \log_a M.$$

(The logarithm of a power of M is the exponent times the logarithm of M.)

To better understand the power rule, note that

$$\log_a M^3 = \log_a(M \cdot M \cdot M) = \log_a M + \log_a M + \log_a M = 3\log_a M.$$

EXAMPLE 3   Use the power rule for logarithms to write an equivalent expression that is a product: **(a)** $\log_a 9^{-5}$; **(b)** $\log_7 \sqrt[3]{x}$.

SOLUTION

a) $\log_a 9^{-5} = -5\log_a 9$ Using the power rule for logarithms

b) $\log_7 \sqrt[3]{x} = \log_7 x^{1/3}$ Writing exponential notation

             $= \frac{1}{3}\log_7 x$ Using the power rule for logarithms

> TRY EXERCISE 17

A Proof of the Power Rule. Let $x = \log_a M$. We then write the equivalent exponential equation, $a^x = M$. Raising both sides to the pth power, we get

$$(a^x)^p = M^p, \quad \text{or} \quad a^{xp} = M^p. \quad \text{Multiplying exponents}$$

Converting back to a logarithmic equation gives us

$$\log_a M^p = xp.$$

STUDENT NOTES ———

Without understanding and *remembering* the rules of this section, it will be extremely difficult to solve the equations of Section 9.6.

But $x = \log_a M$, so substituting, we have

$$\log_a M^p = (\log_a M)p = p \cdot \log_a M.$$

Logarithms of Quotients

The third property that we study is similar to the quotient rule for exponents: $a^m/a^n = a^{m-n}$. Its proof follows Example 5.

> ### The Quotient Rule for Logarithms
>
> For any positive numbers M, N, and a ($a \neq 1$),
>
> $$\log_a \frac{M}{N} = \log_a M - \log_a N.$$
>
> (The logarithm of a quotient is the logarithm of the dividend minus the logarithm of the divisor.)

To better understand the quotient rule, note that

$$\log_2 \tfrac{8}{32} = \log_2 \tfrac{1}{4} = -2$$

and $\quad \log_2 8 - \log_2 32 = 3 - 5 = -2.$

EXAMPLE 4 Express as an equivalent expression that is a difference of logarithms: $\log_t (6/U)$.

SOLUTION

$$\log_t \frac{6}{U} = \log_t 6 - \log_t U \qquad \text{Using the quotient rule for logarithms}$$

> TRY EXERCISE 23

EXAMPLE 5 Express as an equivalent expression that is a single logarithm:

$$\log_b 17 - \log_b 27.$$

SOLUTION

$$\log_b 17 - \log_b 27 = \log_b \frac{17}{27} \qquad \begin{array}{l}\text{Using the quotient rule for}\\ \text{logarithms "in reverse"}\end{array}$$

> TRY EXERCISE 27

A Proof of the Quotient Rule. Our proof uses both the product rule and the power rule:

$$\log_a \frac{M}{N} = \log_a (MN^{-1}) \qquad\qquad \text{Rewriting } \frac{M}{N} \text{ as } MN^{-1}$$

$$= \log_a M + \log_a N^{-1} \qquad \text{Using the product rule for logarithms}$$

$$= \log_a M + (-1)\log_a N \qquad \text{Using the power rule for logarithms}$$

$$= \log_a M - \log_a N.$$

Using the Properties Together

EXAMPLE 6 Express as an equivalent expression, using the individual logarithms of x, y, and z.

a) $\log_b \dfrac{x^3}{yz}$
b) $\log_a \sqrt[4]{\dfrac{xy}{z^3}}$

SOLUTION

a) $\log_b \dfrac{x^3}{yz} = \log_b x^3 - \log_b yz \qquad \begin{array}{l}\text{Using the quotient rule for}\\ \text{logarithms}\end{array}$

$\qquad\qquad\quad = 3\log_b x - \log_b yz \qquad \text{Using the power rule for logarithms}$

$\qquad\qquad\quad = 3\log_b x - (\log_b y + \log_b z) \qquad \begin{array}{l}\text{Using the product rule for}\\ \text{logarithms. Because of the}\\ \text{subtraction, parentheses are}\\ \text{essential.}\end{array}$

$\qquad\qquad\quad = 3\log_b x - \log_b y - \log_b z \qquad \text{Using the distributive law}$

b) $\log_a \sqrt[4]{\dfrac{xy}{z^3}} = \log_a \left(\dfrac{xy}{z^3}\right)^{1/4}$ — Writing exponential notation

$\qquad = \dfrac{1}{4} \cdot \log_a \dfrac{xy}{z^3}$ — Using the power rule for logarithms

$\qquad = \dfrac{1}{4} \left(\log_a xy - \log_a z^3\right)$ — Using the quotient rule for logarithms. Parentheses are important.

$\qquad = \dfrac{1}{4} \left(\log_a x + \log_a y - 3\log_a z\right)$ — Using the product rule and the power rule for logarithms

> TRY EXERCISE 37

> *CAUTION!*   Because the product and quotient rules replace one term with two, it is often essential to apply the rules within parentheses, as in Example 6.

EXAMPLE 7   Express as an equivalent expression that is a single logarithm.

a) $\dfrac{1}{2}\log_a x - 7\log_a y + \log_a z$ **b)** $\log_a \dfrac{b}{\sqrt{x}} + \log_a \sqrt{bx}$

SOLUTION

a) $\dfrac{1}{2}\log_a x - 7\log_a y + \log_a z$

$\qquad = \log_a x^{1/2} - \log_a y^7 + \log_a z$ — Using the power rule for logarithms

$\qquad = \left(\log_a \sqrt{x} - \log_a y^7\right) + \log_a z$ — Using parentheses to emphasize the order of operations; $x^{1/2} = \sqrt{x}$

$\qquad = \log_a \dfrac{\sqrt{x}}{y^7} + \log_a z$ — Using the quotient rule for logarithms. Note that all terms have the same base.

$\qquad = \log_a \dfrac{z\sqrt{x}}{y^7}$ — Using the product rule for logarithms

b) $\log_a \dfrac{b}{\sqrt{x}} + \log_a \sqrt{bx} = \log_a \dfrac{b \cdot \sqrt{bx}}{\sqrt{x}}$ — Using the product rule for logarithms

$\qquad = \log_a b\sqrt{b}$ — Removing a factor equal to 1: $\dfrac{\sqrt{x}}{\sqrt{x}} = 1$

$\qquad = \log_a b^{3/2}, \text{ or } \dfrac{3}{2}\log_a b$ — Since $b\sqrt{b} = b^1 \cdot b^{1/2}$

> TRY EXERCISE 49

If we know the logarithms of two different numbers (with the same base), the properties allow us to calculate other logarithms.

EXAMPLE 8   Given $\log_a 2 = 0.431$ and $\log_a 3 = 0.683$, use the properties of logarithms to calculate a value for each of the following. If this is not possible, state so.

a) $\log_a 6$ **b)** $\log_a \dfrac{2}{3}$ **c)** $\log_a 81$

d) $\log_a \dfrac{1}{3}$ **e)** $\log_a (2a)$ **f)** $\log_a 5$

SOLUTION

a) $\log_a 6 = \log_a(2 \cdot 3) = \log_a 2 + \log_a 3$ Using the product rule for logarithms

$$= 0.431 + 0.683 = 1.114$$

Check: $a^{1.114} = a^{0.431} \cdot a^{0.683} = 2 \cdot 3 = 6$

b) $\log_a \frac{2}{3} = \log_a 2 - \log_a 3$ Using the quotient rule for logarithms

$$= 0.431 - 0.683 = -0.252$$

c) $\log_a 81 = \log_a 3^4 = 4 \log_a 3$ Using the power rule for logarithms

$$= 4(0.683) = 2.732$$

d) $\log_a \frac{1}{3} = \log_a 1 - \log_a 3$ Using the quotient rule for logarithms

$$= 0 - 0.683 = -0.683$$

e) $\log_a (2a) = \log_a 2 + \log_a a$ Using the product rule for logarithms

$$= 0.431 + 1 = 1.431$$

f) $\log_a 5$ *cannot be found using these properties.* $(\log_a 5 \neq \log_a 2 + \log_a 3)$

> **TRY EXERCISE** 55

A final property follows from the product rule: Since $\log_a a^k = k \log_a a$, and $\log_a a = 1$, we have $\log_a a^k = k$.

> ### The Logarithm of the Base to an Exponent
>
> For any base a,
>
> $$\log_a a^k = k.$$
>
> (The logarithm, base a, of a to an exponent is the exponent.)

This property also follows from the definition of logarithm: k is the exponent to which you raise a in order to get a^k.

EXAMPLE **9**   Simplify: **(a)** $\log_3 3^7$; **(b)** $\log_{10} 10^{-5.2}$.

SOLUTION

a) $\log_3 3^7 = 7$ 7 is the exponent to which you raise 3 in order to get 3^7.

b) $\log_{10} 10^{-5.2} = -5.2$ > **TRY EXERCISE** 65

We summarize the properties of logarithms as follows.

> For any positive numbers M, N, and a $(a \neq 1)$:
>
> $$\log_a (MN) = \log_a M + \log_a N; \qquad \log_a M^p = p \cdot \log_a M;$$
>
> $$\log_a \frac{M}{N} = \log_a M - \log_a N; \qquad \log_a a^k = k.$$

> *CAUTION!* Keep in mind that, in general,
>
> $\log_a (M + N) \neq \log_a M + \log_a N$, $\quad \log_a (MN) \neq (\log_a M)(\log_a N)$,
>
> $\log_a (M - N) \neq \log_a M - \log_a N$, $\quad \log_a \dfrac{M}{N} \neq \dfrac{\log_a M}{\log_a N}$.

9.4 EXERCISE SET

↪ **Concept Reinforcement** *In each of Exercises 1–6, match the expression with an equivalent expression from the column on the right.*

1. ____ $\log_7 20$

2. ____ $\log_7 5^4$

3. ____ $\log_7 \frac{5}{4}$

4. ____ $\log_7 7$

5. ____ $\log_7 1$

6. ____ $\log_7 5 + \log_7 6$

a) $\log_7 5 - \log_7 4$

b) 1

c) 0

d) $\log_7 30$

e) $\log_7 5 + \log_7 4$

f) $4 \log_7 5$

Express as an equivalent expression that is a sum of logarithms.

7. $\log_3 (81 \cdot 27)$

8. $\log_2 (16 \cdot 32)$

9. $\log_4 (64 \cdot 16)$

10. $\log_5 (25 \cdot 125)$

11. $\log_c (rst)$

12. $\log_t (3ab)$

Express as an equivalent expression that is a single logarithm.

13. $\log_a 2 + \log_a 10$

14. $\log_b 5 + \log_b 9$

15. $\log_c t + \log_c y$

16. $\log_t H + \log_t M$

Express as an equivalent expression that is a product.

17. $\log_a r^8$

18. $\log_b t^5$

19. $\log_2 y^{1/3}$

20. $\log_{10} y^{1/2}$

21. $\log_b C^{-3}$

22. $\log_c M^{-5}$

Express as an equivalent expression that is a difference of two logarithms.

23. $\log_2 \frac{5}{11}$

24. $\log_3 \frac{29}{13}$

25. $\log_b \dfrac{m}{n}$

26. $\log_a \dfrac{y}{x}$

Express as an equivalent expression that is a single logarithm.

27. $\log_a 19 - \log_a 2$

28. $\log_b 3 - \log_b 32$

29. $\log_b 36 - \log_b 4$

30. $\log_a 26 - \log_a 2$

31. $\log_a x - \log_a y$

32. $\log_b c - \log_b d$

Express as an equivalent expression, using the individual logarithms of w, x, y, and z.

33. $\log_a (xyz)$

34. $\log_a (wxy)$

35. $\log_a (x^3 z^4)$

36. $\log_a (x^2 y^5)$

37. $\log_a (w^2 x^{-2} y)$

38. $\log_a (xy^2 z^{-3})$

39. $\log_a \dfrac{x^5}{y^3 z}$

40. $\log_a \dfrac{x^4}{yz^2}$

41. $\log_b \dfrac{xy^2}{wz^3}$

42. $\log_b \dfrac{w^2 x}{y^3 z}$

43. $\log_a \sqrt{\dfrac{x^7}{y^5 z^8}}$

44. $\log_c \sqrt{\dfrac{x^4}{y^3 z^2}}$

45. $\log_a \sqrt[3]{\dfrac{x^6 y^3}{a^2 z^7}}$

46. $\log_a \sqrt[4]{\dfrac{x^8 y^{12}}{a^3 z^5}}$

Express as an equivalent expression that is a single logarithm and, if possible, simplify.

47. $8 \log_a x + 3 \log_a z$

48. $2 \log_b m + \frac{1}{2} \log_b n$

49. $\log_a x^2 - 2 \log_a \sqrt{x}$

50. $\log_a \dfrac{a}{\sqrt{x}} - \log_a \sqrt{ax}$

51. $\frac{1}{2} \log_a x + 5 \log_a y - 2 \log_a x$

52. $\log_a 2x + 3(\log_a x - \log_a y)$

53. $\log_a (x^2 - 9) - \log_a (x + 3)$

54. $\log_a (2x + 10) - \log_a (x^2 - 25)$

Given $\log_b 3 = 0.792$ *and* $\log_b 5 = 1.161$. *If possible, use the properties of logarithms to calculate values for each of the following.*

55. $\log_b 15$

56. $\log_b \frac{5}{3}$

57. $\log_b \frac{3}{5}$

58. $\log_b \frac{1}{3}$

59. $\log_b \frac{1}{5}$

60. $\log_b \sqrt{b}$

61. $\log_b \sqrt{b^3}$

62. $\log_b 3b$

63. $\log_b 8$

64. $\log_b 45$

Simplify.

Aha! **65.** $\log_t t^{10}$

66. $\log_p p^{-5}$

67. $\log_e e^m$

68. $\log_Q Q^t$

69. Explain the difference between the phrases "the logarithm of a quotient" and "a quotient of logarithms."

70. How could you convince someone that
$$\log_a c \neq \log_c a?$$

Skill Review

To prepare for Section 9.5, review graphing functions and finding domains of functions.

Graph.

71. $f(x) = \sqrt{x} - 3$ [7.1]

72. $g(x) = \sqrt[3]{x} + 1$ [7.1]

73. $g(x) = x^3 + 2$ [2.2]

74. $f(x) = 1 - x^2$ [8.7]

Find the domain of each function.

75. $f(x) = \dfrac{x - 3}{x + 7}$ [2.1]

76. $f(x) = \dfrac{x}{(x - 2)(x + 3)}$ [2.1]

77. $g(x) = \sqrt{10 - x}$ [7.1]

78. $g(x) = |x^2 - 6x + 7|$ [2.1]

Synthesis

79. A student *incorrectly* reasons that
$$\log_b \frac{1}{x} = \log_b \frac{x}{xx}$$
$$= \log_b x - \log_b x + \log_b x = \log_b x.$$
What mistake has the student made?

80. Why are properties of logarithms related to properties of exponents?

Express as an equivalent expression that is a single logarithm and, if possible, simplify.

81. $\log_a (x^8 - y^8) - \log_a (x^2 + y^2)$

82. $\log_a (x + y) + \log_a (x^2 - xy + y^2)$

Express as an equivalent expression that is a sum or a difference of logarithms and, if possible, simplify.

83. $\log_a \sqrt{1 - s^2}$

84. $\log_a \dfrac{c - d}{\sqrt{c^2 - d^2}}$

85. If $\log_a x = 2$, $\log_a y = 3$, and $\log_a z = 4$, what is
$$\log_a \frac{\sqrt[3]{x^2 z}}{\sqrt[3]{y^2 z^{-2}}}?$$

86. If $\log_a x = 2$, what is $\log_a (1/x)$?

87. If $\log_a x = 2$, what is $\log_{1/a} x$?

Solve.

88. $\log_{10} 2000 - \log_{10} x = 3$

89. $\log_2 80 + \log_2 x = 5$

Classify each of the following as true or false. Assume a, x, P, and Q > 0, a ≠ 1.

90. $\log_a \left(\dfrac{P}{Q}\right)^x = x \log_a P - \log_a Q$

91. $\log_a (Q + Q^2) = \log_a Q + \log_a (Q + 1)$

92. Use graphs to show that
$$\log x^2 \neq \log x \cdot \log x.$$
(*Note*: log means $\log_{10}$.)

9.5 Common and Natural Logarithms

Common Logarithms on a Calculator   •   The Base *e* and Natural Logarithms on a Calculator   •
Changing Logarithmic Bases   •   Graphs of Exponential and Logarithmic Functions, Base *e*

Any positive number other than 1 can serve as the base of a logarithmic function. However, some numbers are easier to use than others, and there are logarithmic bases that fit into certain applications more naturally than others.

Base-10 logarithms, called **common logarithms**, are useful because they have the same base as our "commonly" used decimal system. Before calculators became widely available, common logarithms helped with tedious calculations. In fact, that is why logarithms were devised.

The logarithmic base most widely used today is an irrational number named *e*. We will consider *e* and base *e*, or *natural*, logarithms later in this section. First we examine common logarithms.

Common Logarithms on a Calculator

Before the advent of scientific calculators, printed tables listed common logarithms. Today we find common logarithms using calculators.

Here, and in most books, the abbreviation **log**, with no base written, is understood to mean logarithm base 10, that is, a common logarithm. Thus,

$$\log 17 \quad \text{means} \quad \log_{10} 17. \qquad \text{It is important to remember this abbreviation.}$$

The key for common logarithms is usually marked **LOG**. To find the common logarithm of a number, we key in that number and press **LOG**. With most graphing calculators, we press **LOG**, the number, and then **ENTER**.

EXAMPLE 1   Use a calculator to approximate each number to four decimal places.

a) $\log 5312$

b) $\dfrac{\log 6500}{\log 0.007}$

SOLUTION

a) We enter 5312 and then press **LOG**. On most graphing calculators, we press **LOG**, followed by 5312 and **ENTER**. We find that

$$\log 5312 \approx 3.7253. \qquad \text{Rounded to four decimal places}$$

b) We enter 6500 and then press **LOG**. Next, we press **÷**, enter 0.007, and then press **LOG** **=**. On most graphing calculators, we press **LOG**, key in 6500, press **)** **÷** **LOG**, key in 0.007, and then press **)** **ENTER**. Be careful not to round until the end:

$$\frac{\log 6500}{\log 0.007} \approx -1.7694. \qquad \text{Rounded to four decimal places}$$

TRY EXERCISE ▸ 11

The inverse of a logarithmic function is an exponential function. Because of this, on many calculators the **LOG** key doubles as the **10ˣ** key after a **2ND** or SHIFT key is pressed. Calculators lacking a **10ˣ** key may have a key labeled x^y, a^x, or ⏷. Such a key can raise any positive real number to any real-numbered exponent.

EXAMPLE **2** Use a calculator to approximate $10^{3.417}$ to four decimal places.

SOLUTION We enter 3.417 and then press $\boxed{10^x}$. On most graphing calculators, $\boxed{10^x}$ is pressed first, followed by 3.417 and $\boxed{\text{ENTER}}$. Rounding to four decimal places, we have

$$10^{3.417} \approx 2612.1614.$$

TRY EXERCISE 21

The Base *e* and Natural Logarithms on a Calculator

When interest is compounded *n* times a year, the compound interest formula is

$$A = P\left(1 + \frac{r}{n}\right)^{nt},$$

where *A* is the amount that an initial investment *P* is worth after *t* years at interest rate *r*. Suppose that $1 is invested at 100% interest for 1 year (no bank would pay this). The preceding formula becomes a function *A* defined in terms of the number of compounding periods *n*:

$$A(n) = \left(1 + \frac{1}{n}\right)^n.$$

Let's find some function values. We use a calculator and round to six decimal places.

n	$A(n) = \left(1 + \dfrac{1}{n}\right)^n$
1 (compounded annually)	$2.00
2 (compounded semiannually)	2.25
3	2.370370
4 (compounded quarterly)	2.441406
12 (compounded monthly)	2.613035
100	2.704814
365 (compounded daily)	2.714567
8760 (compounded hourly)	2.718127

The numbers in this table approach a very important number in mathematics, called *e*. Because *e* is irrational, its decimal representation does not terminate or repeat.

The Number *e*

$e \approx 2.7182818284\ldots$

Logarithms base *e* are called **natural logarithms**, or **Napierian logarithms**, in honor of John Napier (1550–1617), the "inventor" of logarithms.

The abbreviation "ln" is generally used with natural logarithms. Thus,

$\ln 53$ means $\log_e 53$. It is important to remember this abbreviation.

On most calculators, the key for natural logarithms is marked $\boxed{\text{LN}}$.

EXAMPLE 3   Use a calculator to approximate ln 4568 to four decimal places.

SOLUTION   We enter 4568 and then press **LN**. On most graphing calculators, we press **LN** first, followed by 4568 and **ENTER**. We find that

$$\ln 4568 \approx 8.4268. \qquad \text{Rounded to four decimal places}$$

TRY EXERCISE 25

On many calculators, the **LN** key doubles as the **eˣ** key after a **2ND** or SHIFT key has been pressed.

EXAMPLE 4   Use a calculator to approximate $e^{-1.524}$ to four decimal places.

SOLUTION   We enter -1.524 and then press **eˣ**. On most graphing calculators, **eˣ** is pressed first, followed by -1.524 and **ENTER**. Since $e^{-1.524}$ is irrational, our answer is approximate:

$$e^{-1.524} \approx 0.2178. \qquad \text{Rounded to four decimal places}$$

TRY EXERCISE 31

Changing Logarithmic Bases

Most calculators can find both common and natural logarithms. To find a logarithm with some other base, a conversion formula is often used.

The Change-of-Base Formula

For any logarithmic bases a and b, and any positive number M,

$$\log_b M = \frac{\log_a M}{\log_a b}.$$

(To find the log, base b, of M, we typically compute $\log M / \log b$ or $\ln M / \ln b$.)

Proof.   Let $x = \log_b M$. Then,

$$b^x = M \qquad \log_b M = x \text{ is equivalent to } b^x = M.$$

$$\log_a b^x = \log_a M \qquad \text{Taking the logarithm, base } a, \text{ on both sides}$$

$$x \log_a b = \log_a M \qquad \text{Using the power rule for logarithms}$$

$$x = \frac{\log_a M}{\log_a b}. \qquad \text{Dividing both sides by } \log_a b$$

But at the outset we stated that $x = \log_b M$. Thus, by substitution, we have

$$\log_b M = \frac{\log_a M}{\log_a b}. \qquad \text{This is the change-of-base formula.}$$

EXAMPLE **5**   Find $\log_5 8$ using the change-of-base formula.

SOLUTION   We use the change-of-base formula with $a = 10$, $b = 5$, and $M = 8$:

$$\log_5 8 = \frac{\log_{10} 8}{\log_{10} 5}$$ Substituting into $\log_b M = \dfrac{\log_a M}{\log_a b}$

$$\approx \frac{0.903089987}{0.6989700043}$$ Using **LOG** twice

$$\approx 1.2920.$$ When using a calculator, it is best not to round until the end.

To check, note that $\ln 8 / \ln 5 \approx 1.2920$. We can also use a calculator to verify that $5^{1.2920} \approx 8$.

TRY EXERCISE ▶ 35

EXAMPLE **6**   Find $\log_4 31$.

SOLUTION   As shown in the check of Example 5, base e can also be used.

STUDENT NOTES

The choice of the logarithm base a in the change-of-base formula should be either 10 or e so that the logarithms can be found using a calculator. Either choice will yield the same end result.

$$\log_4 31 = \frac{\log_e 31}{\log_e 4}$$ Substituting into $\log_b M = \dfrac{\log_a M}{\log_a b}$

$$= \frac{\ln 31}{\ln 4} \approx \frac{3.433987204}{1.386294361}$$ Using **LN** twice

$$\approx 2.4771.$$ *Check:* $4^{2.4771} \approx 31$

TRY EXERCISE ▶ 41

Graphs of Exponential and Logarithmic Functions, Base e

EXAMPLE **7**   Graph $f(x) = e^x$ and $g(x) = e^{-x}$ and state the domain and the range of f and g.

SOLUTION   We use a calculator with an e^x key to find approximate values of e^x and e^{-x}. Using these values, we can graph the functions.

x	e^x	e^{-x}
0	1	1
1	2.7	0.4
2	7.4	0.1
−1	0.4	2.7
−2	0.1	7.4

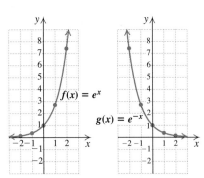

The domain of each function is $\mathbb{R}$ and the range of each function is $(0, \infty)$.

TRY EXERCISE ▶ 61

EXAMPLE **8**   Graph $f(x) = e^{-x} + 2$ and state the domain and the range of f.

SOLUTION   We find some solutions with a calculator, plot them, and then draw the graph. For example, $f(2) = e^{-2} + 2 \approx 0.1 + 2 \approx 2.1$. The graph is exactly like the graph of $g(x) = e^{-x}$, but is translated up 2 units.

x	$e^{-x} + 2$
0	3
1	2.4
2	2.1
-1	4.7
-2	9.4

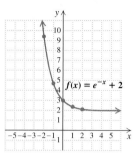

The domain of f is $\mathbb{R}$ and the range is $(2, \infty)$.   **TRY EXERCISE** 49

EXAMPLE **9**   Graph and state the domain and the range of each function.

a) $g(x) = \ln x$ **b)** $f(x) = \ln(x + 3)$

SOLUTION

a) We find some solutions with a calculator and then draw the graph. As expected, the graph is a reflection across the line $y = x$ of the graph of $y = e^x$.

x	$\ln x$
1	0
4	1.4
7	1.9
0.5	-0.7

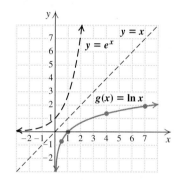

The domain of g is $(0, \infty)$ and the range is $\mathbb{R}$.

b) We find some solutions with a calculator, plot them, and draw the graph.

x	$\ln(x + 3)$
0	1.1
1	1.4
2	1.6
3	1.8
4	1.9
-1	0.7
-2	0
-2.5	-0.7

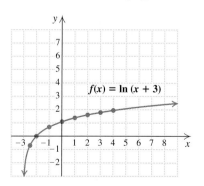

The graph of $y = \ln(x + 3)$ is the graph of $y = \ln x$ translated 3 units to the left. Since $x + 3$ must be positive, the domain is $(-3, \infty)$ and the range is $\mathbb{R}$.

TRY EXERCISE 63

TECHNOLOGY CONNECTION

Logarithmic functions with bases other than 10 or e can be drawn using the change-of-base formula. For example, $y = \log_5 x$ can be written $y = \ln x / \ln 5$.

1. Graph $y = \log_7 x$.
2. Graph $y = \log_5(x + 2)$.
3. Graph $y = \log_7 x + 2$.

Visualizing for Success

Match each function with its graph.

A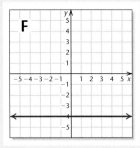

1. $f(x) = 2x - 3$

2. $f(x) = 2x^2 + 1$

3. $f(x) = \sqrt{x + 5}$

4. $f(x) = |x - 4|$

5. $f(x) = \ln x$

6. $f(x) = 2^{-x}$

7. $f(x) = -4$

8. $f(x) = \log x + 3$

9. $f(x) = 2^x$

10. $f(x) = 4 - x^2$

Answers on page A-45

An additional, animated version of this activity appears in MyMathLab. To use MyMathLab, you need a course ID and a student access code. Contact your instructor for more information.

F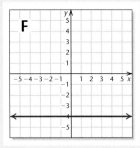

G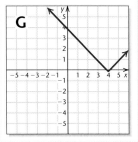

H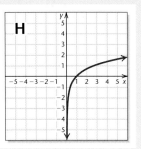

I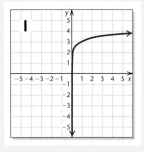

J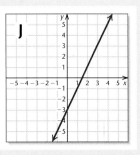

B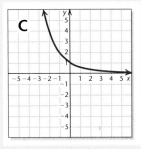

C

D

E

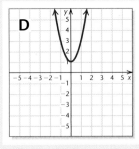

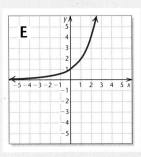

9.5 EXERCISE SET

Concept Reinforcement *Classify each statement as either true or false.*

1. The expression log 23 means $\log_{10} 23$.

2. The expression ln 7 means $\log_e 7$.

3. The number e is approximately 2.7.

4. The expressions log 9 and log 18/log 2 are equivalent.

5. The expressions log 9 and log 18 − log 2 are equivalent.

6. The expressions $\log_2 9$ and ln 9/ln 2 are equivalent.

7. The expressions ln 81 and 2 ln 9 are equivalent.

8. The domain of the function given by $f(x) = \ln(x + 2)$ is $(-2, \infty)$.

9. The range of the function given by $g(x) = e^x$ is $(0, \infty)$.

10. The range of the function given by $f(x) = \ln x$ is $(-\infty, \infty)$.

Use a calculator to find each of the following to four decimal places.

11. log 7

12. log 2

13. log 13.7

14. log 98.3 Aha!

15. log 1000

16. log 100

17. log 0.75

18. log 0.25

19. $\dfrac{\log 8200}{\log 2}$

20. $\dfrac{\log 5700}{\log 5}$

21. $10^{1.7}$

22. $10^{0.59}$

23. $10^{-2.9523}$

24. $10^{-3.2046}$

25. ln 9

26. ln 13

27. ln 0.0062

28. ln 0.00073

29. $\dfrac{\ln 2300}{0.08}$

30. $\dfrac{\ln 1900}{0.07}$

31. $e^{2.71}$

32. $e^{3.06}$

33. $e^{-3.49}$

34. $e^{-2.64}$

Find each of the following logarithms using the change-of-base formula. Round answers to four decimal places.

35. $\log_3 28$

36. $\log_6 37$

37. $\log_2 100$

38. $\log_7 100$

39. $\log_4 5$

40. $\log_8 7$

41. $\log_{0.1} 2$

42. $\log_{0.25} 25$

43. $\log_2 0.1$

44. $\log_{25} 0.25$

45. $\log_\pi 10$

46. $\log_\pi 100$

Graph and state the domain and the range of each function.

47. $f(x) = e^x$

48. $f(x) = e^{-x}$

49. $f(x) = e^x + 3$

50. $f(x) = e^x + 2$

51. $f(x) = e^x - 2$

52. $f(x) = e^x - 3$

53. $f(x) = 0.5e^x$

54. $f(x) = 2e^x$

55. $f(x) = 0.5e^{2x}$

56. $f(x) = 2e^{-0.5x}$

57. $f(x) = e^{x-3}$

58. $f(x) = e^{x-2}$

59. $f(x) = e^{x+2}$

60. $f(x) = e^{x+3}$

61. $f(x) = -e^x$

62. $f(x) = -e^{-x}$

63. $g(x) = \ln x + 1$

64. $g(x) = \ln x + 3$

65. $g(x) = \ln x - 2$

66. $g(x) = \ln x - 1$

67. $g(x) = 2 \ln x$

68. $g(x) = 3 \ln x$

69. $g(x) = -2 \ln x$

70. $g(x) = -\ln x$

71. $g(x) = \ln(x + 2)$

72. $g(x) = \ln(x + 1)$

73. $g(x) = \ln(x - 1)$

74. $g(x) = \ln(x - 3)$

75. Using a calculator, Adan gives an *incorrect* approximation for log 79 that is between 4 and 5. How could you convince him, without using a calculator, that he is mistaken?

76. Examine Exercise 75. What mistake do you believe Adan made?

Skill Review

To prepare for Section 9.6, review solving equations.

Solve.

77. $x^2 - 3x - 28 = 0$ [5.8]   **78.** $5x^2 - 7x = 0$ [5.8]

79. $17x - 15 = 0$ [1.3]   **80.** $\frac{5}{3} = 2t$ [1.3]

81. $(x - 5) \cdot 9 = 11$ [1.3]   **82.** $\frac{x + 3}{x - 3} = 7$ [6.4]

83. $x^{1/2} - 6x^{1/4} + 8 = 0$ [8.5]

84. $2y - 7\sqrt{y} + 3 = 0$ [8.5]

Synthesis

85. Explain how the graph of $f(x) = e^x$ could be used to graph the function given by $g(x) = 1 + \ln x$.

86. How would you explain to a classmate why $\log_2 5 = \log 5 / \log 2$ *and* $\log_2 5 = \ln 5 / \ln 2$?

Knowing only that $\log 2 \approx 0.301$ *and* $\log 3 \approx 0.477$, *approximate each of the following to three decimal places.*

87. $\log_6 81$ **88.** $\log_9 16$ **89.** $\log_{12} 36$

90. Find a formula for converting common logarithms to natural logarithms.

91. Find a formula for converting natural logarithms to common logarithms.

Solve for x. Give an approximation to four decimal places.

92. $\log (275x^2) = 38$ **93.** $\log (492x) = 5.728$

94. $\frac{3.01}{\ln x} = \frac{28}{4.31}$

95. $\log 692 + \log x = \log 3450$

For each function given below, (a) determine the domain and the range, (b) set an appropriate window, and (c) draw the graph. Graphs may vary, depending on the scale used.

96. $f(x) = 7.4e^x \ln x$

97. $f(x) = 3.4 \ln x - 0.25e^x$

98. $f(x) = x \ln (x - 2.1)$

99. $f(x) = 2x^3 \ln x$

100. Use a graphing calculator to check your answers to Exercises 49, 57, and 71.

101. Use a graphing calculator to check your answers to Exercises 48, 54, and 64.

102. In an attempt to solve $\ln x = 1.5$, Emma gets the following graph. How can Emma tell at a glance that she has made a mistake?

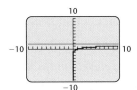

CONNECTING the CONCEPTS

It is important to distinguish between *simplifying* an exponential or logarithmic *expression* and *solving* an exponential or logarithmic *equation*. We use the following properties to simplify expressions and to rewrite equivalent logarithmic and exponential equations.

$\log_a x = m$ means $x = a^m$. $\log_a a^k = k$

$\log_a (MN) = \log_a M + \log_a N$ $\log_a a = 1$

$\log_a \dfrac{M}{N} = \log_a M - \log_a N$ $\log_a 1 = 0$

$\log_a M^p = p \cdot \log_a M$ $\log x = \log_{10} x$

$\log_b M = \dfrac{\log_a M}{\log_a b}$ $\ln x = \log_e x$

MIXED REVIEW

Simplify.

1. $\log_4 16$

2. $\log_5 \frac{1}{5}$

3. $\log_{100} 10$

4. $\log_{10} 100$

5. $\log 10$

6. $\ln 1$

7. $\log 10^4$

8. $\ln e^8$

9. $e^{\ln 7}$

10. $10^{\log 3}$

Rewrite each of the following as an equivalent exponential equation.

11. $\log_x 3 = m$

12. $\log_2 1024 = 10$

Rewrite each of the following as an equivalent logarithmic equation.

13. $e^t = x$

14. $64^{2/3} = 16$

Solve.

15. $\log_x 64 = 3$

16. $\log_3 x = -1$

17. Express as an equivalent expression using $\log x$, $\log y$, and $\log z$:
$$\log \sqrt{\frac{x^2}{yz^3}}.$$

18. Express as an equivalent expression that is a single logarithm: $\log a - 2 \log b - \log c$.

Find each of the following logarithms using the change-of-base formula. Round answers to four decimal places where appropriate.

19. $\log_4 8$

20. $\log_5 100$

9.6   Solving Exponential and Logarithmic Equations

Solving Exponential Equations   ■   Solving Logarithmic Equations

Solving Exponential Equations

Equations with variables in exponents, such as $5^x = 12$ and $2^{7x} = 64$, are called **exponential equations**. In Section 9.3, we solved certain exponential equations by using the principle of exponential equality. We restate that principle below.

The Principle of Exponential Equality

For any real number b, where $b \neq -1$, 0, or 1,

$$b^x = b^y \quad \text{is equivalent to} \quad x = y.$$

(Powers of the same base are equal if and only if the exponents are equal.)

EXAMPLE **1**   Solve: $4^{3x} = 16$.

SOLUTION   Note that $16 = 4^2$. Thus we can write each side as a power of the same base:

$$4^{3x} = 4^2 \qquad \text{Rewriting 16 as a power of 4}$$
$$3x = 2 \qquad \text{Since the base on each side is 4, the exponents are equal.}$$
$$x = \tfrac{2}{3}. \qquad \text{Solving for } x$$

Since $4^{3x} = 4^{3(2/3)} = 4^2 = 16$, the answer checks. The solution is $\tfrac{2}{3}$.

> TRY EXERCISE   9

In Example 1, we wrote both sides of the equation as powers of 4. When it seems impossible to write both sides of an equation as powers of the same base, we use the following principle and write an equivalent logarithmic equation.

The Principle of Logarithmic Equality

For any logarithmic base a, and for $x, y > 0$,

$$x = y \quad \text{is equivalent to} \quad \log_a x = \log_a y.$$

(Two expressions are equal if and only if the logarithms of those expressions are equal.)

The principle of logarithmic equality, used together with the power rule for logarithms, allows us to solve equations in which the variable is an exponent.

EXAMPLE **2**   Solve: $7^{x-2} = 60$.

SOLUTION   We have

Take the logarithm of both sides.

$$7^{x-2} = 60$$
$$\log 7^{x-2} = \log 60 \qquad \begin{array}{l}\text{Using the principle of logarithmic} \\ \text{equality to take the common} \\ \text{logarithm on both sides. Natural} \\ \text{logarithms also would work.}\end{array}$$

Use the power rule for logarithms.

$$(x - 2) \log 7 = \log 60 \qquad \text{Using the power rule for logarithms}$$
$$x - 2 = \frac{\log 60}{\log 7} \qquad \longleftarrow \boxed{\textit{CAUTION!} \text{ This is } not \log 60 - \log 7.}$$
$$x = \frac{\log 60}{\log 7} + 2 \qquad \text{Adding 2 to both sides}$$

Solve for x.

$$x \approx 4.1041. \qquad \begin{array}{l}\text{Using a calculator and rounding to four} \\ \text{decimal places}\end{array}$$

Check.   Since $7^{4.1041-2} \approx 60.0027$, we have a check. We can also note that since $7^{4-2} = 49$, we expect a solution greater than 4. The solution is $\dfrac{\log 60}{\log 7} + 2$, or approximately 4.1041.

> TRY EXERCISE   17

ALGEBRAIC–GRAPHICAL CONNECTION

The solution of $4^x = 16$ can be visualized by graphing $y = 4^x$ and $y = 16$ on the same set of axes. The y-values for both equations are the same where the graphs intersect. The x-value at that point is the solution of the equation. That x-value appears to be 2. Since $4^2 = 16$, we see that this is indeed the case.

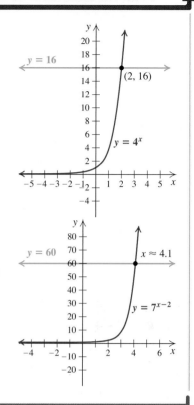

Similarly, the solution of Example 2 can be visualized by graphing $y = 7^{x-2}$ and $y = 60$ and identifying the x-value at the point of intersection. As expected, this value appears to be approximately 4.1.

EXAMPLE   3   Solve: $e^{0.06t} = 1500$.

SOLUTION   Since one side is a power of e, it is easiest to take the *natural logarithm* on both sides:

$$\ln e^{0.06t} = \ln 1500 \qquad \text{Taking the natural logarithm on both sides}$$

$$0.06t = \ln 1500 \qquad \text{Finding the logarithm of the base to a power:} \ \log_a a^k = k. \text{ Logarithmic and exponential functions are inverses of each other.}$$

$$t = \frac{\ln 1500}{0.06} \qquad \text{Dividing both sides by 0.06}$$

$$\approx 121.887. \qquad \text{Using a calculator and rounding to three decimal places}$$

> TRY EXERCISE 21

To Solve an Equation of the Form $a^t = b$ for t

1. Take the logarithm (either natural or common) of both sides.
2. Use the power rule for logarithms so that the variable is no longer written as an exponent.
3. Divide both sides by the coefficient of the variable to isolate the variable.
4. If appropriate, use a calculator to find an approximate solution.

Solving Logarithmic Equations

Recall from Section 9.3 that certain logarithmic equations can be solved by writing an equivalent exponential equation.

EXAMPLE   **4**   Solve: **(a)** $\log_4 (8x - 6) = 3$; **(b)** $\ln (5x) = 27$.

SOLUTION

a) $\log_4 (8x - 6) = 3$

$\qquad\qquad 4^3 = 8x - 6$ Writing the equivalent exponential equation

$\qquad\qquad 64 = 8x - 6$

$\qquad\qquad 70 = 8x$ Adding 6 to both sides

$\qquad\qquad x = \frac{70}{8}$, or $\frac{35}{4}$.

Check:

$$\begin{array}{c|c} \log_4 (8x - 6) = 3 & \\ \hline \log_4 (8 \cdot \frac{35}{4} - 6) & 3 \\ \log_4 (2 \cdot 35 - 6) & \\ \log_4 64 & \\ & 3 \overset{?}{=} 3 \quad \text{TRUE} \end{array}$$

The solution is $\frac{35}{4}$.

STUDENT NOTES

It is essential that you remember the properties of logarithms from Section 9.4. Consider reviewing the properties before attempting to solve equations similar to those in Example 5.

b) $\ln (5x) - 27$ Remember: $\ln (5x)$ means $\log_e (5x)$.

$\qquad e^{27} = 5x$ Writing the equivalent exponential equation

$\qquad \dfrac{e^{27}}{5} = x$ This is a very large number.

The solution is $\dfrac{e^{27}}{5}$. The check is left to the student.   **TRY EXERCISE** ▶ 45

Often the properties for logarithms are needed in order to solve a logarithmic equation. The goal is to first write an equivalent equation in which the variable appears in just one logarithmic expression. We then isolate that expression and solve as in Example 4.

EXAMPLE   **5**   Solve.

a) $\log x + \log (x - 3) = 1$

b) $\log_2 (x + 7) - \log_2 (x - 7) = 3$

c) $\log_7 (x + 1) + \log_7 (x - 1) = \log_7 8$

SOLUTION

a) To increase understanding, we write in the base, 10.

Find a single logarithm.

$\qquad\qquad \log_{10} x + \log_{10} (x - 3) = 1$

$\qquad\qquad\qquad \log_{10} [x(x - 3)] = 1$ Using the product rule for logarithms to obtain a single logarithm

Write an equivalent exponential equation.

$\qquad\qquad\qquad\qquad x(x - 3) = 10^1$ Writing an equivalent exponential equation

$\qquad\qquad\qquad\qquad x^2 - 3x = 10$

$\qquad\qquad\qquad x^2 - 3x - 10 = 0$

$\qquad\qquad\quad (x + 2)(x - 5) = 0$ Factoring

$\qquad\quad x + 2 = 0 \quad or \quad x - 5 = 0$ Using the principle of zero products

Solve.

$\qquad\qquad x = -2 \quad or \qquad x = 5$

Check.

Check:

For -2:

$$\frac{\log x + \log (x - 3) = 1}{\log (-2) + \log (-2 - 3) \stackrel{?}{=} 1} \quad \text{FALSE}$$

For 5:

$$\frac{\log x + \log (x - 3) = 1}{\begin{array}{c|c} \log 5 + \log (5 - 3) & 1 \\ \log 5 + \log 2 & \\ \log 10 & \\ & 1 \stackrel{?}{=} 1 \quad \text{TRUE} \end{array}}$$

The number -2 *does not check* because the logarithm of a negative number is undefined. The solution is 5.

b) We have

$$\log_2 (x + 7) - \log_2 (x - 7) = 3$$

$$\log_2 \frac{x + 7}{x - 7} = 3 \qquad \text{Using the quotient rule for logarithms to obtain a single logarithm}$$

$$\frac{x + 7}{x - 7} = 2^3 \qquad \text{Writing an equivalent exponential equation}$$

$$\frac{x + 7}{x - 7} = 8$$

$$x + 7 = 8(x - 7) \qquad \text{Multiplying by the LCD, } x - 7$$

$$x + 7 = 8x - 56 \qquad \text{Using the distributive law}$$

$$63 = 7x$$

$$9 = x. \qquad \text{Dividing by 7}$$

Check:

$$\frac{\log_2 (x + 7) - \log_2 (x - 7) = 3}{\begin{array}{c|c} \log_2 (9 + 7) - \log_2 (9 - 7) & 3 \\ \log_2 16 - \log_2 2 & \\ 4 - 1 & \\ & 3 \stackrel{?}{=} 3 \quad \text{TRUE} \end{array}}$$

The solution is 9.

c) We have

$$\log_7 (x + 1) + \log_7 (x - 1) = \log_7 8$$

$$\log_7 [(x + 1)(x - 1)] = \log_7 8 \qquad \text{Using the product rule for logarithms}$$

$$\log_7 (x^2 - 1) = \log_7 8 \qquad \text{Multiplying. Note that both sides are base-7 logarithms.}$$

$$x^2 - 1 = 8 \qquad \text{Using the principle of logarithmic equality. Study this step carefully.}$$

$$x^2 - 9 = 0$$

$$(x - 3)(x + 3) = 0 \qquad \text{Solving the quadratic equation}$$

$$x = 3 \quad or \quad x = -3.$$

We leave it to the student to show that 3 checks but -3 does not. The solution is 3.

> **TRY EXERCISE** 55

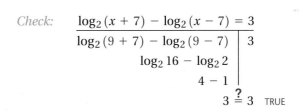

9.6 EXERCISE SET

For Extra Help

↪ *Concept Reinforcement* *In each of Exercises 1–8, match the equation with an equivalent equation from the column on the right that could be the next step in the solution process.*

1. ____ $5^x = 3$

2. ____ $e^{5x} = 3$

3. ____ $\ln x = 3$

4. ____ $\log_x 5 = 3$

5. ____ $\log_5 x + \log_5 (x - 2) = 3$

6. ____ $\log_5 x - \log_5 (x - 2) = 3$

7. ____ $\ln x - \ln (x - 2) = 3$

8. ____ $\log x + \log (x - 2) = 3$

a) $\ln e^{5x} = \ln 3$

b) $\log_5 (x^2 - 2x) = 3$

c) $\log (x^2 - 2x) = 3$

d) $\log_5 \dfrac{x}{x - 2} = 3$

e) $\log 5^x = \log 3$

f) $e^3 = x$

g) $\ln \dfrac{x}{x - 2} = 3$

h) $x^3 = 5$

Solve. Where appropriate, include approximations to three decimal places.

9. $3^{2x} = 81$

10. $2^{3x} = 64$

11. $4^x = 32$

12. $9^x = 27$

13. $2^x = 10$

14. $2^x = 24$

15. $2^{x+5} = 16$

16. $2^{x-1} = 8$

17. $8^{x-3} = 19$

18. $5^{x+2} = 15$

19. $e^t = 50$

20. $e^t = 20$

21. $e^{-0.02t} = 8$

22. $e^{-0.01t} = 100$

23. $4.9^x - 87 = 0$

24. $7.2^x - 65 = 0$

25. $19 = 2e^{4x}$

26. $29 = 3e^{2x}$

27. $7 + 3e^{-x} = 13$

28. $4 + 5e^{-x} = 9$

Aha! 29. $\log_3 x = 4$

30. $\log_2 x = 6$

31. $\log_4 x = -2$

32. $\log_5 x = -3$

33. $\ln x = 5$

34. $\ln x = 4$

35. $\ln (4x) = 3$

36. $\ln (3x) = 2$

37. $\log x = 1.2$

38. $\log x = 0.6$

39. $\ln (2x + 1) = 4$

40. $\ln (4x - 2) = 3$

Aha! 41. $\ln x = 1$

42. $\log x = 1$

43. $5 \ln x = -15$

44. $3 \ln x = -3$

45. $\log_2 (8 - 6x) = 5$

46. $\log_5 (7 - 2x) = 3$

47. $\log (x - 9) + \log x = 1$

48. $\log (x + 9) + \log x = 1$

49. $\log x - \log (x + 3) = 1$

50. $\log x - \log (x + 7) = -1$

Aha! 51. $\log (2x + 1) = \log 5$

52. $\log (x + 1) - \log x = 0$

53. $\log_4 (x + 3) = 2 + \log_4 (x - 5)$

54. $\log_2 (x + 3) = 4 + \log_2 (x - 3)$

55. $\log_7 (x + 1) + \log_7 (x + 2) = \log_7 6$

56. $\log_6 (x + 3) + \log_6 (x + 2) = \log_6 20$

57. $\log_5 (x + 4) + \log_5 (x - 4) = \log_5 20$

58. $\log_4 (x + 2) + \log_4 (x - 7) = \log_4 10$

59. $\ln (x + 5) + \ln (x + 1) = \ln 12$

60. $\ln (x - 6) + \ln (x + 3) = \ln 22$

61. $\log_2 (x - 3) + \log_2 (x + 3) = 4$

62. $\log_3 (x - 4) + \log_3 (x + 4) = 2$

63. $\log_{12} (x + 5) - \log_{12} (x - 4) = \log_{12} 3$

64. $\log_6 (x + 7) - \log_6 (x - 2) = \log_6 5$

65. $\log_2 (x - 2) + \log_2 x = 3$

66. $\log_4 (x + 6) - \log_4 x = 2$

67. Madison finds that the solution of $\log_3 (x + 4) = 1$ is -1, but rejects -1 as an answer. What mistake do you suspect she is making?

68. Could Example 2 have been solved by taking the natural logarithm on both sides? Why or why not?

Skill Review

To prepare for Section 9.7, review using the five-step problem-solving strategy.

Solve.

69. A rectangle is 6 ft longer than it is wide. Its perimeter is 26 ft. Find the length and the width. [1.4]

70. Under one health insurance plan offered in California, the maximum co-pay for an individual is $3000 per calendar year. The co-pay for each visit to a specialist is $40, and the co-pay for a hospitalization is $1000. With hospitalizations and specialist visits, Marguerite reached the maximum co-pay in 2008. If she was hospitalized twice, how many visits to specialists did she make? [4.1]
Source: ehealthinsurance.com

71. Joanna wants to mix Golden Days bird seed containing 25% sunflower seeds with Snowy Friends bird seed containing 40% sunflower seeds. She wants 50 lb of a mixture containing 33% sunflower seeds. How much of each type should she use? [3.3]

72. The outside edge of a picture frame measures 12 cm by 19 cm, and $144 \, cm^2$ of picture shows. Find the width of the frame. [5.8]

73. Max can key in a musical score in 2 hr. Miles takes 3 hr to key in the same score. How long would it take them, working together, to key in the score? [6.5]

74. A sign is in the shape of a right triangle. The hypotenuse is 3 ft long, and the base and the height of the triangle are equal. Find the length of the base and the height. Round to the nearest tenth of a foot. [7.7]

Synthesis

75. Can the principle of logarithmic equality be expanded to include all functions? That is, is the statement "$m = n$ is equivalent to $f(m) = f(n)$" true for any function f? Why or why not?

76. Explain how Exercises 37 and 38 could be solved using the graph of $f(x) = \log x$.

Solve. If no solution exists, state this.

77. $8^x = 16^{3x+9}$

78. $27^x = 81^{2x-3}$

79. $\log_6 (\log_2 x) = 0$

80. $\log_x (\log_3 27) = 3$

81. $\log_5 \sqrt{x^2 - 9} = 1$

82. $x \log \frac{1}{8} = \log 8$

83. $2^{x^2+4x} = \frac{1}{8}$

84. $\log (\log x) = 5$

85. $\log_5 |x| = 4$

86. $\log x^2 = (\log x)^2$

87. $\log \sqrt{2x} = \sqrt{\log 2x}$

88. $1000^{2x+1} = 100^{3x}$

89. $3^{x^2} \cdot 3^{4x} = \frac{1}{27}$

90. $3^{3x} \cdot 3^{x^2} = 81$

91. $\log x^{\log x} = 25$

92. $3^{2x} - 8 \cdot 3^x + 15 = 0$

93. $(81^{x-2})(27^{x+1}) = 9^{2x-3}$

94. $3^{2x} - 3^{2x-1} = 18$

95. Given that $2^y = 16^{x-3}$ and $3^{y+2} = 27^x$, find the value of $x + y$.

96. If $x = (\log_{125} 5)^{\log_5 125}$, what is the value of $\log_3 x$?

97. Find the value of x for which the natural logarithm is the same as the common logarithm.

98. Use a graphing calculator to check your answers to Exercises 11, 31, 41, and 59.

9.7 Applications of Exponential and Logarithmic Functions

Applications of Logarithmic Functions ● Applications of Exponential Functions

We now consider applications of exponential and logarithmic functions.

Applications of Logarithmic Functions

EXAMPLE **1**

Sound levels. To measure the volume, or "loudness," of a sound, the *decibel* scale is used. The loudness L, in decibels (dB), of a sound is given by

$$L = 10 \cdot \log \frac{I}{I_0},$$

where I is the intensity of the sound, in watts per square meter (W/m^2), and $I_0 = 10^{-12}$ W/m^2. (I_0 is approximately the intensity of the softest sound that can be heard by the human ear.)

a) The average maximum intensity of sound in a New York subway car is about 3.2×10^{-3} W/m^2. How loud, in decibels, is the sound level?

Source: Columbia University Mailman School of Public Health

b) The Occupational Safety and Health Administration (OSHA) considers sustained sound levels of 90 dB and above unsafe. What is the intensity of such sounds?

SOLUTION

a) To find the loudness, in decibels, we use the above formula:

$$L = 10 \cdot \log \frac{I}{I_0}$$

$= 10 \cdot \log \dfrac{3.2 \times 10^{-3}}{10^{-12}}$	Substituting
$= 10 \cdot \log \left(3.2 \times 10^{9}\right)$	Subtracting exponents
$= 10 \left(\log 3.2 + \log 10^{9}\right)$	$\log MN = \log M + \log N$
$= 10 \left(\log 3.2 + 9\right)$	$\log_{10} 10^{9} = 9$
$\approx 10 \left(0.5051 + 9\right)$	Approximating log 3.2
$= 10 \left(9.5051\right)$	Adding within the parentheses
$\approx 95.$	Multiplying and rounding

The volume of the sound in a subway car is about 95 decibels.

b) We substitute and solve for I:

$$L = 10 \cdot \log \frac{I}{I_0}$$

$$90 = 10 \cdot \log \frac{I}{10^{-12}} \qquad \text{Substituting}$$

$$9 = \log \frac{I}{10^{-12}} \qquad \text{Dividing both sides by 10}$$

$$9 = \log I - \log 10^{-12} \qquad \text{Using the quotient rule for logarithms}$$

$$9 = \log I - (-12) \qquad \log 10^a = a$$

$$-3 = \log I \qquad \text{Adding } -12 \text{ to both sides}$$

$$10^{-3} = I. \qquad \text{Converting to an exponential equation}$$

Sustained sounds with intensities exceeding 10^{-3} W/m^2 are considered unsafe.

TRY EXERCISE ▶ 15

EXAMPLE 2

Chemistry: pH of liquids. In chemistry, the pH of a liquid is a measure of its acidity. We calculate pH as follows:

$$\text{pH} = -\log [\text{H}^+],$$

where $[\text{H}^+]$ is the hydrogen ion concentration in moles per liter.

a) The hydrogen ion concentration of human blood is normally about 3.98×10^{-8} moles per liter. Find the pH.

Source: www.merck.com

b) The average pH of seawater is about 8.2. Find the hydrogen ion concentration.

Source: www.seafriends.org.nz

SOLUTION

a) To find the pH of blood, we use the above formula:

$$\text{pH} = -\log [\text{H}^+]$$

$$= -\log [3.98 \times 10^{-8}]$$

$$\approx -(-7.400117) \qquad \text{Using a calculator}$$

$$\approx 7.4.$$

The pH of human blood is normally about 7.4.

b) We substitute and solve for $[\text{H}^+]$:

$$8.2 = -\log [\text{H}^+] \qquad \text{Using pH} = -\log [\text{H}^+]$$

$$-8.2 = \log [\text{H}^+] \qquad \text{Dividing both sides by } -1$$

$$10^{-8.2} = [\text{H}^+] \qquad \text{Converting to an exponential equation}$$

$$6.31 \times 10^{-9} \approx [\text{H}^+]. \qquad \text{Using a calculator; writing scientific notation}$$

The hydrogen ion concentration of seawater is about 6.31×10^{-9} moles per liter.

TRY EXERCISE ▶ 11

Applications of Exponential Functions

> **EXAMPLE 3**   *Interest compounded annually.*   Suppose that $25,000 is invested at 4% interest, compounded annually. In t years, it will grow to the amount A given by
>
> $$A(t) = 25{,}000(1.04)^t.$$
>
> (See Example 5 in Section 9.2.)
>
> **a)** How long will it take to accumulate $80,000 in the account?
>
> **b)** Find the amount of time it takes for the $25,000 to double itself.

SOLUTION

a) We set $A(t) = 80{,}000$ and solve for t:

$$80{,}000 = 25{,}000(1.04)^t$$

$$\frac{80{,}000}{25{,}000} = 1.04^t \qquad \text{Dividing both sides by 25,000}$$

$$3.2 = 1.04^t$$

$$\log 3.2 = \log 1.04^t \qquad \text{Taking the common logarithm on both sides}$$

$$\log 3.2 = t \log 1.04 \qquad \text{Using the power rule for logarithms}$$

$$\frac{\log 3.2}{\log 1.04} = t \qquad \text{Dividing both sides by log 1.04}$$

$$29.7 \approx t. \qquad \text{Using a calculator}$$

Remember that when doing a calculation like this on a calculator, it is best to wait until the end to round. At an interest rate of 4% per year, it will take about 29.7 yr for $25,000 to grow to $80,000.

STUDENT NOTES

Study the different steps in the solution of Example 3(b). Note that if 50,000 and 25,000 are replaced with 6000 and 3000, the doubling time is unchanged.

b) To find the *doubling time*, we replace $A(t)$ with 50,000 and solve for t:

$$50{,}000 = 25{,}000(1.04)^t$$

$$2 = (1.04)^t \qquad \text{Dividing both sides by 25,000}$$

$$\log 2 = \log (1.04)^t \qquad \begin{array}{l}\text{Taking the common logarithm}\\\text{on both sides}\end{array}$$

$$\log 2 = t \log 1.04 \qquad \text{Using the power rule for logarithms}$$

$$t = \frac{\log 2}{\log 1.04} \approx 17.7. \qquad \begin{array}{l}\text{Dividing both sides by log 1.04 and using}\\\text{a calculator}\end{array}$$

At an interest rate of 4% per year, the doubling time is about 17.7 yr.

> **TRY EXERCISE** 21

Like investments, populations often grow exponentially.

Exponential Growth

An **exponential growth model** is a function of the form

$$P(t) = P_0 e^{kt}, \quad k > 0,$$

where P_0 is the population at time 0, $P(t)$ is the population at time t, and k is the **exponential growth rate** for the situation. The **doubling time** is the amount of time necessary for the population to double in size.

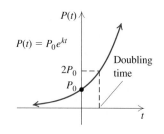

The exponential growth rate is the rate of growth of a population or other quantity at any *instant* in time. Since the population is continually growing, the percent of total growth after one year will exceed the exponential growth rate.

EXAMPLE 4

Text messaging.   In 2000, there were approximately 12 million text messages sent each month in the United States. This number has increased exponentially at an average rate of 108% per year.

Source: Based on information from www.cellulist.com

a) Find the exponential growth function that models the data.

b) Estimate the number of text messages sent each month in 2009.

SOLUTION

a) In 2000, at $t = 0$, the number of messages was 12 million per month. We substitute 12 for P_0 and 108%, or 1.08, for k. This gives the exponential growth function

$$P(t) = 12e^{1.08t}.$$

b) In 2009, we have $t = 9$ (since 9 yr have passed since 2000). To determine the number of messages in 2009, we compute $P(9)$:

$$P(9) = 12e^{1.08(9)} \qquad \text{Using } P(t) = 12e^{1.08t} \text{ from part (a)}$$
$$= 12e^{9.72} \approx 200{,}000. \qquad \text{Using a calculator}$$

In 2009, the number of text messages sent in the United States each month will reach approximately 200,000 million, or 200 billion.

> TRY EXERCISE 23

EXAMPLE 5

Cruise ship passengers.   In 1970, cruise lines carried approximately 500,000 passengers. This number has increased exponentially to 12.1 million in 2006.

Source: Cruise Lines International Association

a) Find the exponential growth rate and the exponential growth function.

b) Estimate the year in which cruise lines will carry 20 million passengers.

SOLUTION

a) We use $S(t) = S_0 e^{kt}$, where t is the number of years since 1970 and $S(t)$ is the number of passengers, in millions. Since 500,000 is half a million, we substitute 0.5 for S_0:

$$S(t) = 0.5\,e^{kt}.$$

To find the exponential growth rate k, note that after 36 yr (2006 − 1970 = 36), there were 12.1 million passengers:

$$S(36) = 0.5e^{k \cdot 36}$$

$$12.1 = 0.5e^{36k} \qquad \text{Substituting}$$

$$24.2 = e^{36k} \qquad \text{Dividing both sides by 0.5}$$

$$\ln 24.2 = \ln e^{36k} \qquad \text{Taking the natural logarithm on both sides}$$

$$\ln 24.2 = 36k \qquad \ln e^{36k} = \log_e e^{36k} = 36k$$

$$\frac{\ln 24.2}{36} = k \qquad \text{Dividing both sides by 36}$$

$$0.089 \approx k. \qquad \text{Using a calculator and rounding}$$

The exponential growth rate is 8.9% and the exponential growth function is given by $S(t) = 0.5e^{0.089t}$.

A visualization of Example 5

b) To estimate the year in which cruise lines will carry 20 million passengers, we replace $S(t)$ with 20 and solve for t:

$$20 = 0.5e^{0.089t}$$

$$40 = e^{0.089t} \qquad \text{Dividing both sides by 0.5}$$

$$\ln 40 = \ln e^{0.089t} \qquad \text{Taking the natural logarithm on both sides}$$

$$\ln 40 = 0.089t \qquad \ln e^a = a$$

$$\frac{\ln 40}{0.089} = t \qquad \text{Dividing both sides by 0.089}$$

$$41.4 \approx t. \qquad \text{Using a calculator}$$

Rounding to 41, we see that, according to this model, cruise lines will carry 20 million passengers 41 yr after 1970, or in 2011. **TRY EXERCISE** 31

EXAMPLE **6** *Interest compounded continuously.* When an amount of money P_0 is invested at interest rate k, compounded *continuously*, interest is computed every "instant" and added to the original amount. The balance $P(t)$, after t years, is given by the exponential growth model

$$P(t) = P_0 e^{kt}.$$

a) Suppose that \$30,000 is invested and grows to \$44,754.75 in 5 yr. Find the exponential growth function.

b) What is the doubling time?

SOLUTION

a) We have $P(0) = 30,000$. Thus the exponential growth function is

$$P(t) = 30,000e^{kt}, \quad \text{where } k \text{ must still be determined.}$$

Knowing that for $t = 5$ we have $P(5) = 44,754.75$, it is possible to solve for k:

$$44,754.75 = 30,000e^{k(5)}$$

$$44,754.75 = 30,000e^{5k}$$

$$\frac{44,754.75}{30,000} = e^{5k} \qquad \text{Dividing both sides by 30,000}$$

$$1.491825 = e^{5k}$$

$$\ln 1.491825 = \ln e^{5k} \qquad \text{Taking the natural logarithm on both sides}$$

$$\ln 1.491825 = 5k \qquad \ln e^a = a$$

$$\frac{\ln 1.491825}{5} = k \qquad \text{Dividing both sides by 5}$$

$$0.08 \approx k. \qquad \text{Using a calculator and rounding}$$

The interest rate is about 0.08, or 8%, compounded continuously. Because interest is being compounded continuously, the yearly interest rate is a bit more than 8%. The exponential growth function is

$$P(t) = 30,000e^{0.08t}.$$

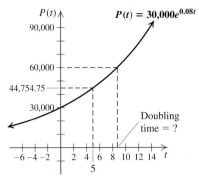

A visualization of Example 6

b) To find the doubling time T, we replace $P(T)$ with 60,000 and solve for T:

$$60{,}000 = 30{,}000e^{0.08T}$$

$$2 = e^{0.08T} \qquad \text{Dividing both sides by 30,000}$$

$$\ln 2 = \ln e^{0.08T} \qquad \text{Taking the natural logarithm on both sides}$$

$$\ln 2 = 0.08T \qquad \ln e^a = a$$

$$\frac{\ln 2}{0.08} = T \qquad \text{Dividing both sides by 0.08}$$

$$8.7 \approx T. \qquad \text{Using a calculator and rounding}$$

Thus the original investment of $30,000 will double in about 8.7 yr.

TRY EXERCISE 41

For any specified interest rate, continuous compounding gives the highest yield and the shortest doubling time.

In some real-life situations, a quantity or population is *decreasing* or *decaying* exponentially.

Exponential Decay

An **exponential decay model** is a function of the form

$$P(t) = P_0 e^{-kt}, \quad k > 0,$$

where P_0 is the quantity present at time 0, $P(t)$ is the amount present at time t, and k is the **decay rate**. The **half-life** is the amount of time necessary for half of the quantity to decay.

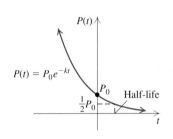

EXAMPLE 7

Chaco Canyon, New Mexico

Carbon dating. The radioactive element carbon-14 has a half-life of 5750 yr. The percentage of carbon-14 in the remains of organic matter can be used to determine the age of that material. Recently, while digging in Chaco Canyon, New Mexico, archaeologists found corn pollen that had lost 38.1% of its carbon-14. The age of this corn pollen was evidence that Indians had been cultivating crops in the Southwest centuries earlier than scientists had thought. What was the age of the pollen?

Source: *American Anthropologist*

SOLUTION   We first find k. To do so, we use the concept of half-life. When $t = 5750$ (the half-life), $P(t)$ is half of P_0. Then

$$0.5P_0 = P_0 e^{-k(5750)} \qquad \text{Substituting in } P(t) = P_0 e^{-kt}$$

$$0.5 = e^{-5750k} \qquad \text{Dividing both sides by } P_0$$

$$\ln 0.5 = \ln e^{-5750k} \qquad \text{Taking the natural logarithm on both sides}$$

$$\ln 0.5 = -5750k \qquad \ln e^a = a$$

$$\frac{\ln 0.5}{-5750} = k \qquad \text{Dividing}$$

$$0.00012 \approx k. \qquad \text{Using a calculator and rounding}$$

Now we have a function for the decay of carbon-14:

$$P(t) = P_0 e^{-0.00012t}.$$ This completes the first part of our solution.

(*Note*: This equation can be used for subsequent carbon-dating problems.) If the corn pollen has lost 38.1% of its carbon-14 from an initial amount P_0, then $100\% - 38.1\%$, or 61.9%, of P_0 is still present. To find the age t of the pollen, we solve this equation for t:

$$0.619 P_0 = P_0 e^{-0.00012t} \qquad \text{We want to find } t \text{ for which } P(t) = 0.619 P_0.$$

$$0.619 = e^{-0.00012t} \qquad \text{Dividing both sides by } P_0$$

$$\ln 0.619 = \ln e^{-0.00012t} \qquad \text{Taking the natural logarithm on both sides}$$

$$\ln 0.619 = -0.00012t \qquad \ln e^a = a$$

$$\frac{\ln 0.619}{-0.00012} = t \qquad \text{Dividing both sides by } -0.00012$$

$$4000 \approx t. \qquad \text{Using a calculator}$$

The pollen is about 4000 yr old.

TRY EXERCISE 35

9.7 EXERCISE SET

Solve.

1. *Asteroids.* The total number $A(t)$ of known asteroids t years after 1990 can be estimated by

$$A(t) = 77(1.283)^t.$$

Source: Based on data from NASA

a) Determine the year in which the number of known asteroids first reached 4000.
b) What is the doubling time for the number of known asteroids?

2. *Social networking.* The number of unique (different) visitors per month to Facebook t months after April 2006 can be estimated by

$$F(t) = 11.755(1.109)^t,$$

where $F(t)$ is in millions.

Source: Based on data from comScore World Metrix

a) In what month will the number of Facebook visitors first reach 1 billion?
b) What is the doubling time for the number of unique Facebook visitors per month?

3. *Health.* The rate of number of deaths due to stroke in the United States can be estimated by

$$S(t) = 180(0.97)^t,$$

where $S(t)$ is the number of deaths per 100,000 people and t is the number of years since 1960.

Source: Based on data from Centers for Disease Control and Prevention

a) In what year was the death rate due to stroke 100 per 100,000 people?
b) In what year will the death rate due to stroke be 25 per 100,000 people?

4. *Alternative fuels.* The number of gallons of ethanol produced in the United States can be estimated by

$$E(t) = 0.18(1.137)^t,$$

where $E(t)$ is the annual production, in billions of gallons, t years after 1980.

a) In what year did the United States produce 5 billion gal of ethanol?
b) In what year will the United States produce 25 billion gal of ethanol?

5. *Student loan repayment.* A college loan of $29,000 is made at 3% interest, compounded annually. After t years, the amount due, A, is given by the function

$$A(t) = 29,000(1.03)^t.$$

a) After what amount of time will the amount due reach $35,000?
b) Find the doubling time.

6. *Spread of a rumor.* The number of people who have heard a rumor increases exponentially. If all who hear a rumor repeat it to two people a day, and if 20 people start the rumor, the number of people N who have heard the rumor after t days is given by

$$N(t) = 20(3)^t.$$

a) After what amount of time will 1000 people have heard the rumor?

b) What is the doubling time for the number of people who have heard the rumor?

7. *Health insurance.*   The percentage of workers covered by a conventional health plan is decreasing exponentially. The percentage of covered workers $W(t)$ enrolled in conventional plans t years after 1988 can be estimated by

$$W(t) = 89(0.837)^t.$$

Sources: Based on data from Kaiser and HRET

a) According to this model, in what year did the percentage of covered workers enrolled in conventional plans drop below 50%?

b) In what year will the percentage of covered workers enrolled in conventional plans drop below 1%?

8. *Smoking.*   The percentage of smokers who received telephone counseling and had successfully quit smoking for t months is given by

$$P(t) = 21.4(0.914)^t.$$

Sources: *New England Journal of Medicine:* data from California's Smoker's Hotline

a) In what month will 15% of those who quit and used telephone counseling still be smoke-free?

b) In what month will 5% of those who quit and used telephone counseling still be smoke-free?

9. *Marine biology.*   As a result of preservation efforts in countries in which whaling was once common, the humpback whale population has grown since the 1970s. The worldwide population $P(t)$, in thousands, t years after 1982 can be estimated by

$$P(t) = 5.5(1.047)^t.$$

a) In what year will the humpback whale population reach 30,000?

b) Find the doubling time.

10. *World population.*   The world population $P(t)$, in billions, t years after 2000 can be approximated by

$$P(t) = 4.553(1.014)^t.$$

Sources: Based on data from U.S. Census Bureau; International Data Base

a) In what year will the world population reach 10 billion?

b) Find the doubling time.

Use the pH formula given in Example 2 for Exercises 11–14.

11. *Chemistry.*   The hydrogen ion concentration of fresh-brewed coffee is about 1.3×10^{-5} moles per liter. Find the pH.

12. *Chemistry.*   The hydrogen ion concentration of milk is about 1.6×10^{-7} moles per liter. Find the pH.

13. *Medicine.*   When the pH of a patient's blood drops below 7.4, a condition called *acidosis* sets in. Acidosis can be deadly when the patient's pH reaches 7.0. What would the hydrogen ion concentration of the patient's blood be at that point?

14. *Medicine.*   When the pH of a patient's blood rises above 7.4, a condition called *alkalosis* sets in. Alkalosis can be deadly when the patient's pH reaches 7.8. What would the hydrogen ion concentration of the patient's blood be at that point?

Use the formula in Example 1 for Exercises 15–18.

15. *Racing.*   The intensity of sound from a race car in full throttle is about 10 W/m². How loud in decibels is this sound level?

Source: nascar.about.com

16. *Audiology.*   The intensity of sound in normal conversation is about 3.2×10^{-6} W/m². How loud in decibels is this sound level?

17. *Concerts.* The crowd at a Hearsay concert at Wembley Arena in London cheered at a sound level of 128.8 dB. What is the intensity of such a sound?
Source: www.peterborough.gov.uk

18. *City ordinances.* In Albuquerque, New Mexico, the maximum allowable sound level from a car's exhaust is 96 dB. What is the intensity of such a sound?
Source: www.cabq.gov

19. *E-mail volume.* The SenderBase® Security Network ranks e-mail volume using a logarithmic scale. The magnitude *M* of a network's daily e-mail volume is given by

$$M = \log \frac{v}{1.34},$$

where *v* is the number of e-mail messages sent each day. How many e-mail messages are sent each day by a network that has a magnitude of 7.5?
Source: forum.spamcop.net

20. *Richter scale.* The Richter scale, developed in 1935, has been used for years to measure earthquake magnitude. The Richter magnitude *m* of an earthquake is given by the formula

$$m = \log \frac{A}{A_0},$$

where *A* is the maximum amplitude of the earthquake and A_0 is a constant. What is the magnitude on the Richter scale of an earthquake with an amplitude that is a million times A_0?

Use the compound-interest formula in Example 6 for Exercises 21 and 22.

21. *Interest compounded continuously.* Suppose that P_0 is invested in a savings account where interest is compounded continuously at 2.5% per year.

   a) Express $P(t)$ in terms of P_0 and 0.025.
   b) Suppose that $5000 is invested. What is the balance after 1 yr? after 2 yr?
   c) When will an investment of $5000 double itself?

22. *Interest compounded continuously.* Suppose that P_0 is invested in a savings account where interest is compounded continuously at 3.1% per year.

   a) Express $P(t)$ in terms of P_0 and 0.031.
   b) Suppose that $1000 is invested. What is the balance after 1 yr? after 2 yr?
   c) When will an investment of $1000 double itself?

23. *Population growth.* In 2008, the population of the United States was 304 million and the exponential growth rate was 0.9% per year.
Source: U.S. Census Bureau

   a) Find the exponential growth function.
   b) Predict the U.S. population in 2012.
   c) When will the U.S. population reach 325 million?

24. *World population growth.* In 2008, the world population was 6.7 billion and the exponential growth rate was 1.14% per year.
Source: U.S. Census Bureau

   a) Find the exponential growth function.
   b) Predict the world population in 2014.
   c) When will the world population be 8.0 billion?

25. *Zebra mussels.* The number of zebra mussels in a river grows at an exponential growth rate of 340% per year. What is the doubling time for zebra mussels?

26. *Population growth.* The exponential growth rate of the population of United Arab Emirates is 4.4% per year (one of the highest in the world). What is the doubling time?
Sources: Based on data from U.S. Census Bureau; International Data Base 2007

27. *World population.* The function

$$Y(x) = 71.41 \ln \frac{x}{4.6}$$

can be used to estimate the number of years $Y(x)$ after 2000 required for the world population to reach *x* billion people.
Sources: Based on data from U.S. Census Bureau; International Data Base

   a) In what year will the world population reach 10 billion?
   b) In what year will the world population reach 12 billion?
   c) Graph the function.

28. *Marine biology.* The function

$$Y(x) = 21.77 \ln \frac{x}{5.5}$$

can be used to estimate the number of years $Y(x)$ after 1982 required for the world's humpback whale population to reach *x* thousand whales.

   a) In what year will the whale population reach 15,000?
   b) In what year will the whale population reach 25,000?
   c) Graph the function.

29. *Forgetting.* Students in an English class took a final exam. They took equivalent forms of the exam at monthly intervals thereafter. The average score $S(t)$, in percent, after t months was found to be given by

$$S(t) = 68 - 20 \log (t + 1), \quad t \geq 0.$$

a) What was the average score when they initially took the test, $t = 0$?

b) What was the average score after 4 months? after 24 months?

c) Graph the function.

d) After what time t was the average score 50%?

30. *Health insurance.* The amount spent each year by the U.S. government for health insurance for low-income children can be estimated by

$$h(t) = 2.6 \ln t,$$

where $h(t)$ is in billions of dollars and t is the number of years after 1998.

Source: Based on data from the Congressional Budget Office

a) How much was spent on health insurance for low-income children in 2007?

b) Graph the function.

c) In what year will $7 billion be spent on health insurance for low-income children?

31. *Wind power.* U.S. wind-power capacity has grown exponentially from about 2000 megawatts in 1990 to 17,000 megawatts in 2007.

Source: American Wind Energy Association.

a) Find the exponential growth rate k and write an equation for an exponential function that can be used to predict U.S. wind-power capacity t years after 1990.

b) Estimate the year in which wind-power capacity will reach 50,000 megawatts.

32. *Spread of a computer virus.* The number of computers infected by a virus t hours after it first appears usually increases exponentially. In 2004, the "MyDoom" worm spread from 100 computers to about 100,000 computers in 24 hr.

Source: Based on data from IDG News Service

a) Find the exponential growth rate k and write an equation for an exponential function that can be used to predict the number of computers infected t hours after the virus first appeared in 100 computers.

b) Assuming exponential growth, estimate how long it took the MyDoom worm to infect 9000 computers.

33. *Cable costs.* In 1997, the cost to construct communication cables under the ocean was approximately $8200 per gigabit per second per mile. This cost for subsea cables dropped exponentially to $500 by 2007.

Source: Based on information from TeleGeography

a) Find the exponential growth rate k, and write an equation for an exponential function that can be used to predict the cost of subsea cables t years after 1997.

b) Estimate the cost of subsea cables in 2010.

c) In what year (theoretically) will it cost only $1 per gigabit per second per mile to construct subsea cables?

34. *Decline in farmland.* The number of acres of farmland in the United States has decreased from 945 million acres in 2000 to 932 million acres in 2006. Assume the number of acres of farmland is decreasing exponentially.

Source: Statistical Abstract of the United States

a) Find the value k, and write an equation for an exponential function that can predict the number of acres of U.S. farmland t years after 2000.

b) Predict the number of acres of farmland in 2015.

c) In what year (theoretically) will there be only 800 million acres of U.S. farmland remaining?

35. *Archaeology.* A date palm seedling is growing in Kibbutz Ketura, Israel, from a seed found in King Herod's palace at Masada. The seed had lost 21% of its carbon-14. How old was the seed? (See Example 7.)

Source: Based on information from www.sfgate.com

36. *Archaeology.* Soil from beneath the Kish Church in Azerbaijan was found to have lost 12% of its carbon-14. How old was the soil? (See Example 7.)

Source: Based on information from www.azer.com

37. *Chemistry.* The exponential decay rate of iodine-131 is 9.6% per day. What is its half-life?

38. *Chemistry.* The decay rate of krypton-85 is 6.3% per year. What is its half-life?

39. *Caffeine.* The half-life of caffeine in the human body for a healthy adult is approximately 5 hr.

a) What is the exponential decay rate?

b) How long will it take 95% of the caffeine consumed to leave the body?

40. *Home construction.* The chemical urea formaldehyde was found in some insulation used in houses built during the mid to late 1960s. Unknown at the time was the fact that urea formaldehyde emitted toxic fumes as it decayed. The half-life of urea formaldehyde is 1 yr.

a) What is its decay rate?

b) How long will it take 95% of the urea formaldehyde present to decay?

41. *Value of a sports card.* Legend has it that because he objected to teenagers smoking, and because his first baseball card was issued in cigarette packs, the great shortstop Honus Wagner halted production of his card before many were produced. One of these cards was purchased in 1991 by hockey great Wayne Gretzky (and a partner) for $451,000. The same card was sold in 2007 for $2.8 million. For the following questions, assume that the card's value increases exponentially, as it has for many years.

WAGNER, PITTSBURG

a) Find the exponential growth rate k, and determine an exponential function that can be used to estimate the dollar value, $V(t)$, of the card t years after 1991.
b) Predict the value of the card in 2012.
c) What is the doubling time for the value of the card?
d) In what year will the value of the card first exceed $4,000,000?

42. *Art masterpieces.* As of April 2008, the highest auction price for a contemporary painting was $72.8 million, paid in 2007 for Rothko's *White Center* (*Yellow, Pink and Lavender on Rose*). The same painting sold for about $10,000 in 1960.

a) Find the exponential growth rate k, and determine the exponential growth function that can be used to estimate $V(t)$, the painting's value, in millions of dollars, t years after 1960.
b) Estimate the value of the painting in 2012.
c) What is the doubling time for the value of the painting?
d) How long after 1960 will the value of the painting be $1 billion?

43. Write a problem for a classmate to solve in which information is provided and the classmate is asked to find an exponential growth function. Make the problem as realistic as possible.

44. Examine the restriction on t in Exercise 29.
  a) What upper limit might be placed on t?
  b) In practice, would this upper limit ever be enforced? Why or why not?

Skill Review

To prepare for Section 10.1, review the distance and midpoint formulas, completing the square, and graphing parabolas (Sections 7.7, 8.1, and 8.7).

Find the distance between each pair of points. [7.7]
45. $(-3, 7)$ and $(-2, 6)$ **46.** $(1, 5)$ and $(4, 1)$

Find the coordinates of the midpoint of the segment connecting each pair of points. [7.7]
47. $(3, -8)$ and $(5, -6)$

48. $(2, -11)$ and $(-9, -8)$

Solve by completing the square. [8.1]
49. $x^2 + 8x = 1$

50. $x^2 - 10x = 15$

Graph. [8.7]
51. $y = x^2 - 5x - 6$

52. $g(x) = 2x^2 - 6x + 3$

Synthesis

53. Will the model used in Example 4 to predict the number of text messages still be realistic in 2030? Why or why not?

54. *Atmospheric pressure.* Atmospheric pressure P at an elevation of a feet above sea level is given by

$$P = P_0 e^{-0.00004a},$$

where P_0 is the pressure at sea level, which is approximately 29.9 in inches of mercury (Hg). Explain how a barometer, or some other device for measuring atmospheric pressure, can be used to find the height of a skyscraper.

55. *Sports salaries.*   As of April 2008, Alex Rodriguez of the New York Yankees had the largest contract in sports history. As part of the 10-year $275-million deal, he will receive $20 million in 2016. How much money would need to be invested in 2008, at 4% interest compounded continuously, in order to have $20 million for Rodriguez in 2016? (This is much like determining what $20 million in 2016 is worth in 2008 dollars.)
Source: *The San Francisco Chronicle*

56. *Supply and demand.*   The supply and demand for the sale of stereos by Sound Ideas are given by

$$S(x) = e^x \quad \text{and} \quad D(x) = 162{,}755e^{-x},$$

where $S(x)$ is the price at which the company is willing to supply x stereos and $D(x)$ is the demand price for a quantity of x stereos. Find the equilibrium point. (For reference, see Section 3.8.)

57. *Stellar magnitude.*   The apparent stellar magnitude m of a star with received intensity I is given by

$$m(I) = -(19 + 2.5 \cdot \log I),$$

where I is in watts per square meter (W/m^2). The smaller the apparent stellar magnitude, the brighter the star appears.
Source: The Columbus Optical SETI Observatory

a) The intensity of light received from the sun is 1390 W/m^2. What is the apparent stellar magnitude of the sun?
b) The 5-m diameter Hale telescope on Mt. Palomar can detect a star with magnitude +23. What is the received intensity of light from such a star?

58. *Growth of bacteria.*   The bacteria *Escherichia coli* (*E. coli*) are commonly found in the human bladder. Suppose that 3000 of the bacteria are present at time $t = 0$. Then t minutes later, the number of bacteria present is

$$N(t) = 3000(2)^{t/20}.$$

If 100,000,000 bacteria accumulate, a bladder infection can occur. If, at 11:00 A.M., a patient's bladder contains 25,000 *E. coli* bacteria, at what time can infection occur?

59. Show that for exponential growth at rate k, the doubling time T is given by $T = \dfrac{\ln 2}{k}$.

60. Show that for exponential decay at rate k, the half-life T is given by $T = \dfrac{\ln 2}{k}$.

61. *Generic drugs.*   Largely because of budget constraints, the Food and Drug Administration (FDA) cannot keep up with the rapidly increasing number of applications for approval of generic drugs. The following table shows the number of applications and the number of approvals for generic drugs for recent years.

Year	Number of New Applications for Generic Drugs	Number of Approvals of Generic Drugs
2001	300	310
2002	361	364
2003	449	373
2004	563	413
2005	766	467
2006	810	525

Source: U.S. Food and Drug Administration

a) Graph the data for applications submitted to the FDA, and determine which would be a better fit for the data: an exponential function or a linear function. Explain your reasoning.
b) Graph the data for approvals from the FDA, and determine which would be a better fit for the data: an exponential function or a linear function. Explain your reasoning.
c) Use regression to fit a function to each set of data.
d) If the trends continue, in what year will there be only half as many approvals as applications?

CORNER

COLLABORATIVE

Safe Listening

Focus: Logarithmic models

Time: 30 minutes

Group size: 2–4

Materials: MP3 players (one per group) with music

The *decibel* scale is used to measure the volume, or "loudness," of a sound. Listening to music at a high volume can lead to damaged hearing, because the power required to produce louder volumes increases exponentially.

ACTIVITY

1. Group members should work together to answer the following questions.

   a) The volume V, in decibels (dB), of sound on an MP3 player is given by

   $$V = 10 \cdot \log \frac{P}{P_0},$$

   where P is the power needed to produce the sound, in milliwatts (mW), and $P_0 = 0.1$ mW is the power needed to produce the lowest volume that registers on the MP3 player. Find V_0, the volume when the lowest power setting is used $(P = P_0)$.

   b) On a typical MP3 player, each time the volume is stepped up by pressing the volume button, the volume increases by 1.5 decibels (dB). If the volume is stepped up 20 times from V_0, by how many decibels does the volume increase?

   c) If the volume is stepped up 20 times from V_0, how much power is required to produce the sound?

   d) By how much does the power level increase every time the volume is stepped up?

   e) Solve the formula for P. What type of function is this?

2. One group member should begin listening to a song on the MP3 player. If possible, the song should be recent and downloaded from a music site.

   a) Most MP3 songs are designed to play at 100 dB. Beyond this volume, the music begins to sound distorted. Starting at the minimum volume level, the listener should increase the volume until the music begins to sound distorted. Calculate the approximate decibel increase per step for this MP3 player by dividing 100 by the number of times the volume was stepped up. How much power would be needed to produce this volume?

   b) The Occupational Safety and Health Administration (OSHA) considers sustained sound levels of 90 dB and above unsafe. Using the decibel increase per step from part 2(a), calculate the number of steps needed to increase the volume from the minimum volume level to 90 dB.

   c) Starting again at the minimum volume level, the listener should increase the volume the calculated number of steps until the volume reaches approximately 90 dB. How much power is required to produce this volume?

   d) How many times as much power is used to produce 100 dB as to produce 90 dB?

Thanks to Greg Massey, Embedded Software Engineer, for suggesting this application.

Study Summary

KEY TERMS AND CONCEPTS	EXAMPLES

SECTION 9.1: COMPOSITE AND INVERSE FUNCTIONS

The **composition** of f and g is defined as
$$(f \circ g)(x) = f(g(x)).$$

If $f(x) = \sqrt{x}$ and $g(x) = 2x - 5$, then
$$(f \circ g)(x) = f(g(x)) = f(2x - 5)$$
$$= \sqrt{2x - 5}.$$

A function f is **one-to-one** if different inputs have different outputs. The graph of a one-to-one function passes the **horizontal-line test**.

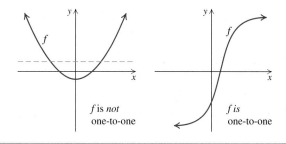

f is *not* one-to-one f is one-to-one

If f is one-to-one, it is possible to find its inverse:

1. Replace $f(x)$ with y.
2. Interchange x and y.
3. Solve for y.

4. Replace y with $f^{-1}(x)$.

If $f(x) = 2x - 3$, find $f^{-1}(x)$.

1. $y = 2x - 3$
2. $x = 2y - 3$
3. $x + 3 = 2y$
   $$\frac{x + 3}{2} = y$$

4. $\dfrac{x + 3}{2} = f^{-1}(x)$

SECTION 9.2: EXPONENTIAL FUNCTIONS
SECTION 9.3: LOGARITHMIC FUNCTIONS

For an **exponential function** f:
$f(x) = a^x$;
$a > 0,\ a \neq 1$;
Domain: $\mathbb{R}$;
$f^{-1}(x) = \log_a x$.

For a **logarithmic function** g:
$g(x) = \log_a x$;
$a > 0,\ a \neq 1$;
Domain: $(0, \infty)$;
$g^{-1}(x) = a^x$.

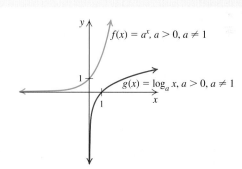

$\log_a x = m$ means $a^m = x$.

$\log_8 x = 2$
$\quad 8^2 = x$ Rewriting as an exponential equation
$\quad 64 = x$

SECTION 9.4: PROPERTIES OF LOGARITHMIC FUNCTIONS
SECTION 9.5: COMMON AND NATURAL LOGARITHMS

Properties of Logarithms

$\log_a (MN) = \log_a M + \log_a N$

$\log_a \dfrac{M}{N} = \log_a M - \log_a N$

$\log_a M^p = p \cdot \log_a M$

$\log_a 1 = 0$

$\log_a a = 1$

$\log_a a^k = k$

$\log M = \log_{10} M$

$\ln M = \log_e M$

$\log_b M = \dfrac{\log_a M}{\log_a b}$

$\log_7 10 = \log_7 5 + \log_7 2$

$\log_5 \dfrac{14}{3} = \log_5 14 - \log_5 3$

$\log_8 5^{12} = 12 \log_8 5$

$\log_9 1 = 0$

$\log_4 4 = 1$

$\log_3 3^8 = 8$

$\log 43 = \log_{10} 43$

$\ln 37 = \log_e 37$

$\log_6 31 = \dfrac{\log 31}{\log 6} = \dfrac{\ln 31}{\ln 6}$

SECTION 9.6: SOLVING EXPONENTIAL AND LOGARITHMIC EQUATIONS

The Principle of Exponential Equality

For any real number b, $b \neq -1, 0,$ or 1:

$$b^x = b^y \quad \text{is equivalent to} \quad x = y.$$

$25 = 5^x$

$5^2 = 5^x$

$2 = x$

The Principle of Logarithmic Equality

For any logarithm base a, and for
$x, y > 0$:

$x = y$ is equivalent to $\log_a x = \log_a y.$

$83 = 7^x$

$\log 83 = \log 7^x$

$\log 83 = x \log 7$

$\dfrac{\log 83}{\log 7} = x$

SECTION 9.7: APPLICATIONS OF EXPONENTIAL AND LOGARITHMIC FUNCTIONS

Exponential Growth Model

$$P(t) = P_0 e^{kt}, \quad k > 0$$

P_0 is the population at time 0.

$P(t)$ is the population at time t.

k is the **exponential growth rate**.

The **doubling time** is the amount of time necessary for the population to double in size.

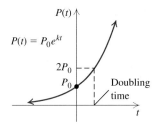

Exponential Decay Model

$$P(t) = P_0 e^{-kt}, \quad k > 0$$

P_0 is the quantity present at time 0.

$P(t)$ is the amount present at time t.

k is the **exponential decay rate**.

The **half-life** is the amount of time necessary for half of the quantity to decay.

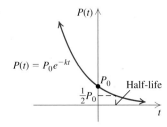

Review Exercises: Chapter 9

Concept Reinforcement   *In each of Exercises 1–10, classify the statement as either true or false.*

1. The functions given by $f(x) = 3^x$ and $g(x) = \log_3 x$ are inverses of each other.   [9.3]

2. A function's doubling time is the amount of time t for which $f(t) = 2 \cdot f(0)$.   [9.7]

3. A radioactive isotope's half-life is the amount of time t for which $f(t) = \frac{1}{2} \cdot f(0)$.   [9.7]

4. $\ln(ab) = \ln a - \ln b$ [9.4]

5. $\log x^a = x \ln a$ [9.4]

6. $\log_a \dfrac{m}{n} = \log_a m - \log_a n$ [9.4]

7. For $f(x) = 3^x$, the domain of f is $[0, \infty)$.   [9.2]

8. For $g(x) = \log_2 x$, the domain of g is $[0, \infty)$.   [9.3]

9. The function F is not one-to-one if $F(-2) = F(5)$.   [9.1]

10. The function g is one-to-one if it passes the vertical-line test.   [9.1]

---

11. Find $(f \circ g)(x)$ and $(g \circ f)(x)$ if $f(x) = x^2 + 1$ and $g(x) = 2x - 3$.   [9.1]

12. If $h(x) = \sqrt{3 - x}$, find $f(x)$ and $g(x)$ such that $h(x) = (f \circ g)(x)$. Answers may vary.   [9.1]

13. Determine whether $f(x) = 4 - x^2$ is one-to-one.   [9.1]

Find a formula for the inverse of each function.   [9.1]

14. $f(x) = x - 10$

15. $g(x) = \dfrac{3x + 1}{2}$

16. $f(x) = 27x^3$

Graph.

17. $f(x) = 3^x + 1$ [9.2]

18. $x = \left(\frac{1}{4}\right)^y$ [9.2]

19. $y = \log_5 x$ [9.3]

Simplify.   [9.3]

20. $\log_9 81$

21. $\log_3 \frac{1}{9}$

22. $\log_2 2^{11}$

23. $\log_{16} 4$

Rewrite as an equivalent logarithmic equation.   [9.3]

24. $2^{-3} = \frac{1}{8}$

25. $25^{1/2} = 5$

Rewrite as an equivalent exponential equation.   [9.3]

26. $\log_4 16 = x$

27. $\log_8 1 = 0$

Express as an equivalent expression using the individual logarithms of x, y, and z.   [9.4]

28. $\log_a x^4 y^2 z^3$

29. $\log_a \dfrac{x^5}{yz^2}$

30. $\log \sqrt[4]{\dfrac{z^2}{x^3 y}}$

Express as an equivalent expression that is a single logarithm and, if possible, simplify.   [9.4]

31. $\log_a 5 + \log_a 8$

32. $\log_a 48 - \log_a 12$

33. $\frac{1}{2} \log a - \log b - 2 \log c$

34. $\frac{1}{3}[\log_a x - 2 \log_a y]$

Simplify.   [9.4]

35. $\log_m m$

36. $\log_m 1$

37. $\log_m m^{17}$

Given $\log_a 2 = 1.8301$ and $\log_a 7 = 5.0999$, find each of the following.   [9.4]

38. $\log_a 14$

39. $\log_a \frac{2}{7}$

40. $\log_a 28$

41. $\log_a 3.5$

42. $\log_a \sqrt{7}$

43. $\log_a \frac{1}{4}$

Use a calculator to find each of the following to four decimal places.   [9.5]

44. $\log 75$

45. $10^{1.789}$

46. $\ln 0.3$

47. $e^{-0.98}$

Find each of the following logarithms using the change-of-base formula. Round answers to four decimal places.   [9.5]

48. $\log_5 50$

49. $\log_6 5$

Graph and state the domain and the range of each function.   [9.5]

50. $f(x) = e^x - 1$

51. $g(x) = 0.6 \ln x$

Solve. Where appropriate, include approximations to four decimal places. [9.6]

52. $5^x = 125$

53. $3^{2x} = \frac{1}{9}$

54. $\log_3 x = -4$

55. $\log_x 16 = 4$

56. $\log x = -3$

57. $6 \ln x = 18$

58. $4^{2x-5} = 19$

59. $2^x = 12$

60. $e^{-0.1t} = 0.03$

61. $2 \ln x = -6$

62. $\log (2x - 5) = 1$

63. $\log_4 x - \log_4 (x - 15) = 2$

64. $\log_3 (x - 4) = 2 - \log_3 (x + 4)$

65. In a business class, students were tested at the end of the course with a final exam. They were then tested again 6 months later. The forgetting formula was determined to be
$$S(t) = 82 - 18 \log (t + 1),$$
where $S(t)$ was the average student grade t months after taking the final exam. [9.7]

   a) Determine the average score when they first took the exam (when $t = 0$).

   b) What was the average score after 6 months?

   c) After what time was the average score 54?

66. A laptop computer is purchased for $1500. Its value each year is about 80% of its value in the preceding year. Its value in dollars after t years is given by the exponential function
$$V(t) = 1500(0.8)^t. \quad [9.7]$$

   a) After what amount of time will the computer's value be $900?

   b) After what amount of time will the computer's value be half the original value?

67. U.S. companies spent $885 million in e-mail marketing in 2005. This amount was predicted to grow exponentially to $1.1 billion in 2010. [9.7]
Source: Jupiter Research

   a) Find the exponential growth rate k, and write a function that describes the amount $A(t)$, in millions of dollars, spent on e-mail marketing t years after 2005.

   b) Estimate the amount spent on e-mail marketing in 2008.

   c) In what year will U.S. companies spend $2 billion on e-mail marketing?

   d) Find the doubling time.

68. In 2005, consumers received, on average, 3253 spam messages. The volume of spam messages per consumer is decreasing exponentially with an exponential decay rate of 13.7% per year. [9.7]

   a) Find the exponential decay function that can be used to predict the average number of spam messages, $M(t)$, t years after 2005.

   b) Predict the number of spam messages received per consumer in 2010.

   c) In what year, theoretically, will the average consumer receive 100 spam messages?

69. The value of Aret's stock market portfolio doubled in 6 yr. What was the exponential growth rate? [9.7]

70. How long will it take $7600 to double if it is invested at 4.2%, compounded continuously? [9.7]

71. How old is a skull that has lost 34% of its carbon-14? (Use $P(t) = P_0 e^{-0.00012t}$.) [9.7]

72. What is the pH of coffee if its hydrogen ion concentration is 7.9×10^{-6} moles per liter? (Use $\text{pH} = -\log [\text{H}^+]$.) [9.7]

73. The roar of a lion can reach a sound intensity of $2.5 \times 10^{-1} \text{ W/m}^2$. How loud in decibels is this sound level? $\left(\text{Use } L = 10 \cdot \log \dfrac{I}{10^{-12} \text{ W/m}^2}. \right)$ [9.7]

Source: en.allexperts.com

Synthesis

74. Explain why negative numbers do not have logarithms. [9.3]

75. Explain why $f(x) = e^x$ and $g(x) = \ln x$ are inverse functions. [9.5]

Solve. [9.6]

76. $\ln (\ln x) = 3$

77. $2^{x^2 + 4x} = \frac{1}{8}$

78. Solve the system:
$$5^{x+y} = 25,$$
$$2^{2x-y} = 64. \quad [9.6]$$

Test: Chapter 9

1. Find $(f \circ g)(x)$ and $(g \circ f)(x)$ if $f(x) = x + x^2$ and $g(x) = 2x + 1$.

2. If
$$h(x) = \frac{1}{2x^2 + 1},$$
find $f(x)$ and $g(x)$ such that $h(x) = (f \circ g)(x)$. Answers may vary.

3. Determine whether $f(x) = x^2 + 3$ is one-to-one.

Find a formula for the inverse of each function.

4. $f(x) = 3x + 4$

5. $g(x) = (x + 1)^3$

Graph.

6. $f(x) = 2^x - 3$

7. $g(x) = \log_7 x$

Simplify.

8. $\log_5 125$

9. $\log_{100} 10$

10. $3^{\log_3 18}$

11. $\log_n n$

12. $\log_c 1$

13. $\log_a a^{19}$

14. Rewrite as an equivalent logarithmic equation: $5^{-4} = \frac{1}{625}$.

15. Rewrite as an equivalent exponential equation: $m = \log_2 \frac{1}{2}$.

16. Express as an equivalent expression using the individual logarithms of a, b, and c:
$$\log \frac{a^3 b^{1/2}}{c^2}.$$

17. Express as an equivalent expression that is a single logarithm:
$$\tfrac{1}{3} \log_a x + 2 \log_a z.$$

Given $\log_a 2 = 0.301$, $\log_a 6 = 0.778$, and $\log_a 7 = 0.845$, find each of the following.

18. $\log_a 14$

19. $\log_a 3$

20. $\log_a 16$

Use a calculator to find each of the following to four decimal places.

21. $\log 25$

22. $10^{-0.8}$

23. $\ln 0.4$

24. $e^{4.8}$

25. Find $\log_3 14$ using the change-of-base formula. Round to four decimal places.

Graph and state the domain and the range of each function.

26. $f(x) = e^x + 3$

27. $g(x) = \ln(x - 4)$

Solve. Where appropriate, include approximations to four decimal places.

28. $2^x = \frac{1}{32}$

29. $\log_4 x = \frac{1}{2}$

30. $\log x = -2$

31. $5^{4-3x} = 87$

32. $7^x = 1.2$

33. $\ln x = 3$

34. $\log(x - 3) + \log(x + 1) = \log 5$

35. The average walking speed R of people living in a city of population P is given by $R = 0.37 \ln P + 0.05$, where R is in feet per second and P is in thousands.

   a) The population of Tulsa, Oklahoma, is 383,000. Find the average walking speed.

   b) San Diego, California, has an average walking speed of about 3 ft/sec. Find the population.

36. The population of Nigeria was about 140 million in 2008 and the exponential growth rate was 2.4% per year.

   a) Write an exponential function describing the population of Nigeria.

   b) What will the population be in 2012? in 2016?

   c) When will the population be 200 million?

   d) What is the doubling time?

37. The average cost of a year at a private four-year college grew exponentially from $21,855 in 2001 to $27,317 in 2006.
   Source: National Center for Education Statistics

   a) Find the exponential growth rate k, and write a function that approximates the cost $C(t)$ of a year of college t years after 2001.

   b) Predict the cost of a year of college in 2012.

   c) In what year will the average cost of college be $50,000?

38. An investment with interest compounded continuously doubled itself in 16 yr. What is the interest rate?

39. How old is an animal bone that has lost 43% of its carbon-14? (Use $P(t) = P_0 e^{-0.00012t}$.)

40. Blue whales and fin whales are the loudest animals, with sound levels up to 188 dB. What is the intensity of such a sound? $\left(\text{Use } L = 10 \cdot \log \dfrac{I}{10^{-12} \text{ W/m}^2}. \right)$
   Source: Guinness World Records

41. The hydrogen ion concentration of water is 1.0×10^{-7} moles per liter. What is the pH? (Use $\text{pH} = -\log[\text{H}^+]$.)

Synthesis

42. Solve: $\log_5 |2x - 7| = 4$.

43. If $\log_a x = 2$, $\log_a y = 3$, and $\log_a z = 4$, find
$$\log_a \frac{\sqrt[3]{x^2 z}}{\sqrt[3]{y^2 z^{-1}}}.$$

Cumulative Review: Chapters 1–9

1. Evaluate $\dfrac{x^0 + y}{-z}$ for $x = 6$, $y = 9$, and $z = -5$.
   [1.2], [1.6]

Simplify.

2. $(-2x^2y^{-3})^{-4}$ [1.6]

3. $(-5x^4y^{-3}z^2)(-4x^2y^2)$ [1.6]

4. $\dfrac{3x^4y^6z^{-2}}{-9x^4y^2z^3}$ [1.6]

5. $(1.5 \times 10^{-3})(4.2 \times 10^{-12})$ [1.7]

6. $3^3 + 2^2 - (32 \div 4 - 16 \div 8)$ [1.1]

Solve.

7. $3(2x - 3) = 9 - 5(2 - x)$ [1.3]

8. $4x - 3y = 15$,
   $3x + 5y = 4$ [3.2]

9. $\quad x + y - 3z = -1$,
   $\quad 2x - y + z = 4$,
   $\quad -x - y + z = 1$ [3.4]

10. $x(x - 3) = 70$ [5.8]

11. $\dfrac{7}{x^2 - 5x} - \dfrac{2}{x - 5} = \dfrac{4}{x}$ [6.4]

12. $\sqrt{4 - 5x} = 2x - 1$ [7.6]

13. $\sqrt[3]{2x} = 1$ [7.6]

14. $3x^2 + 48 = 0$ [8.1]

15. $x^4 - 13x^2 + 36 = 0$ [8.5]

16. $\log_x 81 = 2$ [9.3]

▦ 17. $3^{5x} = 7$ [9.6]

18. $\ln x - \ln (x - 8) = 1$ [9.6]

19. $x^2 + 4x > 5$ [8.9]

20. If $f(x) = x^2 + 6x$, find a such that $f(a) = 11$. [8.2]

21. If $f(x) = |2x - 3|$, find all x for which $f(x) \geq 7$.
    [4.3]

Solve.

22. $D = \dfrac{ab}{b + a}$, for a [6.8]

23. $d = ax^2 + vx$, for x [8.4]

24. Find the domain of the function f given by
    $$f(x) = \dfrac{x + 4}{3x^2 - 5x - 2}.$$ [5.8]

Perform the indicated operations and simplify.

25. $(5p^2q^3 + 6pq - p^2 + p) -$
    $(2p^2q^3 + p^2 - 5pq - 9)$ [5.1]

26. $(3x^2 - z^3)^2$ [5.2]

27. $\dfrac{1 + \dfrac{3}{x}}{x - 1 - \dfrac{12}{x}}$ [6.3]

28. $\dfrac{a^2 - a - 6}{a^3 - 27} \cdot \dfrac{a^2 + 3a + 9}{6}$ [6.1]

29. $\dfrac{3}{x + 6} - \dfrac{2}{x^2 - 36} + \dfrac{4}{x - 6}$ [6.2]

30. $\dfrac{\sqrt[3]{24xy^8}}{\sqrt[3]{3xy}}$ [7.4]

31. $\sqrt{x + 5} \; \sqrt[5]{x + 5}$ [7.5]

32. $(2 - i\sqrt{3})(6 + i\sqrt{3})$ [7.8]

33. $(x^4 - 8x^3 + 15x^2 + x - 3) \div (x - 3)$ [6.6]

Factor.

34. $27 + 64n^3$ [5.6]

35. $6x^2 + 8xy - 8y^2$ [5.4]

36. $x^4 - 4x^3 + 7x - 28$ [5.3]

37. $2m^2 + 12mn + 18n^2$ [5.5]

38. $x^4 - 16y^4$ [5.5]

39. Rationalize the denominator:
    $$\dfrac{3 - \sqrt{y}}{2 - \sqrt{y}}.$$ [7.5]

40. Find the inverse of f if $f(x) = 9 - 2x$. [9.1]

41. Find a linear function with a graph that contains the points $(0, -8)$ and $(-1, 2)$. [2.5]

42. Find an equation of the line whose graph has a y-intercept of $(0, 5)$ and is perpendicular to the line given $2x + y = 6$. [2.5]

Graph.

43. $5x = 15 + 3y$ [2.4]

44. $y = \log_3 x$ [9.3]

45. $-2x - 3y \leq 12$ [4.4]

46. Graph: $f(x) = 2x^2 + 12x + 19$. [8.7]
    a) Label the vertex.
    b) Draw the axis of symmetry.
    c) Find the maximum or minimum value.

47. Graph $f(x) = 2e^x$ and determine the domain and the range. [9.5]

48. Express as a single logarithm:

$3 \log x - \frac{1}{2} \log y - 2 \log z$. [9.4]

Solve.

49. *Colorado River.* The Colorado River delivers 1.5 million acre-feet of water to Mexico each year. This is only 10% of the volume of the river; the remainder is diverted at an earlier time for agricultural use. How much water is diverted each year from the Colorado River? [1.4]
Source: www.sierraclub.org

50. *Desalination.* More cities are supplying some of their fresh water through desalination, the process of removing the salt from ocean water. The worldwide desalination capacity has grown exponentially from 15 million m^3 per day in 1990 to 55 million m^3 per day in 2007. [9.7]
Source: Global Water Intelligence

a) Find the exponential growth rate k, and write an equation for an exponential function that can be used to predict the worldwide desalination capacity $D(t)$, in millions of cubic meters per day, t years after 1990.

b) Predict the worldwide desalination capacity in 2012.

c) In what year will the worldwide desalination capacity reach 100 million m^3 per day?

51. *Gasoline consumption.* The number of barrels of gasoline consumed per day in the United States has increased from 8.5 million in 2000 to 9.3 million in 2006.
Source: U.S. Department of Energy, Energy Information Administration

a) At what rate did gasoline consumption increase from 2000 to 2006? [2.3]

b) Find a linear function g that fits the data. Let t represent the number of years since 2000. [2.5]

c) Find an exponential function G that fits the data. Let t represent the number of years since 2000. [9.7]

52. Good's Candies of Indiana makes all their chocolates by hand. It takes Anne 10 min to coat a tray of candies in chocolate. It takes Clay 12 min to coat a tray of candies. How long would it take Anne and Clay, working together, to coat the candies? [6.5]

53. Joe's Thick and Tasty salad dressing gets 45% of its calories from fat. The Light and Lean dressing gets 20% of its calories from fat. How many ounces of each should be mixed in order to get 15 oz of dressing that gets 30% of its calories from fat? [3.3]

54. A fishing boat with a trolling motor can move at a speed of 5 km/h in still water. The boat travels 42 km downriver in the same time that it takes to travel 12 km upriver. What is the speed of the river? [6.5]

55. What is the minimum product of two numbers whose difference is 14? What are the numbers that yield this product? [8.8]

56. Students in a biology class just took a final exam. A formula for predicting the average exam grade on a similar test t months later is

$S(t) = 78 - 15 \log (t + 1)$.

a) Find the students' average score when they first took the final exam. [9.7]

b) What would the expected average score be on a retest after 4 months? [9.7]

Synthesis

Solve.

57. $\dfrac{5}{3x - 3} + \dfrac{10}{3x + 6} = \dfrac{5x}{x^2 + x - 2}$ [6.4]

58. $\log \sqrt{3x} = \sqrt{\log 3x}$ [9.6]

59. The Danville Express travels 280 mi at a certain speed. If the speed had been increased by 5 mph, the trip could have been made in 1 hr less time. Find the actual speed. [8.4]

Conic Sections

RAGHVENDRA SAHAI
ASTRONOMER
Pasadena, California

As an astronomer, I study stars and interstellar matter, that is, the gas and dust that lie in the vastness of space between the stars in our galaxy. My research includes obtaining images of the clouds of gas and dust ejected by dying sun-like stars. Calculus allows me to solve the equations governing these processes, and thus understand the physical properties of the dust and gas. Trigonometry helps me in figuring out the complex motions and velocities of the ejected matter.

AN APPLICATION

The maximum distance of the planet Mars from the sun is 2.48×10^8 mi. The minimum distance is 3.46×10^7 mi. The sun is at one focus of the elliptical orbit. Find the distance from the sun to the other focus.

This problem appears as Exercise 49 in Section 10.2.

The ellipse described on the preceding page is one example of a *conic section*, meaning that it can be regarded as a cross section of a cone. This chapter presents a variety of applications and equations with graphs that are conic sections. We have already worked with two conic sections, *lines* and *parabolas*, in Chapters 2 and 8.

10.1   Conic Sections: Parabolas and Circles

Parabolas   •   Circles

This section and the next two examine curves formed by cross sections of cones. These curves are all graphs of $Ax^2 + By^2 + Cxy + Dx + Ey + F = 0$. The constants $A, B, C, D, E,$ and F determine which of the following shapes will serve as the graph.

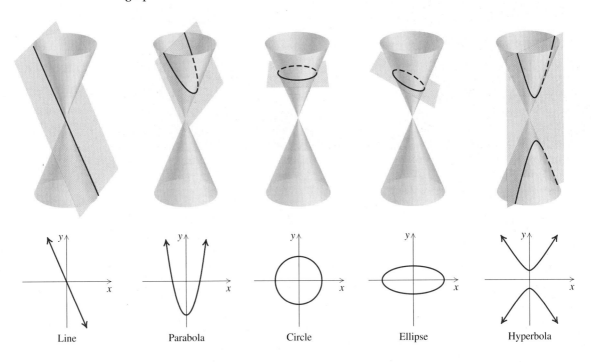

Line   Parabola   Circle   Ellipse   Hyperbola

Parabolas

When a cone is cut as shown in the second figure above, the conic section formed is a **parabola**. Parabolas have many applications in electricity, mechanics, and optics. A cross section of a contact lens or a satellite dish is a parabola, and arches that support certain bridges are parabolas.

> ## Equation of a Parabola
>
> A parabola with a vertical axis of symmetry opens upward or downward and has an equation that can be written in the form
>
> $$y = ax^2 + bx + c.$$
>
> A parabola with a horizontal axis of symmetry opens to the right or to the left and has an equation that can be written in the form
>
> $$x = ay^2 + by + c.$$

Parabolas with equations of the form $f(x) = ax^2 + bx + c$ were graphed in Chapter 8.

EXAMPLE 1

Graph: $y = x^2 - 4x + 9$.

SOLUTION   To locate the vertex, we can use either of two approaches. One way is to complete the square:

$$y = (x^2 - 4x) + 9$$ Note that half of -4 is -2, and $(-2)^2 = 4$.

$$= (x^2 - 4x + 4 - 4) + 9$$ Adding and subtracting 4

$$= (x^2 - 4x + 4) + (-4 + 9)$$ Regrouping

$$= (x - 2)^2 + 5.$$ Factoring and simplifying

The vertex is $(2, 5)$.

A second way to find the vertex is to recall that the x-coordinate of the vertex of the parabola given by $y = ax^2 + bx + c$ is $-b/(2a)$:

$$x = -\frac{b}{2a} = -\frac{-4}{2(1)} = 2.$$

To find the y-coordinate of the vertex, we substitute 2 for x:

$$y = x^2 - 4x + 9 = 2^2 - 4(2) + 9 = 5.$$

Either way, the vertex is $(2, 5)$. Next, we calculate and plot some points on each side of the vertex. As expected for a positive coefficient of x^2, the graph opens upward.

x	y
2	5	← Vertex
0	9	← y-intercept
1	6
3	6
4	9

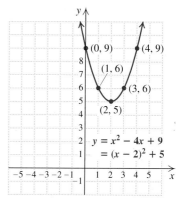

TRY EXERCISE 11

> **To Graph an Equation of the Form** $y = ax^2 + bx + c$
>
> **1.** Find the vertex (h, k) either by completing the square to find an equivalent equation
>
> $$y = a(x - h)^2 + k,$$
>
> or by using $-b/(2a)$ to find the x-coordinate and substituting to find the y-coordinate.
> **2.** Choose other values for x on each side of the vertex, and compute the corresponding y-values.
> **3.** The graph opens upward for $a > 0$ and downward for $a < 0$.

Any equation of the form $x = ay^2 + by + c$ represents a horizontal parabola that opens to the right for $a > 0$, opens to the left for $a < 0$, and has an axis of symmetry parallel to the x-axis.

EXAMPLE 2   Graph: $x = y^2 - 4y + 9$.

SOLUTION   This equation is like that in Example 1 but with x and y interchanged. The vertex is $(5, 2)$ instead of $(2, 5)$. To find ordered pairs, we choose values for y on each side of the vertex. Then we compute values for x. Note that the x- and y-values of the table in Example 1 are now switched. You should confirm that, by completing the square, we get $x = (y - 2)^2 + 5$.

x	y	
5	2	←Vertex
9	0	←x-intercept
6	1	
6	3	
9	4	

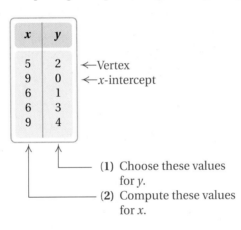

    (1) Choose these values for y.
    (2) Compute these values for x.

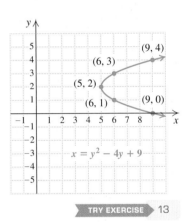

TRY EXERCISE 13

> **To Graph an Equation of the Form** $x = ay^2 + by + c$
>
> **1.** Find the vertex (h, k) either by completing the square to find an equivalent equation
>
> $$x = a(y - k)^2 + h,$$
>
> or by using $-b/(2a)$ to find the y-coordinate and substituting to find the x-coordinate.
> **2.** Choose other values for y that are on either side of k and compute the corresponding x-values.
> **3.** The graph opens to the right if $a > 0$ and to the left if $a < 0$.

EXAMPLE **3**   Graph: $x = -2y^2 + 10y - 7$.

SOLUTION   We find the vertex by completing the square:

$$x = -2y^2 + 10y - 7$$
$$= -2(y^2 - 5y \qquad) - 7$$
$$= -2\left(y^2 - 5y + \tfrac{25}{4}\right) - 7 - (-2)\tfrac{25}{4} \qquad \tfrac{1}{2}(-5) = \tfrac{-5}{2}; \left(\tfrac{-5}{2}\right)^2 = \tfrac{25}{4}; \text{ we}$$
$$\qquad\qquad\qquad\qquad\qquad\qquad\qquad\qquad \text{add and subtract } (-2)\tfrac{25}{4}.$$
$$= -2\left(y - \tfrac{5}{2}\right)^2 + \tfrac{11}{2}. \qquad \text{Factoring and simplifying}$$

The vertex is $\left(\tfrac{11}{2}, \tfrac{5}{2}\right)$.

For practice, we also find the vertex by first computing its y-coordinate, $-b/(2a)$, and then substituting to find the x-coordinate:

$$y = -\frac{b}{2a} = -\frac{10}{2(-2)} = \frac{5}{2}$$
$$x = -2y^2 + 10y - 7 = -2\left(\tfrac{5}{2}\right)^2 + 10\left(\tfrac{5}{2}\right) - 7$$
$$= \tfrac{11}{2}.$$

To find ordered pairs, we choose values for y on each side of the vertex and then compute values for x. A table is shown below, together with the graph. The graph opens to the left because the y^2-coefficient, -2, is negative.

x	y	
$\tfrac{11}{2}$	$\tfrac{5}{2}$	←Vertex
-7	0	←x-intercept
5	2	
5	3	
1	1	
1	4	
-7	5	

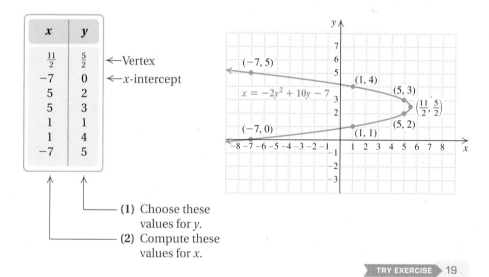

(1) Choose these values for y.
(2) Compute these values for x.

TRY EXERCISE 19.

Circles

Another conic section, the **circle**, is the set of points in a plane that are a fixed distance r, called the **radius** (plural, **radii**), from a fixed point (h, k), called the **center**. Note that the word radius can mean either any segment connecting a point on a circle to the center or the length of such a segment. Using the idea of a fixed distance r and the distance formula,

$$d = \sqrt{(x_2 - x_1)^2 + (y_2 - y_1)^2},$$

we can find the equation of a circle.

If (x, y) is on a circle of radius r, centered at (h, k), then by the definition of a circle and the distance formula, it follows that

$$r = \sqrt{(x - h)^2 + (y - k)^2}.$$

Squaring both sides gives the equation of a circle in standard form.

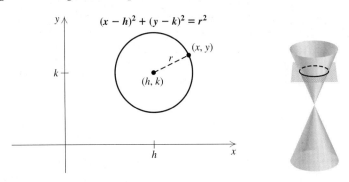

Equation of a Circle (Standard Form)

The equation of a circle, centered at (h, k), with radius r, is given by

$$(x - h)^2 + (y - k)^2 = r^2.$$

Note that for $h = 0$ and $k = 0$, the circle is centered at the origin. Otherwise, the circle is translated $|h|$ units horizontally and $|k|$ units vertically.

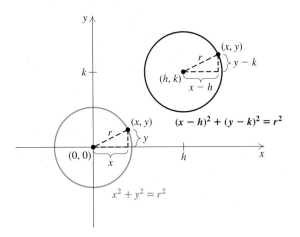

EXAMPLE **4**   Find an equation of the circle centered at $(4, -5)$ with radius 6.

SOLUTION   Using the standard form, we obtain

$$(x - 4)^2 + (y - (-5))^2 = 6^2, \quad \text{Using } (x - h)^2 + (y - k)^2 = r^2$$

or

$$(x - 4)^2 + (y + 5)^2 = 36.$$

TRY EXERCISE 31

EXAMPLE **5**   Find the center and the radius and then graph each circle.

a) $(x - 2)^2 + (y + 3)^2 = 4^2$

b) $x^2 + y^2 + 8x - 2y + 15 = 0$

SOLUTION

a) We write standard form:

$$(x - 2)^2 + [y - (-3)]^2 = 4^2.$$

The center is $(2, -3)$ and the radius is 4. To graph, we plot the points $(2, 1)$, $(2, -7)$, $(-2, -3)$, and $(6, -3)$, which are, respectively, 4 units above, below, left, and right of $(2, -3)$. We then either sketch a circle by hand or use a compass.

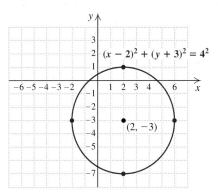

b) To write the equation $x^2 + y^2 + 8x - 2y + 15 = 0$ in standard form, we complete the square twice, once with $x^2 + 8x$ and once with $y^2 - 2y$:

$$x^2 + y^2 + 8x - 2y + 15 = 0$$
$$x^2 + 8x \quad\quad + y^2 - 2y \quad\quad = -15$$

Grouping the x-terms and the y-terms; subtracting 15 from both sides

$$x^2 + 8x + 16 + y^2 - 2y + 1 = -15 + 16 + 1$$

Adding $\left(\frac{8}{2}\right)^2$, or 16, and $\left(-\frac{2}{2}\right)^2$, or 1, to both sides to get standard form

$$(x + 4)^2 + (y - 1)^2 = 2$$

Factoring

$$[x - (-4)]^2 + (y - 1)^2 = (\sqrt{2})^2.$$

Writing standard form

The center is $(-4, 1)$ and the radius is $\sqrt{2}$.

$x^2 + y^2 + 8x - 2y + 15 = 0$
$[x - (-4)]^2 + (y - 1)^2 = (\sqrt{2})^2$
$(-4, 1)$

TRY EXERCISE 51

TECHNOLOGY CONNECTION

Most graphing calculators graph only functions, so graphing the equation of a circle usually requires two steps:

1. Solve the equation for y. The result will include a $\pm$ sign in front of a radical.
2. Graph two functions, one for the $+$ sign and the other for the $-$ sign, on the same set of axes.

For example, to graph $(x - 3)^2 + (y + 1)^2 = 16$, solve for $y + 1$ and then y:

$$(y + 1)^2 = 16 - (x - 3)^2$$
$$y + 1 = \pm\sqrt{16 - (x - 3)^2}$$
$$y = -1 \pm \sqrt{16 - (x - 3)^2},$$

or

$$y_1 = -1 + \sqrt{16 - (x - 3)^2}$$

and

$$y_2 = -1 - \sqrt{16 - (x - 3)^2}.$$

When both functions are graphed (in a "squared" window to eliminate distortion), the result is as follows.

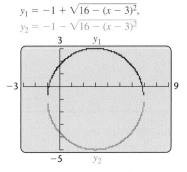

$y_1 = -1 + \sqrt{16 - (x - 3)^2}$,
$y_2 = -1 - \sqrt{16 - (x - 3)^2}$

On many calculators, pressing **APPS** and selecting Conics and then Circle accesses a program in which equations in standard form can be graphed directly and then Traced.

Graph each of the following equations.

1. $x^2 + y^2 - 16 = 0$
2. $(x - 1)^2 + (y - 2)^2 = 25$
3. $(x + 3)^2 + (y - 5)^2 = 16$
4. $(x - 5)^2 + (y + 6)^2 = 49$

10.1 EXERCISE SET

For Extra Help
MyMathLab
Math XL
PRACTICE

WATCH
DOWNLOAD

 Concept Reinforcement *In each of Exercises 1–8, match the equation with the graph of that equation from those shown.*

1. _____ $(x - 2)^2 + (y + 5)^2 = 9$

2. _____ $(x + 2)^2 + (y - 5)^2 = 9$

3. _____ $(x - 5)^2 + (y + 2)^2 = 9$

4. _____ $(x + 5)^2 + (y - 2)^2 = 9$

5. _____ $y = (x - 2)^2 - 5$

6. _____ $y = (x - 5)^2 - 2$

7. _____ $x = (y - 2)^2 - 5$

8. _____ $x = (y - 5)^2 - 2$

a)
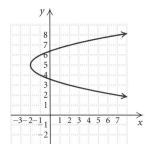

b)
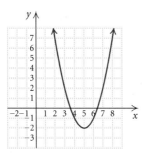

c)
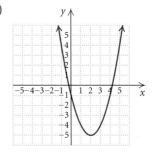

d)
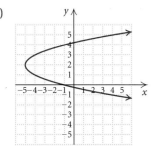

e)
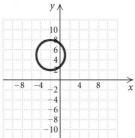

f)
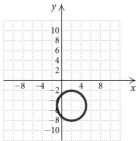

g)
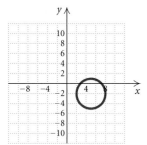

h)
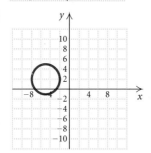

Graph. Be sure to label each vertex.

9. $y = -x^2$

10. $y = 2x^2$

11. $y = -x^2 + 4x - 5$

12. $x = 4 - 3y - y^2$

13. $x = y^2 - 4y + 2$

14. $y = x^2 + 2x + 3$

15. $x = y^2 + 3$

16. $x = -y^2$

17. $x = 2y^2$

18. $x = y^2 - 1$

19. $x = -y^2 - 4y$

20. $x = y^2 + 3y$

21. $y = x^2 - 2x + 1$

22. $y = x^2 + 2x + 1$

23. $x = -\frac{1}{2}y^2$

24. $y = -\frac{1}{2}x^2$

25. $x = -y^2 + 2y - 1$

26. $x = -y^2 - 2y + 3$

27. $x = -2y^2 - 4y + 1$

28. $x = 2y^2 + 4y - 1$

Find an equation of the circle satisfying the given conditions.

29. Center $(0, 0)$, radius 8

30. Center $(0, 0)$, radius 11

31. Center $(7, 3)$, radius $\sqrt{6}$

32. Center $(5, 6)$, radius $\sqrt{11}$

33. Center $(-4, 3)$, radius $3\sqrt{2}$

34. Center $(-2, 7)$, radius $2\sqrt{5}$

35. Center $(-5, -8)$, radius $10\sqrt{3}$

36. Center $(-7, -2)$, radius $5\sqrt{2}$

Aha! **37.** Center $(0, 0)$, passing through $(-3, 4)$

38. Center $(0, 0)$, passing through $(11, -10)$

39. Center $(-4, 1)$, passing through $(-2, 5)$

40. Center $(-1, -3)$, passing through $(-4, 2)$

Find the center and the radius of each circle. Then graph the circle.

41. $x^2 + y^2 = 1$

42. $x^2 + y^2 = 25$

43. $(x + 1)^2 + (y + 3)^2 = 49$

44. $(x - 2)^2 + (y + 3)^2 = 100$

45. $(x - 4)^2 + (y + 3)^2 = 10$

46. $(x + 5)^2 + (y - 1)^2 = 15$

47. $x^2 + y^2 = 8$

48. $x^2 + y^2 = 20$

49. $(x - 5)^2 + y^2 = \frac{1}{4}$

50. $x^2 + (y - 1)^2 = \frac{1}{25}$

51. $x^2 + y^2 + 8x - 6y - 15 = 0$

52. $x^2 + y^2 + 6x - 4y - 15 = 0$

53. $x^2 + y^2 - 8x + 2y + 13 = 0$

54. $x^2 + y^2 + 6x + 4y + 12 = 0$

55. $x^2 + y^2 + 10y - 75 = 0$

56. $x^2 + y^2 - 8x - 84 = 0$

57. $x^2 + y^2 + 7x - 3y - 10 = 0$

58. $x^2 + y^2 - 21x - 33y + 17 = 0$

59. $36x^2 + 36y^2 = 1$

60. $4x^2 + 4y^2 = 1$

61. Does the graph of an equation of a circle include the point that is the center? Why or why not?

62. Is a point a conic section? Why or why not?

Skill Review

To prepare for Section 10.2, review solving quadratic equations (Section 8.1).

Solve. [8.1]

63. $\dfrac{y^2}{16} = 1$

64. $\dfrac{x^2}{a^2} = 1$

65. $\dfrac{(x - 1)^2}{25} = 1$

66. $\dfrac{(y + 5)^2}{12} = 1$

67. $\dfrac{1}{4} + \dfrac{(y + 3)^2}{36} = 1$

68. $\dfrac{1}{9} + \dfrac{(x - 2)^2}{4} = 1$

Synthesis

69. On a piece of graph paper, draw a line and a point not on the line. Then plot several points that are the same distance from the point and from the line. What shape do the points appear to form? How is this set of points different from a circle?

70. If an equation has two terms with the same degree, can its graph be a parabola? Why or why not?

Find an equation of a circle satisfying the given conditions.

71. Center $(3, -5)$ and tangent to (touching at one point) the *y*-axis

72. Center $(-7, -4)$ and tangent to the *x*-axis

73. The endpoints of a diameter are $(7, 3)$ and $(-1, -3)$.

74. Center $(-3, 5)$ with a circumference of 8π units

75. Find the point on the *y*-axis that is equidistant from $(2, 10)$ and $(6, 2)$.

76. Find the point on the *x*-axis that is equidistant from $(-1, 3)$ and $(-8, -4)$.

77. *Wrestling.* The equation $x^2 + y^2 = \frac{81}{4}$, where *x* and *y* represent the number of meters from the center, can be used to draw the outer circle on a wrestling mat used in International, Olympic, and World Championship wrestling. The equation $x^2 + y^2 = 16$ can be used to draw the inner edge of the red zone. Find the area of the red zone.

Source: Based on data from the Government of Western Australia

78. *Snowboarding.* Each side edge of the Burton *X8* 155 snowboard is an arc of a circle with a "running length" of 1180 mm and a "sidecut depth" of 23 mm (see the figure below).
Source: evogear.com

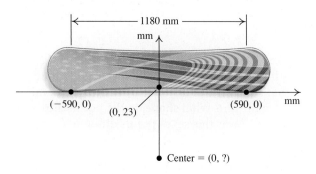

a) Using the coordinates shown, locate the center of the circle. (*Hint*: Equate distances.)
b) What radius is used for the edge of the board?

79. *Snowboarding.* The Never Summer Infinity 149 snowboard has a running length of 1160 mm and a sidecut depth of 23.5 mm (see Exercise 78). What radius is used for the edge of this snowboard?
Source: neversummer.com

80. *Skiing.* The Rossignol Blast ski, when lying flat and viewed from above, has edges that are arcs of a circle. (Actually, each edge is made of two arcs of slightly different radii. The arc for the rear half of the ski edge has a slightly larger radius.)
Source: evogear.com

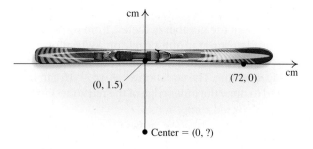

a) Using the coordinates shown, locate the center of the circle. (*Hint*: Equate distances.)
b) What radius is used for the arc passing through (0, 1.5) and (72, 0)?

81. *Doorway construction.* Engle Carpentry needs to cut an arch for the top of an entranceway. The arch needs to be 8 ft wide and 2 ft high. To draw the arch, the carpenters will use as a compass a stretched string with chalk attached at an end.

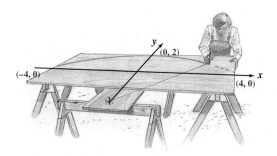

a) Using a coordinate system, locate the center of the circle.
b) What radius should the carpenters use to draw the arch?

82. *Archaeology.* During an archaeological dig, Estella finds the bowl fragment shown below. What was the original diameter of the bowl?

83. *Ferris wheel design.* A ferris wheel has a radius of 24.3 ft. Assuming that the center is 30.6 ft off the ground and that the origin is below the center, as in the following figure, find an equation of the circle.

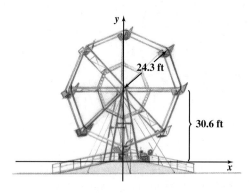

84. Use a graph of the equation $x = y^2 - y - 6$ to approximate to the nearest tenth the solutions of each of the following equations.
a) $y^2 - y - 6 = 2$
b) $y^2 - y - 6 = -3$

85. *Power of a motor.*   The horsepower of a certain kind of engine is given by the formula

$$H = \frac{D^2 N}{2.5},$$

where N is the number of cylinders and D is the diameter, in inches, of each piston. Graph this equation, assuming that $N = 6$ (a six-cylinder engine). Let D run from 2.5 to 8. Then use the graph to estimate the diameter of each piston in a six-cylinder 120-horsepower engine.

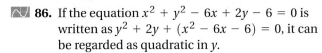 **86.** If the equation $x^2 + y^2 - 6x + 2y - 6 = 0$ is written as $y^2 + 2y + (x^2 - 6x - 6) = 0$, it can be regarded as quadratic in y.

a) Use the quadratic formula to solve for y.

b) Show that the graph of your answer to part (a) coincides with the graph in the Technology Connection on p. 655.

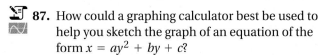 **87.** How could a graphing calculator best be used to help you sketch the graph of an equation of the form $x = ay^2 + by + c$?

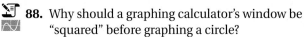 **88.** Why should a graphing calculator's window be "squared" before graphing a circle?

10.2   Conic Sections: Ellipses

Ellipses Centered at $(0, 0)$ ● Ellipses Centered at (h, k)

When a cone is cut at an angle, as shown below, the conic section formed is an *ellipse*. To draw an ellipse, stick two tacks in a piece of cardboard. Then tie a loose string to the tacks, place a pencil as shown, and draw an oval by moving the pencil while stretching the string tight.

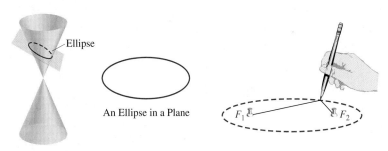

Ellipse

An Ellipse in a Plane

Ellipses Centered at $(0, 0)$

An **ellipse** is defined as the set of all points in a plane for which the sum of the distances from two fixed points F_1 and F_2 is constant. The points F_1 and F_2 are called **foci** (pronounced fō-sī), the plural of focus. In the figure above, the tacks are at the foci and the length of the string is the constant sum of the distances from the tacks to the pencil. The midpoint of the segment $F_1 F_2$ is the **center**. The equation of an ellipse follows. Its derivation is outlined in Exercise 51.

> ### Equation of an Ellipse Centered at the Origin
>
> The equation of an ellipse centered at the origin and symmetric with respect to both axes is
>
> $$\frac{x^2}{a^2} + \frac{y^2}{b^2} = 1, \quad a, b > 0. \quad \text{(Standard form)}$$

To graph an ellipse centered at the origin, it helps to first find the intercepts. If we replace x with 0, we can find the y-intercepts:

$$\frac{0^2}{a^2} + \frac{y^2}{b^2} = 1$$

$$\frac{y^2}{b^2} = 1$$

$$y^2 = b^2 \quad \text{or} \quad y = \pm b.$$

Thus the y-intercepts are $(0, b)$ and $(0, -b)$. Similarly, the x-intercepts are $(a, 0)$ and $(-a, 0)$. If $a > b$, the ellipse is said to be horizontal and $(-a, 0)$ and $(a, 0)$ are referred to as the **vertices** (singular, **vertex**). If $b > a$, the ellipse is said to be vertical and $(0, -b)$ and $(0, b)$ are then the vertices.

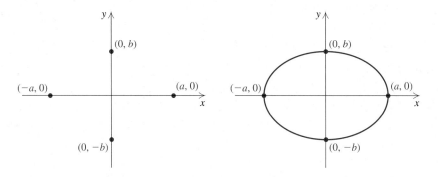

Plotting these four points and drawing an oval-shaped curve, we graph the ellipse. If a more precise graph is desired, we can plot more points.

> ### Using a and b to Graph an Ellipse
>
> For the ellipse
>
> $$\frac{x^2}{a^2} + \frac{y^2}{b^2} = 1,$$
>
> the x-intercepts are $(-a, 0)$ and $(a, 0)$. The y-intercepts are $(0, -b)$ and $(0, b)$. For $a^2 > b^2$, the ellipse is horizontal. For $b^2 > a^2$, the ellipse is vertical.

EXAMPLE **1**   Graph the ellipse

$$\frac{x^2}{4} + \frac{y^2}{9} = 1.$$

SOLUTION   Note that

$$\frac{x^2}{4} + \frac{y^2}{9} = \frac{x^2}{2^2} + \frac{y^2}{3^2}. \qquad \text{Identifying } a \text{ and } b. \text{ Since } b^2 > a^2, \\ \text{the ellipse is vertical.}$$

Since $a = 2$ and $b = 3$, the x-intercepts are $(-2, 0)$ and $(2, 0)$, and the y-intercepts are $(0, -3)$ and $(0, 3)$. We plot these points and connect them with an oval-shaped curve. To plot two other points, we let $x = 1$ and solve for y:

$$\frac{1^2}{4} + \frac{y^2}{9} = 1$$

$$36\left(\frac{1}{4} + \frac{y^2}{9}\right) = 36 \cdot 1$$

$$36 \cdot \frac{1}{4} + 36 \cdot \frac{y^2}{9} = 36$$

$$9 + 4y^2 = 36$$

$$4y^2 = 27$$

$$y^2 = \frac{27}{4}$$

$$y = \pm\sqrt{\frac{27}{4}}$$

$$y \approx \pm 2.6.$$

Thus, $(1, 2.6)$ and $(1, -2.6)$ can also be used to draw the graph. Similarly, the points $(-1, 2.6)$ and $(-1, -2.6)$ should appear on the graph.

TRY EXERCISE   9

EXAMPLE   2

Graph: $4x^2 + 25y^2 = 100$.

SOLUTION   To write the equation in standard form, we divide both sides by 100 to get 1 on the right side:

$$\frac{4x^2 + 25y^2}{100} = \frac{100}{100} \qquad \text{Dividing by 100 to get 1 on the right side}$$

$$\left.\begin{array}{c} \dfrac{4x^2}{100} + \dfrac{25y^2}{100} = 1 \\[2mm] \dfrac{x^2}{25} + \dfrac{y^2}{4} = 1 \end{array}\right\} \qquad \text{Simplifying}$$

$$\frac{x^2}{5^2} + \frac{y^2}{2^2} = 1. \qquad a = 5, b = 2$$

STUDENT NOTES

Note that any equation of the form $Ax^2 + By^2 = C$ (with $A \ne B$ and $A, B > 0$) can be rewritten as an equivalent equation in standard form. The graph is an ellipse.

The x-intercepts are $(-5, 0)$ and $(5, 0)$, and the y-intercepts are $(0, -2)$ and $(0, 2)$. We plot the intercepts and connect them with an oval-shaped curve. Other points can also be computed and plotted.

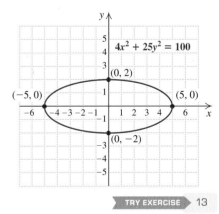

TRY EXERCISE   13

Ellipses Centered at (h, k)

Horizontal and vertical translations, similar to those used in Chapter 8, can be used to graph ellipses that are not centered at the origin.

Equation of an Ellipse Centered at (h, k)

The standard form of a horizontal or vertical ellipse centered at (h, k) is

$$\frac{(x - h)^2}{a^2} + \frac{(y - k)^2}{b^2} = 1.$$

The vertices are $(h + a, k)$ and $(h - a, k)$ if horizontal; $(h, k + b)$ and $(h, k - b)$ if vertical.

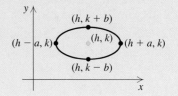

EXAMPLE **3**

Graph the ellipse

$$\frac{(x - 1)^2}{4} + \frac{(y + 5)^2}{9} = 1.$$

SOLUTION Note that

$$\frac{(x - 1)^2}{4} + \frac{(y + 5)^2}{9} = \frac{(x - 1)^2}{2^2} + \frac{(y + 5)^2}{3^2}.$$

Thus, $a = 2$ and $b = 3$. To determine the center of the ellipse, (h, k), note that

$$\frac{(x - 1)^2}{2^2} + \frac{(y + 5)^2}{3^2} = \frac{(x - 1)^2}{2^2} + \frac{(y - (-5))^2}{3^2}.$$

Thus the center is $(1, -5)$. We plot points 2 units to the left and right of center, as well as 3 units above and below center. These are the points $(3, -5)$, $(-1, -5)$, $(1, -2)$, and $(1, -8)$. The graph of the ellipse is shown at left.

Note that this ellipse is the same as the ellipse in Example 1 but translated 1 unit to the right and 5 units down.

> **TRY EXERCISE** 27

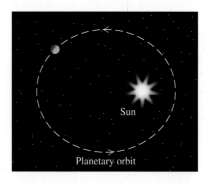

Ellipses have many applications. Communications satellites move in elliptical orbits with the earth as a focus while the earth itself follows an elliptical path around the sun. A medical instrument, the lithotripter, uses shock waves originating at one focus to crush a kidney stone located at the other focus.

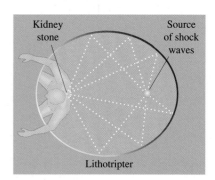

In some buildings, an ellipsoidal ceiling creates a "whispering gallery" in which a person at one focus can whisper and still be heard clearly at the other focus. This happens because sound waves coming from one focus are all reflected to the other focus. Similarly, light waves bouncing off an ellipsoidal mirror are used in a dentist's or surgeon's reflector light. The light source is located at one focus while the patient's mouth or surgical field is at the other.

TECHNOLOGY CONNECTION

To graph an ellipse on a graphing calculator, we solve for y and graph two functions.

To illustrate, let's check Example 2:

$$4x^2 + 25y^2 = 100$$
$$25y^2 = 100 - 4x^2$$
$$y^2 = 4 - \frac{4}{25}x^2$$
$$y = \pm\sqrt{4 - \frac{4}{25}x^2}.$$

Using a squared window, we have our check:

$$y_1 = -\sqrt{4 - \frac{4}{25}x^2}, \quad y_2 = \sqrt{4 - \frac{4}{25}x^2}$$

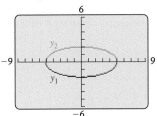

On many calculators, pressing and selecting Conics and then Ellipse accesses a program in which equations in Standard Form can be graphed directly.

10.2 EXERCISE SET

Concept Reinforcement Classify each statement as either true or false.

1. The graph of $\dfrac{x^2}{25} + \dfrac{y^2}{50} = 1$ is a vertical ellipse.

2. The graph of $\dfrac{x^2}{30} + \dfrac{y^2}{20} = 1$ is a vertical ellipse.

3. The graph of $\dfrac{x^2}{25} - \dfrac{y^2}{9} = 1$ is a horizontal ellipse.

4. The graph of $\dfrac{-x^2}{20} + \dfrac{y^2}{16} = 1$ is a horizontal ellipse.

5. The graph of $\dfrac{x^2}{9} + \dfrac{y^2}{25} = 1$ includes the points $(-3, 0)$ and $(3, 0)$.

6. The graph of $\dfrac{x^2}{36} + \dfrac{y^2}{25} = 1$ includes the points $(0, -5)$ and $(0, 5)$.

7. The graph of $\dfrac{(x + 3)^2}{25} + \dfrac{(y - 2)^2}{36} = 1$ is an ellipse centered at $(-3, 2)$.

8. The graph of $\dfrac{(x - 2)^2}{49} + \dfrac{(y + 5)^2}{9} = 1$ is an ellipse centered at $(2, -5)$.

Graph each of the following equations.

9. $\dfrac{x^2}{1} + \dfrac{y^2}{4} = 1$

10. $\dfrac{x^2}{4} + \dfrac{y^2}{1} = 1$

11. $\dfrac{x^2}{25} + \dfrac{y^2}{9} = 1$

12. $\dfrac{x^2}{16} + \dfrac{y^2}{25} = 1$

13. $4x^2 + 9y^2 = 36$

14. $9x^2 + 4y^2 = 36$

15. $16x^2 + 9y^2 = 144$

16. $9x^2 + 16y^2 = 144$

17. $2x^2 + 3y^2 = 6$

18. $5x^2 + 7y^2 = 35$

Aha! 19. $5x^2 + 5y^2 = 125$

20. $8x^2 + 5y^2 = 80$

21. $3x^2 + 7y^2 - 63 = 0$

22. $3x^2 + 3y^2 - 48 = 0$

23. $16x^2 = 16 - y^2$

24. $9y^2 = 9 - x^2$

25. $16x^2 + 25y^2 = 1$

26. $9x^2 + 4y^2 = 1$

27. $\dfrac{(x - 3)^2}{9} + \dfrac{(y - 2)^2}{25} = 1$

28. $\dfrac{(x - 2)^2}{25} + \dfrac{(y - 4)^2}{9} = 1$

29. $\dfrac{(x + 4)^2}{16} + \dfrac{(y - 3)^2}{49} = 1$

30. $\dfrac{(x + 5)^2}{4} + \dfrac{(y - 2)^2}{36} = 1$

31. $12(x - 1)^2 + 3(y + 4)^2 = 48$
(*Hint*: Divide both sides by 48.)

32. $4(x - 6)^2 + 9(y + 2)^2 = 36$

Aha! **33.** $4(x + 3)^2 + 4(y + 1)^2 - 10 = 90$

34. $9(x + 6)^2 + (y + 2)^2 - 20 = 61$

35. Explain how you can tell from the equation of an ellipse whether the graph will be horizontal or vertical.

36. Can an ellipse ever be the graph of a function? Why or why not?

Skill Review

Review solving equations.

Solve.

37. $x^2 - 5x + 3 = 0$ [8.2]   **38.** $\log_x 81 = 4$ [9.6]

39. $\dfrac{4}{x + 2} + \dfrac{3}{2x - 1} = 2$ [6.4]

40. $3 - \sqrt{2x - 1} = 1$ [7.6]

41. $x^2 = 11$ [8.1]   **42.** $x^2 + 4x = 60$ [5.7]

Synthesis

43. Explain how it is possible to recognize that the graph of $9x^2 + 18x + y^2 - 4y + 4 = 0$ is an ellipse.

44. As the foci get closer to the center of an ellipse, what shape does the graph begin to resemble? Explain why this happens.

Find an equation of an ellipse that contains the following points.

45. $(-9, 0)$, $(9, 0)$, $(0, -11)$, and $(0, 11)$

46. $(-7, 0)$, $(7, 0)$, $(0, -10)$, and $(0, 10)$

47. $(-2, -1)$, $(6, -1)$, $(2, -4)$, and $(2, 2)$

48. $(4, 3)$, $(-6, 3)$, $(-1, -1)$, and $(-1, 7)$

49. *Astronomy.*   The maximum distance of the planet Mars from the sun is 2.48×10^8 mi. The minimum distance is 3.46×10^7 mi. The sun is at one focus of the elliptical orbit. Find the distance from the sun to the other focus.

50. *Theatrical lighting.*   The spotlight on a violin soloist casts an ellipse of light on the floor below her that is 6 ft wide and 10 ft long. Find an equation of that ellipse if the performer is in its center, x is the distance from the performer to the side of the ellipse, and y is the distance from the performer to the top of the ellipse.

51. Let $(-c, 0)$ and $(c, 0)$ be the foci of an ellipse. Any point $P(x, y)$ is on the ellipse if the sum of the distances from the foci to P is some constant. Use $2a$ to represent this constant.

a) Show that an equation for the ellipse is given by

$$\frac{x^2}{a^2} + \frac{y^2}{a^2 - c^2} = 1.$$

b) Substitute b^2 for $a^2 - c^2$ to get standard form.

52. *President's office.*   The Oval Office of the President of the United States is an ellipse 31 ft wide and 38 ft long. Show in a sketch precisely where the President and an adviser could sit to best hear each other using the room's acoustics. (*Hint*: See Exercise 51(b) and the discussion following Example 3.)

53. *Dentistry.*   The light source in some dental lamps shines against a reflector that is shaped like a portion of an ellipse in which the light source is

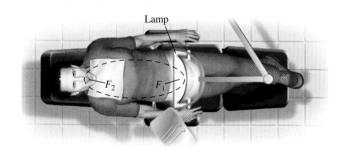

one focus of the ellipse. Reflected light enters a patient's mouth at the other focus of the ellipse. If the ellipse from which the reflector was formed is 2 ft wide and 6 ft long, how far should the patient's mouth be from the light source? (*Hint*: See Exercise 51(b).)

54. *Firefighting.*   The size and shape of certain forest fires can be approximated as the union of two "half-ellipses." For the blaze modeled below, the equation of the smaller ellipse—the part of the fire moving *into* the wind—is

$$\frac{x^2}{40,000} + \frac{y^2}{10,000} = 1.$$

The equation of the other ellipse—the part moving *with* the wind—is

$$\frac{x^2}{250,000} + \frac{y^2}{10,000} = 1.$$

Determine the width and the length of the fire.

Source for figure: "Predicting Wind-Driven Wild Land Fire Size and Shape," Hal E. Anderson, Research Paper INT-305, U.S. Department of Agriculture, Forest Service, February 1983

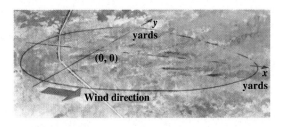

For each of the following equations, complete the square as needed and find an equivalent equation in standard form. Then graph the ellipse.

55. $x^2 - 4x + 4y^2 + 8y - 8 = 0$

56. $4x^2 + 24x + y^2 - 2y - 63 = 0$

57. Use a graphing calculator to check your answers to Exercises 11, 25, 29, and 33.

COLLABORATIVE CORNER

A Cosmic Path

Focus: Ellipses
Time: 20–30 minutes
Group size: 2
Materials: Scientific calculators

On May 4, 2007, Comet 17P/Holmes was at the point closest to the sun in its orbit. Comet 17P is traveling in an elliptical orbit with the sun as one focus, and one orbit takes about 6.88 yr. One astronomical unit (AU) is 93,000,000 mi. One group member should do the following calculations in AU and the other in millions of miles.
Source: Harvard-Smithsonian Center for Astrophysics

ACTIVITY

1. At its *perihelion*, a comet with an elliptical orbit is at the point in its orbit closest to the sun. At its *aphelion*, the comet is at the point farthest from the sun. The perihelion distance for Comet 17P is 2.053218 AU, and the aphelion distance is 5.183610 AU. Use these distances to find a. (See the following diagram.)

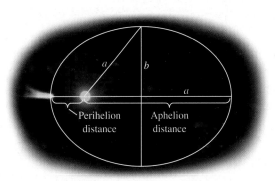

2. Using the figure above, express b^2 as a function of a. Then find b using the value found for a in part (1).

3. One formula for approximating the perimeter of an ellipse is

$$P = \pi\left(3a + 3b - \sqrt{(3a + b)(a + 3b)}\right),$$

developed by the Indian mathematician S. Ramanujan in 1914. How far does Comet 17P travel in one orbit?

4. What is the speed of the comet? Find the answer in AU per year and in miles per hour.

10.3   Conic Sections: Hyperbolas

Hyperbolas   ▪   Hyperbolas (Nonstandard Form)   ▪   Classifying Graphs of Equations

Hyperbolas

A **hyperbola** looks like a pair of parabolas, but the shapes are not quite parabolic. A hyperbola has two **vertices** and the line through the vertices is known as the **axis**. The point halfway between the vertices is called the **center**. The two curves that comprise a hyperbola are called **branches**.

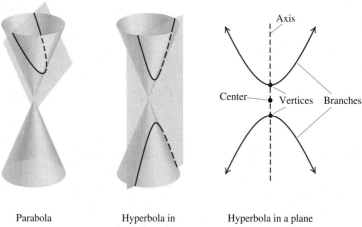

Parabola   Hyperbola in   Hyperbola in a plane
                        three dimensions

Equation of a Hyperbola Centered at the Origin

A hyperbola with its center at the origin* has its equation as follows:

$$\frac{x^2}{a^2} - \frac{y^2}{b^2} = 1 \qquad \text{(Horizontal axis)};$$

$$\frac{y^2}{b^2} - \frac{x^2}{a^2} = 1 \qquad \text{(Vertical axis)}.$$

Note that both equations have 1 on the right-hand side and subtraction between the terms. For the discussion that follows, we assume $a, b > 0$.

To graph a hyperbola, it helps to begin by graphing two lines called **asymptotes**. Although the asymptotes themselves are not part of the graph, they serve as guidelines for an accurate sketch.

As a hyperbola gets farther away from the origin, it gets closer and closer to its asymptotes. The larger $|x|$ gets, the closer the graph gets to an asymptote. The asymptotes act to "constrain" the graph of a hyperbola. Parabolas are *not* constrained by any asymptotes.

---

*Hyperbolas with horizontal or vertical axes and centers *not* at the origin are discussed in Exercises 59–64.

Asymptotes of a Hyperbola

For hyperbolas with equations as shown below, the asymptotes are the lines

$$y = \frac{b}{a}x \quad \text{and} \quad y = -\frac{b}{a}x.$$

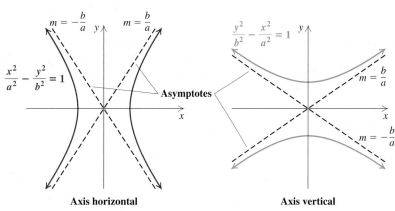

Axis horizontal   Axis vertical

In Section 10.2, we used a and b to determine the width and the length of an ellipse. For hyperbolas, a and b are used to determine the base and the height of a rectangle that can be used as an aid in sketching asymptotes and locating vertices. This is illustrated in the following example.

EXAMPLE **1**   Graph: $\dfrac{x^2}{4} - \dfrac{y^2}{9} = 1$.

SOLUTION   Note that

$$\frac{x^2}{4} - \frac{y^2}{9} = \frac{x^2}{2^2} - \frac{y^2}{3^2}, \quad \text{Identifying } a \text{ and } b$$

so $a = 2$ and $b = 3$. The asymptotes are thus

$$y = \frac{3}{2}x \quad \text{and} \quad y = -\frac{3}{2}x.$$

To help us sketch asymptotes and locate vertices, we use a and b—in this case, 2 and 3—to form the pairs $(-2, 3)$, $(2, 3)$, $(2, -3)$, and $(-2, -3)$. We plot these pairs and lightly sketch a rectangle. The asymptotes pass through the corners and, since this is a horizontal hyperbola, the vertices are where the rectangle intersects the x-axis. Finally, we draw the hyperbola, as shown below.

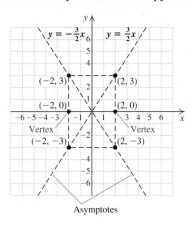

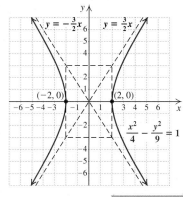

TRY EXERCISE ▶ 11

EXAMPLE   2   Graph: $\dfrac{y^2}{36} - \dfrac{x^2}{4} = 1$.

SOLUTION   Note that

$$\dfrac{y^2}{36} - \dfrac{x^2}{4} = \dfrac{y^2}{6^2} - \dfrac{x^2}{2^2} = 1.$$

> Whether the hyperbola is horizontal or vertical is determined by which term is nonnegative. Here the y^2-term is nonnegative, so the hyperbola is vertical.

STUDENT NOTES

Regarding the orientation of a hyperbola, you may find it helpful to think as follows: "The axis is parallel to the x-axis if $\dfrac{x^2}{a^2}$ is the positive term. The axis is parallel to the y-axis if $\dfrac{y^2}{b^2}$ is the positive term."

Using ± 2 as x-coordinates and ± 6 as y-coordinates, we plot $(2, 6)$, $(2, -6)$, $(-2, 6)$, and $(-2, -6)$, and lightly sketch a rectangle through them. The asymptotes pass through the corners (see the figure on the left below). Since the hyperbola is vertical, its vertices are $(0, 6)$ and $(0, -6)$. Finally, we draw curves through the vertices toward the asymptotes, as shown below.

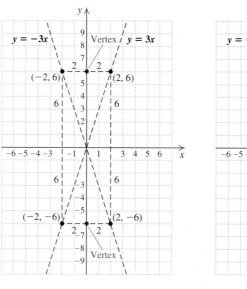
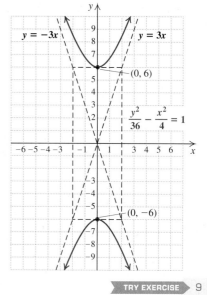

TRY EXERCISE   9

Hyperbolas (Nonstandard Form)

The equations for hyperbolas just examined are the standard ones, but there are other hyperbolas. We consider some of them.

Equation of a Hyperbola in Nonstandard Form

Hyperbolas having the x- and y-axes as asymptotes have equations as follows:

$$xy = c, \quad \text{where } c \text{ is a nonzero constant.}$$

EXAMPLE   3   Graph: $xy = -8$.

SOLUTION   We first solve for y:

$$y = -\frac{8}{x}.$$ Dividing both sides by x. Note that $x \neq 0$.

Next, we find some solutions and form a table. Note that x cannot be 0 and that for large values of $|x|$, the value of y is close to 0. Thus the x- and y-axes serve as asymptotes. We plot the points and draw two curves.

x	y
2	-4
-2	4
4	-2
-4	2
1	-8
-1	8
8	-1
-8	1

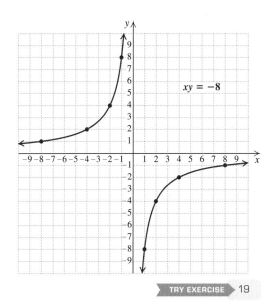

$xy = -8$

TRY EXERCISE   19

Hyperbolas have many applications. A jet breaking the sound barrier creates a sonic boom with a wave front the shape of a cone. The intersection of the cone with the ground is one branch of a hyperbola. Some comets travel in hyperbolic orbits, and a cross section of many lenses is hyperbolic in shape.

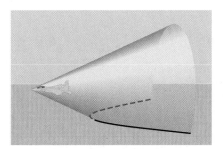

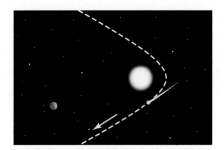

The procedure used to graph a hyperbola in standard form is similar to that used to draw a circle or an ellipse. Consider the graph of

$$\frac{x^2}{25} - \frac{y^2}{49} = 1.$$

The student should confirm that solving for y yields

$$y_1 = \frac{\sqrt{49x^2 - 1225}}{5} = \frac{7}{5}\sqrt{x^2 - 25}$$

and

$$y_2 = \frac{-\sqrt{49x^2 - 1225}}{5} = -\frac{7}{5}\sqrt{x^2 - 25},$$

or $y_2 = -y_1.$

When the two pieces are drawn on the same squared window, the result is as shown. The gaps occur where the graph is nearly vertical.

$$y_1 = \frac{7}{5}\sqrt{x^2 - 25},$$
$$y_2 = -\frac{7}{5}\sqrt{x^2 - 25}$$

On many calculators, pressing **APPS** and selecting Conics and then Hyperbola accesses a program in which hyperbolas in standard form can be graphed directly. Graph each of the following.

1. $\dfrac{x^2}{16} - \dfrac{y^2}{60} = 1$ **2.** $16x^2 - 3y^2 = 64$

3. $\dfrac{y^2}{20} - \dfrac{x^2}{64} = 1$ **4.** $45y^2 - 9x^2 = 441$

Classifying Graphs of Equations

By writing an equation of a conic section in a standard form, we can classify its graph as a parabola, a circle, an ellipse, or a hyperbola. Every conic section can also be represented by an equation of the form

$$Ax^2 + By^2 + Cxy + Dx + Ey + F = 0.$$

We can also classify graphs using values of A and B.

Graph	Standard Form		$Ax^2 + By^2 + Cxy + Dx + Ey + F = 0$
Parabola	$y = ax^2 + bx + c$;	Vertical parabola	Either $A = 0$ or $B = 0$, but not both.
	$x = ay^2 + by + c$	Horizontal parabola	
Circle	$x^2 + y^2 = r^2$;	Center at the origin	$A = B$
	$(x - h)^2 + (y - k)^2 = r^2$	Center at (h, k)	
Ellipse	$\dfrac{x^2}{a^2} + \dfrac{y^2}{b^2} = 1$;	Center at the origin	$A \neq B$, and A and B have the same sign.
	$\dfrac{(x - h)^2}{a^2} + \dfrac{(y - k)^2}{b^2} = 1$	Center at (h, k)	
Hyperbola	$\dfrac{x^2}{a^2} - \dfrac{y^2}{b^2} = 1$;	Horizontal hyperbola	A and B have opposite signs.
	$\dfrac{y^2}{b^2} - \dfrac{x^2}{a^2} = 1$	Vertical hyperbola	
	$xy = c$	Asymptotes are axes	Only C and F are nonzero.

Algebraic manipulations may be needed to express an equation in one of the preceding forms.

EXAMPLE 4 Classify the graph of each equation as a circle, an ellipse, a parabola, or a hyperbola. Refer to the above table as needed.

a) $5x^2 = 20 - 5y^2$

b) $x + 3 + 8y = y^2$

c) $x^2 = y^2 + 4$

d) $x^2 = 16 - 4y^2$

SOLUTION

a) We get the terms with variables on one side by adding $5y^2$ to both sides:

$$5x^2 + 5y^2 = 20.$$

Since x and y are *both* squared, we do not have a parabola. The fact that the squared terms are *added* tells us that we do not have a hyperbola. Do we have a circle? We factor the 5 out of both terms on the left and then divide by 5:

$$5(x^2 + y^2) = 20 \qquad \text{Factoring out 5}$$
$$x^2 + y^2 = 4 \qquad \text{Dividing both sides by 5}$$
$$x^2 + y^2 = 2^2. \qquad \text{This is an equation for a circle.}$$

We see that the graph is a circle centered at the origin with radius 2.

We can also write the equation in the form

$$5x^2 + 5y^2 - 20 = 0. \qquad A = 5, B = 5$$

Since $A = B$, the graph is a circle.

b) The equation $x + 3 + 8y = y^2$ has only one variable that is squared, so we solve for the other variable:

$$x = y^2 - 8y - 3. \qquad \text{This is an equation for a parabola.}$$

The graph is a horizontal parabola that opens to the right.

We can also write the equation in the form

$$y^2 - x - 8y - 3 = 0. \qquad A = 0, B = 1$$

Since $A = 0$ and $B \neq 0$, the graph is a parabola.

c) In $x^2 = y^2 + 4$, both variables are squared, so the graph is not a parabola. We subtract y^2 on both sides and divide by 4 to obtain

$$\frac{x^2}{2^2} - \frac{y^2}{2^2} = 1. \qquad \text{This is an equation for a hyperbola.}$$

The minus sign here indicates that the graph is a hyperbola. Because it is the x^2-term that is nonnegative, the hyperbola is horizontal.

We can also write the equation in the form

$$x^2 - y^2 - 4 = 0. \qquad A = 1, B = -1$$

Since A and B have opposite signs, the graph is a hyperbola.

d) In $x^2 = 16 - 4y^2$, both variables are squared, so the graph cannot be a parabola. We obtain the following equivalent equation:

$$x^2 + 4y^2 = 16. \qquad \text{Adding } 4y^2 \text{ to both sides}$$

If the coefficients of the terms were the same, we would have the graph of a circle, as in part (a), but they are not. Dividing both sides by 16 yields

$$\frac{x^2}{16} + \frac{y^2}{4} = 1.$$ This is an equation for an ellipse.

The graph of this equation is a horizontal ellipse.
We can also write the equation in the form

$$x^2 + 4y^2 - 16 = 0.$$ $A = 1, B = 4$

Since $A \neq B$ and both A and B are positive, the graph is an ellipse.

▶ TRY EXERCISES ▶ 27 and 29

10.3   EXERCISE SET

🖐 **Concept Reinforcement**   *In each of Exercises 1–8, match the conic section with the equation in the column on the right that represents that type of conic section.*

1. ____ A hyperbola with a horizontal axis

2. ____ A hyperbola with a vertical axis

3. ____ An ellipse with its center not at the origin

4. ____ An ellipse with its center at the origin

5. ____ A circle with its center at the origin

6. ____ A circle with its center not at the origin

7. ____ A parabola opening upward or downward

8. ____ A parabola opening to the right or left

a) $\dfrac{x^2}{10} + \dfrac{y^2}{12} = 1$

b) $(x + 1)^2 + (y - 3)^2 = 30$

c) $y - x^2 = 5$

d) $\dfrac{x^2}{9} - \dfrac{y^2}{10} = 1$

e) $x - 2y^2 = 3$

f) $\dfrac{y^2}{20} - \dfrac{x^2}{35} = 1$

g) $3x^2 + 3y^2 = 75$

h) $\dfrac{(x - 1)^2}{10} + \dfrac{(y - 4)^2}{8} = 1$

Graph each hyperbola. Label all vertices and sketch all asymptotes.

9. $\dfrac{y^2}{16} - \dfrac{x^2}{16} = 1$

10. $\dfrac{x^2}{9} - \dfrac{y^2}{9} = 1$

11. $\dfrac{x^2}{4} - \dfrac{y^2}{25} = 1$

12. $\dfrac{y^2}{16} - \dfrac{x^2}{9} = 1$

13. $\dfrac{y^2}{36} - \dfrac{x^2}{9} = 1$

14. $\dfrac{x^2}{25} - \dfrac{y^2}{36} = 1$

15. $y^2 - x^2 = 25$

16. $x^2 - y^2 = 4$

17. $25x^2 - 16y^2 = 400$

18. $4y^2 - 9x^2 = 36$

Graph.

19. $xy = -6$

20. $xy = 8$

21. $xy = 4$

22. $xy = -9$

23. $xy = -2$

24. $xy = -1$

25. $xy = 1$

26. $xy = 2$

Classify each of the following as the equation of a circle, an ellipse, a parabola, or a hyperbola.

27. $x^2 + y^2 - 6x + 10y - 40 = 0$

28. $y - 4 = 2x^2$

29. $9x^2 + 4y^2 - 36 = 0$

30. $x + 3y = 2y^2 - 1$

31. $4x^2 - 9y^2 - 72 = 0$

32. $y^2 + x^2 = 8$

33. $y^2 = 20 - x^2$

34. $2y + 13 + x^2 = 8x - y^2$

35. $x - 10 = y^2 - 6y$

36. $y = \dfrac{5}{x}$

37. $x - \dfrac{3}{y} = 0$

38. $9x^2 = 9 - y^2$

39. $y + 6x = x^2 + 5$

40. $x^2 = 49 + y^2$

41. $25y^2 = 100 + 4x^2$

42. $3x^2 + 5y^2 + x^2 = y^2 + 49$

43. $3x^2 + y^2 - x = 2x^2 - 9x + 10y + 40$

44. $4y^2 + 20x^2 + 1 = 8y - 5x^2$

45. $16x^2 + 5y^2 - 12x^2 + 8y^2 - 3x + 4y = 568$

46. $56x^2 - 17y^2 = 234 - 13x^2 - 38y^2$

47. Explain how the equation of a hyperbola differs from the equation of an ellipse.

48. Is it possible for a hyperbola to represent the graph of a function? Why or why not?

Skill Review

To prepare for Section 10.4, review solving systems of equations and solving quadratic equations (Sections 3.2 and 8.2).

Solve.

49. $5x + 2y = -3,$
$\quad 2x + 3y = 12$ [3.2]

50. $4x - 2y = 5,$
$\quad 3x + 5y = -6$ [3.2]

51. $\frac{3}{4}x^2 + x^2 = 7$ [8.2]

52. $3x^2 + 10x - 8 = 0$ [8.2]

53. $x^2 - 3x - 1 = 0$ [8.2]

54. $x^2 + \dfrac{25}{x^2} = 26$ [8.5]

Synthesis

55. What is it in the equation of a hyperbola that controls how wide open the branches are? Explain your reasoning.

56. If, in

$$\frac{x^2}{a^2} - \frac{y^2}{b^2} = 1,$$

$a = b$, what are the asymptotes of the graph? Why?

Find an equation of a hyperbola satisfying the given conditions.

57. Having intercepts $(0, 6)$ and $(0, -6)$ and asymptotes $y = 3x$ and $y = -3x$

58. Having intercepts $(8, 0)$ and $(-8, 0)$ and asymptotes $y = 4x$ and $y = -4x$

The standard form for equations of horizontal or vertical hyperbolas centered at (h, k) are as follows:

$$\frac{(x - h)^2}{a^2} - \frac{(y - k)^2}{b^2} = 1$$

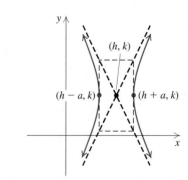

$$\frac{(y - k)^2}{b^2} - \frac{(x - h)^2}{a^2} = 1$$

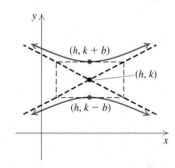

The vertices are as labeled and the asymptotes are

$$y - k = \frac{b}{a}(x - h) \quad and \quad y - k = -\frac{b}{a}(x - h).$$

For each of the following equations of hyperbolas, complete the square, if necessary, and write in standard form. Find the center, the vertices, and the asymptotes. Then graph the hyperbola.

59. $\dfrac{(x - 5)^2}{36} - \dfrac{(y - 2)^2}{25} = 1$

60. $\dfrac{(x - 2)^2}{9} - \dfrac{(y - 1)^2}{4} = 1$

61. $8(y + 3)^2 - 2(x - 4)^2 = 32$

62. $25(x - 4)^2 - 4(y + 5)^2 = 100$

63. $4x^2 - y^2 + 24x + 4y + 28 = 0$

64. $4y^2 - 25x^2 - 8y - 100x - 196 = 0$

65. Use a graphing calculator to check your answers to Exercises 13, 25, 31, and 59.

CONNECTING the CONCEPTS

When graphing equations of conic sections, it is usually helpful to first determine what type of graph the equation represents. We then find the coordinates of key points and equations of lines that determine the shape and the location of the graph.

Graph	Equation	Key Points	Equations of Lines
Parabola	$y = a(x - h)^2 + k$ $x = a(y - k)^2 + h$	Vertex: (h, k) Vertex: (h, k)	Axis of symmetry: $x = h$ Axis of symmetry: $y = k$
Circle	$(x - h)^2 + (y - k)^2 = r^2$	Center: (h, k)	
Ellipse	$\dfrac{x^2}{a^2} + \dfrac{y^2}{b^2} = 1$	x-intercepts: $(-a, 0), (a, 0)$; y-intercepts: $(0, -b), (0, b)$	
Hyperbola	$\dfrac{x^2}{a^2} - \dfrac{y^2}{b^2} = 1$ $\dfrac{y^2}{b^2} - \dfrac{x^2}{a^2} = 1$	Vertices: $(-a, 0), (a, 0)$ Vertices: $(0, -b), (0, b)$	Asymptotes (for both equations): $y = \dfrac{b}{a}x, y = -\dfrac{b}{a}x$
	$xy = c$		Asymptotes: $x = 0, y = 0$

MIXED REVIEW

1. Find the vertex and the axis of symmetry of the graph of $y = 3(x - 4)^2 + 1$.

2. Find the vertex and the axis of symmetry of the graph of $x = y^2 + 2y + 3$.

3. Find the center of the graph of $(x - 3)^2 + (y - 2)^2 = 5$.

4. Find the center of the graph of $x^2 + 6x + y^2 + 10y = 12$.

5. Find the x-intercepts and the y-intercepts of the graph of $\dfrac{x^2}{144} + \dfrac{y^2}{81} = 1$.

6. Find the vertices of the graph of $\dfrac{x^2}{9} - \dfrac{y^2}{121} = 1$.

7. Find the vertices of the graph of $4y^2 - x^2 = 4$.

8. Find the asymptotes of the graph of $\dfrac{y^2}{9} - \dfrac{x^2}{4} = 1$.

Classify each of the following as the graph of a parabola, a circle, an ellipse, or a hyperbola. Then graph.

9. $x^2 + y^2 = 36$

10. $y = x^2 - 5$

11. $\dfrac{x^2}{25} + \dfrac{y^2}{49} = 1$

12. $\dfrac{x^2}{25} - \dfrac{y^2}{49} = 1$

13. $x = (y + 3)^2 + 2$

14. $4x^2 + 9y^2 = 36$

15. $xy = -4$

16. $(x + 2)^2 + (y - 3)^2 = 1$

17. $x^2 + y^2 - 8y - 20 = 0$

18. $x = y^2 + 2y$

19. $16y^2 - x^2 = 16$

20. $x = \dfrac{9}{y}$

10.4 Nonlinear Systems of Equations

Systems Involving One Nonlinear Equation • Systems of Two Nonlinear Equations • Problem Solving

The equations appearing in systems of two equations have thus far in our discussion always been linear. We now consider systems of two equations in which at least one equation is nonlinear.

Systems Involving One Nonlinear Equation

Suppose that a system consists of an equation of a circle and an equation of a line. In what ways can the circle and the line intersect? The figures below represent three ways in which the situation can occur. We see that such a system will have 0, 1, or 2 real solutions.

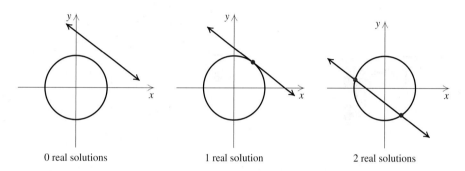

0 real solutions   1 real solution   2 real solutions

Recall that graphing, *elimination*, and *substitution* were all used to solve systems of linear equations. To solve systems in which one equation is of first degree and one is of second degree, it is preferable to use the *substitution* method.

EXAMPLE 1   Solve the system

$$x^2 + y^2 = 25, \quad (1) \qquad \text{(The graph is a circle.)}$$
$$3x - 4y = 0. \quad (2) \qquad \text{(The graph is a line.)}$$

SOLUTION   First, we solve the linear equation, (2), for x:

$$x = \tfrac{4}{3}y. \quad (3) \qquad \text{We could have solved for } y \text{ instead.}$$

Then we substitute $\tfrac{4}{3}y$ for x in equation (1) and solve for y:

$$\left(\tfrac{4}{3}y\right)^2 + y^2 = 25$$
$$\tfrac{16}{9}y^2 + y^2 = 25$$
$$\tfrac{25}{9}y^2 = 25$$
$$y^2 = 9 \qquad \text{Multiplying both sides by } \tfrac{9}{25}$$
$$y = \pm 3. \qquad \text{Using the principle of square roots}$$

Now we substitute these numbers for y in equation (3) and solve for x:

for $y = 3$, $\quad x = \tfrac{4}{3}(3) = 4;$ $\qquad$ The ordered pair is $(4, 3)$.
for $y = -3$, $\quad x = \tfrac{4}{3}(-3) = -4.$ $\qquad$ The ordered pair is $(-4, -3)$.

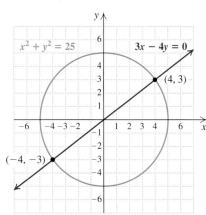

Check: For (4, 3):

$$\frac{x^2 + y^2 = 25}{4^2 + 3^2 \;\Big|\; 25}$$
$$16 + 9$$
$$25 \overset{?}{=} 25 \quad \text{TRUE}$$

$$\frac{3x - 4y = 0}{3(4) - 4(3) \;\Big|\; 0}$$
$$12 - 12$$
$$0 \overset{?}{=} 0 \quad \text{TRUE}$$

It is left to the student to confirm that $(-4, -3)$ also checks in both equations. The pairs $(4, 3)$ and $(-4, -3)$ check, so they are solutions. The graph at left serves as a check. Intersections occur at $(4, 3)$ and $(-4, -3)$.

> **TRY EXERCISE** 7

Even if we do not know what the graph of each equation in a system looks like, the algebraic approach of Example 1 can still be used.

EXAMPLE   2   Solve the system

$$y + 3 = 2x, \qquad (1) \qquad \text{(A first-degree equation)}$$
$$x^2 + 2xy = -1. \qquad (2) \qquad \text{(A second-degree equation)}$$

SOLUTION   First, we solve the linear equation (1) for y:

$$y = 2x - 3. \qquad (3)$$

Then we substitute $2x - 3$ for y in equation (2) and solve for x:

$$x^2 + 2x(2x - 3) = -1$$
$$x^2 + 4x^2 - 6x = -1$$
$$5x^2 - 6x + 1 = 0$$
$$(5x - 1)(x - 1) = 0 \qquad \text{Factoring}$$
$$5x - 1 = 0 \quad or \quad x - 1 = 0 \qquad \text{Using the principle of zero products}$$
$$x = \tfrac{1}{5} \quad or \qquad x = 1.$$

Now we substitute these numbers for x in equation (3) and solve for y:

$$\text{for } x = \tfrac{1}{5}, \quad y = 2\left(\tfrac{1}{5}\right) - 3 = -\tfrac{13}{5}; \qquad \text{The ordered pair is } \left(\tfrac{1}{5}, -\tfrac{13}{5}\right).$$
$$\text{for } x = 1, \quad y = 2(1) - 3 = -1. \qquad \text{The ordered pair is } (1, -1).$$

You can confirm that $\left(\tfrac{1}{5}, -\tfrac{13}{5}\right)$ and $(1, -1)$ check, so they are both solutions.

> **TRY EXERCISE** 13

EXAMPLE   3   Solve the system

$$x + y = 5, \qquad (1) \qquad \text{(The graph is a line.)}$$
$$y = 3 - x^2. \qquad (2) \qquad \text{(The graph is a parabola.)}$$

SOLUTION   We substitute $3 - x^2$ for y in the first equation:

$$x + 3 - x^2 = 5$$
$$-x^2 + x - 2 = 0 \qquad \text{Adding } -5 \text{ to both sides and rearranging}$$
$$x^2 - x + 2 = 0. \qquad \text{Multiplying both sides by } -1$$

Real-number solutions of systems of equations can be found using the INTERSECT option of $\boxed{\text{CALC}}$.

To solve Example 2,

$$y + 3 = 2x,$$
$$x^2 + 2xy = -1,$$

we solve each equation for y and then graph:

$$\left.\begin{array}{l} y_1 = 2x - 3, \\[2mm] y_2 = \dfrac{-1 - x^2}{2x}. \end{array}\right\} \quad \begin{array}{l} \text{Note that} \\ x, y \neq 0. \end{array}$$

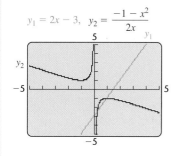

Using INTERSECT, we find the solutions to be $(0.2, -2.6)$ and $(1, -1)$.

Solve each system. Round all values to two decimal places.

1. $4xy - 7 = 0,$
      $x - 3y - 2 = 0$
2. $x^2 + y^2 = 14,$
      $16x + 7y^2 = 0$

Since $x^2 - x + 2$ does not factor, we need the quadratic formula:

$$\begin{aligned} x &= \frac{-b \pm \sqrt{b^2 - 4ac}}{2a} \\[2mm] &= \frac{-(-1) \pm \sqrt{(-1)^2 - 4 \cdot 1 \cdot 2}}{2(1)} \qquad \text{Substituting} \\[2mm] &= \frac{1 \pm \sqrt{1 - 8}}{2} = \frac{1 \pm \sqrt{-7}}{2} = \frac{1}{2} \pm \frac{\sqrt{7}}{2}i. \end{aligned}$$

Solving equation (1) for y gives us $y = 5 - x$. Substituting values for x gives

$$y = 5 - \left(\frac{1}{2} + \frac{\sqrt{7}}{2}i\right) = \frac{9}{2} - \frac{\sqrt{7}}{2}i \quad \text{and}$$

$$y = 5 - \left(\frac{1}{2} - \frac{\sqrt{7}}{2}i\right) = \frac{9}{2} + \frac{\sqrt{7}}{2}i.$$

The solutions are

$$\left(\frac{1}{2} + \frac{\sqrt{7}}{2}i, \frac{9}{2} - \frac{\sqrt{7}}{2}i\right) \quad \text{and} \quad \left(\frac{1}{2} - \frac{\sqrt{7}}{2}i, \frac{9}{2} + \frac{\sqrt{7}}{2}i\right).$$

There are no real-number solutions. Note in the figure at right that the graphs do not intersect. Getting only nonreal solutions tells us that the graphs do not intersect.

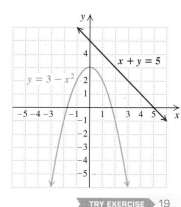

TRY EXERCISE ▶ 19

Systems of Two Nonlinear Equations

We now consider systems of two second-degree equations. Graphs of such systems can involve any two conic sections. The following figure shows some ways in which a circle and a hyperbola can intersect.

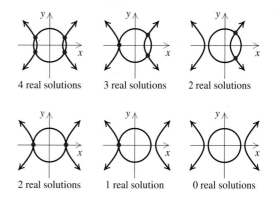

To solve systems of two second-degree equations, we either substitute or eliminate. The elimination method is generally better when both equations are of the form $Ax^2 + By^2 = C$. Then we can eliminate an x^2-term or a y^2-term in a manner similar to the procedure used in Chapter 3.

EXAMPLE **4**

Solve the system

$$2x^2 + 5y^2 = 22, \quad (1) \quad \text{(The graph is an ellipse.)}$$
$$3x^2 - y^2 = -1. \quad (2) \quad \text{(The graph is a hyperbola.)}$$

SOLUTION Here we multiply equation (2) by 5 and then add:

$$\begin{array}{ll} 2x^2 + 5y^2 = 22 & \\ \underline{15x^2 - 5y^2 = -5} & \text{Multiplying both sides of equation (2) by 5} \\ 17x^2 \quad\quad = 17 & \text{Adding} \\ x^2 = 1 & \\ x = \pm 1. & \end{array}$$

There is no x-term, and whether x is -1 or 1, we have $x^2 = 1$. Thus we can simultaneously substitute 1 and -1 for x in equation (2):

$$\left.\begin{array}{r} 3 \cdot (\pm 1)^2 - y^2 = -1 \\ 3 - y^2 = -1 \\ -y^2 = -4 \\ y^2 = 4 \quad \text{or} \quad y = \pm 2. \end{array}\right\} \quad \begin{array}{l} \text{Since } (-1)^2 = 1^2, \text{ we can evaluate for} \\ x = -1 \text{ and } x = 1 \text{ simultaneously.} \end{array}$$

Thus, if $x = 1$, then $y = 2$ or $y = -2$; and if $x = -1$, then $y = 2$ or $y = -2$. The four possible solutions are $(1, 2)$, $(1, -2)$, $(-1, 2)$, and $(-1, -2)$.

Check: Since $(2)^2 = (-2)^2$ and $(1)^2 = (-1)^2$, we can check all four pairs at once.

$$\begin{array}{c|c} 2x^2 + 5y^2 = 22 & \\ \hline 2(\pm 1)^2 + 5(\pm 2)^2 & 22 \\ 2 + 20 & \\ 22 \overset{?}{=} 22 \quad \text{TRUE} & \end{array} \qquad \begin{array}{c|c} 3x^2 - y^2 = -1 & \\ \hline 3(\pm 1)^2 - (\pm 2)^2 & -1 \\ 3 - 4 & \\ -1 \overset{?}{=} -1 \quad \text{TRUE} & \end{array}$$

The solutions are $(1, 2)$, $(1, -2)$, $(-1, 2)$, and $(-1, -2)$. **TRY EXERCISE** 29

When a product of variables is in one equation and the other equation is of the form $Ax^2 + By^2 = C$, we often solve for a variable in the equation with the product and then use substitution.

EXAMPLE **5**

Solve the system

$$x^2 + 4y^2 = 20, \quad (1) \quad \text{(The graph is an ellipse.)}$$
$$xy = 4. \quad\quad (2) \quad \text{(The graph is a hyperbola.)}$$

SOLUTION First, we solve equation (2) for y:

$$y = \frac{4}{x}. \quad \text{Dividing both sides by } x. \text{ Note that } x \neq 0.$$

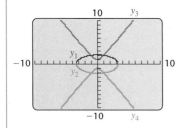

Then we substitute $4/x$ for y in equation (1) and solve for x:

$$x^2 + 4\left(\frac{4}{x}\right)^2 = 20$$

$$x^2 + \frac{64}{x^2} = 20$$

$$x^4 + 64 = 20x^2 \qquad \text{Multiplying by } x^2$$

$$x^4 - 20x^2 + 64 = 0 \qquad \text{Obtaining standard form. This equation is reducible to quadratic.}$$

$$(x^2 - 4)(x^2 - 16) = 0 \qquad \text{Factoring. If you prefer, let } u = x^2 \text{ and substitute.}$$

$$(x - 2)(x + 2)(x - 4)(x + 4) = 0 \qquad \text{Factoring again}$$

$$x = 2 \quad or \quad x = -2 \quad or \quad x = 4 \quad or \quad x = -4. \qquad \text{Using the principle of zero products}$$

Since $y = 4/x$, for $x = 2$, we have $y = 4/2$, or 2. Thus, $(2, 2)$ is a solution. Similarly, $(-2, -2)$, $(4, 1)$, and $(-4, -1)$ are solutions. You can show that all four pairs check.

TRY EXERCISE 37

Problem Solving

We now consider applications that can be modeled by a system of equations in which at least one equation is not linear.

EXAMPLE 6

Architecture.   For a college fitness center, an architect wants to lay out a rectangular piece of land that has a perimeter of 204 m and an area of 2565 m^2. Find the dimensions of the piece of land.

SOLUTION

1. Familiarize. We draw and label a sketch, letting $l =$ the length and $w =$ the width, both in meters.

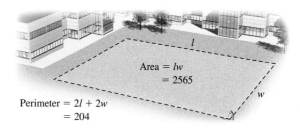

Area $= lw$
   $= 2565$

l

w

Perimeter $= 2l + 2w$
   $= 204$

2. Translate. We then have the following translation:

Perimeter: $2w + 2l = 204$;

Area: $lw = 2565$.

3. Carry out. We solve the system

$$2w + 2l = 204,$$

$$lw = 2565.$$

Solving the second equation for l gives us $l = 2565/w$. Then we substitute $2565/w$ for l in the first equation and solve for w:

$$2w + 2\left(\frac{2565}{w}\right) = 204$$

$$2w^2 + 2(2565) = 204w \qquad \text{Multiplying both sides by } w$$

$$2w^2 - 204w + 2(2565) = 0 \qquad \text{Standard form}$$

$$w^2 - 102w + 2565 = 0 \qquad \text{Multiplying by } \tfrac{1}{2}$$

> Factoring could be used instead of the quadratic formula, but the numbers are quite large.

$$w = \frac{-(-102) \pm \sqrt{(-102)^2 - 4 \cdot 1 \cdot 2565}}{2 \cdot 1}$$

$$w = \frac{102 \pm \sqrt{144}}{2} = \frac{102 \pm 12}{2}$$

$$w = 57 \quad or \quad w = 45.$$

If $w = 57$, then $l = 2565/w = 2565/57 = 45$. If $w = 45$, then $l = 2565/w = 2565/45 = 57$. Since length is usually considered to be longer than width, we have the solution $l = 57$ and $w = 45$, or $(57, 45)$.

4. **Check.** If $l = 57$ and $w = 45$, the perimeter is $2 \cdot 57 + 2 \cdot 45$, or 204. The area is $57 \cdot 45$, or 2565. The numbers check.

5. **State.** The length is 57 m and the width is 45 m.   `TRY EXERCISE` 47

EXAMPLE **7**   *Laptop dimensions.*   The screen on Tara's new laptop has an area of 90 in^2 and a $\sqrt{200.25}$-in. diagonal. Find the width and the length of the screen.

SOLUTION

1. **Familiarize.** We make a drawing and label it. Note that the width, the length, and the diagonal form a right triangle. We let $l =$ the length and $w =$ the width, both in inches.

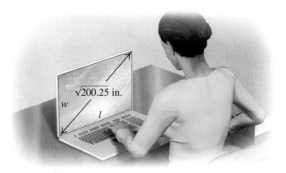

2. **Translate.** We translate to a system of equations:

$$l^2 + w^2 = (\sqrt{200.25})^2, \qquad \text{Using the Pythagorean theorem}$$

$$lw = 90. \qquad \text{Using the formula for the area of a rectangle}$$

3. **Carry out.** We solve the system

$$\left.\begin{array}{l} l^2 + w^2 = (\sqrt{200.25})^2, \\ lw = 90. \end{array}\right\} \qquad \begin{array}{l}\text{You should complete the solution of}\\ \text{this system.}\end{array}$$

We get $(12, 7.5)$, $(7.5, 12)$, $(-12, -7.5)$, and $(-7.5, -12)$.

4. **Check.** Since measurements must be positive and length is usually greater than width, we check only $(12, 7.5)$. In the right triangle, $12^2 + 7.5^2 = 144 + 56.25 = 200.25$. The area is $12(7.5) = 90$, so our answer checks.

5. **State.** The length is 12 in. and the width is 7.5 in.   `TRY EXERCISE` 51

A

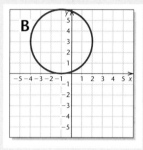

B

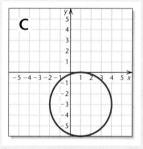

C

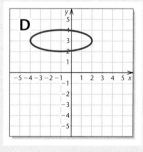

D

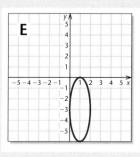

E

Visualizing for Success

Match each equation with its graph

1. $(x - 1)^2 + (y + 3)^2 = 9$

2. $\dfrac{x^2}{9} - \dfrac{y^2}{1} = 1$

3. $y = (x - 1)^2 - 3$

4. $(x + 1)^2 + (y - 3)^2 = 9$

5. $x = (y - 1)^2 - 3$

6. $\dfrac{(x + 1)^2}{9} + \dfrac{(y - 3)^2}{1} = 1$

7. $xy = 3$

8. $y = -(x + 1)^2 + 3$

9. $\dfrac{y^2}{9} - \dfrac{x^2}{1} = 1$

10. $\dfrac{(x - 1)^2}{1} + \dfrac{(y + 3)^2}{9} = 1$

Answers on page A-53

An additional, animated version of this activity appears in MyMathLab. To use MyMathLab, you need a course ID and a student access code. Contact your instructor for more information.

F

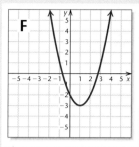

G

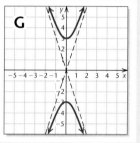

H

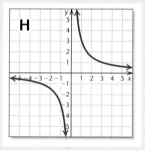

I

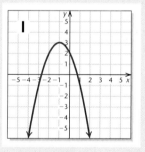

J

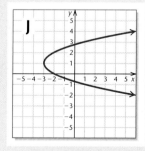

↪ *Concept Reinforcement* *Classify each statement as either true or false.*

1. A system of equations that represent a line and an ellipse can have 0, 1, or 2 solutions.

2. A system of equations that represent a parabola and a circle can have up to 4 solutions.

3. A system of equations representing a hyperbola and a circle can have no fewer than 2 solutions.

4. A system of equations representing an ellipse and a line has either 0 or 2 solutions.

5. Systems containing one first-degree equation and one second-degree equation are most easily solved using the substitution method.

6. Systems containing two second-degree equations of the form $Ax^2 + By^2 = C$ are most easily solved using the elimination method.

Solve. Remember that graphs can be used to confirm all real solutions.

7. $x^2 + y^2 = 41$,
   $y - x = 1$

8. $x^2 + y^2 = 45$,
   $y - x = 3$

9. $4x^2 + 9y^2 = 36$,
   $3y + 2x = 6$

10. $9x^2 + 4y^2 = 36$,
    $3x + 2y = 6$

11. $y^2 = x + 3$,
    $2y = x + 4$

12. $y = x^2$,
    $3x = y + 2$

13. $x^2 - xy + 3y^2 = 27$,
    $x - y = 2$

14. $2y^2 + xy + x^2 = 7$,
    $x - 2y = 5$

15. $x^2 + 4y^2 = 25$,
    $x + 2y = 7$

16. $x^2 - y^2 = 16$,
    $x - 2y = 1$

17. $x^2 - xy + 3y^2 = 5$,
    $x - y = 2$

18. $m^2 + 3n^2 = 10$,
    $m - n = 2$

19. $3x + y = 7$,
    $4x^2 + 5y = 24$

20. $2y^2 + xy = 5$,
    $4y + x = 7$

21. $a + b = 6$,
    $ab = 8$

22. $p + q = -1$,
    $pq = -12$

23. $2a + b = 1$,
    $b = 4 - a^2$

24. $4x^2 + 9y^2 = 36$,
    $x + 3y = 3$

25. $a^2 + b^2 = 89$,
    $a - b = 3$

26. $xy = 10$,
    $x + y = 7$

27. $y = x^2$,
    $x = y^2$

28. $x^2 + y^2 = 25$,
    $y^2 = x + 5$

Aha! 29. $x^2 + y^2 = 16$,
    $x^2 - y^2 = 16$

30. $y^2 - 4x^2 = 25$,
    $4x^2 + y^2 = 25$

31. $x^2 + y^2 = 25$,
    $xy = 12$

32. $x^2 - y^2 = 16$,
    $x + y^2 = 4$

33. $x^2 + y^2 = 9$,
    $25x^2 + 16y^2 = 400$

34. $x^2 + y^2 = 4$,
    $9x^2 + 16y^2 = 144$

35. $x^2 + y^2 = 14$,
    $x^2 - y^2 = 4$

36. $x^2 + y^2 = 16$,
    $y^2 - 2x^2 = 10$

37. $x^2 + y^2 = 10$,
    $xy = 3$

38. $x^2 + y^2 = 5$,
    $xy = 2$

39. $x^2 + 4y^2 = 20$,
    $xy = 4$

40. $x^2 + y^2 = 13$,
    $xy = 6$

41. $2xy + 3y^2 = 7$,
    $3xy - 2y^2 = 4$

42. $3xy + x^2 = 34$,
    $2xy - 3x^2 = 8$

43. $4a^2 - 25b^2 = 0$,
    $2a^2 - 10b^2 = 3b + 4$

44. $xy - y^2 = 2$,
    $2xy - 3y^2 = 0$

45. $ab - b^2 = -4$,
    $ab - 2b^2 = -6$

46. $x^2 - y = 5$,
    $x^2 + y^2 = 25$

Solve.

47. *Art.* Elliot is designing a rectangular stained glass miniature that has a perimeter of 28 cm and a diagonal of length 10 cm. What should the dimensions of the glass be?

48. *Geometry.* A rectangle has an area of 2 yd² and a perimeter of 6 yd. Find its dimensions.

49. *Tile design.* The Clay Works tile company wants to make a new rectangular tile that has a perimeter of 6 in. and a diagonal of length $\sqrt{5}$ in. What should the dimensions of the tile be?

50. *Geometry.* A rectangle has an area of 20 in² and a perimeter of 18 in. Find its dimensions.

51. *Design of a van.* The cargo area of a delivery van must be 60 ft², and the length of a diagonal must accommodate a 13-ft board. Find the dimensions of the cargo area.

52. *Dimensions of a rug.* The diagonal of a Persian rug is 25 ft. The area of the rug is 300 ft². Find the length and the width of the rug.

53. The product of two numbers is 90. The sum of their squares is 261. Find the numbers.

54. *Investments.* A certain amount of money saved for 1 yr at a certain interest rate yielded $125 in simple interest. If $625 more had been invested and the rate had been 1% less, the interest would have been the same. Find the principal and the rate.

55. *Garden design.* A garden contains two square flower beds. Find the length of each bed if the sum of their areas is 832 ft² and the difference of their areas is 320 ft².

56. *TV dimensions.* The Kaplans' new LCD screen has an area of 1100 in² and has a $\sqrt{2561}$-in. diagonal. Find the width and the length of the screen.

57. The area of a rectangle is $\sqrt{3}$ m², and the length of a diagonal is 2 m. Find the dimensions.

58. The area of a rectangle is $\sqrt{2}$ m², and the length of a diagonal is $\sqrt{3}$ m. Find the dimensions.

59. How can an understanding of conic sections be helpful when a system of nonlinear equations is being solved algebraically?

60. Suppose a system of equations is comprised of one linear equation and one nonlinear equation. Is it possible for such a system to have three solutions? Why or why not?

Skill Review

To prepare for Section 11.1, review evaluating expressions (Section 1.2).

Simplify. [1.2]

61. $(-1)^9(-3)^2$

62. $(-1)^{10}(-3)^3$

Evaluate each of the following. [1.2]

63. $\dfrac{(-1)^k}{k-6}$, for $k = 7$

64. $\dfrac{(-1)^k}{k-5}$, for $k = 10$

65. $\dfrac{n}{2}(3 + n)$, for $n = 11$

66. $\dfrac{7(1 - r^2)}{1 - r}$, for $r = \frac{1}{2}$

Synthesis

67. Write a problem that translates to a system of two equations. Design the problem so that at least one equation is nonlinear and so that no real solution exists.

68. Write a problem for a classmate to solve. Devise the problem so that a system of two nonlinear equations with exactly one real solution is solved.

69. Find the equation of a circle that passes through $(-2, 3)$ and $(-4, 1)$ and whose center is on the line $5x + 8y = -2$.

70. Find the equation of an ellipse centered at the origin that passes through the points $(2, -3)$ and $(1, \sqrt{13})$.

Solve.

71. $p^2 + q^2 = 13$,
$\dfrac{1}{pq} = -\dfrac{1}{6}$

72. $a + b = \dfrac{5}{6}$,
$\dfrac{a}{b} + \dfrac{b}{a} = \dfrac{13}{6}$

73. *Fence design.* A roll of chain-link fencing contains 100 ft of fence. The fencing is bent at a 90° angle to enclose a rectangular work area of 2475 ft², as shown. Determine the length and the width of the rectangle.

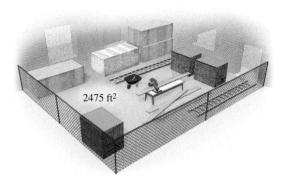

74. A piece of wire 100 cm long is to be cut into two pieces and those pieces are each to be bent to make a square. The area of one square is to be 144 cm^2 greater than that of the other. How should the wire be cut?

75. *Box design.* Four squares with sides 5 in. long are cut from the corners of a rectangular metal sheet that has an area of 340 in^2. The edges are bent up to form an open box with a volume of 350 in^3. Find the dimensions of the box.

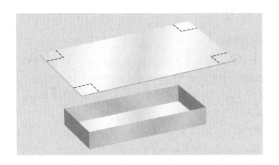

76. *Computer screens.* The ratio of the length to the height of the screen on a computer monitor is 4 to 3. A Dell Inspiron notebook has a 15-in. diagonal screen. Find the dimensions of the screen.

77. *HDTV screens.* The ratio of the length to the height of an HDTV screen is 16 to 9. The Sollar Lounge has an HDTV screen with a 73-in. diagonal screen. Find the dimensions of the screen.

78. *Railing sales.* Fireside Castings finds that the total revenue R from the sale of x units of railing is given by

$$R = 100x + x^2.$$

Fireside also finds that the total cost C of producing x units of the same product is given by

$$C = 80x + 1500.$$

A break-even point is a value of x for which total revenue is the same as total cost; that is, $R = C$. How many units must be sold to break even?

79. Use a graphing calculator to check your answers to Exercises 13, 25, and 47.

Study Summary

KEY TERMS AND CONCEPTS	EXAMPLES

SECTION 10.1: CONIC SECTIONS: PARABOLAS AND CIRCLES

Parabola

$y = ax^2 + bx + c$ Opens upward ($a > 0$) or downward ($a < 0$)

$\quad = a(x - h)^2 + k$;   Vertex: (h, k)

$x = ay^2 + by + c$ Opens right ($a > 0$) or left ($a < 0$)

$\quad = a(y - k)^2 + h$ Vertex: (h, k)

$x = -y^2 + 4y - 1$
$= -(y^2 - 4y \quad\quad) - 1$
$= -(y^2 - 4y + 4) - 1 - (-1)(4)$
$= -(y - 2)^2 + 3 \quad a = -1$; parabola opens left

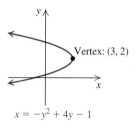

Vertex: $(3, 2)$

$x = -y^2 + 4y - 1$

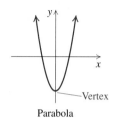

Parabola

Circle

$x^2 + y^2 = r^2$;   Radius: r Center: $(0, 0)$

$(x - h)^2 + (y - k)^2 = r^2$ Radius: r Center: (h, k)

$x^2 + y^2 + 2x - 6y + 6 = 0$
$x^2 + 2x + \quad\quad y^2 - 6y \quad\quad = -6$
$x^2 + 2x + 1 + y^2 - 6y + 9 = -6 + 1 + 9$
$\quad (x + 1)^2 + (y - 3)^2 = 4$
$\quad [x - (-1)]^2 + (y - 3)^2 = 2^2$ Radius: 2
   Center: $(-1, 3)$

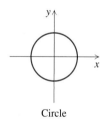

Circle

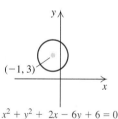

$(-1, 3)$

$x^2 + y^2 + 2x - 6y + 6 = 0$

SECTION 10.2: CONIC SECTIONS: ELLIPSES

Ellipse

$\dfrac{x^2}{a^2} + \dfrac{y^2}{b^2} = 1$;   Center: $(0, 0)$

$\dfrac{(x - h)^2}{a^2} + \dfrac{(y - k)^2}{b^2} = 1$ Center: (h, k)

$\dfrac{(x - 4)^2}{4} + \dfrac{(y + 1)^2}{9} = 1$

$\dfrac{(x - 4)^2}{2^2} + \dfrac{[y - (-1)]^2}{3^2} = 1$ $3 > 2$; ellipse is vertical
   Center: $(4, -1)$

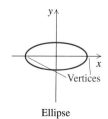

Ellipse

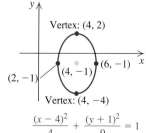

Vertex: $(4, 2)$

$(2, -1)$ $(4, -1)$ $(6, -1)$

Vertex: $(4, -4)$

$\dfrac{(x - 4)^2}{4} + \dfrac{(y + 1)^2}{9} = 1$

SECTION 10.3: CONIC SECTIONS: HYPERBOLAS

Hyperbola

$\dfrac{x^2}{a^2} - \dfrac{y^2}{b^2} = 1;$ Two branches opening right and left

$\dfrac{y^2}{b^2} - \dfrac{x^2}{a^2} = 1$ Two branches opening upward and downward

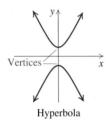

Hyperbola

$\dfrac{x^2}{4} - \dfrac{y^2}{1} = 1$

$\dfrac{x^2}{2^2} - \dfrac{y^2}{1^2} = 1$ Opens right and left

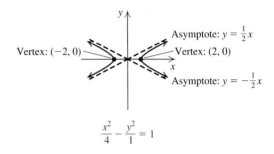

$\dfrac{x^2}{4} - \dfrac{y^2}{1} = 1$

SECTION 10.4: NONLINEAR SYSTEMS OF EQUATIONS

We can solve a system containing at least one nonlinear equation using substitution or elimination.

Solve:

$x^2 - y = -1,$ (1)   (The graph is a parabola.)
$x + 2y = 3.$ (2)   (The graph is a line.)

$x = 3 - 2y$ Solving for x in equation (2)

$(3 - 2y)^2 - y = -1$ Substituting for x in equation (1)
$9 - 12y + 4y^2 - y = -1$
$4y^2 - 13y + 10 = 0$
$(4y - 5)(y - 2) = 0$
$4y - 5 = 0 \quad or \quad y - 2 = 0$
$y = \frac{5}{4} \quad or \qquad y = 2$ Solving for y

If $y = \frac{5}{4}$, then $x = 3 - 2\left(\frac{5}{4}\right) = \frac{1}{2}$. $\left(\frac{1}{2}, \frac{5}{4}\right)$ is a solution.
If $y = 2$, then $x = 3 - 2(2) = -1$. $(-1, 2)$ is a solution.

The solutions are $\left(\frac{1}{2}, \frac{5}{4}\right)$ and $(-1, 2)$.

Review Exercises: Chapter 10

❧ **Concept Reinforcement** *Classify each statement as either true or false.*

1. Every parabola that opens upward or downward can represent the graph of a function. [10.1]

2. The center of a circle is part of the circle itself. [10.1]

3. The foci of an ellipse are part of the ellipse itself. [10.2]

4. It is possible for a hyperbola to represent the graph of a function. [10.3]

5. If an equation of a conic section has only one term of degree 2, its graph cannot be a circle, an ellipse, or a hyperbola. [10.3]

6. Two nonlinear graphs can intersect in more than one point. [10.4]

7. Every system of nonlinear equations has at least one real solution. [10.4]

8. Both substitution and elimination can be used as methods for solving a system of nonlinear equations. [10.4]

Find the center and the radius of each circle. [10.1]

9. $(x + 3)^2 + (y - 2)^2 = 16$

10. $(x - 5)^2 + y^2 = 11$

11. $x^2 + y^2 - 6x - 2y + 1 = 0$

12. $x^2 + y^2 + 8x - 6y = 20$

13. Find an equation of the circle with center $(-4, 3)$ and radius 4. [10.1]

14. Find an equation of the circle with center $(7, -2)$ and radius $2\sqrt{5}$. [10.1]

Classify each equation as a circle, an ellipse, a parabola, or a hyperbola. Then graph.

15. $5x^2 + 5y^2 = 80$ [10.1], [10.3]

16. $9x^2 + 2y^2 = 18$ [10.2], [10.3]

17. $y = -x^2 + 2x - 3$ [10.1], [10.3]

18. $\dfrac{y^2}{9} - \dfrac{x^2}{4} = 1$ [10.3]

19. $xy = 9$ [10.3]

20. $x = y^2 + 2y - 2$ [10.1], [10.3]

21. $\dfrac{(x + 1)^2}{3} + (y - 3)^2 = 1$ [10.2], [10.3]

22. $x^2 + y^2 + 6x - 8y - 39 = 0$ [10.1], [10.3]

Solve. [10.4]

23. $x^2 - y^2 = 21,$
    $x + y = 3$

24. $x^2 - 2x + 2y^2 = 8,$
    $2x + y = 6$

25. $x^2 - y = 5,$
    $2x - y = 5$

26. $x^2 + y^2 = 25,$
    $x^2 - y^2 = 7$

27. $x^2 - y^2 = 3,$
    $y = x^2 - 3$

28. $x^2 + y^2 = 18,$
    $2x + y = 3$

29. $x^2 + y^2 = 100,$
    $2x^2 - 3y^2 = -120$

30. $x^2 + 2y^2 = 12,$
    $xy = 4$

31. A rectangular bandstand has a perimeter of 38 m and an area of 84 m². What are the dimensions of the bandstand? [10.4]

32. One type of carton used by tableproducts.com exactly fits both a rectangular plate of area 108 in² and chopsticks of length 15 in., laid diagonally on top of the plate. Find the length and the width of the carton. [10.4]

15 in.

33. The perimeter of a square mirror is 12 cm more than the perimeter of another square mirror. Its area exceeds the area of the other by 39 cm². Find the perimeter of each mirror. [10.4]

34. The sum of the areas of two circles is 130π ft². The difference of the circumferences is 16π ft. Find the radius of each circle. [10.4]

Synthesis

35. How does the graph of a hyperbola differ from the graph of a parabola? [10.1], [10.3]

36. Explain why function notation rarely appears in this chapter, and list the graphs discussed for which function notation could be used. [10.1], [10.2], [10.3]

37. Solve: [10.4]
    $$4x^2 - x - 3y^2 = 9,$$
    $$-x^2 + x + \quad y^2 = 2.$$

38. Find the points whose distance from $(8, 0)$ and from $(-8, 0)$ is 10. [10.1]

39. Find an equation of the circle that passes through $(-2, -4)$, $(5, -5)$, and $(6, 2)$. [10.1], [10.4]

40. Find an equation of the ellipse with the following intercepts: $(-10, 0)$, $(10, 0)$, $(0, -1)$, and $(0, 1)$. [10.2]

41. Find the point on the x-axis that is equidistant from $(-3, 4)$ and $(5, 6)$. [10.1]

Test: Chapter 10

CHAPTER
Test Prep
VIDEO CD

Step-by-step test solutions are found on the video CD in the front of this book.

1. Find an equation of the circle with center $(3, -4)$ and radius $2\sqrt{3}$.

Find the center and the radius of each circle.

2. $(x - 4)^2 + (y + 1)^2 = 5$

3. $x^2 + y^2 + 4x - 6y + 4 = 0$

Classify the equation as a circle, an ellipse, a parabola, or a hyperbola. Then graph.

4. $y = x^2 - 4x - 1$

5. $x^2 + y^2 + 2x + 6y + 6 = 0$

6. $\dfrac{x^2}{16} - \dfrac{y^2}{9} = 1$

7. $16x^2 + 4y^2 = 64$

8. $xy = -5$

9. $x = -y^2 + 4y$

Solve.

10. $x^2 + y^2 = 36,$
$3x + 4y = 24$

11. $x^2 - y = 3,$
$2x + y = 5$

12. $x^2 - 2y^2 = 1,$
$xy = 6$

13. $x^2 + y^2 = 10,$
$x^2 = y^2 + 2$

14. A rectangular bookmark with diagonal of length $5\sqrt{5}$ has an area of 22. Find the dimensions of the bookmark.

15. Two squares are such that the sum of their areas is 8 m^2 and the difference of their areas is 2 m^2. Find the length of a side of each square.

16. A rectangular dance floor has a diagonal of length 40 ft and a perimeter of 112 ft. Find the dimensions of the dance floor.

17. Brett invested a certain amount of money for 1 yr and earned $72 in interest. Erin invested $240 more than Brett at an interest rate that was $\frac{5}{6}$ of the rate given to Brett, but she earned the same amount of interest. Find the principal and the interest rate for Brett's investment.

Synthesis

18. Find an equation of the ellipse passing through $(6, 0)$ and $(6, 6)$ with vertices at $(1, 3)$ and $(11, 3)$.

19. Find the point on the y-axis that is equidistant from $(-3, -5)$ and $(4, -7)$.

20. The sum of two numbers is 36, and the product is 4. Find the sum of the reciprocals of the numbers.

21. *Theatrical production.* An E.T.C. spotlight for a college's production of *Hamlet* projects an ellipse of light on a stage that is 8 ft wide and 14 ft long. Find an equation of that ellipse if an actor is in its center and x represents the number of feet, horizontally, from the actor to the edge of the ellipse and y represents the number of feet, vertically, from the actor to the edge of the ellipse.

Cumulative Review: Chapters 1–10

Simplify.

1. $(4t^2 - 5s)^2$ [5.2]

2. $\dfrac{1}{3t} + \dfrac{1}{t-3}$ [6.2]

3. $\dfrac{x - \dfrac{1}{a}}{a - \dfrac{1}{x}}$ [6.3]

4. $\sqrt{6t}\,\sqrt{15t^3w}$ [7.3]

5. $(81a^{2/3}b^{1/4})^{3/4}$ [7.2]

6. $\log_2 \dfrac{1}{16}$ [9.3]

7. $(4 + 3i)(4 - 3i)$ [7.8]

8. $\log_m 1$ [9.4]

9. -8^{-2} [1.6]

10. $\sqrt{8} - 2\sqrt{2} + \sqrt{12}$ [7.5]

Factor.

11. $100x^2 - 60xy + 9y^2$ [5.5]

12. $3m^6 - 24$ [5.6]

13. $ax + by - ay - bx$ [5.3]

14. $32x^2 - 20x - 3$ [5.4]

Solve. Where appropriate, give an approximation to four decimal places.

15. $3(x - 5) - 4x \geq 2(x + 5)$ [4.1]

16. $16x^2 - 18x = 0$ [5.8]

17. $\dfrac{2}{x} + \dfrac{1}{x-2} = 1$ [6.4]

18. $5x^2 + 5 = 0$ [8.2]

19. $\log_x 64 = 3$ [9.6]

20. $3^x = 1.5$ [9.6]

21. $x = \sqrt{2x - 5} + 4$ [7.6]

22. $x^2 + 2y^2 = 5,$
    $2x^2 + y^2 = 7$ [10.4]

Graph.

23. $3x - y = 9$ [2.4]

24. $y = \log_5 x$ [9.3]

25. $\dfrac{x^2}{25} + \dfrac{y^2}{1} = 1$ [10.2]

26. $f(x) = 2^{x-1}$ [9.2]

27. $x^2 + (y - 3)^2 = 4$ [10.1]

28. $x < 2y + 1$ [4.4]

29. Graph: $f(x) = -(x + 2)^2 + 3$. [8.7]
    a) Label the vertex.
    b) Draw the axis of symmetry.
    c) Find the maximum or minimum value.

30. Find the domain of the function given by
    $$f(x) = \sqrt{5 - 3x}.\ [4.1]$$

31. Solve $t = \dfrac{ab}{c^2}$ for c. [8.4]

32. Find the slope–intercept equation of the line containing the points $(-3, 6)$ and $(1, 2)$. [2.3]

33. Write a quadratic equation having the solutions $\sqrt{3}$ and $-\sqrt{3}$. Answers may vary. [8.3]

34. Write an equivalent exponential equation:
    $$\log_t 16 = m.\ [9.4]$$

Solve.

35. *Aviation.* BlueAir owns two types of airplanes. One type flies 60 mph faster than the other. Laura often rents a plane from BlueAir to visit her parents. The flight takes 4 hr with the faster plane and 4 hr 24 min with the slower plane. What distance does she fly? [3.3]

36. *Aviation.* It takes Greg 21 hr longer than it takes Kyle to service a Cessna 350. Together they can service the plane in 10 hr. How long would it take each of them, working alone, to service the plane? [6.5]

37. *Employment.* The average supermarket employee worked 32.3 hr per week in 2003. This number fell to 29.5 hr per week in 2007. [2.5]

Source: *The Wall Street Journal*, 3/8/08

   a) Find a linear function that fits the data. Let t represent the number of years since 2003.
   b) Use the function from part (a) to predict the number of hours the average supermarket employee will work in 2010.
   c) Assuming the trend continues, in what year will the average supermarket employee work 20 hr per week?

38. *Picture frames.* The outside edge of a rectangular picture frame measures 18 in. by 11 in., and 120 in^2 of picture shows. How wide is the frame? [5.8]

39. *Population.* The population of Latvia was 2.26 million in 2007 and was decreasing exponentially at a rate of 0.95% per year. [9.7]

Source: Based on information from *CIA World Factbook*

   a) Write an exponential function describing the population of Latvia t years after 2007.
   b) Predict what the population will be in 2025.
   c) What is the half life of the population?

40. *Geometry.* In triangle *ABC*, the measure of angle *B* is three times the measure of angle *A*. The measure of angle *C* is 105° greater than the measure of angle *A*. Find the angle measures. [1.4]

41. *Art.* Elyse is designing a rectangular tray. She wants to put a row of beads around the tray, and has enough beads to make an edge that is 32 in. long. What dimensions of the tray will give it the greatest area? [8.8]

42. *Geometry.* In right triangle *ABC*, the hypotenuse is 8 cm long and one leg is 3 cm long. Find the length of the other leg. Give an exact answer and an approximation to three decimal places. [7.7]

Synthesis

43. Find a and b if the graph of
$$ax - 6y = 3x + by + 2$$
is a vertical line passing through $(-1, 0)$. [2.4]

44. Solve:
$$
\begin{aligned}
w - x + 3y - z &= -1, \\
2w + x + y + 2z &= 8, \\
-w - 2x + 5y - z &= -1, \\
2w + 3x - 4y + z &= 0.
\end{aligned}
$$
[3.4]

45. If y varies inversely as the square root of x and x is multiplied by 100, what is the effect on y? [6.8], [7.1]

46. For $f(x) = x - \dfrac{1}{x^2}$, find all x-values for which $f(x) \le 0$. [8.9]

Sequences, Series, and the Binomial Theorem

**ALEXIS SPENCER-BYERS;
LEE HARPER**
COFFEE SHOP OWNERS
Jackson, Mississippi

At our coffee house, we make many different beverages. To satisfy our customers, each beverage must include the properly measured and combined ingredients to obtain the desired flavor, look, and aroma. Meanwhile, as owners, we need to know exactly how much it costs us to make each item in order to determine our prices. Behind each tasty treat is a series of calculations for determining recipes, managing cash flow, and evaluating profit margins.

AN APPLICATION

Approximately 17 billion espresso-based coffees were sold in the United States in 2007. This number is expected to grow by 4% each year. How many espresso-based coffees will be sold from 2007 through 2015?

Source: Based on data in the *Indianapolis Star,* 11/22/07

This problem appears as Exercise 65 in Exercise Set 11.3.

The first three sections of this chapter are devoted to *sequences* and *series*. A sequence is simply an ordered list. For example, when a baseball coach writes a batting order, a sequence is being formed. When the members of a sequence are numbers, they can be added. Such a sum is called a *series*. Section 11.4 presents the *binomial theorem*, which is used to expand expressions of the form $(a + b)^n$. Such an expansion is itself a series.

11.1   Sequences and Series

Sequences • Finding the General Term • Sums and Series • Sigma Notation

Sequences

Suppose that $10,000 is invested at 5%, compounded annually. The value of the account at the start of years 1, 2, 3, 4, and so on, is

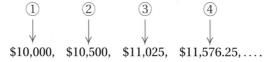

$10,000, $10,500, $11,025, $11,576.25, ...

We can regard this as a function that pairs 1 with $10,000, 2 with $10,500, 3 with $11,025, and so on. This is an example of a **sequence** (or **progression**). The domain of a sequence is a set of consecutive positive integers beginning with 1, and the range varies from sequence to sequence.

If we stop after a certain number of years, we obtain a **finite sequence**:

$10,000, $10,500, $11,025, $11,576.25.

If we continue listing the amounts in the account, we obtain an **infinite sequence**:

$10,000, $10,500, $11,025, $11,576.25, $12,155.06, ...

The three dots near the end indicate that the sequence goes on without stopping.

> **Sequences**
>
> An *infinite sequence* is a function having for its domain the set of natural numbers: $\{1, 2, 3, 4, 5, . . .\}$.
>
> A *finite sequence* is a function having for its domain a set of natural numbers: $\{1, 2, 3, 4, 5, . . . , n\}$, for some natural number n.

As another example, consider the sequence given by

$$a(n) = 2^n, \quad \text{or} \quad a_n = 2^n.$$

The notation a_n means $a(n)$ but is used more commonly with sequences. Some function values (also called *terms* of the sequence) follow:

$$a_1 = 2^1 = 2,$$
$$a_2 = 2^2 = 4,$$
$$a_3 = 2^3 = 8,$$
$$a_6 = 2^6 = 64.$$

The first term of the sequence is a_1, the fifth term is a_5, and the nth term, or **general term**, is a_n. This sequence can also be denoted in the following ways:

$$2, 4, 8, \ldots;$$

or $2, 4, 8, \ldots, 2^n, \ldots.$ The 2^n emphasizes that the nth term of this sequence is found by raising 2 to the nth power.

EXAMPLE 1   Find the first four terms and the 57th term of the sequence for which the general term is given by $a_n = (-1)^n/(n + 1)$.

SOLUTION   We have

$$a_1 = \frac{(-1)^1}{1 + 1} = -\frac{1}{2}, \quad \text{Substituting in } a_n = \frac{(-1)^n}{n + 1}$$

$$a_2 = \frac{(-1)^2}{2 + 1} = \frac{1}{3},$$

$$a_3 = \frac{(-1)^3}{3 + 1} = -\frac{1}{4},$$

$$a_4 = \frac{(-1)^4}{4 + 1} = \frac{1}{5},$$

$$a_{57} = \frac{(-1)^{57}}{57 + 1} = -\frac{1}{58}.$$

Note that the expression $(-1)^n$ causes the signs of the terms to alternate between positive and negative, depending on whether n is even or odd.

TRY EXERCISE   17

TECHNOLOGY CONNECTION

Sequences are entered and graphed much like functions. The difference is that the SEQUENCE MODE must be selected. You can then enter U_n or V_n using n as the variable. Use this approach to check Example 1 with a table of values for the sequence.

Finding the General Term

By looking for a pattern, we can often write an expression for the general term of a sequence. When only a few terms are given, more than one pattern may fit.

EXAMPLE 2   For each sequence, predict the general term.

a) $1, 4, 9, 16, 25, \ldots$

b) $2, 4, 8, \ldots$

c) $-1, 2, -4, 8, -16, \ldots$

SOLUTION

a) $1, 4, 9, 16, 25 \ldots$

These are squares of consecutive positive integers, so the general term could be n^2.

b) $2, 4, 8, \ldots$

We regard the pattern as powers of 2, in which case 16 would be the next term and 2^n the general term. The sequence could then be written with more terms as

$$2, 4, 8, 16, 32, 64, 128, \ldots.$$

c) $-1, 2, -4, 8, -16, \ldots$

These are powers of 2 with alternating signs, so the general term may be

$$(-1)^n[2^{n-1}].$$

Making sure the signs of the terms alternate

Raising 2 to a power that is 1 less than the term's position

To check, note that -4 is the third term, and $(-1)^3[2^{3-1}] = -1 \cdot 2^2 = -4$.

 TRY EXERCISE 29

In part (b) above, suppose that the second term is found by adding 2, the third term by adding 4, the next term by adding 6, and so on. In this case, 14 would be the next term and the sequence would be

$$2, 4, 8, 14, 22, 32, 44, 58, \ldots.$$

This illustrates that the fewer terms we are given, the greater the uncertainty about determining the nth term.

Sums and Series

Series

Given the infinite sequence

$$a_1, a_2, a_3, a_4, \ldots, a_n, \ldots,$$

the sum of the terms

$$a_1 + a_2 + a_3 + \cdots + a_n + \cdots$$

is called an *infinite series* and is denoted S_∞. A *partial sum* is the sum of the first n terms:

$$a_1 + a_2 + a_3 + \cdots + a_n.$$

A partial sum is also called a *finite series* and is denoted S_n.

EXAMPLE 3 For the sequence $-2, 4, -6, 8, -10, 12, -14$, find: **(a)** S_2; **(b)** S_3; **(c)** S_7.

SOLUTION

a) $S_2 = -2 + 4 = 2$ This is the sum of the first 2 terms.

b) $S_3 = -2 + 4 + (-6) = -4$ This is the sum of the first 3 terms.

c) $S_7 = -2 + 4 + (-6) + 8 + (-10) + 12 + (-14) = -8$ This is the sum of the first 7 terms.

TRY EXERCISE 45

Sigma Notation

When the general term of a sequence is known, the Greek letter Σ (upper-case sigma) can be used to write a series. For example, the sum of the first four terms of the sequence $3, 5, 7, 9, 11, \ldots, 2k + 1, \ldots$ can be named as follows, using *sigma notation*, or *summation notation*:

$$\sum_{k=1}^{4} (2k + 1).$$ This represents
$(2 \cdot 1 + 1) + (2 \cdot 2 + 1) + (2 \cdot 3 + 1) + (2 \cdot 4 + 1).$

This is read "the sum as k goes from 1 to 4 of $(2k + 1)$." The letter k is called the *index of summation*. The index need not always start at 1.

EXAMPLE **4**   Write out and evaluate each sum.

a) $\sum_{k=1}^{5} k^2$

b) $\sum_{k=4}^{6} (-1)^k (2k)$

c) $\sum_{k=0}^{3} (2^k + 5)$

STUDENT NOTES

A great deal of information is condensed into sigma notation. Be careful to pay attention to what values the index of summation will take on. Evaluate the expression following sigma, the general term, for each value and then add the results.

SOLUTION

a) $\sum_{k=1}^{5} k^2 = 1^2 + 2^2 + 3^2 + 4^2 + 5^2 = 1 + 4 + 9 + 16 + 25 = 55$

Evaluate k^2 for all integers from 1 through 5. Then add.

b) $\sum_{k=4}^{6} (-1)^k (2k) = (-1)^4 (2 \cdot 4) + (-1)^5 (2 \cdot 5) + (-1)^6 (2 \cdot 6)$

$= 8 - 10 + 12 = 10$

c) $\sum_{k=0}^{3} (2^k + 5) = (2^0 + 5) + (2^1 + 5) + (2^2 + 5) + (2^3 + 5)$

$= 6 + 7 + 9 + 13 = 35$

TRY EXERCISE 49

EXAMPLE **5**   Write sigma notation for each sum.

a) $1 + 4 + 9 + 16 + 25$

b) $3 + 9 + 27 + 81 + \cdots$

c) $-1 + 3 - 5 + 7$

SOLUTION

a) $1 + 4 + 9 + 16 + 25$

Note that this is a sum of squares, $1^2 + 2^2 + 3^2 + 4^2 + 5^2$, so the general term is k^2. Sigma notation is

$$\sum_{k=1}^{5} k^2.$$ The sum starts with 1^2 and ends with 5^2.

Answers can vary here. For example, another—perhaps less obvious—way of writing $1 + 4 + 9 + 16 + 25$ is

$$\sum_{k=2}^{6} (k - 1)^2.$$

b) $3 + 9 + 27 + 81 + \cdots$

This is a sum of powers of 3, and it is also an infinite series. We use the symbol ∞ for infinity and write the series using sigma notation:

$$\sum_{k=1}^{\infty} 3^k.$$

c) $-1 + 3 - 5 + 7$

Except for the alternating signs, this is the sum of the first four positive odd numbers. It is useful to remember that $2k - 1$ is a formula for the kth positive odd number. It is also important to remember that the factor $(-1)^k$ can be used to create the alternating signs. The general term is thus $(-1)^k(2k - 1)$, beginning with $k = 1$. Sigma notation is

$$\sum_{k=1}^{4} (-1)^k(2k - 1).$$

To check, we can evaluate $(-1)^k(2k - 1)$ using 1, 2, 3, and 4. Then we can write the sum of the four terms. We leave this to the student.

> TRY EXERCISE ▶ 61

11.1   EXERCISE SET

For Extra Help MyMathLab Math XL PRACTICE WATCH DOWNLOAD

🖐 **Concept Reinforcement** *In each of Exercises 1–6, match the expression with the most appropriate expression from the column on the right.*

1. ____ $\sum_{k=1}^{4} k^2$

    a) $-1 + 1 + (-1) + 1$

    b) $a_2 = 25$

2. ____ $\sum_{k=3}^{6} (-1)^k$

    c) $a_2 = 8$

3. ____ $5 + 10 + 15 + 20$

    d) $\sum_{k=1}^{4} 5k$

4. ____ $a_n = 5^n$

    e) S_3

5. ____ $a_n = 3n + 2$

    f) $1 + 4 + 9 + 16$

6. ____ $a_1 + a_2 + a_3$

Find the indicated term of each sequence.

7. $a_n = 5n + 3$; a_8

8. $a_n = 3n - 4$; a_8

9. $a_n = (3n + 1)(2n - 5)$; a_9

10. $a_n = (3n + 2)^2$; a_6

11. $a_n = (-1)^{n-1}(3.4n - 17.3)$; a_{12}

12. $a_n = (-2)^{n-2}(45.68 - 1.2n)$; a_{23}

13. $a_n = 3n^2(9n - 100)$; a_{11}

14. $a_n = 4n^2(2n - 39)$; a_{22}

15. $a_n = \left(1 + \dfrac{1}{n}\right)^2$; a_{20}

16. $a_n = \left(1 - \dfrac{1}{n}\right)^3$; a_{15}

In each of the following, the nth term of a sequence is given. Find the first 4 terms; the 10th term, a_{10}; and the 15th term, a_{15}, of the sequence.

17. $a_n = 3n - 1$ 18. $a_n = 2n + 1$

19. $a_n = n^2 + 2$ 20. $a_n = n^2 - 2n$

21. $a_n = \dfrac{n}{n + 1}$ 22. $a_n = \dfrac{n^2 - 1}{n^2 + 1}$

23. $a_n = \left(-\dfrac{1}{2}\right)^{n-1}$ 24. $a_n = (-2)^{n+1}$

25. $a_n = (-1)^n/n$ 26. $a_n = (-1)^n n^2$

27. $a_n = (-1)^n(n^3 - 1)$

28. $a_n = (-1)^{n+1}(3n - 5)$

Look for a pattern and then write an expression for the general term, or nth term, a_n, of each sequence. Answers may vary.

29. $2, 4, 6, 8, 10, \ldots$

30. $1, 3, 5, 7, \ldots$

31. $-1, 1, -1, 1, \ldots$

32. $1, -1, 1, -1, \ldots$

33. $1, -2, 3, -4, \ldots$

34. $-1, 2, -3, 4, \ldots$

35. $3, 5, 7, 9, \ldots$

36. $4, 6, 8, 10, \ldots$

37. $0, 3, 8, 15, 24, \ldots$

38. $2, 6, 12, 20, 30, \ldots$

39. $\frac{1}{2}, \frac{2}{3}, \frac{3}{4}, \frac{4}{5}, \frac{5}{6}, \ldots$

40. $1 \cdot 3, 2 \cdot 4, 3 \cdot 5, 4 \cdot 6, \ldots$

41. $0.1, 0.01, 0.001, 0.0001, \ldots$

42. $\frac{1}{2}, \frac{1}{4}, \frac{1}{8}, \frac{1}{16}, \ldots$

43. $-1, 4, -9, 16, \ldots$

44. $1, -4, 9, -16, \ldots$

Find the indicated partial sum for each sequence.

45. $-1, 2, -3, 4, -5, 6, \ldots;\ S_{10}$

46. $2, -4, 6, -8, 10, -12, \ldots;\ S_{10}$

47. $1, \frac{1}{10}, \frac{1}{100}, \frac{1}{1000}, \ldots;\ S_6$

48. $3, 6, 9, 12, 15, \ldots;\ S_6$

Write out and evaluate each sum.

49. $\displaystyle\sum_{k=1}^{5} \frac{1}{2k}$

50. $\displaystyle\sum_{k=1}^{6} \frac{1}{2k-1}$

51. $\displaystyle\sum_{k=0}^{4} 10^k$

52. $\displaystyle\sum_{k=2}^{6} \sqrt{5k-1}$

53. $\displaystyle\sum_{k=2}^{8} \frac{k}{k-1}$

54. $\displaystyle\sum_{k=2}^{5} \frac{k-1}{k+1}$

55. $\displaystyle\sum_{k=1}^{8} (-1)^{k+1} 2^k$

56. $\displaystyle\sum_{k=1}^{7} (-1)^k 4^{k+1}$

57. $\displaystyle\sum_{k=0}^{5} (k^2 - 2k + 3)$

58. $\displaystyle\sum_{k=0}^{5} (k^2 - 3k + 4)$

59. $\displaystyle\sum_{k=3}^{5} \frac{(-1)^k}{k(k+1)}$

60. $\displaystyle\sum_{k=3}^{7} \frac{k}{2^k}$

Rewrite each sum using sigma notation. Answers may vary.

61. $\dfrac{2}{3} + \dfrac{3}{4} + \dfrac{4}{5} + \dfrac{5}{6} + \dfrac{6}{7}$

62. $\dfrac{1}{1^2} + \dfrac{1}{2^2} + \dfrac{1}{3^2} + \dfrac{1}{4^2} + \dfrac{1}{5^2}$

63. $1 + 4 + 9 + 16 + 25 + 36$

64. $1 + \sqrt{2} + \sqrt{3} + 2 + \sqrt{5} + \sqrt{6}$

65. $4 - 9 + 16 - 25 + \cdots + (-1)^n n^2$

66. $9 - 16 + 25 + \cdots + (-1)^{n+1} n^2$

67. $6 + 12 + 18 + 24 + \cdots$

68. $11 + 22 + 33 + 44 + \cdots$

69. $\dfrac{1}{1 \cdot 2} + \dfrac{1}{2 \cdot 3} + \dfrac{1}{3 \cdot 4} + \dfrac{1}{4 \cdot 5} + \cdots$

70. $\dfrac{1}{1 \cdot 2^2} + \dfrac{1}{2 \cdot 3^2} + \dfrac{1}{3 \cdot 4^2} + \dfrac{1}{4 \cdot 5^2} + \cdots$

71. The sequence $1, 4, 9, 16, \ldots$ can be written as $f(x) = x^2$ with the domain the set of all positive integers. Explain how the graph of f would compare with the graph of $y = x^2$.

72. Consider the sums

$$\sum_{k=1}^{5} 3k^2 \quad \text{and} \quad 3\sum_{k=1}^{5} k^2.$$

a) Which is easier to evaluate and why?

b) Is it true that

$$\sum_{k=1}^{n} ca_k = c \sum_{k=1}^{n} a_k?$$

Why or why not?

Skill Review

To prepare for Section 11.2, review evaluating expressions and simplifying expressions (Sections 1.1 and 1.3).

Evaluate. [1.1]

73. $\dfrac{7}{2}(a_1 + a_7)$, for $a_1 = 8$ and $a_7 = 20$

74. $a_1 + (n - 1)d$, for $a_1 = 3$, $n = 10$, and $d = -2$

Simplify. [1.3]

75. $(a_1 + 3d) + d$

76. $(a_1 + 5d) + (a_n - 5d)$

77. $(a_1 + a_n) + (a_1 + a_n) + (a_1 + a_n)$

78. $(a_1 + 8d) - (a_1 + 7d)$

Synthesis

79. Explain why the equation

$$\sum_{k=1}^{n}(a_k + b_k) = \sum_{k=1}^{n}a_k + \sum_{k=1}^{n}b_k$$

is true for any positive integer n. What laws are used to justify this result?

80. Can a finite series be formed from an infinite sequence? Can an infinite series be formed from a finite sequence? Why or why not?

Some sequences are given by a recursive definition. The value of the first term, a_1, is given, and then we are told how to find any subsequent term from the term preceding it. Find the first six terms of each of the following recursively defined sequences.

81. $a_1 = 1$, $a_{n+1} = 5a_n - 2$

82. $a_1 = 0$, $a_{n+1} = (a_n)^2 + 3$

83. *Value of a projector.* The value of an LCD projector is $2500. Its scrap value each year is 80% of its value the year before. Write a sequence listing the scrap value of the machine at the start of each year for a 10-yr period.

84. *Cell biology.* A single cell of bacterium divides into two every 15 min. Suppose that the same rate of division is maintained for 4 hr. Write a sequence listing the number of cells after successive 15-min periods.

85. Find S_{100} and S_{101} for the sequence in which $a_n = (-1)^n$.

Find the first five terms of each sequence; then find S_5.

86. $a_n = \dfrac{1}{2^n}\log 1000^n$

87. $a_n = i^n$, $i = \sqrt{-1}$

88. Find all values for x that solve the following:

$$\sum_{k=1}^{x} i^k = -1.$$

89. The nth term of a sequence is given by

$$a_n = n^5 - 14n^4 + 6n^3 + 416n^2 - 655n - 1050.$$

Use a graphing calculator with a TABLE feature to determine which term in the sequence is 6144.

90. To define a sequence recursively on a graphing calculator (see Exercises 81 and 82), we use the SEQ MODE. The general term U_n or V_n can often be expressed in terms of U_{n-1} or V_{n-1} by pressing **2ND** ⑦ or **2ND** ⑧. The starting values of U_n, V_n, and n are set as one of the WINDOW variables.

Use recursion to determine how many different handshakes occur when 50 people shake hands with one another. To develop the recursion formula, begin with a group of 2 and determine how many additional handshakes occur with the arrival of each new person. (See the Collaborative Corner following Exercise Set 5.1 on p. 289.)

11.2 Arithmetic Sequences and Series

Arithmetic Sequences ▪ Sum of the First n Terms of an Arithmetic Sequence ▪ Problem Solving

In this section, we concentrate on sequences and series that are said to be arithmetic (pronounced ar-ith-MET-ik).

Arithmetic Sequences

In an **arithmetic sequence** (or **progression**), adding the same number to any term gives the next term in the sequence. For example, the sequence 2, 5, 8, 11, 14, 17, ... is arithmetic because adding 3 to any term produces the next term.

> **Arithmetic Sequence**
>
> A sequence is *arithmetic* if there exists a number d, called the *common difference*, such that $a_{n+1} = a_n + d$ for any integer $n \geq 1$.

EXAMPLE **1**   For each arithmetic sequence, identify the first term, a_1, and the common difference, d.

a) $4, 9, 14, 19, 24, \ldots$ **b)** $27, 20, 13, 6, -1, -8, \ldots$

SOLUTION   To find a_1, we simply use the first term listed. To find d, we choose any term other than a_1 and subtract the preceding term from it.

Sequence	First Term, a_1	Common Difference, d
a) $4, 9, 14, 19, 24, \ldots$	4	$5 \leftarrow 9 - 4 = 5$
b) $27, 20, 13, 6, -1, -8, \ldots$	27	$-7 \leftarrow 20 - 27 = -7$

To find the common difference, we subtracted a_1 from a_2. Had we subtracted a_2 from a_3 or a_3 from a_4, we would have found the same values for d.

Check: As a check, note that when d is added to each term, the result is the next term in the sequence.

TRY EXERCISE 11

To develop a formula for the general, nth, term of any arithmetic sequence, we denote the common difference by d and write out the first few terms:

$$a_1,$$
$$a_2 = a_1 + d,$$
$$a_3 = a_2 + d = (a_1 + d) + d = a_1 + 2d, \quad \text{Substituting } a_1 + d \text{ for } a_2$$
$$a_4 = a_3 + d = (a_1 + 2d) + d = a_1 + 3d. \quad \text{Substituting } a_1 + 2d \text{ for } a_3$$

Note that the coefficient of d in each case is 1 less than the subscript.

Generalizing, we obtain the following formula.

> **To Find a_n for an Arithmetic Sequence**
>
> The nth term of an arithmetic sequence with common difference d is
> $$a_n = a_1 + (n - 1)d, \quad \text{for any integer } n \geq 1.$$

EXAMPLE **2**   Find the 14th term of the arithmetic sequence $6, 9, 12, 15, \ldots$.

SOLUTION   First we note that $a_1 = 6$, $d = 3$, and $n = 14$. Using the formula for the nth term of an arithmetic sequence, we have

$$a_n = a_1 + (n - 1)d$$
$$a_{14} = 6 + (14 - 1) \cdot 3 = 6 + 13 \cdot 3 = 6 + 39 = 45.$$

The 14th term is 45.

TRY EXERCISE 17

EXAMPLE **3** For the sequence in Example 2, which term is 300? That is, find n if $a_n = 300$.

SOLUTION   We substitute into the formula for the nth term of an arithmetic sequence and solve for n:

$$a_n = a_1 + (n - 1)d$$
$$300 = 6 + (n - 1) \cdot 3$$
$$300 = 6 + 3n - 3$$
$$297 = 3n$$
$$99 = n.$$

The term 300 is the 99th term of the sequence.

TRY EXERCISE 23

Given two terms and their places in an arithmetic sequence, we can construct the sequence.

EXAMPLE **4** The 3rd term of an arithmetic sequence is 14, and the 16th term is 79. Find a_1 and d and construct the sequence.

SOLUTION   We know that $a_3 = 14$ and $a_{16} = 79$. Thus we would have to add d a total of 13 times to get from 14 to 79. That is,

$$14 + 13d = 79.$$ a_3 and a_{16} are 13 terms apart; $16 - 3 = 13$

Solving $14 + 13d = 79$, we obtain

$$13d = 65$$ Subtracting 14 from both sides
$$d = 5.$$ Dividing both sides by 13

We subtract d twice from a_3 to get to a_1. Thus,

$$a_1 = 14 - 2 \cdot 5 = 4.$$ a_1 and a_3 are 2 terms apart; $3 - 1 = 2$

The sequence is 4, 9, 14, 19, ... Note that we could have subtracted d a total of 15 times from a_{16} in order to find a_1.

TRY EXERCISE 33

In general, d should be subtracted $(n - 1)$ times from a_n in order to find a_1.

Sum of the First *n* Terms of an Arithmetic Sequence

When the terms of an arithmetic sequence are added, an **arithmetic series** is formed. To develop a formula for computing S_n when the series is arithmetic, we list the first n terms of the sequence as follows:

This is the next-to-last term. If you add d to this term, the result is a_n.

$$a_1, (a_1 + d), (a_1 + 2d), \ldots, (a_n - 2d), (a_n - d), a_n$$

This term is two terms back from the end. If you add d to this term, you get the next-to-last term, $a_n - d$.

Thus, S_n is given by

$$S_n = a_1 + (a_1 + d) + (a_1 + 2d) + \cdots + (a_n - 2d) + (a_n - d) + a_n.$$

Using a commutative law, we have a second equation:

$$S_n = a_n + (a_n - d) + (a_n - 2d) + \cdots + (a_1 + 2d) + (a_1 + d) + a_1.$$

Adding corresponding terms on each side of the two equations above, we get

$$2S_n = [a_1 + a_n] + [(a_1 + d) + (a_n - d)] + [(a_1 + 2d) + (a_n - 2d)]$$
$$+ \cdots + [(a_n - 2d) + (a_1 + 2d)] + [(a_n - d) + (a_1 + d)] + [a_n + a_1].$$

This simplifies to

$$2S_n = [a_1 + a_n] + [a_1 + a_n] + [a_1 + a_n]$$
$$+ \cdots + [a_n + a_1] + [a_n + a_1] + [a_n + a_1].$$ There are n bracketed sums.

Since $[a_1 + a_n]$ is being added n times, it follows that

$$2S_n = n[a_1 + a_n].$$

STUDENT NOTES

The formula for the sum of an arithmetic sequence is very useful, but remember that it does not work for sequences that are not arithmetic.

Dividing both sides by 2 leads to the following formula.

To Find S_n for an Arithmetic Sequence

The sum of the first n terms of an arithmetic sequence is given by

$$S_n = \frac{n}{2}(a_1 + a_n).$$

EXAMPLE 5 Find the sum of the first 100 positive even numbers.

SOLUTION   The sum is

$$2 + 4 + 6 + \cdots + 198 + 200.$$

This is the sum of the first 100 terms of the arithmetic sequence for which

$$a_1 = 2, \quad n = 100, \quad \text{and} \quad a_n = 200.$$

Substituting in the formula

$$S_n = \frac{n}{2}(a_1 + a_n),$$

we get

$$S_{100} = \frac{100}{2}(2 + 200) = 50(202) = 10{,}100.$$ **TRY EXERCISE** 39

The above formula is useful when we know the first and last terms, a_1 and a_n. To find S_n when a_n is unknown, but a_1, n, and d are known, we can use $a_n = a_1 + (n - 1)d$ to calculate a_n and then proceed as in Example 5.

EXAMPLE 6 Find the sum of the first 15 terms of the arithmetic sequence 4, 7, 10, 13, ...

SOLUTION   Note that

$$a_1 = 4, \quad n = 15, \quad \text{and} \quad d = 3.$$

Before using the formula for S_n, we find a_{15}:

$$a_{15} = 4 + (15 - 1)3 \quad \text{Substituting into the formula for } a_n$$
$$= 4 + 14 \cdot 3 = 46.$$

Thus, knowing that $a_{15} = 46$, we have

$$S_{15} = \tfrac{15}{2}(4 + 46) \quad \text{Using the formula for } S_n$$
$$= \tfrac{15}{2}(50) = 375.$$ **TRY EXERCISE** 37

Problem Solving

In problem-solving situations, translation may involve sequences or series. As always, there is often a variety of ways in which a problem can be solved. You should use the approach that is best or easiest for you. In this chapter, however, we will try to emphasize sequences and series and their related formulas.

EXAMPLE **7**

Hourly wages.   Chris accepts a job managing a music store, starting with an hourly wage of $14.60, and is promised a raise of 25¢ per hour every 2 months for 5 years. After 5 years of work, what will be Chris's hourly wage?

SOLUTION

1. **Familiarize.**   It helps to write down the hourly wage for several two-month time periods.

   Beginning: 14.60,
   After two months: 14.85,
   After four months: 15.10,

   and so on.

   What appears is a sequence of numbers: 14.60, 14.85, 15.10, ... Since the same amount is added each time, the sequence is arithmetic.

   We list what we know about arithmetic sequences. The pertinent formulas are

   $$a_n = a_1 + (n - 1)d$$

   and

   $$S_n = \frac{n}{2}(a_1 + a_n).$$

   In this case, we are not looking for a sum, so it is probably the first formula that will give us our answer. We want to determine the last term in a sequence. To do so, we need to know a_1, n, and d. From our list above, we see that

   $$a_1 = 14.60 \quad \text{and} \quad d = 0.25.$$

   What is n? That is, how many terms are in the sequence? After 1 year, there have been 6 raises, since Chris gets a raise every 2 months. There are 5 years, so the total number of raises will be $5 \cdot 6$, or 30. Altogether, there will be 31 terms: the original wage and 30 increased rates.

2. **Translate.**   We want to find a_n for the arithmetic sequence in which $a_1 = 14.60$, $n = 31$, and $d = 0.25$.

3. **Carry out.**   Substituting in the formula for a_n gives us

   $$a_{31} = 14.60 + (31 - 1) \cdot 0.25$$
   $$= 22.10.$$

4. **Check.**   We can check by redoing the calculations or we can calculate in a slightly different way for another check. For example, at the end of a year, there will be 6 raises, for a total raise of $1.50. At the end of 5 years, the total raise will be $5 \times \$1.50$, or $7.50. If we add that to the original wage of $14.60, we obtain $22.10. The answer checks.

5. **State.**   After 5 years, Chris's hourly wage will be $22.10.

TRY EXERCISE 47

EXAMPLE 8   *Telephone pole storage.*   A stack of telephone poles has 30 poles in the bottom row. There are 29 poles in the second row, 28 in the next row, and so on. How many poles are in the stack if there are 5 poles in the top row?

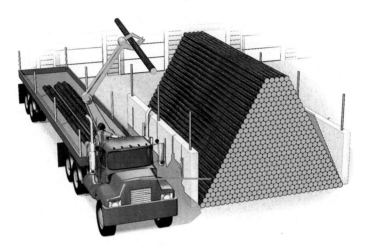

SOLUTION

1. **Familiarize.**   The following figure shows the ends of the poles. There are 30 poles on the bottom and one fewer in each successive row. How many rows will there be?

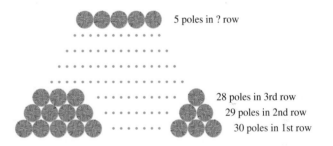

5 poles in ? row

28 poles in 3rd row
29 poles in 2nd row
30 poles in 1st row

   Note that there are $30 - 1 = 29$ poles in the 2nd row, $30 - 2 = 28$ poles in the 3rd row, $30 - 3 = 27$ poles in the 4th row, and so on. The pattern leads to $30 - 25 = 5$ poles in the 26th row.
   The situation is represented by the equation

$$30 + 29 + 28 + \cdots + 5.$$ There are 26 terms in this series.

Thus we have an arithmetic series. We recall the formula

$$S_n = \frac{n}{2}(a_1 + a_n).$$

2. **Translate.**   We want to find the sum of the first 26 terms of an arithmetic sequence in which $a_1 = 30$ and $a_{26} = 5$.

3. **Carry out.**   Substituting into the above formula gives us

$$S_{26} = \frac{26}{2}(30 + 5)$$
$$= 13 \cdot 35 = 455.$$

4. **Check.**   In this case, we can check the calculations by doing them again. A longer, more difficult way would be to do the entire addition:

$$30 + 29 + 28 + \cdots + 5.$$

5. **State.**   There are 455 poles in the stack.

TRY EXERCISE 49

11.2 EXERCISE SET

⤶ *Concept Reinforcement   Classify each statement as either true or false.*

1. In an arithmetic sequence, the difference between any two consecutive terms is always the same.

2. In an arithmetic sequence, if $a_9 - a_8 = 4$, then $a_{13} - a_{12} = 4$ as well.

3. In an arithmetic sequence containing 17 terms, the common difference is $a_{17} - a_1$.

4. To find a_{20} in an arithmetic sequence, add the common difference to a_1 a total of 20 times.

5. The sum of the first 20 terms of an arithmetic sequence can be found by knowing just a_1 and a_{20}.

6. The sum of the first 30 terms of an arithmetic sequence can be found by knowing just a_1 and d, the common difference.

7. The notation S_5 means $a_1 + a_5$.

8. For any arithmetic sequence, $S_9 = S_8 + d$, where d is the common difference.

Find the first term and the common difference.

9. 8, 13, 18, 23, ...

10. 2.5, 3, 3.5, 4, ...

11. 7, 3, −1, −5, ...

12. −8, −5, −2, 1, ...

13. $\frac{3}{2}, \frac{9}{4}, 3, \frac{15}{4}, \ldots$

14. $\frac{3}{5}, \frac{1}{10}, -\frac{2}{5}, \ldots$

15. $8.16, $8.46, $8.76, $9.06, ...

16. $825, $804, $783, $762, ...

17. Find the 19th term of the arithmetic sequence 10, 18, 26, ...

18. Find the 23rd term of the arithmetic sequence 10, 16, 22, ...

19. Find the 18th term of the arithmetic sequence 8, 2, −4, ...

20. Find the 14th term of the arithmetic sequence $3, \frac{7}{3}, \frac{5}{3}, \ldots$.

21. Find the 13th term of the arithmetic sequence $1200, $964.32, $728.64, ...

22. Find the 10th term of the arithmetic sequence $2345.78, $2967.54, $3589.30, ...

23. In the sequence of Exercise 17, what term is 210?

24. In the sequence of Exercise 18, what term is 208?

25. In the sequence of Exercise 19, what term is −328?

26. In the sequence of Exercise 20, what term is −27?

27. Find a_{18} when $a_1 = 8$ and $d = 10$.

28. Find a_{20} when $a_1 = 12$ and $d = -5$.

29. Find a_1 when $d = 4$ and $a_8 = 33$.

30. Find a_1 when $d = 8$ and $a_{11} = 26$.

31. Find n when $a_1 = 5$, $d = -3$, and $a_n = -76$.

32. Find n when $a_1 = 25$, $d = -14$, and $a_n = -507$.

33. For an arithmetic sequence in which $a_{17} = -40$ and $a_{28} = -73$, find a_1 and d. Write the first five terms of the sequence.

34. In an arithmetic sequence, $a_{17} = \frac{25}{3}$ and $a_{32} = \frac{95}{6}$. Find a_1 and d. Write the first five terms of the sequence.

Aha! 35. Find a_1 and d if $a_{13} = 13$ and $a_{54} = 54$.

36. Find a_1 and d if $a_{12} = 24$ and $a_{25} = 50$.

37. Find the sum of the first 20 terms of the arithmetic series $1 + 5 + 9 + 13 + \cdots$.

38. Find the sum of the first 14 terms of the arithmetic series $11 + 7 + 3 + \cdots$.

39. Find the sum of the first 250 natural numbers.

40. Find the sum of the first 400 natural numbers.

41. Find the sum of the even numbers from 2 to 100, inclusive.

42. Find the sum of the odd numbers from 1 to 99, inclusive.

43. Find the sum of all multiples of 6 from 6 to 102, inclusive.

44. Find the sum of all multiples of 4 that are between 15 and 521.

45. An arithmetic series has $a_1 = 4$ and $d = 5$. Find S_{20}.

46. An arithmetic series has $a_1 = 9$ and $d = -3$. Find S_{32}.

Solve.

47. *Band formations.*   The South Brighton Drum and Bugle Corps has 7 musicians in the front row, 9 in the second row, 11 in the third row, and so on, for 15 rows. How many musicians are in the last row? How many musicians are there altogether?

48. *Gardening.*   A gardener is planting tulip bulbs at the entrance to a college. She puts 50 bulbs in the first row, 46 in the second row, 42 in the third row, and so on, for 13 rows. How many bulbs will be in the last row? How many bulbs will she plant altogether?

49. *Archaeology.* Many ancient Mayan pyramids were constructed over a span of several generations. Each layer of the pyramid has a stone perimeter, enclosing a layer of dirt or debris on which a structure once stood. One drawing of such a pyramid indicates that the perimeter of the bottom layer contains 36 stones, the next level up contains 32 stones, and so on, up to the top row, which contains 4 stones. How many stones are in the pyramid?

50. *Telephone pole piles.*   How many poles will be in a pile of telephone poles if there are 50 in the first layer, 49 in the second, and so on, until there are 6 in the top layer?

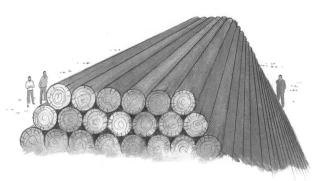

51. *Accumulated savings.*   If 10¢ is saved on October 1, another 20¢ on October 2, another 30¢ on October 3, and so on, how much is saved during October? (October has 31 days.)

52. *Accumulated savings.*   Carrie saves money in an arithmetic sequence: $700 for the first year, another $850 the second, and so on, for 20 yr. How much does she save in all (disregarding interest)?

53. *Auditorium design.*   Theaters are often built with more seats per row as the rows move toward the back. The Community Theater has 20 seats in the first row, 22 in the second, 24 in the third, and so on, for 16 rows. How many seats are in the theater?

54. *Accumulated savings.*   Shirley sets up an investment so that it yields $5000 the first year, $6125 the second year, $7250 the third year, and so on, for 25 yr. What is the total yield from the investment?

55. It is said that as a young child, the mathematician Karl F. Gauss (1777–1855) was able to compute the sum $1 + 2 + 3 + \cdots + 100$ very quickly in his head. Explain how Gauss might have done this and present a formula for the sum of the first n natural numbers. (*Hint*: $1 + 99 = 100$.)

56. Write a problem for a classmate to solve. Devise the problem so that its solution requires computing S_{17} for an arithmetic sequence.

Skill Review

Review finding equations.

Find an equation of the line satisfying the given conditions.

57. Slope $\frac{1}{3}$, *y*-intercept $(0, 10)$ [2.3]

58. Containing the points $(2, 3)$ and $(4, -5)$ [2.5]

59. Containing the point $(5, 0)$ and parallel to the line given by $2x + y = 8$ [2.5]

60. Containing the point $(-1, -4)$ and perpendicular to the line given by $3x - 4y = 7$ [2.5]

Find an equation of the circle satisfying the given conditions. [10.1]

61. Center $(0, 0)$, radius 4

62. Center $(-2, 1)$, radius $2\sqrt{5}$

Synthesis

63. When every term in an arithmetic sequence is an integer, S_n must also be an integer. Given that n, a_1, and a_n may each, at times, be even or odd, explain why $\frac{n}{2}(a_1 + a_n)$ is always an integer.

64. The sum of the first n terms of an arithmetic sequence is also given by

$$S_n = \frac{n}{2}\big[2a_1 + (n - 1)d\big].$$

Use the earlier formulas for a_n and S_n to explain how this equation was developed.

65. A frog is at the bottom of a 100-ft well. With each jump, the frog climbs 4 ft, but then slips back 1 ft. How many jumps does it take for the frog to reach the top of the hole?

66. Find a formula for the sum of the first n consecutive odd numbers starting with 1:

$$1 + 3 + 5 + \cdots + (2n - 1).$$

67. Prove that if p, m, and q are consecutive terms in an arithmetic sequence, then

$$m = \frac{p + q}{2}.$$

68. *Straight-line depreciation.* A company buys a color laser printer for $5200 on January 1 of a given year. The machine is expected to last for 8 yr, at the end of which time its *trade-in*, or *salvage, value* will be $1100. If the company figures the decline in value to be the same each year, then the trade-in values, after t years, $0 \le t \le 8$, form an arithmetic sequence given by

$$a_t = C - t\left(\frac{C - S}{N}\right),$$

where C is the original cost of the item, N the years of expected life, and S the salvage value.

a) Find the formula for a_t for the straight-line depreciation of the printer.
b) Find the trade-in value after 0 yr, 1 yr, 2 yr, 3 yr, 4 yr, 7 yr, and 8 yr.
c) Find a formula that expresses a_t recursively.

69. Use your answer to Exercise 39 to find the sum of all integers from 501 through 750.

11.3 # Geometric Sequences and Series

Geometric Sequences ▪ Sum of the First n Terms of a Geometric Sequence ▪
Infinite Geometric Series ▪ Problem Solving

In an arithmetic sequence, a certain number is added to each term to get the next term. When each term in a sequence is *multiplied* by a certain fixed number to get the next term, the sequence is **geometric**. In this section, we examine both geometric sequences (or progressions) and geometric series.

Geometric Sequences

Consider the sequence

$$2, 6, 18, 54, 162, \ldots$$

If we multiply each term by 3, we obtain the next term. The multiplier is called the *common ratio* because it is found by dividing any term by the preceding term.

> ### Geometric Sequence
>
> A sequence is *geometric* if there exists a number r, called the *common ratio*, for which
>
> $$\frac{a_{n+1}}{a_n} = r, \quad \text{or} \quad a_{n+1} = a_n \cdot r \quad \text{for any integer } n \geq 1.$$

EXAMPLE 1   For each geometric sequence, find the common ratio.

a) $4, 20, 100, 500, 2500, \ldots$

b) $3, -6, 12, -24, 48, -96, \ldots$

c) $\$5200, \$3900, \$2925, \$2193.75, \ldots$

SOLUTION

Sequence	*Common Ratio*	
a) $4, 20, 100, 500, 2500, \ldots$	5	$\frac{20}{4} = 5, \frac{100}{20} = 5$, and so on
b) $3, -6, 12, -24, 48, -96, \ldots$	-2	$\frac{-6}{3} = -2, \frac{12}{-6} = -2$, and so on
c) $\$5200, \$3900, \$2925, \$2193.75, \ldots$	0.75	$\dfrac{\$3900}{\$5200} = 0.75, \dfrac{\$2925}{\$3900} = 0.75$

TRY EXERCISE 11

Note that when the signs of the terms alternate, the common ratio is negative.

To develop a formula for the general, or *n*th, term of a geometric sequence, let a_1 be the first term and let r be the common ratio. We write out the first few terms as follows:

$$a_1,$$
$$a_2 = a_1 r,$$
$$a_3 = a_2 r = (a_1 r)r = a_1 r^2, \qquad \text{Substituting } a_1 r \text{ for } a_2$$
$$a_4 = a_3 r = (a_1 r^2)r = a_1 r^3. \qquad \text{Substituting } a_1 r^2 \text{ for } a_3$$

Note that the exponent is 1 less than the subscript.

Generalizing, we obtain the following.

> ### To Find a_n for a Geometric Sequence
>
> The *n*th term of a geometric sequence with common ratio r is given by
>
> $$a_n = a_1 r^{n-1}, \quad \text{for any integer } n \geq 1.$$

EXAMPLE 2   Find the 7th term of the geometric sequence $4, 20, 100, \ldots$.

SOLUTION   First, we note that

$$a_1 = 4 \quad \text{and} \quad n = 7.$$

To find the common ratio, we can divide any term (other than the first) by the term preceding it. Since the second term is 20 and the first is 4,

$$r = \frac{20}{4}, \quad \text{or } 5.$$

The formula

$$a_n = a_1 r^{n-1}$$

gives us

$$a_7 = 4 \cdot 5^{7-1} = 4 \cdot 5^6 = 4 \cdot 15{,}625 = 62{,}500.$$ **TRY EXERCISE** 19

EXAMPLE 3   Find the 10th term of the geometric sequence

$$64, -32, 16, -8, \dots.$$

SOLUTION   First, we note that

$$a_1 = 64, \quad n = 10, \quad \text{and} \quad r = \frac{-32}{64} = -\frac{1}{2}.$$

Then, using the formula for the nth term of a geometric sequence, we have

$$a_{10} = 64 \cdot \left(-\frac{1}{2}\right)^{10-1} = 64 \cdot \left(-\frac{1}{2}\right)^9 = 2^6 \cdot \left(-\frac{1}{2^9}\right) = -\frac{1}{2^3} = -\frac{1}{8}.$$

The 10th term is $-\frac{1}{8}$.   **TRY EXERCISE** 23

Sum of the First n Terms of a Geometric Sequence

We next develop a formula for S_n when a sequence is geometric:

$$a_1, \ a_1 r, \ a_1 r^2, \ a_1 r^3, \dots, a_1 r^{n-1}, \dots.$$

The **geometric series** S_n is given by

$$S_n = a_1 + a_1 r + a_1 r^2 + \cdots + a_1 r^{n-2} + a_1 r^{n-1}. \tag{1}$$

Multiplying both sides by r gives us

$$r S_n = a_1 r + a_1 r^2 + a_1 r^3 + \cdots + a_1 r^{n-1} + a_1 r^n. \tag{2}$$

STUDENT NOTES

The three determining characteristics of a geometric sequence or series are the first term (a_1), the number of terms (n), and the common ratio (r). Be sure you understand how to use these characteristics to write out a sequence or a series.

When we subtract corresponding sides of equation (2) from equation (1), the color terms drop out, leaving

$$S_n - r S_n = a_1 - a_1 r^n$$
$$S_n(1 - r) = a_1(1 - r^n), \qquad \text{Factoring}$$

or

$$S_n = \frac{a_1(1 - r^n)}{1 - r}. \qquad \text{Dividing both sides by } 1 - r$$

To Find S_n for a Geometric Sequence

The sum of the first n terms of a geometric sequence with common ratio r is given by

$$S_n = \frac{a_1(1 - r^n)}{1 - r}, \quad \text{for any } r \neq 1.$$

EXAMPLE 4   Find the sum of the first 7 terms of the geometric sequence $3, 15, 75, 375, \dots.$

SOLUTION   First, we note that

$$a_1 = 3, \quad n = 7, \quad \text{and} \quad r = \frac{15}{3} = 5.$$

Then, substituting in the formula $S_n = \dfrac{a_1(1 - r^n)}{1 - r}$, we have

$$S_7 = \frac{3(1 - 5^7)}{1 - 5} = \frac{3(1 - 78{,}125)}{-4}$$

$$= \frac{3(-78{,}124)}{-4}$$

$$= 58{,}593.$$

 **TRY EXERCISE** 33

Infinite Geometric Series

Suppose we consider the sum of the terms of an infinite geometric sequence, such as 3, 6, 12, 24, 48, ... We get what is called an **infinite geometric series**:

$$3 + 6 + 12 + 24 + 48 + \cdots.$$

Here, as n increases, the sum of the first n terms, S_n, increases without bound. There are also infinite series that get closer and closer to some specific number. Here is an example:

$$\frac{1}{2} + \frac{1}{4} + \frac{1}{8} + \frac{1}{16} + \cdots + \frac{1}{2^n} + \cdots.$$

Let's consider S_n for the first four values of n:

$$
\begin{aligned}
S_1 &= \tfrac{1}{2} & &= \tfrac{1}{2} = 0.5, \\
S_2 &= \tfrac{1}{2} + \tfrac{1}{4} & &= \tfrac{3}{4} = 0.75, \\
S_3 &= \tfrac{1}{2} + \tfrac{1}{4} + \tfrac{1}{8} & &= \tfrac{7}{8} = 0.875, \\
S_4 &= \tfrac{1}{2} + \tfrac{1}{4} + \tfrac{1}{8} + \tfrac{1}{16} & &= \tfrac{15}{16} = 0.9375.
\end{aligned}
$$

> The denominator of each sum is 2^n, where n is the subscript of S. The numerator is $2^n - 1$.

Thus, for this particular series, we have

$$S_n = \frac{2^n - 1}{2^n} = \frac{2^n}{2^n} - \frac{1}{2^n} = 1 - \frac{1}{2^n}.$$

Note that the value of S_n is less than 1 for any value of n, but as n gets larger and larger, the value of $1/2^n$ gets closer to 0 and the value of S_n gets closer to 1. We can visualize S_n by considering a square with area 1. For S_1, we shade half the square. For S_2, we shade half the square plus half the remaining part, or $\tfrac{1}{4}$. For S_3, we shade the parts shaded in S_2 plus half the remaining part. Again we see that the values of S_n will continue to get close to 1 (shading the complete square).

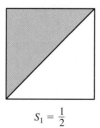

$$S_1 = \frac{1}{2}$$

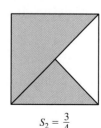

$$S_2 = \frac{3}{4}$$

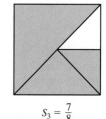

$$S_3 = \frac{7}{8}$$

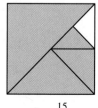

$$S_4 = \frac{15}{16}$$

We say that 1 is the **limit** of S_n and that 1 is the sum of this infinite geometric series. An infinite geometric series is denoted S_∞. It can be shown (but we will not do so here) that the sum of the terms of an infinite geometric sequence exists if and only if $|r| < 1$ (that is, the common ratio's absolute value is less than 1).

To find a formula for the sum of an infinite geometric series, we first consider the sum of the first n terms:

$$S_n = \frac{a_1(1 - r^n)}{1 - r} = \frac{a_1 - a_1 r^n}{1 - r}. \qquad \text{Using the distributive law}$$

For $|r| < 1$, it follows that the value of r^n gets closer to 0 as n gets larger. (Check this by selecting a number between -1 and 1 and finding larger and larger powers on a calculator.) As r^n gets closer to 0, so too does $a_1 r^n$. Thus, S_n gets closer to $a_1/(1 - r)$.

The Limit of an Infinite Geometric Series

For $|r| < 1$, the limit of an infinite geometric series is given by

$$S_\infty = \frac{a_1}{1 - r}. \qquad \text{(For } |r| \geq 1\text{, no limit exists.)}$$

EXAMPLE 5   Determine whether each series has a limit. If a limit exists, find it.

a) $1 + 3 + 9 + 27 + \cdots$ **b)** $-35 + 7 - \frac{7}{5} + \frac{7}{25} + \cdots$

SOLUTION

a) Here $r = 3$, so $|r| = |3| = 3$. Since $|r| \not< 1$, the series does *not* have a limit.

b) Here $r = -\frac{1}{5}$, so $|r| = |-\frac{1}{5}| = \frac{1}{5}$. Since $|r| < 1$, the series *does* have a limit. We find the limit by substituting into the formula for S_∞:

$$S_\infty = \frac{-35}{1 - \left(-\frac{1}{5}\right)} = \frac{-35}{\frac{6}{5}} = -35 \cdot \frac{5}{6} = \frac{-175}{6} = -29\frac{1}{6}.$$

TRY EXERCISE 41

EXAMPLE 6   Find fraction notation for $0.63636363\ldots$.

SOLUTION   We can express this as

$$0.63 + 0.0063 + 0.000063 + \cdots.$$

This is an infinite geometric series, where $a_1 = 0.63$ and $r = 0.01$. Since $|r| < 1$, this series has a limit:

$$S_\infty = \frac{a_1}{1 - r} = \frac{0.63}{1 - 0.01} = \frac{0.63}{0.99} = \frac{63}{99}.$$

Thus fraction notation for $0.63636363\ldots$ is $\frac{63}{99}$, or $\frac{7}{11}$.

TRY EXERCISE 53

Problem Solving

For some problem-solving situations, the translation may involve geometric sequences or series.

EXAMPLE 7   *Daily wages.*   Suppose you were offered a job for the month of September (30 days) under the following conditions. You will be paid $0.01 for the first day, $0.02 for the second, $0.04 for the third, and so on, doubling your previous day's salary each day. How much would you earn? (Would you take the job? Make a guess before reading further.)

SOLUTION

1. **Familiarize.** You earn $0.01 the first day, $0.01(2) the second day, $0.01(2)(2) the third day, and so on. Since each day's wages are a constant multiple of the previous day's wages, a geometric sequence is formed.

2. **Translate.**   The amount earned is the geometric series

$$\$0.01 + \$0.01(2) + \$0.01(2^2) + \$0.01(2^3) + \cdots + \$0.01(2^{29}),$$

where $a_1 = \$0.01$, $n = 30$, and $r = 2$.

3. **Carry out.**   Using the formula

$$S_n = \frac{a_1(1 - r^n)}{1 - r},$$

we have

$$S_{30} = \frac{\$0.01(1 - 2^{30})}{1 - 2}$$

$$= \frac{\$0.01(-1,073,741,823)}{-1} \quad \text{Using a calculator}$$

$$= \$10,737,418.23.$$

4. **Check.**   The calculations can be repeated as a check.

5. **State.**   The pay exceeds $10.7 million for the month. Most people would probably take the job!   **TRY EXERCISE** 69

EXAMPLE 8   *Loan repayment.*   Francine's student loan is in the amount of $6000. Interest is 9% compounded annually, and the entire amount is to be paid after 10 yr. How much is to be paid back?

SOLUTION

1. **Familiarize.**   Suppose we let P represent any principal amount. At the end of one year, the amount owed will be $P + 0.09P$, or $1.09P$. That amount will be the principal for the second year. The amount owed at the end of the second year will be $1.09 \times$ New principal $= 1.09(1.09P)$, or 1.09^2P. Thus the amount owed at the beginning of successive years is as follows:

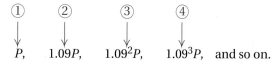

$P, \quad 1.09P, \quad 1.09^2P, \quad 1.09^3P, \quad \text{and so on.}$

We have a geometric sequence. The amount owed at the beginning of the 11th year will be the amount owed at the end of the 10th year.

2. **Translate.**   We have a geometric sequence with $a_1 = 6000$, $r = 1.09$, and $n = 11$. The appropriate formula is

$$a_n = a_1r^{n-1}.$$

3. Carry out.   We substitute and calculate:

$$a_{11} = \$6000(1.09)^{11-1} = \$6000(1.09)^{10}$$

$$\approx \$14{,}204.18.$$ Using a calculator and rounding to the nearest hundredth

4. Check.   A check, by repeating the calculations, is left to the student.

5. State.   Francine will owe \$14,204.18 at the end of 10 yr.

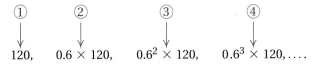 **TRY EXERCISE** 61

EXAMPLE   9   *Bungee jumping.*   A bungee jumper rebounds 60% of the height jumped. Clyde's bungee jump is made using a cord that stretches to 200 ft.

a) After jumping and then rebounding 9 times, how far has Clyde traveled upward (the total rebound distance)?

b) Theoretically, how far will Clyde travel upward (bounce) before coming to rest?

SOLUTION

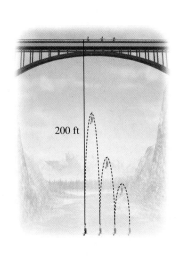

200 ft

1. Familiarize.   Let's do some calculations and look for a pattern.

First fall:	200 ft
First rebound:	0.6 × 200, or 120 ft
Second fall:	120 ft, or 0.6 × 200
Second rebound:	0.6 × 120, or 0.6(0.6 × 200), which is 72 ft
Third fall:	72 ft, or 0.6(0.6 × 200)
Third rebound:	0.6 × 72, or 0.6(0.6(0.6 × 200)), which is 43.2 ft

The rebound distances form a geometric sequence:

①   ②   ③   ④

$120, \quad 0.6 \times 120, \quad 0.6^2 \times 120, \quad 0.6^3 \times 120, \ldots.$

2. Translate.

a) The total rebound distance after 9 bounces is the sum of a geometric sequence. The first term is 120 and the common ratio is 0.6. There will be 9 terms, so we can use the formula

$$S_n = \frac{a_1(1 - r^n)}{1 - r}.$$

b) Theoretically, Clyde will never stop bouncing. Realistically, the bouncing will eventually stop. To approximate the actual distance bounced, we consider an infinite number of bounces and use the formula

$$S_\infty = \frac{a_1}{1 - r}.$$ Since $r = 0.6$ and $|0.6| < 1$, we know that S_∞ exists.

3. Carry out.

a) We substitute into the formula and calculate:

$$S_9 = \frac{120[1 - (0.6)^9]}{1 - 0.6} \approx 297.$$ Using a calculator

b) We substitute and calculate:

$$S_\infty = \frac{120}{1 - 0.6} = 300.$$

4. **Check.** We can do the calculations again.

5. **State.**

   a) In 9 bounces, Clyde will have traveled upward a total distance of about 297 ft.

   b) Theoretically, Clyde will travel upward a total of 300 ft before coming to rest.

TRY EXERCISE 67

11.3 EXERCISE SET

🐍 *Concept Reinforcement* *Classify each of the following as an arithmetic sequence, a geometric sequence, an arithmetic series, a geometric series, or none of these.*

1. $3, 6, 12, 24, \ldots$

2. $-2, 3, 8, 13, \ldots$

3. $10, 7, 4, 1, -2, \ldots$

4. $1000, 500, 250, 125, \ldots$

5. $4 + 20 + 100 + 500 + 2500 + 12{,}500$

6. $10 + 12 + 14 + 16 + 18 + 20$

7. $3 - \frac{3}{2} + \frac{3}{4} - \frac{3}{8} + \frac{3}{16} - \cdots$

8. $1 + \frac{1}{2} + \frac{1}{3} + \frac{1}{4} + \frac{1}{5} + \frac{1}{6} + \cdots$

Find the common ratio for each geometric sequence.

9. $10, 20, 40, 80, \ldots$

10. $5, 20, 80, 320, \ldots$

11. $6, -0.6, 0.06, -0.006, \ldots$

12. $-5, -0.5, -0.05, -0.005, \ldots$

13. $\frac{1}{2}, -\frac{1}{4}, \frac{1}{8}, -\frac{1}{16}, \ldots$

14. $\frac{2}{3}, -\frac{4}{3}, \frac{8}{3}, -\frac{16}{3}, \ldots$

15. $75, 15, 3, \frac{3}{5}, \ldots$

16. $12, -4, \frac{4}{3}, -\frac{4}{9}, \ldots$

17. $\frac{1}{m}, \frac{6}{m^2}, \frac{36}{m^3}, \frac{216}{m^4}, \ldots$

18. $4, \frac{4m}{5}, \frac{4m^2}{25}, \frac{4m^3}{125}, \ldots$

Find the indicated term for each geometric sequence.

19. $2, 6, 18, \ldots$; the 7th term

20. $2, 8, 32, \ldots$; the 9th term

21. $\sqrt{3}, 3, 3\sqrt{3}, \ldots$; the 10th term

22. $2, 2\sqrt{2}, 4, \ldots$; the 8th term

23. $-\frac{8}{243}, \frac{8}{81}, -\frac{8}{27}, \ldots$; the 14th term

24. $\frac{7}{625}, \frac{-7}{125}, \frac{7}{25}, \ldots$; the 13th term

25. $\$1000, \$1040, \$1081.60, \ldots$; the 10th term

26. $\$1000, \$1050, \$1102.50, \ldots$; the 12th term

Find the nth, or general, term for each geometric sequence.

27. $1, 5, 25, 125, \ldots$

28. $2, 4, 8, \ldots$

29. $1, -1, 1, -1, \ldots$

30. $\frac{1}{4}, \frac{1}{16}, \frac{1}{64}, \ldots$

31. $\frac{1}{x}, \frac{1}{x^2}, \frac{1}{x^3}, \ldots$

32. $5, \frac{5m}{2}, \frac{5m^2}{4}, \ldots$

For Exercises 33–40, use the formula for S_n to find the indicated sum for each geometric series.

33. S_9 for $6 + 12 + 24 + \cdots$

34. S_6 for $16 - 8 + 4 - \cdots$

35. S_7 for $\frac{1}{18} - \frac{1}{6} + \frac{1}{2} - \cdots$

Aha! 36. S_5 for $7 + 0.7 + 0.07 + \cdots$

37. S_8 for $1 + x + x^2 + x^3 + \cdots$

38. S_{10} for $1 + x^2 + x^4 + x^6 + \cdots$

39. S_{16} for $\$200 + \$200(1.06) + \$200(1.06)^2 + \cdots$

40. S_{23} for $\$1000 + \$1000(1.08) + \$1000(1.08)^2 + \cdots$

Determine whether each infinite geometric series has a limit. If a limit exists, find it.

41. $18 + 6 + 2 + \cdots$

42. $80 + 20 + 5 + \cdots$

43. $7 + 3 + \frac{9}{7} + \cdots$

44. $12 + 9 + \frac{27}{4} + \cdots$

45. $3 + 15 + 75 + \cdots$

46. $2 + 3 + \frac{9}{2} + \cdots$

47. $4 - 6 + 9 - \frac{27}{2} + \cdots$

48. $-6 + 3 - \frac{3}{2} + \frac{3}{4} - \cdots$

49. $0.43 + 0.0043 + 0.000043 + \cdots$

50. $0.37 + 0.0037 + 0.000037 + \cdots$

51. $\$500(1.02)^{-1} + \$500(1.02)^{-2} + \$500(1.02)^{-3} + \cdots$

52. $\$1000(1.08)^{-1} + \$1000(1.08)^{-2} + \$1000(1.08)^{-3} + \cdots$

Find fraction notation for each repeating decimal.

53. $0.5555\ldots$

54. $0.8888\ldots$

55. $3.4646\ldots$

56. $1.2323\ldots$

57. $0.15151515\ldots$

58. $0.12121212\ldots$

⌨ *Solve. Use a calculator as needed for evaluating formulas.*

59. *Rebound distance.* A ping-pong ball is dropped from a height of 20 ft and always rebounds one-fourth of the distance fallen. How high does it rebound the 6th time?

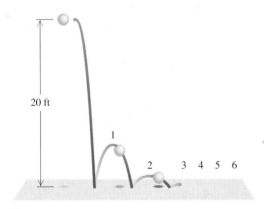

60. *Rebound distance.* Approximate the total of the rebound heights of the ball in Exercise 59.

61. *Population growth.* Yorktown has a current population of 100,000, and the population is increasing by 3% each year. What will the population be in 15 yr?

62. *Amount owed.* Gilberto borrows $15,000. The loan is to be repaid in 13 yr at 5.5% interest, compounded annually. How much will be repaid at the end of 13 yr?

63. *Shrinking population.* A population of 5000 fruit flies is dying off at a rate of 4% per minute. How many flies will be alive after 15 min?

64. *Shrinking population.* For the population of fruit flies in Exercise 63, how long will it take for only 1800 fruit flies to remain alive? (*Hint*: Use logarithms.) Round to the nearest minute.

65. *Food service.* Approximately 17 billion espresso-based coffees were sold in the United States in 2007. This number is expected to grow by 4% each year. How many espresso-based coffees will be sold from 2007 through 2015?
Source: Based on data in the *Indianapolis Star*, 11/22/07

66. *Text messaging.* Approximately 160 billion text messages were sent worldwide in 2000. Since then, the number of text messages sent each year has grown by about 140% per year. How many text messages were sent worldwide from 2000 through 2010?
Source: Based on data from mobilesmsmarketing.com

67. *Rebound distance.* A superball dropped from the top of the Washington Monument (556 ft high) rebounds three-fourths of the distance fallen. How far (up and down) will the ball have traveled when it hits the ground for the 6th time?

68. *Rebound distance.* Approximate the total distance that the ball of Exercise 67 will have traveled when it comes to rest.

69. *Stacking paper.* Construction paper is about 0.02 in. thick. Beginning with just one piece, a stack is doubled again and again 10 times. Find the height of the final stack.

70. *Monthly earnings.*   Suppose you accepted a job for the month of February (28 days) under the following conditions. You will be paid $0.01 the first day, $0.02 the second, $0.04 the third, and so on, doubling your previous day's salary each day. How much would you earn?

Aha! **71.** Under what circumstances is it possible for the 5th term of a geometric sequence to be greater than the 4th term but less than the 7th term?

72. When r is negative, a series is said to be *alternating*. Why do you suppose this terminology is used?

Skill Review

To prepare for Section 11.4, review products of binomials (Section 5.2).

Multiply. [5.2]

73. $(x + y)^2$

74. $(x + y)^3$

75. $(x - y)^3$

76. $(x - y)^4$

77. $(2x + y)^3$

78. $(2x - y)^3$

Synthesis

79. Write a problem for a classmate to solve. Devise the problem so that a geometric series is involved and the solution is "The total amount in the bank is $900(1.08)^{40}$, or about $19,550."

80. The infinite series

$$S_\infty = 2 + \frac{1}{2} + \frac{1}{2 \cdot 3} + \frac{1}{2 \cdot 3 \cdot 4} + \frac{1}{2 \cdot 3 \cdot 4 \cdot 5}$$
$$+ \frac{1}{2 \cdot 3 \cdot 4 \cdot 5 \cdot 6} + \cdots$$

is not geometric, but it does have a sum. Using $S_1, S_2, S_3, S_4, S_5,$ and S_6, make a conjecture about the value of S_∞ and explain your reasoning.

Calculate each of the following sums.

81. $\displaystyle\sum_{k=1}^{\infty} 6(0.9)^k$

82. $\displaystyle\sum_{k=1}^{\infty} 5(-0.7)^k$

83. Find the sum of the first n terms of
$$x^2 - x^3 + x^4 - x^5 + \cdots.$$

84. Find the sum of the first n terms of
$$1 + x + x^2 + x^3 + \cdots.$$

85. The sides of a square are each 16 cm long. A second square is inscribed by joining the midpoints of the sides, successively. In the second square we repeat the process, inscribing a third square. If this process is continued indefinitely, what is the sum of all of the areas of all the squares? (*Hint*: Use an infinite geometric series.)

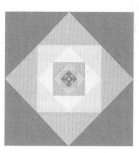

86. Show that $0.999\ldots$ is 1.

87. Using Example 5 and Exercises 41–52, explain how the graph of a geometric sequence can be used to determine whether a geometric series has a limit.

88. To compare the *graphs* of an arithmetic and a geometric sequence, we plot n on the horizontal axis and a_n on the vertical axis. Graph Example 1(a) of Section 11.2 and Example 1(a) of Section 11.3 on the same set of axes. How do the graphs of geometric sequences differ from the graphs of arithmetic sequences?

CONNECTING the CONCEPTS

A *sequence* is simply an ordered list. A *series* is a sum of consecutive terms in a sequence. Some sequences of numbers have patterns and a formula can be found for a general term. If each pair of consecutive terms has a common difference, the sequence is *arithmetic*. If each pair of consecutive terms has a common ratio, the sequence is *geometric*. Arithmetic and geometric sequences have formulas for general terms and for sums.

Arithmetic Sequences	Geometric Sequences
$a_n = a_1 + (n - 1)d$	$a_n = a_1 r^{n-1}$
$S_n = \dfrac{n}{2}(a_1 + a_n)$	$S_n = \dfrac{a_1(1 - r^n)}{1 - r};$ $S_\infty = \dfrac{a_1}{1 - r},\ \|r\| < 1$

MIXED REVIEW

1. Find a_{20} if $a_n = n^2 - 5n$.

2. Write an expression for the general term a_n of the sequence $\frac{1}{2}, \frac{1}{3}, \frac{1}{4}, \frac{1}{5}, \ldots$.

3. Find S_{12} for the sequence $1, 2, 3, 4, \ldots$.

4. Write out and evaluate the sum
$$\sum_{k=2}^{5} k^2.$$

5. Rewrite using sigma notation:
$1 - 2 + 3 - 4 + 5 - 6$.

6. Find the common difference for the arithmetic sequence $115, 112, 109, 106, \ldots$.

7. Find the 21st term of the arithmetic sequence $10, 15, 20, 25, \ldots$.

8. Which term is 22 in the arithmetic sequence $10, 10.2, 10.4, 10.6, \ldots$?

9. For an arithmetic sequence, find a_{25} when $a_1 = 9$ and $d = -2$.

10. For an arithmetic sequence, find a_1 when $d = 11$ and $a_5 = 65$.

11. For an arithmetic sequence, find n when $a_1 = 5$, $d = -\frac{1}{2}$, and $a_n = 0$.

12. Find S_{30} for the arithmetic series $2 + 12 + 22 + 32 + \cdots$.

13. Find the common ratio for the geometric sequence $\frac{1}{3}, -\frac{1}{6}, \frac{1}{12}, -\frac{1}{24}, \ldots$.

14. Find the 8th term of the geometric sequence $5, 10, 20, 40, \ldots$.

15. Find the nth, or general, term for the geometric sequence $2, -2, 2, -2, \ldots$.

16. Find S_{10} for the geometric series $\$100 + \$100(1.03) + \$100(1.03)^2 + \cdots$.

17. Determine whether the infinite geometric series $0.9 + 0.09 + 0.009 + \cdots$ has a limit. If a limit exists, find it.

18. Determine whether the infinite geometric series $0.9 + 9 + 90 + \cdots$ has a limit. If a limit exists, find it.

19. Renata earns \$1 on June 1, another \$2 on June 2, another \$3 on June 3, another \$4 on June 4, and so on. How much does she earn during the 30 days of June?

20. Dwight earns \$1 on June 1, another \$2 on June 2, another \$4 on June 3, another \$8 on June 4, and so on. How much does he earn during the 30 days of June?

CORNER

Bargaining for a Used Car

Focus: Geometric series

Time: 30 minutes

Group size: 2

Materials: Graphing calculators are optional.

ACTIVITY *

1. One group member ("the seller") has a car for sale and is asking $3500. The second ("the buyer") offers $1500. The seller splits the difference ($3500 − $1500 = $2000, and $2000 ÷ 2 = $1000) and lowers the price to $2500. The buyer then splits the difference again ($2500 − $1500 = $1000, and $1000 ÷ 2 = $500) and counters with $2000. Continue in this manner and stop when you are able to agree on the car's selling price to the nearest penny.

2. What should the buyer's initial offer be in order to achieve a purchase price of $2000 or less? (Check several guesses to find the appropriate initial offer.)

---

*This activity is based on the article "Bargaining Theory, or Zeno's Used Cars," by James C. Kirby, *The College Mathematics Journal*, **27**(4), September 1996.

3. The seller's price in the bargaining above can be modeled recursively (see Exercises 81, 82, and 90 in Section 11.1) by the sequence

$$a_1 = 3500, \qquad a_n = a_{n-1} - \frac{d}{2^{2n-3}},$$

where d is the difference between the initial price and the first offer. Use this recursively defined sequence to solve parts (1) and (2) above either manually or by using the SEQ MODE and the TABLE feature of a graphing calculator.

4. The first four terms in the sequence in part (3) can be written as

$$a_1, \quad a_1 - \frac{d}{2}, \quad a_1 - \frac{d}{2} - \frac{d}{8},$$
$$a_1 - \frac{d}{2} - \frac{d}{8} - \frac{d}{32}.$$

Use the formula for the limit of an infinite geometric series to find a simple algebraic formula for the eventual sale price, P, when the bargaining process from above is followed. Verify the formula by using it to solve parts (1) and (2) above.

11.4   The Binomial Theorem

Binomial Expansion Using Pascal's Triangle   ▪   Binomial Expansion Using Factorial Notation

The expression $(x + y)^2$ may be regarded as a series: $x^2 + 2xy + y^2$. This sum of terms is the *expansion* of $(x + y)^2$. For powers greater than 2, finding the expansion of $(x + y)^n$ can be time-consuming. In this section, we look at two methods of streamlining binomial expansion.

Binomial Expansion Using Pascal's Triangle

Consider the following expanded powers of $(a + b)^n$.

$$(a + b)^0 = 1$$
$$(a + b)^1 = a + b$$
$$(a + b)^2 = a^2 + 2a^1b^1 + b^2$$
$$(a + b)^3 = a^3 + 3a^2b^1 + 3a^1b^2 + b^3$$
$$(a + b)^4 = a^4 + 4a^3b^1 + 6a^2b^2 + 4a^1b^3 + b^4$$
$$(a + b)^5 = a^5 + 5a^4b^1 + 10a^3b^2 + 10a^2b^3 + 5a^1b^4 + b^5$$

Each expansion is a polynomial. There are some patterns worth noting:

1. There is one more term than the power of the binomial, n. That is, there are $n + 1$ terms in the expansion of $(a + b)^n$.

2. In each term, the sum of the exponents is the power to which the binomial is raised.

3. The exponents of a start with n, the power of the binomial, and decrease to 0 (since $a^0 = 1$, the last term has no factor of a). The first term has no factor of b, so powers of b start with 0 and increase to n.

4. The coefficients start at 1, increase through certain values, and then decrease through these same values back to 1.

Let's study the coefficients further. Suppose we wish to expand $(a + b)^8$. The patterns we noticed above indicate 9 terms in the expansion:

$$a^8 + c_1 a^7 b + c_2 a^6 b^2 + c_3 a^5 b^3 + c_4 a^4 b^4 + c_5 a^3 b^5 + c_6 a^2 b^6 + c_7 a b^7 + b^8.$$

How can we determine the values for the c's? One method seems very simple, but it has some drawbacks. It involves writing down the coefficients in a triangular array as follows. We form what is known as **Pascal's triangle**:

$$
\begin{array}{cccccccccccc}
(a+b)^0: & & & & & & 1 & & & & & \\
(a+b)^1: & & & & & 1 & & 1 & & & & \\
(a+b)^2: & & & & 1 & & 2 & & 1 & & & \\
(a+b)^3: & & & 1 & & 3 & & 3 & & 1 & & \\
(a+b)^4: & & 1 & & 4 & & 6 & & 4 & & 1 & \\
(a+b)^5: & 1 & & 5 & & 10 & & 10 & & 5 & & 1
\end{array}
$$

There are many patterns in the triangle. Find as many as you can.

Perhaps you discovered a way to write the next row of numbers, given the numbers in the row above it. There are always 1's on the outside. Each remaining number is the sum of the two numbers above:

$$
\begin{array}{ccccccccccccc}
& & & & & & 1 & & & & & & \\
& & & & & 1 & & 1 & & & & & \\
& & & & 1 & & 2 & & 1 & & & & \\
& & & 1 & & 3 & & 3 & & 1 & & & \\
& & 1 & & 4 & & 6 & & 4 & & 1 & & \\
& 1 & & 5 & & 10 & & 10 & & 5 & & 1 & \\
1 & & 6 & & 15 & & 20 & & 15 & & 6 & & 1
\end{array}
$$

We see that in the bottom (seventh) row

the 1st and last numbers are 1;

the 2nd number is $1 + 5$, or 6;

the 3rd number is $5 + 10$, or 15;

the 4th number is $10 + 10$, or 20;

the 5th number is $10 + 5$, or 15; and

the 6th number is $5 + 1$, or 6.

Thus the expansion of $(a + b)^6$ is

$$(a + b)^6 = 1a^6 + 6a^5b + 15a^4b^2 + 20a^3b^3 + 15a^2b^4 + 6ab^5 + 1b^6.$$

To expand $(a + b)^8$, we complete two more rows of Pascal's triangle:

$$
\begin{array}{ccccccccccccccccc}
&&&&&&&& 1 &&&&&&&& \\
&&&&&&& 1 && 1 &&&&&&& \\
&&&&&& 1 && 2 && 1 &&&&&& \\
&&&&& 1 && 3 && 3 && 1 &&&&& \\
&&&& 1 && 4 && 6 && 4 && 1 &&&& \\
&&& 1 && 5 && 10 && 10 && 5 && 1 &&& \\
&& 1 && 6 && 15 && 20 && 15 && 6 && 1 && \\
& 1 && 7 && 21 && 35 && 35 && 21 && 7 && 1 & \\
1 && 8 && 28 && 56 && 70 && 56 && 28 && 8 && 1
\end{array}
$$

The expansion of $(a + b)^8$ has coefficients found in the 9th row above:

$$(a + b)^8 = 1a^8 + 8a^7b + 28a^6b^2 + 56a^5b^3 + 70a^4b^4 + 56a^3b^5 + 28a^2b^6 + 8ab^7 + 1b^8.$$

We can generalize our results as follows:

> ## The Binomial Theorem (Form 1)
>
> For any binomial $a + b$ and any natural number n,
>
> $$(a + b)^n = c_0a^nb^0 + c_1a^{n-1}b^1 + c_2a^{n-2}b^2$$
> $$+ \cdots + c_{n-1}a^1b^{n-1} + c_na^0b^n,$$
>
> where the numbers $c_0, c_1, c_2, \ldots, c_n$ are from the $(n + 1)$st row of Pascal's triangle.

A proof of the binomial theorem is beyond the scope of this text.

EXAMPLE **1**   Expand: $(u - v)^5$.

SOLUTION   Using the binomial theorem, we have $a = u$, $b = -v$, and $n = 5$. We use the 6th row of Pascal's triangle: 1 5 10 10 5 1. Thus,

$$
\begin{aligned}
(u - v)^5 &= [u + (-v)]^5 \qquad \text{Rewriting } u - v \text{ as a sum} \\
&= 1(u)^5 + 5(u)^4(-v)^1 + 10(u)^3(-v)^2 + 10(u)^2(-v)^3 \\
&\quad + 5(u)^1(-v)^4 + 1(-v)^5 \\
&= u^5 - 5u^4v + 10u^3v^2 - 10u^2v^3 + 5uv^4 - v^5.
\end{aligned}
$$

Note that the signs of the terms alternate between $+$ and $-$. When $-v$ is raised to an odd power, the sign is $-$; when the power is even, the sign is $+$.

EXAMPLE **2**   Expand: $\left(2t + \dfrac{3}{t}\right)^6$.

SOLUTION   Note that $a = 2t$, $b = 3/t$, and $n = 6$. We use the 7th row of Pascal's triangle: 1 6 15 20 15 6 1. Thus,

$$\left(2t + \frac{3}{t}\right)^6 = 1(2t)^6 + 6(2t)^5\left(\frac{3}{t}\right)^1 + 15(2t)^4\left(\frac{3}{t}\right)^2 + 20(2t)^3\left(\frac{3}{t}\right)^3$$

$$+ 15(2t)^2\left(\frac{3}{t}\right)^4 + 6(2t)^1\left(\frac{3}{t}\right)^5 + 1\left(\frac{3}{t}\right)^6$$

$$= 64t^6 + 6\left(32t^5\right)\left(\frac{3}{t}\right) + 15(16t^4)\left(\frac{9}{t^2}\right) + 20(8t^3)\left(\frac{27}{t^3}\right)$$

$$+ 15(4t^2)\left(\frac{81}{t^4}\right) + 6(2t)\left(\frac{243}{t^5}\right) + \frac{729}{t^6}$$

$$= 64t^6 + 576t^4 + 2160t^2 + 4320 + 4860t^{-2} + 2916t^{-4} + 729t^{-6}.$$

Binomial Expansion Using Factorial Notation

The drawback to using Pascal's triangle is that we must compute all the preceding rows in the table to obtain the row we need. The following method avoids this difficulty. It will also enable us to find a specific term—say, the 8th term—without computing all the other terms in the expansion. This method is useful in such courses as finite mathematics, calculus, and statistics.

To develop the method, we need some new notation. Products of successive natural numbers, such as $6 \cdot 5 \cdot 4 \cdot 3 \cdot 2 \cdot 1$ and $8 \cdot 7 \cdot 6 \cdot 5 \cdot 4 \cdot 3 \cdot 2 \cdot 1$, have a special notation. For the product $6 \cdot 5 \cdot 4 \cdot 3 \cdot 2 \cdot 1$, we write 6!, read "6 factorial."

Factorial Notation

For any natural number n,

$$n! = n(n - 1)(n - 2) \cdots (3)(2)(1).$$

Here are some examples:

$$6! = 6 \cdot 5 \cdot 4 \cdot 3 \cdot 2 \cdot 1 = 720,$$
$$5! = 5 \cdot 4 \cdot 3 \cdot 2 \cdot 1 = 120,$$
$$4! = 4 \cdot 3 \cdot 2 \cdot 1 = 24,$$
$$3! = 3 \cdot 2 \cdot 1 = 6,$$
$$2! = 2 \cdot 1 = 2,$$
$$1! = 1 = 1.$$

We also define 0! to be 1 for reasons explained shortly.

To simplify expressions like

$$\frac{8!}{5!3!},$$

note that

$$8! = 8 \cdot 7 \cdot 6 \cdot 5 \cdot 4 \cdot 3 \cdot 2 \cdot 1 = 8 \cdot 7! = 8 \cdot 7 \cdot 6! = 8 \cdot 7 \cdot 6 \cdot 5!$$

and so on.

CAUTION! $\dfrac{6!}{3!} \neq 2!$ To see this, note that

$$\frac{6!}{3!} = \frac{6 \cdot 5 \cdot 4 \cdot \cancel{3} \cdot \cancel{2} \cdot \cancel{1}}{\cancel{3} \cdot \cancel{2} \cdot \cancel{1}} = 6 \cdot 5 \cdot 4.$$

EXAMPLE   3

Simplify: $\dfrac{8!}{5!3!}$.

SOLUTION

STUDENT NOTES

It is important to recognize factorial notation as representing a product with descending factors. Thus, 7!, 7 · 6!, and 7 · 6 · 5! all represent the same product.

$$\frac{8!}{5!3!} = \frac{8 \cdot 7 \cdot 6 \cdot 5!}{5! \cdot 3 \cdot 2 \cdot 1} = 8 \cdot 7 \qquad \text{Removing a factor equal to 1: } \frac{6 \cdot 5!}{5! \cdot 3 \cdot 2} = 1$$

$$= 56$$

TRY EXERCISE ▶ 15

The following notation is used in our second formulation of the binomial theorem.

$\dbinom{n}{r}$ **Notation**

For n and r nonnegative integers with $n \geq r$,

$$\binom{n}{r}, \quad \text{read "}n\text{ choose }r\text{,"} \quad \text{means} \quad \frac{n!}{(n-r)!\,r!}.^{*}$$

EXAMPLE   4

Simplify: **(a)** $\dbinom{7}{2}$; **(b)** $\dbinom{9}{6}$; **(c)** $\dbinom{6}{6}$.

SOLUTION

a) $\dbinom{7}{2} = \dfrac{7!}{(7-2)!\,2!}$

$= \dfrac{7!}{5!\,2!} = \dfrac{7 \cdot 6 \cdot 5!}{5! \cdot 2 \cdot 1} = \dfrac{7 \cdot 6}{2}$ We can write 7! as 7 · 6 · 5! to aid our simplification.

$= 7 \cdot 3$

$= 21$

TECHNOLOGY CONNECTION

The PRB option of the MATH menu provides access to both factorial calculations and nCr. In both cases, a number must be entered first. To find $\dbinom{7}{2}$, we press ⑦ **MATH**, select PRB and nCr, and press ② **ENTER**.

```
7 nCr 2
                    21
```

1. Find 12!.
2. Find $\dbinom{8}{3}$ and $\dbinom{12}{5}$.

---

*In many books and for many calculators, the notation $_nC_r$ is used instead of $\dbinom{n}{r}$.

b) $\binom{9}{6} = \dfrac{9!}{3!6!}$

$= \dfrac{9 \cdot 8 \cdot 7 \cdot 6!}{3 \cdot 2 \cdot 1 \cdot 6!} = \dfrac{9 \cdot 8 \cdot 7}{3 \cdot 2}$ Writing 9! as $9 \cdot 8 \cdot 7 \cdot 6!$ to help with simplification

$= 3 \cdot 4 \cdot 7$

$= 84$

c) $\binom{6}{6} = \dfrac{6!}{0!6!} = \dfrac{6!}{1 \cdot 6!}$ Since $0! = 1$

$= \dfrac{6!}{6!}$

$= 1$

> **TRY EXERCISE** 17

Now we can restate the binomial theorem using our new notation.

The Binomial Theorem (Form 2)

For any binomial $a + b$ and any natural number n,

$$(a + b)^n = \binom{n}{0}a^n + \binom{n}{1}a^{n-1}b + \binom{n}{2}a^{n-2}b^2 + \cdots + \binom{n}{n}b^n.$$

EXAMPLE **5**   Expand: $(3x + y)^4$.

SOLUTION   We use the binomial theorem (Form 2) with $a = 3x$, $b = y$, and $n = 4$:

$$(3x + y)^4 = \binom{4}{0}(3x)^4 + \binom{4}{1}(3x)^3 y + \binom{4}{2}(3x)^2 y^2 + \binom{4}{3}(3x)y^3 + \binom{4}{4}y^4$$

$$= \frac{4!}{4!0!}3^4 x^4 + \frac{4!}{3!1!}3^3 x^3 y + \frac{4!}{2!2!}3^2 x^2 y^2 + \frac{4!}{1!3!}3xy^3 + \frac{4!}{0!4!}y^4$$

$$= 1 \cdot 81x^4 + 4 \cdot 27x^3 y + 6 \cdot 9x^2 y^2 + 4 \cdot 3xy^3 + y^4 \left.\vphantom{\begin{array}{c}a\\a\end{array}}\right\}$$ Simplifying

$$= 81x^4 + 108x^3 y + 54x^2 y^2 + 12xy^3 + y^4.$$

> **TRY EXERCISE** 29

EXAMPLE **6**   Expand: $(x^2 - 2y)^{5}$.

SOLUTION   In this case, $a = x^2$, $b = -2y$, and $n = 5$:

$$(x^2 - 2y)^5 = \binom{5}{0}(x^2)^5 + \binom{5}{1}(x^2)^4(-2y) + \binom{5}{2}(x^2)^3(-2y)^2$$

$$+ \binom{5}{3}(x^2)^2(-2y)^3 + \binom{5}{4}(x^2)(-2y)^4 + \binom{5}{5}(-2y)^5$$

$$= \frac{5!}{5!0}x^{10} + \frac{5!}{4!1!}x^8(-2y) + \frac{5!}{3!2!}x^6(-2y)^2$$

$$+ \frac{5!}{2!3!}x^4(-2y)^3 + \frac{5!}{1!4!}x^2(-2y)^4 + \frac{5!}{0!5!}(-2y)^5$$

$$= x^{10} - 10x^8 y + 40x^6 y^2 - 80x^4 y^3 + 80x^2 y^4 - 32y^5.$$

> **TRY EXERCISE** 35

Note that in the binomial theorem (Form 2), $\binom{n}{0}a^n b^0$ gives us the first term, $\binom{n}{1}a^{n-1}b^1$ gives us the second term, $\binom{n}{2}a^{n-2}b^2$ gives us the third term, and so on. This can be generalized to give a method for finding a specific term without writing the entire expansion.

Finding a Specific Term

When $(a + b)^n$ is expanded and written in descending powers of a, the $(r + 1)$st term is

$$\binom{n}{r}a^{n-r}b^r.$$

EXAMPLE   7   Find the 5th term in the expansion of $(2x - 3y)^7$.

SOLUTION   To find the 5th term, we note that $5 = 4 + 1$. Thus, $r = 4, a = 2x, b = -3y$, and $n = 7$. Using the above formula, we have

$$\binom{n}{r}a^{n-r}b^r = \binom{7}{4}(2x)^{7-4}(-3y)^4, \text{ or } \frac{7!}{3!\,4!}(2x)^3(-3y)^4, \text{ or } 22{,}680x^3y^4.$$

> TRY EXERCISE ▶ 45

It is because of the binomial theorem that $\binom{n}{r}$ is called a *binomial coefficient*. We can now explain why 0! is defined to be 1. In the binomial theorem,

$$\binom{n}{0} \text{ must equal 1 when using the definition } \binom{n}{r} = \frac{n!}{(n-r)!\,r!}.$$

Thus we must have

$$\binom{n}{0} = \frac{n!}{(n-0)!\,0!} = \frac{n!}{n!\,0!} = 1.$$

This is satisfied only if 0! is defined to be 1.

Visualizing for Success

A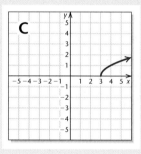

B

C

D

E

Match each equation with its graph.

1. $xy = 2$

2. $y = \log_2 x$

3. $y = x - 3$

4. $(x - 3)^2 + y^2 = 4$

5. $\dfrac{(x - 3)^2}{1} + \dfrac{y^2}{4} = 1$

6. $y = |x - 3|$

7. $y = (x - 3)^2$

8. $y = \dfrac{1}{x - 3}$

9. $y = 2^x$

10. $y = \sqrt{x - 3}$

Answers on page A-56

An additional, animated version of this activity appears in MyMathLab. To use MyMathLab, you need a course ID and a student access code. Contact your instructor for more information.

F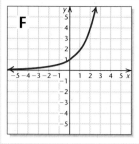

G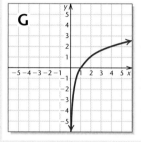

H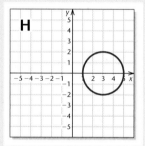

I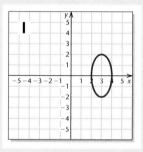

J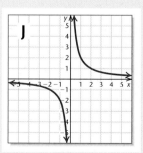

11.4 EXERCISE SET

⮕ *Concept Reinforcement* *Complete each of the following.*

1. The last term in the expansion of $(x + 2)^5$ is _____.

2. The expansion of $(x + y)^7$, when simplified, contains a total of _____ terms.

3. In the expansion of $(a + b)^9$, the exponents in each term add to _____.

4. The expression _____ represents $4 \cdot 3 \cdot 2 \cdot 1$.

5. The expression _____ represents $\dfrac{8!}{3!\,5!}$.

6. In the expansion of $(a + b)^{10}$, the coefficient of $a^8 b^2$ is the same as the coefficient of _____.

7. In the expansion of $(x + y)^9$, the coefficient of y^9 is _____.

8. The notation $\binom{9}{5}$ is read _____.

Simplify.

9. $4!$

10. $9!$

11. $10!$

12. $12!$

13. $\dfrac{10!}{8!}$

14. $\dfrac{12!}{10!}$

15. $\dfrac{9!}{4!\,5!}$

16. $\dfrac{10!}{6!\,4!}$

17. $\binom{10}{4}$

18. $\binom{8}{5}$

Aha! 19. $\binom{9}{9}$

20. $\binom{7}{7}$

21. $\binom{30}{2}$

22. $\binom{51}{49}$

23. $\binom{40}{38}$

24. $\binom{35}{2}$

Expand. Use both of the methods shown in this section.

25. $(a - b)^4$

26. $(m + n)^5$

27. $(p + q)^7$

28. $(x - y)^6$

29. $(3c - d)^7$

30. $(x^2 - 3y)^5$

31. $(t^{-2} + 2)^6$

32. $(3c - d)^6$

33. $(x - y)^5$

34. $(x - y)^3$

35. $\left(3s + \dfrac{1}{t}\right)^9$

36. $\left(x + \dfrac{2}{y}\right)^9$

37. $(x^3 - 2y)^5$

38. $(a^2 - b^3)^5$

39. $(\sqrt{5} + t)^6$

40. $(\sqrt{3} - t)^4$

41. $\left(\dfrac{1}{\sqrt{x}} - \sqrt{x}\right)^6$

42. $(x^{-2} + x^2)^4$

Find the indicated term for each binomial expression.

43. 3rd, $(a + b)^6$

44. 6th, $(x + y)^7$

45. 12th, $(a - 3)^{14}$

46. 11th, $(x - 2)^{12}$

47. 5th, $(2x^3 + \sqrt{y})^8$

48. 4th, $\left(\dfrac{1}{b^2} + c\right)^7$

49. Middle, $(2u + 3v^2)^{10}$

50. Middle two, $(\sqrt{x} + \sqrt{3})^5$

Aha! 51. 9th, $(x - y)^8$

52. 13th, $(a - \sqrt{b})^{12}$

📝 53. Maya claims that she can calculate mentally the first two and the last two terms of the expansion of $(a + b)^n$ for any whole number n. How do you think she does this?

📝 54. Without performing any calculations, explain why the expansions of $(x - y)^8$ and $(y - x)^8$ must be equal.

Skill Review

Review graphing equations and inequalities.

Graph.

55. $y = x^2 - 5$ [8.7]

56. $y = x - 5$ [2.3]

57. $y \geq x - 5$ [4.4]

58. $y = 5^x$ [9.2]

59. $f(x) = \log_5 x$ [9.3]

60. $x^2 + y^2 = 5$ [10.1]

Synthesis

61. Explain how someone can determine the x^2-term of the expansion of $\left(x - \dfrac{3}{x} \right)^{10}$ without calculating any other terms.

62. Devise two problems requiring the use of the binomial theorem. Design the problems so that one is solved more easily using Form 1 and the other is solved more easily using Form 2. Then explain what makes one form easier to use than the other in each case.

63. Show that there are exactly $\dbinom{5}{3}$ ways of choosing a subset of size 3 from $\{a, b, c, d, e\}$.

64. *Baseball.* During the 2007 season, Matt Holliday of the Colorado Rockies had a batting average of 0.340. In that season, if someone were to randomly select 5 of his "at-bats," the probability of Holliday getting exactly 3 hits would be the 3rd term of the binomial expansion of $(0.340 + 0.660)^5$. Find that term and use a calculator to estimate the probability.
Source: www.mlb.com

65. *Widows or divorcees.* The probability that a woman will be either widowed or divorced is 85%. If 8 women are randomly selected, the probability that exactly 5 of them will be either widowed or divorced is the 6th term of the binomial expansion of $(0.15 + 0.85)^8$. Use a calculator to estimate that probability.

66. *Baseball.* In reference to Exercise 64, the probability that Holliday will get *at most* 3 hits is found by adding the last 4 terms of the binomial expansion of $(0.340 + 0.660)^5$. Find these terms and use a calculator to estimate the probability.

67. *Widows or divorcees.* In reference to Exercise 65, the probability that *at least* 6 of the women will be widowed or divorced is found by adding the last three terms of the binomial expansion of $(0.15 + 0.85)^8$. Find these terms and use a calculator to estimate the probability.

68. Find the term of
$$\left(\dfrac{3x^2}{2} - \dfrac{1}{3x} \right)^{12}$$
that does not contain x.

69. Prove that
$$\binom{n}{r} = \binom{n}{n-r}$$
for any whole numbers n and r. Assume $r \le n$.

70. Find the middle term of $(x^2 - 6y^{3/2})^6$.

71. Find the ratio of the 4th term of
$$\left(p^2 - \dfrac{1}{2}p\sqrt[3]{q} \right)^5$$
to the 3rd term.

72. Find the term containing $\dfrac{1}{x^{1/6}}$ of
$$\left(\sqrt[3]{x} - \dfrac{1}{\sqrt{x}} \right)^7.$$

Aha! 73. Multiply: $(x^2 + 2xy + y^2)(x^2 + 2xy + y^2)^2(x + y)$.

74. What is the degree of $(x^3 + 2)^4$?

Study Summary

KEY TERMS AND CONCEPTS	EXAMPLES

SECTION 11.1: SEQUENCES AND SERIES

An ordered list of numbers that ends is a **finite sequence**.

An ordered list of numbers that does not end is an **infinite sequence**.

A **series** is a sum of terms of a sequence.

5, 7, 8, 11, 17 is a finite sequence.

6, 9, 12, 15, ... is an infinite sequence.

6 + 9 + 12 + 15 + ⋯ is an infinite series.

Sigma or **Summation Notation**

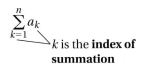

k is the **index of summation**

$$\sum_{k=3}^{5}(-1)^k(k^2) = (-1)^3(3^2) + (-1)^4(4^2) + (-1)^5(5^2)$$
$$= -1 \cdot 9 + 1 \cdot 16 + (-1) \cdot 25$$
$$= -9 + 16 - 25 = -18$$

SECTION 11.2: ARITHMETIC SEQUENCES AND SERIES

Arithmetic Sequences and Series

$a_{n+1} = a_n + d$ d is the **common difference**.

$a_n = a_1 + (n-1)d$ The nth term

$S_n = \dfrac{n}{2}(a_1 + a_n)$ The sum of the first n terms

For the arithmetic sequence 10, 7, 4, 1, ...:

$d = -3;$

$a_7 = 10 + (7-1)(-3) = 10 - 18 = -8;$

$S_7 = \dfrac{7}{2}(10 + (-8)) = \dfrac{7}{2}(2) = 7.$

SECTION 11.3: GEOMETRIC SEQUENCES AND SERIES

Geometric Sequences and Series

$a_{n+1} = a_n \cdot r$ r is the **common ratio**.

$a_n = a_1 r^{n-1}$ The nth term

$S_n = \dfrac{a_1(1 - r^n)}{1 - r}, r \neq 1$ The sum of the first n terms

$S_\infty = \dfrac{a_1}{1 - r}, |r| < 1$ Limit of an infinite geometric series

For the geometric sequence $25, -5, 1, -\frac{1}{5}, \ldots$:

$r = -\dfrac{1}{5};$

$a_7 = 25\left(-\dfrac{1}{5}\right)^{7-1} = 5^2 \cdot \dfrac{1}{5^6} = \dfrac{1}{625};$

$S_7 = \dfrac{25\left(1 - \left(-\frac{1}{5}\right)^7\right)}{1 - \left(-\frac{1}{5}\right)} = \dfrac{5^2\left(\frac{78,126}{5^7}\right)}{\frac{6}{5}} = \dfrac{13,021}{625};$

$S_\infty = \dfrac{25}{1 - \left(-\frac{1}{5}\right)} = \dfrac{125}{6}.$

SECTION 11.4: THE BINOMIAL THEOREM

Factorial Notation

$n! = n(n-1)(n-2)\cdots 3 \cdot 2 \cdot 1$

$7! = 7 \cdot 6 \cdot 5 \cdot 4 \cdot 3 \cdot 2 \cdot 1 = 5040$

Binomial Coefficient

$\dbinom{n}{r} = {}_nC_r = \dfrac{n!}{(n-r)!\,r!}$

$\dbinom{10}{3} = {}_{10}C_3 = \dfrac{10!}{7!\,3!} = \dfrac{10 \cdot 9 \cdot 8 \cdot 7!}{7! \cdot 3 \cdot 2 \cdot 1} = 120$

Binomial Theorem

$$(a + b)^n = \binom{n}{0}a^n + \binom{n}{1}a^{n-1}b + \cdots + \binom{n}{n}b^n$$

$$(r + 1)\text{st term of } (a + b)^n: \binom{n}{r}a^{n-r}b^r$$

$$(1 - 2x)^3 = \binom{3}{0}1^3 + \binom{3}{1}1^2(-2x)$$
$$+ \binom{3}{2}1(-2x)^2 + \binom{3}{3}(-2x)^3$$
$$= 1 \cdot 1 + 3 \cdot 1(-2x) + 3 \cdot 1 \cdot (4x^2) + 1 \cdot (-8x^3)$$
$$= 1 - 6x + 12x^2 - 8x^3$$

$$3\text{rd term of } (1 - 2x)^3: \binom{3}{2}(1)^1(-2x)^2 = 12x^2 \quad r = 2$$

Review Exercises: Chapter 11

✎ *Concept Reinforcement* *Classify each statement as either true or false.*

1. The next term in the arithmetic sequence $10, 15, 20, \ldots$ is 35. [11.2]

2. The next term in the geometric sequence $2, 6, 18, 54, \ldots$ is 162. [11.3]

3. $\sum\limits_{k=1}^{3} k^2$ means $1^2 + 2^2 + 3^3$. [11.1]

4. If $a_n = 3n - 1$, then $a_{17} = 19$. [11.1]

5. A geometric sequence has a common difference. [11.3]

6. The infinite geometric series $10 - 5 + \frac{5}{2} - \cdots$ has a limit. [11.3]

7. For any natural number n, $n! = n(n - 1)$. [11.4]

8. When simplified, the expansion of $(x + y)^{17}$ has 19 terms. [11.4]

Find the first four terms; the 8th term, a_8; and the 12th term, a_{12}. [11.1]

9. $a_n = 10n - 9$

10. $a_n = \dfrac{n - 1}{n^2 + 1}$

Write an expression for the general term of each sequence. Answers may vary. [11.1]

11. $-5, -10, -15, -20, \ldots$

12. $-1, 3, -5, 7, -9, \ldots$

Write out and evaluate each sum. [11.1]

13. $\sum\limits_{k=1}^{5} (-2)^k$

14. $\sum\limits_{k=2}^{7} (1 - 2k)$

Rewrite using sigma notation. [11.1]

15. $7 + 14 + 21 + 28 + 35 + 42$

16. $\dfrac{-1}{2} + \dfrac{1}{4} + \dfrac{-1}{8} + \dfrac{1}{16} + \dfrac{-1}{32}$

17. Find the 14th term of the arithmetic sequence $-3, -7, -11, \ldots$ [11.2]

18. An arithmetic sequence has $a_1 = 11$ and $a_{16} = 14$. Find the common difference, d. [11.2]

19. An arithmetic sequence has $a_8 = 20$ and $a_{24} = 100$. Find the first term, a_1, and the common difference, d. [11.2]

20. Find the sum of the first 17 terms of the arithmetic series $-8 + (-11) + (-14) + \cdots$. [11.2]

21. Find the sum of all the multiples of 5 from 5 to 500, inclusive. [11.2]

22. Find the 20th term of the geometric sequence $2, 2\sqrt{2}, 4, \ldots$ [11.3]

23. Find the common ratio of the geometric sequence $40, 30, \frac{45}{2}, \ldots$ [11.3]

24. Find the nth term of the geometric sequence $-2, 2, -2, \ldots$ [11.3]

25. Find the nth term of the geometric sequence $3, \frac{3}{4}x, \frac{3}{16}x^2, \ldots .$ [11.3]

26. Find S_6 for the geometric series
$$3 + 15 + 75 + \cdots .$$
[11.3]

27. Find S_{12} for the geometric series
$$3x - 6x + 12x - \cdots .\ [11.3]$$

Determine whether each infinite geometric series has a limit. If a limit exists, find it. [11.3]

28. $6 + 3 + 1.5 + 0.75 + \cdots$

29. $7 - 4 + \frac{16}{7} - \cdots$

30. $-\frac{1}{2} + \frac{1}{2} + \left(-\frac{1}{2}\right) + \frac{1}{2} + \cdots$

31. $0.04 + 0.08 + 0.16 + 0.32 + \cdots$

32. $\$2000 + \$1900 + \$1805 + \$1714.75 + \cdots$

33. Find fraction notation for $0.555555\ldots .$ [11.3]

34. Find fraction notation for $1.454545\ldots .$ [11.3]

35. Tyrone took a job working in a convenience store starting with an hourly wage of \$11.50. He was promised a raise of 40¢ per hour every 3 mos for 8 yr. After 8 yr, what will be his hourly wage? [11.2]

36. A stack of poles has 42 poles in the bottom row. There are 41 poles in the second row, 40 poles in the third row, and so on, ending with 1 pole in the top row. How many poles are in the stack? [11.2]

37. Janine's student loan is for \$12,000 at 4%, compounded annually. The total amount is to be paid off in 7 yr. How much will she then owe? [11.3]

38. Find the total rebound distance of a ball, given that it is dropped from a height of 12 m and each rebound is one-third of the preceding one. [11.3]

Simplify. [11.4]

39. $7!$

40. $\begin{pmatrix} 10 \\ 3 \end{pmatrix}$

41. Find the 3rd term of $(a + b)^{20}$. [11.4]

42. Expand: $(x - 2y)^4$. [11.4]

Synthesis

43. What happens to a_n in a geometric sequence with $|r| < 1$, as n gets larger? Why? [11.3]

44. Compare the two forms of the binomial theorem given in the text. Under what circumstances would one be more useful than the other? [11.4]

45. Find the sum of the first n terms of the geometric series $1 - x + x^2 - x^3 + \cdots .$ [11.3]

46. Expand: $(x^{-3} + x^3)^5$. [11.4]

Test: Chapter 11

1. Find the first five terms and the 12th term of a sequence with general term $a_n = \dfrac{1}{n^2 + 1}$.

2. Write an expression for the general term of the sequence $\frac{4}{3}, \frac{4}{9}, \frac{4}{27}, \ldots$.

3. Write out and evaluate:
$$\sum_{k=2}^{5} (1 - 2^k).$$

4. Rewrite using sigma notation:
$$1 + (-8) + 27 + (-64) + 125.$$

5. Find the 13th term, a_{13}, of the arithmetic sequence $\frac{1}{2}, 1, \frac{3}{2}, 2, \ldots$.

6. Find the common difference d of an arithmetic sequence when $a_1 = 7$ and $a_7 = -11$.

7. Find a_1 and d of an arithmetic sequence when $a_5 = 16$ and $a_{10} = -3$.

8. Find the sum of all the multiples of 12 from 24 to 240, inclusive.

9. Find the 10th term of the geometric sequence $-3, 6, -12, \ldots$.

10. Find the common ratio of the geometric sequence $22\frac{1}{2}, 15, 10, \ldots$.

11. Find the nth term of the geometric sequence $3, 9, 27, \ldots$.

12. Find S_9 for the geometric series
$$11 + 22 + 44 + \cdots.$$

Determine whether each infinite geometric series has a limit. If a limit exists, find it.

13. $0.5 + 0.25 + 0.125 + \cdots$

14. $0.5 + 1 + 2 + 4 + \cdots$

15. $\$1000 + \$80 + \$6.40 + \cdots$

16. Find fraction notation for $0.85858585\ldots$.

17. An auditorium has 31 seats in the first row, 33 seats in the second row, 35 seats in the third row, and so on, for 18 rows. How many seats are in the 17th row?

18. Alyssa's uncle Ken gave her $100 for her first birthday, $200 for her second birthday, $300 for her third birthday, and so on, until her eighteenth birthday. How much did he give her in all?

19. Each week the price of a $10,000 boat will be reduced 5% of the previous week's price. If we assume that it is not sold, what will be the price after 10 weeks?

20. Find the total rebound distance of a ball that is dropped from a height of 18 m, with each rebound two-thirds of the preceding one.

21. Simplify: $\dbinom{12}{9}$.

22. Expand: $(x - 3y)^5$.

23. Find the 4th term in the expansion of $(a + x)^{12}$.

Synthesis

24. Find a formula for the sum of the first n even natural numbers:
$$2 + 4 + 6 + \cdots + 2n.$$

25. Find the sum of the first n terms of
$$1 + \frac{1}{x} + \frac{1}{x^2} + \frac{1}{x^3} + \cdots.$$

Simplify.

1. $\left| -\dfrac{2}{3} + \dfrac{1}{5} \right|$ [1.2]

2. $y - [3 - 4(5 - 2y) - 3y]$ [11.3]

3. $(10 \cdot 8 - 9 \cdot 7)^2 - 54 \div 9 - 3$ [1.2]

4. $(2.7 \times 10^{-24})(3.1 \times 10^9)$ [1.7]

5. Evaluate
$$\frac{ab - ac}{bc}$$
for $a = -2$, $b = 3$, and $c = -4$. [1.1], [1.2]

Perform the indicated operations to create an equivalent expression. Be sure to simplify your result if possible.

6. $(5a^2 - 3ab - 7b^2) - (2a^2 + 5ab + 8b^2)$ [5.1]

7. $(2a - 1)(2a + 1)$ [5.2]

8. $(3a^2 - 5y)^2$ [5.2]

9. $\dfrac{1}{x - 2} - \dfrac{4}{x^2 - 4} + \dfrac{3}{x + 2}$ [6.2]

10. $\dfrac{x^2 - 6x + 8}{4x + 12} \cdot \dfrac{x + 3}{x^2 - 4}$ [6.1]

11. $\dfrac{3x + 3y}{5x - 5y} \div \dfrac{3x^2 + 3y^2}{5x^3 - 5y^3}$ [6.1]

12. $\dfrac{x - \dfrac{a^2}{x}}{1 + \dfrac{a}{x}}$ [6.3]

13. $\sqrt{12a}\,\sqrt{12a^3 b}$ [7.3]

14. $(-9x^2 y^5)(3x^8 y^{-7})$ [1.6]

15. $(125x^6 y^{1/2})^{2/3}$ [7.2]

16. $\dfrac{\sqrt[3]{x^2 y^5}}{\sqrt[4]{xy^2}}$ [7.5]

17. $(4 + 6i)(2 - i)$, where $i = \sqrt{-1}$ [7.8]

Factor, if possible, to form an equivalent expression.

18. $4x^2 - 12x + 9$ [5.5]

19. $27a^3 - 8$ [5.6]

20. $12s^4 - 48t^2$ [5.5]

21. $15y^4 + 33y^2 - 36$ [5.7]

22. Divide:
$$(7x^4 - 5x^3 + x^2 - 4) \div (x - 2). [6.6]$$

23. For the function described by
$$f(x) = 3x^2 - 4x,$$
find $f(-2)$. [2.2]

Find the domain of each function.

24. $f(x) = \sqrt{2x - 8}$ [7.1]

25. $g(x) = \dfrac{x - 4}{x^2 - 10x + 25}$ [5.8]

26. Write an equivalent expression by rationalizing the denominator:
$$\frac{1 - \sqrt{x}}{1 + \sqrt{x}}. [7.5]$$

27. Find a linear equation whose graph has a y-intercept of $(0, -8)$ and is parallel to the line whose equation is $3x - y = 6$. [2.5]

28. Write a quadratic equation whose solutions are $5\sqrt{2}$ and $-5\sqrt{2}$. [8.3]

29. Find the center and the radius of the circle given by
$$x^2 + y^2 - 4x + 6y - 23 = 0. [10.1]$$

30. Write an equivalent expression that is a single logarithm:
$$\tfrac{2}{3}\log_a x - \tfrac{1}{2}\log_a y + 5\log_a z. [9.4]$$

31. Write an equivalent exponential equation:
$$\log_a c = 5. [9.3]$$

Use a calculator to find each of the following. Round to four decimal places. [9.5]

32. $\log 120$

33. $\log_5 3$

34. Find the distance between the points $(-1, -5)$ and $(2, -1)$. [7.7]

35. Find the 21st term of the arithmetic sequence $19, 12, 5, \ldots$ [11.2]

36. Find the sum of the first 25 terms of the arithmetic series $-1 + 2 + 5 + \cdots$. [11.2]

37. Write an expression for the general term of the geometric sequence $16, 4, 1, \ldots$ [11.3]

38. Find the 7th term of $(a - 2b)^{10}$. [11.4]

39. Find the sum of the first nine terms of the geometric series $4 + 6 + 9 + \cdots$. [11.3]

Solve.

40. $8(x - 1) - 3(x - 2) = 1$ [1.3]

41. $\dfrac{6}{x} + \dfrac{6}{x + 2} = \dfrac{5}{2}$ [6.4]

42. $2x + 1 > 5 \text{ or } x - 7 \le 3$ [4.2]

43. $5x + 6y = -2,$
$3x + 10y = 2$ [3.2]

44. $x + y - z = 0,$
$3x + y + z = 6,$
$x - y + 2z = 5$ [3.4]

45. $3\sqrt{x - 1} = 5 - x$ [7.6]

46. $x^4 - 29x^2 + 100 = 0$ [8.5]

47. $x^2 + y^2 = 8,$
$x^2 - y^2 = 2$ [10.4]

48. $4^x = 12$ [9.6]

49. $\log(x^2 - 25) - \log(x + 5) = 3$ [9.6]

50. $\log_5 x = -2$ [9.6]

51. $7^{2x+3} = 49$ [9.6]

52. $|2x - 1| \le 5$ [4.3]

53. $15x^2 + 45 = 0$ [8.1]

54. $x^2 + 4x = 3$ [8.2]

55. $y^2 + 3y > 10$ [8.9]

56. Let $f(x) = x^2 - 2x$. Find a such that $f(a) = 80$. [5.8]

57. If $f(x) = \sqrt{-x + 4} + 3$ and $g(x) = \sqrt{x - 2} + 3$, find a such that $f(a) = g(a)$. [7.6]

58. Solve $V = P - Prt$ for r. [1.5]

59. Solve $I = \dfrac{R}{R + r}$ for R. [6.8]

Graph.

60. $3x - y = 7$ [2.4]

61. $x^2 + y^2 = 100$ [10.1]

62. $\dfrac{x^2}{36} - \dfrac{y^2}{9} = 1$ [10.3]

63. $y = \log_2 x$ [9.3]

64. $f(x) = 2^x - 3$ [9.2]

65. $2x - 3y < -6$ [4.4]

66. Graph: $f(x) = -2(x - 3)^2 + 1$. [8.7]
   a) Label the vertex.
   b) Draw the axis of symmetry.
   c) Find the maximum or minimum value.

Solve.

67. The Brighton recreation department plans to fence in a rectangular park next to a river. (Note that no fence will be needed along the river.) What is the area of the largest region that can be fenced in with 200 ft of fencing? [8.8]

68. The perimeter of a rectangular sign is 34 ft. The length of a diagonal is 13 ft. Find the dimensions of the sign. [10.4]

69. A movie club offers two types of membership. Limited members pay a fee of $40 a year and can rent movies for $2.45 each. Preferred members pay $60 a year and can rent movies for $1.65 each. For what numbers of annual movie rentals would it be less expensive to be a preferred member? [4.1]

70. Find three consecutive odd integers whose sum is 177. [1.4]

71. Cosmos Tastes mixes herbs that cost $2.68 an ounce with herbs that cost $4.60 an ounce to create a seasoning that costs $3.80 an ounce. How many ounces of each herb should be mixed together to make 24 oz of the seasoning? [3.3]

72. An airplane can fly 190 mi with the wind in the same time it takes to fly 160 mi against the wind. The speed of the wind is 30 mph. How fast can the plane fly in still air? [6.5]

73. Jared can tap the sugar maple trees in Southway Farm in 21 hr. Delia can tap the trees in 14 hr. How long would it take them, working together, to tap the trees? [6.5]

74. The centripetal force F of an object moving in a circle varies directly as the square of the velocity v and inversely as the radius r of the circle. If $F = 8$ when $v = 1$ and $r = 10$, what is F when $v = 2$ and $r = 16$? [6.8]

75. *Mortgages.*   The loan-to-value ratio of a mortgage is the ratio of the amount owed on the loan to the value of the home. In 2002, the average homeowner owed 80% of the value of the home. By 2007, largely due to falling home prices, this amount had risen to 87%.
Source: UBS Mortgage Strategy Group

  a) What was the average rate of change? [2.3]
  b) Find a linear function that fits the data. Let t represent the number of years since 2000. [2.5]
  c) Use the function from part (b) to predict the average loan-to-value ratio in 2010. [2.5]
  d) Assuming the trend continues, in what year will the average U.S. homeowner owe 95% of the value of the home? [2.5]

76. *Mortgages.*   In a reverse mortgage, the lender makes payments to the borrower and the borrower keeps control of the house. The loan is repaid when the house is sold. The number of reverse mortgages in the United States has increased exponentially from approximately 160 in 1990 to 108,000 in 2007. [9.7]
Source: U.S. Department of Housing and Urban Development

  a) Find the exponential growth rate k, and write an equation for an exponential function that can be used to predict the number of reverse mortgages t years after 1990.
  b) Predict the number of reverse mortgages in 2012.
  c) In what year will there be 1 million reverse mortgages?

77. *Retirement.*   Sarita invested $2000 in a retirement account on her 22nd birthday. If the account earns 5% interest, compounded annually, how much will this investment be worth on her 62nd birthday? [11.3]

Synthesis

Solve.

78. $\dfrac{9}{x} - \dfrac{9}{x + 12} = \dfrac{108}{x^2 + 12x}$ [6.4]

79. $\log_2 (\log_3 x) = 2$ [9.6]

80. y varies directly as the cube of x and x is multiplied by 0.5. What is the effect on y? [6.8]

81. Diaphantos, a famous mathematician, spent $\frac{1}{6}$ of his life as a child, $\frac{1}{12}$ as an adolescent, and $\frac{1}{7}$ as a bachelor. Five years after he was married, he had a son who died 4 years before his father at half his father's final age. How long did Diaphantos live? [3.5]

Appendix

The Graphing Calculator

Different calculators may require different keystrokes to use the same feature. In this appendix, exact keystrokes referenced and screens shown apply to the Texas Instruments TI-83® or TI-84® Plus calculator. If your calculator does not have a key by the name shown, consult the user's manual for your calculator to learn how to perform the procedure. Often it is helpful to simply try different combinations of keystrokes.

Introduction to the Graphing Calculator

A typical screen on a graphing calculator holds up to 8 lines of text and is 16 columns wide. Screens are composed of *pixels*, or dots. After the calculator has been turned on, a blinking rectangle or line appears on the screen. This rectangle or line is called the **cursor**. It indicates your current position on the screen. In the graphing mode, the cursor may appear as a cross. The four arrow keys are used to move the cursor.

KEYPAD

A diagram of the keypad of a graphing calculator appears at the back of this text. Note that there are options written above keys as well as on the keys. To access the options shown above the keys, press **2ND** or **ALPHA** and then the key below the desired option.

CONTRAST

The **contrast** controls how dark the characters appear on the screen. If you do not see anything on the screen after turning the calculator on, try adjusting the contrast by pressing **2ND** and either the up or down arrow key.

HOME SCREEN

To perform computations, you should be in the **home screen**. Pressing (QUIT) (the 2nd option associated with the **MODE** key) will return you to the home screen.

PERFORMING OPERATIONS

To perform addition, subtraction, multiplication, and division using a graphing calculator, type in the expression as it is written. A multiplication symbol will appear on the screen as *, and division is usually shown by the symbol /. Check your typing, and if the expression appears to be correct, press **ENTER**. At that time, the calculator will evaluate the expression and display the result on the screen.

CATALOG

A graphing calculator's catalog lists all the functions of the calculator in alphabetical order. To copy an item from the catalog to the screen, press **2ND** **0** or **CATALOG** and then scroll through the list using the up and down arrow keys until the desired item is indicated. To move through the list more quickly, press the key associated with the first letter of the item. The indicator will move to the first item beginning with that letter. When the desired item is indicated, press **ENTER**.

MENUS

A menu is a list of options that appears when a key is pressed. To select an item from the menu, highlight its number using the up or down arrow keys and press **ENTER** or simply press the number of the item. For example, pressing **MATH** results in a screen like the following. Four menu titles, or **submenus**, are listed across the top of the screen. Use the left and right arrow keys to highlight the desired submenu.

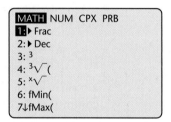

In the screen shown above, the MATH submenu, referred to as the MATH MATH menu, is highlighted. The options in that menu appear on the screen. Note that the submenu contains more options than can fit on a screen, as indicated by the arrow in entry 7. The remaining options will appear as the down arrow is pressed.

ERROR MESSAGES

When a calculator cannot complete an instruction, a message similar to the one shown below appears on the screen.

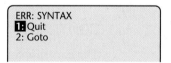

Press **1** to return to the home screen, and press **2** to go to the instruction that caused the error. Not all errors are operator errors; ERR: OVERFLOW indicates a result too large for the calculator to handle.

Entering Expressions

The following list describes how to enter various types of expressions in a graphing calculator.

TYPE OF EXPRESSION	EXAMPLE

Negative Numbers

Graphing calculators have different keys for writing negatives and for subtracting. The key labeled (–) is used to create a negative sign, whereas the one labeled (−) is used for subtraction. Using the wrong key may result in an ERR: SYNTAX message.

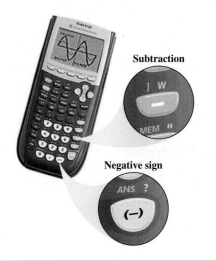

Subtraction

Negative sign

Fraction Notation

To convert a number to fraction notation, enter the number on the home screen and choose the FRAC option from the MATH MATH menu. After the notation ▶FRAC is copied on the home screen, press **ENTER**.

To find fraction notation for $\frac{2}{15} + \frac{7}{12}$, use the keystrokes (2)(÷)(1)(5)(+)(7)(÷) (1)(2) **MATH** (1) **ENTER**.

```
2/15+7/12▶Frac
                    43/60
```

Absolute-Value Notation

The notation "abs" with parentheses indicates absolute value. Thus, abs(2) = |2| = 2. The abs notation is found in the MATH NUM menu. Many calculators automatically supply the left parenthesis.

To find |−3|, we press **MATH** (▷)(1)(−)(3) (▷) **ENTER**.

```
abs(−3)
                        3
```

Exponents

To enter an exponential expression, enter the base and then press **⌃** and enter the exponent. If the exponent is 2, we can also use the **x²** key. If it is −1, we can use the **x⁻¹** key. If the exponent is a single number, not an expression, we need not enclose it in parentheses.

To calculate $(-4.7)^2$, we press ((−)(4) (.)(7)( ) ⌃ (2) **ENTER** or ((−)(4) (.)(7)( ) **x²** **ENTER**. To calculate $(-8)^{-1}$, we press ((−)(8)( ) **x⁻¹** **ENTER**.

```
(−4.7)^2
                    22.09
(−4.7)²
                    22.09
(−8)⁻¹
                    −.125
```

Scientific Notation

Graphing calculators will accept entries using scientific notation, and will normally automatically write very large or very small numbers using scientific notation. The ⟨EE⟩ key, the 2nd option associated with the ⟨,⟩ key, is used to enter scientific notation. On the calculator screen, a notation like E22 represents $\times 10^{22}$.

To calculate $(7.5 \times 10^8)(1.2 \times 10^{-14})$, we press ⟨7⟩ ⟨.⟩ ⟨5⟩ ⟨EE⟩ ⟨8⟩ ⟨×⟩ ⟨1⟩ ⟨.⟩ ⟨2⟩ ⟨EE⟩ ⟨(-)⟩ ⟨1⟩ ⟨4⟩ ⟨ENTER⟩. The result shown is then read as 9×10^{-6}.

```
7.5E8*1.2E-14
                    9E-6
```

Grouping Symbols

Grouping symbols such as brackets or braces are entered as parentheses. For fractions, write parentheses around the numerator and around the denominator.

To calculate $\dfrac{12(9-7)+4\cdot 5}{2^4+3^2}$, we write an equivalent expression with parentheses:

$$(12(9-7)+4\cdot 5) \div (2^4+3^2).$$

```
(12(9-7)+4*5)/(2
^4+3²)►Frac
                   44/25
```

Storing a Value to a Variable

The ⟨STO⟩ key is used to store a value to a variable name. Then the value can be substituted for the variable in an algebraic expression.

To let $X = -5$, press ⟨(-)⟩ ⟨5⟩ ⟨STO⟩ ⟨X,T,θ,n⟩ ⟨ENTER⟩. Any letter can be used as a variable by pressing ⟨ALPHA⟩ and the key associated with that letter.

To see the value of X, press ⟨X,T,θ,n⟩ ⟨ENTER⟩. To evaluate the expression $2X + 1$ for $X = -5$, enter the expression as written and press ⟨ENTER⟩. The value of the expression is -9.

```
-5 → X
                    -5
X
                    -5
2X+1
                    -9
```

Radical Expressions

When entering radical expressions on a graphing calculator, care must be taken to place parentheses properly. Square roots are entered by first pressing ⟨√⟩, the 2nd option associated with the ⟨x²⟩ key. Cube roots can be entered using the $\sqrt[3]{\ }($ option in the MATH MATH menu. For an index other than 2 or 3, first enter the index, then choose the $\sqrt[x]{\ }$ option from the MATH MATH menu, and then enter the radicand. The $\sqrt[x]{\ }$ option may not supply the left parenthesis.

```
MATH NUM CPX PRB
1: ► Frac
2: ► Dec
3: 3
4: 3√(
5: x√
6: fMin(
7↓fMax(
```

To enter $\dfrac{3.5 + \sqrt{4.5 - 2}}{5}$, we press ⟨(⟩ ⟨3⟩ ⟨.⟩ ⟨5⟩ ⟨+⟩ ⟨√⟩ ⟨4⟩ ⟨.⟩ ⟨5⟩ ⟨-⟩ ⟨2⟩ ⟨)⟩ ⟨)⟩ ⟨÷⟩ ⟨5⟩. The outer parentheses enclose the numerator of the expression. The first right parenthesis) indicates the end of the radicand; the left parenthesis of the radicand is supplied by the calculator when ⟨√⟩ is pressed. The result, 1.016227766, is an approximation.

```
(3.5+√(4.5-2))15
                1.016227766
```

Complex Numbers

To perform operations with complex numbers, first choose the a + bi setting in the MODE screen.

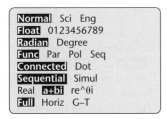

Now complex numbers can be entered as they are written. Press ⓘ for the complex number i. (ⓘ is the 2nd option associated with the ⊙ key.)

Common Logarithms

To find the common logarithm of a number on most graphing calculators, press **LOG**, the number, and then **ENTER**. Some calculators automatically supply the left parenthesis when **LOG** is pressed. Of these, some require that a right parenthesis be entered and some assume one is present at the end of the expression. Even if a calculator does not require an ending parenthesis, it is a good idea to include one.

Natural Logarithms and *e*

We use the **LN** key to find natural logarithms on a graphing calculator. The **LN** key serves as the ⓔˣ key after the **2ND** key is pressed. Some calculators automatically supply a left parenthesis.

Factorials and $\binom{n}{r}$

The PRB submenu of the MATH menu provides access to both factorial calculations and $\binom{n}{r}$, or ${}_nC_r$. In both cases, a number is entered on the home screen before the option is accessed.

```
MATH NUM CPX PRB
1: rand
2: nPr
3: nCr
4: !
5: randInt(
6: randNorm(
7: randBin(
```

To calculate $(-1 + i)(2 - i)$, press ((-) 1 + ⓘ) (2 − ⓘ) **ENTER**. The result is $-1 + 3i$.

To calculate $\dfrac{2 - i}{3 + i}$ and write the result using fraction notation, press (2 − ⓘ) ÷ (3 + ⓘ) **MATH** 1 **ENTER**. Note the need for parentheses around the numerator and the denominator of the expression. The result appears in the form $a + bi$; i is not in the denominator of the fraction. The result is $\frac{1}{2} - \frac{1}{2}i$.

```
(-1+i)(2-i)
                        -1+3i
(2-i)/(3+i)▶Frac
                        1/2-1/2i
```

If we are using a calculator that supplies the left parenthesis, log 5 can be found by pressing **LOG** 5) **ENTER**. Note from the screen that $\log 5 \approx 0.6990$.

```
log(5)
             .6989700043
```

If we are using a calculator that supplies the left parenthesis, ln 8 can be found by pressing **LN** 8) **ENTER**. Note from the screen that $\ln 8 \approx 2.0794$. If such a calculator is used, $e^{2.079441542}$ is found by pressing ⓔˣ 2 . 0 7 9 4 4 1 5 4 2) **ENTER**. As shown on the screen, $e^{2.079441542} \approx 8$.

```
ln(8)
                  2.079441542
e^(2.079441542)
                  8.000000003
```

To find 7!, press 7, select option 4: ! from the MATH PRB menu, and press **ENTER**.

To find $\binom{9}{2}$, press 9, select option 3: nCr from the MATH PRB menu, press 2, and then press **ENTER**.

```
7!
                        5040
9 nCr 2
                          36
```

Matrices

Matrix operations are accessed by pressing (MATRIX). ((MATRIX) is the 2nd feature associated with the (x⁻¹) key.) The MATRIX menu has three submenus: NAMES, MATH, and EDIT. The MATRIX EDIT menu allows matrices to be entered.

After entering the matrix, exit the matrix editor by pressing (QUIT). To access a matrix once it has been entered, use the MATRIX NAMES menu. The brackets around *A* indicate that it is a matrix.

```
NAMES  MATH  EDIT
1: [A] 2x3
2: [B]
3: [C]
4: [D]
5: [E]
6: [F]
7↓[G]
```

The MATRIX MATH menu lists operations that can be performed on matrices.

To enter

$$A = \begin{bmatrix} 1 & 2 & 5 \\ -3 & 0 & 4 \end{bmatrix},$$

choose option [A] from the MATRIX EDIT menu. Enter the **dimensions**, or number of rows and columns, of the matrix first. The dimensions of *A* are 2×3, read "2 by 3," so we press (2) (ENTER) (3) (ENTER). Then enter each element of *A* by pressing the number and (ENTER). The notation $2, 3 = 4$ indicates that the 3rd entry of the 2nd row is 4. Matrices are entered row by row rather than column by column.

```
MATRIX[A] 2 x3
[1        2        5       ]
[-3       0        4       ]



2, 3 = 4
```

Determinants

Determinants can be evaluated using the DET(option of the MATRIX MATH menu. After entering the matrix, go to the home screen and select the determinant operation. Then enter the name of the matrix using the MATRIX NAMES menu. The graphing calculator will return the value of the determinant of the matrix.

If $\mathbf{A} = \begin{bmatrix} 1 & 6 & -1 \\ -3 & -5 & 3 \\ 0 & 4 & 2 \end{bmatrix}$, we have

```
det ([A])
            26
```

Editing Entries

It is possible to correct an error or change a value in an expression by using the *arrow, insert,* and *delete* keys.

As the expression is being typed, use the arrow keys to move the cursor to the character you want to change. Pressing (INS) (the 2nd option associated with the (DEL) key) changes the calculator between the INSERT and OVERWRITE modes. The OVERWRITE mode is often indicated by a rectangular cursor and the INSERT mode by an underscore cursor. The following table shows how to make changes.

Insert a character in front of the cursor.	In the INSERT mode, press the character you wish to insert.
Replace the character under the cursor.	In the OVERWRITE mode, press the character you want as the replacement.
Delete the character under the cursor.	Press (DEL).

After you have evaluated an expression by pressing **ENTER**, the expression can be recalled to the screen by pressing (ENTRY). ((ENTRY) is the 2nd option associated with the **ENTER** key.) Then it can be edited as described above. Pressing **ENTER** will then evaluate the edited expression.

The most recent result of a calculation is stored in the calculator's memory as Ans (short for Answer). To recall this value, press (ANS). ((ANS) is the 2nd option associated with the (-) key.)

Entering Data and Equations

ENTERING DATA

Coordinates of ordered pairs are entered as *data* in lists, using the EDIT option of the **STAT** menu. If there are already numbers in the lists, clear them by moving the cursor to the title of the list (L_1, L_2, and so on) and pressing **CLEAR** **ENTER**.

To enter a number in a list, move the cursor to the correct position, type in the number, and press **ENTER**. Enter the first coordinates of the ordered pairs as one list and the second coordinates as another list. The coordinates of each point should be at the same position on both lists. Note that a DIM MISMATCH error may occur if there is not the same number of items in each list.

To plot the points, first press (STAT PLOT). ((STAT PLOT) is the 2nd option associated with the (Y=) key.) The calculator allows several different sets of points to be displayed at once. See the screen on the left below. Choose Plot1 by pressing (1). Then turn Plot1 on by positioning the cursor over On and pressing **ENTER**, as shown on the right below.

The remaining items on the Plot1 screen define the plot. Use the down arrow key to move to the next item.

The points entered can be used to make a line graph or a bar graph as well as to graph points. The screen on the right below shows six available types of graphs. To plot points, choose the first type of graph shown, a scatter diagram or scatterplot. The second option in the list is a line graph, in which the points are connected. The third type is a bar graph. The last three types will not be discussed in this course.

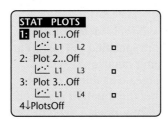
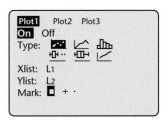

The next item on the screen, Xlist, should be the list in which the first coordinates were entered, probably L_1, and Ylist should be the list in which the second coordinates were entered, probably L_2. List names can be selected by pressing (LIST). ((LIST) is the 2nd option associated with the **STAT** key.) They can also be selected directly from the keyboard by pressing **2ND** (1) for L_1, **2ND** (2) for L_2, and so on.

The last choice on the screen is the type of mark used to plot the points. Different marks can be used to distinguish among several sets of data.

When the STAT PLOT feature has been set correctly, choose window dimensions (see p. 743) that will allow all the points to be seen and press (GRAPH). The Zoom-Stat option of the ZOOM menu will choose an appropriate window automatically. When you no longer wish to plot data, turn off the plot. One way to do this is to move the cursor to the highlighted PLOT name at the top of the equation-editor screen and press **ENTER**. To turn off all the plots, choose option 4: PlotsOff on the STAT PLOTS screen and press **ENTER**.

ENTERING EQUATIONS

Equations are entered using the equation-editor screen, accessed by pressing
(Y=). The first part of each equation, "Y=," is supplied by the calculator, including a subscript that identifies the equation. An equation can be cleared by positioning the cursor on the equation and pressing CLEAR. A symbol before the Y indicates the graph style, and a highlighted = indicates that the equation selected is to be graphed. An equation can be selected or deselected by positioning the cursor on the = and pressing ENTER. Selected equations are then graphed by pressing (GRAPH). *Note:* If the window is not set appropriately, the graph may not appear on the screen at all. Window settings are discussed on p. 743.

Tables

A TABLE feature lists ordered pairs that are solutions of an equation.

Since the value of *y depends* on the choice of the value for *x*, we say that *y* is the **dependent variable** and *x* is the **independent variable**.

After entering an equation on the equation-editor screen, we can view a table of solutions. A table is set up by pressing (TBLSET). ((TBLSET) is the 2nd feature associated with the (WINDOW) key.)

If we want to choose the values for the independent variable, we set Indpnt to Ask, as shown on the left below. Then we can create a table of values. To create a table of values for $y = -\frac{1}{2}x$, we first enter the equation $Y_1 = -(1/2)x$, and then press (TABLE). ((TABLE) is the 2nd feature associated with the (GRAPH) key.) Entering the *x*-values 4, −6, 0, and 2 gives us the table shown on the right below.

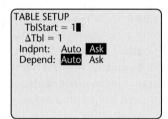

If Indpnt is set to Auto, the calculator will provide values for *x*, beginning with the value specified as TblStart and continuing by adding the value of ΔTbl to the preceding value for *x*.

To create a table of ordered pairs that are solutions of the equation $y = -\frac{1}{2}x$, first enter the equation $Y_1 = -(1/2)x$ and then set up the table, as shown on the left below. Then press (TABLE) to view the table. We see that some solutions of the equation are (−3, 1.5), (−2, 1), (−1, 0.5), and so on. Pressing the up and down arrow keys allows us to scroll through the table.

Graphs

WINDOWS

On a graphing calculator, the rectangular portion of the screen in which a graph appears is called the **viewing window**. Windows are described by four numbers of the form [L, R, B, T], representing the **L**eft and **R**ight endpoints of the x-axis and the **B**ottom and **T**op endpoints of the y-axis. The **standard viewing window** is the window determined by the settings $[-10, 10, -10, 10]$.

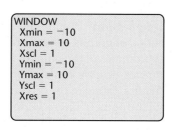

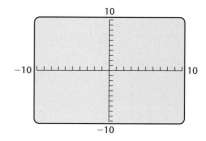

We press (WINDOW) to set the window dimensions and (GRAPH) to display the graph. Xmin is the smallest x-value that will be displayed on the screen, and Xmax is the largest. Similarly, the values of Ymin and Ymax determine the bottom and top endpoints of the vertical axis shown. The scales for the axes are set using Xscl and Yscl. Xres indicates the pixel resolution, which we generally set as Xres = 1. In this text, the window dimensions are written outside the graphs.

Inappropriate window dimensions may result in an error. For example, the message ERR: WINDOW RANGE occurs when Xmin is not less than Xmax.

A standard viewing window is not always the best window to use.

Choosing an appropriate viewing window for a graph can be challenging. There is generally no one "correct" window; the choice can vary according to personal preference and can also be dictated by the portion of the graph that you need to see.

The following screens show the equation $y = -x + 20$ graphed using various viewing windows. Note that the standard viewing window, shown on the left below, does not contain any portion of the line.

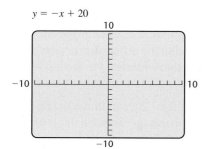

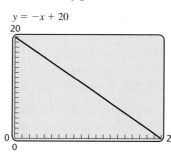

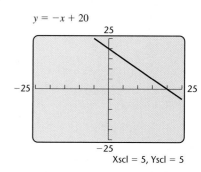

The windows in the middle and on the right above are both appropriate choices for a viewing window.

Pressing (ZOOM) can help in setting window dimensions; for example, choosing ZStandard from the menu will graph the selected equations using the standard viewing window.

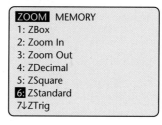

Since the screen on a graphing calculator is rectangular, the 20 units on the x-axis in a standard viewing window are longer than the 20 units on the y-axis in that window. Finding a viewing window with units the same length on both axes is called **squaring** the viewing window.

Windows can be squared by choosing the ZSQUARE option in the ZOOM menu. They can be squared manually by choosing the portions of the axes shown in the same proportion as the length of the sides of the calculator screen.

THE CALC MENU

Point of intersection   A graphing calculator can be used to determine the point of intersection of graphs. We can trace along the graph of one of the functions to find the coordinates of the point of intersection, enlarging the graph using a ZOOM feature if necessary. We can also find the point of intersection directly using an INTERSECT feature, found in the CALC menu.

To find the point of intersection of two graphs, press **2ND** (TRACE) or (CALC) and choose the INTERSECT option. Because more than two equations may be graphed, the questions FIRST CURVE? and SECOND CURVE? are used to identify the graphs with which we are concerned. Position the cursor on each graph, in turn, and press **ENTER**, using the up and down arrow keys if necessary.

The calculator then asks a third question, GUESS?. Since graphs may intersect at more than one point, we must visually identify the point of intersection in which we are interested and indicate the general location of that point. Enter a guess either by moving the cursor near the point of intersection and pressing **ENTER** or by typing a guess for the x-value and pressing **ENTER**. The calculator then returns the coordinates of the point of intersection. At this point, the calculator variables X and Y contain the coordinates of the point of intersection. These values can be used to check an answer, and often they can be converted to fraction notation from the home screen using the FRAC option of the MATH menu.

Zeros of functions   We can determine any zeros of a function using the ZERO option in the CALC menu of a graphing calculator. After graphing the function and choosing the ZERO option, we will be prompted for a Left Bound, a Right Bound, and a Guess. Since there may be more than one zero of a function, the left and right bounds indicate which zero we are currently finding. We examine the graph to find any places where it appears that the graph touches or crosses the x-axis, and then find those x-values one at a time. By using the arrow keys or entering a value on the keyboard, we choose an x-value less than the zero for the left bound, an x-value more than the zero for the right bound, and a value close to the zero for the guess.

Maximum or minimum function values   We can find a maximum or minimum function value for any given interval using the MAXIMUM or MINIMUM feature found in the CALC menu.

To find a maximum or a minimum, enter and graph the function, choosing a viewing window that will show the maximum or minimum point. Next, press (CALC) and choose either the MAXIMUM or MINIMUM option in the menu. The graphing calculator will find the maximum or minimum function value over a specified interval, so the left and right endpoints, or bounds, of the interval must be entered as well as a guess near where the maximum or minimum occurs. The calculator will return the coordinates of the point for which the function value is a maximum or minimum within the interval.

GRAPHING INEQUALITIES

On most graphing calculators, an inequality like $y < \frac{6}{5}x + 3.49$ can be drawn by entering $(6/5)x + 3.49$ as Y1, moving the cursor to the GraphStyle icon just to the left of Y1, pressing **ENTER** until ◣ appears, and then pressing **GRAPH**.

Many calculators have an INEQUALZ program that is accessed using the **APPS** key. Running this program allows us to write inequalities at the **Y=** screen by pressing **ALPHA** and then one of the five keys just below the screen, as shown on the left below.

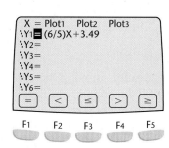

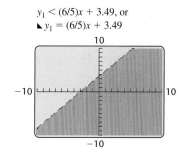

Although the graphs should be identical regardless of the method used, on the newer calculators the boundary line appears dashed when < or > is selected. The graph of the inequality is shown on the right above. Most calculators will shade this area with black vertical lines. Inequalities containing ≤ or ≥ are handled in a similar manner.

When you are finished with the INEQUALZ application, select it again from the **APPS** menu and quit the program.

Regression

Fitting a curve to a set of data is done using the STAT menu. Choose EDIT from the STAT EDIT menu, and enter the values of the independent variable in one list and the corresponding values of the dependent variable in another list. Then press **STAT** again, and choose the CALC menu. To fit a line to the data, choose the LINREG option. After copying LinReg(ax + b) to the home screen, enter the list names containing the data separated by commas, with the independent values first. (The list names are 2nd options associated with the number keys ① through ⑥.)

To copy the equation found to the equation-editor screen, next enter a comma and the function name, found in the VARS Y-VARS menu. To execute the command, press **ENTER**.

The command below indicates that L1 contains the values for the independent variable, L2 contains the values for the dependent variable, and the equation is to be copied to Y1. If no list names are entered, the calculator assumes that the first list is L1 and the second is L2.

```
LinReg (ax+b) L1,
L2, Y1
```

Using CATALOG you can turn DiagnosticOn or DiagnosticOff. If diagnostics are turned on, values of r^2 and r will appear on the screen along with the regression equation. These give an indication of how well the regression line fits the data. When r^2 is close to 1, the line is a good fit. We call r the *coefficient of correlation*.

Answers

The complete step-by-step solutions for the exercises listed below can be found in the *Student's Solutions Manual,* ISBN 0-321-58874-6/978-0-321-58874-6, which can be purchased online or at your bookstore.

CHAPTER 1

Exercise Set 1.1, pp. 10–12

1. Constant   **2.** Variable   **3.** Factors   **4.** Base; exponent   **5.** Evaluating   **6.** Division   **7.** Rational
8. Irrational   **9.** Terminating   **10.** Repeating
11. Let n represent the number; $n - 5$
13. Let x represent the number; $2x$
15. Let x represent the number; $0.29x$, or $\frac{29}{100}x$
17. Let y represent the number; $\frac{1}{2}y - 6$
19. Let s represent the number; $0.1s + 7$, or $\frac{10}{100}s + 7$
21. Let m and n represent the numbers; $mn - 1$
23. $90 \div 4$, or $\frac{90}{4}$ **25.** 36 sq ft, or 36 ft^2
27. 0.25 sq m, or 0.25 m^2 **29.** 7.5 sq ft, or 7.5 ft^2
31. 3.6 sq m, or 3.6 m^2 **33.** 11   **35.** 35   **37.** 8   **39.** 0
41. 5   **43.** 25   **45.** 225   **47.** 0   **49.** 18
51. {a, l, g, e, b, r}   **53.** $\{1, 3, 5, 7, \ldots\}$
55. $\{10, 20, 30, 40, \ldots\}$
57. $\{x \mid x \text{ is an even number between 9 and 99}\}$
59. $\{x \mid x \text{ is a whole number less than 5}\}$
61. $\{x \mid x \text{ is an odd number between 10 and 20}\}$
63. (a) 0, 6;   **(b)** $-3, 0, 6$;   **(c)** $-8.7, -3, 0, \frac{2}{3}, 6$;
(d) $\sqrt{7}$;   **(e)** $-8.7, -3, 0, \frac{2}{3}, \sqrt{7}, 6$ **65. (a)** 0, 8;
(b) $-17, 0, 8$;   **(c)** $-17, -0.01, 0, \frac{5}{4}, 8$;   **(d)** $\sqrt{77}$;
(e) $-17, -0.01, 0, \frac{5}{4}, 8, \sqrt{77}$
67. True   **69.** True   **71.** False   **73.** True   **75.** True
77. False   **79.** ✍   **81.** ✍
83. Let a and b represent the numbers; $\dfrac{a + b}{a - b}$
85. Let r and s represent the numbers; $\dfrac{1}{2}(r^2 - s^2)$, or $\dfrac{r^2 - s^2}{2}$
87. {0}   **89.** $\{5, 10, 15, 20, \ldots\}$ **91.** $\{1, 3, 5, 7, \ldots\}$
93.

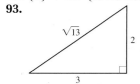

Technology Connection, p. 15

1. ◹   **2.** 0.97

Exercise Set 1.2, pp. 20–22

1. True   **2.** False   **3.** True   **4.** False   **5.** False
6. True   **7.** False   **8.** True   **9.** True   **10.** False
11. 10   **13.** 7   **15.** 46.8   **17.** 0   **19.** $1\frac{7}{8}$ **21.** 4.21
23. -5 is less than or equal to -4; true   **25.** -9 is greater than 1; false   **27.** 0 is greater than or equal to -5; true
29. -8 is less than -3; true   **31.** -4 is greater than or equal to -4; true   **33.** -5 is less than -5; false   **35.** 12
37. -12 **39.** -2.5 **41.** $-\frac{11}{35}$ **43.** -9.06 **45.** $\frac{5}{9}$
47. -6.25 **49.** 0   **51.** 3.8   **53.** -2.37 **55.** 56
57. 0   **59.** -8 **61.** 15   **63.** 4.67   **65.** 0   **67.** 6
69. -6 **71.** 7   **73.** -19 **75.** -3.1 **77.** $-\frac{11}{10}$
79. 5.37   **81.** -24 **83.** 22   **85.** -21 **87.** $-\frac{3}{7}$
89. 0   **91.** $-\frac{1}{2}$ **93.** 4   **95.** -4 **97.** -73 **99.** 0
101. $\frac{1}{8}$ **103.** $-\frac{7}{5}$ **105.** Does not exist   **107.** $\frac{7}{10}$
109. $-\frac{6}{5}$ **111.** $-\frac{1}{36}$ **113.** 1   **115.** -100 **117.** -9
119. 9   **121.** 25   **123.** $-\frac{6}{11}$ **125.** Undefined
127. $\frac{11}{43}$ **129.** 31   **131.** -3 **133.** $xy + 6; 6 + yx$
135. $(ab)(-9); -9(ba)$ **137.** $3(xy)$ **139.** $3y + (4 + 10)$
141. $7x + 7$ **143.** $5m - 5n$ **145.** $-10a - 15b$
147. $9ab - 9ac + 9ad$ **149.** $8(a + b)$ **151.** $3(3p - 1)$
153. $7(x - 3y + 2z)$ **155.** $17(15 - 2b)$ **157.** ✍
159. 16; 16   **160.** 11; 11   **161.** ✍
163. $(8 - 5)^3 + 9 = 36$ **165.** $5 \cdot 2^3 \div (3 - 4)^4 = 40$
167. 15   **169.** -6.2 **171.** ✍

Exercise Set 1.3, pp. 27–28

1. Equivalent expressions   **2.** Equivalent equations
3. Equivalent equations   **4.** Equivalent expressions
5. Equivalent equations   **6.** Equivalent expressions
7. Equivalent expressions   **8.** Equivalent equations
9. Equivalent equations   **10.** Equivalent expressions
11. Equivalent   **13.** Not equivalent   **15.** Not equivalent
17. 16.3   **19.** 9   **21.** 45   **23.** -4 **25.** $\frac{17}{2}$ **27.** $10x$
29. $10t^2$ **31.** $15a$ **33.** $-7n$ **35.** $10x$ **37.** $7x - 2x^2$
39. $21p - 4$ **41.** $-5t^2 + 2t + 4t^3$ **43.** $17x - 21$
45. $5a - 5$ **47.** $-5m + 2$ **49.** $5d - 12$
51. $-2x + 22$ **53.** $p - 16$ **55.** $4a - 12$
57. $-310x - 30$ **59.** $14y + 42$ **61.** 7   **63.** -12
65. 6   **67.** 3   **69.** 3   **71.** -3 **73.** 5   **75.** $\frac{49}{9}$
77. $\frac{4}{5}$ **79.** $\frac{19}{5}$ **81.** $-\frac{4}{11}$ **83.** $\frac{23}{8}$ **85.** $\varnothing$; contradiction
87. {0}; conditional   **89.** $\varnothing$; contradiction

91. $\mathbb{R}$; identity   **93.** ✍   **95.** Let n represent the number; $2n - 9$ **96.** Let n represent the number; $5 + \frac{1}{2}n$ **97.** ✍   **99.** 0.2140224409   **101.** 4   **103.** $\frac{19}{46}$ **105.** ✍

Connecting the Concepts, p. 29

1. Expression; $2x + 7$ **2.** Equation; -3 **3.** Expression; $t + 1$ **4.** Equation; 1   **5.** Equation; $-\frac{3}{8}$ **6.** Expression; $8x + 29$ **7.** Expression; $8x + 2$ **8.** Equation; $-\frac{4}{7}$ **9.** Equation; -11 **10.** Equation; 2   **11.** Expression; $-2p + 10$ **12.** Expression; $18a - 27$ **13.** Equation; 0   **14.** Expression; $-7y - 3$ **15.** Equation; 15   **16.** Expression; $2x - 14$ **17.** Expression; $4t$ **18.** Equation; $\frac{1}{3}$ **19.** Equation; $\frac{19}{3}$ **20.** Expression; $y - 15$

Translating for Success, p. 37

1. F   **2.** D   **3.** I   **4.** C   **5.** E   **6.** J   **7.** O   **8.** M   **9.** B   **10.** L

Exercise Set 1.4, pp. 38–40

1. Let x and $x + 9$ represent the numbers; $x + (x + 9) = 91$ **3.** Let t represent the time, in hours, that it will take Stella to make the trip; $6 = (3.5 - 1.9)t$ **5.** Let x, $x + 1$, and $x + 2$ represent the angle measures, in degrees; $x + (x + 1) + (x + 2) = 180$ **7.** Let t represent the time, in minutes, that it will take Dominik to reach the top of the escalator; $230 = (100 + 90)t$ **9.** Let w represent the wholesale price; $w + 0.5w + 1.50 = 22.50$ **11.** Let t represent the number of minutes spent climbing; $8000 + 3500t = 29{,}000$ **13.** Let x represent the measure of the second angle, in degrees; $4x + x + (2x + 5) = 180$ **15.** Let n represent the first odd number; $n + 2(n + 2) + 3(n + 4) = 70$ **17.** Let s represent the length, in centimeters, of a side of the smaller triangle; $3s + 3 \cdot 2s = 90$ **19.** Let c represent Cody's calls on his next shift; $\frac{5 + 2 + 1 + 3 + c}{5} = 3$ **21.** \$97   **23.** \$396 per month   **25.** 16 seniors   **27.** Length: 45 cm; width: 15 cm   **29.** Length: 52 m; width: 13 m   **31.** 3.75 hr   **33.** $100°, 25°, 55°$ **35.** \$150   **37.** \$14.00   **39.** ✍   **41.** 1   **42.** $\frac{7}{5}$ **43.** -6 **44.** $\frac{1}{2}$ **45.** ✍   **47.** 10 points   **49.** \$110,000

Exercise Set 1.5, pp. 46–49

1. Equation   **2.** Area   **3.** Circumference   **4.** $P = 2l + 2w$ **5.** $A = bh$ **6.** Length   **7.** Subscripts   **8.** Factor   **9.** $A = \dfrac{E}{w}$ **11.** $r = \dfrac{d}{t}$ **13.** $h = \dfrac{V}{lw}$ **15.** $k = Ld^2$ **17.** $n = \dfrac{G - w}{150}$ **19.** $l = p - 2w - 2h$ **21.** $y = \dfrac{4 - 2x}{3}$ **23.** $y = \dfrac{C - Ax}{B}$ **25.** $F = \dfrac{9}{5}C + 32$ **27.** $r^3 = \dfrac{3V}{4\pi}$ **29.** $n = \dfrac{t}{p + m}$ **31.** $v = \dfrac{x}{u + w}$

33. $n = \dfrac{q_1 + q_2 + q_3}{A}$ **35.** $t = \dfrac{d_2 - d_1}{v}$ **37.** $d_1 = d_2 - vt$ **39.** $b = \dfrac{c}{d - a}$ **41.** $w = \dfrac{v}{uv + 1}$ **43.** $m = \dfrac{n}{t^2 + k}$ **45.** 8%   **47.** 16 cm   **49.** About 235 lb   **51.** About 1504.6 g   **53.** 9 ft   **55.** 1 yr   **57.** 816 mL   **59.** 1205   **61.** 5 ft 7 in.   **63.** About 88 kg   **65.** 512 visits per day   **67.** 34 appointments   **69.** About 8.5 cm   **71.** ✍   **73.** $7c - 42$ **74.** $-18t + 39$ **75.** ✍   **77.** About 10.9 g   **79.** 10 times   **81.** $a = \dfrac{2s - 2v_it}{t^2}$ **83.** $w = \dfrac{h + p - b(a + p + f)}{b - 1}$ **85.** $b = \dfrac{ac}{1 + c}$ **87.** $t = \dfrac{1}{s}$ **89.** ✍

Technology Connection, p. 54

1. Answers may vary; (2)(xʸ)(5)(−)(=), (2)(^)((−))(5)(ENTER), (2)(xʸ)(-x)(5)(ENTER)   **2.** Compute $1 \div (2 \times 2 \times 2 \times 2 \times 2)$, or $1 \div 2 \div 2 \div 2 \div 2 \div 2$.

Exercise Set 1.6, pp. 57–59

1. The power rule   **2.** Raising a quotient to a power   **3.** Raising a product to a power   **4.** The quotient rule   **5.** The product rule   **6.** The power rule   **7.** Raising a quotient to a power   **8.** Raising a product to a power   **9.** The quotient rule   **10.** The product rule   **11.** 6^{11} **13.** m^8 **15.** $20x^7$ **17.** $24a^8$ **19.** m^8n^3 **21.** t^5 **23.** $5a^5$ **25.** m^5n^4 **27.** $4x^6y^4$ **29.** $-4x^8y^6z^6$ **31.** -1 **33.** 1   **35.** $\dfrac{1}{t^9}$ **37.** $\dfrac{1}{6^2} = \dfrac{1}{36}$ **39.** $\dfrac{1}{(-3)^2} = \dfrac{1}{9}$ **41.** $-\dfrac{1}{3^2} = -\dfrac{1}{9}$ **43.** $-\dfrac{1}{1^{10}} = -1$ **45.** $10^3 = 1000$ **47.** $\dfrac{6}{x}$ **49.** $\dfrac{3a^8}{b^6}$ **51.** $\dfrac{2}{x^5z^3}$ **53.** $3y^2z^4$ **55.** $\dfrac{ac}{b}$ **57.** $\dfrac{pv^4}{2q^2r^3u^5}$ **59.** x^{-3} **61.** $(-10)^{-3}$ **63.** $\dfrac{1}{8^{-10}}$ **65.** $\dfrac{4}{x^{-2}}$ **67.** $(5y)^{-3}$ **69.** $\dfrac{y^{-4}}{3}$ **71.** 6^{-8}, or $\dfrac{1}{6^8}$ **73.** a^{-7}, or $\dfrac{1}{a^7}$ **75.** 1   **77.** $-8m^4n^5$ **79.** $35x^{-2}y^3$, or $\dfrac{35y^3}{x^2}$ **81.** $10a^{-6}b^{-2}$, or $\dfrac{10}{a^6b^2}$ **83.** 10^{-9}, or $\dfrac{1}{10^9}$ **85.** 2^{-2}, or $\dfrac{1}{2^2}$, or $\dfrac{1}{4}$ **87.** y^9 **89.** $-3ab^2$ **91.** $\dfrac{3}{2}m^{-5}n^7$, or $\dfrac{3n^7}{2m^5}$ **93.** $\dfrac{1}{4}x^3y^{-2}z^{11}$, or $\dfrac{x^3z^{11}}{4y^2}$ **95.** x^{12} **97.** 9^{-12}, or $\dfrac{1}{9^{12}}$ **99.** t^{40} **101.** $25x^2y^2$ **103.** $(-2)^{-3}a^6b^{-3}$, or $-\dfrac{a^6}{8b^3}$ **105.** $\dfrac{m^6n^{-3}}{64}$, or $\dfrac{m^6}{64n^3}$

107. $32a^{-4}$, or $\dfrac{32}{a^4}$ **109.** 1   **111.** $\dfrac{5a^4b}{2}$ **113.** $\dfrac{8x^9y^3}{27}$

115. 1   **117.** $\dfrac{4}{25}x^{-4}y^{22}$, or $\dfrac{4y^{22}}{25x^4}$ **119.** ✍   **121.** 35.1

122. 44   **123.** ✍   **125.** $4a^{-x-4}$ **127.** $2^{-2a-2b+ab}$

129. 3^{a^2+2a} **131.** $2x^{a+2}y^{b-2}$ **133.** $\dfrac{2}{27}$ **135.** $\dfrac{a^{-14ac}}{b^{27ac}}$

Exercise Set 1.7, pp. 64–66

1. Positive power of 10   **2.** Negative power of 10
3. Negative power of 10   **4.** Positive power of 10
5. Positive power of 10   **6.** Negative power of 10
7. 6.4×10^{10} **9.** 1.091×10^9 **11.** 1.3×10^{-6}
13. 9×10^{-5} **15.** 8.03×10^{11} **17.** 9.04×10^{-7}
19. 4.317×10^{11} **21.** 400,000   **23.** 0.00012
25. 0.00000000376   **27.** 8,056,000,000,000
29. 0.00007001   **31.** 9,060,000,000   **33.** 8.8×10^7
35. 3.3×10^{-5} **37.** 1.4×10^{11} **39.** 4.6×10^{-11}
41. 6.0   **43.** 2.5×10^{11} **45.** 2.0×10^{-7}
47. 4.0×10^{-16} **49.** 3.00×10^{-22} **51.** 2.00×10^{26}
53. 7.8×10^{-9} **55.** 1.2×10^{24} **57.** 3.1×10^2 kg
59. Approximately 4.5×10^{-16} in^3 **61.** 4.50×10^{-3} kg,
or 4.50 g   **63.** 1.00×10^5 light years   **65.** 3.08×10^{26} Å
67. 1×10^{22} cu Å, or 1×10^{-8} m^3 **69.** 7.9×10^7 bacteria
71. 4.49×10^4 km/h   **73.** ✍   **75.** 12   **76.** -11
77. ✍   **79.** Approximately 5.53 g/cm^3 **81.** $8 \cdot 10^{-90}$ is
larger by 7.1×10^{-90}.   **83.** 8   **85.** 8×10^{18} grains

Review Exercises: Chapter 1, pp. 70–71

1. (e)   **2.** (g)   **3.** (j)   **4.** (a)   **5.** (i)   **6.** (b)   **7.** (f)
8. (c)   **9.** (d)   **10.** (h)   **11.** Let x and y represent the
numbers; $\dfrac{x}{y} - 8$ **12.** 22   **13.** $\{1, 3, 5, 7, 9\}$; $\{x\mid x$ is an odd
natural number less than $10\}$ **14.** 1750 sq cm   **15.** 19
16. 0   **17.** 6.08   **18.** -11 **19.** $-\dfrac{1}{15}$ **20.** $\dfrac{1}{20}$
21. -25 **22.** 4.4   **23.** -3.8 **24.** 96   **25.** $-\dfrac{5}{12}$
26. -4.8 **27.** 4   **28.** -9.1 **29.** $-\dfrac{21}{4}$ **30.** 6.28
31. $x + 12$ **32.** $y \cdot 7$ **33.** $x \cdot 5 + y$, or $y + 5x$
34. $4 + (a + b)$ **35.** $(xy)z$ **36.** $4(3m + n - 2)$
37. $4x^3 - 6x^2 + 5$ **38.** $47x - 60$ **39.** $\dfrac{1}{2}$ **40.** $\dfrac{21}{4}$
41. $-\dfrac{4}{11}$ **42.** $\mathbb{R}$; identity   **43.** $\varnothing$; contradiction
44. Let x represent the number; $2x + 15 = 21$ **45.** 48
46. $90°, 30°, 60°$ **47.** $c = \dfrac{xt}{b}$ **48.** $x = \dfrac{c}{m - r}$
49. 14 cm   **50.** $-28m^4n^{10}$ **51.** $4xy^6$ **52.** $1, 64, -64$
53. 3^2, or 9   **54.** $8t^{12}$ **55.** $-\dfrac{a^9}{125b^6}$ **56.** $\dfrac{z^8}{x^4y^6}$
57. $\dfrac{n^{12}}{81m^{28}}$ **58.** $\dfrac{3}{7}$ **59.** 0   **60.** 3.07×10^{-4}
61. 3.086×10^{13} **62.** 3.7×10^7 **63.** 2.0×10^{-6}
64. 1.4×10^4 mm^3, or 1.4×10^{-5} m^3 **65.** ✍ To write an
equation that has no solution, begin with a simple equation
that is false for any value of x, such as $x = x + 1$. Then add
or multiply by the same quantities on both sides of the
equation to construct a more complicated equation with no
solution.   **66.** ✍ (a) $-(-x)$ is positive when x is positive;
the opposite of the opposite of a number is the number

itself; (b) $-x^2$ is never positive; x^2 is always nonnegative, so
the opposite of x^2 is always nonpositive;   (c) $-x^3$ is positive
when x is negative; x^3 is negative when x is negative, and
the opposite of a negative number is positive;   (d) $(-x)^2$ is
positive when $x \neq 0$; the square of any nonzero number is
positive;   (e) x^{-2} is positive when $x \neq 0$; $x^{-2} = \dfrac{1}{x^2}$, and x^2 is
positive when x is nonzero.   **67.** 0.0000003%   **68.** $\dfrac{25}{24}$
69. The 17-in. pizza is a better deal. It costs about 5¢ per
square inch; the 13-in. pizza costs about 6¢ per square inch.

70. 729 cm^3 **71.** $z = y - \dfrac{x}{m}$, or $\dfrac{my - x}{m}$
72. $3^{-2a+2b-8ab}$ **73.** $88.\overline{3}$ **74.** -39 **75.** $-40x$
76. $a \cdot 2 + cb + cd + ad = ad + a \cdot 2 + cb + cd = a(d + 2) + c(b + d)$ **77.** $\sqrt{5}/4$; answers may vary

Test: Chapter 1, p. 72

1. [1.1] Let m and n represent the numbers; $mn - 4$
2. [1.1], [1.2] -47 **3.** [1.1] 181.35 sq m   **4.** [1.2] -31
5. [1.2] -3.7 **6.** [1.2] $-\dfrac{1}{6}$ **7.** [1.2] -14.2
8. [1.2] -43.2 **9.** [1.2] -33.92 **10.** [1.2] $\dfrac{1}{12}$
11. [1.2] $\dfrac{5}{49}$ **12.** [1.2] 6   **13.** [1.2] $-\dfrac{4}{3}$ **14.** [1.2] $-\dfrac{5}{2}$
15. [1.2] $x + 3$ **16.** [1.3] $-y$ **17.** [1.3] $a^2b - 4ab^2 + 2$
18. [1.3] -2 **19.** [1.3] $\mathbb{R}$; identity   **20.** [1.5] $p = \dfrac{t}{2 - s}$
21. [1.4] 94   **22.** [1.4] 17, 19, 21   **23.** [1.3] $8x - 11$
24. [1.3] $24b - 9$ **25.** [1.6] $-\dfrac{42}{x^{10}y^6}$
26. [1.6] $-\dfrac{1}{6^2}$, or $-\dfrac{1}{36}$ **27.** [1.6] $-\dfrac{125y^9}{x^3}$ **28.** [1.6] $\dfrac{4y^8}{x^6}$
29. [1.6] 1   **30.** [1.7] 2.01×10^{-7} **31.** [1.7] 3.8×10^2
32. [1.7] 2.0×10^9 neutrinos   **33.** [1.6] $8^{c}x^{9ac}y^{3bc+3c}$
34. [1.6] $-9a^3$ **35.** [1.6] $\dfrac{4}{7y^2}$ **36.** [1.3] -2

CHAPTER 2

Technology Connection, p. 77

1. $y = -4x + 3$ $(-1.5, 9), (1, -1)$

Technology Connection, p. 79

1. $y = 5x - 3$

2. $y = x^2 - 4x + 3$

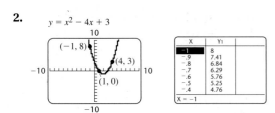

3. $y = (x + 4)^2$

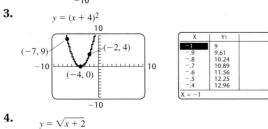

4. $y = \sqrt{x + 2}$

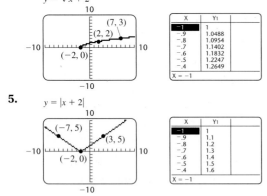

5. $y = |x + 2|$

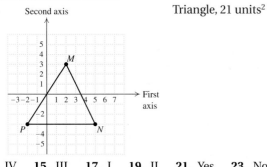

Exercise Set 2.1, pp. 80–82

1. Axes   **2.** Ordered   **3.** Third   **4.** Negative
5. Solutions   **6.** Linear
7. $(5, 3), (-4, 3), (0, 2), (-2, -3), (4, -2)$, and $(-5, 0)$
9.

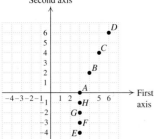

11.   Triangle, 21 units^2

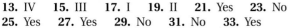

13. IV   **15.** III   **17.** I   **19.** II   **21.** Yes   **23.** No
25. Yes   **27.** Yes   **29.** No   **31.** No   **33.** Yes

35.

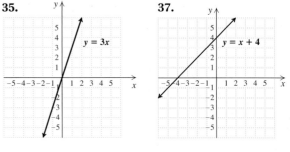

$y = 3x$

37.

$y = x + 4$

39.

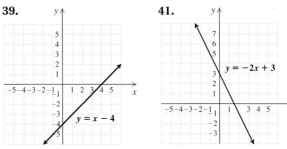

$y = x - 4$

41.

$y = -2x + 3$

43.

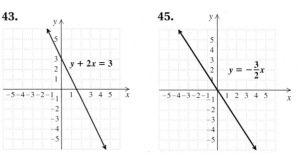

$y + 2x = 3$

45.

$y = -\frac{3}{2}x$

47.

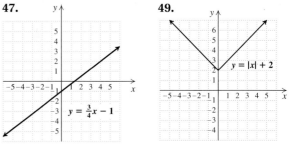

$y = \frac{3}{4}x - 1$

49.

$y = |x| + 2$

51.

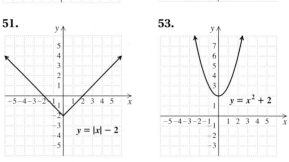

$y = |x| - 2$

53.

$y = x^2 + 2$

55.

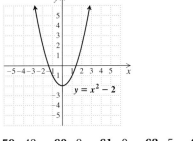

$y = x^2 - 2$

57.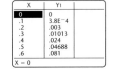

59. 43   **60.** 9   **61.** 0   **62.** 5   **63.** $-\frac{3}{4}$ **64.** 0
65. -4 **66.** 5   **67.** $\frac{1}{2}$ **68.** $-\frac{3}{5}$ **69.** 🖎   **71.** 🖎
73. (a) III; (b) II; (c) I; (d) IV
75. (a) III; (b) II; (c) IV; (d) I
77. $(-1, -2)$, $(-19, -2)$, and $(13, 10)$

79.

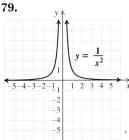

$y = \dfrac{1}{x^2}$

81.

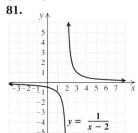

$y = \dfrac{1}{x-2}$

83.

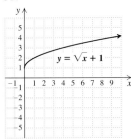

$y = \sqrt{x} + 1$

85.

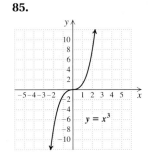

$y = x^3$

87.

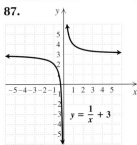

$y = \dfrac{1}{x} + 3$

89. (a)

$y = 0.375x^3$

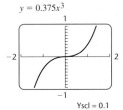

Yscl = 0.1

(b)

$y = -3.5x^2 + 6x - 8$

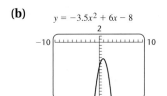

X	Y₁
0	-8
.1	-7.435
.2	-6.94
.3	-6.515
.4	-6.16
.5	-5.875
.6	-5.66
X = 0	

(c)

$y = (x - 3.4)^3 + 5.6$

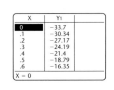

X	Y₁
0	-33.7
.1	-30.34
.2	-27.17
.3	-24.19
.4	-21.4
.5	-18.79
.6	-16.35
X = 0	

Exercise Set 2.2, pp. 91–96

1. Correspondence   **2.** Exactly   **3.** Domain
4. Range   **5.** Horizontal   **6.** Vertical   **7.** "f of 3," "f at
3," or "the value of f at 3"   **8.** Vertical   **9.** Yes   **11.** Yes
13. No   **15.** Yes   **17.** Function   **19.** Function
21. (a) $\{-3, -2, 0, 4\}$; (b) $\{-10, 3, 5, 9\}$; (c) yes
23. (a) $\{1, 2, 3, 4, 5\}$; (b) $\{1\}$; (c) yes
25. (a) $\{-2, 3, 4\}$; (b) $\{-8, -2, 4, 5\}$; (c) no
27. (a) -2; (b) $\{x \mid -2 \le x \le 5\}$; (c) 4; (d) $\{y \mid -3 \le y \le 4\}$
29. (a) -2; (b) $\{x \mid -4 \le x \le 2\}$; (c) -2;
(d) $\{y \mid -3 \le y \le 3\}$ **31.** (a) 3; (b) $\{x \mid -4 \le x \le 3\}$;
(c) -3; (d) $\{y \mid -2 \le y \le 5\}$ **33.** (a) 3;
(b) $\{-4, -3, -2, -1, 0, 1, 2\}$; (c) $-2, 0$; (d) $\{1, 2, 3, 4\}$
35. (a) 4; (b) $\{x \mid -3 \le x \le 4\}$; (c) $-1, 3$;
(d) $\{y \mid -4 \le y \le 5\}$ **37.** (a) 2; (b) $\{x \mid -4 \le x \le 4\}$;
(c) $\{x \mid 0 < x \le 2\}$; (d) $\{1, 2, 3, 4\}$
39. Domain: $\mathbb{R}$; range: $\mathbb{R}$ **41.** Domain: $\mathbb{R}$; range: $\{4\}$
43. Domain: $\mathbb{R}$; range: $\{y \mid y \ge 1\}$
45. Domain: $\{x \mid x \text{ is a real number } and \ x \ne -2\}$;
range: $\{y \mid y \text{ is a real number } and \ y \ne -4\}$
47. Domain: $\{x \mid x \ge 0\}$; range: $\{y \mid y \ge 0\}$ **49.** Yes
51. Yes   **53.** No   **55.** (a) 5; (b) -3; (c) -9; (d) 21;
(e) $2a + 9$; (f) $2a + 7$ **57.** (a) 0; (b) 1; (c) 57;
(d) $5t^2 + 4t$; (e) $20a^2 + 8a$; (f) 48   **59.** (a) $\frac{3}{5}$; (b) $\frac{1}{3}$; (c) $\frac{4}{7}$;
(d) 0; (e) $\dfrac{x - 1}{2x - 1}$; (f) $\dfrac{a + h - 3}{2a + 2h - 5}$
61. $\{x \mid x \text{ is a real number } and \ x \ne 3\}$ **63.** $\mathbb{R}$ **65.** $\mathbb{R}$
67. $\left\{x \mid x \text{ is a real number } and \ x \ne \frac{8}{5}\right\}$
69. $\{x \mid x \text{ is a real number } and \ x \ne -1\}$ **71.** $\mathbb{R}$
73. $\{x \mid x \text{ is a real number } and \ x \ne 0\}$
75. $4\sqrt{3}$ cm² ≈ 6.93 cm²   **77.** 36π in² ≈ 113.10 in²
79. 164.98 cm   **81.** 23°F   **83.** 75 heart attacks per
10,000 men   **85.** 500 movies
87. 19 watts; 30 watts

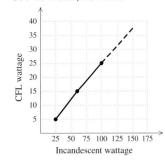

89. 3.5 drinks; 6 drinks

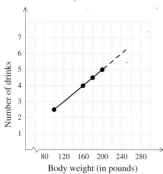

91. $257,000; $306,000   **93.**

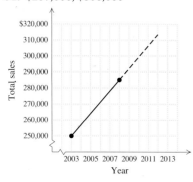

95. $-\frac{1}{3}$ **96.** $-\frac{2}{3}$ **97.** 0   **98.** -1 **99.** $y = 2x - 8$
100. $y = -x + 2$ **101.** $y = -\frac{2}{3}x + 2$ **102.** $y = \frac{5}{4}x - 2$
103.   **105.** 26; 99   **107.** Worm
109. About 2 min 50 sec   **111.** 1 every 3 min
113.

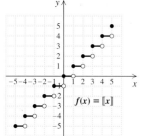

Exercise Set 2.3, pp. 104–108

1. (f)   **2.** (c)   **3.** (e)   **4.** (d)   **5.** (a)   **6.** (b)
7.

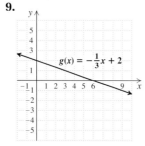

9.

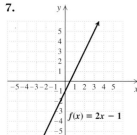

11.

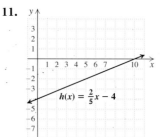

13. $(0, 3)$ **15.** $(0, -1)$ **17.** $(0, -4.5)$ **19.** $\left(0, -\frac{1}{4}\right)$
21. $(0, 138)$ **23.** 4   **25.** -2 **27.** $\frac{1}{3}$ **29.** $-\frac{5}{2}$ **31.** 0
33. Slope: $\frac{5}{2}$;   **35.** Slope: $-\frac{5}{2}$;
     y-intercept: $(0, -3)$ y-intercept: $(0, 2)$

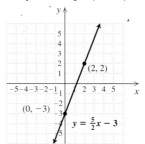   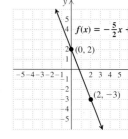

37. Slope: 2;   **39.** Slope: -4;
y-intercept: $(0, 1)$ y-intercept: $(0, 3)$

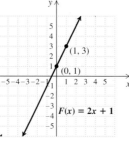   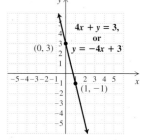

41. Slope: $-\frac{1}{6}$;   **43.** Slope: -0.25;
     y-intercept: $(0, 1)$ y-intercept: $(0, 0)$

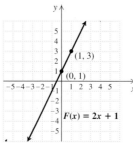   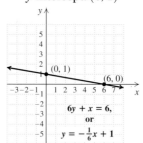

45. Slope: $\frac{4}{5}$;   **47.** Slope: $-\frac{2}{3}$;
     y-intercept: $(0, -2)$ y-intercept: $(0, 2)$

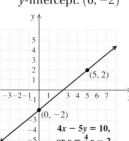   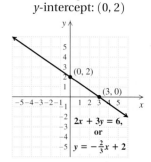

49. Slope: -3; y-intercept: $(0, 5)$

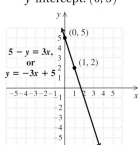

51. Slope: 0; y-intercept: $(0, 4.5)$

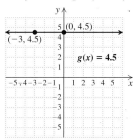

53. $f(x) = 2x + 5$ **55.** $f(x) = -\frac{2}{3}x - 2$
57. $f(x) = -7x + \frac{1}{3}$
59. The distance from home is increasing at a rate of 0.25 km per minute.
61. The distance from the finish line is decreasing at a rate of $6\frac{2}{3}$ m per second.
63. The number of bookcases stained is increasing at a rate of $\frac{2}{3}$ bookcase per quart of stain used.
65. The average SAT math score is increasing at a rate of 1 point per thousand dollars of family income.
67. (a) II; **(b)** IV; **(c)** I; **(d)** III   **69.** 12 km/h   **71.** $\frac{5}{96}$ of the house per hour   **73.** 175,000 hits/yr   **75.** 0.75 signifies that the cost per mile of renting the truck is \$0.75; 30 signifies that the minimum cost is \$30.   **77.** $\frac{1}{2}$ signifies that Lauren's hair grows $\frac{1}{2}$ in. per month; 5 signifies that her hair was 5 in. long when cut.   **79.** $\frac{1}{8}$ signifies that the life expectancy of American women increases $\frac{1}{8}$ yr per year, for years after 1970; 75.5 signifies that the life expectancy in 1970 was 75.5 yr.   **81.** 0.89 signifies that the average price of a ticket increases \$0.89 per year, for years after 2000; 16.63 signifies that the cost of a ticket was \$16.63 in 2000.
83. 849 signifies that the number of acres of organic cotton increases 849 acres per year, for years after 2006; 5960 signifies that 5960 acres were planted with organic cotton in 2006.   **85. (a)** -5000 signifies that the depreciation is \$5000 per year; 90,000 signifies that the original value of the truck was \$90,000; **(b)** 18 yr; **(c)** $\{t | 0 \le t \le 18\}$
87. (a) -200 signifies that the depreciation is \$200 per year; 1800 signifies that the original value of the bike was \$1800; **(b)** after 6 years of use; **(c)** $\{n | 0 \le n \le 9\}$ **89.** ✍
91. 0   **92.** Undefined   **93.** $-\frac{9}{2}$ **94.** $\frac{3}{4}$ **95.** -7
96. $\frac{7}{2}$ **97.** ✍   **99. (a)** III; **(b)** IV; **(c)** I; **(d)** II
101. Sienna to Castellina in Chianti   **103.** Castellina in Chianti   **105.** Slope: $-\dfrac{r}{r+p}$; y-intercept: $\left(0, \dfrac{s}{r+p}\right)$
107. Since (x_1, y_1) and (x_2, y_2) are two points on the graph of $y = mx + b$, then $y_1 = mx_1 + b$ and $y_2 = mx_2 + b$. Using the definition of slope, we have

$$\begin{aligned} \text{Slope} &= \frac{y_2 - y_1}{x_2 - x_1} \\ &= \frac{(mx_2 + b) - (mx_1 + b)}{x_2 - x_1} \\ &= \frac{m(x_2 - x_1)}{x_2 - x_1} \\ &= m. \end{aligned}$$

109. False   **111.** False   **113. (a)** $-\dfrac{5c}{4b}$; **(b)** undefined; **(c)** $\dfrac{a+d}{f}$ **115.** $y_1 = 1.4x + 2$, $y_2 = 0.6x + 2$, $y_3 = 1.4x + 5$, $y_4 = 0.6x + 5$ **117.** ✍

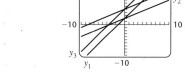

Exercise Set 2.4, pp. 116–118

1. Horizontal   **2.** y-axis   **3.** Undefined   **4.** Vertical
5. 0; x **6.** 0; y **7.** Intersection   **8.** Standard
9. Linear   **10.** Slope   **11.** 0 **13.** Undefined
15. 0 **17.** Undefined   **19.** Undefined   **21.** 0 **23.** 0
25. Undefined   **27.** $-\frac{2}{3}$
29.

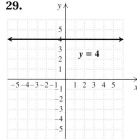

31.

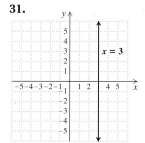

33.

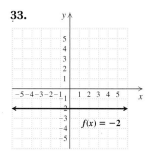

35.

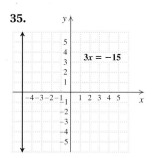

37.

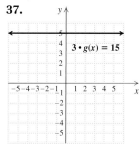

39.

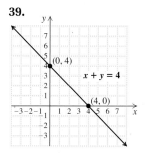

41.

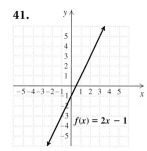

43.

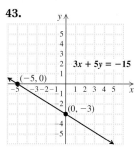

45.

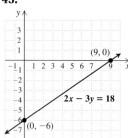

47.

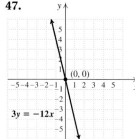

3.

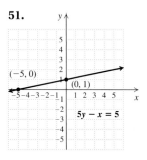

$y_1 = \frac{31}{40}x + 2; \; y_2 = -\frac{40}{30}x - 1$ No: $-\frac{40}{30} \neq -\frac{1}{\frac{31}{40}}$

Although the lines appear to be perpendicular, they are not, because the product of their slopes is not -1:

$$\frac{31}{40}\left(-\frac{40}{30}\right) = -\frac{1240}{1200} \neq -1.$$

Visualizing for Success, p. 123

1. C   **2.** G   **3.** F   **4.** B   **5.** D   **6.** A   **7.** I
8. H   **9.** J   **10.** E

49.

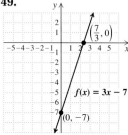

51.

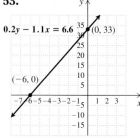

Exercise Set 2.5, pp. 124–127

1. False   **2.** True   **3.** False   **4.** True   **5.** True
6. False   **7.** False   **8.** True   **9.** True   **10.** True
11.   **13.**

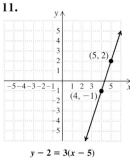

$y - 2 = 3(x - 5)$ $y - 2 = -4(x - 1)$

53.

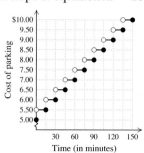

55. 1   **57.** -2 **59.** 4   **61.** $-\frac{3}{2}$ **63.** 2
65. 4 months   **67.** $1350 over $250   **69.** 2 hr 15 min
71. 65 lb   **73.** Linear; $\frac{5}{3}$ **75.** Linear; line is vertical
77. Not linear   **79.** Linear; $\frac{14}{3}$ **81.** Not linear
83. Linear; 3   **85.** Not linear   **87.** ✍   **89.** -1
90. -1 **91.** $-3x - 3$ **92.** $-10x - 70$ **93.** $\frac{2}{3}x - \frac{2}{3}$
94. $-\frac{3}{2}x - \frac{12}{5}$ **95.** ✍   **97.** $4x - 5y = 20$
99. Linear   **101.** Linear   **103.** The slope of equation B is $\frac{1}{2}$ the slope of equation A.   **105.** $a = 7, b = -3$

15.   **17.**

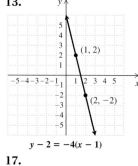

$y - (-4) = \frac{1}{2}(x - (-2))$, or

$y + 4 = \frac{1}{2}(x + 2)$

107.

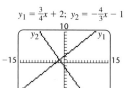

109. $0.\overline{6}$, or $\frac{2}{3}$
111. 2.6, or $\frac{13}{5}$
113. 149 shirts

19. $\frac{1}{4}$; $(5, 3)$ **21.** -7; $(2, -1)$ **23.** $-\frac{10}{3}$; $(-4, 6)$
25. 5; $(0, 0)$
27. $f(x) = 2x - 6$ **29.** $f(x) = -\frac{3}{5}x + \frac{28}{5}$

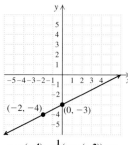

$f(x) = 2x - 6$ $f(x) = -\frac{3}{5}x + \frac{28}{5}$

Technology Connection, p. 122

1.

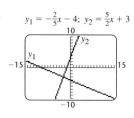

$y_1 = \frac{3}{4}x + 2; \; y_2 = -\frac{4}{3}x - 1$

2.

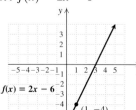

$y_1 = -\frac{2}{5}x - 4; \; y_2 = \frac{5}{2}x + 3$

31. $f(x) = -0.6x - 5.8$

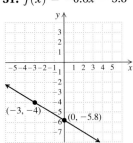

$f(x) = -0.6x - 5.8$

33. $f(x) = \frac{2}{7}x - 6$

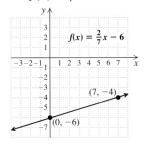

35. $f(x) = \frac{3}{5}x + \frac{42}{5}$

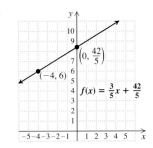

37. $f(x) = 4x - 5$ **39.** $f(x) = 4.5x - 9.4$
41. $f(x) = -2x - 1$ **43.** $f(x) = \frac{5}{3}x$
45. (a) $a(t) = 0.9t + 12.7$; **(b)** 24.4 million cars;
(c) about 2014   **47. (a)** $E(t) = 0.14t + 78.44$; **(b)** 81.52 yr
49. (a) $A(t) = 22.525t + 236.95$; **(b)** \$462.2 million
51. (a) $N(t) = 1.14t + 52.7$; **(b)** 66.38 million tons
53. (a) $N(t) = 9.4t + 16$; **(b)** 110 million Americans;
(c) 2015   **55. (a)** $A(t) = \frac{41}{110}t + \frac{1615}{22}$; **(b)** about
80.9 million acres   **57.** Yes   **59.** Yes   **61.** No
63. $y = \frac{1}{2}x + 4$ **65.** $y = -x - 1$ **67.** $y = 4x - 5$
69. $y = -\frac{2}{3}x - \frac{13}{3}$ **71.** $y = \frac{1}{3}x + 4$ **73.** $x = 5$
75. Yes   **77.** No   **79.** $y = -\frac{3}{2}x + \frac{11}{2}$ **81.** $y = x + 6$
83. $y = -\frac{1}{3}x - \frac{8}{3}$ **85.** $y = -\frac{5}{3}x - \frac{41}{3}$ **87.** $y = -\frac{1}{2}x + 6$
89. $x = -3$ **91.** ✍   **93.** $2x^2 + 2x - 5$
94. $-2t - 4$ **95.** $t + 2$ **96.** $-4x^2 + 7x - 4$
97. $\{x \,|\, x$ is a real number $and\, x \neq 3\}$
98. $\mathbb{R}$ **99.** $\mathbb{R}$ **100.** $\{x \,|\, x$ is a real number $and\, x \neq 0\}$
101. ✍   **103.** 21.1°C   **105.** \$60   **107.** \$8.33 per pound
109. $\{p \,|\, p > 5.5\}$ **111.** $-\frac{40}{9}$
113. (a) $f(x) = 0.256x - 1.746$;
(b) approximately 17 W; ✍   **115.** ◺

Connecting the Concepts, pp. 128–129

1. Standard form   **2.** Slope–intercept form
3. None of these   **4.** Point–slope form
5. Standard form   **6.** Slope–intercept form
7. $2x - 5y = -5$ **8.** $2x + y = 13$ **9.** $y = \frac{3}{5}x - 2$
10. $y = \frac{1}{2}x - \frac{7}{2}$
11.

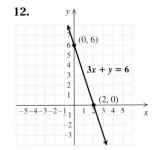

$y = 2x - 1$

12.

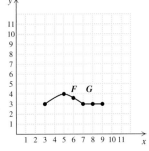

$3x + y = 6$

13.

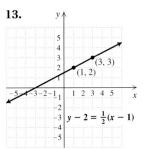

$y - 2 = \frac{1}{2}(x - 1)$

14.

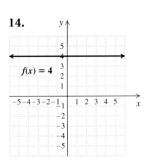

$f(x) = 4$

15.

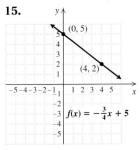

$f(x) = -\frac{3}{4}x + 5$

16. Slope: $\frac{1}{3}$; y-intercept: $\left(0, -\frac{1}{3}\right)$ **17.** $f(x) = -3x + 7$
18. $y - 7 = 5(x - (-3))$ **19.** $y = \frac{2}{3}x - 8$
20. $y = \frac{2}{3}x - \frac{11}{3}$

Exercise Set 2.6, pp. 134–137

1. Sum   **2.** Subtract   **3.** Evaluate   **4.** Domains
5. Excluding   **6.** Sum   **7.** 1   **9.** 5   **11.** -7 **13.** 1
15. -5 **17.** $x^2 - 2x - 2$ **19.** $x^2 - x + 3$ **21.** 5
23. 56   **25.** $\frac{x^2 - 2}{5 - x}, x \neq 5$ **27.** $\frac{7}{2}$ **29.** -2

31. $1.2 + 2.9 = 4.1$ million   **33.** 4%   **35.** About 95 million; the number of tons of municipal solid waste that was composted or recycled in 2005   **37.** About 215 million; the number of tons of municipal solid waste in 1996
39. About 230 million; the number of tons of municipal solid waste that was not composted in 2004   **41.** $\mathbb{R}$
43. $\{x \,|\, x$ is a real number $and\, x \neq -5\}$
45. $\{x \,|\, x$ is a real number $and\, x \neq 0\}$
47. $\{x \,|\, x$ is a real number $and\, x \neq 1\}$
49. $\{x \,|\, x$ is a real number $and\, x \neq -\frac{9}{2}\, and\, x \neq 1\}$
51. $\{x \,|\, x$ is a real number $and\, x \neq 3\}$
53. $\{x \,|\, x$ is a real number $and\, x \neq -4\}$
55. $\{x \,|\, x$ is a real number $and\, x \neq 4\, and\, x \neq 5\}$
57. $\{x \,|\, x$ is a real number $and\, x \neq -1\, and\, x \neq -\frac{5}{2}\}$
59. 4; 3   **61.** 5; -1 **63.** $\{x \,|\, 0 \leq x \leq 9\}$; $\{x \,|\, 3 \leq x \leq 10\}$; $\{x \,|\, 3 \leq x \leq 9\}$; $\{x \,|\, 3 \leq x \leq 9\}$
65.

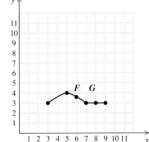

67. 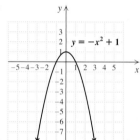   **69.** $y = \frac{1}{6}x - \frac{1}{2}$ **70.** $y = \frac{3}{8}x - \frac{5}{8}$
71. $y = -\frac{5}{2}x - \frac{3}{2}$ **72.** $y = -\frac{1}{8}x + \frac{1}{2}$ **73.** Let n represent
the number; $2n + 5 = 49$ **74.** Let x represent the
number; $\frac{1}{2}x - 3 = 57$ **75.** Let x represent the number;
$x + (x + 1) = 145$ **76.** Let n represent the number;
$n - (-n) = 20$ **77.**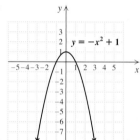
79. $\{x \mid x$ is a real number $and\ x \neq 4\ and\ x \neq 3\ and\ x \neq 2$
$and\ x \neq -2\}$
81. Answers may vary.

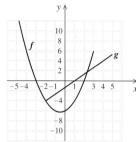

83. $\left\{x \mid x$ is a real number $and -1 < x < 5\ and\ x \neq \frac{3}{2}\right\}$
85. Answers may vary. $f(x) = \dfrac{1}{x + 2}$, $g(x) = \dfrac{1}{x - 5}$

87.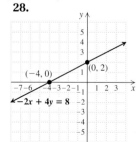

Review Exercises: Chapter 2, pp. 141–143

1. False   **2.** False   **3.** True   **4.** False   **5.** True
6. True   **7.** True   **8.** True   **9.** False   **10.** True
11. No   **12.** Yes   **13.** II
14.

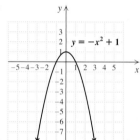

15. The value of the apartment is increasing at a rate of
$7500 per year.   **16.** 72,600 homes per month   **17.** $\frac{4}{7}$
18. Undefined   **19.** $-\frac{1}{4}$ **20.** 0   **21.** Slope: -5;
y-intercept: $(0, -11)$ **22.** Slope: $\frac{5}{6}$; y-intercept: $\left(0, -\frac{5}{3}\right)$
23. 11 signifies that the number of calories consumed each
day increases by 11 per year, for years after 1971; 1542
signifies that the number of calories consumed each day in
1971 was 1542.   **24.** 0   **25.** Undefined
26. x-intercept: $\left(\frac{8}{3}, 0\right)$, y-intercept: $(0, -4)$
27.

28.

29.

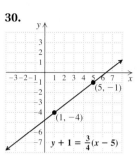

30.

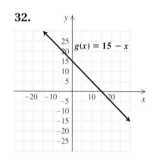

31.

32.

33.

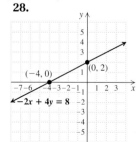

34.

35. -1 **36.** 10 tee shirts   **37.** Perpendicular
38. Parallel   **39.** $f(x) = \frac{2}{9}x - 4$
40. $y - 10 = -5(x - 1)$ **41.** $f(x) = -\frac{1}{4}x + \frac{11}{2}$
42. $y = \frac{3}{5}x - \frac{31}{5}$ **43.** $y = -\frac{5}{3}x - \frac{5}{3}$
44. About $1.50;

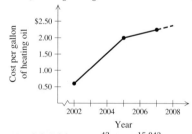

45. About $2.40   **46.** **(a)** $R(t) = -\frac{43}{2400}t + \frac{15,843}{800}$;
(b) about 19.21 sec; about 19.09 sec   **47.** Yes   **48.** Yes
49. No   **50.** No   **51.** **(a)** 3; **(b)** $\{x \mid -2 \leq x \leq 4\}$;
(c) -1; **(d)** $\{y \mid 1 \leq y \leq 5\}$ **52.** Domain: $\mathbb{R}$; range:
$\{y \mid y \geq 0\}$ **53.** Yes   **54.** No   **55.** -6 **56.** 26
57. $3a + 9$ **58.** 102   **59.** $-\frac{9}{2}$ **60.** $x^2 + 3x - 5$
61. $\mathbb{R}$ **62.** $\mathbb{R}$ **63.** $\{x \mid x$ is a real number $and\ x \neq 2\}$
64. 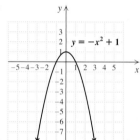 For a function, every member of the domain corresponds to *exactly one* member of the range. Thus, for any
function, each member of the domain corresponds to *at
least one* member of the range. Therefore, a function is a
relation. In a relation, every member of the domain corresponds to *at least one*, but not necessarily *exactly one*,
member of the range. Therefore, a relation may or may not

be a function.   **65.** 📄 The slope of a line is the rise between two points on the line divided by the run between those points. For a vertical line, there is no run between any two points, and division by 0 is undefined; therefore, the slope is undefined. For a horizontal line, there is no rise between any two points, so the slope is 0/run, or 0. **66.** -9 **67.** $-\frac{9}{2}$ **68.** $f(x) = 10.94x + 20$ **69. (a)** III; **(b)** IV; **(c)** I; **(d)** II

Test: Chapter 2, pp. 144–145

1. [2.1] No   **2.** [2.1], [2.2]

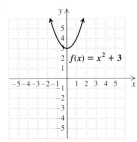

3. [2.3] The number of people on the National Do Not Call Registry is increasing at a rate of 25 million people/year. **4.** [2.3] $\frac{5}{8}$ **5.** [2.3] 0   **6.** [2.3] Slope: $-\frac{3}{5}$; y-intercept: $(0, 12)$ **7.** [2.3] Slope: $-\frac{2}{5}$; y-intercept: $\left(0, -\frac{7}{5}\right)$ **8.** [2.4] 0   **9.** [2.4] Undefined **10.** [2.4] x-intercept: $(3, 0)$; y-intercept: $(0, -15)$ **11.** [2.1], [2.3]   **12.** [2.5]

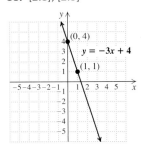   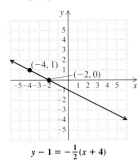

13. [2.4]   **14.** [2.4]

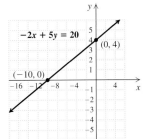   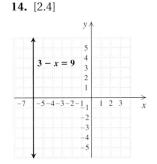

15. [2.4] 3   **16.** [2.4] About 49 million international visitors **17.** [2.4] (a), (c)   **18.** [2.5] Parallel **19.** [2.5] Perpendicular   **20.** [2.3] $f(x) = -5x - 1$ **21.** [2.5] $y - (-4) = 4(x - (-2))$, or $y + 4 = 4(x + 2)$ **22.** [2.5] $f(x) = -x + 2$ **23.** [2.5] $y = \frac{2}{5}x + \frac{16}{5}$ **24.** [2.5] $y = -\frac{5}{2}x - \frac{11}{2}$ **25.** [2.5] **(a)** $C(m) = 0.3m + 25$; **(b)** \$175   **26.** [2.2] **(a)** 1; **(b)** $\{x | -3 \le x \le 4\}$; **(c)** 3;

(d) $\{y | -1 \le y \le 2\}$ **27.** [2.2] -9 **28.** [2.6] $\frac{1}{x} + 2x + 1$ **29.** [2.2] $\{x | x \text{ is a real number } and\, x \ne 0\}$ **30.** [2.6] $\{x | x \text{ is a real number } and\, x \ne 0\}$ **31.** [2.6] $\{x | x \text{ is a real number } and\, x \ne 0 \,and\, x \ne -\frac{1}{2}\}$ **32.** [2.2], [2.3] **(a)** 30 mi; **(b)** 15 mph **33.** [2.5] $s = -\frac{3}{2}r + \frac{27}{2}$, or $\frac{27 - 3r}{2}$

34. [2.6] $h(x) = 7x - 2$

Cumulative Review: Chapters 1–2, pp. 145–146

1. 7   **2.** $\frac{1}{2}$ **3.** -1.83 **4.** -3 **5.** 9   **6.** $-2x + 37$ **7.** $\varnothing$; contradiction   **8.** $\frac{13}{8}$ **9.** $y = \frac{8}{3}x - 4$ **10.** $\frac{y}{3x^2}$ **11.** $-\frac{1}{8}$ **12.** 1   **13.** $\frac{x^6}{25y^2}$ **14.** -5 **15.** 8   **16.** 0 **17.** $f(x) = -x + \frac{1}{5}$ **18.** $f(x) = 4x + 7$ **19.** $y = -x + 3$ **20.** 99   **21.** 0   **22.** $\frac{x^2 - 1}{x + 5}$ **23.** $\{x | x \text{ is a real number } and\, x \ne -6\}$ **24.** Domain: $\{x | -4 \le x \le 3\}$; range: $\{y | -3 \le y \le 0 \,or\, y = 2\}$ **25.** About 60 million; the number of passengers using Newark Liberty and LaGuardia in 2005   **26.** About 112 million; the number of passengers using the three airports in 2007   **27.** About 61 million; the number of passengers using LaGuardia and Newark Liberty in 2006

28.   **29.**

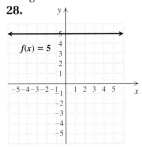   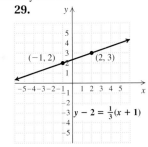

30.   **31.**

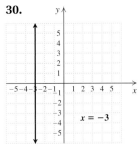   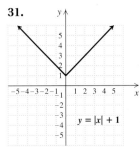

32.   **33.**

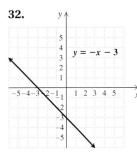   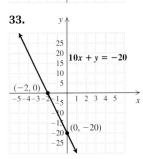

34. Full-time students: 6.8 million; part-time students: 1.7 million   **35. (a)** $c(t) = 153.5t + 4574$; **(b)** $6416; **(c)** about 2019   **36.** $1.09 billion   **37. (a)** 130.2 million payments; **(b)** -8.3 signifies that the number of Social Security payments delivered by paper check is decreasing at a rate of 8.3 million per year; 180 signifies that 180 million Social Security payments were delivered by paper check in 2001.   **38.** 1.1×10^{16} bytes   **39.** 28 pages   **40.** Length: 19.5 in.; width: 6.5 in.   **41.** Let x and y represent the numbers; $x^2 - y^2$ **42.** Let x and y represent the numbers; $(x + y)(x - y)$ **43.** $9^a x^{2a^2} y^{2ac+2a^2}$ **44.** 2   **45.** $f(x) = \frac{1}{2}x + 3$

CHAPTER 3

Technology Connection, p. 152

1. $(1.53, 2.58)$ **2.** $(-0.26, 57.06)$ **3.** $(2.23, 1.14)$ **4.** $(0.87, -0.32)$

Visualizing for Success, p. 154

1. C   **2.** H   **3.** J   **4.** G   **5.** D   **6.** I   **7.** A   **8.** F   **9.** E   **10.** B

Exercise Set 3.1, pp. 155–157

1. False   **2.** True   **3.** True   **4.** True   **5.** True   **6.** False   **7.** False   **8.** True   **9.** Yes   **11.** No   **13.** Yes   **15.** Yes   **17.** $(3, 2)$ **19.** $(2, -1)$ **21.** $(1, 4)$ **23.** $(-3, -2)$ **25.** $(3, -1)$ **27.** $(3, -7)$ **29.** $(7, 2)$ **31.** $(4, 0)$ **33.** No solution   **35.** $\{(x, y) | y = 3 - x\}$ **37.** All except Exercise 33   **39.** Exercise 35   **41.** Let x represent the first number and y the second number; $x + y = 10, x = \frac{2}{3}y$ **43.** Let p represent the number of personal e-mails and b the number of business e-mails; $p + b = 578, b = p + 30$ **45.** Let x and y represent the angles; $x + y = 180, x = 2y - 3$ **47.** Let x represent the number of two-point shots and y the number of foul shots; $x + y = 64, 2x + y = 100$ **49.** Let x represent the number of hats sold and y the number of tee shirts sold; $x + y = 45, 14.50x + 19.50y = 697.50$ **51.** Let h represent the number of vials of Humalog sold and n the number of vials of Lantus; $h + n = 50, 83.29h + 76.76n = 3981.66$ **53.** Let l represent the length, in yards, and w the width, in yards; $2l + 2w = 340; l = w + 50$ **55.** ✏   **57.** $\frac{8}{13}$ **58.** -1 **59.** $-\frac{1}{10}$ **60.** 11   **61.** $y = 3x - 4$ **62.** $x = \frac{5}{2}y - \frac{7}{2}$ **63.** ✏   **65.** Answers may vary. **(a)** $x + y = 6, x - y = 4$; **(b)** $x + y = 1, 2x + 2y = 3$; **(c)** $x + y = 1, 2x + 2y = 2$ **67.** $A = -\frac{17}{4}, B = -\frac{12}{5}$ **69.** Let x and y represent the number of years that Dell and Juanita have taught at the university, respectively; $x + y = 46, x - 2 = 2.5(y - 2)$ **71.** Let s and v represent the number of ounces of baking soda and vinegar needed, respectively; $s = 4v, s + v = 16$ **73.** Mineral oil: 12 oz; vinegar: 4 oz   **75.** $(0, 0), (1, 1)$ **77.** $(0.07, -7.95)$ **79.** $(0.00, 1.25)$

Exercise Set 3.2, pp. 163–165

1. (d)   **2.** (e)   **3.** (a)   **4.** (f)   **5.** (c)   **6.** (b)   **7.** $(2, -1)$ **9.** $(-4, 3)$ **11.** $(2, -2)$ **13.** $\{(x, y) | 2x - 3 = y\}$ **15.** $(-2, 1)$ **17.** $\left(\frac{1}{2}, \frac{1}{2}\right)$ **19.** $(2, 0)$ **21.** No solution   **23.** $(1, 2)$ **25.** $(7, -2)$ **27.** $(-1, 2)$ **29.** $\left(\frac{49}{11}, -\frac{12}{11}\right)$ **31.** $(6, 2)$ **33.** No solution   **35.** $(20, 0)$ **37.** $(3, -1)$ **39.** $\{(x, y) | -4x + 2y = 5\}$ **41.** $\left(2, -\frac{3}{2}\right)$ **43.** $(-2, -9)$ **45.** $(30, 6)$ **47.** $\{(x, y) | 4x - 2y = 2\}$ **49.** No solution   **51.** $(140, 60)$ **53.** $\left(\frac{1}{3}, -\frac{2}{3}\right)$ **55.** ✏   **57.** Toaster oven: 3 kWh; convection oven: 12 kWh   **58.** 90   **59.** $105,000   **60.** 290 mi   **61.** First: 30 in.; second: 60 in.; third: 6 in.   **62.** 165 min   **63.** ✏   **65.** $m = -\frac{1}{2}, b = \frac{5}{2}$ **67.** $a = 5, b = 2$ **69.** $\left(-\frac{32}{17}, \frac{38}{17}\right)$ **71.** $\left(-\frac{1}{5}, \frac{1}{10}\right)$ **73.** ✏

Connecting the Concepts, pp. 165–166

1. $(1, 1)$ **2.** $(9, 1)$ **3.** $(4, 3)$ **4.** $(5, 7)$ **5.** $(5, 10)$ **6.** $\left(2, \frac{2}{5}\right)$ **7.** No solution   **8.** $\{(x, y) | x = 2 - y\}$ **9.** $(1, 1)$ **10.** $(0, 0)$ **11.** $(6, -1)$ **12.** No solution   **13.** $(3, 1)$ **14.** $\left(\frac{95}{71}, -\frac{1}{142}\right)$ **15.** $(1, 1)$ **16.** $(11, -3)$ **17.** $\{(x, y) | x - 2y = 5\}$ **18.** $\left(1, -\frac{1}{19}\right)$ **19.** $\left(\frac{201}{23}, -\frac{18}{23}\right)$ **20.** $\left(\frac{40}{9}, \frac{10}{3}\right)$

Exercise Set 3.3, pp. 174–178

1. 4, 6   **3.** Personal e-mails: 274; business e-mails: 304   **5.** $119°, 61°$ **7.** Two-point shots: 36; foul shots: 28   **9.** Hats: 36; tee shirts: 9   **11.** Humalog vials: 22; Lantus vials: 28   **13.** Length: 110 yd; width: 60 yd   **15.** Regular paper: 32 reams; recycled paper: 84 reams   **17.** 13-watt bulbs: 60; 18-watt bulbs: 140   **19.** HP C7115A cartridges: 180; M3908GA cartridges: 270   **21.** Mexican: 14 lb; Peruvian: 14 lb   **23.** Custom-printed M&Ms: 64 oz; bulk M&Ms: 256 oz   **25.** 50%-chocolate: 7.5 lb; 10%-chocolate: 12.5 lb   **27.** Deep Thought: 12 lb; Oat Dream: 8 lb   **29.** $7500 at 6.5%; $4500 at 7.2%   **31.** Steady State: 12.5 L; Even Flow: 7.5 L   **33.** 87-octane: 2.5 gal; 95-octane: 7.5 gal   **35.** Whole milk: $169\frac{3}{13}$ lb; cream: $30\frac{10}{13}$ lb   **37.** 375 km   **39.** 14 km/h   **41.** About 1489 mi   **43.** Length: 265 ft; width: 165 ft   **45.** Wii game machines: 3.63 million; PlayStation 3 consoles: 1.21 million   **47.** $8.99 plans: 182; $4.99 plans: 68   **49.** Quarters: 17; fifty-cent pieces: 13   **51.** ✏   **53.** 1   **54.** $\frac{1}{2}$ **55.** -13 **56.** 17   **57.** 7   **58.** $\frac{13}{4}$ **59.** ✏   **61.** 0%: 20 reams; 30%: 40 reams   **63.** $10\frac{2}{3}$ oz   **65.** 12 sets   **67.** Brown: 0.8 gal; neutral: 0.2 gal   **69.** City: 261 mi; highway: 204 mi   **71.** $P(x) = \dfrac{0.1 + x}{1.5}$ (This expresses the percent as a decimal quantity.)

Exercise Set 3.4, pp. 185–186

1. True   **2.** False   **3.** False   **4.** True   **5.** True   **6.** False   **7.** Yes   **9.** $(3, 1, 2)$ **11.** $(1, -2, 2)$ **13.** $(2, -5, -6)$ **15.** No solution   **17.** $(-2, 0, 5)$ **19.** $(21, -14, -2)$ **21.** The equations are dependent.

23. $\left(3, \frac{1}{2}, -4\right)$ **25.** $\left(\frac{1}{2}, \frac{1}{3}, \frac{1}{6}\right)$ **27.** $\left(\frac{1}{2}, \frac{2}{3}, -\frac{5}{6}\right)$
29. $(15, 33, 9)$ **31.** $(3, 4, -1)$ **33.** $(10, 23, 50)$
35. No solution   **37.** The equations are dependent.
39. 🖫   **41.** Let x and y represent the numbers: $x = \frac{1}{2}y$
42. Let x and y represent the numbers; $x - y = 2x$
43. Let x represent the first number;
$x + (x + 1) + (x + 2) = 100$
44. Let x, y, and z represent the numbers; $x + y + z = 100$
45. Let x, y, and z represent the numbers; $xy = 5z$
46. Let x and y represent the numbers; $xy = 2(x + y)$
47. 🖫   **49.** $(1, -1, 2)$ **51.** $(1, -2, 4, -1)$
53. $\left(-1, \frac{1}{5}, -\frac{1}{2}\right)$ **55.** 14   **57.** $z = 8 - 2x - 4y$
59. 🖫

Exercise Set 3.5, pp. 190–193

1. 8, 15, 62   **3.** 8, 21, −3   **5.** 32°, 96°, 52°
7. Reading: 502; mathematics: 515; writing: 494
9. Bran muffin: 1.5 g; banana: 3 g; 1 cup of Wheaties: 3 g
11. Basic price: \$30,610; tow package: \$205; camera: \$750
13. 12-oz cups: 17; 16-oz cups: 25; 20-oz cups: 13
15. Bank loan: \$15,000; small-business loan: \$35,000;
mortgage: \$70,000
17. Gold: \$30/g; silver: \$3/g; copper: \$0.02/g
19. Roast beef: 2 servings; baked potato: 1 serving;
broccoli: 2 servings
21. First mezzanine: 8 tickets; main floor: 12 tickets;
second mezzanine: 20 tickets
23. Asia: 5.5 billion; Africa: 2.0 billion; rest of the world:
1.9 billion   **25.** 🖫   **27.** $-4x + 6y$ **28.** $-x + 6y$
29. $7y$ **30.** $11a$ **31.** $-2a + b + 6c$
32. $-50a - 30b + 10c$ **33.** $-12x + 5y - 8z$
34. $23x - 13z$ **35.** 🖫   **37.** Applicant: \$87; spouse: \$87;
first child: \$47; second child: \$42   **39.** 20 yr   **41.** 35 tickets

Exercise Set 3.6, pp. 197–198

1. Matrix   **2.** Horizontal; columns   **3.** Entry
4. Matrices   **5.** Rows   **6.** First   **7.** $(3, 4)$ **9.** $(-2, 5)$
11. $\left(\frac{3}{2}, \frac{5}{2}\right)$ **13.** $\left(2, \frac{1}{2}, -2\right)$ **15.** $(2, -2, 1)$ **17.** $\left(4, \frac{1}{2}, -\frac{1}{2}\right)$
19. $(1, -3, -2, -1)$ **21.** Dimes: 18; nickels: 24
23. Dried fruit: 9 lb; macadamia nuts: 6 lb
25. \$400 at 7%; \$500 at 8%; \$1600 at 9%   **27.** 🖫
29. 17   **30.** −19   **31.** 37   **32.** 422   **33.** 🖫   **35.** 1324

Exercise Set 3.7, pp. 202–203

1. True   **2.** True   **3.** True   **4.** False   **5.** False
6. False   **7.** 4   **9.** −50   **11.** 27   **13.** −3   **15.** −5
17. $(-3, 2)$ **19.** $\left(\frac{9}{19}, \frac{51}{38}\right)$ **21.** $\left(-1, -\frac{6}{7}, \frac{11}{7}\right)$
23. $(2, -1, 4)$ **25.** $(1, 2, 3)$ **27.** 🖫   **29.** 9700
30. $70x - 2500$ **31.** −1800   **32.** 4500   **33.** $\frac{250}{7}$
34. $\frac{250}{7}$ **35.** 🖫   **37.** 12   **39.** 10

Exercise Set 3.8, pp. 207–209

1. (b)   **2.** (f)   **3.** (h)   **4.** (a)   **5.** (e)   **6.** (d)
7. (c)   **8.** (g)
9. (a) $P(x) = 20x - 200,000$;   (b) (10,000 units, \$550,000)

11. (a) $P(x) = 25x - 3100$;   (b) (124 units, \$4960)
13. (a) $P(x) = 45x - 22,500$;   (b) (500 units, \$42,500)
15. (a) $P(x) = 16x - 50,000$;   (b) (3125 units, \$125,000)
17. (a) $P(x) = 50x - 100,000$;   (b) (2000 units, \$250,000)
19. (\$60, 1100)   **21.** (\$22, 474)   **23.** (\$50, 6250)
25. (\$10, 1070)   **27.** (a) $C(x) = 45,000 + 40x$;
(b) $R(x) = 130x$;   (c) $P(x) = 90x - 45,000$;
(d) \$225,000 profit, \$9000 loss   (e) (500 phones, \$65,000)
29. (a) $C(x) = 10,000 + 30x$; (b) $R(x) = 80x$;
(c) $P(x) = 50x - 10,000$; (d) \$90,000 profit, \$7500 loss;
(e) (200 seats, \$16,000)   **31.** 🖫   **33.** 6   **34.** −2
35. $-\frac{11}{9}$ **36.** $\frac{3}{4}$ **37.** −6   **38.** −5   **39.** 🖫
41. (\$5, 300 yo-yo's)   **43.** (a) \$8.74;   (b) 24,509 units

Review Exercises: Chapter 3, pp. 213–214

1. Substitution   **2.** Elimination   **3.** Graphical
4. Dependent   **5.** Inconsistent   **6.** Contradiction
7. Parallel   **8.** Square   **9.** Determinant   **10.** Zero
11. $(4, 1)$ **12.** $(3, -2)$ **13.** $\left(\frac{8}{3}, \frac{14}{3}\right)$ **14.** No solution
15. $\left(-\frac{4}{5}, \frac{2}{5}\right)$ **16.** $\left(\frac{9}{4}, \frac{7}{10}\right)$ **17.** $\left(\frac{76}{17}, -\frac{2}{119}\right)$ **18.** $(-2, -3)$
19. $\{(x, y) \mid 3x + 4y = 6\}$
20. Melon: \$2.49; pineapple: \$3.98   **21.** 4 hr
22. 8% juice: 10 L; 15% juice: 4 L   **23.** $(4, -8, 10)$
24. The equations are dependent.   **25.** $(2, 0, 4)$
26. $\left(\frac{8}{9}, -\frac{2}{3}, \frac{10}{9}\right)$ **27.** A: 90°; B: 67.5°; C: 22.5°
28. Oil: $21\frac{1}{3}$ oz; lemon juice: $10\frac{2}{3}$ oz
29. Man: 1.4; woman: 5.3; one-year-old child: 50
30. $\left(55, -\frac{89}{2}\right)$ **31.** $(-1, 1, 3)$ **32.** −5   **33.** 9
34. $(6, -2)$ **35.** $(-3, 0, 4)$ **36.** (\$3, 81)
37. (a) $C(x) = 4.75x + 54,000$;   (b) $R(x) = 9.25x$;
(c) $P(x) = 4.5x - 54,000$;   (d) \$31,500 loss, \$13,500 profit;
(e) (12,000 pints of honey, \$111,000)
38. 🖫 To solve a problem involving four variables, go
through the *Familiarize* and *Translate* steps as usual. The
resulting system of equations can be solved using the elimination method just as for three variables but likely with
more steps.   **39.** 🖫 A system of equations can be both
dependent and inconsistent if it is equivalent to a system
with fewer equations that has no solution. An example is a
system of three equations in three unknowns in which two
of the equations represent the same plane, and the third
represents a parallel plane.   **40.** 20,000 pints
41. Round Stic: 15 packs; Matic Grip: 9 packs
42. $(0, 2), (1, 3)$ **43.** $a = -\frac{2}{3}, b = -\frac{4}{3}, c = 3$;
$f(x) = -\frac{2}{3}x^2 - \frac{4}{3}x + 3$

Test: Chapter 3, pp. 214–215

1. [3.1] $(2, 4)$ **2.** [3.2] $\left(3, -\frac{11}{3}\right)$
3. [3.2] $\{(x, y) \mid x = 2y - 3\}$ **4.** [3.2] $(2, -1)$
5. [3.2] No solution   **6.** [3.2] $\left(-\frac{3}{2}, -\frac{3}{2}\right)$
7. [3.3] Length: 94 ft; width: 50 ft
8. [3.3] Pepperidge Farm Goldfish: 120 g;
Rold Gold Pretzels: 500 g   **9.** [3.3] 5.5 hr   **10.** [3.4] The
equations are dependent.   **11.** [3.4] $\left(2, -\frac{1}{2}, -1\right)$
12. [3.4] No solution   **13.** [3.4] $(0, 1, 0)$
14. [3.6] $\left(\frac{22}{5}, -\frac{28}{5}\right)$ **15.** [3.6] $(3, 1, -2)$ **16.** [3.7] −14

17. [3.7] -59 **18.** [3.7] $\left(\frac{7}{13}, -\frac{17}{26}\right)$ **19.** [3.5] Electrician: 3.5 hr; carpenter: 8 hr; plumber: 10 hr   **20.** [3.8] ($3, 55)
21. [3.8] **(a)** $C(x) = 25x + 44,000$; **(b)** $R(x) = 80x$; **(c)** $P(x) = 55x - 44,000$; **(d)** $27,500 loss, $5500 profit; **(e)** (800 hammocks, $64,000)
22. [2.3], [3.3] $m = 7, b = 10$ **23.** [3.3] $\frac{120}{7}$ lb

Cumulative Review: Chapters 1–3, pp. 216–217

1. x^{11} **2.** $-\dfrac{9x^4}{2y^6}$ **3.** $-\dfrac{2a^{11}}{5b^{33}}$ **4.** $\dfrac{81x^{36}}{256y^8}$
5. 1.12×10^6 **6.** 4.00×10^6
7. $b = \dfrac{2A}{h} - t$, or $\dfrac{2A - ht}{h}$ **8.** Yes   **9.** -22 **10.** 20
11. -56 **12.** 6   **13.** -5 **14.** $\frac{10}{9}$ **15.** 1   **16.** $(1, 1)$
17. $(-2, 3)$ **18.** $\left(-3, \frac{2}{5}\right)$ **19.** $(-3, 2, -4)$
20. $(0, -1, 2)$

21.
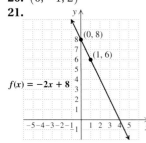

22.
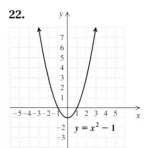

23.
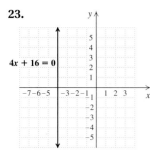

24.
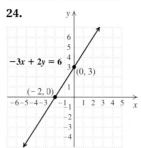

25. Slope: $\frac{9}{4}$; y-intercept: $(0, -3)$ **26.** $\frac{4}{3}$
27. $y = 4x - 19$ **28.** $y = -\frac{1}{10}x + \frac{12}{5}$ **29.** Parallel
30. $y = -2x + 5$
31. $\{-5, -3, -1, 1, 3\}; \{-3, -2, 1, 4, 5\}; -2; 3$
32. $\{x \mid x \text{ is a real number } and\, x \neq -10\}$ **33.** -31
34. 3   **35.** 7   **36.** $2a^2 + 4a - 4$ **37.** 14   **38.** 0
39. **(a)** $1.78; **(b)** 0.15 signifies that the average fee is increasing at a rate of $0.15 per year; 1.03 signifies that the average fee was $1.03 in 2003.   **40.** **(a)** $A(t) = \frac{87}{4}t + 52$; **(b)** 530,500 ATMs; **(c)** about 2013   **41.** 68 professionals; 82 students   **42.** Sea Spray: 45 oz; Ocean Mist: 75 oz
43. 17, 19, 21   **44.** Length: 100 yd; width: 60 yd
45. Adults' tickets: 1346; senior citizens' tickets: 335; children's tickets: 1651   **46.** 86   **47.** $-12x^{2a}y^{b+y+3}$
48. $151,000   **49.** $m = -\frac{5}{9}, b = -\frac{2}{9}$

CHAPTER 4

Exercise Set 4.1, pp. 228–231

1. Equivalent equations   **2.** Equivalent expressions
3. Equivalent inequalities   **4.** Not equivalent
5. Not equivalent   **6.** Equivalent equations
7. Equivalent expressions   **8.** Not equivalent
9. Not equivalent   **10.** Equivalent inequalities
11. **(a)** No; **(b)** no; **(c)** yes; **(d)** yes   **13.** **(a)** Yes; **(b)** no; **(c)** yes; **(d)** no
15. $\{y \mid y < 6\}, (-\infty, 6)$
17. $\{x \mid x \geq -4\}, [-4, \infty)$
19. $\{t \mid t > -3\}, (-3, \infty)$
21. $\{x \mid x \leq -7\}, (-\infty, -7]$
23. $\{x \mid x > -1\}$, or $(-1, \infty)$
25. $\{t \mid t \leq 10\}$, or $(-\infty, 10]$
27. $\{x \mid x \geq 1\}$, or $[1, \infty)$
29. $\{t \mid t < -9\}$, or $(-\infty, -9)$
31. $\{x \mid x < 50\}$, or $(-\infty, 50)$
33. $\{x \mid x \leq -0.9\}$, or $(-\infty, -0.9]$
35. $\left\{y \mid y \geq -\frac{5}{6}\right\}$, or $\left[-\frac{5}{6}, \infty\right)$
37. $\{x \mid x < 2\}$, or $(-\infty, 2)$
39. $\{x \mid x \leq -9\}$, or $(-\infty, -9]$
41. $\{x \mid x < -26\}$, or $(-\infty, -26)$
43. $\left\{t \mid t \geq -\frac{13}{3}\right\}$, or $\left[-\frac{13}{3}, \infty\right)$
45. $\{x \mid x \geq -3\}$, or $[-3, \infty)$
47. $\{x \mid x \geq 2\}$, or $[2, \infty)$
49. $\left\{x \mid x > \frac{2}{3}\right\}$, or $\left(\frac{2}{3}, \infty\right)$
51. $\left\{x \mid x \geq \frac{1}{2}\right\}$, or $\left[\frac{1}{2}, \infty\right)$
53. $\left\{y \mid y \leq -\frac{3}{2}\right\}$, or $\left(-\infty, -\frac{3}{2}\right]$ **55.** $\left\{t \mid t < \frac{29}{5}\right\}$, or $\left(-\infty, \frac{29}{5}\right)$
57. $\left\{m \mid m > \frac{7}{3}\right\}$, or $\left(\frac{7}{3}, \infty\right)$ **59.** $\{x \mid x \geq 2\}$, or $[2, \infty)$
61. $\{y \mid y < 5\}$, or $(-\infty, 5)$ **63.** $\left\{x \mid x \leq \frac{4}{7}\right\}$, or $\left(-\infty, \frac{4}{7}\right]$
65. $\{x \mid x \geq 10\}$, or $[10, \infty)$ **67.** $\{x \mid x \leq 3\}$, or $(-\infty, 3]$
69. $\left\{x \mid x \geq -\frac{7}{2}\right\}$, or $\left[-\frac{7}{2}, \infty\right)$ **71.** $\{x \mid x \leq 4\}$, or $(-\infty, 4]$
73. Lengths of time less than $7\frac{1}{2}$ hr   **75.** At least 56 questions correct   **77.** For 2800 min or more   **79.** For more than 8 transactions   **81.** Gross sales greater than $7000
83. For more than $6000   **85.** Years after 2010
87. **(a)** Body densities less than $\frac{99}{95}$ kg/L, or about 1.04 kg/L; **(b)** body densities less than $\frac{495}{482}$ kg/L, or about 1.03 kg/L
89. **(a)** $\left\{x \mid x < 3913\frac{1}{23}\right\}$, or $\{x \mid x \leq 3913\}$;
(b) $\left\{x \mid x > 3913\frac{1}{23}\right\}$, or $\{x \mid x \geq 3914\}$

91. 📕   **93.** $\{x | x \text{ is a real number } and\ x \neq 0\}$
94. $\{x | x \text{ is a real number } and\ x \neq 6\}$
95. $\{x | x \text{ is a real number } and\ x \neq -\frac{1}{2}\}$
96. $\{x | x \text{ is a real number } and\ x \neq \frac{7}{5}\}$
97. $\mathbb{R}$ **98.** $\{x | x \text{ is a real number } and\ x \neq 0\}$ **99.** 📕
101. $\left\{ x | x \leq \dfrac{2}{a-1} \right\}$ **103.** $\left\{ y | y \geq \dfrac{2a+5b}{b(a-2)} \right\}$
105. $\left\{ x | x > \dfrac{4m-2c}{d-(5c+2m)} \right\}$ **107.** False; $2 < 3$ and
$4 < 5$, but $2 - 4 = 3 - 5$.   **109.** 📕
111. $\mathbb{R}$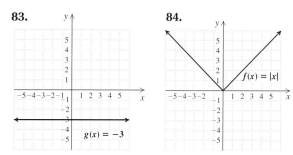
113. $\{x | x \text{ is a real number } and\ x \neq 0\}$ **115.** 📈

Exercise Set 4.2, pp. 238–241

1. (h)   **2.** (j)   **3.** (f)   **4.** (a)   **5.** (e)   **6.** (d)
7. (b)   **8.** (g)   **9.** (c)   **10.** (i)   **11.** $\{4, 16\}$
13. $\{0, 5, 10, 15, 20\}$ **15.** $\{b, d, f\}$ **17.** $\{u, v, x, y, z\}$
19. $\varnothing$ **21.** $\{1, 3, 5\}$
23. ⟨number line⟩ $(1, 3)$
25. ⟨number line⟩ $[-6, 0]$
27. ⟨number line⟩ $(-\infty, -1) \cup (4, \infty)$
29. ⟨number line⟩ $(-\infty, -2] \cup (1, \infty)$
31. ⟨number line⟩ $(-2, 4]$
33. ⟨number line⟩ $(-2, 4)$
35. ⟨number line⟩ $(-\infty, 5) \cup (7, \infty)$
37. ⟨number line⟩ $(-\infty, -4] \cup [5, \infty)$
39. ⟨number line⟩ $[-3, 7)$
41. ⟨number line⟩ $(-7, 0]$
43. ⟨number line⟩ $(-\infty, 5)$
45. $\{x | -5 \leq x < 7\}$, or $[-5, 7)$ ⟨number line⟩
47. $\{t | 4 < t \leq 8\}$, or $(4, 8]$ ⟨number line⟩
49. $\{a | -2 \leq a < 2\}$, or $[-2, 2)$ ⟨number line⟩
51. $\mathbb{R}$, or $(-\infty, \infty)$ ⟨number line⟩
53. $\{x | -3 \leq x \leq 2\}$, or $[-3, 2]$ ⟨number line⟩
55. $\{x | 7 < x < 23\}$, or $(7, 23)$ ⟨number line⟩
57. $\{x | -32 \leq x \leq 8\}$, or $[-32, 8]$
⟨number line⟩
59. $\{x | 1 \leq x \leq 3\}$, or $[1, 3]$ ⟨number line⟩
61. $\{x | -\frac{7}{2} < x \leq 7\}$, or $\left(-\frac{7}{2}, 7\right]$ ⟨number line⟩
63. $\{t | t < 0 \ or\ t > 1\}$, or $(-\infty, 0) \cup (1, \infty)$
⟨number line⟩

65. $\{a | a < \frac{7}{2}\}$, or $\left(-\infty, \frac{7}{2}\right)$
67. $\{a | a < -5\}$, or $(-\infty, -5)$ ⟨number line⟩
69. $\varnothing$
71. $\{t | t \leq 6\}$, or $(-\infty, 6]$ ⟨number line⟩
73. $(-\infty, -6) \cup (-6, \infty)$ **75.** $(-\infty, 0) \cup (0, \infty)$
77. $(-\infty, 4) \cup (4, \infty)$ **79.** 📕
81.   **82.**

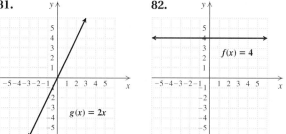

83.   **84.**

85. -1 **86.** -4 **87.** 📕   **89.** $(-1, 6)$
91. Between 2003 and 2009   **93.** Sizes between 6 and 13
95. Densities between 1.03 kg/L and 1.04 kg/L
97. More than 12 and fewer than 125 trips
99. $\{m | m < \frac{6}{5}\}$, or $\left(-\infty, \frac{6}{5}\right)$ ⟨number line⟩

101. $\{x | -\frac{1}{8} < x < \frac{1}{2}\}$, or $\left(-\frac{1}{8}, \frac{1}{2}\right)$ ⟨number line⟩

103. False   **105.** True
107. $(-\infty, -7) \cup \left(-7, \frac{3}{4}\right]$ **109.** 📈   **111.** 📈

Technology Connection, p. 247

1. The x-values on the graph of $y_1 = |4x + 2|$ that are *below* the line $y = 6$ solve the inequality $|4x + 2| < 6$.
2. The x-values on the graph of $y_1 = |3x - 2|$ that are below the line $y = 4$ are in the interval $\left(-\frac{2}{3}, 2\right)$.
3. The graphs of $y_1 = \text{abs}(4x + 2)$ and $y_2 = -6$ do not intersect.

Exercise Set 4.3, pp. 248–250

1. True   **2.** False   **3.** True   **4.** True   **5.** True
6. True   **7.** False   **8.** False   **9.** (g)   **10.** (h)   **11.** (d)
12. (a)   **13.** (a)   **14.** (b)   **15.** $\{-10, 10\}$ **17.** $\varnothing$
19. $\{0\}$ **21.** $\left\{-\frac{1}{2}, \frac{7}{2}\right\}$ **23.** $\varnothing$ **25.** $\{-4, 8\}$
27. $\{6, 8\}$ **29.** $\{-5.5, 5.5\}$ **31.** $\{-8, 8\}$ **33.** $\{-1, 1\}$
35. $\left\{-\frac{11}{2}, \frac{13}{2}\right\}$ **37.** $\{-2, 12\}$ **39.** $\left\{-\frac{1}{3}, 3\right\}$ **41.** $\{-7, 1\}$

43. $\{-8.7, 8.7\}$ **45.** $\left\{-\frac{9}{2}, \frac{11}{2}\right\}$ **47.** $\{-8, 2\}$ **49.** $\left\{-\frac{1}{2}\right\}$
51. $\left\{-\frac{3}{5}, 5\right\}$ **53.** $\mathbb{R}$ **55.** $\left\{\frac{11}{4}\right\}$
57. $\{a \mid -3 \le a \le 3\}$, or $[-3, 3]$

59. $\{t \mid t < 0 \text{ or } t > 0\}$, or $(-\infty, 0) \cup (0, \infty)$

61. $\{x \mid -3 < x < 5\}$, or $(-3, 5)$

63. $\{n \mid -8 \le n \le 4\}$, or $[-8, 4]$

65. $\{x \mid x < -2 \text{ or } x > 8\}$, or $(-\infty, -2) \cup (8, \infty)$

67. $\mathbb{R}$, or $(-\infty, \infty)$

69. $\left\{a \mid a \le -\frac{10}{3} \text{ or } a \ge \frac{2}{3}\right\}$, or $\left(-\infty, -\frac{10}{3}\right] \cup \left[\frac{2}{3}, \infty\right)$

71. $\{y \mid -9 < y < 15\}$, or $(-9, 15)$

73. $\{x \mid x \le -8 \text{ or } x \ge 0\}$, or $(-\infty, -8] \cup [0, \infty)$

75. $\left\{x \mid x < -\frac{1}{2} \text{ or } x > \frac{7}{2}\right\}$, or $\left(-\infty, -\frac{1}{2}\right) \cup \left(\frac{7}{2}, \infty\right)$

77. $\varnothing$
79. $\left\{x \mid x < -\frac{43}{24} \text{ or } x > \frac{9}{8}\right\}$, or $\left(-\infty, -\frac{43}{24}\right) \cup \left(\frac{9}{8}, \infty\right)$

81. $\{m \mid -9 \le m \le 3\}$, or $[-9, 3]$

83. $\{a \mid -6 < a < 0\}$, or $(-6, 0)$

85. $\left\{x \mid -\frac{1}{2} \le x \le \frac{7}{2}\right\}$, or $\left[-\frac{1}{2}, \frac{7}{2}\right]$

87. $\left\{x \mid x \le -\frac{7}{3} \text{ or } x \ge 5\right\}$, or $\left(-\infty, -\frac{7}{3}\right] \cup [5, \infty)$

89. $\{x \mid -4 < x < 5\}$, or $(-4, 5)$
91.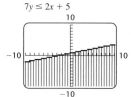
93.

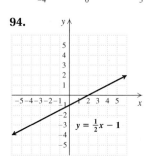

94.

$y = \frac{1}{2}x - 1$

95.

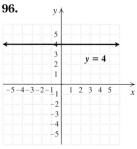

$x = -2$

96.

$y = 4$

97. $\left(4, -\frac{4}{3}\right)$ **98.** $(5, 1)$ **99.** $\left(\frac{5}{7}, -\frac{18}{7}\right)$ **100.** $(-2, -5)$
101. 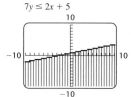 **103.** $\left\{t \mid t \ge \frac{5}{3}\right\}$, or $\left[\frac{5}{3}, \infty\right)$
105. $\mathbb{R}$, or $(-\infty, \infty)$ **107.** $\left\{-\frac{1}{7}, \frac{7}{3}\right\}$ **109.** $|x| < 3$
111. $|x| \ge 6$ **113.** $|x + 3| > 5$
115. $|x - 7| < 2$, or $|7 - x| < 2$ **117.** $|x - 3| \le 4$
119. $|x + 4| < 3$ **121.** Between 80 ft and 100 ft
123. $\{x \mid 1 \le x \le 5\}$, or $[1, 5]$ **125.** 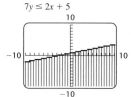,

Connecting the Concepts, pp. 250–251

1. 2 **2.** $\{x \mid x > 3\}$, or $(3, \infty)$ **3.** $\frac{9}{2}$ **4.** $\{-4, 3\}$
5. $\{x \mid x \ge -6\}$, or $[-6, \infty)$ **6.** $\{t \mid -4 < t < 4\}$, or
$(-4, 4)$ **7.** $-\frac{11}{8}$ **8.** $\frac{62}{3}$ **9.** $\left\{-\frac{8}{3}, \frac{8}{3}\right\}$
10. $\{x \mid -7 \le x \le 13\}$, or $[-7, 13]$ **11.** 31 **12.** $\varnothing$
13. $\left\{x \mid x \le -\frac{17}{2} \text{ or } x \ge \frac{7}{2}\right\}$, or $\left(-\infty, -\frac{17}{2}\right] \cup \left[\frac{7}{2}, \infty\right)$
14. $\varnothing$ **15.** $\{m \mid -24 < m < 12\}$, or $(-24, 12)$
16. $\{-42, 38\}$ **17.** $\{t \mid t \le 4 \text{ or } t \ge 10\}$, or
$(-\infty, 4] \cup [10, \infty)$ **18.** $\frac{7}{2}$ **19.** $\{a \mid -7 < a < -5\}$, or
$(-7, -5)$ **20.** $\mathbb{R}$, or $(-\infty, \infty)$

Technology Connection, p. 255

1. $y > x + 3.5$

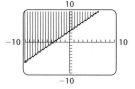

2. $7y \le 2x + 5$

3. $8x - 2y < 11$

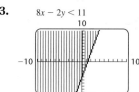

4. $11x + 13y + 4 \ge 0$

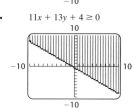

Technology Connection, p. 257

1. $y_1 \le 4 - x, \ y_2 > x - 4$

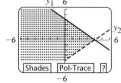

Visualizing for Success, p. 258

1. B **2.** F **3.** J **4.** A **5.** E **6.** G **7.** C **8.** D
9. I **10.** H

Exercise Set 4.4, pp. 259–261

1. (e) **2.** (c) **3.** (d) **4.** (a) **5.** (b) **6.** (f)
7. No **9.** Yes

11.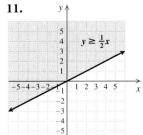
13.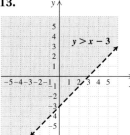

15.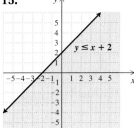
17.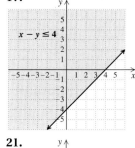

19.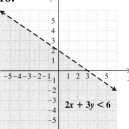
21.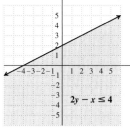

23.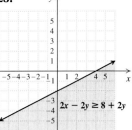
25.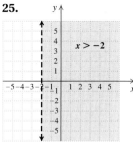

27.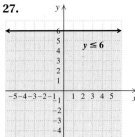
29.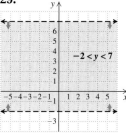

31.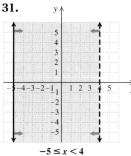
33.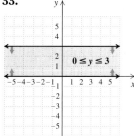

35.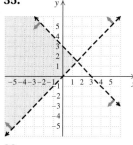
37.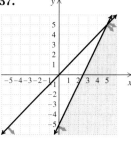

39.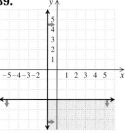
41.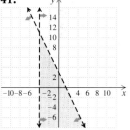

43.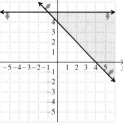
45.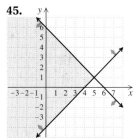

47.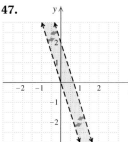
49.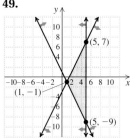

51.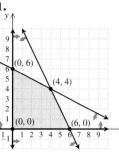
53.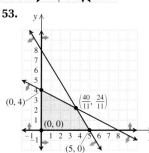

55.

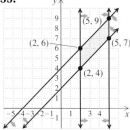

57.   **59.** 3.25%   **60.** $228
61. 3%: $3600; 5%: $6400   **62.** Carrots: 12 lb;
broccoli: 8 lb   **63.** Student tickets: 62; adult tickets: 108
64. Corn: 240 acres; soybeans: 160 acres   **65.**
67.   **69.**

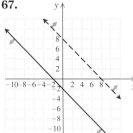   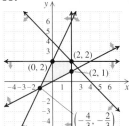

71.
$$w > 0,$$
$$h > 0,$$
$$w + h + 30 \le 62, \text{ or}$$
$$w + h \le 32,$$
$$2w + 2h + 30 \le 130, \text{ or}$$
$$w + h \le 50$$

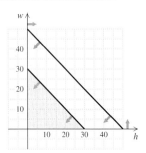

73. $q + v \ge 1150,$
$q \ge 700,$
$q \le 800,$
$v \ge 400,$
$v \le 800$

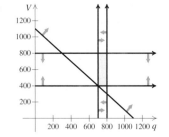

75. $35c + 75a > 1000,$
$c \ge 0,$
$a \ge 0$

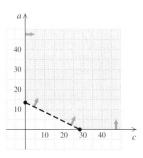

77. $h < 2w,$
$w \le 1.5h,$
$h \le 3200,$
$h \ge 0,$
$w \ge 0$

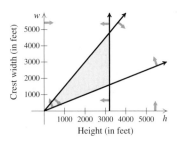

79. (a) $3x + 6y > 2$ **(b)** $x - 5y \le 10$

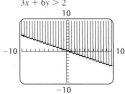   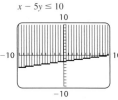

(c) $13x - 25y + 10 \le 0$ **(d)** $2x + 5y > 0$

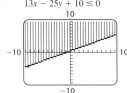   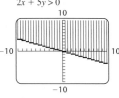

Connecting the Concepts, pp. 262–263

1.   0 ——— 5

2.   0 ——— 5

3.   0 ——— 5

4.   0 ——— 13

5.   −1 0

6.   −2 0

7.   **8.**

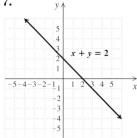   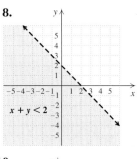

9.   **10.**

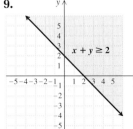   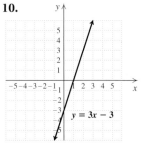

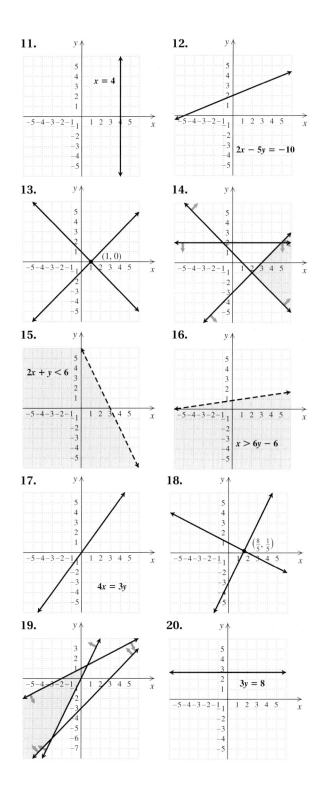

11. y $x = 4$

12. y $2x - 5y = -10$

13. y $(1, 0)$

14. y

15. y $2x + y < 6$

16. y $x > 6y - 6$

17. y $4x = 3y$

18. y $\left(\frac{8}{5}, \frac{1}{5}\right)$

19. y

20. y $3y = 8$

Exercise Set 4.5, pp. 267–269

1. Objective   **2.** Constraints   **3.** Corner
4. Feasible   **5.** Vertices   **6.** Vertex   **7.** Maximum
84 when $x = 0$, $y = 6$; minimum 0 when $x = 0$, $y = 0$
9. Maximum 76 when $x = 7$, $y = 0$; minimum 16 when
$x = 0$, $y = 4$ **11.** Maximum 5 when $x = 3$, $y = 7$;

minimum -15 when $x = 3$, $y = -3$ **13.** Gumbo: 40
orders; sandwiches: 50 orders   **15.** 4-photo pages: 5;
6-photo pages: 15; 110 photos   **17.** Corporate bonds:
$22,000; municipal bonds: $18,000; maximum: $3110
19. Short-answer questions: 12; essay questions: 4
21. Merlot: 80 acres; Cabernet: 160 acres
23. 2.5 servings of each   **25.** 🖩   **27.** 7   **28.** 22
29. $21t - 16$ **30.** $22x + 30$ **31.** t **32.** $12x - 15$
33. 🖩   **35.** T3's: 30; S5's:10   **37.** Chairs: 25; sofas: 9

Review Exercises: Chapter 4, pp. 272–273

1. True   **2.** False   **3.** True   **4.** False   **5.** True
6. True   **7.** True   **8.** False   **9.** False   **10.** False
11. $\{x | x \leq -1\}$, or $(-\infty, -1]$;

12. $\{a | a \leq 4\}$, or $(-\infty, 4]$;

13. $\left\{y | y > -\frac{15}{4}\right\}$, or $\left(-\frac{15}{4}, \infty\right)$;

14. $\{y | y > -30\}$, or $(-30, \infty)$;

15. $\left\{x | x > -\frac{3}{2}\right\}$, or $\left(-\frac{3}{2}, \infty\right)$;

16. $\{x | x < -3\}$, or $(-\infty, -3)$;

17. $\left\{y | y > -\frac{220}{23}\right\}$, or $\left(-\frac{220}{23}, \infty\right)$;

18. $\left\{x | x \leq -\frac{5}{2}\right\}$, or $\left(-\infty, -\frac{5}{2}\right]$;

19. $\{x | x \leq 2\}$, or $(-\infty, 2]$ **20.** More than 125 hr
21. $3000   **22.** $\{a, c\}$ **23.** $\{a, b, c, d, e, f, g\}$
24. $(-3, 2]$

25. $(-\infty, \infty)$

26. $\{x | -8 < x \leq 0\}$, or $(-8, 0]$

27. $\left\{x | -\frac{5}{4} < x < \frac{5}{2}\right\}$, or $\left(-\frac{5}{4}, \frac{5}{2}\right)$

28. $\{x | x < -3 \text{ or } x > 1\}$, or $(-\infty, -3) \cup (1, \infty)$

29. $\{x | x < -11 \text{ or } x \geq -6\}$, or $(-\infty, -11) \cup [-6, \infty)$

30. $\{x | x \leq -6 \text{ or } x \geq 8\}$, or $(-\infty, -6] \cup [8, \infty)$

31. $\left\{x | x < -\frac{2}{5} \text{ or } x > \frac{8}{5}\right\}$, or $\left(-\infty, -\frac{2}{5}\right) \cup \left(\frac{8}{5}, \infty\right)$

32. $(-\infty, -3) \cup (-3, \infty)$ **33.** $[2, \infty)$ **34.** $\left(-\infty, \frac{1}{4}\right]$
35. $\{-11, 11\}$ **36.** $\{t | t \leq -21 \text{ or } t \geq 21\}$, or
$(-\infty, -21] \cup [21, \infty)$ **37.** $\{5, 11\}$
38. $\left\{a | -\frac{7}{2} < a < 2\right\}$, or $\left(-\frac{7}{2}, 2\right)$
39. $\left\{x | x \leq -\frac{11}{3} \text{ or } x \geq \frac{19}{3}\right\}$, or $\left(-\infty, -\frac{11}{3}\right] \cup \left[\frac{19}{3}, \infty\right)$
40. $\left\{-14, \frac{4}{3}\right\}$ **41.** $\varnothing$ **42.** $\{x | -16 \leq x \leq 8\}$, or $[-16, 8]$
43. $\{x | x < 0 \text{ or } x > 10\}$, or $(-\infty, 0) \cup (10, \infty)$
44. $\{x | -6 \leq x \geq 4\}$, or $[-6, 4]$ **45.** $\varnothing$

46.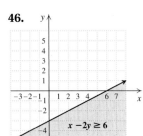

47.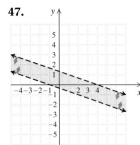

48.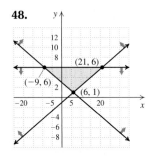

49. Maximum 40 when $x = 7$, $y = 15$; minimum 10 when $x = 1$, $y = 3$ **50.** Ohio plant: 120 computers; Oregon plant: 40 computers   **51.** 📝 The equation $|X| = p$ has two solutions when p is positive because X can be either p or $-p$. The same equation has no solution when p is negative because no number has a negative absolute value.
52. 📝 The solution set of a system of inequalities is all ordered pairs that make *all* the individual inequalities true. This consists of ordered pairs that are common to all the individual solution sets, or the intersection of the graphs.
53. $\left\{x \mid -\frac{8}{3} \le x \le -2\right\}$, or $\left[-\frac{8}{3}, -2\right]$ **54.** False: $-4 < 3$ is true, but $(-4)^2 < 9$ is false.   **55.** $|d - 2.5| \le 0.003$

Test Chapter 4, p. 274

1. [4.1] $\{x \mid x < 11\}$, or $(-\infty, 11)$

2. [4.1] $\{t \mid t > -24\}$, or $(-24, \infty)$

3. [4.1] $\{y \mid y \le -2\}$, or $(-\infty, -2]$

4. [4.1] $\left\{a \mid a \le \frac{11}{5}\right\}$, or $\left(-\infty, \frac{11}{5}\right]$

5. [4.1] $\left\{x \mid x > \frac{16}{5}\right\}$, or $\left(\frac{16}{5}, \infty\right)$

6. [4.1] $\left\{x \mid x \le \frac{9}{16}\right\}$, or $\left(-\infty, \frac{9}{16}\right]$

7. [4.1] $\{x \mid x > 1\}$, or $(1, \infty)$ **8.** [4.1] More than $187\frac{1}{2}$ mi
9. [4.1] Less than or equal to 2.5 hr   **10.** [4.2] $\{a, e\}$
11. [4.2] $\{a, b, c, d, e, i, o, u\}$ **12.** [4.2] $(-\infty, 2]$
13. [4.2] $(-\infty, 7) \cup (7, \infty)$
14. [4.2] $\left\{x \mid -\frac{3}{2} < x \le \frac{1}{2}\right\}$, or $\left(-\frac{3}{2}, \frac{1}{2}\right]$

15. [4.2] $\{x \mid x < 3 \text{ or } x > 6\}$, or $(-\infty, 3) \cup (6, \infty)$

16. [4.2] $\left\{x \mid x < -4 \text{ or } x \ge -\frac{5}{2}\right\}$, or $(-\infty, -4) \cup \left[-\frac{5}{2}, \infty\right)$

17. [4.2] $\{x \mid -3 \le x \le 1\}$, or $[-3, 1]$

18. [4.3] $\{-15, 15\}$

19. [4.3] $\{a \mid a < -5 \text{ or } a > 5\}$, or $(-\infty, -5) \cup (5, \infty)$

20. [4.3] $\left\{x \mid -2 < x < \frac{8}{3}\right\}$, or $\left(-2, \frac{8}{3}\right)$

21. [4.3] $\left\{t \mid t \le -\frac{13}{5} \text{ or } t \ge \frac{7}{5}\right\}$, or $\left(-\infty, -\frac{13}{5}\right] \cup \left[\frac{7}{5}, \infty\right)$

22. [4.3] $\varnothing$
23. [4.2] $\left\{x \mid x < \frac{1}{2} \text{ or } x > \frac{7}{2}\right\}$, or $\left(-\infty, \frac{1}{2}\right) \cup \left(\frac{7}{2}, \infty\right)$

24. [4.3] $\left\{-\frac{3}{2}\right\}$
25. [4.4]   **26.** [4.4]

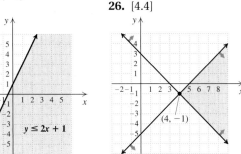

27. [4.4]

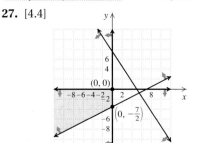

28. [4.5] Maximum 57 when $x = 6$, $y = 9$; minimum 5 when $x = 1$, $y = 0$ **29.** [4.5] Manicures: 35; haircuts: 15; maximum: \$690   **30.** [4.3] $[-1, 0] \cup [4, 6]$
31. [4.2] $\left(\frac{1}{5}, \frac{4}{5}\right)$ **32.** [4.3] $|x + 3| \le 5$

Cumulative Review: Chapters 1–4, pp. 275–276

1. $\frac{5}{9}$ **2.** 22   **3.** $c - 6$ **4.** $-\frac{1}{100}$ **5.** $-\frac{6x^4}{y^3}$ **6.** $\frac{9a^6}{4b^4}$

7. $-\frac{2}{3}$ **8.** 5   **9.** $\mathbb{R}$ **10.** $\left(\frac{22}{17}, -\frac{2}{17}\right)$ **11.** No solution
12. $(5, 1)$ **13.** $\left\{-\frac{7}{2}, \frac{9}{2}\right\}$ **14.** $\left\{x \mid x < \frac{13}{2}\right\}$, or $\left(-\infty, \frac{13}{2}\right)$
15. $\{t \mid t < -3 \text{ or } t > 3\}$, or $(-\infty, -3) \cup (3, \infty)$
16. $\left\{x \mid -2 \le x \le \frac{10}{3}\right\}$, or $\left[-2, \frac{10}{3}\right]$

17.

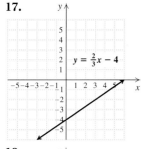

$y = \frac{2}{3}x - 4$

18.

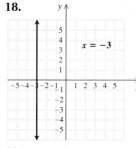

$x = -3$

19.

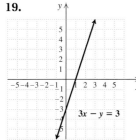

$3x - y = 3$

20.

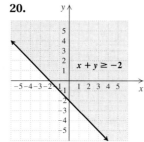

$x + y \geq -2$

21.

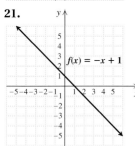

$f(x) = -x + 1$

22.

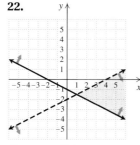

23. Slope: $\frac{4}{9}$; y-intercept: $(0, -2)$ **24.** $f(x) = -7x - 25$
25. $y = \frac{2}{3}x + 4$ **26.** Domain: $\mathbb{R}$; range: $\{y \mid y \geq -2\}$,
or $[-2, \infty)$ **27.** $\{x \mid x$ is a real number $and\, x \neq -\frac{5}{2}\}$,
or $\left(-\infty, -\frac{5}{2}\right) \cup \left(-\frac{5}{2}, \infty\right)$ **28.** 22   **29.** $x^2 + 6x - 9$
30.

$$\xleftarrow{\quad\underset{-1\ \ 0}{\rule{1.2cm}{0pt}}\underset{4}{\rule[0.3ex]{1.4cm}{0.4pt}}\quad}$$

31. $\{x \mid x$ is a real number $and\, x \neq 0\, and\, x \neq \frac{1}{3}\}$

32. $t = \dfrac{c}{a - d}$ **33.** 4.5×10^{10} gal   **34.** Beef: 9300 gal;
wheat: 2300 gal   **35. (a)** $b(t) = \frac{34}{15}t + 34.6$; **(b)** \$55 billion;
(c) 2008   **36. (a)** More than \$40; **(b)** costs greater than \$30
37. Length: 10 cm; width: 6 cm   **38.** \$640
39. \$50 billion   **40.** Vegetarian dinners: 12; steak
dinners: 16   **41.** $m = \frac{1}{3}$, $b = \frac{16}{3}$ **42.** $\frac{64}{3}$
43. $[-4, 0) \cup (0, \infty)$ **44.** 2^{11a-16}

CHAPTER 5

Technology Connection, p. 282

1. Correct   **2.** Incorrect   **3.** Correct   **4.** Incorrect

Exercise Set 5.1, pp. 284–288

1. (g)   **2.** (d)   **3.** (a)   **4.** (h)   **5.** (b)   **6.** (c)   **7.** (j)
8. (e)   **9.** (f)   **10.** (i)   **11. (a)** 5; **(b)** 6, 4, 3, 1, 0; **(c)** 6;
(d) $-5x^6$; **(e)** -5 **13. (a)** 4; **(b)** 4, 5, 3, 0; **(c)** 5; **(d)** a^3b^2;
(e) 1   **15.** 5   **17.** 6   **19.** $-15t^4 + 2t^3 + 5t^2 - 8t + 4$;
$-15t^4$; -15 **21.** $-x^6 + 6x^5 + 7x^2 + 3x - 5$; $-x^6$; -1

23. $-9 + 4x + 5x^3 - x^6$ **25.** $8y + 5xy^3 + 2x^2y - x^3$
27. -38 **29.** -16 **31.** -13; 11   **33.** 282; -9
35. About 28 ft   **37.** 6840   **39.** About 250 horsepower
41. About 20 W   **43.** 150   **45.** 14; 55 oranges
47. About 2.3 mcg/mL   **49.** 2.3   **51.** 56.5 in^2
53. \$18,750   **55.** \$8375   **57.** $3x^3 - x + 1$
59. $-6a^2b - 3b^2$ **61.** $10x^2 + 2xy + 15y^2$
63. $4t^4 - 3t^3 + 6t^2 + t$ **65.** $-2x^2 + x - 3xy + 2y^2 - 1$
67. $6x^2y - 4xy^2 + 5xy$ **69.** $9r^2 + 9r - 9$
71. $-\frac{5}{24}xy - \frac{27}{20}x^3y^2 + 1.4y^3$ **73.** $-(3t^4 + 8t^2 - 7t - 1)$,
$-3t^4 - 8t^2 + 7t + 1$ **75.** $-(-12y^5 + 4ay^4 - 7by^2)$,
$12y^5 - 4ay^4 + 7by^2$ **77.** $7x - 8$ **79.** $-4x^2 - 3x + 13$
81. $6a - 6b + 5c$ **83.** $-2a^2 + 12ab - 7b^2$
85. $8a^2b + 16ab + 3ab^2$ **87.** $x^4 - x^2 - 1$
89. $5t^2 + t + 4$ **91.** $13r^2 - 8r - 1$ **93.** $3x^2 - 9$
95. \$9700   **97.** 📄   **99.** x^8 **100.** y^{10} **101.** a^6b^4
102. t^8 **103.** $25y^6$ **104.** $4x^{10}y^2$ **105.** 📄
107. $68x^5 - 81x^4 - 22x^3 + 52x^2 + 2x + 250$
109. $45x^5 - 8x^4 + 208x^3 - 176x^2 + 116x - 25$
111. 494.55 cm^3 **113.** $5x^2 - 8x$ **115.** $8x^{2a} + 7x^a + 7$
117. $x^{5b} + 4x^{4b} + x^{3b} - 6x^{2b} - 9x^b$ **119.** 〰️

Technology Connection, p. 294

1. $y_1 = x^2 - 9 - (x - 3)(x + 3)$

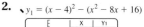

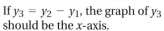

2. $y_1 = (x - 4)^2 - (x^2 - 8x + 16)$

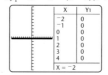

3. $y_1 = x^2 - 4$,
$y_2 = (x + 2)(x - 2)$

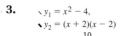
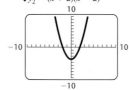

If $y_3 = y_2 - y_1$, the graph of y_3
should be the x-axis.

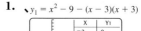

Exercise Set 5.2, pp. 296–298

1. False   **2.** True   **3.** True   **4.** True   **5.** False
6. False   **7.** True   **8.** True   **9.** $15x^5$ **11.** $-48a^3b^2$
13. $36x^5y^6$ **15.** $21x - 7x^2$ **17.** $20c^3d^2 - 25c^2d^3$
19. $x^2 + 8x + 15$ **21.** $8a^2 + 10a - 3$
23. $x^3 - x^2 - 5x + 2$ **25.** $t^3 - 3t^2 - 13t + 15$
27. $a^4 + 5a^3 - 2a^2 - 9a + 5$ **29.** $x^3 + 27$
31. $a^3 - b^3$ **33.** $t^2 - t - 6$ **35.** $20x^2 + 13xy + 2y^2$
37. $t^2 - \frac{7}{12}t + \frac{1}{12}$ **39.** $3t^2 + 1.5st - 15s^2$
41. $r^3 + 4r^2 + r - 6$ **43.** $x^2 + 10x + 25$
45. $4y^2 - 28y + 49$ **47.** $25c^2 - 20cd + 4d^2$
49. $9a^6 - 60a^3b^2 + 100b^4$ **51.** $x^6y^8 + 10x^3y^4 + 25$
53. $12x^4 - 21x^3 - 17x^2 + 35x - 5$ **55.** $25x^2 - 20x + 4$
57. $4x^2 - \frac{4}{3}x + \frac{1}{9}$ **59.** $c^2 - 49$ **61.** $1 - 16x^2$
63. $9m^2 - \frac{1}{4}n^2$ **65.** $x^6 - y^2z^2$
67. $-m^2n^2 + 9m^4$, or $9m^4 - m^2n^2$ **69.** $14x + 58$
71. $3m^2 + 4mn - 5n^2$ **73.** $a^2 + 2ab + b^2 - 1$
75. $4x^2 + 12xy + 9y^2 - 16$ **77.** $A = P + 2Pr + Pr^2$

79. (a) $t^2 - 2t + 6$; **(b)** $2ah + h^2$; **(c)** $2ah - h^2$
81. (a) $2a^2$; **(b)** $a^2 + 2ah + h^2 + a + h$; **(c)** $2ah + h^2 + h$
83.   **85.** $5(x + 3y - 1)$ **86.** $8(x + 1 - 5y)$
87. $16(t - 4)$ **88.** $6(3a + 5b)$ **89.** $x(a + b - c)$
90. $b(x + y - 1)$ **91.**   **93.** $x^4 - y^{2n}$
95. $5x^{n+2}y^3 + 4x^2y^{n+3}$ **97.** $x^{3n} - x^{2n} - 14x^n + 8$
99. $a^2 + 2ac + c^2 - b^2 - 2bd - d^2$ **101.** $x^4 + x^2 + 25$
103. $x^6 - 1$ **105.** $x^{a^2-b^2}$ **107.** 0 **109.** $2a + h$
111.

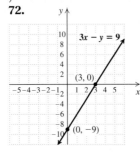

113. (b) and (c) are identities.

Technology Connection, p. 301

1. A table should show that $y_3 = 0$ for any value of x.

Exercise Set 5.3, pp. 302–305

1. True   **2.** False   **3.** True   **4.** True   **5.** True
6. False   **7.** True   **8.** False   **9.** $5(2x^2 + 7)$
11. $2y(y - 9)$ **13.** $5(t^3 - 3t + 1)$
15. $a^3(a^3 + 2a - 1)$ **17.** $6x(2x^3 - 5x^2 + 7)$
19. $b(6a^2 - 2a - 9)$ **21.** $5m^3n(3m + 6m^2n + 5n^2)$
23. $3x^2y^4z^2(3xy^2 - 4x^2z^2 + 5yz)$ **25.** $-5(x + 8)$
27. $-16(t^2 - 6)$ **29.** $-2(x^2 - 6x - 20)$
31. $-5(-1 + 2y)$, or $-5(2y - 1)$ **33.** $-4d(-2d + 3c)$,
or $-4d(3c - 2d)$ **35.** $-1(m^3 - 8)$
37. $-1(p^3 + 2p^2 + 5p - 2)$ **39.** $(b - 5)(a + c)$
41. $(x + 7)(2x - 3)$ **43.** $(x - y)(a^2 - 5)$
45. $(y + z)(x + w)$ **47.** $(y - 1)(y^2 + 3)$
49. $(t + 6)(t^2 - 2)$ **51.** $3a^2(4a^2 - 7a - 3)$
53. $(y - 1)(y^7 + 1)$ **55.** $(x - 2)(3 - xy)$,
or $(xy - 3)(2 - x)$ **57. (a)** $h(t) = -8t(2t - 9)$;
(b) $h(1) = 56$ ft   **59.** $\pi r(2h + r)$ **61.** $P(t) = t(t - 5)$
63. $R(x) = 0.4x(700 - x)$ **65.** $P(n) = \frac{1}{2}(n^2 - 3n)$
67. $N(x) = \frac{1}{6}(x^3 + 3x^2 + 2x)$ **69.**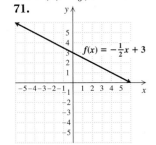
71.

72.

73.

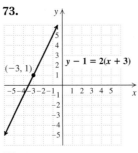

74.

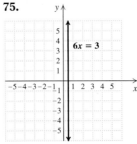

75.

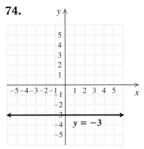

76.

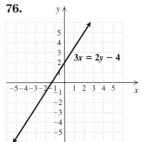

77. 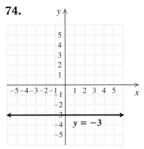   **79.** $x^5y^4 + x^4y^6 = x^3y(x^2y^3 + xy^5)$
81. $(x^2 - x + 5)(r + s)$
83. $(x^4 + x^2 + 5)(a^4 + a^2 + 5)$
85. $x^{-9}(x^3 + 1 + x^6)$ **87.** $x^{1/3}(1 - 5x^{1/6} + 3x^{5/12})$
89. $x^{-5/2}(1 + x)$ **91.** $x^{-7/5}(x^{3/5} - 1 + x^{16/15})$
93. $3a^n(a + 2 - 5a^2)$ **95.** $y^{a+b}(7y^a - 5 + 3y^b)$
97.

Connecting the Concepts, p. 305

1. $4t^3 + 8t^2 - 13t - 1$ **2.** $2a + 2b + 2c$
3. $12x^3y^2 - 8x^5y + 24x^2y^3$ **4.** $10(3x^2 - 4)$
5. $9n^2 - n$ **6.** $8t(2t^2 - 5t + 1)$ **7.** $x^2 - 8y^2$
8. $(x + y)(2 + a)$ **9.** $x^2 + 8x + 7$
10. $10x^2 - 17x + 3$ **11.** $(p - 1)(p^2 + 7)$
12. $\frac{7}{6}x^2 - \frac{1}{6}x - \frac{11}{6}$ **13.** $9m^2 - 60m + 100$
14. $-1.6x^2 - 2.7x - 1.4$ **15.** $a^3 + a^2 - 8a - 12$
16. $(t - 1)(3t^2 + 1)$ **17.** $c^2 - 81$ **18.** $-s^2 + 36t^2$,
or $36t^2 - s^2$ **19.** $4x^2y(2y^2z + 3xy - 4z^3)$
20. $(x - 3)(x^2 - 5)$

Technology Connection, p. 308

1. They should coincide.   **2.** The x-axis
3. Let $y_1 = x^3 - x^2 - 30x$, $y_2 = x(x + 5)(x - 6)$, and
$y_3 = y_2 - y_1$. The graphs of y_1 and y_2 should coincide; the
graph of y_3 should be the x-axis.
4. Let $y_1 = 2x^2 + x - 15$, $y_2 = (2x + 5)(x - 3)$, and
$y_3 = y_2 - y_1$. The graphs of y_1 and y_2 do not coincide; the
graph of y_3 is not the x-axis.

Exercise Set 5.4, pp. 314–315

1. True   **2.** True   **3.** False   **4.** False   **5.** False
6. True   **7.** True   **8.** False   **9.** $(x + 1)(x + 4)$
11. $(y - 3)(y - 9)$ **13.** $(t - 4)(t + 2)$
15. $(a - 1)(a + 2)$ **17.** $2(x - 6)(x + 9)$
19. $(a + 5)(a + 9)$ **21.** $p(p - 9)(p + 8)$

23. $(a - 4)(a - 7)$ **25.** $(x + 3)(x - 2)$
27. $5(y + 1)(y + 7)$ **29.** $(8 - y)(4 + y)$
31. $x(8 - x)(7 + x)$ **33.** $y^2(y + 12)(y - 7)$
35. Prime   **37.** $(x + 3y)(x + 9y)$
39. $(x - 7y)(x - 7y)$, or $(x - 7y)^2$
41. $n^3(n - 1)(n - 79)$ **43.** $x^4(x - 7)(x + 9)$
45. $(x - 2)(3x + 2)$ **47.** $(2t - 3)(3t + 5)$
49. $2(p - 2)(3p - 4)$ **51.** $(3a + 2)(3a + 4)$
53. $2y(3y - 1)(5y + 3)$ **55.** $6(3x - 4)(x + 1)$
57. $t^6(t + 7)(t - 2)$ **59.** $2x^2(5x - 2)(7x - 4)$
61. $(3y - 4)(6y + 5)$ **63.** $(4x + 1)(4x + 5)$
65. $(x + 4)(5x + 4)$ **67.** $-2(2t - 3)(2t + 5)$
69. $xy(6y + 5)(3y - 2)$ **71.** $(24x + 1)(x - 2)$
73. $3x(7x + 3)(3x + 4)$ **75.** $2x^2(6x + 5)(4x - 3)$
77. $(4a - 3b)(3a - 2b)$ **79.** $(2x - 3y)(x + 2y)$
81. $(2x - 7y)(3x - 4y)$ **83.** $(3x - 5y)(3x - 5y)$, or
$(3x - 5y)^2$ **85.** $(9xy - 4)(xy + 1)$ **87.** ✍   **89.** $25a^2$
90. $9x^8$ **91.** $x^2 + 6x + 9$ **92.** $4t^2 - 20t + 25$
93. $y^2 - 1$ **94.** $16x^2 - 9y^2$ **95.** ✍
97. $5(4x^4y^3 + 1)(3x^4y^3 + 1)$ **99.** $\left(y + \frac{4}{7}\right)\left(y - \frac{2}{7}\right)$
101. $ab^2(4ab^2 + 1)(5ab^2 - 2)$ **103.** $(x^a + 8)(x^a - 3)$
105. $a(2r + s)(r + s)$ **107.** $(x - 4)(x + 8)$
109. $76, -76, 28, -28, 20, -20$
111. Since $ax^2 + bx + c = (mx + r)(nx + s)$, from FOIL
we know that $a = mn$, $c = rs$, and $b = ms + rn$. If $P = ms$
and $Q = rn$, then $b = P + Q$. Since $ac = mnrs = msrn$, we
have $ac = PQ$.   **113.** ◿   **115.** ✍

Exercise Set 5.5, pp. 319–321

1. Difference of two squares   **2.** Perfect-square trinomial
3. Perfect-square trinomial   **4.** Difference of two squares
5. None of these   **6.** Polynomial having a common factor
7. Polynomial having a common factor   **8.** None of these
9. Perfect-square trinomial   **10.** Polynomial having a
common factor   **11.** $(x + 10)^2$ **13.** $(t - 1)^2$
15. $4(a - 3)^2$ **17.** $(y + 6)^2$ **19.** $y(y - 9)^2$
21. $2(x - 10)^2$ **23.** $(1 - 4d)^2$ **25.** $-y(y + 4)^2$
27. $(0.5x + 0.3)^2$ **29.** $(p - q)^2$ **31.** $(5a + 3b)^2$
33. $5(a + b)^2$ **35.** $(x + 5)(x - 5)$
37. $(m + 8)(m - 8)$ **39.** $(2a + 9)(2a - 9)$
41. $12(c + d)(c - d)$ **43.** $7x(y^2 + z^2)(y + z)(y - z)$
45. $a(2a + 7)(2a - 7)$
47. $3(x^4 + y^4)(x^2 + y^2)(x + y)(x - y)$
49. $(pq + 10)(pq - 10)$ **51.** $a^2(3a + 5b^2)(3a - 5b^2)$
53. $\left(y + \frac{1}{2}\right)\left(y - \frac{1}{2}\right)$ **55.** $\left(\frac{1}{10} + x\right)\left(\frac{1}{10} - x\right)$
57. $(a + b + 6)(a + b - 6)$ **59.** $(x - 3 + y)(x - 3 - y)$
61. $(t + 8)(t + 1)(t - 1)$ **63.** $(r - 3)^2(r + 3)$
65. $(m - n + 5)(m - n - 5)$
67. $(9 + x + y)(9 - x - y)$
69. $(r - 1 + 2s)(r - 1 - 2s)$
71. $(4 + a + b)(4 - a - b)$ **73.** $(x + 5)(x + 2)(x - 2)$
75. $(a - 2)(a + b)(a - b)$ **77.** ✍   **79.** $8x^6y^{12}$
80. $27a^3b^{18}$ **81.** $-1000x^{30}$ **82.** $-125x^6y^3$
83. $x^3 + 3x^2 + 3x + 1$ **84.** $x^3 - 3x^2 + 3x - 1$
85. $p^3 + 3p^2q + 3pq^2 + q^3$ **86.** $p^3 - 3p^2q + 3pq^2 - q^3$
87. ✍   **89.** $-\frac{1}{54}(4r + 3s)^2$ **91.** $(0.3x^4 + 0.8)^2$, or
$\frac{1}{100}(3x^4 + 8)^2$ **93.** $(r + s + 1)(r - s - 9)$

95. $(x^{2a} + y^b)(x^{2a} - y^b)$
97. $(5y^a + x^b - 1)(5y^a - x^b + 1)$ **99.** $3(x + 1 + 2)^2$,
or $3(x + 3)^2$ **101.** $(s - 2t + 2)^2$
103. $(3x^n - 1)^2$ **105.** $h(2a + h)$
107. (a) $\pi h(R + r)(R - r)$; **(b)** $3{,}014{,}400$ cm^3 **109.** ◿

Exercise Set 5.6, pp. 324–325

1. Difference of cubes   **2.** Sum of cubes   **3.** Difference
of squares   **4.** None of these   **5.** Sum of cubes
6. Difference of cubes   **7.** None of these   **8.** Difference
of squares   **9.** Difference of cubes   **10.** None of these
11. $(x - 4)(x^2 + 4x + 16)$ **13.** $(z + 1)(z^2 - z + 1)$
15. $(t - 10)(t^2 + 10t + 100)$
17. $(3x + 1)(9x^2 - 3x + 1)$
19. $(4 - 5x)(16 + 20x + 25x^2)$
21. $(x - y)(x^2 + xy + y^2)$ **23.** $\left(a + \frac{1}{2}\right)\left(a^2 - \frac{1}{2}a + \frac{1}{4}\right)$
25. $8(t - 1)(t^2 + t + 1)$ **27.** $2(3x + 1)(9x^2 - 3x + 1)$
29. $rs(s + 4)(s^2 - 4s + 16)$
31. $5(x - 2z)(x^2 + 2xz + 4z^2)$
33. $\left(y - \frac{1}{10}\right)\left(y^2 + \frac{1}{10}y + \frac{1}{100}\right)$
35. $(x + 0.1)(x^2 - 0.1x + 0.01)$
37. $8(2x^2 - t^2)(4x^4 + 2x^2t^2 + t^4)$
39. $2y(3y - 4)(9y^2 + 12y + 16)$
41. $(z + 1)(z^2 - z + 1)(z - 1)(z^2 + z + 1)$
43. $(t^2 + 4y^2)(t^4 - 4t^2y^2 + 16y^4)$
45. $(x^4 - yz^4)(x^8 + x^4yz^4 + y^2z^8)$ **47.** ✍
49. 35 cm, 36 cm, 37 cm   **50.** 20 cm
51. Dimes: 3 rolls; nickels: 1 roll; quarters: 6 rolls
52. 65 min or more   **53.** Ken: 54 nests; Kathy: 46 nests
54. 4 weekdays   **55.** ✍
57. $(x^{2a} - y^b)(x^{4a} + x^{2a}y^b + y^{2b})$ **59.** $2x(x^2 + 75)$
61. $5\left(xy^2 - \frac{1}{2}\right)\left(x^2y^4 + \frac{1}{2}xy^2 + \frac{1}{4}\right)$
63. $-(3x^{4a} + 3x^{2a} + 1)$ **65.** $(t - 8)(t - 1)(t^2 + t + 1)$
67. $h(2a + h)(a^2 + ah + h^2)(3a^2 + 3ah + h^2)$ **69.** ◿

Exercise Set 5.7, pp. 328–329

1. (b)   **2.** (a)   **3.** (f)   **4.** (d)   **5.** (c)   **6.** (e)   **7.** (a)
8. (f)   **9.** Factor a trinomial; $(x + 1)(x - 4)$
10. Factor a difference of squares; factor a sum of cubes
and a difference of cubes;
$(x + 1)(x^2 - x + 1)(x - 1)(x^2 + x + 1)$
11. Factor by grouping; factor a difference of squares;
$(x + 1)(x - 1)(2x - 5)$
12. Factor a perfect-square trinomial; $(t - 10)^2$
13. Factor out a common factor; factor a trinomial;
$4(2a + 1)(3a - 2)$
14. Factor out a common factor; $6(x^2 + x + 2)$
15. $(x + 9)(x - 9)$ **17.** $9(m^2 + 10)(m^2 - 10)$
19. $2x(x + 2)(x + 4)$ **21.** $(a + 5)^2$
23. $(2y - 3)(y - 4)$ **25.** $3(x + 12)(x - 7)$
27. $(5x + 3y)(5x - 3y)$ **29.** $(t^2 + 1)(t^4 - t^2 + 1)$
31. $(x + y + 3)(x - y + 3)$
33. $2(4a + 5b)(16a^2 - 20ab + 25b^2)$
35. $7x(x + 3)(x - 5)$ **37.** $t^2(4t + 3)(4t - 3)$
39. $(m^3 + 10)(m^3 - 2)$ **41.** $(a + d)(c - b)$

43. $(2c - d)^2$ **45.** $(5x + y)(8x - y)$
47. $(2a - 5)(2 + a^2)$ **49.** $2(x + 3)(x + 2)(x - 2)$
51. $2(3a - 2b)(9a^2 + 6ab + 4b^2)$ **53.** $(6y - 5)(6y + 7)$
55. $4(m^2 + 4n^2)(m + 2n)(m - 2n)$
57. $ab(a^2 + 4b^2)(a + 2b)(a - 2b)$ **59.** $2t(17t^2 - 3)$
61. $2(a - 3)(a + 3)$ **63.** $7a(a^3 - 2a^2 + 3a - 1)$
65. $(9ab + 2)(3ab + 4)$ **67.** $-5t(2t^2 - 3)$
69. $-2x(3x^3 - 4x^2 + 6)$ **71.** $p(1 - 4p)(1 + 4p + 16p^2)$
73. $(a - b - 3)(a + b + 3)$ **75.** ✒   **77.** -2 **78.** $\frac{5}{2}$
79. 0 **80.** 0 **81.** $\left\{x | x \text{ is a real number } and \, x \neq \frac{2}{3}\right\}$, or
$\left(-\infty, \frac{2}{3}\right) \cup \left(\frac{2}{3}, \infty\right)$ **82.** $\left\{x | x \text{ is a real number } and \, x \neq -\frac{1}{2}\right\}$,
or $\left(-\infty, -\frac{1}{2}\right) \cup \left(-\frac{1}{2}, \infty\right)$ **83.** ✒
85. $a(7a - bc)(4a - 3bc)$ **87.** $x(x - 2p)$
89. $y(y - 1)^2(y - 2)$
91. $(2x + y - r + 3s)(2x + y + r - 3s)$
93. $\left(\dfrac{x^9}{10} - 1\right)\left(\dfrac{x^{18}}{100} + \dfrac{x^9}{10} + 1\right)$ **95.** $3(x - 3)(x + 2)$
97. $3(a + 7)^2$ **99.** $(3x^{2s} + 4y^t)(9x^{4s} - 12x^{2s}y^t + 16y^{2t})$
101. $a(a^w + 1)^2$

Connecting the Concepts, p. 330

1. $(t - 1)^2$ **2.** $2(x - 3)(x - 5)$ **3.** $x(x + 8)(x - 8)$
4. $(2a - 1)(3a + 1)$ **5.** $5t(t^2 + 100)$
6. $(x - 4)(x^2 + 4x + 16)$ **7.** $4x(x + 5)^2$
8. $(2x^2 + 1)(x - 3)$ **9.** $y(4y + 3)(3y - 2)$
10. $(n^3 - 6)(n^3 - 4)$ **11.** $7(t + 1)(t^2 - t + 1)$
12. $6m^2(m + 8)^2$ **13.** $(x + 1)(x - 1)(x + 3)$
14. $\left(a + \frac{1}{3}\right)\left(a - \frac{1}{3}\right)$ **15.** $(0.5 + y)(0.5 - y)$
16. $3n(n - 7)$ **17.** $(x + 1)(x - 1)(x + 2)(x - 2)$
18. $-5c^2(2c - 5)$
19. $(1 + 2t)(1 - 2t + 4t^2)(1 - 2t)(1 + 2t + 4t^2)$
20. $3x(2x^4 - 5x^3 + 6x + 3)$ **21.** $x(2x^3 + 3)(x + 3)$
22. $(2x + y)(z - t)$ **23.** $2(5ab + 4c^2)(5ab - 4c^2)$
24. $(x + 5 + y)(x + 5 - y)$ **25.** $2(x + 2y)(x - 8y)$
26. $(2xy^3 + 5)^2$ **27.** $(m + n - 6)(m - n + 6)$
28. $3b(2a + 5)(a - 4)$ **29.** $(p - 11q)^2$
30. $4ab^2c(2ab - 10c^2 + 1)$

Technology Connection, pp. 334–335

1. The graphs intersect at $(-2, 36)$, $(0, 0)$, and $(5, 225)$, so
the solutions are -2, 0, and 5.
2. The zeros are -2, 0, and 5.

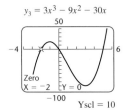

$y_3 = 3x^3 - 9x^2 - 30x$

Visualizing for Success, p. 338

1. D   **2.** J   **3.** A   **4.** E   **5.** B   **6.** C   **7.** I   **8.** F
9. G   **10.** H

1. True   **2.** True   **3.** False   **4.** True   **5.** False
6. True   **7.** $\{2, 5\}$ **9.** $\{-7, -1\}$ **11.** $\left\{-\frac{1}{2}, 0\right\}$
13. $\left\{0, \frac{4}{5}\right\}$ **15.** $\left\{-\frac{5}{2}, 7\right\}$ **17.** $\{-3, 6\}$ **19.** $\{0, 10\}$
21. $\left\{\frac{1}{3}, \frac{5}{4}\right\}$ **23.** $\left\{0, \frac{5}{2}\right\}$ **25.** $\{-2, 8\}$ **27.** $\{-4, 7\}$
29. $\{4\}$ **31.** $\{-10\}$ **33.** $\{-5, -3\}$ **35.** $\{-9, 9\}$
37. $\{-7, 0, 9\}$ **39.** $\{-5, 5\}$ **41.** $\{-6, 6\}$ **43.** $\left\{\frac{1}{3}, \frac{4}{3}\right\}$
45. $\left\{-\frac{3}{4}, -\frac{1}{2}, 0\right\}$ **47.** $\{-3, 1\}$ **49.** $\left\{-\frac{7}{4}, \frac{4}{3}\right\}$
51. $\left\{-\frac{1}{10}, \frac{1}{10}\right\}$ **53.** $\{-5, -1, 1, 5\}$ **55.** $\{-8, -4\}$
57. $\left\{-4, \frac{3}{2}\right\}$ **59.** $\{-9, -3\}$ **61.** $\left\{\frac{1}{4}, \frac{5}{3}\right\}$ **63.** $\left\{-5, 0, \frac{3}{2}\right\}$
65. $\{x | x \text{ is a real number } and \, x \neq -1 \, and \, x \neq 4\}$, or
$(-\infty, -1) \cup (-1, 4) \cup (4, \infty)$
67. $\{x | x \text{ is a real number } and \, x \neq -3 \, and \, x \neq 3\}$, or
$(-\infty, -3) \cup (-3, 3) \cup (3, \infty)$
69. $\left\{x | x \text{ is a real number } and \, x \neq 0 \, and \, x \neq \frac{1}{2}\right\}$, or
$(-\infty, 0) \cup \left(0, \frac{1}{2}\right) \cup \left(\frac{1}{2}, \infty\right)$
71. $\{x | x \text{ is a real number } and \, x \neq 0 \, and \, x \neq 2 \, and \, x \neq 5\}$,
or $(-\infty, 0) \cup (0, 2) \cup (2, 5) \cup (5, \infty)$
73. Length: 15 in.; width: 12 in.   **75.** 3 m   **77.** 3 cm
79. 10 in.   **81.** 16, 18, 20   **83.** Height: 30 in.; base: 50 in.
85. 60 ft; 65 ft   **87.** 41 ft   **89.** Length: 100 m; width: 75 m
91. 2 sets   **93.** 2 sec   **95.** 5 sec   **97.** 2005   **99.** ✒
101. $\frac{3}{2}$ **102.** $-\frac{1}{2}$ **103.** $-\frac{75}{32}$ **104.** $-\frac{2}{27}$ **105.** -2
106. $\frac{4}{3}$ **107.** $\frac{6}{7}$ **108.** $-\frac{11}{6}$ **109.** ✒   **111.** $\left\{-\frac{11}{8}, -\frac{1}{4}, \frac{2}{3}\right\}$
113. $\{-3, 1\}$; $\{x | -4 \leq x \leq 2\}$, or $[-4, 2]$
115. Answers may vary. $f(x) = 5x^3 - 20x^2 + 5x + 30$
117. Length: 28 cm; width: 14 cm   **119.** About 5.7 sec
121. 📈   **123.** $\{6.90\}$ **125.** $\{3.48\}$

Connecting the Concepts, p. 343

1. $(x + 2)(x + 3)$ **2.** $\{-3, -2\}$ **3.** $\{2, 3\}$
4. $4x^2 - x - 5$ **5.** $2x^2 - x + 5$ **6.** $(a + 1)(a - 1)$
7. $a^2 - 1$ **8.** $\{-5, 5\}$ **9.** $\{0, 19\}$
10. $3x^4 + 5x^2 - 7x - 1$ **11.** $4x^2 + 3x - 14$
12. $\left\{-2, \frac{7}{2}\right\}$ **13.** $(t + 1)(t^2 - t + 1)$
14. $(5m + 3)(2m - 7)$ **15.** $25n^2 - 60n + 36$
16. $\left\{-\frac{2}{9}\right\}$ **17.** $2(x + 7)(x - 3)$ **18.** $40y^2 + 7y - 3$
19. $-3x^2 - 9$ **20.** $\{-8, 8\}$

Review Exercises: Chapter 5, pp. 346–347

1. (g)   **2.** (b)   **3.** (a)   **4.** (d)   **5.** (e)   **6.** (j)   **7.** (h)
8. (c)   **9.** (i)   **10.** (f)   **11.** 11
12. $-5x^3 + 2x^2 + 3x + 9$; $-5x^3$; -5
13. $-3x^2 + 2x^3 + 8x^6y - 7x^8y^3$ **14.** 0; -6
15. $2ah + h^2 + 10h$ **16.** $-2a^3 + a^2 - 3a - 4$
17. $-x^2y - 2xy^2$ **18.** $-2x^3 + 2x^2 + 5x + 3$
19. $-2n^3 + 2n^2 - 2n + 11$ **20.** $-5xy^2 - 2xy - 11x^2y$
21. $14x - 7$ **22.** $-2a + 6b$ **23.** $6x^2 - 4xy + 4y^2 + 9y$
24. $-18x^3y^4$ **25.** $x^8 - x^6 + 5x^2 - 3$
26. $8a^2b^2 + 2abc - 3c^2$ **27.** $49t^2 - 1$
28. $9x^2 - 24xy + 16y^2$ **29.** $2x^2 + 5x - 3$
30. $x^4 + 8x^2y^3 + 16y^6$ **31.** $5t^2 - 42t + 16$
32. $x^2 - \frac{1}{2}x + \frac{1}{18}$ **33.** $x(7x + 6)$ **34.** $-3y(y^3 + 3y - 4)$
35. $3x(5x^3 - 6x^2 + 7x - 1)$ **36.** $(a - 9)(a - 3)$

37. $(3m + 2)(m - 4)$ **38.** $(5x + 2)^2$
39. $4(y + 2)(y - 2)$ **40.** $x(x - 2)(x + 7)$
41. $(a + 2b)(x - y)$ **42.** $(y + 2)(3y^2 - 5)$
43. $(a^2 + 9)(a + 3)(a - 3)$ **44.** $4(x^4 + x^2 + 5)$
45. $(3x + 2)(9x^2 - 6x + 4)$
46. $\left(\frac{1}{5}b - \frac{1}{2}c^2\right)\left(\frac{1}{25}b^2 + \frac{1}{10}bc^2 + \frac{1}{4}c^4\right)$
47. $10t^2(7 + 8t)(7 - 8t)$ **48.** $n^5(n^2 + 1)$
49. $(0.1x^2 + 1.2y^3)(0.1x^2 - 1.2y^3)$ **50.** $4y(x - 5)^2$
51. $(3t + p)(2t + 5p)$ **52.** $(x + 3)(x - 3)(x + 2)$
53. $(a - b + 2t)(a - b - 2t)$ **54.** $\{6\}$ **55.** $\left\{\frac{2}{3}, \frac{3}{2}\right\}$
56. $\left\{0, \frac{7}{4}\right\}$ **57.** $\{-2, 2\}$ **58.** $\{-3, 0, 7\}$ **59.** $\{-1, 6\}$
60. $\{-4, 11\}$
61. $\{x \mid x$ is a real number $and\ x \neq -7\ and\ x \neq 8\}$, or
$(-\infty, -7) \cup (-7, 8) \cup (8, \infty)$
62. Height: 16 ft; base: 7 ft   **63.** 5, 7, 9 or $-9, -7, -5$
64. Length: 8 in.; width: 5 in.   **65.** 15 ft; 17 ft   **66.** 2007
67. ✒ When multiplying polynomials, we begin with a product and carry out the multiplication to write a sum of terms. When factoring a polynomial, we write an equivalent expression that is a product.   **68.** ✒ The principle of zero products states that if a product is equal to 0, at least one of the factors must be 0. If a product is nonzero, we cannot conclude that any one of the factors is a particular value.
69. $2(2x - y)(4x^2 + 2xy + y^2)(2x + y)(4x^2 - 2xy + y^2)$
70. $-2(3x^2 + 1)$ **71.** $a^3 - b^3 + 3b^2 - 3b + 1$
72. z^{5n^5} **73.** $\left\{-1, -\frac{1}{2}\right\}$

Test: Chapter 5, p. 348

1. [5.1] 9   **2.** [5.1] $5x^5y^4 - 9x^4y - 14x^2y + 8xy^3$
3. [5.1] $-5a^3$ **4.** [5.1] 4; 2   **5.** [5.2] $2ah + h^2 - 3h$
6. [5.1] $4xy + 3xy^2$ **7.** [5.1] $-y^3 + 6y^2 - 10y - 7$
8. [5.1] $10m^3 - 4m^2n - 3mn^2 + n^2$ **9.** [5.1] $5a - 8b$
10. [5.1] $5y^2 + y^3$ **11.** [5.2] $64x^3y^8$
12. [5.2] $12a^2 - 4ab - 5b^2$ **13.** [5.2] $x^3 - 2x^2y + y^3$
14. [5.2] $16t^2 - 24t + 9$ **15.** [5.2] $25a^6 + 90a^3 + 81$
16. [5.2] $x^2 - 4y^2$ **17.** [5.3] $5x^2(9 + x^2)$
18. [5.5] $(y + 5)(y + 2)(y - 2)$
19. [5.4] $(p - 14)(p + 2)$ **20.** [5.4] $(6m + 1)(2m + 3)$
21. [5.5] $(3y + 5)(3y - 5)$
22. [5.6] $3(r - 1)(r^2 + r + 1)$ **23.** [5.5] $(3x - 5)^2$
24. [5.5] $(x^4 + y^4)(x^2 + y^2)(x + y)(x - y)$
25. [5.5] $(y + 4 + 10t)(y + 4 - 10t)$
26. [5.5] $5(2a - b)(2a + b)$ **27.** [5.4] $2(4x - 1)(3x - 5)$
28. [5.6] $2ab(2a^2 + 3b^2)(4a^4 - 6a^2b^2 + 9b^4)$
29. [5.3] $4xy(y^3 + 9x + 2x^2y + 1)$ **30.** [5.8] $\{-4, 10\}$
31. [5.8] $\{-5, 5\}$ **32.** [5.8] $\left\{-7, -\frac{3}{2}, 0\right\}$
33. [5.8] $\left\{-\frac{1}{2}, 0\right\}$ **34.** [5.8] $\{0, 5\}$
35. [5.8] $\{x \mid x$ is a real number $and\ x \neq -1\}$, or
$(-\infty, -1) \cup (-1, \infty)$ **36.** [5.8] 2.5 in.
37. [5.8] 10 cm; 24 cm   **38.** [5.8] $4\frac{1}{2}$ sec
39. (a) [5.2] $x^5 + x + 1$;
(b) [5.2], [5.7] $(x^2 + x + 1)(x^3 - x^2 + 1)$
40. [5.4] $(3x^n + 4)(2x^n - 5)$

Cumulative Review: Chapters 1–5, pp. 349–350

1. -297 **2.** $x^3 + x^2 - 8x - 7$ **3.** $-8a^2 + 6a^2b + 3b^2$
4. $4x^2 - 81$ **5.** $4x^2 + 36xy + 81y^2$
6. $10m^5 - 5m^3n + 2m^2n - n^2$ **7.** $3(t + 4)(t - 4)$
8. $(a - 7)^2$ **9.** $9x^2y(4xy - 3x^2 + 5y^2)$
10. $(5a + 4b)(25a^2 - 20ab + 16b^2)$
11. $n(m - 7)(m + 3)$
12. $x^2(x + 1)(x^2 - x + 1)(x - 1)(x^2 + x + 1)$
13. $(3y - 2)(4y + 5)$ **14.** $(d + a - b)(d - a + b)$
15. $\varnothing$ **16.** $\left(14, \frac{7}{2}\right)$ **17.** $(-2, -3, 4)$
18. $\left(-\infty, -\frac{1}{4}\right]$, or $\left\{x \mid x \leq -\frac{1}{4}\right\}$
19. $(-8, -6)$, or $\{x \mid -8 < x < -6\}$
20. $[-2, 3]$, or $\{x \mid -2 \leq x \leq 3\}$
21. $\{-2, 12\}$ **22.** $\{0, 6\}$ **23.** $b = \dfrac{a - 2}{2}$
24. Slope: $-\frac{1}{2}$; y-intercept: $(0, 4)$ **25.** $\frac{1}{3}$
26. $f(x) = \frac{1}{3}x - \frac{1}{4}$ **27.** $f(x) = -9x - 13$
28.

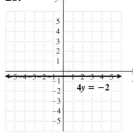

29.

30.

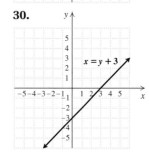

31.

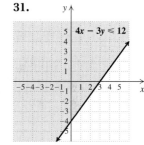

32. -6 **33.** $\{x \mid x$ is a real number $and\ x \neq 1\ and\ x \neq 2\}$, or $(-\infty, 1) \cup (1, 2) \cup (2, \infty)$ **34.** $[-3, \infty)$
35. $x^3 - 2x^2 + x - 2$ **36.** About \$90,909.09
37. Between 850 kWh and 950 kWh   **38.** Length: 8 cm; width: 6 cm   **39.** Length: 13.5 cm; width: 11.5 cm
40. Single-topping pizzas: 29; two-topping pizzas: 19
41. (a) $p(t) = 4t + 13$; **(b)** \$57 billion; **(c)** in 2013
42. 12 ft, 16 ft   **43.** Window-cleaning solution: 10 oz; razor-cleaning solution: 80 oz
44. $d = \dfrac{ct - c}{t + 2}$ **45.** $-\frac{15}{2}$ **46.** $|x + 3| < 4$
47. $(2x^{3n} - 5)(2x^{3n} + 1)$

CHAPTER 6

Technology Connection, p. 356

1. Let $y_1 = (7x^2 + 21x)/(14x)$, $y_2 = (x + 3)/2$, and $y_3 = y_1 - y_2$ (or $y_2 - y_1$). A table or the TRACE feature can be

used to show that, except when $x = 0$, y_3 is always 0. As an alternative, let $y_1 = (7x^2 + 21x)/(14x) - (x + 3)/2$ and show that, except when $x = 0$, y_1 is always 0.
2. Let $y_1 = (x + 3)/x$, $y_2 = 3$, and $y_3 = y_1 - y_2$ (or $y_2 - y_1$). Use a table or the TRACE feature to show that y_3 is not always 0. As an alternative, let $y_1 = (x + 3)/x - 3$ and show that y_1 is not always 0.

Visualizing for Success, p. 359

1. A   **2.** D   **3.** J   **4.** H   **5.** I   **6.** B   **7.** E   **8.** F
9. C   **10.** G

Exercise Set 6.1, pp. 360–363

1. (e)   **2.** (j)   **3.** (g)   **4.** (c)   **5.** (i)   **6.** (h)   **7.** (a)
8. (f)   **9.** (d)   **10.** (b)   **11. (a)** 5; **(b)** 1; **(c)** 5
13. (a) $\frac{9}{4}$; **(b)** does not exist; **(c)** 0   **15.** $\frac{30}{11}$ hr, or $2\frac{8}{11}$ hr
17. $\dfrac{9x(x + 2)}{9x(x - 5)}$ **19.** $\dfrac{(t - 2)(-1)}{(t + 3)(-1)}$ **21.** $\dfrac{t^3}{5}$ **23.** $\dfrac{4}{5x^2y^7}$
25. $a - 5$ **27.** $\dfrac{1}{5y - 6}$ **29.** $\dfrac{x - 4}{x + 5}$
31. $f(x) = \dfrac{5}{x}, x \neq -6, 0$ **33.** $g(x) = \dfrac{x - 3}{5}, x \neq -3$
35. $h(x) = -\dfrac{1}{7}, x \neq 2$ **37.** $f(t) = \dfrac{t + 4}{t - 4}, t \neq 4$
39. $g(t) = -\dfrac{7}{3}, t \neq 3$ **41.** $h(t) = \dfrac{t + 4}{t - 9}, t \neq -1, 9$
43. $f(x) = 3x + 2, x \neq \frac{2}{3}$ **45.** $g(t) = \dfrac{4 + t}{4 - t}, t \neq 4$
47. $\dfrac{6z^3}{7y^3}$ **49.** $\dfrac{8x^2}{25}$ **51.** $\dfrac{(y + 3)(y - 3)}{y(y + 2)}$ **53.** $-\dfrac{a + 1}{2 + a}$
55. 1   **57.** $c(c - 2)$ **59.** $\dfrac{a^2 + ab + b^2}{3(a + 2b)}$ **61.** $\dfrac{9a}{b}$
63. $\dfrac{5}{x^4}$ **65.** $-\dfrac{5x + 2}{x - 3}$ **67.** $-\dfrac{1}{y^3}$ **69.** $\dfrac{(y + 6)(y + 3)}{3(y - 4)}$
71. $\dfrac{x^2 + 4x + 16}{(x + 4)^2}$ **73.** $f(t) = \dfrac{t + 10}{5}, t \neq -4, 10$
75. $g(x) = \dfrac{(x + 5)(2x + 3)}{7x}, x \neq 0, \dfrac{3}{2}, 7$
77. $f(x) = \dfrac{(x + 2)(x + 4)}{x^7}, x \neq -4, 0, 2$
79. $h(n) = \dfrac{n(n^2 + 3)}{(n + 3)(n - 2)}, n \neq -7, -3, 2, 3$
81. $\dfrac{3(x - 3y)}{2(2x - y)(2x - 3y)}$ **83.** $\dfrac{(2a - b)(a - 1)}{(a - b)(a + 1)}$ **85.**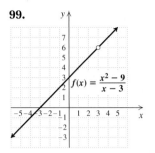
87. $\frac{9}{20}$ **88.** $\frac{8}{15}$ **89.** $-\frac{27}{100}$ **90.** $-\frac{13}{70}$ **91.** $4x^2 - 2$
92. $-t^2 + 9t - 2$ **93.** $y + 13$ **94.** $-x^2 + 29x - 4$
95. 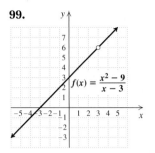   **97.** $2a + h$

99.

101. $\dfrac{(d - 1)(d - 4)(d - 5)^2}{25d^4(d + 5)}$ **103.** $\dfrac{m - t}{m + t + 1}$
105. $\dfrac{x^2 + xy + y^2 + x + y}{x - y}$ **107.** $-\dfrac{2x}{x - 1}$
109. (a) $\dfrac{16(x + 1)}{(x - 1)^2(x^2 + x + 1)}$; **(b)** $\dfrac{x^2 + x + 1}{(x + 1)^3}$;
(c) $\dfrac{(x + 1)^3}{x^2 + x + 1}$ **111.**   **113.** ,

Exercise Set 6.2, pp. 369–372

1. True   **2.** True   **3.** False   **4.** False   **5.** False
6. True   **7.** False   **8.** True   **9.** $\dfrac{5}{a}$ **11.** $\dfrac{1}{3m^2n^2}$
13. 2   **15.** $\dfrac{2t + 4}{t - 4}$ **17.** $\dfrac{1}{x - 5}$ **19.** $\dfrac{-1}{a + 5}$ **21.** $24x^5$
23. $(x + 3)(x - 3)^2$ **25.** $f(x) = \dfrac{3x - 1}{x^2 + 6x + 5}, x \neq -5, -1$
27. $f(x) = \dfrac{-x - 5}{x^2 - 1}, x \neq -1, 1$ **29.** $\dfrac{9x + 2}{15x^2}$ **31.** $\dfrac{y + 3}{2(y - 2)}$
33. $\dfrac{x + y}{x - y}$ **35.** $\dfrac{3x^2 + 7x + 14}{(2x - 5)(x - 1)(x + 2)}$
37. $\dfrac{-a^2 + 7ab - b^2}{(a - b)(a + b)}$ **39.** $\dfrac{x - 5}{(x + 5)(x + 3)}$ **41.** $\dfrac{9}{t}$
43. $-(s + r)$ **45.** $\dfrac{2a^2 - a + 14}{(a - 4)(a + 3)}$ **47.** $\dfrac{5x + 1}{x + 1}$
49. $\dfrac{-x + 34}{20(x + 2)}$ **51.** $\dfrac{8x + 1}{(x + 1)(x - 1)}$ **53.** $-\dfrac{1}{y + 5}$
55. $\dfrac{1}{y^2 + 9}$ **57.** $\dfrac{1}{r^2 + rs + s^2}$ **59.** $\dfrac{y}{(y - 2)(y - 3)}$
61. $\dfrac{7x + 1}{x - y}$ **63.** $\dfrac{-y}{(y + 3)(y - 1)}$ **65.** $-\dfrac{2y}{2y + 1}$
67. $f(x) = \dfrac{3(x + 4)}{x + 3}, x \neq -3, 3$
69. $f(x) = \dfrac{(x - 7)(2x - 1)}{(x - 4)(x - 1)(x + 3)}, x \neq -4, -3, 1, 4$
71. $f(x) = \dfrac{-2}{(x + 1)(x + 2)}, x \neq -3, -2, -1$ **73.**
75. $\dfrac{2}{x}$ **76.** $\dfrac{4}{x^2}$ **77.** $\dfrac{ab}{(a + b)^2}$ **78.** $\dfrac{3p^2}{3 - p}$ **79.** $9x - 6$
80. $3ab^3 + 2a^3$ **81.**   **83.** 420 days   **85.** 12 parts
87. $x^4(x^2 + 1)(x + 1)(x - 1)(x^2 + x + 1)(x^2 - x + 1)$

89. $8a^4, 8a^4b, 8a^4b^2, 8a^4b^3, 8a^4b^4, 8a^4b^5, 8a^4b^6, 8a^4b^7$

91. $\dfrac{x^4 + 6x^3 + 2x^2}{(x+2)(x-2)(x+5)}$ **93.** $\dfrac{x^5}{(x^2-4)(x^2+3x-10)}$

95. $\dfrac{2x+1}{x^2}$ **97.** $\dfrac{9x^2+28x+15}{(x-3)(x+3)^2}$ **99.** $\dfrac{1}{2x(x-5)}$

101. $-4t^4$ **103.** ☄

Technology Connection, p. 376

1. $-1, 0, \frac{2}{3}, 2$

Exercise Set 6.3, pp. 378–380

1. (b)   **2.** (a)   **3.** (f)   **4.** (c)   **5.** (d)   **6.** (e)

7. $\dfrac{(x+1)(x+5)}{(x-3)(x-2)}$ **9.** $\dfrac{3x^2+2}{x(5x-3)}$ **11.** $\dfrac{3(m+n)}{5(m-n)}$

13. $\dfrac{6s-r}{2s+3r}$ **15.** $\dfrac{y(3y+2z)}{z(4-yz)}$ **17.** $\dfrac{a+b}{a}$ **19.** $\dfrac{3}{3x+2}$

21. $\dfrac{1}{x+y}$ **23.** $-\dfrac{1}{x(x+h)}$ **25.** $\dfrac{(a-2)(a-7)}{(a+1)(a-6)}$

27. $\dfrac{x+2}{x+3}$ **29.** $\dfrac{1+2y}{1-3y}$ **31.** $\dfrac{(y+1)(y^2-y+1)}{(y-1)(y^2+y+1)}$

33. $\dfrac{4x-7}{7x-9}$ **35.** $\dfrac{a^2-3a-6}{a^2-2a-3}$ **37.** $\dfrac{a+1}{2a+5}$

39. $\dfrac{-1-3x}{8-2x}$, or $\dfrac{3x+1}{2x-8}$ **41.** $y^2+5y+25$ **43.** $-y$

45. $\dfrac{6a^2+30a+60}{3a^2+2a+4}$ **47.** $\dfrac{(2x+1)(x+2)}{2x(x-1)}$

49. $\dfrac{(2a-3)(a+5)}{2(a-3)(a+2)}$ **51.** -1 **53.** $\dfrac{2x^2-11x-27}{2x^2+21x+13}$

55. ☄   **57.** $\frac{11}{13}$ **58.** $\frac{23}{7}$ **59.** $-5, 5$ **60.** $2, 4$ **61.** $\frac{1}{2}$

62. 28 **63.** ☄   **65.** $\dfrac{5(y+x)}{3(y-x)}$ **67.** $\dfrac{8c}{17}$

69. $\dfrac{-3}{x(x+h)}$ **71.** $\dfrac{2}{(1+x+h)(1+x)}$

73. $\{x \mid x$ is a real number *and* $x \neq \pm1$ *and* $x \neq \pm4$ *and* $x \neq \pm5\}$ **75.** $\dfrac{2+a}{3+a}, a \neq -2, -3$ **77.** $\dfrac{x^4}{81}; \{x \mid x$ is a real number *and* $x \neq 3\}$ **79.** \$168.61

Exercise Set 6.4, pp. 385–387

1. Equation   **2.** Expression   **3.** Expression
4. Equation   **5.** Equation   **6.** Equation   **7.** Equation
8. Expression   **9.** Expression   **10.** Equation   **11.** 6
13. -8 **15.** $\frac{3}{2}$ **17.** $\frac{8}{3}$ **19.** $-\frac{2}{3}$ **21.** No solution
23. $-4, -1$ **25.** $-2, 6$ **27.** $\frac{10}{9}$ **29.** No solution
31. -5 **33.** $-\frac{10}{3}$ **35.** -1 **37.** $2, 3$ **39.** -145
41. -15 **43.** No solution   **45.** -1 **47.** $-4, 1$ **49.** 4
51. $-2, \frac{7}{8}$ **53.** No solution   **55.** $-\frac{3}{2}, 5$ **57.** 14
59. $\frac{3}{4}$ **61.** $\frac{5}{14}$ **63.** $\frac{3}{5}$ **65.** ☄   **67.** $\frac{4}{3}$ hr, or $1\frac{1}{3}$ hr
68. 10 m per minute   **69.** Kylie: 82 lb; Brenton: 41 lb
70. Length: 60 in.; width: 15 in.   **71.** 1.5 in.

72. -12 and -10, 10 and 12   **73.** ☄   **75.** $\frac{1}{5}$ **77.** $-\frac{7}{2}$
79. 0.0854697   **81.** Yes   **83.** ◿   **85.** ◿

Connecting the Concepts, pp. 388–389

1. $\dfrac{x-2}{x+1}$ **2.** $\dfrac{13t-5}{3t(2t-1)}$ **3.** 6   **4.** $\frac{1}{3}$ **5.** $\dfrac{z}{1-z}$

6. $\dfrac{4a}{3(a-1)}$ **7.** $-\frac{1}{2}$ **8.** $\dfrac{1}{a-b}$ **9.** $\dfrac{45a^3}{32}$ **10.** $-6, 6$

11. $\dfrac{(n-1)(n+7)}{(n-2)(n+1)}$ **12.** $\dfrac{5(n^2-n+1)}{3n^2}$

13. $\dfrac{8(t+1)}{(t-1)(2t-1)}$ **14.** $-5, 3$ **15.** $\frac{15}{4}$ **16.** $\dfrac{2a}{a-1}$

17. $-\frac{1}{3}, \frac{1}{2}$ **18.** 0   **19.** $\dfrac{x-6}{(x-8)(x-3)}$ **20.** No solution

Exercise Set 6.5, pp. 395–397

1. $\frac{1}{2}$ cake per hour   **2.** $\frac{1}{3}$ cake per hour
3. $\frac{5}{6}$ cake per hour   **4.** 1 lawn per hour
5. $\frac{1}{3}$ lawn per hour   **6.** $\frac{2}{3}$ lawn per hour
7. 2   **9.** $-3, -2$ **11.** -10 and -9, 9 and 10
13. $3\frac{3}{7}$ hr   **15.** $8\frac{4}{7}$ hr   **17.** $19\frac{1}{11}$ min
19. H470: 135 min; K5400: $67\frac{1}{2}$ min
21. Austin: 15 min; Airgle: 30 min   **23.** Elliot: $\frac{4}{3}$ hr; Sara: 4 hr
25. Zeno: 6 hr; Lia: 3 hr   **27.** 300 min, or 5 hr   **29.** 7 mph
31. 5.2 ft/sec   **33.** Freight: 66 mph; passenger: 80 mph
35. Express: 45 mph; local: 38 mph   **37.** 9 km/h
39. 2 km/h   **41.** 40 mph   **43.** 20 mph   **45.** ☄

47. $6x^6y^8$ **48.** $-5a$ **49.** $-\dfrac{4x}{y}$ **50.** $2x^4 - x^3 - 6x$

51. $5x^2 - 4x - 7$ **52.** $-x - 5$ **53.** ☄   **55.** $49\frac{1}{2}$ hr
57. 30 min   **59.** 2250 people per hour   **61.** $14\frac{7}{8}$ mi
63. Page 278   **65.** $8\frac{2}{11}$ min after 10:30   **67.** $51\frac{3}{7}$ mph

Exercise Set 6.6, pp. 402–403

1. $x - 3$ **2.** $x^2 - x - 1$ **3.** $x + 2$ **4.** 5   **5.** 1

6. 0   **7.** $4x^4 + 2x^3 - 3$ **9.** $-3a^2 - a + \dfrac{3}{7} + \dfrac{2}{a}$

11. $-4z + 2y^2z^3 - 3y^4z^2$ **13.** $8y - \dfrac{9}{2} - \dfrac{4}{y}$

15. $-5x^5 + 7x^2 + 1$ **17.** $1 - ab^2 - a^3b^4$ **19.** $x + 3$

21. $y - 5 + \dfrac{-50}{y-5}$ **23.** $x - 5 + \dfrac{1}{x-4}$ **25.** $y - 5$

27. $y^2 - 2y - 1 + \dfrac{-8}{y-2}$ **29.** $2x^2 - x + 1 + \dfrac{-5}{x+2}$

31. $a^2 - 4a + 6$ **33.** $2y^2 + 2y - 1 + \dfrac{8}{5y-2}$

35. $3x^2 + x + 1$ **37.** $2x^2 - x - 9 + \dfrac{3x+12}{x^2+2}$

39. $2x - 5, x \neq -\frac{2}{3}$ **41.** $4x^2 + 6x + 9, x \neq \frac{3}{2}$
43. $x^2 + 1, x \neq -5, x \neq 5$
45. $2x^3 - 3x^2 + 5, x \neq -1, x \neq 1$ **47.** ☄

49.

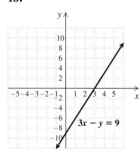

50.

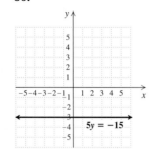

51.

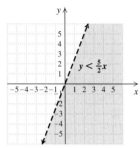

52.

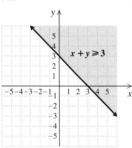

53.

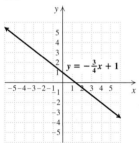

54.

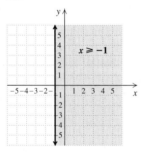

55. ✍   **57.** $a^2 + ab$
59. $a^6 - a^5b + a^4b^2 - a^3b^3 + a^2b^4 - ab^5 + b^6$
61. $-\frac{3}{2}$ **63.** ✍   **65.** 📈

Exercise Set 6.7, pp. 407–408

1. True   **2.** True   **3.** True   **4.** False   **5.** True
6. False   **7.** $x^2 - 3x - 5$ **9.** $a + 5 + \dfrac{-4}{a + 3}$
11. $2x^2 - 5x + 3 + \dfrac{8}{x + 2}$ **13.** $a^2 + 2a - 6$
15. $3y^2 + 2y + 6 + \dfrac{-2}{y - 3}$
17. $x^4 + 2x^3 + 4x^2 + 8x + 16$
19. $3x^2 + 6x - 3 + \dfrac{2}{x + \frac{1}{3}}$ **21.** 6   **23.** 125   **25.** 0
27. ✍   **29.** $c = \dfrac{b}{a}$ **30.** $w = \dfrac{x - y}{z}$ **31.** $q = \dfrac{st}{p - r}$
32. $b = \dfrac{d}{a + c}$ **33.** $b = \dfrac{cd + d}{a - 3}$ **34.** $d = \dfrac{ab - 3b}{c + 1}$
35. ✍

37. (a) The degree of R must be less than 1, the degree of
$x - r$; **(b)** Let $x = r$. Then
$$\begin{aligned} P(r) &= (r - r) \cdot Q(r) + R \\ &= 0 \cdot Q(r) + R \\ &= R. \end{aligned}$$
39. $0; -3, -\frac{5}{2}, \frac{3}{2}$ **41.** 📈   **43.** 0

Exercise Set 6.8, pp. 415–419

1. (d)   **2.** (f)   **3.** (e)   **4.** (b)   **5.** (a)   **6.** (c)
7. Inverse   **8.** Direct   **9.** Direct   **10.** Inverse
11. Inverse   **12.** Direct   **13.** $d = \dfrac{L}{f}$
15. $v_1 = \dfrac{2s}{t} - v_2$, or $\dfrac{2s - tv_2}{t}$ **17.** $b = \dfrac{at}{a - t}$
19. $g = \dfrac{Rs}{s - R}$ **21.** $n = \dfrac{IR}{E - Ir}$ **23.** $q = \dfrac{pf}{p - f}$
25. $t_1 = \dfrac{H}{Sm} + t_2$, or $\dfrac{H + Smt_2}{Sm}$ **27.** $r = \dfrac{Re}{E - e}$
29. $r = 1 - \dfrac{a}{S}$, or $\dfrac{S - a}{S}$ **31.** $a + b = \dfrac{f}{c^2}$
33. $r = \dfrac{A}{P} - 1$, or $\dfrac{A - P}{P}$
35. $t_1 = t_2 - \dfrac{d_2 - d_1}{v}$, or $\dfrac{vt_2 - d_2 + d_1}{v}$ **37.** $t = \dfrac{ab}{b + a}$
39. $Q = \dfrac{2Tt - 2AT}{A - q}$ **41.** $w = \dfrac{4.15c - 98.42}{p + 0.082}$
43. $k = 6; y = 6x$ **45.** $k = 1.7; y = 1.7x$
47. $k = 10; y = 10x$ **49.** $k = 100; y = \dfrac{100}{x}$
51. $k = 44; y = \dfrac{44}{x}$ **53.** $k = 9; y = \dfrac{9}{x}$ **55.** $33\frac{1}{3}$ cm
57. 3.5 hr   **59.** 56 in.   **61.** 32 kg   **63.** 286 Hz
65. About 21 min   **67.** About 110,000,000 tons
69. $y = \frac{1}{2}x^2$ **71.** $y = \dfrac{5000}{x^2}$ **73.** $y = 1.5xz$
75. $y = \dfrac{4wx^2}{z}$ **77.** 61.3 ft   **79.** 308 cm^3
81. About 57 mph   **83.** ✍   **85.** $4a - 7 + h$
86. $4a + 4h - 7$ **87.** $\{x \mid x \text{ is a real number } and\ x \neq -\frac{1}{2}\}$,
or $\left(-\infty, -\frac{1}{2}\right) \cup \left(-\frac{1}{2}, \infty\right)$ **88.** $\mathbb{R}$
89. $\{x \mid x \geq -4\}$ or $[-4, \infty)$
90. $\{x \mid x \text{ is a real number } and\ x \neq -1\ and\ x \neq 1\}$, or
$(-\infty, -1) \cup (-1, 1) \cup (1, \infty)$ **91.** ✍   **93.** 567 mi
95. Ratio is $\dfrac{a + 12}{a + 6}$; percent increase is $\dfrac{6}{a + 6} \cdot 100\%$, or
$\dfrac{600}{a + 6}\%$ **97.** $t_1 = t_2 + \dfrac{(d_2 - d_1)(t_4 - t_3)}{a(t_4 - t_2)(t_4 - t_3) + d_3 - d_4}$
99. The intensity is halved.   **101.** About 1.7 m
103. $d(s) = \dfrac{28}{s}$; 70 yd

Review Exercises: Chapter 6, pp. 424–425

1. True   **2.** True   **3.** False   **4.** False   **5.** False
6. True   **7.** True   **8.** False   **9.** True   **10.** True
11. (a) $-\frac{2}{9}$; (b) $-\frac{3}{4}$; (c) 0   **12.** $120x^3$
13. $(x + 10)(x - 2)(x - 3)$ **14.** $x + 8$ **15.** $\dfrac{20b^2c^6d^2}{3a^5}$
16. $\dfrac{15np + 14m}{18m^2n^4p^2}$ **17.** $\dfrac{(x - 2)(x + 5)}{x - 5}$
18. $\dfrac{(x^2 + 4x + 16)(x - 6)}{(x - 2)^2}$ **19.** $\dfrac{x - 3}{(x + 1)(x + 3)}$
20. $\dfrac{x - y}{x + y}$ **21.** $5(a + b)$ **22.** $\dfrac{-y}{(y + 4)(y - 1)}$
23. $f(x) = \dfrac{1}{x - 1}, x \neq 1, 4$
24. $f(x) = \dfrac{2}{x - 8}, x \neq -8, -5, 8$
25. $f(x) = \dfrac{3x - 1}{x - 3}, x \neq -3, -\frac{1}{3}, 3$
26. $\frac{4}{9}$ **27.** $\dfrac{a^2b^2}{2(b^2 - ba + a^2)}$
28. $\dfrac{(y + 11)(y + 5)}{(y - 5)(y + 2)}$ **29.** $\dfrac{(14 - 3x)(x + 3)}{2x^2 + 16x + 6}$
30. 2   **31.** 6   **32.** No solution   **33.** 0   **34.** $\frac{1}{2}, 5$
35. $-1, 4$ **36.** $5\frac{1}{7}$ hr
37. Core 2 Duo: 45 sec; Core 2 Quad: 30 sec   **38.** 24 mph
39. Motorcycle: 62 mph; car: 70 mph
40. $3s^2 + \frac{5}{2}s - 2rs^2$
41. $y^2 - 2y + 4$ **42.** $4x + 3 + \dfrac{-9x - 5}{x^2 + 1}$
43. $x^2 + 6x + 20 + \dfrac{54}{x - 3}$ **44.** 341
45. $r = \dfrac{2V - IR}{2I}$, or $\dfrac{V}{I} - \dfrac{R}{2}$ **46.** $m = \dfrac{H}{S(t_1 - t_2)}$
47. $c = \dfrac{b + 3a}{2}$ **48.** $t_1 = \dfrac{-A}{vT} + t_2$, or $\dfrac{-A + vTt_2}{vT}$
49. 15 min   **50.** 22.5 lb   **51.** About 2.9 sec
52. ✒ The least common denominator was used to add
and subtract rational expressions, to simplify complex rational expressions, and to solve rational equations.
53. ✒ A rational *expression* is a quotient of two polynomials. Expressions can be simplified, multiplied, or added, but
they cannot be solved for a variable. A rational *equation* is
an equation containing rational expressions. In a rational
equation, we often can solve for a variable.
54. All real numbers except 0 and 13
55. 45   **56.** $9\frac{9}{19}$ sec

Test: Chapter 6, p. 426

1. [6.1] $\dfrac{5}{4(t - 1)}$ **2.** [6.1] $\dfrac{x^2 - 3x + 9}{x + 4}$
3. [6.2] $\dfrac{25x + x^3}{x + 5}$ **4.** [6.2] $3(a - b)$
5. [6.2] $\dfrac{a^3 - a^2b + 4ab + ab^2 - b^3}{(a - b)(a + b)}$

6. [6.2] $\dfrac{-2(2x^2 + 5x + 20)}{(x - 4)(x + 4)(x^2 + 4x + 16)}$
7. [6.2] $f(x) = \dfrac{x - 4}{(x + 3)(x - 2)}, x \neq -3, 2$
8. [6.1] $f(x) = \dfrac{(x - 1)^2(x + 1)}{x(x - 2)}, x \neq -2, 0, 1, 2$
9. [6.3] $\dfrac{a(2b + 3a)}{5a + b}$ **10.** [6.3] $\dfrac{(x - 9)(x - 6)}{(x + 6)(x - 3)}$
11. [6.3] $\dfrac{2x^2 - 7x + 1}{3x^2 + 10x - 17}$ **12.** [6.4] $\frac{6}{5}$ **13.** [6.4] 15
14. [6.4] $-1, \frac{7}{2}$ **15.** [6.1] $-5; -\frac{1}{2}$ **16.** [6.4] $\frac{5}{3}$
17. [6.5] $2\frac{2}{9}$ hr   **18.** [6.6] $4a^2b^2c - \frac{5}{2}a^3bc^2 + 3bc$
19. [6.6] $y - 14 + \dfrac{-20}{y - 6}$ **20.** [6.6] $6x^2 - 9 + \dfrac{5x + 22}{x^2 + 2}$
21. [6.7] $x^2 + 7x + 18 + \dfrac{29}{x - 2}$ **22.** [6.7] 449
23. [6.8] $s = \dfrac{Rg}{g - R}$ **24.** [6.5] -11 and -10, 10 and 11
25. [6.5] $3\frac{3}{11}$ mph   **26.** [6.8] 30 workers   **27.** [6.8] 637 in^2
28. [6.4] $-\frac{19}{3}$ **29.** [6.4] $\{x \mid x$ is a real number *and* $x \neq 0$
and $x \neq 15\}$ **30.** [6.3], [6.4] x-intercept: $(11, 0)$; y-intercept:
$\left(0, -\frac{33}{5}\right)$ **31.** [6.5] Alex: 56 lawns; Ryan: 42 lawns

Cumulative Review: Chapters 1–6, pp. 427–428

1. 10   **2.** 3.91×10^8 **3.** Slope: $\frac{7}{4}$; y-intercept: $(0, -3)$
4. $y = -2x + 5$ **5.** (a) 0; (b) $\{x \mid x$ is a real number *and*
$x \neq 5$ *and* $x \neq 6\}$ **6.** $[9, \infty)$

7.
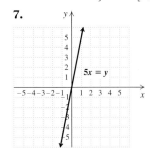

8.
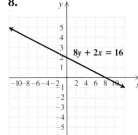

9.
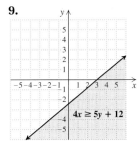

10.
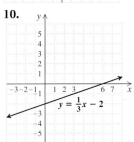

11. $-24x^4y^4$
12. $15x^4 - x^3 - 9x^2 + 5x - 2$
13. $9x^4 + 6x^2y + y^2$ **14.** $4x^4 - 81$
15. $-m^3n^2 - m^2n^2 - 5mn^3$ **16.** $\dfrac{y - 6}{2}$
17. $x - 1$ **18.** $\dfrac{a^2 + 7ab + b^2}{(a - b)(a + b)}$ **19.** $\dfrac{-m^2 + 5m - 6}{(m + 1)(m - 5)}$
20. $\dfrac{3y^2 - 2}{3y}$ **21.** $\dfrac{y - x}{xy(x + y)}$

22. $9x^2 - 13x + 26 + \dfrac{-50}{x+2}$ **23.** $4x(x^2 + 100)$

24. $(x-6)(x+14)$ **25.** $(4y-5)(4y+5)$

26. $8(2x+1)(4x^2 - 2x + 1)$ **27.** $(t-8)^2$

28. $x^2(x-1)(x+1)(x^2+1)$

29. $\left(\frac{1}{2}b - c\right)\left(\frac{1}{4}b^2 + \frac{1}{2}bc + c^2\right)$ **30.** $(3t-4)(t+7)$

31. $(x^2 - y)(x^3 + y)$ **32.** $\frac{1}{4}$ **33.** $-12, 12$

34. $\{x | x \geq -1\}$, or $[-1, \infty)$

35. $\{x | -5 < x < -1\}$, or $(-5, -1)$

36. $\{x | x < -\frac{4}{3} \, or \, x > 6\}$, or $(-\infty, -\frac{4}{3}) \cup (6, \infty)$

37. $\{x | x < -6.4 \, or \, x > 6.4\}$, or $(-\infty, -6.4) \cup (6.4, \infty)$

38. $\{x | -4 \leq x \leq \frac{16}{3}\}$, or $\left[-4, \frac{16}{3}\right]$ **39.** -1

40. No solution   **41.** $(-3, 4)$ **42.** $(-2, -3, 1)$

43. $a = \dfrac{Pb}{4-P}$ **44. (a)** $r(t) = 45.95t + 20;$

(b) \$1076.85 million; **(c)** in 2016–2017

45. 34 performances per month, or 408 performances per year   **46.** Himalayan Diamonds: $6\frac{2}{3}$ lb; Alpine Gold: $13\frac{1}{3}$ lb

47. 72 in.   **48.** IQAir HealthPro: 10 min; Austin Healthmate: 15 min   **49.** 12 billion hr   **50.** $22\frac{1}{2}$ min

51. Magic Kingdom: 41; Disneyland: 57; California Adventure: 35   **52.** $x^3 - 12x^2 + 48x - 64$ **53.** $-3, 3, -5, 5$

54. $\{x | -3 \leq x \leq -1 \, or \, 7 \leq x \leq 9\}$, or $[-3, -1] \cup [7, 9]$

55. All real numbers except 9 and -5 **56.** $-\frac{1}{4}, 0, \frac{1}{4}$

CHAPTER 7

Technology Connection, p. 432

1. False   **2.** True   **3.** False

Visualizing for Success, p. 436

1. B   **2.** H   **3.** C   **4.** I   **5.** D   **6.** A   **7.** F   **8.** J
9. G   **10.** E

Exercise Set 7.1, pp. 437–438

1. Two   **2.** Negative   **3.** Positive   **4.** Negative
5. Irrational   **6.** Real   **7.** Nonnegative   **8.** Negative
9. $8, -8$ **11.** $10, -10$ **13.** $20, -20$ **15.** $25, -25$
17. 7   **19.** -4 **21.** $\frac{6}{7}$ **23.** -13 **25.** $-\frac{4}{9}$ **27.** 0.2

29. 0.09 **31.** $p^2 + 4; 2$ **33.** $\dfrac{x}{y+4}; 5$ **35.** $\sqrt{5}; 0;$

does not exist; does not exist   **37.** -7; does not exist; -1;
does not exist   **39.** $1; \sqrt{2}; \sqrt{101}$ **41.** $10|x|$
43. $|8 - t|$ **45.** $|y + 8|$ **47.** $|2x + 7|$ **49.** -4
51. -1 **53.** $\frac{2}{3}$ **55.** $|x|$ **57.** t **59.** $6|a|$ **61.** 6
63. $|a + b|$ **65.** $|a^{11}|$ **67.** Cannot be simplified
69. $4x$ **71.** $-3t$ **73.** $5b$ **75.** $a + 1$ **77.** 3   **79.** $2x$
81. $x - 1$ **83.** $5y$ **85.** t^9 **87.** $(x-2)^4$
89. $2; 3; -2; -4$ **91.** 2; does not exist; does not exist; 3
93. $\{x | x \geq 6\}$, or $[6, \infty)$ **95.** $\{t | t \geq -8\}$, or $[-8, \infty)$
97. $\{x | x \leq 5\}$, or $(-\infty, 5]$ **99.** $\mathbb{R}$ **101.** $\{z | z \geq -\frac{2}{5}\}$, or $\left[-\frac{2}{5}, \infty\right)$ **103.** $\mathbb{R}$ **105.** 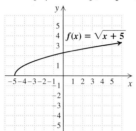   **107.** a^6b^2 **108.** $15x^3y^9$

109. $\dfrac{125x^6}{y^9}$ **110.** $\dfrac{a^3}{8b^6c^3}$ **111.** $\dfrac{x^3}{2y^6}$ **112.** $\dfrac{y^4z^8}{16x^4}$

113. 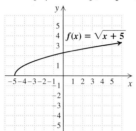   **115.** About 1404 species
117. $\{x | x \geq -5\}$, or $[-5, \infty)$ **119.** $\{x | x \geq 0\}$, or $[0, \infty)$

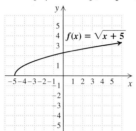

121. $\{x | -3 \leq x < 2\}$, or $[-3, 2)$
123. $\{x | x < -1 \, or \, x > 6\}$, or $(-\infty, -1) \cup (6, \infty)$
125.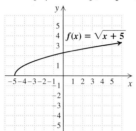

Technology Connection, p. 440

1. Without parentheses, the expression entered would be $\dfrac{7^2}{3}$.

2. For $x = 0$ or $x = 1$, $y_1 = y_2 = y_3$; on $(0, 1)$, $y_1 > y_2 > y_3$; on $(1, \infty)$, $y_1 < y_2 < y_3$.

Technology Connection, p. 442

1. Many graphing calculators do not have keys for radicals of index 3 or higher. On those graphing calculators that offer $\sqrt[x]{}$ in a MATH menu, rational exponents still require fewer keystrokes.

Exercise Set 7.2, pp. 443–445

1. (g)   **2.** (c)   **3.** (e)   **4.** (h)   **5.** (a)   **6.** (d)   **7.** (b)
8. (f)   **9.** $\sqrt[3]{y}$ **11.** 6   **13.** 2   **15.** 8   **17.** $\sqrt{xyz}$
19. $\sqrt[5]{a^2b^2}$ **21.** $\sqrt[6]{t^5}$ **23.** 8   **25.** 625   **27.** $27\sqrt[4]{x^3}$
29. $125x^6$ **31.** $18^{1/3}$ **33.** $30^{1/2}$ **35.** $x^{7/2}$ **37.** $m^{2/5}$
39. $(pq)^{1/4}$ **41.** $(xyz)^{1/5}$ **43.** $(3mn)^{3/2}$

45. $(8x^2y)^{5/7}$ **47.** $\dfrac{2x}{z^{2/3}}$ **49.** $\dfrac{1}{a^{1/4}}$ **51.** $\dfrac{1}{(2rs)^{3/4}}$

53. 8   **55.** $8a^{3/5}c$ **57.** $\dfrac{2a^{3/4}c^{2/3}}{b^{1/2}}$ **59.** $\dfrac{a^3}{3^{5/2}b^{7/3}}$

61. $\left(\dfrac{3c}{2ab}\right)^{5/6}$ **63.** $\dfrac{6a}{b^{1/4}}$ **65.** $11^{5/6}$ **67.** $3^{3/4}$

69. $4.3^{1/2}$ **71.** $10^{6/25}$ **73.** $a^{23/12}$ **75.** 64   **77.** $\dfrac{m^{1/3}}{n^{1/8}}$

79. $\sqrt[3]{x}$ **81.** y^5 **83.** $\sqrt{a}$ **85.** x^2y^2 **87.** $\sqrt{7a}$
89. $\sqrt[4]{8x^3}$ **91.** $\sqrt[10]{m}$ **93.** x^3y^3 **95.** a^6b^{12}
97. $\sqrt[12]{xy}$ **99.** 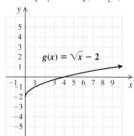   **101.** $x^2 - 25$ **102.** $x^3 - 8$
103. $(2x+5)^2$ **104.** $(3a-4)^2$ **105.** $5(t-1)^2$
106. $3(n+2)^2$ **107.**   **109.** $\sqrt[6]{x^5}$
111. $\sqrt[7]{c-d}, c \geq d$ **113.** $2^{7/12} \approx 1.498 \approx 1.5$
115. (a) 1.8 m; **(b)** 3.1 m; **(c)** 1.5 m; **(d)** 5.3 m
117. 338 cubic feet   **119.**

Technology Connection, p. 447

1. The graphs differ in appearance because the domain of y_1 is the intersection of $[-3, \infty)$ and $[3, \infty)$, or $[3, \infty)$. The domain of y_2 is $(-\infty, -3] \cup [3, \infty)$.

Exercise Set 7.3, pp. 450-451

1. True **2.** False **3.** False **4.** False **5.** True
6. True **7.** $\sqrt{30}$ **9.** $\sqrt[3]{35}$ **11.** $\sqrt[4]{54}$ **13.** $\sqrt{26xy}$
15. $\sqrt[5]{80y^4}$ **17.** $\sqrt{y^2 - b^2}$ **19.** $\sqrt[3]{0.21y^2}$
21. $\sqrt[5]{(x-2)^3}$ **23.** $\sqrt{\dfrac{6s}{11t}}$ **25.** $\sqrt[7]{\dfrac{5x-15}{4x+8}}$ **27.** $2\sqrt{3}$
29. $3\sqrt{5}$ **31.** $2x^4\sqrt{2x}$ **33.** $2\sqrt{30}$ **35.** $6a^2\sqrt{b}$
37. $2x\sqrt[3]{y^2}$ **39.** $-2x^2\sqrt[3]{2}$ **41.** $f(x) = 2x^2\sqrt[3]{5}$
43. $f(x) = |7(x-3)|$, or $7|x-3|$
45. $f(x) = |x-1|\sqrt{5}$ **47.** $a^5b^5\sqrt{b}$ **49.** $xy^2z^3\sqrt[3]{x^2z}$
51. $2xy^2\sqrt[4]{xy^3}$ **53.** $x^2yz^3\sqrt[5]{x^3y^3z^2}$ **55.** $-2a^4\sqrt[3]{10a^2}$
57. $5\sqrt{2}$ **59.** $3\sqrt{22}$ **61.** 3 **63.** $24y^5$ **65.** $a\sqrt[3]{10}$
67. $2x^3\sqrt{5x}$ **69.** $s^2t^3\sqrt[3]{t}$ **71.** $(x-y)^4$
73. $2ab^3\sqrt[4]{5a}$ **75.** $x(y+z)^2\sqrt[5]{x}$ **77.** ⌨ **79.** $9abx^2$
80. $\dfrac{(x-1)^2}{(x-2)^2}$ **81.** $\dfrac{x+1}{2(x+5)}$ **82.** $\dfrac{3(4x^2+5y^3)}{50xy^2}$
83. $\dfrac{b+a}{a^2b^2}$ **84.** $\dfrac{-x-2}{4x+3}$ **85.** ⌨ **87.** 175.6 mi
89. (a) −3.3°C; **(b)** −16.6°C; **(c)** −25.5°C; **(d)** −54.0°C
91. $25x^5\sqrt[3]{25x}$ **93.** $a^{10}b^{17}\sqrt{ab}$
95.

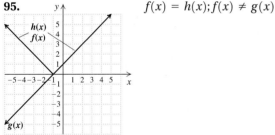

$f(x) = h(x); f(x) \neq g(x)$

97. $\{x \mid x \le 2 \text{ or } x \ge 4\}$, or $(-\infty, 2] \cup [4, \infty)$ **99.** 6
101. ▨, ⌨

Exercise Set 7.4, pp. 456-457

1. (g) **2.** (b) **3.** (f) **4.** (c) **5.** (h) **6.** (d) **7.** (a)
8. (e) **9.** $\dfrac{7}{10}$ **11.** $\dfrac{5}{2}$ **13.** $\dfrac{11}{t}$ **15.** $\dfrac{6y\sqrt{y}}{x^2}$ **17.** $\dfrac{3a\sqrt[3]{a}}{2b}$
19. $\dfrac{2a}{bc^2}$ **21.** $\dfrac{ab^2}{c^2}\sqrt[4]{\dfrac{a}{c^2}}$ **23.** $\dfrac{2x}{y^2}\sqrt[5]{\dfrac{x}{y}}$ **25.** $\dfrac{xy}{z^2}\sqrt[6]{\dfrac{y^2}{z^3}}$
27. 3 **29.** $\sqrt[3]{2}$ **31.** $y\sqrt{5y}$ **33.** $2\sqrt[3]{a^2b}$ **35.** $\sqrt{2ab}$
37. $2x^2y^3\sqrt[4]{y^3}$ **39.** $\sqrt[3]{x^2+xy+y^2}$ **41.** $\dfrac{\sqrt{10}}{5}$
43. $\dfrac{2\sqrt{15}}{21}$ **45.** $\dfrac{\sqrt[3]{10}}{2}$ **47.** $\dfrac{\sqrt[3]{75ac^2}}{5c}$ **49.** $\dfrac{y\sqrt[4]{45y^2x^3}}{3x}$
51. $\dfrac{\sqrt[3]{2xy^2}}{xy}$ **53.** $\dfrac{\sqrt{14a}}{6}$ **55.** $\dfrac{\sqrt[5]{9y^4}}{2xy}$ **57.** $\dfrac{\sqrt{5b}}{6a}$
59. $\dfrac{5}{\sqrt{55}}$ **61.** $\dfrac{12}{5\sqrt{42}}$ **63.** $\dfrac{2}{\sqrt{6x}}$ **65.** $\dfrac{7}{\sqrt[3]{98}}$

67. $\dfrac{7x}{\sqrt{21xy}}$ **69.** $\dfrac{2a^2}{\sqrt[3]{20ab}}$ **71.** $\dfrac{x^2y}{\sqrt{2xy}}$ **73.** ⌨
75. $x(3 - 8y + 2z)$ **76.** $ac(4a + 9 - 3a^2)$
77. $a^2 - b^2$ **78.** $a^4 - 4y^2$ **79.** $56 - 11x - 12x^2$
80. $6ay - 2cy - 3ax + cx$ **81.** ⌨ **83. (a)** 1.62 sec;
(b) 1.99 sec; **(c)** 2.20 sec **85.** $9\sqrt[3]{9n^2}$ **87.** $\dfrac{-3\sqrt{a^2-3}}{a^2-3}$,
or $\dfrac{-3}{\sqrt{a^2-3}}$ **89.** Step 1: $\sqrt[n]{a} = a^{1/n}$, by definition;
Step 2: $\left(\dfrac{a}{b}\right)^n = \dfrac{a^n}{b^n}$, raising a quotient to a power;
Step 3: $a^{1/n} = \sqrt[n]{a}$, by definition **91.** $(f/g)(x) = 3x$,
where x is a real number and $x > 0$
93. $(f/g)(x) = \sqrt{x+3}$, where x is a real number and $x > 3$

Exercise Set 7.5, pp. 462-464

1. Radicands; indices **2.** Indices **3.** Bases
4. Denominators **5.** Numerator; conjugate **6.** Bases
7. $11\sqrt{3}$ **9.** $2\sqrt[3]{4}$ **11.** $10\sqrt[3]{y}$ **13.** $12\sqrt{2}$
15. $13\sqrt[3]{7} + \sqrt{3}$ **17.** $9\sqrt{3}$ **19.** $-7\sqrt{5}$ **21.** $9\sqrt[3]{2}$
23. $(1 + 12a)\sqrt{a}$ **25.** $(x-2)\sqrt[3]{6x}$ **27.** $3\sqrt{a-1}$
29. $(x+3)\sqrt{x-1}$ **31.** $5\sqrt{2} + 2$ **33.** $3\sqrt{30} - 3\sqrt{35}$
35. $6\sqrt{5} - 4$ **37.** $3 - 4\sqrt[3]{63}$ **39.** $a + 2a\sqrt[3]{3}$
41. $4 + 3\sqrt{6}$ **43.** $\sqrt{6} - \sqrt{14} + \sqrt{21} - 7$ **45.** 1
47. -5 **49.** $2 - 8\sqrt{35}$ **51.** $23 + 8\sqrt{7}$ **53.** $5 - 2\sqrt{6}$
55. $2t + 5 + 2\sqrt{10t}$ **57.** $14 + x - 6\sqrt{x+5}$
59. $6\sqrt[4]{63} + 4\sqrt[4]{35} - 3\sqrt[4]{54} - 2\sqrt[4]{30}$ **61.** $\dfrac{18 + 6\sqrt{2}}{7}$
63. $\dfrac{12 - 2\sqrt{3} + 6\sqrt{5} - \sqrt{15}}{33}$ **65.** $\dfrac{a - \sqrt{ab}}{a-b}$ **67.** -1
69. $\dfrac{12 - 3\sqrt{10} - 2\sqrt{14} + \sqrt{35}}{6}$ **71.** $\dfrac{1}{\sqrt{5}-1}$
73. $\dfrac{2}{14 + 2\sqrt{3} + 3\sqrt{2} + 7\sqrt{6}}$ **75.** $\dfrac{x - y}{x + 2\sqrt{xy} + y}$
77. $\dfrac{1}{\sqrt{a+h} + \sqrt{a}}$ **79.** $\sqrt{a}$ **81.** $b^2\sqrt[10]{b^3}$
83. $xy\sqrt[6]{xy^5}$ **85.** $3a^2b\sqrt[4]{ab}$ **87.** $a^2b^2c^2\sqrt[6]{a^2bc^2}$
89. $\sqrt[12]{a^5}$ **91.** $\sqrt[12]{x^2y^5}$ **93.** $\sqrt[10]{ab^9}$ **95.** $\sqrt[6]{(7-y)^5}$
97. $\sqrt[12]{5} + 3x$ **99.** $x\sqrt[6]{xy^5} - \sqrt[15]{x^{13}y^{14}}$
101. $2m^2 + m\sqrt[4]{n} + 2m^3\sqrt{n^2} + \sqrt[12]{n^{11}}$
103. $2\sqrt[4]{x^3} - \sqrt[12]{x^{11}}$ **105.** $x^2 - 7$ **107.** $11 - 6\sqrt{2}$
109. $27 + 6\sqrt{14}$ **111.** ⌨ **113.** 42 **114.** $-\dfrac{1}{3}$
115. $-7, 3$ **116.** $-\dfrac{2}{5}, \dfrac{3}{2}$ **117.** -3 **118.** $-6, 1$
119. ⌨ **121.** $f(x) = 2x\sqrt{x-1}$
123. $f(x) = (x + 3x^2)\sqrt[4]{x-1}$ **125.** $(7x^2 - 2y^2)\sqrt{x+y}$
127. $4x(y+z)^3\sqrt[6]{2x(y+z)}$ **129.** $1 - \sqrt{w}$
131. $(\sqrt{x} + \sqrt{5})(\sqrt{x} - \sqrt{5})$
133. $(\sqrt{x} + \sqrt{a})(\sqrt{x} - \sqrt{a})$ **135.** $2x - 2\sqrt{x^2 - 4}$

Connecting the Concepts, pp. 464-465

1. $t + 5$ **2.** $-3a^4$ **3.** $3x\sqrt{10}$ **4.** $\dfrac{2}{3}$ **5.** $5\sqrt{15t}$
6. $ab^2c^2\sqrt[5]{c}$ **7.** $2\sqrt{15} - 3\sqrt{22}$ **8.** $-2b\sqrt[4]{a^3b^3}$
9. $\sqrt[8]{t}$ **10.** $\dfrac{a^2}{2}$ **11.** $-8\sqrt{3}$ **12.** -4 **13.** $25 + 10\sqrt{6}$

14. $2\sqrt{x-1}$ **15.** $xy\sqrt[10]{x^7y^3}$ **16.** $15\sqrt[3]{5}$ **17.** $\sqrt[5]{x}$
18. $(x+1)\sqrt{3}$ **19.** $ab\sqrt{b}$ **20.** $6x^3y^2$

Technology Connection, p. 467

1. The x-coordinates of the points of intersection should approximate the solutions of the examples.

Exercise Set 7.6, pp. 469–471

1. False   **2.** True   **3.** True   **4.** False   **5.** True
6. True   **7.** 3   **9.** $\frac{16}{3}$ **11.** 20   **13.** -1 **15.** 5
17. 91   **19.** $0, 36$ **21.** 100   **23.** -125 **25.** 16
27. No solution   **29.** $\frac{80}{3}$ **31.** 45   **33.** $-\frac{5}{3}$ **35.** 1
37. $\frac{106}{27}$ **39.** 4   **41.** $3, 7$ **43.** $\frac{80}{9}$ **45.** -1
47. No solution   **49.** $2, 6$ **51.** 2   **53.** 4   **55.**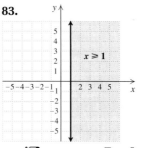
57. Length: 200 ft; width: 15 ft   **58.** Base: 34 in.; height: 15 in.
59. Length: 14 in.; width: 10 in.
60. Length: 30 yd; width: 16 yd   **61.** 6, 8, 10   **62.** 13 cm
63.   **65.** About 68 psi   **67.** About 278 Hz
69. 524.8°C   **71.** $t = \dfrac{1}{9}\left(\dfrac{S^2 \cdot 2457}{1087.7^2} - 2617\right)$
73. 4480 rpm   **75.** $r = \dfrac{v^2 h}{2gh - v^2}$ **77.** $-\frac{8}{9}$ **79.** $-8, 8$
81. $1, 8$ **83.** $\left(\frac{1}{36}, 0\right), (36, 0)$ **85.**   **87.**

Exercise Set 7.7, pp. 479–483

1. (d)   **2.** (c)   **3.** (e)   **4.** (b)   **5.** (f)   **6.** (a)
7. $\sqrt{34}; 5.831$ **9.** $9\sqrt{2}; 12.728$ **11.** 8   **13.** 4 m
15. $\sqrt{19}$ in.; 4.359 in.   **17.** 1 m   **19.** 250 ft
21. $\sqrt{8450}$, or $65\sqrt{2}$ ft; 91.924 ft   **23.** 24 in.
25. $(\sqrt{340} + 8)$ ft; 26.439 ft
27. $(110 - \sqrt{6500})$ paces; 29.377 paces
29. Leg = 5; hypotenuse = $5\sqrt{2} \approx 7.071$
31. Shorter leg = 7; longer leg = $7\sqrt{3} \approx 12.124$
33. Leg = $5\sqrt{3} \approx 8.660$; hypotenuse = $10\sqrt{3} \approx 17.321$
35. Both legs = $\dfrac{13\sqrt{2}}{2} \approx 9.192$
37. Leg = $14\sqrt{3} \approx 24.249$; hypotenuse = 28
39. $5\sqrt{3} \approx 8.660$ **41.** $7\sqrt{2} \approx 9.899$
43. $\dfrac{15\sqrt{2}}{2} \approx 10.607$ **45.** $\sqrt{10,561}$ ft ≈ 102.767 ft
47. $\dfrac{1089}{4}\sqrt{3}$ ft$^2 \approx 471.551$ ft^2 **49.** $(0, -4), (0, 4)$
51. 5   **53.** $\sqrt{10} \approx 3.162$ **55.** $\sqrt{200} \approx 14.142$
57. 17.8
59. $\dfrac{\sqrt{13}}{6} \approx 0.601$ **61.** $\sqrt{12} \approx 3.464$
63. $\sqrt{101} \approx 10.050$ **65.** $(3, 4)$ **67.** $\left(\frac{7}{2}, \frac{7}{2}\right)$
69. $(-1, -3)$ **71.** $(0.7, 0)$ **73.** $\left(-\frac{1}{12}, \frac{1}{24}\right)$
75. $\left(\dfrac{\sqrt{2}+\sqrt{3}}{2}, \dfrac{3}{2}\right)$ **77.**

79.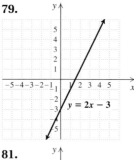

80.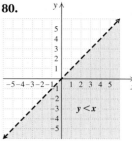

81.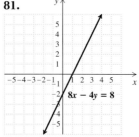

82.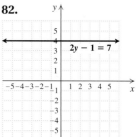

83.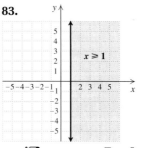

84.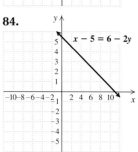

85. 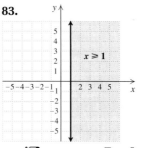   **87.** $36\sqrt{3}$ cm^2; 62.354 cm^2 **89.** $d = s + s\sqrt{2}$
91. 5 gal. The total area of the doors and windows is 134 ft^2 or more.   **93.** 60.28 ft by 60.28 ft   **95.** $\sqrt{75}$ cm

Exercise Set 7.8, pp. 489–490

1. False   **2.** False   **3.** True   **4.** True   **5.** True
6. True   **7.** False   **8.** True   **9.** $10i$ **11.** $i\sqrt{5}$, or $\sqrt{5}i$
13. $2i\sqrt{2}$, or $2\sqrt{2}i$ **15.** $-i\sqrt{11}$, or $-\sqrt{11}i$ **17.** $-7i$
19. $-10i\sqrt{3}$, or $-10\sqrt{3}i$ **21.** $6 - 2i\sqrt{21}$, or $6 - 2\sqrt{21}i$
23. $(-2\sqrt{19} + 5\sqrt{5})i$ **25.** $(3\sqrt{2} - 8)i$ **27.** $5 - 3i$
29. $7 + 2i$ **31.** $2 - i$ **33.** $-12 - 5i$ **35.** -40
37. -24 **39.** -18 **41.** $-\sqrt{30}$ **43.** $-3\sqrt{14}$
45. $-30 + 10i$ **47.** $28 - 21i$ **49.** $1 + 5i$ **51.** $38 + 9i$
53. $2 - 46i$ **55.** 73   **57.** 50   **59.** $12 - 16i$
61. $-5 + 12i$ **63.** $-5 - 12i$ **65.** $3 - i$ **67.** $\frac{6}{13} + \frac{4}{13}i$
69. $\frac{3}{17} + \frac{5}{17}i$ **71.** $-\frac{5}{6}i$ **73.** $-\frac{3}{4} - \frac{5}{4}i$ **75.** $1 - 2i$
77. $-\frac{23}{58} + \frac{43}{58}i$ **79.** $\frac{19}{29} - \frac{4}{29}i$ **81.** $\frac{6}{25} - \frac{17}{25}i$ **83.** 1
85. $-i$ **87.** -1 **89.** i **91.** -1 **93.** $-125i$
95. 0   **97.**   **99.** $-2, 3$ **100.** 5   **101.** $-10, 10$
102. $-5, 5$ **103.** $-\frac{2}{5}, \frac{4}{3}$ **104.** $-\frac{2}{3}, \frac{3}{2}$ **105.**

107.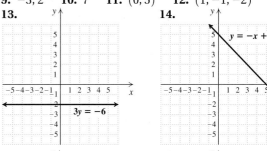

109. 5   **111.** $\sqrt{2}$

113. $-9 - 27i$ **115.** $50 - 120i$ **117.** $\frac{250}{41} + \frac{200}{41}i$
119. 8   **121.** $\frac{3}{5} + \frac{9}{5}i$ **123.** 1

Review Exercises: Chapter 7, pp. 494–495

1. True   **2.** False   **3.** False   **4.** True   **5.** True
6. True   **7.** True   **8.** False   **9.** $\frac{10}{11}$ **10.** -0.6
11. 5   **12.** $\{x | x \geq -10\}$, or $[-10, \infty)$ **13.** $8|t|$
14. $|c + 7|$ **15.** $|2x + 1|$ **16.** -2 **17.** $(5ab)^{4/3}$
18. $8a^4\sqrt{a}$ **19.** x^3y^5 **20.** $\sqrt[3]{x^2y}$ **21.** $\frac{1}{x^{2/5}}$
22. $7^{1/6}$ **23.** $f(x) = 5|x - 6|$ **24.** $2x^5y^2$
25. $5xy\sqrt{10x}$ **26.** $\sqrt{35ab}$ **27.** $3xb\sqrt[3]{x^2}$
28. $-6x^5y^4\sqrt[3]{2x^2}$ **29.** $-\frac{3y^4}{4}$ **30.** $y\sqrt[3]{6}$ **31.** $\frac{5\sqrt{x}}{2}$
32. $\frac{2a^2\sqrt[4]{3a^3}}{c^2}$ **33.** $7\sqrt[3]{4y}$ **34.** $\sqrt{3}$ **35.** $(2x + y^2)\sqrt[3]{x}$
36. $15\sqrt{2}$ **37.** -1 **38.** $\sqrt{15} + 4\sqrt{6} - 6\sqrt{10} - 48$
39. $\sqrt[4]{x^3}$ **40.** $\sqrt[12]{x^5}$ **41.** $4 - 4\sqrt{a} + a$
42. $-4\sqrt{10} + 4\sqrt{15}$ **43.** $\frac{20}{\sqrt{10} + \sqrt{15}}$ **44.** 19
45. -126 **46.** 4   **47.** 2   **48.** $5\sqrt{2}$ cm; 7.071 cm
49. $\sqrt{32}$ ft; 5.657 ft
50. Short leg = 10; long leg = $10\sqrt{3} \approx 17.321$
51. $\sqrt{26} \approx 5.099$ **52.** $\left(-2, -\frac{3}{2}\right)$ **53.** $3i\sqrt{5}$, or $3\sqrt{5}i$
54. $-2 - 9i$ **55.** $6 + i$ **56.** 29   **57.** -1 **58.** $9 - 12i$
59. $\frac{13}{25} - \frac{34}{25}i$ **60.** 📝 A complex number $a + bi$ is real
when $b = 0$. It is imaginary when $b \neq 0$.   **61.** 📝 An
absolute-value sign must be used to simplify $\sqrt[n]{x^n}$ when n
is even, since x may be negative. If x is negative while n is
even, the radical expression cannot be simplified to x, since
$\sqrt[n]{x^n}$ represents the principal, or nonnegative, root. When n
is odd, there is only one root, and it will be positive or negative depending on the sign of x. Thus there is no absolute-
value sign when n is odd.   **62.** $\frac{2i}{3i}$; answers may vary
63. 3   **64.** $-\frac{2}{5} + \frac{9}{10}i$ **65.** The isosceles right triangle is
larger by about 1.206 ft^2.

Test: Chapter 7, p. 496

1. [7.3] $5\sqrt{2}$ **2.** [7.4] $-\frac{2}{x^2}$ **3.** [7.1] $9|a|$
4. [7.1] $|x - 4|$ **5.** [7.2] $(7xy)^{1/2}$ **6.** [7.2] $\sqrt[6]{(4a^3b)^5}$
7. [7.1] $\{x | x \geq 5\}$, or $[5, \infty)$ **8.** [7.5] $27 + 10\sqrt{2}$
9. [7.3] $2x^3y^2\sqrt[5]{x}$ **10.** [7.3] $2\sqrt[3]{2wv^2}$ **11.** [7.4] $\frac{10a^2}{3b^3}$

12. [7.4] $\sqrt[5]{3x^4y}$ **13.** [7.5] $x\sqrt[4]{x}$ **14.** [7.5] $\sqrt[5]{y^2}$
15. [7.5] $6\sqrt{2}$ **16.** [7.5] $(x^2 + 3y)\sqrt{y}$
17. [7.5] $14 - 19\sqrt{x} - 3x$ **18.** [7.5] $\frac{5\sqrt{3} - \sqrt{6}}{23}$
19. [7.6] 4   **20.** [7.6] $-1, 2$ **21.** [7.6] 8
22. [7.7] $\sqrt{10,600}$ ft ≈ 102.956 ft   **23.** [7.7] 5 cm;
$5\sqrt{3}$ cm ≈ 8.660 cm   **24.** [7.7] $\sqrt{17} \approx 4.123$
25. [7.7] $\left(\frac{3}{2}, -6\right)$ **26.** [7.8] $5i\sqrt{2}$, or $5\sqrt{2}i$
27. [7.8] $12 + 2i$ **28.** [7.8] $15 - 8i$ **29.** [7.8] $-\frac{11}{34} - \frac{7}{34}i$
30. [7.8] i **31.** [7.6] 3   **32.** [7.8] $-\frac{17}{4}i$ **33.** [7.6] 22,500 ft

Cumulative Review: Chapters 1–7, pp. 497–498

1. -7 **2.** $-7, 5$ **3.** $-5, 5$ **4.** $\frac{5}{2}$ **5.** -1 **6.** $\frac{1}{5}$
7. $\{x | -3 \leq x \leq 7\}$, or $[-3, 7]$ **8.** $\mathbb{R}$, or $(-\infty, \infty)$
9. $-3, 2$ **10.** 7   **11.** $(0, 5)$ **12.** $(1, -1, -2)$
13. **14.**

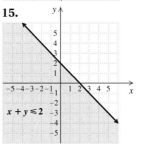

15. **16.**

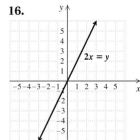

17. Slope: -1; y-intercept: $(0, -6)$
18. $y = 7x - 11$ **19.** 6   **20.** $-8xy^2$
21. $4a^2 - 20ab + 25b^2$ **22.** $c^4 - 9d^2$ **23.** $\frac{2x + 1}{x(x + 1)}$
24. $\frac{x + 13}{(x - 2)(x + 1)}$ **25.** $\frac{(a + 2)(2a + 1)}{(a - 3)(a - 1)}$ **26.** $\frac{2x + 1}{x^2}$
27. 0   **28.** $-1 + 3\sqrt{5}$ **29.** $5ab^2\sqrt{6a}$ **30.** $\sqrt[15]{y^8}$
31. $(x - 7)(x + 2)$ **32.** $4y^5(y - 1)(y^2 + y + 1)$
33. $25(2c + d)(2c - d)$ **34.** $(3t - 8)(t + 1)$
35. $3(x^2 - 2x - 7)$ **36.** $(y - x)(t - z^2)$
37. $\{x | x$ is a real number $and\ x \neq 3\}$, or $(-\infty, 3) \cup (3, \infty)$
38. $\left\{x | x \geq \frac{11}{2}\right\}$, or $\left[\frac{11}{2}, \infty\right)$ **39.** 5   **40.** $6 - 2\sqrt{5}$
41. $(f + g)(x) = x^2 + \sqrt{2x - 3}$ **42.** 4 mph
43. $2\sqrt{3}$ ft ≈ 3.464 ft   **44.** **(a)** $m(t) = 0.08t + 25.02$;
(b) 26.62; **(c)** 2037   **45.** \$36,650   **46.** Swiss chocolate:
45 oz; whipping cream: 20 oz   **47.** 4×10^5 programs
48. \$1.54 per year   **49.** 5 ft   **50.** $5x - 3y = -15$
51. $(0, -1), (1, 0)$ **52.** -2 **53.** 10

CHAPTER 8

Technology Connection, p. 508

1. The right-hand x-intercept should be an approximation of $4 + \sqrt{23}$.   **2.** x-intercepts should be approximations of $(-5 + \sqrt{37})/2$ and $(-5 - \sqrt{37})/2$.
3. Most graphing calculators can give only rational-number approximations of the two irrational solutions. An *exact* solution cannot be found with a graphing calculator.
4. The graph of $y = x^2 - 6x + 11$ has no x-intercepts.

Exercise Set 8.1, pp. 508–510

1. $\sqrt{k}; -\sqrt{k}$ **2.** $7; -7$ **3.** $t + 3; t + 3$ **4.** 16
5. $25; 5$ **6.** $9; 3$ **7.** ± 10 **9.** $\pm 5\sqrt{2}$ **11.** $\pm\sqrt{6}$
13. $\pm\frac{7}{3}$ **15.** $\pm\sqrt{\frac{5}{6}}$, or $\pm\frac{\sqrt{30}}{6}$ **17.** $\pm i$ **19.** $\pm\frac{9}{2}i$
21. $-1, 7$ **23.** $-5 \pm 2\sqrt{3}$ **25.** $-1 \pm 3i$
27. $-\frac{3}{4} \pm \frac{\sqrt{17}}{4}$, or $\frac{-3 \pm \sqrt{17}}{4}$ **29.** $-3, 13$ **31.** $\pm\sqrt{19}$
33. $1, 9$ **35.** $-4 \pm \sqrt{13}$ **37.** $-14, 0$
39. $x^2 + 16x + 64 = (x + 8)^2$
41. $t^2 - 10t + 25 = (t - 5)^2$
43. $t^2 - 2t + 1 = (t - 1)^2$ **45.** $x^2 + 3x + \frac{9}{4} = \left(x + \frac{3}{2}\right)^2$
47. $x^2 + \frac{2}{5}x + \frac{1}{25} = \left(x + \frac{1}{5}\right)^2$
49. $t^2 - \frac{5}{6}t + \frac{25}{144} = \left(t - \frac{5}{12}\right)^2$ **51.** $-7, 1$ **53.** $5 \pm \sqrt{2}$
55. $-8, -4$ **57.** $-4 \pm \sqrt{19}$
59. $(-3 - \sqrt{2}, 0), (-3 + \sqrt{2}, 0)$
61. $\left(-\frac{9}{2} - \frac{\sqrt{181}}{2}, 0\right), \left(-\frac{9}{2} + \frac{\sqrt{181}}{2}, 0\right)$, or $\left(\frac{-9 - \sqrt{181}}{2}, 0\right), \left(\frac{-9 + \sqrt{181}}{2}, 0\right)$
63. $(5 - \sqrt{47}, 0), (5 + \sqrt{47}, 0)$ **65.** $-\frac{4}{3}, -\frac{2}{3}$
67. $-\frac{1}{3}, 2$ **69.** $-\frac{2}{5} \pm \frac{\sqrt{19}}{5}$, or $\frac{-2 \pm \sqrt{19}}{5}$
71. $\left(-\frac{1}{4} - \frac{\sqrt{13}}{4}, 0\right), \left(-\frac{1}{4} + \frac{\sqrt{13}}{4}, 0\right)$, or $\left(\frac{-1 - \sqrt{13}}{4}, 0\right),$ $\left(\frac{-1 + \sqrt{13}}{4}, 0\right)$ **73.** $\left(\frac{3}{4} - \frac{\sqrt{17}}{4}, 0\right), \left(\frac{3}{4} + \frac{\sqrt{17}}{4}, 0\right),$ or $\left(\frac{3 - \sqrt{17}}{4}, 0\right), \left(\frac{3 + \sqrt{17}}{4}, 0\right)$ **75.** 10% **77.** 4%
79. About 4.3 sec   **81.** About 11.4 sec   **83.** ⟰
85. 64 **86.** -15 **87.** $10\sqrt{2}$ **88.** $4\sqrt{6}$ **89.** $2i$
90. $5i$ **91.** $2i\sqrt{2}$, or $2\sqrt{2}i$ **92.** $2i\sqrt{6}$ or $2\sqrt{6}i$
93. ⟰   **95.** ± 18 **97.** $-\frac{7}{2}, -\sqrt{5}, 0, \sqrt{5}, 8$
99. Barge: 8 km/h; fishing boat: 15 km/h   **101.** ⬚
103. ⟰, ⬚

Exercise Set 8.2, pp. 515–516

1. True   **2.** True   **3.** False   **4.** False   **5.** False
6. True   **7.** $-\frac{5}{2}, 1$ **9.** $-1 \pm \sqrt{5}$ **11.** $3 \pm \sqrt{6}$
13. $\frac{3}{2} \pm \frac{\sqrt{29}}{2}$ **15.** $-1 \pm \frac{2\sqrt{3}}{3}$ **17.** $-\frac{4}{3} \pm \frac{\sqrt{19}}{3}$
19. $3 \pm i$ **21.** $\frac{1}{2} \pm \frac{\sqrt{3}}{2}i$ **23.** $-2 \pm \sqrt{2}i$ **25.** $-\frac{8}{3}, \frac{5}{4}$

27. $\frac{2}{5}$ **29.** $-\frac{11}{8} \pm \frac{\sqrt{41}}{8}$ **31.** $5, 10$ **33.** $\frac{3}{2}, 24$
35. $2 \pm \sqrt{5}i$ **37.** $2, -1 \pm \sqrt{3}i$
39. $-\frac{4}{3}, \frac{5}{2}$ **41.** $5 \pm \sqrt{53}$ **43.** $\frac{3}{2} \pm \frac{\sqrt{5}}{2}$
45. $-5.317, 1.317$ **47.** $0.764, 5.236$ **49.** $-1.266, 2.766$
51. ⟰   **53.** $x^2 + 4$ **54.** $x^2 - 180$ **55.** $x^2 - 4x - 3$
56. $x^2 + 6x + 34$ **57.** $-\frac{3}{2}$ **58.** $\frac{1}{6} \pm \frac{\sqrt{6}}{3}i$ **59.** ⟰
61. $(-2, 0), (1, 0)$ **63.** $4 - 2\sqrt{2}, 4 + 2\sqrt{2}$
65. $-1.179, 0.339$ **67.** $\frac{-5\sqrt{2} \pm \sqrt{34}}{4}$ **69.** $\frac{1}{2}$ **71.** ⬚

Technology Connection, p. 519

1. $(-0.4, 0)$ is the other x-intercept of $y = 5x^2 - 13x - 6$.
2. The x-intercepts of $y = x^2 - 175$ are $(-13.22875656, 0)$ and $(13.22875656, 0)$, or $(-5\sqrt{7}, 0)$ and $(5\sqrt{7}, 0)$.
3. The x-intercepts of $y = x^3 + 3x^2 - 4x$ are $(-4, 0), (0, 0)$, and $(1, 0)$.

Exercise Set 8.3, pp. 519–521

1. (b)   **2.** (a)   **3.** (d)   **4.** (b)   **5.** (c)   **6.** (c)
7. Two irrational   **9.** Two imaginary   **11.** Two irrational
13. Two rational   **15.** Two imaginary   **17.** One rational
19. Two rational   **21.** Two irrational
23. Two imaginary   **25.** Two rational
27. Two irrational   **29.** $x^2 + x - 20 = 0$
31. $x^2 - 6x + 9 = 0$ **33.** $x^2 + 4x + 3 = 0$
35. $4x^2 - 23x + 15 = 0$ **37.** $8x^2 + 6x + 1 = 0$
39. $x^2 - 2x - 0.96 = 0$ **41.** $x^2 - 3 = 0$
43. $x^2 - 20 = 0$ **45.** $x^2 + 16 = 0$
47. $x^2 - 4x + 53 = 0$ **49.** $x^2 - 6x - 5 = 0$
51. $3x^2 - 6x - 4 = 0$ **53.** $x^3 - 4x^2 - 7x + 10 = 0$
55. $x^3 - 2x^2 - 3x = 0$ **57.** ⟰   **59.** $c = \frac{d^2}{1 - d}$
60. $b = \frac{aq}{p - q}$ **61.** $y = \frac{x - 3}{x}$, or $1 - \frac{3}{x}$
62. 10 mph   **63.** Jamal: 3.5 mph; Kade: 2 mph
64. 20 mph   **65.** ⟰   **67.** $a = 1, b = 2, c = -3$
69. (a) $-\frac{3}{5}$; (b) $-\frac{1}{3}$ **71.** (a) $9 + 9i$; (b) $3 + 3i$
73. The solutions of $ax^2 + bx + c = 0$ are $x = \frac{-b \pm \sqrt{b^2 - 4ac}}{2a}$. When there is just one solution, $b^2 - 4ac$ must be 0, so $x = \frac{-b \pm 0}{2a} = \frac{-b}{2a}$.
75. $a = 8, b = 20, c = -12$ **77.** $x^2 - 2 = 0$
79. $x^4 - 8x^3 + 21x^2 - 2x - 52 = 0$ **81.** ⟰, ⬚

Exercise Set 8.4, pp. 525–528

1. First part: 60 mph; second part: 50 mph   **3.** 40 mph
5. Cessna: 150 mph, Beechcraft: 200 mph; or Cessna: 200 mph, Beechcraft: 250 mph   **7.** To Hillsboro: 12 mph; return trip: 9 mph   **9.** About 14 mph   **11.** 12 hr
13. About 3.24 mph   **15.** $r = \frac{1}{2}\sqrt{\frac{A}{\pi}}$

17. $r = \dfrac{-\pi h + \sqrt{\pi^2 h^2 + 2\pi A}}{2\pi}$ **19.** $r = \dfrac{\sqrt{Gm_1 m_2}}{F}$

21. $H = \dfrac{c^2}{g}$ **23.** $b = \sqrt{c^2 - a^2}$

25. $t = \dfrac{-v_0 + \sqrt{(v_0)^2 + 2gs}}{g}$ **27.** $n = \dfrac{1 + \sqrt{1 + 8N}}{2}$

29. $g = \dfrac{4\pi^2 l}{T^2}$ **31.** $t = \dfrac{-b \pm \sqrt{b^2 - 4ac}}{2a}$

33. (a) 10.1 sec; (b) 7.49 sec; (c) 272.5 m   **35.** 2.9 sec
37. 0.890 sec   **39.** 2.5 m/sec   **41.** 4.5%   **43.**
45. m^{-2}, or $\dfrac{1}{m^2}$ **46.** $t^{2/3}$ **47.** $y^{1/3}$ **48.** $z^{1/2}$ **49.** 2
50. 81   **51.**
53. $t = \dfrac{-10.2 + 6\sqrt{-A^2 + 13A - 39.36}}{A - 6.5}$ **55.** $\pm\sqrt{2}$

57. $l = \dfrac{w + w\sqrt{5}}{2}$

59. $n = \pm\sqrt{\dfrac{r^2 \pm \sqrt{r^4 + 4m^4 r^2 p - 4mp}}{2m}}$

61. $A(S) = \dfrac{\pi S}{6}$

Exercise Set 8.5, pp. 533–534

1. (f)   **2.** (d)   **3.** (h)   **4.** (b)   **5.** (g)   **6.** (a)
7. (e)   **8.** (c)   **9.** $\sqrt{p}$ **10.** $x^{1/4}$ **11.** $x^2 + 3$ **12.** t^{-3}
13. $(1 + t)^2$ **14.** $w^{1/6}$ **15.** $\pm 2, \pm 3$ **17.** $\pm\sqrt{3}, \pm 2$
19. $\pm\dfrac{\sqrt{5}}{2}, \pm 1$ **21.** 4   **23.** $\pm 2\sqrt{2}, \pm 3$
25. $8 + 2\sqrt{7}$ **27.** No solution   **29.** $-\frac{1}{2}, \frac{1}{3}$ **31.** $-4, 1$
33. $-27, 8$ **35.** 729   **37.** 1   **39.** No solution   **41.** $\frac{12}{5}$
43. $\pm 2, \pm 3i$ **45.** $\pm i, \pm 2i$ **47.** $\left(\dfrac{4}{25}, 0\right)$
49. $\left(\dfrac{3}{2} + \dfrac{\sqrt{33}}{2}, 0\right), \left(\dfrac{3}{2} - \dfrac{\sqrt{33}}{2}, 0\right), (4, 0), (-1, 0)$
51. $(-243, 0), (32, 0)$ **53.** No x-intercepts   **55.**

57.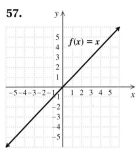

58.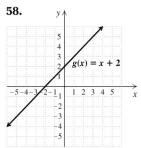

59.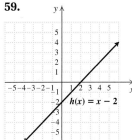

60.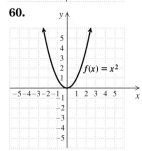

61.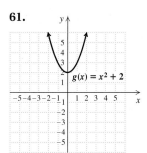

62.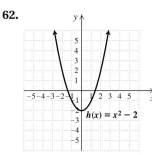

63. 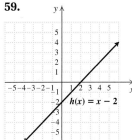   **65.** $\pm\sqrt{\dfrac{-5 \pm \sqrt{37}}{6}}$ **67.** $-2, -1, 6, 7$

69. $\dfrac{100}{99}$ **71.** $-5, -3, -2, 0, 2, 3, 5$ **73.** $1, 3, -\dfrac{1}{2} + \dfrac{\sqrt{3}}{2}i,$

$-\dfrac{1}{2} - \dfrac{\sqrt{3}}{2}i, -\dfrac{3}{2} + \dfrac{3\sqrt{3}}{2}i, -\dfrac{3}{2} - \dfrac{3\sqrt{3}}{2}i$ **75.**

77. ,

Connecting the Concepts, pp. 535–536

1. $-2, 5$ **2.** ± 11 **3.** $-3 \pm \sqrt{19}$ **4.** $-\dfrac{1}{2} \pm \dfrac{\sqrt{13}}{2}$

5. $-1 \pm \sqrt{2}$ **6.** 5   **7.** $\dfrac{1}{2} \pm \dfrac{\sqrt{5}}{2}$ **8.** $1 \pm \sqrt{7}$

9. $\pm\dfrac{\sqrt{11}}{2}$ **10.** $\frac{1}{2}, 1$ **11.** $-\dfrac{1}{2} \pm \dfrac{\sqrt{3}}{2}i$ **12.** $0, \frac{7}{16}$

13. $-\frac{5}{6}, 2$ **14.** $1 \pm \sqrt{7}i$ **15.** $\pm 1, \pm 3$ **16.** $\pm 3, \pm i$
17. $-5, 0$ **18.** $-6, 5$ **19.** $\pm\sqrt{2}, \pm 2i$
20. $\pm\dfrac{\sqrt{3}}{3}, \pm\dfrac{\sqrt{2}}{2}$

Technology Connection, p. 537

1. The graphs of y_1, y_2, and y_3 open upward. The graphs of y_4, y_5, and y_6 open downward. The graph of y_1 is wider than the graph of y_2. The graph of y_3 is narrower than the graph of y_2. Similarly, the graph of y_4 is wider than the graph of y_5, and the graph of y_6 is narrower than the graph of y_5.
2. If A is positive, the graph opens upward. If A is negative, the graph opens downward. Compared with the graph of $y = x^2$, the graph of $y = Ax^2$ is wider if $|A| < 1$ and narrower if $|A| > 1$.

Technology Connection, p. 538

1. Compared with the graph of $y = ax^2$, the graph of $y = a(x - h)^2$ is shifted left or right. It is shifted left if h is negative and right if h is positive.   **2.** The value of A makes the graph wider or narrower, and makes the graph open downward if A is negative. The value of B shifts the graph left or right.

Technology Connection, p. 540

1. The graph of y_2 looks like the graph of y_1 shifted up 2 units, and the graph of y_3 looks like the graph of y_1 shifted down 4 units.   **2.** Compared with the graph of

$y = a(x - h)^2$, the graph of $y = a(x - h)^2 + k$ is shifted up or down. It is shifted down if k is negative and up if k is positive. **3.** The value of A makes the graph wider or narrower, and makes the graph open downward if A is negative. The value of B shifts the graph left or right. The value of C shifts the graph up or down.

Exercise Set 8.6, pp. 542–544

1. (h)   **2.** (g)   **3.** (f)   **4.** (d)   **5.** (b)   **6.** (c)   **7.** (e)
8. (a)
9.

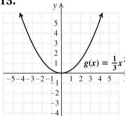

11.

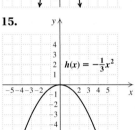

13.

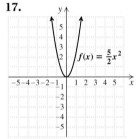

15.

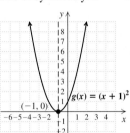

17.

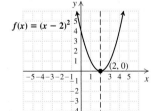

19. Vertex: $(-1, 0)$;
axis of symmetry: $x = -1$

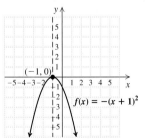

21. Vertex $(2, 0)$;
axis of symmetry: $x = 2$

23. Vertex: $(-1, 0)$;
axis of symmetry: $x = -1$

25. Vertex: $(2, 0)$;
axis of symmetry: $x = 2$

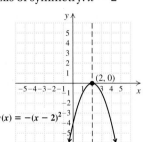

27. Vertex: $(-1, 0)$;
axis of symmetry: $x = -1$

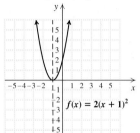

29. Vertex: $(4, 0)$;
axis of symmetry: $x = 4$

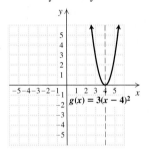

31. Vertex: $(4, 0)$;
axis of symmetry: $x = 4$

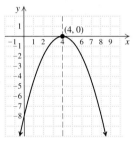

33. Vertex: $(1, 0)$;
axis of symmetry: $x = 1$

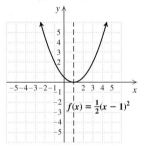

35. Vertex: $(-5, 0)$;
axis of symmetry: $x = -5$

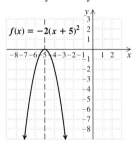

37. Vertex: $\left(\frac{1}{2}, 0\right)$;
axis of symmetry: $x = \frac{1}{2}$

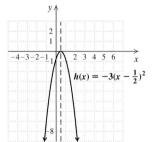

39. Vertex: $(5, 2)$;
axis of symmetry: $x = 5$;
minimum: 2

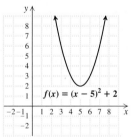

41. Vertex: $(-1, -3)$;
axis of symmetry: $x = -1$;
minimum: -3

43. Vertex: $(-4, 1)$;
axis of symmetry: $x = -4$;
minimum: 1

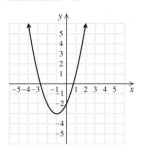

$f(x) = (x + 1)^2 - 3$

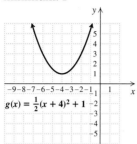

$g(x) = \frac{1}{2}(x + 4)^2 + 1$

45. Vertex: $(1, -3)$;
axis of symmetry: $x = 1$;
maximum: -3

47. Vertex: $(-3, 1)$;
axis of symmetry: $x = -3$
minimum: 1

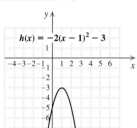
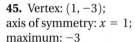

$h(x) = -2(x - 1)^2 - 3$

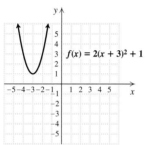

$f(x) = 2(x + 3)^2 + 1$

49. Vertex: $(2, 4)$;
axis of symmetry: $x = 2$;
maximum: 4

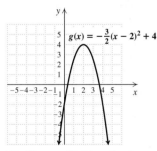

$g(x) = -\frac{3}{2}(x - 2)^2 + 4$

51. Vertex: $(3, 9)$; axis of symmetry: $x = 3$; minimum: 9
53. Vertex: $(-8, 2)$; axis of symmetry: $x = -8$; maximum: 2
55. Vertex: $\left(\frac{7}{2}, -\frac{29}{4}\right)$; axis of symmetry: $x = \frac{7}{2}$;
minimum: $-\frac{29}{4}$ **57.** Vertex: $(-2.25, -\pi)$; axis of
symmetry: $x = -2.25$; maximum: $-\pi$ **59.** 📄
61. x-intercept: $(3, 0)$; y-intercept: $(0, -4)$
62. x-intercept: $\left(\frac{8}{3}, 0\right)$; y-intercept: $(0, 2)$
63. $(-5, 0), (-3, 0)$ **64.** $(-1, 0), \left(\frac{3}{2}, 0\right)$
65. $x^2 - 14x + 49 = (x - 7)^2$
66. $x^2 + 7x + \frac{49}{4} = \left(x + \frac{7}{2}\right)^2$ **67.** 📄
69. $f(x) = \frac{3}{5}(x - 1)^2 + 3$ **71.** $f(x) = \frac{3}{5}(x - 4)^2 - 7$
73. $f(x) = \frac{3}{5}(x + 2)^2 - 5$ **75.** $f(x) = 2(x - 2)^2$
77. $g(x) = -2x^2 - 5$ **79.** The graph will move to the
right.   **81.** The graph will be reflected across the x-axis.
83. $F(x) = 3(x - 5)^2 + 1$

85.

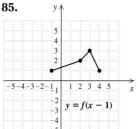

$y = f(x - 1)$

87.

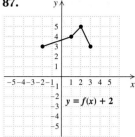

$y = f(x) + 2$

89.

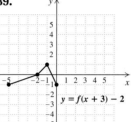

$y = f(x + 3) - 2$

91. 📊   **93.** 📄, 📊

Visualizing for Success, p. 549

1. B   **2.** E   **3.** A   **4.** H   **5.** C   **6.** J   **7.** F
8. G   **9.** I   **10.** D

Exercise Set 8.7, pp. 550–551

1. True   **2.** False   **3.** True   **4.** True   **5.** False
6. True   **7.** False   **8.** True
9. $f(x) = (x - 4)^2 + (-14)$
11. $f(x) = \left(x - \left(-\frac{3}{2}\right)\right)^2 + \left(-\frac{29}{4}\right)$
13. $f(x) = 3(x - (-1))^2 + (-5)$
15. $f(x) = -(x - (-2))^2 + (-3)$
17. $f(x) = 2\left(x - \frac{5}{4}\right)^2 + \frac{55}{8}$
19. (a) Vertex: $(-2, 1)$;
axis of symmetry: $x = -2$;
(b)

21. (a) Vertex: $(-4, 4)$;
axis of symmetry: $x = -4$;
(b)

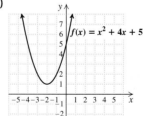

$f(x) = x^2 + 4x + 5$

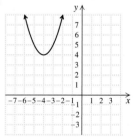

$f(x) = x^2 + 8x + 20$

23. (a) Vertex: $(4, -7)$; axis of symmetry: $x = 4$;
(b)

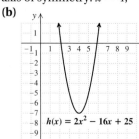

$h(x) = 2x^2 - 16x + 25$

25. (a) Vertex: $(1, 6)$; axis of symmetry: $x = 1$;
(b)

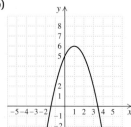

$f(x) = -x^2 + 2x + 5$

39. (a) Vertex: $\left(\frac{5}{6}, \frac{1}{12}\right)$; axis of symmetry: $x = \frac{5}{6}$; maximum: $\frac{1}{12}$;
(b)

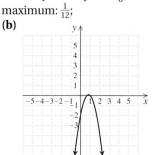

$f(x) = -3x^2 + 5x - 2$

41. (a) Vertex: $\left(-4, -\frac{5}{3}\right)$; axis of symmetry: $x = -4$; minimum: $-\frac{5}{3}$;
(b)

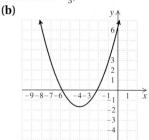

$h(x) = \frac{1}{2}x^2 + 4x + \frac{19}{3}$

27. (a) Vertex: $\left(-\frac{3}{2}, -\frac{49}{4}\right)$; axis of symmetry: $x = -\frac{3}{2}$;
(b)

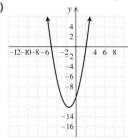

$g(x) = x^2 + 3x - 10$

29. (a) Vertex: $\left(-\frac{7}{2}, -\frac{49}{4}\right)$; axis of symmetry: $x = -\frac{7}{2}$;
(b)

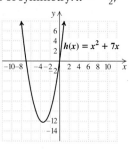

$h(x) = x^2 + 7x$

43. $(3 - \sqrt{6}, 0)$, $(3 + \sqrt{6}, 0)$; $(0, 3)$ **45.** $(-1, 0)$, $(3, 0)$; $(0, 3)$ **47.** $(0, 0)$, $(9, 0)$; $(0, 0)$ **49.** $(2, 0)$; $(0, -4)$
51. $\left(-\frac{1}{2} - \frac{\sqrt{21}}{2}, 0\right)$, $\left(-\frac{1}{2} + \frac{\sqrt{21}}{2}, 0\right)$; $(0, -5)$
53. No x-intercept; $(0, 6)$ **55.** 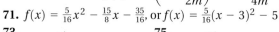   **57.** $(1, 1, 1)$
58. $(-2, 5, 1)$ **59.** $(10, 5, 8)$ **60.** $(-3, 6, -5)$
61. $(2.4, -1.8, 1.5)$ **62.** $\left(\frac{1}{3}, \frac{1}{6}, \frac{1}{2}\right)$ **63.** ◪
65. (a) Minimum: -6.953660714; **(b)** $(-1.056433682, 0)$, $(2.413576539, 0)$; $(0, -5.89)$ **67. (a)** $-2.4, 3.4$;
(b) $-1.3, 2.3$ **69.** $f(x) = m\left(x - \frac{n}{2m}\right)^2 + \frac{4mp - n^2}{4m}$
71. $f(x) = \frac{5}{16}x^2 - \frac{15}{8}x - \frac{35}{16}$, or $f(x) = \frac{5}{16}(x - 3)^2 - 5$

73.

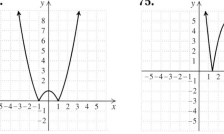

$f(x) = |x^2 - 1|$

75.

$f(x) = |2(x - 3)^2 - 5|$

31. (a) Vertex: $(-1, -4)$; axis of symmetry: $x = -1$;
(b)

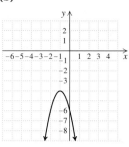

$f(x) = -2x^2 - 4x - 6$

33. (a) Vertex: $(3, 4)$; axis of symmetry: $x = 3$; minimum: 4;
(b)

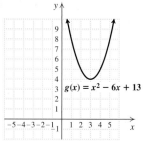

$g(x) = x^2 - 6x + 13$

35. (a) Vertex: $(2, -5)$; axis of symmetry: $x = 2$; minimum: -5;
(b)

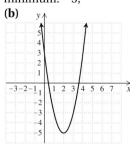

$g(x) = 2x^2 - 8x + 3$

37. (a) Vertex: $(4, 2)$; axis of symmetry: $x = 4$; minimum: 2;
(b)

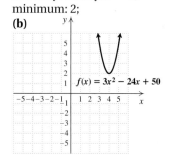

$f(x) = 3x^2 - 24x + 50$

Technology Connection, p. 555

1. About 607 million CDs

Exercise Set 8.8, pp. 556–561

1. (e)   **2.** (b)   **3.** (c)   **4.** (a)   **5.** (d)   **6.** (f)
7. $3\frac{1}{4}$ weeks; 8.3 lb of milk per day   **9.** \$120/dulcimer; 350 dulcimers   **11.** 180 ft by 180 ft   **13.** 450 ft²; 15 ft by 30 ft (The house serves as a 30-ft side.)   **15.** 3.5 in.
17. 81; 9 and 9   **19.** -16; 4 and -4 **21.** 25; -5 and -5
23. $f(x) = ax^2 + bx + c, a < 0$ **25.** $f(x) = mx + b$
27. Neither quadratic nor linear
29. $f(x) = ax^2 + bx + c, a > 0$
31. $f(x) = ax^2 + bx + c, a > 0$ **33.** $f(x) = mx + b$
35. $f(x) = 2x^2 + 3x - 1$ **37.** $f(x) = -\frac{1}{4}x^2 + 3x - 5$
39. (a) $A(s) = \frac{3}{16}s^2 - \frac{135}{4}s + 1750$; **(b)** about 531 accidents
41. $h(d) = -0.0068d^2 + 0.8571d$ **43.** ◪
45. $\{x | x > 4\}$, or $(4, \infty)$ **46.** $\{x | x \geq -3\}$, or $[-3, \infty)$

47. $\{x \mid x \le 7 \ or \ x \ge 11\}$, or $(-\infty, 7] \cup [11, \infty)$
48. $\{x \mid -3 < x < \frac{5}{2}\}$, or $\left(-3, \frac{5}{2}\right)$
49. $f(x) = \dfrac{-4x - 23}{x + 4}, x \ne -4$
50. $f(x) = \dfrac{1}{x - 1}, x \ne 1$ **51.** $-\frac{23}{4}$ **52.** No solution
53. 0 **54.** $-6, 9$ **55.** ✍ **57.** 158 ft **59.** \$15
61. The radius of the circular portion of the window and
the height of the rectangular portion should each be $\dfrac{24}{\pi + 4}$ ft.
63. (a) $h(x) = 11{,}090.60714x^2 - 29{,}069.62143x + 39{,}983.8$;
(b) 858,348 vehicles

Technology Connection, p. 566

1. $\{x \mid -0.78 \le x \le 1.59\}$, or $[-0.78, 1.59]$
2. $\{x \mid x \le -0.21 \ or \ x \ge 2.47\}$, or $(-\infty, -0.21] \cup [2.47, \infty)$
3. $\{x \mid x < -1.26 \ or \ x > 2.33\}$, or $(-\infty, -1.26) \cup (2.33, \infty)$
4. $\{x \mid x > -1.37\}$, or $(-1.37, \infty)$

Exercise Set 8.9, pp. 568–570

1. True **2.** False **3.** True **4.** True **5.** False
6. True **7.** $\left[-4, \frac{3}{2}\right]$, or $\left\{x \mid -4 \le x \le \frac{3}{2}\right\}$
9. $(-\infty, -2) \cup (0, 2) \cup (3, \infty)$, or
$\{x \mid x < -2 \ or \ 0 < x < 2 \ or \ x > 3\}$
11. $\left(-\infty, -\frac{7}{2}\right) \cup (-2, \infty)$, or $\left\{x \mid x < -\frac{7}{2} \ or \ x > -2\right\}$
13. $(5, 6)$, or $\{x \mid 5 < x < 6\}$
15. $(-\infty, -7] \cup [2, \infty)$, or $\{x \mid x \le -7 \ or \ x \ge 2\}$
17. $(-\infty, -1) \cup (2, \infty)$, or $\{x \mid x < -1 \ or \ x > 2\}$
19. $\varnothing$ **21.** $[2 - \sqrt{7}, 2 + \sqrt{7}]$, or
$\{x \mid 2 - \sqrt{7} \le x \le 2 + \sqrt{7}\}$ **23.** $(-\infty, -2) \cup (0, 2)$, or
$\{x \mid x < -2 \ or \ 0 < x < 2\}$ **25.** $[-2, 1] \cup [4, \infty)$, or
$\{x \mid -2 \le x \le 1 \ or \ x \ge 4\}$ **27.** $[-2, 2]$, or
$\{x \mid -2 \le x \le 2\}$ **29.** $(-1, 2) \cup (3, \infty)$, or
$\{x \mid -1 < x < 2 \ or \ x > 3\}$ **31.** $(-\infty, 0] \cup [2, 5]$, or
$\{x \mid x \le 0 \ or \ 2 \le x \le 5\}$ **33.** $(-\infty, 5)$, or $\{x \mid x < 5\}$
35. $(-\infty, -1] \cup (3, \infty)$, or $\{x \mid x \le -1 \ or \ x > 3\}$
37. $(-\infty, -6)$, or $\{x \mid x < -6\}$ **39.** $(-\infty, -1] \cup [2, 5)$, or
$\{x \mid x \le -1 \ or \ 2 \le x < 5\}$ **41.** $(-\infty, -3) \cup [0, \infty)$, or
$\{x \mid x < -3 \ or \ x \ge 0\}$ **43.** $(0, \infty)$, or $\{x \mid x > 0\}$
45. $(-\infty, -4) \cup [1, 3)$, or $\{x \mid x < -4 \ or \ 1 \le x < 3\}$
47. $\left(-\frac{3}{4}, \frac{5}{2}\right]$, or $\left\{x \mid -\frac{3}{4} < x \le \frac{5}{2}\right\}$ **49.** $(-\infty, 2) \cup [3, \infty)$, or
$\{x \mid x < 2 \ or \ x \ge 3\}$ **51.** ✍
53.
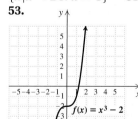
54.
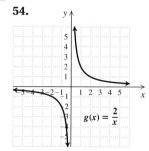
55. $\dfrac{1}{a^2} + 7$ **56.** $a - 8$ **57.** $4a^2 + 20a + 27$

58. $\sqrt{12a - 19}$ **59.** ✍ **61.** $(-1 - \sqrt{6}, -1 + \sqrt{6})$, or
$\{x \mid -1 - \sqrt{6} < x < -1 + \sqrt{6}\}$ **63.** $\{0\}$
65. (a) $(10, 200)$, or $\{x \mid 10 < x < 200\}$;
(b) $[0, 10) \cup (200, \infty)$, or $\{x \mid 0 \le x < 10 \ or \ x > 200\}$
67. $\{n \mid n \text{ is an integer } and \ 12 \le n \le 25\}$
69. $f(x) = 0$ for $x = -2, 1, 3$; $f(x) < 0$ for
$(-\infty, -2) \cup (1, 3)$, or $\{x \mid x < -2 \ or \ 1 < x < 3\}$; $f(x) > 0$
for $(-2, 1) \cup (3, \infty)$, or $\{x \mid -2 < x < 1 \ or \ x > 3\}$
71. $f(x)$ has no zeros; $f(x) < 0$ for $(-\infty, 0)$, or $\{x \mid x < 0\}$;
$f(x) > 0$ for $(0, \infty)$, or $\{x \mid x > 0\}$ **73.** $f(x) = 0$ for
$x = -1, 0$; $f(x) < 0$ for $(-\infty, -3) \cup (-1, 0)$, or
$\{x \mid x < -3 \ or \ -1 < x < 0\}$; $f(x) > 0$ for
$(-3, -1) \cup (0, 2) \cup (2, \infty)$, or
$\{x \mid -3 < x < -1 \ or \ 0 < x < 2 \ or \ x > 2\}$
75. $(-\infty, -5] \cup [9, \infty)$, or $\{x \mid x \le -5 \ or \ x \ge 9\}$
77. $(-\infty, -8] \cup [0, \infty)$, or $\{x \mid x \le -8 \ or \ x \ge 0\}$
79. ✍

Review Exercises: Chapter 8, pp. 574–575

1. False **2.** True **3.** True **4.** True **5.** False
6. True **7.** True **8.** True **9.** False **10.** True
11. $\pm \dfrac{\sqrt{2}}{3}$ **12.** $0, -\frac{3}{4}$ **13.** $3, 9$ **14.** $2 \pm 2i$ **15.** $3, 5$
16. $-\dfrac{9}{2} \pm \dfrac{\sqrt{85}}{2}$ **17.** $-0.372, 5.372$ **18.** $-\frac{1}{4}, 1$
19. $x^2 - 18x + 81 = (x - 9)^2$
20. $x^2 + \frac{3}{5}x + \frac{9}{100} = \left(x + \frac{3}{10}\right)^2$ **21.** $3 + 2\sqrt{2}$
22. 4% **23.** 8.0 sec **24.** Two irrational real numbers
25. Two imaginary numbers **26.** $x^2 + 9 = 0$
27. $x^2 + 10x + 25 = 0$ **28.** About 153 mph **29.** 6 hr
30. $(-3, 0), (-2, 0), (2, 0), (3, 0)$ **31.** $-5, 3$
32. $\pm\sqrt{2}, \pm\sqrt{7}$
33.
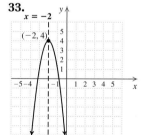
$f(x) = -3(x + 2)^2 + 4$
Maximum: 4
34. (a) Vertex: $(3, 5)$; axis of symmetry: $x = 3$;
(b)
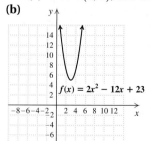
35. $(2, 0), (7, 0); (0, 14)$ **36.** $p = \dfrac{9\pi^2}{N^2}$

37. $T = \dfrac{1 \pm \sqrt{1 + 24A}}{6}$　**38.** Quadratic　**39.** Linear
40. 225 ft^2; 15 ft by 15 ft　**41.** **(a)** $f(x) = -\frac{3}{8}x^2 + \frac{9}{4}x + 8$;
(b) about 10%　**42.** $(-1, 0) \cup (3, \infty)$, or
$\{x | -1 < x < 0 \text{ or } x > 3\}$　**43.** $(-3, 5]$, or
$\{x | -3 < x \le 5\}$　**44.** ✍ The x-coordinate of the maximum or minimum point lies halfway between the
x-coordinates of the x-intercepts.　**45.** ✍ Yes; if the discriminant is a perfect square, then the solutions are rational
numbers, p/q and r/s. (Note that if the discriminant is 0,
then $p/q = r/s$.) Then the equation can be written in
factored form, $(qx - p)(sx - r) = 0$.　**46.** ✍ Four; let
$u = x^2$. Then $au^2 + bu + c = 0$ has at most two solutions,
$u = m$ and $u = n$. Now substitute x^2 for u and obtain
$x^2 = m$ or $x^2 = n$. These equations yield the solutions
$x = \pm\sqrt{m}$ and $x = \pm\sqrt{n}$. When $m \ne n$, the maximum
number of solutions, four, occurs.　**47.** ✍ Completing
the square was used to solve quadratic equations and to
graph quadratic functions by rewriting the function in the
form $f(x) = a(x - h)^2 + k$.　**48.** $f(x) = \frac{7}{15}x^2 - \frac{14}{15}x - 7$
49. $h = 60, k = 60$　**50.** 18, 324

Test: Chapter 8, p. 576

1. [8.1] $\pm\dfrac{\sqrt{7}}{5}$　**2.** [8.1] 2, 9　**3.** [8.2] $-1 \pm \sqrt{2}i$

4. [8.2] $1 \pm \sqrt{6}$　**5.** [8.5] $-2, \frac{2}{3}$
6. [8.2] $-4.193, 1.193$　**7.** [8.2] $-\frac{3}{4}, \frac{7}{3}$
8. [8.1] $x^2 - 20x + 100 = (x - 10)^2$
9. [8.1] $x^2 + \frac{2}{7}x + \frac{1}{49} = \left(x + \frac{1}{7}\right)^2$　**10.** [8.1] $-5 \pm \sqrt{10}$
11. [8.3] Two imaginary numbers　**12.** [8.3] $x^2 - 11 = 0$
13. [8.4] 16 km/h　**14.** [8.4] 2 hr　**15.** [8.5] $(-4, 0), (4, 0)$
16. [8.6]　**17.** [8.7] **(a)** $(-1, -8), x = -1$;
(b)

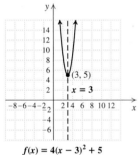
$f(x) = 4(x - 3)^2 + 5$
Minimum: 5

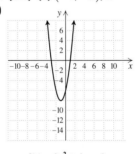
$f(x) = 2x^2 + 4x - 6$

18. [8.7] $(-2, 0), (3, 0); (0, -6)$　**19.** [8.4] $r = \sqrt{\dfrac{3V}{\pi} - R^2}$

20. [8.8] Quadratic　**21.** [8.8] Minimum: $129/cabinet
when 325 cabinets are built　**22.** [8.8] $f(x) = \frac{1}{5}x^2 - \frac{3}{5}x$
23. [8.9] $(-6, 1)$, or $\{x | -6 < x < 1\}$
24. [8.9] $[-1, 0) \cup [1, \infty)$, or $\{x | -1 \le x < 0 \text{ or } x \ge 1\}$
25. [8.3] $\frac{1}{2}$　**26.** [8.3] $x^4 + x^2 - 12 = 0$; answers may vary
27. [8.5] $\pm\sqrt{\sqrt{5} + 2}, \pm\sqrt{\sqrt{5} - 2}i$

Cumulative Review: Chapters 1–8, pp. 577–578

1. 24　**2.** $\dfrac{3a^{10}c^7}{4}$　**3.** $14x^2y - 10xy - 9xy^2$

4. $81p^4q^2 - 64t^2$　**5.** $\dfrac{t(t + 1)(t + 5)}{(3t + 4)^3}$　**6.** $-\dfrac{3}{x + 4}$
7. $3x^2\sqrt[3]{4y^2}$　**8.** $13 - \sqrt{2}i$
9. $3(2x^2 + 5y^2)(2x^2 - 5y^2)$　**10.** $x(x - 20)(x - 4)$
11. $100(m + 1)(m^2 - m + 1)(m - 1)(m^2 + m + 1)$
12. $(2t + 9)(3t + 4)$　**13.** 7　**14.** $\{x | x < 7\}$, or $(-\infty, 7)$
15. $\left(3, \frac{1}{2}\right)$　**16.** $-6, 11$　**17.** $\frac{1}{2}, 2$　**18.** 4　**19.** $-5 \pm \sqrt{2}$
20. $\dfrac{1}{6} \pm \dfrac{\sqrt{11}}{6}i$

21.

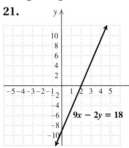
$9x - 2y = 18$

22.

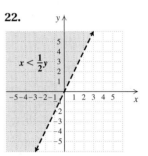
$x < \frac{1}{2}y$

23.

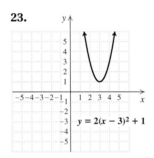
$y = 2(x - 3)^2 + 1$

24.

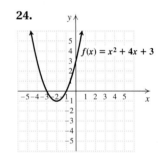
$f(x) = x^2 + 4x + 3$

25. $y = -5x + \frac{1}{2}$　**26.** $-\frac{7}{10}$　**27.** 21
28. $(-\infty, 10]$, or $\{x | x \le 10\}$
29. $\{x | x \text{ is a real number } and x \ne 4\}$, or $(-\infty, 4) \cup (4, \infty)$
30. $a = \dfrac{c}{2b - 1}$　**31.** $t = \dfrac{4r}{3p^2}$　**32.** **(a)** $4.53 billion;
(b) 2015　**33.** **(a)** $h(t) = 33t - 47$; **(b)** 283,000 hotspots;
(c) about 2017　**34.** **(a)** 1.74 oz; **(b)** $600 per ounce;
(c) 75%　**35.** Number tiles: 26 sets; alphabet tiles: 10 sets
36. $125/bunk bed; 400 bunk beds
37. Deanna: 12 hr; Donna: 6 hr
38. 9 km/h　**39.** Mileages no greater than 50 mi
40. $\dfrac{1}{3} \pm \dfrac{\sqrt{2}}{6}i$　**41.** $\{0\}$　**42.** $f(x) = x + 1$
43. $(1 - \sqrt{6}, 16 - 10\sqrt{6}), (1 + \sqrt{6}, 16 + 10\sqrt{6})$

CHAPTER 9

Technology Connection, p. 582

1. To check $(f \circ g)(x)$, we let $y_1 = \sqrt{x}, y_2 = x - 1$,
$y_3 = \sqrt{x - 1}$, and $y_4 = y_1(y_2)$. A table shows that we have
$y_3 = y_4$. The check for $(g \circ f)(x)$ is similar. A graph can also
be used.

Technology Connection, p. 587

1. Graph each pair of functions in a square window along with the line $y = x$ and determine whether the first two functions are reflections of each other across $y = x$. For further verification, examine a table of values for each pair of functions.   **2.** Yes; most graphing calculators do not require that the inverse relation be a function.

Exercise Set 9.1, pp. 588–590

1. True   **2.** True   **3.** False   **4.** False   **5.** False
6. False   **7.** True   **8.** True   **9.** (a) $(f \circ g)(1) = 5$;
(b) $(g \circ f)(1) = -1$; **(c)** $(f \circ g)(x) = x^2 - 6x + 10$;
(d) $(g \circ f)(x) = x^2 - 2$ **11.** (a) $(f \circ g)(1) = -24$;
(b) $(g \circ f)(1) = 65$; **(c)** $(f \circ g)(x) = 10x^2 - 34$;
(d) $(g \circ f)(x) = 50x^2 + 20x - 5$
13. (a) $(f \circ g)(1) = 8$; **(b)** $(g \circ f)(1) = \frac{1}{64}$;
(c) $(f \circ g)(x) = \dfrac{1}{x^2} + 7$; **(d)** $(g \circ f)(x) = \dfrac{1}{(x+7)^2}$
15. (a) $(f \circ g)(1) = 2$; **(b)** $(g \circ f)(1) = 4$;
(c) $(f \circ g)(x) = \sqrt{x+3}$; **(d)** $(g \circ f)(x) = \sqrt{x}+3$
17. (a) $(f \circ g)(1) = 2$; **(b)** $(g \circ f)(1) = \frac{1}{2}$;
(c) $(f \circ g)(x) = \sqrt{\dfrac{4}{x}}$; **(d)** $(g \circ f)(x) = \dfrac{1}{\sqrt{4x}}$
19. (a) $(f \circ g)(1) = 4$; **(b)** $(g \circ f)(1) = 2$;
(c) $(f \circ g)(x) = x + 3$; **(d)** $(g \circ f)(x) = \sqrt{x^2+3}$
21. $f(x) = x^4; g(x) = 3x - 5$ **23.** $f(x) = \sqrt{x}$;
$g(x) = 9x + 1$ **25.** $f(x) = \dfrac{6}{x}; g(x) = 5x - 2$ **27.** Yes
29. No   **31.** Yes   **33.** No   **35.** (a) Yes;
(b) $f^{-1}(x) = x - 3$ **37.** (a) Yes; **(b)** $f^{-1}(x) = \dfrac{x}{2}$
39. (a) Yes; **(b)** $g^{-1}(x) = \dfrac{x+1}{3}$ **41.** (a) Yes;
(b) $f^{-1}(x) = 2x - 2$ **43.** (a) No   **45.** (a) Yes;
(b) $h^{-1}(x) = 10 - x$ **47.** (a) Yes; **(b)** $f^{-1}(x) = \dfrac{1}{x}$
49. (a) No   **51.** (a) Yes; **(b)** $f^{-1}(x) = \dfrac{3x-1}{2}$
53. (a) Yes; **(b)** $f^{-1}(x) = \sqrt[3]{x} - 5$ **55.** (a) Yes;
(b) $g^{-1}(x) = \sqrt[3]{x} + 2$ **57.** (a) Yes;
(b) $f^{-1}(x) = x^2, x \geq 0$
59.

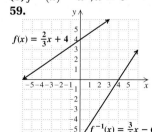

61.

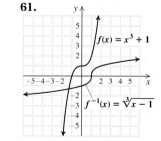

63.

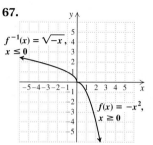

65.

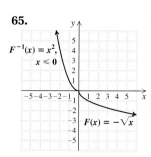

67.

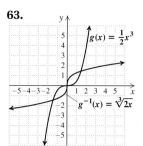

69. (1) $(f^{-1} \circ f)(x) = f^{-1}(f(x))$
$= f^{-1}\left(\sqrt[3]{x-4}\right) = \left(\sqrt[3]{x-4}\right)^3 + 4$
$= x - 4 + 4 = x$;
(2) $(f \circ f^{-1})(x) = f(f^{-1}(x))$
$= f(x^3 + 4) = \sqrt[3]{x^3 + 4 - 4}$
$= \sqrt[3]{x^3} = x$

71. (1) $(f^{-1} \circ f)(x) = f^{-1}(f(x)) = f^{-1}\left(\dfrac{1-x}{x}\right)$

$= \dfrac{1}{\left(\dfrac{1-x}{x}\right) + 1}$

$= \dfrac{1}{\dfrac{1-x+x}{x}}$

$= x$;

(2) $(f \circ f^{-1})(x) = f(f^{-1}(x)) = f\left(\dfrac{1}{x+1}\right)$

$= \dfrac{1 - \left(\dfrac{1}{x+1}\right)}{\left(\dfrac{1}{x+1}\right)}$

$= \dfrac{\dfrac{x+1-1}{x+1}}{\dfrac{1}{x+1}} = x$

73. (a) 40, 44, 52, 60; **(b)** $f^{-1}(x) = (x - 24)/2$, or $\dfrac{x}{2} - 12$
(c) 8, 10, 14, 18   **75.** 🖩   **77.** $\frac{1}{8}$ **78.** $\frac{1}{25}$ **79.** 32
80. Approximately 2.1577
81.

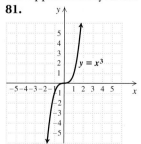

82.

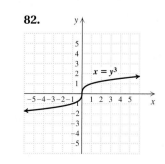

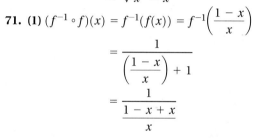

83.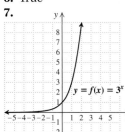
85.

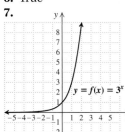

87. $g(x) = \dfrac{x}{2} + 20$ **89.**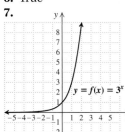

91. Suppose that $h(x) = (f \circ g)(x)$. First, note that for $I(x) = x, (f \circ I)(x) = f(I(x)) = f(x)$ for any function f.

(i) $((g^{-1} \circ f^{-1}) \circ h)(x) = ((g^{-1} \circ f^{-1}) \circ (f \circ g))(x)$
$= ((g^{-1} \circ (f^{-1} \circ f)) \circ g)(x)$
$= ((g^{-1} \circ I) \circ g)(x)$
$= (g^{-1} \circ g)(x) = x$

(ii) $(h \circ (g^{-1} \circ f^{-1}))(x) = ((f \circ g) \circ (g^{-1} \circ f^{-1}))(x)$
$= ((f \circ (g \circ g^{-1})) \circ f^{-1})(x)$
$= ((f \circ I) \circ f^{-1})(x)$
$= (f \circ f^{-1})(x) = x.$

Therefore, $(g^{-1} \circ f^{-1})(x) = h^{-1}(x)$.

93. Yes   **95.** No   **97.** **(1)** C; **(2)** A; **(3)** B; **(4)** D
99.

Technology Connection, p. 593

1. $y_1 = \left(\dfrac{5}{2}\right)^x$; $y_2 = \left(\dfrac{2}{5}\right)^x$

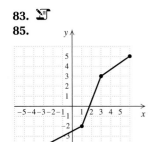

2. $y_1 = 3.2^x$; $y_2 = 3.2^{-x}$

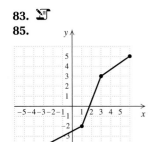

3. $y_1 = \left(\dfrac{3}{7}\right)^x$; $y_2 = \left(\dfrac{7}{3}\right)^x$

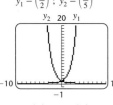

4. $y_1 = 5000(1.08)^x$; $y_2 = 5000(1.08)^{x-3}$

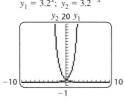

Xscl = 5, Yscl = 1000

Exercise Set 9.2, pp. 596–598

1. True   **2.** True   **3.** True   **4.** False   **5.** False
6. True
7.

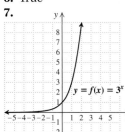

$y = f(x) = 3^x$

9.

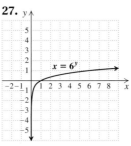

$y = 6^x$

11.

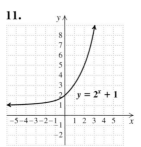

$y = 2^x + 1$

13.

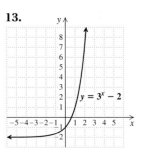

$y = 3^x - 2$

15.

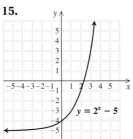

$y = 2^x - 5$

17.

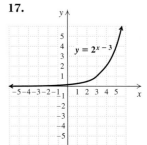

$y = 2^{x-3}$

19.

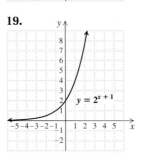

$y = 2^x + 1$

21.

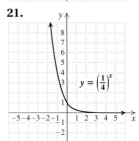

$y = \left(\dfrac{1}{4}\right)^x$

23.

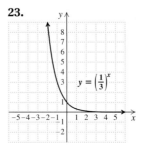

$y = \left(\dfrac{1}{3}\right)^x$

25.

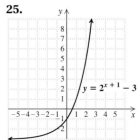

$y = 2^{x+1} - 3$

27.

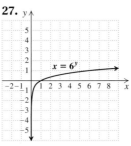

$x = 6^y$

29.

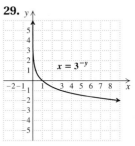

$x = 3^{-y}$

31.

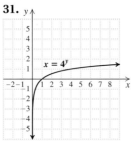

$x = 4^y$

33.

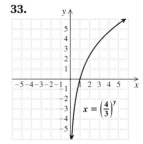

$x = \left(\dfrac{4}{3}\right)^y$

35.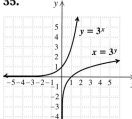

37.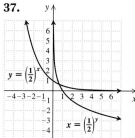

39. (a) About 0.68 billion tracks; about 1.052 billion tracks; about 2.519 billion tracks;

(b)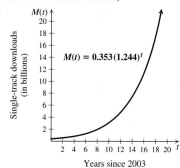

41. (a) 19.6%; 16.3%; 7.3%

(b)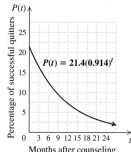

43. (a) About 44,079 whales; about 12,953 whales;

(b)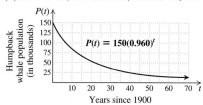

45. (a) About 8706 whales; about 15,107 whales;

(b)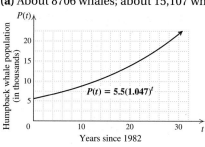

47. (a) 454,354,240 cm^2; 525,233,501,400 cm^2;

(b)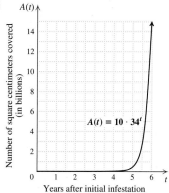

49. 📊   **51.** $3(x + 4)(x - 4)$ **52.** $(x - 10)^2$

53. $(2x + 3)(3x - 4)$

54. $8(x^2 - 2y^2)(x^4 + 2x^2y^2 + 4y^4)$

55. $(t - y + 1)(t + y - 1)$ **56.** $x(x - 2)(5x^2 - 3)$

57. 📊   **59.** $\pi^{2.4}$

61.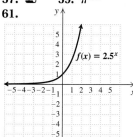

63.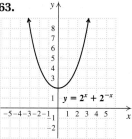

65.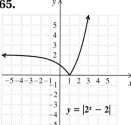

67.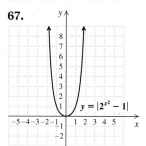

69.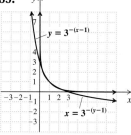

71. $N(t) = 0.464(1.778)^t$; about 464 million devices

73. 📊   **75.** 📊

Exercise Set 9.3, pp. 604–605

1. (g)   **2.** (d)   **3.** (a)   **4.** (h)   **5.** (b)   **6.** (c)
7. (e)   **8.** (f)   **9.** 3   **11.** 2   **13.** 4   **15.** -2
17. -1 **19.** 4   **21.** 1   **23.** 0   **25.** 5   **27.** -2
29. $\frac{1}{2}$ **31.** $\frac{3}{2}$ **33.** $\frac{2}{3}$ **35.** 29

37.
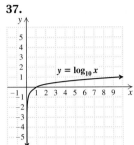

39.
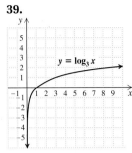

41.
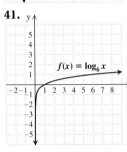

43.
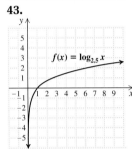

45.
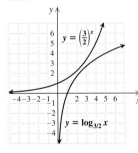

47. $10^x = 8$ **49.** $9^1 = 9$ **51.** $10^{-1} = 0.1$
53. $10^{0.845} = 7$ **55.** $c^8 = m$ **57.** $r^t = C$
59. $e^{-1.3863} = 0.25$ **61.** $r^{-x} = T$ **63.** $2 = \log_{10} 100$
65. $-3 = \log_5 \frac{1}{125}$ **67.** $\frac{1}{4} = \log_{16} 2$
69. $0.4771 = \log_{10} 3$ **71.** $m = \log_z 6$ **73.** $t = \log_p q$
75. $3 = \log_e 20.0855$ **77.** $-4 = \log_e 0.0183$ **79.** 36
81. 5   **83.** 9   **85.** 49   **87.** $\frac{1}{9}$ **89.** 4   **91.** 📙
93. $30a^2b^4$ **94.** $12 - 2\sqrt{30} + 2\sqrt{15} - 5\sqrt{2}$
95. $3\sqrt{3x}$ **96.** $\sqrt[12]{x}$ **97.** $\dfrac{x(3y - 2)}{2y + x}$ **98.** $\dfrac{x + 2}{x + 1}$
99. 📙
101.

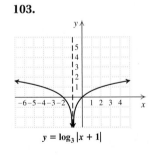

103.

105. 6   **107.** $-25, 4$ **109.** -2 **111.** 0   **113.** Let $b = 0$, and suppose that $x_1 = 1$ and $x_2 = 2$. Then $0^1 = 0^2$, but $1 \neq 2$. Then let $b = 1$, and suppose that $x_1 = 1$ and $x_2 = 2$. Then $1^1 = 1^2$, but $1 \neq 2$.

Exercise Set 9.4, pp. 611–612

1. (e)   **2.** (f)   **3.** (a)   **4.** (b)   **5.** (c)   **6.** (d)
7. $\log_3 81 + \log_3 27$ **9.** $\log_4 64 + \log_4 16$
11. $\log_c r + \log_c s + \log_c t$ **13.** $\log_a (2 \cdot 10)$, or $\log_a 20$
15. $\log_c (t \cdot y)$ **17.** $8 \log_a r$ **19.** $\frac{1}{3} \log_2 y$
21. $-3 \log_b C$ **23.** $\log_2 5 - \log_2 11$
25. $\log_b m - \log_b n$ **27.** $\log_a \frac{19}{2}$ **29.** $\log_b \frac{36}{4}$, or $\log_b 9$
31. $\log_a \dfrac{x}{y}$ **33.** $\log_a x + \log_a y + \log_a z$
35. $3 \log_a x + 4 \log_a z$ **37.** $2 \log_a w - 2 \log_a x + \log_a y$
39. $5 \log_a x - 3 \log_a y - \log_a z$
41. $\log_b x + 2 \log_b y - \log_b w - 3 \log_b z$
43. $\frac{1}{2}(7 \log_a x - 5 \log_a y - 8 \log_a z)$
45. $\frac{1}{3}(6 \log_a x + 3 \log_a y - 2 - 7 \log_a z)$ **47.** $\log_a (x^8 z^3)$
49. $\log_a x$ **51.** $\log_a \dfrac{y^5}{x^{3/2}}$ **53.** $\log_a (x - 3)$ **55.** 1.953
57. -0.369 **59.** -1.161 **61.** $\frac{3}{2}$ **63.** Cannot be found
65. 10   **67.** m **69.** 📙
71.

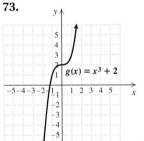

72.

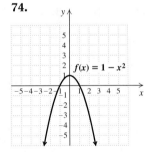

73.

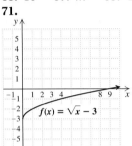

74.

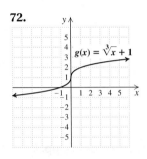

75. $(-\infty, -7) \cup (-7, \infty)$, or $\{x \mid x \text{ is a real number } and \ x \neq -7\}$
76. $(-\infty, -3) \cup (-3, 2) \cup (2, \infty)$, or $\{x \mid x \text{ is a real number } and \ x \neq -3 \ and \ x \neq 2\}$
77. $(-\infty, 10]$, or $\{x \mid x \leq 10\}$ **78.** $(-\infty, \infty)$, or $\mathbb{R}$
79. 📙   **81.** $\log_a (x^6 - x^4 y^2 + x^2 y^4 - y^6)$
83. $\frac{1}{2} \log_a (1 - s) + \frac{1}{2} \log_a (1 + s)$ **85.** $\frac{10}{3}$ **87.** -2
89. $\frac{2}{5}$ **91.** True

Technology Connection, p. 613

1. ⬤LOG ⬤7 ⬤) ⬤÷ ⬤LOG ⬤3 ⬤) ⬤ENTER

Technology Connection, p. 614

1. As x gets larger, the value of y_1 approaches $2.7182818284\ldots$ **2.** For large values of x, the graphs of y_1 and y_2 will be very close or appear to be the same curve,

depending on the window chosen.   **3.** Using ⬭TRACE⬭, no
y-value is given for $x = 0$. Using a table, an error message
appears for y_1 when $x = 0$. The domain does not include 0
because division by 0 is undefined.

Technology Connection, p. 617

1. $y = \log x/\log 7$
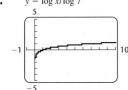

2. $y = \log (x+2)/\log 5$
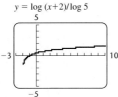

3. $y = \log x/\log 7 + 2$
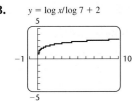

Visualizing for Success, p. 618

1. J   **2.** D   **3.** B   **4.** G   **5.** H   **6.** C   **7.** F   **8.** I
9. E   **10.** A

Exercise Set 9.5, pp. 619–620

1. True   **2.** True   **3.** True   **4.** False   **5.** True
6. True   **7.** True   **8.** True   **9.** True   **10.** True
11. 0.8451   **13.** 1.1367   **15.** 3   **17.** −0.1249
19. 13.0014   **21.** 50.1187   **23.** 0.0011   **25.** 2.1972
27. −5.0832   **29.** 96.7583   **31.** 15.0293   **33.** 0.0305
35. 3.0331   **37.** 6.6439   **39.** 1.1610   **41.** −0.3010
43. −3.3219   **45.** 2.0115

47. Domain: $\mathbb{R}$;
range: $(0, \infty)$
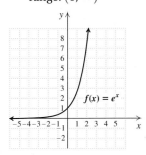

49. Domain: $\mathbb{R}$;
range: $(3, \infty)$
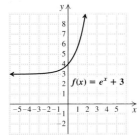

51. Domain: $\mathbb{R}$;
range: $(-2, \infty)$
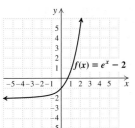

53. Domain: $\mathbb{R}$;
range: $(0, \infty)$
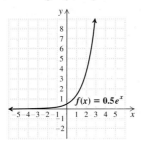

55. Domain: $\mathbb{R}$;
range: $(0, \infty)$
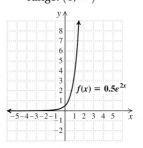

57. Domain: $\mathbb{R}$;
range: $(0, \infty)$
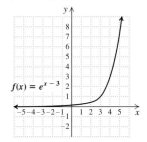

59. Domain: $\mathbb{R}$;
range: $(0, \infty)$
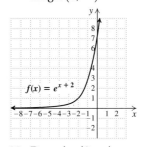

61. Domain: $\mathbb{R}$;
range: $(-\infty, 0)$
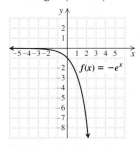

63. Domain: $(0, \infty)$;
range: $\mathbb{R}$
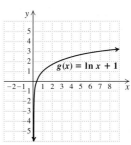

65. Domain: $(0, \infty)$;
range: $\mathbb{R}$
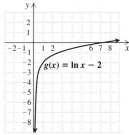

67. Domain: $(0, \infty)$;
range: $\mathbb{R}$
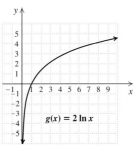

69. Domain: $(0, \infty)$;
range: $\mathbb{R}$
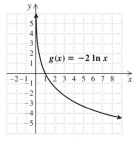

71. Domain: $(-2, \infty)$;
range: $\mathbb{R}$
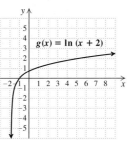

73. Domain: $(1, \infty)$;
range: $\mathbb{R}$
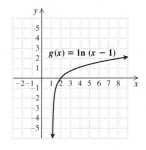

75. 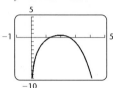   **77.** $-4, 7$ **78.** $0, \frac{7}{5}$ **79.** $\frac{15}{17}$ **80.** $\frac{5}{6}$ **81.** $\frac{56}{9}$
82. 4   **83.** 16, 256   **84.** $\frac{1}{4}, 9$ **85.** 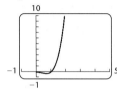   **87.** 2.452

89. 1.442   **91.** $\log M = \dfrac{\ln M}{\ln 10}$ **93.** 1086.5129

95. 4.9855   **97. (a)** Domain: $\{x \mid x > 0\}$, or $(0, \infty)$;
range: $\{y \mid y < 0.5135\}$, or $(-\infty, 0.5135)$;
(b) $[-1, 5, -10, 5]$;   **(c)**

$y = 3.4 \ln x - 0.25 e^x$

99. (a) Domain: $\{x \mid x > 0\}$, or $(0, \infty)$;
range: $\{y \mid y > -0.2453\}$, or $(-0.2453, \infty)$;
(b) $[-1, 5, -1, 10]$;   **(c)**

$y = 2x^3 \ln x$

101.

101. [figure]

Connecting the Concepts, pp. 620-621

1. 2   **2.** -1 **3.** $\frac{1}{2}$ **4.** 2   **5.** 1   **6.** 0   **7.** 4   **8.** 8
9. 7   **10.** 3   **11.** $x^m = 3$ **12.** $2^{10} = 1024$
13. $t = \ln x$ **14.** $\frac{2}{3} = \log_{64} 16$ **15.** 4   **16.** $\frac{1}{3}$

17. $\log x - \frac{1}{2}\log y - \frac{3}{2}\log z$ **18.** $\log \dfrac{a}{b^2 c}$ **19.** 1.5

20. 2.8614

Technology Connection, p. 625

1. 0.38   **2.** -1.96 **3.** 0.90   **4.** -1.53 **5.** 0.13, 8.47
6. $-0.75, 0.75$

Exercise Set 9.6, pp. 626-627

1. (e)   **2.** (a)   **3.** (f)   **4.** (h)   **5.** (b)   **6.** (d)

7. (g)   **8.** (c)   **9.** 2   **11.** $\frac{5}{2}$ **13.** $\dfrac{\log 10}{\log 2} \approx 3.322$

15. -1 **17.** $\dfrac{\log 19}{\log 8} + 3 \approx 4.416$ **19.** $\ln 50 \approx 3.912$

21. $\dfrac{\ln 8}{-0.02} \approx -103.972$ **23.** $\dfrac{\log 87}{\log 4.9} \approx 2.810$

25. $\dfrac{\ln\left(\frac{19}{2}\right)}{4} \approx 0.563$ **27.** $\dfrac{\ln 2}{-1} \approx -0.693$ **29.** 81

31. $\frac{1}{16}$ **33.** $e^5 \approx 148.413$ **35.** $\dfrac{e^3}{4} \approx 5.021$

37. $10^{1.2} \approx 15.849$ **39.** $\dfrac{e^4 - 1}{2} \approx 26.799$

41. $e \approx 2.718$ **43.** $e^{-3} \approx 0.050$ **45.** -4 **47.** 10

49. No solution   **51.** 2   **53.** $\frac{83}{15}$ **55.** 1   **57.** 6
59. 1   **61.** 5   **63.** $\frac{17}{2}$ **65.** 4   **67.**
69. Length: 9.5 ft; width: 3.5 ft   **70.** 25 visits or more
71. Golden Days; $23\frac{1}{3}$ lb; Snowy Friends: $26\frac{2}{3}$ lb
72. 1.5 cm   **73.** $1\frac{1}{5}$ hr   **74.** Approximately 2.1 ft
75.   **77.** -4 **79.** 2   **81.** $\pm\sqrt{34}$ **83.** $-3, -1$
85. $-625, 625$ **87.** $\frac{1}{2}, 5000$ **89.** $-3, -1$
91. $\frac{1}{100{,}000}, 100{,}000$ **93.** $-\frac{1}{3}$ **95.** 38   **97.** 1

Exercise Set 9.7, pp. 634-639

1. (a) Approximately 2006;   **(b)** 2.8 yr
3. (a) Approximately 1979;   **(b)** approximately 2025
5. (a) 6.4 yr;   **(b)** 23.4 yr   **7. (a)** 1991;   **(b)** 2013
9. (a) 2018;   **(b)** 15.1 yr   **11.** 4.9
13. 10^{-7} moles per liter   **15.** 130 dB   **17.** $7.6\ \text{W/m}^2$
19. Approximately 42.4 million messages per day
21. (a) $P(t) = P_0 e^{0.025t}$;   **(b)** \$5126.58; \$5256.36;   **(c)** 27.7 yr
23. (a) $P(t) = 304 e^{0.009t}$;   **(b)** 315 million;   **(c)** about 2015
25. 0.2 yr   **27. (a)** About 2055;   **(b)** about 2068;

(c)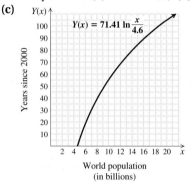

29. (a) 68%;   **(b)** 54%, 40%

(c) 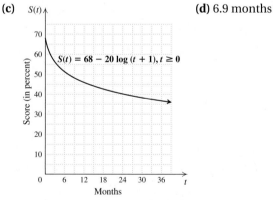   **(d)** 6.9 months

31. (a) $k \approx 0.126$; $P(t) = 2000 e^{0.126t}$;   **(b)** 2015
33. (a) $k \approx 0.280$; $P(t) = 8200 e^{-0.280t}$;   **(b)** \$215 per gigabit
per second per mile;   **(c)** 2029   **35.** About 1964 yr
37. 7.2 days   **39. (a)** 13.9% per hour;   **(b)** 21.6 hr
41. (a) $k \approx 0.114$; $V(t) = 451{,}000 e^{0.114t}$;   **(b)** \$4.9 million;
(c) 6.1 yr;   **(d)** 2010   **43.**   **45.** $\sqrt{2}$ **46.** 5
47. $(4, -7)$ **48.** $\left(-\frac{7}{2}, -\frac{19}{2}\right)$ **49.** $-4 \pm \sqrt{17}$
50. $5 \pm 2\sqrt{10}$

51. 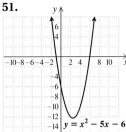   **52.**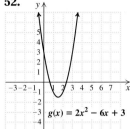

$y = x^2 - 5x - 6$ $g(x) = 2x^2 - 6x + 3$

53. 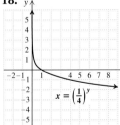   **55.** \$14.5 million   **57. (a)** -26.9;
(b) $1.58 \times 10^{-17}\,\mathrm{W/m^2}$ **59.** Consider an exponential
growth function $P(t) = P_0 e^{kt}$. At time T, $P(T) = 2P_0$.
Solve for T:

$$2P_0 = P_0 e^{kt}$$
$$2 = e^{kt}$$
$$\ln 2 = kT$$
$$\frac{\ln 2}{k} = T.$$

61.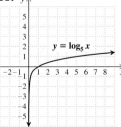

Review Exercises: Chapter 9, pp. 643–644

1. True   **2.** True   **3.** True   **4.** False   **5.** False
6. True   **7.** False   **8.** False   **9.** True   **10.** False
11. $(f \circ g)(x) = 4x^2 - 12x + 10$; $(g \circ f)(x) = 2x^2 - 1$
12. $f(x) = \sqrt{x}$; $g(x) = 3 - x$ **13.** No
14. $f^{-1}(x) = x + 10$ **15.** $g^{-1}(x) = \dfrac{2x - 1}{3}$
16. $f^{-1}(x) = \dfrac{\sqrt[3]{x}}{3}$ **17.**

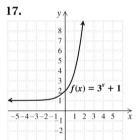

$f(x) = 3^x + 1$

18. 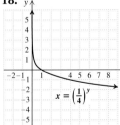   **19.**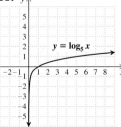

$x = \left(\frac{1}{4}\right)^y$ $y = \log_5 x$

20. 2   **21.** -2 **22.** 11   **23.** $\frac{1}{2}$ **24.** $\log_2 \frac{1}{8} = -3$
25. $\log_{25} 5 = \frac{1}{2}$ **26.** $16 = 4^x$ **27.** $1 = 8^0$
28. $4\log_a x + 2\log_a y + 3\log_a z$
29. $5\log_a x - (\log_a y + 2\log_a z)$, or
$5\log_a x - \log_a y - 2\log_a z$
30. $\frac{1}{4}(2\log z - 3\log x - \log y)$ **31.** $\log_a(5 \cdot 8)$, or $\log_a 40$
32. $\log_a \frac{48}{12}$, or $\log_a 4$ **33.** $\log \dfrac{a^{1/2}}{bc^2}$ **34.** $\log_a \sqrt[3]{\dfrac{x}{y^2}}$
35. 1   **36.** 0   **37.** 17   **38.** 6.93   **39.** -3.2698
40. 8.7601   **41.** 3.2698   **42.** 2.54995   **43.** -3.6602

44. 1.8751   **45.** 61.5177   **46.** -1.2040 **47.** 0.3753
48. 2.4307   **49.** 0.8982
50. Domain: $\mathbb{R}$;   **51.** Domain: $(0, \infty)$;
range: $(-1, \infty)$ range: $\mathbb{R}$

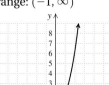   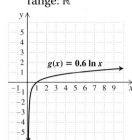

$f(x) = e^x - 1$ $g(x) = 0.6 \ln x$

52. 3   **53.** -1 **54.** $\frac{1}{81}$ **55.** 2   **56.** $\frac{1}{1000}$
57. $e^3 \approx 20.0855$ **58.** $\frac{1}{2}\left(\dfrac{\log 19}{\log 4} + 5\right) \approx 3.5620$
59. $\dfrac{\log 12}{\log 2} \approx 3.5850$ **60.** $\dfrac{\ln 0.03}{-0.1} \approx 35.0656$
61. $e^{-3} \approx 0.0498$ **62.** $\frac{15}{2}$ **63.** 16   **64.** 5
65. (a) 82; **(b)** 66.8; **(c)** 35 months   **66. (a)** 2.3 yr;
(b) 3.1 yr   **67. (a)** $k \approx 0.043$; $A(t) = 885e^{0.043t}$;
(b) \$1.0 billion; **(c)** 2023; **(d)** 16.1 yr
68. (a) $M(t) = 3253e^{-0.137t}$;   **(b)** 1640 spam messages per
consumer; **(c)** 2030   **69.** 11.553% per year   **70.** 16.5 yr
71. 3463 yr   **72.** 5.1   **73.** About 114 dB
74. 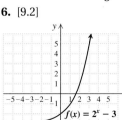 Negative numbers do not have logarithms because
logarithm bases are positive, and there is no exponent to
which a positive number can be raised to yield a negative
number.   **75.** If $f(x) = e^x$, then to find the inverse
function, we let $y = e^x$ and interchange x and y: $x = e^y$. If
$x = e^y$, then $\log_e x = y$ by the definition of logarithms.
Since $\log_e x = \ln x$, we have $y = \ln x$ or $f^{-1}(x) = \ln x$. Thus,
$g(x) = \ln x$ is the inverse of $f(x) = e^x$. Another approach is
to find $(f \circ g)(x)$ and $(g \circ f)(x)$:

$$(f \circ g)(x) = e^{\ln x} = x, \text{ and}$$
$$(g \circ f)(x) = \ln e^x = x.$$

Thus, g and f are inverse functions.
76. e^{e^3} **77.** $-3, -1$ **78.** $\left(\frac{8}{3}, -\frac{2}{3}\right)$

Test: Chapter 9, p. 645

1. [9.1] $(f \circ g)(x) = 2 + 6x + 4x^2$;
$(g \circ f)(x) = 2x^2 + 2x + 1$
2. [9.1] $f(x) = \dfrac{1}{x}$; $g(x) = 2x^2 + 1$ **3.** [9.1] No
4. [9.1] $f^{-1}(x) = \dfrac{x - 4}{3}$ **5.** [9.1] $g^{-1}(x) = \sqrt[3]{x} - 1$
6. [9.2]   **7.** [9.3]

    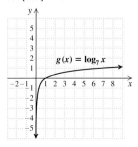

$f(x) = 2^x - 3$ $g(x) = \log_7 x$

8. [9.3] 3   **9.** [9.3] $\frac{1}{2}$ **10.** [9.3] 18   **11.** [9.4] 1
12. [9.4] 0   **13.** [9.4] 19   **14.** [9.3] $\log_5 \frac{1}{625} = -4$
15. [9.3] $2^m = \frac{1}{2}$ **16.** [9.4] $3 \log a + \frac{1}{2} \log b - 2 \log c$
17. [9.4] $\log_a\left(z^2 \sqrt[3]{x}\right)$ **18.** [9.4] 1.146   **19.** [9.4] 0.477
20. [9.4] 1.204   **21.** [9.5] 1.3979   **22.** [9.5] 0.1585
23. [9.5] -0.9163 **24.** [9.5] 121.5104   **25.** [9.5] 2.4022
26. [9.5]

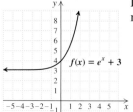

Domain: $\mathbb{R}$;
range: $(3, \infty)$

$f(x) = e^x + 3$

27. [9.5]

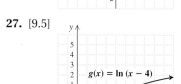

Domain: $(4, \infty)$;
range: $\mathbb{R}$

$g(x) = \ln(x - 4)$

28. [9.6] -5 **29.** [9.6] 2   **30.** [9.6] $\frac{1}{100}$
31. [9.6] $-\frac{1}{3}\left(\frac{\log 87}{\log 5} - 4\right) \approx 0.4084$
32. [9.6] $\frac{\log 1.2}{\log 7} \approx 0.0937$ **33.** [9.6] $e^3 \approx 20.0855$
34. [9.6] 4   **35.** [9.7] **(a)** 2.25 ft/sec; **(b)** 2,901,000
36. [9.7] **(a)** $P(t) = 140e^{0.024t}$, where t is the number of
years after 2008 and $P(t)$ is in millions; **(b)** 154 million;
170 million; **(c)** 2023; **(d)** 28.9 yr
37. [9.7] **(a)** $k \approx 0.045$; $C(t) = 21,855e^{0.045t}$; **(b)** \$35,853;
(c) 2019   **38.** [9.7] 4.3%   **39.** [9.7] 4684 yr
40. [9.7] $6.3 \times 10^6 \, \text{W/m}^2$ **41.** [9.7] 7.0
42. [9.6] $-309, 316$ **43.** [9.4] 2

Cumulative Review: Chapters 1–9, pp. 646–647

1. 2   **2.** $\frac{y^{12}}{16x^8}$ **3.** $\frac{20x^6z^2}{y}$ **4.** $-\frac{y^4}{3z^5}$ **5.** 6.3×10^{-15}
6. 25   **7.** 8   **8.** $(3, -1)$ **9.** $(1, -2, 0)$ **10.** $-7, 10$
11. $\frac{9}{2}$ **12.** $\frac{3}{4}$ **13.** $\frac{1}{2}$ **14.** $\pm 4i$ **15.** $\pm 2, \pm 3$ **16.** 9
17. $\frac{\log 7}{5 \log 3} \approx 0.3542$ **18.** $\frac{8e}{e - 1} \approx 12.6558$
19. $(-\infty, -5) \cup (1, \infty)$, or $\{x | x < -5 \text{ or } x > 1\}$
20. $-3 \pm 2\sqrt{5}$ **21.** $\{x | x \leq -2 \text{ or } x \geq 5\}$,
or $(-\infty, -2] \cup [5, \infty)$ **22.** $a = \frac{Db}{b - D}$
23. $x = \frac{-v \pm \sqrt{v^2 + 4ad}}{2a}$

24. $\{x | x \text{ is a real number } and \; x \neq -\frac{1}{3} \; and \; x \neq 2\}$, or
$\left(-\infty, -\frac{1}{3}\right) \cup \left(-\frac{1}{3}, 2\right) \cup (2, \infty)$
25. $3p^2q^3 + 11pq - 2p^2 + p + 9$ **26.** $9x^4 - 6x^2z^3 + z^6$
27. $\frac{1}{x - 4}$ **28.** $\frac{a + 2}{6}$ **29.** $\frac{7x + 4}{(x + 6)(x - 6)}$
30. $2y^2 \sqrt[3]{y}$ **31.** $\sqrt[10]{(x + 5)^7}$ **32.** $15 - 4\sqrt{3}i$
33. $x^3 - 5x^2 + 1$ **34.** $(3 + 4n)(9 - 12n + 16n^2)$
35. $2(3x - 2y)(x + 2y)$ **36.** $(x - 4)(x^3 + 7)$
37. $2(m + 3n)^2$ **38.** $(x - 2y)(x + 2y)(x^2 + 4y^2)$
39. $\frac{6 + \sqrt{y} - y}{4 - y}$ **40.** $f^{-1}(x) = \frac{x - 9}{-2}$, or $f^{-1}(x) = \frac{9 - x}{2}$
41. $f(x) = -10x - 8$ **42.** $y = \frac{1}{2}x + 5$
43.

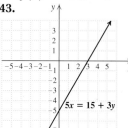

$5x = 15 + 3y$

44.

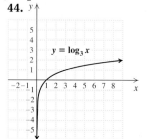

$y = \log_3 x$

45.

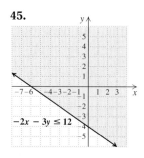

$-2x - 3y \leq 12$

46.

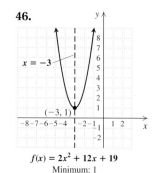

$x = -3$

$(-3, 1)$

$f(x) = 2x^2 + 12x + 19$
Minimum: 1

47.

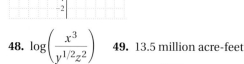

$f(x) = 2e^x$

Domain: $\mathbb{R}$;
range: $(0, \infty)$

48. $\log\left(\frac{x^3}{y^{1/2}z^2}\right)$ **49.** 13.5 million acre-feet
50. **(a)** $k \approx 0.076$; $D(t) = 15e^{0.076t}$; **(b)** 79.8 million cubic
meters per day; **(c)** 2015   **51.** **(a)** $\frac{2}{15}$ million barrels per
day per year; **(b)** $g(t) = \frac{2}{15}t + 8.5$; **(c)** $G(t) = 8.5e^{0.015t}$
52. $5\frac{5}{11}$ min   **53.** Thick and Tasty: 6 oz; Light and Lean:
9 oz   **54.** $2\frac{7}{9}$ km/h   **55.** -49; -7 and 7
56. **(a)** 78; **(b)** 67.5   **57.** All real numbers except 1
and -2 **58.** $\frac{1}{3}, \frac{10,000}{3}$ **59.** 35 mph

CHAPTER 10

Technology Connection, p. 655

1. $x^2 + y^2 - 16 = 0$

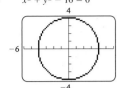

2. $(x - 1)^2 + (y - 2)^2 = 25$

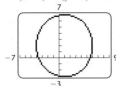

3. $(x + 3)^2 + (y - 5)^2 = 16$

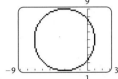

4. $(x - 5)^2 + (y + 6)^2 = 49$

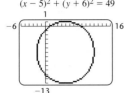

Exercise Set 10.1, pp. 656–659

1. (f)   **2.** (e)   **3.** (g)   **4.** (h)   **5.** (c)   **6.** (b)
7. (d)   **8.** (a)

9.

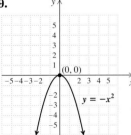

11.

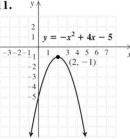

13.

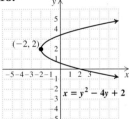

15.

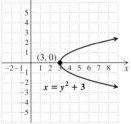

17.

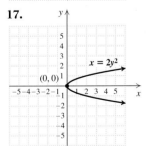

19.

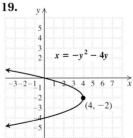

21.

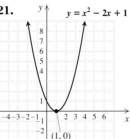

23.

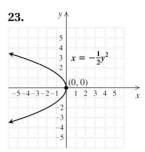

25.

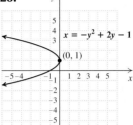

27.

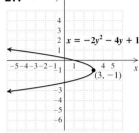

29. $x^2 + y^2 = 64$ **31.** $(x - 7)^2 + (y - 3)^2 = 6$
33. $(x + 4)^2 + (y - 3)^2 = 18$
35. $(x + 5)^2 + (y + 8)^2 = 300$
37. $x^2 + y^2 = 25$ **39.** $(x + 4)^2 + (y - 1)^2 = 20$
41. $(0, 0)$; 1   **43.** $(-1, -3)$; 7

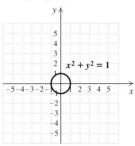

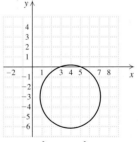

$(x + 1)^2 + (y + 3)^2 = 49$

45. $(4, -3)$; $\sqrt{10}$ **47.** $(0, 0)$; $2\sqrt{2}$

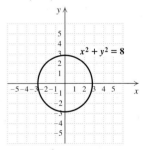

$(x - 4)^2 + (y + 3)^2 = 10$

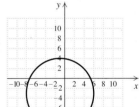

49. $(5, 0)$; $\frac{1}{2}$ **51.** $(-4, 3)$; $\sqrt{40}$, or $2\sqrt{10}$

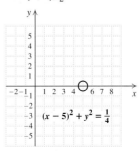

$(x - 5)^2 + y^2 = \frac{1}{4}$

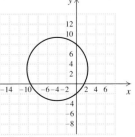

$x^2 + y^2 + 8x - 6y - 15 = 0$

53. $(4, -1); 2$

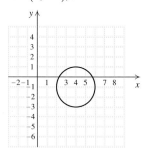

$$x^2 + y^2 - 8x + 2y + 13 = 0$$

55. $(0, -5); 10$

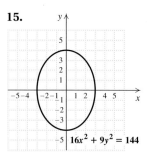

$$x^2 + y^2 + 10y - 75 = 0$$

57. $\left(-\dfrac{7}{2}, \dfrac{3}{2}\right); \sqrt{\dfrac{98}{4}}$, or $\dfrac{7\sqrt{2}}{2}$ **59.** $(0, 0); \dfrac{1}{6}$

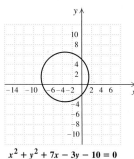

$$x^2 + y^2 + 7x - 3y - 10 = 0$$

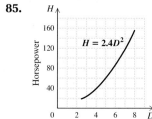

61.   **63.** ± 4 **64.** $\pm a$ **65.** $-4, 6$
66. $-5 \pm 2\sqrt{3}$ **67.** $-3 \pm 3\sqrt{3}$ **68.** $2 \pm \dfrac{4\sqrt{2}}{3}$
69. 🖩   **71.** $(x - 3)^2 + (y + 5)^2 = 9$
73. $(x - 3)^2 + y^2 = 25$ **75.** $(0, 4)$ **77.** $\dfrac{17}{4}\pi$ m², or
approximately 13.4 m²   **79.** 7169 mm
81. (a) $(0, -3)$; **(b)** 5 ft   **83.** $x^2 + (y - 30.6)^2 = 590.49$
85.

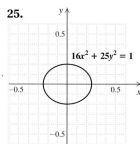

7 in.   **87.** 🖩, 〰

Diameter of piston
(in inches)

Exercise Set 10.2, pp. 663–665

1. True   **2.** False   **3.** False   **4.** False   **5.** True
6. True   **7.** True   **8.** True
9.

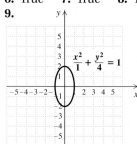

$$\dfrac{x^2}{1} + \dfrac{y^2}{4} = 1$$

11.

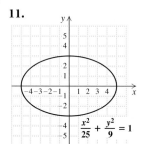

$$\dfrac{x^2}{25} + \dfrac{y^2}{9} = 1$$

13.

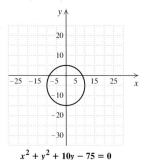

$$4x^2 + 9y^2 = 36$$

15.

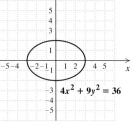

$$16x^2 + 9y^2 = 144$$

17.

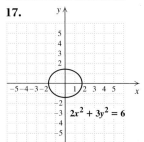

$$2x^2 + 3y^2 = 6$$

19.

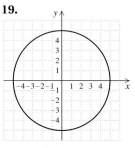

$$5x^2 + 5y^2 = 125$$

21.

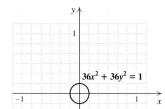

$$3x^2 + 7y^2 - 63 = 0$$

23.

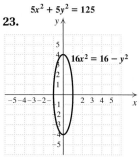

$$16x^2 = 16 - y^2$$

25.

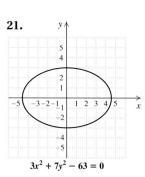

$$16x^2 + 25y^2 = 1$$

27.

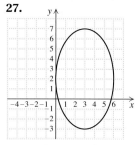

$$\dfrac{(x - 3)^2}{9} + \dfrac{(y - 2)^2}{25} = 1$$

29.

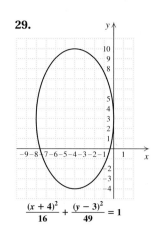

$$\dfrac{(x + 4)^2}{16} + \dfrac{(y - 3)^2}{49} = 1$$

31.

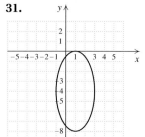

$$12(x - 1)^2 + 3(y + 4)^2 = 48$$

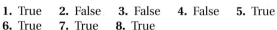

33.

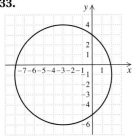

$4(x + 3)^2 + 4(y + 1)^2 - 10 = 90$

35.   **37.** $\dfrac{5}{2} \pm \dfrac{\sqrt{13}}{2}$ **38.** 3   **39.** $-\dfrac{3}{4}, 2$ **40.** $\dfrac{5}{2}$

41. $-\sqrt{11}, \sqrt{11}$ **42.** $-10, 6$ **43.**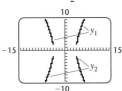

45. $\dfrac{x^2}{81} + \dfrac{y^2}{121} = 1$ **47.** $\dfrac{(x - 2)^2}{16} + \dfrac{(y + 1)^2}{9} = 1$

49. 2.134×10^8 mi   **51. (a)** Let $F_1 = (-c, 0)$ and $F_2 = (c, 0)$. Then the sum of the distances from the foci to P is $2a$. By the distance formula,

$$\sqrt{(x + c)^2 + y^2} + \sqrt{(x - c)^2 + y^2} = 2a, \text{ or}$$
$$\sqrt{(x + c)^2 + y^2} = 2a - \sqrt{(x - c)^2 + y^2}.$$

Squaring, we get

$$(x + c)^2 + y^2 = 4a^2 - 4a\sqrt{(x - c)^2 + y^2} + (x - c)^2 + y^2,$$

or

$$x^2 + 2cx + c^2 + y^2 = 4a^2 - 4a\sqrt{(x - c)^2 + y^2} + x^2 - 2cx + c^2 + y^2.$$

Thus,

$$-4a^2 + 4cx = -4a\sqrt{(x - c)^2 + y^2}$$
$$a^2 - cx = a\sqrt{(x - c)^2 + y^2}.$$

Squaring again, we get

$$a^4 - 2a^2cx + c^2x^2 = a^2(x^2 - 2cx + c^2 + y^2)$$
$$a^4 - 2a^2cx + c^2x^2 = a^2x^2 - 2a^2cx + a^2c^2 + a^2y^2,$$

or

$$x^2(a^2 - c^2) + a^2y^2 = a^2(a^2 - c^2)$$
$$\dfrac{x^2}{a^2} + \dfrac{y^2}{a^2 - c^2} = 1.$$

(b) When P is at $(0, b)$, it follows that $b^2 = a^2 - c^2$. Substituting, we have

$$\dfrac{x^2}{a^2} + \dfrac{y^2}{b^2} = 1.$$

53. 5.66 ft   **55.**   **57.**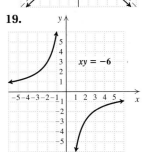

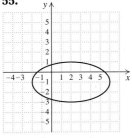

$$\dfrac{(x - 2)^2}{16} + \dfrac{(y + 1)^2}{4} = 1$$

Technology Connection, p. 670

1.
$$y_1 = \dfrac{\sqrt{15x^2 - 240}}{2};$$
$$y_2 = -\dfrac{\sqrt{15x^2 - 240}}{2}$$

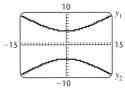

2.
$$y_1 = \sqrt{\dfrac{16x^2 - 64}{3}};$$
$$y_2 = -\sqrt{\dfrac{16x^2 - 64}{3}}$$

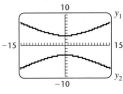

3.
$$y_1 = \dfrac{\sqrt{5x^2 + 320}}{4};$$
$$y_2 = -\dfrac{\sqrt{5x^2 + 320}}{4}$$

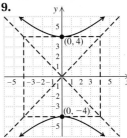

4.
$$y_1 = \sqrt{\dfrac{9x^2 + 441}{45}};$$
$$y_2 = -\sqrt{\dfrac{9x^2 + 441}{45}}$$

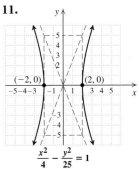

Exercise Set 10.3, pp. 672-673

1. (d)   **2.** (f)   **3.** (h)   **4.** (a)   **5.** (g)   **6.** (b)
7. (c)   **8.** (e)

9.

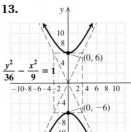

$$\dfrac{y^2}{16} - \dfrac{x^2}{16} = 1$$

11.

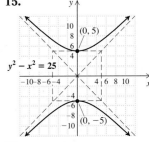

$$\dfrac{x^2}{4} - \dfrac{y^2}{25} = 1$$

13.

$$\dfrac{y^2}{36} - \dfrac{x^2}{9} = 1$$

15.

$$y^2 - x^2 = 25$$

17.

$$25x^2 - 16y^2 = 400$$

19.

$$xy = -6$$

21.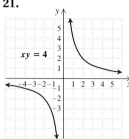

23.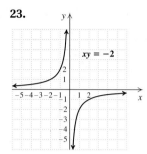

25.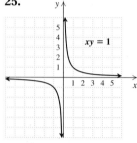

63. $\dfrac{(x+3)^2}{1} - \dfrac{(y-2)^2}{4} = 1$; C: $(-3, 2)$; V: $(-4, 2)$, $(-2, 2)$; asymptotes: $y - 2 = 2(x + 3)$, $y - 2 = -2(x + 3)$

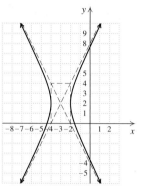

$4x^2 - y^2 + 24x + 4y + 28 = 0$

65.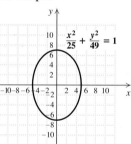

27. Circle   **29.** Ellipse   **31.** Hyperbola   **33.** Circle
35. Parabola   **37.** Hyperbola   **39.** Parabola
41. Hyperbola   **43.** Circle   **45.** Ellipse   **47.** ✍
49. $(-3, 6)$ **50.** $\left(\frac{1}{2}, -\frac{3}{2}\right)$ **51.** $-2, 2$ **52.** $-4, \frac{2}{3}$
53. $\dfrac{3}{2} \pm \dfrac{\sqrt{13}}{2}$ **54.** $\pm 1, \pm 5$ **55.** ✍   **57.** $\dfrac{y^2}{36} - \dfrac{x^2}{4} = 1$
59. C: $(5, 2)$; V: $(-1, 2)$, $(11, 2)$; asymptotes: $y - 2 = \frac{5}{6}(x - 5)$, $y - 2 = -\frac{5}{6}(x - 5)$

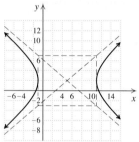

$\dfrac{(x-5)^2}{36} - \dfrac{(y-2)^2}{25} = 1$

61. $\dfrac{(y+3)^2}{4} - \dfrac{(x-4)^2}{16} = 1$; C: $(4, -3)$; V: $(4, -5)$, $(4, -1)$; asymptotes: $y + 3 = \frac{1}{2}(x - 4)$, $y + 3 = -\frac{1}{2}(x - 4)$

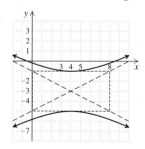

$8(y+3)^2 - 2(x-4)^2 = 32$

Connecting the Concepts, p. 674

1. $(4, 1)$; $x = 4$ **2.** $(2, -1)$; $y = -1$ **3.** $(3, 2)$
4. $(-3, -5)$ **5.** $(-12, 0)$, $(12, 0)$, $(0, -9)$, $(0, 9)$
6. $(-3, 0)$, $(3, 0)$ **7.** $(0, -1)$, $(0, 1)$ **8.** $y = \frac{3}{2}x$, $y = -\frac{3}{2}x$
9. Circle

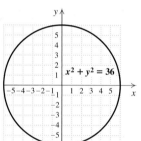

10. Parabola

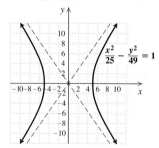

11. Ellipse

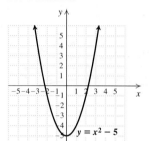

12. Hyperbola

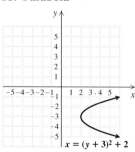

13. Parabola

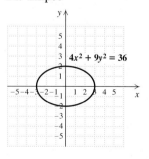

14. Ellipse

15. Hyperbola

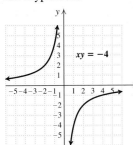

$xy = -4$

16. Circle

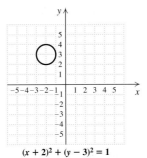

$(x + 2)^2 + (y - 3)^2 = 1$

17. Circle

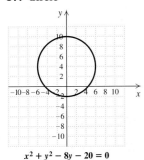

$x^2 + y^2 - 8y - 20 = 0$

18. Parabola

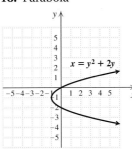

$x = y^2 + 2y$

19. Hyperbola

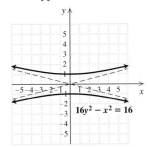

$16y^2 - x^2 = 16$

20. Hyperbola

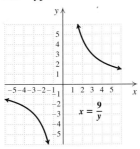

$x = \dfrac{9}{y}$

Technology Connection, p. 677

1. $(-1.50, -1.17); (3.50, 0.50)$
2. $(-2.77, 2.52); (-2.77, -2.52)$

Technology Connection, p. 678

1.

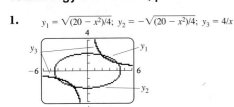

$y_1 = \sqrt{(20 - x^2)/4}; \ y_2 = -\sqrt{(20 - x^2)/4}; \ y_3 = 4/x$

Visualizing for Success, p. 681

1. C   **2.** A   **3.** F   **4.** B   **5.** J   **6.** D   **7.** H   **8.** I
9. G   **10.** E

Exercise Set 10.4, pp. 682–684

1. True   **2.** True   **3.** False   **4.** False   **5.** True

6. True   **7.** $(-5, -4), (4, 5)$ **9.** $(0, 2), (3, 0)$
11. $(-2, 1)$
13. $\left(\dfrac{5 + \sqrt{70}}{3}, \dfrac{-1 + \sqrt{70}}{3}\right), \left(\dfrac{5 - \sqrt{70}}{3}, \dfrac{-1 - \sqrt{70}}{30}\right)$
15. $\left(4, \dfrac{3}{2}\right), (3, 2)$ **17.** $\left(\dfrac{7}{3}, \dfrac{1}{3}\right), (1, -1)$ **19.** $\left(\dfrac{11}{4}, -\dfrac{5}{4}\right), (1, 4)$
21. $(2, 4), (4, 2)$ **23.** $(3, -5), (-1, 3)$
25. $(-5, -8), (8, 5)$ **27.** $(0, 0), (1, 1),$
$\left(-\dfrac{1}{2} + \dfrac{\sqrt{3}}{2}i, -\dfrac{1}{2} - \dfrac{\sqrt{3}}{2}i\right), \left(-\dfrac{1}{2} - \dfrac{\sqrt{3}}{2}i, -\dfrac{1}{2} + \dfrac{\sqrt{3}}{2}i\right)$
29. $(-4, 0), (4, 0)$ **31.** $(-4, -3), (-3, -4), (3, 4), (4, 3)$
33. $\left(\dfrac{16}{3}, \dfrac{5\sqrt{7}}{3}i\right), \left(\dfrac{16}{3}, -\dfrac{5\sqrt{7}}{3}i\right), \left(-\dfrac{16}{3}, \dfrac{5\sqrt{7}}{3}i\right),$
$\left(-\dfrac{16}{3}, -\dfrac{5\sqrt{7}}{3}i\right)$ **35.** $(-3, -\sqrt{5}), (-3, \sqrt{5}), (3, -\sqrt{5}),$
$(3, \sqrt{5})$ **37.** $(-3, -1), (-1, -3), (1, 3), (3, 1)$
39. $(4, 1), (-4, -1), (2, 2), (-2, -2)$ **41.** $(2, 1), (-2, -1)$
43. $\left(2, -\dfrac{4}{5}\right), \left(-2, -\dfrac{4}{5}\right), (5, 2), (-5, 2)$ **45.** $(-\sqrt{2}, \sqrt{2}),$
$(\sqrt{2}, -\sqrt{2})$ **47.** Length: 8 cm; width: 6 cm
49. Length: 2 in.; width: 1 in.   **51.** Length: 12 ft; width: 5 ft
53. 6 and 15; −6 and −15   **55.** 24 ft, 16 ft   **57.** Length:
$\sqrt{3}$ m; width: 1 m   **59.**   **61.** −9   **62.** −27
63. −1   **64.** $\dfrac{1}{5}$ **65.** 77   **66.** $\dfrac{21}{2}$ **67.**
69. $(x + 2)^2 + (y - 1)^2 = 4$ **71.** $(-2, 3), (2, -3),$
$(-3, 2), (3, -2)$ **73.** Length: 55 ft; width: 45 ft
75. 10 in. by 7 in. by 5 in.   **77.** Length: 63.6 in.;
height: 35.8 in.   **79.**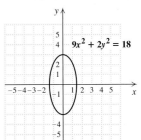

Review Exercises: Chapter 10, pp. 687–688

1. True   **2.** False   **3.** False   **4.** True   **5.** True
6. True   **7.** False   **8.** True   **9.** $(-3, 2), 4$
10. $(5, 0), \sqrt{11}$ **11.** $(3, 1), 3$ **12.** $(-4, 3), 3\sqrt{5}$
13. $(x + 4)^2 + (y - 3)^2 = 16$
14. $(x - 7)^2 + (y + 2)^2 = 20$
15. Circle

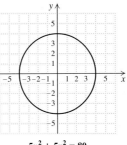

$5x^2 + 5y^2 = 80$

16. Ellipse

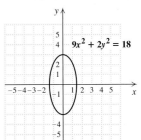

$9x^2 + 2y^2 = 18$

17. Parabola

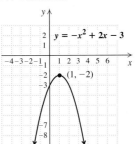

$y = -x^2 + 2x - 3$
$(1, -2)$

18. Hyperbola

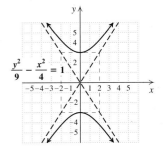

$\dfrac{y^2}{9} - \dfrac{x^2}{4} = 1$

19. Hyperbola

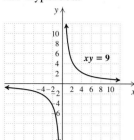

$xy = 9$

20. Parabola

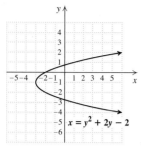

$x = y^2 + 2y - 2$

6. [10.3] Hyperbola

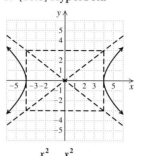

$$\frac{x^2}{16} - \frac{y^2}{9} = 1$$

7. [10.2], [10.3] Ellipse

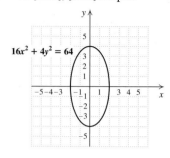

$16x^2 + 4y^2 = 64$

21. Ellipse

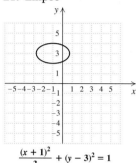

$$\frac{(x + 1)^2}{3} + (y - 3)^2 = 1$$

22. Circle

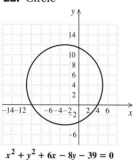

$x^2 + y^2 + 6x - 8y - 39 = 0$

8. [10.3] Hyperbola

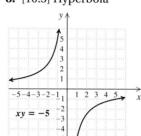

$xy = -5$

9. [10.1], [10.3] Parabola

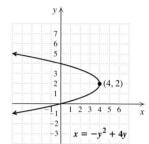

$(4, 2)$

$x = -y^2 + 4y$

23. $(5, -2)$ **24.** $(2, 2), \left(\frac{32}{9}, -\frac{10}{9}\right)$ **25.** $(0, -5), (2, -1)$
26. $(4, 3), (4, -3), (-4, 3), (-4, -3)$ **27.** $(2, 1), (\sqrt{3}, 0),$
$(-2, 1), (-\sqrt{3}, 0)$ **28.** $(3, -3), \left(-\frac{3}{5}, \frac{21}{5}\right)$ **29.** $(6, 8),$
$(6, -8), (-6, 8), (-6, -8)$ **30.** $(2, 2), (-2, -2),$
$(2\sqrt{2}, \sqrt{2}), (-2\sqrt{2}, -\sqrt{2})$ **31.** Length: 12 m; width: 7 m
32. Length: 12 in.; width: 9 in. **33.** 32 cm, 20 cm
34. 3 ft, 11 ft **35.** ✍ The graph of a parabola has one
branch whereas the graph of a hyperbola has two branches.
A hyperbola has asymptotes, but a parabola does not.
36. ✍ Function notation rarely appears in this chapter
because many of the relations are not functions. Function
notation could be used for vertical parabolas and for hyperbolas that have the axes as asymptotes.
37. $(-5, -4\sqrt{2}), (-5, 4\sqrt{2}), (3, -2\sqrt{2}), (3, 2\sqrt{2})$
38. $(0, 6), (0, -6)$ **39.** $(x - 2)^2 + (y + 1)^2 = 25$
40. $\dfrac{x^2}{100} + \dfrac{y^2}{1} = 1$ **41.** $\left(\frac{9}{4}, 0\right)$

10. [10.4] $(0, 6), \left(\frac{144}{25}, \frac{42}{25}\right)$ **11.** [10.4] $(-4, 13), (2, 1)$
12. [10.4] $(3, 2), (-3, -2), \left(-2\sqrt{2}i, \dfrac{3\sqrt{2}}{2}i\right), \left(2\sqrt{2}i, -\dfrac{3\sqrt{2}}{2}i\right)$
13. [10.4] $(\sqrt{6}, 2), (\sqrt{6}, -2), (-\sqrt{6}, 2), (-\sqrt{6}, -2)$
14. [10.4] 2 by 11 **15.** [10.4] $\sqrt{5}$ m, $\sqrt{3}$ m
16. [10.4] Length: 32 ft; width: 24 ft **17.** [10.4] $1200, 6%
18. [10.2] $\dfrac{(x - 6)^2}{25} + \dfrac{(y - 3)^2}{9} = 1$ **19.** [10.1] $\left(0, -\frac{31}{4}\right)$
20. [10.4] 9 **21.** [10.2] $\dfrac{x^2}{16} + \dfrac{y^2}{49} = 1$

Test: Chapter 10, p. 688

1. [10.1] $(x - 3)^2 + (y + 4)^2 = 12$ **2.** [10.1] $(4, -1), \sqrt{5}$
3. [10.1] $(-2, 3), 3$
4. [10.1], [10.3] Parabola

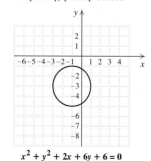

$y = x^2 - 4x - 1$

5. [10.1], [10.3] Circle

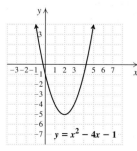

$x^2 + y^2 + 2x + 6y + 6 = 0$

Cumulative Review: Chapters 1–10, pp. 689–690

1. $16t^4 - 40t^2s + 25s^2$ **2.** $\dfrac{4t - 3}{3t(t - 3)}$ **3.** $\dfrac{x}{a}$
4. $3t^2\sqrt{10w}$ **5.** $27a^{1/2}b^{3/16}$ **6.** -4 **7.** 25 **8.** 0
9. $-\frac{1}{64}$ **10.** $2\sqrt{3}$ **11.** $(10x - 3y)^2$
12. $3(m^2 - 2)(m^4 + 2m^2 + 4)$ **13.** $(x - y)(a - b)$
14. $(4x - 3)(8x + 1)$ **15.** $\left(-\infty, -\frac{25}{3}\right], $ or $\left\{x \,\middle|\, x \le -\frac{25}{3}\right\}$
16. $0, \frac{9}{8}$ **17.** $1, 4$ **18.** $\pm i$ **19.** 4
20. $\dfrac{\log 1.5}{\log 3} \approx 0.3691$ **21.** 7
22. $(-\sqrt{3}, -1), (-\sqrt{3}, 1), (\sqrt{3}, -1), (\sqrt{3}, 1)$

23.

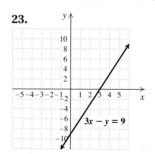

$3x - y = 9$

24.

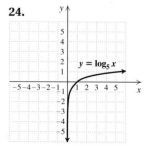

$y = \log_5 x$

25.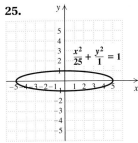

26.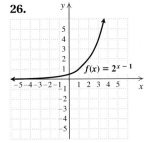

27.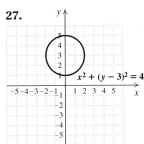

28.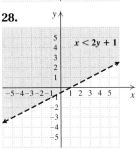

29.

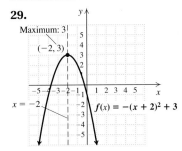

30. $\left(-\infty, \frac{5}{3}\right]$, or $\left\{x \mid x \leq \frac{5}{3}\right\}$　**31.** $c = \pm\sqrt{\dfrac{ab}{t}}$

32. $y = -x + 3$　**33.** $x^2 - 3 = 0$　**34.** $t^m = 16$

35. 2640 mi　**36.** Greg: 35 hr; Kyle: 14 hr

37. (a) $h(t) = -0.7t + 32.3$; **(b)** 27.4 hr per week;

(c) approximately 2021　**38.** 1.5 in.

39. (a) $P(t) = 2.26e^{-0.0095t}$; **(b)** 1.90 million;

(c) 73 yr　**40.** A: 15°; B: 45°; C: 120°　**41.** 8 in. by 8 in.

42. $\sqrt{55}$ cm $\approx$ 7.416 cm　**43.** $a = 1, b = -6$

44. $(1, -2, 0, 4)$　**45.** y is divided by 10.

46. $(-\infty, 0) \cup (0, 1]$, or $\{x \mid x < 0 \, or \, 0 < x \leq 1\}$

CHAPTER 11

Exercise Set 11.1, pp. 696–698

1. (f)　**2.** (a)　**3.** (d)　**4.** (b)　**5.** (c)　**6.** (e)　**7.** 43

9. 364　**11.** −23.5　**13.** −363　**15.** $\frac{441}{400}$

17. 2, 5, 8, 11; 29; 44　**19.** 3, 6, 11, 18; 102; 227

21. $\frac{1}{2}, \frac{2}{3}, \frac{3}{4}, \frac{4}{5}, \frac{10}{11}; \frac{15}{16}$　**23.** $1, -\frac{1}{2}, \frac{1}{4}, -\frac{1}{8}; -\frac{1}{512}; \frac{1}{16,384}$

25. $-1, \frac{1}{2}, -\frac{1}{3}, \frac{1}{4}; \frac{1}{10}; -\frac{1}{15}$　**27.** 0, 7, −26, 63; 999; −3374

29. $2n$　**31.** $(-1)^n$　**33.** $(-1)^{n+1} \cdot n$　**35.** $2n + 1$

37. $n^2 - 1$, or $(n + 1)(n - 1)$　**39.** $\dfrac{n}{n + 1}$

41. $(0.1)^n$, or 10^{-n}　**43.** $(-1)^n \cdot n^2$　**45.** 5

47. 1.11111, or $1\frac{11,111}{100,000}$　**49.** $\dfrac{1}{2} + \dfrac{1}{4} + \dfrac{1}{6} + \dfrac{1}{8} + \dfrac{1}{10} = \dfrac{137}{120}$

51. $10^0 + 10^1 + 10^2 + 10^3 + 10^4 = 11,111$

53. $2 + \dfrac{3}{2} + \dfrac{4}{3} + \dfrac{5}{4} + \dfrac{6}{5} + \dfrac{7}{6} + \dfrac{8}{7} = \dfrac{1343}{140}$

55. $(-1)^2 2^1 + (-1)^3 2^2 + (-1)^4 2^3 + (-1)^5 2^4 +$
$(-1)^6 2^5 + (-1)^7 2^6 + (-1)^8 2^7 + (-1)^9 2^8 = -170$

57. $(0^2 - 2 \cdot 0 + 3) + (1^2 - 2 \cdot 1 + 3) +$
$(2^2 - 2 \cdot 2 + 3) + (3^2 - 2 \cdot 3 + 3) +$
$(4^2 - 2 \cdot 4 + 3) + (5^2 - 2 \cdot 5 + 3) = 43$

59. $\dfrac{(-1)^3}{3 \cdot 4} + \dfrac{(-1)^4}{4 \cdot 5} + \dfrac{(-1)^5}{5 \cdot 6} = -\dfrac{1}{15}$　**61.** $\displaystyle\sum_{k=1}^{5} \dfrac{k+1}{k+2}$

63. $\displaystyle\sum_{k=1}^{6} k^2$　**65.** $\displaystyle\sum_{k=2}^{n} (-1)^k k^2$　**67.** $\displaystyle\sum_{k=1}^{\infty} 6k$

69. $\displaystyle\sum_{k=1}^{\infty} \dfrac{1}{k(k+1)}$　**71.** ✍　**73.** 98　**74.** −15

75. $a_1 + 4d$　**76.** $a_1 + a_n$　**77.** $3(a_1 + a_n)$, or $3a_1 + 3a_n$

78. d　**79.** ✍　**81.** 1, 3, 13, 63, 313, 1563　**83.** \$2500,
\$2000, \$1600, \$1280, \$1024, \$819.20, \$655.36, \$524.29,
\$419.43, \$335.54　**85.** $S_{100} = 0$; $S_{101} = -1$

87. $i, -1, -i, 1, i; i$　**89.** 11th term

Exercise Set 11.2, pp. 704–706

1. True　**2.** True　**3.** False　**4.** False　**5.** True

6. True　**7.** False　**8.** False　**9.** $a_1 = 8, d = 5$

11. $a_1 = 7, d = -4$　**13.** $a_1 = \frac{3}{2}, d = \frac{3}{4}$

15. $a_1 = \$8.16, d = \$0.30$　**17.** 154　**19.** −94

21. −\$1628.16　**23.** 26th　**25.** 57th　**27.** 178

29. 5　**31.** 28　**33.** $a_1 = 8$; $d = -3$; 8, 5, 2, −1, −4

35. $a_1 = 1$; $d = 1$　**37.** 780　**39.** 31,375　**41.** 2550

43. 918　**45.** 1030　**47.** 35 musicians; 315 musicians

49. 180 stones　**51.** \$49.60　**53.** 560 seats　**55.** ✍

57. $y = \frac{1}{3}x + 10$　**58.** $y = -4x + 11$　**59.** $y = -2x + 10$

60. $y = -\frac{4}{3}x - \frac{16}{3}$　**61.** $x^2 + y^2 = 16$

62. $(x + 2)^2 + (y - 1)^2 = 20$　**63.** ✍　**65.** 33 jumps

67. Let $d =$ the common difference. Since p, m, and q form
an arithmetic sequence, $m = p + d$ and $q = p + 2d$.

Then $\dfrac{p + q}{2} = \dfrac{p + (p + 2d)}{2} = p + d = m$.　**69.** 156,375

Exercise Set 11.3, pp. 713–715

1. Geometric sequence　**2.** Arithmetic sequence

3. Arithmetic sequence　**4.** Geometric sequence

5. Geometric series　**6.** Arithmetic series

7. Geometric series　**8.** None of these　**9.** 2　**11.** −0.1

13. $-\frac{1}{2}$　**15.** $\frac{1}{5}$　**17.** $\dfrac{6}{m}$　**19.** 1458　**21.** 243

23. 52,488　**25.** \$1423.31　**27.** $a_n = 5^{n-1}$

29. $a_n = (-1)^{n-1}$, or $a_n = (-1)^{n+1}$

31. $a_n = \dfrac{1}{x^n}$, or $a_n = x^{-n}$　**33.** 3066　**35.** $\frac{547}{18}$

37. $\dfrac{1 - x^8}{1 - x}$, or $(1 + x)(1 + x^2)(1 + x^4)$　**39.** \$5134.51

41. 27　**43.** $\frac{49}{4}$　**45.** No　**47.** No　**49.** $\frac{43}{99}$

51. \$25,000　**53.** $\frac{5}{9}$　**55.** $\frac{343}{99}$　**57.** $\frac{5}{33}$　**59.** $\frac{5}{1024}$ ft

61. 155,797   **63.** 2710 flies   **65.** Approximately 179.9 billion coffees   **67.** 3100.35 ft   **69.** 20.48 in.   **71.**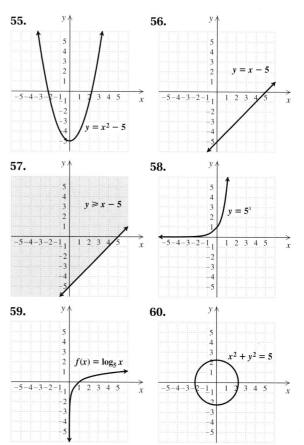
73. $x^2 + 2xy + y^2$ **74.** $x^3 + 3x^2y + 3xy^2 + y^3$
75. $x^3 - 3x^2y + 3xy^2 - y^3$
76. $x^4 - 4x^3y + 6x^2y^2 - 4xy^3 + y^4$
77. $8x^3 + 12x^2y + 6xy^2 + y^3$
78. $8x^3 - 12x^2y + 6xy^2 - y^3$ **79.**   **81.** 54
83. $\dfrac{x^2[1 - (-x)^n]}{1 + x}$ **85.** 512 cm^2 **87.** ,

Connecting the Concepts, p. 716

1. 300   **2.** $\dfrac{1}{n + 1}$ **3.** 78   **4.** $2^2 + 3^2 + 4^2 + 5^2 = 54$

5. $\displaystyle\sum_{k=1}^{6} (-1)^{k+1} \cdot k$ **6.** -3 **7.** 110   **8.** 61st   **9.** -39
10. 21   **11.** 11   **12.** 4410   **13.** $-\frac{1}{2}$ **14.** 640
15. $2(-1)^{n+1}$ **16.** $1146.39   **17.** 1   **18.** No
19. $465   **20.** $1,073,741,823

Technology Connection, p. 721

1. 479,001,600   **2.** 56; 792

Visualizing for Success, p. 724

1. J   **2.** G   **3.** A   **4.** H   **5.** I   **6.** B   **7.** E   **8.** D
9. F   **10.** C

Exercise Set 11.4, pp. 725–726

1. 2^5, or 32   **2.** 8   **3.** 9   **4.** 4!   **5.** $\binom{8}{5}$, or $\binom{8}{3}$
6. a^2b^8 **7.** 1   **8.** 9 choose 5   **9.** 24   **11.** 3,628,800
13. 90   **15.** 126   **17.** 210   **19.** 1   **21.** 435   **23.** 780
25. $a^4 - 4a^3b + 6a^2b^2 - 4ab^3 + b^4$
27. $p^7 + 7p^6q + 21p^5q^2 + 35p^4q^3 + 35p^3q^4 + 21p^2q^5 + 7pq^6 + q^7$
29. $2187c^7 - 5103c^6d + 5103c^5d^2 - 2835c^4d^3 + 945c^3d^4 - 189c^2d^5 + 21cd^6 - d^7$
31. $t^{-12} + 12t^{-10} + 60t^{-8} + 160t^{-6} + 240t^{-4} + 192t^{-2} + 64$
33. $x^5 - 5x^4y + 10x^3y^2 - 10x^2y^3 + 5xy^4 - y^5$
35. $19{,}683s^9 + \dfrac{59{,}049s^8}{t} + \dfrac{78{,}732s^7}{t^2} + \dfrac{61{,}236s^6}{t^3} + \dfrac{30{,}618s^5}{t^4} + \dfrac{10{,}206s^4}{t^5} + \dfrac{2268s^3}{t^6} + \dfrac{324s^2}{t^7} + \dfrac{27s}{t^8} + \dfrac{1}{t^9}$
37. $x^{15} - 10x^{12}y + 40x^9y^2 - 80x^6y^3 + 80x^3y^4 - 32y^5$
39. $125 + 150\sqrt{5}\,t + 375t^2 + 100\sqrt{5}\,t^3 + 75t^4 + 6\sqrt{5}\,t^5 + t^6$
41. $x^{-3} - 6x^{-2} + 15x^{-1} - 20 + 15x - 6x^2 + x^3$
43. $15a^4b^2$ **45.** $-64{,}481{,}508a^3$ **47.** $1120x^{12}y^2$
49. $1{,}959{,}552u^5v^{10}$ **51.** y^8 **53.**

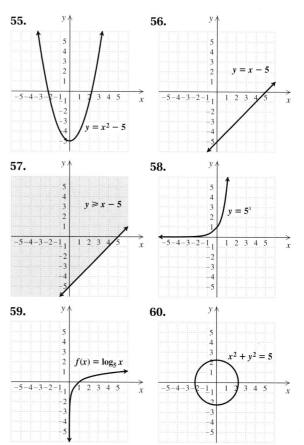

55. $y = x^2 - 5$
56. $y = x - 5$
57. $y \geq x - 5$
58. $y = 5^x$
59. $f(x) = \log_5 x$
60. $x^2 + y^2 = 5$

61.   **63.** List all the subsets of size 3: $\{a, b, c\}, \{a, b, d\},$ $\{a, b, e\}, \{a, c, d\}, \{a, c, e\}, \{a, d, e\}, \{b, c, d\}, \{b, c, e\},$ $\{b, d, e\}, \{c, d, e\}$. There are exactly 10 subsets of size 3 and $\binom{5}{3} = 10$, so there are exactly $\binom{5}{3}$ ways of forming a subset of size 3 from $\{a, b, c, d, e\}$.

65. $\binom{8}{5}(0.15)^3(0.85)^5 \approx 0.084$

67. $\binom{8}{6}(0.15)^2(0.85)^6 + \binom{8}{7}(0.15)(0.85)^7 + \binom{8}{8}(0.85)^8 \approx 0.89$

69. $\binom{n}{n - r} = \dfrac{n!}{[n - (n - r)]!(n - r)!}$
$= \dfrac{n!}{r!(n - r)!} = \binom{n}{r}$

71. $\dfrac{-\sqrt[3]{q}}{2p}$ **73.** $x^7 + 7x^6y + 21x^5y^2 + 35x^4y^3 + 35x^3y^4 + 21x^2y^5 + 7xy^6 + y^7$

Review Exercises: Chapter 11, pp. 728–729

1. False   **2.** True   **3.** True   **4.** False   **5.** False
6. True   **7.** False   **8.** False   **9.** 1, 11, 21, 31; 71; 111
10. $0, \frac{1}{5}, \frac{1}{5}, \frac{3}{17}, \frac{7}{65}, \frac{11}{145}$ **11.** $a_n = -5n$
12. $a_n = (-1)^n(2n - 1)$
13. $-2 + 4 + (-8) + 16 + (-32) = -22$
14. $-3 + (-5) + (-7) + (-9) + (-11) + (-13) = -48$

15. $\displaystyle\sum_{k=1}^{6} 7k$ **16.** $\displaystyle\sum_{k=1}^{5} \frac{1}{(-2)^k}$ **17.** -55 **18.** $\frac{1}{5}$
19. $a_1 = -15, d = 5$ **20.** -544 **21.** $25,250$
22. $1024\sqrt{2}$ **23.** $\frac{3}{4}$ **24.** $a_n = 2(-1)^n$

25. $a_n = 3\left(\dfrac{x}{4}\right)^{n-1}$ **26.** $11,718$ **27.** $-4095x$ **28.** 12

29. $\frac{49}{11}$ **30.** No   **31.** No   **32.** $40,000   **33.** $\frac{5}{9}$ **34.** $\frac{16}{11}$
35. $24.30   **36.** 903 poles   **37.** $15,791.18   **38.** 6 m
39. 5040 **40.** 120 **41.** $190a^{18}b^2$
42. $x^4 - 8x^3y + 24x^2y^2 - 32xy^3 + 16y^4$
43. 🖋 For a geometric sequence with $|r| < 1$, as n gets larger, the absolute value of the terms gets smaller, since $|r^n|$ gets smaller.   **44.** 🖋 The first form of the binomial theorem draws the coefficients from Pascal's triangle; the second form uses factorial notation. The second form avoids the need to compute all preceding rows of Pascal's triangle, and is generally easier to use when only one term of an expansion is needed. When several terms of an expansion are needed and n is not large (say, $n \leq 8$), it is often easier to use Pascal's triangle.   **45.** $\dfrac{1 - (-x)^n}{x + 1}$
46. $x^{-15} + 5x^{-9} + 10x^{-3} + 10x^3 + 5x^9 + x^{15}$

Test: Chapter 11, p. 730

1. [11.1] $\frac{1}{2}, \frac{1}{5}, \frac{1}{10}, \frac{1}{17}, \frac{1}{26}; \frac{1}{145}$ **2.** [11.1] $a_n = 4\left(\frac{1}{3}\right)^n$
3. [11.1] $-3 + (-7) + (-15) + (-31) = -56$
4. [11.1] $\displaystyle\sum_{k=1}^{5} (-1)^{k+1} k^3$ **5.** [11.2] $\frac{13}{2}$ **6.** [11.2] -3
7. [11.2] $a_1 = 31.2; d = -3.8$ **8.** [11.2] 2508
9. [11.3] 1536 **10.** [11.3] $\frac{2}{3}$ **11.** [11.3] 3^n
12. [11.3] 5621 **13.** [11.3] 1 **14.** [11.3] No
15. [11.3] $\frac{25,000}{23} \approx 1086.96 **16.** [11.3] $\frac{85}{99}$
17. [11.2] 63 seats   **18.** [11.2] $17,100
19. [11.3] $5987.37   **20.** [11.3] 36 m   **21.** [11.4] 220
22. [11.4] $x^5 - 15x^4y + 90x^3y^2 - 270x^2y^3 + 405xy^4 - 243y^5$ **23.** [11.4] $220a^9x^3$ **24.** [11.2] $n(n+1)$

25. [11.3] $\dfrac{1 - \left(\dfrac{1}{x}\right)^n}{1 - \dfrac{1}{x}}$, or $\dfrac{x^n - 1}{x^{n-1}(x-1)}$

Cumulative Review/Final Exam: Chapters 1–11, pp. 731–733

1. $\frac{7}{15}$ **2.** $-4y + 17$ **3.** 280 **4.** 8.4×10^{-15}
5. $\frac{7}{6}$ **6.** $3a^2 - 8ab - 15b^2$ **7.** $4a^2 - 1$
8. $9a^4 - 30a^2y + 25y^2$ **9.** $\dfrac{4}{x+2}$ **10.** $\dfrac{x-4}{4(x+2)}$
11. $\dfrac{(x+y)(x^2 + xy + y^2)}{x^2 + y^2}$ **12.** $x - a$ **13.** $12a^2\sqrt{b}$
14. $-27x^{10}y^{-2}$, or $-\dfrac{27x^{10}}{y^2}$ **15.** $25x^4y^{1/3}$
16. $y\sqrt[12]{x^5y^2}, y \geq 0$ **17.** $14 + 8i$
18. $(2x - 3)^2$ **19.** $(3a - 2)(9a^2 + 6a + 4)$

20. $12(s^2 + 2t)(s^2 - 2t)$ **21.** $3(y^2 + 3)(5y^2 - 4)$
22. $7x^3 + 9x^2 + 19x + 38 + \dfrac{72}{x - 2}$ **23.** 20
24. $[4, \infty)$, or $\{x | x \geq 4\}$
25. $(-\infty, 5) \cup (5, \infty)$, or $\{x | x < 5 \text{ or } x > 5\}$
26. $\dfrac{1 - 2\sqrt{x} + x}{1 - x}$ **27.** $y = 3x - 8$ **28.** $x^2 - 50 = 0$
29. $(2, -3); 6$ **30.** $\log_a \dfrac{\sqrt[3]{x^2} \cdot z^5}{\sqrt{y}}$ **31.** $a^5 = c$
32. 2.0792 **33.** 0.6826 **34.** 5 **35.** -121 **36.** 875
37. $16\left(\frac{1}{4}\right)^{n-1}$ **38.** $13,440a^4b^6$ **39.** $\frac{19,171}{64}$, or 299.546875
40. $\frac{3}{5}$ **41.** $-\frac{6}{5}, 4$ **42.** $\mathbb{R}$, or $(-\infty, \infty)$ **43.** $\left(-1, \frac{1}{2}\right)$
44. $(2, -1, 1)$ **45.** 2 **46.** $\pm 2, \pm 5$
47. $(\sqrt{5}, \sqrt{3}), (\sqrt{5}, -\sqrt{3}), (-\sqrt{5}, \sqrt{3}), (-\sqrt{5}, -\sqrt{3})$
48. 1.7925 **49.** 1005 **50.** $\frac{1}{25}$ **51.** $-\frac{1}{2}$
52. $\{x | -2 \leq x < 3\}$, or $[-2, 3]$ **53.** $\pm i\sqrt{3}$
54. $-2 \pm \sqrt{7}$ **55.** $\{y | y < -5 \text{ or } y > 2\}$, or $(-\infty, -5) \cup (2, \infty)$ **56.** $-8, 10$ **57.** 3
58. $r = \dfrac{V - P}{-Pt}$, or $\dfrac{P - V}{Pt}$ **59.** $R = \dfrac{lr}{1 - I}$

60.
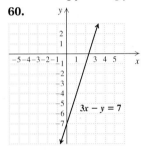
$3x - y = 7$

61.
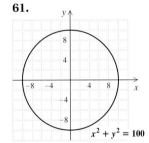
$x^2 + y^2 = 100$

62.
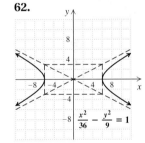
$\dfrac{x^2}{36} - \dfrac{y^2}{9} = 1$

63.
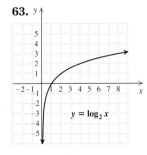
$y = \log_2 x$

64.
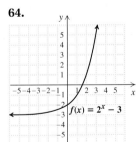
$f(x) = 2^x - 3$

65.

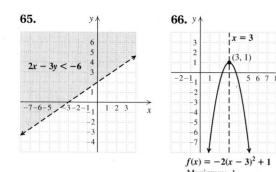

$2x - 3y < -6$

66.

$f(x) = -2(x - 3)^2 + 1$
Maximum: 1

67. 5000 ft^2 **68.** 5 ft by 12 ft   **69.** More than 25 rentals
70. 57, 59, 61   **71.** $2.68 herb: 10 oz; $4.60 herb: 14 oz
72. 350 mph   **73.** $8\frac{2}{5}$ hr, or 8 hr 24 min   **74.** 20
75. (a) The loan-to-value ratio increased 1.4% per year;
(b) $f(t) = 1.4t + 77.2$, where $f(t)$ is the loan-to-value ratio,
in percent;   **(c)** 91.2%;   **(d)** about 2013
76. (a) $k \approx 0.383$; $P(t) = 160e^{0.383t}$;   **(b)** 730,273 reverse
mortgages;   **(c)** about 2013   **77.** $14,079.98
78. All real numbers except 0 and -12 **79.** 81
80. y gets divided by 8   **81.** 84 yr

Glossary

A

Absolute value [1.2]   The distance that a number is from 0 on the number line.

Additive inverse [1.2]   A number's opposite. Two numbers are additive inverses of each other if their sum is zero.

Algebraic expression [1.1]   A collection of numbers and variables on which the operations $+, -, \cdot, \div, (\;)^n$, or $\sqrt[n]{(\;)}$ are performed.

Arithmetic sequence [11.2]   A sequence in which the difference between any two successive terms is constant.

Arithmetic series [11.2]   A series for which the associated sequence is arithmetic.

Ascending order [5.1]   A polynomial in one variable written with the terms arranged according to degree, from least to greatest.

Associative law of addition [1.2]   The statement that when three numbers are added, regrouping the addends gives the same sum.

Associative law of multiplication [1.2]   The statement that when three numbers are multiplied, regrouping the factors gives the same product.

Asymptote [9.1], [10.3]   A line that a graph approaches more and more closely as x increases or as x decreases.

Average [1.4]   Most commonly, the mean of a set of numbers.

Axes [2.1] (singular, axis)   Two perpendicular number lines used to identify points in a plane.

Axis of symmetry [8.6]   A line that can be drawn through a graph such that the part of the graph on one side of the line is an exact reflection of the part on the opposite side.

B

Base [1.1]   In exponential notation, the number being raised to a power.

Binomial [5.1]   A polynomial composed of two terms.

Branches [6.1], [10.3]   The two or more curves that comprise the graph of some relations.

Break-even point [3.8]   In business, the point of intersection of the revenue function and the cost function.

C

Circle [10.1]   A set of points in a plane that are a fixed distance r, called the radius, from a fixed point (h, k), called the center.

Circumference [1.5]   The distance around a circle.

Closed interval $[a, b]$ [4.1]   The set of all numbers x for which $a \le x \le b$. Thus, $[a, b] = \{x | a \le x \le b\}$.

Coefficient [5.1]   The numerical multiplier of a variable.

Combined variation [6.8]   A mathematical relationship in which a variable varies directly and/or inversely, at the same time, with more than one other variable.

Common logarithm [9.5]   A logarithm with base 10.

Commutative law of addition [1.2]   The statement that when two numbers are added, changing the order in which the numbers are added does not affect the sum.

Commutative law of multiplication [1.2]   The statement that when two numbers are multiplied, changing the order in which the numbers are multiplied does not affect the product.

Completing the square [8.1]   Adding a particular constant to an expression so that the resulting sum is a perfect square.

Complex number [7.8]   Any number that can be written as $a + bi$, where a and b are real numbers and $i = \sqrt{-1}$.

Complex rational expression [6.3]   A rational expression that has one or more rational expressions within its numerator and/or denominator.

Complex-number system [7.8]   A number system that contains the real-number system and is designed so that negative numbers have defined square roots.

Composite function [9.1]   A function in which a quantity depends on a variable that, in turn, depends on another variable.

Composite number   A natural number, other than 1, that is not prime.

Compound inequality [4.2]   A statement in which two or more inequalities are combined using the word *and* or the word *or*.

Compound interest [8.1]   Interest computed on the sum of an original principal and the interest previously accrued by that principal.

Conditional equation [1.3]   An equation that is true for some replacements of a variable and false for others.

Conic section [10.1]   A curve formed by the intersection of a plane and a cone.

Conjugates [7.5], [7.8]   Pairs of radical expressions, like $a\sqrt{b} + c\sqrt{d}$ and $a\sqrt{b} - c\sqrt{d}$, for which the product does not have a radical term, or pairs of imaginary numbers, like $a + bi$ and $a - bi$, for which the product is real.

Conjunction [4.2]   A sentence in which two statements are joined by the word *and*.

Consecutive numbers [1.4]   Integers that are one unit apart.

Consistent system of equations [3.1], [3.4]   A system of equations that has at least one solution.

Constant [1.1]   A known number.

Constant function [2.4]   A function given by an equation of the form $f(x) = b$, where b is a real number.

Constant of proportionality [6.8]   The constant in an equation of variation.

Constraint [4.5]   A requirement imposed on a problem.

Contradiction [1.3]   An equation that is never true.

Coordinates [2.1]   The numbers in an ordered pair.

Cube root [7.1]   The number c is called the cube root of a if $c^3 = a$.

D

Data point [8.8]   A given ordered pair of a function, usually found experimentally.

Degree of a polynomial [5.1]   The degree of the term of highest degree in a polynomial.

Degree of a term [5.1]   The number of variable factors in a term.

Demand function [3.8]   A function modeling the relationship between the price of a good and the quantity of that good demanded.

Denominator   The number below the fraction bar in a fraction, or the expression below the fraction bar in a rational expression.

Dependent equations [3.1], [3.4]   Equations in a system from which one equation can be removed without changing the solution set.

Descending order [5.1]   A polynomial in one variable written with the terms arranged according to degree, from greatest to least.

Determinant [3.7]   A descriptor of a matrix. The determinant of a two-by-two matrix $\begin{bmatrix} a & c \\ b & d \end{bmatrix}$ is denoted $\begin{vmatrix} a & c \\ b & d \end{vmatrix}$ and represents $ad - bc$. Determinants are defined for all square matrices.

Difference of (two) squares [5.2], [5.5]   An expression that can be written in the form $a^2 - b^2$.

Direct variation [6.8]   A situation that translates to an equation of the form $y = kx$, with k a constant.

Discriminant [8.3]   The expression $b^2 - 4ac$ from the quadratic formula.

Disjunction [4.2]   A sentence in which two statements are joined by the word *or*.

Distributive law [1.2]   The statement that multiplying a factor by the sum of two numbers gives the same result as multiplying the factor by each of the two numbers and then adding.

Domain [2.2]   The set of all first coordinates of the ordered pairs in a function.

Doubling time [9.7]   The time necessary for a population to double in size.

E

Element [1.1], [3.6]   A member of a set or an entry in a matrix.

Elimination method [3.2]   An algebraic method that uses the addition principle to solve a system of equations.

Ellipse [10.2]   The set of all points in a plane for which the sum of the distances from two fixed points F_1 and F_2, called foci, is constant.

Empty set [1.3]   The set containing no elements, denoted $\varnothing$ or { }.

Equation [1.1]   A number sentence with the verb =.

Equation of variation [6.8]   An equation used to represent direct, inverse, or combined variation.

Equilibrium point [3.8]   The point of intersection between the demand function and the supply function.

Equivalent equations [1.3]   Equations with the same solutions.

Equivalent expressions [1.2]   Expressions that have the same value for all allowable replacements.

Equivalent inequalities [4.1]   Inequalities that have the same solution set.

Evaluate [1.1]   To substitute a value for each occurrence of a variable in an expression.

Exponent [1.1]   In expressions of the form a^n, the number n is an exponent. For n a natural number, a^n represents n factors of a.

Exponential decay [9.7]   A decrease in quantity over time that can be modeled by an exponential equation of the form $P(t) = P_0 e^{-kt}$, $k > 0$.

Exponential equation [9.6]   An equation in which a variable appears as an exponent.

Exponential function [9.2]   A function that can be described by an exponential equation.

Exponential growth [9.7]   An increase in quantity over time that can be modeled by an exponential function of the form $P(t) = P_0 e^{kt}$, $k > 0$.

Exponential notation [1.1]   A representation of a number using a base raised to a power.

Extrapolation [2.2]   The process of predicting a future value on the basis of given data.

F

Factor [1.2], [5.3]   *Verb:* to write an equivalent expression that is a product. *Noun:* a multiplier.

Finite sequence [11.1]   A function having for its domain a set of natural numbers: $\{1, 2, 3, 4, 5, \ldots, n\}$, for some natural number n.

Fixed costs [3.8]   In business, costs that are incurred whether or not a product is produced.

Focus [10.2] (plural, foci)   One of two fixed points that determine the points of an ellipse.

FOIL [5.2]   To multiply two binomials by multiplying the First terms, the Outer terms, the Inner terms, and the Last terms, and then adding the results.

Formula [1.5]   An equation that uses numbers and/or letters to represent a relationship between two or more quantities.

Fraction notation [1.1]   A number written using a numerator and a denominator.

Function [2.2]   A correspondence that assigns to each member of a set called the domain exactly one member of a set called the range.

G

General term of a sequence [11.1] The nth term, denoted a_n.

Geometric sequence [11.3] A sequence in which the ratio of every pair of successive terms is constant.

Geometric series [11.3] A series for which the associated sequence is geometric.

Graph [2.1] A picture or diagram of the data in a table, or a line, a curve, a plane, or a collection of points, etc., that represent all the solutions of an equation.

Greatest common factor [5.3] The common factor of a polynomial with the largest possible coefficient and the largest possible exponent(s).

H

Half-life [9.7] The amount of time necessary for half of a quantity to decay.

Half-open interval [4.1] An interval that includes exactly one of two endpoints.

Horizontal-line test [9.1] The statement that if it is impossible to draw a horizontal line that intersects the graph of a function more than once, then that function is one-to-one.

Hyperbola [10.3] The set of all points P in the plane such that the difference of the distance from P to two fixed points called foci is constant.

Hypotenuse [5.8] In a right triangle, the side opposite the right angle.

I

i [7.8] The square root of -1. That is, $i = \sqrt{-1}$ and $i^2 = -1$.

Identity [1.3] An equation that is always true.

Identity property of 0 The statement that the sum of a number and 0 is always the original number.

Identity property of 1 The statement that the product of a number and 1 is always the original number.

Imaginary number [7.8] A number that can be written in the form $a + bi$, where a and b are real numbers and $b \neq 0$.

Inconsistent system of equations [3.1] A system of equations for which there is no solution.

Independent equations [3.1] Equations that are not dependent.

Index [7.1] In the radical $\sqrt[n]{a}$, the number n is called the index.

Inequality [1.2] A mathematical sentence using $<, >, \leq, \geq,$ or $\neq$.

Infinite geometric series [11.3] The sum of the terms of an infinite geometric sequence.

Infinite sequence [11.1] A function having for its domain the set of natural numbers: $\{1, 2, 3, 4, 5, \dots\}$.

Input [2.2] A member of the domain of a function.

Integers [1.1] The whole numbers and their opposites.

Interpolation [2.2] The process of estimating a value between given values.

Intersection of two sets [4.2] The set of all elements that are common to both sets.

Interval notation [4.1] The use of a pair of numbers inside parentheses and/or brackets to represent the set of all numbers between those two numbers. *See also* Closed, Open, and Half-open intervals.

Inverse relation [9.1] The relation formed by interchanging the members of the domain and the range of a relation.

Inverse variation [6.8] A situation that translates to an equation of the form $y = k/x$, with k a constant.

Irrational number [1.1] A real number that cannot be named as a ratio of two integers.

Isosceles right triangle [7.7] A right triangle in which both legs have the same length.

J

Joint variation [6.8] A situation that translates to an equation of the form $y = kxz$, with k a constant.

L

Leading coefficient [5.1] The coefficient of the term of highest degree in a polynomial.

Leading term [5.1] The term of highest degree in a polynomial.

Least common denominator [6.2] The least common multiple of the denominators of two or more rational expressions.

Legs [5.8] In a right triangle, the two sides that form the right angle.

Like radicals [7.5] Radical expressions that have a common radical factor.

Like terms [1.3], [5.1] Terms that have exactly the same variable factors.

Linear equation [1.3], [2.1], [2.3], [3.4] In two variables, any equation that can be written in the form $y = mx + b$, or $Ax + By = C$, where x and y are variables. In three variables, an equation that is equivalent to one of the form $Ax + By + Cz = D$, where x, y, and z are variables.

Linear function [2.3] A function that can be described by an equation of the form $f(x) = mx + b$, where m and b are constants.

Linear inequality [4.4] An inequality whose related equation is a linear equation.

Linear programming [4.5] A branch of mathematics involving graphs of inequalities and their constraints.

Logarithmic equation [9.6] An equation containing a logarithmic expression.

Logarithmic function, base a [9.3] The inverse of an exponential function with base a.

M

Matrix [3.6] (plural, matrices) A rectangular array of numbers.

Maximum value [8.6] The greatest function value (output) achieved by a function.

Minimum value [8.6] The least function value (output) achieved by a function.

Monomial [5.1] A constant, a variable, or a product of a constant and one or more variables.

Motion problem [3.3], [6.5] A problem that deals with distance, speed, and time.

Multiplicative inverses [1.2] Reciprocals; two numbers whose product is 1.

Multiplicative property of zero The statement that the product of 0 and any real number is 0.

N

Natural logarithm [9.5] A logarithm with base e.

Natural numbers [1.1] The counting numbers: 1, 2, 3, 4, 5, ...

Nonlinear equation [2.1] An equation whose graph is not a straight line.

Numerator The expression above the fraction bar in a fraction or in a rational expression.

O

Objective function [4.5] In linear programming, the function in which the expression being maximized or minimized appears.

One-to-one function [9.1] A function for which different inputs have different outputs.

Open interval (a, b) [4.1] The set of all numbers x for which $a < x < b$. Thus, $(a, b) = \{x | a < x < b\}$.

Opposite [1.2] The opposite, or additive inverse, of a number a is written $-a$. Opposites are the same distance from 0 on the number line but on different sides of 0.

Ordered pair [2.1] A pair of numbers of the form (h, k) for which the order in which the numbers are listed is important.

Origin [2.1] The point on a coordinate plane where the two axes intersect.

Output [2.2] A member of the range of a function.

P

Parabola [8.1], [8.6], [10.1] A graph of a second-degree polynomial in one variable.

Parallel lines [2.5] Lines that extend indefinitely in the same plane without intersecting.

Pascal's triangle [11.4] A triangular array of coefficients of the expansion $(a + b)^n$ for $n = 0, 1, 2, \ldots$.

Perfect square [7.1] A rational number for which there exists a number a for which $a^2 = p$.

Perfect-square trinomial [5.2], [5.5] A trinomial that is the square of a binomial.

Perpendicular lines [2.5] Lines that form a right angle.

Point–slope equation [2.5] An equation of the type $y - y_1 = m(x - x_1)$, where x and y are variables.

Polynomial [5.1] A monomial or a sum of monomials.

Polynomial equation [5.8] An equation in which two polynomials are set equal to each other.

Polynomial inequality [8.9] An inequality that is equivalent to an inequality with a polynomial as one side and 0 as the other.

Price [3.8] The amount a purchaser pays for an item.

Prime factorization [6.2] The factorization of a whole number into a product of its prime factors.

Prime number A natural number that has exactly two different factors: the number itself and 1.

Principal square root [7.1] The nonnegative square root of a number.

Pure imaginary number [7.8] A complex number of the form $a + bi$, with $a = 0$ and $b \neq 0$.

Pythagorean theorem [5.8] The theorem that states that in any right triangle, if a and b are the lengths of the legs and c is the length of the hypotenuse, then $a^2 + b^2 = c^2$.

Q

Quadrants [2.1] The four regions into which the axes divide a plane.

Quadratic equation [5.8] An equation equivalent to one of the form $ax^2 + bx + c = 0$, where $a \neq 0$.

Quadratic formula [8.2] $x = \dfrac{-b \pm \sqrt{b^2 - 4ac}}{2a}$, which gives the solutions of $ax^2 + bx + c = 0$, where $a \neq 0$.

Quadratic function [8.1] A second-degree polynomial function in one variable.

Quadratic inequality [8.9] A second-degree polynomial inequality in one variable.

R

Radical equation [7.6] An equation in which a variable appears in a radicand.

Radical expression [7.1] An algebraic expression in which a radical sign appears.

Radical sign [7.1] The symbol $\sqrt{}$.

Radical term [7.5] A term in which a radical sign appears.

Radicand [7.1] The expression under the radical sign.

Radius [10.1] The distance from the center of a circle to a point on the circle. Also, a segment connecting the center to a point on the circle.

Range [2.2] The set of all second coordinates of the ordered pairs in a function.

Ratio [2.3] The ratio of a to b is a/b, also written $a:b$.

Rational equation [6.4] An equation containing one or more rational expressions.

Rational expression [6.1] A quotient of two polynomials.

Rational inequality [8.9] An inequality containing a rational expression.

Rational number [1.1] A number that can be written in the form $\dfrac{a}{b}$, where a and b are integers and $b \neq 0$.

Rationalizing the denominator [7.4] A procedure for finding an equivalent expression without a radical in the denominator.

Rationalizing the numerator [7.4]   A procedure for finding an equivalent expression without a radical in the numerator.

Real number [1.1]   Any number that is either rational or irrational.

Reciprocal [1.2]   A multiplicative inverse. Two numbers are reciprocals if their product is 1.

Reflection [8.6]   The mirror image of a graph.

Relation [2.2]   A correspondence between the domain and the range such that each member of the domain corresponds to at least one member of the range.

Repeating decimal [1.1]   A decimal in which a block of digits repeats indefinitely.

Right triangle [5.8]   A triangle that includes a right angle.

Row-equivalent operations [3.6]   Operations used to produce equivalent systems of equations.

S

Scientific notation [1.7]   A number written in the form $N \times 10^m$, where m is an integer, $1 \le N < 10$, and N is expressed in decimal notation.

Sequence [11.1]   A function for which the domain is a set of consecutive positive integers beginning with 1.

Series [11.1]   The sum of specified terms in a sequence.

Set [1.1]   A collection of objects.

Set-builder notation [1.1]   The naming of a set by describing basic characteristics of the elements in the set.

Sigma notation [11.1]   The naming of a sum using the Greek letter Σ (sigma) as part of an abbreviated form.

Similar triangles   Triangles in which corresponding sides are proportional.

Simplify [1.3]   To rewrite an expression in an equivalent, abbreviated, form.

Slope [2.3]   The ratio of the rise to the run for any two points on a line.

Slope–intercept equation [2.3]   An equation of the form $y = mx + b$, where x and y are variables.

Solution [1.1], [3.1], [4.1]   A replacement or substitution that makes an equation or inequality or a system of equations or inequalities true.

Solution set [1.3], [3.1], [4.1]   The set of all solutions of an equation, an inequality, or a system of equations or inequalities.

Solve [1.1], [3.1], [4.1]   To find all solutions of an equation, an inequality, or a system of equations or inequalities; to find the solution(s) of a problem.

Speed [3.3]   The ratio of distance traveled to the time required to travel that distance.

Square matrix [3.7]   A matrix with the same number of rows and columns.

Square root [7.1]   The number c is a square root of a if $c^2 = a$.

Subset [1.1]   A collection of objects entirely contained within a given set.

Substitute [1.1]   To replace a variable with a number.

Substitution method [3.2]   An algebraic method for solving systems of equations.

Supply function [3.8]   A function modeling the relationship between the price of a good and the quantity of that good supplied.

System of equations [3.1]   A set of two or more equations that are to be solved simultaneously.

T

Term [1.3], [5.1]   A number, a variable, or a product or a quotient of numbers and/or variables.

Terminating decimal [1.1]   A decimal that can be written using a finite number of decimal places.

Total cost [3.8]   The amount spent to produce a product.

Total profit [3.8]   The amount taken in less the amount spent, or total revenue minus total cost.

Total revenue [3.8]   The amount taken in from the sale of a product.

Trinomial [5.1]   A polynomial that is composed of three terms.

U

Union of A and B [4.2]   The set of all elements belonging to either A or B.

V

Value   The numerical result after a number has been substituted into an expression.

Variable [1.1]   A letter that represents an unknown number.

Variable costs [3.8]   In business, costs that vary according to the quantity of products produced.

Vertex [8.6], [10.1], [10.2], [10.3] (plural, vertices)   The point at which the graph of a parabola, an ellipse, or a hyperbola crosses its axis of symmetry.

Vertical-line test [2.2]   The statement that a graph represents a function if it is impossible to draw a vertical line that intersects the graph more than once.

W

Whole numbers [1.1]   The natural numbers and 0: $0, 1, 2, 3, \ldots$

X

x-intercept [2.4]   A point at which a graph crosses the x-axis.

Y

y-intercept [2.4]   A point at which a graph crosses the y-axis.

Z

Zeros [8.9]   The x-values for which $f(x)$ is 0, for any function f.

Index

Index of Applications

Selected Keys of the Scientific Calculator

This secondary function takes the square root of number displayed.

Squares number displayed.

Activates secondary functions printed above certain keys. Also denoted (INV) or (2nd).

Used when entering numbers in scientific notation. Also denoted (EXP).

Finds reciprocal of number displayed.

Used to raise any base to a power. Also denoted (y^x), (a^x), or (∧).

Stores number displayed in memory. Also denoted (MIN) or (M).

Recalls number stored in memory. Also denoted (MR).

This secondary function raises 10 to any power entered.

Clears all preceding numbers and operations. Also used to turn calculator on.

Used as an approximation for pi.

Used to perform indicated operation.

Used to control order in which certain operations are performed.

Clears last number displayed but not preceding operations.

Used when entering decimal notation.

Used to change sign of number displayed.

3.141592654